P9-CKL-861

Biotechnology

Second Edition

Volume 9

Enzymes, Biomass, Food and Feed

Biotechnology

Second Edition

Fundamentals

Volume 1
Biological Fundamentals

Volume 2
Genetic Fundamentals and
Genetic Engineering

Volume 3
Bioprocessing

Volume 4
Measuring, Modelling, and Control

Products

Volume 5
Genetically Engineered Proteins and
Monoclonal Antibodies

Volume 6
Products of Primary Metabolism

Volume 7
Products of Secondary Metabolism

Volume 8
Biotransformations

Special Topics

Volume 9
Enzymes, Biomass, Food and Feed

Volume 10
Special Processes

Volume 11
Environmental Processes

Volume 12
Legal, Economic and
Ethical Dimensions

Distribution:

VCH, P. O. Box 101161, D-69451 Weinheim (Federal Republic of Germany)

Switzerland: VCH, P. O. Box, CH-4020 Basel (Switzerland)

United Kingdom and Ireland: VCH (UK) Ltd., 8 Wellington Court, Cambridge CB1 1HZ (England)

USA and Canada: VCH, 220 East 23rd Street, New York, NY 10010–4606 (USA)

Japan: VCH, Eikow Building, 10-9 Hongo 1-chome, Bunkyo-ku, Tokyo 113 (Japan)

ISBN 3-527-28319-6 (VCH, Weinheim)
Set ISBN 3-527-28310-2 (VCH, Weinheim)

A Multi-Volume Comprehensive Treatise

Biotechnology

Second, Completely Revised Edition

Edited by
H.-J. Rehm and G. Reed
in cooperation with
A. Pühler and P. Stadler

Volume 9

Enzymes, Biomass, Food and Feed

Edited by
G. Reed and T. W. Nagodawithana

Weinheim · New York
Basel · Cambridge · Tokyo

Series Editors:
Prof. Dr. H.-J. Rehm
Institut für Mikrobiologie
Universität Münster
Correnssstraße 3
D-48149 Münster

Dr. G. Reed
2131 N. Summit Ave.
Appartment #304
Milwaukee, WI 53202-1347
USA

Prof. Dr. A. Pühler
Biologie VI (Genetik)
Universität Bielefeld
P.O. Box 100131
D-33501 Bielefeld

Dr. P. J. W. Stadler
Bayer AG
Verfahrensentwicklung Biochemie
Leitung
Friedrich-Ebert-Straße 217
D-42096 Wuppertal

Volume Editors:
Dr. G. Reed
2131 N. Summit Ave.
Appartment #304
Milwaukee, WI 53202-1347
USA

Dr. T. W. Nagodawithana
Universal Foods Corp.
6143 N. 60th Street
Milwaukee, WI 53218
USA

Published jointly by
VCH Verlagsgesellschaft mbH, Weinheim (Federal Republic of Germany)
VCH Publishers Inc., New York, NY (USA)

Editorial Director: Dr. Hans-Joachim Kraus
Editorial Manager: Christa Maria Schultz
Copy Editor: Karin Dembowsky
Production Manager: Dipl. Wirt.-Ing. (FH) Hans-Jochen Schmitt

Library of Congress Card No.: applied for

British Library Cataloguing-in-Publication Data:
A catalogue record for this book is available from the British Library

Die Deutsche Bibliothek – CIP-Einheitsaufnahme
Biotechnology : a multi volume comprehensive treatise / ed. by
H.-J. Rehm and G. Reed. In cooperation with A. Pühler and
P. Stadler. – 2., completely rev. ed. – Weinheim ; New York ;
Basel ; Cambridge ; Tokyo : VCH.
ISBN 3-527-28310-2 (Weinheim ...)
ISBN 1-56081-602-3 (New York)
NE: Rehm, Hans J. [Hrsg.]

2., completely rev. ed.
Vol. 9. Enzymes, biomass, food and feed / ed. by G. Reed and
T. W. Nagodawithana. – 1995
ISBN 3-527-28319-6
NE: Reed, Gerald [Hrsg.]

Printed on acid-free and chlorine-free paper.

Composition and Printing: Zechnersche Buchdruckerei, D-67330 Speyer.
Bookbinding: J. Schäffer, D-67269 Grünstadt.
Printed in the Federal Republic of Germany

**In memory of Professor Anthony H. Rose,
scholar, gentleman and friend**

In memory of Professor Anthony H. Haines,
scholar, gentleman and friend

Preface

In recognition of the enormous advances in biotechnology in recent years, we are pleased to present this Second Edition of "Biotechnology" relatively soon after the introduction of the First Edition of this multi-volume comprehensive treatise. Since this series was extremely well accepted by the scientific community, we have maintained the overall goal of creating a number of volumes, each devoted to a certain topic, which provide scientists in academia, industry, and public institutions with a well-balanced and comprehensive overview of this growing field. We have fully revised the Second Edition and expanded it from ten to twelve volumes in order to take all recent developments into account.

These twelve volumes are organized into three sections. The first four volumes consider the fundamentals of biotechnology from biological, biochemical, molecular biological, and chemical engineering perspectives. The next four volumes are devoted to products of industrial relevance. Special attention is given here to products derived from genetically engineered microorganisms and mammalian cells. The last four volumes are dedicated to the description of special topics.

The new "Biotechnology" is a reference work, a comprehensive description of the state-of-the-art, and a guide to the original literature. It is specifically directed to microbiologists, biochemists, molecular biologists, bioengineers, chemical engineers, and food and pharmaceutical chemists working in industry, at universities or at public institutions.

A carefully selected and distinguished Scientific Advisory Board stands behind the series. Its members come from key institutions representing scientific input from about twenty countries.

The volume editors and the authors of the individual chapters have been chosen for their recognized expertise and their contributions to the various fields of biotechnology. Their willingness to impart this knowledge to their colleagues forms the basis of "Biotechnology" and is gratefully acknowledged. Moreover, this work could not have been brought to fruition without the foresight and the constant and diligent support of the publisher. We are grateful to VCH for publishing "Biotechnology" with their customary excellence. Special thanks are due to Dr. Hans-Joachim Kraus and Christa Schultz, without whose constant efforts the series could not be published. Finally, the editors wish to thank the members of the Scientific Advisory Board for their encouragement, their helpful suggestions, and their constructive criticism.

H.-J. Rehm
G. Reed
A. Pühler
P. Stadler

Scientific Advisory Board

Contents

Contributors

Keshab K. Batajoo, M.S.
Department of Dairy Science
266 Animal Sciences Building
University of Wisconsin
1675 Observatory Drive
Madison, WI 53706-1284, USA
Chapter 20

Manuel Brenes Balbuena
Instituto de la Grasa y sus Derivados
Avenida Padre Garcia Tejero 4
E-41042 Sevilla, Spain
Chapter 16

Prof. Dr. Larry R. Beuchat
Center for Food Safety and
Quality Enhancement
Department of Food Science and Technology
University of Georgia
Griffin, GA 30223-1797, USA
Chapter 13

Prof. Dr. Jürgen-Michael Brümmer
Bundesforschungsanstalt für Getreide-
und Kartoffelforschung
Schützenberg 12
D-32756 Detmold, Germany
Chapter 7

Hélène Boze
Chaire de Microbiologie Industrielle et
de Génétique des Microorganismes
E.N.S.A.-I.N.R.A.
2, Place Pierre Viala
F-34060 Montpellier, Cedex 1
France
Chapter 5

Clifford Caron
Lallemand Inc.
1620 Prefontaine
Montreal PQ H1W 2N8
Canada
Chapter 8

Dr. Frederick Breidt, Jr.
Department of Food Science
322 Schaub Hall
N.C. State University
Raleigh, NC 27695-7624, USA
Chapter 17

Dr. Ramesh C. Chandan
World Class Dairy Foods Consultants
3257 Rice Creek Road
New Brighton, MN 55112, USA
Chapter 10

Prof. Dr. Athel Cornish-Bowden
Laboratoire de Chimie Bactérienne
Centre National de la Recherche Scientifique
31, Chemin Joseph Aigueir, B.P. 71
F-13402 Marseille, Cedex 20
France
and
Departamento de Biologia
Facultad de Ciencias
Universidad de Chile
Santiago, Chile
Chapter 3

Dr. Heinrich Follmann
Heinrich Frings GmbH & Co KG
Jonas-Cahn-Straße 9
D-53115 Bonn, Germany
Chapter 15

Prof. Dr. Paul S. Dimick
Department of Food Science
Pennsylvania State University
116 Borland Laboratory
University Park, PA 16802, USA
Chapter 14

Karen A. Foster
Pharmacia P.L. Biochemicals Inc.
2202 N. Bartlett Avenue
Milwaukee, WI 53202, USA
Chapter 2

Prof. Dr. Helmut H. Dittrich
Kreuzweg 19
D-65366 Geisenheim, Germany
Chapter 12

Susan Frackman
Pharmacia P.L. Biochemicals Inc.
2202 N. Bartlett Avenue
Milwaukee, WI 53202, USA
Chapter 2

Dr.-Ing. Heinrich Ebner
Piringerhofstraße 13
A-4020 Linz, Austria
Chapter 15

Prof. Dr. Pierre Galzy
Chaire de Microbiologie Industrielle et
de Génétique des Microorganismes
E.N.S.A.-I.N.R.A.
2, Place Pierre Viala
F-34060 Montpellier, France
Chapter 5

Dr. Antonio Garrido Fernández
Instituto de la Grasa y sus Derivados
Avenida Padre Garcia Tejero 4
E-41042 Sevilla, Spain
Chapter 16

Pedro García García
Instituto de la Grasa y sus Derivados
Avenida Padre Garcia Tejero 4
E-41042 Sevilla, Spain
Chapter 16

Prof. Dr. Henry P. Fleming
USDA-ARS
322 Schaub Hall
N.C. State University
Raleigh, NC 27695-7624, USA
Chapter 17

Dr. Ronald E. Hebeda
CPC International Inc.
Moffett Technical Center
6500 S. Archer Road
Summit-Argo, IL 60501-0345, USA
Chapter 19

Dr. James F. Jolly
Pharmacia P.L. Biochemicals Inc.
2202 N. Bartlett Avenue
Milwaukee, WI 53202, USA
Chapter 2

Prof. Dr. Georg-Burkhard Kresse
Boehringer Mannheim Therapeutics
Biotechnology, Dept. of Biochemistry
Nonnenwald 2
D-82372 Penzberg, Germany
Chapter 4

Dr. Kyu Hang Kyung
Department of Food Science
Kunja-dong, Sungdong-ku
Seoul 133-747, Korea
Chapter 17

Dr. Alex S. Lopez
Cocoa Research Center
CEPLAC/CEPEC/SETEA
Itabuna, Bahia 45600-000
Brazil
and
Department of Food Science
Pennsylvania State University
212 Boreland Laboratory
University Park, PA 16802, USA
Chapter 14

Prof. Dr. Guy Moulin
Chaire de Microbiologie Industrielle et de Génétique des Microorganismes
E.N.S.A.-I.N.R.A.
2, Place Pierre Viala
F-34060 Montpellier, Cedex 1
France
Chapter 5

Edwina B. Murray
United Nations University
Charles St. Sta., P.O. Box 500
Boston, MA 02114-0500, USA
Chapter 6

Dr. Hans Sejr Olsen
Manager Industrial Technology
Novo Nordisk A/S
DK-2880 Bagsvaerd
Denmark
Chapter 18

Prof. Dr. Norman F. Olson
Department of Dairy Technology
University of Wisconsin – Madison
Madison, WI 53706, USA
Chapter 9

Dr. Ingeborg Russell
John Labatt Ltd.
150 Simcoe Street
London, Ontario, N6A 4M3
Canada
Chapter 11

Prof. Dr. Nevin S. Scrimshaw
United Nations University
Charles St. Sta., P.O. Box 500
Boston, MA 02114-0500, USA
Chapter 6

Dr. Sylvia Sellmer
Heinrich Frings GmbH & Co KG
Jonas-Cahn-Straße 9
D-53115 Bonn, Germany
Chapter 15

Dr. Khem Shahani
Department of Food Science and Technology
University of Nebraska
116 H.C. Filley Hall
Lincoln, NE 68583-0919, USA
Chapter 10

Prof. Dr. Randy D. Shaver
Department of Dairy Science
266 Animal Sciences Building
1675 Observatory Drive
Madison, WI 53706-1284, USA
Chapter 20

Dr. Gottfried Spicher
Hohe Straße 13
D-32756 Detmold, Germany
Chapter 7

Prof. Dr. Gary M. Smith
Department of Food Science and Technology
University of California in Davis
Davis, CA 95616, USA
Chapter 1

Graham G. Stewart
John Labatt Ltd.
150 Simcoe Street
London, Ontario, N6A 4M3
Canada
Chapter 11

Introduction

GERALD REED
TILAK W. NAGODAWITHANA
Milwaukee, WI 53202, USA

The present volume combines four distinct, but related, sections: Enzymes, Biomass Production, Food Fermentations and Feed Fermentations. The section on *enzymes* is introduced by a general description of the properties of enzymes. This is followed by a comprehensive chapter on the production of enzymes as fine chemicals, a subject which has not previously been treated in the literature in such detail. The section also includes a challenging chapter on the function of multienzyme systems. Finally, the analytical uses of enzymes are treated in detail. Additional chapters on enzymes, on their modification by genetic methods (Vol. 5), on their use in biotransformations (Vol. 8) and on microbial enzyme inhibitors (Vol. 7) should be consulted. Indeed there is no volume of this series which does not deal extensively with the innumerable aspects of biocatalysis.

The section on *biomass* deals in one chapter with the production of fungal, bacterial and yeast biomass for use in human foods and in feed. A second chapter treats the nutritional properties of microbial biomass.

The third section of this volume treats *food fermentations* on a world-wide basis. The chapters deal with the staples of our diet, the yeast-raised baked goods, the production of yeasts, and the cheeses, yogurts and other fermented dairy products. Beer and wine may also be considered staples because of their major contribution to the diets in various countries. Other chapters deal with the production of cocoa, vinegar, olives, and fermented vegetables such as pickles, sauerkraut and the Korean kimchi. The chapters on the use of enzymes in food processing and the specific chapter on the use of enzymes for the production of syrups and sugars from starches are included in this section because the raw materials and end products are foods. A chapter on distilled beverages will be included in Vol. 10 of *Biotechnology*.

The fourth section deals in a single chapter with the *fermentation of feed stuffs*.

Biotechnology has been defined in many ways. For this volume it may be defined as an application of biological principles for the purpose of converting foodstuffs into more palatable, nutritious or stable foods. Biotechnology then, is not a new science. On the contrary it originated with indigenous food fermentations and has been practiced for millenia (paraphrased from L. R. BEUCHAT, 1995: Application of biotechnology to indigenous fermented foods. *Food Technol.* **49** (1), 97–99).

In many respects the treatment of food fermentations differs from that of primary and secondary products of microbial metabolism. Food fermentations still involve a good deal of art or craftsmanship. They are never carried out as pure culture fermentations because the starting material cannot be sterilized (flour, milk, etc.) or sterilization would be too costly (ultrafiltration of must).

Many food fermentations are characterized by the sequential action of various microorganisms, often by a succession of yeasts and lactic acid bacteria as in the production of sour dough bread, soda crackers, and some wines. It is not surprising that food fermentations show the traditional aspects of their development from prehistoric times, and they differ in their scientific and practical aspects from country to country. An attempt has been made to stress those aspects which are common features of these various fermentations.

The editors are grateful to the authors who made it possible to publish this volume which deals largely with traditional aspects of biotechnology. We also wish to thank our editorial colleagues and the staff of VCH Publ. Co., especially Dr. Achim Kraus and the Managing Editor of Biotechnology, Mrs. Christa Schultz.

Milwaukee, June 1995

Gerald Reed
Tilak W. Nagodawithana

I. Enzymes

1 The Nature of Enzymes

GARY M. SMITH

Davis, CA 95616, USA

Introduction

The original literature of enzymology is immense. Multi-volume reviews of enzymes and closely related subjects appear in various series including *The Enzymes, Advances in Enzymology, Methods in Enzymology, Advances in Protein Chemistry,* and a previous edition of *Biotechnology.* Recent editions of these series are recommended as further reading. The current state of the art in structural biochemistry including X-ray crystallography, NMR spectroscopy and molecular dynamics/molecular graphics is sufficiently advanced to support the emergence of new journals and the shift of the emphasis of existing journals to highlight protein structure (e.g., *Protein Structure, Current Opinion in Structural Biology, Protein Engineering, The Journal of Protein Chemistry*). The pace of new publications has been heightened by the techniques of modern molecular genetics, which allow rapid "protein" sequencing at the gene level, as well as overproduction of enzymes for study in the laboratory and for industrial uses. This wealth of information, together with the spectacular improvements in computer and networking technology have fostered the birth and use of databases that contain readily accessible data and analysis (DOOLITLE, 1990). Having acknowledged that it is impossible, the reviewer will attempt to capture the essence of an entire science, the structure and function of enzymes, in a brief review.

1 Nomenclature: Enzymes as Catalysts

The word "enzyme", meaning "from yeast" was reportedly coined by KÜHNE in 1887 to denote a catalytic substance derived from yeast. Not a precise definition, to be sure. After SUMNER crystallized urease in 1926 and showed that the material that formed the crystals had catalytic activity, the inference was drawn that all enzymes are proteins. The simplest definition may therefore be that enzymes are proteins that have catalytic activity. Despite the recent well-deserved commotion over the existence of "ribozymes", segments of RNA that participate in excision or rearrangement of mRNA (CECH and BASS, 1986), ribozymes and all other non-protein catalysts will be excluded from this discussion.

In addition to its proteinaceous nature, the essence of an enzyme is its catalytic activity. Activity is characterized by the enzyme's substrates and products, the relationship among which, in turn, defines the nature of the reaction catalyzed by the enzyme. It is eminently reasonable, then, to classify enzymes according to the nature of the reaction they catalyze (e.g., oxidation/reduction, hydrolysis, etc.) and sub-classify them according to the exact identity of their substrates and products. A nomenclature scheme employing this framework was set forth in 1961 by the Enzyme Commission, an *ad hoc* committee of the International Union of Biochemistry (IUB, 1964). The scheme, updated and reissued periodically (approximately deciennially) by a standing committee, is currently the most concise and extensive classification and nomenclature system in use. The 1992 compilation is now current (IUB, 1992). The system has at least two uses: to give structure to comparative enzymology in much the same way as microbial taxonomy provides a framework for comparison of species, and to facilitate effective communication among scientists.

The Enzyme Commission system consists of a numerical classification hierarchy of the form "EC i.j.k.l", in which "i" represents the class of reaction catalyzed (see classes below), "j" denotes the sub-class, "k" denotes the sub-subclass, and "l" is usually the serial number of the enzyme within its sub-subclass. The criteria used to assign "j" and "k" depend on the class and represent details useful to distinguish one activity from another. The Enzyme Commission's report gives a list of guidelines to aid in assigning an enzyme to its proper category. In addition, a systematic name with a logical form (defined for each class of reaction) is given together with a rational common name. All enzymes are placed into one of the following classes, which are discussed more fully below:

1. Oxidoreductases
2. Transferases
3. Hydrolases
4. Lyases
5. Isomerases
6. Ligases

1. *Oxidoreductases* catalyze oxidation–reduction reactions. Their systematic names have the form "donor:acceptor oxidoreductase", where the donor is the molecule becoming oxidized (donating a hydrogen or electron). Their recommended common names have the form "donor dehydrogenase", unless O_2 is the acceptor, in which case "donor oxidase" is permitted. The subclass describes the chemical group on the donor that actually becomes oxidized (e.g., an alcohol, keto- or aldo-group). Sub-subclasses generally, but not always, distinguish among acceptors (e.g., NAD(P)H, cytochromes, O_2, etc.).

2. *Transferases* catalyze group transfers from one molecule to another. Systematic names logically have the form "donor:acceptor grouptransferase". Recommended common names are "donor grouptransferase" or "acceptor grouptransferase", but "acceptor-kinase" (e.g., hexokinase) is used for many phosphotransferases. The subclasses distinguish in a general way among the various groups that are transferred (e.g., one-carbon transfers, acyl transfers, glycosyl transfers) and the sub-subclasses employ greater detail in distinguishing among the groups transferred. It is noteworthy that transamination reactions between an amine and a ketone are classified as group transfer reactions even though the ketone becomes reduced to an amine and the amine becomes oxidized to a ketone.

3. *Hydrolases* catalyze hydrolytic cleavage of C—C, C—N, C—O or O—P bonds. They are essentially group transfer reactions but the acceptor is always water. For this reason, and probably because of the ubiquity and importance of hydrolases, they are awarded their own class. Since the reactions are comparatively simple (two substrates, one of which is always water), the systematic names are also simple: "substratehydrolase". Common names may simply be "substratease". However, since a substrate may have more than one hydrolytically labile bond, it is useful to include the kind of group being transferred to water (e.g., methylesterase, *O*-glycosidases). The subclass number reflects the need to specify the bond being hydrolyzed, and the sub-subclass further defines the nature of the substrate. In the case of proteases (peptidyl-peptide hydrolases), the sub-subclass reflects properties of the enzyme itself (e.g., metalloproteases, serine proteases, etc.), rather than the substrate.

4. *Lyases* catalyze elimination reactions resulting in the cleavage of C—C, C—O, C—N or a few other bonds, or the addition that constitutes the reverse of these reactions. Examples from this category include decarboxylases, aldolases and dehydratases. The systematic names are written as "substrate group-lyase", in which the hyphen is not optional. If the reverse (addition) reaction is more important than the elimination reaction, the name "product synthase" may be used. Subclasses contain enzymes that break different bonds (C—C, C—N, etc.), and sub-subclasses distinguish among enzymes on the basis of the identity of the group eliminated.

5. *Isomerases* catalyze structural rearrangements. Their recommended names correspond to the kind of isomerizations carried out by members of the different subclasses: racemases and epimerases, *cis-trans*-isomerases, tautomerases, mutases and cyclo-isomerases. The sub-subclasses depend upon the nature of the substrate.

6. *Ligases* catalyze bond formation coupled with the hydrolysis of a high-energy phosphate bond. The systematic names are written "A:B ligase", and may specify "ADP-forming", etc., depending on the coupled energy source. Common names often include the term synthetase, which the Commission discourages because of confusion with the name synthase (which does not involve ATP hydrolysis). Subclasses are created on the basis of the kind of bond formed (C—C, C—O, etc.); sub-subclasses exist only for C—N ligases.

All enzymes possess both a systematic name and a number. But, like microbial taxonomy, the Enzyme Commission system is fundamentally a classification system, rather than a naming system. The categories, like taxa, are not unique names, but categories that contain groups of elements which can be further distinguished from one another using other criteria. The inability of the EC system to provide a unique name for each different enzyme is that it defines an enzyme solely in terms of its activity, rather than by its chemical identity, i.e., its structure. Genuinely different proteins may catalyze the same reaction, hence, they are *classified* as the "same enzyme". So, some information in addition to the EC number, generally the source of the enzyme, is required in order to identify an enzyme unambiguously. For instance, yeast aldolase and rabbit muscle aldolase have quite different properties, structures and mechanisms, although both are EC 4.1.2.13, D-fructose-1,6-bisphosphate D-glyceraldehyde-3-phosphate lyase. Even within the same organism there may exist isozymes, different proteins (i.e., proteins arising from different genes) that have the same classification, such as heart type and muscle type lactate dehydrogenase (EC 1.1.1.28), which are truly different proteins, separable by electrophoresis. Isoforms of enzymes, which arise from the same gene (or identical copies) but have different kinds or extent of post-translational modification, also exist. Isoforms may be separable and may have different properties and stabilities, although they arise from the same DNA sequence.

Another problem encountered during the designing of the EC system is that, in order to classify an enzyme, the identity of its "true" substrate must be known or assumed. Many enzymes catalyze more than one reaction and accept more than one substrate. Sometimes the substrate for which the enzyme has the highest V_{max} (see below) *in vitro*, is not present in sufficient concentration *in vivo* to be the physiological substrate. Consequently the assignments to categories may be somewhat arbitrary and, on occasion, enzymes assigned different EC numbers are later found to be the same protein.

In short, although the EC nomenclature system is useful for identifying enzymes, it is not entirely complete. To determine whether two enzymes are really the same requires at least some degree of structure determination.

2 Enzymes as Proteins

2.1 Structure

Proteins have several "levels" of structure. They are composed of amino acids, and hence, the amino acid composition might be termed its zero-order structure. The structures of the amino acids commonly found in proteins are shown in Tab. 1 along with their standard abbreviations. The amino acid composition is partially responsible for a protein's net charge, solubility and nutritional value. The amino acids are strung together via amide bonds (peptide bonds, see Fig. 1), and the order of amino acids, the protein's "sequence", is termed its "primary structure". The primary structure is at least indirectly responsible for the higher levels of structure, and therefore, for all properties of the protein, including enzymatic activity.

2.1.1 The Effect of Primary Structure on Three-Dimensional Structure

Although proteins are generally linear polymers, the active ("native") form of an enzyme is folded into a globular structure. The primary structure of a protein places several constraints on how it can fold to produce its three-dimensional structure. The backbone of a protein consists of three kinds of bonds: peptide C—N, N—C_α, and C_α—C (see Fig. 1). First, the peptide bond itself has two conformations that are lower in energy than all others. The carbonyl carbon is sp^2 hybridized to form a double bond with the carbonyl oxygen atom. The amide nitrogen, bonded to the α-carbon of the next amino acid in the sequence (usually drawn as sp^3 hybridized) contains some sp^2 character so that it can participate in a partial π-bond with the carbonyl carbon to gain stabilization through electron delocalization (Fig. 1).

The result of this "amide resonance" is that the system is more stable if the dihedral angle between the C_α—carbonyl C bond and the N—C_α bond is either 0° or 180°, independent

Tab. 1. Side Chains of the Common Amino Acids

Amino Acid		Abbrev.	Side Chain Structure
Small, neutral	Glycine	Gly, G	$-H$
	Alanine	Ala, A	$-CH_3$
Polar	Serine	Ser, S	$-CH_2OH$
	Threonine	Thr, T	$-CH(OH)CH_3$
	Cysteine	Cys, C	$-CH_2CH_2SH$
	Asparagine	Asn, N	$-CH_2CONH_2$
	Glutamine	Gln, Q	$-CH_2CH_2CONH_2$
	Methionine	Met, M	$-CH_2CH_2SCH_3$
	Histidine	His, H	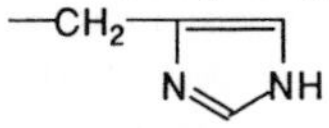 $-CH_2-$ N NH
Anionic	Aspartate	Asp, D	$-CH_2COO-$
	Glutamate	Glu, E	$-CH_2CH_2COO-$
Cationic	Lysine	Lys, K	$-CH_2CH_2CH_2CH_2NH_3^+$
	Arginine	Arg, R	$-CH_2NHC(NH_2)_2^+$
Hydrophobic	Leucine	Leu, L	$-CH_2CH(CH_3)_2$
	Isoleucine	Ile, I	$-CH(CH_3)CH_2CH_3$
	Valine	Val, V	$-CH(CH_3)_2$
Aromatic	Phenylalanine	Phe, F	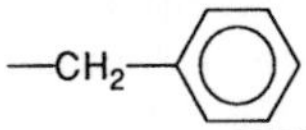 $-CH_2-$
	Tyrosine	Tyr, Y	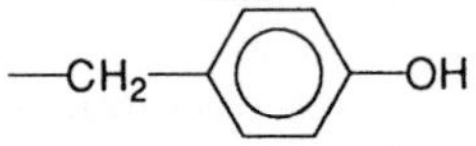 $-CH_2-$ OH
	Tryptophan	Trp, W	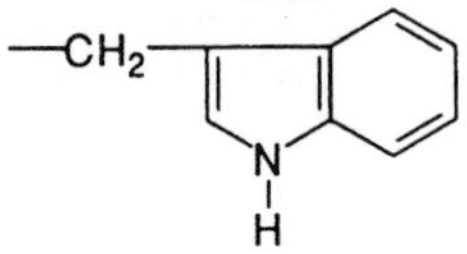 $-CH_2-$ N H

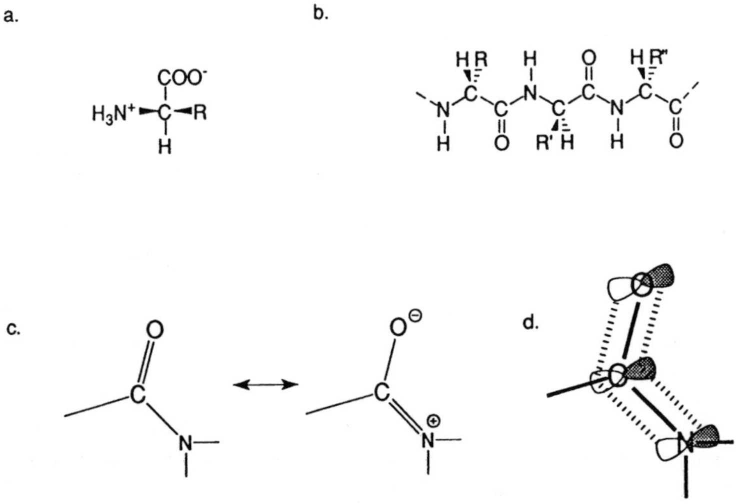

Fig. 1. Amino acids and the peptide bond. (a) The general structure of an α-amino carboxylic acid like those found in proteins. The structures of the R-groups are found in Tab. 1. (b) A stretch of polypeptide showing peptide bonds between the C=O and N. (c) "Amide resonance" which leads to partial double-bond character in the peptide bond. (d) The p-orbitals involved in the π-bonds shown in (c).

of the sequence of the protein. Since this energy barrier is not large, the two forms interconvert readily at room temperature, and the angles can be deformed by rotation if such distortion allows stabilization to be gained elsewhere in the molecule. Nevertheless, amide resonance provides a constraint on protein structure.

The remaining two types of bonds in the backbone are pure single bonds and are free to rotate, subject to steric interactions. These interactions can be appreciable and depend partly on the bulk of the R-group of the amino acids involved in the bond. The conformation about the N$-$C$_\alpha$ bond is termed the ϕ torsion angle and is defined as the dihedral angle between the C$_\alpha$$-$C bond and the peptide C$-$N bond of the next (toward the N-terminus) amino acid (see Fig. 2). The conformation about the C$_\alpha$$-$C bond is termed the ψ torsion angle which is defined as the dihedral angle between the N$-$C$_\alpha$ bond and the next (toward the C-terminus) peptide C$-$N bond. Certain combinations of ϕ and ψ would place adjacent carbonyl oxygens in unpleasant proximity, i.e., within each others' van der Waals radii, and are thus not allowed. The beta-carbon of the R-group provides an additional steric barrier: rotation about ψ can bring C$_\beta$ into van der Waals contact with the carbonyl oxygen, and rotation about ϕ can bring it into van der Waals contact with the N$-$H. The magnitude of this unfavorable interaction depends on the size and character of the R-group and is thus responsive to the sequence of the peptide.

Ramachandran et al. (1963) demonstrated this effect graphically by plotting the energy of these steric interactions against all possible ϕ and ψ "coordinates". This rudimentary contour plot, called a Ramachandran plot (Fig. 3), vividly shows that only certain combinations of ϕ and ψ produce energetically favorable conformations. (The details of the energetic considerations and methodology were described thoroughly by Ramachandran and Sasisekharan (1968). (It should be mentioned that the convention for what constitutes a torsion angle of zero has changed at least twice since the inception of the Ramachandran plot.) The steric interactions are least unfavorable for ϕ angles between about $-50°$ and $-160°$. There are two ranges of ψ angles that lead to energetic minima, 90° to 180°, and $-45°$ to $-80°$, so there are two regions of conformational stability in

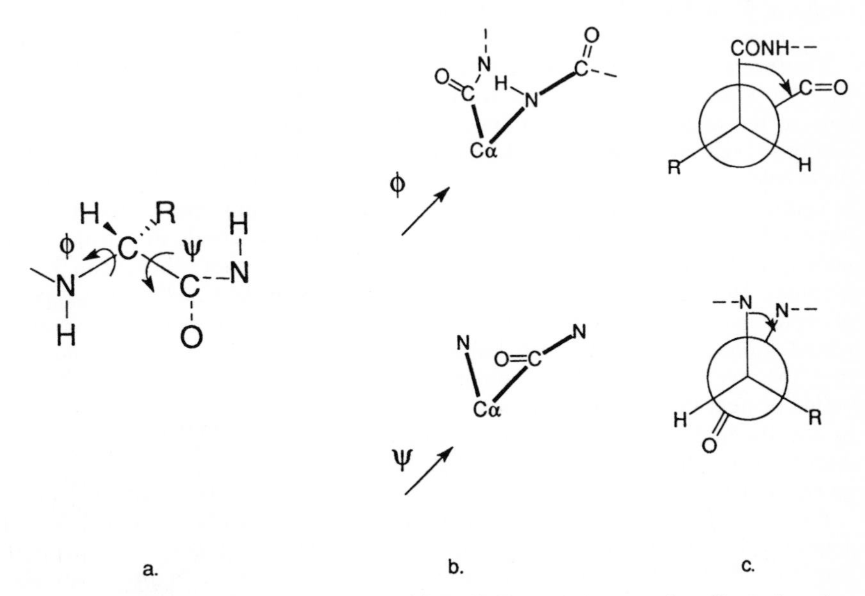

Fig. 2. The torsion angles defining protein conformation. (a) Definition of the ϕ and ψ dihedral angles. (b) and (c) Two representations of how the torsion angles are measured.

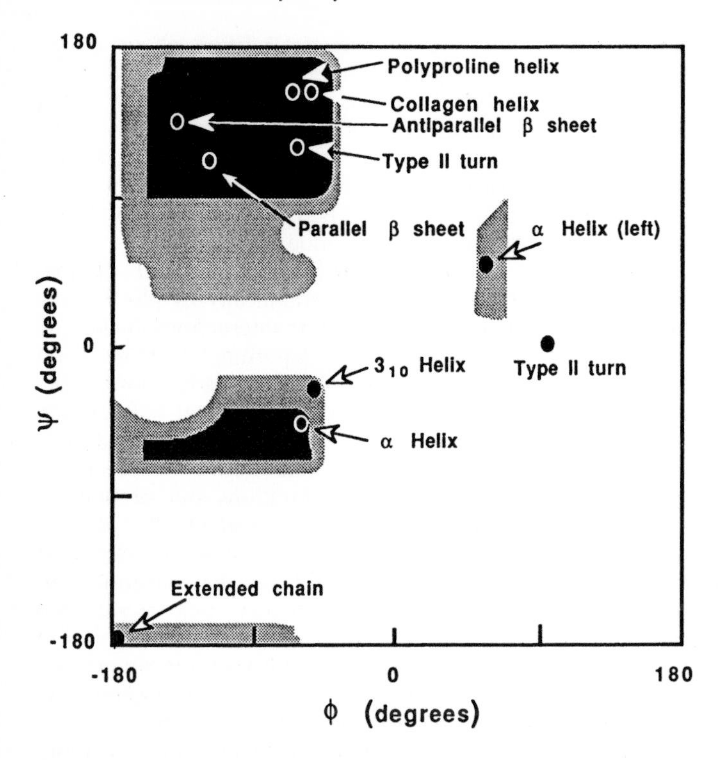

Fig. 3. A Ramachandran diagram. Torsion angles of various kinds of secondary structure are shown. The figure was sketched from those in RAMACHANDRAN and SASISEKHARAN (1968) and DICKERSON and GEIS (1969), but with the newer convention for measurement of ϕ and ψ.

the ϕ-ψ coordinate space (Fig. 3). The character of the side chain can determine which range of ψ angles yields the conformation of lower energy.

Conformations in which there is only slight van der Waals contact are not truly forbidden, so the regions in which allowed conformations exist are fairly broad. Of course, these rules may be completely violated if other stabilizing factors elsewhere in the protein outweigh the destabilization that arises from unfavorable contacts at a particular pair of ϕ-ψ values. Nonetheless, plots of the frequency of occurrence of the torsion angles in proteins whose atomic coordinates are known show that the majority of ϕ-ψ combinations fall within or near the energetic minima predicted by the Ramachandran plot. A more detailed discussion of these considerations has been given by RICHARDSON (1981).

Proline is a bit of an exception to these general principles because its α-amino nitrogen is bonded to its δ-carbon to form a ring. There are thus additional constraints arising from the bulk and relative rigidity of the 5-membered ring. Peptides having a peptide bond to the nitrogen atom of proline exist in one of two conformations, termed *cis*- and *trans*-, although they actually represent rotations about the peptide bond, which has only partial double bond character. Even for small peptides, these two conformational isomers interconvert slowly; both ^{13}C- and ^{15}N-NMR spectra show distinct peaks for the *cis*- and *trans*-conformers. In proteins, most peptidyl proline bonds are in the *trans*-conformation. In addition, there can be no free rotation about the C_α—N bond because of the ring, so the ϕ angle is fixed at about $-50°$.

2.1.2 Secondary Structure

If strands of amino acids are arranged to have (nearly) identical ϕ angles and (nearly) identical ψ angles corresponding to the energy minima mapped by RAMACHANDRAN, regular structures are produced. Other interactions (e.g., hydrogen bonding, described below), render certain of these structures more stable than most of the others, so they occur frequently in many proteins. One of these structures, with $\phi = \sim -57°$ and $\psi = \sim -47°$, is the α-helix (right-handed) as predicted by PAULING and COREY (1951). The two other most common regular structures, with $\phi = \sim -119°$, $\psi = \sim +113°$, or with $\phi = \sim -139°$, $\psi = +135°$, are extended strands called β-structure. There are many variations on these schemes: more extended or compact helices are reasonably stable (e.g., the ω and 3_{10} helices), and the similar polyproline and collagen helices, which have torsion angles more like β-structure than α (Fig. 4). Proline, with its constrained ϕ angle, cannot participate in an α-helix, and thus often occurs at the end of an α-helix. Counter-clockwise (left-handed) helices ($\phi = \sim 60°$, $\psi = 45°$) such as the α_L and ω_L, are also reasonably stable, although they appear in a lesser valley of the Ramachandran plot completely separate from the α- and β-structures.

Not only are the structures described above relatively stable from steric considerations, there are other stabilizing interactions that strengthen them further. Hydrogen bonds, weak interactions between an electronegative atom such as oxygen or nitrogen and a proton bonded to another electronegative atom, have bond energies of 1–5 kcal/mol (4.2–21 kJ/mol). (The hydrogen bond can be thought of as a dipole-dipole interaction in which a proton forms the positive end of one of the dipoles by virtue of having its electrons withdrawn by the electronegative atom to which it is bonded (JENCKS, 1987). This interaction is unique to hydrogen because it is the only element that uses all of its electrons in a single bond, rendering its entire electron cloud susceptible to partial withdrawal by its partner.) Whereas a single hydrogen bond can provide only modest stabilization to a structure, the stabilization afforded by many such bonds operating in concert can be substantial. The α-helix has a pitch of 3.6 residues per turn, which places the carbonyl oxygen of each residue in proximity to the NH group of the fourth residue along the chain. Thus, every nth residue pairs with its $n+4$th neighbor to lock the structure into the sterically allowed helical form.

A β-strand has no such intra-chain hydrogen bonding, but if two β-strands occur side-by-side, they can form inter-chain hydrogen bonds that provide mutual stabilization. The side-by-side interaction can occur in two forms, with the two strands either parallel ($\phi = \sim -119°$ and $\psi = \sim +113°$) or antiparallel ($\phi = -139°$ and $\psi = +135°$) (see Fig. 4). Furthermore, only half of the groups involved in intra-chain hydrogen bonds are utilized in the interaction between a pair of chains, leaving the remaining groups to hydrogen bond with additional chains to form a sheet. Because of the zig-zag-arrangement of the extended $-C_\alpha-CO-N-$ atoms in the strands, the sheets appear pleated, hence, the name β-pleated sheet. The sheets also twist slightly. The sheets can be formed of parallel strands or of antiparallel strands; mixed forms occur, but are more rare. Antiparallel sheets may be formed from adjacent runs of the sequence connected by a "hairpin loop', usually of 1–5 residues. Both parallel and antiparallel sheets can also be formed from portions of the polypeptide that are not adjacent in the sequence. The occurrence of recognizable, well-defined structures within a protein such as α-helix, and β-sheet are termed secondary structure.

A further feature of secondary structure is the loop or turn (RICHARDSON, 1981; ROSE et al., 1985). Turns are regions in which the peptide backbone reverses its overall direction (ROSE et al., 1985). Loops are turns, such as those connecting adjacent runs of polypeptide that form antiparallel β-sheet, in which the ends of the turns are somehow fused. The distinction between the terms, if one exists, is vague.

2.1.3 Tertiary Structure and Structural Motifs

The term tertiary structure of a protein refers to the overall folding pattern, since proteins are not all α-helix or β-sheet, but helices and sheets, connected by loops and turns and regions of less well-defined structure. Detailed studies by numerous investigators have shown that certain combinations of helices or sheet together with turns or loops occur over and over in many proteins. These often-oc-

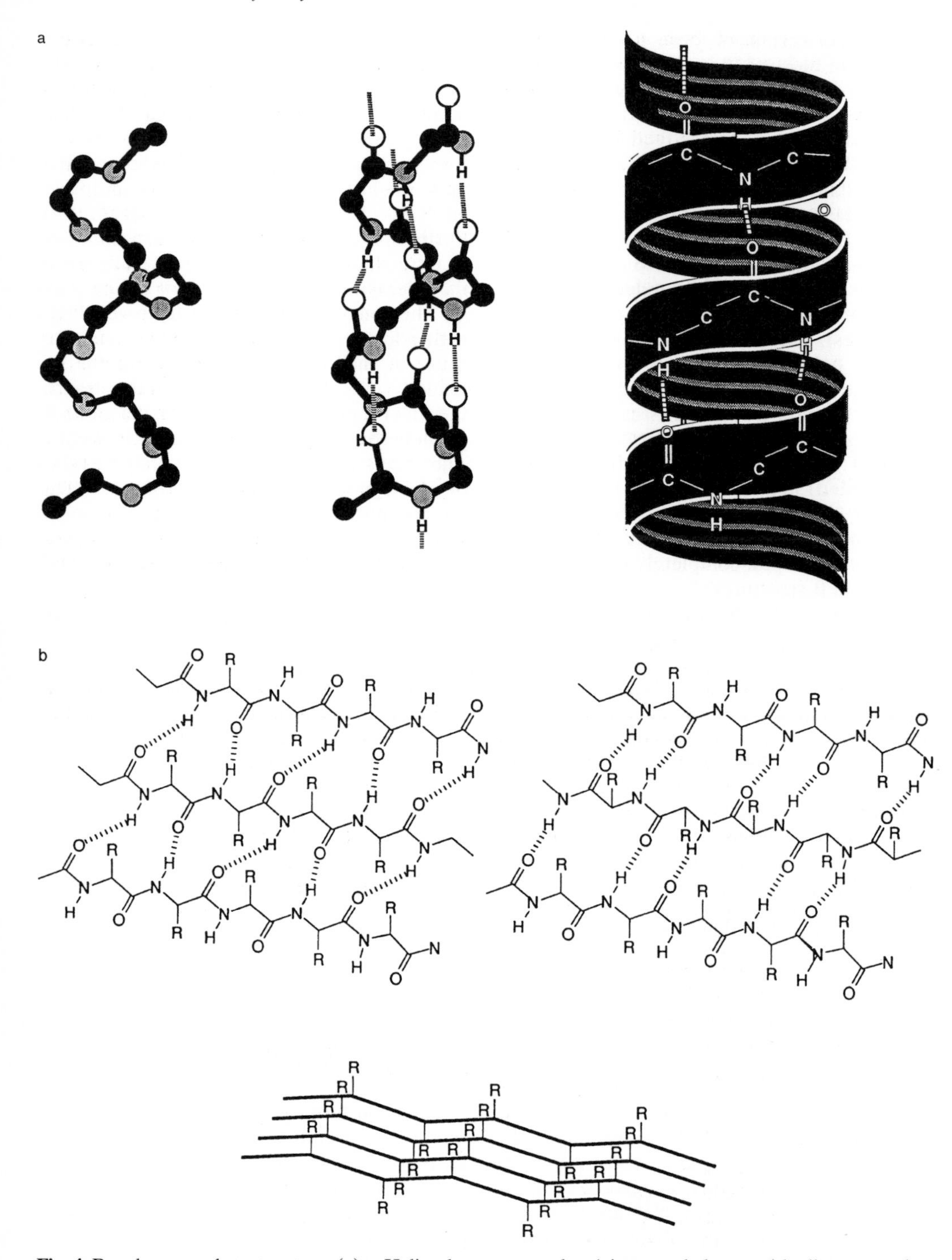

Fig. 4. Regular secondary structure. (a) α-Helix, shown as a carbon/nitrogen skeleton, with all atoms and as a ribbon to dramatize the helicity. (b) Hydrogen bonding patterns in parallel (left) and antiparallel (right) β-sheet and a schematic representation of the pleated appearence of the sheet.

curring structures have been termed motifs, or super-secondary structure. Some motifs indicate a related function in different proteins, such as the calcium-binding "E-F hand" motif (so named because it is formed from helices labeled "E" and "F" in parvalbumin, the protein in which it was first observed). An early example of a structural motif was the "nucleotide fold" proposed by EVENTOFF and ROSSMAN (1975). Other motifs have no apparent function other than holding the protein in a particular conformation and may arise simply because they are stable or by evolutionary processes. Indeed, some E-F hands do not bind calcium. CHOTHIA and FINKELSTEIN (1990) have contributed greatly to the classification of folding patterns containing structural motifs. A relatively complete and colorful discussion of motifs is given in the text by BRANDÉN and TOOZE (1991). Some motifs and examples of proteins in which they occur are given in Tab. 2.

Another term describing tertiary structure, the structural "domain", has arisen since the mid 1970s. It is used differently in different contexts, but it usually refers to a unit of structure that is separate in some way from other regions of the protein. One of the first examples was the Bence–Jones proteins. These are globular dimers held together by a disulfide bridge that are found in the urine of patients with multiple myeloma. They were found to be fragments of immunoglobulins (i.e., two light chains), and their existence as independent globular structures helped EDELMAN (1970) formulate his conception of IgG structure. A more current usage of the term "domain" is to denote structures that fold independently of other regions of the same polypeptide chain, or of a larger aggregate. Domains are especially evident when different regions of the protein have decidedly different structures, such as transmembrane helices or helical coil regions attached

Tab. 2. Examples of Folding Motifs Found in Enzymes and Other Proteins

Name	Function/Description	Example of a Protein in Which it Occurs
β-α-β	Common structural element; two β-strands forming parallel β-sheet connected by an α-helix	Triosephosphate isomerase
β-Hairpin	Common structural element; two β-strands involved in an antiparallel sheet connected by a 2–~5 residue loop	Many proteins, e.g., bovine pancreatic trypsin inhibitor
Greek key	Common structural element; four β-strands in a contiguous sequence associated in antiparallel β-sheet: strand 1 pairs with strand 2, which pairs with strand 3, strand 4 pairs with strand 1.	Staphylococcal nuclease
Jelly roll	A "barrel" of β-sheets formed from two Greek key motifs	γ-Crystallin
Helix-turn-helix	DNA binding	DNA-binding proteins
E-F hand	A helix-turn-helix motif sometimes involved in Ca^{2+} binding	Parvalbumin
Leucine zipper	Reversible dimer formation; two parallel α-helices with leucine residues at contact points between the two helices (every 7th residue in each helix)	Eukaryotic transcription factors
Zinc finger	DNA binding; four side chains (2 Cys, 2 His; 3 Cys, 1 His; or 4 Cys) bind one Zn^{2+} ion to form a loop of 4–14 residues, which constitutes the "finger"	DNA binding proteins (transcription factors)

to truly globular domains. Like small proteins (i.e., single-domain proteins), domains may be composed of secondary structure and identifiable motifs and therefore represent a higher level of structure than motifs. LEVITT and CHOTHIA (1976) have classified domains of globular proteins according to their structures into four categories that depend on the predominance of α- or β-structure or a mixture of the two.

It is not surprising that domains are not only separate structural elements of proteins but may have separate functional roles as well. Furthermore, domains having a similar function in different proteins may have considerable structural similarity (although perhaps low sequence homology). Thus, domains are often named and compared according to their function (e.g., the flavin mononucleotide-binding domain (XIA and MATHEWS, 1990). Conversely, proteins are also commonly classified, by crystallographers, at least, according to the structures of the domain that comprises the recognizable portion of their structures or according to the motifs that comprise the domain. Examples are β-barrel proteins (composed of antiparallel β-sheet), α/β-barrel proteins (composed of β-α-β motifs) (BRANDÉN and TOOZE, 1991; CHOTHIA, 1984).

Because the folding together of helices, β-sheet and other structures may occasionally bring charged side chains to the interior of the protein, another interaction is important. Since there is insufficient water on the interior of a protein to solvate charged groups, the occurrence of such side chains within a protein is highly unfavorable unless side chains of opposite charge are able to pair with one another. This interaction is called a salt bridge and has a strength of about 10–20 kcal/mol (42–84 kJ/mol). Salt bridges may also occur on the surface of a protein or between subunits.

One final consideration in the folding pattern of globular proteins is that they may contain non-protein species, including solvent, cofactors or metal ions. Water molecules are contained within the structure of essentially all soluble proteins large enough to surround them, and such water molecules are clearly visible in electron density maps. Even a miniprotein, such as the bovine pancreatic trypsin inhibitor, contains water molecules bound sufficiently tightly that they might be considered part of the protein's structure (BRUNNE et al., 1993; OTTING et al., 1991; WÜTHRICH et al., 1992). Structural water molecules occur in polar regions within the protein and are probably hydrogen-bonded to specific groups to provide stabilization of the structure in their vicinity. Although clefts and active sites are often represented as voids, they are generally occupied by some component, usually water.

2.1.4 The Driving Force

One curious point that arises from a Ramachandran plot is that the conformation defined by $\phi = \psi = -180°$, corresponding essentially to an extended chain, appears to be reasonably stable (see Fig. 3). A naive question would be whether the stabilization afforded by hydrogen bonding and salt bridging and the decrease in steric interaction sufficient to cause a protein to fold into a globular state if the extended form is reasonably stable. A great deal of information about the energetics of protein structures has been provided by calorimetric studies of protein denaturation. The reader is referred to numerous detailed reviews for more information (BALDWIN, 1986; BECKTEL and SCHELLMAN, 1987; PRIVALOV, 1979, 1982; PRIVALOV and GILL, 1988; PRIVALOV and POTHKIN, 1986). One of the somewhat surprising results of denaturation studies is that folded (native) forms of proteins are not adamantly more stable than their denatured or unfolded forms. The drive toward the folded structure is therefore not strong and is usually attributed to an indirect source: the hydrophobic effect.

2.1.4.1 The Hydrophobic Effect

A thoughtful, qualitative description of hydrophobic "forces" has been given by JENCKS (1987), and is summarized briefly here. A more theoretical description together with experimental support has been given by PRIVALOV and GILL (1988), and a detailed treatise has been provided by TANFORD (1980). Liquid water is highly self-cohesive because of the ability of water molecules to form extensive, though irregular, hydrogen bonds with each other. Placing a hydrophobic molecule in aqueous medium causes some of these hydrogen bonds to be broken to make room for the foreign molecule, as though a hole were created in the water. Stabilizing bonds are lost, and the enthalpy of the system increases. If the solute is not hydrophobic, some of the energy may be regained by hydrogen bonding or solvation interactions between water and charged, polar or polarizable groups of the molecule. For hydrophobic molecules, no such interactions are available, and water molecules at the interface maximize their hydrogen bonding to minimize enthalpy by orienting their dipoles toward other water molecules and away from the hydrophobic molecule. Entropy is therefore decreased because of the loss of randomness in orientation. Hydrophobic molecules are thus surrounded by a "cage" of water molecules, and the increase in energy has both enthalpic and entropic components. The total energy is smallest if the hydrophobic molecule presents the smallest possible surface area to the water. Thus, molecules of oil coalesce into droplets, detergents and phospholipids form micelles, and polypeptide chains fold into globular structures. The folding of polypeptides should favor hydrophobic residues on the inside of the folded molecule, and hydrophilic groups (i.e., side chains with charged or polar groups) on the outside, although it is clear that some nonpolar groups are in contact with solvent. The exterior polar groups allow establishment of enthalpically favorable interactions with water that do not force an increase in order of the solvent (i.e., cage formation). There is no "hydrophobic bond" but the interactions among hydrophobic groups (van der Waals–London forces, reviewed by BURLEY and PETSKO, 1988) within the folded protein are more favorable than the order they would otherwise impose on water molecules surrounding them. The same considerations hold for the folding of hydrophobic molecules, except that they possess insufficient numbers of polar groups to surround them, and seek interactions with other hydrophobic molecules, (e.g., imbed themselves in a membrane) in order to escape from the aqueous phase.

2.1.4.2 Additional Stabilization: The Disulfide Bond

Once proteins have folded or as they fold, oxidizable groups, the —SH groups of cysteine residues, may come into contact. Two —SH groups oxidize readily in the presence of oxygen to form the only covalent inter-residue crosslink commonly found in proteins, the disulfide bond or bridge. Indeed, enzymes with free sulfhydryl groups are relatively rare unless the —SH groups are protected from interaction with other —SH groups (e.g., buried in active sites), or they exist in regions of the cell from which O_2 is excluded or scavenged. Otherwise, inappropriate intermolecular crosslinking would occur. Hence, extracellular enzymes contain free sulfhydryl groups much less frequently than intracellular enzymes. As a covalent bond (bond energy 30–100 kcal/mol or 126–420 kJ/mol) the disulfide bridge confers significant stability to the folded structure. In fact, the stability of a protein, measured by temperature of denaturation, can generally be related directly to the number of disulfide bonds it possesses (MATSUMURA et al., 1989), though there are certainly examples of extraordinarily stable proteins that lack disulfide bonds. It may not be obvious how the formation of a single covalent bond, strong though it may be compared to H-bonding, etc., reinforces the structure of an entire protein. A simple explanation is that a major drive toward unfolding lies in the increase in conformational entropy of the unfolded state compared to the native conformation. If the conformations available to the unfolded state are constrained by a covalent bond fusing distant parts of the polypeptide, the gain in conformational entropy would be decreased significantly; multiple linkages would provide significant additional stabilization.

2.1.5 Multisubunited Enzymes

It has been estimated that enzymes that contain more than 30% nonpolar residues cannot possibly fold in such a way as to cover themselves in their hydrophilic residues (VANHOLDE, 1966). The same energetic considerations that drive proteins to fold, therefore drive molecules with surface hydrophobic residues to associate with other hydrophobic molecules. The options available to these proteins are to bind lipophilic molecules such as lipids to form lipoproteins, to become sunken to some degree in a phospholipid bilayer (or coat the hydrophobic side of a monolayer), or to associate with other hydrophobic proteins. The multipolypeptide-chain complexes may consist of subunits (protomers) that are identical or non-identical or of various numbers of polypeptides (e.g., for three types of polypeptide chain of stoichiometry 2:2:4, $a_2b_2g_4$). The pyruvate dehydrogenase complex, for instance, consists of fifty-six subunits in all with the stoichiometry $a_8b_{24}g_{24}$ (REED and COX, 1966). These subunits catalyze separate but sequential reactions and may also be thought of as a multienzyme complex or particle. There are also examples of disulfide bridges between subunits of multisubunited proteins.

2.1.6 Modulating the Hydrophobic Effect: Protein Solubility, Stability and Other Solutes

It was shown by HOFMEISTER in 1888 that other species in the solvent affect the solubility of proteins. A ten-year-old review of the Hofmeister effect (COLLINS and WASHBAUGH, 1985) contained a thousand references, more than one hundred of which were themselves reviews, so a detailed treatment cannot be included here. Risking profound oversimplification as well as overgeneralization, suffice it to say that some solutes act as kosmotropes (producing order) and others act as chaotropes (producing disorder). The effect is usually interpreted in terms of the effect of the solute on the structure of liquid water, since it is the entropy of the solvent that appears to be affected by the presence of a protein in solution. The idea is that kosmotropes increase the structure of water so that the adverse effect caused by the presence of protein in the water would be amplified. Chaotropes have the opposite effect; they decrease the structure of water so that it plays a less important role. Kosmotropes therefore tend to force proteins to expose the least possible disruptive surface area to the solvent. Proteins therefore fold into a globular state, or in more extreme cases, associate with one another and precipitate. Chaotropes allow proteins to unfold or dissociate. Kosmotropes build the structure of both water and of proteins; chaotropes destroy both water and protein structures. Sodium chloride appears to be essentially neutral and is neither a chaotrope nor a kosmotrope.

ARAKAWA and TIMASHEFF (1982) have proposed a variation on this description with applications to solubility (ARAKAWA and TIMASHEFF, 1985b) and to stability of enzymes. The premise put forth by TIMASHEFF is that components in the solvent that bind to the surface of proteins cause destabilization (and solubilization), while components that are specifically excluded from the hydration sphere of proteins (i.e., produce "preferential hydration") confer stability (TIMASHEFF, 1992, 1993). Although the physical mechanism of preferential exclusion seems obscure, the theory is supported by numerous studies using densimetry to determine the partial specific volume of proteins in the presence of salts, amino acids, etc., and from temperatures of denaturation. There are exceptions to this theory, such as molecules (e.g., substrates and certain ions) that bind specifically to sites on the native (folded) protein. In this case, the protein could be stabilized thermodynamically, essentially by creation of an alternate enzyme state in equilibrium with the folded protein. The binding energy would contribute to the stability of the folded protein essentially by "pulling the (folding) reaction through".

A major piece of information with which all theories of protein stability must deal is the decrease in stability afforded by such solutes as urea and guanidine hydrochloride. Indeed, hundreds of papers have been published that make use of these reagents to denature proteins. Such experiments have yielded a great deal of information about the energetics of protein unfolding. Some experimental aspects have been explored by PACE (1986).

In any case, it can be said that small solutes can alter the stability of proteins. These small solutes can be employed in the laboratory to stabilize proteins during purification or to aid in purification (i.e., as "salting out" agents). Solutes called "osmolytes" or "compatible solutes" may also be enlisted by nature to stabilize proteins during periods of environmental stress (YANCEY et al., 1982).

2.2 Protein Folding

Proteins are, of course, synthesized one residue at a time on the ribosome. It is a significant question to ask how the protein assumes its active three-dimensional conformation. It should be clear from the foregoing discussion that there is a driving force for polypeptides to fold, a few steric interactions to select particular conformations, and a few relatively weak polar interactions that stabilize the structure. The question of how they fold – the order of the folding processes, the occurrence of intermediates and the kinetics of these processes – has been under active investigation for some time. It is not the aim of this chapter to review this field in depth, but some considerations are presented below. A concise review is found in FISCHER and SCHMIDT (1990).

2.2.1 Folding of Cytoplasmic Proteins

In bacteria, which have no internal membranes, the folding of an enzyme has always been thought to occur spontaneously, during and after synthesis on the ribosome. In bacteria, there is relatively little posttranslational processing or modification, except that the proteolytic removal of formyl methionine, coded for at initiation, plus some other N-terminal amino acids is common. This would seem to be the simplest case of protein folding, since all the folding information is contained in the sequence, and the medium, which provides most of the thermodynamic drive, is uniform. There is, however, evidence for the involvement of other molecules, such as the chaperonins, in folding of cytoplasmic proteins (see below).

An authoritative review by ANFINSEN and SCHERAGA (1975) detailed the relevant knowledge about protein folding in 1975. At that time, "folding" referred to the nature and energetics of the three-dimensional structure, as there was relatively little knowledge about the mechanism of the process. The folding mechanism was envisioned as beginning with one or several nucleation steps that form areas of local secondary structure, folding of these sites into an approximately correct structure, and minor modification ("energy minimization") of the trial structure to produce the final fold. This idea suggests the existence of folding pathways or groups of convergent pathways. Whether or not there exist relatively stable folding intermediates was not directly addressed by this model, because the process could be cooperative to the extent that once begun, it proceeded directly to completion with energetic minima too shallow for intermediates to accumulate. Intermediates might be labeled by chemical modification agents or by proteolysis of unfolded regions, but these techniques would alter the protein and pervert the folding process. In 1989, BALDWIN discussed three viable theories, hydrophobic collapse, formation of secondary structure and formation of specific interactions, as the possible initial nucleation event in protein folding. He also indicated the possibility that there exist multiple pathways of folding from the denatured state.

Since the mid 1970s, a spectacular amount of information has emerged concerning mechanisms of protein folding from two separate lines of investigation: NMR spectroscopy and trapping of folding intermediates. In addition, the ability to assess the effect of the replacement of individual amino acids and short sequences upon protein folding has allowed more detailed information to be obtained (LECOMPTE and MATHEWS, 1993). Yet, these approaches generally deal with denaturation or renaturation of proteins *in vitro* and, therefore, represent cases different from the folding of nascent proteins *in vivo*.

One of the useful NMR approaches was developed by BALDWIN (KUWAJIMA and BALDWIN, 1983) and KUWAJIMA et al. (1983), extended by RODER and WÜTHRICH (1986), and recently reviewed by BALDWIN and RODER (1991). A review of somewhat broader context is provided by GREGORY and ROSENBERG (1986). The method involves the measurement of deuterium exchange into or out of proteins. Only protons bonded to nitrogen, oxygen or sulfur are able to exchange readily with solvent protons; exchange of $-OH$ and $-SH$ protons is usually rapid if the groups are in contact with solvent water. Amide $-NH$ exchange is slower and pH-dependent (e.g., WÜTHRICH, 1986). The rate of exchange can be quite slow if the amide groups are held in a hydrogen bond and/or sequestered from solvent on the interior of the protein. Such buried groups can exchange readily only in denatured protein, provided that aggregation does not occur. If a protein is denatured to a given extent by addition of urea then allowed to renature by rapid dilution of the urea, refolding will occur. If ^2H_2O is added at the time that the urea is diluted or shortly thereafter, some of the solvent deuterium will exchange into sites on the protein that are not accessible in the folded protein. Since ^2H resonates at much lower frequency than ^1H at the same magnetic field, the resonances from the exchanged groups disappear from the proton spectrum. Rapid-mixing experiments carried out at different times after renaturation is initiated by dilution give an indication of which groups were exposed in the denatured protein (i.e., the extent of denaturation) and the order in which the same groups become protected from exchange (i.e., the order of folding). The experiment can also be done by exchanging all exchangeable protons for deuterons in the denatured protein to simplify the spectrum, then observing exchange of protons into exposed sites in various intermediates of the protein as it renatures.

Low temperature enzymology, the chief proponent of which has been DOUZOU (DOUZOU, 1973; DOUZOU and PETSKO, 1984), involves decreasing the rate of a reaction by chilling the sample to temperatures much below the freezing point of water. This is accomplished by using mixed solvents to prevent freezing. At such low temperatures, relatively few intermediates have sufficient thermal energy to attain the activation energy on the path leading to the next intermediate, and become trapped. Depending on the temperature, a particular intermediate can account for a substantial fraction of the total and can be characterized as though it were a stable compound. In the present context, the reaction is not catalysis, but folding. FINK (e.g., BIRINGER and FINK, 1982) and others have examined folding using the techniques of low-temperature enzymology. Intermediates in the refolding pathway of chemically denatured proteins have been trapped and examined using circular dichroism, fluorescence of tryptophan or tyrosine and NMR spectroscopy. A review of practical aspects of such experiments has been provided by FINK (1986).

One common folding intermediate found in studies of numerous proteins has been called the molten globule. Besides being a folding intermediate, this protein state can be produced by mild denaturation by acid, base or in the presence of denaturants. Circular dichroism (CD) has been used profitably to study the nature of the molten globule. CD spectra in the far UV (200–240 nm, covering the absorption of the peptide bond) monitor the backbone structure of a protein and can be used to calculate the amounts of various types of secondary structure. CD spectra in the region of the absorption of aromatic and sulfhydryl side chains monitor tertiary structure since these groups are affected by the nature of their environment in the folded protein but are not directly affected by differences in ϕ or ψ angles. Comparison of CD spectra of the molten globule form of several proteins in the far and near UV shows that the secondary structure remains relatively intact, but a unique tertiary structure is absent. The molten globule is still quite fluid, as judged by its calorimetric similarity to the denatured state (KUWAJIMA, 1989). It is also compact or partially folded as evidenced by size exclusion chromatography studies. Further discussion and characterization of the molten globule states of several proteins was recently provided by FINK et al. (1994).

The folding and association of oligomeric enzymes contains an additional level of structural assembly. Fundamental considerations and a discussion of experimental approaches have been summarized by JAENICKE and RUDOLPH (1986).

2.2.2 Targeting, Excretion and Misfolding of Proteins

In Gram negative bacteria, some enzymes appear in the periplasm, the space between the cytoplasmic membrane and the cell wall, which means that they must pass through the cytoplasmic membrane. Likewise, in eukaryotic cells proteins are synthesized in one location, but are often "targeted" to appear in another cellular compartment (e.g., mitochondria), which also requires passage through a membrane. Other enzymes in bacteria and eukaryotes enter and remain firmly ensconced in a membrane. The question in these cases (other than how proteins are transported and how the signaling polypeptide sequence targets the protein toward a given location) is how and when they fold into an active conformation. There is some indication that, for at least some proteins, it is the molten globule or other "pre-folded" form that passes across a membrane (FISCHER and SCHMIDT, 1990; KUMAMOTO, 1991) and completes its folding after transport.

It is certainly possible that enzymes subjected, either *in vitro* or *in vivo*, to stress such as that provided by denaturants, high (or low) temperature or ionic composition of the medium might unfold partially or completely. Upon a change in conditions, if the cell is to remain viable, these proteins must either refold or be cleared by proteolysis to make way for newly synthesized enzymes. Since proteases are much more effective on unfolded proteins than on native structures, clearing by proteolysis would seem simple, so long as the proteases themselves remain active. Refolding, however, is another matter. Since the primary drive for folding is provided by hydrophobic forces, an alternative to refolding is simply aggregation to form disorganized complexes which may precipitate. *In vitro*, a partial solution to this problem is to keep the protein concentration low and to alter conditions slowly to allow folding rather than aggregation.

A modern strategy for laboratory or industrial production of proteins in large amounts is by overproduction in a microorganism. Overproduction is accomplished by inserting the gene that codes for the protein of interest behind a strong, perhaps inducible, promotor in a microorganism such as *Escherichia coli* and express the protein using the bacterial machinery. Overproduction of a foreign protein in a microorganism raises the local concentration of protein and may favor aggregation. The desired protein may therefore precipitate in a denatured form as protein bodies (inclusion bodies, refractile bodies) so that it cannot be recovered. This undesirable event might even be predicted if the protein of interest is one that is normally targeted toward a specific cellular compartment and contains a signaling peptide sequence that is necessary for transport across a membrane and may be necessary for folding. Some degree of processing, such as clipping of the leader sequence, may also be necessary before or as folding occurs. Since *E. coli* does not contain the machinery to accommodate these possibilities, protein bodies may be formed. A common strategy in this case is to extract the protein, solubilize the inclusion bodies by denaturation, then remove the denaturant slowly and await refolding into a native state. In at least one case (HATTORI et al., 1993), protein refolded in this manner appears native by most criteria, but lacks some epitopes recognized by antibodies to the native protein. Apparently, misfolding occurs.

Misfolding may also occur upon renaturation of any denatured protein. Such misfolding could occur because of slow *cis-trans* interconversion of peptide bonds involving proline, or from inappropriate association of hydrophobic regions of the polypeptide that occur more rapidly than correct folding patterns. If the surrounding environment is oxidizing in nature and if more than two cysteine residues are present, the incorrect conformation can be locked in by formation of incorrect disulfide bonds. In the laboratory, misfolding and aggregation may be an inconvenience; *in vivo*, they can be fatal. Biological systems are resilient and provide enzymes that protect cellular proteins from these perils.

2.2.3 Catalysis of Folding or Refolding: Molecular Chaperones, Disulfide Isomerases and Peptidylproline Isomerases

As recently as 1987, it was stated that enzymes that catalyze the folding of proteins were not known (CORNISH-BOWDEN and CÁRDENAS, 1987). Since then, at least three classes of enzymes have been shown to catalyze folding or refolding. One class was found in a group of "heat-shock proteins" or HSPs (BECKER and CRAIG, 1994; CRAIG et al., 1993; HARTL et al., 1994; HORVICH and WILLISON, 1993; JAKOB et al., 1993; WELCH 1991, 1993), which are synthesized by many kinds of cells in response to heat stress (LINDQUIST, 1985). Such proteins are also expressed in response to other forms of environmental stress, and appear to form part of a generalized stress response (WELCH, 1993). Dozens of these proteins do indeed enhance the rate of the refolding of unfolded proteins at the expense of ATP, and thus catalyze a true, energy-coupled reaction, rather than simply providing a template or nucleating the folding process (BUCHBERGER et al., 1994; LUND, 1994; MARTIN et al., 1993; NADEAU et al., 1993; SCHMID et al., 1994). These folding enzymes have been called molecular chaperones or chaperonins. The most famous of these are HSP 70 (or Cpn 60, or *E. coli* GroEL) and its helper, HSP 20 (Cpn 20, or *E. coli* GroES) (AZEM et al., 1994; BOCHKAREVA and GIRSHOVICH, 1992; LUND, 1994; NADEAU et al., 1993; SCHMID et al., 1994). There are several families of chaperonins, distinguished by structural similarities (CRAIG et al., 1993; HORWICH and WILLISON, 1993; WELCH, 1991; WYNN et al., 1994). Not all chaperones are stress-inducible; some are constitutive (WYNN et al., 1994). Some specifically catalyze folding of nascent proteins but they may also be important for transport and targeting (HARDY and RANDALL, 1993; HEEB and GABRIEL, 1984; KUMAMOTO, 1991; NEUPERT and PFANNER, 1993; PFANNER et al., 1994; STUART et al., 1994). It has been clearly demonstrated that chaperonins can aid in the refolding of denatured proteins *in vitro* (e.g., HOBSON et al., 1993; KUBO et al., 1993; PERALTA et al., 1994) and prevent aggregation (EDGERTON et al., 1993; HARTL et al., 1994). They can confer heat stability to proteins (SCHRÖDER et al., 1993; TAGUCHI and YOSHIDA, 1993) and thermotolerance (ZIMMERMANN and COHILL, 1991) or osmotolerance (MEURY et al., 1993) to organisms. Clearly, molecular chaperones are ubiquitous and essential, not an inconsequential biochemical curiosity.

The importance of chaperonins appears to be far-reaching; they also play a role in gene expression and regulation. While only a few such cases have been demonstrated, and these are generally related to expression of other stress response proteins (NADEAU et al., 1993; ZYLICZ, 1993), it seems reasonable to believe that they may control the activity of many protein factors important in gene expression such as sigma factors, repressors, etc. (GOVEZENSKY et al., 1994).

Chaperones generally do not bind native proteins, but associate with unfolded or partially unfolded proteins, probably via the same hydrophobic interactions that would otherwise cause non-specific aggregation (RICHARME and KOHIYAMA, 1993, 1994; ROSENBERG et al., 1993). There is even evidence that some chaperonins specifically recognize certain folding intermediates (e.g., a pre-folded form, perhaps a molten globule) but not others (HAYER-HARTL et al., 1994; KUMAMOTO and FRANCETIC, 1993; MELKI and COWAN, 1994; PERALTA et al., 1994). So, the prodigious work done on uncatalyzed protein folding pathways *in vitro* is relevant, even in light of these recent discoveries.

Molecular chaperones also appear to be useful as a laboratory tool. They can be used to refold denatured enzymes and even untangle aggregated proteins *in vitro*. *In vivo*, coexpression (by molecular genetics) of a chaperone and an enzyme one wishes to study can lead to enhanced recovery of the active enzyme and a reduction in unfolded or aggregated product (DUENAS et al., 1994; FERREYRA et al., 1993).

Folding enzymes of the second class, protein disulfide isomerases (PDIs), establish the formation of proper disulfide bonds (FREEDMAN et al., 1994; KAJI and LODISH, 1993; PUIG et al., 1994; WANG and TSOU, 1993). Although there is evidence that at least one chaperonin is capable of rearranging mispaired disulfide bonds, the PDIs form a separate class of folding-catalyzing enzymes. Although protein disulfide isomerases have been known for at least 20 years, they were perhaps thought of more as maintenance enzymes rather than as catalysts for proper folding. BULLEID and FREEDMAN (1988) have presented evidence that PDIs are required for proper folding of nascent proteins *in vivo*. The importance of PDIs should not be underestimated; in a protein containing 4 pairs of cysteines, 105 different disulfide bonding patterns can exist (ANFINSEN and SCHERAGA, 1975), 104 of which are wrong. Like chaperonins, these enzymes are ubiquitous, judging from comparison of cDNA sequences in organisms as diverse as trypanosomes, yeast, alfalfa and mammals (FREEDMAN et al., 1994). The PDI from the endoplasmic reticulum lumen is found to contain a sequence homologous to the active site domain of thioredoxin, which has led to a proposal of a possible catalytic mechanism (FREEDMAN et al., 1994).

The third class of enzymes known to be involved in protein folding or refolding are proline isomerases, which interconvert the *cis*- and *trans*-forms, of proline peptides (e.g., LANG et al., 1987). Since this isomerization is much slower than simple rotation about single bonds, establishing the proper isomer (which is usually the *trans*-isomer) can be the rate-limiting step of protein folding (SCHMIDT and BALDWIN, 1978).

Certainly an additional group of enzymes that could be thought of as catalyzing the formation of an active enzyme structure includes posttranslational processing enzymes. These enzymes include methylating enzymes, glycosylating enzymes, kinases, and proteases, among others. These enzymes are so diverse that it is perhaps misleading to place them together as a class and certainly misleading to say that they catalyze folding *per se*. Nonetheless, they are required in order to obtain active enzyme. For instance, proteases are often necessary to remove leader (targeting) sequences or to activate enzymes synthesized in an inactive form, such as the zymogens (HUBER and BODE, 1978). Some of these processes are autocatalytic, such as the chymotryptic cleavage required in the conversion of chymotrypsinogen to chymotrypsin or the autocatalytic conversion of serine to the active site pyruvoyl residue of *Lactobacillus* 30a histidine decarboxylase (RECSEI and SNELL, 1970).

2.3 Determination of Protein Structure

While a number of methods can be used to obtain information about molecular shape (e.g., viscometry, ultracentrifugation, size exclusion chromatography), amount and type of secondary structure (e.g., circular dichroism, Fourier transform infrared spectroscopy, etc.) only two methods are currently available that yield protein structure to atomic resolution, X-ray crystallography (see, for instance, chapter 17 in BRANDÉN and TOOZE, 1991) and NMR spectroscopy (for brief reviews, see KAPTEIN et al., 1988; WÜTHRICH, 1990). The two methods, actually fields in themselves, have been compared recently (WAGNER et al., 1992). Although NMR spectroscopy has the advantage that it allows the determination of the structure in solution rather than in a crystal, most enzymes are too large to be studied, even by multidimensional techniques. X-ray crystallography requires the availability of high-quality crystals, which may be impossible to obtain for many proteins.

The structures of so many proteins have now been determined that methods used to predict protein structure can be evaluated in detail by comparison to experimental results. FASMAN, in many ways the founder of secondary structure prediction methodology, reviewed the approaches available in 1989. It is now possible to obtain secondary structure predictions for a protein directly from the nucleotide sequence (i.e., without even having the protein!) using various algorithms or by comparison to databases. FASMAN (1989) pointed out that the methods are not entirely reliable, and do not substitute for actual structure determination.

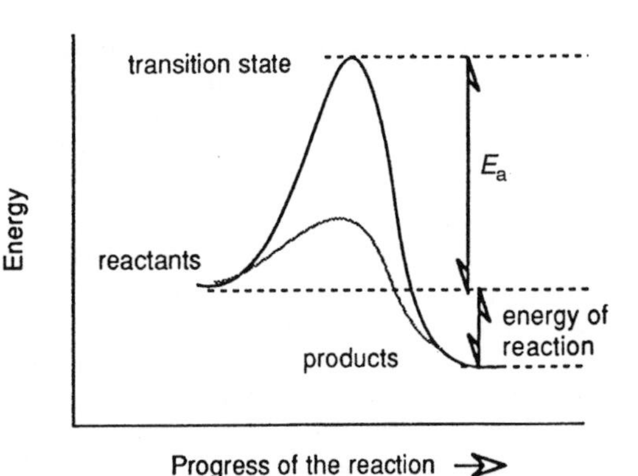

Fig. 5. Schematic representation of energy as a function of the progress of a reaction. The difference between energies of products and reactants is the overall thermodynamic energy change of the reaction. The difference between the energies of the reactants and the transition state is the energy of activation E_a. A similar diagram could be drawn for free energy ΔG and free energy of activation $\Delta G^{\neq}$. An actual reaction would likely have more than one energy barrier, with chemical intermediates residing in the valleys between them. The dotted line shows the energy path lowered by a catalyst.

3 Catalysis and Mechanism

Detailed discussions of the mechanism of enzyme action can be found in many texts and monographs (e.g., FERSHT, 1977; JENCKS, 1987). One of the more complete collections of relevant articles can be found in volumes 1, 2, 3 and 19 of the third edition of the series *The Enzymes*) (BOYER, 1970–1990). Some important considerations are summarized here.

3.1 Substrate Binding

Enzymes enhance the rates of reactions by several distinct mechanisms. According to the Arrhenius equation $k = A_0 e^{-E_a/RT}$, the rate constant k for the reaction depends on an activation energy E_a, and the temperature, which supplies the energy for reactants to attain the activation energy, and the gas law constant. This equation is largely phenomenological, but indicates the need to overcome an energy barrier, as shown schematically in Fig. 5. Indeed, the form is the same as the Boltzmann equation, with the exponential term representing the fraction of reactants having adequate energy for the reaction to proceed. All catalysts in some way decrease the energy barrier so that the reaction proceeds more rapidly than it otherwise would at a given temperature. In absolute rate theory, the Eyring equation shows the dependence of the rate constant upon the free energy of activation, $\Delta G^{\neq}$, and further identifies the contribution of enthalpic and entropic terms (GLASSTONE et al., 1941):

$$k = \frac{k_B T}{h} e^{-\frac{\Delta G^{\neq}}{RT}}$$

or

$$k = \frac{k_B T}{h} e^{-\frac{\Delta H^{\neq}}{RT} + \frac{\Delta S^{\neq}}{R}}$$

where $\Delta H^{\neq}$ and $\Delta S^{\neq}$ represent the enthalpic and entropic difference, respectively, between the reactants and the transition state, h is the Planck's constant and k_B the Boltzmann constant. By analogy to chemical thermodynamics, it is apparent that reaction rates are governed by the thermodynamics of the reactants and the excited state. $k_B T/h$ is a vibration frequency that is said to represent the breakdown of the transition state complex into product. Catalysis occurs by decreasing the free energy of activation $\Delta G^{\neq}$, which can occur via changes in enthalpy or entropy or both. An enzyme thus provides a lower energy path between substrate and product. It does so by lowering the transition state energy, a saddle point on an energy surface, which means that it provides more stable intermediates (e.g., ES complexes) and lower energy paths between them. Such energy lowering can come from several sources which have been discussed in some detail by JENCKS (1987), LIPSCOMB (1983), and many others.

Catalysis can occur via binding of two substrates in proximity to one another, which increases the local concentration of reactants, and favors interaction (JENCKS, 1975). Binding can also reduce orientational entropy by holding the substrates in the proper position for reaction to occur. Binding can introduce strain in bonds that are to be broken and force proximity between nuclei between which new bonds will form. More details of the effect of binding energy on catalysis were recently reviewed by HACKNEY (1990) and by HANSEN and RAINES (1990). From these arguments, it is reasonable to state that binding of a substrate by an enzyme favors the transition state. If true, this idea suggests that enzymes should have higher affinity for the transition state than for either reactant or product. It also suggests that the equilibrium constant for the reaction of bound substrates, for instance, between ES and EP, should be closer to unity than for the overall reaction, S to P. There is evidence for both suppositions. Compounds that resemble the transition state of an enzyme-catalyzed reaction, transition state analogs, have proven to be potent enzyme inhibitors that often have dissociation constants much lower than those of substrate or product (e.g., WOLFENDEN, 1988), supporting the idea that enzymes bind structures intermediate to those of substrate and product. ^{31}P-NMR studies of enzyme-bound intermediates of kinase reactions, which are generally nearly irreversible, have shown that ES and EP complexes are present in similar amounts, indicating the equilibrium constant for interconversion to be near unity (COHN and RAO, 1980).

Binding interactions involve the same kinds of interactions that are responsible for maintaining protein structure: hydrogen bonding, charge–charge interactions, dipole–dipole interactions, etc. These substrate-specific interactions are provided by amino acid side chains, in general, but even peptide C=O and =NH– groups are capable of hydrogen bonding and other dipole–dipole interactions. Thus, even glycine may participate in catalysis despite its otherwise undistinguished chemistry.

To summarize, enzymes use binding energy to stabilize the transition state. The binding interactions are the same as the interactions that stabilize protein structure: the hydrophobic effect, salt bridging, hydrogen bonding. Transient covalent interactions also occur, but play a more intimate role in catalysis (see below).

3.2 General Acid/Base Catalysis

In addition to pure binding effects, enzymes provide catalytic groups of several different types. In solution, acidic compounds can catalyze reactions by supplying protons at specific locations to stabilize an intermediate. For instance, acid-catalyzed amide hydrolysis proceeds by protonation of the carbonyl oxygen atom of the amide to render the carbonyl carbon more electrophilic for attack by water, and to stabilize the resulting tetrahedral intermediate. This function could be carried out by "acid", H_3O^+ in aqueous solvent. But it could equally well be performed by any proton-donating group, (i.e., any Brønstead acid). The acidic group might be called a general acid, because it is its proton-donating capacity, not its identity, that is important. Similarly, hydroxide ion can act as a catalyst by removing a proton from an amine, alcohol, etc., to render it more nucleophilic. This function could also be performed by any Brønstead base. Catalysis in this case would be said to be general base catalysis. In both general acid and general base catalysis, the proton being donated or accepted, need not be fully transferred between groups. Simple hydrogen bonding may be sufficient to stabilize the transition state or intermediate. In isomerases that move protons from one atom to another, the same group may act first as a general base, then as a general acid.

Side chains of amino acid residues that can function as general acids include lysine ($-NH_3^+$), histidine ($=NH^+-$), serine and threonine ($-OH$), cysteine ($-SH$), and, if their pK_a is such that they are protonated at ambient pH, the carboxyl groups of aspartic and glutamic acids. Side chains that can function as general bases include histidine ($=\ddot{N}-$), and the carboxylate groups of aspartate and glutamate. Tab. 3 lists side chains of enzymes that are known or thought to function as general acids or bases.

Tab. 3. Active Site Residues

Residue Side Chain	Function During Catalysis	Example
Aspartate	a) General acid or base b) Binding charged groups of substrate, metal ions	b) Trypsin (EC 3.4.21.4)
Threonine	a) H-bonding to "activate" substrate b) Binding to polar regions of substrates via H-bonding	a) Carbonic anhydrase (EC 4.2.1.1) b) Elastase (EC 3.4.21.36), tyrosyl-tRNA synthetase (EC 6.1.1.1)
Serine	a) Nucleophile b) Binding to polar regions of substrates via H-bonding	a) Serine proteases (chymotrypsin, EC 3.4.21.1) b) Liver alcohol dehydrogenase (EC 1.1.1.1)
Glutamate	a) Binding charged groups of substrate b) Binding metal ions c) General acid or base	a) Various b) Carboxypeptidase A (EC 3.4.17.1) c) Triosephosphate isomerase (EC 5.3.1.1)
Glutamine	Hydrogen-bond donor (N—H) or acceptor (C=O)	Tyrosyl-tRNA synthetase (EC 6.1.1.1)
Valine	Line hydrophobic sites	Elastase (EC 3.4.21.36)
Methionine	Hydrogen bond acceptor, provides weak polar interactions	
Cysteine	a) Zinc binding b) Nucleophile c) Redox mediator d) General acid/base e) Polar/hydrogen-bonding interactions	a) *All* zinc proteins b) Thiol proteases (e.g., papain, EC 3.4.22.2) c) Protein disulfide isomerases (EC 5.3.4.1) d) β-Galactosidase (EC 3.2.1.23)
Isoleucine	Line hydrophobic sites	
Leucine	Line hydrophobic sites	
Lysine	a) Nucleophile b) Binding of anionic substrates	a) Muscle aldolases (EC 4.1.2.13) b) Tyrosyl-tRNA synthetase (EC 6.1.1.1)
Histidine	a) H-bonding, e.g., catalytic triad of serine proteases b) General acid/base c) Zinc binding	a) Serine protease (e.g., subtilisin, EC 3.4.21.14) b) Triosephosphate isomerase (EC 5.3.1.1) c) All zinc enzymes (most zinc proteins)
Arginine	H-bonding to stabilize intermediates	Staphylococcal nuclease (EC 3.1.31.1)
Tryptophan	Potential H-bonding, $\pi-\pi$-interaction	
Tyrosine	H-bonding to substrates	Carboxypeptidase A (EC 3.4.17.1) Tyrosyl-tRNA synthetase (EC 6.1.1.1)
Phenylalanine	Line hydrophobic sites; $\pi-\pi$-interaction	

3.3 Covalent Catalysis, Nonprotein Catalytic Groups and Metal Ions

The side chains of active site residues can also effect catalysis by covalent catalysis, i.e., by forming covalent intermediates with substrates. Most examples of covalent catalysis involve nucleophilic attack by a side chain nitrogen, oxygen or sulfur atom. Many hydrolases employ covalent catalysis in which an enzyme-bonded intermediate is formed with the release of one product. The covalent enzyme intermediate must be hydrolyzed in a second step. A key feature of this variety of catalysis is that there is an initial burst of release of the first product in amounts stoichiometric with the enzyme concentration. Hydrolysis of the enzyme intermediate bond is rate-limiting, so further turnover requires release of the second product as well, before more substrate can bind. The serine proteases constitute a classic example of a covalent catalysis. The sequence of events is shown in Fig. 6.

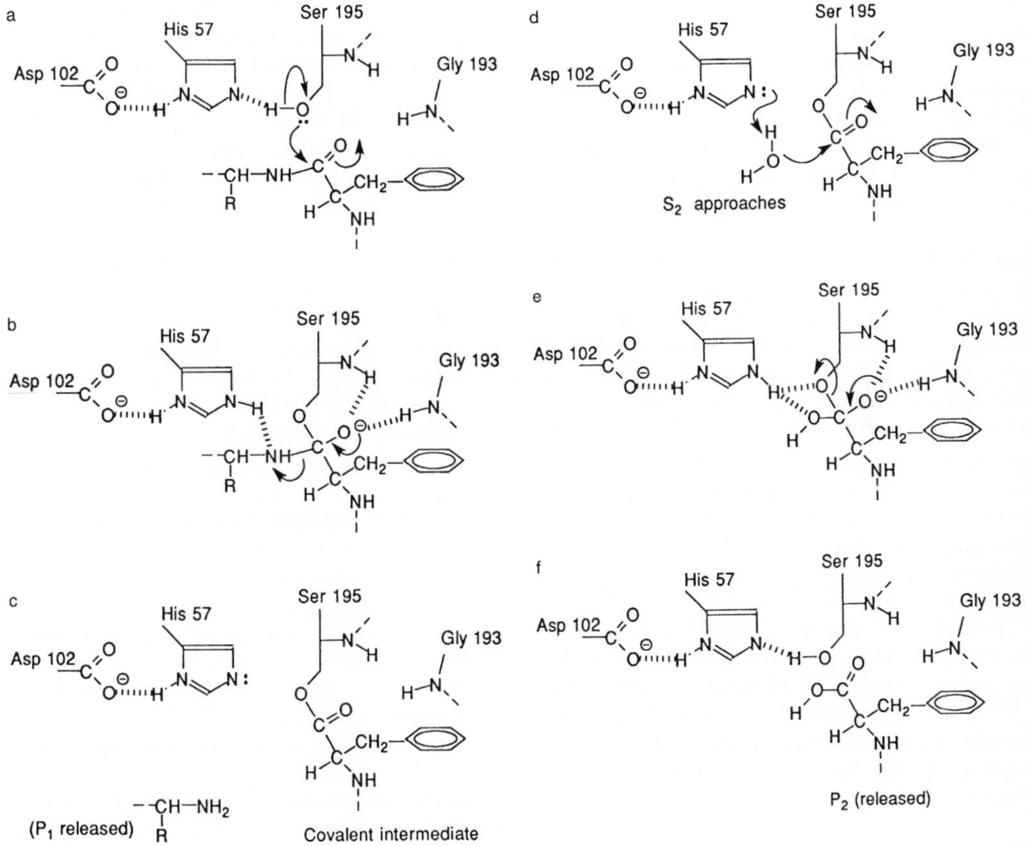

Fig. 6. Sequence of steps by which chymotrypsin catalyzes the hydrolysis of a peptide bond in which phenylalanine provides the carbonyl group. (a) Nucleophilic attack by serine 195 to produce a tetrahedral intermediate. (b) The intermediate, stabilized by the Ser 195 and Gly 193 amide NH groups collapses to form an enzyme ester as His 57 acts as a general acid to protonate the leaving group, P_1. (c) Release of first product, P_1, leaving a covalent intermediate. (d) Hydrolysis of the ester intermediate begins as His 57 acts as a general base to deprotonate an entering water molecule. (e), (f) The tetrahedral intermediate collapses, breaking the bond to Ser 195, and the product P_2 is released.

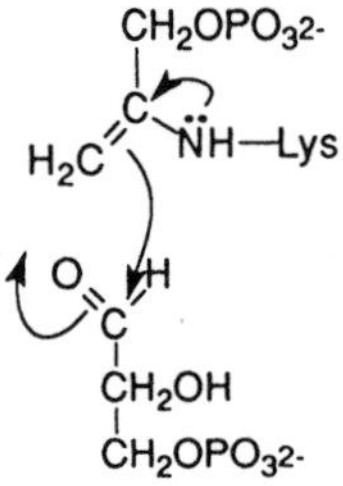

Fig. 7. Enamine intermediate in the muscle aldolase reaction.

Active-site cysteine residues, as well as a few other compounds (see below) can provide a nucleophilic sulfur atom. Perhaps the best-known example of a nucleophilic active-site cysteine involved in covalent catalysis is papain, a protease from the papaya fruit (DRENTH et al., 1968).

Nitrogen can also act as a nucleophile. A classic example is HORECKER'S mechanistic study of muscle aldolase (HORECKER et al., 1963), which catalyzes the aldol condensation (or the reverse of the condensation, in the forward direction of glycolysis) that interconverts fructose-1,6-bisphosphate and the trioses, dihydroxyacetone phosphate and glyceraldehyde-3-phosphate. After binding of the keto substrate, the side-chain nitrogen atom of an active site lysine attacks the keto group to form an imine (Schiff base). This form of the bound substrate more easily loses a proton (to a general base) or accepts the electron pair from the carbon–carbon bond (depending on the direction of the reaction) to form an enamine that can attack the aldose substrate. The enamine structure (Fig. 7) is more stable than the corresponding enol that would be obtained without the intervention of the nitrogen atom. Release of the product occurs after hydrolysis of the enamine which occurs by the reverse of the reaction that created it.

3.4 Cofactors, Coenzymes and Prosthetic Groups

Non-protein groups can also be used by enzymes to effect catalysis (see Tab. 4 and Fig. 8). These groups, called cofactors, can be organic or inorganic and are divided into three classes: coenzymes, prosthetic groups, and metal ion cofactors. Prosthetic groups are tightly bound to an enzyme and therefore are considered to be an integral, though non-protein, part of the enzyme. Coenzymes are organic cofactors. Some coenzymes associate, become altered, and dissociate from the enzyme with each turnover, and may more precisely be thought of as cosubstrates rather than as integral parts of an enzyme. This group includes the coenzymes $NAD(P)^+$, NAD(P)H, coenzyme A, coenzyme Q, and folic acid derivatives. Flavin-containing coenzymes flavin mononucleotide (FMN) and flavin adenine dinucleotide (FAD), on the other hand, are generally tightly bound to their enzymes. Metal ions such as Zn^{2+}, Mg^{2+}, Mn^{2+}, Fe^{2+}, Cu^{2+}, and Mo may be cofactors, held in the protein by simple coordination with electron-donating atoms of amino acid side chains (e.g., histidine N, cysteine S, carboxylate O^-). (Some metal ions, particularly Mg^{2+}, are often associated with the substrate rather than the enzyme. Mg-ATP, for instance, is the true substrate of kinase reactions.) Alternatively, they may form part of a prosthetic group in which they are bound by coordinate bonds (e.g., heme), in addition to side-chain groups, or in combination with other inorganic components (e.g., iron–sulfur clusters, molybdenum–iron clusters). Some of these ions participate in reactions in which there is net electron transfer (i.e., in oxidation–reduction reactions). Metal ions are also used by a number of enzymes in order to bind substrate which coordinates to the metal ion or to polarize a bond prior to attack using such coordination (see Fig. 8).

Many cases are known in which monovalent cations, usually K^+ or NH_4^+ ion, activate enzymes. This activation is usually non-essential in that there is at least some catalytic activity in their absence, unlike the essential activation one observes with many divalent cations. Monovalent cations must act either by altering the enzyme conformation in some subtle way or by providing additional beneficial interactions at the active site. One exception to the non-essential function of monovalent cations in enzyme systems includes transmembrane transport proteins that require

Tab. 4. Cofactors: Catalytic Groups Other Than Amino Acid Side Chains

	Function	Example	Structure[a]
Cosubstrates			
NAD^+/NADH	2-Electron oxidation/reduction reactions, i.e., hydride transfer, usually in catabolic reactions or energy metabolism	Alcohol dehydrogenase (EC 1.1.1.1) Lactate dehydrogenase (EC 1.1.1.27) NADH dehydrogenase (EC 1.6.5.3)	a
$NADP^+$/NADPH	2-Electron oxidation/reduction reactions, usually in biosynthesis	Glyceraldehyde-3-phosphate dehydrogenase (EC 1.2.1.9)	a
Coenzyme A (CoA)	Acyl group carrier, especially of acetate	Citrate synthase (EC 4.1.3.7)	b
Coenzyme Q	Electron transport	NADH dehydrogenase (EC 1.6.5.3)	c
Ascorbate	Electron donor in oxidation/reduction reactions	Lysine hydroxylase (EC 1.14.11.4)	d
Tetrahydrofolate	1-Carbon transfer other than methyl groups	Thymidylate synthase (EC 2.1.1.45) Glycinamide	e
Tetrahydrobiopterin	Hydroxylation reactions	Phenylalanine-, tyrosine- and tryptophan-monooxygenases (EC 1.14.16.1, EC 1.14.16.2, EC 1.4.16.4, respectively)	e
S-Adenosyl methionine	Methyl group carrier		f
Prosthetic Groups			
FAD (usually tightly bound to enzyme)	1- or 2-Electron oxidation/reduction reactions, hydrogenation/dehydrogenation of carbon–carbon double bonds	Dihydrolipoamide dehydrogenase (EC 1.8.1.4) Succinate dehydrogenase (EC 1.3.5.1) Fatty acyl CoA dehydrogenase (EC 1.3.99.3)	g
FMN	1- or 2-Electron oxidation/reduction reactions, hydrogenation/dehydrogenation of carbon–carbon double bonds	NAD(P)H dehydrogenase (EC 1.6.8.1)	g
TPP	Decarboxylation reactions	Pyruvate dehydrogenase (EC 1.2.4.1)	h
Biotin (biocytin)	Carboxylation or decarboxylation reactions	Pyruvate carboxylase (EC 6.4.1.1)	i
Lipoamide	Acyl group transfer reactions	Dihydrolipoamide acetyltransferase (EC 2.3.1.12)	j
4-Phosphopantetheine	Acyl-carrying group	Acyl carrier protein	k
Pyridoxal-5′-phosphate	Decarboxylation, group elimination, group transfer reactions	Amino acid decarboxylases, e.g., histidine decarboxylase (EC 4.1.1.22)	l
PQQ	Oxidation reactions in bacteria, e.g., sugar oxidation in energy metabolism	Glucose dehydrogenase (EC 1.1.99.17) Methanol dehydrogenase (EC 1.1.99.8)	m

Tab. 4. Cofactors: Catalytic Groups Other Than Amino Acid Side Chains (Continued)

	Function	Example	Structure[a]
Amino Acid Derivatives			
Pyruvate	Pyridoxal-like reactions (rare)	Histidine decarboxylase of *Lactobacillus* 30a (EC 4.1.1.22)	n
Topaquinone	Oxidative deamination of amines	Bovine serum amine oxidase (EC 1.4.3.13)	o
Tryptophylquinone (TTQ)	Oxidative deamination of methylamine to formaldehyde	Methylamine dehydrogenase from methylotrophic bacteria (EC 1.4.99.3)	p
Selenocysteine	Oxidation/reduction reactions	Glutathione peroxidase (EC 1.11.1.9)	
Iron protoporphyrin (heme)		Catalase (EC 1.11.1.6) Cytochrome P-450 (EC 1.14.15.6)	q
Deoxyadenosyl cobalamin (a vitamin B_{12} derivative)	Intramolecular rearrangements	Methylmalonyl-CoA mutase (EC 5.4.99.2)	
Metal Ions			
Fe		Nitrogenase (EC 1.18.6.1) Aconitase (EC 4.2.1.3)	
Mo		Nitrogenase (EC 1.18.6.1) Nitrite reductase from *Micrococcus halodenitrificans* (EC 1.7.99.4)	
Zn^{2+}		RNA polymerase (EC 2.7.7.6) Alcohol dehydrogenase (EC 1.1.1.1) Carbonic anhydrase (EC 4.2.1.1)	s
Cu^{2+}		Mitochondrial cytochrome oxidase (EC 1.9.31) Superoxide dismutase (EC 1.15.1.1) Polyphenol oxidase (EC 1.10.3.1)	
Mn^{2+}		Arginase (EC 3.5.3.1)	

[a] refers to structures in Fig. 8

symport or antiport of a monovalent cation (perhaps a proton) in its mechanism of transport of another compound (KABACK, 1970). In these cases, a transmembrane concentration gradient of the monovalent cation (often ^1H) may supply the thermodynamic drive for transport.

3.5 Kinetics of Enzyme-Catalyzed Reactions

The field of enzyme kinetics has been strongly influenced by the work of CLELAND, who has written ample reviews of the subject (CLELAND, 1967, 1970, 1979, 1990). Another useful review is found in the detailed textbook by SEGEL (1975).

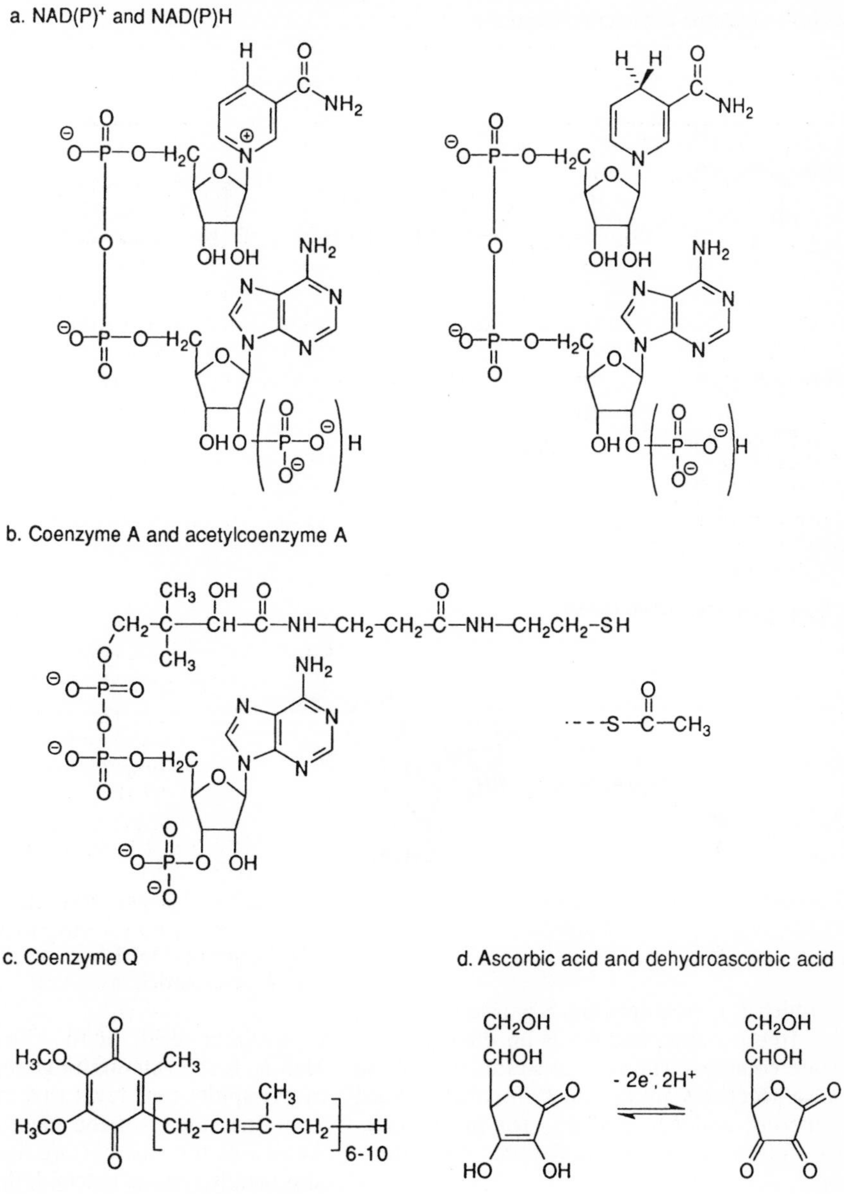

a. NAD(P)$^+$ and NAD(P)H

b. Coenzyme A and acetylcoenzyme A

c. Coenzyme Q

d. Ascorbic acid and dehydroascorbic acid

Fig. 8. Structures of compounds listed in Tab. 4.

3.5.1 Simple Cases

The kinetics of enzyme-catalyzed reactions are usually described in terms of the treatment set forth by MICHAELIS and MENTEN in 1913, before the nature of enzymes as proteins was known. The treatment is based on a number of assumptions that are valid for most enzyme systems, providing that the experiments used to determine reaction rates are designed properly. The reaction is modeled as a two-step reaction:

e. Tetrahydrofolate and tetrahydrobiopterin

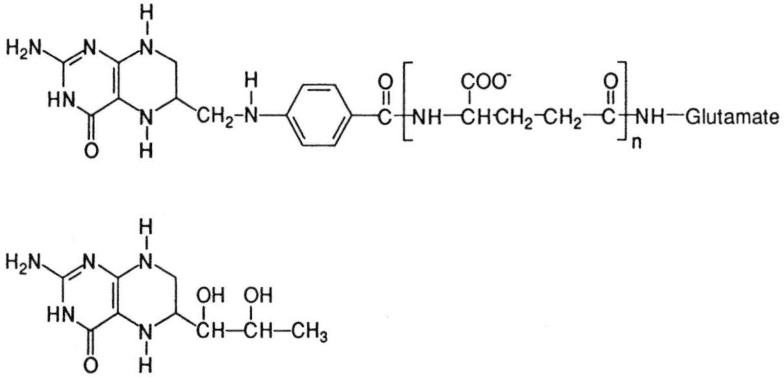

f. S-Adenosyl methionine (SAM)

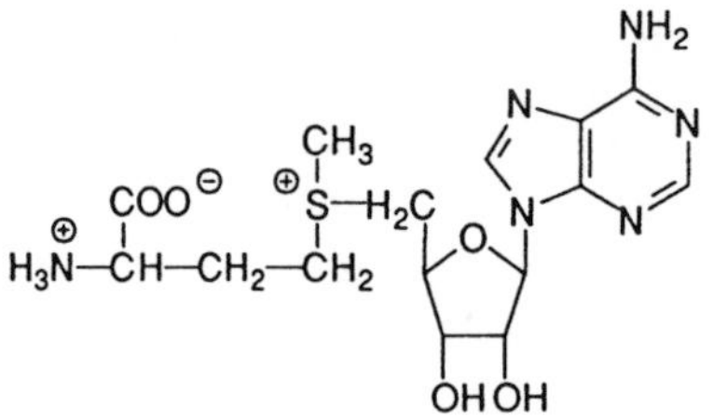

Fig. 8e–f.

$$E + S \rightleftharpoons ES \rightleftharpoons E + P$$

in which E is free enzyme, S is free substrate, P is free product, and ES is an enzyme–substrate complex. The rate constants describing these processes are k_1 and k_2 for the forward reactions, and k_{-1} and k_{-2} for the reverse. The following derivation assumes the following:

(1) The enzyme is present only in catalytic amounts, i.e., $[E]_t$ the total enzyme concentration is much lower than [S].
(2) Initial rates are used, i.e., [S] is sufficiently large and the time over which rate determinations are made is sufficiently short that [S] is essentially constant, and [P] is negligible. If [P] is negligible, the reverse reaction, $k_{-2}[E][P]$, can be omitted.
(3) Formation of ES is rapid and reversible, i.e., $k_2 \ll k_1$ and k_{-1}.

A direct result of the third assumption is that k_2 is the rate-limiting step, and that [ES] rises rapidly to a level that remains constant over the course of the measurement. This is known as the steady-state assumption. When this steady state is reached, the rate of change of [ES] is zero, despite the fact that it is continuously being formed by the k_1 reaction and destroyed by both the k_{-1} and k_2 reactions:

$$\frac{d[ES]}{dt} = 0 = k_1[E][S] - [ES](k_{-1} + k_2)$$

The total enzyme present, $[E]_t$, must be equal to the sum of the concentrations of all en-

g. Flavin adenine dinucleotide (FAD), flavin mononucleotide (FMN) and reduced form of FMN

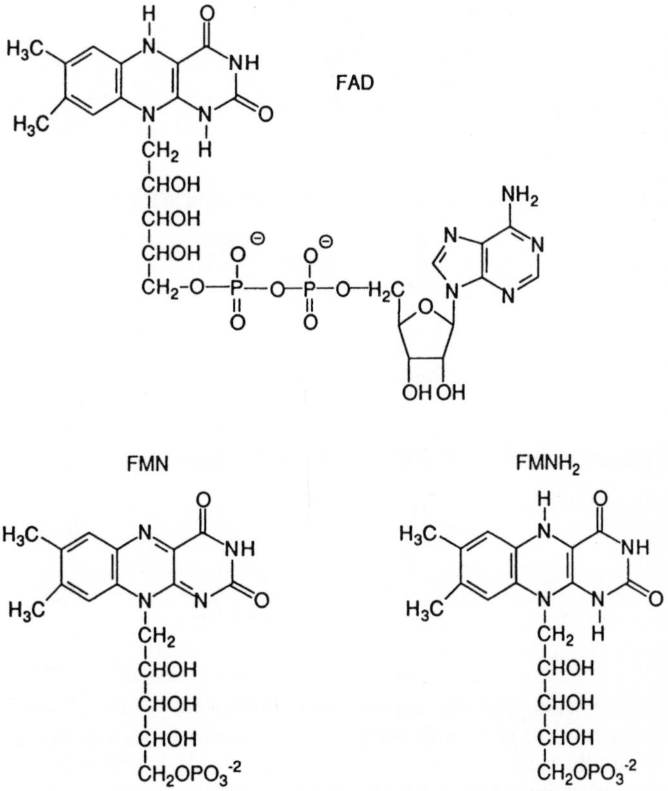

FAD

FMN

$FMNH_2$

Fig. 8g.

zyme forms: $[E]_t = [E] + [ES]$. So, in terms of [ES], the equation becomes

$$0 = k_1([E]_t - [ES])[S] - [ES](k_{-1} + k_2)$$

Hence, at steady state,

$$[ES] = \frac{[E]_t S}{\dfrac{k_{-1}+k_2}{k_1} + S}$$

The overall reaction rate v, measured either as appearance of product or disappearance of substrate, is $k_2[ES]$, the first-order decomposition of the enzyme–substrate complex, ES:

$$v = k_2[ES] = \frac{k_2[E]_t S}{\dfrac{k_{-1}+k_2}{k_1} + S}$$

The reaction proceeds at the fastest rate possible when the enzyme is saturated, i.e., all the enzyme exists as ES, so that $[E]_t = [ES]$. The term $k_2[E]_t$ is consequently given the name V_{max}. One further simplification is usually applied: the term $(k_{-1} + k_2/k_1)$ is called K_m, the Michaelis constant, so the full equation, the Michaelis–Menten equation, is

$$v = \frac{V_{max} S}{K_m + S}$$

k_2 is a first-order rate constant and therefore has the dimension time^{-1}. V_{max} $(= k_2[E]_t)$ contains units of concentration, as well it should, since rates are in units of concentration/time. If it is to be used to compare the results of different experiments or different enzymes, V_{max} must be reported per unit

h. Thiamine pyrophosphate (TPP)

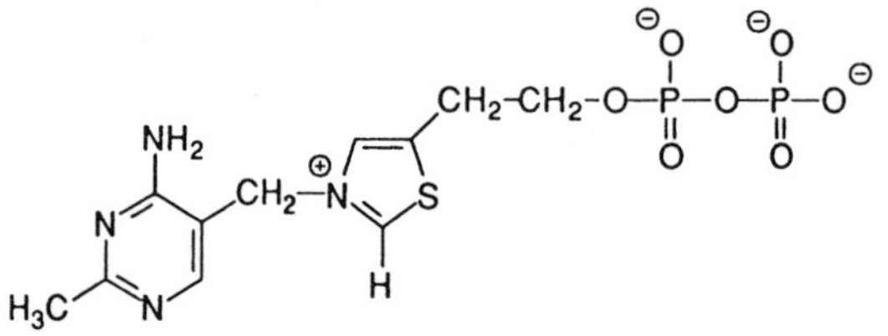

i. Biocytin (bound form of biotin) forms an amide bond to the protein

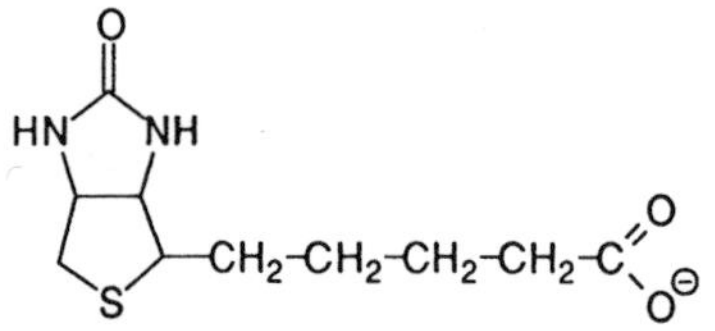

j. Lipoamide (bound form of lipoic acid) and dihydrolipoamide

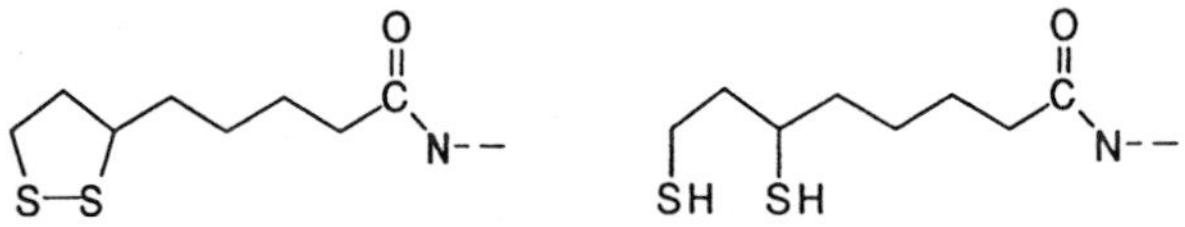

Fig. 8h–j.

enzyme (mg protein, moles, etc). However, when reported per unit enzyme, the quantity is usually referred to as "turnover number". The most enlightening form of the turnover number is when it is reported as the ratio of moles of substrate turned over per unit time per mole of enzyme. In this form, it truly represents catalytic efficiency.

K_m, the ratio of the sum of first-order rate constants (dimension: time^{-1}) to a second-order rate constant (dimensions: concentration^{-1}·time^{-1}), has the units of concentration, and is independent of enzyme concentration, so long as assumption (1), above, is met. It represents an aggregate dissociation constant for ES in that it consists of the ratio of rate constants for dissociation of ES to that for its formation. Provided that $k_2 \ll k_{-1}$ (assumption (3), above), it is equal to the equilibrium constant for formation of ES.

A plot of v versus [S] for an enzyme obeying the Michaelis–Menten equation, where the experimental conditions conform to its implicit assumptions, is shown in Fig. 9. The rate at low [S] is first-order (i.e., is proportional to) in [S], but eventually becomes zero-order in [S] when the enzyme becomes saturated, and $v = V_{max}$. At constant $[E]_t$, the dependence of rate on enzyme concentration is not apparent, but it is clearly always first-order in enzyme, so long as assumption (1), above, is met. Some algebra will show that the concentration of substrate that produces a rate equal to half of V_{max} is $[S] = K_m$. Thus the values of V_{max} and K_m can be obtained from inspection of a plot of v versus [S]. Also, from this plot, it is clear that, in order to obtain reliable values, data points should be taken at values of [S] that are in the range of K_m. Rates obtained at $[S] > 2K_m$ are likely to be nearly the same, and differences among them dominated by experimental error.

It is rather difficult to judge the quality of data from a direct v vs. [S] plot, especially in the past, before desktop and on-line computers. Several methods for linearizing the data were therefore developed. The most common of these is the double-reciprocal or Lineweav-

k. 4-Phosphopantetheine

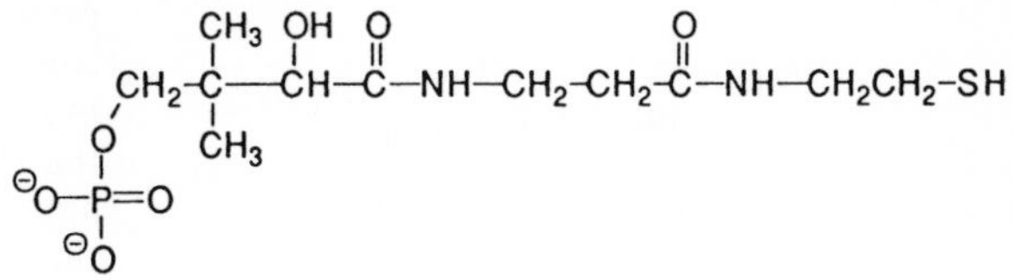

l. Pyridoxal-5'-phosphate

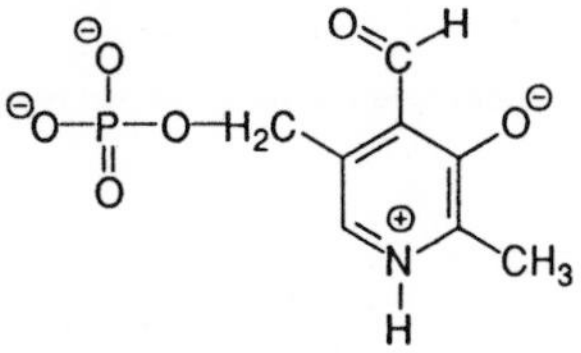

m. Pyrroloquinoline quinone (PQQ)

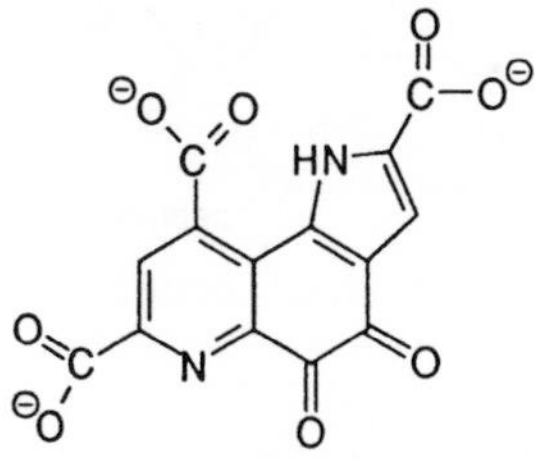

n. Pyruvate

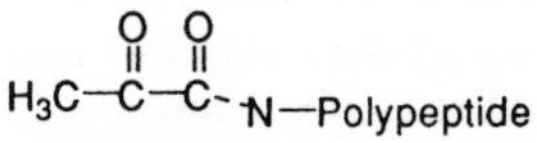

Fig. 8 k–n.

er–Burk plot. If the Michaelis–Menten equation is inverted, it is easily seen that

$$\frac{1}{v} = \frac{K_m}{V_{max}} \frac{1}{[S]} + \frac{1}{V_{max}}$$

A double reciprocal plot ($1/v$ vs. $1/[S]$) therefore yields a slope of K_m/V_{max}, a y-intercept of $1/V_{max}$, and an x-intercept of $-1/K_m$ (Fig. 9). The linearity is easy to judge, and both kinetic parameters can be evaluated graphically, or simple least-squares fitting can be employed. It is worth noting, however, that the most reliable data points are those taken at relatively high [S] (but $<2K_m$, or so), because the rates are high and easily measurable. Data taken at very low [S] are less well-determined. Unfortunately, the lowest rates play a major role in determining the slope of the line in a double reciprocal plot. The higher rates are often weighted more heavily in the analysis by employing a weighted least-squares routine.

An alternative graphical method that is often used is the Eadie–Hofstee plot (Fig. 9). It is most commonly employed in impure systems such as by microbiologists working with whole cells or crude extracts, where more than one enzyme might be expected to catalyze the same reaction. It is widely used in such systems probably because two activities with appreciably different K_m's, can be re-

o. Topaquinone (TPQ)

p. Tryptophylquinone, a derivative involving two Trp residues

Fig. 8o–p.

solved easily. In this treatment, v is plotted against $v/[S]$ to yield a linear plot with a negative slope. The y-intercept in this case is V_{max} and the slope, $-K_m$.

Catalysis is efficient if the substrate has a high V_{max} (i.e., is turned over rapidly) and a low K_m (i.e., is bound readily). These properties are therefore indicative of the specificity of an enzyme for a substrate. Indeed, the ratio V_{max}/K_m has been termed the specificity factor (e.g., WHITAKER, 1994).

3.5.2 Multi-Substrate Reactions: Cleland's Notation, and the King-Altman Method

In the common situation in which more than one substrate is involved, the simple Michaelis–Menten treatment does not suffice. Multisubstrate systems include pyridine–nucleotide-linked dehydrogenases, ligases, kinases and many others. CLELAND (1967, 1970) has formulated a clear description of such systems, which consists of a line showing the progress of the reaction, punctuated by arrows showing explicitly when substrates are bound and products are released. Beneath the lines are shown the interconversions among intermediates and a representation of the complexes. An example might be the following scheme in which two substrates bind in a particular order, then two products are released in a particular order:

$$\begin{array}{ccccc} & S_1\downarrow & S_2\downarrow & & P_1\uparrow \quad P_2\uparrow & \\ \hline E & ES_1 & ES_1S_2 \rightleftharpoons EP_1P_2 & EP_2 & E \end{array}$$

Although the addition and release processes are shown as single-headed arrows, they are assumed to be reversible, exactly as in rapid-equilibrium steady-state kinetics.

The binding of two substrates and/or the release of the products might also occur in random order, in which case an additional branch of the kinetic pathway must be specified explicitly. Both are shown as random in the example below:

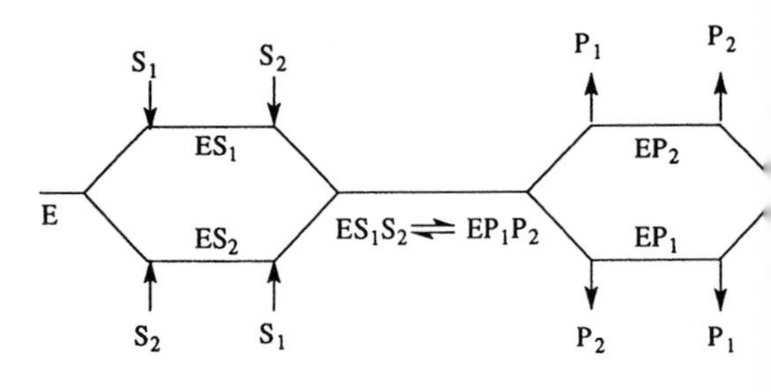

This kinetic mechanism contains an enzyme-substrate complex that contains two bound substrate molecules (a "bi" complex) before any reaction occurs and two bound products (another "bi" complex) immediately after the reaction, so the mechanism may be called a "random bi bi" mechanism.

Sometimes, especially in cases in which the binding sites of the two substrates overlap or are mutually exclusive, the first substrate S_1 must dissociate after engaging in some reaction that leaves the enzyme in a modified form, E′, before the second substrate S_2 can bind:

q. Heme (iron protoporphyrin IX)

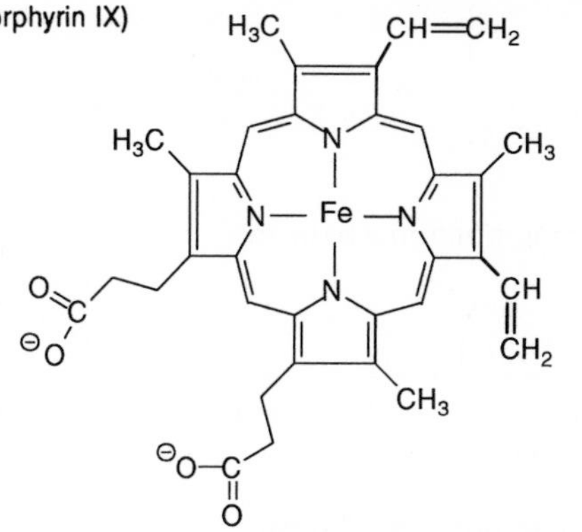

r. Four-iron and two-iron clusters found in iron-sulfur proteins ("non-heme iron" proteins)

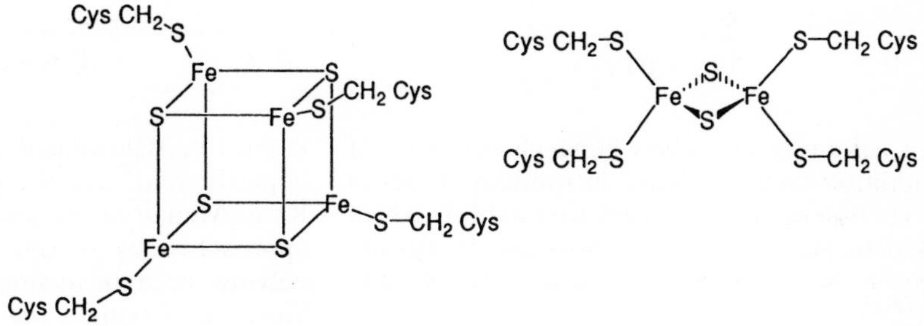

s. A tetrahedral Zn^{2+} center (alcohol dehydrogenase)

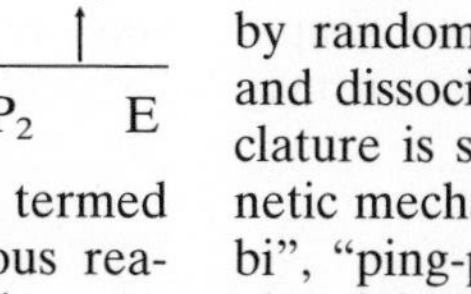

Fig. 8 q–s.

$$\begin{array}{ccccc} S_1\downarrow & & P_1\uparrow \quad S_2\downarrow & & P_2\uparrow \\ \hline E & ES_1 \rightleftharpoons E'P_1 & E' & E'S_2 \rightleftharpoons EP_2 & E \end{array}$$

This kinetic mechanism has been termed "ping-pong" by CLELAND, for obvious reasons. This kinetic mechanism contains two enzyme–substrate complexes that contain only one bound substrate or product, so it may also be called a uni uni ping-pong reaction.

For reactions involving more than two substrates, some complexes may be bimolecular and others termolecular, and may be formed by random, ordered, or ping-pong addition and dissociation. In these cases, the nomenclature is straightforward, if entertaining; kinetic mechanisms may be called "ordered ter bi", "ping-pong uni uni", etc. A good example might be a system having three substrates and two products, such as a coenzyme A-linked dehydrogenase. The substrates are aldehyde (for instance), NAD^+ and coenzyme A; the products are NADH and an acyl CoA. Such a reaction could proceed by an ordered ter bi mechanism,

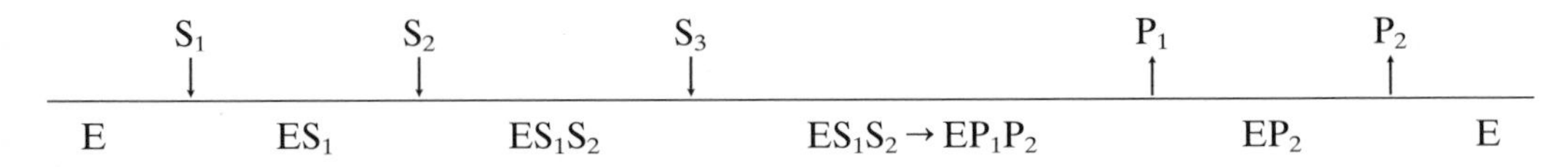

a bi uni uni uni ping-pong mechanism,

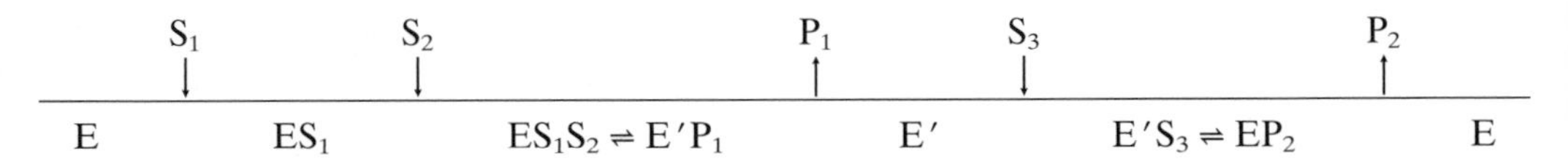

a uni uni bi uni ping-pong mechanism,

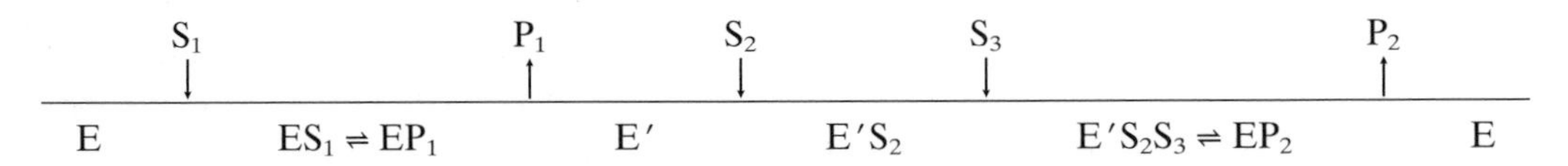

or various bi variations in which the order of addition and/or release is random. Kinetics experiments can be used to establish which kinetic mechanism is operative and to identity S_1, S_2, S_3, P_1 and P_2 (CLELAND, 1990; SEGEL, 1975).

KING and ALTMAN (1956) have devised a straightforward graphical method for writing Michaelis–Menten-like equations. In the standard derivation of the simple single-substrate reaction, the key element was to establish how much of the enzyme existed as ES, because the overall rate is the first-order decomposition of ES, given by k_2[ES]. The term $[E]_t$ was introduced, and the remaining enzyme form was eliminated from the equations so that it did not have to be quantitated. In a slightly more complicated case such as one in which the dissociation of product might not be immediate, the enzyme is divided among three species, E, ES, and EP, so our simple derivation is inadequate. The situation may be depicted as follows:

$$E + S \underset{k_{-1}}{\overset{k_1}{\rightleftharpoons}} ES \underset{k_{-2}}{\overset{k_2}{\rightleftharpoons}} EP \underset{k_{-3}}{\overset{k_3}{\rightleftharpoons}} E + P$$

which could also be written in Cleland's notation as

S ↓ P ↑

E — ES ⇌ EP — E

It must be determined how the total enzyme is partitioned over these three forms. In the King–Altman formalism, the partitioning is approached by writing all the reactions involving interconversion of enzyme forms. Since the enzymatic reaction is necessarily cyclical if free enzyme is to be regenerated for the next turnover, a convenient way to keep track of the interconversions is to draw a plane polyhedron with the enzyme forms at the apices. Rates of interconversion are first-order in the enzyme form, but may also be first-order in substrate or product, so the rates are kept track of by listing the associated rate constants and kinetic contributions from the ligands (substrate or product). For the above case, which has three enzyme forms, the figure would be a triangle. The rate expressions for the forward reaction are shown on the outside, while those for the reverse reaction are shown on the inside:

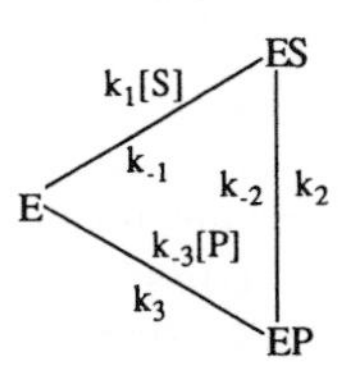

The next step involves listing separately all the possible pathways that lead to production

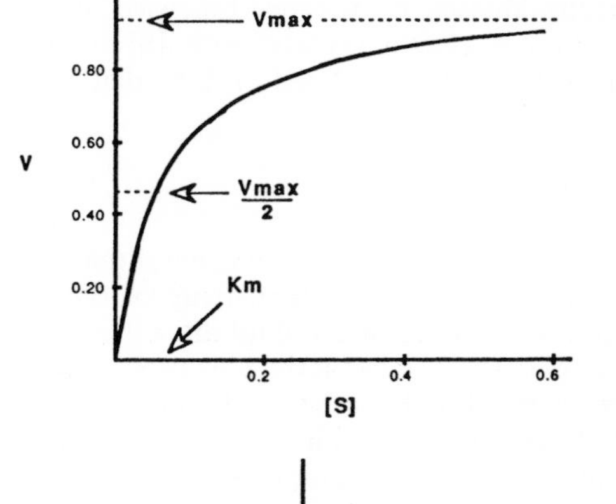

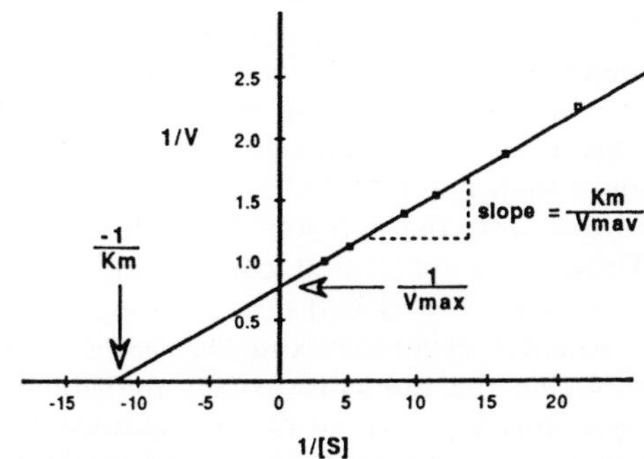

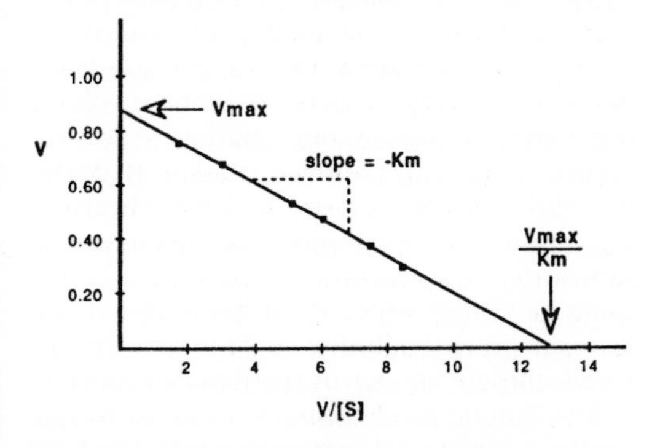

Fig. 9. Three common plots for examining initial velocity data from an enzyme-catalyzed reaction: top, V versus [S]; center, double-reciprocal or Lineweaver–Burk plot; bottom, Eadie–Hofstee plot. Relevant parameters are labeled on each plot.

or loss of each enzyme form. For an n-sided closed figure, these patterns would be a series of $n-1$-sided open figures:

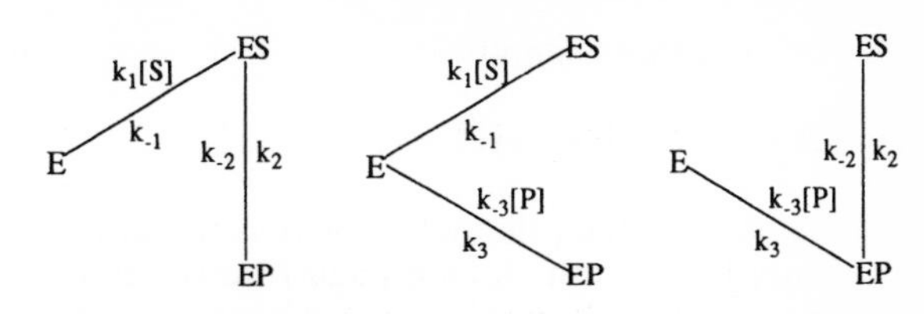

The fraction of total enzyme represented by each of the three enzyme forms can be obtained by taking the product of rate constants (or rate constant times ligand concentration) for all reactions forming a particular enzyme form in each of the $n-1$-sided figures, summing the contributions from each such figure, and dividing by the denominator D described below. For example, from the leftmost two-sided figure the reactions that create free enzyme are from EP to ES to E, yielding a term $k_{-2}k_{-1}$. From the center figure, the non-sequential reactions from EP to E and from ES to E are involved in formation of free enzyme, yielding the term $k_{-1}k_3$. The rightmost figure yields the term k_2k_3. $[E]/[E]_t$ is obtained from

$$\frac{[E]}{[E]_t} = \frac{k_{-2}k_{-1}+k_{-1}k_3+k_2k_3}{D}$$

A similar process is carried out for ES and for EP:

$$\frac{[ES]}{[E]_t} = \frac{k_{-2}k_1[S]+k_3k_1[S]+k_{-2}k_{-3}[P]}{D}$$

$$\frac{[EP]}{[E]_t} = \frac{k_2k_1[S]+k_{-1}k_{-3}[P]+k_2k_{-3}[P]}{D}$$

As might be expected for a partition equation, the denominator is equal to the sum of all possible processes, i.e., all nine terms:

$$\begin{aligned} D = {} & k_{-2}k_{-1}+k_{-1}k_3+k_2k_3+k_{-2}k_1[S] \\ & +k_3k_1[S]+k_{-2}k_{-3}[P]+k_2k_1[S] \\ & +k_{-1}k_{-3}[P]+k_2k_{-3}[P] \end{aligned}$$

Terms containing [P] can be collected, as can terms in [S] to give

$$\begin{aligned} D = {} & (k_{-2}k_{-1}+k_{-1}k_3+k_2k_3) \\ & +k_1(k_{-2}+k_2+k_3)[S] \\ & +k_{-3}(k_{-1}+k_{-2}+k_2)[P] \end{aligned}$$

or even more simply

$$D = C + C_S[S] + C_P[P]$$

At steady state, the net velocity is the difference between the forward and reverse velocities at any step (because the flux must be uniform through all parts of the reaction at steady state):

$$v = k_2[ES] - k_{-2}[EP]$$
$$= k_2\left(\frac{k_{-2}k_1[S] + k_3k_1[S] + k_{-2}k_{-3}[P]}{D}\right)[E]_t$$
$$- k_{-2}\left(\frac{k_2k_1[S] + k_{-1}k_{-3}[P] + k_2k_{-3}[P]}{D}\right)[E]_t$$

Collecting terms in [S] and [P], this equation can be written:

$$v = \frac{k_1k_2k_3[S] - k_{-1}k_{-2}k_{-3}[P]}{C + C_S[S] + C_P[P]}E_t$$

which has the intuitively satisfying form of products of forward rate constants times [S] minus products of reverse rate constants times [P], all over an equally orderly denominator. But the expression can be simplified further, if we make the following substitutions:

$$K_{m_S} = \frac{C}{C_S},\ K_{m_P} = \frac{C}{C_P},$$

$$V_{max_f} = \frac{k_1k_2k_3}{C_S},\ V_{max_r} = \frac{k_{-1}k_{-2}k_{-3}}{C_P},\ \text{and}$$

$$K_{eq} = \frac{k_1k_2k_3}{k_{-1}k_{-2}k_{-3}}$$

After a substantial amount of algebra, the expression is reduced to

$$v = \frac{V_{max_f}V_{max_r}\left([S] - \frac{[P]}{K_{eq}}\right)}{K_{m_S}V_{max_r} + V_{max_r}[S] + \frac{V_{max_f}[P]}{K_{eq}}}$$

If the measurements are initial rate measurements ([P] = 0), terms in [P] drop out, V_{max_r}'s cancel, and we are left with the standard Michaelis–Menten expression. Likewise, if the reaction is run in reverse, with [S] = 0, the equation can be simplified to show that

$$v = \frac{-V_{max_r}[P]}{K_{m_P} + [P]}$$

This development is more general than the standard derivation in that many of the assumptions are not made until after the equation is derived. The derivation procedure is very general and can be used almost exactly as described above for systems with more than one substrate and/or product, which is its major value. A notable exception is the case in which enzyme-bound intermediates interconvert in more than one way (i.e., there is more than one path between enzyme forms). In this case, the polygons have closed loops. These cases are treated logically and in detail in SEGEL's text (1975).

Kinetics experiments in the case of multisubstrate reactions are usually performed at saturating levels of all but one substrate, the varied substrate. Since the varied substrate is present at levels near its K_m, the enzyme is never saturated with the varied substrate, which effectively renders the steps involving the varied substrate rate-limiting processes. Hence, the kinetic parameters associated with the varied substrate (i.e., K_m) are observed. V_{max}, however, is a property of the enzyme at saturation of all substrates and should be the same no matter which substrate is varied, unless substrate inhibition is operative at the levels chosen for one of the fixed substrates.

The kinetic mechanism of a multisubstrate reaction, which includes determination of the identities of S_1, S_2 ... and P_1, P_2 ..., can be investigated by examining the form of the rate equation that describes the data (CLELAND, 1970; VIOLA and CLELAND, 1982), from studies using alternate substrates, and from inhibition studies (CLELAND, 1970; SEGEL, 1975; VIOLA and CLELAND, 1982). The simplest of these are probably inhibition studies; a few considerations are presented here. It should be clear that, if an inhibitor engages in mutually exclusive binding the same form of the enzyme as one of the substrates, it exhibits competitive inhibition toward that substrate (see below). If it is competitive toward one substrate, inhibition toward another of the substrates is indirect and will appear to be "mixed type". In this case, a Lineweaver–Burk plot shows lines that intersect above the 1/[S] axis, to the left of the $1/V$ axis. Product inhibition studies are also useful. For a more detailed discussion, see SEGEL (1975). Substrate inhibition, as well, can produce unusual patterns in double reciprocal plots. In ordered reactions, for instance, a substrate can bind out of order and compete with another substrate, or it may simply bind to the wrong intermediate enzyme form and act as an uncompetitive inhibitor (see below).

3.5.3 Enzyme Inhibitors and Inactivators

Enzyme inhibitors are compounds or ions that reduce the velocity of an enzyme-catalyzed reaction. Some examples of enzyme inhibitors are given in Tab. 5. By using a number of different criteria, inhibitors can be classified into several categories; the broadest categories depend on whether inhibition is reversible or irreversible. Reversible inhibitors fall into different classes according to how they affect the kinetics of catalysis. Inhibitors that block catalytic activity irreversibly are often termed inactivators. Agents that cause denaturation (i.e., unfolding) of the enzyme are generally not included in this category since their action is trivial from a mechanistic point of view. Those inactivators that modify a particular kind of amino acid side chain or are selective for a certain binding site on an enzyme can be used to obtain information of structural and mechanistic relevance. The kinetic consequences of irreversible inhibition are simpler than those of reversible inhibitors, so they are discussed first.

3.5.3.1 Irreversible Inhibitors, Affinity Reagents, Photoaffinity Labels and Suicide Reagents

Inactivators include agents that chemically modify an active site residue to block catalysis or modify a nearby residue to prevent substrate binding. Residue-selective reagents (e.g., thiol-selective agents, histidine-selective reagents) have been widely used in studying the nature of the active site of an enzyme. These reagents are not selective for active-site residues, *per se*, but a substantial reduction in activity caused by the chemical modification of a given type of amino acid is often taken as evidence that an active-site residue has been modified. The possibility exists that modification of one or more extra-active-site residues might perturb the enzyme's structure enough to decrease activity. It is therefore difficult to interpret residue-selective chemical modification data with certainty.

Three classes of active-site-directed chemical modification reagents have been developed for the identification of amino acid residues located in the active site. Members of all three classes are reagents that bind to the active site, thus increasing the local concentration of the reagent in that vicinity, then react with a side-chain group. Members of the simplest class, the affinity reagents (COLMAN, 1990), bear structural homology to a substrate, product or transition state, but contain a reactive group that is labile to substitution. The most common such affinity labels contain a good leaving group such as iodide that is easily displaced by a nearby nucleophilic group. Since they mimic substrate etc., the rate of inactivation by such reagents exhibits saturation; they are competitive against substrate and/or product; they have a characteristic binding constant toward the enzyme. These properties are to be distinguished from non-specific reagents, such as thiol reagents (e.g., Ellman's reagent (ELLMAN, 1959), *p*-chloromercuribenzoate (BOYER, 1954), and methyl methanethiolsulfonate (SMITH et al., 1975), which modify groups solely on the basis of their exposure and reactivity. Substrate may protect against even non-specific reagents by decreasing exposure.

One of the problems often associated with affinity reagents is the possibility of non-specific labeling of exposed, non-active site groups. This and other considerations lead to the development of similar reagents that bind to enzyme at the active site, but contain a photoactivatable group rather than an intrinsically reactive group. Using these reagents,

Tab. 5. Examples of Enzyme Inhibitors

Irreversible Inhibitors[a]	Enzyme or Class of Enzyme Inhibited	Site of Attachment
Side-Chain Selective Reagents		
5,5′-Dithiobis(2-nitrobenzoic acid (DTNB)	Many enzymes with $-SH$ groups in or near their active sites	Cys $-CH_2-S-$
p-Chloromercuribenzoate (PCMB)	Many enzymes with $-SH$ groups in or near their active sites	Cys $-CH_2-S-$
Methylmethanethiol sulfonate (MMTS)	Enzymes requiring free Cys SH groups	Cys $-CH_2-S-$
Diethyl pyrocarbonate (DPEC)	Enzymes with active-site histidine residues	His imidazole $=N-$
N-Ethyl maleimide (NEM)	Enzymes with active-site nucleophiles, especially $-SH$	Cys $-CH_2-S-$ Lys $-CH_2-NH_2$
Iodoacetate (IAA)	Enzymes with active-site nucleophiles, especially $-SH$	
N-Bromo succinimide	Tryptophan	Oxidation of indole ring
Affinity Label		
Tosylphenylalanine chloromethylketone (TPCK)	Serine proteases	Active-site serine
Glycidol phosphate	Triosephosphate isomerase, enolase	Labels nucleophiles or general base in active site
S-(4-Bromo-2,3-dioxybutyl)coenzyme A	Citrate synthase	Lys $-CH_2-NH_2$ or Glu $-COO^-$
4-Chloropropionyl CoA	Fatty acyl synthase	
2′,3′-Dialdehyde derivative of ADP	Pyruvate carboxylase	Labels Lys $-CH_2-NH_2$ in the presence of a reducing agent
5′-*p*-Fluorosulfonylbenzoyladenosine	Several dehydrogenases, e.g., glutamate dehydrogenase	NAD(P)H site residues
Photoaffinity Label		
8-Azido-ATP	a) F1-ATPase b) recA protein	b) Tyrosine
8-Azido-cAMP	Bovine pancreatin Ribonuclease A	Threonine
5-Azido-UTP	UTP-glucose pyrophosphorylase	
3′-*O*(4-Benzoyl)benzoyl-ATP	F1-ATPase	
Suicide Reagents		
Isatoic anhydride	Chymotrypsin	Active-site serine
β-D-Galactopyranosylmethyl(*p*-nitrophenyl)triazine	β-Galactosidase	Methionine
Cyclopropyl benzylamine	Monamine oxidase	Flavin
Adenosine 2′,3′-riboepoxide-5′-triphosphate	DNA polymerase I	Unknown

Tab. 5. Examples of Enzyme Inhibitors (Continued)

Reversible Inhibitors	Enzyme or Class of Enzyme Inhibited
Cyano NAD^+	Pyridine-nucleotide-linked dehydrogenases (e.g., lactic dehydrogenase)
Desulfo coenzyme A	CoA-requiring enzymes
Malonate	Succinate dehydrogenase

[a] Examples compiled from ATOR and ORTIZ DE MONTELLANO (1990), COLMAN (1990), MEANS and FEENEY (1990), WALSH (1984)

an excess of enzyme is first loaded with the reagent, which is subsequently made reactive by irradiation with ultraviolet light. These reagents are termed photoaffinity labels (COLMAN, 1990). Because they bind reversibly to the enzyme before photoactivation, there is always the possibility that some unbound reagent exists at the time of photoactivation, which can lead to non-specific labeling. For this reason, it is usually necessary to include some scavenger molecules to the reaction mixture that do not compete for the binding site to trap unbound photoactivated label. Photoaffinity reagents have the added advantage over affinity labels that they are highly reactive and are capable of bonding to the less reactive side chains, such as aromatic and hydrophobic amino acids. They do not rely so strongly on nucleophilic or electrophilic properties of the side chains and can give a more complete map of the active site.

The third class of affinity reagents are actually alternate substrates for the enzyme. However, the product of turnover is an unstable molecule that is capable of bonding covalently to active site residues (ATOR and ORTIZ DE MONTELLANO, 1990; WALSH, 1984). The possibility of non-specific labeling is greatly reduced. These reagents are called mechanism-based, turnover-based or suicide inhibitors.

3.5.3.2 Reversible Inhibitors

The most common types of reversible inhibitors are competitive, non-competitive and uncompetitive inhibitors. The term competitive refers to competition with the substrate for binding to the enzyme. For multisubstrate reactions, the classification is more complex, because the inhibitor may compete with any of the substrates, or with all.

Simple *competitive inhibitors* are compounds that bind to free enzyme and prevent binding of substrate. Conversely, binding of substrate prevents inhibitor binding. The derivation of the steady-state equation under initial rate conditions follows that of the simple Michaelis–Menten equation, except that enzyme can exist in three forms, E, ES and EI, whose concentrations sum to equal $[E]_t$ and are further related by the equilibrium equation:

$$K_I = \frac{[E][I]}{[EI]}$$

The Michaelis–Menten equation becomes

$$v = \frac{V_{max} S}{K_m \left(1 + \frac{[I]}{K_I}\right) + S}$$

In words, at fixed [I], the behavior is the same as in the absence of inhibitor, except that the apparent K_m is $K_m(1+[I]/K_I)$. V_{max} is ultimately unaffected, but is reached only at higher [S], since [ES] is reduced by the competition and must be restored by the addition of more substrate. Since V_{max} is constant, a double-reciprocal plot of data acquired at several levels of inhibitor (but fixed for a set of substrate concentrations) should intersect the $1/v$ axis at $1/V_{max}$, regardless of [I]. However, the intersection with the 1/[S] axis occurs at $-(1/K_m)(1+[I]/K_I)$. The slope of the line (K_m/V_{max}, without inhibitor) must also change by the same factor as K_m, i.e., $(1+[I]/K_I)$. Thus, the value of K_I can be obtained from replots of either the slopes or the apparent K_m's of double-reciprocal lines obtained at different values of [I] (see Fig. 10). In the case of slope replots, the slope of the replot is $K_m/K_I V_{max}$, and its x-intercept, $-K_I$. In the case of replots of apparent K_m, the slope is K_m/K_I, and the x-intercept, $-K_I$. Clearly, experiments designed to determine K_I must be carried out at combinations of [S] and [I] near their respective dissociation constants or balanced with respect to one another. If, for instance, saturating levels of S were used, little, if any inhibition would be observed unless [I] were also much higher than K_I. This merely restates the premise that substrate and inhibitor compete with one another.

Linear competitive inhibition, in which the slope- or intercept-replots of inhibited reactions are linear, is the most straightforward variety of competitive inhibition. But, nonlinear varieties of competitive and other inhibition patterns are not uncommon. The reader is referred to SEGEL (1975) for details.

It is tempting to infer from the observation of linear competitive inhibition that the substrate and inhibitor compete for the same site on the enzyme. However, the kinetic formulation requires only that binding be mutually exclusive. Conjecture upon whether the sites are different or the same (or different, but overlapping) depends on structural similarities beween the inhibitor and substrate. However, KOSHLAND's principle of induced fit (KOSHLAND and NEET, 1968) argues that binding of a substrate, inhibitor or other effector at a distant site can change the conformation of the enzyme enough to alter or prevent binding of other species at the active site. (The validity of the induced fit theory is vividly illustrated by hexokinase (ANDERSON et al., 1979).) Thus, the possibility exists that inhibitor and substrate bind at different, but mutually-exclusive sites. Nonetheless, observation of competitive inhibition, together with structural similarity to the substrate is usually taken as evidence in support (not proof) of binding of the inhibitor at the active site.

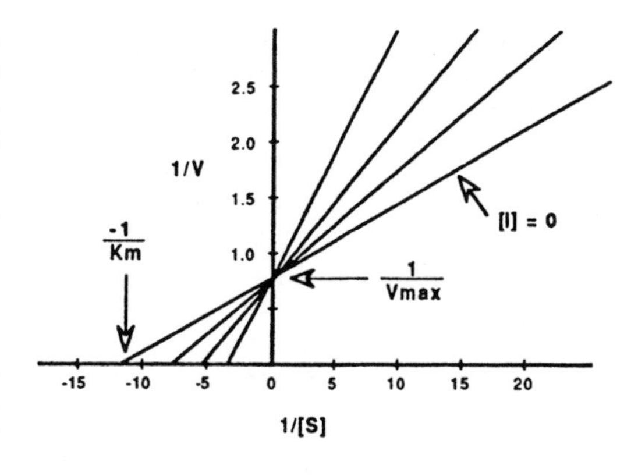

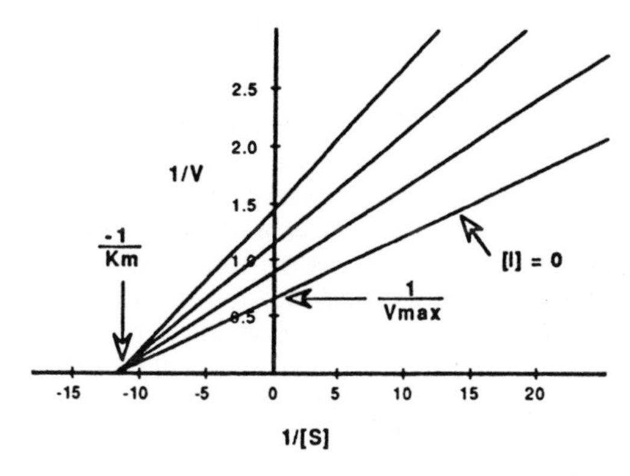

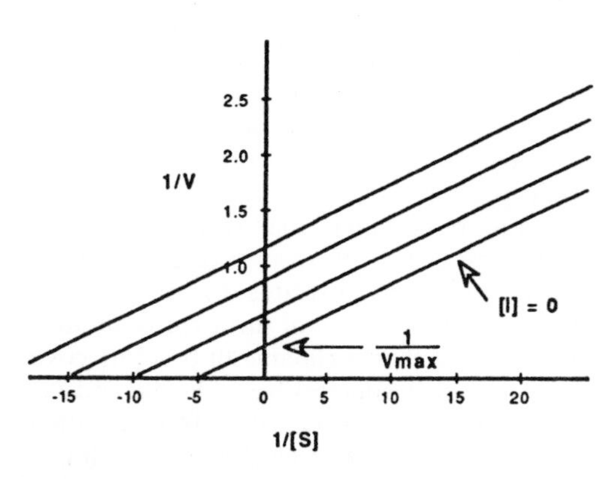

Fig. 10. Modes of inhibition shown in double-reciprocal plots of initial velocity data: top, competitive; center, non-competitve; bottom, uncompetitive.

If the site at which an inhibitor binds to an enzyme does not overlap with the active site, it is possible that both substrate and inhibitor can bind at the same time. For inhibition to occur, there must be some alteration of the catalytic groups at the active site that prevents turnover. This variety of inhibition is termed "*non-competitive*". In non-competitive inhibition, the enzyme can exist in four forms, E, ES, EI and EIS, which sum to E_t. The complex formation is described by an inhibition constant, $K_I = [E][I]/[EI]$, just as in competitive inhibition. Since inhibitor and substrate bind independently, the relation also holds that $K_I = [ES][I]/[EIS]$. In this situation it is useful to separate the dissociation of ES to E + S from catalytic turnover, k_2, so the equilibrium constant $K_S = [E][S]/[ES]$ is often used and K_m does not appear in any of the equations. The overall rate is still $k_2[ES]$, so [ES] is calculated just as in the absence of inhibitor, but using K_S and K_I to remove the terms in EI and EIS from the equation for $[E]_t$ (i.e., $[E]_t = [E] + [ES] + [EI] + [EIS]$), so that [ES] can be solved in terms of $[E]_t$ and the dissociation constants. The resulting Michaelis–Menten equation is

$$v = \frac{V_{max} S}{\left(1 + \frac{[I]}{K_I}\right)(K_S + S)}$$

In double-reciprocal form, the $1 + [I]/K_I$ term is associated with V_{max}, not K_S; the K_S is unaffected by inhibitor binding, but the apparent V_{max}, V'_{max}, equals

$$\frac{V_{max}}{1 + \frac{[I]}{K_I}}$$

and depends upon [I]. The slope and $1/v$-intercept of a double reciprocal plot of data taken at different levels of inhibitor should vary with [I], but the $1/[S]$-intercept is constant at $1/K_S$ (see Fig. 10). The value of K_I can be found from either slope or $1/v$-intercept replots.

Inactivation of an enzyme by irreversible inhibitors (i.e., K_I approaches zero) may fit the pattern of non-competitive inhibition if only catalysis, not binding, is affected by the modification. Inactivators can be distinguished from reversible non-competitive inhibitors by varying $[E]_t$ in different runs. Since $V_{max} = k_2[E]_t$, the apparent V_{max} should increase with increasing total enzyme and should extrapolate back to the true V_{max} at zero inhibitor, if the inhibitor is reversible. If the inhibitor is an inactivator, it prevents a stoichiometric amount of enzyme from participating in catalysis of the reaction. A plot of apparent V_{max} vs. $[E]_t$ would therefore intersect the $[E]_t$ axis at $[E]_t = [I]$, since V_{max} is zero until the amount of enzyme exceeds the amount of irreversible inhibitor.

One additional case will be considered, that of *uncompetitive inhibition*. Simple uncompetitive inhibition occurs if the inhibitor binds to ES but binds negligibly to E. This condition seems unlikely and is probably uncommon in unireactant systems. But, in bireactant systems (or higher), it is easy to visualize if the inhibitor were to mimic the second substrate in an ordered sequential reaction. Thus, it would bind competitively with the second substrate, but would appear uncompetitive with respect to the first. (The effect might not be observed if the second substrate were present at saturating levels, which is usually the case in experiments in which the first substrate is varied.)

In the case of uncompetitive inhibition, the enzyme is partitioned over all possible forms: $[E]_t = [E] + [ES] + [ESI]$. The observed rate is still $k_2[ES]$, and the velocity equation is formulated by the same steps as before to yield

$$v = \frac{V_{max}[S]}{K_S + [S]\left(1 + \frac{[I]}{K_I}\right)}$$

which can also be written

$$v = \frac{\frac{V_{max}}{1 + \frac{[I]}{K_I}}[S]}{\frac{K_S}{1 + \frac{[I]}{K_I}} + [S]}$$

Thus, in double-reciprocal plots, the apparent values of both K_S and V_{max} are modulated to the same extent, by the factor $1/(1 + [I]/K_I)$. The slope of the plots would remain K_S/V_{max}, but both intercepts would be multiplied by $1+[I]/K_I$. A series of parallel lines would therefore be obtained (see Fig. 10).

It is not always guaranteed that an inhibitor will fit precisely into one of the three categories of reversible inhibitors. For instance, it could form an EI complex competitively with substrate and also bind ES to form a non-productive ESI complex with a dissociation constant similar to that for EI. Double reciprocal plots for varied substrate concentrations at different fixed inhibitor concentrations would show characteristics of both competitive and uncompetitive inhibition. Lines would intersect, but in the upper left quadrant rather than on the 1/[S] or $1/V$ axes. This type of inhibition is an example of mixed-type inhibition. Mixed-type inhibition can result from other scenarios, and lines obtained from different fixed levels of inhibitor can intersect essentially anywhere or they can be parallel. For a collection of different mixed inhibition patterns and the mechanisms that cause them, see SEGEL (1975).

3.5.3.3 Substrate Inhibition

The kinetic development presented here assumes binding of substrate at a single site or at multiple, kinetically identical, independent sites. It also assumes that there is only one kind of ES complex that is always capable of turnover at a rate described by k_2. Of course, in many cases these expectations are not met. The most common exceptions are substrate inhibition and cooperativity of binding.

The case of substrate inhibition is relatively simple. It arises from binding of substrate in a non-productive fashion or from binding at a remote site that causes decreased binding or slower turnover at the active site. In bireactant systems, substrate inhibition can also arise if one substrate binds to the site for (or overlapping with) the other substrate. Substrate inhibition might be expected in ping-pong reactions because the site for the second substrate S_2 may be essentially the same as that for the first, S_1. If S_1 rather than S_2 binds to the site immediately after P_1 has vacated it, S_1 will act as an inhibitor. The non-productive binding is usually weaker than proper binding at the active site, or there would be negligible activity, hence the inhibition is observed only at higher substrate concentrations. A double reciprocal plot involving substrate inhibition typically shows linear behavior at low [S], and the line can be extrapolated (or fit by computer, taking the inhibition into account) to yield the K_m and V_{max}. However, at high [S], the data points rise at decreasing 1/[S]. CLELAND has generated a graphical method that can be used to determine the inhibition constant (CLELAND, 1970).

3.5.3.4 Biological Roles of Inhibitors

Inhibitors are not only important tools for the enzymologist, but are often critical metabolic regulators or protective agents in the cell. Examples of protection of cellular components include protease inhibitors that block the action of proteases until they are delivered to the site at which they are intended to function. For instance, the bovine pancreatic trypsin inhibitor presumably protects the pancreas and its system of ducts from proteolytic digestion until the enzyme is delivered to the large intestine.

Simple inhibition by product usually decreases the rate of the forward reaction (S and P generally bind to the same enzyme form and are therefore competitive) to prevent excessive buildup of the product. A more striking example of metabolic regulation by inhibitors is the process which includes feedback inhibition. Feedback inhibitors are end products (or late intermediates, or byproducts) of metabolic pathways that inhibit one of the early enzymes in the pathway. The pathway is thus blocked when its ultimate product is present in quantities adequate to fulfill cellular requirements. The feedback system of regulation works most effectively if the enzyme inhibited catalyzes the "committed step", the reaction that defines the beginning of a pathway, converting a common metabolite to one found exclusively in the pathway. In other cases, high concentrations of other molecules that provide a measure of metabolic status (e.g., ATP, ADP,

AMP), may activate or inhibit pathways. The structural resemblance between a late intermediate in a pathway and the common metabolite at the committed step of the pathway may be slight. It is not reasonable to expect simple competitive inhibition (for instance) to be an effective mechanism for control of such a pathway. Indeed, the mechanism involves interactions at more than one site and communication among different regions of the enzyme. These concepts are brought into play in two intimately related phenomena, cooperativity and allostery.

3.5.4 Cooperativity and Allostery

In monomeric enzymes, the binding of a substrate to one molecule of enzyme does not depend on whether other molecules of enzyme have bound substrate or not; binding occurs at independent sites. In oligomeric enzymes that contain more than one active site, the case may be different. In the mid 1960s, KOSHLAND proposed that the binding of substrate, inhibitor or effector can alter the conformation of an enzyme such that binding of other ligands might be altered (KOSHLAND et al., 1966). This theory of "induced fit" also suggests that subunits in an oligomeric complex can communicate with one another, since subunit–subunit interactions could also be altered when one subunit binds a ligand. Thus, binding of a ligand by one subunit can alter the binding constant of the ligand at one of the other sites, as had been indicated by MONOD et al. (1965). The result is that the binding constant (or, in principle, the turnover number) can change with changing substrate concentration: Michaelis–Menten kinetics are not obeyed. The effect can be put into quantitative terms by comparing the results of kinetic measurements to those expected from the Michaelis–Menten treatment. According to Michaelis–Menten, the substrate concentration at which $v/V_{max}=0.9$ is $[S]_{0.9}=9\ K_m$; that at which $v/V_{max}=0.1$ is $[S]_{0.1}=K_m/9$. Thus, for any enzyme, the ratio of substrate concentrations producing these rates, R_S, is 81. However, if binding of substrate enhances the binding of additional substrate (presumably at other sites of the same oligomer), $R_S>81$. This condition is called positive cooperativity. Conversely, negative cooperativity occurs if binding of substrate decreases the affinity for additional substrate, and $R_S<81$. Graphically, the v vs. [S] plot is distorted and appears sigmoidal, but the data can be fit (loosely) to theory using a modification of the Michaelis–Menten equation analogous to the Hill equation (ATKINSON, 1966; HILL, 1913):

$$v = \frac{V_{max}[S]^n}{K+[S]^n}$$

In this equation, K represents the average of all K_m or K_S operative during catalysis, and n represents the number of sites involved in substrate binding. The plot can be linearized by putting it in the form

$$\log \frac{v}{V_{max}-v} = n \log[S] - \log K$$

and the value of n can be determined from the slope. The value of [S] that yields $v=0.5\ V_{max}$ can be referred to as $S_{0.5}$, and is roughly analogous to K_m in non-cooperative systems (i.e., exactly equal to K_m if $n=1$). The exact relation between $S_{0.5}$ and K depends on the value of n. The extent of cooperativity depends on the strength of the interaction among sites and has been discussed more fully by ATKINSON (1966).

Allostery involves the regulation of activity by agents other than substrate, by mechanisms other than simple inhibition. The term allostery itself comes from the Greek for "other shape". It usually involves changes in subunit interactions or association–dissociation equilibria, sometimes caused by covalent modifications, such as phosphorylation or adenylation. The fundamental requirements for and properties of an allosteric system were defined by MONOD, WYMAN and CHANGEUX who proposed the "symmetry model" (MONOD et al., 1965). The flexible theories of KOSHLAND produced the "ligand-induced model" (KOSHLAND et al., 1966). Both models indicate that a change in the structure of one subunit of a multi-subunited enzyme can alter the properties of the other subunits.

Binding of an effector can alter the structure of one subunit, which therefore affects all subunits. In kinetics experiments, allostery is silent unless conditions are changed during the experiment. Details have been discussed by SEGEL (1975).

3.5.5 Binding of Ligands to Enzymes: The Scatchard Plot

It is often worthwhile to measure the binding of ligands such as metal ions, inhibitors, activators, or cosubstrates to enzymes in order to determine stoichiometry and affinity. Binding studies are usually carried out as stepwise titrations monitored using essentially any observable parameter that allows the investigator to measure either the residual free ligand or the enzyme–ligand complex. Such observable properties may include UV/VIS absorption, fluorescence or its quenching, epr signals, NMR linewidth or other relaxation parameters, or even turnover. If the complex cannot be detected, equilibrium dialysis may be employed to differentiate bound from free ligand. Titrations may be done varying the concentration of the ligand, holding the enzyme concentration constant, or the reverse, or variations in which both concentrations change. In order to interpret the data, it is usually cast in the form of a Scatchard plot (Fig. 11) in which the vertical axis is the ratio of bound to free ligand, usually normalized to total enzyme concentration if known, or to total protein, if not. The horizontal axis is the concentration of bound ligand normalized in the same way as the vertical axis. For a single set of n identical, independent sites with binding constants K_a (association constant), a plot of [bound ligand]/[free ligand] versus [bound ligand], normalized per unit enzyme, yields a straight line that intersects the vertical (Bound/Free) axis at a value equal to $n\ K_a$ and intersects the horizontal (B) axis at n, the number of binding sites. The slope of the line is $-n\ K_a$.

If the plot is not linear, however, the data are difficult to interpret, and in fact, are often interpreted incorrectly (BIETH, 1974; KLOTZ, 1993; WOFSY and GOLDSTEIN, 1992; ZIERLER, 1989). Some difficulties arise in the way in which the curve is resolved into two separate lines, some difficulties arise from experimental design, and some problems arise because the asumptions used are incorrect. It should be clear that the appropriate lines are not obtained by extrapolating linear regions of the curve until they intersect the axes: the

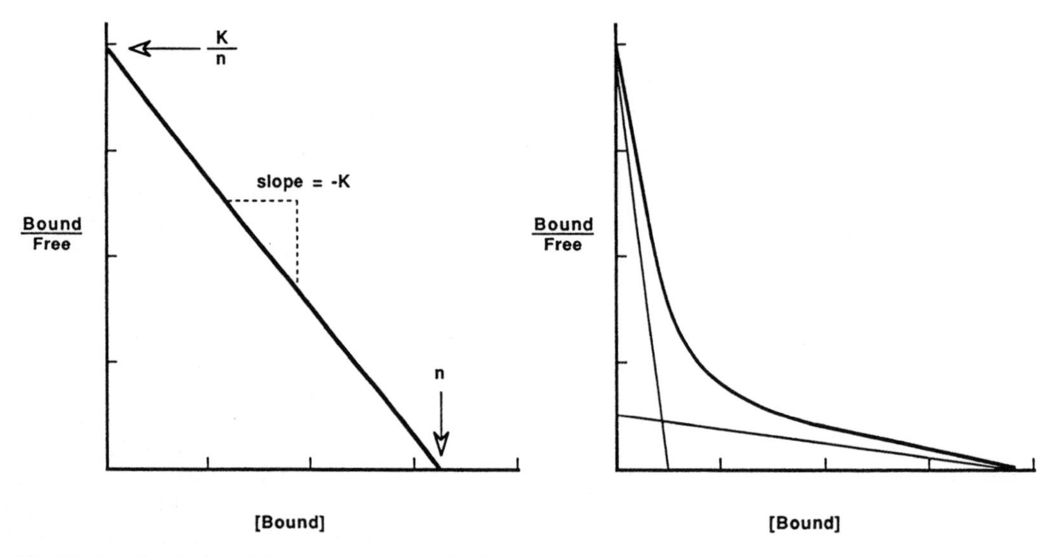

Fig. 11. Scatchard plots: left, one group of n binding sites per mole of enzyme and binding with association constant K; right, a more difficult example (see text).

two binding behaviors are additive, so that both individual lines should fall below the curve. The fit is easily obtained using non-linear least squares, or other computer programs that take this into account, although a graphical method is available (ROSENTHAL, 1967).

Experimental problems are also abundant. The titration must be performed over a range covering 10 to 90% of saturation of the enzyme. Lower levels of saturation tend to produce curvature near the B axis which can be misinterpreted as a weak class of binding site. The failure to approach saturation causes errors in the extrapolated y-intercept as well as the slope. The enzyme and ligand must both be present at substantial concentration because, depending on whether bound or free ligand is measured, the other form of the ligand is usually obtained by subtraction from the total ligand concentration. This caveat suggests that some systems cannot be studied using a Scatchard plot. If the dissociation constant is very low, essentially all ligand may be bound by the enzyme at low ligand concentration, so that it is impossible to measure or calculate the concentration of free ligand accurately. Since it is bound/free ligand that is plotted, errors in the concentration of free ligand are serious.

The assumption that sites are independent might also be false (e.g., cooperative binding) so that the behavior will appear to represent multiple classes of different sites. Lastly, it is not possible to determine how many classes of different sites are responsible for the observed behavior unless there is only one and the plot is linear. If the curve can be fit by m classes of n_m sites each, it can also be fit by $m+1$ classes. It is therefore unwise to attempt to resolve the data into more than two classes of sites unless there is clear information from other sources to verify the existence of more classes of sites. In addition, although the Scatchard plot can be and is often used in impure systems (e.g., whole cells, crude extracts, etc.), it is not possible to determine the binding stoichiometry unless the enzyme is pure and its molar concentration known.

3.5.6 Dependence of the Reaction Rate on pH

The pH dependence of an enzyme-catalyzed reaction has been used effectively to infer the identity of active site groups. Clearly, general acids and general bases functional in catalysis must be in their correct ionic forms for turnover to occur. Nucleophiles must possess a non-bonded (unprotonated) electron pair. Binding may also depend upon the charge of groups in the active site. Since pH is a logarithmic parameter and concentrations of protonated and deprotonated species (hence, observed rates) change by an order of magnitude per pH unit near their pK_a, it is common to plot logarithms of the kinetic parameters versus pH. The pH optimum of a reaction and the apparent pK_a's associated with it are often determined as part of the characterization of an enzyme, but sufficient information to interpret the rate versus pH profiles is often lacking. It is impossible to derive definite information from a v vs. pH profile unless the value of [S] used in the experiments relative to K_m is known. If [S] is known to be saturating at all pH values used, then all enzyme is present as ES, thus K_m is irrelevant. Any pH effect must come from groups involved in catalysis (or from substrate). Also, it is obviously but tacitly assumed (by the requirement that [S] is large compared to K_m) that the enzyme remains undenatured over the entire pH range used. If [S] is subsaturating, pH-dependent changes in K_m arising from groups involved in binding could also be responsible for changes in the rate. Therefore, it is essential that two levels of substrate be used in order to obtain meaningful information concerning mechanism. An alternative approach is to measure the rate of reaction at several values of [S] for each pH value, then plot $\log(V/K)$ (i.e., the reciprocal of the slope of a Lineweaver–Burk plot) versus pH.

The effect of pH on reaction rate can also originate from ionization of functional groups of the substrate. This possibility can be checked by using alternate substrates having different pK_a's or simply by measuring the substrate pK_a independently. Ionizations that can alter substrate binding can also alter inhibitor binding. Therefore, $\log K_I$ can also be plotted against pH to determine the pK_I for the reaction. Just as in the case of substrate, pH-dependent behavior can originate from ionizing groups of the inhibitor. If the inhibitor has a different pK_a than substrates, the origin of the pH effect on rate can be ascribed to the enzyme or to the ligand by comparing pK_a to the observed pK_a for the reaction.

A concise discussion of pH dependence of reaction rates is provided by WHITAKER (1994), and a somewhat more detailed discussion is presented by CLELAND (1982, 1990).

4 Practical Enzymology: Purification, Estimation of Purity

4.1 Laboratory-Scale Purification of Enzymes

Enzymology has definite requirements for pure enzymes and pure substrates for an isolated study of their fundamental interactions. The techniques for the laboratory-scale purification of enzymes have not changed dramatically since the development of affinity chromatography in the mid 1970s, but many significant improvements have been made. A recent summary of current methods for enzyme purification is found in DEUTSCHER (1990).

Organic chemists would not consider performing syntheses, recrystallization or kinetic analyses in impure or undefined solvents, lest the effect of a small contaminant exert some untoward effect. Water for use in enzymology (GANZI, 1984) must be purified by distillation, deionization or reverse osmosis, etc., for exactly the same reasons.

4.1.1 Extraction

The first step in an enzyme purification is nearly always the extraction of proteins from cells or tissue. Techniques to liberate the enzymes include disruption using a French pressure cell (FRENCH and MILNER, 1955), sonication, or tissue homogenization (POTTER, 1955), disruption or milling with glass beads, or a few other physical methods. The methods employed depend to some extent on the nature of the material from which the enzymes are to be extracted, e.g., bacteria (GUNSALUS, 1955), plants (NASON, 1955), cultured mammalian cells (LIEBERMAN and GINGOLD, 1962). Some older chemical methods, such as preparation of an acetone powder or cell rupture using dry ice and toluene are occasionally used (GUNSALUS, 1955; NASON, 1955), especially for walled cells, such as yeast and higher plants.

4.1.1.1 Stability of Crude Extracts

In general, cell disruption is performed at 4°C to prevent denaturation and to decrease protease activity. A few enzymes are cold-labile, and purifications of these enzymes are best done at room temperature. Cell or tissue disruption also liberates proteases. Nucleophilic proteases such as serine proteases can be inhibited by addition of inactivating agents such as diisopropylfluorophosphate (DFP) or phenylmethylsulfonyl fluoride (PMSF). Both of these reagents become covalently attached to the active-site nucleophilic group and cause permanent inactivation. Both are toxic, but the solid PMSF is easier to deal with than the volatile DFP. Also, because of the lability of freshly extracted enzymes, enzymes in various states of purification are generally stored frozen, except perhaps for ammonium sulfate precipitates. Frozen enzyme preparations should be kept at −20°C or −70°C and thawed rather quickly by placing the tube in room-temperature water. If the frozen solution is allowed to thaw slowly, ice surrounding protein molecules melts quickly, yielding isolated regions of very high local concentrations of proteins, which are subject to aggregation or proteolysis.

Care is also taken to prevent foaming during extraction. Proteins are easily denatured at the interface between air and water in the foam. Also, unpaired sulfhydryl groups oxidize easily in air-saturated solutions. Consequently, reducing agents such as 2-mercaptoethanol or Cleland's reagents (dithiothreitol or dithioerythritol) may be included (CLELAND, 1964).

4.1.1.2 Buffers for Enzyme Purification

Enzymes are usually extracted into pH-buffered solutions. Sodium or potassium phosphate or acetate buffers have always been popular (e.g., GREEN and HUGHES, 1985) for buffering near physiological pH, though buffers based on many other acid–base systems were employed historically (CLELAND, 1964). However, in cases where the phosphate ion may affect enzyme activity or interfere with assays, or when other pH values are required, "Good buffers" (GOOD et al., 1966) and related compounds (BLANCHARD, 1984) may be used. An array of these organic, usually zwitterionic, buffers is now available for buffering within most pH ranges. Tris(hydroxymethyl)aminomethane ("Tris"), commonly used to buffer near pH 8, HEPES, near pH 7, and MES near pH 6 are common choices.

Membrane proteins (PENEFSKY and TZAGOLOFF, 1972) present their own special set of problems. Most are not water-soluble and are denatured by organic solvents. A useful approach has been to include chaotropes, non-ionic detergents (e.g., Triton X-100) and emulsifiers (e.g., bile salts, such as deoxycholate) in the extraction buffers (HJELMELAND and CHRAMBACH, 1984).

4.1.2 Salting-Out and Other Precipitation Steps

One of the oldest methods of purification, salting-out, is still a useful procedure, either as a crude purification step or as a method to concentrate the extracted proteins before the next purification step (GREEN and HUGHES, 1955). Crude purification or concentration of extracts uses ammonium sulfate fractionation. As mentioned earlier, the addition of ammonium sulfate to a solution of protein lowers the solubility of the protein, and it lowers the solubility of different proteins to different extents. Solid enzyme-grade ammonium sulfate is pulverized and added slowly, with stirring, in the cold, to a crude extract of protein to a particular final concentration which is usually expressed as a percent of saturation. The precipitate is allowed to flocculate and is removed by centrifugation. The supernatant solutions and resuspended pellet are assayed for total activity and total protein. If necessary, more ammonium sulfate is added to the supernatant solution, using a table or nomogram (GREEN and HUGHES, 1955) to determine the amount of ammonium sulfate required to reach the desired concentration. With luck, one of the fractions (e.g., the precipitate that occurred between 40 and 60% of saturation with ammonium sulfate) may contain the bulk of the activity of interest or may be enriched in the activity. This fraction is used in further purification steps.

It is well-known that the solubility of a protein is lowest at its isoelectric point. It may be useful to adjust the pH of the extract to the isoelectric point of the enzyme if it is known. However, ammonium sulfate is reasonably acidic, and most buffers are completely overpowered by the high concentrations of ammonium sulfate used in the precipitation. It is therefore more effective to adjust the pH after the addition of the salt. Since most proteins are acidic, the ambient pH after addition of ammonium sulfate may be near the isoelectric point and adjustment may not be necessary.

If an ammonium sulfate fractionation is to be reproducible in other laboratories, the total protein concentration of the extract should be reported. Less protein precipitates from solutions of lower protein concentration.

Other precipitants have been used, including metal salts and organic solvents (GREEN and HUGHES, 1955; KAUFMAN, 1971) or compounds such as polyethylene glycol (FRIED and CHUN, 1971; INGHAM, 1984). In addition, many older procedures have heat precipitation steps in which proteins other than the one of interest are precipitated or denatured.

Precipitation using ammonium sulfate is also used as a concentration step prior to gel filtration chromatography, crystallization, etc., in which either small volumes or concentrated solutions are required. The protein is collected by centrifugation, or by filtration if the protein is reluctant to flocculate. Filter aids packed into large chromatography columns can be used to entrap suspended particles which can be "eluted" by slowly washing the column with buffer. Other methods for concentrating protein solutions include lyophilization (freeze-drying) (EVERSE and STOLZENBACH, 1971) and various forms of ultrafiltration (BLATT, 1971) using vacuum or pressure, or dialysis versus an osmotically active polymer or absorbant material.

4.1.3 Desalting

Before an ion-exchange chromatographic step or after an ammonium sulfate fractionation, it is usually necessary to remove the salt from the solution of protein. Desalting is accomplished in one of two ways: dialysis (MCPHIE, 1971) or gel filtration (PORATH and FLODIN, 1959).

4.1.3.1 Dialysis

Dialysis is performed by filling a section of dialysis tubing (a semipermeable membrane) having a sufficiently small molecular weight "cut-off", with the protein solution, and placing the filled tubing in a large volume of buffer. The decrease in salt concentration can be calculated easily from the ratio of the volumes inside and outside of the bag. Dialysis requires a few hours, after which the bag may be transferred to fresh buffer if the reduction in salt concentration effected by one cycle is deemed to be insufficient.

In dialysis, all small molecules, including salt ions, metal ions and cofactors, pass through the membrane, which retains only macromolecules. The equilibrium is subject, however, to specific binding effects and to the Donnan equilibrium. Neither tightly bound metal ions and cofactors, nor counterions to the macromolecule are effectively removed.

Since the initial solution in the bag is of much greater osmotic strength than the surrounding buffer, the bag generally increases in volume. The volume of the contents of the bag must be measured after dialysis if either total protein or total enzyme units are to be calculated.

4.1.3.2 Size Exclusion Chromatography for Desalting

Gel filtration (also called size exclusion) is a form of chromatography in which solutes are separated from one another based upon the degree to which the solutes fit into pores on the stationary phase (PORATH and FLODIN, 1959). Several theories exist, and four useful treatments were compared experimentally by SIEGEL and MONTY (1966). Gel filtration media can be obtained with many different pore sizes, so that the technique may be tailored to the separation needed (REILAND 1971). For desalting, usually a gel of rather small pore size is employed so that only very small solutes ($M_r < 5000$) fit easily into the pores, and enzymes are totally excluded. For bulk desalting of protein solutions, usually a porous polysaccharide such as Sephadex G-25 or BioGel P10 is selected. The elution behavior in gel filtration follows the standard rules of partition chromatography:

$$V_e = V_0 + K_{av} V_s$$

in which V_e is the elution volume, V_0 is the volume of the mobile phase between particles (the "void volume"), K_{av} is a partition coefficient of sorts, and V_s is the volume of the stationary phase. In gel filtration chromatography, the quantity V_s includes the volume of pores plus the volume of the beads themselves, a departure from the true partition chromatography theory. K_{av} correlates well with the Stokes radius of the solute and varies between 0 (for excluded solutes) and 1 (for solutes that are much smaller than the pores). Thus the void volume can be measured easily by determining the elution volume of any excluded molecule, typically blue dextran. The volume of the stationary phase can be calculated simply by measuring the bed volume, V_{bed}, the total volume occupied by the packing (which includes the void spaces between particles) and subtracting the void volume. The void volume is typically about a third of the bed volume. Thus K_{av} can be determined for any solute from its V_e by the relation

$$K_{av} = \frac{V_e - V_0}{V_{bed} - V_0}.$$

Brief inspection of this equation yields the perhaps surprising prediction that large solutes elute earlier than small ones. So, gel filtration is not a filter at all.

It should be clear that K_{av} for salts is essentially 1, while that for proteins is essentially zero, hence separation is maximally efficient for desalting. Because of the unusually large separation, a desalting column can be loaded with a volume of sample that is up to 30% of the bed volume to maximize throughput.

Gel filtration is not without artifacts. Some aromatic compounds and strongly charged proteins adsorb to the packing itself and are retained by this second mechanism. This problem can usually be resolved by using a buffer (i.e., mobile phase) of rather high ionic strength (>50 mM). It is also worth mentioning that gel filtration is a rather inefficient variety of chromatography. (The efficiency of a column is quantified as the number of theoretical plates experienced by a solute during chromatography. The number of plates, N, which represents the number of repetitions of the partitioning process, is calculated from the equation $N=16(V_e/W)^2$, where V_e is the elution volume (or retention time) and W is the width of the peak at its base, measured in the same units at V_e.) Efficiency is maximized by using low flow rates and small diameter media. For desalting, high throughput is more valuable than high resolution, so larger media (e.g., coarse rather than fine or superfine) are used, but the flow rates should be kept low to prevent band broadening.

4.1.4 Ion Exchange Chromatography

Since proteins have different net charge and charge distribution, ion exchange chromatography can be an effective purification tool (HIMMELHOCH, 1971; HIRS, 1955). For bench-top preparations, usually gravity-flow columns are employed, but HPLC (REGNIER, 1984) and automated HPLC-like systems have grown in popularity. For gravity flow or for use with low pressure peristaltic pumps, ion exchange media are usually carbohydrate-based. Charged groups are attached to solid supports ("inert phase") such as sepharose, Sephadex and cellulose. Since these carbohydrates are compressible, they are not used in higher-pressure systems, and more rigid inert phases such as TSK (a polyether-coated gel) are used. For higher pressures, reinforced polysaccharides, and organically coated silica (e.g., TSK) are used. The resins, especially poly(styrenedivinylbenzene) described by HIRS (1955) for use with enzymes were used by MOORE and STEIN (1952) in their famous amino acid analyzer. They are commonly employed for ion exchange chromatography of small molecules, but have given way to the ion exchange polysaccharides (e.g., PETERSON and SOBER, 1956) for preparative applications in enzymology. The charged groups used with the solid supports depend to some extent on the chemistry of the support material itself, but are remarkably similar. Groups containing charged nitrogen atoms are almost universally used for anion exchange media. These include, from strong to weak, quaternary amino methyl or ethyl (QAE), tertiary amino (diethylaminoethyl, DEAE, or diethylaminomethyl) and secondary plus tertiary nitrogens (polyethylenimine, PEI). The quaternary amino compounds are positively charged at any pH, but the others must be used at a pH below the pK_a of the protonated form (~10, for DEAE). The conjugate base of the strongly acidic sulfonic acid (i.e., alkyl or aryl sulfonate) and the weakly acidic carboxylic acids (e.g., carboxymethyl, CM) are the most common charged groups employed in cation exchangers. The carboxymethyl packings must be used at a pH above their pK_a (~4). Methods for determining the optimal pH for separation of proteins depends, of course, on the proteins. Since most proteins are acidic, they are negatively charged at pH 7–8. They therefore adsorb to a positively charged stationary phase to which they act as counterions, providing that other anions are not available to play the role of counterion. The cationic stationary phase is known as an anion exchanger because it functions by exchanging one anionic counterion for another. Anionic proteins may bind more tightly to anion-exchange stationary phases than simple salts because they possess more negative charges than a simple anion. However, it is not the total charge on a protein, but the charge density that determines the affinity. More precisely, it is the charge distribution. Since a protein may interact with a stationary phase on one side at a time, proteins with densely charged patches may be bound more tightly (DRAGER and REGNIER, 1986, 1987).

At pH values below the isoelectric point of a protein, the net charge is positive, so negatively charged stationary phases (cation exchange phases) are used. If a protein has an isoelectric point near neutrality, either a cation exchange or an anion exchange system can be used, depending on the pH employed.

The important considerations in choosing an optimal pH for separation of enzymes by ion exchange chromatography have been reviewed by RICHEY (1984).

Protein solutions are generally desalted, then loaded onto a column packed with a stationary phase having the appropriate charge. Loading can often be done as rapidly as the columns will flow without undue pressure; proteins that adsorb are retained at the top of the column. As long as the capacity of the column is not exceeded, liters of a (desalted, buffered) crude extract can be loaded onto a column of modest size, so that a pre-chromatography concentration step is not needed. After loading, the column is washed with the loading buffer to remove unadsorbed and weakly adsorbed proteins. The adsorbed proteins are then eluted by washing the column with buffers of increasing salt concentration (e.g., NaCl), which corresponds to increasing solvent strength. This method of elution using a series of isocratic (constant strength) elutions of progressively increasing strength is known as batch elution. The ion having a charge of the same sign as the protein can act as a displacing ion by competing for charged sites on the stationary phase. (A physical model and a mathematical model with a simple interpretation has been devised by DRAGER and REGNIER (1986, 1987). At some concentration, the eluting ion competes effectively with the protein, which accordingly, spends a larger fraction of its time in the mobile phase, leading to elution. This concentration would be ideal to purify the protein of interest providing that more loosely bound proteins were removed first, because it affords the maximum discrimination among the charge densities of the proteins still on the column. However, the protein might elute as a broad, dilute band. Also, it is not possible to know *a priori* what the salt concentration of this solvent should be. A simple and common solution to elution is to employ a linear concentration gradient of salt, using an easily constructed siphon apparatus. Such a gradient can cover a range from 0 to 1 M NaCl over the volume of a few hundred mL to a few liters, depending on the dimensions of the column and the steepness of the gradient desired. A major advantage of gradient elution is that proteins having a wide range of affinities for the column can be eluted in a single run. The information obtained from a gradient elution may be used to determine an optimum salt concentration to be used in isocratic elution, but the procedure is not straightforward. The theory of gradient elution is messy, even in the simplest case (DOLAN et al., 1979; SNYDER et al., 1979). One egregious misstatement appears in numerous papers on enzyme purification: "the enzyme elutes at such and such a concentration of sodium chloride". Because the gradient travels much more rapidly in the column than the protein (the protein is retained to some extent), the concentration of sodium chloride in which the enzyme actually appears at the bottom of the column is much higher than the concentration at which it began to elute appreciably. Thus, the concentration in which it appears to elute (concentration of sodium chloride in the fraction in which the activity appears) is much too strong for use as an isocratic eluent. In addition, the concentration in which the enzyme appears varies with the dimensions of the column; longer columns cause the enzyme to appear to elute in a higher salt concentration, simply because the gradient progresses as the enzyme moves down the column.

To exercise maximum control over the system, it is useful to separate the effects of pH from those of ionic strength during ion exchange chromatography. One of the ions involved in the buffering system bears the same charge as the protein and can therefore act as a displacing ion. The concentration of this ion should not change with pH, so it should not be the one involved in the equilibrium with solvent protons. Buffering ions selected for use in ion exchange chromatography should have the same charge as the column, i.e., cations for an anion exchange column, anions for cation exchange. Hence, phosphate buffers are used for cation exchange chromatography, and Tris (for instance) buffers are used for anion exchange.

For the sake of performance and reproducibility, it is necessary for the column to be completely equilibrated with the starting solvent. Equilibration can be checked by measurement of both pH and ionic strength (e.g., by conductivity) prior to loading the column.

Elution from an ion-exchange column could also be accomplished using a change in pH. Stepwise pH changes are sometimes employed, but do not generally produce high resolution of complex mixtures. Reproducible continuous pH gradients are difficult to obtain because so many of the components in the system engage in acid–base equilibria. A workable system along these lines has been devised using a proprietary mixed-bed packing and a multi-component buffer system to elute proteins at their isoelectric pH (FÄGERSTAM, 1983). The process is called chromatofocusing because of a loose analogy to isoelectric focusing gel electrophoresis.

4.1.5 Hydroxyapatite Chromatography

Although size exclusion and ion exchange chromatography have been used often since the 1960s, a few other chromatographic techniques are occasionally employed. A useful stationary phase has been hydroxyapatite (BERNARDI, 1971; SWINGLE and TISELIUS, 1951). Like other calcium phosphate gels employed for chromatography, the adsorbant has had a checkered and mystical past; it always seemed necessary to indicate the age of the adsorbant and how it had been aged (COLOWICK, 1955). GORBUNOFF and TIMASHEFF (1984) have reported a systematic study and can explain many of the anomalies involved in the use of hydroxyapatite. It appears that there are two different interactions that are fundamentally opposite. The gel, which has the formula $Ca^{2+}_{(10\text{-}X)}(H_3O^+)_{2X}(PO_4^{3-})_6(OH^-)_2$, is basically a crosslinked series of Ca^{2+} and PO_4^{3-} ions. Since the surface of the gel is largely negative, owing to the presence of the phosphate oxygens, positive groups on the proteins can bind to the gel. On the other hand, there are some Ca^{2+} sites near the surface of the gel which can attract carboxylate groups of proteins. Thus, the gel binds acidic proteins because they have negative groups on the surface, and binds basic proteins because they have positive groups on the surface. Either interaction can be enhanced by changing the equilibration buffer. If the mobile phase is phosphate buffer, the Ca^{2+} sites tend to be covered and more phosphate sites exposed. If the mobile phase is $CaCl_2$, there are more cationic sites and fewer partially negative sites. This kind of dynamic flux probably plays a role in the ageing process.

In order to explain the elution of proteins from hydroxyapatite, GORBUNOFF classifies proteins into three loose groups: (1) basic proteins ($pI>8$), which are eluted by 1–3 mM Mg^{2+}, (2) neutral proteins ($5 \geq pI \geq 8$), which are eluted by 1–2 M Mg^{2+}, and (3) acidic proteins, which are not eluted by Mg^{2+}, even at 3 M. For basic proteins, there are two desorption mechanisms. The first is competition between the proteins' positive charges and the divalent cation in the buffer for the phosphate portion of the gel. These interactions are important at very low divalent cation concentration because divalent cations form tight complexes with the phosphate. The second effect is simple charge screening, which is non-specific and requires much higher concentrations. F^-, Cl^-, SCN^- and phosphate are all equally effective in this mechanism. For acidic proteins, desorption occurs by competition between anions that bind the calcium portion of the column and the carboxylate groups of the protein. F^- and phosphate are effective but Cl^-, which does not complex with calcium, is not.

4.1.6 Size Exclusion Chromatography

In addition to its use in desalting procedures, size exclusion chromatography can be used for separation of proteins from one another. As described above, the method discriminates among proteins on the basis of their effective hydrodynamic radius, their Stokes radius (SIEGEL and MONTY, 1966), and several models exist that purport to explain the mechanism of separation. For globular proteins, there is a reasonable correlation between the logarithm of molecular weight and K_{av}, although there is a strong depencence upon molecular shape: rods and elipsoids with large axial ratios behave as though they have an inflated molecular weight. Nonetheless, the stationary phase for size exclusion chromatography is chosen almost solely on the basis of the expected molecular weight of the protein of interest. It is obvious that all molecules having Stokes' radii larger than the pores have $K_{av}=0$ and cannot be separated from one another. It is perhaps not as obvious that molecules much smaller than the pores have $K_{av} \sim 1$ and cannot be separated from one another, either. Size exclusion media are therefore sold according to the useful range of "molecular weights" they are capable of separating. Since the correlation is really according to Stokes' radius, the discrepancy is often accounted for by listing two molecular weight ranges, one for globular proteins and one for dextrans, which have large axial ratios. The HPLC version of size exclusion chromatography has been treated by UNGER (1984).

4.1.7 HPLC

High pressure (or high performance) liquid chromatography (HPLC) was developed because small particle (<20 μm) column packings are more efficient (i.e., yield more theoretical plates) than larger-diameter packings (GIDDINGS, 1965), but require high pressure to obtain adequate flow rates. HPLC has not traditionally been used to purify proteins because it is engineered to produce sharp peaks when loaded with small volume (generally <25 μL) samples. However, some of its principles can be applied profitably to preparative techniques. There is virtue in automation because it leads to reproducibility. And, the possibility of autosampling and microprocessor-controlled systems suggests that repeated, automated runs of column chromatography can lead to relatively high throughput even though relatively small samples are used in each run. In addition, the development of macroporous adsorbants has increased the capacity and efficiency, rendering HPLC a useful method. Furthermore, several companies have developed lower pressure automated liquid chromatography systems that mimic the features of HPLC, including programmed solvent gradients and detection systems with low mixing volumes. One system bears the trademarked name of FPLC (Pharmacia), for fast protein liquid chromatography. The systems often have programmable fraction collectors, which are necessary for preparative applications and rarely found on HPLC chromatographs. In general, such systems employ larger diameter columns than those used in HPLC (for increased throughput), larger particles, and lower pressure. Essentially any kind of chromatography commonly used for enzymes can be used in automated systems.

4.1.8 Affinity Chromatography

One of the few quantum leaps in protein purification was developed during the 1970s. True affinity chromatography is a system based of the binding of an enzyme or other protein to its natural ligand, such as a substrate, product, or effector, or of any compound resembling the substrate, product, intermediate or transition state. There are cases in which the resemblance between the stationary phase and the natural ligand is not clear or is purely coincidental. The ligands are covalently attached to a solid support, often sepharose, via a linker or spacer arm. The spacer arm fixes the ligand to the solid support, but at some distance, to prevent steric interaction between the protein and the solid support.

Almost all low pressure affinity systems employ either agarose or sepharose as the solid support ("inert phase"), although glass beads and polyacrylamide beads are sometimes used. In HPLC, where more mechanical stability is required, derivatized silicas are used (LARSSON, 1984). The porous polysaccharide gels contain many OH groups which can be "activated" using a number of common reagents. Cyanogen bromide, CNBr, for instance, is used to convert the alcohol groups to more reactive groups. Following activation of the gel, the spacer arm is attached. A common one is 1,6-diaminohexane. The sequence of reactions seems to be different for sepharose and agarose. Sepharose is an α-1,6-linked polymer of glucose with α-1,3-crosslinks; agarose is a linear polymer of D-galactosyl-3,6-anhydro-L-galactosyl units. For such crosslinked dextrans the activated gel with an attached 1,6-diaminohexane spacer arm is thought to have one of the following structures or a mixture of the two.

$O-C(=NH(CH_2)_6NH_2)-O$ (cyclic imidocarbonate)

OH, $O-C(=NH)NH(CH_2)_6NH_2$

This spacer arm contains a primary amine, which is a relatively good nucleophile and can add to double bonds (especially C=O and C=N), participate in nucleophilic substitutions and form amide bonds.

Purine-containing compounds (e.g., adenosine, AMP, ADP, ATP, NAD(P)(H), coenzyme A, aminoacyl adenylates) are of great importance in biochemistry and are commonly attached to solid supports. One way to make the attachment is to use the 8-bromo purine (e.g., 8-bromo AMP) analog of the compound. Under mildly basic conditions, the spacer arm nitrogen displaces bromide to yield 8-(aminohexylamino-sepharose)-AMP (or NADH ...). In principle, this reaction can be used for any water-soluble ligand that has a displaceable bromine. If the ligand contains a carboxyl group or if such a derivative can be made easily, it can be covalently attached to the side arm by forming an amide bond with the aid of a water-soluble carbodiimide activating agent such as 1-ethyl-3-(3-dimethylaminopropyl) carbodiimide hydrochloride (EDAC or EDC) and 1-cyclohexyl-3-(2-morpholinoethyl) carbodiimide metho-*p*-toluene sulfonate (CMC). Carbodiimides promote the formation of amides and can be used to link primary amines to carboxylic acids of ligands or proteins.

Bis epoxides, such as 1,4-*bis*(2,3-epoxypropoxy-)butane and N-hydroxysuccinimide esters are also used as spacer arms. Cyanogen bromide-activated sepharose is commercially available and may be purchased as may agarose with choice of several spacer arms attached, including N-hydroxysuccinimide esters.

For HPLC applications, epoxy-, glyceryl-, propyl-, aldehydo-, and tresyl silicas are used. They are all prepared from silica by initial reaction of silica with γ-glycidoxypropyltrimethysiloxane, followed by further derivatizations (LARSSON, 1984).

4.1.8.1 Choice of the Affinity Ligand

There is no way to know *a priori* what affinity ligand will be bound by a particular enzyme. One or more may be prepared or purchased, then screened for binding capacity (number of activity units adsorbed from a crude, desalted extract). Affinity adsorbants that have useable binding capacity must also be screened for the ability to release the native enzyme in response to a wash with buffer solutions containing high salt or a specific eluting molecule (competitive inhibitor, etc.).

For use, affinity adsorbants are packed into a small column and loaded with a desalted extract. Loading is sometimes continued until the capacity of the column is exceeded, when appreciable activity begins to elute from the column. The loaded column is washed with loading buffer until all unbound or loosely bound proteins are removed. Elution is begun using either high-salt buffers or a specific eluting agent, such as a competitive inhibitor or analog. The purification obtained from a single affinity chromatography step can be spectacular and the yield impressive.

It is perhaps obvious that affinity chromatography offers two opportunities to obtain specificity: loading and elution. Even if loading cannot be accomplished with great specificity, by judicious choice of eluting agent (LARSSON, 1984; WILCHEK et al., 1984), the chromatographer may be able to obtain specific elution.

The presence of substrates, cosubstrates, products, etc., in the loading buffer can have an appreciable effect on adsorption to the affinity column. If these compete with the affinity ligand, the effect is obvious and undesirable. However, especially in the case of multisubstrate enzymes, the presence of cosubstrates can have a favorable effect on binding. A subtle corollary to this statement is that it may be possible to employ a very common (hence, commercially available) affinity column, and enhance the specificity by adding cosubstrates to the loading buffer. For example, AMP-sepharose might be considered a generic affinity column, because all pyridine-nucleotide-linked and FAD-linked dehydrogenases, kinases and a host of other enzymes might be expected to recognize it as a portion of their cosubstrate. Yet, in some cases, the presence of the partner substrate enhances binding specifically of the enzyme of interest. An artful example is the affinity purification of methionyl-tRNA synthetase on an AMP column in the presence of methionol and magnesium (FAYAT et al., 1977). The combination of the alcohol and AMP apparently mimic a reaction intermediate, an aminoacyl adenylate, so that the synthetase binds in preference to all other AMP-binding enzymes. The enzyme was eluted by either ATP or methionine, both of which bind at sites that overlap with the aminoacyl adenylate site.

Proteins can be attached to solid supports via the formation of amide bonds to a linker molecule with either the carboxylate or amino groups of amino acid side chains using water-soluble carbodiimides. Thus, immobilized antibodies could be used to bind a protein toward which they had been raised or another protein with identical epitopes, even if the protein were present only in trace quantities. This approach is not useful unless the protein or a close relative has previously been prepared in sufficient purity and quantity to obtain antibodies. Lectins can also be used to chromatograph oligosaccharides or glycoproteins containing the sugars recognized by the lectin. Concanavalin A attached to sepharose is probably the most common.

4.1.8.2 Pseudoaffinity Chromatography

The resemblance between an enzyme's natural ligand and the affinity ligand is ultimately judged by the enzyme, not by the scientist. It has been found that a number of enzymes bind to immobilized organic dyes in a manner mimicking affinity chromatography (reviewed by LOWE and PEARSON, 1984). It has been said that aromatic heterocycles resemble the adenine ring, so pyridine–nucleotide-linked dehydrogenases bind to Cibacron Blue™. But, serum albumin also binds, perhaps by mistaking the organic compound for a natural lipid. Nonetheless, sepharose-bound dyes have been used widely for purification of proteins. Several varieties (and colors) exist (see catalogs of companies such as Pharmacia and Biorad, or LOWE and PEARSON, 1984), each with a different range of proteins that bind to it. Companies offer kits containing samples of their products which can be screened for binding capacity for the enzyme of interest in much the same way as putative affinity adsorbants are screened.

4.1.9 Hydrophobic Interaction Chromatography

When affinity chromatography was being developed, journal reviewers were concerned that what was being reported as affinity chromatography might actually be binding of the enzyme to sites on the adsorbant other than the putative affinity ligand (i.e., to the "inert phase"). They required authors to include in their papers control experiments in which the capacity of an affinity adsorbant was compared to those of the solid support and support with spacer arms attached. Ironically, some enzymes adsorbed to sepharose-bound hexylamine or other spacer arms rather well, and the adsorption was reversible. These observations led to the development of another branch of chromatography, now called hydrophobic interaction chromatography (HIC) (SHALTIEL, 1984). The key feature of HIC stationary phases is that relatively nonpolar organic ligands are affixed to a polar-surfaced support. The ligands are commonly $-NH-(CH_2)_n-X$, where X is NH_2, OH, COOH or even H. The length of the chain, n, varies typically from 1 to 8. In addition, phenyl sepharose has become quite popular. The nonpolar ligands are attached relatively sparsely, compared to reversed-phase chromatography stationary phases. This feature allows the ligands to penetrate slightly into the interior of a protein by hydrophobic interaction, while the polar surface of the protein makes contact with the polar surface of the polar support. These interactions should be encouraged by the presence of structure-building solutes and discouraged by chaotropes. For this reason, proteins are sometimes loaded in buffers containing ammonium sulfate, then eluted using a decreasing ammonium sulfate gradient, sometimes in combination with an increasing concentration gradient of ethylene glycol, or of "deforming buffers", such as imidazole/mercaptoethanol. The number of theoretical plates experienced by proteins in HIC appears to be greater than in gel filtration. HIC has therefore become a popular adjunct to ion exchange and gel filtration chromatography.

4.1.10 Covalent Chromatography

Chromatography depends on reversible interactions between a solute and a stationary phase. Any kind of interaction can be used in principle, as long as it leads to reversible binding. Reversible covalent bonding has been employed in several cases. Reversible disulfide bond formation has been used most successfully. One of the earliest applications of covalent chromatography was the purification of papain from papaya latex (BROCKLEHURST et al., 1973). One of the first commercially available disulfide columns was simply an affinity-like column containing glutathione directly linked to cyanogen bromide-activated sepharose. The $-S$-group of glutathione was engaged in a disulfide bond with 2-dipyridylsulfide. SH-containing proteins bind by forming a mixed disulfide and releasing 2-thiopyridone. The enzyme can be eluted using cysteine, which replaces the enzyme in the disulfide linkage.

4.1.11 Molecular Genetics

Probably the most significant boon to enzymology since the development of chromatography is the advent of cloning of enzymes. This subject is of such a large scope that several other articles (Vol. 5) in this series have been devoted to it. It will therefore be essentially ignored, here.

4.2 Assessment of Purity

4.2.1 Specific Activity

The ability to assess the purity of an enzyme is essential to detailed enzymological studies. Probably the most fundamental measure of purity, though certainly not the most sensitive to contaminants, is the specific activity. The specific activity of an enzyme is the number of enzyme activity units divided by the amount of protein, usually expressed in milligrams. This measurement of purity only expresses purity in terms of contaminating (or inactive) protein. Thus, an enzyme may be deemed $>90\%$ pure and still contain massive amounts of other components, such as buffers, salts or other non-protein material. For specific activity measurements, the rate of enzymatic reaction is measured at saturating levels of substrate (unless substrate inhibition complicates matters), so that it is V_{max} that is determined. Activity units are calculated in Δmoles/time rather than in Δconcentration/time, because a fixed amount of enzyme turns over the same number of moles of substrate no matter what the volume, so long as the enzyme remains saturated. The standard activity unit is the International Unit of enzyme activity (IU), which is the number of μmoles of substrate (or product) removed (or formed) per minute. The SI unit of activity, the katal or kat, expresses the rate in $mol \cdot s^{-1}$, so 1 μkat = 1 IU.

4.2.1.1 Coupled Enzyme Assays

It is often difficult to monitor the rate of an enzymatic reaction because neither the substrate nor product has a measurable property (e.g., absorbance, fluorescence, optical rotation) that would allow the enzymologist to distinguish between them. One popular solution is to include in the assay another enzyme that is capable of converting the product to a second product that has a distinguishable property. A very common system involves the addition of a pyridine–nucleotide-linked dehydrogenase and the appropriate form of the coenzyme (NAD(P)H or $NAD(P)^+$). NAD(P)H has a strong absorbance at 340 nm with an extinction coefficient of $6.22 \times 10^3\ M^{-1}$, whereas the oxidized form of the coenzyme absorbs negligibly at 340 nm. Thus, the production of glucose 6-phosphate by hexokinase could be monitored by observing the reduction of NAD^+ to NADH (by absorbance at 340 nm) coupled to the oxidation of glucose 6-phosphate to 6-phosphogluconate by glucose 6-phosphate dehydrogenase. Such an assay requires that the coupling enzyme and its substrates be present at sufficient level so that the second reaction is not rate-limiting. A corollary to this requirement is that, if a coupled enzyme assay is used to determine other kinetic properties of an enzyme such as temperature or pH dependence, the activity of the coupling enzyme must not become limiting due to the effect of changing conditions.

Since enzymic reactions are quite sensitive to pH, salt concentration, temperature and sometimes specific components in the assay mixture (e.g., K^+, Ca^{2+}, Mg^{2+}), these factors must be controlled and reported. These effects have been described earlier, except for that of temperature. Even though the enzyme exerts its catalytic effect by lowering the activation energy of a reaction, and thus, its temperature dependence, the reaction is still determined by the fraction of reactants that have enough thermal energy to exceed the activation energy, lowered though it may be. Thus, the Arrhenius equation still holds, and the reaction rate increases with temperature proportional to $e^{-\frac{E_a}{RT}}$. However, at some temperature, the enzyme denatures, by a process with its own activation energy, and the increase in rate reverses, the subsequent decrease in rate corresponding to a loss in active enzyme. Even at temperatures below the denaturation temperature, the enzyme may lose activity with time.

4.2.1.2 Measurement of Protein Concentration

The other element of the specific activity is the protein concentration. There are about a dozen methods for measuring protein, each of which responds to a different property of proteins. For this reason, it is not surprising that different methods sometimes give different answers. The most common methods are described below.

Because of the UV absorbance of tryptophan, and to a lesser degree, tyrosine, phenylalanine and cysteine, proteins absorb UV light near 280 nm. On the average, a 1 mg/mL solution of protein has an absorbance of 1 absorbance unit. The total protein concentration of a solution containing many proteins (e.g., a crude extract not containing nucleic acids) can therefore be measured quite accurately from a simple absorbance measurement. However, the amino acid composition of individual proteins varies sufficiently that their extinction coefficients range between about 0.4 and 2 absorbance units per mg/mL. The nearly universal primary protein standard used to calibrate other methods, bovine serum albumin, has an extinction coefficient of 0.66. To minimize the variation in extinction coefficients among different proteins, some researchers use the absorbance of the peptide bond, which occurs below 240 nm. The wavelength of maximum absorbance (λ_{max}) occurs near 190 nm, but most species containing double bonds (including nucleic acids, carboxylic acids, phosphate, sulfate) absorb strongly below about 210 nm, so the method is limited in usefulness.

In addition to variability in extinction coefficients, UV absorbance measurements are highly susceptible to interference, especially by nucleic acids. Thus crude extracts that may contain DNA and RNA may give inappropriately high readings. A traditional solution to the problem is to measure absorbance at two wavelengths, 280 and 260 nm (the wavelength of maximum absorbance of nucleic acids) and use simultaneous equations or a table to calculate the protein (LAYNE, 1957). The difference in absorbance at 280 and 215 nm can also be used (WHITAKER and GRANUM, 1980).

Another problem associated with UV absorbance at 280 nm is that it is not very sensitive. Colorimetric reactions have therefore been devised to enhance sensitivity. The biuret method (ROBINSON and HODGEN, 1940) is much more sensitive and depends upon the number of peptide bonds, so that it gives relatively uniform detection of different proteins. Peptides with more than two peptide bonds form a blue-colored complex with Cu^{2+} salts (citrate or tartrate) in alkaline solution which can conveniently be detected at 540 nm. This method is quite reliable, and protein determinations used for commercial enzyme preparations are usually performed using the biuret procedure. It is relatively insensitive to interference. The color yield is not precisely reproducible from day to day or in different laboratories, so accuracy is increased by running standard curves, usually using known concentrations of bovine serum albumin, alongside the unknowns.

The biuret procedure is not sensitive enough to be used in all cases, and a procedure >100 times more sensitive, developed in LOWRY's laboratory, was published in 1951 (LOWRY et al., 1951). The paper by LOWRY, ROSEBROUGH, FARR and RANDALL rapidly became the most referenced scientific publication in history, and remains so today. The method depends on the Cu^{2+}-catalyzed oxidation of aromatic amino acids by phosphotungstic and phosphomolybdic acid that produces a blue color which is measured at 500, 560, 650 or 700 nm (the absorbance is greater near 700 nm) (HARTREE, 1972). Like the biuret method, standard curves are generally run along with the unknowns. The method has two drawbacks: it relies on the presence of aromatic residues in a protein, and it is subject to interference by a few common components of enzyme solutions, including reducing agents (e.g., 2-mercaptoethanol) and the common buffer, Tris.

To obtain uniform color yields for different proteins and avoid interference by reducing agents, BRADFORD (1976) developed a protein assay utilizing Coomassie brilliant blue G-250. The Bradford reagent is dissolved in acidic ethanol solution. It is brownish (red protonated form plus green neutral form) in solution, but its pK_a changes upon binding to proteins and the resulting anionic form is blue. The blue color is measured at 595 nm and compared to a standard curve. The method is about 10 times more sensitive than LOWRY's method.

The newest popular method of protein determination is marketed by Pierce Chemical Company and is termed the Pierce or BCA method. Like the Lowry method, it depends on the reduction of Cu^{2+} to Cu^{+} (SMITH et al., 1985). The Cu^{+} produced is complexed by bicinchoninic acid (BCA); the $Cu(I) \cdot BCA_2$ complex has a high absorbance at 562 nm. Reducing agents and chelating agents interfere with the BCA assay, but the color yield is relatively constant.

KJELDAHL's method for determination of protein (BRADSTREET, 1965) is inconvenient and insensitive. It involves the metal ion-catalyzed digestion of protein by mineral acid to yield free ammonium ion, which is determined by Nessler's reagent or distilled into an acid trap containing a known excess of nonvolatile acid. The remaining acid is titrated with base to determine how much ammonia was trapped. This method measures nitrogen rather than protein and the weight percent of nitrogen calculated from the titration or Nesslerization must be multiplied by a conversion constant to calculate the protein concentration. Most proteins contain about 16% nitrogen, so the common conversion factor is 1/0.16 = 6.25. The obvious advantage of this method over other methods of protein determination is that the sample need not be soluble. The Kjeldahl method is rarely used in enzymology.

Occasionally, proteins that are poor in aromatic residues are determined by precipitation using trichloroacetic acid or another precipitant, followed by measurement of turbidity (LAYNE, 1957; TAPPAN, 1966).

4.2.2 Polyacrylamide Gel Electrophoresis

Polyacrylamide gel electrophoresis (PAGE) has been widely used to assess the purity of a protein. Electrophoresis is a method for separation of proteins, thus the appearance of a single protein band on a gel is evidence for, though not proof of purity. There are a number of variations commonly employed for gel electrophoresis (DEYL, 1979).

4.2.2.1 Native Gels

The simplest variety of PAGE is known as a native gel electrophoresis, referring to the state of the protein. In this variation, a polyacrylamide gel is formed from a mixture of acrylamide and N,N′-methylene-bisacrylamide via a free radical reaction catalyzed by a free radical form of N,N,N′,N′-tetramethylethylenediamine (TEMED) (DAVIS, 1964; GABRIEL, 1971a; ORNSTEIN, 1964). The reaction is actually initiated by the addition of ammonium persulfate, which abstracts a hydrogen atom from the TEMED. The total acrylamide (acrylamide plus bisacrylamide), known as % T, determines the porosity of the gel. The fraction of bisacrylamide, known as % C, determines the extent of crosslinking, which alters the physical properties of the gel and is usually held constant between 2 and 5%. The solutions containing the monomers are buffered at the desired pH. The gels can be polymerized in tubes of ~5 mm diameter or between glass plates to form slabs. For slab gels, a teflon "comb" is inserted into the top of the slab before it polymerizes to make small wells into which the sample is later layered. After polymerization, the top and bottom of the gel are immersed in buffered electrolyte solutions and samples of buffered protein solution, usually containing a small charged dye to report the progress of the electrophoretic motion, are layered into the wells or onto a gel in a tube. Since most proteins are acidic, buffers are used to hold the pH above neutrality, so that all of the proteins in a sample have a net negative charge. The anode (positive electrode) is placed in the bottom reservoir and the cathode is placed in the top. For basic proteins, the gel buffer and the sample buffer maintain a pH below the p*I* of the protein, and the positions of the anode and cathode are reversed. When a DC potential of ~100 volts is applied to the system, ion currents flow in the gel (usually a few milliamps per cm^2 (cross-section) of gel. A protein molecule (actually, a protein ion) is also accelerated by the electric field by an amount proportional to its charge. However, as a protein moves through the gel, it experiences a decelerating force, or frictional drag, that is proportional to its velocity. The frictional coefficient, the proportionality constant between velocity and frictional force, depends on the size (and shape) of the protein. It also depends on the apparent viscosity of the gel, which in turn depends on the % T used in the gel. When the protein is accelerated to a velocity that causes its frictional drag to equal the accelerating force provided by the electric field, acceleration ceases, and the protein travels at a constant velocity that depends upon its charge, its mass, and the % T of the gel. After electrophoresis, the gel is removed and stained for protein (see below). The relative mobility, R_m, (or R_f, retardation factor) is calculated as the ratio between the distance traveled by the protein to that traveled by the tracking dye, which moves with the ion front.

The mobility and separation in native PAGE depends upon both the charge and size of the protein. To improve band separation or to distinguish the effect of charge from the effect of size, there are two options: changing pH and changing the amount of acrylamide in the gel. For very large proteins a % T between 5 and 7% might be used, and for smaller proteins a % T up to or even above 15% might be used to increase the apparent viscosity of the gel. If the same set of proteins are run in gels of different total acrylamide concentration but under otherwise identical conditions, all of the proteins will likely have different R_m's in the different gels. If the log R_m for each protein is plotted against the % T at which it was obtained (a Ferguson plot; FERGUSON, 1964), an approximately straight line is obtained. The slope of the line depends only upon molecular size (HEDRICK and SMITH, 1968), because it is only the frictional factor that differs from gel to gel. When a line is extrapolated to zero acrylamide concentration, the intercept is a measure of the charge on the protein at that pH, since there is negligible friction in the absence of acrylamide. Obviously, lines for proteins having different charges and different molecular sizes may well cross, so that *either* higher or lower % T can yield greater resolution. The resolution can also be altered by changing the pH of the gel and sample to change the net charge on the proteins, presumably to different extents.

Resolution depends on both the separation of band centers and the width of the bands. Although it is not always employed, the original native PAGE procedure (DAVIS, 1964; ORNSTEIN, 1964) involves discontinuous gels, which yield narrowed bands. The gels are poured in two stages, a lower "resolving gel" identical to that described above, and an upper "stacking gel" which is of low % T and contains a different buffer system than the resolving gel. The buffer in the stacking gel contains an ion (often glycine) that bears only a small net charge at the pH employed and therefore moves slowly. The stacking gel becomes ion-poor at the onset of electrophoresis because the mobile ions move onto the resolving gel, but the trailing ions lag behind. The electric current must be the same at all stages of the gel, which implies that the electric field in the stacking gel must be much higher in order for the trailing ion to maintain the current. Proteins in the stacking gel therefore move very rapidly to the top of the resolving gel, where they slow down because the field is smaller. Thus the proteins become focused at the very top of the resolving gel and begin their treck across the resolving gel as a very narrow band. This common form of PAGE is called discontinuous gel electrophoresis, which gives rise to the confusing term "disc gels", which is sometimes even more confusingly spelled "disk" (DAVIS, 1964; ORNSTEIN, 1964).

4.2.2.2 SDS Gels

In native PAGE, two proteins of very different size may co-electrophorese if their charge densities are such that they produce the same steady-state velocity. A common method to separate proteins solely on the basis of their molecular weight is to treat them with the ionic detergent sodium dodecyl sulfate (SDS) and heat, then submit the denatured samples to electrophoresis (LAEMMLI, 1970). The proteins become unfolded and coated with the surfactant, which has a negative charge. The total charge on a protein is proportional to the number of detergent molecules bound, which is proportional to molecular weight. Thus, the charge to mass ratio for all proteins is approximately identical. In addition, any possible effects of molecular shape are removed, since the proteins are denatured. Multimeric proteins are converted to denatured monomers, so SDS PAGE can be used to detect the presence of non-identical subunits. A minor complication is that disulfide bonds, which can covalently link subunits, are not broken. Furthermore, free sulfhydryl groups may become exposed and could form disulfide bonds during the experiment. For this reason, the original SDS PAGE procedures called for performic acid oxidation of sulfhydryl groups. More commonly, the samples are prepared under reducing conditions, i.e., in the presence of 2-mercaptoethanol. The final result is that the relative mobility is proportinal to the logarithm of molecular weight. By the use of standards of known molecular weight, molecular weight can be calculated directly from mobility in a single SDS gel.

There are, of course, exceptions to this behavior. A large fraction of the molecular weight of glycoproteins can reside in the attached oligosacharides, which do not bind SDS. Thus, glycoproteins have a mobility smaller than would be expected for their molecular weights. The anomalous behavior can be reduced by using an alternate buffer system (PODUSLO, 1981) or using gradient gels (ROTHE and MAURER, 1986) (see below). A simple staining procedure exists to identify glycoproteins in a gel (GANDER, 1984).

If a mixture of proteins having a large range of molecular weights must be resolved in a single gel, gradient gels can be used. These are poured using a gradient mixing apparatus that has two reservoirs, one containing a higher and one containing a lower % T mixture of monomers. The reagents of higher % T are siphoned into the tube or slab holder. The reagents of lower % T are allowed to siphon into the reservoir of higher % T, diluting it uniformly as flow proceeds. The resulting gel may vary from 20 % T at the bottom to 5 % T at the top. Proteins travel during electrophoresis until they reach a level in the gel through which they can pass only with great difficulty. There is a modest degree of focusing in a gradient gel, because the molecules at the trailing end of the band have higher mobility than molecules at the leading edge. Since it is ultimately frictional drag that determines the final mobility of the proteins, gradient gels can be used to estimate molecular size of enzymes, if standards are employed. Gradient procedures may be used either for native gels or for SDS gels (HAMES, 1990).

4.2.2.3 Isoelectric Focusing Gels

Another major modification of PAGE is gel isoelectric focusing (IEF), which is a direct adaptation of a preparative technique developed for use in density gradient solutions (VESTERBERG, 1971; VESTERBERG and SVENSSEN, 1966). In gels (WRIGLEY, 1971), the technique involves the use of buffers of different pH in the two buffer reservoirs. If the positive electrode (cathode) is placed in the lower reservoir and the buffer is more basic than that in the upper reservoir, amphoteric anionic molecules will move toward the cathode until they encounter more alkaline pH and lose their charge. If a number of amphoteric molecules of different isoelectric point (called ampholytes) are included in the gel when it is poured, they lose charge (and mobility) at different levels in the gel. They also tend to buffer at this pH, since they can either gain or lose protons in response to a change in pH. Thus, a stable pH gradient is created within the gel during electrophoresis. Proteins loaded onto such a gel travel until they reach their own isoelectric point, then stop and focus to a width that depends on the steepness of the pH gradient. The gels employed are generally of low % T so as to remove the effect of molecular size and to avoid retardation of focusing. The pH gradient can be measured using a needle-tipped pH electrode, by grinding the gel in pure water and measuring the pH, or by running standards of known isoelectric point. This process resolves protein solely on the basis of isoelectric point, not net charge.

If proteins can be separated on the basis of isoelectric point using IEF or on the basis of molecular weight using SDS PAGE, the two techniques can be combined in two-dimensional gels to produce exquisite resolution (e.g., O'FARRELL, 1975).

4.2.2.4 Staining Gels for Proteins and for Enzymes

There are several ways to detect the proteins on electrophoresis gels. The simplest and most common is to stain the gels with a dye that binds to protein. A solution of Coomassie blue G in methanol/acetic acid is usually used (BLACKSHEAR, 1984). This staining solution also causes the proteins to precipitate inside the gel so that the bands do not broaden by diffusion. Stain not bound to protein must be removed from the gels, so there is a destaining step using methanol/acetic acid. The proteins can be visualized by eye or scanned using a spectrophotometer equipped with a device to move the gel in the sample compartment, or scanned using a laser densitometer. A more rapid, though less sensitive stain involves the use of Coomassie blue R (REISNER, 1984), which does not require destaining. A newer method that is more sensitive than Coomassie blue R involves staining with silver (MERRIL et al., 1984), or with colloidal gold.

Some proteins contain non-protein components such as phosphate or attached sugars. There are specific stains for phosphorylated proteins (CUTTING, 1984), and for glycoproteins (GANDER, 1984). Proteins that have been labeled with radioactive nuclides can be detected fluorographically (BONNER, 1984).

Native, gradient and IEF gels contain active enzymes after electrophoresis. If the enzymes catalyze a reaction that can be detected colorimetrically etc., substrates can be diffused into the gels, and the product (or the absence of substrate) can be detected by scanning the gel in a spectrophotometer. The same gel can then be stained for protein, so that the enzyme under study can be identified among other bands. GABRIEL has provided two compendia of such activity stains (GABRIEL, 1971b; HEEB and GABRIEL, 1984). This procedure has occasionally led to the embarrassing conclusion that the enzyme in a supposedly nearly-pure preparation actually corresponds to a minor band (e.g., SMITH and KAPLAN, 1979).

Antibodies can also be used to identify a protein from among many bands on an electrophoresis gel. To prevent problems associated with diffusion of an immunoglobulin into a gel, the protein is usually transferred to derivatized paper or to a nitrocellulose membrane before staining (RENART and SANDOVAL, 1984). The transfer can be done by diffusion, using a vacuum, or electrophoretically. The antibodies may be radioactively labeled, in which case they can be visualized using an autoradiogram, or they may be labeled with a fluorescent compound.

It should also be recognized that the various kinds of gel electrophoresis can be adapted to small-scale purification of enzymes. Before the advent of polyacrylamide gels, electrophoresis was used as a preparative method in starch or agarose gels and in density-gradient-stabilized liquids.

4.2.3 HPLC

Because of the development of high efficiency ion exchange, hydrophobic interaction and reverse phase columns, HPLC has become a good method for analysis of proteins. The use of IEC and HIC columns has already been discussed. The reverse-phase columns are generally hydrocarbons covalently attached to silica particles of diameters of 10, 5 or 3 μm. Octyl silane ("C8") is probably the most common, although octadecyl silane (ODS, a variety of C18) is more common for small molecules. Resin-based versions of the same columns also exist, as do silica columns in which the silica is coated with an organic polymer to protect it from the solvent (e.g., TSK).

The retention mechanism in reverse-phase chromatography can be viewed in at least two ways: the solvophobic model promoted by HORVÁTH (MELANDER and HORVÁTH, 1980), and the stoichiometric displacement model proposed by REGNIER (GENG and REGNIER, 1985). The solvophobic model envisions the binding to occur because of energetics related to the hydrophobic effect. Thus stronger solvents are those that contain a higher proportion of organic component in an aqueous solvent. A more hydrophobic solvent is a better solvent for hydrophobic solutes. REGNIER's model describes the octyl or octadecyl groups of the stationary phase as a hydrophobic surface stacked on end. Modeling the phase as a surface is reminiscent of adsorption chromatography, which suggests that the important event is competition between the binding of solute and the strong component of the solvent for a limited amount of surface area. For small organic molecules, the mobile phase is generally a binary mixture of water and one or more of the following: acetonitrile, methanol or tetrahydrofuran, though many other systems exist. For proteins, the mobile phases are generally harsher, and usually contain other agents such as formic acid or trifluoroacetic acid (HEARN, 1984; STEFFENSEN and ANDERSON, 1987).

Since proteins have a reasonable absorbance at 280 nm and a strong absorbance below 210 nm, detectors using UV absorbance are the most common.

Acknowledgement

The author wishes to thank Dr. LINDA TOMBRAS SMITH, a scientific granddaughter of FRITZ LIPMANN, for her help and patience.

5 References

ANDERSON, C., ZUCKER, F., STEITZ, T. (1979), Space-filling models of kinase clefts and conformational changes, *Science* **204,** 375–380.

ANFINSEN, C., SCHERAGA, H. (1975), Experimental and theoretical aspects of protein folding, *Adv. Protein Chem.* **29,** 205–300.

ARAKAWA, T., TIMASHEFF, S. (1982), Stabilization of protein structure by sugars, *Biochemistry* **21,** 6536–6544.

ARAKAWA, T., TIMASHEFF, S. (1985a), Stabilization of proteins by osmolytes, *Biophys. J.* **47,** 411–414.

ARAKAWA, T., TIMASHEFF, S. (1985b), Theory of protein solubility, *Methods Enzymol.* **114,** 49–77.

ATKINSON, D. (1966), Regulation of enzyme activity, *Annu. Rev. Biochem.* **35,** 85–124.

ATOR, M., ORTIZ DE MONTELLANO, P. (1990), Mechanism-based (suicide) enzyme inactivation, in *The Enzymes* (SIGMAN, D., BOYER, P., Eds.), pp. 214–282, New York: Academic Press.

AZEM, A., KESSEL, M., GOLOUBINOFF, P. (1994), Characterization of a functional GroEL14 $(GroES7)_2$ chaperonin hetero-oligomer, *Science* **265,** 653–656.

BALDWIN, R. (1986), Temperature dependence of hydrophobic interactions in protein folding, *Proc. Natl. Acad. Sci. USA* **83,** 8069–8072.

BALDWIN, R. (1989), How does protein folding get started?, *Trends Biochem. Sci.* **14,** 291–294.

BALDWIN, R. L., RODER, H. (1991), Characterizing protein folding intermediates, *Curr. Biol.* **1,** 218–220.

BECKER, J., CRAIG, E. A. (1994), Heat-shock proteins as molecular chaperones, *Eur. J. Biochem.* **219,** 11–23.

BECKTEL, W., SCHELLMAN, J. (1987), Protein stability curves, *Biopolymers* **26,** 1859–1877.

BERNARDI, G. (1971), Chromatography of proteins on hydroxyapatite, *Methods Enzymol.* **22,** 325–339.

BIETH, J. (1974), Some kinetic consequences of the tight binding of protein-proteinase-inhibitors to the determination of dissociation constants, in: *Proteinase Inhibitors,* pp. 463–469, Heidelberg: Springer-Verlag.

BIRINGER, R., FINK, A. (1982), Methanol-stabilized intermediates in the thermal unfolding of ribonuclease A, *J. Mol. Biol.* **160,** 87–116.

BLACKSHEAR, P. (1984), Systems for polyacrylamide gel electrophoresis, *Methods Enzymol.* **104,** 237–255.

BLANCHARD, J. (1984), Buffers for enzymes, *Methods Enzymol.* **104,** 404–414.

BLATT, W. (1971), Ultrafiltration for enzyme concentration, *Methods Enzymol.* **22,** 39–49.

BOCHKAREVA, E. S., GIRSHOVICH, A. S. (1992), A newly synthesized protein interacts with GroES on the surface of chaperonin GroEL, *J. Biol. Chem.* **267,** 25672–25675.

BONNER, W. (1984), Fluorography for the detection of radioactivity in gels, *Methods Enzymol.* **104,** 460–465.

BOYER, P. (1970–1990), *The Enzymes,* New York: Academic Press.

BOYER, P. (1954), Spectrophotometric study of the reaction of protein sulfhydryl groups with organic mercurials, *J. Am. Chem. Soc.* **76,** 4331–4337.

BRADFORD, M. (1976), A rapid and sensitive method for the quantitation of microgram quantities of protein using the principle of protein-dye binding, *Anal. Biochem.* **72,** 248–254.

BRADSTREET, R. (1965), *The Kjeldahl Method for Organic Nitrogen,* New York: Academic Press.

BRANDÉN, C., TOOZE, J. (1991), *Introduction to Protein Structure,* New York: Garland Publishing, Inc.

BROCKLEHURST, K., CARLSSON, J., KIERSTAN, M. (1973), Covalent chromatography. Preparation of fully active papain from dried papaya latex, *Biochem. J.* **133,** 573–584.

BRUNNE, R. M., LIEPINSH, E., OTTING, G., WÜTHRICH, K., VAN GUNSTEREN, W. F. (1993), Hydration of proteins. A comparison of experimental residence times of water molecules solvating the bovine pancreatic trypsin inhibitor with theoretical model calculations, *J. Mol. Biol.* **231,** 1040–1048.

BUCHBERGER, A., VALENCIA, A., MCMACKEN, R., SANDER, C., BUKAU, B. (1994), The chaperone function of DnaK requires the coupling of ATPase activity with substrate binding through residue E171, *EMBO J.* **13,** 1687–1695.

BULLEID, N., FREEDMAN, R. (1988), Defective cotranslational formation of disulphide bonds in protein-disulphide-isomerase-deficient microsomes, *Nature* **335,** 649–651.

BURLEY, S., PETSKO, G. (1988), Weakly polar interactions in proteins, *Adv. Protein Chem.* **38,** 125–189.

CECH, T., BASS, B. (1986), Biological catalysis by RNA, *Annu. Rev. Biochem.* **55,** 599–629.

CHOTHIA, C. (1984), Principles that determine the structure of proteins, *Annu. Rev. Biochem.* **53,** 537–572.
CHOTHIA, C., FINKELSTEIN, A. (1990), Classification and origins of protein folding patterns, *Annu. Rev. Biochem.* **59,** 1007–1039.
CLELAND, W. (1964), Dithiothreitol, a new protective reagent for SH groups, *Biochemistry* **3,** 480–482.
CLELAND, W. (1967), Steady-state kinetics, *Adv. Enzymol.* **29,** 1.
CLELAND, W. (1970), Steady-state kinetics, in: *The Enzymes* (SIGMAN, D., BOYER, P., Eds.), pp. 1–65, San Diego: Academic Press.
CLELAND, W. (1979), Steady-state kinetics, *Methods Enzymol.* **63,** 103.
CLELAND, W. (1982), The use of pH studies to determine chemical mechanisms of enzyme-catalyzed reactions, *Methods Enzymol.* **87,** 390–405.
CLELAND, W. (1990), Steady-state kinetics, in: *The Enzymes* (SIGMAN, D., BOYER, P., Eds.), pp. 99–158, San Diego: Academic Press.
COHN, M., RAO, B. N. (1980), ^{31}P NMR studies of enzymatic reactions, *Bull. Magn. Reson.* **1,** 38–60.
COLLINS, K., WASHBAUGH, M. (1985), The Hofmeister effect and the behaviour of water at interfaces, *Quart. Rev. Biophys.* **18,** 323–422.
COLMAN, R. (1990), Site-specific modification of enzyme sites, in: *The Enzymes* (SIGMAN, D., BOYER, P., Eds.), pp. 283–321, San Diego: Academic Press.
COLOWICK, S. (1955), Separation of proteins by use of adsorbants, *Methods Enzymol.* **1,** 90–98.
CORNISH-BOWDEN, A., CÁRDENAS, M. (1987), Chemistry of enzymes, in: *Biotechnology* (REHM, H.-J., REED, G., Eds.), Vol. 7a, pp. 3–33, Weinheim: VCH Verlagsgesellschaft.
CRAIG, E. A., GAMBILL, B. D., NELSON, R. J. (1993), Heat shock proteins: molecular chaperones of protein biogenesis, *Microbiol. Rev.* **57,** 402–414.
CUTTING, J. (1984), Gel protein stains: Phosphoproteins, *Methods Enzymol.* **104,** 451–455.
DAVIS, B. (1964), Disk electrophoresis II. Method and application to serum proteins, *Ann. NY Acad. Sci.* **121,** 404–427.
DEUTSCHER, M. P. (Ed.) (1990), Guide to enzyme purification, *Methods Enzymol.* **182**.
DEYL, Z. (1979), Electrophoresis – A survey of techniques and applications, in: *Journal of Chromatography Library,* Amsterdam: Elsevier.
DICKERSON, R. E., GEIS, I. (1969), *The Structure and Action of Proteins,* New York: Harper and Row.
DOLAN, J., GANT, J., SNYDER, L. (1979), Gradient elution in HPLC: II. Practical application to reversed-phase systems, *J. Chromatogr.* **165,** 31–58.
DOOLITTLE, R. F. (Ed.) (1990), Molecular evolution: Computer analysis of protein and nucleic acid structure, *Methods Enzymol.* **183**.
DOUZOU, P. (1973), Enzymology at subzero temperatures, *Mol. Cell. Biochem.* **1,** 15–27.
DOUZOU, P., PETSKO, G. (1984), Proteins at work: "Stop-Action" pictures at subzero temperatures, *Adv. Protein Chem.* **36,** 246–361.
DRAGER, R., REGNIER, F. (1986), Application of the stoichiometric displacement model of retention to anion-exchange chromatography of nucleic acids, *J. Chromatogr.* **359,** 147–155.
DRAGER, R., REGNIER, F. (1987), Retention mechanism of lactate dehydrogenase in anion-exchange chromatography, *J. Chromatogr.* **406,** 237–246.
DRENTH, J., JANSONIUS, J., KOEKOEK, R., SWEN, H., WOLTHERS, B. (1968), Structure of papain, *Nature* **218,** 929–932.
DUENAS, M., VAZQUEZ, J., AYALA, M., SODERLIND, E., OHLIN, M., PEREZ, L., BORREBAECK, C. A., GAVILONDO, J. V. (1994), Intra- and extracellular expression of an scFv antibody fragment in *E. coli:* effect of bacterial strains and pathway engineering using GroES/L chaperonins, *Biotechniques* **16,** 476–477, 480–483.
EDELMAN, G. (1970), The covalent structure of a human γG-immunoglobulin. XI. Functional implications, *Biochemistry* **9,** 3197–3205.
EDGERTON, M. D., SANTOS, M. O., JONES, A. M. (1993), *In vivo* suppression of phytochrome aggregation by the GroE chaperonins in *Escherichia coli, Plant Mol. Biol.* **21,** 1191–1194.
ELLMAN, G. (1959), Tissue sulfhydryl groups, *Arch. Biochem. Biophys.* **82,** 70–77.
EVENTOFF, W., ROSSMAN, M. (1975), The evolution of dehydrogenases and kinases, *CRC Crit. Rev. Biochem.* **3,** 111–140.
EVERSE, J., STOLZENBACH, F. (1971), Lyophilization, *Methods Enzymol.* **22,** 33–39.
FÄGERSTAM, L. (1983), Fast chromatofocussing of human serum proteins with special reference to α-1-antitrypsin and Ge-globulin, *J. Chromatogr.* **266,** 523–532.
FASMAN, G. (1989), Protein conformational prediction, *Trends Biochem. Sci.* **14,** 295–299.
FAYAT, G., FROMANT, M., KAHN, D., BLANQUET, S. (1977), Methionyl-tRNA synthetase, *Eur. J. Biochem.* **78,** 333.
FERGUSON, K. (1964), Starch gel electrophoresis – application to the classification of pituitary proteins and polypeptides, *Metabolism* **13,** 985–1001.
FERREYRA, R. G., SONCINI, F. C., VIALE, A. M. (1993), Cloning, characterization, and functional

expression in *Escherichia coli* of chaperonin (groESL) genes from the phototrophic sulfur bacterium *Chromatium vinosum, J. Bacteriol.* **175,** 1514–1523.

FERSHT, A. (1977), *Enzyme Structure and Mechanism,* San Francisco: W. H. Freeman and Company.

FINK, A. (1986), Protein folding in cryosolvents and at subzero temperatures, *Methods Enzymol.* **131,** 173–185.

FINK, A., CALCIANO, L., GOTO, Y., KUROTSU, T., PALLEROS, D. (1994), Classification of acid denaturation of proteins: Intermediates and unfolded states, *Biochemistry* **33,** 12504–12511.

FISCHER, G., SCHMIDT, F. (1990), The mechanism of protein folding. Implications of *in vitro* refolding models for *de novo* protein folding and translocation in the cell, *Biochemistry* **29,** 2205–2212.

FREEDMAN, R., HIRST, T., TUITE, M. (1994), Protein disulfide isomerase: Building bridges in protein folding, *Trends Biochem. Sci.* **19,** 331–336.

FRENCH, C., MILNER, H. (1955), Disintegration of bacteria and small particles by high-pressure extrusion, *Methods Emzymol.* **1,** 64–67.

FRIED, M., CHUN, P. (1971), Water-soluble nonionic polymers in protein purification, *Methods Enzymol.* **22,** 238–248.

GABRIEL, O. (1971a), Analytical disc gel electrophoresis, *Methods Enzymol.* **22,** 565–578.

GABRIEL, O. (1971b), Locating enzymes on gels, *Methods Enzymol.* **22,** 578–604.

GANDER, J. (1984), Gel protein stains: Glycoproteins, *Methods Enzymol.* **104,** 447–451.

GANZI, G. (1984), Preparation of high-purity laboratory water, *Methods Enzymol.* **104,** 391–403.

GENG, X., REGNIER, F. (1985), Stoichiometric displacement of solvent by nonpolar solutes in reversed-phase liquid chromatography, *J. Chromatogr.* **406,** 147–168.

GIDDINGS, J. (1965), *Dynamics of Chromatography,* Part 1: *Principles and Theory,* New York: Marcel Dekker.

GLASSTONE, S., LAIDLER, K., EYRING, H. (1941), *The Theory of Rate Processes,* New York: McGraw Hill.

GOOD, N., WINGET, G., WINTER, W., CONNOLLY, T., IZAWA, S., SINGH, R. (1966), Hydrogen ion buffers for biological research, *Biochemistry* **5,** 467–477.

GORBUNOFF, M., TIMASHEFF, S. (1984), The interaction of proteins with hydroxyapatite. III. Mechanism, *Anal. Biochem.* **136,** 440–445.

GOVEZENSKY, D., BOCHKAREVA, E. S., ZAMIR, A., GIRSHOVICH, A. S. (19945), Chaperonins as potential gene regulatory factors. *In vitro* interaction and solubilization of NifA, the nif transcriptional activator, with GroEL, *J. Biol. Chem.* **269,** 14003–14006.

GREEN, A., HUGHES, W. (1955), Protein fractionation on the basis of solubility in aqueous salts and organic solvents, *Methods Enzymol.* **1,** 67–90.

GREGORY, R., ROSENBERG, A. (1986), Protein conformational dynamics measured by hydrogen exchange techniques, *Methods Enzymol.* **131,** 448–508.

GUNSALUS, I. (1955), Extraction of enzymes from microorganisms, *Methods Enzymol.* **1,** 51–62.

HACKNEY, D. (1990), Binding energy and catalysis, in: *The Enzymes* (SIGMAN, D., BOYER, P., Eds.), pp. 1–36. San Diego: Academic Press.

HAMES, D. (1990), One-dimensional polyacrylamide gel electrophoresis, in: *Gel Electrophoresis of Proteins. A Practical Approach* (HAMES, B., RICKWOOD, D., Eds.), pp. 1–148, Oxford: IRL Press.

HANSEN, D., RAINES, R. (1990), Binding energy and enzymatic catalysis, *J. Chem. Educ.* **67,** 483–489.

HARDY, S. J., RANDALL, L. L. (1993), Recognition of ligands by SecB, a molecular chaperone involved in bacterial protein export, *Philos. Trans. R. Soc. Lond. Biol.* **339,** 343–352; discussion 352–354.

HARTL, F. U., HLODAN, R., LANGER, T. (1994), Molecular chaperones in protein folding: the art of avoiding sticky situations, *Trends Biochem. Sci.* **19,** 20–25.

HARTREE, E. (1972), Determination of protein: A modification of the Lowry method that gives linear photometric response, *Anal. Biochem.* **48,** 422–427.

HATTORI, M., AMETANI, A., KATAKURA, Y., SHIMIZU, M., KAMINOGAWA, S. (1993), Unfolding/refolding studies on bovine β-lactoglobulin with monoclonal antibodies as probes, *J. Biol. Chem.* **268,** 22414–22419.

HAYER-HARTL, M. K., EWBANK, J. J., CREIGHTON, T. E., HARTL, F. U. (1994), Conformational specificity of the chaperonin GroEL for the compact folding intermediates of alpha-lactalbumin, *EMBO J.* **13,** 3192–3202.

HEARN, M. (1984), Reversed-phase high-performance liquid chromatography, *Methods Enzymol.* **104,** 190–212.

HEDRICK, J., SMITH, A. (1968), Size and charge isomer separation and estimation of molecular weights of proteins by disc gel electrophoresis, *Arch. Biochem. Biophys.* **126,** 155–164.

HEEB, M., GABRIEL, O. (1984), Enzyme localization in gels, *Methods Enzymol.* **104,** 416–439.

HENDRICK, J. P., HARTL, F. U. (1993), Molecular chaperone functions of heat-shock proteins, *Annu. Rev. Biochem.* **62,** 349–384.

HILL, A. (1913), The combinations of haemoglobin with oxygen and with carbon monoxide. I., *Biochem. J.* **7,** 471–480.

HIMMELHOCH, S. (1971), Chromatography of proteins on ion-exchange adsorbents, *Methods Enzymol.* **22,** 273–286.

HIRS, C. (1955), Chromatography of enzymes on ion exchange resins, *Methods Enzymol.* **1,** 113–125.

HJELMELAND, L., CHRAMBACH, A. (1984), Solubilization of functional membrane proteins, *Methods Enzymol.* **104,** 305–318.

HOBSON, A. H., BUCKLEY, C. M., AAMAND, J. L., JORGENSEN, S. T., DIDERICHSEN, B., MCCONNELL, D. J. (1993), Activation of a bacterial lipase by its chaperone, *Proc. Natl. Acad. Sci. USA* **90,** 5682–5686.

HORECKER, B., ROWLEY, P., GRAZI, E., CHING, T., TCHOLA, O. (1963), The mechanism of action of aldolases IV. Lysine as the substrate binding site, *Biochem. Z.* **338,** 36–51.

HORWICH, A. L., WILLISON, K. R. (1993), Protein folding in the cell: functions of two families of molecular chaperone, hsp 60 and TF55-TCP1, *Philos. Trans. R. Soc. Lond. Biol.* **339,** 313–325; discussion 325–326.

HUBER, R., BODE, W. (1978), Structural basis for the activation of trypsin, *Acc. Chem. Res.* **11,** 114–122.

INGHAM, K. (1984), Protein precipitation with polyethylene glycol, *Methods Enzymol.* **104,** 351–356.

IUB (International Union for Biochemistry and Molecular Biology) (1964), *Enzyme Nomenclature: Recommendations of the International Union of Biochemistry,* Amsterdam: Elsevier.

IUB (International Union of Biochemistry and Molecular Biology) (1992), *Enzyme Nomenclature: Recommendations,* San Diego: Academic Press.

JAENICKE, R., RUDOLPH, R. (1986), Refolding and association of oligomeric proteins, *Methods Enzymol.* **131,** 218–250.

JAKOB, U., GAESTEL, M., ENGEL, K., BUCHNER, J. (1993), Small heat shock proteins are molecular chaperones, *J. Biol. Chem.* **268,** 1517–1520.

JENCKS, W. (1975), Binding energy, specificity, and enzymatic catalysis: the Circe effect, *Adv. Enzymol.* **43,** 219–410.

JENCKS, W. (1987), *Catalysis in Chemistry and Enzymology,* New York: Dover Publications, Inc.

KABACK, H. (1970), Transport, *Annu. Rev. Biochem.* **39,** 561–598.

KAJI, E. H., LODISH, H. F. (1993), *In vitro* unfolding of retinol-binding protein by dithiothreitol. Endoplasmic reticulum-associated factors, *J. Biol. Chem.* **268,** 22195–22202.

KAPTEIN, R., BOELENS, R., SHEEK, R., VAN GUNSTEREN, W. (1988), Protein structures from NMR, *Biochemistry* **27,** 5389–5395.

KAUFMAN, S. (1971), Fractionation of protein mixtures with organic solvents, *Methods Enzymol.* **22,** 233–238.

KING, E., ALTMAN, C. (1956), A schematic method of deriving the rate laws for enzyme-catalyzed reactions, *J. Phys. Chem.* **60,** 1375–1378.

KLOTZ, I. M. (1993), Biogenesis: number mysticism in protein thinking, *Faseb J.* **7,** 1219–1225.

KOSHLAND JR, D., NEET, K. (1968), The catalytic and regulatory properties of enzymes, *Annu. Rev. Biochem.* **37,** 359–410.

KOSHLAND JR, D., NÉMETHY, G., FILMER, D. (1966), Comparison of experimental binding data and theoretical models in proteins containing subunits, *Biochemistry* **5,** 365–385.

KUBO, T., MIZOBATA, T., KAWATA, Y. (1993), Refolding of yeast enolase in the presence of the chaperonin GroE. The nucleotide specificity of GroE and the role of GroES, *J. Biol. Chem.* **268,** 19346–19351.

KUMAMOTO, C. A. (1991), Molecular chaperones and protein translocation across the *Escherichia coli* inner membrane, *Mol. Microbiol.* **5,** 19–22.

KUMAMOTO, C. A., FRANCETIC, O. (1993), Highly selective binding of nascent polypeptides by an *Escherichia coli* chaperone protein *in vivo*, *J. Bacteriol.* **175,** 2184–2188.

KUWAJIMA, K. (1989), The molten globule state as a clue for understanding the folding and cooperativity of globular-protein structure, *Proteins Struct. Funct. Genet.* **6,** 87–103.

KUWAJIMA, K., BALDWIN, R. L. (1983), Exchange behavior of the H-bonded amide protons in the 3 to 13 helix of ribonuclease S, *J. Mol. Biol.* **169,** 299–323.

KUWAJIMA, K., KIM, P., BALDWIN, R. L. (1983), Strategy for trapping intermediates in the folding of ribonuclease and for using ^{1}H NMR to determine their structures, *Biopolymers* **22,** 59–67.

LAEMMLI, U. (1970), Cleavage of structural proteins during the assembly of the head of bacteriophage T4, *Nature* **227,** 680–685.

LANG, K., SCHMIDT, F., FISCHER, G. (1987), Catalysis of protein folding by prolyl isomerase, *Nature* **329,** 268–270.

LARSSON, P.-O. (1984), High-performance liquid affinity chromatography, *Methods Enzymol.* **104,** 212–223.

LAYNE, E. (1957), Spectrophotometric and turbidimetric methods for measuring proteins, *Methods Enzymol.* **3,** 447–454.

LECOMPTE, J., MATTHEWS, C. (1993), Unraveling the mechanism of protein folding: new tricks for an old problem, *Protein Eng.* **6,** 1–10.

LEVITT, M., CHOTHIA, C. (1976), Structural patterns in globular proteins, *Nature* **261,** 552–558.

LIEBERMAN, I., GINGOLD, J. (1962), Techniques for study of mammalian cell cultures: Extraction of enzymes, *Methods Enzymol.* **5,** 119–122.

LINDQUIST, S. (1985), The heat-shock response, *Annu. Rev. Biochem.* **55,** 1151–1191.

LIPSCOMB, W. (1983), Structure and catalysis of enzymes, *Annu. Rev. Biochem.* **52,** 17–34.

LOWE, C., PEARSON, J, (1984), Affinity chromatography on immobilized dyes, *Methods Enzymol.* **104,** 97–113.

LOWRY, O., ROSEBROUGH, N., FARR, A., RANDALL, R. (1951), Protein measurement with the Folin phenol reagent, *J. Biol. Chem.* **193,** 265–275.

LUND, P. (1994), The chaperonin cycle and protein folding, *Bioessays* **16,** 229–231.

MARTIN, J., GEROMANOS, S., TEMPST, P., HARTL, F. U. (1993), Identification of nucleotide-binding regions in the chaperonin proteins GroEL and GroES, *Nature* **366,** 279–282.

MATSUMURA, M., SIGNOR, G., MATTHEWS, B. (1989), Substantial increase in protein stability by multiple disulfide bonds, *Nature* **342,** 291–293.

MCPHIE, P. (1971), Dialysis, *Methods Enzymol.* **22,** 23–32.

MEANS, G., FEENEY, R. (1990), Chemical modifications of proteins: History and applications, *Bioconjugate Chem.* **1,** 2–12.

MELANDER, W., HORVÁTH, C. (1980), Reversed-phase chromatography, in: *High Performance Liquid Chromatography: Advances and Perspectives* (HORVÁTH, C., Ed.), pp. 114–319, New York: Academic Press.

MELKI, R., COWAN, N. J. (1994), Facilitated folding of actins and tubulins occurs via a nucleotide-dependent interaction between cytoplasmic chaperonin and distinctive folding intermediates, *Mol. Cell. Biol.* **14,** 2895–2904.

MERRIL, C., GOLDMAN, D., VANKEUREN, M. (1984), Gel protein stains: Silver stain, *Methods Enzymol.* **104,** 441–447.

MEURY, J., BAHLOUL, A., KOHIYAMA, M. (1993), Impairment of nucleoid segregation and cell division at high osmolarity in a strain of *Escherichia coli* overproducing the chaperone DnaK, *FEMS Microbiol. Lett.* **113,** 93–99.

MICHAELIS, L., MENTEN, M. (1913), Die Kinetik der Invertinwirkung, *Biochem. Z.* **49,** 333–369.

MONOD, J., WYMAN, J., CHANGEUX, J. (1965), On the nature of allosteric transitions: a plausible model, *J. Mol. Biol.* **12,** 88–118.

MOORE, S., STEIN, W. (1952), Chromatography, *Annu. Rev. Biochem.* **21,** 521–546.

NADEAU, K., DAS, A., WALSH, C. T. (1993), Hsp90 chaperonins possess ATPase activity and bind heat shock transcription factors and peptidyl prolyl isomerases, *J. Biol. Chem.* **268,** 1479–1487.

NASON, A. (1955), Extraction of enzymes from higher plants, *Methods Enzymol.* **1,** 62–64.

NEUPERT, W., PFANNER, N. (1993), Roles of molecular chaperones in protein targeting to mitochondria, *Philos. Trans. R. Soc. Lond. Biol.* **339,** 355–361; discussion 361—362.

O'FARRELL, P. (1975), High resolution two dimensional electrophoresis of proteins, *J. Biol. Chem.* **250,** 4007–4021.

ORNSTEIN, L. (1964), Disk electrophoresis I. Background and theory, *N Y Acad. Sci.* **121,** 321–349.

OTTING, G., LIEPINSH, E., WÜTHRICH, K. (1991), Protein hydration in aqueous solution, *Science* **254,** 974–980.

PACE, C. (1986), Determination and analysis of urea and guanidine hydrochloride denaturation curves, *Methods Enzymol.* **131,** 266–280.

PAULING, L., COREY, R. (1951), Atomic coordinates and structure factors for two helical configurations of polypeptide chains, *Proc. Natl. Acad. Sci. USA* **37,** 235–240.

PENEFSKY, H., TZAGOLOFF, A. (1971), Extraction of water-soluble enzymes and proteins from membranes, *Methods Enzymol.* **22,** 204–218.

PERALTA, D., HARTMAN, D. J., HOOGENRAAD, N. J., HOJ, P. B. (1994), Generation of a stable folding intermediate which can be rescued by the chaperonins GroEL and GroES, *FEBS Lett.* **339,** 45–49.

PETERSON, E., SOBER, H. (1956), Chromatography of proteins I. Cellulose ion-exchange adsorbents, *J. Am. Chem. Soc.* **78,** 751–755.

PFANNER, N., CRAIG, E., MEIJER, M. (1994), The protein import machinery of the mitochondrial inner membrane, *Trends Biochem. Res.* **19,** 368–372.

PODUSLO, J. (1981), Glycoprotein molecular-weight estimation using sodium dodecyl sulfate-pore gradient electrophoresis: comparison of Tris-glycine and Tris-borate-EDTA buffer systems, *Anal. Biochem.* **114,** 131–139.

PORATH, J., FLODIN, P. (1959), Gel filtration: A method for desalting and group separation, *Nature* **183,** 1657–1659.

POTTER, V. (1955), Tissue homogenates, *Methods Enzymol.* **1,** 10–15.

PRIVALOV, P. (1979), Stability of proteins: Small globular proteins, *Adv. Protein Chem.* **33,** 167–241.

PRIVALOV, P. (1982), Stability of proteins: Proteins which do not present a single cooperative system, *Adv. Protein Chem.* **35,** 1–104.

PRIVALOV, P., GILL, S. (1988), Stability of protein structure and hydrophobic interaction, *Adv. Protein Chem.* **39,** 191–234.

PRIVALOV, P., POTHKHIN, S. (1986), Scanning microcalorimetry in studying temperature-induced changes in proteins, *Methods Enzymol.* **131,** 4–51.

PUIG, A., LYLES, M. M., NOIVA, R., GILBERT, H. F. (1994), The role of the thiol/disulfide centers and peptide binding site in the chaperone and anti-chaperone activities of protein disulfide isomerase, *J. Biol. Chem.* **269,** 19128–19135.

RAMACHANDRAN, G., SASISEKHARAN, V. (1968), Conformation of polypeptides and proteins, *Adv. Protein Chem.* **23,** 283–437.

RAMACHANDRAN, G., RAMAKRISHNAN, C., SASISEKHARAN, V. (1963), Stereochemistry of polypeptide chain configurations, *J. Mol. Biol.* **7,** 95–99.

RECSEI, P., SNELL, E. (1970), Histidine decarboxylase of *Lactobacillus* 30a. VI. Mechanism of action and kinetic properties, *Biochemistry* **9,** 1492–1497.

REED, L., COX, D. (1966), Macromolecular organization of enzyme systems, *Annu. Rev. Biochem.* **35,** 57–84.

REGNIER, F. (1984), High-performance ion-exchange chromatography, *Methods Enzymol.* **104,** 170–189.

REILAND, J. (1971), Gel filtration, *Methods Enzymol.* **22,** 287–321.

REISNER, A. (1984), Gel protein stains: A rapid procedure, *Methods Enzymol.* **104,** 439–441.

RENART, J., SANDOVAL, I. (1984), Western blots, *Methods Enzymol.* **104,** 455–460.

RICHARDSON, J. (1981), The anatomy and taxonomy of protein structure, *Adv. Protein Chem.* **34,** 167–339.

RICHARME, G., KOHIYAMA, M. (1993), Specificity of the *Escherichia coli* chaperone DnaK (70-kDa heat shock protein) for hydrophobic amino acids, *J. Biol. Chem.* **268,** 24074–24077.

RICHARME, G., KOHIYAMA, M. (1994), Amino acid specificity of the *Escherichia coli* chaperone GroEL (heat shock protein 60), *J. Biol. Chem.* **269,** 7095–7098.

RICHEY, J. (1984), Optimal pH conditions for ion exchangers on macroporous supports, *Methods Enzymol.* **104,** 223–233.

ROBINSON, H., HODGEN, C. (1940), The biuret reaction in the determination of serum protein. I. A study of the conditions necessary for the production of the stable color which bears a quantitative relationship to the protein concentration, *J. Biol. Chem.* **135,** 707–725.

RODER, H., WÜTHRICH, K. (1986), Protein folding kinetics by combined use of rapid mixing techniques and NMR observation of individual amide protons, *Proteins* **1,** 34–42.

ROSE, G., GIERASCH, L., SMITH, J. (1985), Turns in peptides and proteins, *Adv. Protein Chem.* **37,** 1–109.

ROSENBERG, H. F., ACKERMAN, S. J., TENEN, D. G. (1993), Characterization of a distinct binding site for the prokaryotic chaperone GroEL, on a human granulocyte ribonuclease, *J. Biol. Chem.* **268,** 4499–4503.

ROSENTHAL, H. (1967), A graphic method for the determination and presentation of binding parameters in a complex system, *Anal. Biochem.* **20,** 525–532.

ROTHE, G., MAURER, W. (1986), in: *Gel Electrophoresis of Proteins,* pp. 312–322, Bristol: Wright.

SCHMID, D., BAICI, A., GEHRING, H., CHRISTEN, P. (1994), Kinetics of molecular chaperone action, *Science* **263,** 971–973.

SCHMIDT, F., BALDWIN, R. (1978), Acid catalysis of the formation of the slow-folding species of RNase A: evidence that the reaction is proline isomerization, *Proc. Natl. Acad. Sci. USA* **75,** 4764–4768.

SCHRÖDER, H., LANGER, T., HARTL, F. U., BUKAU, B. (1993), DnaK, DnaJ and GrpE form a cellular chaperone machinery capable of repairing heat-induced protein damage, *EMBO J.* **12,** 4137–4144.

SEGEL, I. (1975), *Enzyme Kinetics,* New York: John Wiley & Sons.

SHALTIEL, S. (1984), Hydrophobic chromatography, *Methods Enzymol.* **104,** 69–96.

SIEGEL, L., MONTY, K. (1966), Dermination of molecular weights and frictional ratios of proteins in impure systems by use of gel filtration and density gradient centrifugation. Application to crude preparations of sulfite and hydroxylamine reductases, *Biochim. Biophys. Acta* **112,** 346–362.

SMITH, D., MAGGIO, E., KENYON, G. (1975), Simple alkanethiol groups for temporary blocking of sulfhydryl groups of enzymes, *Biochemistry* **14,** 766–771.

SMITH, L., KAPLAN, N. (1979), Purification of phosphotransacetylase by affinity chromatography, *Anal. Biochem.* **95,** 2–7.

SMITH, P., KROHN, R., HERMANSON, G., et al. (1985), Measurement of protein using bicinchoninic acid, *Anal. Biochem.* **150,** 76–85.

SNYDER, L., DOLAN, J., GANT, J. (1979), Gradient elution in HPLC: I. Theoretical basis for reversed-phase systems, *J. Chromatogr.* **165,** 3–30.

STEFFENSEN, R., ANDERSON JR, J. (1987), Reversed-phase separation of proteins, *BioChromatography* **2,** 85.

STUART, R. A., CYR, D. M., CRAIG, E. A., NEUPERT, W. (1994), Mitochondrial molecular chaperones: their role in protein translocation, *Trends Biochem. Sci.* **19,** 87–92.

SUMNER, J. (1926), The isolation and crystallization of the enzyme urease, *J. Biol. Chem.* **69,** 435–441.

SWINGLE, S., TISELIUS, A. (1951), Tricalcium phosphate as an adsorbent in the chromatography of proteins, *Biochem. J.* **48,** 171–174.

TAGUCHI, H., YOSHIDA, M. (1993), Chaperonin from *Thermus thermophilus* can protect several enzymes from irreversible heat denaturation by capturing denaturation intermediate, *J. Biol. Chem.* **268,** 5371–5375.

TANFORD, C. (1980), *The Hydrophobic Effect,* New York: Wiley.

TAPPAN, D. (1966), A light-scattering method for measuring protein concentration, *Anal. Biochem.* **14,** 171–182.

TIMASHEFF, S. N. (1992), Water as ligand: preferential binding and exclusion of denaturants in protein unfolding, *Biochemistry* **31,** 9857–9864.

TIMASHEFF, S. N. (1993), The control of protein stability and association by weak interactions with water: how do solvents affect these processes?, *Annu. Rev. Biophys. Biomol. Struct.* **22,** 67–97.

TODD, M. J., VIITANEN, P. V., LORIMER, G. H. (1994), Dynamics of the chaperonin ATPase cycle: implications for facilitated protein folding, *Science* **265,** 659–666.

UNGER, K. (1984), High-performance size-exclusion chromatography, *Methods Enzymol.* **104,** 154–169.

VANHOLDE, K. (1966), in: *Molecular Architecture in Cell Physiology* (HAYASHI, T., SZENT-GYÖRGI, A. G., Eds.), p. 81, Englewood Cliffs, NJ: Prentice Hall.

VESTERBERG, O. (1971), Isoelectric focusing of proteins, *Methods Enzymol.* **22,** 389–412.

VESTERBERG, O., SVENSSEN, H. (1966), Isoelectric fractionation, analysis, and characterization of ampholytes in natural pH gradients, *Acta Chem. Scand.* **20,** 820–834.

VIOLA, R., CLELAND, W. (1982), Initial velocity analysis for terreactant mechanisms, *Methods Enzymol.* **87,** 353–366.

WAGNER, G., HYBERTS, S., HAVEL, T. (1992), NMR structure determination in solution: A critique and comparison with X-ray crystallography, *Annu. Rev. Biophys. Biomol. Struct.* **21,** 167–198.

WALSH, C. (1984), Suicide substrates, mechanism-based inactivators: Recent developments, *Annu. Rev. Biochem.* **53,** 493–535.

WANG, C. C., TSOU, C. L. (1993), Protein disulfide isomerase is both an enzyme and a chaperone, *Faseb J.* **7,** 1515–1517.

WELCH, W. J. (1991), The role of heat-shock proteins as molecular chaperones, *Curr. Opin. Cell. Biol.* **3,** 1033–1038.

WELCH, W. J. (1993), Heat shock proteins functioning as molecular chaperones: their roles in normal and stressed cells, *Philos. Trans. R. Soc. Lond. Biol.* **339,** 327–333.

WHITAKER, J. (1994), *Principles of Enzymology for the Food Sciences,* New York: Marcel Dekker, Inc.

WHITAKER, J., GRANUM, P. (1980), An absolute method for protein determination based on differences in absorbance at 235 and 280 nm, *Anal. Biochem.* **109,** 156–159.

WILCHEK, M., MIRON, T., KOHN, J. (1984), Affinity chromatography, *Methods Enzymol.* **104,** 3–55.

WOFSY, C., GOLDSTEIN, B. (1992), Interpretation of Scatchard plots for aggregating receptor systems, *Math. Biosci.* **112,** 115–154.

WOLFENDEN, R. (1988), Analog approaches to the structure of the transition state of enzyme reactions, *Acc. Chem. Res.* **5,** 10–18.

WRIGLEY, C. (1971), Gel electrofocusing, *Methods Enzymol.* **22,** 559–564.

WÜTHRICH, K. (1986), *NMR of Proteins and Nucleic Acids,* New York: John Wiley and Sons.

WÜTHRICH, K. (1990), Protein structure determination in solution by NMR spectroscopy, *J. Biol. Chem.* **265,** 22059–22062.

WÜTHRICH, K., OTTING, G., LIEPINSH, E. (1992), Protein hydration in aqueous solution, *Faraday Discuss.* **93,** 35–45.

WYNN, R. M., DAVIE, J. R., COX, R. P., CHUANG, D. T. (1994), Molecular chaperones: heat-shock proteins, foldases, and matchmakers, *J. Lab. Clin. Med.* **124,** 31–36.

XIA, Z., MATTHEWS, F. (1990), Molecular structure of flavocytochrome b2 and 2.4 Å resolution, *J. Mol. Biol.* **212,** 837–863.

YANCEY, P., CLARK, M., HAND, S., BOWLUS, R., SOMERO, G. (1982), Living with water stress: evolution of osmolyte systems, *Science* **217,** 1214–1222.

ZIERLER, K. (1989), Misuse of nonlinear Scatchard plots, *Trends Biochem. Sci.* **14,** 314–317.

ZIMMERMAN, J. L., COHILL, P. R. (1991), Heat shock and thermotolerance in plant and animal embryogenesis, *New Biol.* **3,** 641–650.

ZYLICZ, M. (1993), The *Escherichia coli* chaperones involved in DNA replication, *Philos. Trans. R. Soc. Lond. Biol.* **339,** 271–277; discussion 277–278.

2 Production of Enzymes as Fine Chemicals

KAREN A. FOSTER
SUSAN FRACKMAN
JAMES F. JOLLY

Milwaukee, WI 53202, USA

1 Enzyme Overproduction

1.1 Introduction

In considering the starting material for protein purification there are two basic options: using the wild-type organism as a source of the protein of interest or expressing that protein in a recombinant expression system. Until the advent of recombinant technology all proteins were purified from the organism from which they originated. The major advantage to this approach is that all of the various posttranslational modifications which may be necessary for the production of a functional protein will occur. These include: phosphorylation, N-linked and O-linked glycosylation, signal peptide cleavage, proteolytic processing, palmitylation, myristylation, farnesylation, carboxyl methylation, disulfide formation and multimeric assembly. In many cases the protein of interest is present only at low levels in its natural host resulting in very low yields of purified protein. Difficulty in obtaining and breeding the organism of interest is another often encountered problem. These problems have led many researchers to develop systems for the abundant expression of heterologous proteins. All of the heterologous systems require that the gene for the protein of interest be cloned and present on an expression vector. These systems have all been developed with the goal of abundant expression of functional protein. Unfortunately, despite the advanced state of knowledge concerning regulation of gene expression in many of the organisms that are being used for heterologous protein expression, one single expression system that works for all heterologous proteins has not been developed. Some of the factors that affect the expression of individual proteins in heterologous systems are: mRNA and protein stability, codon usage, presence of appropriate chaparonins, posttranslational modification systems, and the toxicity of the protein to the host. Each protein behaves differently and the best system for expression must be determined on an individual empirical basis.

1.2 Expression in Bacteria

The organism that is, by far, the most commonly used for the expression of heterologous proteins is *Escherichia coli*. Other popular bacterial expression systems for the extracellular production of heterologous proteins have been developed using *Bacillus subtilis* and *Bacillus brevis* hosts. The major drawback to the use of a bacterial system for heterologous protein expression is that the posttranscriptional processing required for the proper functioning of many eukaryotic proteins does not occur.

1.2.1 Expression in *Escherichia coli*

The extensive information that has been accumulated on the genetics, physiology and molecular biology of *E. coli* has made it possible to develop a number of well characterized systems for the expression of foreign proteins. The general requirements for an *E. coli* expression vector are: an origin of replication, mechanism for propagation and maintenance, transcriptional and translational initiation sites, and transcriptional and translational termination signals. Numerous *E. coli* expression systems have been developed that utilize either naturally occurring plasmids or bacteriophages.

1.2.1.1 Vectors

Most *E. coli* plasmid expression systems are based on the high-copy number pBR322 or pUC18/19 plasmids. The high-copy number of these plasmids can be beneficial for foreign protein expression by increasing the gene dosage. However, a high-copy number can be also detrimental for the expression of a toxic protein. One solution to this problem is the use of an adjustable-copy-number plasmid. These plasmids can be maintained at low-copy number during growth and can be induced to high-copy number at the time of or just preceding the induction of recombinant protein production. There are some temperature-sensitive copy number mutants whose copy number can be increased by altering the temperature of the growth culture (BITTNER and VAPNEK, 1981; KISHIMOTO et al., 1986). Another type of adjustable-copy-number plasmid contains two different origins of replication (*oris*) (CAULCOTT, 1985; FURMAN et al., 1987; CHEW and TACON, 1990). The low copy-number *ori* functions during growth of the culture. The high-copy-number *ori* is controlled by an inducible promoter.

While not as frequently used as plasmid expression systems, expression systems that utilize bacteriophage λ (YOUNG and DAVIS, 1983; PARK et al., 1991; PADUKONE et al., 1990; PANASENKO et al., 1977) and bacteriophage Mu (WEINBERG et al., 1993; GROISMAN et al., 1991; GRAMAJO et al., 1988) as vectors have been developed. In these systems the bacteriophage can be maintained in a lysogenic state during growth. During growth there is a single copy of the heterologous gene integrated into the bacterial chromosome. The presence of a temperature-sensitive repressor in either system makes it possible to shift the bacteriophage to lytic growth by increasing the temperature at the time that the recombinant protein is induced. During lytic growth these bacteriophages can replicate to 50–100 copies/cell effectively amplifying the foreign gene. An expression system that utilizes the bacteriophage M13 as a vector has recently been described (TAN et al., 1994). In this system the heterologous gene is introduced by infection with M13 following growth of the host strain.

1.2.1.2 *E. coli* Promoters

E. coli promoters contain three DNA elements that are necessary for transcriptional initiation. There are two hexanucleotide sequences located at −10 and −35 with respect to the start of transcription and a spacer between them. The most useful promoters for heterologous gene expression are regulated. The regulation can be either positive or negative and is generally mediated by the binding of regulatory proteins to specific sequences (operators) close to or within the promoter. This makes it possible to grow a culture under conditions in which the heterologous protein is not produced and then turn on its production. The ability to separate the growth phase from the production phase is especially important for the production of proteins that may adversely affect the growth of the host. One of the drawbacks of many of the regulated promoters in *E. coli* is that they are never completely silent but rather have a basal level of expression that may be problematic for expressing toxic genes. There are several very well characterized promoters that have been routinely used for heterologous protein expression in *E. coli*. They are promoters for the genes involved in tryptophan (*trp*) biosynthesis, lactose (*lac*) catabolism, an outer membrane protein (*lpp*) and repression of bacteriophage λ (P_L and P_R). A tightly regulated system utilizing the bacteriophage T7 promoter has been developed that facilitates the expression of toxic proteins.

lac Promoter

The *lac* genes of *E. coli* are required for lactose utilization. The three structural genes (*lac*Z, *lac*Y and *lac*A) are regulated by a repressor protein that is the product of the *lac*I gene. In the absence of lactose the repressor binds the operator site and the *lac* genes are not transcribed. When lactose is present in the culture medium, the repressor binds to the allolactose that has been formed from lactose and cannot bind to its operator site and transcription occurs. Many expression vectors are engineered so that a foreign gene can be inserted downstream of the *lac* promoter which makes the expression of that gene dependent upon lactose or a lactose analog such as isopropyl β-D-thiogalactoside (IPTG) in the culture medium. When the *lac* promoter is present on a high-copy-number vector, the amount of repressor made by a wild-type host strain may not be sufficient to ensure repression of the *lac* promoter. Strains with a mutation in the *lac*I gene, *lac*I^q, are frequently used to maintain repression when the *lac* promoter is present in high-copy number. These strains make more Lac repressor protein than wild-type strains.

The *lac* promoter is subject to another type of regulation known as catabolite repression. Catabolite repression causes the repression of the *lac* genes in the presence of glucose. To alleviate any problems caused by catabolite repression, a mutant of the *lac* promoter, *lac*UV5, is frequently used for expression of heterologous proteins. The *lac*UV5 promoter is still controlled by the Lac repressor but is not subject to catabolite repression.

trp Promoter

The *trp* promoter directs the transcription of five genes that are involved in tryptophan biosynthesis. In the absence of tryptophan the *trp* genes are actively transcribed. When tryptophan is present in the growth medium, it binds to the TrpR repressor; this complex then binds to its operator site, thereby inhibiting transcription. When the *trp* promoter is used for recombinant protein expression, it is generally induced by starvation for tryptophan and the addition of a tryptophan analog, indole acrylic acid (IAA). IAA has a higher affinity for the Trp repressor than tryptophan and lowers its operator-binding affinity 30-fold. Heterologous genes expressed from the *trp* promoter can accumulate to levels up to 30% of total cell protein (YANSURA and HENNER, 1990).

The *tac* promoter is a hybrid *trp-lac* promoter that has been useful for expression of recombinant proteins (AMMAN et al., 1983; DEBOER et al., 1983). It contains the −10 region of the *lac*UV5 promoter and the −35 region of the *trp* promoter and the *lac* operator site. The *tac* promoter is a very efficient promoter that can be repressed by overexpression of the Lac repressor and induced by IPTG.

lpp Promoter

The Lpp protein is an abundantly expressed outer membrane protein in *E. coli*. Both constitutive and regulated expression systems have been developed using the *lpp* promoter (NAKAMURA and INOUYE, 1982; MASUI et al., 1984). The constitutive system places the foreign gene downstream of the *lpp* promoter, while the regulated system places the *lac*UV5 promoter just downstream of the *lpp* promoter making expression dependent upon the presence of a *lac* inducer.

Bacteriophage λ Promoters

Two regulated promoters from bacteriophage λ, the right and left promoters (P_L and P_R), have been extensively used for heterologous expression. In the lysogenic mode of replication, λ is integrated into the *E. coli* chromosome, and all of its lytic genes are repressed by the phage encoded repressor protein, cI. cI represses by binding to its operator sites on the two phage promoters P_L and P_R. The switch from lysogenic growth to lytic growth occurs, when the repressor protein is inactivated and transcription initiates at P_L and P_R. A mutation in the *c*I gene that renders the repressor temperature-sensitive has been used extensively as a heterologous expression system. In this system a foreign gene is inserted downstream of P_L or P_R and is controlled by the the temperature-sensitive repressor, *c*I857, which is supplied either on a lysogen or on a plasmid.

Bacteriophage T7 Promoters

An expression system based on the bacteriophage T7 promoters has been developed that provides tight regulation of heterologous genes and abundant expression upon induction. This is a useful system for the expression of extremely toxic proteins. This system relies on the fact that the bacteriophage T7 produces an RNA polymerase that is specific for its own promoters. It is also a very active, stable, and processive enzyme. Recombinant proteins have been produced at levels of up to 50% of total intracellular protein using this system (STUDIER and MOFFAT, 1986). In this system the heterologous gene is cloned downstream of a T7 promoter on an expression vector. T7 polymerase under the control of an inducible promoter is provided on another plasmid (STUDIER and MOFFATT, 1986; TABOR and RICHARDSON, 1985) or a bacteriophage (MIAO et al., 1993). Two modifications to this system have been made to ensure that the T7 RNA polymerase is not produced until induction. One modification relies upon the fact that T7 lysozyme inhibits T7 RNA polymerase. In this system T7 lysozyme is produced at low levels from a plasmid (STUDIER, 1991). The low level of T7 lysozyme inactivates the small amount of T7 RNA polymerase that is made before induction. In the other system, a *lac* operator is placed downstream of the T7 promoter (GIORDANO et al., 1989; DUBENDORFF and STUDIER, 1991). This places the control of the foreign gene under the dual control of T7 RNA polymerase and the Lac repressor. A recent paper describes a system in which the heterologous gene under the control of a T7 promoter is present on an M13 bacteriophage chromosome (TAN et al., 1994). The host strain is infected at the same time that the T7 polymerase is induced alleviating the problem of low level of expression of toxic proteins before induction.

1.2.1.3 Transcriptional Termination

Transcriptional termination in *E. coli* requires a number of factors: a specific nucleotide sequence, mRNA secondary structure, and termination proteins. Appropriate termination of mRNA synthesis of a heterologous gene is necessary because transcription that is not limited to the heterologous gene can proceed through the *ori* region and destabilize a plasmid. A number of transcription termination sequences have been used in vectors for the expression of foreign genes. These include: the termination sequences from the *trp* operon, the bacteriophage λ terminators, and the *rrn*B ribosome RNA terminator.

1.2.1.4 Translational Initiation

There are specific requirements for the initiation of translation that must be provided for high-level expression of heterologous genes (GOLD and STORMO, 1990). Translational initiation requires a ribosome binding site (RBS) known as a Shine-Delgarno site, which is a short nucleotide sequence that is complementary to 16S rRNA. An initiation codon located 6–12 nucleotides downstream of the RBS is required. Translational initiation is also influenced by the nucleotide sequence and the potential for secondary structure around the initiation site. A variety of translational initiation sites from highly expressed *E. coli* and bacteriophage genes have been included in expression vectors.

1.2.1.5 Export of Heterologous Proteins

A number of vectors for heterologous expression have been constructed that direct the protein of interest to the periplasmic space of *E. coli*. These vectors have the signal sequence that is required for protein export positioned so that it is fused to the amino-terminus of the heterologous protein. The signal sequences from the genes for the membrane proteins, *omp*A, *omp*F and *lam*B have been used for the export of foreign proteins (GHRAYEB et al., 1984; MARULLO et al., 1989; SHIBUI et al., 1989). Signal sequences for genes from *Bacillus subtilis* have also been used for expression of heterologous proteins in *E. coli* (SMITH et al., 1988). A number of host strains that overexpress the components of the secretory pathway in *E. coli* have been found to increase the yield of foreign proteins exported to the periplasm (VAN DIJIL et al., 1991; PEREZ-PEREZ et al., 1994).

In many cases it is advantageous to express heterologous proteins so that they are secreted into the culture medium. This has generally been accomplished by treatments to destabilize the outer membrane and cause leakage of proteins (HSIUNG et al., 1989; AONO, 1989). Another system used for secretion to the culture medium utilized fusions to the staphylococcal protein A. Proteins containing two of the IgG binding domains of that protein are secreted in *E. coli* (JOSEPHSON and BISHOP, 1988).

1.2.1.6 Fusion Proteins

Heterologous proteins are often expressed as fusion proteins in *E. coli*. In these systems the protein of interest is expressed with another protein or part of a protein fused to its N- or C-terminus. The fusion partner can generally be removed either *in vivo* or *in vitro* to produce the protein of interest. There are a number of reasons to express a heterologous protein as a fusion. The fusion partner can provide a convenient method for purification. It may provide protection against proteolysis or may enhance the solubility of the foreign protein. It can also be used to direct protein export.

1.2.1.7 Inclusion Bodies

Heterologous proteins that are expressed in *E. coli* often form inclusion bodies which are insoluble protein aggregates. Proteins found within inclusion bodies are generally non-functional due to improper folding. Some proteins can be denatured and then allowed to refold to produce functional proteins. In these cases the presence of inclusion bodies is not a problem. In the vast majority of cases, however, it is very difficult or impossible to renature proteins found in inclusion bodies. Therefore, there has been some effort to develop methods to decrease the formation of inclusion bodies. Some success has been achieved by using low growth temperature, rich growth media and limiting the extent of induction of the promoter for heterologous expression. In some instances the overexpression of the chaperonins, GroESL have been useful in decreasing inclusion body formation in *E. coli* (BROSS et al., 1993).

1.2.1.8 Proteolysis of Heterologous Proteins

Frequently, one of the major obstacles to high-yielding heterologous protein systems is proteolysis. Many foreign proteins are unstable when produced in *E. coli*. Several methods have been utilized to minimize the effect of proteolysis on heterologous protein expression. The expression of a heterologous protein as a fusion to a stable protein has been used to increase yields (SCHULZ et al., 1987; HAMMARBERG et al., 1989). Expression systems that target the heterologous protein into the periplasm of *E. coli* or to the culture medium have been used to avoid proteolysis. Limited success in eliminating the problem of proteolysis has been achieved by using host strains that carry mutations in one or more of the protease genes (NAKANO et al., 1994; OBUKOWICZ et al., 1988; BUELL et al., 1985).

1.2.2 Expression in *Bacillus* spp.

B. subtilis and *B. brevis* heterologous expression systems have been developed for extracellular protein production. These organisms share two properties that make them attractive as systems for heterologous protein expression. They are able to secrete proteins directly into the growth medium and they are non-pathogenic.

B. subtilis secretion vectors have been developed using naturally occurring plasmids (JANNIERE et al., 1990), bacteriophages that infect *B. subtilis* (EAST and ERRINGTON, 1989) or integrating vectors (PETIT et al., 1990). Temporally regulated promoters that become active in the stationary phase have been used for recombinant gene expression (MOUNTAIN, 1989). The sucrose inducible promoter for the *sac*B gene which encodes levansucrase and inducible promoters from the bacteriophage ϕ105 have been used for foreign gene expression (JOLIFF et al., 1989; GIBSON and ERRINGTON, 1992; THORNEWELL et al., 1993). Several hybrid promoters have been developed for expression in *B. subtilis*. The *B. subtilis spo1* promoter has been fused to the *lac* operator from *E. coli* for inducible expression (YANSURA and HENNER, 1990). Another hybrid inducible promoter couples the *lac* regulatory elements with the bacteriophage T5 promoter (LE GRICE et al., 1987; LE GRICE, 1990). All of these vectors also contain a signal sequence from a *Bacillus* exoenzyme that efficiently directs the secretion of proteins that are fused to them. A major drawback of this system is the production by *B. subtilis* of a number of secreted proteases that can adversely affect the accumulation of heterologous proteins.

The major advantage of the *B. brevis* expression system is that only low levels of protease activity are secreted by this strain so that extracellular heterologous proteins are usually stable. An expression system that uses the promoter and signal peptide for the abundantly expressed cell-wall protein gene (*cwp*) has been developed using both high- and low-copy-number plasmids (UDAKA and YAMAGATA, 1933). This protein is secreted during the stationary phase and can accumulate to very high levels in the growth medium. The major problem of this system is that frequently plasmids carrying foreign genes are unstable in this system.

1.3 Yeast Expression Systems

Yeast expression systems are extensively utilized for the production of heterologous proteins (ROMANOS et al., 1992). They provide a genetically well characterized eukaryotic system that is easy to manipulate. While there is an enormous wealth of information on heterologous protein expression in the intensively studied yeast, *Saccharomyces cerevisiae*, increasingly other yeasts are being used for foreign gene expression. These include: *Pichia pastoris, Hansenula polymorpha, Kluyveromyces lactis* and *Yarrowia lipolytica*.

1.3.1 Expression in *Saccharomyces cerevisiae*

S. cerevisiae has a number of properties that make it an extremely useful system for the expression of heterologous proteins. It can be rapidly grown on simple media to high cell density; it is a food organism which makes it highly acceptable for expression of pharmaceutical proteins. Furthermore, the classical and molecular genetics of *S. cerevisiae* is more advanced than that of any other eukaryote making it possible to develop sophisticated expression systems for the production of foreign proteins.

1.3.1.1 Vectors

There are a number of vectors that can be used for expression of foreign proteins in *S. cerevisiae*. These include: yeast-episomal plasmids (YEp), yeast-replicating plasmids (YRp), yeast-centromeric plasmids (YCp) and yeast-integrating plasmids (YIp). The first three types of vector are extrachromosomally maintained while the *YIp* vectors integrate into the yeast genome. The most commonly used expression vectors are the *YEp* vectors that are based on the naturally occurring 2-μm plasmid. Most expression vectors are *E. coli*–yeast shuttle vectors that can be maintained and propagated in *E. coli* and *S. cerevisiae*.

2-μm Plasmid and Vectors

The 2-μm plasmid is a 6.3 kb plasmid present in most *Saccharomyces* strains at approximately 100 copies per haploid strain. It is very stably maintained with plasmid-free cells arising at a rate of 1 in 10^4 generations (FUTCHER and COX, 1983). It replicates once per cell cycle; its stability is due to a plasmid partitioning system and an amplification system. There are four open reading frames which are necessary for the plasmid partitioning system and the amplification system. The 2-μm plasmid also contains an origin of replication (*ORI*), a *cis*-acting element (*STB*) required for stabilization, and two 599 bp inverted repeat sequences. The products of two plasmid encoded genes *REP1* and *REP2* are required for plasmid stabilization. The plasmid encoded FLP protein is a site-specific recombinase that acts on a sequence within the inverted repeats (*FRT*) resulting in the presence of two forms of the 2-μm plasmid within the cell. This site-specific recombination is the basis of the 2-μm amplification system.

A wide variety of vectors containing various amounts of the 2-μm plasmid have been developed. The simplest of these vectors contains the 2-μm *ORI-STB* sequences, a yeast selectable marker and bacterial plasmid sequences. The resident 2-μm plasmid supplies the REP1 and REP2 proteins for the stabilization of these vectors; they cannot amplify. These vectors are small and easily manipulated. They are maintained at a copy number of 10–40 copies per cell and are lost in 1–3% of cells per generation under non-selective conditions. More complex vectors contain increasing amounts of the 2-μm plasmid. Overexpression of the FLP recombinase results in very high copy numbers of 2-μm vectors that are capable of amplification (ARMSTRONG et al., 1989).

ARS and *CEN* Vectors

Yeast-replicating vectors contain autonomous replication sequences (*ARS*) which function as origins of replication. They are present in 1–20 copies per cell but they are unsuitable for foreign gene expression because of their instability. In the absence of selection, plasmid-free cells accumulate at a rate of up to 20% per generation. These vectors can be stabilized by the addition of centromeric (*CEN*) sequences but the copy number is reduced to 1–2 copies per cell. These vectors are only used for heterologous expression when low-level expression is desired.

Vectors for Integration, Transplacement and Transposition

Yeast-integrating vectors are useful for foreign gene expression primarily because of their stability. They are lost at <1% per generation in the absence of selection (HINNEN et al., 1978). They generally contain yeast chromosomal DNA to target integration, a selectable marker, and a bacterial replicon. Integration occurs by homologous recombination resulting in the duplication of the chromosomal target site. Digestion of the plasmid at a unique restriction site in the homologous sequences increases the transformation frequency and targets the plasmid to the desired integration site.

Yeast-integrating vectors have been used to generate stable multicopy integrants. Tandem multi-copy inserts can result when high concentrations of integrating vector DNA are used (ORR-WEAVER and SZOSTAK, 1983). Another method targets the integrating vector into the ribosomal DNA (rDNA) cluster. This gene cluster contains about 140 tandem repeats. The targeting of an integration vector to a non-transcribed spacer of the rDNA cluster in conjunction with the use of a promoter-defective marker can result in transformants with 100–200 integrated copies (LOPES et al., 1989, 1990). Other strategies for generating multicopy integrants are targeting of the integrating vector to a transposable element (*Ty*) that is present in 30–40 copies per genome in most strains or to the δ element that is found both alone and as part of the *Ty* element (KINGSMAN et al., 1985; SHUSTER et al., 1990).

Another method for introducing foreign DNA into yeast chromosomes is *transplacement*. In transplacement there is no duplication of the target chromosomal site because the foreign DNA is introduced into the chromosome by a double homologous recombination event. Transplacement vectors contain the heterologous DNA and a selectable marker flanked by yeast DNA that is homologous to the 3′ and 5′ regions of the yeast chromosomal DNA targeted for replacement. Prior to transformation the vector is digested to produce a fragment whose ends are within the flanking 3′ and 5′ DNA. This is a useful type of vector when a single-copy transformant lacking bacterial DNA sequences is required.

Ty-based *transposition* vectors have been constructed to produce multi-copy integrants (BOEKE et al., 1988). These vectors are analogous to retroviral vectors in mammalian systems. They are constructed so that all of the DNA to be integrated is transcribed as a single unit. Integration occurs by way of a full-length transcript that is encapsidated into virus-like particles and reverse transcribed to DNA and integrated into the genome.

1.3.1.2 Selectable Markers

Commonly, the selection for transformants in yeast is achieved by using an auxotrophic host strain with a defect in a biosynthetic enzyme and a vector that expresses that enzyme. *LEU2*, *TRP1*, *HIS3*, and *URA3* are frequently used markers for the selection of transformants. A variation of this approach is used to produce transformants with very high copy numbers. Alleles of *URA3* or *LEU2* with a defective or truncated promoter that results in low-level expression of the selectable marker have been introduced into vectors (ERHART and HOLLENBERG, 1983; LOISON et al., 1989). Under selective conditions the plasmid must be in high copy number in order to produce enough enzyme for complementation of the host *leu2* or *ura3* mutation.

There are some dominant markers that have been used for selection of transformants in *S. cerevisiae*. A number of antibiotic resistance genes have been successfully used as dominant markers (GRITZ and DAVIES, 1883; HADFIELD et al., 1986; JIMINEZ and DAVIES, 1980; WEBSTER and DICKSON, 1983). Vectors which produce resistance to G418, hygromycin B and chloramphenicol have been constructed. Copper resistance, which is conferred by the presence of multiple copies of the yeast *CUP1* gene, has also been used as a dominant marker for selection (HENDERSON et al., 1985). The advantages of using a dominant selectable marker are that they increase the range of hosts that can be used to include prototrophic strains of *S. cerevisiae*, and it is possible to maintain selection while growing the cultures on rich media.

A number of systems, called autoselection systems, have been developed to maintain plasmid selection under any culture conditions. Vectors carrying the cDNAs for killer toxin and immunity are examples of an autoselection system (BUSSEY and MEADEN, 1985). In this case the vector is maintained because its loss results in the death of the cell due to the killer toxin produced by the other cells in the culture. Another autoselection system employs a *ura3 fur1* which cannot utilize exogenous uracil and therefore absolutely requires the presence of the vector encoded *URA3* gene (LOISON et al., 1986). An *fbal* mutant host strain is deficient in 1,6-diphosphate aldolase, a glucolytic enzyme that is essential for both fermentative and gluconeogenic growth; it requires the vector encoded *FBA1* gene under all culture conditions (COMPAGNO et al., 1993). LUDWIG et al. (1993) have developed a system for heterologous gene expression in which a 2 μ vector carries a selectable marker only until the plasmid has been established in the cell; site-specific recombination mediated by the Flp recombinase results in the excision of the selectable marker gene. The plasmid carrying the heterologous gene is stably maintained in the absence of selection.

1.3.1.3 *Saccharomyces cerevisiae* Promoters

The structure and regulation of promoters in *S. cerevisiae* has been intensively studied. *S. cerevisiae* promoters may be very complex, extending over 500 bp (STRUHL, 1989). They consist of at least three elements that regulate transcriptional initiation: upstream activation sequence (UAS), TATA element, and initiator elements. Furthermore, elements involved in repression of transcription (BRENT, 1985) and downstream activation sequence (DAS) elements (MELLOR et al., 1987; LIAO et al., 1987; ZAMAN et al., 1992) are involved in the regulation of expression of some *S. cerevisiae* genes. The UAS is a binding site for regulatory proteins that activate transcription from that promoter. UASs can act at variable distances 5′ from the site of initiation. A promoter can have multiple UASs. TATA elements are found 40 to 120 bp upstream of the initiation site. A promoter can have multiple TATA elements which direct initiation from different sites. The initiator element is a poorly defined sequence element that directs the initiation of transcription. The presence of DAS elements is particularly problematic for heterologous gene expression, because they are located downstream of the start of translation and are required for maximal transcription.

An important aspect of a heterologous expression system is the selection of an appropriate endogenous promoter to drive expression of the foreign protein. A large number of promoters have been used for high-level expression of foreign proteins in *S. cerevisiae*. These include: the promoters from the abundantly expressed but poorly regulated glycolytic enzymes, tightly regulated promoters, and hybrid promoters.

Constitutive Promoters

The promoters for the genes encoding the glycolytic enzymes were utilized for much of the initial experimentation on heterologous gene expression (SHUSTER, 1989). These genes include *ADH1* encoding alcohol dehydrogenase, *GADH* encoding glyceraldehyde-3-phosphate dehydrogenase, *PGK* encoding phosphoglycerate kinase, *ENO* encoding enolase, and *TPI* encoding triose-phosphate isomerase. These genes are abundantly and constitutively expressed on glucose-containing media. However, there are several drawbacks to the use of these promoters. The expression of heterologous proteins can be toxic or inhibitory to cell growth limiting the usefulness of these constitutive promoters. The *PGK* gene has been shown to contain a DAS element (MELLOR et al., 1987). It is frequently advantageous, especially for large-scale applications to have maximum expression at high cell density. Since the glycolytic promoters are down-regulated in the absence of glucose, transcription will be reduced in the late stages of growth when glucose is limiting and cell density is highest.

Regulated Promoters

The use of a regulated promoter for the expression of heterologous genes overcomes many of the limitations of constitutively expressed promoters. Most significantly, tightly regulated promoters make it possible to separate the growth phase from the protein production phase, thereby minimizing the problems associated with continuous heterologous protein production.

The promoters for the galactose-regulated genes that are involved in galactose utilization have been successfully used for foreign protein expression. Transcription of the *GAL1*, *GAL7*, and *GAL10* genes is rapidly induced (>1000-fold) by galactose to approximately 1% of total mRNA. These genes are also subject to catabolite repression by glucose. The molecular basis for galactose activation has been extensively studied. The basis of galactose regulation lies in the interactions between the GAL4 transcriptional activator protein, the *GAL* UAS and the GAL80 repressor protein. In the absence of galactose the GAL80 protein binds the GAL4 protein preventing the interaction between the GAL4 protein and the *GAL* UAS. In the presence of galactose the GAL4 protein binds to the *GAL* UAS to activate transcription. The GAL4 activator can be limiting in the induction of a heterologous gene on a high-copy-number plasmid. This problem has been overcome by the construction of strains that overexpress the GAL4 protein (BAKER et al., 1987; HASHIMOTO et al., 1983; SCHULTZ et al., 1987). The most useful of these strains regulates the expression of GAL4 protein so that it is only overexpressed during induction thereby alleviating the problem of gratuitous induction in the absence of galactose caused by titration of the GAL80 protein (SCHULTZ et al., 1987).

The phosphate-regulated promoter for the acid phosphatase gene, *PHO5*, has been used for foreign gene expression (HINNEN et al., 1989). Transcription from the *PHO5* gene is induced in low-phosphate medium. The PHO4 protein is a transcriptional activator, the PHO80 protein is a repressor, and there are two UAS elements upstream of the *PHO5* gene that are necessary for regulation (VOGEL and HINNEN, 1990). A host strain has been constructed for phosphate-independent induction of the *PHO5* gene by temperature (KRAMER et al., 1984). This strain is defective in *PHO80* and produces a temperature-sensitive PHO4 protein. Lowering the growth temperature of this strain results in the induction of the *PHO5* gene. The major drawback to using the *PHO5* promoter is that it is not a very strong promoter.

There are a number of other regulated promoters that have been used to express recombinant proteins in *S. cerevisiae*. The promoter for the glucose-repressible gene for alcohol dehydrogenase (*ADH2*) has been used for foreign gene expression (PRICE et al., 1990). The promoter for α-factor has been used for heterologous gene expression in a strain that has a temperature-sensitive *sir* mutation which effectively renders the promoter for α-factor temperature sensitive (BRAKE et al., 1984). A copper-regulated promoter from the *CUP1* gene, encoding copper metallothionein, has also been utilized for expression of heterologous genes (ETCHEVERRY, 1990).

Hybrid Promoters

Hybrid promoters have been constructed for the expression of heterologous proteins. With the construction of these promoters investigators have attempted to link the promoters for the highly expressed genes for the glycolytic enzymes to a UAS element from a tightly regulated gene. Some examples of these hybrid promoters are the replacement of the glycolytic UAS element with a *GAL* UAS element in the *GAP* promoter (BITTER and EGAN, 1988) and the *PGK* promoter (COUSENS et al., 1990). Hybrids of the *GAP* gene have been constructed with both the *PHO5* UAS (HINNEN et al., 1989) and the *ADH2* UAS (COUSENS et al., 1987) elements. Hybrid promoters have also been constructed that exploit the ability of mammalian steroid hormone receptors to function as transcriptional activators in yeast. Promoters that fuse the glucocorticoid response elements to the *CYC1* promoter (SCHENA et al., 1991) and the androgen response elements to the *PGK* promoter (PURVIS et al., 1991) have been constructed.

1.3.1.4 Transcriptional Terminators

Transcriptional terminators from higher eukaryotes or prokaryotes are not active in *S. cerevisiae*. It is therefore necessary that expression vectors contain a *S. cerevisiae* specific terminator. Transcription terminators from a number of genes have been used for heterologous expression including: *TRP1*, *ADH1*, *GAP*, *MF1*, and *FLP*.

1.3.1.5 Intracellular Posttranslational Processing

The intracellular posttranslational processing of proteins in *S. cerevisiae* include: amino-terminal modification, phosphorylation, myristylation, and isoprenylation. The specificity of the enzymes involved in acetylation of the N-terminal residue and in removal of the N-terminal amino acid appears to be conserved among eukaryotes (CHANG et al., 1990; HALLEWELL et al., 1987).

1.3.1.6 Secretion of Foreign Proteins

Expression systems in *S. cerevisiae* have been developed for the secretion of heterologous proteins. This is important for the expression of proteins that are normally secreted since the correct folding and posttranslational processing is associated with the secretory process. It can also be advantageous to engineer the secretion of heterologous proteins that are not normally secreted. The removal of the heterologous protein from the intracellular environment may lessen its toxic effect or decrease its accessibility to intracellular proteolysis. Since the secreted protein is cleaved from a fusion with an amino terminal signal sequence, it is possible to engineer a recombinant protein with the native amino terminus. Furthermore, purification of a heterologous protein from the culture medium is generally simpler than purification of an intracellularly expressed protein. Some problems that have been associated with extracellular expression of heterologous proteins are low yield, incomplete or incorrect signal peptide cleavage and inappropriate glycosylation.

1.3.1.7 Signal Sequences

Many of the promoter systems previously described have been used in systems designed for extracellular expression. In these systems, the signal sequence of a yeast-secreted protein is present downstream of the promoter such that the heterologous protein is expressed with a yeast signal sequence at its amino terminus. Although there have been some successful attempts to utilize heterologous signal sequences in *S. cerevisiae* (DE BAETSELIER et al., 1991; RUOHONEN et al., 1987; SLEEP et al., 1990; SOGAARD and SVENSSON, 1990), most often the heterologous protein accumulates intracellularly. *S. cerevisiae* signal sequences that have been used for extracellular expression of heterologous proteins include those from acid phosphatase (PHO5) (HINNEN et al., 1989), invertase (SUC2) (CHANG, 1986; HITZEMAN et al., 1983; HORWITZ et al., 1988) and α-factor (MFa1) (BRAKE, 1989).

1.3.1.8 Glycosylation

Glycosylation of proteins may be essential for their appropriate function and immunogenicity. Both N- and O-linked glycosylation of foreign proteins has been reported in *S. cerevisiae* (KUKURUZINSKA et al., 1987). The amino acid recognition site for N-linked glycosylation is conserved in yeast and higher eukaryotes; however, the type of oligosaccharide added and the extent of glycosylation differs. *S. cerevisiae* is not capable of adding the oligosaccharides that are found in the complex glycosylation of higher eukaryotes. Many *S. cerevisiae* strains have been found to hyperglycosylate foreign proteins by the addition of multiple mannose residues. *S. cerevisiae* strains with mutations that affect the glycosylation pathway have been used for expression in order to avoid hyperglycosylation of foreign proteins. The O-linked oligosaccharides in yeast are very different than those in higher eukaryotes.

1.3.2 Expression in *Pichia pastoris*

P. pastoris is a methylotrophic yeast that has been used to express high levels of many proteins. Techniques for efficient, high-yield, cultivation using a defined medium containing methanol have been developed. This fermentation technology has been adapted for the expression of heterologous proteins from highly expressed and tightly regulated *P. pastoris* promoters for the methanol-inducible gene *AOX1* (CREGG et al., 1987). *AOX1* encodes alcohol oxidase which is the first step in the pathway for methanol assimilation; it can be induced by methanol to 30% of total cell protein (COUDERC and BARATTI, 1980). The selective markers used in *P. pastoris* vectors are the *HIS4* (CREGG et al., 1987), *ARG4* (CREGG and MADDEN, 1987), and *SUC2* (SREEKRISHNA et al., 1987) genes from *S. cerevisiae*; the dominant G418 resistance gene (SCORER et al., 1993) from *E. coli* Tn90 has also been used. *P. pastoris* expression systems utilized vectors that are designed for chromosomal integration or transplacement. Multiple integration and transplacement events can occur during transformation of *P. pastoris* in many cases resulting in higher levels of foreign gene expression than from strains carrying a single copy (CLARE et al., 1991a; ROMANOS et al., 1991). Secretion of foreign proteins from *P. pastoris* has been achieved using the signal sequences from invertase and α-factor from *S. cerevisiae* (CLARE et al., 1991b; TSCHOPP et al., 1987). Hyperglycosylation of some foreign proteins expressed in *P. pastoris* has been observed.

1.3.3 Expression in Other Yeasts

Expression systems in other yeasts have been developed to a much lesser extent. An expression system for the methylotrophic yeast, *Hansenula polymorpha* that is similar to the *P. pastoris* expression system has been developed (GLEESON and SUDBERY, 1988). It utilizes the methanol-inducible promoter for the *MOX* gene that encodes methanol oxidase. The large-scale cultivation of *Kluyveromyces lactis* has been developed in the food industry for the production of β-galactosidase. The advantages of *K. lactis* as an expression system lie in the fact that it grows on inexpensive substrates such as lactose and whey, and it is an accepted system for the production of proteins for human consumption. The major advantages of *Yarrowia lipolytica* for foreign gene expression are that it naturally secretes a variety of high-molecular-weight proteins at very high levels and it can grow to high cell density at industrial scale.

1.4 Baculovirus Expression

A wide variety of proteins from many sources have been expressed in insect cells using baculovirus vectors (LUCKOW, 1992;

LUCKOW and SUMMERS, 1988). It is possible in many cases to achieve high levels of heterologous protein expression in insect cells from a regulated and highly expressed promoter for a non-essential baculovirus gene. The advantages of the baculovirus expression system include: (1) the vectors are not dependent on helper virus; (2) standard virological methods are utilized; (3) baculoviruses are not pathogenic to vertebrates or plants; (4) baculoviruses can be grown and maintained *in vitro* using insect cell lines. Recombinant proteins produced in baculovirus-infected insect cells undergo many of the post-translational modifications of higher eukaryotes. A number of baculovirus-expressed recombinant proteins have been shown to be transported to their natural cellular location and undergo appropriate oligomerization. For some but not all proteins, baculovirus mediated expression in insect cells has resulted in 20–250 times more expression than in mammalian expression systems (LUCKOW, 1992). These features have made the baculovirus expression system particularly useful for expression of proteins that have been difficult to express in lower eukaryotic or prokaryotic expression systems.

1.4.1 Baculovirus Biology

Baculoviruses have been divided into three morphologically distinct subgroups: nuclear polyhedrosis viruses (NPV), granulosis viruses and non-occluded viruses. The baculovirus genome is a covalently closed, circular, double-stranded DNA molecule of 80–220 kb. The virus particle is composed of an outer lipid membrane containing rod-shaped nucleocapsids into which the DNA has been packaged. Only NPVs have been utilized as expression vectors. There are two infectious forms of NPVs which result from a biphasic replication cycle within infected insect cells (LUCKOW, 1992; MILLER, 1988; ATKINSON et al., 1990). These forms are the extracellular budded virus (EV) and the occluded virus (OV). In OVs the virus particles are embedded in a proteinaceous crystalline matrix called polyhedra. Polyhedrin, a 29 kD protein, is the major component of the polyhedra.

The virus known as AcNPV is the most intensively studied baculovirus; it is also the one that has been most extensively developed as a vector for foreign gene expression. It was originally isolated from *Autographa californica* (alfalfa looper) but has a wide host range infecting over 30 insect species (GRONER, 1986). Baculoviruses isolated from *Bombyx mori* (silkworm) and *Helliothis zea* have been utilized to a lesser extent as a vector for heterologous gene expression (MAEDA et al., 1985; FRASER, 1989).

1.4.2 Gene Expression

Four phases of viral gene expression in AcNPV infected cultured insect cells have been described (BLISSARD and ROHRMANN, 1990; MILLER, 1988; LUCKOW, 1992). Immediate early genes, 0–4 hours post-infection (hpi), and delayed early (5–7 hpi) are expressed prior to DNA replication. The late genes (8–18 hpi) are expressed with the onset of viral DNA replication; they encode structural proteins. The very late genes are expressed 18 hpi and beyond. It is during this phase that polyhedrin synthesis and OV formation occurs. Polyhedrin is expressed at very high levels, reaching 25–50% of the total stainable protein of infected cells. Another abundantly expressed, very late protein, p10, is expressed slightly before the polyhedrin promoter (WEYER et al., 1990).

1.4.3 Recombinant Virus Construction

There are a number of methods that have recently been developed to facilitate the isolation of recombinant baculovirus (DAVIS, 1994). Most of these methods are based on *in vivo* homologous recombination between a transplacement vector containing the gene of interest and a wild-type baculovirus genome.

The first system that was developed for the construction of recombinant baculoviruses involved the transfection of cultured insect cells with a transplacement vector and wild-type baculovirus DNA. The transplacement vectors are bacterial plasmids that carry the heterologous gene downstream of a baculovirus promoter and varying amounts of 3′ and 5′ viral sequences flanking the gene normally expressed from that promoter. A certain fraction of the viral genomes undergoes recombination with the transplacement plasmid resulting in allelic replacement of the viral gene with the heterologous gene. The frequency of recovery of recombinant viruses has been reported to be between 0.1 and 5 percent (LUCKOW, 1992; CORSARO and FRASER, 1989; SUMMERS and SMITH, 1987).

Modifications of this method have been developed to increase the yield of recombinant virus. These include ultraviolet irradiation just prior to transfection (PEAKMAN et al., 1989) and linearizing the baculovirus genome prior to transfection (KITTS et al., 1990). KITTS and POSSEE (1993) developed vectors in which the baculovirus genome is deleted for an essential gene. The transplacement vector is designed so that recombination with the baculovirus chromosome results in the transfer of both the heterologous gene and essential baculovirus gene to the viral chromosome. Since only the recombinant viruses are viable, this strategy has resulted in yields of 85–99% of recombinant viruses from a coinfection.

Methods that employ heterologous hosts for the isolation of recombinant baculovirus have been developed (PATEL et al., 1991; LUCKOW et al., 1993). In one of these methods the baculovirus genome is modified so that it can be maintained in the yeast *S. cerevisiae*. Recombination between the baculovirus and a transplacement vector results in the loss of a gene that produces canavanine sensitivity in the host. In the second method the virus is maintained in *E. coli* by the introduction of a mini-F replicon into the viral chromosome. This large plasmid also carries the target site for the Tn7 transposon. The transplacement plasmid contains the left and right arms of Tn7 flanking the gene of interest. When the proteins for Tn7 transposition are supplied in *trans*-site, specific recombination between the bacmid and the transplacement plasmid results in a recombinant baculovirus.

A method that employs *in vitro* recombination for the construction of recombinant baculovirus has been described by PEAKMAN et al. (1992a). This method uses the *cre/lox* site specific recombination system of bacteriophage P1. Both the baculovirus chromosome and the transplacement plasmid contain *lox* sites which can be acted upon *in vitro* by the Cre recombinase to form the appropriate recombinant baculovirus. The proportion of recombinant viruses obtained by this method has been as high as 50%.

1.4.4 Promoters for Heterologous Expression

The promoters that are utilized for foreign gene expression are very active and result in very high levels of the natural viral protein that is normally expressed from them. When foreign genes replaced the viral genes, the levels of expression have been variable. Heterologous proteins have been produced as fusion and nonfusion proteins at levels ranging from 1 to 500 mg/liter (LUCKOW and SUMMERS, 1988; LUCKOW, 1992).

Generally, the regulated, non-essential, and highly expressed very late promoters for the polyhedrin gene and the *p10* gene from AcNPV are used for the expression of foreign proteins. A vector that uses the promoter for the AcNPV basic protein has also been constructed (HILL-PERKINS and POSSEE, 1990). This promoter is active during late times in infection (8–24 hpi). There are transplacement vectors that carry the promoter for two different baculovirus genes making it possible to produce viruses that express two foreign proteins within the same cell (WEYER and POSSEE, 1991). This is useful for studies involving multisubunit proteins and protein interactions.

1.4.5 Posttranslational Processing

A major reason for selecting a eukaryotic expression system for protein production is the requirement for posttranslational processing by many eukaryotic proteins in order to be biologically active. There are many posttranslation processing steps. These steps include: phosphorylation, N-linked and O-linked glycosylation, signal peptide cleavage, proteolytic processing, palmitylation, myristylation, farnesylation and carboxyl methylation (LUCKOW, 1992; LUCKOW and SUMMERS, 1988; MILLER, 1988; ATKINSON et al., 1990).

1.4.6 Growing and Expressing Recombinant Viruses

The cell lines used most frequently to grow recombinant baculoviruses are derived from *Spodoptera frugipera* (fall army worm) and *Trichoplusia ni* (cabbage looper). Some studies have indicated that *T. ni* cells may be useful in producing higher levels of proteins particularly those that are glycosylated and secreted than other cell lines (WICKHAM et al., 1992, WICKHAM and NIMEROW, 1993). There has been some interest in the expression of foreign genes in insect larvae (LUCKOW, 1992). This is considered a potentially low cost way to produce large quantities of a protein.

2 Enzyme Purification

2.1 Introduction

It is nearly impossible for the inexperienced researcher to find detailed guidelines to protein purification problems in the literature. Numerous methods of protein purification are available, and the successful purification process will require the use of several of these methods in combination. The growth of biotechnology as an industry and the confidentiality of methods that coincides with such development have made it difficult to find published information on new purification developments. Manufacturers of purification equipment and materials provide little application data because of confidentiality agreements they have with their clients. While many publications regarding protein purification dwell on the theory and mathematical modeling of the processes, this chapter will focus on some practical considerations based on years of the authors' experience purifiying proteins as molecular biology reagents.

With the exception of perfusion chromatography, streamline chromatography, some membrane technology and automated systems, industries selling purification equipment and resins have introduced little new purification technology in the past decade. Developing novel or improved purification schemes has become a challenge. Complicating this issue is the demand for purer protein preparations for both clinical and research use.

As one sets out to purify proteins as fine chemicals there is much to consider. From the practical view point the budgetary constraints and required yield will have a major impact on the plan. Will the purification costs be supported by the final use? Is a protein purified to the homogeneity desired, or may the protein be functionally pure? Will the final use of the product be FDA regulated? In such cases not only is purity an issue, but one must choose purification steps that are easily validated. The ultimate scale of the purification process plays a major role in the purification scheme developed. The ultimate process should contain as few steps as possible in order to maximize yield and to minimize the required investment of labor, materials and equipment. In general, one can expect a 20% loss of yield with each critical purification step.

Due to all of the parameters involved in the process, such as starting material, fermentation processes, cell lysis techniques, clarification techniques and the wide range of purification options, developing a purification design that produces the optimal yield of the purest protein in the least amount of time for the lowest cost is likely to take a great deal of effort. In fact, the high number of options leaves designing the optimal scheme a virtual impossibility. As the design is developed, the endpoint will most likely be based on reaching an acceptable yield of sufficient purity that supports the cost of required labor and materials. The purpose of this chapter is to provide a review of current purification techniques that have practical applications to designing such a system.

2.2 Starting Material: Extracellular versus Intracellular

The purification scheme can be significantly different depending on whether the protein is expressed intracellularly or extracellularly. If the protein of interest is excreted from the cell, several advantages are presented. It is not necessary to rupture the cells, a process which requires an investment in equipment and/or time. Cell rupturing techniques can generate heat and aeration, both of which can be detrimental to the target protein. Extracellular proteins tend to be more stable as they are not offered the protection of the cell and they are often produced in larger amounts because of the greater volume outside the cell (WANG, 1987). Lastly, because few other proteins are secreted, the purification steps can be minimized. However, a concentration step is usually necessary for processing of extracellular proteins. If the cells require washing in order to maximize yield, the washing will further contribute to the dilution of the material for downstream processing. If presented with a process that requires the concentration of an extracellular protein, the best solution is to choose a concentration step that also serves as a purification step. This can be achieved in a number of ways, such as ammonium sulfate precipitation, adsorption onto a resin or by ultrafiltration – all of which are discussed in the remainder of this chapter.

The processing of intracellular proteins has the disadvantage of requiring cell disruption, creating protein loss and contamination with all other intracellular proteins. However, when working with intracellular proteins, one generally works with much smaller initial volumes.

2.3 Initial Processing Steps

2.3.1 Fermentation

Enzyme fermentations are difficult to optimize and are difficult to control to the point of yielding reproducible batches. The fermentation process that delivers the highest cell mass yield will not necessarily deliver the highest yield of pertinent protein. For example, the temperature optimum for the synthesis of a particular enzyme may be different from that for the growth of the cells. In many cases the optimum pH for growth is different than the optimum pH for enzyme stability. The ultimate fermentation is one that delivers the highest yield of protein while minimizing the functional contaminants. The optimum fermentation parameters should be determined experimentally. Because of the parameters that must be controlled and because extensive protein purification may be required to judge the success of a fermentation, developing such a process is difficult and time-consuming.

Enzyme fermentations are typically carried out in stirred tank reactors ranging in size from 0.1 cubic meters to 250 cubic meters. It is best to have fermentation facilities capable of monitoring and controlling all important parameters when optimizing a fermentation. These parameters include temperature, pH, head space pressure, liquid volume, agitator speed, air flow, liquid feed rates, and the presence of foam, dissolved gasses and ionic levels. Any change in these parameters can significantly affect the outcome of the fermentation.

Fermentations typically consist of three stages: the strain maintenance and seed preparation stage, the inoculum growth stage and the fermentation stage. Two methods of strain maintenance are recommended. The cells can be lyophilized or frozen as a glycerol stock. Both sources can then be used to prepare agar slopes or plates that can then be used for inoculum growth. The inoculum growth is typically done in a shake flask. The volume of this growth should be 1–10% of the final fermentation volume. A larger inoculum (10%), called transfer rate, will lessen the chance that a contaminant will outgrow the production strain, will reduce lag times and will minimize the cycle time in subsequent inoculum and fermentation stages. Smaller transfer rates (1%) may reduce the size of equipment necessary; i.e., a 4 liter shake flask will suffice over a 12 liter stir jar. The inoculum stage may consist of one to four transfers (see Fig. 1).

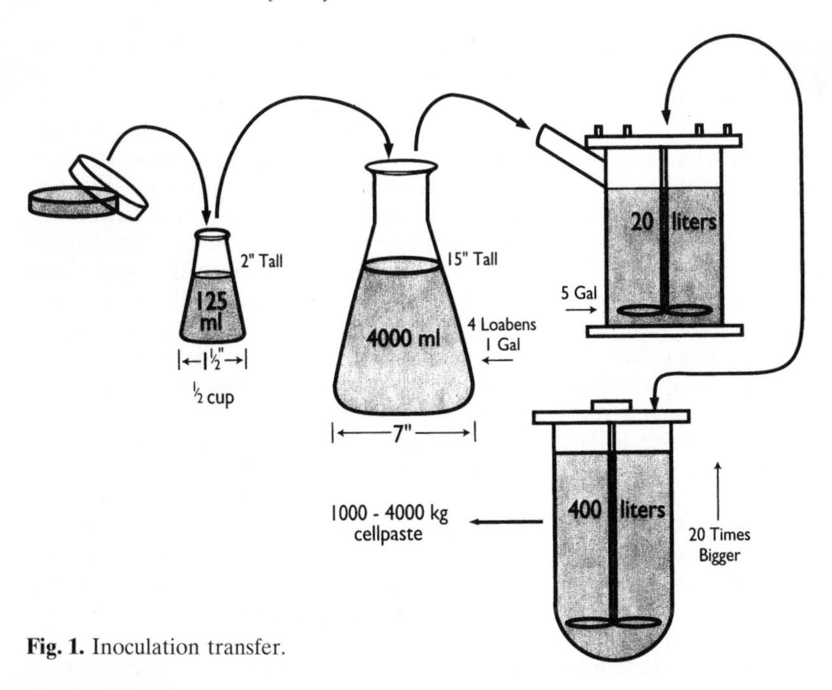

Fig. 1. Inoculation transfer.

There are three ways of performing the fermentation stage: the batch method, the feed-batch method and the continuous cultivation method. When performing the batch method, all necessary components are present at the start of the cultivation. As growth proceeds, the nutrients are consumed, the biomass increases and the activity and yield of an enzyme rises to a maximum level before conditions become detrimental, either because of nutrient exhaustion or the presence of a toxic metabolite. The optimum yield requires optimizing the point of harvest. The typical batch fermentation takes 18–180 hours. This type of fermentation is necessary when the optimal yield occurs during a particular stage in a bacterial growth curve.

During a feed-batch fermentation the fermentation is fed at certain times. This process can be used to extend the fermentation phase of optimal target protein expression. Such interruption of the fermentation process, however, increases the chances of contaminating the tank.

When performing the continuous cultivation method, the fermentor is continuously fed fresh culture medium, while the same volume of the used broth is continuously removed. This process has been used to a limited extent in the biotechnology industry. As with the feed method, there is an increased opportunity for contamination. However, if unique growth conditions are being used, there is a reduced possibility for a contaminant to grow. Also, the yield of extracellular products is often less when using this method. This method encourages the development of a faster growing low-enzyme-producing strain.

When purifying intracellular enzymes, one must choose whether to process the cells immediately after harvest or to store them for future use. The authors recommend storage at 4 °C for no longer than 24 hours. Longer storage should be done at −70 °C; if a −70 °C storage is chosen, storage for no longer than 6 months is recommended. Some enzymatic activities become more difficult to recover in cell paste that has been stored much longer than this. If longer storage is necessary, it is recommended that a stability study be performed with the cell paste, using a quick screening assay for the protein of interest. While cells processed immediately after harvesting may provide the greatest yield, the cells may be easier to rupture after a freeze/thaw cycle. A comprehensive review of fermentation methods has been written by FROST and MOSS (1987) in Volume 7 of this series (*Biotechnology* First Edition).

2.3.2 Cell Rupture

If the protein to be recovered from the cells is intracellular, the cells must first be ruptured. Tissue culture cells are easier to rupture than are bacterial cells. Additional capital investment is usually necessary when working with bacterial cells. Several methods for cell rupture exist and include bead mills, pressure, sonication and chemical treatments. The manner in which cells are disrupted has an effect on the yield and quality of the protein of interest. Protein recovery will increase with increasing exposure to disruption; however, increasing exposure will also increase the release of contaminants, thereby complicating downstream processing. Experimentation is important to define the point of optimum recorvery while minimizing contaminant release. Most methods of rupture create heat and aeration; therefore, a method that will minimize denaturation is desirable. The ultimate scale of a purification process must be kept in mind during the design phase. Some methods of cell rupture are impractical with greater than a kilogram of cell paste. During disruption, the rate of release of the target protein depends on the location of the protein in the cell. The growth conditions of the cells will also affect the disruptability. GRAY et al. (1972) reported that cells grown rapidly on synthetic media disrupt more quickly than those grown slowly on complex media. Cells harvested during the log phase of growth are more easily disrupted than those harvested during the stationary phase. Certainly, the efficiency of cell disruption will vary according to cell type.

2.3.2.1 Sonication

Sonication is an acceptable method for disrupting a fraction of a gram to 500 grams of cells. Any more than 500 grams would be time-consuming. Flow-through sonication probes are available, allowing faster processing of larger volumes. However, the viscosity of a cell suspension can interfere with the performance of such equipment. The cells should be suspended in 2–3 mL of buffer per gram wet weight for sonication, and no more than 300 mL should be sonicated at one time. The efficiency of sonication is not significantly influenced by cell density. It is more efficient to sonicate a large volume for longer times than to sonicate multiple volumes. Pulses of no longer than 3 minutes at most power levels should be applied to the homogenate in order to prevent extended exposure to high temperatures. For example, the temperature of a 300 mL homogenate will rise approximately 10 °C after 3 min of sonication. The use of cooling rosettes will minimize the temperature increase. The progress of the cell rupture should be monitored microscopically and the lysate should be cooled to at least 8 °C before resuming any further pulses. If the probe is not inserted far enough into the solution, air will be introduced to the sonicate, increasing the chance for protein denaturation. Manufacturers recommend the probe to be inserted to a depth of approximately 1 cm. Complete rupture of a 300 mL homogenate by sonication can take 3–15 min for *E. coli*, depending on the strain and the point of harvest. During sonication the homogenate will initially become more viscous as nucleic acids are released from the cells. The viscosity will decrease as sonication is continued.

2.3.2.2 Bead Mills

Bead mills operate by mixing the cell suspension with glass beads and agitating the mixture at a very high rate of speed. The bombardment of the cells with the beads causes cell rupture. Bead mills present a heating problem, although they work well for filamentous organisms. Another risk is the potential for the protein to absorb to the glass beads, which can be minimized by buffer optimization. Typically, increasing the ionic strength of the extraction buffer will minimize this absorption.

2.3.2.3 Pressure

Cell rupture by pressure is the most widely used method. As with sonication and bead mills, pressurized rupture generates heat. APV Gaulin manufactures a few models for varying scales. The model 15 MR works very well for 300–6000 grams of wet cell paste. As with sonication, suspending the cells in 2–3 mL of buffer per gram of cell paste works well. One liter of suspension can be passed once through the unit within 1 min. The unit operates at a maximum pressure of 8000–9000 psi. At this operating pressure the temperature of the lysate will increase 18–20 °C above the starting temperature. Collecting the lysate in a large stainless steel receptacle on wet ice cools the lysate down to 8 °C in approximately 5–15 min. Gram-negative organisms are more susceptible to rupture than are Gram-positive organisms. *E. coli* and *Thermus aquaticus* are easily ruptured in one pass through the unit while *Haemophilus influenzae* may take 2–3 passes. *Bacillus* species require 3–4 passes. Yeasts are more difficult to rupture. Larger cells tend to be easier to process. Some authors report that this process does not work well for filamentous organisms (ASENJO and PATRICK, 1990). As with sonication, the homogenate will initially become more viscous during pressure rupture as nucleic acids are released from the cells; the viscosity will decrease with additional passes. The small size of the cell debris created by this method can make cell debris removal difficult.

2.3.2.4 Chemical Treatment

Chemical treatment can be used to rupture cells, although large-scale use can be limited by cost considerations. Chemicals such as lysozyme, detergents (e.g., SDS, deoxycholate), deoxyribonucleases and alkali have all been used to release intracellular contents. Lysozyme is active only on some bacterial cells and will leave a very viscous solution. The addition of any of these chemicals introduces an additional contaminant to the process that may produce a purification problem downstream. Proper titration of any of these chemicals can lead to selective membrane permeability, which may present a purification advantage.

2.3.3 Lysate Clarification

Clarifying the cell lysate improves downstream processing by removing cellular debris, nucleic acids, and other contaminants that will interfere with centrifugation, chromatography and detection of the pertinent protein. Numerous precipitation and cell debris removal aids are available, and a brief discussion on the practical aspects of some of these follows. Several methods to clarify cellular extracts are available; however, methods generating extremes of temperature, pH, ionic strength, solvents and foams should be avoided. It is important to perform optimization experiments when designing precipitation steps into a purification protocol. During the design phase keep in mind the concentration of precipitants used, salt concentrations and pH of buffers, centrifugation and the ultimate scale of the preparation. The type and amount of precipitating agent along with optimal buffer conditions should be determined by the clarity of the resulting supernatant and by the recovery of pertinent protein. If the protein is prevalent enough to see on an SDS gel, Pharmacia's PhastSystem™ is a rapid method to quantitate recovery (see Fig. 2).

2.3.3.1 Solid Filtering Aids

A solid filter aid such as Solka Floc™ (James River Corp.) works very well as a means to filter mammalian and plant tissue extracts that are too large to clarify by centrifugation. This diatomaceous earth product also works well to remove lipids, but can also bind the protein of interest. A cellulosic filtering agent called Cell Debris Remover (Whatman) works well to remove cell debris, nucleic acids and lipids. Anionic exchangers, such a DE-52 (Whatman) and DEAE Sepharose (Pharmacia) are often added to cell lysates to bind nucleic acids, and if enough salts are added proteins will not bind to the resin. When using an ion exchanger for primary extraction choose an inexpensive, large, low cross-linked

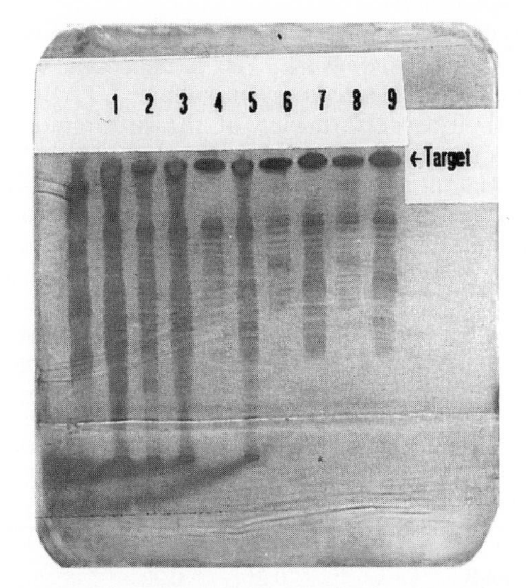

Fig. 2. Use of Pharmacia's PhastSystem to determine optimum cell debris precipitation method. Molecular weight of target protein: 12000 daltons. *Lane 1:* crude supernatant; *Lane 2:* 0.25% PEI; *Lane 3:* 0.25% CTAB; *Lane 4:* 0.50% PEI; *Lane 5:* 0.50% CTAB; *Lane 6:* 0.75% PEI; *Lane 7:* 0.75% CTAB; *Lane 8:* 1.0% PEI; *Lane 9:* 1.0% CTAB. Lane 4, 0.5% PEI precipitation, was deemed optimal.

resin, such as DE-52. The higher cross-linked smaller beaded resins are more suitable for chromatographic applications.

The industrial purification of terminal deoxynucleotide transferase involves extracting hundreds of pounds of calf thymus glands. The tissue is ground and extracted with buffer. The volume of this process is too large for centrifugation and the consistency of the homogenate does not condone filtration. Filtration is aided by the addition of a diatomaceous earth product. The suspension is then filtered using an industrial filter press fitted with filtration cloths.

2.3.3.2 Polyethylenimine

Polyethylenimine (PEI) can be added directly to a cell lysate to precipitate nucleic acids and to aid in cell debris removal. PEI is a long-chain cationic polymer with a molecular weight of 24000 daltons. Proteins that bind to DNA and RNA may be precipitated with the PEI pellet. In such instances the protein can be recovered by extracting the pellet with a buffer of higher ionic strength. Precipitation of the target protein with the nucleic acid pellet could serve as a purification step. If precipitation of the target protein in this fashion is not desirable, it can be prevented by increasing the salt concentration of the lysis buffer. The ionic properties of PEI will interfere with downstream processing, particularly cationic chromatography steps. The large size of the PEI molecule prevents removal by size exclusion chromatography or by dialysis. Therefore, the PEI must be removed by ammonium sulfate precipitation. A PEI concentration of 0.5–2% in the homogenate is sufficient to precipitate nucleic acids. The optimal concentration should be determined by a titration of PEI concentrations relative to clarification, purification and protein recovery (see Fig. 2 for an example).

2.3.3.3 Biocryl Bioprocessing Aids

Biocryl BPA-1000 is manufactured by TosoHaas (Montgomery, Pennsylvania) to serve as a bioprocessing aid by removing nucleic acids, lipids and cell fragments from cell lysates. The material is an aqueous suspension of strongly cationic cross-linked particles. The suspension is simply mixed with the lysate before clarification, resulting in an improved clarification after centrifugation, a reduction in any lipid material present, and potential purification from contaminating proteins.

2.3.3.4 Hexadecyl Trimethylammonium Bromide

Hexadecyl trimethylammonium bromide (CTAB) can be used in the same fashion as PEI. CTAB has been used in our laboratory to improve the preparation of a DNA binding protein. Without CTAB and ammonium sulfate precipitation the preparation of the enzyme required 4–5 chromatographic steps. CTAB precipitation removed the functional

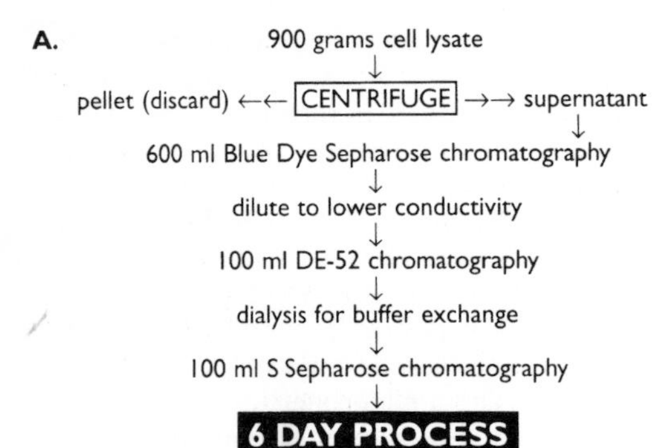

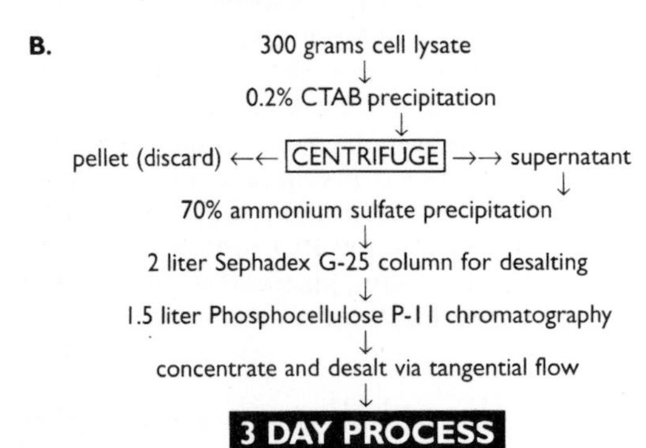

Fig. 3. An example of a process improvement in the purification of a DNA-binding protein by the introduction of a cell debris precipitation step. A. Process without precipitation. B. Process improved by the addition of CTAB precipitation.

contaminant, which resulted in the need for only one chromatographic step to sufficiently purify the protein (see Fig. 3). The success of this process is strongly dependent on the concentration of the CTAB solution, which can be difficult to maintain. A 10% CTAB solution is quickly crystallized at refrigeration temperature (4–10 °C), making a slow addition of a constant concentration of CTAB in the cold difficult.

2.3.3.5 Streptomycin Sulfate

Streptomycin sulfate is often used as a clarification agent but does not appear to work as often as does PEI and CTAB. We performed an experiment to determine the optimum precipitant to clarify a homogenate of *Streptomyces achromogenes*. The homogenate was treated with streptomycin sulfate, CTAB or PEI. The performance of the precipitation was determined by the clarification of the supernatant as determined by the absorbance of the supernatant at 660 nm. The results are summarized in Tab. 1.

Tab. 1. Clarification of a Bacterial Homogenate after Treatment with a Precipitating Agent and Centrifugation – as Measured at 660 nm

Treatment	Duration of Centrifugation		
	30 min	60 min	90 min
PEI	0.08	–	–
CTAB	0.21	0.20	–
Streptomycin sulfate	1.44	0.82	0.85

2.3.3.6 Ammonium Sulfate

Ammonium sulfate is not used as a lysate clarifyer, but is routinely used to concentrate protein solutions, making downstream processing easier. This precipitation technique can result in a crude purification from proteins of differing hydrophobicities. Ammonium sulfate precipitation is usually performed prior to any chromatography, so it is included in this section. As noted earlier, ammonium sulfate precipitation is necessary to remove large soluble precipitation agents that could not otherwise be removed by gel filtration or dialysis. The use of ammonium sulfate to precipitate proteins with molecular weights of less than 15000 daltons is inefficient and recovery may be poor. Precipitations performed with an ultrapure ammonium sulfate can be significantly different from precipitations performed with lower grades. Proteins are relatively stable when stored as ammonium sulfate precipitates. If a crude preparation must be left overnight, the protein is more likely to retain biological activity if stored as a precipitate rather than in a crude solution. An extensive discussion of ammonium sulfate precipitation of proteins has been prepared by ENGLARD and SEIFTER (1990).

2.3.3.7 Ultra- and Micro-Filtration

Ultrafiltration is an effective tool for clarifying, concentrating and purifying protein preparations. Two modes of ultrafiltration are available: stirred cell and tangential flow filtration. Manufacturers of filtration membranes have developed an extensive line of filters for ultrafiltration with the ability to process volumes as small as 50 mL to as large as 12000 liters. Tangential flow filtration differs from conventional dead-end filtration, in that in tangential flow filtration the solution to be processed flows tangentially across the surface of the membrane (see Fig. 4).

The tangential flow method of ultrafiltration is preferable as it performs filtrations faster and prevents fouling of the membranes by lowering concentration polarization at the membrane surface, eliminating the need for frequent membrane changes. These membranes can be used multiple times without flux rate decreases and without cross-contamination, provided the units are cleaned and stored properly after each use. Filtron's Minisette™ system fitted with two 10K membranes will concentrate and desalt 4 liters of a column pool to 500 mL in approximately 4 hours. The rate of salt removal is a function of volume and not of salt concentration. This process eliminates the need for time-consuming overnight dialysis and will reduce the volume necessary to load onto the next column. This step works well between the first and second column, because the target protein typically elutes very dilute from the first column. The second column is generally of very high capacity and will be significantly smaller in volume than the first column. Loading such a large volume onto a small column takes considerable time. The desalted and concentrated preparation can be more quickly loaded onto the next column, saving a day in the processing of the protein that would otherwise require an overnight dialysis and lengthy loading time. In general, the faster a protein is purified, the greater the yield will be. One risk of tangential flow filtration is aeration. Because of the high flow rate and turbulence through the apparatus, any introduction of air easily causes protein denaturation. A book edited by E. L. V. HARRIS and S. ANGAL (1988) provides a good review of the use of this technique.

While the various manufacturers of tangential flow filtration apparatus recommend the use of their products for bacterial lysate clarification, few have application data to share. Clarifying cell lysates in this fashion is referred to as microfiltration rather than ultrafiltration. The only difference between the two processes is the pore size of the membrane used. Attempts by the authors to clarify lysates have provided some promising results; however, in order to maximize the recovery of the intracellular protein the retentate must be continually diluted to "rinse" the protein from the debris. This results in a cumbersome volume of filtrate that makes downstream processing more difficult. While maximum recovery of cell debris clarified by microfiltration may not be as great as recovery from cell debris that has been centrifuged, microfiltration is a promising option when the scale of the clarification exceeds centrifuge capacity. An interesting scenario for downstream processing would be to pump the filtrate directly onto the first column as the clarification is continuing (see Fig. 5). In one case where we clarified lysate with tangential flow filtration the protein preparation was less contaminated with an interfering enzyme than lysate clarified by centrifugation. In another experiment we were able to purify a 12000 dalton protein to 90% physical purity by passing the cell debris over a 30K membrane. However,

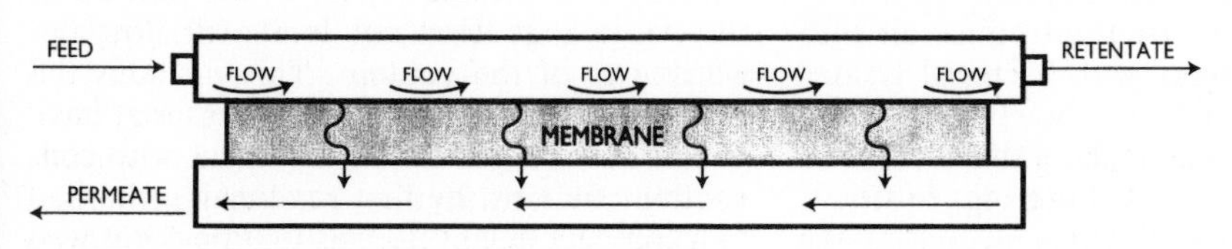

Fig. 4. Tangential-flow filtration.

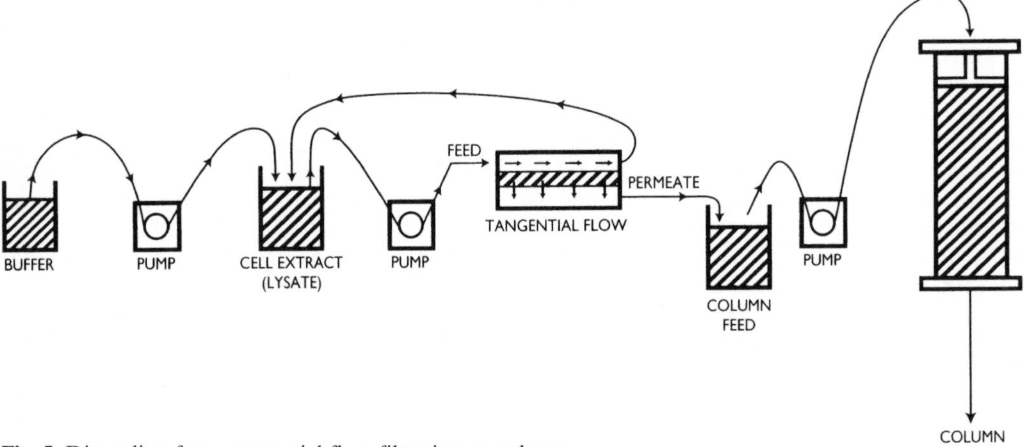

Fig. 5. Direct line from tangential-flow filtration to column.

this required a 25-fold dilution of the cell debris.

2.3.3.8 Centrifugation

When selecting the mode of centrifugation the following should be considered: the size of the particulates to be removed, the volume to be clarified and the time available for clarification.

Depending on the scale of the process, the effectiveness of conventional batch centrifugation can be limited both by gravitational force and by the amount of the preparation that can be processed at one time. The largest laboratory batch centrifuge can process approximately 6 liters at a time. However, with a centrifuge this large, the centrifugal force is often not sufficient to clarify bacterial lysates, particularly those obtained from pressure rupture. Several alternatives to batch centrifugation are available, aside from microfiltration that was discussed previously. Several types of continuous flow centrifuges are available. While larger *g*-forces can be established, these types of rotors often introduce air into the supernatants. Also, with bacterial lysates that are 25–30% solids, the bowl would need frequent cake removal, making this technique labor-intensive. Heinkel Filtering Systems, Inc. (Germany) manufactures a continuous flow centrifuge (Inverting Filter Centrifuge) that delivers pressure to the top of the solution in addition to centrifugal force, increasing the separation power of the unit. In addition, this unit is designed to automatically remove the pelleted cake without operator assistance. However, the maximum *g*-force available is only $1500 \times g$.

Because of limitations of *g*-force with scales larger than 6 liters, the ultimate scale and available *g*-force must be considered during the small-scale design phase. All centrifugations at a small scale must be done with centrifugal forces to mimic the final *g*-force available.

2.3.3.9 Fluidized Bed Technology

When performing fluidized bed (or expanded bed) adsorption, the chromatographic resin is poured into a column designed with excess head space. The cellular lysate is pumped upward into the resin bed, expanding the bed upwards. As the lysate continues to be pumped onto the column, the bed will eventually reach a fixed height. The portion of the lysate that does not bind will flow upwards out of the column. This includes the cell debris (see Fig. 6). The proteins that have bound to the resin can be eluted as with convention columns, by first readjusting the bed to a compact height. Resins that perform well in this method include ion exchangers and

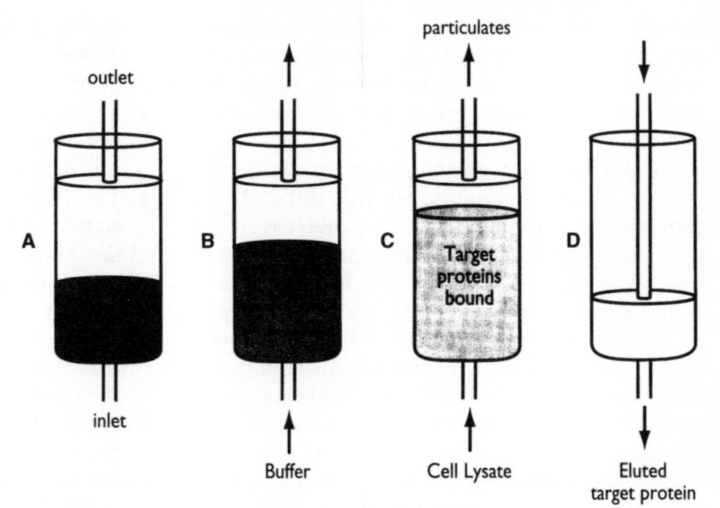

Fig. 6. Expanded bed chromatography. A. Settled bed of resin. B. Equilibration of fluidized bed. C. Application of cell extract, clarification. D. Elution of target protein from settled resin.

some affinity matrices. Pharmacia has prepared columns and resins especially designed for this application (Streamline^{TM}).

2.3.3.10 Depth Filtration

Filtration of 100 liters or more can be performed using a filter press. This unit consists of a series of metal plates, between each of which is placed a piece of cotton cloth. The steel plate/cloth "sandwiches" are pressed together, and the slurry to be filtered is pumped through the series of plates. The debris is filtered by the cloth filters. A filtering aid is usually necessary when performing this type of filtration (see previous discussion of terminal deoxynucleotide transferase purification). Filtering aids are particles of high surface area, such a diatomaceous earth, silica, alumina and carbon. They efficiently absorb nonpolar compounds from polar solutions (Box, 1985). The filtering aid is slurried with the protein solution prior to filtration. While this method works well to clarify large volumes, two risks are presented. The target protein may bind to the filtering aid; this risk can be determined by experimentation. The second risk to the protein solution is the amount of aeration created by the filtration process; therefore, this method may not be a choice for proteins sensitive to denaturation by aeration.

2.4 Purification Methods

Liquid chromatography is generally recognized as the technique that allows the highest degree of purification of biomolecules. In contrast to precipitation, electrophoresis and ultrafiltration, chromatography does not involve heat generation or major shear forces. Conditions close to physiological can be maintained. The first step in designing a chromatographic purification strategy is to characterize as much about the protein sample as possible. This includes characterization of both the protein of interest and of contaminating proteins. Knowledge of the molecular weight (both native and denatured); amino acid sequence; isoelectric point; solubilities in the presence of different salts, organic solvents and temperature; hydrophobicity; location in the cell and binding characteristics to the chromatographic media as a function of pH will maximize the successful design of the chromatography step. Taking advantage of the biochemical differences between the target protein and the contaminants is the key to successful purification. Purification of the target protein from a functional contaminant is a challenge. A functional contaminant is one that interferes with the biological activity of the target protein; often the proteins have similar binding sites and are, therefore, very similar biochemically. For example, purifying a DNA binding protein such as a DNA polymerase from other DNA binding proteins that would interfere with the successful use of a polymerase for molecular biology research (e.g., a restriction nuclease), is an arduous undertaking.

In general, it is easy to remove the first 95% of the contaminants, but it is difficult to remove the remaining 5%. In one case by chromatographically purifying the target protein away from an unacceptable level of functional contaminant and assaying the amount of protein in the fractions containing the functional protein, the functional contaminant was less than 0.0025% of remaining protein. When purifying proteins from wild-type organisms, the pertinent protein can be 0.00001%, or even less, of the total cell mass. This high degree of required purity demands choosing reagents of high purity. Some ACS-grade reagents have been demonstrated in our laboratory to occasionally introduce nucleolytic contaminants into preparations of DNA modifying enzymes. Reagents used should be of ultrapure grade, molecular biology grade, or should be tested for the absence of contaminants of concern. Buffers should be prepared from liquid reagents rather than solid reagent, such as phosphoric acid instead of monobasic and dibasic potassium salts, and 50% NaOH instead of NaOH pellets. Liquid reagents tend to be of higher purity.

During the design of a purification process it is important to keep in mind the stability of the protein both during the purification and as a purified product, the required yield, the ultimate scale, and capital expense of the process.

Other factors to be considered when designing a purification procedure include:

- The initial chromatography steps should be of high capacity and low cost.
- The lower capacity, higher cost steps should be used for the final steps of the purification.
- Since the scale of a purification usually decreases as the process proceeds, the most difficult and expensive steps should be used last.
- Choose purification steps that are very different from each other.
- Enzymes can lose activity when exchanged from one buffer into another. The organization of columns to minimize buffer exchanges is recommended.
- The most plentiful impurity should be removed early and the most selective purification steps should be used first.
- The ideal chromatography step is designed such that the target protein does not bind to the column and the contaminant does, or to have the target protein be the first or last protein to elute, presenting the need to pool away from contaminants on only one side of the peak.
- An inverse relationship exists between the number of purification steps and the final yield of the protein. A typical yield after each step is approximately 80% of that which went into the step. Therefore, it is necessary to limit the number of purification steps.

Manufacturers of chromatographic resins have designed resins to bind proteins using one or several of the following interactions: electrostatic, hydrophobic, hydrogen bonding or van der Waals (JONES, 1990). We have found that the best matrix for protein purification is a cross-linked agarose. Several cellulose-based resins are available, but these fibrous particles present flow rate challenges. Cross-linked acrylamides or styrenes tend to result in non-specific binding to the resin itself, rather than to the functional group.

The size of the column should be 5–20 times the binding capacity of the protein preparation as determined by batch experimentation. The dimensions of the chromatography column can help or hinder the purification and should be seriously considered. When increasing the scale of a column, increase the radius of the column but maintain the height and linear flow rate. MAO et al. (1993) offer mathematical modeling as a means of optimizing purifications as scale increase is performed. The linear flow rate is calculated by dividing the volumetric flow rate by the column cross-sectional area.

Example: A Pharmacia XK 50 column has a cross-sectional area of 19.6 cm^2. If a flow rate of 5 mL/min was being applied to the column, the linear flow rate would be:

linear flow rate = 5 mL/min ÷ 19.6 cm^2
linear flow rate = 0.25 $mL/cm^2/min$

If the column is going to be scaled up into a Pharmacia BP 113, with a cross-sectional area of 100 cm^2, the flow rate needed for proper scale-up would be:

0.25 $mL/cm^2/min$ =
volumetric flow rate ÷ 100 cm^2
volumetric flow rate = 25 mL/min

The resolution of the purification is primarily determined by the elution conditions and not by the length of the column. Columns should be wide and short, although long columns are useful for proteins with weak binding characteristics as the long contact time will maximize binding. One can try to improve the purification from a persistent contaminant by increasing the length of the column, but maintaining the column volume. Columns with a height of much over 25 cm will remove the effect of the resin wall support and will lead to gel compression. When a longer column is desired, try connecting columns together and running them in series.

One column manufacturer, Sepragen, has created an interesting column that allows for maximizing flow rate. The sample is introduced into the column through the inlet port at the top of the column. It is evenly distributed to the outer channel through the radial distributor. The inner wall of this outer channel is a porous tube that allows the sample to pass radially into the chromatographic packing. The sample proteins are bound and eluted radially inward by the passage of eluent through the bed. The inner wall of the bed is another porous tube. The sample passes through this tube into the inner channel to the column outlet. Due to the large surface area of the outer porous tube and the relatively small bed height, separations can be performed at high flow rates with low back pressure. Scale-up just requires a longer tube.

Whatever type of liquid chromatography is performed, the column should be designed to minimize pressure drops, mixing and loss of resolution. The dead space between the end of the column and the fraction should be minimized to prevent mixing and loss of resolution. This requires minimizing the length of tubing between the column and UV-detector and between the detector and the fraction. Because of inertness and durability, any pump used should be fitted with silicone tubing.

The packing of a chromatographic column is critical to the performance of the column. An unevenly packed chromatography bed and entrapped air bubbles will lead to channeling, zone broadening and loss of resolution. Cross-linked resins are simple to pack. It is recommended that any alcohol the resin is shipped in is first rinsed from the resin with water. This can be performed on a sintered glass funnel. Rinsing the alcohol from the resin before it is poured into the column will prevent bubble formation from the degassing due to the heat generated from the alcohol mixing with the aqueous buffer. The water rinse will also prevent any of the buffer components from precipitating in the alcohol. Gel filtration matrices and nonspherical matrices are more difficult to pack into columns and must be poured exactly to the manufacturers' requirements.

The resins can be used to purify proteins in one of three ways. The *batch method* does not utilize a column and is the quickest and most amenable to scale up. The protein solution is mixed into the resin and the resin is washed by filtration or centrifugation. For elution, the protein–resin complex can be either poured into a column and eluted or batch eluted. The batch method is best if the load has a high degree of particulates, which will impede column flow during the load, or when the target protein is at a low concentration and binding will increase if the protein is kept in contact with the resin for a longer period of time. The *isocratic method* elutes a protein, usually weakly bound to the resin, by continually pumping the same buffer, without changing ionic strength or pH, onto the column. This can be a powerful means of purification, but the protein can take a long time to elute, leading to a large volume of a very dilute protein. The third method of elution, *gradient elution*, is not as amenable to scale up, but provides some of the best resolutions. In order to scale up, the gradient needs to reach the same height over the same fluid length. For example, when increasing the scale of a column from 100 mL to 1000 mL with a 0–1 M NaCl gradient, the total gradient volume must also be increased 10-fold.

While most modern liquid chromatographic resins are amenable to multiple uses, frequently a cleaning procedure more complex than a simple salt wash is needed to maintain acceptable performance. Before deciding to regenerate columns for subsequent use, the cost of the labor required to properly clean and store resins should be compared to that of purchasing new resin for future needs. When regenerating a column for multiple uses it is important to consider whether the labor required to clean the column is more or less costly than repacking the column with fresh resin, and whether there is a risk of cross-contaminating future preparations with either the protein previously applied to the column or with one of the cleaning solutions. Most likely, the purchase of new resin is the practical choice. Exceptions would be prepacked columns, such as Pharmacia's Mono bead columns and columns made with resins that are time-consuming to process and pack, such as Sephadex G-25 columns used for routine protein desalting.

The average amount of time required for a purification depends on the scale of the process, the degree of purification needed and the degree of assay difficulty. In general, purification of a protein from a cloned source requires 1–2 precipitation steps followed by 1–3 chromatographic steps. The purification from a non-cloned source requires 1–2 precipitation steps, followed by 3–6 chromatographic steps. One day should be allowed for the disruption and precipitation steps. An additional day should be allowed for each chromatographic step, followed by a last day for final quantitation and measurements for purity. Therefore, the purification of a protein will require 5–10 days of labor. While the purification scheme is designed to yield a protein with sufficient purity, it is also important to design a process that can be completed rapidly. This will certainly reduce the required amount of labor to complete the process and will increase the yield by removing proteolytic activity as soon as possible. In addition, there is some indication from work done in our laboratory, that a preparation that remains in an unpurified state for 24–48 hours is more difficult to purify than the same preparation that is purified quickly. When the cells are ruptured, the release of previously compartmentalized proteins can result in some protein interactions that are difficult to separate.

The following is a summary of the major types of chromatography used for protein purification. Any protein contaminant remaining after using these steps will be the most difficult to remove as they are similar to the target protein in molecular weight, isoelectric point and hydrophobicity.

2.4.1 Ion Exchange Chromatography

The most versatile of chromatographic interactions is ionic exchange chromatography. Ion exchange chromatography is the most frequently used mode of protein purification. Some resins are capable of separating two proteins differing by only one charged amino acid. There are several manufacturers of ion exchange resins, but this chapter will focus on Pharmacia's products as examples. Q- and DEAE Sepharose are anion exchangers, with Q- being a strong exchanger while DEAE is a weak exchanger. S- and Carboxymethyl Sepharose are cation exchangers, with S- being a strong exchanger while Carboxymethyl Sepharose is a weak exchanger. A strong versus weak exchanger refers to the extent of ionization with pH and not to the strength of protein binding. Strong ion exchangers are completely ionized over a wide pH range, whereas with weak ion exchangers, the degree of dissociation and thus exchange capacity varies more markedly with pH.

Isoelectric point titration curve analysis will provide information about the protein solution that will aid in the optimization of an ion exchange step. Pharmacia's PhastSystem™ will generate a titration curve in approximately 100 minutes. This electrophoretic step will indicate the buffer pH that will create the greatest difference in charged properties of the proteins in the solution and whether the proteins will bind better to an anion or cation exchange resin at that pH (see Fig. 7). When performing an IEF titration curve analysis, running an IEF standard marker is recommended, as the pH gradient created by the ampholines may not be linear. If the pertinent protein is easy to detect by assay, a mini-column packed with an ion exchanger can be prepared and a binding experiment can be performed. Alternatively, the protein solution can be batch-mixed with 1 mL of resin and binding experiments can be performed. Information generated from titration curve analysis should be used to define optimal chromatography conditions which are best employed on a strong ion exchanger, which can be utilized over a wider pH range than weak exchangers. It is recommended to choose a pH that is within one pH of the buffer pK_a and one unit of the isoelectric point of the pertinent protein.

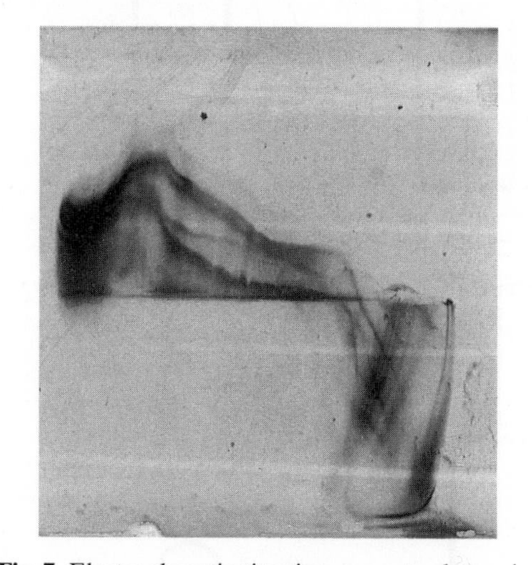

Fig. 7. Electrophoretic titration curve to determine optimum ion exchange chromatography.

The binding capacity of Q- and SP-Sepharose is approximately 50 mg/mL; the binding capacity is protein-dependent. The recommended flow rate is 400 cm/h. The available capacity of a column is different for different proteins and is related to the molecular weight of the protein and to the charge of the protein at the equilibrated pH of the column. The flow rate and temperature of the column also play a role. Regardless of the binding capacity of any column, when performing preparative work a column should be loaded to only 10% of capacity.

While a conventional column may be able to handle the pressures generated by a high flow rate, the higher flow rate may result in proteins not binding to the column that would normally bind at lower flow rates. Pharmacia has designed the FPLC™ system to manage high performance ion exchange chromatography with Mono S and Mono Q high performance resins. FPLC™ operates under moderate pressure, giving higher speed and resolution than conventional chromatography. Mono S and Mono Q are composed of smaller, more rigid beads than that of their Sepharose counterparts. These monobeads are of plastic composition, which make them more hydrophilic than their Sepharose counterparts. The smaller and more rigid the bead size, the better and faster the resulting separation. The FPLC™ has been designed to deliver and withstand the pressures required to deliver flow through columns composed of small beads. All tubing on the FPLC™ is non-metallic so as to prevent metal exposure that may be detrimental to the biological activity of proteins. The rapid processing provided by this system, along with titration curve analysis, will allow for efficient optimization of an ion exchange purification. A salt gradient providing high resolution can be applied to the column in as little as 10 min. With automation, several pH conditions and gradient slopes can be tried within a few hours. While these columns are expensive, the Mono S and Mono Q columns are very rugged and can continue to produce acceptable results after years of multiple uses. We continue to successfully use a Mono Q column for the purification of fine biologically active proteins that was purchased in 1987 and has been used repeatedly since then. Fully utilizing the valving mechanisms of the FPLC™ can allow one to process two or three chromatographic steps with the push of a button (see Fig. 8). Other comparable automated systems are Unicorn (Pharmacia), Trio Bioprocessing (Sepracor) and BioCor (PerSeptive Biosystems).

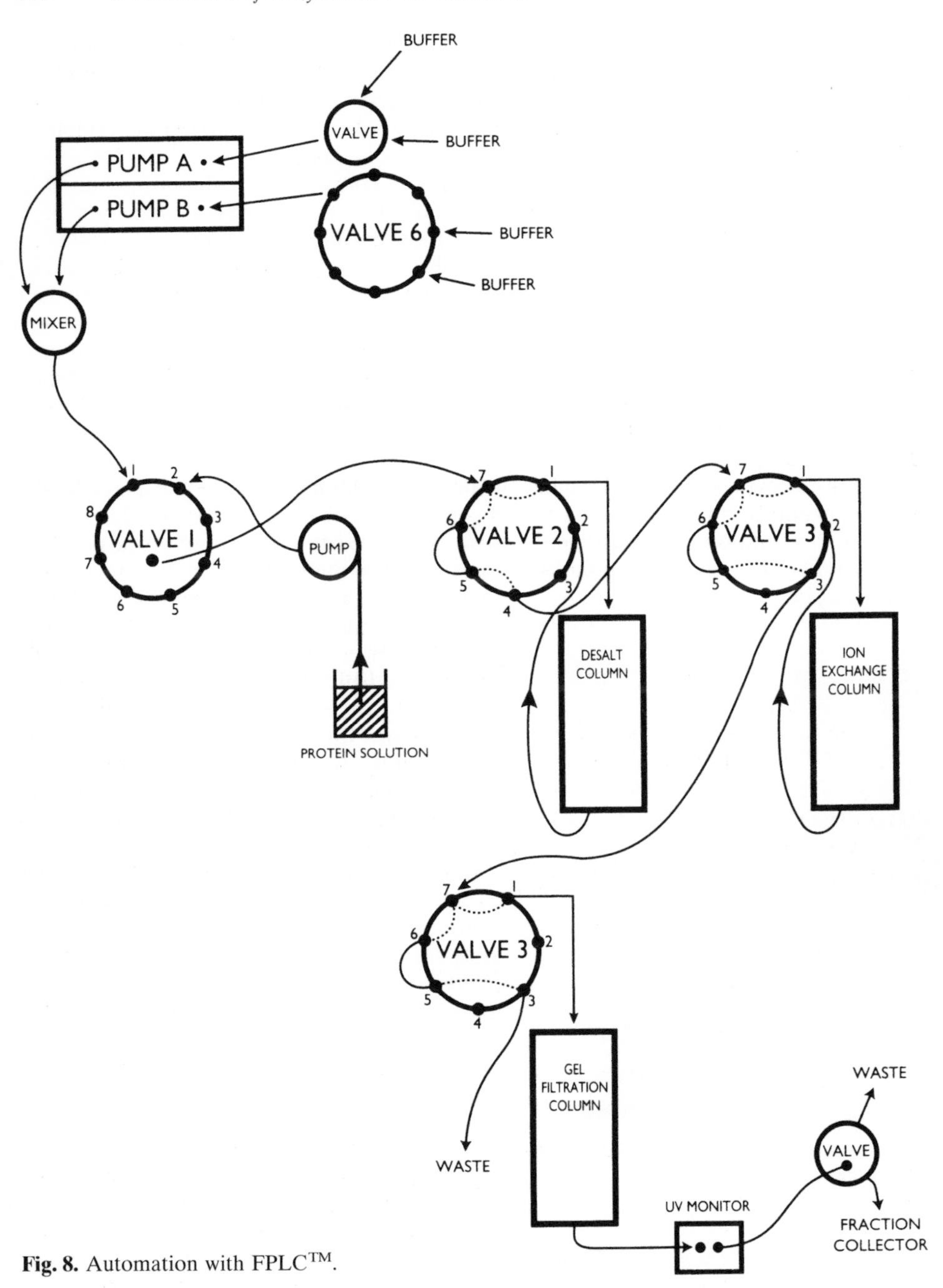

Fig. 8. Automation with FPLC™.

The buffer in which the ion exchange column is run is critical to the performance of the chromatography. The buffer chosen should be used within 1 pH unit of the pK_a. An anionic buffer should be used for anion exchangers and a cationic buffer should be used for a cation exchanger. A cationic buffer will bind to an anion exchanger, occupying valuable binding capacity of the resin. In addition, if an incorrect buffer were chosen, it would take a large volume of the buffer to completely equilibrate the pH of the column as the resin will bind the buffer ions.

Proteins are eluted from ion exchange columns with an increase of ionic strength. Resolution is optimized by applying a gradient of increasing ionic strength. The optimum gradient for a column starts with the highest ionic strength that allows for binding and ends with the lowest ionic strength that causes elution. When designing the parameters at which a column should be run, it is best to first run a gradient from 0–1 M elution salt. When the molarity of the point of elution is determined, continue to modify the gradient to the point where the protein is loaded onto the column at the highest salt concentration that it will bind. Apply a gradient to a final salt concentration at which the protein is completely eluted. The resolution during chromatography is inversely proportional to the slope of the salt gradient applied to the column. The risk of this design, however, is that the volume required to elute the protein will increase with decreasing gradient slope. The ultimate salt molarity at which to load the protein onto the column is one where either the pertinent protein binds and the contaminant does not, or *vice versa*. Generally, the former method is more useful since it allows a greater degree of fractionation and concentration of the target protein.

In some instances where a persistent contaminant cannot be purified from the target protein, the application of a pH gradient to the ion exchange column has led to the required separation. We currently use this mode of separation in our laboratories when we are purifying target proteins from persistant contaminants. In such cases the functional contaminant is a DNA-binding protein and applying a pH gradient, at a low pH, to an anion exchanger has been successful. This alternate method should be used only for difficult purifications as the gradients are difficult to reproduce. A linear pH gradient cannot be prepared by simply mixing buffers of different pH conditions in linear volume ratios, since the buffering capacities of the system produced are pH-dependent.

Ion exchangers that are cross-linked resins, along with Mono Q and Mono S, can be used as an efficient concentration step. Because of the large binding capacity of these columns, a large volume of protein solution can be quickly loaded onto these columns and eluted simply by applying 1–2 volumes of a buffer containing 1 M NaCl or KCl.

2.4.2 Hydrophobic Interaction Chromatography

Hydrophobic interaction chromatography (HIC) takes advantage of the surface hydrophobicity of proteins. This technique is based on a separation principle different from most other modes of chromatographic purification; the use of this technique in combination with ion exchange and gel filtration chromatography will offer a high degree of purification. Generally, this mode of chromatography is useful only when purifying membrane-associated proteins, due to their hydrophobic characteristics. However, this mode may be of particular utility in other instances where ion exchange or gel filtration are ineffective, when the desired protein already exists in a high salt environment or when there is a high susceptibility to denaturation. The protein must be loaded onto a HIC column in a high concentration of salt, such as ammonium sulfate (0.75–2 M) or NaCl (1–4 M), which lowers the solvation of proteins and exposes nonpolar amino acid side chains on the protein. The exposed side chains interact with hydrophobic groups on the resin (polyethylene glycol, octyl- or phenyl-, for example). Pharmacia developed some of the first HIC resins with Octyl- and Phenyl-Sepharose. The bound proteins are eluted in order of increasing hydrophobicity with a gradient of decreasing salt concentration; adding ethylene glycol, glycerol or isopropanol to the eluent, or the addition of detergents. Salt is the most frequent mode of elution; the other methods can be tried if the protein remains bound to the resin after low salt concentration is achieved. The advantage of this step is, when used immediately after an ion exchange column or immediately after an ammonium sulfate precipitation, desalting before the subsequent step may be deemed unnecessary.

Generally, the affinity of proteins for the separation medium surface increases as the chain length of the ligand increases. For proteins with poor solubility in buffers of high salt concentration HIC absorbents with longer ligands are recommended.

The high capacity of commercially available HIC resins makes them suitable for use early in the purification process. In addition, because of the stabilizing effect of the salt, recovery of the target protein from this column may be higher than that from other columns. The disadvantage is that the protein may precipitate at the salt concentration necessary for protein binding. HIC has been particularly useful for the purification of antibodies.

Temperature plays a role in the performance of a HIC column; the hydrophobicity of a protein will increase with temperature. No general trend in the interaction strength and protein p*I* values has been observed (ERIKSSON, 1989).

Reverse phase chromatography (RPC) is a variation of HIC. While HIC depends on the surface hydrophobicity of a protein, reverse phase chromatography depends on the hydrophobicity of the entire protein, as it is usually run under denaturing conditions. These silica resins contain polar side groups (hexyl-, phenyl- and octyl-, for example) and, as does HIC chromatography, separate proteins based on their differences in hydrophobicity. Proteins are applied to the column in a solution containing an organic solvent. Because of the solvent, the method is not amenable for the purification of proteins that are required to maintain biological activity. The use of solvents may make this a useful mode of separating inclusion bodies, which typically must be denatured before restoring biological activity (SOFER and NYSTROM, 1989).

Both HIC and RPC are chromatography steps providing high speed and capacity, while producing high resolution. A thorough account of hydrophobic interaction chromatography has been written by ERIKSSON (1989).

2.4.3 Size Exclusion Chromatography

Size exclusion chromatography (SEC), or gel filtration, separates proteins based on molecular size. This method is also often referred to as gel permeation chromatography. Gel filtration matrices are composed of porous beads that are typically cross-linked dextran (Sephadex, Pharmacia) or cross-linked acrylamide (BioGel, BioRad).

Chromatography suppliers, such as Pharmacia, supply a wide range of gel filtration matrices. The major difference between these matrices is the pore size of the bead. The porosity of the product chosen should be made to maximize the effect of the separation.

The size of the pores in each resin bead is influenced by the degree of cross-linking. A range of matrices are available that can separate compounds with molecular weights within the nominal range of 700–800000 daltons. When these beads are packed into a column, a volume external to the beads and a volume internal comprising the internal volume of the beads is created. Molecules larger than the pore size of the beads will be excluded from the internal volume, while proteins that will fit into the pores will travel through both volumes. When flow is applied to the column, the distance through the column is longer, then, for molecules able to fit into the pores than it is for molecules excluded from the pores. Therefore, larger molecules will elute from the column first. Proteins will not elute solely based on the molecular weight, but also based on the molecular size. An asymmetrical protein will elute before a globular protein of the same molecular weight.

Elution is performed isocratically since there is no binding to the column. A measurement of elution is made in volume, referred to as elution volume, and not in concentration of eluent. The selectivity of the resin is not adjustable by changing the composition of the mobile phase and is only dependent on the dimensions of the resin itself. During gel filtration, the protein is not bound to the resin, and, therefore, labile proteins usually are not inactivated. However, the resolution provided by gel filtration is limited. Typically, fewer than 10 proteins can be resolved from one elution. This technology is most useful when purifying a large protein from a small one, and may work very well for aggregate or degradation product removal. HAGEL (1989) has written a thorough review of gel filtration chromatography.

The size of the column load should be no more than 5–10% of the total volume of the column. A concentration step prior to gel filtration is usually necessary and can be done by ultrafiltration, ammonium sulfate precipitation, or ion exchange chromatography. Another disadvantage of size exclusion chromatography is that the allowable flow rates are very low. These two parameters of SEC make this method rather inefficient. No method is so inefficient, however, if it provides the required purification, especially if other methods have failed. The success of SEC is also influenced by the viscosity of the protein preparation; the sample should not have a protein concentration much greater than 50 mg/mL, and should be free of particulates. Resolution generated by SEC is low but it may be improved by lowering the flow rate, selecting a resin with a smaller bead size, or by selecting a resin with a smaller fractionating range. Too low a flow rate, however, can result in peak diffusion. When there is a choice of bead sizes for a given porosity, the smallest bead size should be used, as this will improve the resolution. Reversing the flow of a column will aid resolution by decreasing diffusion due to gravity. Gel filtration chromatography is best left for later stages of purification because of volume constraints.

The resolution increases and the separation time increases when a resin of very small bead size is chosen. However, the column back pressure will increase with a decrease in particle size. When very small beads are used, shear forces or frictional heat may result in protein denaturation. Such resins may best be used for analytical purposes rather than preparative work.

For best results, the resin should be poured into a long, narrow column. A column can be built too high, creating pressure on the beads within the column. This pressure can crush the beads, impeding flow and changing the pore size of the beads. If a large column is necessary, consider plumbing multiple columns in a series to accommodate your scale needs. Stacked columns will also allow a faster flow rate.

Because proteins in the solution can be loosely associated with each other, increasing the ionic strength of the protein solution and of the column running buffer may improve purification. This higher ionic strength will also minimize binding of the proteins to the gel by electrostatic or van der Waals interactions. A molarity of 200–300 mM works well. Purification, however, may be enhanced by taking advantage of these interactions with the gel. In such instances, the ionic strength should be no higher than 150 mM (HAGEL, 1989).

Performing PAGE prior to optimizing the matrix selection will provide information regarding the protein mixture that will prove useful in designing optimum gel filtration parameters. While SDS-PAGE are generally easier to perform than native PAGE, only a native-PAGE will provide a true representation of the size difference between the proteins in the solution to be purified. Pharmacia's PhastSystem can generate a completed gel in less than 2 hours.

A gel filtration matrix like Pharmacia's Sephadex G-25 provides an excellent means of desalting a protein preparation. Other methods of desalting or buffer exchange include dialysis and ultrafiltration. Unless the protein solution must also be concentrated (in these cases ultrafiltration is recommended) Sephadex G-25 group separation is the method of choice. When choosing gel filtration as a desalting step, the pore size of the resin should completely exclude the protein of interest to maximize the speed of the process and recovery of the protein. When performing such group separation, the loading capacity can be increased to 25% of the total column volume, and the column dimensions can be shorter and squatter than those columns used for protein fractionation. The large bead size allows for flow rates that generate a separation of protein from salt in as little as 20 minutes. This mode of desalting eliminates the need for time-consuming dialysis or dilution, which can be detrimental to a protein and will take a long time to load onto the next column.

Pharmacia manufactures a series of small disposable Sephadex G-25 columns: NAP-5, NAP-10 and NAP-25, which serve as an excellent means of desalting 0.5–1.5 mL of protein solution in minutes. Care must be taken when desalting proteins of low molecular weight in this fashion. In our lab we used Sephadex G-25 to desalt a protein solution containing our target protein of 12000 daltons. This protein did not elute with the bulk of the protein, but in the volume overlapping both the protein eluent and the salt eluent. Recovery and desalting were, therefore, suboptimal.

When performing a desalting step, a protein precipitate often forms. If this precipitate is the target protein, precipitation upon desalting can be minimized by performing the desalting step gradually. Dialysis or dilution may work best in these cases. Often the precipitate can be removed by centrifugation, or just resuspension in buffer, and loading onto a column, if the column will handle the back pressure created by the particulates. The increasing gradient will often redissolve the protein, and the protein will bind to the column. Never filter particulates from the protein solution. The filtration process will introduce foaming and shear force on the protein. These precipitates are best removed by a brief centrifugation.

2.4.4 Affinity Chromatography

Affinity chromatography is considered by many chromatographers as the simplest, yet most powerful, chromatographic method. Affinity chromatography is simply a method that takes advantage of a protein's natural interaction with other biomolecules. The molecule to which the protein binds is referred to as the ligand. A good ligand choice will be one that is involved with the target protein in either an enzyme–substrate interaction, an enzyme–cofactor interaction or an enzyme–inhibitor complex. Affinity chromatography is a high-capacity, high-resolution purification step that is fast. Affinity purifications generally are very fast to perform because no gradient may be necessary. All that is needed is adsorption, washing and desorption. Affinity resins are typically expensive and may best be used late in the purification stage if multiple purification steps are necessary. It may be disadvantageous to use affinity columns as the last step of purification, as the ligand typically leaches from the column. CARLSSON et al. (1989) have written a complete discussion of affinity chromatography.

The blocking (or deactivating) of remaining reactive groups of the affinity resin with a low-molecular-weight substance (Tris, ethanolamine, glycine) after coupling of the ligand should be as complete as possible. This will minimize nonspecific binding of contaminants to the resin and the binding of the target protein to the resin. If nonspecific proteins bind to the resin, they may leach from the resin during elution, contaminating the preparation. The interaction between the target protein and the resin may be stronger than the interaction between the ligand and the target protein. When this happens, the yield will be reduced. Optimal protein–ligand binding constants need to exceed 10^5 M. Interactions with binding constants of 10^{10} or greater make it almost impossible to elute the protein without affecting the biological activity of the protein.

Affinity purification is particularly powerful in instances when the target protein is a small percentage of the total cellular protein. This method is useful for concentrating target proteins in dilute solutions. These biological interactions often rely on the biological activity of the target protein. Therefore, affinity chromatography often serves as a means to separate functionally active from inactive molecules. Proteins are stabilized when they are absorbed to the ligands (JONES, 1990).

One difficulty with affinity chromatography is the purification or synthesis of the ligand, if one is not already commercially available. Another difficulty is selecting an elution condition that will not also elute the ligand from the gel matrix. Many chromatography suppliers provide activated resins that make the coupling of the ligand to the resin a simple process. The cyanogen bromide technique is the most used technique for the preparations of affinity absorbents and works well for protein ligands. The preparation of CNBr-activated resins is a dangerous, toxic process;

the use of a commercially prepared CNBr-activated support is recommended. The matrix providing the best support for affinity chromatography is a cross-linked agarose. Good choices are CNBr Activated Sepharose 4B (Pharmacia) and Affi-Gel (Bio-Rad).

Ligands can be either mono-specific or group-specific. Mono-specific ligands are ligands that bind to a single or a very small number of proteins in a particular extract. An example is an antibody–antigen complex. Mono-specific ligands tend to bind the target protein tighter, creating the need for harsher elution conditions. An example of a group-specific ligand is DNA.

Due to the power of an affinity purification, this technique may be the only one necessary to purify the target protein. In some instances, however, it is recommended to precede the affinity column with a precipitation or ion exchange step to remove particulates, lipids, major contaminants and to reduce the volume of the material to be processed. Affinity matrices are typically expensive, so it is advantageous to keep the amount of resin used to a minimum. Therefore, particulates should be removed to prevent fouling of a column that could otherwise be used repeatedly.

The affinity resin should be prepared to contain 1–10 μg ligand/mL of resin with a low-molecular-weight ligand or 1–10 mg protein ligand/mL of resin. Because the dimensions of the column are not critical to the performance of the matrix, the column can be packed short and squat to allow rapid flow rates. One author (STELLWAGEN, 1990) recommends a column height/diameter ratio of 2 to 5. Lower flow rates will maximize resolution when a gradient is being used to separate several proteins that have been absorbed. The size of the column is dependent on the capacity of the resin. If the target protein is only loosely associated by the ligand, a longer column will improve fractionation. Flow rates can be reduced or stopped or batch adsorption can be used to maximize binding efficiency. Proteins typically elute from affinity columns in broad peaks. This effect can be reduced by reversing the flow of the column.

As with any chromatographic step, it is important to wash the column after the protein load, at least until the A280 absorbance has established a minimum. This typically requires approximately 3 column volumes of wash. In some instances, especially with affinity chromatography, exhaustive washing may increase the purity of the target protein upon elution. We have experience with one protein purified from mammalian tissue that is purified to near homogeneity by a single chromatographic affinity step. The key to the success of the purification is the washing of the column after the load and the use of a preelution wash. While the absorbance at 280 nm reached a minimum in approximately 3 column volumes, a troublesome contaminant eluted with the target protein unless the column was washed with approximately 40 column volumes.

2.4.4.1 Antibody Affinity Methods

Probably no other chromatographic step can be as selective as antibody affinity chromatography (also referred to as immunoaffinity chromatography). It can be difficult, however, to prepare poly- or monoclonal antibodies to the target protein. As a result, this mode of purification is very expensive. However, the resulting savings in labor and additional equipment needs may compensate for the expense of this step. This one purification step can increase the specific activity of a protein preparation several hundred-fold to two thousand-fold. The major limitation of this technique is the strength with which antibodies bind their target proteins. Most methods of elution are very harsh, and proteins isolated in this manner are generally inactive. When using immunoaffinity, antibodies that require harsh elution conditions should be avoided. Typical eluents include high or low pH buffers, chaotropic agents, high salt concentrations, SDS, urea and guanidine-HCl.

2.4.4.2 Dye Ligand Chromatography

Dye molecules often mimic biological compounds, such as coenzymes, nucleotides and polynucleotides, to which the target protein may naturally bind. These reactive dyes can bind proteins by either specific interactions at the protein's active site or by a range of nonspecific interactions (BOYER and HSU, 1993). Dye resins have been prepared to take advantage of this phenomenon. Dye chromatography is well suited for proteins that have an affinity for aromatic compounds, such as those proteins that interact with nucleotides and cofactors. Detailed information regarding the target protein, such as p*I* or molecular weight, is not needed before the separation can be attempted. Therefore, this mode of affinity chromatography is useful for uncharacterized proteins, as well as those that have been characterized. These dye ligand resins are more likely to withstand wear than are the affinity resins prepared with antibodies or enzymes. Three examples of such dye–ligand affinity resins are Matrex Red (Amicon), Blue Sepharose CL-6B (Pharmacia) and Affi-Gel Blue (BioRad). Blue dye, or Cibacron Blue F3G-A, is an analog of adenylyl-containing cofactors. The red dye is Procion Red HE-3B. Lextin Scientific International and Amicon supply a kit of several reusable leak-proof synthetic dye–ligand affinity matrices for analysis. A useful review of performing optimization experiments with such kits has been written by STELLWAGEN (1990).

Dyes typically used for this method of affinity purification are often of crude purity; therefore, care should be taken in choosing the source of dye, as dyes will typically leach from the resin. The use of this method to purify a protein that must meet FDA guidelines should be questioned. A method to remove the dye, and to validate that it has been removed, is needed in these cases. Refer to SANTAMBINE (1992) for an account of detecting the leaching of some dyes used in this type of chromatography.

The binding capacity of these resins ranges from 1 to 15 mg protein/mL. An immobilized dye can simultaneously bind 5–60% of the total protein in a crude cell extract (SCOPES, 1986). As with ion exchanges resins, the bed should be prepared to accommodate 10-fold more protein than is present in the solution to be purified. Proteins can be eluted from a dye ligand column with high salt or with a competing ligand, such as ATP; a gradient is recommended for fractionation. The binding of some proteins to these affinity dyes can be enhanced by the presence of a low concentration of a metallic cation such as Zn^{2+}, Co^{2+}, Mn^{2+}, Ni^{2+}, Cu^{2+}, or Al^{3+} in the buffer. Elution can be achieved by the addition of a chelating agent to the eluting buffer. The affinity of most proteins for immobilized dyes is substantially weakened with an ionic strength of 0.2 M. Many proteins have been purified to homogeneity in a single step using immobilized dye chromatography (STELLWAGEN, 1990).

2.4.4.3 Heparin Sepharose

Heparin, a sulfonated polysaccharide, is frequently used for the purification of plasma proteins, DNA binding proteins, a variety of enzymes, steroid receptors and virus surface antigens (CARLSSON et al., 1989). The biological role of heparin is not understood; it may play a role in several biological interactions. Therefore, several types of proteins have an affinity for the molecule. Heparin is commercially available (Heparin Sepharose, Pharmacia), bound to cross-linked agarose, and is as convenient to use as other ion exchange media. The binding capacity of the resins is very large, and columns can be prepared short and squat to allow for rapid flow properties. Elution can be performed by using a salt gradient. Heparin coupled to cross-linked resins can be used as an efficient concentration step. Because of the large binding capacity of heparin columns, a large volume of protein solution can be quickly loaded onto a heparin column and eluted off by simply applying 1–2 volumes of a Tris-based buffer containing 1 M NaCl or KCl. One author (MURPHY, 1990) states that Heparin Sepharose can have lot to lot variability.

As with affinity chromatography, the heparin may leach from the column during elution. If this is a concern for the purification process, the heparin column should be followed with either ion exchange or gel filtration.

2.4.5 Hydroxyapatite

Hydroxyapatite chromatography is an example of adsorption chromatography, but the mechanism of adsorption is not completely understood. This mode of separation does not fractionate proteins based on differences in isoelectric point, molecular weight or elution molarity. Some chromatographers suspect a hydrophobic interaction between the matrix and the protein. Some suspect that proteins interact with hydroxyapatite via carboxyl groups (GORBUNOFF, 1990). HTP, a form of calcium phosphate, has two absorbing sites: calcium sites and phosphate sites. The calcium will bind acidic groups, carboxyls and phosphates. Elution of these types of interactions should be performed with anions, usually in the form of a phosphate gradient. The phosphate sites will bind basic proteins. Elution of these types of interactions should be performed with cations, usually in the form of sodium ions.

HTP matrices are generally supplied in a crystalline form that must first be defined. This defining procedure can take up to an hour to perform. Since the crystals are not spherical and are prone to breakage, the matrix must be treated with care. The resin packed in a column will demonstrate backpressure problems if the slurry has not been properly defined. Even when defined, flow cannot be as rapid as that obtained with cross-linked beads. Aside from this disadvantage, HTP has a high binding capacity and provides highly resolved fractionation. Sepracor supplies a modified HTP matrix, HA-Ultrogel™, which contains no crystal fines. The matrix is a cross-linked agarose with trapped homogeneous microcrystals of HTP.

As with ion exchangers and heparin columns, HTP columns can be used as a concentration step. HTP can be used to purify proteins from nucleic acids, as nucleic acids will bind to HTP. HTP also works well to remove low-molecular-weight components, as these molecules have a low affinity for the matrix. An informative chapter regarding hydroxyapatite chromatography has been written by GORBUNOFF (1990).

HTP columns are usually equilibrated in a low concentration of phosphate buffer at pH 6.8. In only a few situations does the presence of NaCl or ammonium sulfate play a role in the binding of the protein. Therefore, a protein can often be loaded directly onto HTP from an ion exchange step without first desalting (KARLSSON et al., 1989).

2.4.6 Desalting/Buffer Exchange

Desalting or buffer exchange is generally required after an ammonium sulfate precipitation or between chromatographic steps. Two useful methods for performing these operations have already been discussed: gel filtration chromatography and tangential flow filtration. In many situations, simple dialysis with cellulose dialysis tubing is the most convenient means of performing this task. When preparing dialysis tubing, invest in high quality material (Sepracor), and be sure to boil the tubing in 1 mM EDTA for 10 min to chelate any metals often present in the tubing, as these metals can be detrimental to the biological activity of the target protein. During dialysis the salt concentration of the protein solution reaches equilibrium with the dialysis buffer. Therefore, at least once during the dialysis the tubing must be moved into fresh buffer. Dialyzing against two changes of buffer, each equal to ten-fold the retentate volume, for 8–16 hours works well. The dialysis will occur faster with more frequent buffer changes. Since buffer exchanges can be detrimental to a protein, it is important to organize chromatographic steps to minimize required buffer changes.

2.4.7 Perfusion Chromatography

Perfusion chromatography is one of the newest developments in protein purification by liquid chromatography. Perfusion matrices are beads containing channels, or pores. The design allows the material to flow through the pores and interact with the active sites that line the pores. Therefore, much more of the surface is readily available to the protein sample. Because there is more contact with active sites, higher flow rates can be used without sacrificing resolution and binding efficiency.

The manufacturers of these columns state that separations of acceptable resolution can be performed in minutes using these columns; separations can be performed 100 times faster than with conventional low pressure chromatography. These speeds can allow for the quick optimization of pH, gradient profile, sample chemistry, and column chemistry. The short residence time on the column increases recovery by minimizing proteolysis. These columns are currently manufactured by Pharmacia, Sepracor and PerSeptive Biosystems and are available as cation and anion exchange, hydrophobic interaction, reverse phase and affinity resins.

2.4.8 Membrane-Based Chromatography

Manufacturers such as Millipore, BioRad, and Sepracor have prepared membranes to which have been coupled typical liquid chromatography matrix functional groups. These membranes contain the chromatographic beads permanently enmeshed in the membrane. The objective is to provide fractionation as quickly as it is to pass a liquid through a membrane. At this point the membranes do not provide the degree of separation available from a conventional liquid chromatography column and are most useful for crude, or group, separations. They may serve well as a means to concentrate a dilute protein preparation.

3 Enzyme Stabilization

If one is to study an enzyme or use an enzyme as a reagent over a long period of time, it is important to maintain the enzyme for ideally a number of years without significant loss of activity. After investing weeks in purifying an enzyme, thought should be given to the best method for enzyme storage. All too often one is tempted to simply pool final enzyme fractions and dialyze into a "final storage buffer" containing 50% glycerol and store the enzyme at −20 °C without much thought about the optimal storage buffer or the advantages of other storage options. This shortcut is often followed by significant loss of enzyme activity in a few weeks or months.

3.1 Mechanisms of Enzyme Inactivation

Before reviewing the various options for enzyme storage and stabilization, it is important to have an understanding of how enzymes maintain their native conformation and what forces disrupt in an irreversible way the native structure of enzymes, leading to loss of activity. Protein structure is determined by the amino acid sequence of the protein. In particular, the order of hydrophilic and hydrophobic moieties dictates how the protein is folded. It has been established that hydrophobic amino acids tend to be folded away from the protein surface buried within the globular protein structure, where they can be removed from water and can associate with each other, while hydrophilic amino acids tend to be located on the surface of the protein and often interact by hydrogen bonds or salt bridges to give structure to the globular protein molecule.

It is therefore the unique folding of each protein sequence that not only gives each protein its specific biological activity, but also dictates the stability of the protein. It is now clear that proteins are only just as stable as they need be for biological activity. The free-energy difference between the native state and the unfolded random state for many proteins is only about 12 kcal/mol (50 kJ) (FRANKS, 1993). This is equivalent to 2–3 hydrogen bonds. The first step in protein inactivation is the substantial unfolding of the native protein conformation. This unfolding is spontaneous and reversible (PRIVOLOV, 1979). Once a protein is in the unfolded (denatured) state, additional processes can occur (e.g., hydrolysis, oxidation, deamination, S-S interchange, aggregation and precipitation) which prevent the denatured protein from folding back to the native state, resulting in an irreversible loss of activity.

In order to stabilize an enzyme the process of inactivation must be prevented or at least slowed down. This can be done by either (1) stabilizing the native protein relative to the unfolded (denatured) protein or (2) once a protein is unfolded, prevent irreversible inactivation from occurring. The following methods of enzyme stabilization can be understood in these terms.

3.2 Addition of Substrate

It is well known that the addition of a substrate can stabilize an enzyme. For example, lactate dehydrogenase and malate dehydrogenase are stabilized by the presence of NAD and NADH (CITRI, 1973). Nucleoside phosphates, substrates or effectors for many enzymes, have been shown to stabilize enzymes when present in storage buffer. An example is the stabilization of deoxycytidine kinase by ATP and TTP (KIERDASZUK and ERIKSSON, 1990). Sugars have been found to stabilize enzymes whose substrates are carbohydrates. Examples are the stabilization of glucoamylase with glucose or lactose and the stabilization of β-glucosidase with glucose and fructose (DEAN and ROGERS, 1969). Other examples are the stabilization of ribonuclease with phosphate (PACE, 1990) and the stabilization of asparaginase with aspartate or asparagine (CITRI and ZYK, 1972). This stabilization can be explained by the fact that when an enzyme binds its substrate, a conformational change occurs as a result of interaction between groups not normally in a position to interact in the native form of the enzyme. Recently, the change in conformation of an enzyme when it binds its substrate was observed directly by use of the atomic force microscope (AFM). Individual molecules of lysozyme were seen to change shape over a period of 50 milliseconds when they bound the substrate oligoglycoside. In the presence of the inhibitor chitobiose, these size fluctuations were not observed (RADMACHER et al., 1994). These new interactions result in a stabilization of the native enzyme–substrate complex relative to the native conformation of the enzyme resulting in a shift of the equilibrium from the unfolded conformation to the native conformation.

3.3 Addition of Polyhydroxy Compounds

The addition of hydroxy compounds to enzymes in solution has a large effect on stability. The addition of monohydric alcohols to ribonuclease or lysozyme lowers the transition temperature (T_m) for the reversible denaturation of these proteins (GERLSMA and STRUIR, 1974). The effect is increased by increasing the concentration of the alcohol or by increasing the hydrophobic character of the alcohol. These denaturing properties can be explained by the fact that alcohols stabilize the denatured conformation of proteins by interacting with the exposed hydrophobic groups. In contrast, polyhydroxy compounds (e.g., glycerol, sorbitol, mannitol, inositol, sucrose, lactose and glucose) increase the T_m for denaturation for a number of proteins (ARAKAWA and TIMASHEFF, 1982), thereby stabilizing the native conformation of the proteins. This is observed, for example, by a stabilizing effect which the addition of glycerol has on invertase at 60°C (COMBES et al., 1987). Polyhydroxy compounds are thought to stabilize proteins by reducing the rate of inactivation once a protein is denatured or unwound. It is thought that polyhydric alcohols interact with water through hydrogen bonding. This interaction results in an increase in the structural organization of water, which limits the unfolding of proteins (COMBES et al., 1987).

3.4 Addition of Salts

Many enzymes require the presence of metal ions for stability. The binding sites for metal ions are the negatively charged carboxylate groups of aspartyl and glutamyl residues on the surface of the protein molecule that are brought close together by the globular structure of the native protein. If the metal ions were not present, the presence of negatively charged groups close to each other would be destabilizing to the structure of the native protein. The stabilization effect can be explained by formation of metal ion bridges or cross-links (usually involving calcium ions) which provides rigidity to the protein structure in much the same way as disulfide bond formation (FONTANA et al., 1976). The process of binding metal ions is also thermodynamically favorable, because the water molecules bound to hydrated metal ions are released when the metal ions bind to the protein (SCHMID, 1979). The binding of calcium ions seems especially important for the stability of thermostable enzymes, which exhibit a correlation between the number of bound calcium ions and thermostability (IMANAKA et al., 1986).

At much higher concentrations, ions can either stabilize or de-stabilize native proteins against reversible thermodenaturation. The stabilizing effect of ions is outlined in the following lyotropic series (von HIPPEL and SCHLEICH, 1969):

Cations: $(CH_3)_4N^+ > NH_4^+ > K^+ > Na^+ > Mg^{2+} > Ca^{2+} > Ba^{2+}$

Anions: $SO_4^{2-} > Cl^- > Br^- > NO_3^- > ClO_4^- > SCN^-$

Thus ammonium sulfate (both cation and anion near top of the list) stabilizes many proteins by a process referred to as "salting out" of hydrophobic groups into the interior of the protein structure, that is by precipitation of proteins from solution. Stabilization and precipitation of proteins are related because, in both cases, hydrophobic groups are removed from the solvent: in the case of protein stabilization, the hydrophobic groups are directed within the globular protein structure; in the case of enzyme precipitation, the hydrophobic groups on the surface of native protein molecules associate or aggregate resulting in precipitation. Those ions which increase protein solubility ("salting in") destabilize protein structure. The "salting in" ions stabilize the exposed groups of the unwound, denatured protein structure leading to greater solubility and unwinding (denaturation) of the native protein conformation.

3.5 Addition of Miscellaneous Stabilizers

The following stabilizers have proven useful as additives to enzyme storage buffers:

1. Anti-oxidants, such as the thiol compound dithiothreitol (DTT), prevent the oxidation of sulfur-containing groups (SCHMID, 1979). Sulfhydryl groups at the enzyme active site are very prone to oxidation. Possible reactions include disulfide bond formation, partial oxidation to a sulfinic acid or irreversible oxidation to a sulfonic acid. To prevent oxidation, the compound DTT is usually present at 1–5 mM concentrations in enzyme storage buffers.

2. Disulfide formation and oxidation is greatly accelerated by the presence of divalent metal ions which activate oxygen molecules and complex with sulfhydryls (O'FAGAIN et al., 1988). To prevent such inactivation, trace metal ions can be removed by a complexing agent such as EDTA (usually present at 0.1–0.2 mM).

3. The addition of ammonium ions may suppress the deamination of asparagine and glutamine residues (O'FAGAIN et al., 1988).

4. Trace amounts of protease can be inhibited by the addition of protease inhibitors, such as phenylmethylsulfonyl fluoride (PMSF) at 0.5 mM, which inhibits serine proteases, some thiol proteases and carboxypeptidases. The reagent hydrolyzes rapidly, but proteases once inhibited are inhibited irreversibly.

5. The addition of exogenous proteins, usually BSA, can inhibit proteolytic activity and loss of protein on surfaces, especially glass, if present in large excess over the protein of interest.

3.6 Stabilization by Storage at Low Temperature

Stabilization of enzymes by chilling at low temperature (but not freezing) occurs because of a significant reduction of chemical deterioration rates and inhibition of microbial growth and attack at low temperatures. This is usually accomplished by storing enzymes in storage buffer plus 50% glycerol at −20 °C. We have found that many common DNA/RNA modifying enzymes and restriction enzymes are stable for several years when stored at −20 °C in the appropriate storage buffer.

Alternatively, enzymes have been stored frozen in small aliquots at usually −70 °C. Freezing (separation of pure water in the form of ice) can be damaging and result in loss of activity unless done properly. All the chemical processes that are harmful as a result of freezing are secondary effects of concentration changes that accompany the removal of water from the solution phase as ice. Under freezing conditions, salt crystal nucleation and growth are very slow compared to ice crystal growth. As a result, the crystallization of salts from a freezing solution is not common for salt concentrations <0.5 M. It is therefore not unusual for freezing solutions to become supersaturated. The combination of freeze-concentration of protein, salts and pH shifts often leads to freeze-denaturation.

3.6.1 Stabilization by Undercooling

The formation of ice depends on the generation of crystal nuclei which are capable of growth into macroscopic crystals. The probability of ice formation can be reduced several orders of magnitude at subzero temperatures by dispersing the aqueous sample as microscopic droplets in an immiscible organic phase (MICHELMORE and FRANKS, 1982). The organic phase used must not catalyze ice nucleation nor permit the aqueous droplets to coalesce. The aqueous phase should be readily recoverable after the emulsion is returned to ambient temperature. Suitable dispersion fluids (U-COOL™) are commercially available (HATLEY et al., 1987). Undercooling is especially useful as a method of stabilizing intermediate fractions during a purification procedure. Complete retention of enzyme activity can be achieved with storage at −20 °C for several years. Since the ratio of organic: aqueous phase is optimal at 4:1, the technique is not useful for large volumes of protein. Some advantages listed by FRANKS (1993) include:

- The process is simple and cost-effective, especially when compared to freeze-drying.
- Protecting additives are not usually required.
- Proteins can be processed and supplied over a very wide range of concentrations.
- Aliquots can be taken from a sample without affecting the activity of the remainder.
- Products are stable and not sensitive to medium-term temperature fluctuations, provided the temperature never exceeds about 4 °C.
- The storage temperature is not critical, provided it does not approach the ice nucleation temperature (in the region of −40 °C).
- Products are not susceptible to moisture pick-up.
- The process is essentially the same for all proteins.

3.6.2 Stabilization by Freeze-Drying

Freeze-drying is occasionally attempted on a small scale as a method of stabilizing enzymes. Unfortunately, freeze-drying protocols are often developed haphazardly with little thought toward optimization. The ideal situation is that a product when frozen will be a mixture of completely crystallized non-aqueous material and ice. The ice is then removed by sublimation leaving the pure crystal product. This ideal situation, however, does not occur with protein solutions. Instead, the frozen material is a mixture of freeze-concentrated amorphous phase embedded in ice. At the subzero temperatures of freeze-drying, the vapor pressure of ice is lower than that of the unfrozen water held in the solute matrix. The water in the solute matrix is not frozen because of the high viscosity of the amorphous solid. The ice sublimation rate is lowered by the matrix which imposes resistance to the migration of water vapor from ice crystals. However, it is still possible to calculate the sublimation rate from the fill volume, vial dimensions, total solid content and pressure and to develop a formulation that will provide a product with acceptable shelf life (FRANKS, 1993).

A secondary drying cycle is necessary to remove the water from the matrix, which should be porous after removal of ice crystals during primary drying. The porosity of the matrix is responsible for the ease of rehydration of final product. The final product will be in the form of a highly viscous glassy solid characterized by a glass transition temperature (T_g) below which the product is a glass and above which the product becomes a less viscous "rubber". If the sample is kept below its T_g, the very high viscosity of the glass will prevent enzyme denaturation. However, if the product is stored a few degrees above the T_g, then the viscosity will rapidly drop, the product will collapse and the solid will be much less readily rehydratable. More importantly, the rates of reactions damaging to the protein will be high enough to effect the stability of the enzyme.

The ideal situation is to develop a formulation and freeze-drying protocol that will produce a freeze-dried product with a T_g greater than room temperature. In practice this is rarely obtained and most freeze-dried products are stored at −20 °C. There is always a trade-off between the time and expense of reaching a product with a high T_g and the convenience of room temperature stability. In general, the formulation of product for freeze-drying should contain excipients with high T_g values (e.g., sucrose, T_g of 65 °C), salt content should be minimized and glycerol (T_g of −78 °C) should be avoided or eliminated. If one follows straight forward rules for formulation development and freeze-drying protocols the guess work can be taken out of freeze-drying (FRANKS, 1993).

3.6.3 Stabilization by Permazyme Technology

When excipients that exhibit high T_g values as dry powders (e.g., carbohydrates, carbohydrate polymers) are added to protein solutions in increasing amounts, viscosity increases and crystallization of salts is inhibited. Depending on the excipient concentration the mixture can undergo rapid changes in viscosity over a narrow temperature change. The physical state can change from a fluid, through a syrup, viscoelastic "rubber" to a brittle, solid glass. The change to a glass occurs when the solute concentration reaches about 80% by weight (FRANKS, 1993). The glass formed can be viewed as an undercooled liquid with a very high viscosity. Because of the high viscosity, the rates of reactions responsible for enzyme denaturation and loss of activity are very slow.

Unlike freeze-drying, which removes residual water by freezing and sublimation, the permazyme technology removes residual water by evaporation. The advantage is that water can be removed much faster by evaporation (5 mL of solution can be dried from 5% to 95% total solids in about 1 hour), the process is more easily controlled and the capital costs are minimal. Since the glass state is formed very rapidly, there is little loss of activity even at room temperature for many enzymes. The process can be optimized with respect to pressure, temperature and formulation to produce room temperature stable glass (T_g > 30 °C) as quickly as possible to prevent loss of activity while drying enzyme solutions at temperatures above freezing. We have found at Pharmacia that enzymes and also complex reaction systems containing enzymes, nucleotides, primers and reaction buffer can be stabilized at room temperature for over one year by using the permazyme technology.

4 References

AMMAN, E., BROSIUS, J., PTASHNE, M. (1983), Vectors bearing a hybrid *trp-lac* promoter useful for regulated expression of cloned genes in *Escherichia coli*, *Gene* **25,** 167–178.

AONO, R. (1989), Subcellular distribution of alkalophilic *Bacillus* penicillinase produced by *Escherichia coli* HB101 carrying pEAP31, *Appl. Microbiol. Biotechnol.* **31,** 397–400.

ARAKAWA, T., TIMASHEFF, S. N. (1982), Stabilization of protein structure by sugars, *Biochemistry* **21,** 6536–6544.

ARMSTRONG, K. A., et al. (1989), Propagation and expression of genes in yeast using 2-micron circle vectors, in: *Yeast Genetic Engineering* (BARR, P. J., BRAKE, A. J., VALENZUELA, P., Eds.), pp. 83–108. Boston: Butterworth.

ASENJO, J. A., PATRICK, I. (1990), Large-scale protein purification, in: *Protein Purification Applications. A Practical Approach* (JARRIS, E. L. V., ANGAL, S., Eds.), pp. 1–28, New York: IRL Press at Oxford University Press.

ATKINSON, A. E., et al. (1990), Baculoviruses as vectors for foreign gene expression in insect cells, *Pestic. Sci.* **28,** 215–224.

BAKER, S. M., et al. (1987), Transcription of multiple copies of the yeast *GAL7* gene is limited by specific factors in addition to *GAL4*, *Mol. Gen. Genet.* **208,** 127–134.

BITTER, G. A., EGAN, K. M. (1988), Expression of interferon-gamma from hybrid yeast *GPD* promoters containing upstream regulatory sequences from *GAL7–GAL10* intergenic region, *Gene* **69,** 193–207.

BITTNER, M., VAPNEK, D. (1981), Versatile cloning vectors derived from the run-away-replication plasmid pKN402, *Gene* **15,** 319–329.

BLISSARD, G. W., ROHRMANN, G. F. (1990), Baculovirus diversity and molecular biology, *Annu. Rev. Entomol.* **35,** 127–155.

BOEKE, J. D., XU, H., FINK, G. R. (1988), A general method for the chromosomal amplification of genes in yeast, *Science* **239,** 280–282.

BOX, S. J. (1985), Approaches to the isolation of an unidentified microbial product, in: *Discovery and Isolation of Microbial Products* (VERRAL, M. S., Ed.), pp. 32–51, London: Ellis Horwood Ltd.

BOYER, P. M., HSU, J. T. (1993), Protein purification by dye–ligand chromatography, *Adv. Biochem. Eng. Biotechnol.* **49,** 1–44.

BRAKE, A. J., et al. (1984), α-Factor-directed synthesis and secretion of mature foreign proteins in *Saccharomyces cerevisiae*, *Proc. Natl. Acad. Sci. USA* **81,** 4642–4646.

BRAKE, A. J. (1989), Secretion of heterologous proteins directed by the yeast α-factor leader, in: *Yeast Genetic Engineering* (BARR, P. J., BRAKE, A. J., VALENZUELA, P., Eds.), pp. 269–280, Boston: Butterworth.

BRENT, R. (1985), Repression of transcription in yeast, *Cell* **42,** 3–4.

BROSS, P., et al. (1993), Co-overexpression of bacterial GroESL chaperonins partly overcomes non-productive folding and tetramer assembly of *E.-coli*-expressed human medium-chain acyl-CoA dehydrogenase (MCAD) carrying the prevalent disease-causing K304E mutation, *Biochim. Biophys. Acta* **1182,** 264–274.

BUELL, G., et al. (1985), Optimizing the expression in *E. coli* of a synthetic gene encoding somatomedin-C (IGF-I), *Nucleic Acids Res.* **13,** 1923–1938.

BUSSEY, H., MEADEN, P. (1985), Selection and stability of yeast transformant expressing cDNA of M1 killer toxin-immunity gene, *Curr. Genet.* **9,** 285–291.

CARLSSON, J., et al. (1989), Affinity chromatography, in: *Protein Purification Principles: High Resolution Methods and Applications* (JANSON, J. C., RYDEN, L., Eds.), pp. 63–106, Weinheim–New York: VCH.

CAULCOTT, C. A. (1985), Investigation of the instability of plasmids directing the expression of met-prochyosin in *Escherichia coli*, *J. Gen. Microbiol.* **131,** 3355–3365.

CHANG, C. N. (1986), *Saccharomyces cerevisiae* secretes and correctly processes human interferon hybrid proteins containing yeast invertase signal peptides, *Mol. Cell. Biol.* **6,** 1812–1819.

CHANG, Y.-N., TEICHERT, U., SMITH, J. A. (1990), Purification and characterization of a methionine aminopeptidase from *Saccharomyces cerevisiae*, *J. Biol. Chem.* **265,** 19892–19897.

CHEW, L. C., TACON, W. C. (1990), Simultaneous regulation of plasmid replication and heterologous gene expression in *Escherichia coli*, *J. Biotechnol.* **13,** 47–60.

CITRI, N. (1973), Conformational adaptability in enzymes, *Adv. Enzymol.* **37,** 397–648.

CITRI, N., ZYK, N. (1972), Stereospecific features of the conformative response of L-asparaginase, *Biochemistry* **11,** 2110–2116.

CLARE, J. J., et al. (1991a), High-level expression of tetanus toxin fragment C in *Pichia pastoris* strains containing multiple tandem integrations of the gene, *Bio/Technology* **9,** 455–460.

CLARE, J. J., et al. (1991b), Production of mouse epidermal growth factor in yeast: high-level secretion using *Pichia pastoris* strains containing multiple gene copies, *Gene* **105,** 205–212.

COMBES, D., et al. (1987), Mechanism of enzyme stabilization, *Ann. N.Y. Acad. Sci.* **501,** 59–62.

COMPAGNO, C., et al. (1993), Copy number modulation in an autoselection system for stable plasmid maintenance in *Saccharomyces cerevisiae*, *Biotechnol. Prog.* **9,** 594–599.

CORSARO, B. G., FRASER, M. J. (1989), Transfection of Lepidopteran insect cells with baculovirus DNA, *J. Tissue Culture Meth.* **12,** 7–11.

COUDERC, R., BARATTI, J. (1980), Oxidation of methanol by the yeast *Pichia pastoris:* Purification and properties of alcohol oxidase, *Agric. Biol. Chem.* **44,** 2279–2289.

COUSENS, L. S., et al. (1987), High level expression of proinsulin in the yeast *Saccharomyces cerevisiae*, *Gene* **61,** 265–275.

COUSENS, L. S., WILSON, M. J., HINCHCLIFFE, E. (1990), Construction of a regulated *PGK* expression vector, *Nucleic Acids Res.* **18,** 1308.

CREGG, J. M., MADDEN, K. N. (1987), Development of transformation systems and construction of methanol-utilization-defective mutants of *Pichia pastoris* by gene disruption, in: *Biological Research on Industrial Yeasts*, Vol. II (STEWART, G. G., RUSSELL, I., KLEIN, R. D., HIEBSCH, R. R., Eds.), pp. 1–18, Boca Raton: CRC Press.

CREGG, J., et al. (1987), High-level expression of and efficient assembly of hepatitis B surface antigen in methylotrophic yeast *Pichia pastoris*, *Bio/Technology* **5,** 479–485.

DAVIS, A. H. (1994), Current methods for manipulating baculoviruses, *Bio/Technology* **12,** 47–50.

DEAN, A. C. R., ROGERS, P. J. (1969), Action of urea on the activity of dehydrogenases and alpha glucosidase in *Aerobacter aerogenes* grown in continuous culture, *Nature* **221,** 969–971.

DE BAETSELIER, A., et al. (1991), Fermentation of a yeast producing *A. niger* glucose oxidase: scale-up, purification and characterization of the recombinant enzyme, *Bio/Technology* **9,** 559–561.

DEBOER, H. A., COMSTOCK, L. J., VASSER, M. (1983), The *tac* promoter: A functional hybrid derived from *trp* and *lac* promoters, *Proc. Natl. Acad. Sci. USA* **80,** 21–25.

DUBENDORFF, J. W., STUDIER, F. W. (1991), Controlling basal expression in an inducible T7 expression system by blocking the target T7 promoter with *lac* repressor, *J. Mol. Biol.* **219,** 45–59.

EAST, A. K., ERRINGTON J. (1989), A new bacteriophage vector for cloning in *Bacillus subtilis* and the use of phi105 for protein synthesis in maxicells, *Gene* **81,** 35–43.

ENGLARD, S., SEIFTER, S. (1990), Precipitation techniques, *Meth. Enzymol.* **183,** 285–300.

ERHART, E., HOLLENBERG, C. P. (1983), The presence of a defective *LEU2* gene in 2μm DNA recombinant plasmids of *Saccharomyces cerevisiae* is responsible for curing and high copy number, *J. Bacteriol.* **156,** 625–635.

ERIKSSON, K.-O. (1989), Hydrophobic interaction chromatography, in: *Protein Purification: Principles, High Resolution Methods and Applications* (JANSON, J.-C., RYDEN, L., Eds.), pp. 207–226, Weinheim–New York: VCH.

ETCHEVERRY, T. (1990), Induced expression using copper metallothionein promoter, *Meth. Enzymol.* **185,** 319–329.

FONTANA, A., BOCCU, E., VERONESE, F. M. (1976), in: *Enzymes and Proteins from Thermophilic Microorganisms* (ZUBER, H., Ed.), pp. 55–59, Basel: Birkhäuser Verlag.

FRANKS, F. (1993), Storage stability of proteins, in: *Protein Biotechnology* (FRANKS, F., Ed.), pp. 409–530, Clifton, NJ: The Humana Press, Inc.

FRASER, M. J. (1989), Expression of eukaryotic genes in cell cultures, *In Vitro Cell. Dev. Biol.* **25,** 225–235.

FROST, G. M., MOSS, D. A. (1987), Production of enzymes by fermentation, in: *Biotechnology* (REHM, H.-J., REED, G., Eds.), Vol. 7a, pp. 65–212, Weinheim: VCH.

FURMAN, T. C., et al. (1987), Recombinant human insulin-like growth factor II expressed in *Escherichia coli*, *Bio/Technology* **5,** 1047–1051.

FUTCHER, A. B., COX, B. S. (1983), Maintenance of the 2μm circle plasmid in populations of *Saccharomyces cerevisiae*, *J. Bacteriol.* **154,** 283–290.

GERLSMA, S. Y., STRUIR, E. R. (1974), The effects of combining two different alcohols on the heat-induced reversible denaturation of ribonuclease, *Int. J. Peptide Prot. Res.* **6,** 65–74.

GHRAYEB, J., et al. (1984), Secretion cloning vectors in *Escherichia coli*, *EMBO J.* **3,** 2437–2442.

GIBSON, R. M., ERRINGTON, J. (1992), A novel *Bacillus subtilis* expression vector based on bacteriophage phi 105, *Gene* **121,** 137–142.

GIORDANO, T. J., et al. (1989), Regulation of T3 and T7 RNA polymerases by the *lac* repressor-operator system, *Gene* **84,** 209–219.

GLEESON, M. A., SUDBERY, P. E. (1988), The methylotrophic yeasts, *Yeast* **4,** 1–15.

GOLD, L., STORMO, G. D. (1990), High-level translation initiation, *Meth. Enzymol.* **185,** 89–93.

GORBUNOFF, M. J. (1990), Protein chromatography in hydroxyapatite columns, *Meth. Enzymol.* **183,** 329–339.

GRAMAJO, H. C., VIALE, A. M., DE MENDOZA, D. (1988), Expression of cloned genes by *in vivo* insertion of *tac* promoter using a mini-Mu bacteriophage, *Gene* **65,** 305–314.

GRAY, P. P., et al. (1972), *Proc. IV IFS Ferment. Technol. Today*, 347–351.

GRITZ, L., DAVIES, J. (1983), Plasmid encoded hygromycin B resistance: the sequence of hygromycin B phosphotransferase and its expression in *Escherichia coli* and *Saccharomyces cerevisiae*, *Gene* **25,** 179–188.

GROISMAN, E. A., PAAGRATIS, N., CASADABAN, M. J. (1991), Genome mapping and protein coding region identification using bacteriophage Mu, *Gene* **99,** 1–7.

GRONER, A. (1986), Specificity and safety of baculoviruses, in: *The Biology of Baculoviruses*, Vol. I, *Biological Properties and Molecular Biology* (GRANANDOS, R. R., FEDERICI, B. A., Eds.), pp. 177–202, Boca Rotan: CRC Press.

HADFIELD, C., CASHMORE, A. M., MEACOCK, P. A. (1986), An efficient chloramphenicol-resistance marker for *Saccharomyces cerevisiae* and *Escherichia coli*, *Gene* **45,** 149–158.

HAGEL, L. (1989), Gel filtration, in: *Protein Purification: Principles, High Resolution Methods and Applications* (JANSON, J.-C., RYDEN, L., Eds.), pp. 63–106, Weinheim–New York: VCH.

HALLEWELL, R. A., et al. (1987), Amino terminal acetylation of authentic human Cu, Zn superoxide dismutase produced in yeast, *Bio/Technology* **5,** 363–366.

HAMMARBERG, B., et al. (1989), Dual affinity fusion approach and its use to express recombinant human insulin-like growth factor II, *Proc. Natl. Acad. Sci USA* **86,** 4367–4371.

HARRIS, E. L. V., ANGAL, S. (1988), *Protein Purification Applications, a Practical Approach*, Oxford: IRL Press.

HASHIMOTO, H., et al. (1983), Regulation of expression of the galactose gene cluster in *Saccharomyces cerevisiae*, *Mol. Gen. Genet.* **191,** 31–38.

HATLEY, R. H. M., FRANKS, F., MATHIAS, S. F. (1987), *Process Biochem.* **22,** 169–172.

HENDERSON, R. C. A., COX, B. S., TUBB, R. (1985), The transformation of brewing yeast with a plasmid containing the gene for copper resistance, *Curr. Genet.* **9,** 31–48.

HILL-PERKINS, M. S., POSSEE, R. D. (1990), A baculovirus expression vector from the basic protein promoter of *Autographa californica* nuclear polyhidrosis virus, *J. Gen. Virol.* **71,** 971–976.

HINNEN, A., HICKS, J. B., FUNK, G. R. (1978), Transformation of yeast, *Proc. Natl. Acad. Sci. USA* **75,** 1929–1933.

HINNEN, A., MEYHACK, B., HEIM, J. (1989), Heterologous gene expression in yeast, in: *Yeast Genetic Engineering* (BARR, P. J., BRAKE, A. J., VALENZUELA, P., Eds.), pp. 83–108, Boston: Butterworth.

HITZEMAN, R. A., et al. (1983), Secretion of human interferons by yeast, *Science* **219,** 620–625.

HORWITZ, A. H., et al. (1988), Secretion of functional antibody and Fab fragment from yeast cells, *Proc. Natl. Acad. Sci. USA* **85,** 8678–8682.

HUSING, H. M., et al. (1989), Use of bacteriocin release protein in *E. coli* for secretion of human growth hormone into the culture medium, *Bio/Technology* **7,** 267–271.

IMANAKA, T., SHIBAZAKE, M., TAKAGI, M. (1986), A new way of enhancing the thermostability of proteases, *Nature* **324,** 695–697.

JANNIERE, L., BRUAND, C., EHRLICH, S. D. (1990), Structurally stable *Bacillus subtilis* cloning vectors, *Gene* **87,** 53–61.

JIMENEZ, A., DAVIES, J. (1980), Expression of a transposable antibiotic resistance element in *Saccharomyces*, *Nature* **287,** 869–871.

JOLIFF, G., EDELMAN, A., KLIER, A., RAPOPORT, G. (1989), Inducible secretion of a cellulase from *Clostridium thermocellum* in *Bacillus subtilis*, *Appl. Environ. Microbiol.* **55,** 2739–2744.

JONES, K. (1990), Affinity chromatography: a technology update, *American Biotechnology Laboratory*, October.

JOSEPHSON, S., BISHOP, R. (1988), Secretion of peptides from *E. coli:* A production system for the pharmaceutical industry, *Trends Biotechnol.* **6,** 218–223.

KARLSSON, E., et al. (1989), Ion exchange chromatography, in: *Protein Purification: Principles, High Resolution Methods and Applications* (HANSON, J.-C., RYDEN, L., Eds.), pp. 108–148, Weinheim–New York: VCH.

KIERDASZUK, B., ERIKSSON, S. (1990), Selective inactivation of the deoxyadenosine phosphorylating activity of pure human deoxycytidine kinase: stabilization of different forms of the enzyme by substrates and biological detergents, *Biochemistry* **29,** 4109–4114.

KINGSMAN, S. M., et al. (1985), Heterologous gene expression in *Saccharomyces cerevisiae*, *Biotechnol. Genet. Eng. Rev.* **3,** 377–416.

KISHIMOTO, F., et al. (1986), Direct expression of urogastrone gene in *Escherichia coli*, *Gene* **45,** 311–316.

KITTS, P. A., POSSEE, R. D. (1993), A method for producing recombinant baculovirus expression vectors at high frequency, *Bio/Techniques* **14,** 810–817.

KITTS, P. A., AYRES, M., POSSEE, R. D. (1990), Linearization of baculovirus DNA enhances the recovery of recombinant virus expression vectors, *Nucleic Acids Res.* **18,** 5667–5672.

KRAMER, R. A., et al. (1984), Regulated expression of a human interferon gene in yeast: control by phosphate concentration or temperature, *Proc. Natl. Acad. Sci. USA* **81,** 367–370.

KUKURUZINSKA, M. A., BERGH, M. L. E., JACKSON, B. L. (1987), Protein glycosylation in yeast, *Annu. Rev. Biochem.* **56,** 915–944.

LE GRICE, S. F. J. (1990), Regulated promoter for high-level expression of heterologous genes in *Bacillus subtilis*, *Meth. Enzymol.* **185,** 201–214.

LE GRICE, S. F. J., BEUCK, V., MOUS, J. (1987), Expression of biologically active human T-cell lymphotropic virus type III reverse transcriptase in *Bacillus subtilis*, *Gene* **55,** 95–103.

LIAO, X.-B., CLARE, J. J., FARABAUGH, P. J. (1987), The upstream activation site of a Ty2 element of yeast is necessary but not sufficient to promote maximal transcription of the element, *Proc. Natl. Acad. Sci. USA* **84,** 8520–8524.

LOISON, G., et al. (1986), Plasmid-transformed *URA3 FUR1* double-mutants of *S. cerevisiae:* an autoselection system applicable to the production of foreign proteins, *Bio/Technology* **4,** 433–437.

LOISON, G., et al. (1989), High level of expression of a protective antigen of schistosomes in *Saccharomyces cerevisiae*, *Yeast* **5,** 497–507.

LOPES, T. S., et al. (1989), High-copy-number integration into the ribosomal DNA of *Saccharomyces cerevisiae:* a new vector for high-level expression, *Gene* **79,** 199–206.

LOPES, T. S., et al. (1990), Mechanism of high-copy-number integration of pMIRY-type vectors into the ribosomal DNA of *Saccharomyces cerevisiae*, *Gene* **105,** 83–90.

LUCKOW, V. A. (1992), Cloning and expression of heterologous genes in insect cells with baculovirus vectors, in: *Recombinant DNA Technology and Applications* (PROKOP, A., BAJPAI, R. K., HO, C., Eds.), pp. 97–152, New York: McGraw Hill.

LUCKOW, V. A., SUMMERS, M. D. (1988), Trends in the development of baculovirus expression vectors, *Bio/Technology* **6,** 47–55.

LUCKOW, V. A., et al. (1993), Efficient generation of infectious recombinant baculoviruses by site-specific transposon-mediated insertion of foreign genes into a baculovirus genome propagated in *Escherichia coli*, *J. Virol.* **67,** 4566–4579.

LUDWIG, D., UGOLINI, S., BRUSCHI, C. (1993), High-level heterologous gene expression in *Saccharomyces cerevisiae* from a stable 2μm plasmid system, *Gene* **132,** 33–40.

MAEDA, S., et al. (1985), Production of human α-interferon in silkworm using a baculovirus vector, *Nature* **315,** 592–594.

MAO, Q. M., et al. (1993), High performance liquid chromatography of amino acids, peptides and proteins. Optimisation of operating parameters for protein purification with chromatographic columns, *J. Chromatogr.* **636** (1), 81–89.

MARULLO, S., et al. (1989), Expression of human β1 and β2 adrenergic receptors in *E. coli* as a new tool for ligand screening, *Bio/Technology* **7,** 923–927.

MASUI, Y., MIZUNO, T., INOUYE, M. (1984), Novel high-level expression cloning vehicles: 104-fold amplification of *Escherichia coli* minor protein, *Bio/Technology* **2,** 81–85.

MELLOR, J., et al. (1987), A transcriptional activator is located in the coding region of the yeast *PGK* gene, *Nucleic Acids Res.* **15,** 6243–6259.

MIAO, F., DRAKE, S. K., KOMPALA, D. S. (1993), Characterization of gene expression in recombinant *Escherichia coli* cells infected with phage λ, *Biotechnol. Prog.* **9,** 153–159.

MICHELMORE, R. W., FRANKS, F. (1982), Nucleation rates of ice in undercooled water and aqueous solutions of polyethylene glycol, *Cryobiology* **19,** 163–171.

MILLER, L. (1988), Baculoviruses as gene expression vectors, *Annu. Rev. Microbiol.* **42,** 177–199.

MOUNTAIN, A. (1989), Gene expression system for *Bacillus subtilis*, in: *Bacillus* (HARWOOK, C. R., Ed.), pp. 73–114, New York: Plenum Press.

MURPHY, G. (1990), Purification of connective tissue metalloproteinases, in: *Protein Purification Applications. A Practical Approach* (JARRIS, E. L. V., ANGAL, S., Eds.), pp. 142–147, New York: IRL Press at Oxford University Press.

NAKAMURA, K., INOUYE, M. (1982), Construction of versatile expression cloning vehicles using the lipoprotein gene of *Escherichia coli*, *EMBO J.* **1,** 771–775.

NAKANO, H., et al. (1994), Purification of glutathione S-transferase fusion proteins as a non-degraded form by using a protease-negative *E. coli*, *Nucleic Acids Res.* **22,** 543–544.

OBUKOWICZ, M. G., et al. (1988), Secretion and export of IGF-1 in *Escherichia coli* strain JM101, *Mol. Gen. Genet.* **215,** 19–25.

O'FAGAIN, C., et al. (1988), *Process Biochem.* **23,** 166–171.

ORR-WEAVER, T. L., SZOSTAK, J. W. (1983), Multiple, tandem plasmid integration in *Saccharomyces cerevisiae*, *Mol. Cell. Biol.* **30,** 747–749.

PACE, C. N. (1990), Conformational stability of globular proteins, *Trends Biochem. Sci.* **15,** 14–17.

PADUKONE, N., PERETTI, S. W., OLLIS, D. F. (1990), λ vectors for stable cloned gene expression, *Biotechnol. Prog.* **6,** 277–282.

PANASENKO, S. M., et al. (1977), Five hundredfold overproduction of DNA ligase after induction of a hybrid λ lysogen constructed *in vitro*, *Science* **196,** 188.

PARK, T. H., SEO, J.-H., LIM, H. C. (1991), Two-stage fermentation with bacteriophage λ as an expression vector in *Escherichia coli*, *Biotechnol. Bioeng.* **37,** 297–302.

PATEL, G., NASMYTH, K., JONES, N. (1992), A new method for the isolation of recombinant baculovirus, *Nucleic Acids Res.* **20,** 97–104.

PEAKMAN, T. C., PAGE, M., GERWERT, D. R. (1989), Increased recombinational efficiency in insect cells irradiated with short wavelength ultra-violet light, *Nucleic Acids Res.* **13,** 5403.

PEAKMAN, T. C., HARRIS, R. A., GERWERT, D. R. (1992a), Highly efficient generation of recombinant baculoviruses by enzymatically mediated site-specific *in vitro* recombination, *Nucleic Acids Res.* **20,** 495–500.

PEAKMAN, T. C., et al. (1992b), Enhanced expression of recombinant proteins in insect cells using a baculovirus vector containing a bacterial leader sequence, *Nucleic Acids Res.* **20,** 6111–6112.

PEREZ-PEREZ, J., et al. (1994), Increasing the efficiency of protein export in *Escherichia coli*, *Bio/Technology* **12,** 178–180.

PETIT, M.-A., et al. (1990), Hypersecretion of cellulase from *Clostridium thermocellum* in *Bacillus subtilis* by induction of chromosomal DNA amplification, *Bio/Technology* **8,** 559–563.

PRICE, V. L., et al. (1990), Expression of heterologous proteins in *Saccharomyces cerevisiae* using *ADH2* promoter, *Meth. Enzymol.* **185,** 308–318.

PRIVOLOV, P. L. (1979), Stability of proteins: small globular proteins, *Adv. Prot. Chem.* **33,** 167–241.

PURVIS, I. J., et al. (1991), An androgen-inducible expression system for *Saccharomyces cerevisiae*, *Gene* **106,** 35–42.

RADMACHER, M., et al. (1994), Direct observation of enzyme activity with the atomic force microscope, *Science* **265,** 1577–1579.

ROMANOS, M. A., et al. (1991), Recombinant *Bordetella pertussis* pertactin (P69) from the yeast *Pichia pastoris:* high-level production and immunological properties, *Vaccine* **9,** 901–906.

ROMANOS, M. A., SCORER, C. A., CLARE, J. J. (1992), Foreign gene expression in yeast: a review, *Yeast* **8,** 423–488.

RUOHONEN, L., et al. (1987), Efficient secretion of *Bacillus amyloliquefaciens* α-amylase by its own signal peptide from *Saccharomyces cerevisiae* host cells, *Gene* **59,** 161–170.

SANTAMBINE, P. (1992), Immunoenzyme assay of dyes currently used in affinity chromatography for protein purification, *J. Chromatogr.* **597,** 312–322.

SCHENA, M., PICARD, D., YAMAMOTO, K. R. (1991), Vectors for constitutive and inducible gene expression in yeast, *Meth. Enzymol.* **194,** 389–398.

SCHMID, R. P. (1979), *Adv. Biochem. Eng.* **12,** 42–64.

SCHULTZ, L. D., et al. (1987), Regulated overexpression of the *GAL4* gene product greatly increases expression from galactose-inducible promoters on multi-copy expression vectors in yeast, *Gene* **61,** 123–133.

SCHULZ, M.-F., et al. (1987), Increased expression in *Escherichia coli* of a synthetic gene encoding human somatomedin C after gene duplication and fusion, *J. Bacteriol.* **169,** 5384–5392.

SCOPES, R. K. (1986), *J. Chromatogr.* **376,** 131.

SCORER, C. A., et al. (1993), Rapid selection using G418 of high copy number transformants of *Pichia pastoris* for high-level foreign gene expression, *Bio/Technology* **12,** 181–184.

SHIBUI, T., et al. (1989), Periplasmic production of human pancreatic prokallikrein in *Escherichia coli*, *Appl. Microbiol. Biotechnol.* **31,** 253–258.

SHUSTER, J. R. (1989), Regulated transcriptional systems for the production of proteins in yeast: regulation by carbon source, in: *Yeast Genetic Engineering* (BARR, P. J., BRAKE, A. J., VALENZUELA, P., Eds.), pp. 83–108, Boston: Butterworth.

SHUSTER, J. R., LEE, H., MOYER, D. L. (1990), Integration and amplification of DNA at yeast delta sequences, *Yeast* **6,** 579.

SLEEP, D., et al. (1991), Cloning and characterization of the *Saccharomyces cerevisiae* glycerol-3-phosphate dehydrogenase (*GUT2*) promoter, *Gene* **101,** 89–96.

SMITH, H., et al. (1988), Characterization of signal sequence – coding regions selected from the *Bacillus subtilis* chromosome, *Gene* **70,** 351–361.

SOFER, G. K., NYSTROM, L. E. (1989), *Process Chromatography, a Practical Guide*, New York: Academic Press.

SOGAARD, M., SVENSSON, B. (1990), Expression of cDNAs encoding barley α-amylase 1 and 2 in yeast and characterization of the secreted proteins, *Gene* **94,** 173–179.

SREEKRISHNA, K., TSCHOOP, J. F., FUKE, M. (1987), Invertase gene (*SUC2*) of *Saccharomyces cerevisiae* as a dominant marker for transformation of *Pichia pastoris*, *Gene* **59,** 115–125.

STELLWAGEN, E. (1990), Chromatography on immobilized reactive dyes, *Meth. Enzymol.* **183,** 343–357.

STRUHL, K. (1989), Molecular mechanisms of transcriptional regulation in yeast, *Annu. Rev. Biochem.* **58,** 1051–1077.

STUDIER, F. W. (1991), Use of bacteriophage T7 lysozyme to improve an inducible T7 expression system, *J. Mol. Biol.* **219,** 37–44.

STUDIER, F. W., MOFATT, B. A. (1986), Use of bacteriophage T7 RNA polymerase to direct selective high-level expression of cloned genes, *J. Mol. Biol.* **189,** 113–130.

SUMMERS, M. D., SMITH, G. E. (1987), A manual of methods for baculovirus vectors and insect cell culture procedure, *Texas Agric. Exp. Sta. Bull.*, 1555.

TABOR, S., RICHARDSON, C. C. (1985), A bacteriophage T7 RNA polymerase/promoter system for controlled exclusive expression of specific genes, *Proc. Natl. Acad. Sci. USA* **82,** 1074–1078.

TAN, S., CONAWAY, R. C., CONAWAY, J. W. (1994), A bacteriophage vector suitable for site-directed mutagenesis and high-level expression of multisubunit proteins in *E. coli*, *BioTechniques* **16,** 824–828.

THORNWELL, S. J., EAST, A. K., ERRINGTON, J. (1993), An efficient expression and secretion system based on *Bacillus subtilis* phage phi105 and its use for the production of *B. cereus* β-lactamase, I, *Gene* **133,** 47–53.

TSCHOPP, J. F., et al. (1987), High-level secretion of glycosylated invertase in the methylotrophic yeast, *Bio/Technology* **5,** 1305–1308.

UDAKA, S., YAMAGATA, H. (1993), High-level secretion of heterologous protein by *Bacillus brevis*, *Meth. Enzymol.* **217,** 23–33.

VAN DIJIL, J. M., et al. (1991), Signal peptidase I overproduction results in increased efficiencies of export and maturation of hybrid secretory proteins in *Escherichia coli*, *Mol. Gen. Genet.* **227,** 40–48.

VOGEL, K., HINNEN, A. (1990), The yeast phosphatase system, *Mol. Microbiol.* **4,** 2013–2017.

VON HIPPEL, P. H., SCHLEICH, T. (1969), in: *Structure and Stability of Biological Macromolecules* (TIMASHEFF, S. N., FESMAN, G. D., Eds.), pp. 417–574, New York: Marcel Dekker.

WANG, D. I. C. (1987), Separations for biotechnology, in: *Separations for Biotechnology* (VERRALL, M. S., HUDSON, M. J., Eds.), pp. 30–48, London: Ellis Horwood Ltd.

WEBSTER, T. D., DICKSON, R. C. (1983), Direct selection of *Saccharomyces cerevisiae* resistant to the antibiotic G418 following transformation with a DNA vector carrying the kanamycin resistance gene of Tn 903, *Gene* **26,** 243–252.

WEINBERG, R. A., DE CIECHI, P. A., OBUKOWICZ, M. (1993), A chromosomal expression vector for *Escherichia coli* based on the bacteriophage Mu, *Gene* **126,** 25–33.

WEYER, U., POSSEE, R. D. (1991), A baculovirus dual expression vector from the *Autographa californica* nuclear polyhedrosis virus polyhedrin and p10 promoters: co-expression of two influenza virus genes in insect cells, *J. Gen. Virol.* **72,** 2967–2974.

WEYER, U., KNIGHT, S., POSSEE, R. D. (1990), Analysis of very late gene expression by *Autographa californica* nuclear polyhedrosis virus and the further development of multiple expression vectors, *J. Gen. Virol.* **71,** 1525–1534.

WICKHAM, T. J., NIMEROW, G. R. (1993), Optimization of growth methods and recombinant protein production in BTI-Tn-5B1-4 insect cells using the baculovirus expression system, *Biotechnol. Prog.* **9,** 25–30.

WICKHAM, T. J., et al. (1992), Screening insect cell lines for the production of recombinant proteins and infectious virus in the baculovirus expression system, *Biotechnol. Prog.* **8,** 391–396.

YANSURA, D. G., HENNER, D. J. (1990), Use of *Escherichia coli trp* promoter for direct expression of proteins, *Meth. Enzymol.* **185,** 54–60.

YOUNG, R. A., DAVIS, R. W. (1983), Efficient isolation of genes using antibody probes, *Proc. Natl. Acad. Sci. USA* **80,** 1194–1198.

ZAMAN, Z., BROWN, A. J. P., DAWES, I. W. (1992), A 3′ transcriptional enhancer within the coding sequence of a yeast gene encoding the common subunit of two multienzyme complexes, *Mol. Microbiol.* **6,** 239–246.

3 Kinetics of Multi-Enzyme Systems

ATHEL CORNISH-BOWDEN

Marseille, France and Santiago, Chile

1 The Impact of Molecular Biology on Biotechnology

1.1 Introduction

The development of molecular biology in recent years has enormously increased efforts in biotechnology. The possibility of increasing selected enzyme activities, essentially at will, has apparently opened the door to the application of classical ideas of enzyme regulation to useful industrial objectives. Obtaining a product on an industrial scale from a microorganism might seem to be quite simple and straightforward: first identify the enzyme that catalyzes the rate-limiting step in the normal biosynthetic pathway to the required product; next clone the gene and use the standard techniques of genetic manipulation to produce a greatly increased activity in the organism; finally reap the reward in the form of much higher yields.

The idea seems simple, until one searches the literature for examples of its successful application. There are a few examples of failures, but no examples of increases in production rates large enough to compensate for the added costs of the process. Perhaps when the entire new edition of *Biotechnology* is available it will contain some examples, but it is unlikely. For a while it might, perhaps, have been possible to explain the lack of examples in terms of industrial secrecy, but this is not an explanation that can be credible indefinitely; eventually the possibility will have to be faced that the lack of reported successes reflects a lack of actual successes, and that the simple idea outlined above is misconceived and unlikely ever to work.

This chapter describes the theory of metabolic control that has developed during the past two decades, with a view to indicating what is wrong with the simple approach. Although the message of the chapter will in part be a negative one, it can still fulfill a positive function, if it helps to direct efforts away from methods that cannot work and towards others that have some hope of success.

1.2 Example: Production of Tryptophan in Yeast

A recent study by NIEDERBERGER et al. (1992) of the tryptophan pathway in the yeast *Saccharomyces cerevisiae* provides some experimental evidence for the views expressed in the Introduction. The biosynthetic part of this pathway (Fig. 1) consists of five enzymes: the activities of four of these were increased individually by factors of 10 to 60; in other ex-

(a) Chorismate $\xrightarrow{E_1}$ Anthranilate $\xrightarrow{E_2}\xrightarrow{E_3}\xrightarrow{E_4}\xrightarrow{E_5}$ Tryptophan

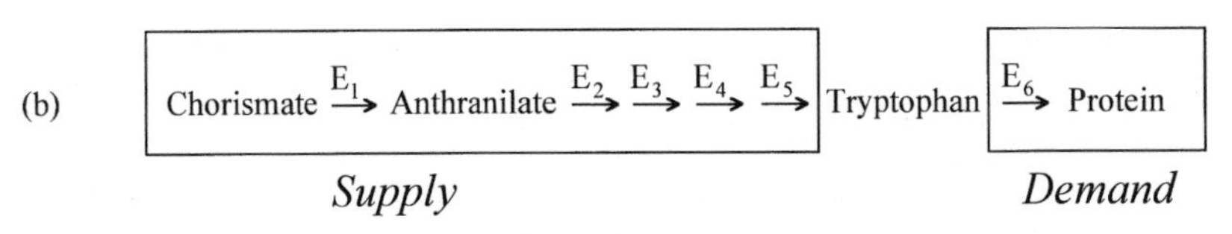

Fig. 1. Biosynthesis of tryptophan in *Saccharomyces cerevisiae*. The identities of the enzymes E_1 to E_5 are as listed in Tab. 1. E_6 represents not only tryptophan tRNA synthetase but also all of the subsequent process of protein synthesis. (a) Biosynthetic pathways are usually drawn in textbooks without explicit indication of the reactions that consume the "end product". (b) However, as discussed in the text, regulation of such a pathway involves communication via the end product between a demand block and a supply block. In this example only the most important use of tryptophan is shown as the demand block; other pathways, such as degradation to tryptophanol, constitute competing demand blocks.

periments all of them were similarly increased in various combinations. As may be seen in Tab. 1, increasing no single enzyme activity had any significant effect on the flux to tryptophan; for example, even a 50-fold increase in the activity of indoleglycerol phosphate synthase left the flux unchanged.

The classical explanation of this would be that anthranilate synthase (E_1 in Fig. 1), the only enzyme whose activity was not increased by itself, was the "key" enzyme in the pathway, i.e., the one whose activity controlled the flux through the whole pathway. This explanation is made plausible by the fact that anthranilate synthase catalyzes the first committed step of the pathway, i.e., the first step after the branch point separating the biosynthesis of tryptophan from that of phenylalanine and tyrosine. Unfortunately for any such argument, however, increasing the activity of anthranilate synthase by a factor of 30, at the same time as two other activities were increased by factors of 30 and 40, produced only a 20% increase in flux. Moreover, decreasing the activity of anthranilate synthase by a factor of 4 had almost no effect (Fig. 1a of NIEDERBERGER et al., 1992), an observation that makes it highly implausible that increasing it would have had a large effect.

Only when all five activities were increased by factors of about 20 (or all five apart from phosphoribosylanthranilate isomerase, which appears to have little importance) was there a substantial increase in flux; even then it was considerably less than the increase in the activity of the pathway enzymes.

This study is the most detailed that can be found in the literature, but it is not the only one, and the results of others were equally disappointing. For example, SCHAAFF et al. (1989) used a similar approach in efforts to improve the yield of ethanol in alcoholic fermentation of yeast, with essentially the same results as those shown in Tab. 1. It is only when one searches for examples of success that one searches in vain.

Tab. 1. Effect of Increasing Activities of Enzymes in the Tryptophan Pathway of *Saccharomyces cerevisiae*

Plasmid	Activities Relative to Wild Type E_1	E_2	E_3	E_4	E_5	Trp Flux
Wild-type	1.0	1.0	1.0	1.0	1.0	0.43
pME530	1.2	**19.9**	1.2	1.6	1.2	0.57
pME526	1.5	1.2	**10.6**	1.6	0.7	0.50
pME550	1.5	0.8	1.8	**51.5**	1.1	0.44
pYAS-1	1.3	0.8	1.7	1.6	**37.1**	0.52
pME529	1.5	0.6	**58.6**	**56.0**	0.5	0.87
pME553	**25.0**	**34.0**	1.7	**47.0**	1.0	1.03
pME557	**29.4**	1.0	**30.6**	**40.9**	1.0	0.53
pME558	**26.5**	**26.4**	**30.3**	**37.4**	0.8	0.91
pME554	**20.0**	**15.2**	1.6	**20.7**	**18.1**	3.54
pME559	**24.0**	**19.6**	**22.1**	**26.9**	**23.9**	3.51
$C^J_{E_i}$...	0.018	0.174	0.013	0.013	0.040	

The table shows data of NIEDERBERGER et al. (1992) for the effects on the biosynthetic flux to tryptophan in yeast of engineering large increases in the activities of the five enzymes in the supply block of the pathway: E_1, anthranilate synthase; E_2, anthranilate phosphoribosyl transferase; E_3, phosphoribosylanthranilate isomerase; E_4, indoleglycerol phosphate synthase; E_5, tryptophan synthase. Tryptophan fluxes are in nmol min^{-1} per mg protein. The bottom line of the table shows flux control coefficients $C^J_{E_i}$ measured in separate experiments in which the activities of the five enzymes were individually decreased by various amounts and curves fitted to the results.

1.3 Need for a Theory of Regulation

For those who have followed the development of metabolic control theory over the past 20 years since its origins in articles of KACSER and BURNS (1973) and of HEINRICH and RAPOPORT (1974), the disappointing results of attempts to improve yields by increasing activities of rate-limiting enzymes can hardly have been a surprise, because a major emphasis in this theory has been that rate-limiting enzymes as classically conceived do not exist; rather, the control of flux in any pathway is shared, in proportions that cannot be predicted *a priori*, among all the enzymes of the organism. WESTERHOFF and KELL (1987) have described how this theory can be applied to biotechnological questions, but it has yet to have a major impact on the practice of biotechnology. Efforts to use genetic engineering to realize commercial objectives have been guided more by a vague optimism than by application of the principles of metabolic control.

Important though it is, however, metabolic control theory as now applied is incomplete, because it has given inadequate attention to the large body of valid information about metabolic regulation that exists. After the discovery of feedback inhibition of the enzyme that catalyzes the first committed step of a pathway by the end product of the pathway (YATES and PARDEE, 1956; UMBARGER, 1956), experimental work in the period 1956–1970 led to detailed information about how the activity of an enzyme in a cell can be modified; and the theoretical work of MONOD et al. (1965) and KOSHLAND et al. (1966) led to a good understanding of the molecular properties that permit the necessary variations in enzyme activity.

It would be an exaggeration to say that current metabolic control theory ignores cooperativity and feedback inhibition, but it does confine them to the periphery of the subject, far away from the central place that they had in the classical studies of metabolic regulation. Instead, metabolic control analysis has decreased the emphasis on mechanisms as a way of emphasizing the importance of treating metabolic control systemically, i.e., treating systems as systems and not just as collections of components. Nonetheless, the time has arrived when one must try to reestablish the importance of the classical mechanisms in achieving regulation.

A problem has been the emphasis on *control* at the expense of *regulation*. Not everyone would agree that these terms have different meanings, but if one examines how they are actually used in the biochemical literature, it appears that regulation conveys much more an idea of biological function than does control. One can quantify metabolic control without specifying what purpose the control serves, but discussion of regulation implies some definition of the biological objectives. We have argued that metabolic regulation needs to be discussed in terms similar to those used of everyday technological devices such as refrigerators (HOFMEYR and CORNISH-BOWDEN, 1991). A domestic refrigerator is considered to be well regulated if it is able to maintain a constant temperature (loosely speaking, a constant "concentration of heat") in the face of variations in the outside temperature and in the heat flux when the door is opened at arbitrary intervals unpredictable by the engineer who designed it. The metabolic analogy would be to define a well regulated system as one capable of maintaining a near-constant concentration of a metabolite in the face of unpredictable variations in the demand by biosynthetic pathways that use it as a starting material. This approach certainly oversimplifies biology, as there may be circumstances where constant concentration in the face of variations in flux is not a desirable objective, but such a criticism misses the point: *whatever* may be the desirable state, it has to be specified before one can analyze the extent to which it is realized.

In summary, our present view is that the classical mechanisms are essential for achieving satisfactory regulation, but that to be properly understood they need to be considered in the framework of a systemic theory of control (HOFMEYR and CORNISH-BOWDEN, 1991). I shall not here enter into a discussion of which systemic theory of control is to be preferred, beyond noting that biochemical systems theory (reviewed by SAVAGEAU,

1990) offers an alternative way of looking at many of the same ideas.

2 Classical Regulatory Mechanisms

2.1 Feedback Inhibition

The major classical regulatory mechanism is *feedback inhibition* of the first committed step of a pathway by its end product. Although this is sometimes presented in textbooks as simple and obvious, there are a number of points of definition that need to be addressed. Virtually the whole of metabolism consists of a single highly branched network, and any reaction can in principle lead eventually to any other, so there are no real beginnings or ends of pathways. When biochemists refer to end products, they hardly ever mean excretion products such as water or CO_2; instead they mean substances that appear somewhere in the middle of a typical chart of metabolic pathways. In the classic review of metabolic regulation (STADTMAN, 1970), the term "end product" or even, with added emphasis but no change of meaning, "ultimate end product", always means a metabolite such as aspartate that is not excreted but serves as the starting point for other pathways. Such metabolites usually constitute "pools", i.e., they occur at concentrations that are relatively high and stable compared with those of the intermediates that connect them; to the extent that we know the true fluxes of pathways in living cells, it appears reasonable to view the steps connecting the pools as constituting coherent pathways because, even though they may contain branches, there is a sufficiently uniform flow from one pool to another that most of the material produced in one step of the pathway is consumed by the next. Thus, although to some extent a metabolic pathway exists as much in the mind of the biochemist as in the cell, it is not wholly arbitrary.

Once the end product of a pathway is defined, the "first committed step" can be understood as the first step after a branch point that leads to only that product. As many pathways have more than two or three steps, it is obvious that the end product of a pathway may differ substantially in structure from the substrate and immediate product of the inhibited enzyme. Consequently the inhibition can hardly be just a chance effect of structural similarity, but must result from an interaction that has evolved specifically. Such inhibition by a molecule structurally distinct from the immediate reactants is known as *allosteric inhibition*. (Allosteric activation may also occur, but this will not be discussed here.)

2.2 Cooperativity

In practice allosteric inhibition is often accompanied by a second property, known as *cooperativity*, i.e., the inhibition is more sensitive than the simplest mechanisms predict. For example, if an enzyme displays ordinary competitive inhibition and has rate

$$v = V_s/[K_m(1+i/K_i)+s]$$

at concentrations s and i of the substrate and inhibitor, respectively, the rate equation may be rearranged to show that the ratio of uninhibited to inhibited rates is

$$v_0/v = 1+(i/K_i)/(1+s/K_m),$$

and a simple calculation shows that to go from 10% to 90% inhibited (i.e., from $v/v_0 = 0.9$ to $v/v_0 = 0.1$, or $v_0/v = 10/9$ to $v_0/v = 10$) it is necessary to increase i by a factor of 81.

This ratio R is the *cooperativity index* (TAKETA and POGELL, 1965), and the value of $R = 81$ is the baseline value for a non-cooperative interaction. It is an enormous factor for a concentration to have to change to produce a reasonable physiological effect, and enzymes that are believed to fulfill important regulatory functions have evolved to respond with much greater sensitivity, with values of R as low as 3, though not usually lower. Cooperativity is usually expressed in terms of the *Hill coefficient* h, a parameter with a less transparent meaning than R, but with a more convenient range of values: the approximate relationship is $R=81^{1/h}$, so a non-cooperative interaction ($R=81$) has $h=1$, whereas a highly cooperative one ($R=3$) has $h=4$.

Although allosteric inhibition and cooperativity are two different properties, they often occur together, and the highly influential paper of MONOD et al. (1965) suggested that they might be two different aspects of the same fundamental mechanism, which they called the *allosteric model.* Although many later workers have preferred the more neutral name *symmetry model*, this history has often caused the two properties to be confused, with the terms *allosteric* and *cooperative* treated as interchangeable. This is unfortunate, however, as they are clearly distinguishable and both need to be considered in discussing regulation.

Before attempting a modern view of the function of these properties in metabolic regulation, we need to digress to discuss the development of metabolic control analysis.

3 Metabolic Control Analysis

3.1 Terminology

Metabolic control theory has developed from landmark papers of KACSER and BURNS (1973) and HEINRICH and RAPOPORT (1974). The initial impetus was a desire to move away from the traditional mechanism-based view of metabolism towards a systemic view. Instead of discussing the kinetic properties of pathways in terms of what mechanisms the individual enzymes display, proponents of metabolic control analysis argue that they need to be discussed in terms of how the whole system responds to a change in the parameters that define its environment. The most fundamental such parameters are the catalytic activities of the component enzymes: if the activity of a particular enzyme increases, how do the flux or fluxes and the concentrations of the intermediates change? If we suppose that a change in some parameter p changes the rate v_i of the ith enzyme E_i when considered in isolation from the pathway, i.e., in the presence of all metabolites that influence it, at the concentrations that occur in the complete system, but in the absence of all other enzymes, then one can express the sensitivity of a flux J through the system in terms of a *flux control coefficient:*

$$C_i^J = \frac{\partial \ln J}{\partial p} \Big/ \frac{\partial \ln v_i}{\partial p} \tag{1}$$

The identity of the parameter p does not have to be specified, because the value of C_i^J does not depend on it. If, as is often but not always the case, the rate v_i is directly proportional to the total concentration e_i of E_i, then it may be convenient to take p as identical to e_i, in which case C_i^J may be defined more simply, as follows:

$$C_i^J = \frac{\partial \ln J}{\partial \ln e_i} \tag{2}$$

This type of definition was used by KACSER and BURNS (1973), but the present trend in metabolic control analysis is towards preferring the one in Eq. (1), which originated with HEINRICH et al. (1977). Partly this is because it is more general, but a more important reason may be that the simpler definition has led to a widespread misunderstanding that metabolic control analysis deals only with effects brought about by changes in enzyme concentrations or changes in limiting rates (V) of enzyme reactions. For example, ATKINSON (1990) has argued that metabolic control analysis has no relevance to regulation in the cell, because enzyme concentrations hardly change in the time scale usually considered in discussions of regulation and most effectors of enzymes alter substrate binding rather than limiting rates. Leaving aside the question of whether this last generalization is factually accurate, it should be clear from the first definition that the whole criticism is misconceived, because metabolic control analysis deals with effects of changes in enzyme activity regardless of the mechanisms that produce them.

A different criticism of describing control in terms of control coefficients is that they are defined as derivatives, and thus refer to infinitesimal changes in parameters: they express the effects of small parameter changes but not of large changes. This criticism is analogous to rejecting any analysis of traffic flow in terms of vehicle velocities, which are likewise defined as derivatives, and is invalid for the same reason. If one were trying to describe the control structure of a system entirely in terms of a set of constants called control coefficients the point might be valid, but in reality control coefficients are just a starting point: they provide a language for discussing control, and allow definition of some ground rules that have to be obeyed.

Control coefficients are defined similarly for other kinds of variables, for example, a *concentration control coefficient* refers to effects on the concentration s_j of an intermediate S_j:

$$C_i^{s_j} = \frac{\partial \ln s_j}{\partial p} \bigg/ \frac{\partial \ln v_i}{\partial p} \tag{3}$$

In the standard mechanistic approach to enzymology the kinetic properties of enzymes are expressed in terms of Michaelis constants, inhibition constants, limiting rates, etc. For purposes of metabolic control analysis, however, it is more convenient to decrease the emphasis on mechanism by using quantities that relate effects to causes in a more phenomenological way, similar in definition to control coefficients:

$$\varepsilon_{s_j}^{v_i} = \frac{\partial \ln v_i}{\partial \ln s_j} \tag{4}$$

This quantity $\varepsilon_{s_j}^{v_i}$ is called the *elasticity* of v_i with respect to s_j. The term was borrowed by KACSER and BURNS (1973) from econometrics, where it may be said, for example, that the demand for cars has an elasticity of 2.5 if it decreases by 2.5% when the price increases by 1% [note that this would be an elasticity of −2.5 if the definition were strictly analogous to Eq. (4)]. However, most biochemists are not trained in econometrics, and the meaning would probably have been less obscure than many people have found it if it had been called the *kinetic order*, the term used by SAVAGEAU (1990) for a similarly defined quantity. [SAVAGEAU himself considers the definitions to be identical, not merely "similar". The important difference is that he uses the kinetic order as the basis for power-law rate equations, and as any law has to be expressed in terms of constants, the kinetic order must be regarded as a constant, at least over the range of applicability of the law; the elasticity, however, is always regarded as a variable in metabolic control analysis (CORNISH-BOWDEN, 1989).]

Unlike a control coefficient, which describes a property of the whole system, an elasticity is a local property, i.e., it expresses a property of an enzyme isolated from the rest of the system. This means that it describes the response of an enzyme to a metabolite, in the presence, at the concentrations that exist in the intact system, of all other metabolites that influence it, but in the absence of the other enzymes of the system. In a sense, measuring elasticities is what enzymologists have been doing since the time of MICHAELIS and MENTEN (1913), but they have usually expressed their results in ways that emphasize their mechanistic significance rather than their contribution to a multi-enzyme system. More important, perhaps, ordinary enzyme kinetic experiments are done under conditions as simple as possible, i.e., metabolic concentrations are often either zero or saturating, and are in any case chosen according to mechanistic criteria rather than to mimic conditions in the cell.

A third kind of coefficient is needed to define the response of a system to a change in a concentration x_j that is defined independently of the system (in contrast to an intermediate concentration that is set by the system itself). This is called a *response coefficient* and is defined similarly:

$$R_{x_j}^{J} = \frac{\partial \ln J}{\partial \ln x_j} \tag{5}$$

In this example the flux J is taken to be the response, but it could equally well be an intermediate concentration or other variable of interest.

For studying a complex pathway it is often convenient to consider it to consist of distinct "blocks" of enzymes, so that a group of enzymes catalyzing consecutive steps can be treated as if it were a single enzyme. The quantity that would be the flux if the block were the complete system becomes the local rate when it is treated as part of the larger system; a metabolite that is an external pool for the small system becomes an intermediate in the larger system. Similarly, a response coefficient in the smaller system is a block elasticity in the larger one.

Some workers (e.g., CRABTREE and NEWSHOLME, 1987) deny the need for so many kinds of coefficient, arguing that they are mathematically equivalent "sensitivities". However, they arise from the need to distinguish between properties of the system (control coefficients, response coefficients) and properties of its components (elasticities), and from the need to distinguish between parameters that are defined independently of the system (kinetic constants, concentrations of enzymes and pool metabolites, including external effectors) and variables that are set by the system (fluxes and concentrations of intermediate metabolites). The fact that changing the limits of the system may convert a response coefficient of the smaller system into a block elasticity when it is considered to be part of a larger system, does not mean that the two terms are equivalent. CRABTREE and NEWSHOLME (1987) attempt to get around this difficulty by defining three different kinds of metabolite, i.e., they define "partially external" metabolites as an intermediate stage between internal (variable) and external (constant) metabolites. However, although a quantity that is constant under one set of conditions may vary under others, it cannot be both constant and variable at the same time. Users of metabolic control analysis accordingly consider three kinds of coefficient to be less confusing than three kinds of metabolite.

The essential relationship between the three kinds of quantity is the *partitioned response property*, which is given here without proof:

$$R^J_{x_j} = \varepsilon^{v_i}_{x_j} C^J_i \qquad (6)$$

This expresses the fact that any response can be regarded as the combination of an effect on an enzyme with the control exercised by that enzyme. It follows that defining control coefficients in relation to enzyme concentrations, as in Eq. (2), does not restrict control analysis to changes in enzyme concentration, because it allows any kind of parameter change to be related to any kind of response. Comparison with Eqs. (1) and (2) shows that the kind of control coefficient expressed by Eq. (2) is really a response coefficient that happens to be equal to a control coefficient because the relevant elasticity is unity, i.e., $\varepsilon^{v_i}_{e_i}=1$ if v_i is proportional to e_i.

3.2 The Summation Relationships

It is obvious that the properties of a system must depend on the properties of its components, but the form of the dependence is less obvious. Accordingly, much effort in metabolic control analysis has been directed towards showing mathematically how the systemic properties derive from the elasticities (for current reviews, see FELL, 1992, or CORNISH-BOWDEN, 1994). However, this has limited direct application to biotechnology, and will not be discussed here. The important relationships for an introductory understanding of control analysis are the *summation relationships*, which will now be discussed.

For any flux J the summation relationship is as follows:

$$\sum_{\substack{\text{all} \\ \text{enzymes}}} C^J_i = 1 \qquad (7)$$

whereas for any concentration s_j it is

$$\sum_{\substack{\text{all} \\ \text{enzymes}}} C^{s_j}_i = 0 \qquad (8)$$

Two important points should be made about these summations. If the pathway is branched, there will be more than one distinct flux through it at steady state, but each control coefficient in Eq. (4) must refer to the same flux, and likewise each one in Eq. (5) must refer to the same concentration. The second point is that the summations are made over all enzymes in the system, not just those in the pathway or branch in which the flux or concentration of interest occurs. Thus, if the system is defined as the whole organism, the summations are made over all the enzymes in the organism; if it is just the glycolytic pathway, then they are made over all the glycolytic enzymes, and so on. Although this implies some arbitrariness in how "the system" is defined, most of this can be avoided by defining the limits of the system as pools of metabolites whose concentrations are stable enough to be treated as parameters, i.e., as quantities with values that are set independently of the system and cannot be changed by the enzymes of the system.

The summation relationships may seem surprising at first sight, but they can be made more plausible (though not of course proved) by reflecting that if all enzyme activities in the system were increased by the same factor the only change in the system would be a change of time scale: this means that every flux would increase by the same factor and every metabolite concentration would be unchanged.

Provided that one accepts that a control coefficient does in some way measure control, i.e., that it is not just a name, Eq. (4) immediately establishes the crucial fact about flux control, that it is shared among all the enzymes in a system, so that each enzyme has on average a flux control coefficient $1/n$ in a system of n enzymes. Although hypothetical pathways of three or four enzymes are often used in model calculations to illustrate metabolic control analysis, most of the pathways of interest to biochemists are longer than this, and in the limit the system is the whole organism, or at least a whole cell, with thousands of different enzyme activities. Even though there is no reason to postulate that this control is divided equally among the enzymes, it is obvious that in a real system most individual enzymes have very little control, and even the exceptions cannot have very much. This is illustrated by the flux control coefficients shown in the bottom line of Tab. 1, where the largest value is only 0.174.

A complicating factor is that although all flux control coefficients in an unbranched pathway are positive, negative flux control coefficients can occur in branched pathways, so that it is not impossible for the total to be 1 even though some individual values are much larger than 1. However, in practice it proves very difficult to imagine realistic systems in which there are enough large negative values to allow this to happen to a significant extent. Thus it is not far from reality to conclude that the typical flux control coefficient is close to zero, and that the largest ones in any system are less than 1. The experimental systems that have been studied to date confirm that this is true (see, for example, FELL, 1992).

All of this leaves little hope for biotechnological methods based on the traditional ideas of "key enzymes", "metabolic bottlenecks", "rate-limiting enzymes", "pacemakers", etc. If rate-limiting steps do not exist, one cannot increase yields by increasing the activities of the enzymes that catalyze them. Thus, the example of tryptophan synthesis discussed at the beginning of this chapter ceases to be a mysterious result, but just illustrates what any student of metabolic control analysis would have predicted.

3.3 Treatment of Large Activity Changes

As noted above, a control coefficient is defined as a derivative, and can thus be treated as a constant only if the parameter change is very small, say a few percent at most. Biotechnologists, however, are normally interested in much larger changes than this, and maybe tempted to see hope in the fact that control coefficients change as conditions change. For a biotechnologist wanting to decrease or eliminate a flux, this hope is not without foundation, because the flux control coefficient of an enzyme for the flux through its own reaction always increases as its catalytic activity decreases, becoming unity when the catalytic activity is zero. However, this effect works in the wrong direction if one is interested in increasing fluxes or yields: the non-constancy of control coefficients makes matters worse, not better. Thus even if one manages to find an enzyme with a flux control coefficient greater than 0.9, one cannot expect that cloning it and increasing its activity by a factor of 100 will results in a more than 63-fold (i.e., $100^{0.9}$-fold) increase in flux; one will be lucky to achieve a 2-fold increase. The problem is that the high control coefficient reflects a shortage of the enzyme in the starting conditions, but as its concentration (or activity) increases it ceases to be in short supply.

3.4 The Role of Cooperative Feedback Inhibition

3.4.1 Theory

KACSER and BURNS (1973) made an interesting analysis of the role of feedback inhibition, but many recent articles in the field of metabolic control analysis give so little attention to the classical regulatory mechanisms that one can easily obtain the impression that they are regarded as having no relevance to modern control theory. Such an impression could certainly explain part of the slowness with which metabolic control analysis has been accepted by biochemists, as anyone who has studied the self-consistent picture of regulation given by the classical approach will not easily believe that it consists of a set of chance properties that have no relevance to how fluxes and concentrations are actually regulated. (A more important explanation, however, may be that some papers on metabolic control analysis suggest that their authors are more interested in matrix algebra than in biology.)

HOFMEYR and CORNISH-BOWDEN (1991) have tried to redress the balance by analyzing the role of cooperative feedback inhibition in the context of metabolic control analysis. They point out that much of metabolism can be categorized in terms of supply and demand, so that almost any pathway can be separated into a supply block and a demand block. For tryptophan biosynthesis, for example, the supply block consists of the five enzymes studied by NIEDERBERGER et al. (1992), and the demand block consists mainly of the consumption of tryptophan by tryptophanyl-tRNA synthetase and the reactions of protein synthesis. Biosynthetic pathways are commonly shown in textbooks as in Fig 1a, however, without any mention of the demand block. This is unfortunate, because the regulatory design, which transfers control from the supply block to the demand block, cannot be understood unless the demand for the end product is made explicit, as in Fig 1b of this chapter or Fig. 2 of NIEDERBERGER et al. (1992) (compare these with, for example, Fig. 24-13 and the schemes on pp. 588–589 of STRYER, 1988). In the specific case of tryptophan biosynthesis, the flux control coefficients shown at the bottom of Tab. 1 have a sum of 0.258, i.e., only a little more than a quarter of the control of the flux through the pathway lies within the pathway itself, if this is represented in the usual way.

An end product such as tryptophan in Fig. 1b acts as the boundary between its supply block and its demand block, and transmits information about changes in demand to the supply block. We can define $\varepsilon_{\mathrm{Trp}}^{\mathrm{supply}}$ and $\varepsilon_{\mathrm{Trp}}^{\mathrm{demand}}$ as the block elasticities expressing how the supply and demand rates in isolation respond respectively to the concentration of tryptophan. If these were separate pathways, these block elasticities would be response coefficients of the two fluxes to tryptophan, which would then be a pool metabolite and not an intermediate, but here we are concerned with the complete system.

As supply elasticities are normally negative, because of product inhibition, whereas demand elasticities are normally positive, it is convenient to consider the ratio $Q = -\varepsilon_{\mathrm{Trp}}^{\mathrm{supply}}/\varepsilon_{\mathrm{Trp}}^{\mathrm{demand}}$, which is normally positive and corresponds to the *loop gain* as used in control engineering. Analysis of systems similar to Fig. 1b shows that the distribution of flux control depends on this ratio as follows:

$$C^{J}_{\mathrm{demand}}/C^{J}_{\mathrm{supply}} = Q = -\varepsilon_{\mathrm{Trp}}^{\mathrm{supply}}/\varepsilon_{\mathrm{Trp}}^{\mathrm{demand}} \tag{9}$$

Thus, flux control can be transferred from supply to demand by increasing the sensitivity of the supply to end product or decreasing the sensitivity of demand to end product. As far as flux is concerned, these are equivalent, but, as emphasized above, regulation is not only a matter of varying fluxes; it is also a matter of keeping concentrations as constant as possible while doing so. The control of the tryptophan concentration does not depend on the ratio of the absolute block elasticities, but on their sum, i.e.,

$$C^{\mathrm{Trp}}_{\mathrm{supply}} = -C^{\mathrm{Trp}}_{\mathrm{demand}} = 1/(-\varepsilon^{\mathrm{supply}}_{\mathrm{Trp}} + \varepsilon^{\mathrm{demand}}_{\mathrm{Trp}}) \quad (10)$$

Thus, although decreasing the demand elasticity may improve flux regulation, it works against stabilizing the end-product concentration. Increasing the absolute value of the supply elasticity, on the other hand, benefits both objectives. One may expect, therefore, that regulation should arise in biological systems by the evolution of mechanisms to increase the supply elasticity, and this is exactly what cooperative feedback inhibition does.

Although ordinary product inhibition back through the supply block will by itself tend to increase the supply elasticity, it is very feeble unless the block is very short, because each intermediate in turn has to accumulate sufficiently to inhibit the enzyme that produces it. In very short pathways, such as serine biosynthesis, where feedback inhibition occurs in bacteria but not in mammals, this may be sufficient (Fell and Snell, 1988), but this is likely to be the exception. By acting directly on the first enzyme of the block, which often has one of the largest flux control coefficients in the block, feedback inhibition goes directly to the target and avoids the need for intermediates to accumulate. The role of cooperativity is simply to increase the elasticity of the interaction between metabolite and enzyme.

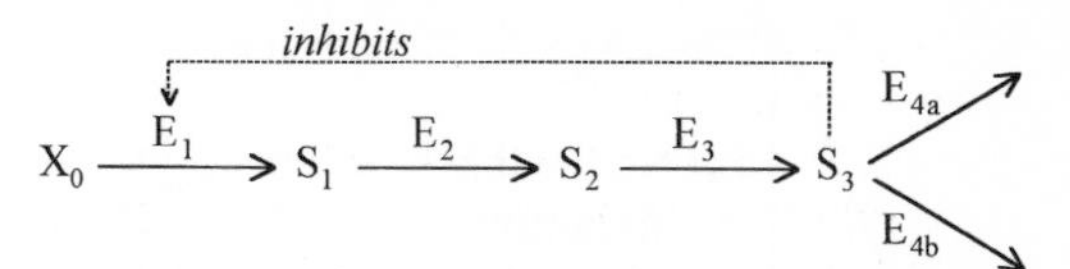

Fig. 2. Simple branched model for studying regulatory effects. There are two demand blocks, represented by the enzymes E_{4a} and E_{4b}, and effects of changes in demand in these two pathways can be simulated by varying the catalytic activities of these enzymes and examining the resulting changes in the steady-state fluxes and concentrations of intermediates. The steady states were calculated using the metabolic modeling program MetaModel (Cornish-Bowden and Hofmeyr, 1991), assuming the following rate equations:

$$v_1 = (10x_0 - s_1)/(1 + x_0 + s_1 + s_3^h)$$

$$v_2 = (10s_1 - 2s_2)/(1 + s_1 + s_2)$$

$$v_3 = (10s_2 - 2s_3)/(1 + s_2 + s_3)$$

$$v_{4a} = V_{4a}s_3/(1 + s_3)$$

$$v_{4b} = V_{4b}s_3/(1 + s_3).$$

The pool concentration x_0 was held constant at a value of 10. When there was no feedback inhibition, the term s_3^h was omitted from the first equation, and when it was present the exponent h had a value of either 1 (non-cooperative) or 4 (cooperative).

3.4.2 Numerical Example

The principles presented in the previous section can best be developed by reference to a numerical example. Hofmeyr and Cornish-Bowden (1991) examined the effect of cooperativity in a simple unbranched pathway, but it is more instructive to consider a pathway in which there are two demand blocks competing for the end product (Fig. 2), and to examine not only the importance of cooperativity but also the effect of having no feedback inhibition at all, as in Cornish-Bowden et al. (1994).

A pathway such as that in Fig. 2 may be regarded as well regulated if it displays the following properties: (1) the flux through either of the demand blocks responds sensitively to changes in the demand in that block, as indicated by the catalytic activity of the demand-block enzymes; (2) the flux through either demand block is hardly affected by changes in demand in the other block; (3) changes in flux occur with almost no changes in the concentration of the end product S_3. Fig. 3 illustrates the extent to which the first two of these are satisfied when there is (a) no feedback inhibition of E_1 by S_3, (b) non-cooperative feedback inhibition of E_1 by S_3, and (c) cooperative feedback inhibition of E_1 by S_3. It is striking that there is almost no difference between these three plots. They are not quite superimposable, but qualitatively they are the same: regardless of whether there is feedback inhibition or not, each demand flux increases sensitively (though not proportionally) when demand for it increases, but is little affected by changes in demand in the other branch. This leads to the surprising conclusion that neither feedback inhibition nor cooperativity are necessary for flux regulation, at least in this sim-

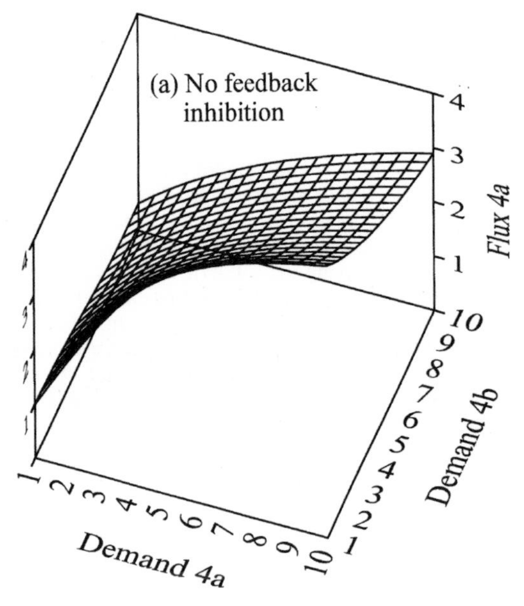

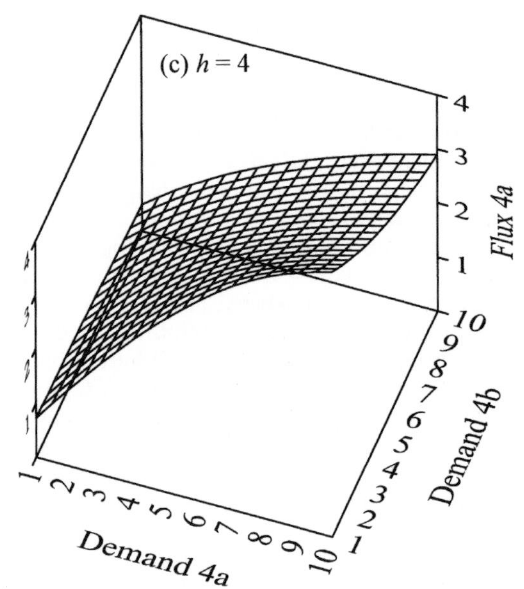

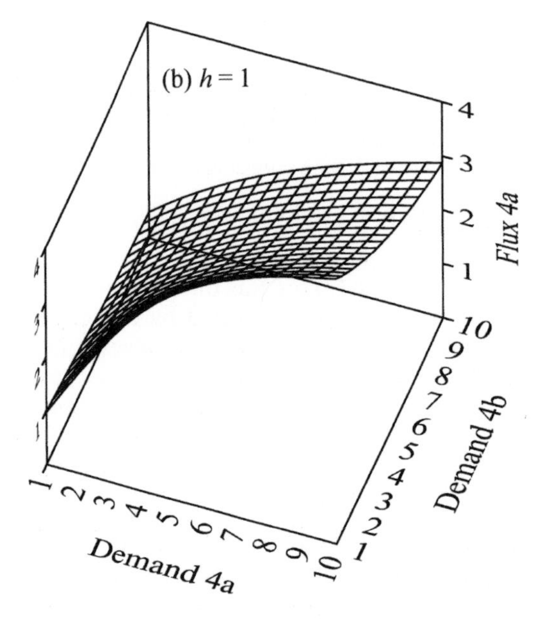

Fig. 3. Effect of feedback inhibition and cooperativity on metabolic fluxes. The model of Fig. 2 was studied with the values of V_{4a} and V_{4b} varied independently in the range 1 to 10, with (a) no feedback inhibition of E_1 by S_3, (b) non-cooperative feedback inhibition, and (c) cooperative feedback inhibition.

ple model: fluxes can be regulated quite satisfactorily without either of the classical regulatory mechanisms, and when they are included in the model, they make hardly any difference to the results!

Biochemists trained before 1975 will be shocked at the suggestion that the classical regulatory mechanisms have nothing to do with regulation, and will probably find it impossible to accept. However, as emphasized earlier in this chapter, regulation is not just a matter of fluxes; it is also a matter of concentrations, and when one examines the behavior of the concentration of end product in the same three cases, the results are very different (Fig. 4). When there is no feedback inhibition (Fig. 4a), the variation in flux when demand in both blocks increases by a factor of 10 is achieved at the expense of a more than 250-fold fall in the concentration of end product; when there is non-cooperative feedback inhibition (Fig. 4b), this fall becomes about 73-fold, a substantial improvement but still far from a stable concentration; when there is cooperative feedback inhibition (Fig. 4c), it becomes less than 7-fold.

These numbers, of course, are specific to the model and the particular conditions con-

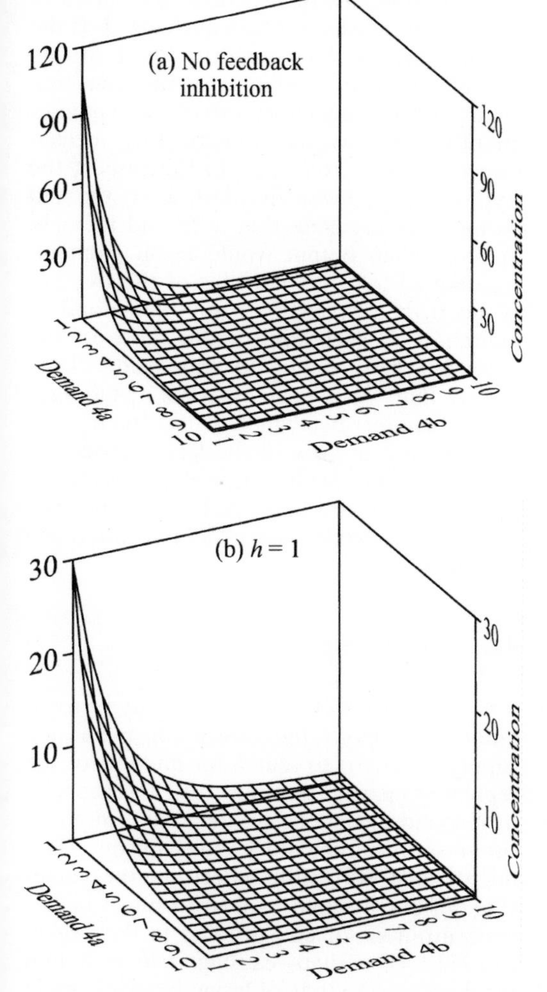

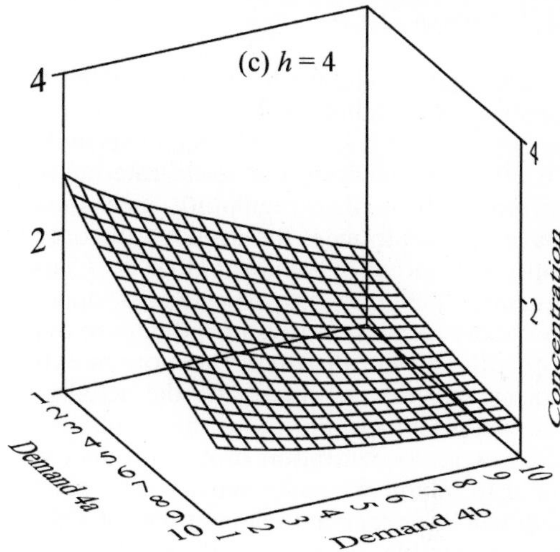

Fig. 4. Effect of feedback inhibition and cooperativity on the end-product concentration. The conditions were the same as in Fig. 3.

sidered, but qualitatively the same conclusions apply generally: neither feedback inhibition nor cooperativity make an important contribution to flux regulation, but they minimize the variations in end-product concentration that accompany the variations in flux. Thus, only by recognizing the duality of regulation, varying one effect while minimizing variations in another, is it possible to understand the role of the classic regulatory mechanisms.

4 Manipulation of Metabolic Fluxes on an Industrial Scale

4.1 Attitudes to Regulatory Mechanisms

The possible ways for the biotechnologist to take account of regulatory mechanisms may be classified as *opposition, oblivion, evasion, suppression*, and *subversion*. The most common approach in practice is opposition, and, as illustrated in Tab. 1, it does not work. Previous suggestions from metabolic control analysis amount to consigning regulation mechanisms to oblivion or evading them by ensuring that they are not triggered by changes in metabolite concentration: although they may work, they may not be the simplest or best. The analysis in this chapter suggests that either suppression or subversion will be preferable, and as subversion takes advantage of existing regulatory mechanisms whereas suppression does not, it is probably the better of the two. The use of politically suggestive terms is not accidental, but is intended to have mnemonic value.

4.2 Opposition

The first conclusion to be drawn from the analysis of metabolic regulation given above is obvious, but it has largely been overlooked in efforts to identify and clone rate-limiting enzymes. It is that regulatory mechanisms have evolved to meet the needs of the organism, not those of biotechnologists. Consequently, they can be expected to oppose, probably very efficiently, any attempt to overcome them by brute force. The more one tries to increase flux by increasing the activity of the supply block, the more the resulting increases in concentration of end product will tend to inhibit the same activity. The net result will be as seen in the upper part of Tab. 1: a considerable amount of effort without significant result.

4.3 Oblivion

The literature of metabolic control analysis has to date offered two suggestions about how to improve yields of commercially valuable products. WESTERHOFF and KELL (1987) suggest that the answer lies in carrying out a full control analysis of the system of interest, followed by inspection of the resulting control matrix to decide which enzyme activities and which elasticities ought to be modified by genetic manipulation. Although this approach appears valid, making a complete control analysis of even a short pathway is a far from trivial task, and presents formidable problems when the pathway contains more than four or five enzymes, though a recently proposed approach (CORNISH-BOWDEN and HOFMEYR, 1994) ought to decrease these. Moreover, as it largely ignores the classical regulation mechanisms, one may doubt whether it is likely to be the easiest or most cost-effective.

4.4 Evasion

NIEDERBERGER et al. (1992) make a similar suggestion, which is developed more fully in KACSER and ACERENZA (1993). The essential idea is that the activities of enzymes leading into and out of branches should be altered by fractions that increase the fluxes of interest but leave all others unchanged. If the calculations are done correctly, there are no changes in any metabolite concentrations that might trigger regulatory effects, and consequently the existence of regulatory mechanisms becomes irrelevant. In the case of the tryptophan pathway in yeast, KACSER and ACERENZA calculate that a 25-fold increase in tryptophan output would result from increasing 25-fold the activities of the five enzymes from chorismate to tryptophan and simultaneously increasing 5-fold those of the eight enzymes leading to chorismate. They believe that "there is no aspect of metabolism which cannot be handled by [their] approach", but in view of the large amount of precise genetic manipulation that it requires, potential users are likely to be cautious until its effectiveness has been demonstrated experimentally.

4.5 Suppression

One way to overcome the capacity of regulation to oppose the effects of increasing supply activity is to search for mutants of the regulated enzyme that have lost the capacity to respond to the end product, i.e., simply to suppress the regulatory mechanisms. This might have very little effect on the fluxes (Fig. 3), but it ought to lead to much higher concentrations of end product in the organism (Fig. 4), perhaps causing it to leak into the medium. A study of lysine biosynthesis in *Escherichia coli* by DAUCE-LE REVEREND et al. (1982) supports this idea. These authors used a plasmid pDA1 to increase about 30-fold the activity of dihydrodipicolinate synthetase, the putative rate-limiting enzyme in lysine biosynthesis, both in a wild-type strain RM4102 and in a mutant TOCR21 that lacked feedback inhibition of aspartokinase III by lysine. Unfortunately, the rate of lysine production was too low in the wild type to be measured, with or without the plasmid, and so it is not possible to estimate how much the increased enzyme activity increased the flux to lysine. Significantly, however, lysine secretion in the mutant increased by about 60% (in large-scale tests) when the plasmid was present. Although this is very small in relation to the increase in enzyme activity, it is much larger than any of the increases in tryptophan flux discussed earlier that could be obtained by increasing activities of single enzymes.

4.6 Subversion

Suppressing the regulatory mechanisms is, however, rather crude; more subtle would be to pervert the regulatory mechanisms to biotechnological ends. If, as suggested above, the primary role of metabolic regulation is to transfer control from the supply block of a biosynthetic pathway to the demand block, the solution is obvious. One should increase demand by manipulating or introducing a non-transformational secretory process, and leave the organism to solve the problems of supply. If the desired end product is actively secreted by the natural strain of the organism, this would be particularly simple, as one could try to increase the activity of the secretion process. It is likely to be exceptional for the desired product to be secreted naturally, but the same idea could in principle be applied to a product that is not ordinarily secreted by seeking mutants that cause it to leak the medium. Active secretion is much better than passive secretion or leaking, of course, because a passive process will never allow the extracellular concentration to rise above the intracellular concentration, which, in this approach, will be the physiological concentration at best; in practice it will be slightly less than the physiological concentration, because of the increase in demand.

Increasing demand cannot work miracles, of course, and in particular it cannot produce a flux greater than the smallest limiting rate of the enzymes in the supply block. That is why, in Fig. 3, the flux cannot increase indefinitely as both demands increase: the largest value of the flux in step 4a of Fig. 3a is about 2.9, with the same flux in step 4b, making a total flux of about 5.8, or 58% of saturation, as each of the three supply enzymes has the same limiting rate of 10. Once this plateau is approached, the flux ceases to be controlled by demand, and one must turn back to manipulation of the supply to increase it further. However, this would require large increases in the activities of all three of the supply enzymes. All in all, it may be realistic to increase fluxes by a factor of ten above their ordinary values, but it would be very optimistic to hope for much more than this.

Acknowledgement

This chapter was written during tenure of the Cátedra Hermann Niemeyer Fernández at the Faculty of Sciences, University of Chile, Santiago.

5 References

Atkinson, D. E. (1990), What should a theory of metabolic control offer to the experimenter? in: *Control of Metabolic Processes* (Cornish-Bowden, A., Cárdenas, M. L., Eds.), pp. 3–11, New York: Plenum Press.

Cornish-Bowden, A. (1989), Metabolic control theory and biochemical systems theory: different objectives, different assumptions, different results, *J. Theor. Biol.* **136,** 365–377.

Cornish-Bowden, A. (1994), Metabolic control analysis in theory and practice, *Adv. Mol. Cell. Biol.* **12,** in press.

Cornish-Bowden, A., Hofmeyr, J.-H. S. (1991), MetaModel: a program for modelling and control analysis of metabolic pathways on the IBM PC and compatibles, *Comp. Appl. Biosci.* **7,** 89–93.

Cornish-Bowden, A., Hofmeyr, J.-H. S. (1994), Determination of control coefficients in intact metabolic systems, *Biochem. J.* **298,** 367–375.

Cornish-Bowden, A., Cárdenas, M. L., Hofmeyr, J.-H. S. (1994), Vers une meilleure compréhension de la régulation métabolique, *Images Rech.* **2,** 179–182.

Crabtree, B., Newsholme, E. A. (1987), A systematic approach to describing and analysing metabolic control systems, *Trends Biochem. Sci.* **12,** 4–12.

Dauce-Le Reverend, B., Boitel, M., Deschamps, A. M., Lebeault, J.-M., Sano, K., Takinami, K., Patte, J.-C. (1982), Improvement of *Escherichia coli* strains overproducing lysine using recombinant DNA techniques, *Eur. J. Appl. Microbiol. Biotechnol.* **15,** 227–231.

Fell, D. A. (1992), Metabolic control analysis: a

survey of its theoretical and experimental development, *Biochem. J.* **286,** 313–330.

Fell, D. A., Snell, K. (1988), Control analysis of mammalian serine biosynthesis: feedback inhibition on the final step, *Biochem. J.* **256,** 97–101.

Heinrich, R., Rapoport, T. A. (1974), A linear steady-state treatment of enzymatic chains: general properties, control and effector strength, *Eur. J. Biochem.* **42,** 89–95.

Heinrich, R., Rapoport, S. M., Rapoport, T. A. (1977), Metabolic regulation and mathematical models, *Prog. Biophys. Mol. Biol.* **32,** 1–82.

Hofmeyr, J.-H. S., Cornish-Bowden, A. (1991), Quantitative assessment of regulation in metabolic systems, *Eur. J. Biochem.* **200,** 223–236.

Kacser, H., Acerenza, L. (1993), A universal method for achieving increases in metabolite production, *Eur. J. Biochem.* **216,** 316–367.

Kacser, H., Burns, J. A. (1973), The control of flux, *Symp. Soc. Exp. Biol.* **27,** 65–104.

Koshland, D. E., Jr., Némethy, G., Filmer, D. (1966), Comparison of experimental binding data and theoretical models in proteins containing subunits, *Biochemistry* **5,** 365–385.

Michaelis, L., Menten, M. L. (1913), Kinetik der Invertinwirkung, *Biochem. Z.* **49,** 333–369.

Monod, J., Wyman, J., Changeux, J.-P. (1965), On the nature of allosteric transitions: a plausible model, *J. Mol. Biol.* **12,** 88–118.

Niederberger, P., Prasad, R., Miozzari, G., Kacser, H. (1992), A strategy for increasing an *in vivo* flux by genetic manipulations: the tryptophan system of yeast, *Biochem. J.* **287,** 473–479.

Savageau, M. A. (1990), Biochemical systems theory: alternative views of metabolic control, in: *Control of Metabolic Processes* (Cornish-Bowden, A., Cárdenas, M. L., Eds.), pp. 69–87, New York: Plenum Press.

Schaaff, I., Heinisch, J., Zimmermann, F. K. (1989), Overproduction of glycolytic enzymes in yeast, *Yeast* **5,** 285–290.

Stadtman, E. R. (1970), Mechanisms of enzyme regulation in metabolism, in: *The Enzymes*, 3rd Ed. (Boyer, P. D., Ed.) Vol. 1, pp. 397–459, New York: Academic Press.

Stryer, L. (1988), *Biochemistry*, 3rd Ed., pp. 575–600, New York: W. H. Freeman and Co.

Taketa, K., Pogell, B. M. (1965), Allosteric inhibition of rat liver fructose 1,6-diphosphatase by adenosine 5′-monophosphate, *J. Biol. Chem.* **240,** 651–662.

Umbarger, H. E. (1956), Evidence for a negative-feedback mechanism in the biosynthesis of isoleucine, *Science* **123,** 848.

Westerhoff, H. V., Kell, D. B. (1987), Matrix method for determining steps most rate-limiting to metabolic fluxes in biotechnological processes, *Biotechnol. Bioeng.* **30,** 101–107.

Yates, R. A., Pardee, A. B. (1956), Control of pyrimidine biosynthesis in *Escherichia coli* by a feed-back mechanism. *J. Biol. Chem.* **221,** 757–770.

4 Analytical Uses of Enzymes

GEORG-BURKHARD KRESSE

Penzberg, Federal Republic of Germany

1 Introduction

Enzymes are proteins with catalytic activity which are highly specific both in the reaction catalyzed as well as in their choice of substrates.

Due to their specificity, enzymes have been used as analytical tools for the selective determination of the concentration of substances for many decades (BERGMEYER and GAWEHN, 1983). These analytes may either be chemically transformed in the presence of an enzyme (i.e., as substrates), or may modulate enzyme activity in a manner related to their concentration (i.e., as activators or inhibitors).

Enzymes also serve as "markers" in assay techniques based on non-enzymatic interactions, especially antigen–antibody binding (enzyme immunoassay) or DNA-oligonucleotide hybridization.

The quantitation of enzyme activities in biological samples is also part of the methodology termed *Enzymatic Analysis* (BERGMEYER et al., 1983–1986).

Enzymes are also widely used in some other applications related to analytical problems, such as in DNA sequencing and gene technology (reviewed by KESSLER, 1987, 1990) and in protein and glycoconjugate sequence determination (reviews: ALLEN, 1989; JACOB and SCUDDER, 1994). These specialized topics will not be dealt with in this chapter.

(Besides antibodies, enzymes are the most specific reagents known. The use of enzymes in analysis therefore has a number of advantages as compared to chemical reagents)

- high specificity and selectivity for the analyte (especially isomeric and enantiomeric specificity)
- thus low effort for tedious sample preparation or pretreatment even from complex mixtures
- mild reaction conditions (aqueous solutions of near-neutral pH, ambient temperature) suitable for labile analytes
- non-hazardous, environmentally acceptable reagents
- reasonable cost due to simple instrumentation and low working time
- assays amendable for automation.

This chapter presents the fundamentals of enzymatic analysis in a concise manner. For more details, also on the special assay systems mentioned as examples, the reader is referred to reference monographs, e.g., BERGMEYER et al. (1983–1986).

2 Fields of Application

The main fields of application of enzymatic analysis today are still biochemistry, clinical chemistry, and food analysis. However, its importance is also growing in organic chemistry, pharmacology, pharmaceutical chemistry, and environmental analysis. In contrast to earlier belief, many enzymes are active not only in aqueous solution, but also in organic solvents (DORDICK, 1989; GUPTA, 1992). This should allow the use of enzymes for enzymatic analysis procedures under low-water conditions although, up to now, only few examples have been reported.

The acceptance and further increase of use of enzymatic assays critically depends on the commercial availability of "analytical grade" enzymes or even complete test kits and systems; otherwise, the convenience and speed of enzymatic procedures would be compromised by the time and labor involved in the multi-step purification of enzymes from biological materials. Fortunately, an increasing number of highly purified enzymes (see Sect. 9.3) as well as test combinations or complete analytical systems, such as biosensors, can be obtained from several suppliers.

3 Techniques of Measurement

In analytical systems designed for measuring substrate concentration, enzymes can be used for analyte recognition as well as for generation of a concentration-dependent signal which can then be detected by a suitable measuring principle. Among these, the most popular and convenient methods include absorption photometry, fluorometry, and electrochemical measurements (biosensors).

Optical methods are often used for enzymatic assays because of the simple technique and the availability of reliable, reasonably priced instruments. Photometric tests can be carried out in a quick and convenient way when a substrate or product absorbs light in the visible or UV range, by following the change of absorbance at a selected wavelength with a spectrophotometer. In many cases, inexpensive spectral-line photometers or filter photometers will be sufficient.

The dependence of light absorption on substrate concentration, in dilute solutions, is given by Beer's law

$$A = \varepsilon \cdot c \cdot d \quad (1)$$

where A is the absorbance (dimensionless), ε the molar absorption coefficient (L mol^{-1} mm^{-1}), c the concentration (mol L^{-1}) and d the path length (mm).

With ε values of frequently used absorbing substances such as NAD(P)H, 4-nitrophenolate, or quinonimine dyes, it is easily possible to quantitate analytes in the 10^{-3} to 10^{-5} mol L^{-1} range.

In recent years, fluorometry has also been widely used in enzymatic analysis. Using similar instrumentation as photometry, fluorescence measurements have the advantage of being more sensitive, in many cases by a factor of more than 10^3 (Tab. 1). Furthermore, since only a limited number of components give a fluorescence signal, the method usually is highly selective. In some cases, naturally occurring enzyme substrates possess fluorescence properties which change during the enzyme reaction, so that the conversion can be detected directly in a fluorometer. On the other hand, for many enzymes, especially hydrolases (proteases, glycohydrolases, etc.), fluorescing substrate analogs are available.

In dilute solutions, fluorescence intensity is linearly proportional to the concentration of the analyte (just as light absorption). However, the measured fluorescence does not only depend on the electronic structure of the fluorescent molecule, but also on quenching effects (radiationless transition from the excited to the ground state) heavily influenced by the surrounding solvent, as well as on instrumental changes such as lamp fluctuation. Therefore, the results of fluorometric assays usually cannot be calculated from tabulated parameters (such as ε in photometry), but require standardization. Luminescence measurements are used increasingly in enzyme immunoassays (see MAYER and NEUENHOFER (1994) and references cited therein).

Electrochemical methods (biosensors) are described in Sect. 7.2.

Tab. 1. Detection Limits of Enzymatic Methods (Modified from BERGMEYER et al. 1983)

Measuring Technique	Approximate Detection Limit (mol/L sample)
Photometry	
– end-point methods	$10^{-6} \ldots 10^{-5}$
– kinetic methods	$10^{-7} \ldots 10^{-6}$
– catalytic methods	$10^{-9} \ldots 10^{-8}$
Fluorometry	
– end-point methods	$10^{-9} \ldots 10^{-8}$
– enzymatic cycling	$10^{-15} \ldots 10^{-14}$
Luminometry	$10^{-13} \ldots 10^{-8}$
Enzyme immunoassay	
– heterogeneous	$10^{-13} \ldots 10^{-10}$
– homogeneous	$10^{-10} \ldots 10^{-5}$

4 Determination of Analyte Concentrations with End-Point Methods

4.1 Simple End-Point Reactions

In the most simple case, the analyte takes part as the substrate in an enzyme-catalyzed reaction and is converted practically completely with the simultaneous and stoichiometric production of a detectable signal (e.g., change in absorbance) allowing the direct determination of the analyte concentration.

Example 1: Determination of uric acid ($\varepsilon_{293\,nm} = 12.6 \cdot 10^2$ L mol^{-1} mm^{-1}) with urate oxidase (uricase). The product allantoin does not absorb at 293 nm wavelength, so the decrease in absorbance at this wavelength, after complete conversion, is directly correlated to the concentration of the urate initially present:

$$\text{Urate} + O_2 + 2H_2O \xrightarrow{\text{urate oxidase}} \text{Allantoin} + H_2O_2 + CO_2 \tag{2}$$

Example 2: Determination of triglycerides by hydrolysis with lipases. Triglyceride is converted into glycerol and free fatty acids which can be titrated directly with base, e.g., NaOH:

$$\text{Triglyceride} + 3H_2O \xrightarrow{\text{Lipase}} \text{Glycerol} + 3CH_3\text{-}(CH_2)_n\text{-}COOH \tag{3}$$

$$3CH_3\text{-}(CH_2)_n\text{-}COOH + 3NaOH \longrightarrow 3CH_3\text{-}(CH_2)_n\text{-}COO^- + 3Na^+ + 3H_2O \tag{4}$$

The same principles, of course, apply if not the analyte conversion by itself but the stoichiometrically linked conversion of a co-substrate or cofactor can be detected. This is the case in the large number of dehydrogenase-catalyzed assays which make use of the absorption of the reduced coenzyme NADH (or NADPH, respectively) at 340 nm (Fig. 1) for detection.

With common laboratory equipment, the detectability of NADH-based photometric assays is in the range of 10^{-3} to 10^{-4} mol L^{-1}. If NADH is measured fluorometrically, the detection limit may be as low as 10^{-8} to 10^{-9} mol L^{-1}.

Example 3: Determination of pyruvate with lactate dehydrogenase. The decrease of absorbance at 340 nm is measured and correlated, by means of the molar extinction coefficient $\varepsilon_{340\,nm} = 6.3 \cdot 10^2$ L mol^{-1} mm^{-1}, to the consumption of NADH which is stoichiometrically equal to the initial pyruvate concentration:

$$\text{Pyruvate} + \text{NADH} + H^+ \xrightarrow{\text{Lactate dehydrogenase}} \text{Lactate} + \text{NAD}^+ \tag{5}$$

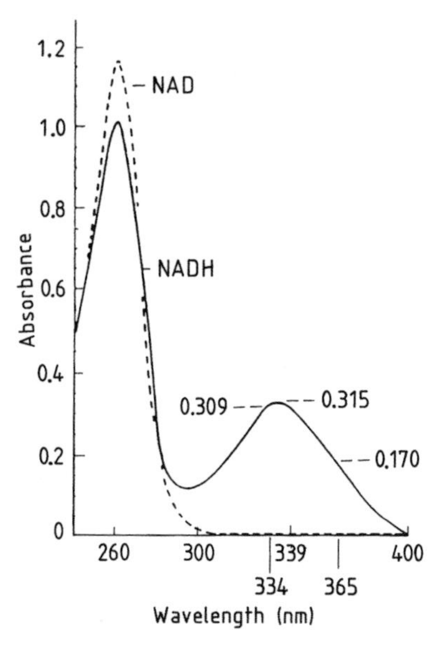

Fig. 1. Absorbance of oxidized and reduced nicotinamide-adenine dinucleotide (NAD^+/NADH). Concentration: $5 \cdot 10^{-5}$ mol^{-1}. From MATTENHEIMER (1982), with permission.

The experimental design of enzymatic analyses of this kind is simple. The reaction is allowed to proceed until completion, and the result can easily be calculated from known physical constants, e.g., the molar absorption coefficient in the case of light-absorbing substances (Fig. 2).

It is desirable that the incubation times during the assay are as short as possible, e.g., several minutes, in order to minimize non-enzymatic side reactions and to enhance convenience. For substrate concentrations well below the numerical value of the Michaelis constant K_m, reaction rate obeys the following equation:

$$v = [S] \cdot \frac{V}{K_m} \tag{6}$$

with v, the reaction rate; [S], the substrate concentration; V, the maximum activity.

Hence, the time required for 'quantitative' (i.e., $\geq 99\%$) substrate conversion will depend critically on the K_m and V values of the enzyme used. If only an enzyme with a high K_m or a low V value is available, this must be

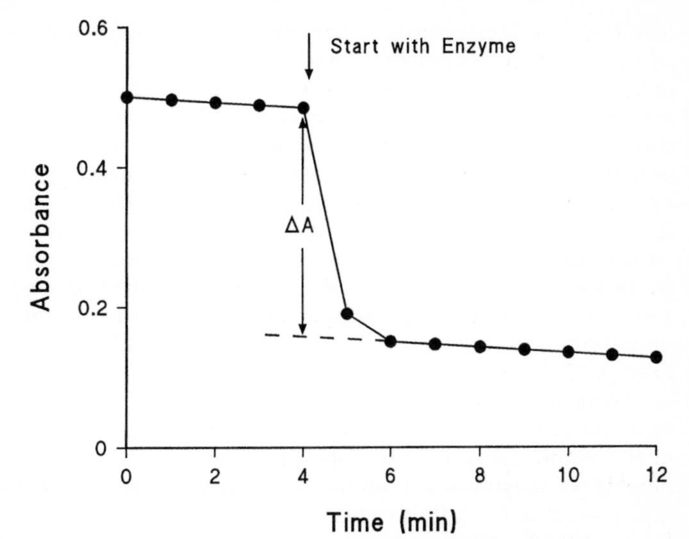

Fig. 2. Schematic time course of the enzymatic determination of a substrate by an end-point method (modified from BERGMEYER, 1983a). The change in absorbance (ΔA) is measured and correlated to substrate conversion according to Beer's law.

Tab. 2. Examples of Practicable Enzyme Concentrations for End-Point Methods with Given K_m Values (Modified from BERGMEYER, 1983a)

Analyte	Enzyme	K_m (mol/L)	V/K_m[a] = 1 mL/min is given by
ADP	Adenylate kinase	1.6×10^{-3}	1600 U/L
Glucose	Hexokinase	1.0×10^{-4}	100 U/L
Glycerol	Glycerol kinase	5.0×10^{-5}	50 U/L
Uric acid	Urate oxidase	1.7×10^{-5}	17 U/L
Fumarate	Fumarase	1.7×10^{-6}	1.7 U/L

[a] The value of V/K_m should be of the order of 1 mL/min if 99% conversion of the substrate is to take only a few minutes.

compensated by applying a higher enzyme concentration (as measured in units of catalytic activity per unit of assay volume); cf. Tab. 2.

4.2 Assay Systems Involving Coupled Reactions; Indicator Systems

In many cases, none of the reactants or products of the enzymatic reaction lend themselves readily to physical or chemical measurement so that no detectable signal is produced. In these cases, the primary (called 'auxiliary') reaction has to be coupled to a stoichiometrically linked 'indicator' reaction (which is very often also enzyme-catalyzed), with one of the products of this second reaction being easily detectable.

Often the auxiliary and indicator reactions can be performed simultaneously in the same assay mixture. The more common sequence is to measure the product of the auxiliary reaction in a succeeding indicator reaction, e.g., in glucose determination with hexokinase and glucose-6-phosphate dehydrogenase.

Example 4: Glucose determination with hexokinase (HK) and glucose-6-phosphate dehydrogenase (G6P-DH). The hexokinase reaction is not easily amenable to detection, but coupling to the G6P-DH reaction allows for stoichiometric NADH formation:

$$\text{Glucose} + \text{ATP} \xrightarrow{\text{HK}} \text{Glucose-6-phosphate} + \text{ADP} \tag{7}$$

$$\text{Glucose-6-phosphate} + \text{NAD}^+ + \text{H}_2\text{O} \xrightarrow{\text{G6P-DH}} \text{6-Phosphogluconate} + \text{NADH} + \text{H}^+ \tag{8}$$

This example also shows one important feature of coupled enzymatic assay systems:

hexokinase is an enzyme with very broad substrate specificity, accepting a large number of hexoses as substrates, whereas G6P-DH is highly specific for glucose-6-phosphate. In the coupled system, it is sufficient to employ an enzyme with narrow substrate specificity in only one of the steps, because the specificity of the whole reaction sequence only depends on the most specific enzyme. This means that, in practice, general (unspecific) indicator reactions can be coupled without individual optimization to various specific auxiliary reactions.

Especially in clinical chemistry, the use of indicator reactions for photometric detection in the visible range is very popular. A large number of diagnostically important metabolites present in body fluids can be determined in a technically simple way by use of specific H_2O_2-forming oxidase reactions coupled to chromogenic indicator systems (see MICHAL et al., 1983). Among these, the most common indicator reactions are those in which either a single chromogen (a leuco-dye, e.g., 3,5,3′,5′-tetramethylbenzidine) is converted to a colored substance by a peroxidase-catalyzed oxidation, or in which two chromogens are oxidatively coupled in the presence of peroxidase as in the reaction first described by TRINDER in 1969 where phenol is coupled with 4-aminophenazone to produce a quinonimine dye absorbing in the visible range at about $\lambda = 500$ nm (see example 5 below).

Due to the low substrate specificity of horseradish peroxidase, phenol can be replaced by a large number of phenolic or anilinic derivatives (MICHAL et al., 1983). The molar extinction coefficient of the resulting dyes depends on the substituents chosen, thus the lower detection limit of assays using these indicator systems is in the range of 10^{-4} to 10^{-6} mol L^{-1}.

Example 5: For determination of serum total cholesterol, i.e., the sum of free cholesterol and cholesterol esterified with fatty acids, cholesterol esters are first hydrolyzed with cholesterol esterase (auxiliary reaction 9). Cholesterol is oxidized with cholesterol oxidase to give Δ^4-cholesten-3-one and hydrogen peroxide (auxiliary reaction II). In the indicator reaction, H_2O_2 is then used for the peroxidase-catalyzed coupling of "Trinder"-type chromogens to form the dye molecule:

$$\text{Cholesterol ester} + H_2O \xrightarrow{\text{Cholesterol esterase}} \text{Cholesterol} + \text{Fatty acid} \tag{9}$$

Cholesterol —(Cholesterol oxidase; O_2 → H_2O_2)→ Δ^5-Cholestenone —(Isomerization)→ Δ^4-Cholestenone (10)

$2\,H_2O_2$ + 4-Aminophenazone + Phenol —(Peroxidase)→ 4-(p-Benzoquinone-monoimino)-phenazone + $4\,H_2O$ (11)

Sometimes, one of the reactants of an auxiliary reaction is produced *in situ* through a preceding indicator reaction rather than being applied in excess as a co-substrate. In this case, this indicator reaction has to be a reversible (equilibrium) reaction, e.g., the malate dehydrogenase reaction in example 6. Whereas the results of assays involving succeeding indicator reactions can be calculated in a straightforward manner due to the simple stoichiometries involved, calculation is somewhat more complicated in the case of preceding indicator reactions, since K_{eq} of the indicator reaction has to be taken in account (for details see BERGMEYER, 1983a).

Example 6: Determination of acetyl-coenzyme A with malate dehydrogenase and citrate synthase. Oxaloacetate which is the co-substrate of the citrate synthase reaction is formed *in situ* through the malate dehydrogenase reaction whose equlibrium concentrations must be reached before citrate synthase is added. Oxaloacetate consumed by condensation with acetyl-CoA is then replenished with simultaneous formation of NADH (measured parameter, NADH absorbance) through the malate dehydrogenase reaction:

$$\text{Malate} + \text{NAD}^+ \xrightarrow{\text{Malate dehydrogenase}} \text{Oxaloacetate} + \text{NADH} + \text{H}^+ \quad (12)$$

$$\text{Acetyl-CoA} + \text{Oxaloacetate} \xrightarrow{\text{Citrate synthase}} \text{Citrate} + \text{CoA-SH} \quad (13)$$

4.3 Reactions with Unfavorable Equilibria

If, in an enzyme-catalyzed reaction, the substrate conversion is incomplete because of an unfavorable equilibrium constant, an endpoint determination is only possible if the resulting substrate and product concentrations can be influenced, either by increasing the initial concentrations of a second substrate or by trapping one of the reaction products by irreversible (or quasi-irreversible) chemical or enzyme-catalyzed sequel processes. This is illustrated by the assay system used for determination of ethanol.

Example 7: In the presence of alcohol dehydrogenase, ethanol is converted into acetaldehyde with concomitant reduction of NAD^+. Due to the unfavorable equilibrium constant ($K_{eq} = 8.0 \cdot 10^{-12}$ mol L^{-1}), complete conversion of the analyte requires an excess of the co-substrate NAD^+, an alkaline pH (9.0) and trapping of acetaldehyde either chemically with, e.g., semicarbazide or tris(hydroxymethyl)aminomethane, or preferably enzymatically using aldehyde dehydrogenase:

$$\text{CH}_3\text{-CH}_2\text{OH} + \text{NAD}^+ \underset{}{\overset{\text{Alcohol dehydrogenase}}{\rightleftharpoons}} \text{CH}_3\text{-CHO} + \text{NADH} + \text{H}^+ \quad (14)$$

$$\text{CH}_3\text{-CHO} + \text{NAD}^+ + \text{H}_2\text{O} \xrightarrow{\text{Aldehyde dehydrogenase}} \text{CH}_3\text{-COOH} + \text{NADH} + \text{H}^+ \quad (15)$$

In this special case, the 'trapping' reaction also leads to the stoichiometric formation of NADH, so that the detected signal intensity ($\Delta A_{340\text{ nm}}$) is doubled.

5 Determination of Analyte Concentrations with Kinetic Methods

5.1 First-Order Reactions

If the substrate concentration [S] is much smaller than the Michaelis constant K_m of the enzyme, $[S] \ll K_m$, it follows from the Michaelis–Menten equation

$$v = \frac{V \cdot [S]}{K_m + [S]} \quad (16)$$

that the observed reaction rate v is linearly proportional to [S]. Thus, by monitoring the reaction kinetics, i.e., either the disappearance of substrate, $-d[S]/dt$, or the formation of product, $+d[P]/dt$, the analyte concentration can be determined with the aid of a calibration curve under these conditions.

Kinetic assays allow a drastic reduction of the time required for analysis and are less sensitive to interferences (e.g., turbidity, intrinsic color of the sample) than end-point assays. On the other hand, for rate measurements the reaction conditions (e.g., temperature, pH, amount of enzyme) must be held constant carefully. Therefore, kinetic assays are especially advantageous for use with automated analyzers rather than for assays performed manually.

Substrate concentration determination by rate measurement requires that the initial substrate concentration is low as compared to the K_m value of the enzyme. In order to increase the dynamic concentration range of the assay, enzymes with high K_m values (low substrate affinity) therefore are preferred in kinetic assays in contrast to end-point methods. If such enzymes are not available, it may be possible to increase K_m by the addition of a competitive inhibitor I; in this case, the apparent K_m value is given by

$$K_m^{app} = K_m \left(\frac{1 + [I]}{K_I} \right) \tag{17}$$

where K_I is the dissociation constant of the enzyme–inhibitor complex) and, thus, may be increased considerably as compared to K_m (SIEDEL et al., 1983).

Example 7: Kinetic determination of glucose using the hexokinase/glucose-6-phosphate dehydrogenase method (cf. example 4) (DEEG et al., 1980). Because of the low K_m value of G6P-DH with respect to glucose-6-phosphate ($6.4 \cdot 10^{-5}$ mol L^{-1}), the method can be only used if the apparent K_m value is increased by the addition of high concentrations ($1.7 \cdot 10^{-2}$ mol L^{-1}) of ATP which is a competitive inhibitor of G6P-DH. Under these conditions, the fixed-time absorbance change is a linear function of the glucose concentration (Fig. 3).

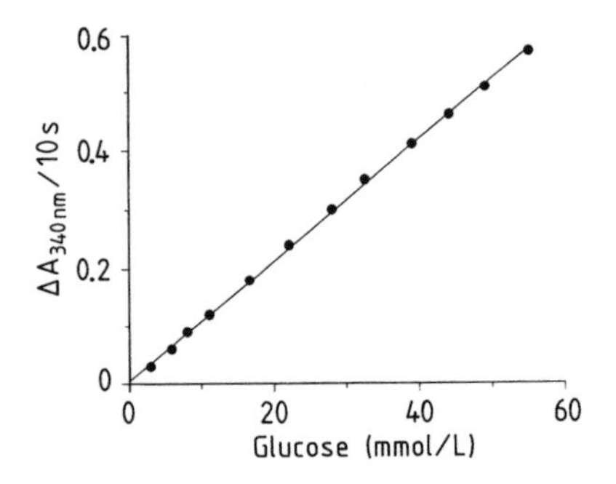

Fig. 3. Kinetic substrate determination: fixed-time absorbance change as a function of glucose concentration measured using an automated analyzer. Measuring interval: 10 s. Modified from DEEG et al. (1980).

5.2 Zero-Order Reactions

Under conditions where the substrate concentration $[S] \gg K_m$ (substrate saturation), it follows from the Michaelis–Menten equation that the observed reaction rate v becomes independent of the substrate concentration (zero-order kinetics). Under these conditions, the enzyme activity cannot be used to determine the substrate concentration directly. However, in zero-order reactions a direct correlation exists between the reaction rate v and V, which corresponds to the total enzyme activity present in the assay mixture. Therefore, if a substance exerts an activating or inhibiting effect on the catalyst in a concentration-dependent manner, it is possible to determine its concentration indirectly by kinetic measurement of the enzyme activity under substrate saturation conditions (SIEDEL et al., 1983).

This principle has been used to design kinetic substrate concentration assays for, e.g., heparin (based on inhibition of thrombin), insecticides (by inhibition of acetylcholinesterase), theophyllin (by inhibition of alkaline phosphatase), lead (by inhibition of δ-aminolaevulinate dehydratase), and a number of ions able to enhance the activity of apoenzymes, e.g., Na^+ with β-galactosidase, Mg^{2+} with isocitrate dehydrogenase or glucokinase, Cu^{2+} with apo-galactose oxidase, or chloride by activation of α-amylase (for references, see SIEDEL et al., 1983; KRESSE and SCHMID, 1995).

Example 8: Determination of K^+ with pyruvate kinase (PK). Muscle PK activity is stimulated by K^+ ions. Thus, to determine K^+ concentration, PK activity is assayed by measuring, with potassium-free reagents, the rate of conversion of phospho-*enol*-pyruvate to pyruvate, which is immediately converted to lactate in an indicator reaction (measured parameter, increase in absorbance at 340 nm):

$$\text{Phospho-enol-pyruvate} + \text{ADP} \xrightarrow{\text{Pyruvate kinase}/K^+} \text{Pyruvate} + \text{ATP} \tag{18}$$

$$\text{Pyruvate} + \text{NADH} + H^+ \xrightarrow{\text{Lactate dehydrogenase}} \text{Lactate} + \text{NAD}^+ \tag{19}$$

PK is also activated by NH_4^+ and Na^+. In order to eliminate interference by these ions, NH_4^+ is removed by enzymatic conversion to glutamate with α-ketoglutarate and glutamate dehydrogenase, and Na^+ is selectively bound by a cryptand so that it is no longer available for PK activation (BERRY et al., 1989).

One potential drawback in these modulator-dependent assays is that for every test, exactly the same amount of enzyme activity has to be applied, so that the reproducibility of the results critically depends on enzyme stability under storage and assay conditions. This may be a problem especially with apoenzymes, since these are frequently not very stable on the shelf.

5.3 Catalytic Assays

The detectability of end point and kinetic substrate determinations is governed by physicochemical constants of the detected species, such as the molar absorption coefficient ε in photometry. In order to lower the detection limit, attempts have been made to design catalytic assays (also performed under zero-order conditions, i.e., with saturating concentrations of all substrates involved except for the target analyte) in which the analyte itself is recycled in a multi-reaction system and, due to its low concentration, is limiting for the overall rate (thus, acts as a catalyst). Examples include, among others, methods for the determination of malonyl-coenzyme A (TAKAMURA et al., 1985), phenylpyruvate (COOPER et al., 1989), carnitine (TAKAHASHI et al., 1994) and coenzyme A (MICHAL and BERGMEYER, 1985).

Example 9: Determination of coenzyme A in a catalytic assay. The assay mixture contains acetyl phosphate, malate, NAD^+, phosphotransacetylase (PTA), citrate synthase (CS), and malate dehydrogenase (MDH). The measured parameter is the rate of NADH formation (measured photometrically or fluorometrically) during malate oxidation which is dependent on the concentration of coenzyme A, since all other reactants and enzymes are present in non-limiting amounts. Calculation of the coenzyme A concentration has to be done with the aid of a calibration curve.

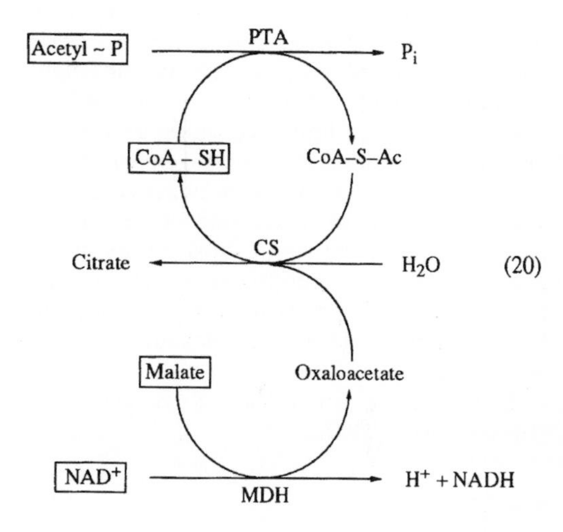

(20)

A number of assay systems based on catalytic cycling have been described for the determination of $NAD(P)^+/NAD(P)H$ or analogs (e.g., LOWRY et al., 1961; SELF, 1985; BERGEL et al., 1989). With these methods, signal amplification by a factor of at least 20, but up to 5000 is possible. Since these techniques are used mainly in connection with enzyme immunoassays, they are described in some more detail in Sect. 8.4.

In catalytic substrate determination systems, enzymes and other reagents of extremely high purity have to be used, since impurities may interfere with the complete reaction system or may even be amplified. Detection limits as low as 10^{-9} to 10^{-13} mol L^{-1} can be reached, but because of their complexity, assays of this type are seldom used in routine clinical chemistry analysis.

6 Determination of the Catalytic Activity of Enzymes

6.1 General Remarks

According to BERGMEYER and GAWEHN (1983), the assay of enzyme activities in biological materials, especially in the field of clinical enzymology, is also part of the methodology called "enzymatic analysis". A detailed overview on enzyme assays has been published (EISENTHAL and DANSON, 1993) where reviews of the various measurement techniques (e.g., photometric, radiometric, high performance liquid chromatography, or electrochemical assays) can also be found.

The catalytic activity of an enzyme is measured as the rate of conversion of a substrate (not necessarily the same substrate as in the living cell!) under optimized *in vitro* conditions in order to approximate the theoretically maximum rate (V in the Michaelis–Menten equation).

The rate of conversion $\dot{\xi}$ is given by (BERGMEYER, 1983c)

$$\dot{\xi} = \frac{d\xi}{dt} \, (\text{mol s}^{-1}) \tag{21}$$

where ξ is the extent of reaction.

The rate of reaction v is the rate of conversion, related to the volume of the analytical system (assay volume V_t):

$$v = \frac{\dot{\xi}}{V_t} = \frac{dc}{dt} \, (\text{mol L}^{-1} \text{ s}^{-1}) \tag{22}$$

with c as the substance concentration. In practice, one always measures time-dependent changes in concentration. Thus, if the assay volume is known, the change in substrate amount per unit time can be calculated.

The International Union of Biochemistry (IUB) has defined the unit of enzyme activity (1 U) as the amount of enzyme catalyzing the conversion of 1 μmol of substrate per minute under specified conditions; the SI unit (1 katal [kat]), equal to the amount of enzyme converting 1 mol of substrate per second, is still rarely used because it is somewhat impractical for most enzymes (1 U = 16,67 nkat [nanokatal]). For many enzymes, historical or arbitrary definitions of activity units have survived in the literature (sometimes even several different unit definitions for the same enzyme) so that, in order to compare activity values from different laboratories, it is necessary to check for the special reaction conditions and unit definition used by the authors. Unfortunately, in many cases these different "units" cannot even be converted by simple recalculation if determined under different assay conditions.

6.2 Factors Influencing Enzyme Activity

Temperature. Due to the dependence of chemical reaction rates on temperature, the reaction mixture has to be thermostated to a constant value, e.g., 25°C (as recommended by IUB in 1961) or 30°C (as recommended in 1964), or 37°C (as common in many laboratories although not recommended by IUB). Enzymes from thermophilic organisms will have to be assayed at higher temperature since their activity, in the 25–30°C range, often is very low.

The temperature coefficient of enzyme-catalyzed reaction rates may be as large as 10% per degree centigrade, i.e., for a temperature rise of 1°C, the value found for the catalytic activity of an enzyme may be about 10% too high (Fig. 4).

Proteins are intrinsically unstable against heat, although the temperature range of thermal stability and the rate of thermal inactivation are very different for different enzymes. Therefore, the measured enzyme activity (at a given temperature) may be influenced by the duration of pre-incubation at this temperature, as well as by the reaction time itself (unless true initial activities are measured): although the initial reaction velocity will increase according to the Arrhenius equation, this may be compensated by time-dependent thermal inactivation (Fig. 5). Therefore, it is not useful to give "temperature optimum"

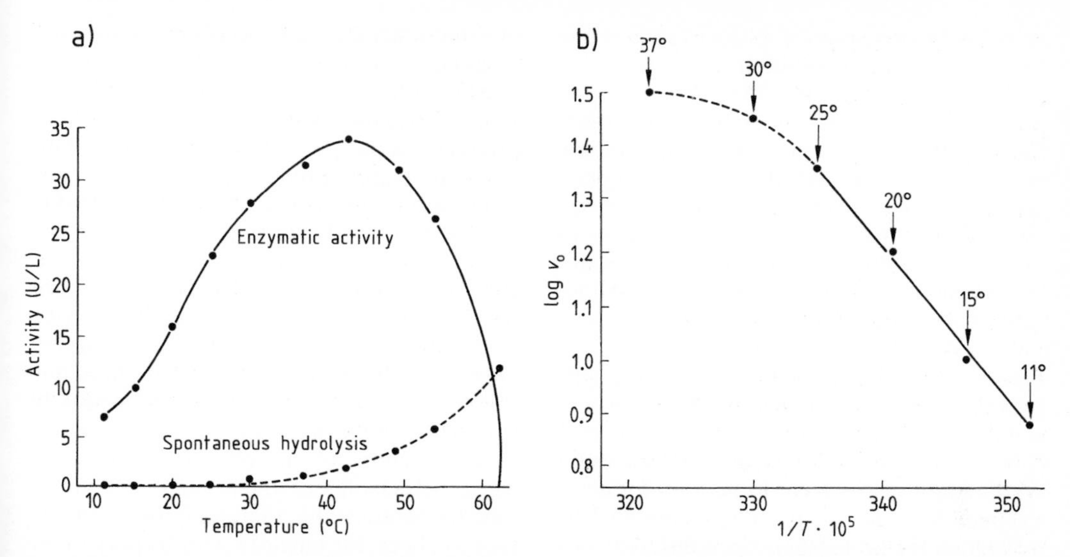

Fig. 4. Dependence of γ-glutamyl transpeptidase activity on temperature. a) Dependence of enzyme activity on assay temperature; b) Arrhenius plot of data from A. v_0, initial reaction velocity; T, absolute temperature. Modified from Cousins (1976).

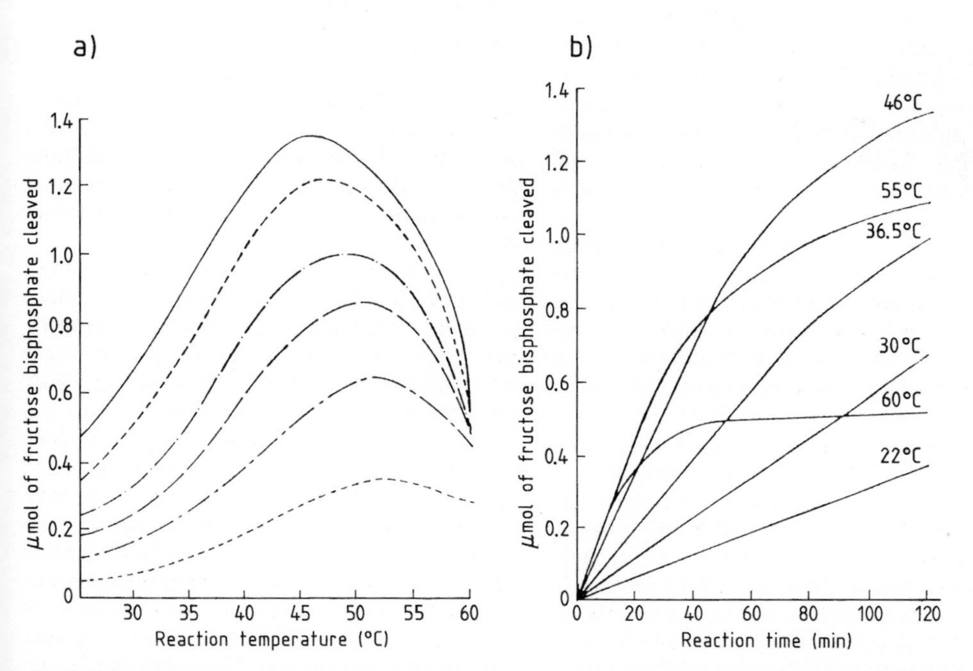

Fig. 5. Dependence of serum aldolase activity on time of incubation at different temperatures.
a) Activity was measured at various reaction temperatures after different preincubation periods at the given temperature (———, 120 min; – – – –, 90 min; —·—, 60 min; — — —, 45 min; ——–, 30 min; ----------, 22 min).
b) Data from A, plotted differently to show the effect of different reaction temperatures on the initial velocity. Modified from Cousins (1976).

values without reference to the exact reaction and pre-incubation conditions.

Buffer and pH values. The optimum pH range for activity varies greatly between different enzymes and may also depend on temperature and choice of substrate. The type of buffer may also affect catalytic activity. Choosing the optimal buffer and pH value is especially important when coupled systems are used for activity determination (see Sect. 6.4).

Choice of substrate. In many cases, the substrate physiologically converted by an enzyme is either not known, not available, or its conversion is not easily detectable. In these cases, synthetic substrates have to be used for activity measurement (e.g., oligopeptide *p*-nitroanilides for proteases). Since different substrates will be converted at different rates, the choice of substrate is of decisive importance for the numerical activity values.

Substrate concentration. In order to approach the maximum activity, V in the Michaelis–Menten equation, Eq. (6), enzyme activity assays have to be run under "substrate saturation" conditions, $[S] \gg K_m$, to achieve pseudo-zero-order kinetics (independence of the reaction rate on remaining substrate concentration).

In practice, the possible range of substrate concentration is limited due to solubility, physicochemical properties (e.g., light absorption, substrate inhibition effects), and in some cases also by the cost of the substrate. From the Michaelis–Menten equation, the initial substrate concentrations which will result in a tolerable deviation from V can be calculated (BERGMEYER, 1983b); e.g., if 1% deviation is accepted ($v = 0.99\ V$), [S] has to be $99 \cdot K_m$ in the simplest case of a one-substrate reaction. For two-substrate reactions, there is not only one single optimum condition with respect to substrate concentration, but the same v/V ratio can be obtained using an infinitely large number of pairs of substrate concentrations (for details, see BERGMEYER, 1983b).

Effectors. Enzyme activity is influenced by inhibitors and activators. Activators may have to be added to the assay mixture to achieve maximum activity (e.g., thiols in the assay of creatine kinase), whereas inhibitors may be present either in the sample or in the reagents or may arise during the reaction in the case of product inhibition. In this latter case, inhibition can be minimized by removal of the inhibiting species in a consecutive (uncatalyzed or enzyme-catalyzed) reaction.

Enzyme concentration. In most enzyme-catalyzed reactions, the initial velocity is proportional to the concentration of the enzyme. However, there are some cases where this simple relationship does not hold; this may be the case when the assay mixture contains small amounts of an enzyme inhibitor or when dissociable enzyme activators or inhibitors are present. It is therefore always important to check for linearity with respect to enzyme concentration.

6.3 Assay Techniques: Direct Assays

Many enzyme-catalyzed reactions result in changes of the properties of the reactants that are relatively easily measured directly and continuously. In these cases, the progress curve for the reaction can be followed directly.

It is generally necessary to measure the initial rate of the enzyme-catalyzed reaction since with proceeding substrate consumption, the conversion rate will no longer be linear with time due to substrate exhaustion, product accumulation, and approximation to the thermodynamic equilibrium (Fig. 6). The reaction therefore must be made to proceed slowly so that only a small portion of the substrate is converted during the measuring period; often, enzyme solutions have to be diluted appropriately. It is not uncommon to observe "blank" reaction rates in the absence of either the enzyme or its substrate, especially with biological samples. These may have several reasons (e.g., settling of particles, adsorption effects on the vessel wall, non-enzymic processes, or presence of contaminants). The sample itself may also contain the target enzyme together with its substrate, but may lack

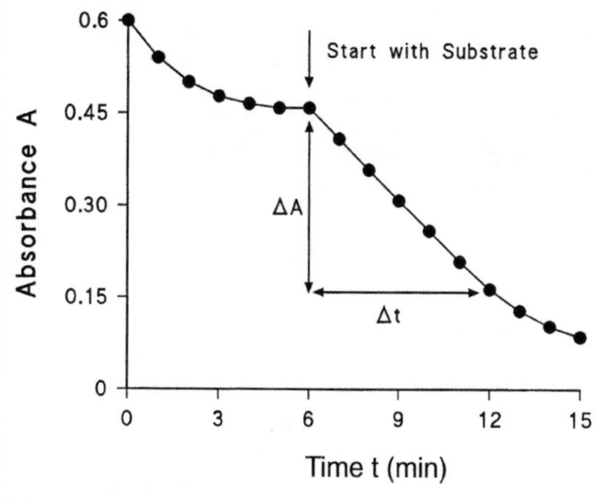

Fig. 6. Determination of the catalytic activity of an enzyme in a direct continuous photometric assay (schematic). Measured parameter, $\Delta A/\Delta t$. Modified from BERGMEYER (1983c).

an essential cofactor. If these blank reactions cannot be completed during a short pre-incubation time (as in Fig. 6), it is necessary to correct for this blank rate in an appropriate way (see EISENTHAL and DANSON, 1993).

6.4 Indirect and Coupled Assay Systems

If the primary enzymatic reaction cannot be measured directly, the reaction mixture has to be treated either to produce a measurable product, or to improve the sensitivity or convenience of the assay procedure. In discontinuous assays, the reaction is stopped after a fixed time, and the amount of product formed is analyzed by any suitable method (e.g., high performance liquid chromatography). In assays of this type, it is especially important to check signal linearity with time, since in many cases the reaction may not run under "initial rate" conditions for the full assay time.

Alternatively, coupled reaction systems can be used where the target enzyme reaction is followed by an uncatalyzed or enzyme-catalyzed indicator reaction. In consecutive reaction systems, the substrate is converted to an intermediate first which will accumulate to some extent before the coupling enzyme has reached its steady-state velocity for formation of the final, detectable product. Coupled reaction systems therefore inevitably show a lag in the rate of formation of the product (Fig. 7). In order to minimize the lag period, the indicator enzyme should have a low K_m value so that it will be able to work efficiently at low substrate (intermediate) concentrations. Furthermore, it is essential that the primary reaction (catalyzed by the target enzyme to be assayed) is rate-limiting, so that for all auxiliary and indicator reactions, v must be much greater. Thus, these other enzymes have to be added in very large excess (BROOKS and SUELTER, 1989).

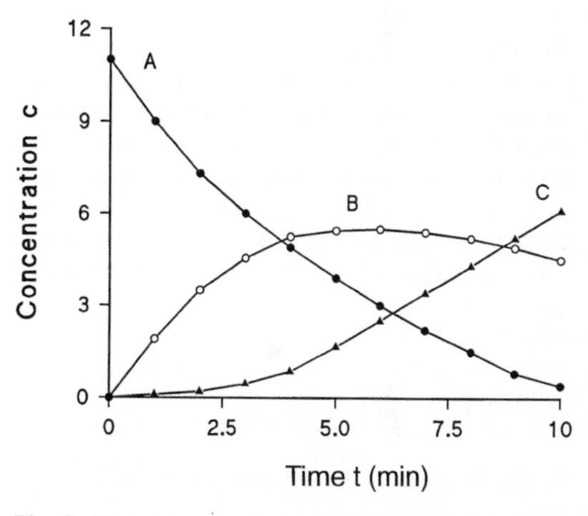

Fig. 7. Time course of a coupled enzyme assay involving a single coupling enzyme. Curve A, concentration of substrate; curve B, concentration of intermediate (product of target enzyme reaction); curve C, concentration of final product of indicator reaction which is detected.

Example 10: Determination of α-amylase activity in a coupled assay with α-glucosidase as the coupling enzyme (RAUSCHER et al., 1985). The endoglycosidase α-amylase cleaves the chromogenic substrate, *p*-nitrophenyl(pNP)-maltoheptaoside (G_7-pNP), at several glycosidic bonds within the chain, giving rise to the formation of free and *p*-nitrophenylated oligosaccharides of different chain length. Due to its substrate specificity, the exoglycosidase α-glucosidase (indicator enzyme) will then release *p*-nitrophenolate with

60% 31% 9%

Maltotriose + pNP-Tetraoside

Maltotetrose + pNP-Trioside

Maltopentaose + pNP-Dioside

α-Glucosidase + H_2O

5 Glucose + 2 p-Nitrophenolate

(23)

detectable velocity only from G_2-pNP and G_3-pNP (which amount to about 40% of the products of the primary reaction). p-Nitrophenolate is the detected species due to its absorption at 410 nm. Even if p-nitrophenol is dissociated, at pH 7.1, by only 50%, the assay has to be run at this pH because it corresponds to the pH optimum of α-amylase.

7 "Dry Reagent Chemistry" Systems and Biosensors

7.1 Test Strips

In order to facilitate handling of the reagents used for substrate concentration as well as enzyme activity determination, attempts have been made to use immobilized enzymes for analytical purposes. Since, as known for a long time, labile enzymes often can be stabilized by immobilization, this will also contribute to enhance the shelf-life of the reagents. Furthermore, in some cases it may be possible to re-use expensive enzymes for series of analyses when immobilized.

An early and still useful approach is to adhere the enzyme non-covalently to paper. This allows the design of the well-known test strips for use in biological samples such as urine or blood. These systems include, in dried state, the enzyme(s) as well as all the necessary substrates and buffer substances so that, after application of a droplet of sample, the reaction can occur resulting in the formation of colored products. These can be detected either by simple comparison with a reference color table obtained with several standard concentrations of the analyte, or (in technologically more advanced systems) quantitatively by reflectance photometry (WERNER and RITTERSDORF, 1983). Commercial systems such as Ektachem^R (Kodak) or Reflotron^R (Boehringer Mannheim) are available which include many of the parameters of interest in clinical chemistry (SONNTAG, 1993). These assays (though developed and optimized for use in human diagnostics) may be conveniently used also for other applications such as fermentation control; this, however, requires evaluation of possible sources of error due to sample ingredients not present in human blood or serum.

7.2 Biosensors

Biosensors are miniaturized devices consisting of a biological substrate recognition system, e.g., an enzyme, antibody, receptor, nucleic acid, or cell, connected to a transducer component which converts the biological (or physicochemical) interaction into an electronic signal which is then amplified and processed (Fig. 8). Transducers can be electro-

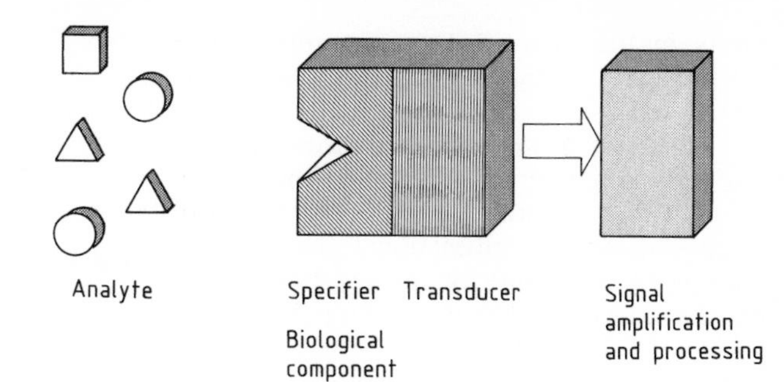

Fig. 8. General concept of a biosensor.

chemical, optical, mechanical, acoustic, or calorimetric, although electrochemical biosensors with amperometric or potentiometric detection are most common.

The first enzyme electrode described by CLARK and LYONS (1962) was a soluble-enzyme amperometric biosensor for the determination of blood glucose employing, as the specifier enzyme, membrane-entrapped glucose oxidase. Hydrogen peroxide, the product of enzyme-catalyzed oxidation of glucose, was detected with a platinum electrode whose potential was set at the appropriate polarity.

A major breakthrough was reached when HIGGINS and HILL (1985) described the use of electrochemical mediators, such as ferrocene, in order to lower the detection potential of amperometric oxidase-based biosensor assays:

$$\text{D-Glucose} + \text{Ferricinium salt} \xrightarrow{\text{Glucose oxidase}} \text{D-}\delta\text{-Gluconolactone} + \text{Ferrocene} \quad (24)$$

This system not only was less sensitive to interference from electrochemically active substances present in the sample (e.g., ascorbic acid), but also resulted in independence of oxygen concentration. Amperometric measurement devices based on disposable ferrocene-mediated glucose oxidase electrodes have been introduced into the U.S. and European market for diabetes home monitoring.

Recent trends and advances in biosensor development have been described in several reviews (SCHELLER and SCHUBERT, 1989; SCHELLER and SCHMID, 1992; TURNER et al., 1990).

Major advantages of biosensors as compared to "conventional" enzymatic assays are based on the rapid progress in electronics, leading to the possibilities of system miniaturization, increase in sensitivity, and cost reduction. Drawbacks, however, are limited stability and reproducibility, and problems concerning large-scale production process technology. Although exploding market sizes for biosensors had been expected, progress towards marketable products has been sluggish up to now, and if biosensor technology is an intellectually stimulating area of scientific research, it is not really clear at present whether it will in fact be of significant functional advance for applications such as laboratory diagnostics.

8 Enzyme Immunoassay

8.1 General Principles

In enzyme immunoassay (EIA), enzymes are used merely as "markers" for the detection of antibody (or hapten)–antigen interactions and thus compete with other labeling principles, such as radioisotopes (radioimmunoassay, RIA) or physicochemical labels (e.g., measurement by fluorescence polarization or luminescence). Since the first description of an immunoassay by YALOW and BERSON (1959), measurements based on immune reactions have become increasingly popular for determination of high (e.g., catalytically inactive proteins) as well as low molecular weight substances. Immunoassay systems are often quick, easy, and sensitive with detection limits in the nanogram to picogram range (OELLERICH, 1983, 1984; MASSEYEFF et al., 1993).

Conjugates of the marker enzyme with an antibody or hapten (depending on the type of assay) can be synthesized by chemical coupling with several bifunctional reagents according to the nature of the reactive groups on the enzyme (O'SULLIVAN et al., 1979). Compared to RIA, the use of enzymes as labels has the advantage that contact with radioactive materials is avoided, the label can be easily detected with commonly available laboratory equipment (e.g., spectrophotometers or fluorometers), and the reagents have a much longer shelf-life.

It should be mentioned here that enzyme-labeled antibodies can be used, similar as in sandwich enzyme immunoassays, for the non-radioactive detection of many other classes of biomolecules, such as specific DNA sequences (using DNA-oligonucleotide hybrid formation as the specific recognition principle), or carbohydrate residues of glycoproteins (using carbohydrate–lectin binding). These techniques have been reviewed by KESSLER (1992).

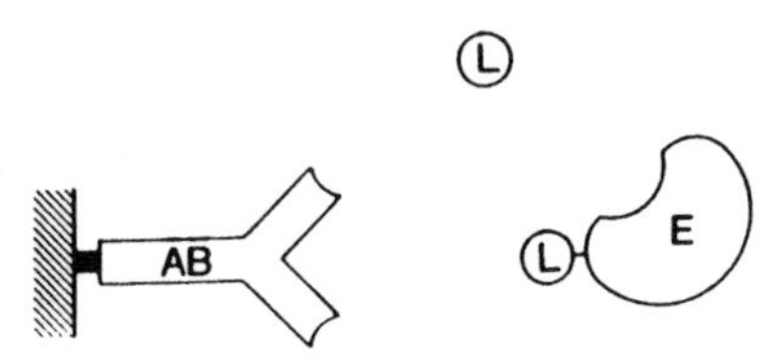

Fig. 9. Competitive enzyme immunoassay. L, ligand (analyte); E-L, ligand–marker enzyme conjugate (added in fixed concentration); AB, antibody against L present in limiting amounts. Depending on the concentration ratio, L and E-L will compete for the binding sites on AB. Solid-phase bound enzyme activity will decrease with increasing concentration of L. From OELLERICH (1983).

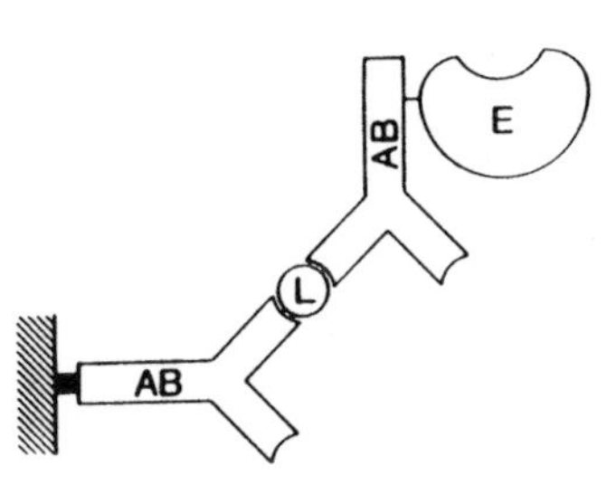

Fig. 10. "Sandwich" antigen ELISA. For abbreviations, see legend to Fig. 9; AB-E, antibody against L conjugated to marker enzyme E. L will be completely bound to immobilized AB (present in excess). In a second step, AB-E is bound to another epitope of AB-bound L. Solid-phase bound enzyme activity will increase with increasing concentration of L. From OELLERICH (1983).

8.2 Heterogeneous Enzyme Immunoassays

Assay systems in which the activity of the marker enzyme conjugate is not influenced by the antigen–antibody reaction require a separation step, usually between a liquid and a solid (vessel bound or precipitated) phase, and are therefore called heterogeneous EIAs. *Competitive* EIA principles (such as in the now classical radioimmunoassay technique of YALOW and BERSON) are often used (Fig. 9), whereas the use of "*sandwich*"-type assays (enzyme-linked immunosorbent assay = ELISA, Fig. 10) is limited to antigens large enough to accomodate binding of two antibodies simultaneously, one of them being labeled with the marker enzyme. The enzymes most often used as labels in heterogeneous immunoassays are calf intestine alkaline phosphatase, *E. coli* β-galactosidase, and horseradish peroxidase, due to their high specific activity, availability of convenient chromogenic substrates, good stability, and reasonable cost.

8.3 Homogeneous Enzyme Immunoassays

A separation step can be avoided if the enzymatic activity is influenced (inhibited or increased) as a consequence of the antigen–antibody binding. The choice of the marker enzyme very strongly depends on the special type of assay so that various enzymes have been used (OELLERICH, 1983, 1984). A large number of different assay principles have been described, of which only two are cited here as examples. As most homogeneous immunoassay techniques, both can only be used for small analytes, but not for large antigens (e.g., proteins).

In the now classical EMIT[R] test (enzyme multiplied immunoassay test, RUBENSTEIN et al., 1972), the activity of the marker enzyme is reduced when the hapten–enzyme conjugate is bound to the antibody (Fig. 11). With the addition of increasing amounts of free hapten from the sample to be analyzed, an increase of enzyme activity is observed due to competition for the limiting amounts of antibody.

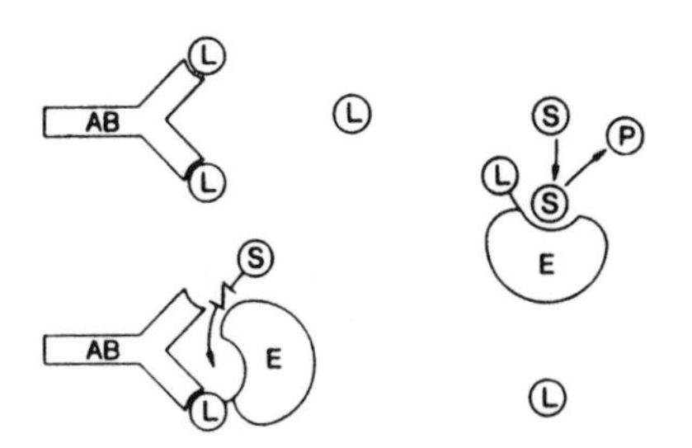

Fig. 11. Homogeneous EIA according to EMIT[R] test. For abbreviations, see legends to Figs. 9 and 10; P, product of marker enzyme reaction. Catalytic activity of E-L is inhibited if bound to AB. With increasing concentration of free analyte L (from sample), enzyme activity will increase. From OELLERICH (1983).

In the more recent CEDIA[R] (cloned enzyme donor immuno test; HENDERSON et al., 1986) test (Fig. 12), hapten or analyte is attached to a small (ED, enzyme donor) fragment of β-galactosidase produced by recombinant DNA technology. The hapten–ED conjugate spontaneously associates with a large (EA, enzyme acceptor) fragment to give enzymatically active β-galactosidase, but this aggregation is inhibited by analyte-specific antibodies present in limiting amounts. So again analyte from the sample will compete with the hapten–ED conjugate to give β-galactosidase activity proportional to the analyte concentration.

8.4 Enzymatic Signal Amplification Systems

In order to enhance the detectability of immunoassays, a number of enzymatic signal amplification systems have been described (BATES, 1987). For example, with a redox cycle based on the cycling of NAD (Fig. 13), the signal from the enzyme alkaline phosphatase can be amplified 100 to 5000-fold giving a detection limit of only 3000 molecules (SELF, 1985; JOHANSSON et al., 1985).

It should be emphasized that enzymatic systems of this or similar types will amplify the signal produced by the marker enzyme, but will not (in contrast to, e.g., the polymerase chain reaction) amplify the target molecule itself. As a consequence, blank or unspecific reaction products may also be amplified, and the signal-to-noise ratio may not be significantly better than with the original unamplified reaction system. In an attempt to improve this ratio, the amplification system was run in two steps; the accumulation of the triggering substance from the primary enzyme reaction was allowed to proceed for a fixed period of time before the components of the amplification system were added. In this way, for each molecule of NAD generated by the first reaction up to 500 molecules of formazan could be produced in about 10 minutes (STANLEY et al., 1985).

9 Enzymes Used for Analytical Purposes

9.1 Quality Requirements

Enzymes which shall be used in substrate assays have to fulfill a number of quality criteria concerning

- *specificity* (i.e., absence of side activities towards other substances which may be present in the sample or the assay mixture),
- *purity* (i.e., absence of contaminating activities or other contaminants interfering with the analytical and detection systems, not

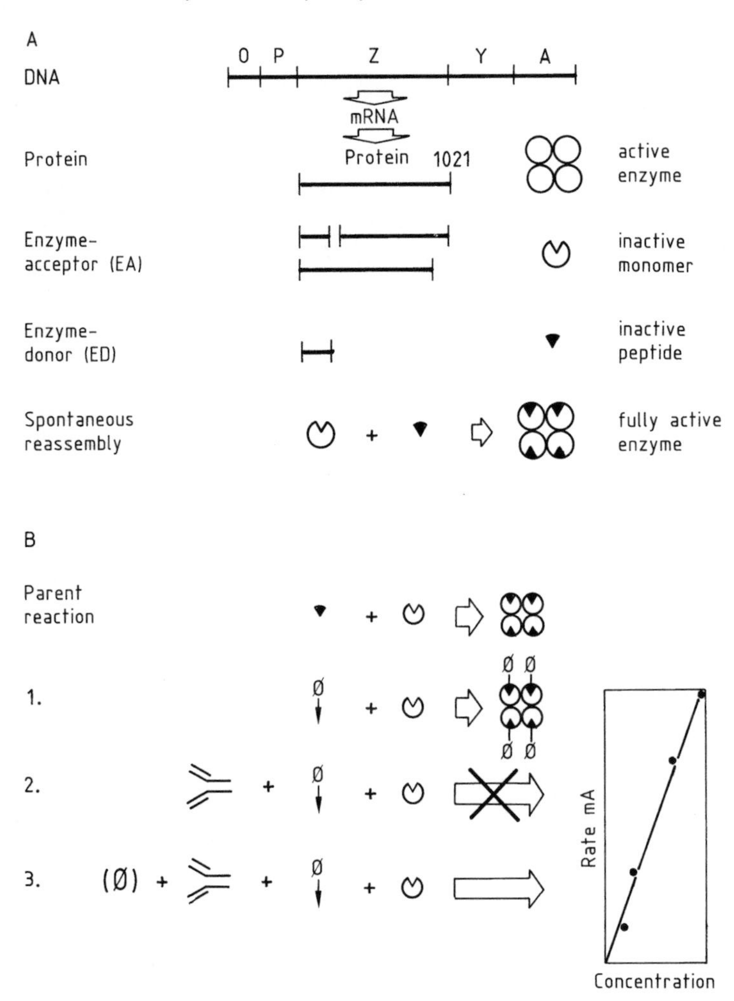

Fig. 12. Principle of the CEDIA^R test.
A: *lac* operon of *E. coli*. The *lac* operon consists of the operator, promotor, *Z*, *Y*, and *A* genes. The *Z* gene encodes β-galactosidase; mutant *Z* genes encode enzyme–acceptor (EA) and enzyme–donor (ED) fragments. Both EA and ED are enzymatically inactive but, when mixed, spontaneously associate to form enzymatically active β-galactosidase.
B: Principle of CEDIA^R immunoassay. An analyte (∅) is attached to ED so that the ∅-ED conjugate is able to recombine with EA to yield actice β-galactosidase (1). The addition of an analyte-specific antibody inhibits this enzyme assembly (2). Analyte in the sample will compete with ∅-ED conjugate for the limiting amount of antibody (3) so that, with increasing analyte concentrations in the sample, increasing amounts of ∅-ED conjugate will be available for active enzyme formation. The signal generated by β-galactosidase turnover therefore is directly proportional to analyte concentration. From HENDERSON et al. (1986), with permission.

necessarily purity with respect to the absence of other inactive proteins),
- *stability* (in the test mixture during the assay reaction as well as on long-term storage),
- *kinetic properties* (suitable K_m and k_{cat} values, no inhibition by substances present in the sample),
- *pH optimum* (suitable for the required assay conditions),

Tab. 3. List of Enzymes Used in Diagnostics and Food Analysis (Modified from KRESSE, 1990). For reference to the enzyme data listed, see KRESSE (1987, 1990)

Enzyme, E.C. number, and CAS registry number	Source	M_r, 10^3 dalton enzyme (subunit)	K_m, mol/L (substrate)	pH optimum	Isoelectric point	Substrate measured
Acetyl-CoA synthetase (acetate–CoA ligase) E.C. 6.2.1.1 [*9012-31-1*]	yeast	151 (78)	2.8 $\times 10^{-4}$ (acetate) 3.5 $\times 10^{-4}$ (CoA)	7.6	–	acetate
Adenylate kinase (myokinase) E.C. 2.7.4.3 [*9013-02-9*]	porcine muscle	21	5 $\times 10^{-4}$ (AMP) 3 $\times 10^{-4}$ (ATP) 1.58 $\times 10^{-3}$ (ADP)	ca. 7.5	6.1	AMP
Alcohol dehydrogenase E.C. 1.1.1.1 [*9031-72-5*]	yeast	141 (35)	1.3 $\times 10^{-2}$ (ethanol) 7.8 $\times 10^{-4}$ (acetaldehyde) 7.4 $\times 10^{-5}$ (NAD^+) 1.08 $\times 10^{-5}$ (NADH)	9.0	5.4–5.8	ethanol, other alcohols, aldehydes
Aldehyde dehydrogenase E.C. 1.2.1.5 [*9013-02-9*]	yeast	207 (57)	9 $\times 10^{-6}$ (acetaldehyde) 1.3 $\times 10^{-4}$ (NAD^+)	9.3	–	acetaldehyde, ethanol
D-Amino acid oxidase E.C. 1.4.3.3 [*9000-88-8*]	porcine kidney	100 (50 or 38–39)	1.8 $\times 10^{-3}$ (D-alanine) 1.8 $\times 10^{-4}$ (oxygen)	9.5	–	penicillin, hormones
Amyloglucosidase (glucamylase, glucan 1,4-α-glucosidase) E.C. 3.2.1.3 [*9032-08-0*]	*Aspergillus niger*	97 (48)	2.2 $\times 10^{-7}$ (amylopectin) 3.2 $\times 10^{-5}$ (amylose) 1.85 $\times 10^{-2}$ (maltose)	ca. 5.0	4.2	starch, glycogen
Ascorbate oxidase E.C. 1.10.3.3 [*9029-44-1*]	*Cucurbita* species	140 (65)	2.4 $\times 10^{-4}$ (L-ascorbate)	5.6–7.0	5.0–5.5	ascorbate; (removal of ascorbate interference)
Batroxobin (*Bothrops atrox* serine proteinase, reptilase) E.C. 3.4.21.29 [*9039-61-6*]	*Bothrops atrox* venom (different zoological varieties)	32–43	1.6–2.9 $\times 10^{-4}$ (B_z^c-Phe-Val-) Arg-*p*-nitroanilide)	7.4–8.2	6.6	fibrinogen
Catalase E.C. 1.11.1.6 [*9001-05-2*]	bovine liver	232 (57.5)	–[a]	6.8–7.0	5.4–5.8	uric acid, cholesterol (indicator enzyme)
Cholesterol esterase E.C. 3.1.1.13 [*9026-00-0*]	*Pseudomonas fluorescens*	ca. 129 (27)	7 $\times 10^{-5}$ (cholesterol oleate) 5.5 $\times 10^{-5}$ (cholesterol linoleate)	7.3–7.6	4.5	cholesterol esters
Cholesterol oxidase E.C. 1.1.3.6 [*9028-76-6*]	*Nocardia erythropolis*	59	1 $\times 10^{-6}$ (cholesterol)	7.5	4.85	cholesterol
Cholesterol oxidase E.C. 1.1.3.6 [*9028-76-6*]	*Streptomyces* species	55	6 $\times 10^{-4}$ (cholesterol)	7.0–7.5	4.5	cholesterol

Tab. 3 (Continued)

Enzyme, E.C. number, and CAS registry number	Source	M_r, 10^3 dalton enzyme (subunit)	K_m, mol/L (substrate)	pH optimum	Isoelectric point	Substrate measured
Choline kinase E.C. 2.7.1.32 [*9026-67-9*]	yeast	67–68	1.5×10^{-5} (choline) 1.4×10^{-4} (ATP)	8.0–9.5	ca. 6	phosphatidylcholine
Choline oxidase E.C. 1.1.3.17 [*9028-67-5*]	*Arthrobacter globiformis*	84 or 71 [b]	1.2×10^{-3} (choline) 8.7×10^{-2} (betaine aldehyde)	7.5	4.5	phospholipids
Citrate lyase (citrate(*pro-3S*)-lyase) E.C. 4.1.2.6 [*9012-83-3*]	*Aerobacter aerogenes*	575 (73.8)	2.1×10^{-4} (citrate)	8.0–9.0	8.0	citrate
Citrate synthase E.C. 4.1.3.7 [*9027-96-7*]	porcine heart	100 (50)	2.5×10^{-4} (citrate) 2.8×10^{-5} (CoA)	8.0	5.05	acetate
Creatinase (creatine amidinohydrolase) E.C. 3.5.3.3 [*37340-58-2*]	*Pseudomonas* species	94 (47)	1×10^{-2} (creatine)	8.0	4.8	creatine, creatinine
Creatininase (creatinine amidohydrolase) E.C. 3.5.2.10 [*9025-13-2*]	*Pseudomonas* species	175 (23)	3×10^{-2} (creatinine) 6×10^{-2} (creatine)	7.8	4.7	creatinine
Creatinine deiminase E.C. 3.5.4.21 [*37289-15-5*]	*Corynebacterium lilium*	200	6×10^{-3} (creatine)	7.5–9.0	4.2	creatine
Dihydrolipoamide dehydrogenase (diaphorase) E.C. 1.8.1.4 [*9001-18-7*]	porcine heart	114 (57)	5×10^{-3} (lipoamide) 2×10^{-3} (lipoate) 2×10^{-4} (NAD^+) 2.7×10^{-4} (ferricyanide)	4.8 (diaphorase) 5.6–6.5 (lipoate reduction)	5.9–7.2 [c]	NADH (indicator enzyme)
Esterase (carboxylesterase) E.C. 3.1.1.1 [*9016-18-6*]	porcine liver	168 (42)	5×10^{-4} (ethyl *n*-butyrate) 4.4×10^{-4} (methyl *n*-butyrate) 1.5×10^{-4} (phenyl *n*-butyrate)	8.6–8.8	5.0	triglycerides
Factor Xa (thrombokinase) E.C. 3.4.21.6 [*9002-05-5*]	bovine plasma	47.2 (Xaα); 44.2 (Xaβ)	5×10^{-4} (B_z^c-Ile-Glu-Gly-Arg-*p*-nitroanilide)	8.3	–	prothrombin
Formaldehyde dehydrogenase E.C. 1.2.1.46 [*68821-75-0*]	*Pseudomonas* species	150 (75)	6.7×10^{-5} (formaldehyde) 1.2×10^{-4} (NAD^+)	7.8	5.25	formaldehyde
Formate dehydrogenase E.C. 1.2.1.2 [*9028-85-7*]	*Candida boidinii*	74 (36)	1.3×10^{-2} (formate) 9×10^{-5} (NAD^+)	7.5–8.5	–	formate, oxalate
β-Fructosidase (invertase, saccharase) E.C. 3.2.1.26 [*9001-57-4*]	yeast	270 (135)	9.1×10^{-3} (sucrose) 2.4×10^{-1} (raffinose)	3.4–4.0	4.02, 4.24 [c]	sucrose
β-D-Galactose dehydrogenase E.C. 1.1.1.48 [*9028-54-0*]	*Pseudomonas fluorescens*	64 (32)	7×10^{-4} (D-galactose) 2.4×10^{-4} (NAD^+) 2.3×10^{-3} ($NADP^+$)	9.1–9.5	5.13	galactose, raffinose

α-Galactosidase E.C. 3.2.1.22 [*9025-35-8*]	*Escherichia coli*	329 (82)	2 × 10^{-2} (galactose)	6.0–7.0	5.1	raffinose
β-Galactosidase E.C. 3.2.1.23 [*9031-11-2*]	*Escherichia coli*	465 (116)	3.85 × 10^{-3} (lactose) 9.5 × 10^{-4} (2-nitrophenyl-β-D-galactoside) 4.45 × 10^{-4} (4-nitrophenyl-β-D-galactoside)	8.0	4.61	lactose; (marker enzyme)
Gluconokinase E.C. 2.7.1.12 [*9030-55-1*]	*Escherichia coli*	67 (34)	7 × 10^{-5} (D-gluconate) 5 × 10^{-4} (ATP)	8.0	–	D-gluconate
Glucose dehydrogenase E.C. 1.1.1.47 [*9028-53-9*]	*Bacillus megaterium*	118 (28)	4.75 × 10^{-2} (glucose) 4.5 × 10^{-3} (NAD$^+$)	8.0 (tris-HCl) 9.0 (acetate)	– –	glucose
Glucose oxidase E.C. 1.1.3.4 [*9001-37-0*]	*Aspergillus niger*	160 (79)	3.3 × 10^{-2} (glucose) 2 × 10^{-4} (oxygen)	5.5–6.5	4.2	glucose
Glucose 6-phosphate dehydrogenase E.C. 1.1.1.49 [*9001-40-5*]	*Leuconostoc mesenteroides*	103.7 (54.8)	6.4 × 10^{-5} (glucose-6-phosphate) 1.15 × 10^{-4} (NAD$^+$) 7 × 10^{-6} (NADP$^+$)	7.8	4.6	glucose-6-phosphate, sucrose, ATP; (marker enzyme)
Glucose 6-phosphate isomerase E.C. 5.3.1.9 [*9001-41-6*]	yeast	120 (60)	0.7–1.5 × 10^{-3} (glucose 6-phosphate) 2.3 × 10^{-4} (fructose 6-phosphate)	7.6	5.0–5.4[c]	fructose
α-Glucosidase (maltase) E.C. 3.2.1.20 [*9001-42-7*]	yeast	68	1.8–2.8 × 10^{-4} (4-nitrophenyl-α-D-glucoside)	7.5–8.0	5.6–5.9[c]	α-amylase
β-Glucosidase E.C. 3.2.1.21 [*9001-22-3*]	*Amygdalae dulces*	135 (65)[c] +90	6 × 10^{-3} (2-nitrophenyl-β-D-glucoside)	4.4–6.0[c]	7.3[c]	α-amylase, amygdalin
Glutamate dehydrogenase E.C. 1.4.1.3 [*9029-12-3*]	bovine liver	oligomers of 332, e.g., 2200 (55)	1.8 × 10^{-3} (L-glutamate) 7 × 10^{-4} (2-oxoglutarate) 3.2 × 10^{-3} (ammonia) 7.0 × 10^{-4} (NAD$^+$) 4.7 × 10^{-5} (NADP$^+$) 2.5 × 10^{-5} (NADH) 2.4 × 10^{-5} (NADPH)	8.5–9.0 (glutamate oxidation) 7.8 (reductive amination)	4.5	glutamate, 2-oxoglutarate, ammonia
Glutamic–oxaloacetic transaminase (aspartate aminotransferase) E.C. 2.6.1.1 [*9000-97-9*]	porcine heart	94 (47)	8.9 × 10^{-3} (L-glutamate) 8.8 × 10^{-5} (oxaloacetate) 3.9 × 10^{-3} (L-aspartate) 4.3 × 10^{-4} (2-oxoglutarate)	8.0–8.5	5.0	malate

Tab. 3 (Continued)

Enzyme, E.C. number, and CAS registry number	Source	M_r, 10^3 dalton enzyme (subunit)	K_m, mol/L (substrate)	pH optimum	Isoelectric point	Substrate measured
Glutamic–pyruvic transaminase (alanine aminotransferase) E.C. 2.6.1.2 [*9000-86-6*]	porcine heart	115	2.5 $\times 10^{-2}$ (L-glutamate) 3 $\times 10^{-4}$ (pyruvate) 2.8 $\times 10^{-2}$ (L-alanine) 4 $\times 10^{-4}$ (2-oxoglutarate)	8.0–8.5	–	lactate
Glycerol dehydrogenase E.C. 1.1.1.6 [*9028-14-2*]	*Enterobacter aerogenes*	340 (56)	1.7 $\times 10^{-2}$ (glycerol) 1.3 $\times 10^{-3}$ (dihydroxyacetone) 1.5 $\times 10^{-4}$ (NAD^+) 1.4 $\times 10^{-5}$ (NADH)	9.0	–	glycerol, triglycerides
Glycerol kinase E.C. 2.7.1.30 [*9030-66-4*]	*Bacillus stearothermophilus*	230 (58)	4.4 $\times 10^{-5}$ (glycerol) 6 $\times 10^{-5}$ (ATP)	10.0–10.5	–	glycerol, triglycerides, lipase
Glycerol 3-phosphate dehydrogenase E.C. 1.1.1.8 [*9075-65-4*]	rabbit muscle	78 (37.5)	1.1 $\times 10^{-4}$ (glycerol 3-phosphate) 3.8 $\times 10^{-4}$ (NAD^+)	7.5–8.6	6.45	glycerol, triglycerides
Glycerol 3-phosphate oxidase E.C. 1.1.3.21 [*9046-28-0*]	*Aerococcus viridans*	75	3.2 $\times 10^{-3}$ (L-glycerol 3-phosphate)	7.5–8.5	4.2	triglycerides
Hexokinase E.C. 2.7.1.1 [*9001-51-8*]	yeast	104 (52)	1.0 $\times 10^{-4}$ (D-glucose) 7.0 $\times 10^{-4}$ (D-fructose) 2.0 $\times 10^{-4}$ (ATP) 5.0 $\times 10^{-5}$ (D-mannose)	8.0–9.0	4.7	glucose, fructose, mannose, ATP; creatine kinase
3-Hydroxybutyrate dehydrogenase E.C. 1.1.1.30 [*9028-38-0*]	*Rhodopseudomonas spheroides*	85	4.1 $\times 10^{-4}$ (D-3-hydroxybutyrate) 8.0 $\times 10^{-5}$ (NAD^+)	6.2–6.9 (reduction) 7–9 (oxidation)	–	3-hydroxybutyrate
Isocitrate dehydrogenase E.C. 1.1.1.42 [*9028-48-2*]	porcine heart	ca. 60	2.6 $\times 10^{-6}$ (isocitrate) 9.2 $\times 10^{-6}$ (NADPH) 1.3 $\times 10^{-4}$ (2-oxoglutarate) 1 $\times 10^{-7}$ ($NADP^+$)	7.0–7.5	7.4	isocitrate
D-Lactate dehydrogenase E.C. 1.1.1.28 [*9028-36-8*]	*Lactobacillus leichmanii*	68 or 80[d]	7 $\times 10^{-2}$ (D-lactate) 1.2 $\times 10^{-3}$ (pyruvate) 7.1 $\times 10^{-5}$ (NADH)	7.0	–	D-lactate; glutamic–oxaloacetic and glutamic–pyruvic transaminases
L-Lactate dehydrogenase E.C. 1.1.1.27 [*9001-60-9*]	porcine heart	140 (35)	1.5 $\times 10^{-4}$ (pyruvate) 3.3 $\times 10^{-3}$ (L-lactate) 1.1 $\times 10^{-5}$ (NADH) 6.7 $\times 10^{-5}$ (NAD^+)	7.0	–[c]	L-lactate, pyruvate, citrate, ADP

Lipase (triacylglycerol acylhydrolase) E.C. 3.1.1.3 [*9001-62-1*]	*Pseudomonas* species	32	9 $\times 10^{-4}$ (tributyrin)	7.0–8.0	4.3	triglycerides
Luciferase, bacterial (alkanal monooxygenase, FMN-linked) E.C. 1.14.14.3 [*9014-00-0*]	*Photobacterium fischeri*	80 (40 + 40)	4–8 $\times 10^{-7}$ ($FMNH_2$)	6.8	–	NADH
Luciferase, firefly (*Photinus*-luciferin 4-monooxygenase) E.C. 1.13.12.7 [*61970-00-1*]	*Photinus pyralis*	100 (50)	5 $\times 10^{-5}$ (ATP)	7.5–7.8	6.2–6.3	ATP
Malate dehydrogenase E.C. 1.1.1.37 [*9001-64-3*]	porcine heart mitochondria	67 (34)	4 $\times 10^{-4}$ (L-malate) 3.3 $\times 10^{-5}$ (oxaloacetate)	7.4–7.5	6.1–6.4	malate, oxaloacetate, acetate, citrate; (marker enzyme)
Mannose 6-phosphate isomerase E.C. 5.3.1.8 [*9023-88-5*]	yeast	45	0.8–1.35 $\times 10^{-3}$ (mannose 6-phosphate)	7.0–7.2	–	mannose
NADH peroxidase E.C. 1.11.1.1 [*9032-24-0*]	*Streptococcus faecalis*	120 (59)	1.7 $\times 10^{-5}$ (NADH) 2.8 $\times 10^{-5}$ (H_2O_2)	6.0	–	H_2O_2 (indicator enzyme)
Nitrate reductase E.C. 1.6.6.2 [*9029-27-0*]	*Aspergillus* species	200 (97)	3.2 $\times 10^{-4}$ (nitrate)	7.5	–	nitrate
Peroxidase E.C. 1.11.1.7 [*9003-99-0*]	horseradish	ca. 40	–[a]	6.0–7.0	–[c]	H_2O_2 (indicator enzyme and marker enzyme)
Phosphatase, alkaline E.C. 3.1.3.1 [*9001-78-9*]	calf intestine	140 (69)	–[a]	9.8	5.7	(marker enzyme)
6-Phosphogluconate dehydrogenase E.C. 1.1.1.44 [*9073-95-4*]	yeast	100	1.6 $\times 10^{-4}$ (6-phosphogluconate) 2.6 $\times 10^{-5}$ ($NADP^+$)	8.0	–	gluconate
Phospholipase C E.C. 3.1.4.3 [*9001-86-9*]	*Bacillus cereus*	23	2 $\times 10^{-2}$ (phosphatidylcholine)	8.0	7.0	phosphatidylcholine
Phospholipase D E.C. 3.1.4.4 [*9001-87-0*]	*Streptomyces chromofuscus*	57 or 50[b]	1.43 $\times 10^{-3}$ (phosphatidylcholine) 5.6 $\times 10^{-2}$ (sphingomyelin)	8.0	5.1	triglycerides, choline phospholipids

Tab. 3 (Continued)

Enzyme, E.C. number, and CAS registry number	Source	M_r, 10^3 dalton enzyme (subunit)	K_m, mol/L (substrate)	pH optimum	Isoelectric point	Substrate measured
Plasmin E.C. 3.4.21.7 [*9001-90-5*]	human or bovine plasma	87–91	1.7 $\times 10^{-4}$ (tosyl-Gly-Pro-Lys-*p*-nitroanilide)	7.4	6.1–8.4[c]	α_2-antiplasmin
Pyruvate kinase E.C. 2.7.1.40 [*9001-59-6*]	rabbit muscle	237 (57)	7.0 $\times 10^{-5}$ (phosphoenolpyruvate) 1.0 $\times 10^{-2}$ (pyruvate) 3.0 $\times 10^{-4}$ (ADP) 8.6 $\times 10^{-4}$ (ATP)	7.0–7.8	5.98	ADP, IDP, creatine kinase, phosphoenol-pyruvate
Pyruvate oxidase E.C. 1.2.3.3 [*9001-96-1*]	*Pediococcus* species	150 (38)	1.7 $\times 10^{-3}$ (pyruvate) 5 $\times 10^{-4}$ (phosphate)	6.5–7.5	4.0	pyruvate, ADP, glutamic–oxaloacetic and glutamic–pyruvic transaminases
Sarcosine oxidase E.C. 1.5.3.1 [*9029-22-5*]	*Pseudomonas* species	174 (110, 44, 21, 10)	4 $\times 10^{-3}$ (sarcosine) 1.2 $\times 10^{-2}$ (*N*-methylalanine) 1.3 $\times 10^{-4}$ (oxygen)	8.0	–	sarcosine, creatinine
Sorbitol dehydrogenase (L-iditol dehydrogenase) E.C. 1.1.1.14 [*9028-21-1*]	sheep liver	115	0.7–1.1 $\times 10^{-3}$ (sorbitol) 1.8 $\times 10^{-4}$ (xylitol)	7.9–8.1	–	sorbitol, xylitol
Succinyl-CoA synthetase (succinate-CoA ligase, GDP-forming) E.C. 6.2.1.4 [*9014-36-2*]	porcine heart	75 (42.5 + 34.5)	4–8 $\times 10^{-4}$ (succinate) 5–10 $\times 10^{-6}$ (GTP) 5–20 $\times 10^{-6}$ (CoA)	8.3	5.8–6.4[c]	succinate
Sulfite oxidase E.C. 1.8.3.1 [*9029-38-3*]	chicken liver	110 (55)	2.5 $\times 10^{-5}$ (sulfite)	8.5	–	sulfite
Thrombin E.C. 3.4.21.5 [*9002-04-4*]	bovine plasma	39 (α), 28 (β)	1.3 $\times 10^{-3}$ (B_z^e-Arg-*p*-nitroanilide) 5.9 $\times 10^{-6}$ (tosyl-Gly-Pro-Arg-*p*-nitroanilide)	9.0	5.3–5.75	antithrombin III, fibrinogen, heparin, coagulation status
Trypsin E.C. 3.4.21.4 [*9002-07-7*]	bovine pancreas	23.3	9.4 $\times 10^{-4}$ (B_z^e-Arg-*p*-nitroanilide) 4.3 $\times 10^{-6}$ (B_z^e-Arg-ethyl ester)	8.0	10.5–10.8	α_1-proteinase inhibitor, α_2-macroglobulin
Urate oxidase (uricase) E.C. 1.7.3.3 [*9002-12-4*]	*Arthrobacter protophormiae*	170 (40)	6.6 $\times 10^{-5}$ (urate)	9.0	–	urate
Urease E.C. 3.5.1.5 [*9002-13-5*]	jack beans	480 (75–83)	1.05 $\times 10^{-2}$ (urea)	7.0	5.0–5.1	urea; (marker enzyme)
Xanthine oxidase E.C. 1.1.3.22 [*9002-17-9*]	cow milk	283 (150)	1.7 $\times 10^{-6}$ (xanthine) 2.4 $\times 10^{-5}$ (oxygen)	8.5–9.0	6.2	xanthine, hypoxanthine, phosphate

[a] Enzyme reaction does not obey Michaelis–Menten kinetics. [b] Depending on method (gel filtration or polyacrylamide gel electrophoresis in the presence of sodium dodecylsulfate). [c] Isoenzymes or multiple species. [d] Depending on method (ultracentrifugation or gel filtration). [e] B_z = benzoyl.

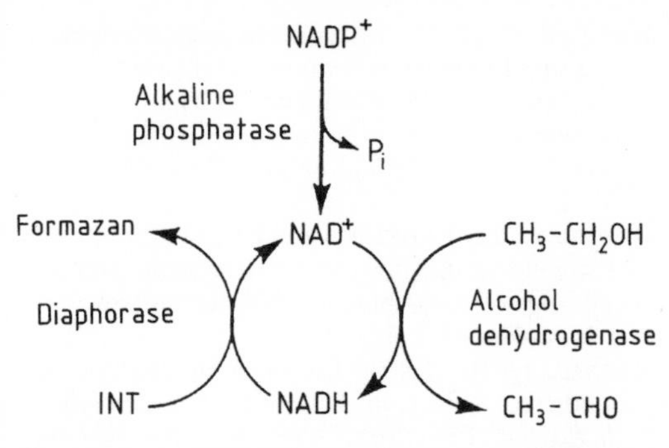

Fig. 13. An amplification system for the enzyme alkaline phosphatase using $NADP^+$ as the primary substrate. The marker enzyme acts on $NADP^+$ which is dephosphorylated to NAD^+. This is used as a trigger for a redox cycle involving alcohol dehydrogenase and diaphorase, as well as the substrates ethanol and INT (2-*p*-iodophenyl-3-[*p*-nitrophenyl]-5-phenyltetrazolium chloride) which is oxidized to give a deeply colored formazan product.

- *solubility* and *surface properties* (no interference by adsorption or aggregation effects), and
- *cost.*

These criteria are not independent of each other (e.g., purity, activity, and cost). So the choice and quality of an enzyme must be optimized in each case with regard to the particular analytical application.

9.2 Use of Recombinant Enzymes for Analytical Purposes

The advances in recombinant DNA technology have made it possible to clone genes of interest and to manipulate bacterial, yeast, insect, or mammalian cells to overproduce the desired protein. During the last decade, many enzymes have been expressed in microorganisms at levels 10 to 100 times higher than those in the natural host cell. This has allowed not only better economics in enzyme production, but also significant reduction of the environmental burden because of the decrease in fermentation volumes (and, thereby, in waste formation).

Furthermore, the techniques of unspecific and, more recently, site-directed mutagenesis have opened the way to improve relevant properties of enzymes for analytical application, e.g., stability under the assay conditions, pH optimum, or solubility. One example is the enzyme pyruvate oxidase used in coupled assays for transaminase activity where three thermostabilized mutants have been obtained by plasmid mutagenesis, followed by construction of a triple mutant with even more enhanced stability (SCHUMACHER et al., 1990).

The achievements and possibilities of recombinant DNA technology as applied to enzymes used in diagnostics have recently been reviewed (KOPETZKI et al., 1994).

9.3 List of Enzymes Used for Analytical Purposes

Tab. 3 is a comprehensive list of important, commercially available enzymes used for analytical purposes.

10 References

ALLEN, G. (1989), *Sequencing of Proteins and Peptides*, pp. 73–104, Amsterdam: Elsevier.

BATES, D. L. (1987), Enzyme amplification in diagnostics, *Trends Biotechnol.* **5,** 204–209.

BERGEL, A., SOUPPE, J., COMTAT, M. (1989), Enzymatic amplification for spectrophotometric and electrochemical assays of NAD^+ and NADH, *Anal. Biochem.* **179,** 382–388.

BERGMEYER, H. U. (1983a), Determination of metabolite concentrations with end-point methods, in: *Methods of Enzymatic Analysis* (BERGMEYER, H. U., GRASSL, M., BERGMEYER, J., Eds.), 3rd Ed., Vol. 1, pp. 163–181, Weinheim: VCH.

BERGMEYER, H. U. (1983b), Optimization of substrate concentrations, in: *Methods of Enzymatic Analysis* (BERGMEYER, H. U., GRASSL, M., BERGMEYER, J., Eds.), 3rd Ed., Vol. 1, pp. 116–142, Weinheim: VCH.

BERGMEYER, H. U. (1983c), Determination of the catalytic activity of enzymes, in: *Methods of Enzymatic Analysis* (BERGMEYER, H. U., GRASSL, M., BERGMEYER, J., Eds.), 3rd Ed., Vol. 1, pp. 105–114, Weinheim: VCH.

BERGMEYER, H. U., GAWEHN, K. (1983), Brief history and definition, in: *Methods of Enzymatic Analysis* (BERGMEYER, H. U., GRASSL, M., BERGMEYER, J., Eds.), 3rd Ed., Vol. 1, pp. 2–7, Weinheim: VCH.

BERGMEYER, H. U., HØRDER, M., MARKOWETZ, D. (1983), Reliability of laboratory results and practicability of procedures, in: *Methods of Enzymatic Analysis* (BERGMEYER, H. U., GRASSL, M., BERGMEYER, J., Eds.), 3rd Ed., Vol. 1, pp. 21–40, Weinheim: VCH.

BERGMEYER, H. U., GRASSL, M., BERGMEYER, J. (Eds.) (1983–1986), *Methods of Enzymatic Analysis*, 3rd Ed., Vols. 1–12, Weinheim: VCH.

BERRY, M. N., MAZZACHI, R. D., PEJAKOVIC, M., PEAKE, M. J. (1989), Enzymatic determination of potassium in serum, *Clin. Chem.* **35,** 817–820.

BROOKS, S. P. J., SUELTER, C. H. (1989), Practical aspects of coupling enzyme theory, *Anal. Biochem.* **176,** 1–14.

CLARK, L., LYONS, L. (1962), Glucose enzyme electrode, *Ann. N. Y. Acad. Sci.* **102,** 29–45.

COOPER, A. J. L., LEUNG, L. K. H., ASANO, Y. (1989), Enzymatic cycling assay for phenylpyruvate, *Anal. Biochem.* **183,** 210–214.

COUSINS, C. L. (1976), Principles of enzymology, *Clin. Biochem.* **9,** 160–164.

DEEG, R., KRAEMER, W., ZIEGENHORN, J. (1980), Kinetic determination of serum glucose by use of the hexokinase/glucose-6-phosphate dehydrogenase method. *J. Clin. Chem. Clin. Biochem.* **18,** 49–52.

DORDICK, J. S. (1989), Enzymatic catalysis in monophasic organic solvents, *Enzyme Microb. Technol.* **11,** 194–211.

EISENTHAL, R., DANSON, M.J. (1993), *Enzyme Assays – A Practical Approach*, Oxford: IRL Press.

GUPTA, M. N. (1992), Enzyme function in organic solvents, *Eur. J. Biochem.* **203,** 25–32.

HENDERSON, D. R., FRIEDMAN, S. B., HARRIS, J. D., MANNING, W. S., ZOCCOLI, M. A. (1986), CEDIA™, a new homogeneous immunoassay system, *Clin. Chem.* **32,** 1637–1641.

HIGGINS, I. J., HILL, H. A. O. (1985), Bioelectrochemistry, *Essays Biochem.* **21,** 119–145.

JACOB, G. S., SCUDDER, P. (1994), Glycosidases in structural analysis, *Methods Enzymol.* **230,** 280–299.

JOHANSSON, A., STANLEY, C. J., SELF, C. H. (1985), A fast highly sensitive colorimetric enzyme immunoassay system demonstrating benefits of enzyme amplification in clinical chemistry, *Clin. Chim. Acta* **148,** 119–124.

KESSLER, C. (1987), Enzymes in genetic engineering, in: *Ullmann's Encyclopedia of Industrial Chemistry*, 5th Ed., Vol. A9, pp. 457–502, Weinheim: VCH.

KESSLER, C. (1990), Enzymes in genetic engineering, in: *Enzymes in Industry* (GERHARTZ, W., ed.), pp. 185–248, Weinheim: VCH.

KESSLER, C. (Ed.) (1992), *Nonradioactive Labeling and Detection of Biomolecules*, Berlin: Springer-Verlag.

KOPETZKI, E., LEHNERT, K., BUCKEL, P. (1994), Enzymes in diagnostics: Achievements and possibilities of recombinant DNA technology, *Clin. Chem.* **40,** 688–704.

KRESSE, G.-B. (1987), Enzymes in analysis and medicine: Survey, in: *Ullmann's Encyclopedia of Industrial Chemistry*, Vol. A9, pp. 434–441, Weinheim: VCH.

KRESSE, G.-B. (1990), Enzymes in analysis and medicine: Survey, in: *Enzymes in Industry* (GERHARTZ, W., Ed.), pp. 154–159, Weinheim: VCH.

KRESSE, G.-B., SCHMID, R. D. (1995), Enzymatic analysis and biosensors, in: *Handbook of Enzyme Catalysis in Organic Solvents* (DRAUZ, K. H., WALDMANN, H., Eds.), Weinheim: VCH, in press.

LOWRY, O. H., PASSONEAU, J. V., SCHULZ, D. W., ROCK, M. R. (1961), The measurement of pyridine nucleotides by enzymatic cycling, *J. Biol. Chem.* **236,** 2746–2755.

MASSEYEFF, R. F., ALBERT, W. A., STAINES, N. A. (Eds.) (1993), *Methods of Immunological Analysis*, Vols. 1–3, Weinheim: VCH.

MATTENHEIMER, H. (1982), *Die Theorie des enzymatischen Tests*, Mannheim: Boehringer Mannheim GmbH.

MAYER, A., NEUENHOFER, S. (1994), Lumineszenzmarker – mehr als nur eine Alternative zu Radioisotopen, *Angew. Chem.* **106,** 1097–1126; *Angew. Chem. Int. Ed. Engl.* **33,** 1044–1072.

MICHAL, G., BERGMEYER, H. U. (1985), Coenzyme A, catalytic method with phosphate acetyltransferase, in: *Methods of Enzymatic Analysis* (BERGMEYER, H. U., GRASSL, M., BERGMEYER, J., Eds.), 3rd Ed., Vol. 7, pp. 169–177, Weinheim: VCH.

MICHAL, G., MÖLLERING, H., SIEDEL, J. (1983), Chemical design of indicator reactions for the visible range, in: *Methods of Enzymatic Analysis* (BERGMEYER, H. U., GRASSL, M., BERGMEYER, J., Eds.), 3rd Ed., Vol. 1, pp. 197–232, Weinheim: VCH.

OELLERICH, M. (1983), Principles of enzyme-immunoassays, in: *Methods of Enzymatic Analysis* (BERGMEYER, H. U., GRASSL, M., BERGMEYER, J., Eds.), 3rd Ed., Vol. 1, pp. 233–260, Weinheim: VCH.

OELLERICH, M. (1984), Enzyme-immunoassay: A review, *J. Clin. Chem. Clin. Biochem.* **22,** 895–904.

O'SULLIVAN, M. J., BRIDGES, J. W., MARKS, V. (1979), Enzyme immunoassay: a review, *Ann. Clin. Biochem.* **16,** 221–240.

RAUSCHER, E., NEUMANN, U., SCHAICH, E., VON BÜLOW, S., WAHLEFELD, A.W. (1985), Optimized conditions for determining activity concentration of α-amylase in serum with 1,4-α-D-4-nitrophenylmaltoheptaoside in serum, *Clin. Chem.* **31,** 14–19.

RUBENSTEIN, K. E., SCHNEIDER, R. S., ULLMAN, E. F. (1972), "Homogeneous" enzyme immunoassay. A new immunochemical technique, *Biochem. Biophys. Res. Commun.* **47,** 846–851.

SCHELLER, F., SCHMID, R. D. (1992), *Biosensors: Fundamentals, Technologies and Applications, GBF Monograph* Vol. **17,** Weinheim: VCH.

SCHELLER, F. , SCHUBERT, F. (1989), *Biosensors,* Basel: Birkhäuser.

SCHUMACHER, G., MÖLLERING, H., GEUSS, U., FISCHER, S., TISCHER, W., DENEKE, U., KRESSE, G.-B. (1990), Properties and application of a recombinant pyruvate oxidase from *Lactobacillus plantarum* and a mutant with enhanced stability, in: *Biochemistry and Physiology of Thiamin Diphosphate Enzymes* (BISSWANGER, H., ULLRICH, J., Eds.), pp. 300–307, Weinheim: VCH.

SELF , C. H. (1985), Enzyme amplification – A general method to provide an immunoassisted assay for placental alkaline phosphatase, *J. Immunol. Methods* **76,** 389–393.

SIEDEL, J., DEEG, R., ZIEGENHORN, J. (1983), Determination of metabolic concentrations by kinetic methods, in: *Methods of Enzymatic Analysis* (BERGMEYER, H. U., GRASSL, M., BERGMEYER, J., Eds.), 3rd Ed., Vol. 1, pp. 182–197, Weinheim: VCH.

SONNTAG, O. (1993), *Dry Chemistry, Laboratory Techniques in Biochemistry and Molecular Biology,* Vol. 25, Amsterdam: Elsevier.

STANLEY, C. J., PARIS, F., PLUMB, A., WEBB, A., JOHANSSON, A. (1985), Enzyme amplification: a new technique for enhancing the speed and sensitivity of enzyme immunoassays, *Am. Biotechnol. Lab.*, May/June Issue.

TAKAHASHI, M., UEDA, S., MISAKI, H., SUGIYAMA, N., MATSUMOTO, K., MATSUO, N., MURAO, S. (1994), Carnitine determination by an enzymatic cycling method with carnitine dehydrogenase, *Clin. Chem.* **40,** 817–821.

TAKAMURA, Y., KITAYAMA, Y., ARAKAWA, A., YAMANAKA, S., TOSAKI, M., OGAWA, Y. (1985), Malonyl-CoA:acetyl-CoA cycling. A new micromethod for determination of acyl-CoAs with malonate decarboxylase, *Biochim. Biophys. Acta* **834,** 1–7.

TRINDER, P. (1969), Determination of glucose in blood using glucose oxidase with an alternative oxygen acceptor, *Ann. Clin. Biochem.* **6,** 24–27.

TURNER, A. P. F., KARUBE, I., WILSON, G. (1990), *Biosensors, Fundamentals and Applications*, 2nd Ed., Oxford: Oxford Science Press.

WERNER, W., RITTERSDORF, W. (1983), Reflectance photometry, in: *Methods of Enzymatic Analysis* (BERGMEYER, H. U., GRASSL, M., BERGMEYER, J., Eds.), 3rd Ed., Vol. 1, pp. 305–326, Weinheim: VCH.

YALOW, R. S., BERSON, S. A. (1959), Assay of plasma insulin in human subjects by immunological methods, *Nature* **184,** 1648–1649.

II. Biomass

5 Production of Microbial Biomass

HÉLÈNE BOZE
GUY MOULIN
PIERRE GALZY

Montpellier, France

1 Historical Background

Human beings have used microorganisms since prehistoric times. The first utilization was accidental and then, after much trial and error, microorganisms were used empirically in the making of beverages, foods, textiles and antibiotics. It was not until 1830 that CAGNIARD DE LATOUR and KUTZING and SCHWANN discovered that the growth of yeasts and other Protista caused the fermentation processes involved in the making of products such as wine.

In about 1850, extensive work by PASTEUR resulted in considerable progress. PASTEUR described bacteria and yeasts at a physiological level, introduced aseptic methods and the notion of minimum medium and defined nutrient and oxygen requirements.

MONOD modeled bacterial growth in 1949. He launched the ideas of growth yield, specific growth rate and the concentration of growth-limiting substrates. The discovery of microorganisms and understanding of growth mechanisms made possible the development of culture methods. The theoretical approach to chemostat culture was performed by MONOD and NOVICK and SZILARD in 1950.

Control of microbial cultures requires knowledge of the metabolic pathways involved, determination of the nutrient requirements of the microorganism and the physicochemical conditions required for optimum growth. Furthermore, development and changes in culture equipment thanks to progress in process engineering enable optimization of the production of microorganisms and their metabolites.

In addition to traditional agrofood processing (wine, beer, dairy products, bread, etc.), applications have been developed in the pharmaceutical sector (antibiotics, vitamins, etc.) and in the production of metabolites (enzymes and amino acids).

Many processes have been developed to produce microorganisms able to use organic material as a source of carbon and energy and to convert inorganic nitrogen into high-food-value proteins. These can be used in human foodstuffs or animal feed to replace traditional plant or animal sources.

Microorganisms such as algae, bacteria, yeasts and fungi have been considered as protein sources. Culture of microorganisms for their nutritional value started at the end of the First World War. The Germans developed yeast culture for use in animal and human diets. The term "fodder yeast" was coined (BRAUDE, 1942; WEITZEL and WINCHEL, 1932; SCHÜLEIN, 1937). After the Second World War, production of fodder yeast using the pentoses in sulfite liquor was developed in the USA, and other processes were developed in the United Kingdom. Carbon substrates such as molasses, pulp of the coffee and seed pulp of the palmyra palm were used.

Interest in the production of microbial biomass as food and fodder intensified after the 1950s and reached a peak towards 1977. The term "Single Cell Protein (SCP)" was invented, and much information on these processes was published in a large number of journals (cf. References). The most important features resulting from research on SCP concerned the following topics:

- diversity of the substrates used and their catabolism (renewable substrate, fossil mass);
- breeding and genetic improvement of a great variety of microorganisms.

The various microorganisms used for biomass production and the various metabolic pathways involved in substrate catabolism are described here. Physiological aspects, growth parameters, energy and nutritional requirements and the influence of the physicochemical parameters are discussed. The different types of microbial culture and examples of processes are described.

2 Microorganisms

Four types of microorganisms are used to produce biomass: bacteria, yeasts, fungi and algae. The choice of a microorganism depends on numerous criteria, the most impor-

Tab. 1. Composition (g per 100 g DW) of Microorganisms of Interest in Biomass Production (LICHTFIELD, 1979)

Microorganism	Substrate	Nitrogen	Protein	Fat	Carbohydrate
Algae					
Chlorella sorokiniana	CO_2	9.6	60	8	22
Spirulina maxima	CO_2	10	62	3	
Spirulina maxima	CO_2	8.5	53	4.8	28
Bacteria and Actinomycetes					
Cellulomonas sp.	Bagasse	14	87	8	
Methylomonas clara	Methanol	12–13	80–85	8–10	
Yeasts					
Candida lipolytica	*n*-Alkanes	10	65	8.1	
Candida utilis	Ethanol	8.3	52	7	
Kluyveromyces fragilis	Whey	9	54	1	
Saccharomyces cerevisiae	Molasses	8.4	53	6.3	
Molds and Higher Fungi					
Aspergillus niger	Molasses	7.7	50		
Morchella crassipes	Glucose	5	31	3.1	
Paecilomyces variotii	Sulfite waste liquor	8.8	55	1.3	25

tant of which is the nature of the raw material available. The other criteria are:

- nutritional: energy value, protein content, amino acid balance;
- technological: type of culture, nutritional requirements, type of separation;
- toxicological.

The ideal microorganism should possess the following technological characteristics:

- high specific growth rate (μ) and biomass yield ($Y_{x/s}$);
- high affinity for the substrate;
- low nutritional requirements, i.e., few indispensable growth factors;
- ability to use complex substrates;
- ability to develop high cell density;
- stability during multiplication;
- capacity for genetic modification;
- good tolerance to temperature and pH.

In addition, it should have a balanced protein and lipid composition. It must have a low nucleic acid content, good digestibility and be non-toxic (Tab. 1).

2.1 Yeasts

Yeasts were the first microorganisms known, the best studied and generally best accepted by consumers. Yeasts are rarely toxic or pathogenic and can be used in human diets. Although their protein content rarely exceeds 60%, their concentration in essential amino acids such as lysine (6 to 9%), tryptophan and threonine is satisfactory. In contrast, they contain small amounts of the sulfur-containing amino acids methionine and cysteine. They are also rich in vitamins (B group), and their nucleic acid content ranges from 4 to 10%.

They are larger than bacteria, facilitating separation. They can be used in a raw state. However, their specific growth rate is relatively slow (generation time 2 to 5 hours).

2.2 Bacteria

The specific growth rate and biomass yield of bacteria are greater than those of the other categories of microorganisms. Total protein content may reach 80%. Their amino acid profile is balanced and their sulfur-containing amino acid and lysine concentrations are high. In contrast, their nucleic acid content (10 to 16%) is greater than that of yeasts.

A limited number of bacterial species can be used in foodstuffs as many are pathogenic. In addition, separation is difficult because of their small size.

2.3 Fungi

The use of fungi as biomass is relatively new. They are more conventionally used for producing enzymes, organic acids and antibiotics. Their generation times (5 to 12 hours) are distinctly longer than those of yeasts and bacteria. This is generally an apparent generation time as they grow through elongation of mycelium; growth is not really exponential. Their protein content (50%) is often smaller than that of yeasts and bacteria, and they are deficient in sulfur-containing amino acids. There are also problems of wall digestibility. However, the nucleic acid content is low (3 to 5%).

The principal merits of fungi are their ability to use a large number of complex growth substances such as cellulose and starch and easy recovery by simple filtration, reducing production costs.

2.4 Algae

The potential merits of algae are related to their ability to multiply with CO_2 as the only carbon source. Sone genera (Cyanophyta) can use atmospheric nitrogen. Algae production takes place in natural waterbodies (ponds, lakes and lagoons). Algae are traditionally a food complement for some populations in Mexico (*Spirulina platensis*) and Chad (*Spirulina maxima*). However, algae have a low sulfur-containing amino acid content. Their nucleic acid content is about 4 to 6%. They are easy to recover, but multiplication is very slow, and investment costs involved in artificial shallow ponds result in low process profitability.

2.5 Selection and Improvement of Strains

2.5.1 Selection of Microorganisms

The strategy for selecting microorganisms can be described schematically in several stages (STEELE and STOWERS, 1991).

The first stage consists of:

- definition of the types of transformation sought (carbon substrate, type of microorganism) and definition and characterization of the product to be obtained (baker's yeast, wine yeast, food yeast);
- identification of the microorganisms capable of the transformations desired;
- selection of microorganisms in collections, the natural environment or a favorable environment (waste water, refuse, etc.);
- definition and development of screening.

The second stage consists of the study and comparison of several pre-selected strains. The main features investigated are the growth parameters (μ, $Y_{x/s}$), the physicochemical parameters (pH, temperature, nutritional requirements) and the characteristics peculiar to each strain (cell composition, pathogenicity, toxicity, food value). All these studies can be performed in Erlenmeyer flasks or in a batch fermenter at laboratory scale.

After selection of the microorganisms displaying the best characteristics, the last stage consists of examining and determining the optimum growth parameters of the microorganism and choosing the most suitable culture procedure (batch, fed-batch, continuous, recycling, etc.). These studies should make it possible to extrapolate the laboratory-designed process for use at industrial scale.

2.5.2 Examples of the Selection and Improvement of Strains

The microorganisms involved in biomass production are mainly obtained by searching for mutants occurring by natural selection, mutation or gene manipulation.

Heat-tolerant yeasts and bacteria have thus been selected for processes using methanol or the *n*-paraffins as the carbon source; the metabolisms of these substrates usually generate considerable heat.

A mutant of *Pichia pastoris* was selected with reduced biotin and thiamine requirements (IFP-Technip methanol process). The protein content has also been increased in molds such as *Trichoderma album* (INRA-Blachère process). A *Candida tropicalis* yeast was isolated for its high protein content and – mainly – for its high lysine content (IFP process on *n*-paraffins). The protein yields of a bacterial strain (*Methylophilus methylotrophus*) were increased after gene manipulation affecting its nitrogen metabolism (ICI process).

Strains with higher specific growth rates than normal have also been isolated. The obtaining of a mutant form of *Pichia pastoris* led to a 20 to 30% saving in production cost (IFP-Technip). The type of mutation sought is always closely related to the carbon substrate and the process used. The search for mutants with specific growth rate or even protein content features can generally be carried out by continuous multiplication and steady increase in the dilution rate. Screening techniques suitable for each case are used for the other characteristics.

2.6 Characteristics of Single-Cell Biomass

The food value of microorganisms is directly related to their protein and amino acid composition and their lipid, vitamin and nucleic acid contents. Various analyses must also be performed:

- overall analysis: water, lipid, protein, fiber and mineral contents;
- lipid analysis: proportions of fatty acids, sterols and phospholipids;
- analysis of nitrogen compounds: total nitrogen, amino acid profile, nucleic acid nitrogen, purine and pyrimidine base, quantification of RNA and DNA;
- analysis of minerals: major elements (Na, K, Mg, Ca, Cl) and trace elements (Mn, Zn, Cu, Fe, Co, Mo, As, Pb, Hg);
- analysis of carbohydrates;
- analysis of vitamins.

2.6.1 Composition

The compositions of various microorganisms used for biomass production are shown in Tab. 1. The protein content is often defined as total Kjeldahl nitrogen × 6.25. The real protein content is found using the sum of amino acids determined by analysis according to the recommendations of the Protein Advisory Group of the United Nations System (PAG) Guideline No 6. Comparison of the protein content of different microorganisms cultured on various carbon substrates with the protein content of egg and soy is shown in Tab. 2.

2.6.2 Nutritional Value and Toxicological Status

Important aspects of the quality of the biomass produced are as follows:

- nutritional value of the product,
- safety of the product,
- production of protein concentrate free of nucleic acid and toxic substances.

Three parameters are used to establish the nutrient value of biomass: digestibility, biological value and protein efficiency ratio. Digestibility (D) is the percentage of total nitrogen consumed in relation to the nitrogen in the food ration: $D=[(I-F)/I]\,100$. The total quantity of microbial protein ingested by animals is measured and the nitrogen content (I) is analyzed. Nitrogen contents in feces (F) and urine (U) are collected and measured. Digestibility of algae and bacteria is 83 to 88% and that of yeasts ranges from 88 to 96% (LITCHFIELD, 1979).

Biological value (BV) is the percentage of total nitrogen assimilated that is retained by the body, taking into account the simultaneous loss of endogenous nitrogen through urinary excretion: $BV=[(I-(F+U))/$

Tab. 2. Essential Amino Acid Content of Microorganisms of Interest for Biomass Production (g per 16 g N) (Lichtfield, 1979; Boze et al., 1992)

Protein Source	Substrate	Cys	Ile	Leu	Lys	Met	Phe	Thr	Try	Val
FAO reference		2.0	4.2	4.8	4.2	2.2	2.8	2.8	1.0	4.2
Soy bean meal		0.7	2.2	3.5	2.8	0.6	2.2	1.9	0.6	2.3
Fish meal		0.7	3.2	5.0	4.9	1.9	2.9	3.0	0.9	3.7
Egg		2.4	6.7	8.9	6.5	5.1	5.8	5.1	1.6	7.3
Algae										
Chlorella sorokiniana	CO_2		3.4	4.0	7.8	1.8	2.7	3.2	1.4	5.1
Spirulina maxima	CO_2	0.4	5.8	7.8	4.8	1.5	4.6	4.6	1.3	6.3
Bacteria and Actinomycetes										
Cellulomonas alcaligenes	Bagasse		5.4	7.4	7.6	2.0	4.7	5.5		7.1
Methylophilus methylotrophus	Methanol	0.6	4.3	6.8	5.9	2.4	3.4	4.6	0.9	5.2
Thermomonospora fusca	Cellulose pulping	0.4	3.2	6.1	3.6	2.0	2.6	4.0		13.0
Yeasts										
Candida lipolytica	*n*-Alkanes	1.1	4.5	7.0	7.0	1.8	4.4	4.9	1.4	5.4
Candida utilis	Ethanol	0.4	4.5	7.1	6.6	1.4	4.1	5.5	1.2	5.7
Kluyveromyces fragilis	Whey		4.0	6.1	6.9	1.9	2.8	5.8	1.4	5.4
Saccharomyces cerevisiae	Molasses	1.6	5.5	7.9	8.2	2.5	4.5	4.8	1.2	5.5
Molds and Higher Fungi										
Aspergillus niger	Molasses	1.1	4.2	5.7	5.9	2.6	3.8	5.0	2.1	5.2
Morchella crassipes	Glucose	0.4	2.9	5.6	3.5	1.0	1.9	3.0	1.5	3.0
Paecilomyces variotii	Sulfite waste liquor	1.1	4.3	6.9	6.4	1.5	3.7	4.6	1.2	5.1

$(I-F)]$ 100. Variation is substantial: from 30 to 90% according to the microorganism (Litchfield, 1979).

The protein efficiency ratio (*PER*) is the proportion of nitrogen retained by the animal in comparison with reference proteins, e.g., egg albumin.

Numerous nutritional and toxicological tests on animals are obligatory before any use in human or animal foodstuffs. These tests were published in a series of Guidelines (IUPAC, 1974) by the Protein Advisory Group (PAG) of FAO/WHO/UNICEF and IUPAC.

2.7 Microorganisms Used for Biomass Production

Numerous species of algae, bacteria, yeasts and fungi are used to produce biomass from the various carbon substrates available (Goldberg, 1985; Atkinson and Mavituna, 1983). A list (not exhaustive) is given in Tab. 3.

In principle, the microorganisms mentioned here are not toxic, and it should be possible to use them all in animal feed and most in human food. However, they are not all officially accepted in all countries. It should also be noted that excessive use of antibiotics in the treatment of certain illnesses has resulted in the appearance of antibiotic-resistant strains (e.g., *Candida tropicalis*) of non-toxic species. Use of resistant strains should be avoided.

3 Carbon Substrates and Metabolic Pathways

3.1 Carbon Sources and Energy Sources

The variety of carbon sources available (CO_2, carbohydrates, hydrocarbons, lipids) makes it necessary to choose microorganisms with specific metabolic pathways. The various

Tab. 3. Microorganisms Used According to Carbon Source

Substrate	Organism
Algae	
CO_2	*Chlorella pyrenoidosa, C. regularis, C. sorokiniana, Oocystis polymorpha, Scenedesmus quadricaula, Spirulina maxima, Spirulina platensis, Dunaliella bardawil*
Bacteria and Actinomycetes	
n-Alkanes	*Acinetobacter cerificans, Achromobacter delvacuate, Mycobacterium phlei* sp., *Nocardia* sp., *Pseudomonas* sp.
Methane	*Corynebacterium hydrocarbonoclastus, Nocardia paraffinica, Acinetobacter* sp., *Flavobacterium* sp., *Hyphomicrobium* sp., *Methylomonas methanica, Methylococcus capsulatus*
Methanol	*Methylomonas methylovora, M. clara, M. methanolica, Flavobacterium* sp., *Methylophilus methylotrophus, Pseudomonas* sp., *Streptomyces* sp., *Xantomonas* sp.
Ethanol	*Acinetobacter calcoaceticus*
Cellulosic wastes	*Thermomonospora fusca*
Sulfite waste liquor	*Pseudomonas denitrificans*
Yeasts	
n-Alkanes, *n*-paraffins	*Candida lipolytica, C. tropicalis, C. guilliermondii, C. maltosa, C. paraffinica, C. oleophila, Yarrowia lipolytica*
Methanol	*Candida utilis, Hanseniaspora* sp., *Pichia pastoris, Hansenula* sp., *Kloeckera* sp.
Ethanol	*Candida enthanothermophilum, C. utilis, C. kruzei*
Whey	*Kluyveromyces fragilis, Candida intermedia*
Cane molasses	*Saccharomyces cerevisiae*
Starch	*Schwanniomyces alluvius, Lipomyces kononenkoe*
Lipids	*Candida rugosa, C. utilis, C. lipolytica, C. blankii, C. curvata, C. deformans, C. parapsilosis*
Cellulose	*Candida utilis*
Sulfite waste liquor	*Candida utilis, C. tropicalis*
Molds and Higher Fungi	
Glucose	*Agaricus blazei, A. campestris*
Malt – molasses	*Agaricus campestris*
Starch	*Aspergillus niger, Fusarium graminearum*
Sulfite waste liquors	*Paecilomyces variotii*
Cellulose	*Chaetomium cellulolyticum, Trichoderma viride*
Brewery waste	*Calvatica gigantea*
Carob bean extract	*Aspergillus niger, Fusarium moniliforme*

microorganisms can be classified according to the carbon and energy sources that they are able to use (ATKINSON and MAVITUNA, 1983) (Tab. 4).

The microorganisms used for biomass production are in the photolithotrophic and chemoorganotrophic categories. Special attention has been paid to the CO_2 metabolism of photolithotrophs and the hydrocarbon C1 compound, ethanol, glycerol, carbohydrate and lipid metabolism in chemoorganotrophs.

3.2 CO_2 Metabolism and Photosynthesis

CO_2 is the simplest carbon source for biomass production. It is present in the atmosphere at about 300 ppm. It is the most oxidized carbon source and cannot be used as an energy source. It can also be used in carbonate form. Energy is provided by light and converted into chemical energy by the photosynthesis mechanism.

Tab. 4. Classification of Microorganisms According to Carbon and Energy Sources

Carbon Source	Energy Source	
	Chemical	Light
Organic (Heterotrophs)	Chemoorganotrophs Bacteria, yeasts, fungi	Photoorganotrophs Nonsulfur purple bacteria
Carbon dioxide (Autotrophs)	Chemolithotrophs Hydrogen, sulfur, iron denitrifing bacteria	Photolithotrophs Blue-green algae, photosynthetic bacteria

The different reactions involved in photosynthesis can be assembled in two categories.

- One category of light-dependent reactions. Light energy is converted into chemical energy through pigments such as chlorophyll. This electron transfer process results in ATP synthesis and the formation of NADPH. Water is oxidized and oxygen released.
- The second reaction category is not light-dependent (dark reactions). The ATP and NADPH formed during the light-dependent reaction serve as energy source enabling conversion of CO_2 into carbohydrate via the carbon-fixation cycle (or Calvin-Benson cycle).

Conversion of CO_2 into cell material can be summarized by the following equation:

$$6\,CO_2 + 18\,ATP + 12\,(NADPH + H^+) \rightarrow \text{fructose-6-P} + 18\,ADP + 12\,NADP^+ + 17\,P_i$$

The assimilation of 1 mole of CO_2, therefore requires 3 moles of ATP and 2 moles of NADPH. Photosynthesis occurs in green algae, cyanobacteria and a few species of bacteria. Water is used as an electron donor for algae, Eq. (1), and oxygen is produced.

$$CO_2 + H_2O + \text{light} \rightarrow (CH_2O) + O_2 + \text{chemical energy} \qquad (1)$$

CH_2O = organic molecule.

Bacterial photosynthetic processes, Eq. (2), are essentially anaerobic. Molecular hydrogen, sulfur compounds (H_2S) and some organic compounds are electron donors.

$$CO_2 + 2\,H_2A + \text{light} \rightarrow (CH_2O) + 2\,A + H_2O \qquad (2)$$

3.3 Hydrocarbon Metabolism

The ability of some microorganisms to use aliphatic and/or aromatic hydrocarbons as sole carbon source has been described by MIYOSHI (1895). Hydrocarbons, *n*-alkanes and paraffins are produced from gas oil and kerosene.

3.3.1 Oxidation of Aliphatic Hydrocarbons

The metabolism of *n*-alkanes can be divided into several stages: (1) incorporation of alkanes in the cell, (2) oxidation of the alkanes into their corresponding fatty acids, and (3) degradation of the fatty acids.

Penetration of alkanes is by (1) direct contact between the alkane and the cell, (2) uptake via an accommodated alkane phase, and (3) penetration by pseudo-solubilized alkanes (TANAKA and FUKUI, 1989).

Linear chains of *n*-alkanes are more easily attacked than ramified chains. Oxidation of alkanes into a fatty acid takes place in several stages (Fig. 1) by two main pathways, the subterminal oxidation and the terminal oxidation pathways (monoterminal and diterminal oxidation).

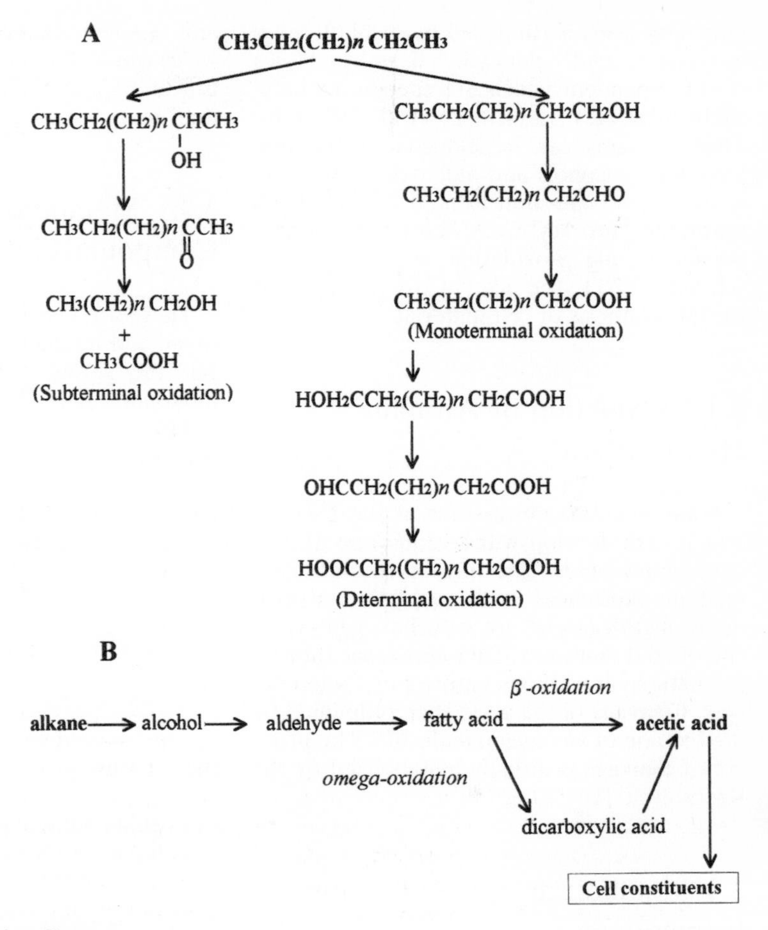

Fig. 1. Microbial oxidation of *n*-alkanes.
A – Subterminal and terminal oxidation of *n*-alkanes (FUKUI and TANAKA, 1981).
B – Schematic representation of terminal oxidation pathways of *n*-alkanes.

The first stages of the subterminal oxidation pathway convert the alkane into a secondary alcohol and then into the corresponding ketone.

In terminal oxidation, the methyl group is attacked and the first intermediary is the corresponding alcohol, which is oxidized into an aldehyde and then a fatty acid. The first stage (transformation of alkane into alcohol) is catalyzed by monooxygenases.

There are several different types of monooxygenase in bacteria. The best-known system is present in *Pseudomonas*; it consists of three proteins: rubredoxin, NADH rubredoxin reductase and ω-hydroxylase (PETERSON et al., 1967). *Corynebacterium* possesses a hydroxylation system consisting of cytochrome P-450 and NADH cytochrome c reductase (CARDINI and JURTSHUK, 1970).

Monoterminal oxidation is dominant in yeasts. Participation of cytochrome P-450 and NADH cytochrome c reductase has also been demonstrated in *Candida tropicalis, Candida guilliermondii*, and *Lodderomyces elongisporus* (LEBEAULT et al., 1971; GALLO et al., 1971; RIEGE et al., 1981). In diterminal oxidation, the fatty acid obtained after monoterminal oxidation is hydroxylated again (ω-oxidation) to form dicarboxylic acid.

Oxidation of the higher alcohols formed

into fatty acids is catalyzed by alcohol dehydrogenase and aldehyde dehydrogenases NAD dependent which are specific to long-chain substrates (LEBEAULT et al., 1970a, b). These systems can be induced by alkanes, long-chain alcohols and aldehydes. The fatty acids formed are transformed by acyl-CoA synthetase into acyl-CoA; the latter is transformed during β-oxidation into acetyl-CoA (cf. Sect. 3.7.1). The acetyl-CoA is then used for the synthesis of cell material.

3.3.2 Oxidation of Aromatic Hydrocarbons

Some microorganisms such as the *Pseudomonas* can develop with assimilation of aromatic compounds. The three principal intermediate substances in the oxidation of aromatic hydrocarbons are catechol, protocatechuate and gentisate. They represent the last aromatic hydrocarbons before ring fission occurs. Cleavage of the cycle is accompanied by the fixation of an oxygen molecule. The product of cleavage is directly metabolized by the Krebs cycle (DOELLE, 1981).

3.4 Metabolism of One-Carbon Compounds

Compounds with one or more carbon atoms but no carbon bonds are referred to as one-carbon compounds (C1). The following compounds are in this category: methane, methanol, methylamine, formaldehyde, formic acid, dimethylether or dimethylamine. Methane and methanol are the most commonly used. Methane is obtained from natural gas, coal or by gasification of wood. Methanol is produced by chemical synthesis.

C1 compounds can be used as energy and carbon source by microorganisms referred to as being methylotrophic. Use of methylotrophic bacteria has been known since the beginning of the century (SÖHNGEN, 1906), whereas that of eukaryotes was first described by OGATA et al. only in 1970. Numerous yeasts (LEE and KOMAGATA, 1980) and several filamentous fungi (GONCHAROVA et al., 1977) can metabolize methanol. The metabolism of C1 compounds in bacteria is different than that of yeasts.

3.4.1 Metabolization of C1 Compounds by Bacteria

There are two pathways for the use of methane and methanol in methylotrophic bacteria, the serine pathway and the ribulose monophosphate (RMP) pathway.

Full oxidation of methane and methanol is relatively simple:

CH_4 (methane) → CH_3OH (methanol) → HCOH (formaldehyde) →

(from HCOH: ↙ Serine pathway; ↘ Ribulose phosphate pathway)

HCOOH (formate) → CO_2 (carbon dioxide)

The formaldehyde (HCOH) formed during these pathways may then be:

- either fully oxidized into CO_2, two moles of $(NADH^+ + H^+)$ are formed in the last two reactions, and reoxidation by the respiratory chain permits the formation of ATP (4 or 6 ATP are formed according to the number of phosphorylation sites) used in the biosynthesis of cell material;
- or taken up in the cell material by the serine pathway (Fig. 2) or the ribulose monophosphate pathway (Fig. 3).

3.4.1.1 Serine Pathway

Carbon uptake occurs at two different degrees of oxidation in the form of CO_2 and of formaldehyde (HCOH) in the serine pathway. Formaldehyde is incorporated in the carbon structure of glycine; the serine thus formed is converted into 3-phosphoglycerate (3-PGA) and phosphoenolpyruvate (PEP) on which CO_2 becomes fixed and gives oxaloacetate (OAA). Glycine regeneration occurs

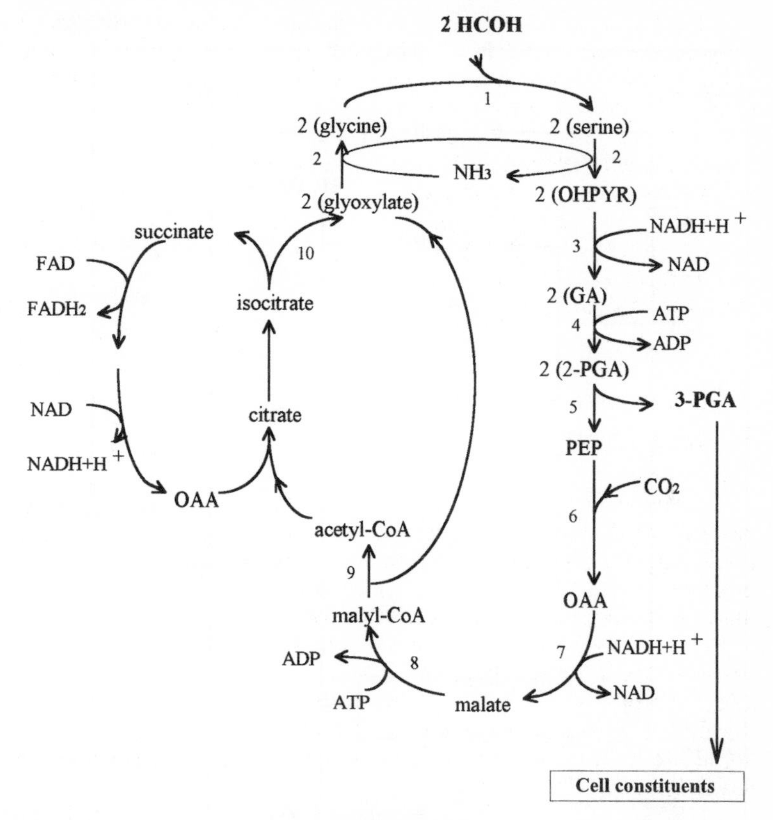

Fig. 2. Serine pathway (COLBY et al., 1979). **1,** serine transhydroxymethylase; **2,** serine glyoxylate amino-transferase; **3,** hydroxypyruvate reductase; **4,** glycerate kinase; **5,** phosphopyruvate hydratase; **6,** phosphoenolpyruvate carboxylase; **7,** malate dehydrogenase; **8,** malate thiokinase; **9,** malyl-CoA lyase; **10,** isocitrate lyase. OHPYR, hydroxypyruvate; GA, glycerate; PGA, phosphoglycerate; PEP, phosphoenolpyruvate; OAA, oxaloacetate.

thanks to a malyl-CoA and to Krebs cycle enzymes. The 3-phosphoglycerate formed enters the EMP pathway (cf. Sect. 3.6.1.1) and is used for synthesis of cell material (Fig. 2).

The overall balance of the reaction is summarized by the following equation (COLBY et al., 1979):

$$2\,HCOH + CO_2 + FAD + 2\,NADH_2 + 3\,ATP \rightarrow CH_2OP\ CHOH\ COO^- + 3\,ADP + 2\,NAD + FADH_2$$

3.4.1.2 Ribulose Monophosphate Pathway (RMP)

Only formaldehyde is taken up in this pathway. The RMP has three stages (Fig. 3) (COLBY et al., 1979).

In stage 1, three molecules of formaldehyde condense with three molecules of ribulose-5-phosphate to form three molecules of fructose-6-phosphate. This stage is common to all methylotrophic bacteria using the pathway.

In stage 2, a molecule of fructose-6-phosphate is divided into two molecules of C3 compounds. Depending on the strain, the reactions are caused either by glycolysis enzymes or by Entner–Doudoroff pathway enzymes (cf. Sect 3.6.1.3).

In the last stage, 3 molecules of ribulose-5-phosphate are regenerated from 2 molecules of fructose-6-phosphate and 1 molecule of glyceraldehyde-3-phosphate produced in the first two stages.

The net balance of the cycle is the formation of one molecule of glyceraldehyde-3-P (GAP) from three molecules of formaldehyde (HCOH). The glyceraldehyde-3-P formed is used in the synthesis of cell material.

$$3\,HCOH + 3\,ATP \rightarrow 1\,GAP + 3\,ADP + 2\,P_i$$

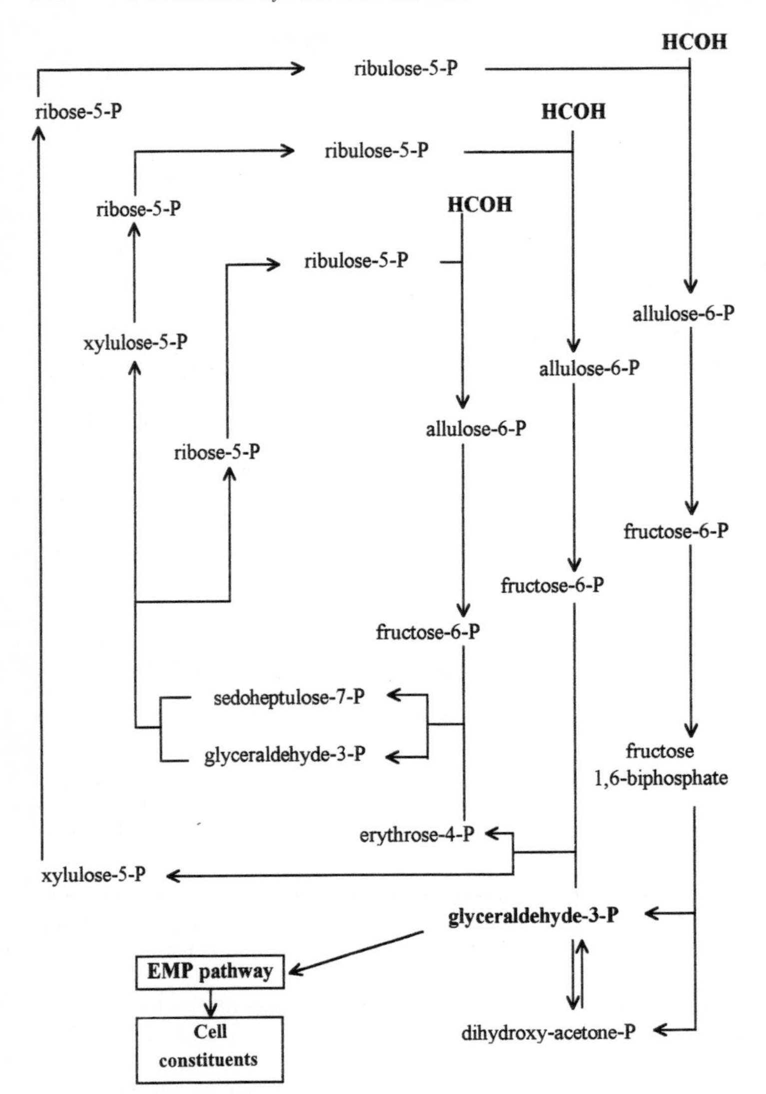

Fig. 3. Ribulose monophosphate pathway of methylotrophic bacteria.

3.4.2 Methanol Metabolism in Yeasts

The methanol metabolism in yeasts is much more complicated than in bacteria. There are many differences; nature and localization of the enzymes involved, cell compartmentation and the role of mitochondria in energy transduction (Fig. 4). Methanol cannot be taken up by the serine pathway because hydroxypyruvate reductase is not present. There is only the hexose monophosphate pathway in yeasts; it is similar to that described in bacteria (COLBY et al., 1979).

The methanol formed is oxidized into formaldehyde by a flavine-dependent dehydrogenase which catalyzes the reaction

$$CH_3OH + O_2 \rightarrow HCOH + H_2O_2$$

(Fig. 4). Oxidation of formate into CO_2 takes place in the cytoplasm, and NAD is regenerated by the NADH dehydrogenases on the

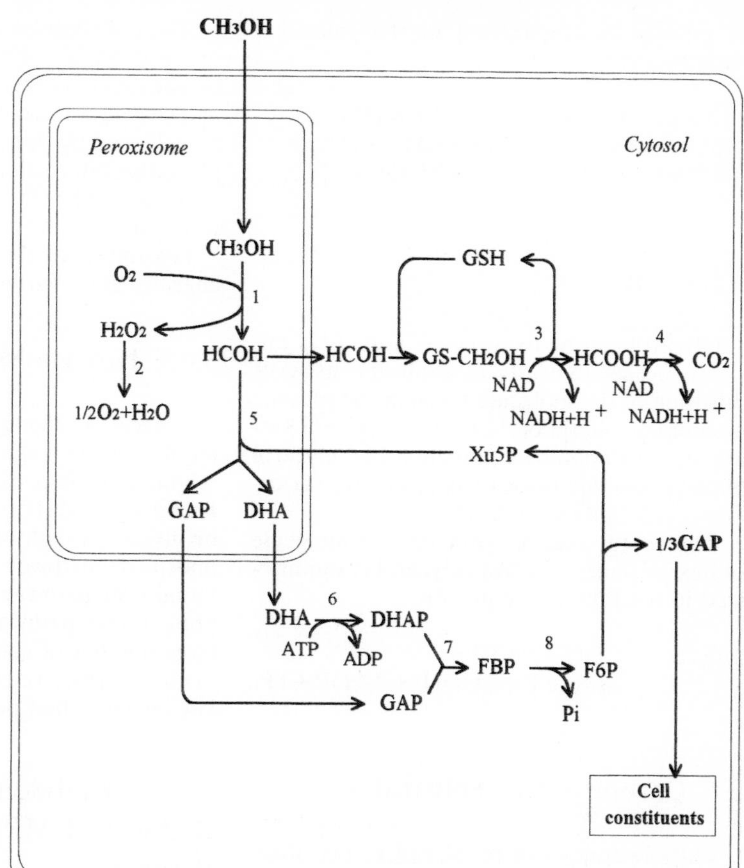

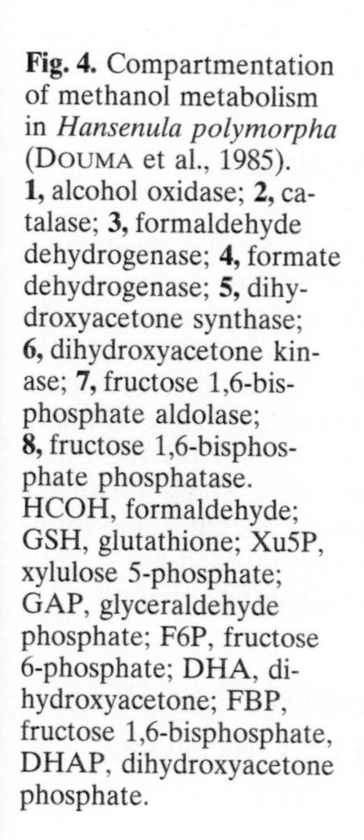
Fig. 4. Compartmentation of methanol metabolism in *Hansenula polymorpha* (DOUMA et al., 1985). **1**, alcohol oxidase; **2**, catalase; **3**, formaldehyde dehydrogenase; **4**, formate dehydrogenase; **5**, dihydroxyacetone synthase; **6**, dihydroxyacetone kinase; **7**, fructose 1,6-bisphosphate aldolase; **8**, fructose 1,6-bisphosphate phosphatase. HCOH, formaldehyde; GSH, glutathione; Xu5P, xylulose 5-phosphate; GAP, glyceraldehyde phosphate; F6P, fructose 6-phosphate; DHA, dihydroxyacetone; FBP, fructose 1,6-bisphosphate, DHAP, dihydroxyacetone phosphate.

outer face of the internal membrane of the mitochondria which are connected to the respiratory chain at the level of ubiquinone, short-circuiting the first phosphorylation site. Oxidation of one molecule of $NADH+H^+$ therefore only produces two ATP (HARDER and VEENHUIS, 1989). this may explain the fact that the yeasts have lower substrate–biomass transformation efficiency than bacteria (ANTONY, 1982).

3.5 Ethanol and Glycerol Metabolism

These two substrates can be used as carbon and energy sources under aerobic conditions.

3.5.1 Ethanol

The first stage of ethanol transformation is oxidation into acetaldehyde which is itself oxidized into acetate. Two moles of $NADH+H^+$ are produced during these reactions. Acetate is either excreted out of the cell or converted into acetyl CoA and incorporated in the Krebs cycle at the level of the citrate or the malate according to the cell energy state. Use of ethanol, a C2 compound, requires the functioning of gluconeogenesis and the glyoxylic shunt in the yeast, enabling synthesis of C3 compounds from C2 compounds.

The energy cost of the formation of a molecule of fructose-6-phosphate from molecules of ethanol is summarized in the following equation:

$$4\,\text{ethanol} + 12\,\text{NAD} + 4\,\text{CoA} + 8\,\text{ATP} + 2\,\text{UQ} \rightarrow \text{hexose-P} + 12(\text{NADH}^+ + \text{H}^+) + 4\,\text{AMP} + 4\,\text{ADP} + 2\,\text{CO}_2 + 2\,\text{UQH}_2 + 11\,\text{P}_i$$

3.5.2 Glycerol

Glycerol can be oxidized by numerous microorganisms. In yeasts, catabolism consists of diffusion of the molecule through the plasmic membrane, phosphorylation by a glycerol kinase and oxidation by a mitochondrial glycerol phosphate-ubiquinone oxidoreductase (GANCEDO and SERRANO, 1989).

The energy cost of synthesis of a molecule of hexose phosphate from glycerol is summarized in the following equation:

$$2\,\text{glycerol} + 2\,\text{UQ} + 2\,\text{ATP} \rightarrow \text{hexose-P} + 2\,\text{UQH}_2 + 2\,\text{ADP} + 1\,\text{P}_i$$

3.6 Carbohydrate Substrates

Carbohydrates can be placed in two main categories: saccharides and polysaccharides. Saccharides include:

- hexoses: glucose, fructose, galactose, mannose;
- disaccharides: sucrose, lactose, maltose;
- pentoses: xylose, arabinose.

These various saccharides are found in corn syrup, molasses, whey, sulfite waste liquor and vegetable and fruit wastes.
Polysaccharides include:

- starch, a glucose polymer;
- lignocellulose consisting of cellulose (glucose polymer), hemicellulose (xylose polymer) and lignin.

Starch is abundant in maize, wheat, potato, cassava and rice. Lignocellulose is found in wood chips, crop residues, forest and mill residues, vegetable and fruit wastes.

These glycosides enter cells as mono-, di- or trisaccharides. Prior hydrolysis (enzymatic or chemical) is required for polysaccharides and certain disaccharides. Sugar transport can be achieved by three different processes, simple diffusion, facilitated diffusion and active transport. The last two processes involve transport systems or permeases.

Hydrolysis of the various di- and polysaccharides gives various assimilable hexoses.

3.6.1 Hexose Metabolism

Glucose is the carbon source most commonly used by microorganisms. It can be metabolized in four different pathways: (1) the Embden–Meyerhof–Parnas pathway (EMP) or glycolysis pathway, (2) the hexose monophosphate pathway (HMP), (3) the Entner–Doudoroff pathway (ED) and (4) the phosphoketolase pathway (PK). These possess a large number of common enzymes and intermediaries (Fig. 5). However, each has specific key enzymes that permit differentiation.

3.6.1.1 Embden–Meyerhof–Parnas Pathway (EMP) or Glycolysis Pathway

This pathway is found in most microorganisms. It permits oxidation of glucose into pyruvate and is anaerobic. Phosphofructokinase is one of the key enzymes of the pathway. The glycolysis balance can be broken down as follows:

$$\text{glucose} + 2\,\text{ATP} \rightarrow 2\,\text{glyceraldehyde-3-P} + 2\,\text{ADP}$$

$$2\,\text{glyceraldehyde-3-P} + 4\,\text{ADP} + 2\,\text{P}_i + 2\,\text{NAD}^+ \rightarrow 2\,\text{pyruvate} + 4\,\text{ATP} + 2(\text{NADH} + \text{H}^+)$$

The overall glycolysis balance is as follows:

$$\text{glucose} + 2\,\text{ATP} + 2\,\text{NAD}^+ \rightarrow 2\,\text{pyruvate} + 4\,\text{ATP} + 2(\text{NADH} + \text{H}^+)$$

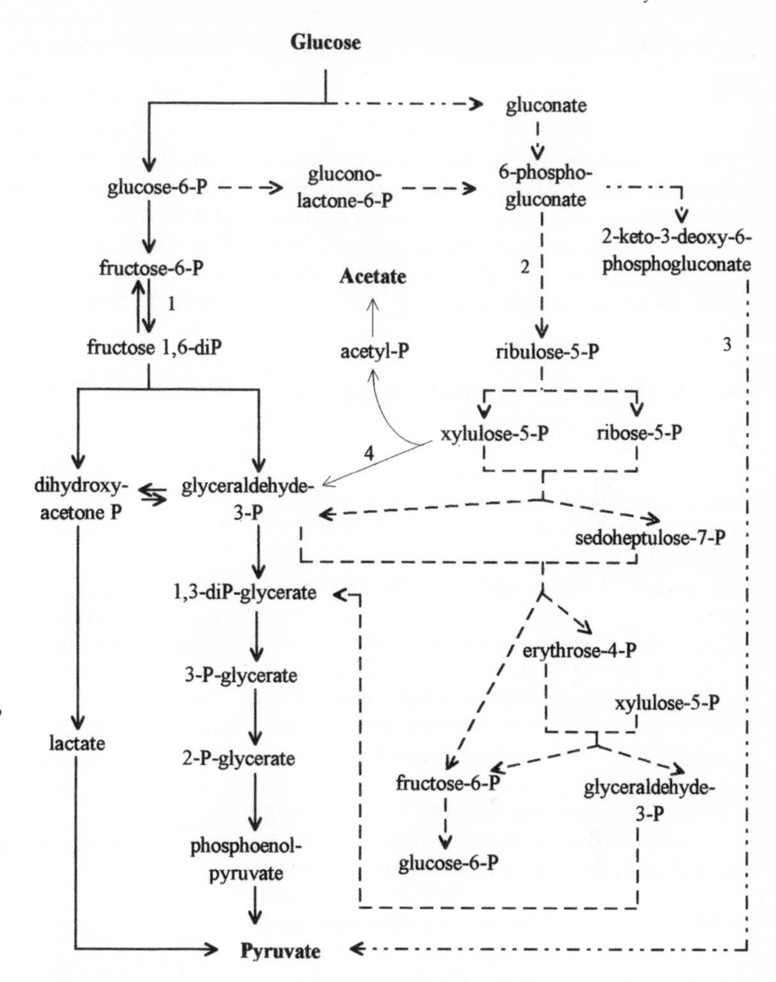

Fig. 5. Interlinkage of EMP (**——**), HMP (----), ED (-··-··) and PK (——). **1,** phosphofructokinase (EMP); **2,** 6-phosphogluconate dehydrogenase (HMP); **3,** 2-keto-3-deoxy-6-phosphogluconate aldolase (ED); **4,** xylulose-5-phosphoketolase (PK).

3.6.1.2 Hexose Monophosphate Pathway (HMP)

This is a strictly aerobic pathway and is found in numerous yeasts and bacteria. Via oxidation of three glucose molecules one glyceraldehyde molecule and two fructose-6-P molecules are formed. Its essential role is to supply the $NADPH+H^+$ required for the biosynthesis reactions and to supply the ribose required for the synthesis of nucleotides and nucleic acids. One of the key enzymes is 6-phosphogluconate dehydrogenase. The balance is:

$$3\,\text{glucose} + 6\,NADP^+ + 3\,ATP \rightarrow 2\,\text{fructose-6-P} + \text{glyceraldehyde-3-P} + 3\,CO_2 + 3\,ADP + 6\,(NADP + H^+)$$

3.6.1.3 Entner–Doudoroff Pathway (ED)

This pathway is only present in bacteria. It permits the aerobic oxidation of glucose in *Pseudomonas* bacteria and oxidation in *Zymomonas mobilis*. It possesses stages in common with glycolysis (Fig. 5). One of the key enzymes is 2-keto-3-deoxy-6-phosphogluconate.

3.6.1.4 Phosphoketolase Pathway (PK)

PK is used solely by heterofermentative lactobacilli and results in the formation of ethanol, lactate, CO_2 and sometimes acetate. Heterofermentative lactobacilli possess only part of glycolysis, the glyceraldehyde phosphate dehydrogenase system (Fig. 5).

3.6.1.5 Pyruvate Oxidation and the Glyoxylic Shunt

In the presence of air, strict or facultative aerobic microorganisms oxidize acetyl-CoA through the Krebs cycle (or the tricarboxylic acid cycle (TCA)) and by the glyoxylic shunt.

TCA also permits oxidation of the acetate produced during glycolysis, by the pentose pathway and the oxidation of fatty acids. The intermediate substances formed serve as the precursors of numerous biosynthesis reactions (amino acids, etc.).

The microorganisms capable of developing on 2- and 4-carbon substrates such as acetate, ethanol and succinic acid possess not only TCA enzymes but also isocitrate lyase and malate synthetase. The first enzyme obtains a C4 compound, succinate, and a C2 compound, glyoxylate, from a C6 compound, isocitrate. Malate synthetase condenses glyoxylate (C2) and acetyl-CoA (C2) into malate (C4). This cyclic mechanism is called the glyoxylate cycle.

The net oxygen balance is as follows:

1. TCA cycle:

$$\text{acetyl-CoA} + 3\,NAD^+ + UQ + ADP + P_i \rightarrow 2\,CO_2 + 3\,(NADH + H^+) + UQH_2 + ATP + CoA$$

2. Glyoxylate cycle:

$$2\,\text{acetyl-CoA} + NAD^+ \rightarrow \text{succinate} + (NADH + H^+) + 2\,CoA$$

In eukaryotes, the electrons are conveyed to the oxygen by the respiratory chain. TCA and respiratory chain enzymes are mitochondrial. If it is considered that oxidation of one $NADH + H^+$ is coupled with the formation of 3 ATP and oxidation of one $FADH + H^+$ with the formation of 2 ATP, the overall oxidation balance of the glucose is 2 ATP during glycolysis and 36 ATP during the Krebs cycle. It is more difficult to know how many ATPs are formed in bacteria. Experimental and theoretical values range from 16 to 36.

As an example, comparison of ATP production by the various glucose oxidation pathways is shown in Tab. 5.

Tab. 5. ATP Production by Different Glucose Oxidation Pathways

Pathway	Moles of ATP
Homolactic fermentation	
glucose → 2 lactate + $2\,H^+$	
via EM	2
via ED	1
Heterolactic fermentation	
glucose → ethanol + CO_2 + lactate + H^+	
via HMP	1
Alcoholic fermentation	
glucose → ethanol + $2\,CO_2$	
via EM	2
Full oxidation of glucose	
glucose + $6\,O_2 \rightarrow 6\,CO_2 + 6\,H_2O$	
via EM + TCA (P/O = 3)	38

3.6.2 Disaccharides

3.6.2.1 Sucrose

Sucrose or glucosyl-α-[β-fructofuranoside] is mainly found in molasses. Beet molasses contains about 50% w/v sucrose, whereas cane molasses contains about 30%. Sucrose is hydrolyzed by a β-D-fructofuranosidase, or invertase, into glucose and fructose. Hydrolysis is generally performed outside the cell.

3.6.2.2 Lactose

The main source of lactose, or galactosyl-β-1-4 glucose, is whey, obtained from milk during cheese manufacture. Few microorganisms metabolize lactose. Less than 12% of yeasts are capable of developing on this substrate (BARNETT, 1981). Lactose is hydrolyzed into galactose and glucose by an intracellular enzyme specific to β(1-4) linkages, β-galactosidase. The galactose released is metabolized by the enzymes of the Leloir Kalckar pathway. Galactose is phosphorylated by galactokinase into galactose-1-phosphate, and uridyl transferase converts galactose-1-P and UDP-glucose into UDP-galactose and glucose-1-P. Epimerase converts UDP-galactose into UDP-glucose.

3.6.3 Polysaccharides

3.6.3.1 Starch

Starch is the main form of reserve carbohydrate in the plant kingdom. It is the main constituent of certain seeds, roots and tubers (maize, wheat, barley, rice, potatoes, etc. ...). Starch is made up of amylose and amylopectin. Amylose consists of a linear chain of glucose molecules connected by α(1-4) glucosidic linkages. Amylopectin consists of ramifications of 20 to 40 glucose units linked by α(1-4) linkages connected to a longer chain by α(1-6) linkages.

Starch is broken down by glycosyl-hydrolases or amylases. These enzymes can be placed in three categories according to their specificity.

- Enzymes specific to α(1-4) linkages:
  - endo α(1-4) glucanases (α-amylase) which attack α(1-4) linkages of amylose and amylopectin and release oligosides from 2 to 7 glucose units. They are found in animals and plants and also in many microorganisms (*Aspergillus oryzae, Bacillus subtilis*, etc.)
  - exo α(1-4) glucanases (β-amylase) hydrolyze chains from their reducing extremities and release maltose. The best-known are from plants and a few are of bacterial origin (*Bacillus* sp., *Pseudomonas* sp., *Streptomyces* sp.) (LELOUP et al., 1991)
- Enzymes specific to α(1-6) linkages or debranching enzymes:
  - endo α(1-6) glucanases: pullulanase and isoamylase, found only in microorganisms (*Aerobacter aerogenes*)
  - exo α(1-6) glucanases: exopullulanase
- Enzymes specific to α(1-4) and α(1-6) linkages or glucoamylases which release glucose. They are mainly produced by microorganisms, and especially molds such as *Aspergillus, Penicillium, Rhizopus* and *Endomycopsis*.

3.6.3.2 Cellulose

Cellulose is the major constituent of wood and textile fibers. It is the most abundantly synthesized organic molecule in the world (in each photosynthesis cycle). It is a glucosidic polymer consisting of some 3000 glucose units bound by β(1-4) linkages. Cellulose is broken down by β(1-4) glucanases in microorganisms (*Trichoderma viride, T. koningii*, etc.). Endo β(1-4) glucanases break the long cellulose chains into several fragments. Exo β(1-4) glucanases allow the release of cellobiose from the extremities of the chains. The cellobiose is hydrolyzed into two glucose molecules by β(1-4) glucosidases.

3.6.3.3 Hemicellulose and Lignin

Hemicellulose is defined as a complex set of very varied polyosides: homopolysaccharides and heteropolysaccharides. The latter are hydrolyzed in an alkaline medium into a mixture of hexose, pentose and uronic acid. The central structure generally consists of a β(1-4) linked xylose or glucose polymer. This chain has ramifications of galactose or galacturonic and glucuronic acid. Average polymerization of hemicellulose is 150 to 200 units and it forms 25 to 40% of the dry weight of plants.

Lignin is not glucidic. It consists of phenylpropane polymers linked by carbon–carbon bonds. Lignin forms 14 to 30% of the dry weight of plants.

These two compounds are rarely broken down directly by microorganisms, and preliminary treatment is necessary.

3.7 Lipids

Lipids form a heterogeneous group of compounds related to the fatty acids. They have the common property of being insoluble in water and soluble in non-polar solvents such as ether, chloroform and benzene. Lipids include fats, oils and waxes.

Lipids can be divided into two main categories:

(1) *simple lipids:* esters of fatty acids and various alcohols,

- fats or acylglycerol: esters of fatty acids and glycerol,
- waxes: esters of fatty acids and high molecular weight monohydric alcohols,

(2) *complex lipids:* esters of fatty acids whose molecule contains various groups in addition to the fatty acid and the alcohol,

- phospholipids: esters of fatty acids and glycerol containing phosphate groups, nitrogenous substances and various other constituents,
- glycolipids: derivatives of fatty acids and sugars.

A large proportion of the lipids available comes from oil plants: rape, sunflower, maize, olive and oil palm. Lipids can also be found in factory effluent, domestic wastes, fish oil and fishery by-products. These substances consist of fatty acids and acylglycerol in varying proportions. Animal fats are not easy to use because of their solid form.

Lipids can be used as an energy and carbon source by yeasts of the genera *Candida, Cryptococcus, Rhodotorula, Torulopsis, Yarrowia* and *Trichosporon*, which take up the fatty acids by facilitated diffusion (low concentrations) and simple diffusion (high concentrations) (RATLEDGE and EVANS, 1989).

3.7.1 Breakdown of Fatty Acids

The first stage in the degradation process is the conversion of the fatty acid into acyl-coenzyme A. One mole of ATP is used by this reaction.

$$\text{R-COOH} + \text{ATP} + \text{CoA-SH} \rightarrow \text{RCO-S-CoA} + \text{AMP} + \text{PP}_i + \text{H}_2\text{O}$$

There are three possible fatty acid breakdown pathways in yeasts: α-, β- and ω-oxidation.

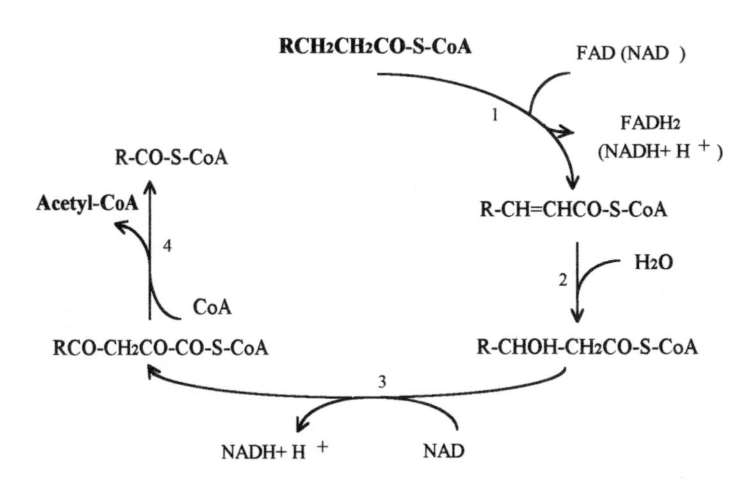

Fig. 6. The β-oxidation cycle for acyl-CoA esters.
1, fatty acyl-CoA oxidase FAD linked or fatty acyl-CoA dehydrogenase-NAD$^+$ linked;
2, enoyl hydratase; **3,** L-3-hydroxyacyl-CoA dehydrogenase; **4,** oxoacyl-CoA thiolase.

Only a few yeasts display α-oxidation. The ω-oxidation pathway is found in common with alkanes (cf. Sect. 3.3). The β-oxidation pathway is the most widespread and most important system.

β-Oxidation is catalyzed by four soluble enzymes (Fig. 6). The first stage is catalyzed either by a mitochondrial acyl-CoA dehydrogenase and reoxidation of the $NADH + H^+$ formed is combined with ATP production, or by a peroxisomal acyl-CoA oxidase; in the latter case reoxidation of the FADH is not related to ATP production.

The second reaction is catalyzed by an enoyl-CoA hydratase permitting the formation of 3-hydroxyacyl-CoA. In the third stage, the compound is converted by a 3-hydroxyl-CoA dehydrogenase with NAD as electron acceptor.

The final stage is catalyzed by acetyl-CoA acetyltransferase resulting in the formation of a C2 compound, acetyl-CoA, and a compound with two carbon atoms less than the initial fatty acid.

3.7.2 Triacylglycerols

Triacylglycerols are hydrolyzed to fatty acid and glycerol by intra- or extracellular lipases. These are reversible reactions. Lipases can be classified in several groups according to their specificity (RATLEDGE and EVANS, 1989; RATLEDGE and TAN, 1989):

Non-specific lipases

$$\begin{array}{l} R_1COOCH_2 \\ \quad\;\; | \\ R_2COOCH + 3\,H_2O \Leftrightarrow R_1COOH + R_2COOH + R_3COOH + \\ \quad\;\; | \\ R_3COOCH_2 \end{array} \begin{array}{r} CH_2OH \\ | \\ CHOH \\ | \\ CH_2OH \\ | \\ R_2COOCH \\ | \\ R_3COOCH \end{array}$$

1,3-Specific lipases

$$\begin{array}{l} R_1COOCH_2 \\ \quad\;\; | \\ R_2COOCH + 2\,H_2O \Leftrightarrow \\ \quad\;\; | \\ R_3COOCH_2 \end{array} \begin{array}{l} CH_2OH \\ \quad | \\ R_2COOCH + R_1COOH + R_3COOH \\ \quad | \\ CH_2OH \end{array}$$

2-Specific lipases

$$\begin{array}{l} R_1COOCH_2 \\ \quad\;\; | \\ R_2COOCH + H_2O \Leftrightarrow \\ \quad\;\; | \\ R_3COOCH_2 \end{array} \begin{array}{l} R_1COOCH_2 \\ \quad\;\; | \\ \qquad CHOH + R_2COOH \\ \quad\;\; | \\ R_3COOCH_2 \end{array}$$

Fatty-acyl specific lipases

$$\begin{array}{l} R_1COOCH_2 \\ \quad\;\; | \\ R_2COOCH + 2\,H_2O + \\ \quad\;\; | \\ R_3COOCH_2 \end{array} \begin{array}{l} R_2COOCH_2 \\ \quad\;\; | \\ R_1COOCH \Leftrightarrow 2\,R_1COOH + \\ \quad\;\; | \\ R_3COOCH_2 \end{array} \begin{array}{r} CH_2OH \\ | \\ R_2COOCH \\ | \\ R_3COOCH \end{array} + \begin{array}{r} R_2COOCH_2 \\ | \\ CHOH \\ | \\ R_3COOCH_2 \end{array}$$

3.7.3 Energy Balance

Using the breakdown of palmitate (C16) as an example, the overall oxidation equation can be written as follows:

$$\text{palmitoyl-S-CoA} + 7\,\text{CoA} + 7\,\text{FAD} + 7\,\text{NAD} + 7\,H_2O \rightarrow 8\,\text{acetyl-CoA} + 7\,\text{FADH}_2 + 7\,\text{NADH} + 7\,H^+$$

Five moles ATP are produced for each of the seven acetyl-CoA molecules formed by β-oxidation. The breakdown of one mole acetyl-CoA permits the formation of twelve moles ATP. The overall balance is thus $(7\times5)+(8\times12)=131$ mol ATP.

3.8 Principal Pathways of the Oxidative Metabolism of Microorganisms

3.8.1 Schematic Recapitulation

Degradation of the various substrates – hydrocarbons, C1 compounds, fatty acids and ethanol – leads to the formation of C3 compounds (triose-P) and/or C2 compounds (acetyl-CoA). Degradation of carbohydrates and glycerol leads to the formation of pyruvate (C3) through glycolysis. These compounds are then incorporated in the cell material.

Oxidation of acetyl-CoA is realized by TCA (cf. Sect. 3.6.1.5). During the cycle, the intermediates serve as precursors for biosynthesis reactions. The NADH formed is reoxidized at the respiratory chain level (Fig. 7).

3.8.2 Respiratory Chain/Energy Conservation

The respiratory chain consists of a multienzyme complex. NADH and succinate are the major substrates. The mechanism of the reaction involves a series of steps in which a component oxidizes its neighbor on the substrate side, itself being reduced. The energy produced by the oxido-reduction reactions is coupled with ATP synthesis. The term oxidative phosphorylation is used to describe the entire process.

The respiratory chain is located in the mitochondria in eukaryotes. It consists of flavoproteins (FP) combined with dehydrogenases, coenzymes Q (CoQ) and cytochromes. Distinction is made between three types of cytochrome: cytochromes b, c and a.

$$\text{NADH}_2 \rightarrow \text{FP} \rightarrow \underset{\downarrow H^+}{\text{CoQ}} \rightarrow \text{cyt b} \rightarrow \underset{\downarrow H^+}{\text{cyt c}_1\text{c}} \rightarrow \text{cyt aa}_3 \rightarrow \underset{\downarrow H^+}{O_2}$$

There are three phosphorylation sites, the first between the flavoproteins and the CoQ, the second between cytochromes b and c_1 and the third at the level of cytochrome oxidase (aa_3). Measurement of the ratio P/O, the number of phosphorylation sites per number of oxygen atoms consumed, makes it possible to calculate the number of functional phosphorylation sites in a cell.

The respiratory chain of the bacteria sited in the plasma membrane is similar to that of eukaryotes. It consists of flavoprotein dehydrogenase, quinones and cytochromes (b, c and a). Cytochromes b are always present, and cytochromes c are absent in a large number of bacteria. There are four types of oxidase: aa_3 system, cytochromes o, a_1 and d (or a_2).

Oxidative phosphorylation differs in bacteria and yeasts. In bacteria, adenine nucleotides cannot cross the plasmic membrane, and it is, therefore, difficult to know the P/O ratio. Nevertheless, a phosphorylation site associated with NAD dehydrogenase and another site associated with the cytochrome b complex have been detected. A third site was found in bacteria possessing site c cytochromes (PREBBLE, 1981).

$$\text{NADH}_2 \rightarrow \text{FP} \rightarrow \underset{\downarrow H^+}{\text{Q}} \rightarrow \text{cyt b} \rightarrow \underset{\downarrow H^+}{\text{(cyt c)}} \rightarrow \underset{\downarrow (H^+)}{\text{cyt oxidase}} \rightarrow$$

A unifying biochemical concept emerges from the summary of all aerobic metabolic processes (Fig. 7), demonstrating the economics of energy production and cellular biosynthesis of microorganisms (DOELLE, 1981).

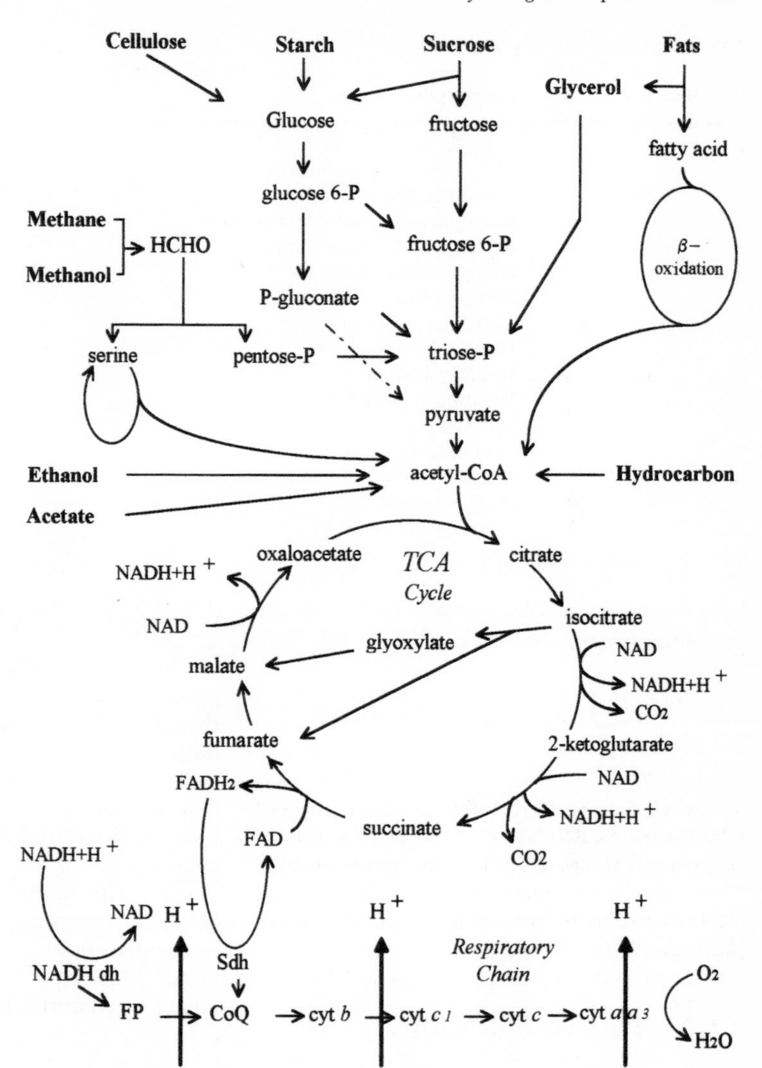

Fig. 7. Schematic representation of aerobic catabolism in microorganisms. FP, flavoprotein; Sdh, succinate dehydrogenase; CoQ, coenzyme Q.

4 Physiological Aspects

The growth parameters of a microorganism depend on the microorganism itself, the carbon source, the environment: physicochemical parameters (pH, temperature and oxygen) and the composition of the growth medium. Growth can be evaluated using several parameters: the specific growth rate (μ) and the doubling time (t_d), the growth yield ($Y_{X/S}$ = g of biomass produced per g of substrate uptake), the metabolic coefficients (qS, qO_2, qCO_2), affinity for a substrate (K_s, K_m) and maximum biomass productivity (P_x). Definitions of the different parameters and the units used are shown in Tab. 6.

4.1 Growth Parameters

4.1.1 Specific Growth Rate

This is the increase in biomass (dx) over a very short period (dt) for total biomass (X), $\mu=(1/X)(dx/dt)$. The following equation can

Tab. 6. Definitions of the Parameters Used

Parameters	Definition	Unit
μ	Specific growth rate	h^{-1}
μ_m	Maximum specific growth rate	h^{-1}
t_d	Doubling time	h
S	Residual substrate concentration	$g L^{-1}$ or mM
Sr	Initial substrate concentration	$g L^{-1}$ or mM
X	Biomass concentration	$g L^{-1}$
D	Dilution rate (F/V)	h^{-1}
D_m	Maximum dilution rate	h^{-1}
F	Input or output flow	$L h^{-1}$
V	Reactor volume	L
K_s	Saturation constant	$g L^{-1}$ or mM
qS	Specific substrate uptake rate	$g g^{-1} h^{-1}$
qO_2	Specific oxygen uptake rate	$g g^{-1} h^{-1}$ or mmol $g^{-1} h^{-1}$
qCO_2	Specific carbon dioxide production rate	$g g^{-1} h^{-1}$ or mmol $g^{-1} h^{-1}$
$Y_{X/S}$	Biomass yield	$g g^{-1}$ or g mol^{-1}
P_x	Biomass productivity	$g L^{-1} h^{-1}$

be written when μ is constant during the exponential growth phase: $X = X_0 \cdot e^{\mu t}$ (cf. Sect. 5.1).

4.1.2 Substrate Uptake

Growth is directly related to penetration of the substrate into the cell. The uptake rate of substrate is the first limiting factor for growth. The solute transport through the plasma membrane can be achieved by three different processes (SHECHTER, 1984).

4.1.2.1 Passive Diffusion

Passive diffusion is governed by the law of mass action. This procedure can be defined as migration of molecules across the plasma membrane from a region of high concentration to one of lower concentration. Ions and organic molecules with low molecular weights can thus cross the plasma membrane. The initial rate of diffusion of a molecule is directly proportional to the external concentration of that molecule. Diffusion stops when the internal and external concentrations are identical. There is no saturation phenomenon. Passive diffusion takes place without specific interaction between the solute and a membrane constituent.

4.1.2.2 Facilitated Diffusion

This involves a membrane carrier, usually a protein. Solutes are diffused more rapidly than would be indicated by their hydrophilic nature or size. The diffusion rate displays saturation when the external concentration increases; transport may be inhibited by structural analogs of the solute. The mechanism occurs spontaneously and does not require energy. The net flux stops when the electrochemical potential is the same on both sides of the membrane. Membrane constituents are either proteins (in the case of organic molecules) or channels (in ions).

4.1.2.3 Active Transport

Like facilitated diffusion, active transport is carrier-mediated. It requires energy and must be linked with an exergonic reaction and functions against a concentration gradient. The nature of the reaction makes it possible to define two main categories, primary and secondary active transport. To these should be added a particular mechanism – transport by group translocation – that involves chemical modification of the solute.

Numerous ions penetrate by a *primary active transport* mechanism. The exergonic reaction involved may be the hydrolysis of ATP by means of an ATPase. Active transport of Ca^{2+}, Na^+, K^+ and H^+ can be mentioned. These transport processes can keep intracellular concentrations (H^+, Na^+) smaller than extracellular concentrations. The energy source may be light in some cases (active transport of H^+ in halophilic bacteria). Finally, many organic compounds (sugars and amino acids) penetrate cells by primary active transport involving specific proteins or permeases.

In *secondary active transport* mechanisms, transport of a solute is linked to the facilitated diffusion of a second solute in the direction of its decreasing electrochemical potential. The second solute is often an ion whose difference in electrochemical potential has been created by primary active transport. This type of transport is that of dicarboxylic acids through the mitochondrial membrane or the transport of lactose in *E. coli*, for example.

Unlike the transport types defined above, *group translocation* includes the formation of a covalent bond between the solute and a membrane constituent. The chemical form of the solute is different after transport. The sugar transport (glucose, fructose and mannose) takes place in this way in *E. coli*.

4.1.2.4 Special Cases of Active Transport (Button, 1985)

The ligand "taxi": In this case, a specific carrier is exuded by the microorganism; it bonds to the substrate and both enter the cell. Iron penetrates by this mechanism in siderophore algae, as do paraffinic hydrocarbons in some bacteria (Käppeli and Finnerty, 1979).

Vectorial partitioning: This mechanism operates for non-polar substrates which cannot bond to the cell membrane. A favorable partition coefficient of lipophilic substrate between phospholipid and water results in amplified hydrocarbon concentrations in the cell membrane. Lipophilic substrates can be hydroxylated, so that their polarity increases. Those which diffuse inward are trapped. Their concentration increases, so that the reaction rate is enhanced.

4.1.3 Saturation Constant

4.1.3.1 Definition

Monod (1949) described growth empirically and related the growth rate (μ) to substrate concentration (S) according to the formula $\mu = \mu_m (S/(K_s + S))$. The value of K_s corresponds to the substrate concentration at which $\mu = \mu_m/2$.

The value of the saturation constant K_s for an element is generally constant for a given microorganism. However, when there are several transport systems according to the substrate concentration, several K_s values may be found for a single element (Höfle, 1983).

Modeling microbial growth requires knowledge of K_s values. When the substrate is the carbon and energy source, the K_s value is an important factor in the efficiency of conversion of substrate into biomass. Knowledge of K_s values is essential in mixed cultures as it permits assessment of flora equilibrium during a fermentation (cf. Sect. 5.5).

4.1.3.2 Determination

The K_s value can be calculated either by measurement of the specific growth rate for known substrate concentrations or by measurement of the concentration of elements for known growth rates. The linear form of Monod's equation is generally used. The following equations are sometimes used:

- Lineweaver-Burk: $1/\mu = 1/\mu_m + K_s/(\mu_m S)$; a plot of $1/\mu$ against $1/S$ is linear with intercepts $1/\mu_m$ and $-1/K_s$.
- Eadie-Hofstee: $\mu/S = \mu_m/K_s - \mu/K_s$; a plot of μ against μ/S is linear with slope $-K_s$ and intercept μ_m.
- Langmuir: $S/\mu = K_s/\mu_m + S/\mu_m$; a plot of S against S/μ is linear with intercept $-K_s$.

4.1.3.3 Calculation Methods

Owens and Legan (1987) described five of the main methods of determination of K_s.

Tab. 7. Saturation Constants (K_s) for Growth of Organisms on Diverse Substrates

Substrate	Organism	K_s (µM)	Method
Glucose	*Pseudomonas* sp. (1)	5.9–105	Direct assay of *S* in chemostat
	Streptococcus	110–2170	Indirect assay of *S* in chemostat
	Achromobacter sp.	30	Direct assay of *S* in chemostat
	Saccharomyces cerevisiae	139	Direct assay of *S* in chemostat
	Aspergillus nidulans	28	Radial growth rate
Glycerol	*Candida utilis*	50	Direct assay of *S* in chemostat
Methanol	*Methylococcus* sp.	42	Respirometry
Methane	*Pseudomonas* (2)	26	
Oxygen	*Candida utilis* (1)	1.3–2.6	Direct assay of *S* in chemostat
		14	Indirect assay of *S* in chemostat

(1) OWENS and LEGAN (1987)
(2) PIRT (1975)

(1) Direct detemination in chemostat culture by measurement of residual substrate for different dilution rates: $K_s = S(\mu_m - D)/D$. In indirect determination for chemostat culture, the residual substrate concentration is calculated using the formula $X = Y_{X/S}(Sr - S)$, $Y_{X/S}$ is determined in batch culture.
(2) Determination in batch culture by measurement of the initial growth rates with known initial concentrations of elements.
(3) Determination during washout in continuous culture by measurement of the growth rate with known concentrations of elements.
(4) Determination in chemostat culture at the maximum dilution rate: $D_m = \mu_m[1 - K_s/\sqrt{(K_s + Sr)}]$, D_m, μ_m, and Sr are known, and K_s can be calculated.
(5) Respirometer determination. This is based on the oxygen uptake rate with different substrate concentrations.

According to OWENS and LEGAN (1987), the best method is the direct assay of growth rate-limiting nutrient in steady-state continuous culture. Several K_s values of several microorganisms for various carbon substrates are shown in Tab. 7.

4.1.4 Metabolic Coefficients

The parameters qS, qO_2 and qCO_2 are the most commonly used. Their variation is directly related to the physiological state of the cells. They vary according to substrate and microorganism.

4.2 Energy and Carbon Source Requirements

The biomass yield ($Y_{X/S}$ = quantity of biomass produced per quantity of substrate uptake) is an important factor in industrial processes. High biomass yields are particularly advantageous as the cost of the raw material has considerable weight in the cost of the end product.

The biomass yield indicates the efficiency of the conversion of carbon substrate into biomass. It depends on several factors:

- the quantity of energy required for the synthesis of cell material (Y_{ATP});
- the quantity of energy required for maintenance functions and the transport of nutrients;
- the quantity of energy (ATP) produced during catabolism, depending mainly on oxidative phosphorylation efficiency (P/O).

These factors are in turn influenced by numerous parameters.

4.2.1 Energy Required for the Formation of Cell Material

The ATP yield (Y_{ATP}) is defined as the quantity of cell material (g) formed per mole of ATP. The theoretical $Y_{ATP,max}$ can be calculated; it is the sum of all the ATP taken up and produced during the reactions enabling formation of biomass (STOUTHAMER, 1973; VERDUYN et al., 1990). The experimental Y_{ATP} is calculated in anaerobic batch culture. Only under anaerobic conditions the ATP yield for substrate breakdown can be calculated exactly, since the catabolic pathways for anaerobic breakdown of substrate are known (STOUTHAMER, 1979).

The ATP yield depends to a considerable extent on the cell composition, the carbon source and the metabolic pathways.

4.2.1.1 Cell Composition

Microorganisms are constituted of four main groups of polymers, protein (40–70%), carbohydrate (10–40%), nucleic acid (3–16%) and lipid (1–20%) whose quantities and proportions vary according to the carbon source and also according to physiological factors such as growth rate and physicochemical characteristics such as temperature. The theoretical ATP requirements for the synthesis of each constituent can be calculated (Tab. 8). The protein content has the main effect on total ATP requirements (VERDUYN, 1991).

Theoretical calculation of Y_{ATP} is, therefore, based on knowledge of the energy requirements for the synthesis of cell compounds.

4.2.1.2 Carbon Source

The theoretical ATP requirements for the synthesis of the various cell compounds vary according to the carbon source (Tab. 8). STOUTHAMER (1973) gives the value of Y_{ATP}=28.8 g of biomass formed per mole of ATP for microorganisms cultured in the presence of glucose and mineral salts. VERDUYN (1991) proposes 25.1 and 9.5 for yeasts cultured in the presence of glucose and ethanol, respectively.

Tab. 8. Calculation of the Theoretical Y_{ATP} for Growth of Yeast on Glucose and Ethanol (cell composition: 52% protein, 28% carbohydrate, 7% lipid, 6% ash) (VERDUYN, 1991)

Substrate	Glucose	Ethanol
	mmol ATP (100 g DW)$^{-1}$	
Amino acid synthesis	200	4264
polymerization	1960	1960
Carbohydrate synthesis	358	2329
Lipid synthesis	179	651
RNA synthesis and polymerization	182	341
Turnover of mRNA	71	71
NADPH generation	88	0
Transport: ammonium	700	700
phosphorus and potassium	240	240
Total	3978	10556
Theoretical Y_{ATP} (g DW (mol ATP)$^{-1}$)	25.1	9.5

Tab. 9. Theoretical Y_{ATP} for Microorganisms Using Different Carbon Assimilation Pathways for Growth on Methane and Methanol (HARDER and VAN DIJKEN, 1976)

Assimilation Pathway	Cell Material g (mol ATP)$^{-1}$
Ribulose monophosphate cycle, fructose diphosphate variant	27.3
Ribulose monophosphate cycle, 2-keto-3-deoxy-6-phosphogluconate aldolase variant	19.4
Serine pathway	12.5
Ribulose monophosphate cycle	6.5

4.2.1.3 Metabolic Pathways

The various nitrogen uptake pathways (according to the nitrogen source available: NH_3, NO_3^- or N_2) and the different carbon source uptake pathways affect ATP requirements for the formation of cell material. Theoretical calculations of Y_{ATP} using different methane and methanol uptake pathways are shown in Tab. 9 (STOUTHAMER, 1979).

4.2.1.4 Transport

The active (energy-requiring) transport involved in the metabolisms of numerous sugars and polyols may increase ATP requirements.

4.2.2 Energy Produced During the Metabolism

This depends on the carbon source, the metabolic pathways used and the P/O ratio. The latter represents the efficiency of oxidative phosphorylation, i.e., the number of phosphorylation sites in the respiratory chain. The P/O ratio varies from 2 to 3 according to the microorganisms; it has an important effect on the number of ATP formed and hence on the biomass yield (VERDUYN et al., 1991; VERDUYN, 1991).

4.2.3 Biomass Yields

Calculation of the quantity of ATP needed for the formation of cell material is not sufficient for the description of biomass yields. Yields are generally expressed in $Y_{X/S}$ (g DW per mol substrate consumed), which shows the efficiency of the transformation of the substrate into biomass, or in Y_{X/O_2} (g DW per mol oxygen consumed). The Y_{X/O_2} expresses the quantity of cell material formed per mole of oxygen consumed. It varies with the carbon source and the P/O ratio. In addition to the carbon substrate and cell composition, other physiological or cell environment factors affect the biomass yield. These are growth rate, pH, temperature, osmotic stress, partial oxygen and carbon dioxide pressures and medium composition.

STOUTHAMER (1979) proposed a method for calculating the theoretical values of $Y_{X/S}$ and Y_{X/O_2} from theoretical values of Y_{ATP}. The important data are the formula and molecular weight of one pseudo-mole of biomass. The substrate uptake reaction can, therefore, be written as follows:

$$C_6H_{12}O_6 + 1.4\,NH_3 + 0.25\,\text{“}H_2\text{”} \rightarrow C_6H_{10.84}N_{1.4}O_{3.07} + 2.93\,H_2O$$

The biomass yield is given by: $Y_{X/S} = Y_{ATP} \times$ (total ATP production per mole of glucose). STOUTHAMER (1979) reported theoretical $Y_{X/S}$ values of 124 and 131.2 for two and three phosphorylation sites. The corresponding Y_{X/O_2} values are 125.7 and 183.3.

Tab. 10. Variation of Yields According to the Type of Carbon Source (PIRT, 1975) ($Y_{X/S}$, $Y_{X/C}$, Y_{X/O_2} represent the amount of cell material formed per g of substrate, carbon or oxygen)

Microorganism	Substrate	$Y_{X/S}$	$Y_{X/C}$	Y_{X/O_2}
Candida utilis	Glucose	0.51	1.28	1.30
	Acetic acid	0.36	0.90	0.62
	Ethanol	0.68	1.30	0.58
C. intermedia	*n*-Alkanes	0.81	0.96	0.35
Methylococcus sp.	Methane	1.01	1.34	0.29
Penicillium chrysogenum	Glucose	0.43	1.08	1.35

The values shown in Tab. 10 permit the comparison of the yields of several microorganisms cultured under different conditions on various carbon substrates.

Biomass yields ($Y_{X/S}$), productivity (P_X) and the final biomass concentration of various microorganisms cultured on different substrates with the aim of producing biomass are shown in Tabs. 11 and 12 (LICHTFIELD, 1979; MOULIN et al., 1983a; SHAY et al., 1987). The examples mentioned were developed either on a laboratory, pilot or industrial scale.

4.3 Oxygen Demand and Transfer

4.3.1 Oxygen Demand

In an aerobic metabolism, oxygen plays the role of the final electron or hydrogen acceptor. The process is mediated by an oxidase enzyme. Oxygen may also be incorporated in carbon substrates (mainly hydrocarbons and aromatic compounds) during catabolism.

Oxygen is required in the presence of hydrocarbons and aromatic compounds not only to produce energy, but also to increase the oxidation of these substrates. Production of biomass from a hydrocarbon requires more oxygen (2.2 g O_2 per g DW) than when a carbohydrate is used (1 g O_2 per g DW) (PIRT, 1975). The amount of oxygen required can be calculated from theoretical equations of transformation of carbon substrate into biomass (EINSELE, 1983).

In the presence of glucose:

$$1.8\text{n}\,CH_2O + 0.8\text{n}\,O_2 + 0.19\text{n}\,NH_4^+ \rightarrow \text{n}\,(CH_{1.7}O_{0.5}N_{0.19}) + 1.5\text{n}\,H_2O + 0.8\text{n}\,CO_2 + \text{n}\,19000\ \text{kJ}$$

In the presence of hydrocarbons:

$$2\text{n}\,CH_2 + 2\text{n}\,O_2 + 0.19\text{n}\,NH_4^+ \rightarrow \text{n}\,(CH_{1.7}O_{0.5}N_{0.19}) + 1.5\text{n}\,H_2O + \text{n}\,CO_2 + \text{n}\,48000\ \text{kJ}$$

The oxygen requirements of a microorganism can be expressed in terms of qO_2; this is the quantity of oxygen consumed per unit of time and per unit of biomass. It varies according to the microorganism in question and, for a given microorganism, it varies with the physiological state of the cells (growth stage), the carbon substrate and the physicochemical parameters (partial pressure of dissolved oxygen). A critical value can be defined for dissolved oxygen beyond which the growth parameters μ, qO_2 and qCO_2 are constant. These values are affected below this critical value, and the growth rate is proportional to the dissolved oxygen concentration. Oxygen becomes a limiting factor. The critical concentration varies from one organism to another and according to the substrate; it is about 0.1 mg L^{-1} for carbohydrates and 1 mg L^{-1} for hydrocarbons.

This Michaelis–Menten type equation can be written:

$$qO_2 = qO_{2\max}\,[C/(C+K_s)],$$

with C being the dissolved oxygen concentration and K_s the affinity constant for oxygen.

The oxygen demand (mmol L^{-1} h^{-1}) of a

Tab. 11. Growth of Selected Algae, Bacteria and Actinomycetes on Various Substrates (LITCHFIELD, 1979)

Microorganism	Substrate	Scale	Temperature (°C)	pH	μ or D (h^{-1})	X	$Y_{X/S}$ or P_x
Algae							
Chlorella pyrenoidosa	CO_2	2.7 L	38	6.4–6.5		8.98 g L^{-1}	36.5 g d^{-1}
Chlorella regularis	Acetic acid	20 L	36	6.8	0.28	15 mL L^{-1}	0.48 g g^{-1}
Scenedesmus quadricauda + *Sc. obliquus*	CO_2	54000 L	34	6.5–7.0		1.5–2.0 g L^{-1}	10 g m^{-2} d^{-1}
Spirulina maxima	CO_2	700 m^2 pond	Ambient				15 g m^{-2} d^{-1}
Spirulina platensis	CO_2	10 L	35–37	8–10.5	0.019	4.2 g L^{-1}	
Bacteria and Actinomycetes							
Achromobacter delvacuate	Diesel oil	6000 L	35–36	7.0–7.2		10–15 g L^{-1}	
Acinetobacter cerificans	*n*-Hexadecane	7.5 L	30	6.8	μ = 1.33		1.20 g g^{-1}
Cellulomonas sp.	Bagasse	7, 14, 530 L	34	6.6–6.8	μ = 0.20–0.29 D = 0.08–0.10	16 g L^{-1} 10 g L^{-1}	0.44–0.50 g g^{-1}
Methylococcus capsulatus	Methane	2.8 L	37	6.9	μ = 0.14	0.4 g L^{-1}	1–1.03 g g^{-1}
Methylomonas clara	Methanol	1000 L	39	6.8	μ = 0.5		0.50 g g^{-1}
Pseudomonas sp.	Fuel oil	6000 L	36–38	7.0	μ = 0.16	16 g L^{-1}	1.00 g g^{-1}
Thermomonospora fusca	Cellulose-pulping fine	10 L	55	7.4			0.35–0.40 g g^{-1}

Tab. 12. Growth of Selected Yeasts, Molds and Higher Fungi on Various Substrates (LITCHFIELD, 1979)

Microorganism	Substrate	Scale	Temperature (°C)	pH	μ or D (h^{-1})	X ($g\ L^{-1}$)	$Y_{X/S}$ ($g\ g^{-1}$)
Yeasts							
Candida enthanothermophilum	Ethanol	30 L	40	3.5	$D=0.20$	8.0	0.95
Candida lipolytica	*n*-Alkanes	1800 L	32	5.5	$D=0.16$	23.6	0.88
Candida tropicalis	*n*-Alkanes	50 m^3	30	3.0	$\mu=0.15$–0.24	10–30	1.0–1.1
Candida utilis	Sulfite waste liquor	Waldhof fermenter	32	5.0	$\mu=0.5$		0.39
Endomycopsis fibuliger *Candida utilis* (1)	Starch	300 m^3	30	4.5	$D=0.10$	10	0.60
Kluyveromyces fragilis (2)	Whey	23 m^3	38	3.5	$D=0.33$	13	0.55
Kluyveromyces fragilis (3)	Whey	1500 L	37	4.5	$\mu=0.23$	90	0.45
Saccharomyces cerevisiae	Cane molasses	75–225 m^3	30	4.5–5.0	$\mu=0.20$–0.25	40–45	0.50
Pichia pastoris (3)	Methanol	1500 L	30	3.5	$\mu=0.11$	105	0.4
Molds and Higher Fungi							
Agaricus campestris	Glucose	20 L	25	4.5		20	
Aspergillus niger	Carob bean extract	3000 L	30–36	3.4	$\mu=0.25$	31.5	45
Geotrichum sp.	Corn and pea waste	37.8 m^3	Ambient	3.7		0.75–1.0	
Morchella deliciosa	Sulfite waste liquor	18.93 L	Ambient	6.0		10	32
Trichoderma viride	Corn and pea waste	37.8 m^3	Ambient	4.6		1.2	

(1) JARL (1969), (2) MOULIN et al. (1983a), (3) SHAY et al. (1987)

microbial culture (qO_2X) must be achieved by the oxygen transfer rate (OTR) into the fermenter. The maximum theoretical demand by a microorganism can be calculated using the formula $(\mu_m/Y_{X/O_2})X$ and determined experimentally using a respirometer ($qO_{2max}X$).

Oxygen demands of some 150 to 380 mmol O_2 L^{-1} h^{-1} have been observed in high cell density cultures (15–20 g DW L^{-1}) in the presence of alkanes (EINSELE et al., 1973).

4.3.2 Oxygen Transfer – K_La

The oxygen transfer rate (OTR) in a fermenter is characteristic of a given system. Oxygen transfer can be written:

$$OTR = dC/dt = K_La(C^* - C_L).$$

K_L is the mass transfer coefficient (cm h^{-1}), a is the spherical exchange area ($cm^2 \cdot cm^{-3}$), C^* is the oxygen solubility (mmol L^{-1}), and C_L is the dissolved oxygen concentration in the solution. The transfer rate is expressed in mmol L^{-1} h^{-1}. This therefore depends on the K_La and C^*.

The oxygen transfer capacity of a culture medium varies mainly according to two parameters: agitation rate and aeration (AIBA et al., 1973). Agitation performs the following functions:

- maintenance of high mass transfer with even distribution of all the reagents,
- dispersion of the gas in small bubbles and hence enhancement of the contact area,
- increasing the gas–liquid contact time through the disturbance and whirl caused,
- increased turbulence and decrease in the consistency of the film of the liquid.

The oxygen solubility is 1.16 mmol L^{-1} at 30 °C in water. It decreases when the temperature increases and also depends on the partial pressure of oxygen in the gas phase, the mineral composition of the culture medium and the presence of antifoam.

The oxygen transfer rate in conventional fermenters varies from 240 to 480 mmol L^{-1} h^{-1}. This gives a biomass productivity of some 5 to 10 g L^{-1} h^{-1}.

4.4 Physical and Chemical Parameters

4.4.1 pH

The optimum pH for the growth of numerous microorganisms is about 7, with pH 5 to 8 being a favorable range (Tabs. 11 and 12). The pH range for yeast growth is 2.5 to 7 with optimum growth at 4 to 5. The optimum pH for most bacteria is between 6.5 and 7.5; bacterial growth is often inhibited at pH values lower than 4 and higher than 8.5, with the exception of acetic acid bacteria oxidizing ethanol into acetate and the *Thiobacillus* oxidizing sulfides into sulfuric acid which can develop at pH values as low as 2.

The optimum pH for fungi is from 5 to 7 and they tolerate pH values of 3 to 8.5. Algae can grow in a pH range of 4.5 to 10.5. The pH of their natural habitat is often over 10.

4.4.2 Temperature

Microorganism can be classified in three categories according to their growth sensitivity to temperature:

- psychrophiles develop at temperatures lower than 20 °C;
- mesophiles develop at between 20 and 45 °C;
- thermophiles develop at between 45 and 60 °C.

Temperature affects the growth rate, nutritional requirements, cell composition and cell permeability (PIRT, 1975). It also affects protein and lipid structure and the temperature coefficients of reaction rates which in turn depend on the activation energies of the reactions. There is an optimum temperature and a lethal temperature for each species and each strain (Tabs. 11 and 12) (ATKINSON and MAVITUNA, 1983).

Optimum growth temperature for bacteria varies from 10 to 40°C (SPECTOR, 1956). Yeasts are essentially mesophilic and are facultative psychrophiles. Maximum growth is around 20 to 30°C. Strictly psychrophilic yeasts (e.g., *Candida gelida, Candida frigida* and *Rhodotorula infirmo-miniata*) are unable to develop at temperatures higher than 20–30°C (STOKES, 1971). Optimum growth of fungi takes place at between 10 and 37°C (SPECTOR, 1956). The highest temperatures in the habitat of algae range from 30 to 80°C. The maximum temperatures enabling growth vary from 70 to 85°C for blue-green algae and 10 to 70°C for green algae (SPECTOR, 1956).

4.5 Nutritional Requirements

The growth of microorganisms is related to the presence in the environment of the substances required for the synthesis of cell material and energy production for biosynthesis processes. Nutritional requirements depend on the microorganism and culture conditions.

4.5.1 Nutritional Requirements of Microorganisms

Composition of the culture medium plays a fundamental role in the processes of production of biomass or metabolites. The availability of growth substances is decisive for optimum performance.

Two aspects of medium composition are examined:

(1) the qualitative aspect, to identify the substances required for growth;
(2) the quantitative aspect, to establish the concentrations which are sufficient but not inhibitory for the growth or type of metabolite required.

The main substances can be placed in four categories: substrate, macro elements, trace elements and growth factors (FIECHTER, 1984; PIRT, 1975).

4.5.1.1 Substrate

Carbon forms approximately 50% (w/w) of dry matter (Tab. 13). Carbon is generally a limiting substrate. Nitrogen or phosphorus may serve as limiting substrate in some cases.

Tab. 13. Elemental Composition of Dry Yeast (mg g^{-1}) (FIECHTER, 1984)

Elements	*Saccharomyces*	*Candida*
C	470	454
H	60	67
N	85	73
O	325	320
P	14.3	16.8
K	17.2	18.8
Mg	2.3	1.3
Na	0.7	0.1
Fe	0.1	0.1
Zn	0.0387	0.0992
Mn	0.0057	0.0387
Ca	1.3	5.7
Co	0.0002	–
Cu	0.033	0.013

4.5.1.2 Macro Elements

These are N, O, H, C, S, P, K, Ca and Mg. They are used in large quantities (g L^{-1} or mg L^{-1}).

Nitrogen

Nitrogen forms 10 to 12% of cell dry weight (Tab. 13). It can be used in various forms: organic (urea, peptides, amino acids, purine, pyrimidine) and inorganic (ammonia, ammonium salts, nitrate).

Phosphorus

Phosphorus forms approximately 1.5% of cell dry weight (Tab. 13). It is provided as phosphoric acid or phosphate salts. Organic molecules such as phytates can also be used as a source of phosphorus (LAMBRECHTS et al., 1992). Phosphorus is incorporated in the nucleic acids, phospholipids and polymers forming cell walls. It may be accumulated in polyphosphate form.

Other macro elements (K, Mg, S)

Potassium plays a major role in the regulation of transport of divalent cations. The presence of potassium is indispensable for $H_2PO_4^-$ penetration. Potassium ions also serve as coenzymes.

Magnesium is an activator for numerous enzymes in glycolysis and membrane ATPases; it enhances fatty acid synthesis and regulates the intracellular ion level. Magnesium is found in chlorophyll in algae.

Sulfur is a component in amino acids and coenzymes (PIRT, 1975; JONES and GREENFIELD, 1984).

4.5.1.3 Trace Elements

These are mainly Fe, Mn, Mo, Zn, Co, Ni, V, B, Cl and Si. They are supplied in form of very small quantities of salts (μg L^{-1} or ng L^{-1}). They may have a specific role in enzymes and coenzymes. They may be growth inhibitors at certain concentrations (JONES and GREENFIELD, 1984).

4.5.1.4 Growth Factors

These are mainly vitamins, organic compounds which cannot be synthesized by cells or which limit growth kinetics. Vitamins are provided in small quantities, on the μg L^{-1} scale. They are mainly enzyme and coenzyme precursors or constituents. The most important vitamins are thiamine, riboflavin, pantothenic acid, pyridoxine, nicotinic acid, biotin, *p*-aminobenzoic acid and folic acid.

4.5.2 Improvement of Medium Composition for Biomass Production

The culture media used in industry are not always clearly defined. Many difficulties may also be encountered as a result of the following factors:

- the interaction between elements;
- the inhibitory effect of certain elements (Fe, Cu, etc.);
- the presence of chelating agents;
- the deficiency in certain elements (nitrogen, phosphorus, iron, zinc, etc.).

There are several possible approaches to the optimization of cultures to achieve optimum growth parameters.

4.5.2.1 Cell Composition

Knowledge of the cell composition makes it possible to develop a basic medium containing the necessary growth substances (see Tab. 13). The resulting medium must be adapted to each strain by successive approaches. The method is time-consuming and empirical.

4.5.2.2 The Pulse and Shift Method

Chemostat culture is one of the most effective ways of optimizing a culture medium. The principles were described by MATELES and BATTAT (1974). A bibliographical summary on the improvement of media by use of these methods was published by GOLDBERG and ER-EL (1981).

The pulse technique consists of injection of suspected growth-limiting substances into the reactor while the culture is in a steady state. The change in biomass is monitored for about two or three renewal times. The types of response expected are shown in Fig. 8. The technique also makes it possible to keep substance concentrations to a minimum.

In contrast, the shift technique consists of modifying the concentration of a nutrient in

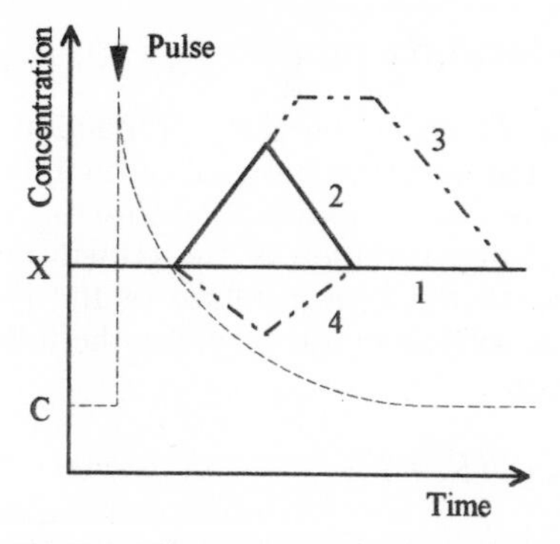

Fig. 8. Examples of pulse of nutrients in chemostat, *C:* nutrient concentration, *X:* biomass concentration (GOLDBERG and ER-EL, 1981).
1, response of a non-limiting nutrient; **2,** response of a limiting nutrient; **3,** response of a second limitation; **4,** response of a toxic nutrient.

the feed tank. The strain responds more slowly, and the culture reaches a new state of equilibrium which can be sustained in time.

Chemostat optimization techniques are faster than batch techniques. They permit a rapid response from cells during growth to the addition of a substance. They are frequently used when strain requirements are high.

4.5.2.3 Experimental Designs

Experimental research methodology (experimental designs) is used to determine the influence of several substances and their interactions and to optimize one or more experimental responses. These objectives are attained in successive stages, each stage being a homogeneous set of experiments forming a body of results. The aim is to collect the largest number of valid observations with as few tests as possible (CHERUY et al., 1989; GOUPY, 1988).

In the experimental design method, the levels of all the factors are varied at the same time in each experiment. The advantages are as follows:

- fewer tests,
- larger number of factors tested,
- detection of interactions between factors,
- more accurate results,
- optimization and modeling of the results.

This methodology is used in many industrial laboratories. It was used for optimizing fungus culture methods (CHERUY et al., 1989). These methods are complementary. It can be observed that the chemostat pulse and shift method gives the fastest qualitative and quantitative responses in growing cells.

5 Types of Biomass Production

Biomass production processes are grouped in three categories: batch, continuous and fed-batch. The basic structure of the fermentation system has been described and discussed in detail by PIRT (1975) and MOSER (1985a, b, c).

5.1 Batch Culture

Batch culture is the most common method in the microorganism production industry. The development of continuous and fed-batch processes always starts with a batch culture. Finally, whatever the subsequent method, numerous culture parameters are first determined in batch culture.

The main characteristic of a batch culture is the variation of the concentration of all the medium components in time. There is usually no addition of substances (except for air) during culturing; the cells develop until there is a deficiency of nutrients or an accumulation of a toxic metabolite which stop growth.

Cell multiplication takes place in several phases (PIRT, 1975): the lag phase, the accelerating phase, the exponential phase, the stationary phase and the decreasing phase. The aim in batch cultures is to shorten the lag phase to maximize the exponential growth phase.

Lag phase

Adding an inoculum to a culture medium means that the cells must adapt to new conditions. Duration of the lag phase varies and is difficult to estimate. It is generally shortest when the cells forming the inoculum are collected during the exponential growth phase after development under the same culture conditions. There is no production of biomass ($dx/dt=0$) during this phase.

Accelerating phase

This generally short phase is between the lag phase and the exponential phase. The value of μ varies from zero to μ_m (μ in the exponential phase).

Exponential phase

Cell multiplication is at its peak during this phase. However, all essential substances must be available in sufficient quantities and there must be no inhibiting factor.

Under these conditions, growth is defined by the following equation:

$$dx/dt=\mu_m X \quad (1)$$

$$\text{hence } \ln X-\ln X_0=\mu_m t \quad (2)$$

$$\text{leading to } \ln(X/X_0)=\mu_m t \quad (3)$$

and hence the fundamental equation for biomass production in batch culture:

$$X=X_0 e \mu_m t \quad (4)$$

The doubling time (t_d) can thus be calculated from Eq. (4) on the assumption that

$$X=2X_0 \text{ and } t=t_d$$

$$t_d=\ln 2/\mu_m=0.693/\mu_m$$

t_d is also defined as the generation time.

Decelerating phase

In most cases, biomass production stops when the concentration of growth-limiting substrate is consumed. MONOD (1942) showed that variation of the growth rate according to the concentration of the growth-limiting substance was given by the following equation:

$$\mu=\mu_m(S/(K_s+S))$$

where S (g L^{-1}) represents the limiting substrate concentration and K_s (g L^{-1}) is the saturation constant (cf. Sect. 4.1.3). Consequently, the period when nutrient limitation affects the growth rate is limited to a small fraction of the last generation, provided that there is no inhibitory product accumulated.

This management method is used in a large number of industrial metabolite production processes (antibiotics, enzymes, amino acids).

5.2 Continuous Culture

Continuous culture is an open system in comparison with batch culture. Fresh medium is fed into the reactor at a steady rate, and the volume is kept constant by drawing off the same amount. Continuous cultures are characterized by constant substrate, biomass and metabolite concentrations. There are two main types of continuous culture:

- the turbidostat, in which the biomass concentration is kept constant in the reactor by adjustment of the feed flow according to biomass variation at equilibrium;
- the chemostat, in which the biomass concentration is kept constant by the addition of a constant concentration of a limiting nutrient factor.

The chemostat is more commonly used and is the only system described here.

5.2.1 The Principle of the Chemostat

Cell growth is limited in a chemostat by the amount of a nutrient (referred to as the limiting substrate); this may be the carbon source, the nitrogen source, phosphate, a trace element or a vitamin. All other components are available in excess.

The theoretical bases of continuous culture were established by MONOD (1950) and by NOVICK and SZILARD (1950). These authors showed that the biomass specific growth rate could be set between zero and a maximum value (μ_m).

The aim of the model is to forecast in steady state the growth rate and biomass and limiting substrate concentrations. Symbols are shown in Tab. 6.

5.2.1.1 Specific Growth Rate

Biomass variation can be established from the following balance:

net increase in biomass = growth − output

for very short periods of time:

$$V\,dx = V\,\mu\,X\,dt - F\,X\,dt$$

which can be written by dividing by $V\,dt$:

$$dx/dt = \mu\,X - F\,X/V$$

giving $dx/dt = X(\mu - D)$

In steady state conditions, when $dx/dt = 0$, $\mu = D$. This shows that the specific growth rate (μ) is directly controlled by the dilution rate (D). The limits of this equality are discussed later.

5.2.1.2 Relation Between Biomass and Limiting Substrate

The substrate balance in the reactor can be defined as follows:

substrate concentration = input − output − quantity used.

For a very short period of time the balance can be written as follows:

$$V\,ds = F\,Sr\,dt - F\,S\,dt - V\,\mu\,X\,dt/Y_{X/S}$$

hence $ds/dt = F\,Sr/V - F\,S/V - \mu\,X/Y_{X/S}$

In steady state conditions, $ds/dt = 0$ and $\mu = D$

hence $X = Y_{X/S}\,(Sr - S)$.

Obtaining the values of X and S requires substitution of μ by D in Monod's equation:

$$\mu = \mu_m\,S/(K_s + S)$$

This gives

$$S = K_s\,D/(\mu_m - D) \tag{5}$$

and

$$X = Y_{X/S}\,[Sr - (K_s\,D)/(\mu_m - D)]$$

In steady state conditions, when $ds/dt = 0$ and $dx/dt = 0$, the residual substrate concentration depends only on the dilution rate (D); the biomass (X) depends only on Sr and D provided that yield $Y_{X/S}$ is constant.

5.2.1.3 Critical Dilution Rate

The critical dilution rate (D_c) shows the limit of functioning of continuous culture in a reactor. It is the theoretical value that can be calculated when $S = Sr$, i.e., when no more substrate is used. The equation

$$\mu = D = \mu_m\,S/(K_s + S)$$

shows that D_c is very close to μ_m if $K_s \ll Sr$.

5.2.2 Discussion

The ideal mode described above requires clear definition of a number of factors when this choice of biomass production is made. The main factor is the nature of the limiting substance. The carbon source is chosen for this in most biomass production. Choice of carbon source must, therefore, take into account the affinity of the strain for the substrate, estimated by measurement of K_s. As is shown by Eq. (5), low K_s values allow increase in dilution rates to levels close to the μ_{max} without resulting in a high residual substrate content (S).

This parameter may be even more important when the limiting substrate is inhibitory from a certain concentration onwards. Under such conditions, values of D close to μ_m may lead to an S value higher than the growth inhibition threshold. This type of problem may be found in the production of biomass from carbon substrates such as methanol or ethanol.

The second factor is medium composition. At this level, limitation by a substance in a medium other than that chosen as the limiting factor often causes deviation from the ideal behavior described. Deviation of the metabolism may cause a fall in yield ($Y_{X/S}$) and/or the specific growth of the strain. This type of production permits development of microorganisms in a constant environment, once the parameters have been defined.

5.2.3 Advantages of Continuous Culture over Batch Fermentation

Continuous culture is particularly well-suited to biomass production for the following reasons:

- multiplication conditions can be maintained indefinitely;
- each parameter (pH, pO_2, medium component concentrations) can be optimized separately;
- productivity of chemostat culture (P_x, expressed in g of DW per unit volume in reactor per unit time) is greater than that of batch culture;
- management of production units is easier; an on-line computer can be used;
- production is more homogeneous and steadier than in batch production.

However, the continuous method can only be used for metabolite production, when synthesis is strictly growth-linked (primary metabolite). Batch culture or fed-batch culture techniques are more often used for secondary metabolites.

Continuous culture also avoids the diauxic growth problems observed in batch culture, when a medium contains two carbon sources. The residual substrate concentration (S) is small in continuous culture, and there is thus no repression of the catabolism enzymes of one substrate by the other.

Cell metabolism is also better controlled in continuous culture (FIECHTER and SEGHEZZI, 1992). Thus, in *Saccharomyces cerevisiae*, respirative glucose catabolism is observed at low dilution rates ($<Dr$) whereas a respiro-fermentative glucose metabolism is triggered for values of $D>Dr$. Dr is the dilution rate at which the cell metabolism changes (SONNLEITNER and KÄPPELI, 1986).

5.3 Fed-Batch Culture

Fed-batch culture is a semi-open method. Fed-batch fermentation processes lie between batch and continuous cultures. Many industrial fermentation processes are varieties of fed-batch techniques. Three parameters may vary in each culture:

- input rate, F;
- culture volume, V;
- substrate concentration, S.

5.3.1 Change of Input Flow

The following working states are possible:

- constant flow, $F(t)=F_0$;
- variable flow with linear flow increase, $F(t)=\alpha_1 t+\beta_1$;
- variable flow with exponential flow increase, $F(t)=\alpha_2 e^{\beta_2 t}$; α_1, α_2, β_1, β_2 are constants.

The specific growth rate of the biomass (μ) varies in a different manner with time according to the type of limiting substrate input chosen (Fig. 9).

With constant inflow, μ decreases with time.

With linearly increased variable inflow, the growth rate also falls.

The specific growth rate remains constant with exponentially increased variable flow.

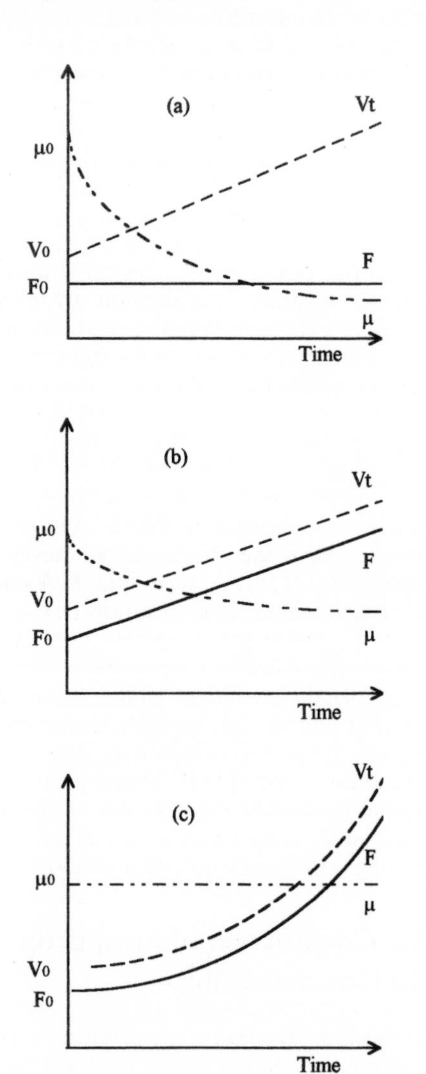

Fig. 9. Schematic representation of evolution of growth rate (μ) and volume (V) in fed-batch culture for different variations of flow rate (F).
(a) $F(t) = F_0$
(b) $F(t) = \alpha_1 t + \beta_1$
(c) $F(t) = \alpha_2 \, e^{\beta_2 t}$

5.3.2 Change in Culture Volume

Some processes are fed with the limiting carbon source alone. Here, all the medium substances except for the carbon source are provided at the start of culture. The limiting substrate is provided as a concentrated solution causing only a small change in volume. This type of fed-batch culture is called „semi-batch“ culture.

Other processes consist of supplying all the components of the medium throughout the multiplication phase. In this case, the initial volume frequently represents only 10% of the final volume. This type of fed-batch causing considerable variation in volume is called "extended" culture.

5.3.3 Change in the Concentration of Limiting Substrate

In most cases of fed-batch culture, whatever the type ("extended" or "semi-batch"), the limiting substrate concentration is kept constant.

5.3.4 Effect on Biomass

In "extended" culture, the biomass concentration is kept constant throughout the cycle as in continuous culture.

In "semi-batch" culture, the biomass concentration increases strongly during the cycle following a small variation in volume caused by the addition of a concentrated solution of limiting substrate.

5.3.5 Advantages

Many microbiological processes are affected by sub-optimal behavior at excessive substrate concentration. In batch culture, concentration may be supra-optimal at the start and then optimal and finally sub-optimal. Thus, excessive initial concentrations of glucose, one of the most important carbon sources in the fermentation industry, may cause various inhibitory effects such as the Crabtree effect (FIECHTER et al., 1981). Likewise, substrates such as ethanol or methanol (MACHEK et al., 1975) may be toxic at high concentrations. These effects can be limited by fed-batch culture. The latter technique is also well-suited to high cell density cultures (MATSUMURA et al., 1982).

Numerous fermentation processes also require oxygen. This is one of the limiting factors in the scaling-up of microbiological processes (EINSELE, 1978; WHITAKER, 1980). Fed-batch culture can adjust oxygen demand by reducing the carbon supply as the cell density increases. This type of production is used in many cases, e.g.,

- production of biomass on methanol (MACHEK et al., 1975);
- production of baker's yeast (cf. Sect. 6.1.1);
- production of heterologous proteins (BLONDEAU et al., 1993; ALBERGHINA et al., 1991; TØTTRUP and CARLSEN, 1990).

5.4 Mixed Cultures

Multiplication of strains of different species is used in the making of many fermented beverages and foodstuffs (wines, cheese, etc.) and in sewage treatment. In spite of this, fermentation industries for the production of metabolites (enzymes, antibiotics, organic acids, etc.) or biomass have developed using monocultures. There are several reasons for the predominance of pure cultures in fermentation industries.

First, the production of a given molecule by a strain requires isolation of the strain and then genetic improvement and optimization of production conditions. Second, it is generally recognized that monoculture is preferable for obtaining new foodstuffs or animal feed of constant quality. However, mixed cultures may have certain advantages, especially when complex carbon sources are used.

The main patterns studied (VELDKAMP, 1972, 1977; FREDRICKSON, 1983) generally refer to a combination of two strains. The following features are most frequently identified among all the mixed culture possibilities:

- competition between strains for the same carbon source;
- a strain develops on the primary carbon source and produces a metabolite (ethanol, organic acid) which is metabolized in turn by a second strain, or a strain develops on a carbon source and another strain develops on another substrate in the medium. Each strain thus occupies a different ecological niche;
- complex interactions between strains.

5.4.1 Competition for a Single Limiting Substrate

When two organisms are in competition for the same substrate, it is possible to forecast the type of competition between strains using the K_s and μ_m of each strain for the substrate. Thus, the evolution of the population can be forecast if strain A has K_sA and μ_mA and strain B has K_sB and μ_mB (Fig. 10).

If $\mu_m A > \mu_m B$ and $K_s A < K_s B$ for all substrate concentrations at equilibrium (S), strain A is dominant in batch culture and eliminates B in the chemostat whatever the dilution rate. If $\mu_m A < \mu_m B$ and $K_s A < K_s B$, there is a substrate concentration in the chemostat at equilibrium at which strains A and B can remain. However, conserving such balances is strictly dependent on the affinity constant (K_s) and the μ_m of each strain for the substrate. These parameters may vary according to many physical (pH, temperature, pO_2) or biotic factors. Mixed cultures of this type are generally considered to be unstable and one strain eliminates the other in most cases.

5.4.2 Coexistence of Strains on Different Substrates

In this type of culture, coexistence becomes possible if each strain uses a different carbon source for growth. Each microorganism uses a different ecological niche. Under these conditions, it is possible to determine a chemostat functioning point using the specific growth rate and the affinity constant of each strain for its respective substrate.

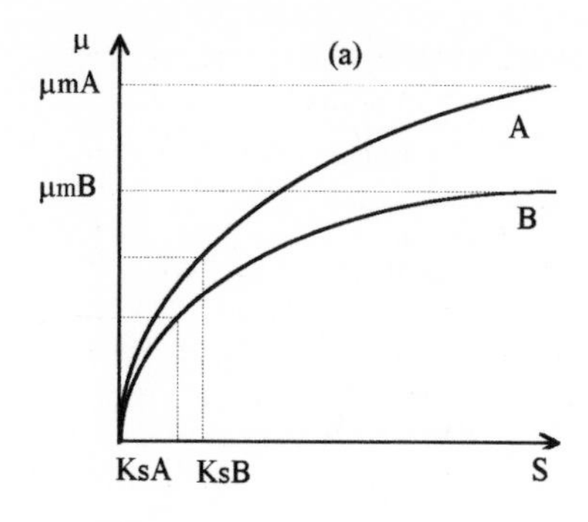

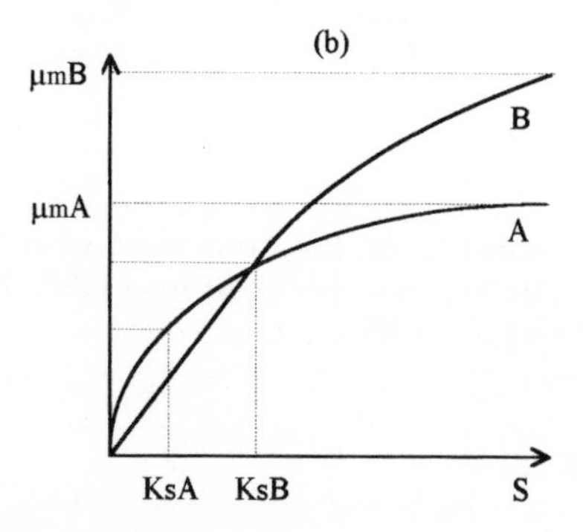

Fig. 10. Competition between two organisms (A and B) for a single growth-limiting substrate (VELDKAMP, 1972). Relation between substrate concentration (S) and growth rate (μ).
(a) $K_s\text{A} < K_s\text{B}$ $\mu_m\text{A} > \mu_m\text{B}$
(b) $K_s\text{A} < K_s\text{B}$ $\mu_m\text{A} < \mu_m\text{B}$

The substrates limiting the growth of each strain may or may not be present in the medium. Thus in some cases a strain may depend on the production by another strain of an organic compound from the primary carbon source. The terms commensalism or mutualism are used here. A case of this type is described (cf. Sect. 6.1.2.1).

5.4.3 Complex Interaction Between Strains

Many authors have described a number of cases of complex interaction (VELDKAMP, 1977; BAZIN, 1981). A strain sometimes produces a metabolite required for the growth of another strain. Likewise, mixed cultures may be developed when the multiplication of a strain causes a variation in pH. In a culture of this type, a strain sensitive to an acid pH may produce an acid and its growth rate falls until it reaches a situation in which $D > \mu$. The population of this strain falls in the fermenter while the acid-tolerant strain develops once again. The pH of the medium may increase, permitting the acid-sensitive strain to develop once again. Oscillating populations of two organisms may thus continue in a chemostat.

6 Examples of Biomass Production

Many raw materials can be considered as potential carbon and energy sources for biomass production (see Tabs. 11 and 12):

- renewable carbon (carbohydrate, starch, cellulose);
- fossil mass consisting of methane, methanol and alkanes;
- CO_2, which is a carbon source for several species of algae (see Tabs. 3 and 11).

As the cost of the carbon source may account for as much as 60–70% of production costs, biomass production processes make considerable use of by-products or co-products from agriculture (starch), the agrofood sector (molasses, whey), industry (sulfite liquor, alkanes, *n*-paraffins, methanol). Many articles contain detailed descriptions of the main biomass production processes (cf. References). A number of examples are given here.

6.1 Carbohydrate Substrate

6.1.1 Molasses

Molasses was long considered as a waste product in the sugar industry, but is now one of the main sources of carbon for fermentation industries. The main reasons are:

Tab. 14. Average Values for Some Constituents inBeet and Cane Molasses at 75% DW (OLBRICH, 1973)

Constituent	Beet Molasses	Cane Molasses
Total sugar (%)	48–52	48–56
Non-sugar organic matter (%)	12–17	9–12
Protein (N × 6.25) (%)	6–10	2–4
Potassium (%)	2.0–7.0	1.5–5.0
Calcium (%)	0.1–0.5	0.4–0.8
Magnesium (%)	ca. 0.09	0.06
Phosphorus (%)	0.02–0.07	0.6–2.0
Biotin (mg kg^{-1})	0.02–0.15	1.0–3.0
Panthothenic acid (mg kg^{-1})	50–100	15–55
Inositol (mg kg^{-1})	5000–8000	2500–6000
Thiamine (mg kg^{-1})	ca. 1.3	1.8

- its relatively low price in comparison with those of other carbon substrates;
- the nature of the substrate (sucrose); this is metabolized by many microorganism species;
- useful mineral, organic and vitamin contents for the development of biomass.

The compositions of beet and cane molasses are different, and molasses samples vary from one sugar-refinery to another. The general composition is shown in Tab. 14. Molasses requires little treatment before use as a culture medium. Preparation is frequently limited to clarification followed by adjustment of the pH, adjustment of the sucrose concentration to the level required and addition of nutrients, nitrogen, phosphorus and trace elements (especially zinc). Baker's yeast is the main biomass product from molasses.

Baker's yeast

World production of baker's yeast is estimated to be 2 million tons of pressed yeast, i.e., approximately 0.6 million tons of dry matter, and is the largest production of microorganisms. Successive improvement to yeast strains and processes since the start of production at the end of the 18th century have led to biomass yields from the carbon source (sucrose) of about 0.5. Many articles have been published on the production of baker's yeast (BURROWS, 1979; REED, 1982; BEUDEKER et al., 1989).

Preparation of inoculum

Fed-batch culture is used in most production processes today. A typical production cycle lasts for 5 to 7 days. An inoculum from a slant of a pure production strain is transferred to a first culture medium. After successive subculturing operations, 300 kg of fresh yeast is obtained in three stages of about 24 hours. After centrifugation, the yeast is stored and used for inoculating industrial fermentation operations.

Production phase

The first phase starts with a fed-batch culture (10 kg of yeast) with a small amount of oxygen, producing in 12 hours 500 kg of fresh yeast well-suited to the subsequent phases. This yeast is then transferred to another tank. Supply with molasses and nutrients is continuous and aeration is increased gradually to 1 vvm. This results in 8 tons of yeast in 16 hours. One can inoculate one or more tanks for producing yeast for sale. In this phase, development of the population is controlled by regulation of the flow of nutrient supply (molasses, ammonium phosphate, ammonia) and aeration. Pressed yeast production is 48 tons in 16 hours. Achieving yeast yield ($Y_{X/S}$) and quality makes this the most delicate phase.

Each installation is computer-controlled. In addition to the traditional regulation parameters (pH, temperature, pO_2), increasing use is made of gas analysis (O_2, CO_2) and on-line assaying of ethanol. Entry and processing of these parameters allows adjustment of the feed flow to obtain the final culture stage required. Changes in the oxygen uptake and carbon dioxide release rates are used to calculate the respiratory quotient. Use of this parameter and the on-line measurements of ethanol makes it possible to maintain an oxidative metabolism and thus achieve maximum biomass production.

Final phase

When the preliminary stages have resulted in acid pH levels (4.0) to limit contamination, the pH can be set at 5 or 6 at the end of production to reduce, for example, the uptake of pigments from molasses at the surface of the yeasts. Likewise, fermentation can be continued for 30 to 60 minutes at the end of the cycle without addition of medium in order to exhaust certain components. The final biomass concentration varies from 40 to 80 g L^{-1} dry weight. It is recovered, washed, concentrated to 200 g L^{-1} by centrifuging and then packaged as pressed yeast containing 30% dry matter or dried yeast with a moisture content of only 6 to 8%.

6.1.2 Biomass from Whey

Whey is a by-product of cheese, casein and butter-making. Whey composition varies depending on its origin (Tab. 15), but it contains an average of 6.5 to 7% dry matter, of which approximately 70% is lactose. The other constituents are protein, lipids, lactic acid, minerals and vitamins which are of interest for biomass production. Biomass can be made from whey or ultrafiltration permeate in three ways:

- direct use of lactose by the microorganism;
- conversion of lactose into glucose and galactose by enzymatic or chemical hydrolysis;
- prior fermentation by lactic bacteria producing a mixture of lactic acid and galactose.

Lactose can be converted directly into biomass by numerous microorganisms (MEY-

Tab. 15. Average Composition of Defatted Whey (% DW), According to FÉVRIER and BOURDIN (1977)

Crude Type	Rennet	Cow Mixed	Lactic	Ewe Rennet	Goat Lactic
Dry, defatted extract	100	100	100	100	100
Lactose	78.76	75.76	69.71	65.88	62.68
Total nitrogen	2.20	2.17	1.88	3.79	2.35
Total protein (N×6.25)	13.75	13.56	11.75	23.69	14.69
Non-protein nitrogen	0.56	0.62	0.83	1.03	1.07
Ammonium nitrogen	0.06	0.13	0.22	0.17	0.28
Urea nitrogen	0.21	0.14	0.11	0.18	0.20
Lactic acid	0.49	3.32	11.64	2.28	13.38
Citric acid	1.97	1.63	0.40	1.33	0.25
Ash	7.98	8.77	11.30	7.31	13.37
Calcium	0.71	0.94	1.93	0.64	2.15
Phosphorus	0.63	0.70	1.00	0.70	1.12
Potassium	2.21	2.22	2.29	1.66	2.90
Sodium	0.77	0.80	0.81	0.80	0.69
Chlorides (as NaCl)	3.34	3.29	3.22	3.06	5.26

RATH and BAYER, 1979; MOULIN and GALZY, 1984). The most commonly used strains belong to the species *Kluyveromyces marxianus* (Hansen) Van der Walt.

6.1.2.1 The Bel Process

The French company Fromageries Bel currently uses a food yeast production process based on deproteinated whey. The process was described by MOULIN et al. (1983a). It permits reduction of the pollutant load (BOD_5) from 50000 mg L^{-1} to 3000 mg L^{-1}. Continuous culture is used. The deproteinated whey is diluted to adjust the lactose concentration to between 20 and 25 g L^{-1} and then complemented with nitrogen, phosphate and trace elements. The pH is set at 3.5, the temperature at 38°C and the dilution rate at 0.33 h^{-1}. Aeration (1.5 vvm) is provided by an air-lift system according to the method proposed by LEFRANÇOIS (1964).

Unlike the Vienna process (MEYRATH and BAYER, 1979) in which a strain of *Candida intermedia* is used, the Bel process (MOULIN et al., 1983a) is based on a balance of three species of flora:

- *Kluyveromyces marxianus* (Hansen) Van der Walt var. *marxianus*, formerly described as *Kluyveromyces fragilis*;
- *Kluyveromyces marxianus* (Hansen) Van der Walt var. *lactis*, formerly described as *Kluyveromyces lactis*;
- *Candida pintolopepsii* (Van Uden) Meyer and Yarrow, formerly described as *Torulopsis bovina*.

The respective proportions of the species are 90%, 9% and 1%.

The balance between the strains of these three species is particularly stable. Each species has a separate ecological niche. The strains of the first group preferentially metabolize lactose, those of the second group metabolize lactic acid and those of the third group metabolize ethanol. MOULIN et al. (1983b) showed that the strains in the first group produce 0.06 mg ethanol (g DW)$^{-1}$ h^{-1} in aerobic culture. This is sufficient ethanol to maintain the strains in the third group which do not metabolize lactose.

This balance ensures complete conversion of the carbon substrates with a 0.55 yield and productivity (P_x) of 4.5 g L^{-1} h^{-1}. The yeasts are separated by centrifugation, washed, plasmolyzed and then dried.

6.1.2.2 Other Processes

Protein-enriched whey

In parallel, different processes have been proposed to enrich whey protein contents without separation of the biomass. HALTER et al. (1981) proposed the enrichment of ultrafiltration permeate with a strain of *Trichosporon cutaneum* in fed-batch culture. FLEURY and HENRIET (1977) proposed the continuous culture of *Penicillium cyclopium* filamentous fungi. In addition to the production of biomass, the process proposes the recovery of amino acids and peptides from the fermentation medium.

MOLINARO et al. (1977) proposed the production of *Kluyveromyces fragilis* yeast on concentrated whey. The product of fermentation is concentrated and then dried. The protein content may reach 45%.

Hydrolyzed whey

Biomass production from whey from which 85 to 95% of the lactose has been hydrolyzed enzymatically to glucose and galactose makes it possible to choose strains which do not metabolize lactose. ARNAUD et al. (1978) described a two-stage process. Glucose is metabolized in the first fermenter by a *Candida utilis* strain which does not metabolize lactose. After separation of yeasts by centrifugation, exhaustion of the carbon substrate in the medium is carried out in the second fermenter. Residual lactose can be metabolized by flora identical to that described in the Bel process.

ARNAUD et al. (1978) also proposed production of baker's yeast. A process of this type was used by the Kroger Company in the USA. Lactose in deproteinated whey is hydrolyzed in a reactor in which cells are immobilized by covalent bonds using the Corning process (DOHAN and BARET, 1980). It consists of the following phases (ANONYMOUS, 1984): baker's yeast is multiplied in a batch fermenter (300 L) so that a 90 m^3 reactor can be inoculated. After initial batch culture in which the substrates are used up successively (glucose and then galactose), the continuous method is used at 30°C in an aerobic environment at pH 4.5–5. Cell concentration is 15 g L^{-1}. At a 90 m^3 reactor output, the material is transferred to finishing fermenters for ripening. They are pressed to 30% dry weight after centrifuging and washing. The compressed yeast displays good fermentative activity and is of a desirable light color.

Fermented whey

Biomass production from fermented whey has also been described. MOEBUS and KIESBYE (1975) proposed production of baker's yeast. After preliminary fermentation by *Lactobacillus bulgaricus* strains which convert lactose into lactic acid, baker's yeast multiplies on the substrate. CHAMPAGNE et al. (1990) proposed the use of a *Streptococcus thermophilus* strain which converts lactose into lactic acid and galactose, thus reducing the inhibitory effect of lactic acid. Baker's yeast is produced by multiplication on the two substrates.

6.1.3 Sulfite Waste Liquor

Like whey and molasses, sulfite waste liquor can be used as a fermentation medium. Some 7 m^3 of sulfite liquor is produced per ton of cellulose processed for pulp. The composition of spent sulfite liquors varies according to the type of wood trees and the reagent used in the cooking process. Reducing substances fermented by yeast from spent sulfite liquor from softwood contain about 80% hexoses and 20% pentoses, but reducing substances in liquors from hardwoods contain only 20% hexoses. In addition, sulfite liquors contain compounds such as SO_2 or sulfites which may inhibit growth. The development of a biomass production process therefore requires various treatments before the material can be used (NOONAI and FLEGEL, 1981).

Most processes use *Candida utilis* strains. Biomass is produced continuously in most cases. Productivity is limited by the low level of metabolizable carbon (RYCHTERA et al., 1977; LITCHFIELD, 1983; SOLOMONS, 1983).

The "Pekilo" process was developed by the Finnish Pulp and Paper Institute. It is a continuous process using a strain of *Paecilomyces variotii*. Developed at the Jansan-Koski pulp mill, it runs with a 0.2 h^{-1} dilution rate in an aerated medium without sterilization. The mycelium concentration in the fermenter attains 17 g L^{-1} with 0.55 yield ($Y_{X/S}$) and productivity (P_x) of approximately 2.8 g L^{-1} h^{-1}. Biomass is separated by filtration. It contains 55–60% proteins and vitamins and minerals and is used as an animal feed supplement (ROMANTSCHUK, 1975, 1976).

However, replacement of the sulfite process by the Kraft of sulfate process in paper manufacturing considerably reduces the quantity of sulfite liquor available for biomass production.

6.1.4 Starch

Many plants (wheat, maize, cassava, potatoes, etc.) have high starch contents and are available in large quantities in many countries. There are also large amounts of starch in waste water produced by factories producing starch products. Starch is thus a useful carbon substrate for biomass production.

Numerous species of yeast (see Tab. 3) and filamentous fungi (*Fusarium graminareum, Aspergillus oryzae, Aspergillus fumigatus, Aspergillus hennebergi* var. *niger*) can develop on starch substrates. Many processes have been proposed for biomass production.

6.1.4.1 The Symba Process

This is the only process which has been used at industrial scale (JARL, 1969; SKOGMAN, 1976). It uses waste water and wastes from a potato processing factory. Two yeasts, *Candida utilis* and *Endomycopsis fibuligera*, are used in symbiosis. The latter hydrolyzes starch to glucose, and *Candida utilis* grows on this substrate. The starch-rich water is complemented with growth substances (especially urea and ammonium sufate). The growth medium is sterilized before inoculation. The pH is set at 6.5, the optimum for the functioning of the amylase system. Culture is performed in two continuously operating air-lift fermenters. The temperature is set at 30°C. Feed dry matter content varies from 2 to 3%. The strain *Endomycopsis fibuligera* (Linder) Kloecker develops in a first, small fermenter (30 m^3) on starch substrate and excretes amylases into the medium. The second, large fermenter (300 m^3) is fed simultaneously with the sterile culture medium and the contents of the first fermenter. It is inoculated with a *Candida utilis* strain which develops on glucose in symbiosis with the strain *Endomycopsis fibuligera*. The dilution rate is relatively low ($D = 0.18\ h^{-1}$) to maintain the amylolytic strain. Conversion yield ($Y_{X/S}$) is close to 0.60. This process permits reduction of the pollutant load (BOD_5) from 10000 to 20000 mg L^{-1} to 2000 mg L^{-1}. Productivity (P_x) of the installations (1.2 g L^{-1} h^{-1}) is limited by enzymatic activity.

6.1.4.2 The Adour-Speichim Process

The chopped and crushed cassava substrate used contains 80% starch which is partially solubilized by heat treatment. The medium is complemented with ammonium phosphate, urea, magnesium and potassium. The culture is managed in such a way as to ensure partial conversion of the starch. A *Candida tropicalis* strain with a parietal amylase is used (AZOULAY et al., 1980). The end product is concentrated and dried at fermenter output. Composition according to fermenter residence time may be 56% carbohydrates and 20% proteins or 32% carbohydrates and 37% proteins (REVUZ and VOISIN, 1980).

6.1.4.3 The Ircha-Orstom Process

The process was proposed by SENEZ et al. (1980) to produce a protein-enriched foodstuff from cassava. An *Aspergillus niger* strain is used. the mycelium has a protein content of about 40%. As in the preceding process, substrate conversion is only partial. The raw material (cassava roots) is heated and wetted for a production with 30 to 35% dry weight. The medium is complemented with nitrogen (urea, ammonium sulfate). The pH is set at pH 4.5 and the temperature at 35°C. The medium is seeded with spores and the moisture content increased to 50%. The protein content of the end product is about 20% after 30 hours of growth. Pilot plant technology consists of a simple apparatus of the baker's kneader type mixer with a capacity of 100 kg.

6.1.4.4 The SETA Process

This was proposed by the Société d'Exploitation de Techniques Alimentaires (KANTER, 1987). It is a continuous process ($D = 0.11\ h^{-1}$) for a *Candida utilis* strain at a pH between 3 and 5 at a temperature of 35°C on starchy substrate previously hydrolyzed by amylases.

6.1.4.5 The ENSA-INRA Process

The aim of the process is total conversion of gelatinized starch without prior hydrolysis. Strains of *Schwanniomyces occidentalis* Klöcker var. *castellii* are used. Culture is continuous at pH 3.5 and 30°C without sterilization. Starch (wheat, cassava, potato or maize flour) is gelatinized and then added continuously. The medium is complemented with nitrogen, phosphate, sulfate and trace elements. Dilution rate is 0.35 h^{-1} and 0.63 biomass yield is obtained with certain strains. Productivity at pilot installation scale is 10 g L^{-1} h^{-1} (BOZE et al., 1987). The biomass produced is separated by centrifugation, washed, plasmolyzed and dried; it contains 40% protein. A variant of the process can use higher starch concentrations (80 g L^{-1}) and results in only partial conversion. This can produce a product whose protein content is fixed beforehand.

6.2 Non-Carbohydrate Substrates

6.2.1 Lipids

Several types of substrate can be considered for biomass production from lipids: triglycerides, soaps and residues with a lipid by-product composition. Yeasts are the most interesting microorganisms (HOTTINGER et al., 1974; BA et al., 1983; KOH et al., 1983). Although multiplication is easier in the presence of oil, use of an emulsifier permits development on triglycerides such as the solid fractions resulting from palm oil production (MARTINET et al., 1982). Most lipases possess an optimum hydrolysis pH of between 6 and 7. The specific growth rate of the strains is therefore always lower at pH 3.5 than at pH 6 (MONTET et al., 1983).

However, some strains have interesting growth characteristics at pH 3.5. *Candida rugosa* and *Geotrichum candidum* on rapeseed oil have generation times of 3 and 2.5 hours, respectively, at pH 3.5 and 1.5, and 2.5 hours at pH 6.5. However, on a solid fraction of palm oil and with an emulsifier, *Candida rugosa* generation time (7 hours) is identical at pH 3.5 and 6.5, whereas it increases from 8 to 18 h for *Geotrichum candidum* (MONTET et al., 1983).

Palm oil contains 95% glycerides which include approximately 50% saturated fatty acids; because of their high melting point (60–70°C) these are responsible for a large proportion of palm oil being solid at ambient temperature. The liquefaction temperature (45°C) reduces the possibility of attack of the substrate by yeasts at their optimal growth temperature (30°C). At 45°C, only thermophiles can develop on palm oil (KOH et al., 1985). LABORBE et al. (1989) showed that it is possible to saponify the free fatty acids (4 to 7%) in raw oil palm and to use the resulting soaps as an endogenous emulsifier. RIAUBLANC et al. (1992) studied biomass production from palm oil after saponification of free fatty acids with ammonia. A strain of *Candida rugosa* Diddens and Lodder was used. Continuous culture at pH 4 after optimization of the medium gave a yield ($Y_{X/S}$) of 1.13; 15 g L^{-1} biomass was obtained at equilibrium when the dilution rate was lower than 0.25 h^{-1}. Productivity (P_x) was 3.4 g L^{-1} h^{-1}. In fed-batch culture on the same substrate, final biomass concentration reached 50 g L^{-1}, and productivity was 6 g L^{-1} h^{-1}. Such values were obtained with a constant growth rate (0.1 h^{-1}). As in cultures on alkanes, the dissolved oxygen content is the main growth-limiting factor.

6.2.2 Methanol

Methanol is a useful carbon source for biomass production. It can be obtained from various sources: natural gas, naphtha, heavy fuel oil, coal and cellulose. Use has been envisaged either for pure yeast cultures, pure bacterial cultures and mixed cultures.

The fermentation process on methanol raises specific problems concerning volatility of the substrate and its toxicity for cells at a low concentration and the low affinity of microorganisms for methanol (HARRISON, 1973).

Growth of yeast on methanol has been the subject of interesting reviews by COONEY and LEVINE (1975) and by SAHM (1977). The main strains used belong to the genera *Candida, Trichosporon, Hansenula, Kloeckera, Pichia* and *Torulopsis*. They take up methanol using the dihydroxyacetone pathway. Optimal growth temperatures range from 37 to 42°C, and biomass yields are close to 0.4.

Many studies have been carried out to develop mixed cultures on methanol substrate (HARRISON et al., 1975). Thus, BALLERINI et al. (1977, 1978) isolated and studied the mixed growth of a combination of bacteria: *Methylomonas, Methylovora, Xanthomonas* sp., *Flavobacterium* sp. and *Pseudomonas pseudoalcaligenes*. Equilibrium can be maintained at between 34 and 41°C. The population growth rate reaches 0.49 h^{-1} with a conversion yield varying from 0.44 at 34°C to 0.36 at 41°C. HARRISON et al. (1975) proposed a combination including the primary bacterium, an obligate methanol utilizer and four heterotrophic strains: two *Pseudomonas* sp. strains, an *Acinetobacter* sp. and a *Curtobacterium* sp. As a combination, the population displays a growth rate of 0.59 h^{-1} and a yield of 0.52 at 42°C. The main difficulty involved in the use of mixed cultures for biomass production lies in the need for good definition of the balance between the different strains in order to keep the final biomass composition constant.

The most interesting of the pure cultures of bacteria was developed by Imperial Chemical Industries (ICI) in England using the strain *Methylophilus methylotrophus* (GOW et al., 1975; LICHTFIELD, 1983; SOLOMONS, 1983). In the ICI "pruteen process", the problems of volatility and toxicity related to the substrate were solved by the development of a new fermenter, the "pressure-cycle fermenter". In the general design, the methanol, ammonia and nutrient solutions are sterilized and fed into the fermenter at a dilution rate of between 0.16 and 0.19 h^{-1}. The residual methanol content is 2 ppm, 3 g L^{-1} biomass is obtained at equilibrium and the yield is 0.65.

The "pruteen" air-lift fermenter is the largest aerobic fermenter in operation in the world. Its capacity is nearly 1500 m^3; useful column height is 42 m and diameter is 7 m. The oxygen transfer capacity of this fermenter is about 10 g O_2 L^{-1} h^{-1}. Air is injected at the base (80000 m^3 h^{-1}) at a pressure of several atmospheres. Cells are separated by flocculation, and the liquid is recycled after sterilization. The thick foam of cells is centrifuged and then dried. The end product contains 64% true proteins.

6.2.3 Alkanes

n-Alkanes are saturated, straight-chain hydrocarbons called paraffins. Among the different types of hydrocarbon in crude oil, the aliphatic (*n*-paraffin) compounds are those most easily attacked by microorganisms. The lower range of liquid alkanes, below *n*-nonane (C9) are not taken up by yeast and rarely taken up by bacteria. The *n*-alkanes with chains longer than 9 carbon atoms can be taken up by various microorganisms on the condition that the molecule is made accessible to the cell.

In most processes for biomass production from *n*-alkanes, the raw materials (gas oil, kerosene range, wide range cut from waxy crude) contain a mixture of *n*-alkanes with chain lengths of between C10 and C20 (LEVI et al., 1979; EINSELE, 1983).

The strain used in the pioneering work carried out by the British Petroleum Group is a *Candida tropicalis* (Castellani) Berkhout. Two processes have been developed in parallel by BP. In France, the process developed by Société Française de Pétrole BP uses a heavy gas oil as substrate; this has a boiling range of 300 to 380°C and contains 15% *n*-paraffin (C15–C30). Fermentation is continuous without sterilization using the strain *Candida tropicalis* (Castellani) Berkhout. The yeasts metabolize only the *n*-paraffins. The presence of gas oil and the pH of about 3 reduce risks of contamination (LAINE, 1975). Conversion yield is 1.1, temperature is about 30°C, biomass production at equilibrium varies from 10 to 30 g L^{-1} and the dilution rates applied vary from 0.15 to 0.24 h^{-1}. A centrifugation process is used to separate the yeast cells, the deparaffined gas oil and the spent growth medium.

A process using pure *n*-paraffins (99%) has been developed by BP in Scotland. A *Saccharomycopsis lipolytica* (Wickerham, Kurtman, Hermann) Yarrow in its imperfect form was chosen. The process is continuous under sterile conditions. The culture is controlled in such a way as to eliminate all the paraffins from the medium. Yeasts are more easily recovered under these conditions than in the preceding process and are not subjected to washing with solvents.

Other yeast strains can metabolize *n*-paraffins and have been proposed: *Candida guilliermondii* (Castellani) Langeron and Guerra, *Candida maltosa* (Komagato, Nakase, Katsuya). Few processes using bacteria have been described.

6.3 Carbon Dioxide

Carbon dioxide is the simplest carbon source for biomass production, but it cannot be used as an energy source. Energy must be supplied as light energy and converted to chemical energy during photosynthesis. Microorganisms which can develop using CO_2 include photosynthetic bacteria and algae.

More research has been carried out on the latter, including *Chlorella, Scenedesmus, Spirulina* and *Euglena*.

The production of algal biomass has been described in detail by BENEMANN et al. (1979), GOLDMAN (1979a, b) and RICHMOND (1983). As in agriculture, extensive land, water, nutrients, light and specific climatic conditions are required for open-system culture.

A temperature of between 20 and 40 °C is required for growth. It is limited to a depth of 8 to 10 cm from the surface due to reduced light penetration through the medium. This phenomenon is limited by agitating the medium sufficiently for all the cells to develop under identical conditions.

The biomass concentration obtained (0.2 to 0.5 g L^{-1}) is often limited by low carbon concentrations in the medium usually caused by the respiration of organic compounds by bacteria. Addition of CO_2 from the combustion of certain motors or from fermentation processes thus increases biomass production (1 to 2 g L^{-1}), with a productivity of 10 g m^{-2} to 30 g m^{-2}.

Harvesting algae is particularly difficult. Numerous techniques including flocculation, sedimentation, filtration and centrifugation can be used to recover biomass. Large-scale production is limited by these difficulties (light energy, low biomass concentration and recovery cost). Nevertheless, microalgae production is an interesting field of investigation for obtaining specific biological molecules.

7 Perspectives and Conclusion

There was a great deal of interest in the production of biomass from 1965 to 1985. Enthusiasm for single cell proteins ran high and many research teams

- bred strains capable of metabolizing a great many carbon sources;
- investigated the metabolic pathways involved in the catabolism of these substrates (alkanes, methane, methanol, pentoses, etc.);
- defined the optimal conditions for development of these strains.

The work led to:

- gaining better knowledge of the biology of microbial species hitherto little-studied or not studied at all;
- development of fermentation technology with the design of various probes (measurement and regulation of pH, temperature and partial pressure of oxygen) and the governing of the production rate by a number of physiological parameters (qO_2, qCO_2, QR) or the concentration of a single component of the medium (substrate, ethanol, etc.).

All this research has resulted in numerous applications in industrial sectors such as manufacture of baker's yeast, antibiotics, enzymes and metabolites and in more traditional industries such as brewing and wine-making.

More recently, the development of methods in molecular biology has enabled the production of various molecules using recombined cells (FLEER, 1992; BUCKHOLZ and GLEESON, 1991; ROMANOS et al., 1992). Biomass is not the only objective in this type of production, although higher cell concentration often results in higher product concentration and hence decreased manufacturing costs. Knowledge acquired concerning the metabolism of strains such as *Kluyveromyces lactis* and *Pichia pastoris* in higher density growth (SHAY et al., 1987) is now used in developments in heterologous protein production (FLEER et al., 1991). In addition, optimization of production yields is not limited to the improvement of strains. Recent examples have shown that control of growth and induction of certain factors can considerably increase the production of heterologous protein (ROMANOS et al., 1992; ALBERGHINA et al., 1991).

8 References

AIBA, S., HUMPHREY, A. E. MILLIS, N. F. (1973), Aeration and agitation, in: *Biochemical Engineering* (AIBA, S., HUMPHREY, A. E., MILLIS, N. F., Eds.), 2nd Ed., pp. 163–194, New York: Academic Press.

ALBERGHINA, L., PORRO, D., MARTEGANI, E., RANZI, B. M. (1991), Efficient production of recombinant DNA proteins in *Saccharomyces cerevisiae* by controlled high-cell density fermentation, *Biotechnol. Appl. Biochem.* **14,** 82–92.

ANONYMOUS (1984), Immobilized enzyme and fermentation technologies combined to produce baker's yeast from whey, *Food Technol.*, June, 26–27.

ANTONY, C. (1982), *The Biochemistry of Methylotrophs*, London: Academic Press.

ARNAUD, M., MALIGE, B., GALZY, P., MOULIN, G. (1978), Perfectionnement à la fabrication de levure lactique, *French Patent* 7830229.

ATKINSON, B., MAVITUNA, F. (1983), Thermodynamic aspects of microbial metabolism, in: *Biochemical Engineering and Biotechnology Handbook,* pp. 71–112, New York: The Nature Press.

AZOULAY, E., JOUANNEAU, F., BERTRAND, J. C., RAPHAEL, A., JANSSENS, J., LEBEAULT, J. M. (1980), Fermentation methods for protein enrichment of cassava and corn with *Candida tropicalis, Appl. Environ. Microbiol.* **39**, 41–47.

BA, A., RATOMAHENINA, R., GRAILLE, J., GALZY, P. (1983), Etude de la croissance de quelques souches de levures sur les sous-produits de raffinage de l'huile d'arachide, *Oléagineux* **36**, 439–445.

BALLERINI, D., PARLOUAR, D., LAPEYRONNIE, M., SRI, K. (1977), Mixed culture of bacteria utilizing methanol for growth. II. Influence of growth rate and temperature on growth yield and composition of mixed MS1 culture of bacteria, *Eur. J. Appl. Microbiol.* **4**, 11–19.

BALLERINI, D., CRÉMIEUX, A., LAPEYRONNIE, M., PARLOUAR, D., WEILL-THEVENET (1978), *French Patent* 2338992.

BARNETT, J. (1981), The utilization of disaccharides and some other sugars by yeasts, *Adv. Carbohydr. Chem. Biochem.* **39**, 347–404.

BAZIN, M. J. (1981), Mixed culture kinetics, in: *Mixed Culture Fermentation* (BUSHELL, M. E., SLATER, J. H., Eds.), pp. 25–51, New York: Academic Press.

BENEMANN, J. R., WEISSMAN, J. C., OSWALD, W. J. (1979), Algae biomass, in: *Economic Microbiology* (ROSE, A. H., Ed.), Vol. 4, pp. 177–206, New York: Academic Press.

BEUDEKER, R. F., VAN DAM, H. W., VAN DER PLAAT, J. B., VELLENGA, K. (1989), Developments in baker's yeast production, in: *Yeast Biotechnology and Biocatalysis* (VERACHTER, H., DE MOT, R., Eds.) pp. 103–146, New York: Marcel Dekker Inc.

BLONDEAU, K., BOUTUR, O., JUNG, G., MOULIN, G., GALZY, P. (1993), Development of high density fermentation for heterologous interleukin 1β production in *Kluyveromyces lactis* controlled by PHO5 promoter, *Appl. Microbiol. Biotechnol.* **4**, 324–329.

BOZE, H., MOULIN, G., GALZY, P. (1987), Influence of culture conditions on the cell yield and amylase biosynthesis in continuous culture by *Schwanniomyces castellii, Arch. Microbiol.* **148**, 162–166.

BOZE, H., MOULIN, G., GALZY, P. (1992), Production of food and fodder yeast, *Crit. Rev. Biotechnol.* **12**, 65–86.

BRAUDE, R. (1942), *J. Inst. Brew.* **48**, 206.

BUCKHOLZ, R. G., GLEESON, A. G. (1991), Yeast systems for the commercial production of heterologous proteins, *Bio/Technology* **9**, 1067–1072.

BURROWS, S. (1979), Baker's yeast, in: *Economic Microbiology* (ROSE, A. H., Ed.) Vol. 4, pp. 31–61, New York: Academic Press.

BUTTON, D. K. (1985), Kinetics of nutrient-limited transport and microbial growth, *Microbiol. Rev.* **49**, 270–297.

CAGNIARD DE LATOUR, C. (1838), Mémoire sur la fermentation vineuse, *Ann. Chim. Phys.* **68**, 206–222.

CARDINI, G., JURTSHUK, P. (1970), The enzymatic hydroxylation of *n*-octane by *Corynebacterium* species strains, *J. Biol. Chem.* **245**, 2789–2796.

CHAMPAGNE, C. P., GOULET, J., LACHANCE, R. A. (1990), Production of baker's yeast in cheese whey ultrafiltrate, *Appl. Environ. Microbiol.* **56**, 425–430.

CHERUY, A., DUMENIL, G., SANGLIER, J. J. (1989), Optimisation et modélisation, in: *Biotechnologie des Antibiotiques* (LARPENT, J. P., SANGLIER, J. J., Eds.), pp. 289–338, Paris: Masson.

COLBY, J., DALTON, H., WHITTENBURY, R. (1979), Biological and biochemical aspects of microbial growth on C1 compounds, *Annu. Rev. Microbiol.* **33**, 481–517.

COONEY, C. L., LEVINE, D. W. (1975), S.C.P. production from methanol by yeast, in: *Single Cell Protein II* (TANNENBAUM, S. R., WANG, D. I. C., Eds.), pp. 402–423, Cambridge, MA: MIT Press.

DOELLE, H. W. (1981), Basic metabolic processes, in: *Biotechnology 1st Edition* (REHM, H.-J., REED, G., Eds.), Vol. 1, pp. 113–210, Weinheim: Verlag Chemie.

DOHAN, L. A., BARET, J. L. (1980), Lactose hydrolysis by immobilized lactase: semi-industrial experience, *Enzyme Eng.* **5**, 279–293.

DOUMA, A. C., VEENHUIS, M., DE KONING, W., EVERS, M., HARDER, W. (1985), Dihydroxyacetone synthetase is localized in the peroximal matrix of methanol grown *Hansenula polymorpha, Arch. Microbiol.* **143**, 237–243.

EINSELE, A. (1978), Scaling up bioreactors, *Process Biochem.* **13** (7), 13–14.

EINSELE, A. (1983), Biomass from higher *n*-alkanes, in: *Biotechnology 1st Edition* (REHM, H.-J., REED, G., Eds.), Vol. 3, pp. 43–81, Weinheim: Verlag Chemie.

EINSELE, A., BLANCH, H. W., FIECHTER, A. (1973), Agitation and aeration in hydrocarbon fermentations, *Biotechnol. Bioeng. Symp.* **4**, 455–466.

FÉVRIER, C., BOURDIN, C. (1977), Utilisation du lactosérum et des produits lactosés par les porcins, in: *Colloque, les lactosérums, une richesse alimentaire,* pp. 129–174, Paris: A.P.R.I.A.

FIECHTER, A. (1984), Physical and chemical parameters of microbial growth, *Adv. Biochem. Eng. Biotechnol.* **30**, 7–60.

FIECHTER, A., SEGHEZZI, W. (1992), Regulation of glucose metabolism in growing yeast cells, *J. Biotechnol.* **27**, 27–45.

FIECHTER, A., FUHRMANN, G. F., KÄPPELI, O. (1981), Regulation of glucose metabolism in growing yeast cells, *Adv. Microbiol. Physiol.* **22**, 123–183.

FLEER, R. (1992), Engineering yeast for high level expression, *Curr. Opin. Biotechnol.* **3**, 486–496.

FLEER, R., YEH, P., AMELLAI, N., MAURY, I., FOURNIER, A., BACHETTA, F., BADUEL, P., JUNG, G., L'HÔTE, H., BECQUART, J., FUKUHARA, H., MAYAUX, I. F. (1991), Stable multicopy vectors for high-level secretion of recombinant human serum albumin by *Kluyveromyces lactis, Bio/Technology* **9**, 968–975.

FLEURY, H., HENRIET, P. (1977), Culture de champignons filamenteux. Traitement du lactosérum par le procédé Caliqua/Sireb, in: *Colloque, les lactosérums, une richesse alimentaire*, pp. 101–116, Paris: A.P.R.I.A.

FREDRICKSON, A. G. (1983), Interactions of microbial populations in mixed culture situations, *ACS Symp. Ser.* **207**, 202–227.

FUKUI, S., TANAKA, A. (1981), Metabolism of alkanes by yeasts, *Adv. Biochem. Eng.* **19**, 217–237.

GALLO, M., BERTRAND, J. C., AZOULAY, E. (1971), Participation of cytochrome P-450 in the oxidation of alkanes by *Candida tropicalis, FEBS Lett.* **19**, 45–49.

GANCEDO, C., SERRANO, R. (1989), Energy-yielding metabolism, in: *The Yeasts, Metabolism and Physiology of Yeasts* (ROSE, A. H., HARRISON, J. S., Eds.), 2nd Ed., Vol. 3, pp. 205–259, London: Academic Press.

GOLDBERG, I. (1985), Single cell protein, in: *Biotechnology Monographs* (AIBA, S., FAN, L. T., FIECHTER, A., SCHÜGERL, K., Eds.), Berlin: Springer-Verlag.

GOLDBERG, I., ER-EL, Z. (1981), The chemostat: an efficient technique for medium optimization, *Process Biochem.*, Oct/Nov., 2–8.

GOLDMAN, J. C. (1979a), Outdoor algal mass cultures: I. Applications, *Water Res.* **13**, 1–20.

GOLDMAN, J. C. (1979b), Outdoor algal mass cultures: II. Photosynthetic yield limitations, *Water Res.* **13**, 119–136.

GONCHAROVA, I. A., BABITSKAYA, V. G., LOBANOJ, A. G. (1977), in: *Proceedings Symposium Microbial Growth on C1-compounds* (SKRYABIN, G. K., IVANOV, H. V., KONDRATIEVA, E. N., ZAVARZIN, G. A., YU, A., TROTSENKO, A., NESTEROV, A. I., Eds.), pp. 187–200, Pushkino, USSR: USSR Academy of Sciences.

GOUPY, J. (1988), La méthode des plans expériences. Optimisation du choix des essais et de l'interprétation des résultats (DUNOD, Ed.), Paris: Masson.

GOW, J. S., LITTLEHAILES, J. D., SMITH, S. R. L., WALTER, R. B. (1975), S.C.P. production from methanol: Bacteria, in: *Single Cell Protein II* (TANNENBAUM, S. R., WANG, D. I. C., Eds.), pp. 370–384, Cambridge, MA: MIT Press.

HALTER, N., PUHAN, Z., KÄPPELI, O. (1981), Upgrading of milk U.F. permeate by yeast fermentation, in: *Advances in Biotechnology* (MOO-YOUNG, M. O., Ed.), Vol. 2, pp. 351–356, New York: Pergamon Press.

HARDER, W., VAN DIJKEN, J. P. (1976), Theoretical calculations on the relation between energy production and growth of methane-utilizing bacteria, in: *Microbial Production and Utilization of Gases* (SCHLEGEL, H. G., PFENNIG, N., GOTTSCHALK, G., Eds.), pp. 403–418, Göttingen: Erich Goltze Verlag.

HARDER, W., VEENHUIS, M. (1989), Metabolism of one-carbon compounds, in: *The Yeasts, Metabolism and Physiology of Yeasts* (ROSE, A. H., HARRISON, J. S., Eds.), 2nd Ed., Vol. 3, pp. 289–315, London: Academic Press.

HARRISON, D. E. F. (1973), Studies on the affinity of methanol and methane-utilizing bacteria for their carbon substrates, *J. Appl. Bacteriol.* **36**, 301–308.

HARRISON, D. E. F., WILKINSON, T. C., WREN, S. J., HARWOOD, J. H. (1975), Mixed bacterial cultures as a basis for continuous production of

SCP from C1 compounds, in: *Continuous Culture: Applications and New Fields* (DEAN, A. C. R., ELLWOOD, D. C., EVAN, C. G. T., MELLING, J., Eds.), pp. 122–134, Chichester, UK: Ellis Horwood.

HÖFLE, M. G. (1983), Long-term changes in chemostat cultures of *Cytophaga johnsonae, Appl. Environ. Microbiol.* **46**, 1045–1053.

HOTTINGER, H. H., RICHARDSON, T., AMUNDSON, C. H., STUIBER, D. A. (1974), Utilization of fish oil by *Candida lipolytica* and *Geotrichum candidum, J. Milk Food Technol.* **37**, 522–528.

IUPAC (International Union of Pure and Applied Chemistry) (1974), *Technical Report No. 12. Proposed Guidelines for Testing S.C.P. Destined as Major Protein Sources for Animal Feed*, AUG.

JARL, K. (1969), Symba yeast process, *Food Technol.* **23**, 23–26.

JONES, R. P., GREENFIELD, P. F. (1984), A review of yeast ionic nutrition. Part I: Growth and fermentation requirements, *Process Biochem.* (April), 48–60.

KANTER, N. (1987), Procédé de production à partir d'un substrat végétal hydrocarboné quelconque d'un produit naturel alimentaire et produit obtenu par ce procédé, *French Patent* 2597723.

KÄPPELI, P., FINNERTY, W. R. (1979), Partition of alkane by an extracellular vesicle derived from *Acinetobacter, J. Bacteriol.* **140**, 707–712.

KOH, J. J., KODAMA, T., MINODA, Y. (1983), Screening of yeast and culture conditions of cell production from palm oil, *Appl. Microbiol. Biotechnol.* **47**, 1207–1212.

KOH, J. J., YAMAKAYA, T., KODOMA, T., MINODA, Y. (1985), Rapid and dense culture of *Acinetobacter calcoaceticus* on palm oil, *Agric. Biol. Chem.* **49**, 1411–1416.

LABORBE, J. M., DWECK, C., RATOMAHENINA, R., PINA, M., GRAILLE, J., GALZY, P. (1989), Production of single cell protein from palm oil using *Candida rugosa, MIRCEN J.* **5**, 517–523.

LAINE, B. M. (1975), Gas-oil as a substrate for single-cell protein, in: *Single Cell Protein II,* (TANNENBAUM, S. R., WANG, D. I. C., Eds.), pp. 424–437, Cambridge, MA: MIT Press.

LAMBRECHTS, C., BOZE, H., MOULIN, G., GALZY, P. (1992), Utilization of phytate by some yeasts, *Biotechnol. Lett.* **14**, 61–66.

LEBEAULT, J. M., ROCHE, B., DUVNJAK, Z., AZOULAY, E. (1970a), Alcool et aldéhyde deshydrogénases particulières de *Candida tropicalis* cultivé sur hydrocarbures, *Biochim. Biophys. Acta* **220**, 373–385.

LEBEAULT, J. M., MEYER, F. ROCHE, B., AZOULAY, E. (1970b), Oxydation des alcools supérieurs chez *Candida tropicalis* cultivé sur hydrocarbures, *Biochim. Biophys. Acta* **220**, 386–395.

LEBEAULT, J. M., LODE, E. T., COON, M. J. (1971), Fatty acid and hydrogen hydroxylation in yeast: role of cytochrome P-450 in *Candida tropicalis, Biochem. Biophys. Res. Commun.* **42**, 413–419.

LEE, J. D., KOMAGATA, K. (1980), Taxonomic study of methanol assimilating yeasts, *J. Gen. Appl. Microbiol.* **26**, 133–158.

LEFRANÇOIS, L. (1964), Problèmes de l'aération et de la circulation dans les cuves de fermentations aérobies, *Ind. Alim. Agric.* **1**, 3–18.

LELOUP, V., COLONNA, P., BULÉON, A. (1991), Les transformations enzymatiques des glucides, in: *Biotransformation des Produits Ceréaliers* (GODON, B., Ed.), pp. 79–128, Paris: Technique et Documentation, Lavoisier.

LEVI, J. D., SHENNAN, J. L., EBBON, G. P. (1979), Biomass from liquid *n*-alkanes, in: *Economic Microbiology* (ROSE, A. H., Ed.), Vol. 4, pp. 361–419, London: Academic Press.

LITCHFIELD, J. H. (1979), Production of single-cell protein for use in food or feed, in: *Microbial Technology* (PEPPLER, H. J., PERLMAN, D., Eds.) Vol. 1, p. 93, New York: Academic Press.

LITCHFIELD, J. H. (1983), Single cell protein, *Science* **219**, 740–746.

MACHEK, F., STROS, F., PROKOP, A., ADAMER, L. (1975), Production and isolation of protein from synthetic ethanol, in: *Continuous Culture: Applications and New Fields* (DEAN, A. C. R., ELLWOOD, D. C., EVAN, C. G. T., MELLING, J., Eds.), pp. 135–148, Chichester, UK: Ellis Horwood.

MARTINET, F., RATOMAHENINA, R., GRAILLE, J., GALZY, P. (1982), Production of food yeast from solid fraction of palm oil, *Biotechnol. Lett.* **4**, 9–14.

MATELES, R. I., BATTAT, E. (1974), Continuous culture used for media optimization, *Appl. Microbiol.* **28**, 901–904.

MATSUMURA, M., UMEMOTO, K., SHINABE, K., KOBAYASHI, J. (1982), Application of pure oxygen in a new gas entraining fermentor, *J. Ferment. Technol.* **60**, 565–571.

MEYRATH, J., BAYER, K. (1979), Biomass from whey, in: *Economic Microbiology* (ROSE, A. H., Ed.) Vol. 4, pp. 207–269, London: Academic Press.

MIYOSHI, M. (1895), *Jahrb. Wiss. Bot.* **28**, 269.

MOEBUS, O., KIESBYE, P. (1975), Continuous process for producing yeast protein and baker's yeast, *G.F.R. Patent Application* 2410349.

MOLINARO, R., HONDERMARCK, J. C., JACQUOT, J. (1977), Production de lactoprotéines levurées, in: *Colloque, les lactosérums, une richesse alimentaire*, pp. 83–93, Paris: A.P.R.I.A.

MONOD, J. (1949), Recherche sur la croissance des cultures bactériennes, *Annu. Rev. Microbiol.* **3**, 371–394.

MONOD, J. (1950), La technique de la culture continue. Théorie et applications, *Ann. Inst. Pasteur* **79**, 390–410.

MONTET, D., RATOMAHENINA, R., BA, A., PINA, M., GRAILLE, J., GALZY, P. (1983), Production of single cell proteins from vegetable oils, *J. Ferment. Technol.* **61**, 417–420.

MOSER, A. (1985a), Continuous cultivation, in: *Biotechnology 1st Edition* (REHM, H.-J., REED, G., Eds.), Vol. 2, pp. 285–309, Weinheim: VCH.

MOSER, A. (1985b), Kinetic of batch fermentations, in: *Biotechnology 1st Edition* (REHM, H.-J., REED, G., Eds.), Vol. 2, pp. 243–283, Weinheim: VCH.

MOSER, A. (1985c), Special cultivation techniques, in: *Biotechnology 1st Edition* (REHM, H.-J., REED, G., Eds.), Vol. 2, pp. 310–347, Weinheim: VCH.

MOULIN, G., GALZY, P. (1984), Whey a potential substrate for biotechnology, in: *Biotechnology and Genetic Engineering Reviews* (RUSSEL, G. E., Ed.), Vol. 1, pp. 347–374, Newcastle Upon-Tyne: Intercept.

MOULIN, G., MALIGNE, B., GALZY, P. (1983a), Balanced flora of an industrial fermenter. Production of yeast from whey, *J. Dairy Sci.* **66**, 21–28.

MOULIN, G., LEGRAND, M., GALZY, P. (1983b), The importance of residual aerobic fermentation in aerated medium for the production of yeast from glucidic substrates, *Process Biochem.* **18** (5), 5–17.

NOONAI, A., FLEGEL, T. W. (1981), Biomass production from acid hydrolysate of spent sulfite liquor, in: *Microbial Utilization of Renewable Resources* (TAGUCHI, H., Ed.), p. 33, Osaka University.

NOVICK, A., SZILARD, L. (1950), *Science* **112**, 715.

OGATA, K., NISHIKAWA, H., OHSUGI, M. (1970), Studies on the production of yeast (I). A yeast utilizing methanol as sole carbon source, *J. Ferment. Technol.* **48**, 389–396.

OLBRICH, H. (1973), Melasse als Rohstoffproblem der Backhefeindustrie, *Branntweinwirtschaft* **113**, 53–68, 270–271.

OWENS, J. D., LEGAN, J. D. (1987), Determination of the Monod substrate saturation constant for microbial growth, *FEMS Microbiol. Rev.* **46**, 419–432.

PASTEUR, L. (1860), Mémoire sur la fermentation alcoolique, *Ann. Chim. Phys.* **58**, 323–426.

PETERSON, J. A., KUSUNOSE, M., KUSUNOSE, E., COON, M. J. (1967), Enzymatic ω-oxydation: II. function of rubredoxin as the electron carrier in ω-hydroxylation. *J. Biol. Chem.* **242**, 4334–4340.

PIRT, S. J. (1975), *Principles of Microbe and Cell Cultivation*, Oxford: Blackwell Scientific Publication.

PREBBLE, J. N. (1981), Bacterial energy transformation, in: *Mitochondria, Chloroplasts and Bacterial Membranes* (PREBBLE, J. N., Ed.), pp. 211–230, New York: Longman Inc.

RATLEDGE, C., EVANS, C. T. (1989), Lipids and their metabolism, in: *The Yeasts, Metabolism and Physiology of Yeasts* (ROSE, A. H., HARRISON, J. S., Eds.), 2nd Ed., Vol. 3, pp. 367–455, London: Academic Press.

RATLEDGE, C., TAN, K. H. (1989), Oils and fats: Production, degradation and utilization by yeast, in: *Yeast Biotechnology and Biocatalysis* (VERACHTER, H., DE MOT, R., Eds.), pp. 223–253, New York: Marcel Dekker Inc.

REED, G. (1982), Production of baker's yeast, in: *Prescott and Dunn's Industrial Microbiology* (REED, G., Ed.), pp. 593–633, Westport: AVI.

REVUZ, B., VOISIN, B. (1980), Le manioc protéiné, *Ind. Alim. Agric.* **10**, 1079–1084.

RIAUBLANC, A., BOZE, H., DEMUYNCK, M., MOULIN, G., RATOMAHENINA, R., GRAILLE, J., GALZY, P. (1992), Optimization of biomass production from palm oil cultures using *Candida rugosa, Fat Sci. Technol.* **2**, 46–51.

RICHMOND, A. (1983), Phototrophic microalgae, in: *Biotechnology 1st Edition* (REHM, H.-J., REED, G, Eds.), Vol. 3, pp. 109–143, Weinheim: Verlag Chemie.

RIEGE, P., SCHUNCK, W. H., HONECK, H., MUELLER, H. G. (1981), Cytochrome P-450 from *Lodderomyces elongisporus:* Its purification and some properties of the highly purified protein, *Biochem. Biophys. Res. Commun.* **98**, 527–534.

ROMANOS, M. A., SCORER, C. A., CLARE, J. J. (1992), Foreign gene expression in yeast: A review, *Yeast* **8**, 423–488.

ROMANTSCHUK, H. (1975), The Pekilo process: protein from spent sulfite liquor, in: *Single Cell Protein* (TANNENBAUM, S. R., WANG, D. I. C., Eds.), pp. 344–356, Cambridge, MA: MIT Press.

ROMANTSCHUK, H. (1976), The Pekilo process: A development project. Applications and new fields, in: *Continuous Culture* (DEAN, A. C. R., ELLWOOD, D. C., EVAN, C. G. T., MELLING, J., Eds.), pp. 116–121, Chichester, UK: Ellis Horwood.

RYCHTERA, M., BARTA, J., FIECHTER, A., EINSELE, A. A. (1977), Several aspects of the yeast cultivation on sulfite waste liquors and synthetic ethanol, *Process Biochem.* **12**, 26–30.

SAHM, H. (1977), Metabolism of methanol by yeast, *Adv. Biochem. Eng.* **6**, 77–103.

SCHÜLEIN, J. (1937), The brewer's yeast as medicine and feeding stuff. Dresden: Verlag Steinkopf.

SENEZ, J. C., RAIMBAULT, M., DESCHAMPS, F. (1980), Protein enrichment of starchy substrates for animal feeds by solid state fermentation, *World Anim. Rev. F.A.A.* **35**, 36–39.

SHAY, L. K., HUNT, H. R., WEGNER, G. H. (1987), High productivity fermentation process for cultivating industrial microorganisms, *J. Ind. Microbiol.* **2**, 79–85.

SHECHTER, E. (1984), Transport, in: *Membranes Biologiques* (MASSON, Ed.), pp. 83–152, Paris: Masson.

SKOGMAN, H. (1976), The symba process. *Stärke* **8**, 278–281.

SÖHNGEN, N. L. (1906), *Zentralbl. Bakteriol. Parasitenkd. Infektionskr. Hyg. Abt.* **15**, 513.

SOLOMONS, G. L. (1983), Single cell protein, *C.R.C. Crit. Rev. Biotechnol.* **1**, 21–58.

SONNLEITNER, B., KÄPPELI, O. (1986), Growth of *Saccharomyces cerevisiae* is controlled by its limited respiratory capacity: formulation and verification of a hypothesis, *Biotechnol. Bioeng.* **28**, 927–937.

SPECTOR, W. S. (1956), *Handbook of Biological Data*, Philadelphia: Saunders.

STEELE, D. B., STOWERS, M. S. (1991), Techniques for selection of industrially important microorganisms, *Annu. Rev. Microbiol.* **45**, 89–106.

STOKES, J. L. (1971), Influence of temperature on the growth and metabolism of yeast, in: *The Yeasts, Metabolism and Physiology of Yeasts* (ROSE, A. H., HARRISON, J. S., Eds.), 2nd Ed., Vol. 2, pp. 119–134, New York: Academic Press.

STOUTHAMER, A. H. (1973), A theoretical study on the amount of ATP required for synthesis of microbial cell material, *Antonie van Leeuwenhoek* **39**, 545–565.

STOUTHAMER, A. H. (1979), The search for correlation between theoretical and experimental growth yields, in: *International Reviews of Biochemistry* (QUAYLE, J. R., Ed.), Vol. 21, pp. 1–47, Baltimore, University Park Press.

TANAKA, A., FUKUI, S. (1989), Metabolism of *n*-alkanes, in: *The Yeasts, Metabolism and Physiology of Yeasts* (ROSE, A. H., HARRISON, J. S., Eds.), 2nd Ed., Vol. 2, pp. 261–287, London: Academic Press.

TØTTRUP, H. V., CARLSEN, S. (1990), Process for the production of human proinsulin in *Saccharomyces cerevisiae, Biotechnol. Bioeng.* **35**, 339–348.

VELDKAMP, H. (1972), Mixed cultures studies with the chemostat, *J. Appl. Chem. Biotechnol.* **22**, 105–123.

VELDKAMP, H. (1977), Ecological studies with the chemostat, *Adv. Microb. Ecol.* **1**, 59–89.

VERDUYN, C. (1991), Physiology of yeasts in relation to biomass yields, *Antonie van Leeuwenhoek* **60**, 325–353.

VERDUYN, C., POSTMA, E., SCHEFFERS, W. A., VAN DIJKEN, J. P. (1990), Energetics of *Saccharomyces cerevisiae* in anaerobic glucose-limited chemostat cultures, *J. Gen. Microbiol.* **136**, 395–403.

VERDUYN, C., STOUTHAMER, A. H., SCHEFFERS, W. A., VAN DIJKEN, J. P. (1991), A theoretical evaluation of growth of yeasts, *Antonie van Leeuwenhoek* **59**, 49–63.

WEITZEL, W., WINCHEL, M. (1932), *The Yeast, Its Nutritive and Therapeutic Value,* Berlin: Verlag Rothgiese und Diesing.

WHITAKER, A. (1980), Fed-batch culture, *Process Biochem.* **15** (4), 10–12, 14–15, 32.

Further Reading: Books on the production of biomass

Economic Microbiology (ROSE, A. H., Ed.), Volume 4: *Microbial Biomass* (1979), *London: Academic Press.*

Biotechnology 1st Edition (REHM, H.-J., REED, G., Eds.), Volume 1: *Microbial Fundamentals* (1981), Weinheim: Verlag Chemie.

Biotechnology 1st Edition (REHM, H.-J., REED, G., Eds.), Volume 3: *Biomass, Microorganisms for Special Applications, Microbial Products I, Energy from Renewable Resources* (1983), Weinheim: Verlag Chemie.

Biotechnology 2nd Edition (REHM, H.-J., REED, G., PÜHLER, A., STADLER, P., Eds.), Volume 1: *Biological Fundamentals* (1993), Weinheim: VCH.

Biochemical Engineering and Biotechnology Handbook (1983) (ATKINSON, B., MAVITUNA, F., Eds.), New York: The Nature Press.

Biotechnology 1st Edition (REHM, H.-J., REED, G., Eds.), Volume 2: *Fundamentals of Biochemical Engineering* (1985), Weinheim: VCH.

Single cell protein (1985), in: *Biotechnology Monographs* (AIBA, S., FIECHTER, A., SCHÜGERL, K., Eds.), Berlin: Springer-Verlag.

6 Nutritional Value and Safety of "Single Cell Protein"

NEVIN S. SCRIMSHAW
EDWINA B. MURRAY

Boston, MA 02114-0500, USA

1 Introduction

The term "Single Cell Protein" was selected in 1967 for the title of the first international conference on microbial biomass produced on relatively pure carbohydrate or hydrocarbon substrates (SCRIMSHAW, 1968, 1975). It offered a neutral term to include bacteria, yeast and filamentous microfungi. This name, usually abbreviated to SCP, soon became the generic term throughout the world. Had it been coined earlier, such public relations disasters as calling it "petro-protein" in Japan or microbial protein might have been avoided. Major substrates for SCP are listed in Tab. 1.

Tab. 1. Major Substrates for Single-Cell Protein (SCP)

From primary plant sources:
- Sulfite Liquor (waste from paper manufacturers)
- Molasses (from sugar cane)
- Ethanol (from carbohydrate fermentation)
- Whey (from milk)

From petroleum hydrocarbons:
- "Gas Oil" (extracted)
- Normal alkanes (extracted)
- Paraffins (extracted)
- Ethanol (synthesized)
- Methanol (synthesized)

When a conference was convened in Guatemala ten years later in 1978 on the *Bioconversion of Organic Residues* (UNITED NATIONS UNIVERSITY, 1979), it was apparent that a name was also needed to designate products based on microbial growth on vegetable and animal wastes in which the residues themselves are a significant part of a final product that has been enriched or transformed by organic residues. The name microbial biomass product (MBP) was adopted. It applies equally to wholly new products and such traditional ones as tempeh, ontjom, and bongkrek (STEINKRAUS, 1983). Major substrates available for MBP are listed in Tab. 2.

Tab. 2. Substrates for Microbial Biomass Products (MBP) (JOHNSON, 1967)

Agriculture	Other
Cereals	Animal manure
Straw	Municipal garbage
Bran	Paper mill effluents
Coffee hulls	Cannery effluents
Cocoa hulls	Fishery effluents
Coconut hulls	Slaughterhouse effluents
Fruit and vegetable wastes	Milk-processing effluents
Bagasse	
Oilseed cakes	
Cotton wastes	
Coffee and tea wastes	
Bark and sawdust	

Once interest was aroused by the 1967 international meeting at MIT on "Single Cell Protein" (MATELES and TANNENBAUM, 1968), several others were held in the next few years including *Proteins from Hydrocarbons* (GOUNELLE DE PONTANEL, 1972), *Single Cell Protein II* (TANNENBAUM and WANG, 1975), *International Symposium on Single Cell Protein* (DAVIS, 1977), *Single Cell Protein: Safety for Animal and Human Feeding* (GARRATINI et al., 1979). These meetings were supplemented during this period by symposia or workshops on the topic at various national and international scientific meetings. Because the economics proved unattractive for a number of reasons, there has been no major SCP conference since, and there is no longer significant commercial interest in SCP products. There was, however, an international symposium on *Single Cell Proteins from Hydrocarbons for Animal Feeding* held in Algiers in 1983 (HAMDAN, 1983). During the period of active interest in the 1970s and early 1980s, sophisticated technologies and products were successfully developed.

2 Nutritional Value

An extensive literature on the nutritional value of food yeast for rats (14 reports), chickens (11 reports), pigs (18 reports) and humans (10 reports) has been listed by KIHLBERG (1972). He also described studies with feeding bacteria to rats (13 reports), chickens (1 study) and humans. For filamentous microfungi, he reported 8 studies in rats, 4 in chickens and 5 in swine, and furthermore, he cited a number of review articles. The results can be generalized below without describing them separately.

2.1 Protein and Amino Acids

As compared with vegetable protein sources, the nitrogen content of SCPs is relatively high (YOUNG and SCRIMSHAW, 1975a). Calculation of the protein value of the SCPs is biased by their nucleic acid content. Any rapidly growing cell has relatively large amounts of ribonucleic acid (RNA) and about half as much deoxyribonucleic acid (DNA). Unless specially processed to reduce the nucleic acid content as discussed below, the conventional calculation of crude protein, as N×6.25, overestimates protein content.

Most nucleic acid molecules consist of equimolar proportions of a pyrimidine and a purine. The nitrogen content of pyrimidines is about 40% of that for purines. Multiplying by 6.25 for protein nitrogen and 9.0 for nucleic acid nitrogen gives corrected values. The conventional calculation, which continues to be used, results in biological values that are lower than if "true protein" were estimated.

Depending on the specific organism, the protein content of SCPs, calculated as N×6.25, ranges from approximately 50% for yeast to over 70% for bacteria. The microfungi have the lowest crude protein concentration, bacteria have the highest, and yeasts are intermediate. If it has been processed without excessive heat destruction, the digestibility and quality of SCP protein is comparable to that of other good vegetable proteins. Digestibility ranges from approximately 50 to 90.

The amino acid composition of a protein determines its biological value as a source of nitrogen. While the amino acid compositions of the cell wall and the cytoplasm of the cell are somewhat different, the overall essential amino acid pattern is quite good. Both SCPs and legumes are limiting in the amino acid methionine. This deficiency is detectable in conventional experimental animal studies to determine net protein utilization (NPU) and in humans fed protein-deficient diets with yeast as the only nitrogen source (YOUNG and SCRIMSHAW, 1975b). More important is the evidence that the amino acid pattern of SCPs, like that of legumes, complements the deficiencies of lysine and threonine or tryptophan in cereal grains.

Information on the average digestibility and protein value of 11 various single cell proteins is given in Tab. 3. The amino acid pattern of a yeast and a filamentous microfungus compared with that of rumen bacteria, soy protein and the latest FAO/WHO reference pattern is given in Tab. 4. It will be noted that except for methionine/cystine all of the essential amino acids are present in the SCP products at levels exceeding their requirement per 100 g protein and are quite comparable to soy protein whose protein value is well recognized (SCRIMSHAW and YOUNG, 1979).

Our own metabolic balance studies in young adults fed microfungi include *Fusarium graminearium* grown on sugar hydrolysate and *Paecilomyces variotii* grown on sulfite waste from paper manufacture. The results for digestibility were 78% and 81%, respectively, biological value 84% and 67%, and from net protein utilization 65% and 54% (UDALL et al., 1984).

Feeds from rapidly growing animals that utilize either legumes or microbial biomass as major protein sources are improved by either the addition of methionine or the incorporation of some animal protein. For humans and adult animals, however, the amino acid pattern is sufficiently similar to that of soy and animal proteins, so that amino acid supplementation is not required. When humans are fed at or above their requirement levels, methionine supplementation no longer significantly improves protein retention (BRESSANI and ELIAS, 1968). As for all foods, excessive heat treatment or other improper processing can reduce protein quality.

Tab. 3. Digestibility and Protein Values of Various Single-Cell Proteins

Genus and Species	No. of Subjects	Protein Level	Digestibility	Biological Value	Reference
Yeasts					
Candida utilis	7	28 g	83±7[b]	70±5[b]	WASLIEN et al., 1970
Candida utilis	7	51 g	87±3[b]	58±6[b]	WASLIEN et al., 1970
Candida utilis	–	–	70–90	52–87	BRESSANI, 1968
Candida utilis	52	20 g	64		MIT[c]
Candida utilis[a]	8	0.35 g/kg	84	58	MIT[c]
Pichia spp.	6	–	84[b]	54[b]	SCRIMSHAW and UDALL, 1981
Pichia spp.[a]	3	–	83[b]	51[b]	SCRIMSHAW, 1983
Bacteria					
Methylomonas clara	4	–	85	48	MIT[c]
Fungi					
Fusarium graminearium	13	0.30 g/kg	78[b]	65[b]	UDALL et al., 1984
Fusarium graminearium	6	0.30 g/kg + methionine	79[b]	73[b]	UDALL et al., 1984
Paecilomyces variotii	6	0.35 g/kg	81[b]	54[b]	UDALL et al., 1984

[a] RNA-reduced
[b] Corrected for endogenous N loss
[c] Unpublished MIT data

Tab. 4. A Comparison of the Essential Amino Acids in Samples of Yeast, Bacteria, and Fungi (g per 100 g protein)

Amino Acid	Soy[a]	Rumen Bacteria[b]	Bacteria *Pseud.*[c]	Bacteria *H. eut.*[d]	Yeast[e]	Micro-fungi[f]	Refer.[g]
Lysine	4.3	9.0	5.3	8.6	7.8	7.6	6.4
Leucine	4.8	8.7	7.0	8.3	7.8	7.6	7.7
Isoleucine	3.2	6.3	3.9	4.6	5.3	4.2	3.1
Valine	3.4	4.7	5.9	7.1	5.8	5.3	3.8
Methionine/cystine	2.0	2.2	2.1	2.7	2.5	2.8	2.7
Tryptophan	0.8	1.4	NA	1.1	1.3	NA	1.2
Threonine	2.7	6.9	4.5	4.5	5.4	5.3	3.7
Phenylalanine/tyrosine	5.6	8.8	NA	NA	8.8	7.6	6.9

[a] RACKIS et al., 1961
[b] BERGEN et al., 1968
[c] *Pseudomonas* grown on methanol, KIHLBERG, 1972
[d] *Hydrogenomonas eutropha*, KIHLBERG, 1972
[e] *Candida lipolytica* grown on *n*-paraffin, YOUNG and SCRIMSHAW, 1975b
[f] *Fusarium graminearium*, from RHM Research Ltd.
[g] FAO/WHO, 1989

Tab. 5. Fatty Acid Composition of Selected SCPs[a] (LASKIN and LECHEVALIER, 1977)

	Saturated		Mono-unsaturated		
	C16	C18	C16	C18	C20
Candida lipolytica	32.6	3.25	17.8	14.3	
Candida utilis	26.0	4.0	14.1	39.0	5.5
Agaricus bisporus					
Sporophore	6.1	2.0	trace	trace	
Mycelium	18.8	3.7	8.5	14.1	

[a] Expressed as a percentage of total fats for the *Candida* species and as a percentage of polar lipids for *Agaricus*.

2.2 Fatty Acids

The crude fat content of SCPs is very low, less than 10% and some in the range of 1–2%. The fatty acid content of selected SCPs is given in Tab. 5.

2.3 Vitamins

Yeasts are an excellent source of B vitamins. The administration of baker's yeast was successful in treating and preventing symptoms of thiamin, riboflavin and niacin deficits among poor families in the southern United States in the 1930s. The vitamin content of a number of yeasts is given in Tab. 6. Bacterial cells are also good sources of B vitamins including the "animal protein factor", vitamin B_{12} (WOLNAK et al., 1967). The concentration of B vitamins in fungal mycelia is usually lower than in yeasts and bacteria (LITCHFIELD, 1967).

2.4 Other Components

After extensive testing in experimental animals as described below, Liquichimica Biosintei began constructing a plant for the large-scale production of *Candida maltosa* ("Liquipron") in Sardinia. The marketers of soy meal and other competing products sought all possible ways of blocking this development and enlisted considerable political support. Issues of plant safety and environmental contamination were dealt with, but the charge that the material contained fatty acids with "odd-numbered carbon chains" was particularly troublesome.

Large sums of money were spent in demonstrating that they were utilized through normal pathways, that rat liver mitochondria utilize even and uneven carbon chain fatty acids at about the same rate, and that liver and heart mitochondria of animals fed uneven carbon chain fatty acids respond in the same manner as those of control animals to a number of chemicals that affect metabolism. They have no effect on the viability and respiratory function of cells and disappear from tissues in a normal manner when their ingestion is terminated (GARRATINI et al., 1979). Moreover, these fatty acids are also found in fish, milk, and a number of other common foods.

It was also suggested that the increased amounts of *n*-paraffin residues in the tissues of animals fed yeasts grown on alkanes might harm humans consuming them. However, *n*-paraffins occur naturally in many vegetable products in general use for animal feeding and are common in food products. They are metabolized through normal channels. Feeding trials were conducted in laying hens, turkeys, rabbits and trout, utilizing different levels of hydrocarbon grown yeast, and their tissues examined. Residues were found to be similar to those of meat products already marketed in Italy (VALFRÉ et al., 1979). The scientific arguments in favor of the safety and nutritional value of "Liquipron" seemed overwhelming, but the political battle was never won, and the partly finished factory was abandoned.

Tab. 6. The Vitamin Content of Some Food Yeasts (in µg per g dried yeast) (From DUNN, 1975, and SHACKLADY and GAMUTEL, 1972)

Food Yeast	Thiamin	Riboflavin	Nicotinic Acid	Pantothenic Acid	Pyri-doxin	Folic Acid	Biotin	*p*-Aminobenzoic Acid	Medium
Baker's compressed	20–40	60–85	200–700	180–330	–	–	0.6–1.8	–	–
Baker's foil	80–150	50–65	180–400	130–160	–	–	0.5–1.5	–	–
Brewer's	104–250	25–80	300–627	72–86	23–40	19.0–30.0	1.1	15–40	–
Candida arborea	32–33	52–70	492–580	–	–	14.8–16.0	–	–	Molasses, beet, cane
Candida arborea	13–33	46–70	301–580	–	–	12.0–26.0	0.24–3.2	11–21	–
Hansenula strain	9	54	590	180	–	1.7	1.7	16	Wood sugar stillage
Mycotorula strain	5	59	600	–	–	3.1	1.8	31	Wood sugar stillage
Oidium lactis	12–29	40–55	186–248	–	–	6.0–15.0	1.30–2.1	–	–
Oidium lactis	20–29	40–55	193–248	–	–	5.6–7.8	–	–	Molasses, beet, cane
Saccharomyces cerevisiae	28–41	39–62	277–568	–	–	19.0–36.0	0.45–3.6	11–62	–
Candida lipolytica	4	180	430	125	25	6.4	–	40	*n*-Alkane
Candida utilis	5	43	417	39	33	21.5	2.3	–	Spent sulfite liquor
Candida utilis	35–38	54–62	511–600	–	–	10.6–15.2	–	–	Molasses, beet, cane
Candida utilis	22	54	440–490	–	–	–	–	–	Molasses, beet, cane

3 Biochemical and Metabolic Parameters

For all of the subjects fed single cell protein in nutritional value and tolerance trials extensive biochemical data were obtained at the beginning and end of the study. Those included liver function tests (SGOT, SGPT, bilirubin, albumin, and alkaline phosphatase), hematologic assessment (hemoglobin, hematocrit, and mean corpuscular volume), and urinary function (urine analysis, blood urea nitrogen, and serum creatinine). Additional tests including serum and urine uric acid, serum amylase, and cholesterol were also obtained.

With the exception of an expected increase in serum and urine uric acid, especially when the material was not RNA reduced, no abnormalities attributed to the feeding were ever observed in these measurements; nor have such changes been reported by others.

4 Food Use

More than 80 years of use of *Candida utilis* as a food additive or vitamin source might be expected to confirm its tolerance in human diets, especially since this applied to cells grown on a wide variety of substrates including molasses, starch and sulfite liquor. Moreover, brewer's yeast and baker's yeast go back to antiquity. As commonly used, food yeast is on the GRAS list of the US Food and Drug Administration and is used in small quantities without restriction in processed foods worldwide.

The most extensive use of yeast as food occurred in Germany and Eastern Europe during World Wars I and II when it became a meat extender and meat substitute (PRESCOTT and DUNN, 1959). Yet, as noted below, processing yeast to reduce its RNA content can render it unsuitable for human consumption, and when other species of yeast and/or when hydrocarbon substrates are used extensive preclinical and clinical testing is required.

Some of the foods that require yeast for their production are listed in Tab. 7, and include wine and beer, breads, and miso. Bacteria are, of course, necessary for the production of many traditional foods, including, yoghurts, some cheeses, sauerkraut, pickles, soy sauce, the holes in Swiss and Emmentaler cheeses, the flavor of Limburger cheese, and such traditional fermented foods as kishk (Egypt), trahanas (Greek), kaumias and kafir (Russia), and many others (STEINKRAUS, 1983). Filamentous microfungi are used in the production of such traditional foods as tempeh and onjom (Indonesia). Some traditional foods such as ogi (Nigeria), oji (Kenya), and koko (Ghana) depend on a mixture of molds and bacteria.

Tab. 7. Examples of Applications of Yeasts in Food Processing

Saccharomyces
- Wine fermentation
- Beer fermentation
- Galactosidase production

Candida
- Xylose fermentation
- B vitamin source

Torulopsis
- Bread leavening

Zygosaccharomyces
- Shoyus, miso

Kluyveromyces
- Lactose fermentation
- Galactosidase production
- Invertase production

Pachysolen, Pichia
- Xylose fermentation

Since yeast functions as a naturally concentrated source of protein and B vitamins, it is often added to foods in small quantities to fortify their protein value and vitamin content. It is used in smaller quantities for purely functional reasons as a flavor enhancer and vehicle for flavorings. Although the protein in SCP is not readily soluble, SCP still has a useful capacity for water and fat absorption. Hence its incorporation in foods based on cereal products. It can also be used in weaning foods, reconstitutable beverages, sandwich spreads, soups, sauces, and the like.

In the 1970s there was major interest and technical success in producing textured protein products simulating animal protein from soy protein, and the methods were shown to be equally applicable to yeast and bacterial protein. For economic reasons this approach has not been pursued. However, filamentous microfungi give texture to foods without the costly purification and extrusion or spinning required for yeast or bacteria. One company, Rank Hovis McDougall, Ltd. (RHM) of the U.K. has developed and is marketing breads and other baked goods containing this product.

5 Safety for Animal and Human Consumption

The safety of microbial products for feed and food use depends on the selection of organisms, the substrate, the process, and the characteristics of the organism. The latter in turn applies not only to the initial organism, but also to the effect of any processing that is involved. From 1956–1973 the Protein Advisory Group (PAG) of the UN System held many meetings to examine issues related to novel protein sources. (The record of its meetings and documents was published in 1975 as *The PAG Compendium* in nine large volumes by Worldmark Press, Ltd., a division of John Wiley & Sons, New York and should be available in some large libraries.) Its guidelines for the evaluation of novel sources of protein were updated and republished by the United Nations University (UNU) in 1983 and are still considered authoritative. From 1969–1973 a PAG "Working Group on Single-Cell Protein" focused further attention on SCP issues. The assurances necessary for approval of SCP or any other novel protein are listed in Tab. 8.

PAG-UNU GUIDELINE NO. 6 for *Preclinical Testing of Novel Sources of Protein* (1983) emphasizes the requirement for toxicological safety of the methods of production, content of microorganisms and their metabolites, effects of laboratory animals and on normal

Tab. 8. Assurances Necessary for Approval of an SCP

For animal use:
- Safety of species
- Safety of substrates
- Safety of products[a]
- Adequate nutritional value

Additional requirements for human use:
- Lack of mutagenicity/carcinogenicity
- Lack of teratogenicity
- Minimal allergenicity
- Favorable organoleptic and/or functional characteristics
- Cultural acceptability

[a] From animals fed SCP

subjects in limited feeding studies, nutritional value, and sanitation. If the results of chemical and microbiological examinations are favorable, they must be followed by multiple feeding studies in rodents and other experimental animals, both short-term and long-term.

PAG-UNU GUIDELINE NO. 7 for *Human Testing of Supplementary Food Mixtures* (1983) was originally developed for the evaluation of weaning foods utilizing novel protein sources but applies to any food use. Once all of the requirements of Guideline No. 6 have been met, human testing falls into four main categories as detailed in Guideline No. 7.

5.1 Acceptability and Tolerance Tests

Even when the original novel protein source is acceptable, it must be tested in the form in which it is processed and consumed. This testing must be done under rigorously controlled conditions on a sufficiently large number of individuals.

5.2 Metabolic Studies

Nitrogen balance is sensitive to variations in protein quality, and such a determination in human subjects is the only definitive way of determining protein value for food use.

5.3 Growth Studies in Children

Growth is also very sensitive to variations in the protein adequacy of a diet when energy and other nutrient needs are met. If a material is to be used in foods for children, particularly weaning foods, it is reassuring to have evidence of good growth.

5.4 Other Criteria

A variety of laboratory measures including serum albumin, creatine excretion, indices of liver function, and immunological parameters can provide further assurance.

Because of the novelty and hence lack of long-term experience with some of the substrates and organisms proposed for SCP, the preceding two guidelines have been supplemented by PAG-UNU GUIDELINE NO. 12 on the *Production of Single Cell Protein for Human Consumption* (1983). This guideline emphasizes that before testing for nutritional value is warranted the following additional considerations must be met.

5.4.1 Type of Organism

It is preferable to select an organism of a species known not to produce pathogenic or toxic variants and evaluate it under the exact conditions proposed for industrial production.

5.4.2 Raw Materials

As listed in Tab. 1, potential carbon sources include carbohydrate-containing materials such as molasses, starch, cellulose, whey, sulfite liquor, etc., various classes of hydrocarbons including methane and longer-chain alkanes, alcohols such as methanol and ethanol. Substrates for either SCP or MBP are required to be free of pathogens and foreign bodies. They must meet standards for pesticide residues, drug residues, and toxic metabolites. The safety of other materials must also be assured including not only buffer salts and nutrients but also such additives as antifoam agents, detergents, and flocculants.

5.4.3 Process Variables

Operating variables of the fermentation such as temperature, air oxygen supply, cell growth rate, and cell concentration must be carefully controlled to ensure product quality and uniformity.

5.4.4 Quality Control

Maintenance of the integrity of the original strain of the organism and lack of contamination must be assured by an appropriate series of ongoing microbiological and biochemical tests.

6 Clinical Tolerance Studies of SCPs

No problems have been encountered in animals fed SCPs that have been processed under well defined and maintained conditions. This includes rats, mice, dogs, swine, monkeys, and chimpanzees tested with proportionately much larger doses than given human subjects. However, no amount of testing in experimental and domestic animals, including primates, can determine whether or not a given product can cause allergic symptoms in humans. The importance of careful attention to the preceding clinical guidelines has been repeatedly demonstrated by practical experience with the actual clinical evaluation of new single cell protein products.

Tab. 9 lists commercial products that, with one exception, have been evaluated in the out-patient facilities of the MIT Clinical Research Center. Processing to reduce the RNA content of the yeasts and bacteria was found to introduce allergic problems in *all* of these cases, although the problem could usually be overcome by process modifications.

Essentially all protein-containing foods are capable of causing an allergic response in some individuals, although the frequency and nature of this response varies widely with the specific food and the population at risk (EASTHAM, 1979). Foods that are major sources of protein in human diets include cow's milk, meat, and fish, as well as fruit, cereals, legumes and oilseeds. Allergic reac-

Tab. 9. Major Industrial Efforts to Develop SCP for Human Consumption in the 1970s

SCP	Organism	Substrate	Proposed Name	RNA-Reduced
Bacteria				
Nestlé/Esso	*Acinetobacter calcoaceticus*	Ethanol	Protecel	Acid wash
Imperial Chemical Industries Ltd.	*Methylophilus methylotrophus*	Methanol	Pruteen	Acid wash
Hoechst	*Methylomonas clara*	Methanol	Probion	Acid wash
Yeast				
Nestlé/Esso	*Candida* sp.	Ethanol	Protocel	Acid wash
Nestlé	*Candida utilis*	*n*-Alkanes	–	Heat shock
Liquiquimica	*Candida maltosa*	*n*-Alkanes	Liquipron	None
Provesta	*Pichia* sp.	Methanol	Provesteen	Acid wash
British Petroleum Grangemouth	*Candida lipolytica*	*n*-Alkanes	Toprina	Alkaline hydrolysis, acid wash
Lavera	*Candida tropicalis*	"Gas Oil"	Toprina[a]	Acid wash
Filamentous microfungi				
Tampella Product Engineering	*Paecilomyces variotii*	Sulfite waste	Pekilo Protein	None
Rank Hovis McDougall	*Fusarium graminearium*	Glucose	–	Acid wash

[a] Product solvent-extracted

tions have been found common with some single-cell protein products, and the problem has been exacerbated by the effects of RNA reduction. It is the frequency and severity of such reactions, rather than their occasional occurrence, that is of concern. Depending on the nature and frequency of the allergic reactions encountered with any food, there are three alternatives: (1) prohibit its sale, (2) clearly identify it so that sensitive individuals can avoid it, (3) process it in such a manner that the responsible allergen is removed or inactivated. A great deal of commercial effort has been directed at this third alternative with the assistance of our research group at MIT, and some of the results are described below.

6.1 Yeast

Historically, most feeding of yeast as food has been free of any adverse gastrointestinal effects. In those few cases in which they were observed, contamination can be suspected (Scrimshaw, 1985). This includes an unpublished feeding trial in the US during World War II, in which severe gastrointestinal reactions developed quite suddenly after 30 days of uneventful consumption. Our MIT experiences began with feeding a commercially available food-grade sulfite-grown *Candida utilis*, from Lakes State Yeast Co., Rhinelander, Wisconsin, to 12 young men for 90 days. No adverse gastrointestinal reactions were observed, but a mild papular rash appeared on the palms and soles of eight of these subjects. It was painless and disappeared spontaneously when the yeast feeding was discontinued. Similar reactions were experienced by three of five subjects consuming 135 g per day and one of 11 subjects consuming 45 g daily. We tried to determine whether a contaminant might have been responsible, but the manufacturer was uncooperative. No such lesions were observed when a glucose-grown *C. utilis* provided by the Nestlé Co. of Switzerland was fed at 90 g per day to 21 subjects for 90 days (Scrimshaw, 1972; Scrimshaw and Udall, 1981).

In 1977 we began a series of double-blind studies with experimental yeasts at a level of 20 g per day and utilized the Lakes States *C. utilis* as the control substance, because it was commercially available and widely consumed in small quantities. For several years we encountered no problems of any kind in feeding this yeast at this level of intake to a total of 90 individuals. However, in the fall of 1979, 5 of 28 control subjects developed some degree of nausea, vomiting and diarrhea, and one individual developed a papular rash on his palms and soles (Scrimshaw and Udall, 1981). It seems clear that in this case some difference in processing must have occurred that presumably was not detected by quality controls. By this time we had experience with other SCPs that apparently minor differences in processing conditions can cause problems when the yeast is fed at low levels.

In 1970 we undertook the evaluation of the first two SCP products grown on a petroleum-derived hydrocarbon substrate, even though the substrate itself, pure ethanol, was not a concern. The products were developed by the Nestlé Co. and consisted of a bacterium, identified as *Acinetobacter calcoaceticus*, and a *Candida* sp. yeast identified as MA100. Both materials had undergone preclinical testing more rigorous than called for in Guideline No. 6. Nine of the 50 volunteers fed the yeast and eight of 50 fed the bacterium, at levels of 20 g per day, developed the syndrome of nausea, vomiting and diarrhea (NVD) quite suddenly after about two weeks. However, the remaining volunteers experienced no problems. Both materials had been *washed in alkali* to reduce their RNA content. It was subsequently demonstrated that the *same* materials when *acid washed* were innocuous, even to the individuals previously sensitized to the alkaline washed materials.

We then tested, in the same double-blind manner, a strain of *Candida lipolytica* grown on *n*-alkanes by the British Petroleum Co. (BP) in Grangemouth that was also RNA-reduced by alkaline hydrolysis (Scrimshaw, 1972). Seven of 106 individuals consuming 20 mg daily of this material stirred into fruit juice of their choice developed mild NVD compared with none of the controls consuming the commercially available non-RNA-reduced *C. utilis* for 90 days. A characteristic of these trials was that most individuals had no adverse symptoms and those developing them did so between 9 and 19 days. A new lot of the BP material that was acid washed was obtained for another 90 day trial and caused no problems for about 45 days. However, it had been specially made by a batch process, and a second batch used mid-way in the trial caused NVD reactions in sufficient numbers that the trial was discontinued. We subsequently learned that one of 20 volunteers fed the second material in Sweden developed an NVD reaction (ABRAHAMSSON et al., 1971).

This experience demonstrated once again that small variations in processing can make a significant difference to safety and acceptability for human consumption. In 1975 we completed an uneventful tolerance study with 106 subjects fed 20 g of *Candida tropicalis* grown by BP in Lavera, France on a "gas oil" substrate and solvent-extracted with hexane (SCRIMSHAW and UDALL, 1981).

We next undertook an evaluation of a *Candida utilis* grown on beet molasses and corn steep liquor and in wide use as a food additive in Europe. It differed only in having been subjected to heat shock to release sufficient of the cell's own ribonuclease to reduce its nucleic acid content. When fed to 8 subjects at a level of 50–60 grams per day in a 15-day nitrogen balance study of its protein value, no problems were encountered (SCRIMSHAW and DILLON, 1979).

We then embarked with confidence on a tolerance trial in which 35 g were given daily in fruit juice. On the third and fifth days for the first cohort of 13 subjects, two males reported a mild pruritic erythema on the inner surface of the elbow, axilla, and back of the knee. Both stopped taking the material and the rash disappeared in two days. On day 8, a 35 year old female noted an erythematous and pruritic eruption on her neck which she did not report. It was a weekend and when she took the next dose the rash extended over her entire body and she experienced chills and fever. She recovered completely in three days. The trial was terminated and no additional cohorts were exposed, but not before a 27 year old female experienced itching and noted a few red papules on the dorsum of her feet and toes.

It turned out that the material used in the nitrogen balance study was RNA-reduced by exposure to 140 °C, but for the tolerance trial, a different sample was supplied that was treated at 80 °C for a slightly longer time in an effort to reduce costs. Subjects who had developed the rash with the lower-temperature treated material showed no adverse reaction when cautiously retested in double-blind fashion with the original higher-temperature processed material.

We were also able to carry out extensive tolerance studies with a *Pichia* sp. grown on methanol by a subsidiary of Phillips Petroleum Co. When adult subjects were fed cookies with or without 20 g of the whole cells in an intended double-blind cross-over study, the trial had to be terminated after 18 days because of adverse cutaneous reactions in some subjects and NVD in others. Because onset was staggered, cohorts of 25 subjects completed 18, 14, and 11 days, respectively. When a yeast sample that had been RNA-reduced by alkaline hydrolysis was tested, 6 of 17 subjects developed a characteristic rash on hands and feet on the second to ninth day of the study.

A sample grown on a modified substrate and RNA-reduced in the same manner was more acceptable at the same level of intake, but 9 of 41 subjects developed very mild rashes on their palms and in some cases also on the soles of their feet during the 5th to 11th day of a 30 day test period that they all completed. A further modification resulted in a material that caused no symptoms in a pilot study of 8 subjects for 30 days. All of these studies with *Pichia* supplied 20 g daily.

6.2 Bacteria

Administration of a few grams of thoroughly washed and boiled cells of *Hydrogenomonas eutropha* caused severe NVD, and in some subjects vertigo and weakness (CALLOWAY and KUMAR, 1969; WASLIEN et al., 1969). *Aerobacter aerogenes* (marketed as *Escherichia coli*) produced comparable symptoms. Our first experience with bacterial protein was with *Acinetobacter calcoaceticus* grown by Nestlé on ethanol with the clinical test results described in Sect. 6.1. We continued with the evaluation of a bacterium, identified as *Methylophilus methylotrophus* produced by Imperial Chemical Industries (ICI) and one identified as *Methylomonas clara* from Hoechst.

For the tolerance study with *Methylophilus methylotrophus*, grown on methanol (Pruteen), of the 48 adult subjects (28 male, 20 female) half consumed 15 g of Pruteen per day and the remainder 15 g of soy protein isolate. The supplements were taken in water, hot bouillion, or fruit juice according to individual preference in addition to their normal diets. Various clinical symptoms (abdominal discomfort,. abdominal pain, gas, nausea, vomiting, loose stools, loss of appetite, dizziness) were observed in 18 of 24 subjects (11 male, 7 female) who consumed the Pruteen protein supplement and 1 of 24 subjects who consumed the soy protein isolate. The symptoms were mild to moderate in severity, no severe reactions were observed. Onset of symptoms ranged from day 1 to day 11 with a median of 3.5 days. The symptoms cleared up in one or two days after intake of material ceased. Due to the large number of subjects developing adverse reactions, the code was broken and the trial terminated. Half of the subjects began the trial one week later than the first group so that when the study was terminated, the first group had completed 14 days and the second group 7 days.

For the trial with *Methylomonas clara*, a *Candida utilis*, grown on beet molasses and supplied by Hoechst, was the control material. For the first cohort of 19 subjects, the trial was stopped on day 15 because of cases of nausea with vomiting and/or diarrhea in 5 subjects on days 5, 6, and 9 and in 2 subjects in cohort 2 of 21 subjects on days 3 and 4 and terminated on day 7. This type or reaction was not seen in the control subjects.

A second material in which the RNA reduction process had been modified was tested in 9 subjects of which 5 developed mild diarrhea and associated symptoms. No further work was done on the Hoechst materials but as noted in Sect. 7.

As noted in Sect. 3, no adverse biochemical changes were observed except in tests for allergenicity. It is noteworthy that it later became evident that the ICI and Hoechst organisms, although given different names were identical and the differences in severity of reactions were due to processing differences.

6.3 Filamentous Microfungi

We have evaluated two filamentous microfungal products, one *Fusarium graminearium* from Rank Hovis McDougall, Ltd. grown in the U.K. by continuous fermentation on a medium of commercial glucose syrup and treated to reduce nucleic acid content. The other was *Paecilomyces variotii* (Pekilo protein) from Tampella Process Engineering Co., Finland which was produced on a substrate of the spent liquor from the sulfite process for pulping wood. Twenty grams of the RHM product was fed in double-blind cross-over studies in cookies to 100 individuals daily for 30 days with no adverse symptoms. Cupcakes with 10 g of the Tampella material were fed in the same manner to 50 individuals. Except for mild transient rashes in two of the subjects, that may or may not have been related to the test material, no adverse symptoms were noted and no significant changes occurred in 17 serum constituents, except for a fall in cholesterol.

7 Laboratory Evidence of Intolerance as an Allergic Phenomenon

In an effort to explore the immunological basis for the intolerance phenomena described above, the radioimmunoabsorption test (RIST) was used to measure total circulating IgE antibodies in serum and the radioallergosorbent test (RAST) to no avail. We then explored the capacity of aqueous extracts of material that had caused NVD or cutaneous allergic symptoms to stimulate the multiplication *in vitro* of leukocytes from individuals that had shown reactions to it. The first material tested in this manner was the Nestlé *Candida utilis* processed at high temperature to reduce its nucleic acid content (SCRIMSHAW and DILLON, 1979). The more the cells respond to the stimulation, the more tritiated thymidine they take up *in vitro*. The cultures are counted in a scintillation counter, and a stimulation index (SI) is calculated by dividing the counts per minute (cpm) in the presence of the stimulant by the cpm without the stimulant.

The results indicated clearly that the SI was very significantly increased in the sensitized lymphocytes exposed to an extract of the yeast responsible for cutaneous allergies. Heating the extract for 30 minutes at 70, 80, and 100 °C did not reduce its capacity for stimulation. It was then possible to separate the protein in the extract by molecular weight using Sephadex elution. The stimulation activity proved to be in a fraction of approximately 50000 molecular weight. Extracts from the material that had not provoked allergic responses had no stimulation capacity when tested with the lymphocytes of subjects who had reacted to the material processed at 80 °C.

The SI approach offered an opportunity to test the effectiveness of variations in processing designed to eliminate allergenicity without having to expose new groups of subjects to tolerance trials as long as sensitized leukocytes are available. The limitation was that the sensitivity slowly decreased and after 4–6 months samples from sensitized individuals were no longer useful for the purpose. When we applied this technique to the lymphocytes of individuals developing either cutaneous or gastrointestinal symptoms when consuming the Nestlé yeast and bacterium, the SIs were positive. However, there was no cross reactivity between products and sensitized individuals. When we determined SIs for the two bacterial samples (from ICI and Hoechst) described previously to cause allergic responses, the SI results were again positive (PHUA, 1981). We were surprised, however, to observe cross reactivity. The leukocytes of individuals with allergic reactions to either material responded to extracts of either material in similar manners. When this finding was pursued with the two companies, it turned out that the organisms, although given different names, were identical.

It must be emphasized that in all of these tolerance trials the original cells did not provoke allergic responses; they did so only when they were processed to reduce their RNA content. It is unfortunate that the loss of commercial interest in the development of SCPs for human consumption led to the abandonment of this promising approach.

8 Nucleic Acid Limitations for Human Use

Any rapidly growing cell contains relatively large amounts of nucleic acids (NA), of which the purine half of the molecule is broken down and excreted as uric acid. Humans lack the enzyme urate oxidase, present in other mammals, that breaks down uric acid to the more soluble allantoin. When expressed on a protein basis, between 8 and 25% of the nitrogen of SCPs comes from NA compared with 4% for liver, 2% for sardines and about 1% for wheat flour (KIHLBERG, 1972). An increased consumption of SCP therefore results in increased blood levels and higher excretion of uric acid.

Since uric acid is sparingly soluble, a rise in blood levels above normal can result in crystals deposited in joints leading to gouty arthritis. The ratio between RNA consumed and uric acid excretion appears to be constant over a range up to excretion values of 1750 mg/day, and there is a proportional rise in plasma uric acid, amounting to 0.9 mg per 100 mL per g RNA (ANONYMOUS, 1975). The prevalence of gout on usual diets is less than 0.4% in the US and Europe. Tab. 10 summarizes the relationship between serum urate concentration and symptoms of gout in males. Similar data are not available for nephropathy and renal stones, but their association with hyperuricemia is also well established (WYNGAARDEN, 1982).

As long as SCPs are used only for animal feeding or are minor constituents in human diets, their nucleic acid content is of no clini-

Tab. 10. Prevalence of Gout in Males Related to Serum Urate Concentration and Age (GIBSON and GRAHAME, 1974; MYERS et al., 1968)

Serum Urate Concentration (mg/100 mL)	Mean at Age 49 (%)	Mean at Age 58 (%)
7.0–7.9	4.7	16
8.0–8.9	11.4	25
9.0–9.9	31.6	90
10 or above	47.6	90

cal significance, but it becomes a limiting factor when they are proposed as dietary protein sources. In men the risk begins with puberty and in women after the menopause. As shown in Tab. 11, the UN Protein Advisory Group has estimated the amount of additional nucleic acid that could come from the use of SCP as a protein source without an unacceptable increase in the risk of gout (ANONYMOUS, 1975). The limit is 20 grams per day for adults on Western diets and more for those in developing countries consuming diets that are lower in nucleic acids from meat and other sources. To meet this limitation, SCP for human consumption as a major protein source must have its nucleic acid content reduced by processing. The common methods are heat shock to release the cell's own ribonuclease for the purpose or either alkaline or acid hydrolysis. This is *not* necessary when SCPs are used as protein sources in feeds, because only humans lack the enzyme to break down uric acid into soluble compounds.

9 Historical Comment

Eventually, each of the companies in Tab. 9 attempting to develop bacterial or yeast SCP as a protein source for human consumption decided that it would be uneconomical to pursue the effort. One by one they therefore ceased to sponsor tolerance studies of their materials. For those using hydrocarbon-based substrates, the increase in petroleum prices associated with the oil shock of 1973 made any hydrocarbon-based product too costly as a protein source in animal feeds. However, the added cost of producing RNA-reduced materials for human use that are free of allergenicity was the final blow. Although the production of the microfungal SCP in Finland on the sulfite waste of paper production seemed economically promising because it cleaned up effluent that could no longer be discharged into streams, the sulfite process itself became obsolete.

In the end only the microfungus produced on sugar by RHM has reached the market. However, the technological problems of

Tab. 11. Permissible Levels of Intake of Additional Nucleic Acid from SCP for Different Ages and Sex (ANONYMOUS, 1975)

Age Group (Years)	Sex	Body Weight (kg)	Intake of Nucleic Acid Purine (N×9 g)
Adult	M	65	2.0
Adult	F	55	1.7
16–19	M	63	1.9
16–19	F	54	1.7
13–15	M	51	1.6
13–15	F	50	1.5
10–12	M	37	1.1
10–12	F	38	1.2
7–9	M, F	28	0.9
4–6	M, F	20	0.6
1–3	M, F	13	0.4

large-scale production of these SCPs on various substrates have been solved and their protein value established. Limitations associated with their relatively high nucleic acid content and allergenicity after processing to lower RNA can be overcome by suitable process modifications and controls. Interest in SCP as a major food source will return, whenever an increase in the prices of currently conventional protein sources as the result of population growth and resource depletion is sufficient to make SCPs economically viable.

10 Summary

SCP products can be produced with yeasts, bacteria, and filamentous microfungi that are good protein supplements to cereal diets for both animals and humans as well as excellent natural sources of B vitamins. Exhaustive toxicological testing has demonstrated that they are safe for animal feeding and that they *can* be produced in a manner suitable for human feeding. SCPs are higher in protein content than cereals and legumes. For human consumption at requirement levels, the quality is comparable to that of legumes. For rapidly growing animals, methionine is the limiting essential amino acid. It must be supplied in their rations from other sources, either methionine itself or the methionine in cereals and fish meal.

As a protein source in human diets, acceptable quantities of yeasts are limited to about 20 g per day for adults unless their high nucleic acid content is reduced. The allowance for bacteria and filamentous microfungi can be adjusted for their actual RNA content in relation to *Candida utilis*. This process can cause the product to provoke cutaneous and/or gastrointestinal allergies in human subjects and, therefore, must be carefully controlled. New SCP products must be screened for allergenicity before they are marketed for human consumption. Yeast and bacteria can be given functional properties through processing, and filamentous microfungi already have useful textures for food use.

11 References

ABRAHAMSSON, L. L., HAMBRAEUS, L., HOFVANDER, Y., VAHLQUIST, B. (1971), Single-cell protein in clinical testing. A tolerance test in healthy adult subjects comprising biochemical, clinical and dietary evaluation, *Nutr. Metab.* **13,** 186–199.

ANONYMOUS (1975), PAG *ad hoc* working group meeting on clinical evaluation and acceptable nucleic acid levels of SCP for human consumption, *PAG Bull.* **5** (3), 17–23.

BERGEN, W. G., PURSER, D. B., CLINE, J. H. (1968), Effect of ration on the nutritive quality of rumen microbial protein, *J. Anim. Sci.* **27,** 1497–1501.

BRESSANI, R. (1968), The use of yeast in human foods, in: *Single-Cell Protein* (MATELES, R. I., TANNENBAUM, S. R., Eds.), pp. 90–121, Cambridge, MA: The MIT Press.

BRESSANI, R., ELIAS, L. G. (1968), Processed vegetable protein mixtures for human consumption in developing countries, in: *Advances in Food Research* (CHICESTER, C. O., MRAK, E. W., STEWART, O. F., Eds.), Vol. 16, pp. 1–103, New York: Academic Press.

CALLOWAY, D. H., KUMAR, A. M. (1969), Protein quality of the bacterium *Hydrogenomonas eutropha, Appl. Microbiol.* **17,** 176–178.

DAVIS, P. (Ed.) (1977), *Single Cell Protein*, London: Academic Press.

DUNN, C. G. (1975), Uses of A. Yeast and yeast-like microorganisms in human nutrition. B. Bacteria and bacteria-like microorganisms in human nutrition, in: *PAG Compendium.* Vol. C2, pp. 2051–2071, New York: Worldmark Press, Ltd.

EASTHAM, E. J. (1979), Clinical gastrointestinal allergy, in: *Single-Cell Protein – Safety for Animal and Human Feeding* (GARATTINI, S., PAGLIALUNGA, S., SCRIMSHAW, N. S., Eds.), pp. 179–185, Oxford: Pergamon Press.

FAO/WHO (1989), *Protein Quality Evaluation.* Report of a Joint FAO/WHO Expert Consultation, Rome: FAO.

GARATTINI, S., PAGLIALUNGA, S., SCRIMSHAW, N. S. (Eds.) (1979), *Single-Cell Protein – Safety for Animal and Human Feeding*, Oxford: Pergamon Press.

GIBSON, T., GRAHAME, R. (1974), Gout and hyperlipidaemia. *Ann. Rheum. Dis.* **33,** 298–303.

GOUNELLE DE PONTANEL, H. (Ed.) (1972), Proteins from Hydrocarbons, *Proceedings of the 1972 Symposium* at Aix-en-Provence, London–New York: Academic Press.

HAMDAN, I. Y. (Ed.) (1983), *Single Cell Proteins from Hydrocarbons for Animal Feeding*, Kuwait: Kuwait Institute for Scientific Research.

JOHNSON, M. (1967), Growth of microbial cells on hydrocarbons, *Science* **15,** 1515–1519.

KIHLBERG, R. (1972), The microbe as a source of food, in: *Annual Review of Microbiology* (CLIFTON, C. E., RAFFEL, S., STARR, M. P., Eds.), pp. 427–466, Palo Alto, CA: Annual Reviews Inc.

LASKIN, A. I., LECHEVALIER, H. A. (Eds.) (1977), *Handbook of Microbiology*, Vol. II: *Microbial Composition*, Cleveland, OH: CRC Press.

LITCHFIELD, J. H. (1967), Submerged culture of mushroom mycelium, in: *Microbial Technology* (PEPPLER, H. J., Ed.), pp. 107–144, New York: Reinhold.

MATELES, R. I., TANNENBAUM, S. R. (Eds.) (1968), *Single-Cell-Protein*, Cambridge, MA: The MIT Press.

MYERS, A., EPSTEIN, F. H., DODGE, H. J., MIKKELSON, W. M. (1968), The relationship of serum uric acid to risk factors in coronary heart disease, *Am. J. Med.* **45,** 520–528.

PAG/UNU GUIDELINE NO. 6: *Preclinical Testing of Novel Sources of Food* (1983), *Food Nutr. Bull.* **5** (1), 60–63.

PAG/UNU GUIDELINE NO. 7: *Human Testing of Novel Foods* (1983), *Food Nutr. Bull.* **5** (2), 77–80.

PAG/UNU GUIDELINE NO. 12: *The Production of Single-Cell Protein for Human Consumption* (1983), *Food Nutr. Bull.* **5** (1), 64–66.

PHUA, C. C. (1981), Safety of Single-Cell Protein for Human Feeding: The Development of Immunological Assays for Allergens in Some Single-Cell Proteins, *PhD Thesis*, Massachusetts Institute of Technology, Cambridge, MA.

PRESCOTT, S. C., DUNN, C. G. (1959), *Industrial Microbiology*, New York: McGraw-Hill.

RACKIS, J. J., ANDERSON, R. L., SASAME, H. A., SMITH, A. K., VANETTEN, C. H. (1961), Amino acids in soy bean hulls and oil mill fraction, *J. Agric. Food Chem.* **9,** 409.

SCRIMSHAW, N. S. (1968), Introduction, in: *Single-Cell Protein* (MATELES, R. I., TANNENBAUM, S. R., Eds.), pp. 3–7, Cambridge, MA: The MIT Press.

SCRIMSHAW, N. S. (1972), The future outlook for feeding the human race. The PAG's recommendations Nos. 6 and 7, in: *Proteins from Hydrocarbons* (GOUNELLE DE PONTANEL, H., Ed.), pp. 215–228, London: Academic Press.

SCRIMSHAW, N. S. (1975), Single-cell protein for human consumption: An overview, in: *Single-Cell Protein II* (TANNENBAUM, S. R., WANG, D. I. C., Eds.), pp. 24–45, Cambridge MA: The MIT Press.

SCRIMSHAW, N. S. (1983), Non-photosynthetic sources of single-cell protein – their safety and nutritional value for human consumption, in: *A Systems Analysis Approach to the Assessment of Non-Conventional Protein Production Technologies, Proc. of a Task Force Meeting*, Sofia, Bulgaria, 1982 (WORGAN, J. T., Ed.), pp. 119–127, Laxenburg, Austria: International Institute of Applied Systems Analysis.

SCRIMSHAW, N. S. (1985), Acceptance of single-cell protein for human food applications, in: *Comprehensive Biotechnology: The Principles, Applications and Regulations of Biotechnology in Industry, Agriculture and Medicine* (MOO-YOUNG, M., ROBINSON, C. W., HOWELL, J. A., Eds.), Vol. 4, *The Practice of Biotechnology: Specialty Products and Service Activities*, pp. 673–684, Oxford: Pergamon Press.

SCRIMSHAW, N. S., DILLON, J.-C. (1979), Allergic responses to some single-cell proteins in human subjects, in: *Single-Cell Protein: Safety for Animal and Human Feeding* (GARATTINI, S., PAGLIALUNGA, S., SCRIMSHAW, N. S., Eds.), pp. 171–178, Oxford: Pergamon Press.

SCRIMSHAW, N. S., UDALL, J. (1981), The nutritional value and safety of single-cell protein for human consumption, in: *Proc. Int. Colloq. on Proteins from Single-Cell Organisms* (SENEZ, J., Ed.), Paris: Association pour la Promotion Industrie-Agriculture (APRIA).

SCRIMSHAW, N. S., YOUNG, V. R. (1979), Soy protein in adult human nutrition: A review with new data, in: *Soy Protein and Human Nutrition* (WILCKE, H. K., Ed.), pp. 121–148, New York: Academic Press.

SHACKLADY, C. A., GATUMEL, E. (1972), The nutritional value of yeast grown on alkanes, in: *Proteins from Hydrocarbons* (GOUNELLE DE PONTANEL, H., Ed.), pp. 27–52, London: Academic Press.

STEINKRAUS, K. H. (Ed.) (1983), *Handbook of Indigenous Fermented Foods*, New York: Marcel Dekker.

TANNENBAUM, S. R., WANG, D. I. C. (Eds.) (1975), *Single-Cell Protein II*, Cambridge, MA: The MIT Press.

UDALL, J. N., LO, C. N., YOUNG, V. R., SCRIMSHAW, N. S. (1984), The tolerance and nutritional value of two microfungal foods in human subjects, *Am. J. Clin. Nutr.* **40,** 285–292.

UNITED NATIONS UNIVERSITY (1979), Bioconversion of organic residues for rural communities, *Food Nutr. Bull. Suppl.* No. 2, Tokyo: UNU.

VALFRÉ, F., BOSI, G., BELLEZZA, O., OLIVIERI, O., MOCA, S. (1979), Effect of feeding *n*-paraffins on animal tissue levels, in: *Single-Cell Protein – Safety for Animal and Human Feeding* (GARATTINI, S., PAGLIALUNGA, S., SCRIMSHAW, N. S., Eds.), pp. 133–147, Oxford: Pergamon Press.

WASLIEN, C. I., CALLOWAY, D. H., MARGEN, S. (1969), Human intolerance to bacteria as food, *Nature* **221,** 84–85.

WASLIEN, C. I., CALLOWAY, D. H., MARGEN, S., COSTA, F. (1970), Uric acid levels in men fed algae and yeast as protein sources, *J. Food Sci.* **35,** 294–298.

WOLNAK, B., ANDREAN, B. H., CHISHOLM, J. A., SAEDEH, M. (1967), Fermentation of methane, *Biotechnol. Bioeng.* **9,** 57–76.

WYNGAARDEN, J. B. (1965), Disorder of purine and pyrimidine metabolism, in: *Cecil's Textbook of Medicine* (WYNGAARDEN, J. B., SMITH, L. H., Eds.), pp. 1107–1118, Philadelphia: Saunders.

YOUNG, V. R., SCRIMSHAW, N. S. (1975a), Clinical studies in the United States on the amino acid fortification of protein foods, in: *Amino Acid Fortification of Protein Foods* (SCRIMSHAW, N. S., ALTSHUL, A., Eds.), pp. 248–265, Cambridge, MA: The MIT Press.

YOUNG, V. R., SCRIMSHAW, N. S. (1975b), Clinical studies on the nutritional value fo SCPs, in: *Single-Cell Protein II* (TANNENBAUM, S. R., WANG, D. I. C., Eds.), pp. 564–586, Cambridge, MA: The MIT Press.

III. Food Fermentations

7 Baked Goods*

GOTTFRIED SPICHER
JÜRGEN-MICHAEL BRÜMMER
Detmold, Federal Republic of Germany

* Updated version of Chapter 1, Volume 5 of the First Edition of *Biotechnology*

1 Introduction

Cereals and bread have been the basic food for individuals and population groups for many centuries. In various countries bread accounts for 18–80% of all nutrients.

The history of bread can be traced back about 6 millenia. It is probable that it developed from a gruel. During the earlier stone age the preparation of a gruel made from rubbed or ground grain and water or milk was known by all civilizations. The flat bread developed from this gruel which was either air-dried or baked on hot stones. The first primitive baking was done by placing formed doughs into hot ashes.

Towards the end of the stone age the flat bread had assumed the shape of a disc. These breads were generally eaten warm, or they could be dried and stored. All types of grain could be used for the preparation of the gruel, but the choice of grains for the preparation of baked flat breads is narrower. Therefore, flat breads were introduced only slowly and only in some localities. Barley and other grains (hardly known today) were used. With the change to the loaf bread the choice of suitable grains is still narrower. Today wheat is grown worldwide for this purpose (Fig. 1, Tab. 1). Beyond this rye is grown in Central and Northern Europe. Wheat and rye are also called "bread" grains because of their suitability for the production of baked goods.

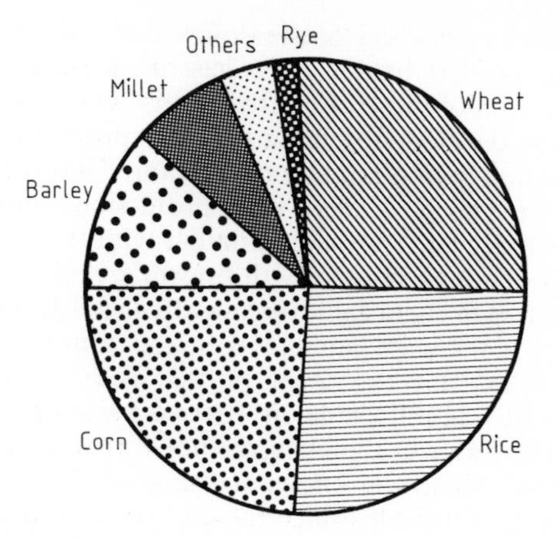

Fig. 1. Production of various cereals worldwide (calculated from data of the FAO Production Year Book by ROTHE, 1980).

The production of a loaf bread presupposes the development of the baking oven

Tab. 1. Wheat and Rye Production of Some Countries in 1977 (in 1000 tons)

Country	Wheat	Rye	Wheat/Rye Ratio
Italy	6330	31	204/1
Bulgaria	3010	20	150/1
Great Britain	5230	38	138/1
USA	55130	432	128/1
Canada	19650	392	50/1
France	17450	376	46/1
Hungary	5310	147	36/1
Argentina	5300	170	31/1
USSR	92040	8470	10.9/1
Czechoslovakia	5240	870	6.0/1
Sweden	1560	364	4.3/1
Austria	1070	351	3.0/1
West Germany	7180	2538	2.8/1
East Germany	3100	1500	2.1/1
Denmark	605	320	1.9/1
Poland	5310	6200	0.9/1
Totals	386600	23770	16 /1

(ROTHE, 1980)

and the discovery of the dough fermentation. The earliest baking oven has been found in the area of the early Babylonians. The Egyptians developed a baking oven approaching the design of a modern oven about 2700 B.C. Since 1750 B.C. there were professional bakers in Egypt. They were familiar with the raising of wheat doughs with brewer's yeast and even with chemical leavening. In 450 B.C. the Egyptians were familiar with the souring of doughs. But the form of this sour dough bread was still similar to that of a flat bread. Discoveries in tombs have shown that a sour dough bread was known in rye growing areas north of the Alps since the end of the bronze age (800 B.C.). In 100 B.C. the sour dough bread was generally known throughout the world (VON STOKAR, 1951). The Greeks developed the art of baking further, and especially the design of baking ovens. Greek bakers brought the art of baking bread and cake to Rome. The Romans in turn developed the earlier techniques of other populations to a high degree of technical and organizational perfection. They also developed the form of the baking oven which remained the model until the 19th century. In Central Europe the preparation of bread in its proper sense stems from the middle of the first millenium A. C. The further development was very fast, and at the beginning of the 8th century bakers are mentioned in the "lex alamanorum" (ROHRLICH, 1976).

The introduction of the stone oven permitted an increase in the heating surface and an increase in production. The oldest dough kneading machines date from the second half of the 18th century. Finally, the industrial production of baker's yeast in the last century was decisive for the technology of baking. Only this development permitted the production of the loaf bread as we know it today. The concentration of the population in towns at the beginning of the 20th century led to technical advances and the rationalization of production. A lack of workers, undesirable working hours, the desire to make the work easier, and to lower production costs contributed to this development. Mainly since 1950 automation has advanced in a broad front and has led to the establishment of some fully automated bakeries. The typical properties of bread, its form, appearance, texture and flavor, were originally produced manually and in conformity with the quality of available grains. Now the baker has to produce bread in the same form and with the same quality by means of rational, technical processes (WASSERMANN, 1981). This change from the manual to the industrial method of producing bread and other baked goods can be observed in almost all industrialized countries. The conditions for this change are particularly favorable in countries in which the assortment of types of breads and baked goods is not very large. Therefore, the development started earlier in countries producing bread almost exclusively from wheat, as in the United States. During the past years the industrial production of rye breads and mixed grain breads has progressed considerably. The production of sweet goods and shelf stable baked goods occurs today already largely in industrial facilities (SEIBEL, 1970).

The development of cereal foods has proceeded through several stages: from roasted grain to gruels to flat breads and finally to leavened bread loaves. But the latter stages have not replaced the earlier ones. Today all stages are still practiced. About 60% of the world population, mainly in Central America, parts of South America, Africa, the Middle East and the Far East, eat gruels made from grain crops and flat breads.

2 The Nutritional Contribution of Bread

The individual grain consists of a starchy endosperm whose outer layers are rich in protein, vitamins, and minerals; a largely indigestible fruit coat (or pericarp); and a germ or embryo. The higher the degree of extraction of the flour, that is, the greater the yield of flour for a given weight of grain, the greater the amount of seed coat and of the outer layer in the flour. Such flours are darker. Short extraction flour (or very white flour) contains almost exclusively portions of the endosperm. Wheat flours generally contain more protein and less fiber than rye flours. The ash content of rye flours is almost twice as high as that of comparable wheat flours.

Grains and products made from milled grains have a favorable combination of nutrients. For 4 billion people they provide more than 50% of the required calories, and in some regions more than 90%. Beyond this they supply considerable percentages of the nutritionally important proteins, B vitamins, and minerals and trace minerals. Bread is a staple food in Germany (as it is in other industrialized countries) and supplies the following percentages of the required nutrients: 20% of the calories, 45–55% of the carbohydrates, 22–25% of the proteins in the form of plant protein. Some specialty breads also contain animal proteins. In addition bread contributes 25–33% of the required vitamins of the B complex (thiamin, riboflavin), 2 grams daily of minerals (among them calcium and iron), as well as a considerable part of the requirement for sodium chloride. If bread is supplemented with foods rich in nutrients such as animal protein, vitamins A, C and D, and some minerals, it can be a suitable basis for a nutritionally optimal diet (CREMER et al., 1969) (Tab. 2).

Carbohydrates occur mainly in the form of starch (on the average about 45–50% of the weight of fresh bread) (Tab. 3). Mono- and disaccharides are not very important since they account for less than 5% of the carbohydrates. The carbohydrates of bread serve mainly as carbon and energy sources. In the form of gelatinized starch they are preferable to excessive amounts of carbohydrates in the form of sugar because of the lesser risk of contributing to arteriosclerosis and diabetes (YUDKIN and RODDY, 1964). In addition carbohydrates counteract the tendency toward an excessive intake of fats in industrialized countries. They also aid in the digestion and complete oxidation of fats in humans because of large inner surface and porosity which distributes the fat finely in the stomach and intestinal tract.

Bread, and particularly bread from high extraction flours and whole grain, contains bulk materials. These are the indigestible, organic substances present in the structural molecules and the cell wall of plants. They consist mainly of cellulose, hemicellulose, pectin, and lignin. These substances increase the mass of food in the intestines and increase the feeling of satiety. They regulate the passage of food through the intestinal tract and may have prophylactic and therapeutic functions with regard to gastrointestinal and cardiovascular diseases.

Bread contains only about 5–9% protein and contributes to protein nutrition only in a limited way (Tab. 3). Nevertheless, the contribution of bread is not negligible. It is the second largest contributor of proteins to the diet after meat protein, and ahead of milk, cheese, and cottage cheese (SEIBEL and ZENTGRAF, 1981). The biological value of the protein of grains is low in comparison with proteins of animal origin, at least for the growing organism. Rye bread has a slightly lower protein content than wheat bread but a slightly higher biological value. For flour of 80% extraction it is 75 for rye bread and 65 for wheat bread

Tab. 2. Contribution of Some Foods to Caloric Supply in West Germany

	Percent of Total Caloric Content		
	1950	1960	1974
1. Bread flour (75% wheat, 25% rye)	32	26	20
2. Pork	3	5	8
3. Beef	2	3	4
4. Sugar	9	11	13
5. Alcohol	2	5	7
Sum of 2 to 5	16	24	32

(MENDEN et al., 1975)

Tab. 3. Nutrient Content and Caloric Content of Various Types of Bread

	Moisture %	Crude Fat %	Protein (N·6.25) %	Total Carbohydrates %	Caloric Value per 100 g kJ	kcal
Cracked grain rye and whole grain rye bread	43.8	1.2	6.8	45.5	935	220
Rye (flour) bread	42.2	0.9	5.2	47.8	935	220
Rye (mixed grain) bread	41.0	1.4	7.0	49.0	1005	237
Wheat (mixed grain) bread	40.0	1.5	7.5	49.0	1018	240
Wheat (flour) bread	36.2	1.8	8.9	51.6	1104	258
Wheat toast bread	35.0	3.9	8.5	49.9	1141	269
Cracked grain wheat and whole grain wheat bread	43.4	1.2	7.2	44.9	931	219
Knäckebrot	6.4	1.7	11.4	76.2	1554	366

(RABE and SEIBEL, 1981)

(CREMER et al., 1969). Prolonged baking lowers the biological value of the protein. This is particularly true if higher concentrations of sugar or sugar containing ingredients are added, as for instance, with milk rolls. The Maillard reaction between the sugar (in this case lactose) and the protein leads to addition compounds which cannot be hydrolyzed by the intestinal enzymes. However, these addition compounds contribute to the aroma of bread.

The limiting amino acid in grain protein is always lysine (2.5% in wheat protein and 3.3% in rye protein), followed by threonine for wheat and presumably by isoleucine and tryptophan for rye (CREMER et al., 1969). The protein of flour of higher extraction and of whole grain has a higher biological value. In countries with a deficiency of animal and vegetable protein sources good results have been obtained with the fortification of wheat flour with synthetic L-lysine and D,L-threonine or with the addition of skimmilk powder. However, the amount of added lysine must be large enough to compensate for the loss of this amino acid through the Maillard reaction (CREMER et al., 1969). The nutritional value of grain proteins is considerably improved by fermentation with sour dough bacteria and with yeast. Yeast contains about 1.3% lysine (as is). Its contribution in straight doughs with 5–7% yeast (based on flour) is considerable (BECKER, 1966).

Bread is also important as a source of essential vitamins of the B complex and of vitamin E (Tab. 4). Normal diets generally contain considerable amounts of riboflavin, niacin, and some other vitamins. However, vitamin B1 occurs in sufficient concentrations only in pork, bread, and a few foods which are not consumed regularly. Therefore, bread is most important for the supply of vitamin B1. Its concentration depends on the degree of extraction of the flour and on the time of baking and the baking temperature. The consumption of light breads covers only about 16% of the daily requirement for vitamin B1 (thiamin). Darker flours and whole grain contain higher concentrations of vitamin B1 as well as niacin, vitamin E, and minerals. Bread which has been baked or roasted for a long time, such as Pumpernickel or Zwieback, shows an almost complete loss of the heat labile vitamin. Yeast contains about 15–30 micrograms of thiamin per gram (as is) and improves the concentration of this vitamin in bread. In several countries in which low extraction flours are used extensively such flours are fortified with a mixture of vitamins (mostly thiamin, riboflavin, and niacin, and occasionally with pyridoxin) and in some countries they are also fortified with calcium and iron (CREMER et al., 1969).

Cereal grains contain only small concentrations of fats but these are not without some importance. The approximately 1.5 to 2.5% of

Tab. 4. Vitamins in 300 g of Bread

Vitamin	Daily Requirement	Vitamin Content as % of the Recommended Daily Allowance						
		White Bread	Mixed Bread	Rye Bread	Whole Wheat Bread	Whole Rye Bread	Knäckebrot	Graham-Bread
A	5000 IU	0	0	0	20	–	–	–
E	5 mg	–	–	–	134	–	240	–
B1	1.6 mg	16	25	26	38	28	38	40
B2	1.8 mg	10	17	17	25	23	23	18
Niacin	12 mg	23	37	25	80	42	50	62
B6	1.5 mg	28	–	44	60	–	60	40
Folic acid	0.5 mg	8	–	12	14	–	–	–
Panthothenic acid	6 mg	20	–	–	31	–	–	–

(KRAUT, 1963)

fat in wheat or rye grain is evenly divided between the endosperm and the germ. Linoleic acid and linolenic acid account for 60% of total fats. These fatty acids play a part as carriers of fat soluble vitamins.

The minerals and trace elements of bread are also important (Tab. 5). Particularly for the darker breads they contribute a considerable percentage of the required elements such as potassium, calcium, iron, magnesium, manganese, and zinc (LUDEWIG, 1975). Rye also contributes fluorine. The major portion of the minerals is found in the outer layers of the grain. Therefore, they are also found in higher concentration in high extraction flours. Wheat flour (type 405) contains 0.4% minerals; cracked wheat contains 1.8%. An average daily intake of 140 g of whole wheat flour provides 46% of the iron, 8% of the calcium and 100% of the zinc requirement of adults. A low extraction flour (type 550) provides only 15% of the iron and 3% of the calcium.

Other bread ingredients such as water, salt, yeast, eggs, dairy products, fruits and dough conditioners enrich the mineral content of bread (LUDEWIG, 1975).

The question of flour fortification has been debated repeatedly. Fortification is the attempt to replace vitamins and minerals lost during milling process. In some countries such as the United States, Great Britain, and the Scandinavian countries this fortification is already practiced. There is, however, no real need for such fortification as long as the consumption of light and dark breads and of cracked grain and whole grain breads provides a reasonable mixture in the daily diet (SEIBEL and ZENTGRAF, 1981).

Tab. 5. Mineral Content in 200 g of Wheat or Rye Whole Meal

Mineral	Daily Requirement (%)
Potassium	60–70
Phosphorus	70–80
Magnesium	70–90
Calcium	10–20
Manganese	30
Iron	50
Copper	50
Zinc	100

(CREMER and ACKER, 1960)

Bread is very important for special diets. The wide variety of types of baked goods permits their use in many such diets. Such baked goods may be the blandest white breads and Zwieback or the whole grain breads which are richest in fiber. The cariogenic activity of bread is only slight because of its low concentration of sugars. A change from a high fat to a high carbohydrate diet lowers the blood cholesterol concentration. This effect is greater for bread than for sugar (CREMER et al., 1969). Rye or wheat bread which is almost free from fat can be used in diets of patients with liver or gall bladder diseases. White bread, Zwieback, Knäckebrot or Graham bread are in low fiber diets. High gluten breads, in which a portion of the carbohydrate has been replaced with protein, has been used in diets for patients with diabetes mellitus. Today whole grain bread, which is absorbed more slowly, is preferred for this purpose. The natural sodium content of flour is very low (2–3 mg per 100 g). Therefore, low salt specialty breads can be used in low salt diets for patients with heart or kidney diseases.

Newer Raw Materials

In past years many new bread varieties have been introduced. Besides the traditional bread grains, wheat and rye, other grains have additionally been used. Wheat meal and in some regions rye meals have also been used.

The nutritional advantages of the consumption of whole grain products have recently been stressed. This has prompted a reevaluation of the technology of their use. The use of premium, whole grain wheat products has resulted in improvements in the production of breads, buns and rolls. The processing of whole meal rye products continues to be characterized by the use of rye meal and coarse rye meal. The use of whole meal rye flour has not brought the same advantages as the use of whole wheat flours.

The use of other grains such as oats, barley, corn, rice or milo has increased. Together with the use of various oil seeds there is now a wide spectrum of specialty breads in trade (BRÜMMER et al., 1988a, b). Apart from the mentioned grains some so-called pseudocereals have been used as additives to traditional grains. Buckwheat (BRÜMMER and MORGENSTERN, 1988) amaranth (*Amaranthus*) and quinea (genus *Chenopodium quinea*) have been considered. In the U.S. a major source of fiber additives are oat fibers, but soluble and insoluble fibers of non-cereals have also been used.

The higher content of phytic acid in whole grain products has been widely discussed. This is no problem with breads since the intensive fermentations, particularly with the additions of sour dough, lead to substantial reductions in phytic acid (FRETZDORFF and BRÜMMER, 1992).

3 Outline of the Technology of Baked Goods

A dough prepared with only flour and water shows variable extensibility and elasticity. Without the use of a leavening agent one obtains a flat bread (tortilla) as it is often prepared by natives of Southeastern Europe, Africa, and South America from corn flour, wheat flour, or millet flour. The very dense flat bread is at best slightly leavened by escaping steam. The production of baked goods with a definite cell structure as it is produced in Western countries requires leavening of the doughs with gases or steam. The dough must then be able to retain the developed gas.

Modern baked goods are produced from the milled products of wheat, rye, or other grains, potable water, salt, leavening, and some optional ingredients (sugar, fat, eggs, fruits, milk, spices, etc.). Rye breads also require the addition of acids. Depending on recipe and process one can distinguish between bread (including rolls and buns) and sweet goods (including shelf stable items). In the German Federal Republic these categories are defined as follows:

- bread contains for each 90 parts of flour or milled grain no more than 10 parts of fat and/or sugar. Rolls and buns are distinguished from bread only by size, form, and weight (less than 250 g). They are produced almost entirely with wheat flour. Some special types contain small amounts of rye flour and/or other grains.
- sweet goods are products which contain for each 90 parts of flour of milled grain at least 10 parts of fat and/or sugar. Sweet goods with a low moisture con-

tent such as cookies, macaroons, honey cakes, waffels, or Zwieback are often called shelf stable baked goods.

Within these two categories of baked goods there is a multiplicity of types which can be distinguished by the kind and relative proportions of raw materials and additives, by particular processes, dough formation and baking, and the resulting form of the baked goods. The availability in Germany of more than 200 different types of breads is unique in the world (KUNKEL, 1966; PELSHENKE, 1949; SCHULZ, 1970; SEIBEL et al., 1978).

The production of baked goods consists basically of the following steps (Fig. 2):

(1) Preparation of raw materials (choice, of preparation and weighing of ingredients)
(2) Dough formation (kneading, maturing)
(3) Dough processing (fermentation and leavening, dividing, molding, and shaping)
(4) Baking
(5) Final preparation (steps for retention of quality, slicing, packaging, sterilization, or pasteurization, etc.)

Various mechanical, physical, chemical, biochemical and microbiological processes occur during the production of baked goods, and these act either at the same time or in succession. These processes may mutually affect each other. They cause chemical and structural changes by swelling, solution, leavening, changes in form, and solidification, as well as by the formation of aroma and flavor. These changes and the following baking process convert the indigestible native starch and gluten of the flour to a digestible and desirable form.

Preparation of raw materials. One can distinguish between a few basic recipes in spite of the great variability in detail. Basically one can distinguish between wheat doughs and rye doughs. Wheat doughs are leavened with yeast. Rye doughs require besides yeast some acidification either by use of a sour dough starter or by addition of an acid. Mixed grain breads which may contain regionally from 20–80% rye flour and 80–20% wheat flour must be acidified according to their content of rye flour.

Dough processing. The various dough components are mixed forming the basis for the development of the structure of the dough and the baked product. Leavening begins at

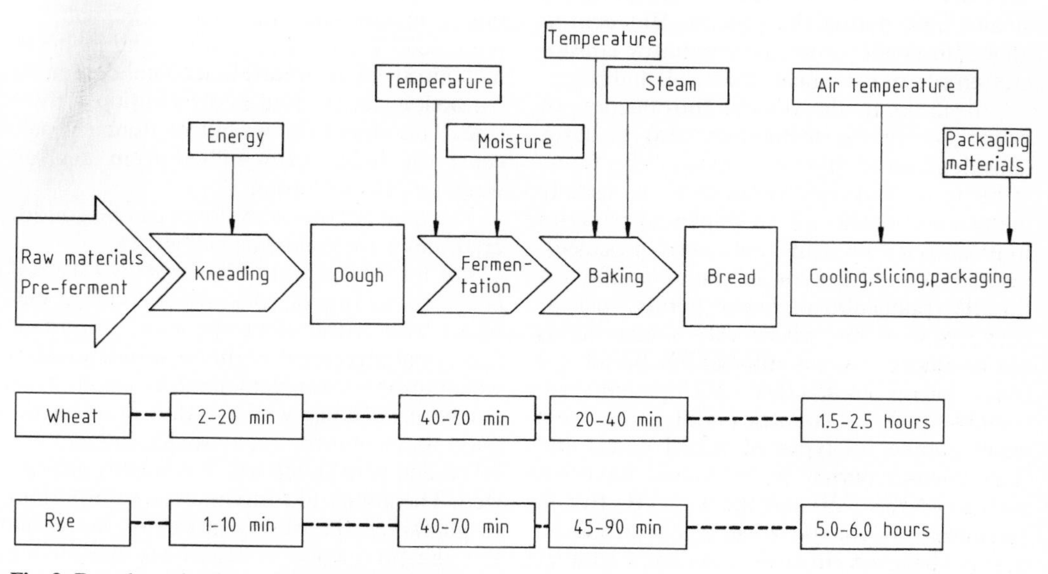

Fig. 2. Bread production – Flow sheet. (SEIBEL, 1979).

the same time. In wheat doughs the retention of the leavening gas depends on the gluten structure. Rye flour doughs retain the gas formed by yeasts and heterofermentative lactic acid bacteria (and the air contained in the flour) to a lesser extent. In rye doughs the presence of rye mucilage and the high viscosity of the doughs are most important.

During the following *dough fermentation* and the proofing of dough pieces the major portion of the leavening gas is formed, and formation of a coherent dough is completed. The external surface of the dough pieces, and consequently that of the baked loaf is increased 2 to 3 fold because of the action of the leavening gas. The matured and raised dough is then divided and formed. This processing of the dough and the subsequent firming of the structure by baking leads to the formation of small, evenly distributed pores in the crumb of baked goods. For doughs leavened by fermentation the primary products of the biological process are carbon dioxide and ethanol (which has a neutral taste). In addition secondary fermentation products have a lesser or greater effect on the taste of bread. Finally, the structure of the dough is fixed by baking, resulting in an edible product.

The various processing steps can be carried out batchwise or continuously. During manual production corrective action can be taken at any time during the process. But continuous, automatic production requires consideration of the biological, chemical, and physical reactions in the dough. Introduction of newer processing techniques also requires consideration of these reactions.

In retail bakeries production is usually manual and batchwise; in wholesale bakeries it is carried out with continuous processes. The best conditions for full automation exist for the production of baked goods, such as shelf stable items, where very similar items are produced in great volume. For bread and other baked goods this development proceeded fastest in countries with a relatively small number of types of baked goods and high volume output in individual bakeries, such as in the U.S. and the U.S.S.R. But in Germany 70% of the bread is still produced in retail bakeries because of the large number of different types of bread.

4 Choice and Preparation of Raw Materials

4.1 Flour

Flour is the major ingredient in most recipes. It has a central function because of its ability to absorb water and to form a cohesive, visco-elastic mass, the dough. Flour is largely responsible for the properties and quality criteria of baked goods, and it requires special consideration in working out the recipe. This is particularly true for the industrial production of baked goods since – in contrast to retail baking – deviations of the quality of flour cannot be corrected during processing.

Each type of baked goods requires flour with particular qualities and good baking properties. The concept of good baking properties is, of course, relative. A flour suitable for a particular application may not perform well in another application. The baking properties of wheat and rye are genetically determined, but they may be affected favorably or unfavorably by the conditions of growth and the climate.

The commercial value of a flour and its quality characteristics are of prime importance. In Germany its commercial value is regulated by its mineral content and its maximal moisture concentration. The degree of extraction can be used as an additional criterion. This shows the weight of flour (in percent) which has been milled from a given weight of cleaned grain.

The quality criteria of flours can be considered under the following categories:

For the production of *biologically leavened baked goods* (bread, rolls, sweet goods, Zwieback) with wheat flour the most important functional ingredient is gluten; a functionality which must be complemented by starch. One needs wheat flours with a gluten which shows good water absorption, elasticity and extensibility, and with starch which is readily gelatinized. The ability to form maltose is important to guarantee the raising power of the yeast. For the production of Zwieback the mixing properties, gas retention, and fermentation tolerance are important to yield a high volume product. Mature flours with a high protein concentration and good gluten quality are required for this use. The same characteristics are required for the production of "toast bread" except that the yield of a good volume is not as critical. The protein concentration may be slightly lower. Flours for the production of rolls require an intermediate yield of volume, and must produce goods with a good break and shred and a persistent crustiness. The production of biscuits or crackers requires flours of a low protein concentration and a weak gluten. The required rheological properties are more difficult to determine. For a given recipe they are related to particular processing conditions (SEIBEL, 1970). The production of whole grain breads, among them graham bread (wheat) or Pumpernickel (rye), requires the use of whole meal or coarse whole meal (Tabs. 6 and 7).

The baking quality of *rye flours* is largely determined by the properties of the pentosans and the starch. Enzymatic activity is also important.

Chemically leavened baked goods (such as cookies, batter cakes, tender pastry, shelf stable items, etc.) require flours which yield a quality generally called shortness. Flours with a low concentration of gluten but with starch of good pasting qualities are required.

Physically leavened baked goods require flours with the following properties: low protein, weak gluten flours for biscuits and coo-

Tab. 6. Quality Criteria for German Wheat Flours

Criteria of Quality	Wheat Flours for the Production of			
	Fluffy Batter	Non-yeasted Dough	Light Yeast Dough	Heavy Yeast Dough
Flour type	405	550	550	550
Protein (% of dry weight)	to 9.0	9.5–11.0	12.0–13.0	13.1–14.0
Moist gluten (%)	to 20.0	21.5–24.0	27.0–29.0	30.0–33.0
Sedimentation value	to 20	25–30	33–39	40–45
Maltose number	to 1.5	1.5–2.0	2.5–1.8	1.5–1.8
Falling number	above 300	200–300	250–300	250–300
Particle size				
>160 micron (%)	0	0–3	4–8	4–8
>125 micron (%)	3–5	3–6	5–10	5–10
<125 micron (%)	95	91	82	82
Absorption (%)	48.0–50.0	51.0–53.0	53.5–55.0	55.5–57.0
Volume of baked goods (mL/100 g)	to 450	475–525	620–660	670–740

(WEILAND, 1976)

Tab. 7. Quality Criteria for French Wheat Flours

Criteria of Quality	Bread	Zwieback	Sweet Goods (a)	Sweet Goods (b)	Sweet Goods (c)	All Purpose Flour
Alveograph value	120–160	160–180	200–300	100–120	60– 80	140–180
Protein (N·5.7)	10– 11	11– 12	13– 14	9– 10	6– 8	10– 12
Sedimentation value	23– 28	33– 38	40– 45	17– 20	12– 17	20– 30
Amylogram units	300–450	400–500	500–800	400–600	500–600	800

(CALVEL, 1972)

kies; high protein flours with an extensible gluten and a high maltose value for puff paste; or flours with very good pasting qualities of their starch (scalded flours); low protein flours with little elasticity of the gluten but with a high water absorption and good starch pasting qualities for the production of waffles (SCHAUZ, 1969; SEIBEL, 1970).

4.2 Water

The quality of water has some importance for the production of baked goods (ANGERMANN and SPICHER, 1964). This is particularly true for doughs made with wheat flour of low extraction. Mineral constituents of the dough water (mainly carbonates and sulfates) give a firmer, more resistant gluten; the doughs do not collapse during fermentation, the gas retention is improved, and with a normal volume the grain is finer and more elastic. Only potable water may be used for the production of baked goods. A water of medium hardness or a hard water is preferred (75–150 ppm hardness). Whole milk or skimmilk may also be used as dough liquids. But such doughs have different properties from water doughs and require different methods of processing.

4.3 Salt

Salt is used in all baked goods to provide flavor and because of its effect on the baking process. It inhibits the hydration of gluten. The gluten becomes "shorter", doughs do not collapse and gas retention is improved. Bread volume and fineness of grain is also improved. Unsalted doughs show a high gas development and a fast extension of the dough. The dough is more moist and runny.

Higher concentrations of salt inhibit enzymatic reactions. This can be used to lessen the degradation of proteins and starches in flours from sprout damaged grains. However, salt concentrations above 1.5% (always based on the weight of flour) also inhibit the fermentation activity of yeast. In general 1.5 to 2% salt are used in bread and roll doughs (slightly higher for wheat doughs than for rye doughs); 0.8 to 1% for "fine" baked goods, and 0.1 to 0.5% for other doughs and batters.

4.4 Other Ingredients

The production of bread requires basically only flour, water, yeast or sour dough, salt and some additives such as dough conditioners. Optional ingredients are fat, sugars, milk, and/or oil seeds, skimmilk powder, eggs, fruits, spices, and other aroma forming compounds. Some of these ingredients also affect the rheological properties of doughs. The following can be distinguished. For softer doughs: fats, sugars, chocolate, egg yolk; for tougher doughs: flour, skimmilk powder, egg white; for moister doughs: fluid milk and fluid eggs; for drier doughs: flour, sugar, skimmilk powder, cocoa. Aroma forming ingredients are sugars, cocoa, fats, eggs, and spices.

Fat makes baked goods "shorter". It increases the shelf life, and produces a finer grain and, if used in small concentrations, a greater volume of baked goods. The crust is more elastic and softer. The shortening effect is due to the formation of a film of fat between the starchy and protein layers of the flour. Surface active materials such as mono- and diglycerides or lecithin promote the formation of this film of fat and have a fat sparing effect. The shortening effect is greater for fats with a lower melting point than for harder fats. Oils are also suitable for the production of baked goods. Hydrogenated vegetable fats with a melting point between 30 and 40 °C are also suitable.

Eggs have diverse functions depending on which part of the egg is used: Leavening through foam (egg white); binding effects (egg white and yolk); shortening effect (egg yolk because of its content of fat and lipid materials); and an emulsifying effect (egg yolk because of its lecithin content). In addition the use of eggs promotes browning, and it affects the color of the crumb as well as the taste. The nutritional quality of the baked goods is also improved.

Sugar promotes the fermentation and browning. In addition it makes the dough more stable, more elastic and shorter, and the baked goods more tender. For increasing additions of sugar and fat the amount of added liquids must be reduced for a given dough consistency. The use of liquid eggs requires 33% more liquid and that of frozen eggs 50% more liquid than when water or milk is used. Somewhat more milk has to be used than water for a given dough consistency. Eggs, fats and sugar make the dough shorter and more elastic and the baked product more tender.

4.5 Leavening Agents

For the production of bread, rolls, and some sweet goods leavening is done by microbial fermentation. Many sweet goods, particularly those with higher concentrations of fats, sugars, eggs or spices, require doughs which have a high osmotic pressure and inhibit yeast fermentation. These as well as low moisture, shelf stable items, and items for which a long fermentation time is undesirable are leavened chemically.

4.5.1 Biological Leavening Agents

Sour doughs have been used traditionally by bakers and for the preparation of bread in the home. Originally sour doughs have been used for the production of all types of bread because yeast was not available. The leavening action of sour doughs is also largely due to its natural yeasts, but also in part to the presence of heterofermentative lactobacilli. The introduction of special baker's yeasts at the beginning of this century has limited the use of sour doughs with few exceptions (e.g., Pannetoni and some ginger bread cookies) to the production of rye and rye mixed grain breads. Sour doughs are used mainly to acidify rye flours (up to 6%) and to produce a slightly sour aromatic taste of the bread. Today even wheat sour doughs are used to enhance the flavor of wheat bread or rolls (BRÜMMER, 1985).

4.5.1.1 Yeast

Industrially produced yeasts are strains of the top fermenting species *Saccharomyces cerevisiae* grown on molasses in an aerobic fed-batch fermentation (baker's yeast, compressed yeast). Such baker's yeast has been introduced at the turn of the last century. Its use has increased greatly since the Second World War since processes for growing yeasts as part of a sour dough process or a multiple stage dough process have been almost abandoned. In addition the trend towards simplified sour dough processes with a major emphasis on acid production has led to a greater demand for baker's yeast. In Germany 58% of the supply of baker's yeast is used for the production of wheat doughs and 42% for the production of acidified doughs at the present time.

Baker's yeast has optimum temperatures for growth and fermentation between 28 and 32 °C. The optimum pH is between 4 and 5. Leavening of doughs requires the addition of 1–6% yeast based on the weight of flour. The exact percentage depends on the recipe, the process, the quality of the flour and of the yeast, as well as on operating considerations. Use of more than 8–9% yeast has an undesirable effect on the taste of baked goods. Baker's yeast is available in several forms:

Yeast cakes. The traditional compressed yeast cake is still the most popular form. After yeast has been grown in fermenters it is separated, washed and recovered, for instance, by vacuum filtration. It is then extruded into strands with a rectangular cross section and cut into blocks of 500 g or 2500 g (in the U.S. 454 g). The yeast cakes have a solids content of 28–32% in Europe and of 30% in the U.S.

Prior to use the yeast press cakes may be slurried in water. This permits an easier and more precise measurement of the amount of yeast added but does not affect the fermentation in any other way (BRÜMMER, 1990). Baker's yeast cakes may be stored at 4 °C for a period of 6–8 days without significant loss of its fermentation activity (SCHULZ, 1968; BRÜMMER and ELSNER, 1982).

Bulk yeast. Bulk yeast (or crumbled yeast in the U.S.) is produced in a similar manner as compressed yeast. However, after removal from the filter it is not extruded but broken into irregular pieces and packed into 25 kg plastic film bags to exclude oxygen. The concentration of yeast solids of the crumbled material is 30–32%. It may also be suspended in the bakery in water and pumped for fully automatic delivery in highly mechanized bakeries.

Yeast cream. Yeast cream or liquid yeast is a centrifuged and washed suspension of baker's yeast with about 18% yeast solids. Yeast cream may be delivered in tank trucks directly to bakeries, particularly to those in the immediate vicinity of the yeast plant. Yeast creams should be stored no longer than one day (BRÜMMER, 1990).

Active dry baker's yeast (ADY). By almost complete removal of water one can produce ADY with a solids content of 92–96%. They are granular or powdered products which are packaged in hermetically sealed containers under vacuum or an inert gas atmosphere. The low moisture content permits storage of this ADY for a one year period without significant loss of bake activity. ADY is mainly used in bakeries in subtropical or tropical countries. It is also used widely in home baking. Frequently it is used (separately packed) in complete baking mixes for home use. It is not widely used in the bakeries of industrialized countries because of its higher price. ADY has to be re-hydrated before use, and the conditions of rehydration affect its fermentation activity and its effect on the dough rheology. During the first phase of rehydration some of the cell content leaches out. The lower the temperature of the rehydration water, the greater the amount of leached solids (PONTE et al., 1960). Therefore, rehydration should be carried out with water of 35–43 °C to prevent excessive leaching.

Compressed yeast must be replaced with at least 40% of its weight of ADY to compensate for differences in strains and methods of production (BACHMANN et al., 1973). If compressed yeast and ADY are compared on an equivalent dry yeast solids basis, one must expect a 25–35% lower fermentation activity for the ADY.

Instant ADY. The use of modern methods of drying, particularly with air lift driers, has permitted the production of an ADY (95% solids) which gives equivalent fermentation activity to compressed yeast (LANGEJAN, 1974). This so-called Instant ADY may be added directly to the flour or the dough during mixing, while the traditional ADY requires reactivation, respectively rehydration (LANGEJAN, 1972, 1974; OSZLANYI, 1980). The activity of the Instant ADY in forming CO_2 is about 87% of that of compressed yeast (Tab. 8). Its moisture content is less than 5%.

Instant ADY provides maximum fermentation activity if it is added to flour immediately prior to mixing of the doughs (BRUINSMA and FINNEY, 1981). Otherwise it loses more than 22% of its activity within 18 hours (Fig. 3). Instant ADY packed under vacuum has a shelf life of 22 months at 18 °C. Once the package has been opened it should be kept at 3 °C with a shelf life of up to 10 weeks.

4.5.1.2 Sour Dough Starter Cultures

Sour dough starters are commercially available under various designations (pure culture sour, baking ferment). They contain from $2 \cdot 10^7$ to $9 \cdot 10^{11}$ "sour dough bacteria" per g and $1.7 \cdot 10^5$ to $8 \cdot 10^6$ yeasts per g. Even starters designated as "pure culture sours" are

Tab. 8. Activity of Instant vs. Compressed Yeast (OSZLANYI, 1980)

Form	Method of Drying	Protein[a]	Gas Production[b]
Compressed	None	52	390
Threads	Fluidized	52	341
Irregular spheres	Drum	43	185

[a] % dry matter basis (N·6.25)
[b] mL CO_2 produced in 165 min per 300 mg of yeast (dry basis) in a dough consisting of flour, water, yeast, and salt

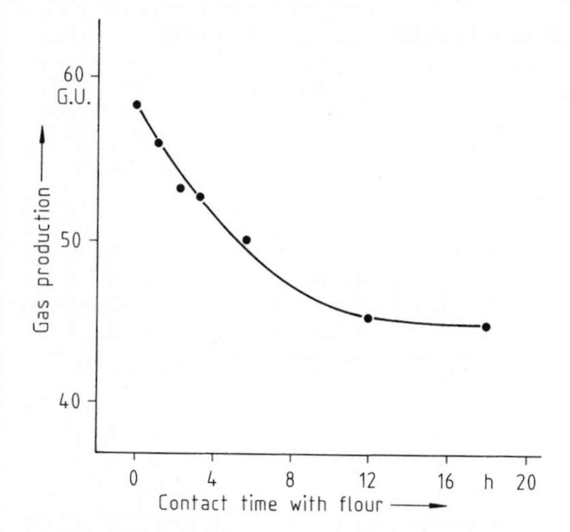

Fig. 3. Gas production as a function of the contact time of wheat flour with Fermipan active dry yeast (2%) at 25 °C. (BRUINSMA and FINNEY, (1981).

merely concentrates of not further defined lactic acid bacteria.

The "sour dough bacteria" are not an independent group of microorganisms occurring only in sour doughs. They are strains specially adapted to doughs as their medium but belong to the lactobacilli which also occur on other products (silage, sauerkraut, mashes). The "sour dough bacteria" belong to the genus *Lactobacillus* (family Lactobacillaceae). They are relatively well characterized. They are Gram positive rods, non-motile, and do not form spores. The "sour dough bacteria" are also characterized by some common physiological properties. They are anaerobes or microaerophiles, acid tolerant, and capable of intensive fermentation of carbohydrates (Tab. 9). The fermentation of glucose results either in the formation of lactic acid (homofermentative lactic acid bacteria) or lactic acid, plus acetic acid, ethanol, and CO_2 (heterofermentative lactic acid bacteria).

Sour dough starters used in German bakeries contain up to 9 different species of lactobacilli (SPICHER and SCHRÖDER, 1978). Some sour dough starters contain a wide spectrum of homo- and heterofermentative species, while others contain only a limited number of species (Tab. 10).

At some time a starter culture has been commercially available which contained propionic acid bacteria apart from the lactic acid bacteria and yeasts (Böcker-Pure Culture Sour). The purpose was to form propionic acid in addition to lactic and acetic acids during the course of the fermentation. The bread contained up to 0.28% propionic acid and had an increased shelf life because of the inhi-

Tab. 9. Lactic Acid Bacteria and Yeasts Occurring in Sour Dough Starters

Designation of Starter	Lactobacilli: homofermentative				Lactobacilli: heterofermentative					Yeasts			
	L. acidophilus	*L. casei*	*L. plantarum*	*L. farciminis*	*L. brevis*	*L. brevis* var. *lindneri*	*L. buchneri*	*L. fermentum*	*L. fructivorans*	*Sacch. cerevisiae*	*Pichia saitoi*	*Candida crusei*	*Torulopsis holmii*
A	−	−	−	−	−	−	−	+	+	+	+	−	−
B	+	+	+	+	+	+	+	+	−	−	+	+	+
C	−	−	+	+	+	−	−	−	−	−	−	−	−
D	−	−	+	−	−	−	−	−	+				
E	−	−	+	−	−	−	−	−	−				
F	+	−	+	−	+	−	−	+	−				
G	−	−	+	+	−	−	−	+	+				
H	+	+	−	−	−	+	+	+	+				

(SPICHER, 1983)

Tab. 10. Sugar Fermentation by Lactic Acid Bacteria of Sour Doughs

Lactobacillus spp.		Monosaccharides							Disaccharides							Tri-saccharides		Sugar Alcohols	
		Pentoses			Hexoses														
		Arabinose	Ribose	Xylose	Fructose	Galactose	Glucose	Mannose	Cellobiose	Lactose	Maltose	Melibiose	Rhamnose	Saccharose	Trehalose	Melizitose	Raffinose	Mannitol	Sorbitol
homofermentative	*Lactobacillus farciminis*	–	±	–	+	+	+	+	+	–	+	–	–	+	+	±	–	–	–
	L. plantarum	–	+	±	+	+	+	+	+	+	+	+	–	+	+	+	+	+	+
	L. casei	–	+	–	+	+	+	+	+	–	+	–	–	+	+	+	–	+	+
	L. acidophilus	–	–	–	+	+	+	+	+	+	+	–	–	+	–	–	–	–	–
	L. delbrueckii	–	–	–	–	+	+	+	–	–	±	–	–	–	–	–	–	–	–
heterofermentative	*L. brevis*	+	+	+	+	+	+	–	–	–	+	+	–	–	–	–	±	±	–
	L. brevis var. *lindneri* I	–	–	–	+	–	+	–	–	–	+	–	–	–	–	–	–	–	–
	L. brevis var. *lindneri* II	–	–	–	–	–	+	–	–	–	+	–	–	–	–	–	–	–	–
	L. buchneri	+	+	+	+	+	+	–	–	–	+	+	–	+	–	+	+	+	–
	L. fermentum	+	+	±	+	+	+	±	–	+	+	+	–	+	–	–	+	–	–
	L. fructivorans	–	(+)	–	+	–	(+)	–	–	–	+	–	–	+	–	–	–	–	–

+ ferments; – does not ferment; (+) ferments weakly; ± some strains ferment, others do not (SPICHER and SCHROEDER, 1978)

bition of molds by propionic acid (SCHULZ, 1947; PELSHENKE, 1950). This effect could only be achieved if the population of propionic acid bacteria exceeded $250 \cdot 10^6$ per g of sour dough (SCHULZ, 1959).

Sour dough starters are characterized (depending on their microflora) by a particular manner of acidification, such as the drop in pH or the ratio of formed lactic acid to acetic acid (hereafter called the fermentation quotient). The fermentation quotient also depends on the conditions of the process itself as shown for instance for the Detmold single stage sour dough process with variations of the fermentation quotient from 1.4 to 2.5 (Fig. 4).

The lactic acid bacteria of sour dough starter do not necessarily have the same technological effect. They are all basically capable of forming acid which may result in bread with a good grain and an elastic crumb. But homofermentative lactobacilli usually do not produce the sensory qualities desired. In contrast acidification by heterofermentative lactic acid bacteria usually leads to the characteristic sensory quality of sour dough bread (Tab. 11).

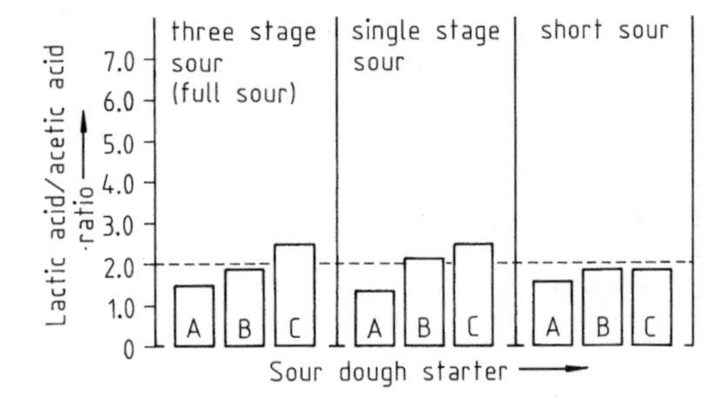

Fig. 4. The effect of starter culture and method of processing on the ratio of lactic acid to acetic acid in sour doughs. (SPICHER, 1982).

Lactobacillus brevis var. *lindneri* can be considered as the representative microorganism for production of sour doughs in Central Europe. The rod-shaped organism is 0.7–1.1 μm wide and 2.5–4.0 μm long. It occurs singly or in chains. They do not grow or grow only very slowly on the common laboratory media. In nutrient media they cause either a

Tab. 11. Quality Criteria of Three-Stage Sour Doughs and Breads as a Function of the Species of Lactobacilli of the Starter (SPICHER et al., 1980)

Lactobacillus spp.	Sour Dough		Bread Quality			
	pH	Acidity	pH	Acidity[b]	Elasticity of Crumb (1 2 3 4 5)	Flavor (0 1 2 3 4 5)
L. acidophilus	4.50 4.20–5.00	7.25 5.00–8.20	4.60 4.35–4.75	6.40 5.80–6.80	2.5	2
L. casei	4.25 4.10–4.30	7.00 6.60–7.50	4.70 4.65–4.80	6.20 6.10–6.50	3	2
L. plantarum	4.05 3.95–4.15	7.90 7.60–8.20	4.50 4.45–4.60	6.50 6.20–6.60	4	2
L. farciminis	3.95 3.95–4.00	8.10 8.00–8.30	4.40 4.35–4.50	6.60 6.40–6.80	4	2
L. brevis	4.20 4.10–4.40	9.95 9.00–11.20	4.40 –	8.00 7.90–8.10	3.5	3
L. brevis var. *lindneri* I	3.90 3.80–4.00	11.50 10.60–12.30	4.30 4.20–4.30	8.90 8.30–9.20	5	4
L. brevis var. *lindneri* II	3.90 3.80–4.00	11.65 11.20–12.00	4.25 4.10–4.30	9.00 8.90–9.20	5	4
L. buchneri	4.35 4.30–4.40	9.10 8.50–9.60	4.60 4.55–4.70	7.70 7.60–7.90	3	3
L. fermentum	4.30 4.20–4.50	8.46 7.50–9.70	5.70 4.65–4.75	6.80 6.40–7.10	2.5	3
L. fructivorans	4.25 4.10–4.50	9.30 8.20–11.10	4.30 4.10–4.50	8.50 8.20–8.70	3.5	2
Starter (B)[a]	3.80 3.70–3.90	12.55 11.60–13.40	4.30 –	8.50 8.10–9.00	5	4.5
Spontaneous sour	4.60	6.10	4.80	6.10	2	0

[a] commercial culture [b] mL 0.1 N NaOH/10 g

flocculation (*L. brevis* var. *lindneri* I) or a uniform haze (*L. brevis* var. *lindneri* II). Characteristically they are limited to the fermentation of glucose, maltose, and partially fructose (*L. brevis lindneri* I) (SPICHER and SCHRÖDER, 1978). They had first been described by KNUDSEN (1924) and called *Betabacterium*. Corresponding lactic acid bacteria have been isolated from beer (ESCHENBECHER, 1968, 1969), from a commercially used culture of *Lactobacillus delbrueckii* and from molasses (RÖCKEN, 1976). According to ESCHENBECHER (1968, 1969) the behavior of this species is the same as that of *Bacillus lindneri* (*Lactobacillus lindneri*) described by HENNEBERG (1903).

Sour doughs of the San Francisco area (U.S.) for the production of "San Francisco sour dough French bread" contain mainly heterofermentative lactobacilli. The predominant species was called *Lactobacillus sanfrancisco*. It grows only on maltose. Xylose, arabinose, glucose, galactose, saccharose, rhamnose, and raffinose are not fermented. In sour doughs the microorganism produces from 70–80% lactic acid and from 20–30% acetic acid (KLINE and SUGIHARA, 1971).

The microflora of the Balady- or Soltani-starter which is used in Egyptian cities for the production of Balady bread consists of lactobacilli (63–64%), yeasts (30–32%), streptococci (1–2.5%), micrococci (1–2%), and bacilli (1–3%) (ABD-EL-MALEK et al., 1974). *Lactobacillus brevis, L. fermentum,* and *L. plantarum* are always present. Occasionally *L. casei* and *L. helveticus* appear. During the dough fermentation *L. brevis* becomes dominant. After a 2 h fermentation it accounts for 75% of the lactic acid bacterial rods (*L. fermentum* 21–23%). Other species developing in the dough are: *Leuconostoc, Enterococcus, Streptococcus bovis, Streptococcus lactis, Staphylo-*

coccus sp., *Micrococcus varians, Micrococcus intermedia* as well as *Bacillus subtilis, B. cereus,* and *B. licheniformis.*

"Torsh" is the sour dough starter used in Iran for the production of the popular Sangak bread. Its microflora consists of *Leuconostoc* sp. (mainly *L. mesenteroides*) and *Lactobacillus* sp. (*L. plantarum, L. brevis*) which together account for 77% of the organisms. A small number of heterofermentative cocci are also present (*Pediococcus cerevisiae*). The number of *Leuconostoc* sp. may account for up to 85% of the bacterial population. Yeasts are only present in small numbers (AZAR et 1977).

Yeasts occur more or less regularly besides the sour dough bacteria in sour dough starters or sour doughs used in the Federal Republic of Germany (SPICHER et al., 1979). These are *Pichia saitoi, Saccharomyces cerevisiae, Candida krusei,* and *Torulopsis holmii,* the imperfect form of *Saccharomyces exiguus.*

In "San Francisco sour dough" the following two yeast species occur: *Torulopsis holmii* which ferments glucose, sucrose, galactose, and raffinose, but not maltose, and *Saccharomyces inusitus* which ferments maltose, glucose, sucrose, and raffinose, but not galactose. *Saccharomyces exiguus* is considered typical of "San Francisco sour dough" since it occurs most frequently and can grow in the presence of lactic acid bacteria (SUGIHARA et al., 1971).

Two groups of yeasts are also found in "Torsh", the starter for Sangak bread. The predominant large, oval cells which ferment glucose, sucrose, and maltose, but not lactose are *Torulopsis colluculosa.* The small, oval yeasts which ferment glucose and sucrose, but not maltose or lactose are described as *Torulopsis candida* (AZAR et al., 1977).

4.5.2 Chemical Leavening Agents

Chemical leavening is generally used with sweet goods and cakes. Carbon dioxide develops by chemical reaction of a carbonate with acid and/or moisture during heating in the oven. A good leavening agent should have the following properties. The amount of CO_2 liberated during preparation of the dough (bench rise) should have the right ratio to the gas developed during baking (oven rise). The total capacity for the development of CO_2 must be sufficient. And, finally, the baked product must not be discolored or show any off-flavor.

Cake batters, pound cake doughs, and other baked goods are leavened with baking powder, which is also used to assist the leavening action of yeast (Zwieback, Streussel cake, doughnuts). Baking powders consist of a carbonate (sodium bicarbonate) and one or a combination of the following leavening acids: potassium acid tartrate, tartaric acid or its Na-, Ca-, or potassium salts, citric acid, dicalcium phosphate, mono sodium phosphate, sodium acid pyrophosphate, or sodium aluminum phosphate. Flour or starch are used as excipients to separate the carbonate from the acids in the baking powder. For some flat shelf stable items such as cookies, crackers, waffles, or honey cakes one can use sodium bicarbonate, sometimes together with ammonium carbonate and without use of a leavening acid. The baking powder is sifted into the flour. It must develop at least 2.35 g of CO_2 but no more than 2.85 g CO_2 (about 1.25 L) per 100 g of flour. About two thirds of the CO_2 should be developed on the bench and one third during the oven rise. Doughs for the production of low moisture products such as cookies, lozenges, or small honey cakes may be leavened with a mixture of ammonium bicarbonate and ammonium carbamate. Ammonium acid carbamate may also be used by itself by addition to the dough or to the dough water. At temperatures exceeding 60°C it is split into carbon dioxide, ammonia, and water. Baked goods leavened with this salt may not contain more than 100 mg NH_3 per 100 g of the final product.

Pottash, potassium carbonate, is the oldest chemical leavening agent. It is used for the leavening of ginger bread and honey cakes. Carbon dioxide is liberated by the action of acids during a period of several weeks or months. The acids are formed by the action of lactic acid bacteria. Potassium carbonate is added to doughs just before dividing together with the spices. It is usually used in conjunction with ammonium carbonate or ammonium carbamate because of its low leavening activity.

4.5.3 Physical Leavening Agents

Cakes and sweet doughs can also be leavened by physical, respectively mechanical means. This can be done by mixing air into the dough. The air expands during baking and leavens the dough. This process presupposes a soft-viscous consistency of the dough which can be enhanced by the addition of eggs, fats, emulsifiers, and stabilizers. The incorporation of air into doughs can also be accomplished by the following: the addition of beaten egg whites or whole eggs beaten to a foam; incorporation of air in the form of small bubbles into the fat phase (batters); the beating of soft fluid doughs (wafer doughs) or by the incorporation of ground ice (as in some Knaeckebrot/crisp bread doughs).

Mechanical leavening can also be achieved by steam which is retained between thinly sheeted dough and intervening fat layers (up to 144 fat layers and 288 dough layers) (TSCHIRPE, 1967). This process is used in the production of baked goods with a characteristic soft leafy structure such as Danish pastry, flaky pastry, or puff paste.

In many instances the desired structure of the baked goods can be achieved by a combination of mechanical, chemical, or biological leavening. Physical (or mechanical) leavening has little effect on taste or aroma of baked goods. Hence the naturally formed aroma substances are more readily perceived than with biological or chemical leavening.

4.6 Additives Affecting the Processing Characteristics of Flour and Doughs (Dough Conditioners)

Various additives such as enzymes, swelling agents, emulsifiers, oxidizing agents and reducing agents have been developed to compensate for variations in the processing characteristics of flours. Such standardized dough conditioners are required in view of the changed conditions of the production of baked goods (COLE, 1973). They permit by enzyme catalysis or physical-chemical means a regulation of the various stages of processing. Even for flours with normal baking properties the machineability of doughs can be improved, processes can be simplified, the quality of the baked goods can be improved and their shelf life extended. Such dough conditioners are either simple compounds with a specific effect or they may contain many compounds with multiple effects on doughs and baked goods. The kind and concentration of dough conditioners permitted in baking is regulated in the different countries. In addition dough conditioners should not change the basic composition of baked goods and should not affect their taste.

Dough conditioners are mainly used in yeast doughs or sour doughs. Their use is generally limited to white, low-ash flours. Darker flours with a higher ash content cannot be readily improved by dough conditioners because of their higher content in enzymes, sugars and their higher absorption. Dough conditioners may be added directly to the flour (powders), dissolved in the dough water (highly viscous dough conditioners) or they may be added to the dough (liquid and paste forms).

4.6.1 Regulation of Water Absorption in Doughs

A certain amount of liquid is required to achieve adequate swelling and gelatinization of starch during baking. But the addition of water, milk, or other liquids may be varied only within narrow limits which are determined by the swelling properties of colloidal components of the flour. If not enough liquid is used the swelling is insufficient, the stiffness of the dough results in poor leavening, and starch does not gelatinize completely during baking. The bread will have a dry, rough crumb or the crumb may even tear. But if too much liquid is used the water cannot be bound by the colloids of the dough which will be soft and sticky. It cannot be worked well and yields bread with a coarse grain and a moist crumb. Flour that does not have enough dough forming elements can be improved by the addition of materials with high water absorbing qualities, such as pre-gelatinized flours or starches. It is assumed that the pre-gelatinized flour releases its bound water during baking so that it is available for the gelatinization of starch.

Pre-gelatinized flour. The use of pre-gelatinized flour, starch from steamed grain or cracked grain or of boiled potatoes has a favorable effect. These materials are produced by gelatinization (boiling or steaming in an autoclave) and subsequent drying in a drum drier, milling and sifting of materials derived from wheat, rye, rice, milo, or corn. Pre-gelatinized starches are produced by the corresponding starches. The gelatinized starch in these materials may bind 4 to 8 times its weight of water. The water absorption of the formula can be adjusted by mixing such gelatinized flours with un-gelatinized flour or by the addition of other water binding substances such as locust bean gum, guar, or alginate (SCHÄFER, 1972).

Gelatinized flours have little acidity and may decrease the acidity of doughs. Therefore, they are often mixed with small amounts of organic acids (lactic acid, tartaric or citric acid, or their acid Na or K salts). They may also be mixed with lecithin.

Gelatinized flours are used mainly in doughs for rye or mixed grain rye breads. Acidified pre-gelatinized flour is also used for the production of mixed grain wheat breads. The addition of these gelatinized flours gives a higher yield of doughs and breads, a better shelf life, and such breads are easier to slice.

Pentosanases. The ability of rye flour to absorb water, the moist and sticky consistency of rye doughs and the properties of the crust and crumb of rye bread depend largely on the concentration and the kind of mucilagenous substances in rye flour. Therefore, the structure of the bread can be improved by a controlled hydrolysis of the soluble pentosans during the fermentation. Use of pentosanases results in a lower viscosity of rye doughs, and the water binding properties during each stage of dough processing become more uniform. This leads to better leavening, an improvement of the volume, and a better shelf life of rye and mixed grain rye breads (ROTSCH and STEPHAN, 1966).

4.6.2 Improvement of the Properties of Doughs and Baked Goods

For the production of white bread and rolls the following properties may be affected by the addition of various additives: gas retention, the machineability of doughs, structure of the crumb, browning and taste of the baked goods and their shelf life.

Dairy products. Browning of the crust and production of a finer grain can be obtained by the addition of dairy products; and within certain limits taste and shelf life may be improved. Such dough conditioners should contain at least 30% of skimmilk, buttermilk, yoghurt, or whey as well as casein in dry, liquid, or paste form. The products may or may not be acidified. The improving action of such additives is often accompanied by an inhibition of the fermentation, excessive browning, and a lower bread volume. Therefore, dough conditioners containing dairy products are usually mixed enzymes, lecithin, sugar, pregelatinized flour, etc.

Soy products. Soy dough conditioners are used for the regulation of the water balance, and partly also for the stabilization of fats to prevent rancidity. They contain at least 30% full fat soy flour. Dough conditioners containing soy flour with its full content of oil and lecithin are used mainly for the production of sweet goods where they act principally as antioxidants. Fat free soy flours which contain no lecithin are used mainly in bread doughs. The proteins and polysaccharides of the soy beans are responsible for the water binding capacity in doughs which is greater than that of wheat flour. Addition of soy flour containing dough conditioners also causes a more intense browning of the crust and a well rounded taste (MENGER et al., 1972). Enzyme active, undebittered soy flour has also been recommended for the production of a lighter bread crumb. The stabilizing effect of soy flour is based on lipoxidase activity on unsaturated fatty acids with the formation of peroxides, which in turn oxidize the $-SH$ groups of proteins. The lighter color of the bread crumb is caused by a coupled oxidation of unsaturated fatty acids and the carotenoids of flour by lipoxidase. The latter reaction requires the presence of oxygen from the air.

Lecithin. A greater bread volume, and a finer grain of mixed grain breads can be obtained by the addition of lecithin. Pure lecithin is usually blended with excipients to facilitate handling, for instance with milled grain products, pre-gelatinized flour, skimmilk powder, oils or sugars. The effect of lecithin is due to its surface active and emulsifying properties. It also reacts directly with flour in the following stages: physical adsorption on the surface of flour particles, swelling and formation of chemical linkages, and condensation to lipoproteins. The decisive reaction is obviously the formation of the lipoprotein complex. This strengthens the protein structure of the dough and reduces the shifting of protein chains; thus creating optimal conditions for water absorption and desirable rheological properties. The gluten (or the dough) becomes more elastic and smoother. Gas retention is improved which results in a shorter fermentation time. Retrogradation of starch is inhibited and the shelf life is extended (SCHÄFER, 1972). The improving action of lecithin applies only to low extraction wheat flours with weak gluten; rye flours are not significantly improved.

Lecithin also has a synergistic effect with other dough improvers, for instance with enzyme preparations and emulsifying agents of various degrees of hydrophilic/hydrophobic balance (SCHÄFER, 1972).

Emulsifying agents. The properties of doughs, their gas retention, their structure (grain, break, and shred), and shelf life can also be improved for wheat flour doughs with synthetic emulsifying agents (see Tab. 12). The use of emulsifying agents also permits the addition of higher concentrations of ingredients such as dairy products or soy flour, which are added to provide nutritional benefits. The most important emulsifying agents are mono- and diglycerides of higher saturated fatty acids which are incorporated into commercial products together with other emulsifying agents. Such mixtures include those of mono- and diglycerides of higher fatty acids and mono- and diglycerides of edible organic acids such as acetic or tartaric acid; "transesterification" products of mono- and diglyceride fatty acids mixed with thermal oxidized plant oils; or mixtures of "transesterification" products of mono- and diglycerides of fatty acids with polyglyceride esters of fatty acids. Optimum effectiveness is related to the so-called HLB value, which means the hydrophilic/lipophilic balance of the emulsifying agent. It expresses the mass ratio of hydrophilic to lipophilic groups of the compound. Emulsifying agents with an HLB ratio of 9–11 are significantly more effective than those with lower HLB ratio, as for instance, glyceryl monostearate with an HLB value of 3.8 (KNIGHTLY, 1968).

The effect of emulsifying agents on doughs and baked goods is based on its reaction with the starch-protein-fat-water system. The molecular basis of this interaction has not yet been clarified. The following mechanisms have been proposed to explain this effect: Finer emulsification of fats in the dough; adsorption on the surface of the starch granule

Tab. 12. Effect of Emulsifiers on Bread Quality

Emulsifier	Volume	Shelf Life	Bread Quality
Sorbitan ester	very good	no effect	good
Calcium stearoyl-2-lactylate	very good	good +	very good
Lactic acid monostearate	good +	good +	very good
Sodium stearoyl fumarate	fair	very good	good
Succinic acid monoglyceride	very good	very good	very good
Ethoxylated mono- and diglycerides	very good	no effect	fair
Sodium stearoyl lactylate	very good	good	very good

(MARNETT, 1977)

and loosening of the bonds between gelatinized starch particles; retention of the soluble amylose in the starch granule; blockage of water penetration into the starch granule and the consequent delay of starch gelatinization; promotion of water uptake by gluten because of the inhibition of starch swelling; and binding with flour proteins, mainly with lipoproteins (COLE, 1973; COPPOCK et al., 1954; HÜTTINGER, 1972; KNIGHTLY, 1968, 1973; SEIBEL et al., 1969). The specific effect of diacetyl tartaric acid monoglycerides is an improvement of gas retention in doughs. These doughs are also less sensitive to mechanical abuse.

The emulsifying agents used for the production of white bread and rolls can be divided into softeners and dough improvers. The softening emulsifying agents presumably reduce the rate of crumb firming (KNIGHTLY, 1973). The dough improvers strengthen the gluten structure of doughs and improve the handling characteristics of doughs as well as their gas retention (BADE, 1974). In the Federal Republic of Germany the use of mono- and diglycerides of natural fatty acid esters is permitted without restrictions. The use of monoglyceride esters of diacetyl tartaric acid derivatives is permitted for rolls and sweet doughs except for doughs containing dairy products. Apart from their emulsifying action which produces shorter crumbs, the monoglycerides act very much like fats. The monoglyceride esters of fatty acids with lower melting points have a shortening effect. The monoglycerides of higher fatty acids impart better gas retention and consequently better leavening and a larger volume. They also impart to doughs a suppleness, lessen their stickiness, and they improve the aerating capacity of cake batters and cremes. Staling is delayed, and toast breads have a better shred.

In the U.S. the widely used continuous mixing process subjects the dough to high stress during the short mixing period. This necessitates the addition of relatively high concentrations of oxidants and emulsifying agents, the dough conditioners and dough "strengtheners". These additives increase the tolerance of doughs to mechanical stress and result in a satisfactory loaf volume. Such additives are, alone or in combination, sorbitan esters, calcium stearoyl-2-lactylate, sodium stearoyl-2-lactylate, the succinic acid ester of monoglyceride, and ethoxylated mono- and diglycerides (MARNETT, 1977).

Oxidants. The use of oxidants results in an improvement of the rheological properties of doughs and of the gas retention. The time of dough maturation is shorter, the oven spring is greater, the volume is large and the quality of the grain is better. This improvement results from the oxidation of $-SH$ groups of proteins to $-SS-$ groups. The bonds thus established within and between protein chains lead to a firmer gluten structure. There are important differences between the action of various oxidants.

The action of one of the more important oxidants, ascorbic acid, is due to its conversion to dehydro-ascorbic acid during dough preparation. This compound oxidizes flour components, in particular the $-SH$ groups of proteins. Ascorbic acid requires the presence of atmospheric oxygen or of bromate for its action. Therefore, it is often used in the U.S. in combination with bromate. In several countries, such as Germany, Belgium, and France, ascorbic acid is the only permitted oxidant. Ascorbic acid acts as a reducing agent in closed mixing systems in which the dough is not exposed to the oxygen of the atmosphere.

Oxidants such as potassium bromate, potassium iodate and chlorine dioxide produce a spongy, somewhat dry and extensible dough which machines well, and which has a short fermentation time because of its good gas retention. The temperature during mixing may be reduced by lowering the rpm of the mixer because of the faster softening of the dough.

In France up to 2% of bean meal has been traditionally added to flour. Particularly during high speed mixing the bean meal facilitates the oxidation of the dough. This leads also to a whitening of the bread crumb. The bean meal also contributes to the amylolytic activity of the flour. This additive also improves the tolerance of the dough but has a certain negative effect on bread aroma (CALVEL, 1972).

Reducing agents. The gluten may be weakened by a reduction of the disulfide bonds which slackens the doughs. This is desirable in the production of cookies, for a reduction of the energy requirements during mixing, and for chemically leavened doughs. The reduced elasticity of such doughs permits better machining. Cysteine in combination with ascorbic acid is well suited for baking purposes (BRÜMMER et al., 1980b). Water absorption is reduced and consequently the time of baking is shorter.

Proteases. The rheological properties of doughs which are determined by the gluten can also be changed by the use of fungal or bacterial proteases. Proteases lead to slacker doughs by splitting peptide bonds and liberating amino acids. Browning and aroma of the baked goods is improved by a higher concentration of the products of the Maillard reaction. The use of proteases is useful for the production of rolls and bread if the maximum of the extensogram exceeds 500 F.E., and for use with cookie flours whose protein content should not exceed 11% or a moist gluten content of 22%. Doughs made from flour with very strong gluten are bucky. After sheeting such dough pieces contract and lead to small poorly leavened baked goods with a fissured or blistered crust.

Fungal proteases are active in the slightly acid pH range. Bacterial proteases are mainly active in the neutral to slightly alkaline range; and they have better heat stability. Optimal concentrations of such enzymes depend on their activity and also on the conditions of use. They are more active at elevated temperatures and for longer time periods in the dough, and less active at higher concentrations of fats and sugars in the dough.

With a joint action of proteases and ascorbic acid it is possible to control the gluten structure as desired for a particular product (ROTSCH, 1966).

4.6.3 Effect of Dough Fermentation

The leavening of doughs by yeast fermentation presupposes the presence of fermentable sugars which become available only during the dough phase, and particularly during the oven spring. The required hydrolysis of starch begins with the action of alpha-amylase on damaged starch. The resulting dextrins are hydrolyzed by beta-amylase to maltose or by amyloglucosidase to glucose. If there is a lack of fermentable sugars, the amount of CO_2 formed by yeast is decreased. The volume of the baked goods is small and the grain is dense. Flours with low enzyme activity produce breads which do not brown well, which have a crumbly crumb and which stale rapidly. Such flours must be supplemented by direct addition of fermentable sugars or by added amylases. The addition of alpha-amylase is most important to start the hydrolysis of starch and to form the substrate for beta-amylase action.

Malt preparations. Malt preparations are used mainly in the production of rolls and buns. Their effectiveness is due to their content of alpha-amylase, beta-amylase, maltose, and glucose. These accelerate dough fermentation, particularly the oven spring. To the extent the resulting sugars are not used by yeast, they lead to the formation of caramelized substances which contribute to the aroma and flavor of bread and to browning and the sheen of the crust. The use of malt preparations can also improve the shelf life of breads (ROTSCH and STEPHAN, 1966). In rolls and buns they improve crispness and break and shred. Malt preparations always contain proteases (besides amylases). Therefore, they are not as well suited for use with flours with a weak gluten, particularly flours from sprouted grain. They weaken the gluten and lead to slack doughs and cause a loss of elasticity of the doughs. Grain is generally coarser. On the other hand, the use of malt preparations may be advantageous with strong gluten flours.

Malt preparations are available in two forms: (1) *Malted flour* (with a moisture content of about 12%) is usually added to flour at the mill, and (2) active *malt extract,* the aqueous extract of malt mashes which is concentrated in vacuo to a solids content of 70–75%. The latter is used exclusively in bakeries. However, the pure malt preparations have been largely replaced by formulated dough conditioners. In such preparations the undesired side effects of malt preparations can be lessened by use of surface active and stabilizing compounds; and the positive effects can be strengthened and complemented.

Malt extracts have a much greater concentration of easily fermentable sugars (about 60–72% maltose) than the malted flours. This accelerates dough fermentation considerably. In contrast the amylolytic activity of malted flours requires some time and their effect is greater on the oven spring.

Microbial alpha-amylases. During the past 30 years amylases of fungal and bacterial origin have been introduced into the baking industry. These have the advantage that their enzymatic activity can be standardized and that they are nearly free from proteases and beta-amylase. They are less heat-stable and, therefore, permit a greater tolerance in their dosage. This permits a fortification of the natural enzymes of flour according to technological requirements.

Fungal alpha-amylases are produced by submerged culture of *Aspergillus oryzae* and *A. niger.* They are available as powdered concentrates. Bacterial alpha-amylases are produced by submerged culture of *Bacillus subtilis.* These have excellent temperature stability. They are inactivated only slowly at 80 °C and often remain active after baking, leading to a moist and slimy crumb. More recently a heat-labile alpha-amylase from *B. subtilis* has become available which is inactivated at 60 °C, that is, at the temperature of starch gelatinization. This preparation can be used as a flour improver without dangerous consequences (SCHULZ and UHLIG, 1972).

Malt enzymes form basically maltose from starches. The microbial amylases differ in the formation of sugars from starch. Fungal amylases form largely maltose and maltotriose, while the liquefying bacterial enzyme forms sugars from glucose to maltohexose. In addition the fungal preparations contain alpha-amyloglucosidase, and the end product is mainly glucose (SPRÖSSLER and UHLIG, 1972). Doughs treated with such fungal preparations are drier and firmer and are, therefore, easier to machine. In contrast to the use of malt preparations, breads show a finer and more uniform grain, a lighter color of the crumb and a more elastic crumb, and the breads have a longer shelf life (SCHULZ and UHLIG, 1972). But addition of enzymatic agents has to be considered in combination with the effects of propionic and sorbic acid salts which are used as preservative. This may decrease crumb elasticity (BRÜMMER and MORGENSTERN, 1975).

Products obtained by starch hydrolysis. Corn syrup, maltose syrup, high fructose corn syrup, or dextrose are used as a source of fermentable sugars either as such or in formulated dough improvers. Their effect is similar to that of malt products. But the acceleration of the fermentation occurs sooner, usually already at the start, because of the presence of readily fermentable sugars such as glucose and maltose. Fermentation slows down as soon as the sugar has been fermented. These additives cause an increased volume of breads and rolls, better grain, and particularly if dextrose (glucose) is used a deeper color of the crust. The consistency of the dough is not affected. These products are best suited as additives to flour with a sufficient concentration of natural enzymes, that is, flours which do not require the addition of amylolytic or proteolytic enzymes. Products derived by starch saccharification are often more suitable as additives to flours with weak gluten than malt products.

These fermentable sugars are principally produced from the starch of grains, mainly from corn, but also from wheat or milo by acid or enzymatic hydrolysis. Apart from dextrose which consists of a single compound (glucose) they contain mixtures of fermentable sugars, and sometimes trisaccharides and various dextrins.

4.6.4 Dough Acidification

Rye flour doughs are usually acidified by use of a sour dough, but they may also be acidified by the addition of acid compounds. Such compounds simplify the process. They are particularly suitable for use with flours from sprouted rye since they cause an immediate drop in pH which inhibits the alpha-amylase activity of such flours. Such dough acidifying compounds are organic edible acids such as lactic, acetic, citric, or tartaric acids or their acid Na or Ca salts, or the acid Na or Ca salts of ortho- or pyrophosphoric acid. They are used as such or in combination with swelling compounds, fats, or other additives. Otherwise, they may be concentrates of sour doughs or grain mashes fermented with lactic acid bacteria.

These acidifying agents may be used to replace sour doughs in the single stage sour dough procedure, or as a partial replacement in multiple stage sour doughs. The required amounts depend on the acid content of the additives. The use of such acids as additives has certain disadvantages, mainly a lessening of the aroma of the bread. Occasionally, they cause a shorter crumb or faster staling of the bread. But they give a higher yield of doughs and breads. This may compensate for the higher ingredient costs in comparison with self fermented sour doughs.

4.7 Formulated Additives

Such improving mixtures are available for use with wheat or mixed wheat breads. They consist usually of mixtures of the compounds described above. Often they contain such basic improvers as malt extract, glucose, lecithin, etc., with the addition of minerals, swelling agents, enzyme concentrates, emulsifiers, and oxidizing or reducing compounds. Ammonium chloride or phosphate are often added as yeast nutrients. Mono calcium phosphate and calcium pyrophosphate are often used to affect the swelling properties of doughs and to strengthen the gluten of soft wheat flours. Calcium phosphates also improve the growth of yeasts and produce a more uniform, finer grain and an increase in volume. Tri calcium phosphate is sometimes added to powdered preparations containing oils or emulsifying agents in order to improve their free flow properties. Occasionally formulated improvers contain preservatives such as Ca propionate, Ca acetate, or sorbic acid to prevent mold growth or ropiness. The uses of additives have to conform to local regulations.

Formulated dough improvers are generally used only in doughs with insignificant additions of compounds other than flour. With doughs containing high concentrations of sugars, of meal or fats, in particular with batters, such additives are less effective since flour plays only a minor role in the formation of the dough.

5 Formation of the Dough

It is important to mix the raw materials and additives thoroughly and to obtain good aeration. The incorporated air is distributed in very small bubbles which are essential for the leavening of the dough and the grain of the bread. During subsequent yeast fermentation the number of bubbles does not increase. The gas bubbles merely increase the volume. The oxygen in the air bubbles stimulates the fermentation by yeast and improves the properties of the flour for dough formation. The choice of the mixer and the conditions during mixing largely determine the final volume of the baked goods and the structure of the crumb.

5.1 Preparation of Raw Materials

The baking quality of flour depends on the variety of the grain, on all agricultural conditions including the harvest and on the milling process. It is hardly possible to produce this raw material in uniform and optimal quality over long periods of time. But the formulation or the processing conditions can be changed appropriately if the properties of the flour are known qualitatively and quantitatively. The important processes for the preparation of the flour are its storage, transport through the plant, sieving, weighing, and mixing. These routine operations can be carried out by automated machinery.

Storage of flour. Immediately after milling the properties of the flour are not optimal. For proper maturation rye flours require a storage time of 1–2 weeks, and wheat flours one of 2–4 weeks. Flour maturation is an oxidative process greatly dependent on the supply of oxygen from the air. It improves the baking properties by formation of a shorter gluten.

Flour cannot be stored indefinitely. It is exposed to environmental conditions because of its large surface area. Flour may be stored for up to half a year at 20°C and with a moisture content of less than 12% (a_w<65%) without a change in its properties and without microbial spoilage or infestation. Wheat flours may be stored longer than rye flours, and short extraction flours longer than high extraction flours. During storage the concentration of free fatty acids increases due to enzymatic hydrolysis. This leads to a change in the rheological properties of the gluten. The oxidation products of unsaturated free fatty acids lessen the baking properties of flour and may produce an off-flavor.

All raw materials for the production of bread may be stored in large tanks or bins. Mixed charges can be stored (based on a given formula involving a short period of mixing of the dry ingredients (60–90 seconds) and a precision of 1:1000) (REIMELT, 1970; STEUB, 1970).

Sieving of flour. Prior to the preparation of doughs the flour is sieved. This removes extraneous material such as fibers from bags, insects, etc., and it aerates the flour. This leads to an increase in the specific volume of the flour (up to 60%). The loosened flour particles swell more readily during the formation of the dough. The time required for formation of the dough is reduced. At the same time the oxygen brought into contact with the flour particles improves the gluten. The increased surface of the flour particles favors adsorption of air oxygen and increases the number of air bubbles or pockets in the dough. An analogous effect is obtained by pneumatic transport of flour in the bakery.

Mixing of flour. Flours of diverse origin and diverse properties are generally mixed to achieve more uniform baking properties.

Temperature. Each dough process requires a given temperature range. Yeast or sour dough processes require a range from 24–30°C. Chemically leavened doughs such as short bread doughs, batters, and puff paste doughs, require lower temperatures. The dough temperature depends on the temperature of the flour and the dough water. That of the dough water should not exceed 45°C or at most 50°C. Otherwise there is a danger of starch gelatinization or damage to yeasts or lactic acid bacteria.

5.2 Doughs Mixing

Formulations consisting largely of flour and water are produced in mixers. Formulae with lesser amounts of flour, such as batters, are produced by stirring or beating. Some yeast leavened doughs are also mixed by stirring (Tab. 13). Processes of swelling and solubilization as well as the effect of mechanical work lead to a body with certain plastic-elastic properties, that is extensible and can be formed into different shapes – the dough. The consistency of wheat doughs and rye doughs differs greatly. Wheat doughs are elastic and

Tab. 13. Distinguished Characteristics of Doughs and Batters

Characteristic	Dough	Batter
Basic ingredients in recipe	Flour, sugar, fat	Sugar, eggs, fat, flour
Processing	Kneading, mixing	Beating, stirring, mixing, short heating
Raising	Biological, chemical, physical	Chemcial, physical
Factors affecting binding of water or consistency	Wheat gluten, pentosans, damaged starch, swelling agents	Eggs, fat, sugar, damaged starch, and in part wheat gluten and swelling agents
Consistency	Elastic to plastic	Foamy, soft-plastic, salve-like to semi-fluid

(MENGER, 1972)

have a dry surface. Rye doughs are plastic and somewhat moist (WEIPERT, 1976). The weight of the dough obtained from 100 parts of flour plus the weight of the other dry ingredients and the dough water is called the dough yield. Dough yields vary for different kinds of baked goods. Doughs for rye or mixed rye breads are softer with dough yields of 160–189; doughs for wheat and mixed wheat breads have dough yields of 155–160. The use of make-up equipment requires firmer doughs. Pan bread may be produced from softer doughs than hearth bread. The dough yield affects the quality of the baked goods. Softer doughs produce generally better leavened, more aromatic bread, while firmer doughs produce a flat taste and quality defects are more likely to occur. There are no general rules which relate to a certain degree of dough development in the mixer with optimal crumb structure. The technology of mixing has to adapt to the demands of consumers in various countries. In England a fine grain but a denser crumb is preferred. In the U.S. and in Sweden a soft, sponge-like crumb is preferred which is obtained by use of a soft "sponge dough". The walls of the pores of doughs mixed under pressure in continuous mixers (Do-Maker, Amflow, Brimec process) are extremely thin and the crumb is less coherent and has a very soft feel. In Australia as well as in France a coarse grain, a so-called aerated crumb is preferred. In Germany there is no definite indication of a preference for a crumb structure since wheat bread is not one of the major bread types. So-called toast bread should have a fine uniform grain with a crumb that is not too dense.

5.2.1 Mixers

A wide variety of mixers based on different principles of construction are available both for batch and for continuous mixing. The more traditional slow speed mixers (revolving arm mixer) turn at 20–40 rpm. The kneading arm imitates the motion of the hand by turning, pushing and pulling the dough. These kneaders can accommodate up to 600 kg of dough. They are used mainly for dough containing high percentages of rye flour.

High speed and high intensity kneaders rotate at 60–240 rpm. These are vertical mixers in which the shaft of the mixer can turn or move up and down and in which ribbons, cones or other mechanical structures knead the dough. Such mixers find use for a great variety of applications.

High speed mixers with a vertical shaft rotate at 1000–3000 rpm. They have a tearing or cutting action rather than the traditional kneading action. Within 0.5–3 min the dough can be fully developed. In some of these mixers the vertical, cylindrical trough can be sealed hermetically. This permits mixing under pressure as in the Brimec process (plus 0.75 atm) or under partial vacuum as in the Tweedy mixer (0.5 atm). Such mixers require strict control of the uniformity of the raw materials, and adherence to a definite formulation and a definite time of mixing. They are used mainly for the production of wheat bread doughs, for sweet goods and for batters.

Some batch mixers consist of a device which can turn 360° around a vertical axis. The device has 4 or 8 fork-like arms to accommodate a corresponding number of kneading troughs. The troughs are rotated past the following stations: Filling, kneading, dough rest, and emptying by tipping. The kneading action is by means of high speed or high intensity mixers. The minimum dwell time at each station is 6 min. With a batch size of 240 kg of dough an hourly throughput of 2.4 tons of dough can be achieved. Such mixers are used mainly for the large scale production of wheat doughs and mixed grain doughs.

Fully continuous mixers are commercially available: Buss-Ko mixer, Continua, Strahmann Mixer, Amflow Developer, Do-Maker Developer. Automatic feeders supply the powdered ingredients (flour, skimmilk powder, dough conditioners) through screw feeders or the liquid ingredients (water, yeast, or salt solutions) through pumps (Fig. 5). Compounds for the acidification of doughs and other salts are usually dissolved in part of the dough water and fed by pumps. Yeast is suspended in water and pumped. Fats and emulsions are warmed to a temperature at which they become pumpable. The mixing chamber

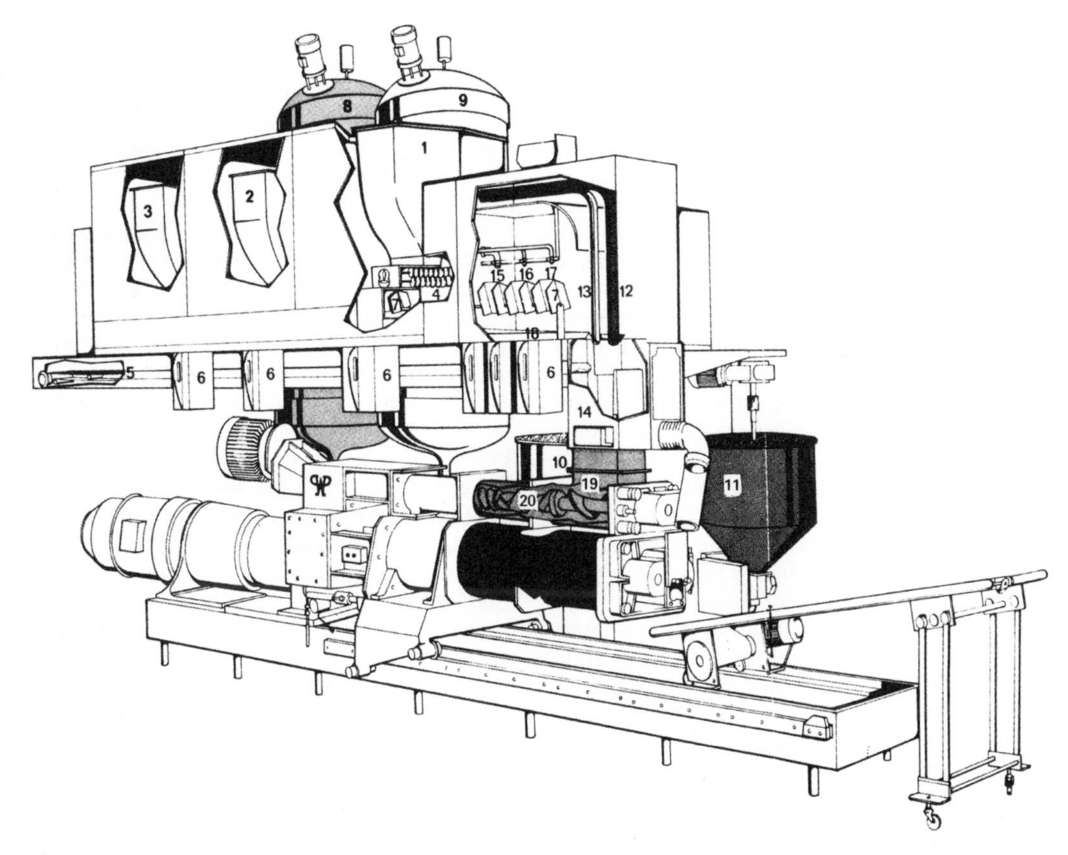

Fig. 5. Equipment for continuous mixing of doughs. – 1 Feeding device for flour; 2 Feeding device for other ingredients; 3 Feeding device for other powdered materials; 4 Feed auger; 5 Conveyor belt; 6 Sampling device; 7 Valves; 8, 9 Ingrediator tanks for yeast, sugar, or salt solutions; 10 Mixer for viscous masses (e. g., fat); 11 Mixer for pre-dough; 12, 13 Piping for viscous materials; 14 Feed tunnel; 15 Water feed pipe; 16 Yeast slurry feed pipe; 17 Feed pipe for other liquids (i. e., sugar solutions); 18 Collecting trough; 19 Feed cone for mixer; 20 Continuous mixer.

is cylindrical or slightly cone-shaped. The mixer shaft has various shapes which may be in the form of screws or wings. In the first part of the mixing chamber the ingredients are mixed intimately and wetted. In the second part the dough is formed and developed. The dwell time in the mixing chamber is from 1–4 min. The dough is extruded from the mixing chamber in the form of a strand with a round, square, or rectangular cross section. The amount of dough that is extruded in a given time period can be regulated by changing the rpm of the mixer (from 25–170 rpm) or in some cases by changing the cross section of the opening through which the dough is extruded. A change in the cross section of this opening also affects the back pressure and changes the intensity of kneading in the mixer chamber. The uniform cross section of the extruded strand permits the cut-off of exact weights, and consequently direct deposition into baking pans. This makes the use of dividers, rounders, intermediate proofers, and molders unnecessary. However, such make-up equipment may be used if desired.

Continuous mixers can process 13 tons of flour per hour. They are particularly suitable for bakeries producing one or only few types of bread. They can be used for all types of baked goods which require a uniform dough consistency and a certain firmness of the dough. Commercially they have been used mainly for the production of white wheat bread and rolls. They may also be used for the production of rye bread and mixed rye bread (HUBER, 1965).

5.2.2 The Mixing Process

The conditions required for optimal dough development depend largely on the formulation, particularly on the quality of the flour and the relative concentration of wheat and rye flours; also on dough temperature and to a limited extent on the weight of the dough. On the other hand, each mixing system has specific effects on the development of the dough and consequently on its properties as they affect subsequent processing. Mixing may be carried out at various pressures either to accelerate dough development or to produce specific properties of the crumb. For instance, mixing under vacuum results in bread with a lesser specific volume but a finer and more uniform grain.

Each dough requires a certain time for development in the mixer. Wheat flour doughs require longer periods than rye flour doughs. Wheat gluten does not swell readily and requires mechanical distribution and stretching. This change in gluten structure is required for the formation of an elastic gluten network. Wheat gluten is also quite tough, which is particularly true for American wheat flour of high gluten concentration and good gluten strength. This strong gluten can only be made elastic by long sponge fermentations and/or by intensive kneading in high speed mixers. Hence wheat flour doughs require a considerable amount of energy for mixing. The amount of free water in the dough decreases with increasing swelling of the gluten. Therefore, the film of water separating the starch granules and the gluten strands becomes thinner. This is responsible for the high energy requirement of about 9–11 watt hours per kg dough. In contrast doughs from rye flours require an energy uptake of only 3.2–3.4 watt hours per kg dough (HUBER, 1965). Dough development is essentially complete at the time the ingredients have been uniformly distributed and wetted. With additional mixing rye doughs become unstable. Bound water is set free, and the dough becomes softer and more sticky. Mixed grain wheat doughs require an energy input of 3.8–4.0 watt per h per kg (HUBER, 1965).

Mixer speeds. With traditional revolving arm mixers the energy requirement is relatively modest, 3–5 watt per h per kg dough, while high speed mixers require 8–12 watt (GASZTONYI, 1970). Wheat doughs which have been mixed in revolving arm mixers are firm, they tear easily and their gas retaining properties are poor. With more intensive kneading the resistance of the dough decreases, it becomes more extensible and acquires better gas retention. Water binding in cool doughs is slower than in warm doughs. Therefore, they have to be mixed longer. On the other hand, higher mixing speeds increase dough temperatures considerably (Tab. 14). In revolving arm mixer the temperature increases during 20 min of mixing by no more

Tab. 14. Average Increase in Dough Temperature During Kneading of Wheat Doughs in Various Kneading Systems

Kneading Equipment	rpm	Optimal Time of Mixing (min)	Dough Temperature After Mixing (°C)	Dough Temperature After Kneading (°C)	Increase in Dough Temp. (°C)/ Mixing Time
Revolving arm mixer	25	20	22	23	1
Fast kneader	60	20	22	24	2
Fast kneader	120	10	22	27	5
High speed mixer	1440	1	22	31	9
High speed mixer	2990	0.75	22	36	14

(HUBER, 1970)

than 1 °C. In high speed mixers it may rise by 8–10 °C within 1–2 min (HUBER and BANSBACH, 1971). In continuous mixers with an average mixing time of 2–4 min the temperature of the dough reaches 38–40 °C. This accelerates dough maturation because of the faster and better swelling of flour components, and requires the addition of up to 3% more dough water. It has recently been found that this warming up of doughs actually increases the volume of the bread. The fast dough formation in such mixers permits a reduction of the time of intermediate proof or its omission (HUBER, 1972). With increased speeds of the mixer the time span between the undermixed stage and the overmixed stage is shorter. There is a definite optimum for dough development which may last only for 30 s with high speed mixers (HUBER, 1966).

Mixing time. Wheat flours require longer mixing times than rye flours. Flours with strong gluten require a longer time than those with weak gluten, and granular flours require a longer time than smooth flours. Firm doughs in which less water is available for swelling require shorter mixing times than softer doughs. The mixing time depends also on the rpm of the mixer and on its specific construction. For instance a wheat flour dough may require a mixing time of 15–30 min in a revolving arm mixer, while optimum results in high speed mixers may require as little as 2 min and occasionally 40–60 s.

A better development of the dough during mixing often includes changes which otherwise occur later during the fermentation time of the dough. In such cases the fermentation time can be shortened or omitted. In general the volume of baked goods increases to an optimal value with increased mixing times. At the same time this leads to a finer grain, a lighter crumb, and facilitates slicing.

The mixing tolerance of doughs depends largely on the formula. With higher mixer speeds differences in the quality of the raw materials become more apparent. Doughs containing sugars or shortenings require longer mixing times. Higher concentrations of fat, emulsifying agents, and salt (up to 2.3%) increase the mixing tolerance of doughs; skimmilk powder, sugars or eggs reduce it.

Dough temperature. The most suitable temperature is not necessarily the best temperature for the fermentation since the rheological properties of the dough are most important. Each blend of flour has a certain optimum temperature. Doughs for wheat bread should have a temperature of 22–24 °C, for rye and mixed rye bread 28 °C and for cracked grain bread 30 °C. Cooler doughs improve the taste and the shelf life (HUBER, 1965, 1970). The use of lower temperatures (2–3 °C below the above given temperatures) is increasing today (TUNGER, 1971). The temperature during mixing is important for both dough formation and for the subsequent processing of the dough, for its fermentation tolerance and the quality of the baked goods. Cool dough temperatures delay swelling and require longer mixing times. Very high dough temperatures have the disadvantage of shortening the time of dough maturation and of the proof time too much. The sensitivity of doughs to higher temperatures increases with increasing percentages of wheat flour in the formula. Wheat breads made from doughs with high temperatures have a smaller loaf volume, are difficult to slice, and stale sooner. This is more pronounced with firm doughs than with soft ones. Doughs with more than 5% fat and sugar are also more sensitive to higher temperatures.

However, each mixing system has its own optimal temperature. High speed mixers do not just increase the temperature. They also produce optimal results at higher temperature levels (Tab. 15). Warm wheat doughs produced in high speed mixers are not as sensitive as doughs of slower mixers at the same temperature. This is due to the short time required for swelling and the higher work input in a shorter time period. On the other hand, some doughs require very low temperatures; for instance, Knäckebrot. This may require the addition of ice water. Rye and rye mixed doughs are mixed at higher temperatures than wheat doughs (28–30 °C).

5.2.3 Dough Formation

The physico-chemical transformations during the early stages of mixing and during the

Tab. 15. Optimal Dough Temperature for Various Types of Kneaders

Kneader	rpm	Optimal Dough Temp. (°C) (After Kneading)
Revolving arm mixer	20–40	22–24
High speed kneader	60–70	24–26
Intensity kneader	80–120	26–28
Mixer with spiral mixing shaft	150–250	26–28
High speed mixer	1000–3000	28–30

(HUBER and BANSBACH, 1971)

subsequent kneading involve largely the proteins, the starch, and the gums. But other flour constituents such as glutathione, lipids, and their associations with carbohydrates and proteins, etc., also play a role, although these constituents are present only in very small concentrations. The mechanical processes during mixing do not just establish contact between the different constituents but determine also the extent of their interactions.

During the first stage the various constituents are brought into contact; the dough water enters the interstices between the flour particles and wets them (blending stage). At the end of this stage the dough is not coherent and its surface is rough, somewhat crumbly, sticky, and moist. During the following stage (development stage) the constituents interact with the dough water. The starch adsorbs up to ⅓ of its weight of water. This hydration water is bound by adsorption as well as by capillary action. Besides, the starch granules have a certain amount of water of crystallization. Another portion of the water reacts with the pentosans. The proteins, particularly the gluten, also swell and the gluten takes up 2–3 times its weight of water. This water is bound (a) by chemical sorption to certain reactive groups of the protein molecule, (b) by chemical binding within the protein molecule (adsorption), (c) by surface binding to molecules of gliadin and glutenin, and (d) by capillary forces. In particular the water bound by capillary forces can readily be given up and is available for other processes during the baking process, for instance for starch gelatinization. The end result is a homogeneous, flowable, aqueous phase which contains the free water with the other soluble and swelled constituents such as sugars, minerals, and phospholipids. The formed dough becomes very sticky and putty-like. It is still unelastic. It consists of a continuous aqueous and proteinaceous phase (about 56% of the dough volume) with imbedded hydrated starch granules. The beginning plasticity of the dough permits ultimately the retention of carbon dioxide produced by the fermentation. About 8% of the dough consists of air bubbles.

With further mechanical input the dough becomes elastic. The kneading process has reached its final stage when the dough becomes elastic, and its surface is smooth, dry, and slightly lighter in color. At this point the dough is picked up from the wall of the mixer. The dough remains in this condition for a longer or shorter time period. The length of this phase is most important since doughs with a lesser mixing tolerance require more careful process control.

With additional mixing the dough structure disintegrates. The dough becomes softer, less elastic and very extensible. It loses its coherence, becomes more fluid, and loses its gas holding ability. This is the breakdown stage of an overmixed dough.

The described processes are based largely on the formation of a structure of swelled gluten caused by the kneading action in the presence of oxygen (RITTER, 1966; ROHRLICH, 1973). The rheological properties of wheat flour doughs are largely determined by the presence of free cysteine-SH groups in the macromolecule and their relation to cystine –SS– groups, which determines the redox potential. According to the lipoprotein model of GROSSKREUTZ (1960) disulfide bridges are formed through oxidation of the –SH groups of cysteine. These bridges connect the polypeptide chains of the protein molecule. The dough is firmer and has greater stability the larger the number of –SS– bridges. The tertiary structure of the gluten molecule is originally spherical. Swelling and shearing action during dough formation results in an opening of the polypeptide chains. The tertiary structure is maintained by binding between the acid and basic groups of the amino acid side chains, by disulfide bridges, by hydrogen bonding and also by van der Waals forces. Changes in structure occur when new active groups are formed during dough formation which lead to new inter- and intramolecular bonds. This results in a structure in which the protein hydrogel forms the continuous phase of the dough, and in which the starch granules are imbedded. This structure permits gas retention in wheat flour doughs. Recent investigations have shown that gummy substances (pentosans) have a considerable effect on the structure and the gas retaining ability of the dough (DREWS, 1971; NEUKOM, 1972). Flour contains only about 1% soluble pentosans. But these bind one third of the added dough water during dough formation and form a firm, gel-like mass which contributes to the firmness of the dough. This is particularly true for rye flour doughs. Differences in the structure of wheat and rye pentosans are slight, but their physical properties and their role during dough formation differ appreciably. Rye generally contains macromolecules of a lesser degree of polymerization, and contains more soluble proteins and sugars, and the starch is of poorer quality. The hemicelluloses and the pentosans of rye are more water soluble, and particularly the pentosans form more viscous solutions. In wheat flour doughs the structure of the gluten is most important. In rye flour, which lacks the gluten quality of wheat flour, the gummy components are decisive for the formation of a highly viscous medium. This medium contains the rye protein and all water soluble constituents of rye flour. The water binding capacity of the gums necessitates the addition of more dough water. Rye doughs do not show the elastic properties of wheat flour doughs but are rather plastic. Hence, rye flour doughs bind more water and require different processing conditions.

6 Leavening of Doughs

The production of bread, buns, rolls, Zwieback, etc. requires generally biological leavening. Yeast is used for wheat flour doughs and sour dough as well as for rye flour doughs. The extent of leavening depends not only on the amount of leavening gas produced but also on the rate of gas retention in the dough. Gas development continues for a few minutes in the oven until the dough has lost its plasticity and the structure of the baked goods is firmly established. A leavened dough absorbs during the same time period 2 to 3 times more heat than an unleavened dough. Heat also penetrates faster into the interior of a leavened dough.

6.1 Processing of Wheat Doughs

With straight doughs and sponge doughs the yeast is added at the beginning. The choice of the process depends not only on technological considerations but also on the quality of the flour. Straight doughs are preferred for flour of poorer gluten quality and consequently poorer gas retention. The total fermentation time should be relatively short since such doughs become too soft if they are fermented for extended time periods. The sponge dough process is preferred for flours with a strong gluten and a low amylolytic activity (for instance for the production of Zwieback). With prolonged fermentation the gluten becomes softer and more extensible and is easier to machine. The sponge dough process is also recommended for formulations with higher concentration of dough ingredients other than flour.

Tab. 16. Use of Compressed Yeast in Various Types of Baked Goods

Bread or Baked Goods	Yeast Usage (%) Average Value	Range
Rye bread	1	0.5– 1.5
Rye mixed grain bread	1.5	1.0– 2.0
Wheat mixed grain bread	2.2	2.0– 2.4
Wheat bread	2	1.5– 2.5
Rolls	4	3.0– 4.0
Zwieback	8	6.0–10.0
Sweet bread	5	4.0– 6.0
Berlin filled doughnuts	7	6.0– 8.0

(VOM STEIN, 1952)

6.1.1 Straight Dough Process

In the straight dough process the flour, water, yeast, salt, and all other ingredients are added at the same time and together formed into a dough. In Germany this process is used almost exclusively for the production of bread and rolls. The required amount of yeast depends on processing conditions, dough acidity and the quality of the flour and the yeast. Bread doughs require from 0.5–4% compressed yeast, and sweet doughs 6–10%. Tab. 16 shows the concentrations used with various types of baked goods.

The dough ferments at temperatures from 26–32°C. Within limits the time of fermentation can be varied by varying the concentration of yeast. The fermentation time is relatively short. Occasionally, for overnight fermentation of 8–12 h the concentration of compressed yeast can be reduced to 0.15–0.3% and the temperature to 18–20°C. The resulting longer enzymatic activity requires the use of strong flours.

The straight dough process has the advantage of saving time. It is simpler because it does not require re-mixing. Fermentation losses are smaller than with other processes.

6.1.2 Sponge Dough Process

In this "indirect" process parts of flour and water and a portion of the yeast are first mixed into a sponge or "pre-dough" (temperature 25°C). When the yeast has multiplied sufficiently or when the pre-dough has fully fermented it is mixed with the remaining flour, water and other ingredients to the final dough (temperature 26–28°C). The size of the pre-dough depends on the quality of the flour. For flours with weak gluten the pre-dough contains 25% of the total flour; for flours with strong gluten 50%. The concentration of compressed yeast depends on the desired fermentation time of the pre-dough; 0.25–1% for short fermentation times and 1–1.5% for longer fermentations. Besides smaller dough masses require higher yeast concentrations (see Tab. 17).

A short wheat dough fermentation (Vienna dough fermentation) is used mainly for the production of rolls and buns. The pre-dough is quite soft with a dough yield of 180–220 (based on the weight of flour). For shorter fermentations sometimes a firmer consistency is preferred (dough yield 160). For overnight fermentations the dough yield should be 160–180.

For doughs high in fat or sugar the fermentation time is often only 20–30 min. In this case all of the yeast is used with one fourth to one third of the flour and water to form a warm, soft pre-dough. This causes better swelling of the flour and acclimatizes the yeast to the conditions of the dough.

The pre-dough (or sponge dough) method requires more time and labor than the straight dough method but permits greater variations in the operations of the process. Improved swelling of the flour components and a more uniform distribution of the yeast

Tab. 17. Conditions for Wheat Flour Sponges

Length of Fermentation	Flour (%) of Total Fluor	Dough Yield	Yeast (%) of Flour in Sponge	Temperature (°C)	Time of Fermentation (h)
Short	50	150–160	6 –10	25–28	0.5– 1
Medium	20–40	160–200	1 – 2	25–28	2 – 4
Long	10–20	150–160	0.1– 0.2	22–25	12 –20

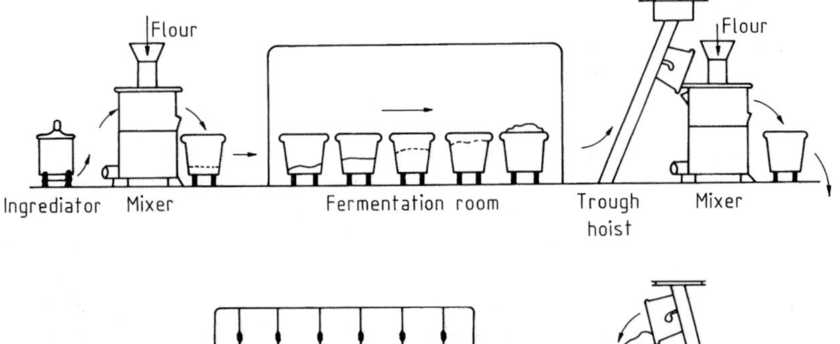

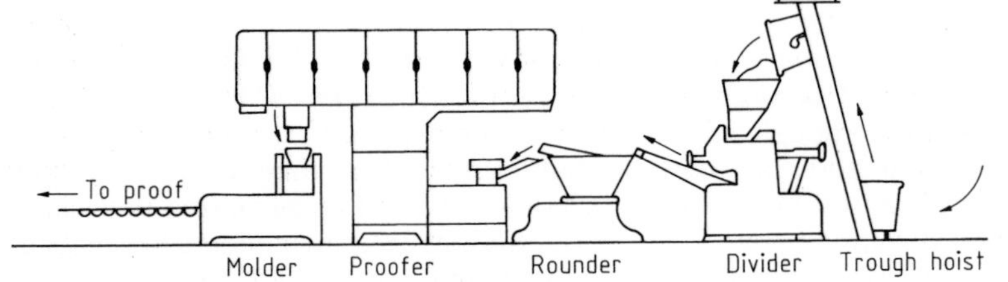

Fig. 6. Flow sheet for the sponge-dough process.

result in a larger volume and a better shelf life. In addition the longer fermentation time improves the aroma.

In Central Europe the pre-dough method has been replaced largely by the straight dough method during the past decades. In contrast flours used in North America require a definite sponge dough method to produce bread with the desired quality. In the U.S. about 80% of all of the white bread is produced by a pre-dough or pre-ferment method such as the sponge dough method, the liquid pre-ferment method, or the continuous mix method. The straight dough method is used mainly in smaller bakeries. In larger bakeries the straight dough method is used for the production of variety breads.

The sponge of the sponge dough process contains from 50–70% of the flour and water and all of the yeast. The fermentation time is 3.5–5 h at 26–28 °C. It is then mixed with the rest of the flour and water and all of the other ingredients to the dough. Fig. 6 shows a scheme of this process.

American white bread (toast) is also distinguished by the use of added ingredients such as sugar, skimmilk powder, yeast nutrients, and oxidants which aid in the production of bread with a very soft, cake-like crumb. Tab. 18 shows the ingredients used in American straight doughs and sponge doughs.

6.1.3 Liquid Pre-Ferment Processes

Liquid, pumpable pre-ferments have been introduced into American bakeries since the 1950s. They simplify the conventional sponge dough process for the production of white bread, rolls and buns, and facilitate the continuous mix process. These pre-ferments contain yeast, sugars, and inorganic salts, lately also up to 70% of the flour of the ingredient

Tab. 18. Straight Dough and Sponge-Dough Formulations (PONTE, 1971)

	Straight Dough	Sponge-Dough	
		Sponge	Dough
Flour[a]	100.0	65.0	35.0
Water (variable)	65.0	40.0	25.0
Yeast	3.0	2.5	–
Yeast food	0.2–0.5	0.2–0.5	–
Salt	2.25	–	2.25
Sweetener (solid)	8–10	–	8–10
Fat	3.0	–	3.0
Nonfat milk solids	3.0	–	3.0
Softener	0.2–0.5	–	0.2–0.5
Rope and mold inihibitor	0.125	–	0.125
Dough improver	0–0.5	–	0–0.5
Enrichment	as needed	–	as needed

[a] Ingredients based on 100 parts flour

formula. They precede the true dough fermentation. The use of pre-ferments reduces labor costs and saves time. It permits better control of the production process and a more uniform quality of the final product. A single pre-ferment may be used for the production of a variety of baked goods such as bread, rolls, buns, sweet goods, doughnuts, and variety breads (CAVALIER, 1963).

The composition of the pre-ferment and the conditions of fermentation are decisive for the quality of the bread. Carbon dioxide, ethanol, organic acids, and other microbial metabolites are formed during the fermentation. Some of these contribute to the aroma of bread or are precursors of bread aroma components. The regulation of the pH is particularly important to assure an optimum dough fermentation, good machineability, and a desirable bread aroma. The production of white pan bread with a pH of 5.1–5.4 requires a pre-ferment with a pH of 4.5–5.2 (REED, 1965). This can be obtained by the use of buffer substances such as mineral salts, skimmilk powder or flour (GROSS et al., 1968). Tab. 19 shows the composition of various pre-ferments.

Fig. 7 shows a flow diagram of the preparation of a liquid pre-ferment. Yeast, sugar, salt, and yeast food are dissolved or suspended with a portion of the required water in the ingredient mix tank. The remaining water and sometimes flour is added at the blending tank. From here the liquid pre-ferment is pumped into one or alternately into two or more fermentation tanks. The temperature of the fermentation is 23–31 °C with a temperature rise of 2–5 °C during the fermentation. The fermenter liquid is stirred slowly to prevent foaming. After the desired fermentation time the liquid is cooled through a heat exchanger to 8–10 °C and kept in a refrigerated storage tank until use. It may be stored for a 24 hour period without loss of activity. The fermentation temperatures may be higher the lower the concentration of flour in the liquid pre-ferment. Fermentation losses are smaller than in comparable pre-doughs in spite of the longer time of the fermentation (ROJTER et al., 1967). Longer fermentation times are needed for liquid pre-ferments with higher concentrations of flour or sugar. The purpose of the liquid pre-ferment is basically an activation of the yeast and its acclimatization to the substrate. With 2% sugar the fermentation time is 1 h 10 min; for 4.5% sugar it is 3 h 20 min (CHOI, 1955). It also depends on the concentration of skimmilk powder, which is between 4 and 6% (SWORTFIGUER, 1955).

A process developed by the Am. Dry Milk Institute in the 1950s is based on the use of up to 6% skimmilk powder as buffer and at least 3% sugar (Tab. 20) (PIRRIE, 1954; PIRRIE and GLABAU, 1954). Today it is mainly of historical interest. The fermentation lasted 6 hours at 35–38 °C (CHOI, 1955). In the mean-

Tab. 19. Basic Differences Between Sponge-Dough Process and Liquid Pre-Ferment Dough Process

Sponge Side	Sponge	Liquid Pre-ferments			Concentrated Brew
Flour levels (%)	50–100	40–70	15–40	0–15	none
Water (%)	30– 42±	59–64	60–65	61–66	25.0
Yeast (%)	2.5	2.5	3.0	3.5	3.5
Yeast nutrients (%)	0.5±	0.5±	0.55±	0.60±	0.60±
Buffers (%)	none	0.05–0.10	0.10–125	0.1875	0.175
Sugar (%)	none	0.5	2.0±	2.5±	2.5
Salt (%)	none	none	0.25	0.25	0.50
Enrichment		Normal enrichment levels			
Weight of ferment (lbs)	variable	107–132	85–105	73–83	32
Set temperature (°C)	23–27	24–27	27–30	28–30	
Fermentation time (h)	3– 5	2– 3	2 or less		
Yeast-nutrient-oxidation (ppm)	15	20–50	25–60	35–75	
Mix reducing time					
Dough mixing time	normal	+10%	+40%	+60%	
Dough temperatures	normal (27°C)	Gradual increase to 35°C			
Dough floor time	normal (20 min)	Gradual decrease in time			

(UHRICH, 1975)

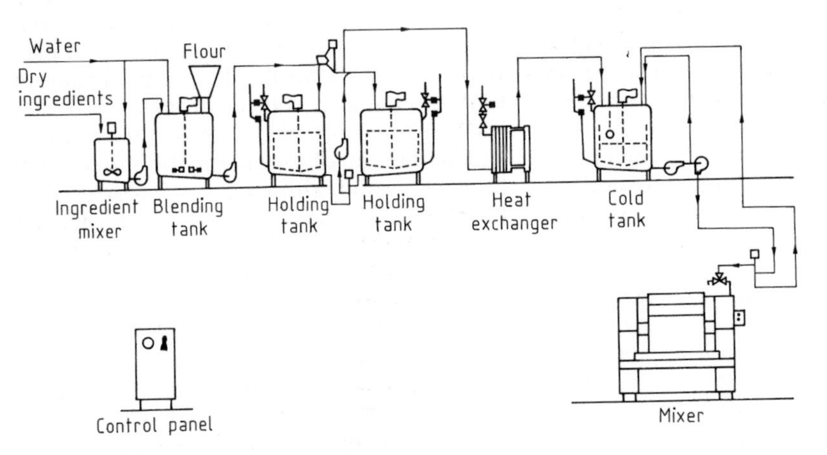

Fig. 7. Flow sheet for the liquid ferment process.

time fermentation periods have been reduced to 2–2.5 h (PYLER, 1970). The pH of the fermented liquid was between 5.14 and 5.18 (SWORTFIGUER, 1955), and it could be stored as a "stable" ferment for up to 48 h at 10°C. The use of skimmilk powder did not merely serve as a buffer. It also contributed soluble nitrogenous and phosphate compounds for yeast metabolism. The yeast does not multiply but cell volume increases by 25% (CHOI, 1955). The bacterial flora of the "stable" fer-

Tab. 20. ADMI Stable Ferment Formula (PYLER, 1970)

Water[a]	70
Yeast	2.0
Yeast food	0.5
Malt	0.4
Sugar	3.0
Nonfat dry milk	6.0
Salt	2.0

[a] Ingredients based on 100 parts flour

ments consists mainly of *Lactobacillus* spp. Several variations of this process have been introduced with lesser concentrations of skimmilk powder, the use of inorganic buffer salts, or the substitution of flour for sugar (SCARBOROUGH, 1955).

In the so-called salt buffer process the buffering action is caused by inorganic salts. A typical composition is 61.5% diammonium phosphate, 15.4% potassium sulfate, 7.7% magnesium sulfate, and 15.4% calcium carbonate (COLE et al., 1962). The liquid pre-ferment consists of ⅔ of the total water and the following ingredients: Compressed yeast, 1.5%; sugar, 2%; salt, 1.25%; dough improver, 0.25%; buffer salts, 0.2% (all concentrations based on the total weight of flour in the formula). Fermentation time is 3.5 h at 26–29 °C with constant stirring. For mixing of the final dough the following ingredients are still added: the remaining ⅓ of the water; all of the flour; yeast, 1.5%; salt, 1%; yeast nutrient, 0.5%; sugar, 6–8%; skimmilk powder, 3–6%; and fat, 2–3%.

A liquid pre-ferment may be made with only half of the total water. A formula for such a concentrated pre-ferment is shown in Tab. 21. The lesser volume saves space for the fermentation and storage tanks (CAVALIER, 1963; PYLER, 1970). Tab. 21 also shows a formula for a flour containing pre-ferment. It contains all of the required water of the dough but only a portion of the flour.

The concentrated pre-ferments are fermented at 30–35 °C for 3 h. They may be stored at 15.5 °C for up to 3 days. In contrast flour containing liquid pre-ferments must be cooled to 7–8 °C for storage since fermentation still proceeds at 15.5 °C.

Microbiological basis of liquid pre-ferments. The optimum fermentation temperature is 35 °C (COLE et al., 1962). At 30 °C the total bacterial count increases during the first two hours, remains at the same level to the 6th hour, and then increases again (Fig. 8). During the early hours the bacterial flora is derived from the flour and the compressed yeast. Thereafter species of *Lactobacillus* (*L. fermentum, L. plantarum,* and others) predominate (MILLER and JOHNSON, 1958). Similar lactic acid bacteria can also be found in conventional predoughs. Inoculation with pure cultures of *L. bulgaricus, L. fermentum, L. casei, L. plantarum, L. leichmannii,* and also buttermilk cultures produce the desired aroma of the bread (ROBINSON et al., 1958). In liquid pre-ferments about 50% of the available sugar is fermented within 1.5 h, and 90% after 5 h. At this time CO_2 formation is maximal. The pH value of pre-ferments buffered with skimmilk powder or inorganic salts remains practically constant at pH 5.2 for

Tab. 21. Concentrated Liquid Ferment and Straight Liquid Ferment Formula (CAVALIER, 1963)

	Concentrated Liquid Ferment (Basic Ferment)	Straight Liquid Ferment (Basic Ferment)
Flour[a]	–	40.0 lbs
Water	35.6 lbs	55.6 lbs
Sugar	2.5 lbs	2.5 lbs
Yeast	2.5 lbs	2.5 lbs
Salt	0.7 lbs	0.7 lbs
Buffer salt	0.2 lbs	0.2 lbs
Total weight	41.5 lbs	101.5 bls
Average fermentation loss	1.5 lbs	1.5 lbs
Final weight	40.0 lbs	100.0 lbs

[a] Formula is based on 100 lbs of flour; 50% of total water is used in basic concentrated liquid ferment, resp. nearly all the water is used in basic straight liquid ferment

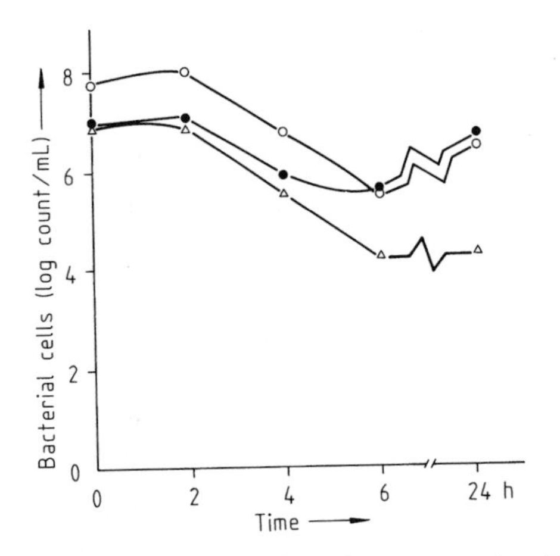

Fig. 8. Bacterial counts of pre-ferments made with yeast at 30°C. (ROBINSON et al., 1958). ○ Admi, ● salt buffered, △ sugar.

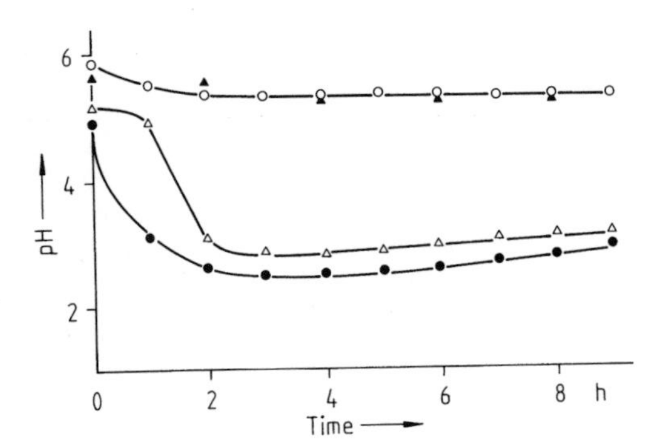

Fig. 9. Effect of fermentation time on pH of different pre-ferments. ○ dry milk buffered pre-ferment, ▲ salt buffered pre-ferment, △ sugar pre-ferment, △ flour pre-ferment.

24 h. In contrast the pH of unbuffered pre-ferments drops to 2.6 within 2 h (Fig. 9). Acid formation is basically due to lactic and acetic acids. It is proportional to the amount of sugar fermented, and reaches its highest value after about 3–5 h (COLE et al., 1962). After 8 hours of fermentation the concentration of acetic acid reaches a maximum of 50 mg/100 mL and that of lactic acid 70–80 mg/100 mL (Fig. 10). After this time the concentration of volatile carbonyl compounds remains constant or decreases while the concentration of non-volatile carbonyl compounds continues to increase.

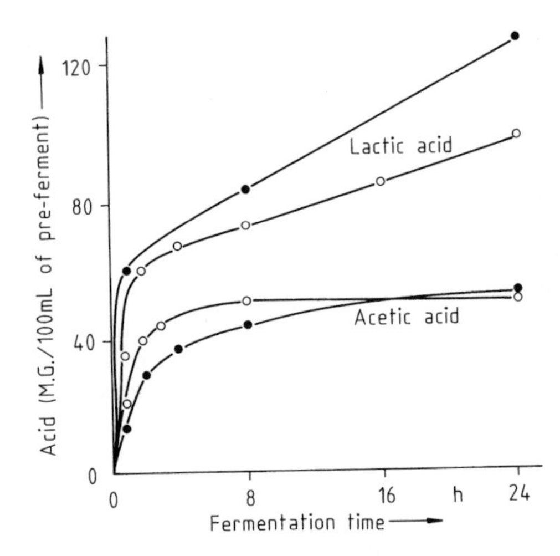

Fig. 10. Effect of fermentation time on concentration of acetic and lactic acids produced in different pre-ferments. (JOHNSON et al., 1961). ● dry milk buffered pre-ferment, ○ salt buffered pre-ferment.

Use of liquid pre-ferments in doughs. The liquid pre-ferments have the same function as sponges or pre-doughs. They are mixed with the remaining ingredients (flour or meal, water, yeast, sugar, yeast nutrients, shortening, monoglycerides or other dough conditioners) to the final dough (Tab. 22). Liquid pre-ferments which do not contain any flour do not yield a satisfactory bread aroma (WUTZEL, 1967). The taste is too strong. Also a longer re-mix time is required and the mixing tolerance is slight (Tab. 23). The addition of flour to liquid pre-ferments improves the rheological properties of the dough, the grain, the slicing properties, the taste, and the shelf life (THOMPSON, 1980; BRÜMMER, 1989).

In the U.S. and Canada the liquid pre-ferment process is used widely for the production of soft rolls (Frankfurter buns), but only few bakeries use it for the production of bread.

6.1.4 Continuous Mix Process

The development of continuous mix processes started in 1926 with the observation of SWANSON and WORKING that the enzymatic and oxidative changes in dough structure during the fermentation may be replaced by in-

Tab. 22. Typical Dough Formulations Using Sponges and Liquid Pre-Ferments

Dough Side	Sponge	Liquid Pre-Ferments			Concentrated Brew
Liquid pre-ferment (lbs)	Sponge	105–130	83–103	71–81	30
Flour (%)	0–50	30–60	60–85	85–100	100
Water (%)	22–34	0.5	0–6	0–4	41
Sugar (%)	8–10	8–10	8–10	8–10	8–10
Shortening (%)	3.0±	3.0±	3.0±	3.0±	3.0±
Hard flakes (%)	none	Some may be required in continuous mixing			
Nonfat dry milk (%)	3–6	3–6	3–6	3–6	3–6
Salt (%)	2.25	2.25	2.0	2.0	1.75
Softeners (%)	0.50±	0.50±	0.50±	0.50±	0.50±
Dough conditioners (%)	none	to 0.50	to 0.50	to 0.60	to 0.60
Enzymatic gluten mellowing agents	none	to 0.25	0.50	1.00	1.00
Additional oxidation	none	10–25 ppm	15–35 ppm	20–50 ppm	50 ppm

(UHRICH, 1975)

Tab. 23. Comparison of Sponge Dough and Liquid Pre-Ferment Processes

	Sponge Dough	Liquid Pre-Ferments		
Flour levels (%)	50 to 100	40 to 70	15 to 40	0 to 15
Yeast conditions	very good	very good	good	fair
Gas development	very good	very good	good	fair
Gluten development	very good	very good	little	none
Mixing tolerance to overmix	critical	gradual improvement to very good		
Mixing tolerance to undermix	good	gradual decrease to poor		
Tolerance to shop delays	fair	good to very good		
Space requirements	considerable	graduated to less space		
Flexibilities for handling adds and cuts	limited	very flexible		

(UHRICH, 1975)

tensive mixing of doughs at high rpm. But the practical use in automated plants for the continuous mixing of doughs began in the U.S. with the introduction of the "Do-Maker" process (BAKER, 1954). Soon thereafter the "Amflow" process was introduced (ANONYMOUS, 1958). It was recognized that intensive, high speed mixing cannot replace the development of CO_2 and of aroma substances during the fermentation. Therefore, the typical high speed mixing was combined with the use of a liquid pre-ferment. This assures a desirable aroma in the final bread. The effect of the fermentation on bread flavor has been widely discussed in the literature (COLLYER, 1966; RUSSELL-EGGITT et al., 1965; KILBORN and TIPPLES, 1968; REDFERN et al., 1968; SNELL et al., 1965).

The Do-Maker and the Amflow process have the following common features: (1) Use of a liquid pre-ferment (with or without flour in the case of the Do-Maker); (2) a premixer in which all of the ingredients of the formula are combined, namely, the liquid pre-ferment, fat, sugar, oxidants, flour, any remaining water, other minor ingredients. The homogeneous mass which has not formed a dough is pumped continuously to the dough developer; (3) a dough developer in which the dough is formed under pressure and with intensive mixing (50–200 rpm). This produces a smooth, soft dough with good gas retaining properties. It is not subjected to the normal make-up steps (dividing, rounding, intermediate proof molding) but is extruded directly into pans at a rate of 4000 to 6000 lbs/h. In the meantime there have been considerable changes in the ingredient formulation and the processing conditions of the pre-ferments. Most continuous mix plants for the production of white pan bread now include 50–60% of the total flour into the pre-ferment (THOMPSON, 1980).

The dough produced by this process is softer and much warmer (38 °C) than a comparable sponge dough. It resembles a batter more than a dough. This is the result of the high pressure and the strong shearing action during mixing, which results in a tender and sensitive dough structure. Continuously mixed doughs are quite sensitive to any manipulation after proofing. Besides, the doughs are so moist that they cannot be worked by hand. The mentioned continuous mix processes save a great deal of time in comparison with conventional processes (Fig. 11).

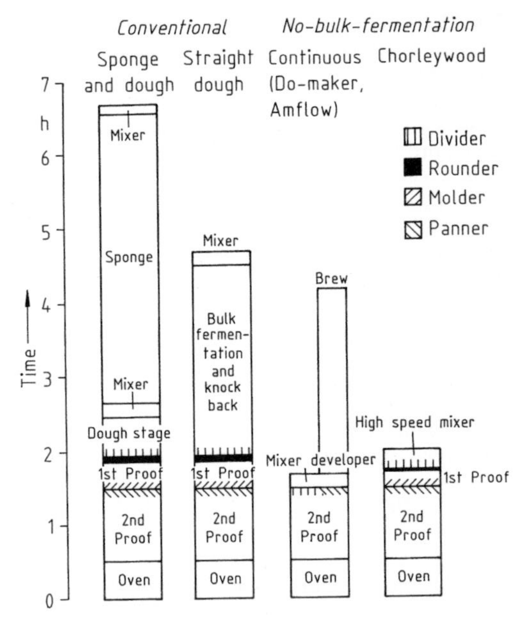

Fig. 11. Comparison of production processes and times for conventional and "no bulk fermentation" doughs. (TIPPLES, 1967).

Do-Maker process. The process has been described by BAKER (1960) and a schematic flow diagram is shown in Fig. 12. The liquid pre-ferment contains all required ingredients except the flour and the shortening (Tab. 24). The fermentation requires 2.5 h at 30–32 °C with constant, slow stirring. During the fermentation the temperature rises by 5–7 °C, and the pH drops to 4.7. The fermentation tank has a volume of 500 liter so that it can supply the mixer for a ½ h period. The liquid pre-ferment is then cooled and mixed in the pre-mixer with the flour, the shortening, and the oxidants. The ingredients are mixed for

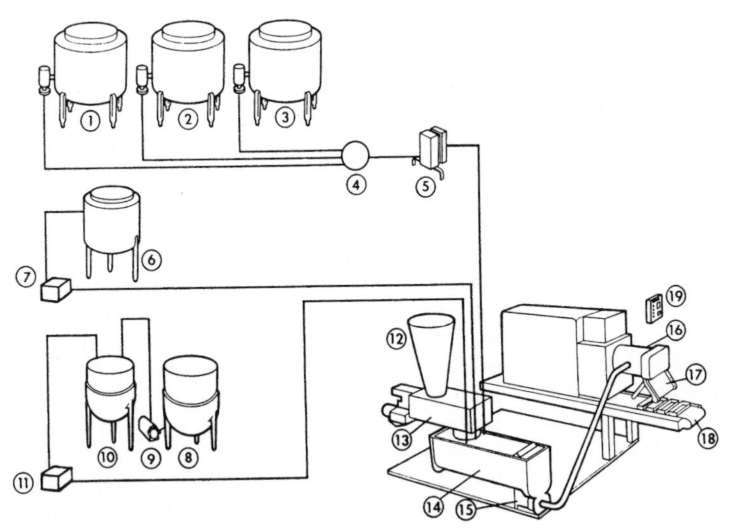

Fig. 12. Continuous mixing process flow sheet – Do-Maker process. (SNYDER, 1963). (1) Broth tank; (2) Broth tank; (3) Broth tank; (4) Broth selector valve; (5) Broth heat exchanger; (6) Oxidation solution tank; (7) Oxidation solution feeder; (8) Shortening blending kettle; (9) Shortening transfer pump; (10) Shortening holding kettle; (11) Shortening feeder; (12) Flour hopper; (13) Flour feeder; (14) Pre-mixer; (15) Dough pump; (16) Developer; (17) Divider; (18) Panner; (19) Control panel.

30–60 s and continuously pumped to the developer. The dwell time in the developer is only about 60 s. Is is continuously extruded into pans.

Amflow process. In contrast to the Do-Maker process the pre-ferment of the Amflow process contains flour. Originally 10–20% of the total flour were added to the pre-ferment (also called a liquid sponge), but at present 50–70% of the total flour are used (TRUM, 1964). A portion of the fermented, liquid sponge is returned to the start of the operation (re-cycle). Occasionally a certain amount of fully developed dough pieces is also recycled. In this respect the process differs also from the Do-Maker process (Fig. 13).

The liquid sponge is produced in two steps (Tab. 25). In the first step water, yeast, yeast nutrient, mold inhibitor (Ca propionate), and a portion of the sugar are blended to a liquid pre-ferment. This ferments for 0.5–1 h in a holding tank. If mono calcium phosphate is used in the formula it contributes to an optimization of the pH (GROSS et al., 1968). In the second step flour, salt, skimmilk powder, and additional sugar is added in a fermentation trough. This is permitted to ferment for another hour with stirring at 30–32 °C. Finally the liquid sponge is pumped into the premixer where the remaining ingredients (flour, water, sugar, fat, and oxidant) are combined. This mixture is then developed in the mixer-

Tab. 24. Continuous Breadmaking Formula – Non-Flour Brew (REDFERN et al., 1964)

Brew Ingredients[a]		Dough Composition	
Water	60.32	Flour (%)	100
Sugar	7.61	Brew (%)	74
Yeast	2.67	Shortening blend (%)	3
Yeast food	0.50	Water (%)	2
Salt	2.10	$KBrO_3$ (ppm)	60
Milk powder	0.33	KIO_3 (ppm)	12
Mold inhib.	0.10		

[a] Ingredients based on 100 parts flour

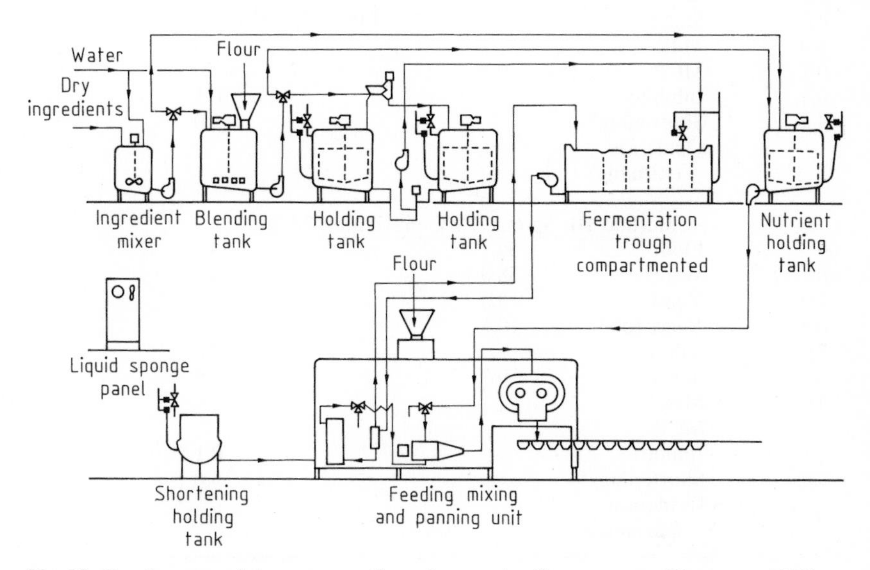

Fig. 13. Continuous mixing process flow sheet – Amflow process. (SEILING, 1969).

developer and directly extruded into baking pans.

The quality of bread produced by the continuous mix method has been the subject of an intensive debate. The grain of the bread is very uniform and fine, although it could be modified by injection of air into the dough (FORTMANN et al., 1966). A modification of the Do-Maker process permits production of a molded loaf (RUSSELL-EGGITT et al., 1965). The taste of the continuously mixed bread differs from that of the conventional sponge

Tab. 25. Three Procedures with Ingredients Addedat Different Phases (TRUM, 1964)

Formula (%)		Phase I (%)	Phase II (%)	Mixing Phase (%)
		Method I: Sponge 28–29 °C, 2.5 h		
100	Flour	20		80.0
68	Water	56	7.0	5.0
3.0	Yeast	3.0		
0.5	Yeast food	0.5		
2.0	Salt		2.0	
6.0	Sugar		1.0	5.0
3.0	Milk		3.0	
0.2	MCP	0.2		
0.1	Inhibitor		0.1	
3.0	Shortening			3.0
	Oxidation (variable)			
		Method II: Sponge 26–28 °C, 2.5 h		
100	Flour	30		70
67	Water	56	4.0	7.0
3.0	Yeast	30		
0.5	Yeast food	0.5		
2.0	Salt		2.0	
6.0	Sugar		1.0	5.0
3.0	Milk			3.0
0.1	MCP	0.1		
0.1	Inhibitor			0.1
3.0	Shortening			3.0
	Oxidation (variable)			
		Method III: Sponge 25 °C maximum, 2.5 h		
100	Flour	50		50
66	Water	58.0		8.0
3.0	Yeast	3.0		
0.5	Yeast food	0.5		
2.0	Salt			2.0
6.0	Sugar	0.5		5.5
3.0	Milk			3.0
	MCP			
0.1	Inhibitor			0.1
3.0	Shortening			3.0
	Oxidation (variable)			

(all values expressed as percent of total amount of flour)

dough bread (BRÜMMER and MORGENSTERN, 1988). The mentioned processes have found their widest application in the U.S. for the production of white pan bread. In Germany these processes have not been successful because consumers have not generally accepted this type of bread (SEIBEL, 1979).

6.1.5 No Time Doughs

Dough fermentation and dough maturing may be accelerated by the addition of reducing and oxidizing substances. This has led to the development of "no time doughs" in which dough maturation is obtained by mechanical and chemical means rather than by prolonged fermentation (TIPPLES, 1967; FINK, 1968). The "no time doughs", more appropriately called "short time doughs" have the advantage of a shorter production period.

Chorleywood bread process. This process was introduced in 1961 by the "British Baking Industries Research Association". In the production of wheat bread it substitutes mechanical and in part chemical development for biological maturation of the dough (CHAMBERLAIN et al., 1962, 1965). STEWART reported in 1972 that 75% of the wheat bread produced in Great Britain was made by this process. The significant process is mixing of the dough intensively in a batch, high speed mixer (Tweedy kneader) within 3–5 min with control of the energy input. The optimum work input is 11 watt hours per kg dough (40 Joule/g). However, the energy requirements differ for flours of varying quality and the type of bread (ELTON, 1965). With increasing amounts of rye flour in the formula the required energy for mixing decreases.

Mechanical dough development is supported by relatively high concentrations of oxidants. A mixture of fast acting and slow acting oxidants is particularly useful so that oxidative reactions can proceed during proofing and even during the first stages of baking (TIPPLES, 1967). Depending on the quality of the flour 10–20 ppm potassium iodate and 45–55 ppm potassium bromate are used. A similar effect can be obtained with 75 ppm ascorbic acid. But this requires some intermediate proof time (Tabs. 26, 27). The addition of shortening, about 0.7%, during mechanical dough development is advantageous (ELTON et al., 1966). The melting point of the shortening, which should be higher than the dough temperature, is more important than the concentration used (ELTON, 1965). The concentration of yeast should be increased by 50–100% (in comparison with traditional formulations) in order to obtain normal fermentation times (CHAMBERLAIN et al., 1965). The process does not include a pre-ferment stage. The fermentation loss of 0.4% for the Chorleywood process is lower than the loss in conventional dough processes (1%) because of the absence of bulk fermentation.

The Tweedy mixer permits mixing of the dough under a vacuum of 380–400 mm Hg (AXFORD et al., 1963). This leads to the formation of a fine, silky grain (DE RUITER, 1968). The Chorleywood process may be carried out with continuous extrusion of dough directly into baking pans. This produces bread with a very fine grain. However, if dough development is followed by traditional make-up procedures (with a 5–8 min intermediate proof) the grain resembles that of conventional bread.

Brimec process. A process for largely mechanical dough development and maturation has been developed by the Bread Research Institute of Australia (MARSTON, 1966). In contrast to the Chorleywood process dough mixing is carried out with high speed or low speed mixers of traditional design. The addition of 1% shortening and of high levels of oxidants is required to produce good bread. Potassium bromate (30 ppm), potassium iodate (20 ppm), ascorbic acid (100 ppm) or L-cysteine are used (BOND, 1967). Yeast is added directly to the dough without activation in a pre-dough or a liquid pre-ferment (Tab. 28).

Tab. 26. Chorleywood Bread Process Formula[a]

Flour	100
Water	61
Yeast	1.8 to 2.0
Salt	1.8 to 2.0
Fat	0.7
Ascorbic acid	75 ppm

[a] Ingredients based on 100 parts flour

Tab. 27. Effect of Ascorbic Acid and Intermediate Proof on the Quality of Bread (DE RUITER, 1968)

Ascorbic Acid (ppm)	0	0	40	40
Intermediate Proof (min)	0	10	0	10
Loaf volume (mL/kg flour)	5300	5770	5590	6440
Browning[a]	5	6	6	6
Shape[a]	3	4	4	6
Crumb texture[a]	3	4	5	7
Crumb color[a]	3	4	5	6

[a] Maximum attainable score is 10 points

Tab. 28. Recipe for Production of White Bread According to the Brimec Process

Ingredients	No Time Dough (ascorbic acid)	Traditional Process (3 h fermentation)
Flour	50 kg	50 kg
Water	28.5 kg	28 kg
Compressed yeast	1 kg	0.75 kg
Salt	1 kg	1 kg
Fat	1 kg	1 kg
Malted flour	0.25 kg	0.25 kg
Ascorbic acid	100 ppm	–
Potassium bromate	10–30 ppm	7–10 ppm
Ammonium chloride	300 ppm	300 ppm

(BOND, 1967)

A special mixer, the Brimec mixer, has been developed for this process. It can knead at two different speeds, 100 rpm and 150 rpm. The mixing chamber may be closed by a piston-like cover. This permits working at a pressure of about 1.5 atm. With this pressure and with direct extrusion of the dough into baking pans a bread with very fine grain may be produced. On the other hand, mixing under atmospheric pressure followed by traditional make-up and proofing produces so-called "aerated bread" with an open grain (BOND, 1967). The energy required for dough development is about 9 watt hours per kg of dough.

The distance of the cover of the mixer from the dough determines the time required and the extent of dough development. If the cover is close to the dough energy can be transferred to the dough more efficiently.

6.1.6 Frozen Yeast Leavened Doughs

The freezing of dough pieces for later use, for instance, over weekends has been practiced by wholesale bakers for some time. It is also useful for in-store bakeries which can be supplied with the frozen dough pieces by a central bakery. In either case the use of frozen doughs allows greater flexibility in operation and permits a greater output in a given time period. But frozen doughs deteriorate during frozen storage. Proof times increase over time, loaf volumes decrease, the texture worsens and the grain is coarser.

These changes are principally due to a decrease in gassing power or to the death of yeast cells, which can be readily determined by thawing dough pieces and determination of their CO_2 production rate. There is also a deleterious change in the rheological properties of the doughs as can be shown by determinations of their extensibility.

The loss of yeast activity is not simply due to the effect of the freezing on the yeast. Compressed yeast can be frozen and thawed again with only a slight loss in activity. But during the frozen storage of a slightly or fully fermented dough piece the loss of yeast activity can be readily observed. At present there is no entirely satisfactory explanation for this effect. The number of variables which affect survival in frozen doughs is very large. Among these variables are the size of the dough piece, freezing rate, temperature of frozen storage, number of freeze/thaw cycles, presence of fermentation metabolites, formulation, and yeast concentration, to mention only the obvious ones.

The storage of doughs for later processing is more successful for small dough pieces used in the production of rolls and buns than for bread doughs. Tab. 29 shows the possibilities of attaining various storage periods by reduction of the yeast concentration, by lowering of the dough temperature (retardation) or by freezing (BRÜMMER, 1993).

As one would expect, the rate of freezing of the dough pieces and the temperature of frozen storage affect storage stability. Freezing of small dough pieces in liquid nitrogen kills all yeast cells. In general slower freezing (test 1: 90 min to −20°C) is preferable to faster freezing (test 2: 20 min to −20°C). The number of surviving yeast cells in tests 1 and 2 was 77% and 62%, respectively, after 1 day storage and 61% and 51%, resp., after one week (BRÜMMER, 1993).

Storage stability is also affected by the formulation. The use of strong flours or the addition of vital wheat gluten is preferred. Yeasted dough pieces which have been frozen before any fermentation takes place are preferred for frozen storage periods exceeding one week. Active dry yeast has sometimes been preferred over compressed yeast because it starts the fermentation more slowly.

In the U.S. the freezing of larger dough pieces for the production of white bread is common practice, mainly for use in in-store bakeries. The technical problems of such operations have been described by REED and NAGODAWITHANA (1991).

6.2 Rye Dough (Sour Dough) Processes

Bread doughs containing more than 20% rye flour require acidification (Fig. 14). This is a necessary step for (1) the technical quality of rye containing breads (swelling and baking; enzymatic activity, grain, elasticity, and suitability for slicing of the crumb, shelf life) and (2) the development of the characteristic flavor and aroma (lactic acid, acetic acid, precursors of aroma substances) and (3) the inhibition of undesirable fermentations caused by bacteria and yeasts of the flour. Further the lactic acid bacteria of the sour dough (4) have a synergistic effect with the yeasts and (5) inhibit the growth of molds and of rope (*Bacillus mesentericus*).

In general a higher percentage of the used rye flour has to be acidified the smaller the

Tab. 29. Storage of Yeasted Doughs Used in the Production of Buns and Rolls

Fermentation	Yeast Concn. (%)	Dough Temp. (°C)	Storage Temp. (°C)	Permissible Storage Time
Long ferm.	1	26	20 to 25	4 h
Long ferm.	4–5	26	4 to 8	8 h
Long ferm.	1	26	4 to 8	15 h
Retard	4–5	26	−4 to 4	24 h
Freeze	3–5	26	−15 to −20	days/weeks/months

(BRÜMMER, 1993)

percentage of rye flour in the formula. Rye breads require acidification of 35–45% of the rye flour; mixed rye breads require acidification of 40–60% and mixed wheat breads 60–100% of the rye flour. The use of very dark rye flour or flour high in maltose concentration requires acidification of 5–10% more of the rye flour. Depending on the type of bread and its content of rye flour a final pH of 4.7–4.2 and a final degree of acidity from 6.0–14.0 (mL 0.1 N NaOH/10 g) should be obtained (Tab. 30).

Acidification is achieved by various sour dough processes. These are distinguished by a particular ratio of flour to water (firmness and yield of the dough), the amount of "seed" dough used, the ratio of flour of the preceding stage to the flour of the following stage (ratio of flour increase), the temperature and the time of fermentation. The traditional sour dough processes are carried out in stages. Two stage and single stage sour dough processes have been developed by omission of one or more stages. Multiple stage sour

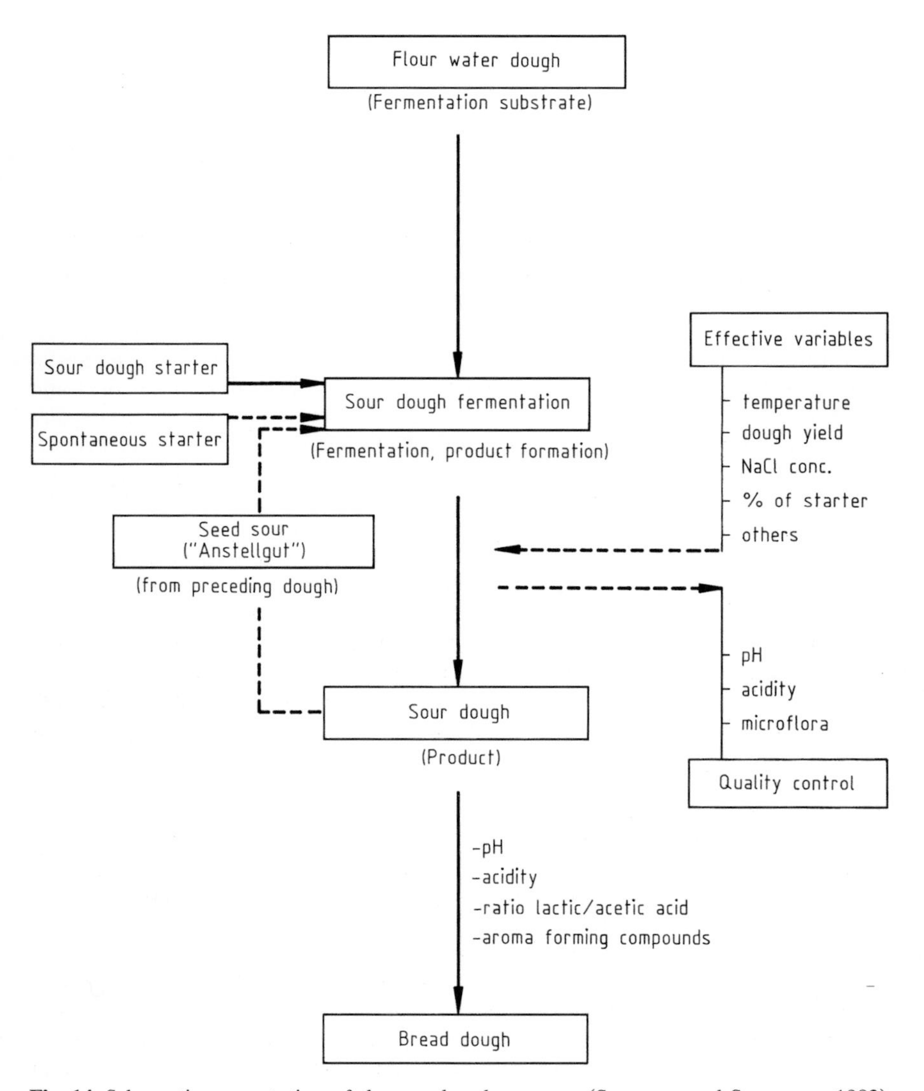

Fig. 14. Schematic presentation of the rye dough process. (SPICHER and STEPHAN, 1982).

Tab. 30. Required pH and Acidity in Breads Made with Rye Flour or Cracked Grain Rye (SPICHER and STEPHAN, 1982)

Bread Type	Recommended Specification pH	Acidity[a]
Rye bread	4.20–4.30	8.0–10.0
Rye mixed grain bread (50–89% rye)	4.30–4.40	7.0– 9.0
Wheat mixed grain bread (10–49% rye)	4.65–4.75	6.0– 8.0
Rye cracked grain bread	4.00–4.60	8.0–14.0
Whole rye bread	4.00–4.60	8.0–14.0

[a] mL 0.1 N NaOH per 10 g

Tab. 31. Percentage of Sour Dough Used with Rye Flour (Type 997) (SPICHER and STEPHAN, 1982)

Sour Dough Process	Total Time of Process (h)	Fermentation Time of Full Sour (h)	Degree of Acidity of Mature Full Sour[a]	pH of Mature Full Sour	Sour Dough Used (%)
Berlin short sour	3– 4	3–4	9.5	4.1	45–50
2 and 3 Stage sour process (basic sour overnight)	15–20	3	10.5	4.0	40–45
3 Stage sour dough process (full sour overnight)	20	8	13	3.9	35–40
Detmold single stage sour dough process	15–20	17	17	3.8	30–35
Monheim salt sour dough process	18–48	48	20	3.6	25–30

[a] mL 0.1 N NaOH/10 g

dough processes require a separate "sour" for each dough. For single stage sour doughs the same "sour" may be used for several doughs. Two stage and single stage sour doughs are more tolerant to variations of the fermentation time. The purpose is merely acidification of the doughs. Baker's yeast has to be added to such doughs because they do not develop their own yeast flora and lack leavening. The schemes of the various sour dough processes may be changed within limits to permit adjustments for the swelling capacity of the flour, the type of bread produced, the conditions in the bake shop and the sour dough equipment (BRÜMMER, 1991).

The final dough may contain variable percentages of the "full sour". This percentage is important for physical characteristics of the bread and for its taste and aroma. The percentage of the "full sour" in the final dough depends on the type of bread but should be at least 30% (Tab. 31).

6.2.1 Seed

For seeding one may use a commercial sour dough starter culture, a spontaneous sour, or a portion of a preceding sour dough (see also Fig. 15).

Sour dough starter culture. The development of such commercial cultures over several decades has led to a certain selection of organisms (see Tab. 9). However, the use of pure cultures as it is common in other branches of the fermentation industry (for instance, baker's yeast) has not been successful.

Spontaneous sour. This is an older method. Today it is used mainly for the production of some specialty breads (ROTSCH and SCHULZ, 1958). A dough prepared from ground grain and water is permitted to stand at 26–35 °C (Tab. 32). The start of the fermentation is evident from gas formation and an acid odor. This dough is then enriched several times with flour and water. After several days a sour dough with a pH of 3.6–3.9 and a degree of acidity of 14.0 to 20.0 develops. The microflora of such spontaneous sours consists of $3.7–7.5 \cdot 10^9$ bacteria and $1.0 \cdot 10^6$ to $3.0 \cdot 10^8$ yeasts per g of dough.

Seed sour. This seed material will be designated as "Anstellgut". It is usually taken from the full sour of a preceding fermentation. It is used at a concentration of 0.5% for the multiple stage sour (overnight) and 20% for the Berlin short sour or the salt sour process. A seed sour may be kept at 4 °C up to 4 days for a short sour process and even longer for a multiple stage process. It may also be kept alive for several days or weeks by kneading with additional flour or cracked grain until it dries out (crumbly sour).

6.2.2 Sour Dough Processes With Several Stages

There are several variations of this process. The multiple stage process serves the growth of both lactic acid bacteria and yeasts. In Germany it is the most widely used process in larger bakeries (Fig. 15).

The required percentage of seed sour (0.5–1.0%) is usually taken from a mature full sour. It is enriched with some flour and water to form the "fresh sour". The fresh sour serves mainly for the growth of yeast and is, therefore, kept cool (25–26 °C) and soft (dough yield 200–240). All dough yields are expressed as weight of dough divided by the weight of flour times 100. It ferments for about 6 hours. Addition of flour and water leads to the basic sour. At a temperature of 26–30 °C and a dough yield of 160–180 lactic acid bacteria develop optimally in 6–8 hours with enzymatically active rye flour and 8–10 hours with enzymatically inactive rye flour. Rye meals need longer time periods for optimal acidification. This is the principal souring stage. Temperature and firmness of the dough determine the ratio of lactic to acetic acid which is in turn decisive for the flavor of the bread. Cold and firm doughs favor the development of acetic acid; soft and warm doughs that of lactic acid. The final stage, the "full sour", matures within 2.5–3 h. It is soft (dough yield 180–200) and warm (30–33 °C). The amount of flour used for each succeeding stage depends on the fermentation times used. As a rule the amount of flour added for the next stage is doubled if the succeeding stage ferments for 2 hours, tripled for a 3 hour fermentation, and so on.

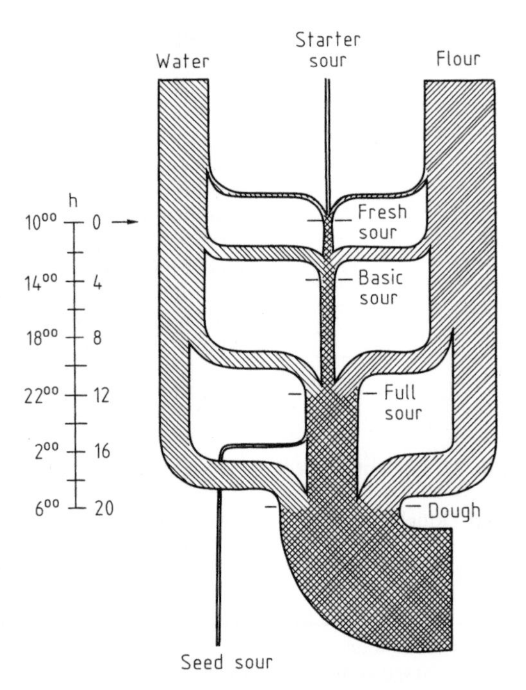

Fig. 15. Multi stage sour dough process flow sheet.

A three stage sour process takes about 16–24 h. Yeasts usually multiply sufficiently, but 0.5–1% compressed yeast are usually added to the final dough.

Basic sour overnight. This is a classical multiple stage process. Tab. 33 shows the details of the processing steps. Bread with a mild flavor is obtained with a soft and warm full sour which has fermented for 3 h.

Tab. 32. Scheme for the Spontaneous Sour Dough Process

	Preceding Sour (g)	Rye Flour or Cracked Grain (g)	Water (mL)	Total Weight (g)	Temp. (°C)	Time of Fermentation (h)
1st Stage	–	500	1000	1500	26–28	24
2nd Stage	500	500	500	1500	26–28	24
3rd Stage	500	500	500	1500	32–35	8
4th Stage	500	500	500	1500	26–28	16

(SPICHER and STEPHAN, 1982)

Tab. 33. Scheme of a Three Stage Sour Dough Process for Production of Rye Mixed Grain Bread (Basic Sour Overnight) Consisting of 70% Rye Flour and 30% Wheat Flour

Stage	Duration Clock Time	Sour Dough (kg)	Rye Flour (kg)	Wheat Flour (kg)	Water (L)	Dough Yield[a]	Sour Dough Temp. (°C)
Seed sour	14–20	0.3	1.0	–	1.0	200	25–26
Basic sour	20–04	2.0	11.0	–	6.0	158	23–27
Full sour	04–07	19.0	24.0	–	26.5	190	28–31
Dough[b]	07	68.5	34.0	34.0	35.5	168	28

[a] In relation to 100 parts of flour
[b] Yeast 1.0 kg; salt 1.8 kg; dough rest 5 min

Full sour overnight process. The fresh sour is fermented at 25–26°C and with a dough yield of 200 for 5 h. The basic sour (dough yield 160, temp. 23–27°C) ferments for 7 h. The full sour which ferments overnight for about 8 h must be kept rather firm (dough yield 160–170) and cool (25–28°C) to prevent excessive souring.

Foamy sour process. This is a five stage sour dough process (LUBIG, 1949). Almost all of the water of the final dough is already used in the preparation of the sour stages which improves swelling of the flour and gelatinization during baking. The amount of sour dough starter is 0.5–1.5% of the total weight of flour. The individual stages are soft (dough yield about 300) and cool (22–24°C). At the start of each stage air is beaten into the dough. The full sour contains 40–50% of the rye flour of the final dough.

6.2.3 Two Stage Sour Dough Process

This is a simplified process (STEPHAN, 1959). Either the fresh sour or the basic sour can be omitted depending on whether the full sour is fermented for 3–4 h over day or 8–24 h over night. The shorter stage is always fermented softer and warmer, the overnight stage firmer and cooler. This permits two variations. The first one with a short "full sour" fermentation is shown in Tab. 34. The second one consists of a fresh sour (dough yield 200, 25–26°C, 6 h) and a "full sour" (dough yield 170, 23–24°C, 8 h). In both cases 0.4–1.2% compressed yeast must be added to the final dough.

Tab. 34. Scheme of a Two Stage Sour Dough Process for the Production of Mixed Grain Bread with a Fermentation Time of 2.5 to 3.5 hours for the "Full Sour"

	Rye Mixed Grain Bread	Wheat Mixed Grain Bread	Process Conditions			
Full sour in dough (% of total flour)	40%	20%				
Rye flour in sour dough	40.0 kg	20.0 kg				
Seed sour 2.5% of rye flour in dough	1.0 kg	0.5 kg				
1st Stage: basic sour						
Seed sour	1.0 kg	0.5 kg	dough yield 150; temperature 22–26 °C; fermentation time 15–24 h			
Rye flour (16×)	16.0 kg	8.0 kg				
Water (8×)	8.0 L	4.0 L				
Total without starter	24.0 kg	12.0 kg				
2nd Stage: full sour						
Basic sour	24.0 kg	12.0 kg	dough yield 180;			
Flour (1×)	24.0 kg	12.0 kg	temperature	ferm. time	taste of bread	
Water (1×)	24.0 kg	12.0 kg	33 °C	2½ h	milder	
Total	72.0 kg	36.0 kg	31 °C	2¾ h	more acid	
			29 °C	3 h		
			27 °C	3½ h		
Dough						
Full sour	72.0 kg	36.0 kg			rye mixed grain bread	wheat mixed grain bread
Rye flour	40.0 kg	20.0 kg				
Wheat flour	20.0 kg	60.0 kg	dough yield (theory)		168	163
Yeast	1.8 kg	2.0 kg	dough temperature		28 °C	26 °C
Salt	1.8 kg	2.0 kg	dough rest time		5 min	5 min
Water, about	36.0 L	47.0 L				
Total	171.6 kg	167.0 kg				

6.2.4 Single Stage Sour Doughs

This process has the advantage of a greater tolerance with regard to the fermentation time of the sour dough (15–24 h). It can be accomplished by proper choice of the amount of seed sour taken from the full sour (Detmold single stage sour dough process) or by the addition of salt which inhibits acid formation (salt sour dough process).

Detmold single stage process. The details of the process are shown in Tab. 35 (STEPHAN, 1957). The temperature and the firmness of the single stage are particularly important because of the long fermentation time. The amount of starter used depends on the temperature. It is 20% at 20–23 °C and as little as 2% at 27–28 °C.

Berlin short sour dough process. This is the simplest and shortest process (PELSHENKE, 1941). The single stage sour can be ready within 3–4 h. This can be achieved by use of a high percentage of seed sour (20% of the amount of flour used), a high temperature (35–36 °C) and a soft consistency (dough yield 190) (Tab. 36). Fermentation losses are only half of those for a multiple stage sour dough process. 1.0 or 1.5% baker's compressed yeast is added to the dough for rye breads and mixed rye breads, respectively. The degree of acidity obtained in the sour is sufficient if the dough contains 50–70% of the sour.

Tab. 35. Scheme of a Detmold One Stage Sour Dough Process for the Production of Rye Mixed Grain Bread (70% Rye Flour, 30% Wheat Flour) and Wheat Mixed Grain Bread (70% Wheat Flour, 30% Rye Flour)

	Rye Mixed Grain Bread	Wheat Mixed Grain Bread	Process Conditions		
Full sour in dough (% of total flour)	25%	15%			
Detmold 1 stage sour					
Rye flour	25.0 kg	15.0 kg	dough yield 180;		
Seed sour	0.5 kg	0.3 kg	temp. at start 31 °C;		
Water	20.0 L	12.0 L	at end 23 °C;		
Total without starter	45.0 kg	27.0 kg	fermentation time 15–24 h		
Dough					
Detmold 1 stage sour	45.0 kg	27.0 kg		rye mixed grain bread	wheat mixed grain bread
Rye flour	45.0 kg	15.0 kg			
Wheat flour	30.0 kg	70.0 kg	dough yield (theory)	168	163
Yeast	1.8 kg	2.0 kg	dough temperature	28 °C	26 °C
Salt	1.8 kg	2.0 kg	dough rest time	5 min	5 min
Water, about	48.0 L	51.0 L			
Total	171.6 kg	167.0 kg			

Tab. 36. Scheme of a Berlin Short (One Stage) Sour Dough Process for the Production of Rye Mixed Grain Bread (70% Rye Flour, 30% Wheat Flour) and Wheat Mixed Grain Bread (70% Wheat Flour, 30% Rye Flour)

	Rye Mixed Grain Bread	Wheat Mixed Grain Bread	Process Conditions		
Full sour in dough (% of total flour)	40%	20%			
Berlin short sour					
Rye flour	40.0 kg	20.0 kg	dough yield 190;		
Seed sour	8.0 kg	4.0 kg	temperature 35 °C;		
Water	36.0 L	18.0 L	ferm. time 3–4 h		
Total without sour starter	76.0 kg	38.0 kg			
Dough					
Berlin short sour	76.0 kg	38.0 kg		rye mixed grain bread	wheat mixed grain bread
Rye flour	30.0 kg	10.0 kg			
Wheat flour	30.0 kg	70.0 kg	dough yield (theory)	168	163
Yeast	2.0 kg	2.0 kg	dough temperature	28 °C	26 °C
Salt	2.0 kg	2.0 kg	dough rest time	5 min	5 min
Water, about	32.0 L	45.0 L			
Total	172.0 kg	167.0 kg			

6.2.5 Salt Sour Processes

In these processes salt is added already to the single stage sour. With a desired concentration of 1–2% of salt in the final dough up to 6% salt may be added to the sour (based on the weight of the flour in the sour). This permits a regulation of the sour dough fermentation. Fermentation time may be varied from 1–80 h without excessive souring or fermentation losses. The soft consistency of the sour (dough yield 200) permits fermentation in a tank. Such sours are also useful for continuous dough processes.

Salt sour process. The process requires use of a starter of one third of the amount of rye flour that must be acidified. All of the salt required later for the dough is already added to the sour. The fermentation temperature is 35–37 °C (VOM STEIN, 1958), and a fermentation time of 12 h is required. The amount of the sour used for preparation of the dough may be reduced from 35% to 25% if the sour is held for longer time periods. Yeasts are also inhibited by the addition of salt and twice the normal amount must be added to the dough (2.5–3.5%).

Monheim salt sour process. This is a variant of the preceding process. With about the same formulation (dough yield 200, seed sour 20%, salt 2%) the sour is fermented for 18–24 h with a decreasing temperature during this time. Details of the process are shown in Tab. 37. Sour doughs made by this process can be kept for several days without a change of pH or the degree of acidity and without a noticeable effect on bread quality (VOM STEIN, 1971).

6.2.6 Long-Time Sour Dough Process

From the traditional single stage sour dough process, which was carried out overnight, the special long-time sour dough processes have been developed. Normally a sour dough is prepared over the weekend, so that it is available for production on Monday morning. At this time acidification has been

Tab. 37. Scheme of a Monheim Salt-Sour One Stage Sour Dough Process for the Production of Rye Mixed Grain Bread (70% Rye Flour, 30% Wheat Flour) and Wheat Mixed Grain Bread (70% Wheat Flour, 30% Rye Flour)

	Rye Mixed Grain Bread	Wheat Mixed Grain Bread	Process Conditions		
Full sour in dough (% of total flour)	30%	20%			
Monheim salt-sour					
Rye flour	30.0 kg	20.0 kg	dough yield	200	
Seed sour	6.0 kg	4.0 kg	temp. at start	30–35 °C	
Salt	0.6 kg	0.4 kg	at end	20–25 °C	
Water	30.0 L	20.0 L	fermentation time	18–24 h	
Total without starter	60.6 kg	40.0 kg			
Dough				rye mixed grain bread	wheat mixed grain bread
Salt-sour	60.6 kg	40.4 kg			
Rye flour	40.0 kg	10.0 kg			
Wheat flour	30.0 kg	70.0 kg	dough yield (theory)	168	163
Yeast	2.0 kg	2.0 kg	dough temperature	28 °C	26 °C
Salt	1.2 kg	1.6 kg	dough rest period	5 min	5 min
Water, about	38.0 L	43.0 L			
Total	171.8 kg	167.0 kg			

largely completed. It has been suggested that such sour doughs can be preserved for up to one week, which avoids the daily preparation of fresh sour doughs. But this creates some problems since undesirable fermentations can take place during too long a period of preservation. These can lead to off-flavors in the bread. Also, such sour doughs have high acid concentrations which means that they are used at lesser percentages of the final dough. This clearly reduces the positive effect on shelf life and aroma of the bread (BRÜMMER, 1990). However, the trade has accepted the idea of the longer preservation of sour doughs; not for a whole week but certainly for two and certainly no more than 3 days. Thereafter a new sour dough is used. Longer storage times than 2 or 3 days require cooling of the sour dough after 2 days of ripening and subsequent storage at 10°C.

6.2.7 Freeze-Dried Sour Dough

This product can be obtained by a normal three stage sour dough process by freeze- or drum-drying. At the end of the fermentation the sour dough is frozen to an interior temperature of −10 to −30°C and dried in vacuo. The freeze-dried preparation is commercially available (Ulmer full sour) with a pH of 3.8 and a degree of acidity of 40. It contains viable lactic acid bacteria and the aromatic substances characteristic of sour doughs. The freeze-dried sour dough replaces the acidified rye flour in the formulation. For instance, for a rye mixed bread with 70% rye flour and 30% wheat flour the following proportions are used: freeze-dried sour dough 20–25%; rye flour type 997 or 1150 50–45%; wheat flour type 812 or 1050 30%. Dried sour doughs need more time for re-activation (6–8 h) than undried ones (2–4 h) (BRÜMMER and STEPHAN, 1986; SPICHER et al., 1990).

6.2.8 Effect of Processing Conditions on the Sour Dough Fermentation

The sour dough is a complex biological system. The factors which act in the system are either processing factors such as the temperature, the gas phase (O_2/CO_2 concentration), or properties of the fermentation substrate such as the flour type, the pH, salt or water concentration, or antimicrobial compounds. To this one must add biological factors of the microflora and the interaction between the different microorganisms (Fig. 16). All of these variables affect the production of acids as well as the ratio of non-volatile lactic acid to volatile acetic acid and the amount of carbon dioxide formed.

Temperature. The optimal temperature for the multiplication of lactic acid bacteria of sour doughs is between 30 and 35°C. Outside of this range the generation time is highly variable (Fig. 17). Optimal temperatures for growth are not necessarily optimal for acid formation (Tab. 38). The temperature also affects the relative amounts of lactic and acetic acid formed. A sour dough produced by a commercial sour dough starter and a total process time of 24 hours usually has 0.25–0.30% acetic acid and 0.80–0.97% lactic acid per 100 g of sour dough (Fig. 18). But there are definite differences between various species of lactic acid bacteria with regard to the ratio of lactic to acetic acid formed (ROHRLICH and ESSNER, 1951) (see also Tab. 39).

The ratio of D- to L-lactic acid formed also depends on the species (SPICHER and RABE, 1980, 1981). In sour doughs produced with a "pure culture starter" a relatively constant ratio of 59:41 at 35°C and of 62:38 at 30°C can be found. The same ratios are also found for *Lactobacillus brevis* spp. *lindneri* (Fig. 19). The lactic acid bacteria occurring in sour dough may be divided into 3 groups: (1) those forming D(+)-lactic acid such as *L. delbrueckii, L. lactis,* and *L. plantarum;* (2) those which form largely L(+)-lactic acid with only small percentages of the D-isomer such as *L. casei* spp. *casei, L. casei* spp. *rhamnosus, L. farciminis,* and (3) those which form DL-lactic acid. The latter do indeed form the racemate but depending on the growth phase one or the other isomer may predominate. *Lactobacillus acidophilus, L. casei* spp. *pseudoplantarum, L. fermentum, L. brevis,* and *L. fructivorans* form an excess of the L(+)-isomer which diminishes with the age of the culture. *L. plantarum* spp. *plantarum* and *L. plantarum*

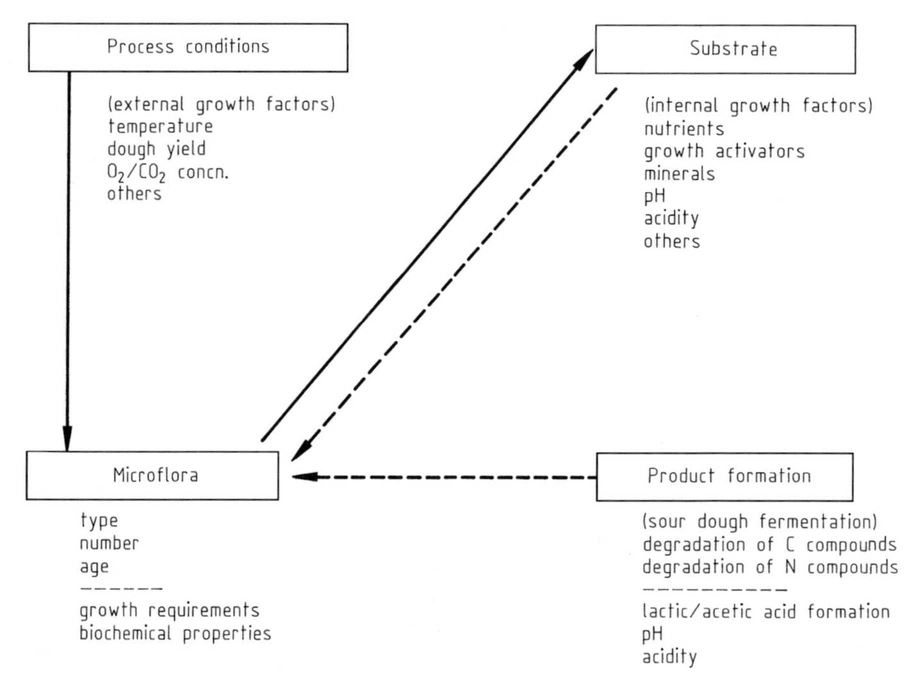

Fig. 16. Variables affecting the sour doughs fermentation.

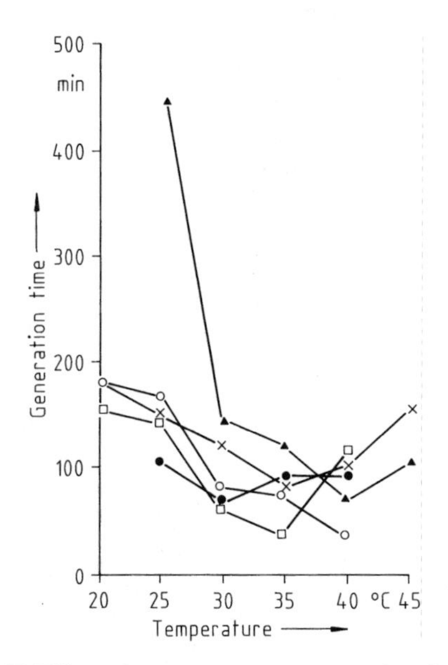

Fig. 17. Effect of temperature on generation time of lactic acid bacteria. (SPICHER et al., 1982). × *Lactobacillus brevis* var. *lindneri;* □ *L. brevis;* ● *L. fructivorans;* ▲ *L. fermentum;* ○ *L. plantarum.*

spp. *arabinosus* form the racemate or only a slight excess of the D(−)-isomer. *Lactobacillus buchneri* is characterized by a considerable excess of the D(−)-isomer during the early growth phase, an excess which diminishes with the age of the culture.

Oxygen. Oxygen is important with regard to the foamy sour process (LUBIG, 1949). The number of viable bacteria is twice as high as in sour doughs of more common processes (Tab. 40). The incorporation of air into the dough also causes a shift in the microflora. *L. brevis* is more prevalent while homofermentative species such as *L. plantarum* are inhibited (PANZER, 1956). *L. brevis* shows a certain increase of acid formation in the presence of oxygen. In contrast lactic acid fermentation by *L. plantarum* is stimulated in substrates saturated with CO_2 or N_2 (ROHRLICH et al., 1959).

Sodium chloride. Addition of the smallest amounts of NaCl (e.g., 0.1% based on flour) affect the sour dough fermentation. Acid formation is inhibited with increasing concentrations of salt. Lactic acid bacteria differ in their

Tab. 38. Temperature Optima for the Multiplication and Acid Formation by Lactic Acid Bacteria of Sour Dough

	Lactobacillus plantarum	*L. brevis*	*L. fermenti*
Fastest growth (shortest generation time)	35°C	35°C	40°C
Highest cell counts	35°C	30–35°C	30–35°C
Fastest production of acid	35°C	30–35°C	40°C
Greatest amount of acid formed	30°C	30°C	30–35°C

(SPICHER, 1968)

Tab. 39. Effect of Temperature and Dough Yield on the Ratio of Lactic to Acetic Acid in Sour Doughs Fermented with Various Species of Lactic Acid Bacteria

Species	20°C Dough Yield			25°C Dough Yield			30°C Dough Yield			35°C Dough Yield		
	150	180	210	150	180	210	150	180	210	150	180	210
Lactobacillus plantarum	8.9	13.3	10.4	8.9	10.4	6.7	10.4	10.4	6.7	8.9	13.3	6.3
Lactobacillus casei	8.9	32.7	32.7	21.6	21.6	21.6	21.6	32.7	32.7	32.7	32.7	32.7
Lactobacillus farciminis	7.9	7.9	15.3	8.9	10.4	13.3	10.4	10.4	15.3	8.9	10.4	13.3
Lactobacillus acidophilus	7.9	7.9	6.7	7.9	7.9	6.7	7.9	7.9	6.7	7.9	7.9	7.9
Lactobacillus brevis var. *lindneri*	2.5	2.8	3.3	2.8	3.3	4.1	2.4	3.0	4.1	2.4	3.0	3.8
Lactobacillus brevis	1.9	2.4	2.8	2.0	2.4	2.8	2.0	2.4	2.7	1.7	2.4	2.7
Lactobacillus fermentum	2.1	2.2	2.1	3.8	4.1	4.1	4.4	4.9	4.9	4.9	5.4	4.9
Lactobacillus buchneri	1.0	2.0	1.1	1.0	1.2	1.3	1.1	1.4	1.6	1.1	1.2	1.5
Lactobacillus fructivorans	0.3	0.2	0.3	1.0	1.1	1.3	2.2	2.8	3.8	2.2	3.3	3.8

(SPICHER and RABE, 1983)

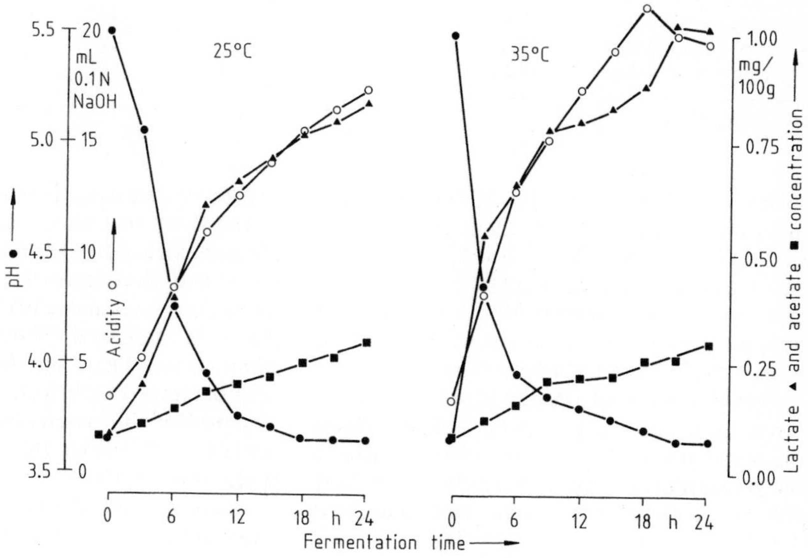

Fig. 18. Effect of temperature on the formation of lactic and acetic acids by a sour dough starter. (SPICHER and RABE, 1980).

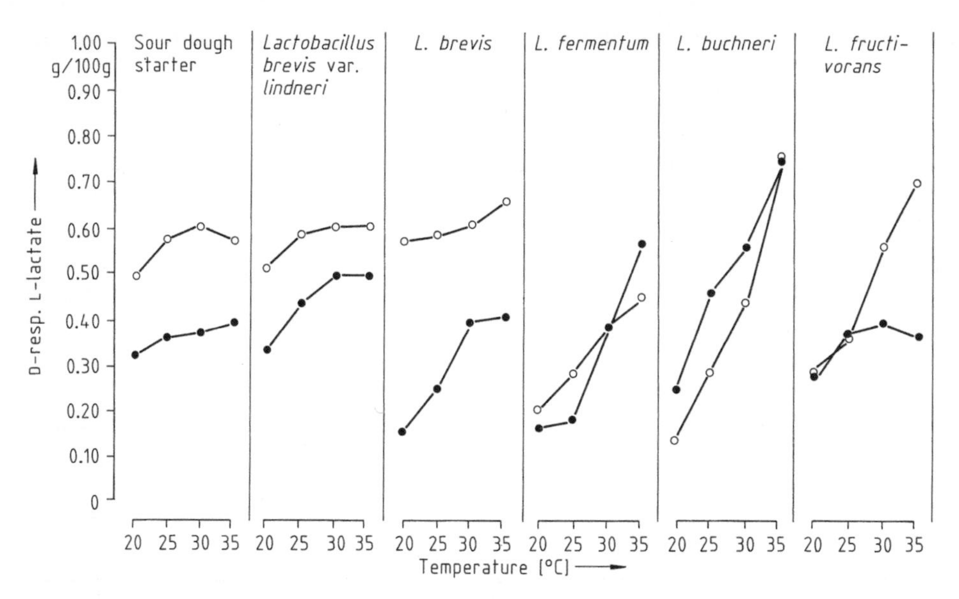

Fig. 19. Effect of temperature on the formation of D- and L-lactate by heterofermentative lactic acid bacteria of sour dough. (SPICHER and RABE, 1983). ○ L-lactate; ● D-lactate.

Tab. 40. Growth of Bacteria and Yeasts in a Foam Sour Dough and a Normal Sour Dough

	Live Cell Counts $\times 10^6$ per g Dough			
	Foam Sour		Normal Sour	
	Bacteria	Yeasts	Bacteria	Yeasts
Sour starter	864	21	318	19
1st Stage sour dough	1082	35	524	54
2nd Stage sour dough	1052	40	552	59
3rd Stage sour dough	1200	77	604	101
4th Stage sour dough	1832	95	1179	95
5th Stage sour dough	1122	49	564	67

(PANZER, 1956)

tolerance of NaCl (SPICHER, 1961). Homofermentative organisms sometimes show a considerable tolerance. With NaCl addition exceeding 4% heterofermentative lactic acid bacteria are not significantly active in sour doughs. This is also apparent in the taste of breads made from such sour doughs.

Dough yield, resp., dough firmness. The ratio of flour and water in sour doughs affects the fermentation time. This ratio also affects the formation of lactic and acetic acids. Soft doughs produce more lactic acid and firm doughs more acetic acid (Tab. 41).

Flour extraction. The degree of extraction of the flour also affects acid formation. With greater extraction the flour contains more Ca, Fe, P and B vitamins (mainly thiamin, riboflavin, and nicotinic acid) and its buffering capacity is increased (THOMAS and TUNGER, 1966). Accordingly the development of lactic acid bacteria which form acids and of yeasts which form CO_2 is greater the higher the ash content of the flour (OPUSZYNSKA and KOWALCZUK, 1967).

Antimicrobial factors. Such effects on the microflora of sour doughs are mainly caused

Tab. 41. Effect of Temperature and Firmness of the Sour Dough on the Formation of Lactic and Acetic Acid (DREWS and STEPHAN, 1956)

	Quality Characteristics of the Bread			
	warm/soft	warm/firm	cold/soft	cold/firm
Rye flour type 1150				
acidity	7.4	5.2	6.4	5.8
Ratio of lactic to acetic acid	89.7/10.3	86.2/13.8	70.1/29.9	71.0/29.0
Taste	fairly strong, aromatic	mild aromatic	bland	bland
Rye flour type 1370				
acidity	7.1	6.8	8.8	7.7
Ratio of lactic to acetic acid	85.6/14.4	84.2/15.8	69.1/30.8	72.9/27.1
Taste	strong aromatic	mild aromatic	mild acetic	bland, acid after taste

warm: sour starter 25–26 °C; basic sour 28–30 °C; full sour 30–32 °C
cold: sour starter 20–23 °C; basic sour 20–23 °C; full sour 20–23 °C

soft: sour starter 200 D.Y.; basic sour 200 D.Y.; full sour 220 D.Y.
firm: sour starter 200 D.Y.; basic sour 160 D.Y.; full sour 160 D.Y.

D.Y.: dough yield

by the fermentation and the feedback effect of the formed lactic and acetic acids. The growth of sour dough yeasts is limited below a pH of 3.7 although the degree of inhibition depends on the type of lactic acid fermentation (SPICHER and SCHRÖDER, 1979). If the drop in pH is the result of a homofermentative fermentation with the formation of lactic acid (0.5–2.0 mL/100 g culture substrate; pH 3.25 to 2.65) then the inhibition of yeasts is slight. However, with heterofermentative fermentations and a concentration of 1 mL acetic acid/100 g substrate (pH 3.6) the yeasts of the sour dough do not grow (Fig. 20).

Increase in sour dough starter bacteria. The increase in the amount of flour for succeeding sour dough stages is of great importance. A doubling of the amount of flour causes a fast increase in acid formation; a 6 fold increase in the next stage causes only a slow increase in acidity. But the increase of the microbial population is faster (shorter generation time) the smaller the amount of a preceding sour dough stage which is present in the following stage (Tab. 42). But the effect of variations in the increase of flour from stage to stage is somewhat limited. At 25–30 °C it is greater than the

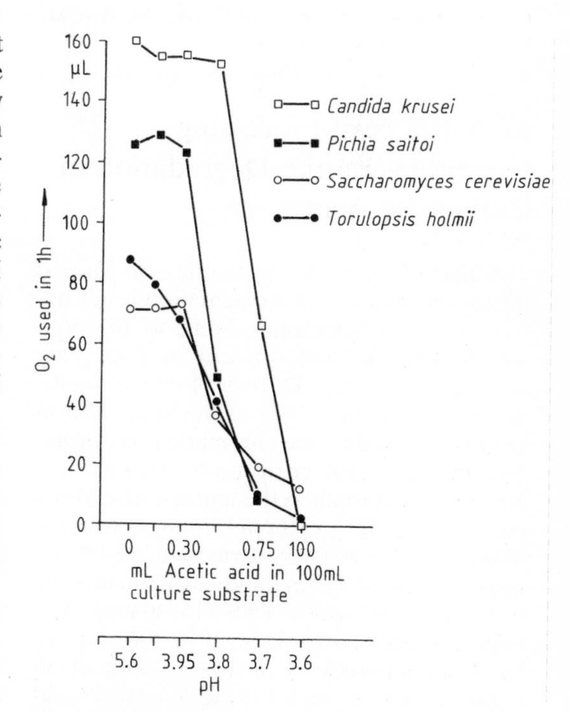

Fig. 20. Effect of pH and acetic acid concentration on the respiration of yeasts in sour dough. (SPICHER and SCHRÖDER, 1979).

Tab. 42. Effect of the Multiplication of Sour Dough Starter Cultures on the Development of the Microflora of Sour Doughs (OPUSZYNSKA, 1965)

	Multiplication of Sour Dough		
	5 fold	10 fold	15 fold
Basic sour			
Generation time			
of yeasts	4 h 7 min	2 h 52 min	2 h 53 min
of bacteria	3 h 19 min	2 h 41 min	2 h 46 min
Ratio yeasts/bacteria	1/7.5	1/6.2	1/5.2
Full sour			
Generation time			
of yeasts	3 h 20 min	3 h 27 min	2 h 3 min
of bacteria	13 h 3 min	6 h 52 min	2 h 22 min
Ratio yeasts/bacteria	1/4.3	1/4.0	1/3.0

basic sour: dough yield 165
full sour: dough yield 190

effect of the temperature. At 30–35 °C the effect of the temperature is greater (SPICHER, 1968; STEPHAN, 1967). The conditions applying to wheat sour doughs and similarities to rye doughs have been summarized by BRÜMMER and LORENZ (1991).

6.2.9 Effect of Processing Conditions on the Degradation of Carboxylic Acids

A part of the lactic and acetic acid formed during the sour dough fermentation is not derived from carbohydrates. Some of the nonvolatile organic acids present in flour may serve as substrates. DREWS (1961) has estimated that about 5–10% of the lactic acid in bread is derived by the enzymatic transformation of malic and citric acids. HERRMANN (1974) has determined the amount of carboxylic acids in 100 g of bread grain as follows: Malic acid 90–200 mg; citric acid 40–90 mg; succinic acid 30–60 mg; and smaller amounts of fumaric and oxalic acids (15–40 mg). The total amount of organic acids in 100 g of rye has been reported by SCHORMÜLLER et al. (1961) as 475 mg; with 143 mg of tartaric acid and 141 mg of succinic acid. Organic acids are found mainly in the outer layers of the grain and in the germ.

Malic acid. In mature sour doughs malic acid is absent or present only in very small quantities. Wheat flour doughs do not contain malic acid after a sour dough fermentation. But in doughs fermented with yeast the concentration of succinic and malic acids increases. In sour doughs the bacteria decrease the amount of malic acid (DREWS, 1958). *Lactobacillus plantarum, L. casei* and *L. brevis* have an enzyme or enzyme system, today designated as malolactic enzyme, which forms lactic acid and CO_2 from malic acid without the release of intermediate compounds such as oxalic or pyruvic acid (SCHÜTZ and RADLER, 1973, 1974; HEGAZI and ABO-ELNAGA, 1980). The high acid production rates of high extraction flours, which contain higher concentrations of malic acid, may be due in part to the high concentration of biodegradable organic acids (DREWS, 1958).

Citric acid. Citric acid concentrations are reduced during the first hours of a sour dough fermentation, but a minimum amount of acid is always detectable. With yeast fermentations this decrease in citric acid concentration cannot be observed (TÄUFEL and BEHNKE, 1957). Citric acid is degraded to lactic acid and acetic acid by the bacterial flora (DREWS, 1961). Both heterofermentative bacteria (*Lactobacillus brevis*) and homofermentative bacteria (*Lactobacillus plantarum, L. casei*) are able to ferment citric acid (WEBB and IN-

GRAHAM, 1969; HEGAZI and ABO-ELNAGA, 1980).

6.2.10 Effect of Processing Conditions on the Hydrolysis of Proteins

Lactic acid bacteria contain proteinases and peptidases which can be found in variable concentrations in species of sour dough organisms. Only 30% of the protein is soluble in flour water doughs which have not been acidified. After a 17 h fermentation 62.1% of the protein is soluble and with "strong" acidification the solubility increases to 82.7% (LEMMERZAHL, 1937). There is also a development of smaller, dialyzable peptides and amino acids during a sour dough fermentation (ROHRLICH and HERTEL, 1966). There is an almost linear increase in free amino acids, particularly of leucine, phenylalanine, methionine, tyrosine, lysine, and isoleucine. *Lactobacillus plantarum* and *L. brevis* spp. *lindneri* form higher concentrations of amino acids than *L. fructivorans* (Fig. 21).

The proteolytic activity is subject to some of the processing conditions. Temperature has a greater effect than dough yield (Fig. 22). *Lactobacillus plantarum, L. brevis* spp. *lindneri,* and *L. fructivorans* form higher concentrations of amino acids at 35°C than at 25°C. This is particularly pronounced for some aliphatic amino acids such as alanine, leucine, and valine, for dicarboxylic amino acids such as aspartic and glutamic acids and their amines, as well as for lysine, phenylalanine, serine, and proline. The temperature effect is greater for *L. plantarum* than for the other two mentioned organisms. The effect of dough yield is of lesser importance.

The substrate also affects the degree of proteolysis. Rye flours with a higher ash content lead to the formation of higher concentrations of amino acids during the sour dough fermentation, particularly leucine, alanine, asparagine, lysine, and proline (Fig. 23).

Amino acids may also be utilized by yeasts during a sour dough fermentation, e.g., alanine, glycine, asparagine, and leucine. This is more pronounced for *S. cerevisiae* than for

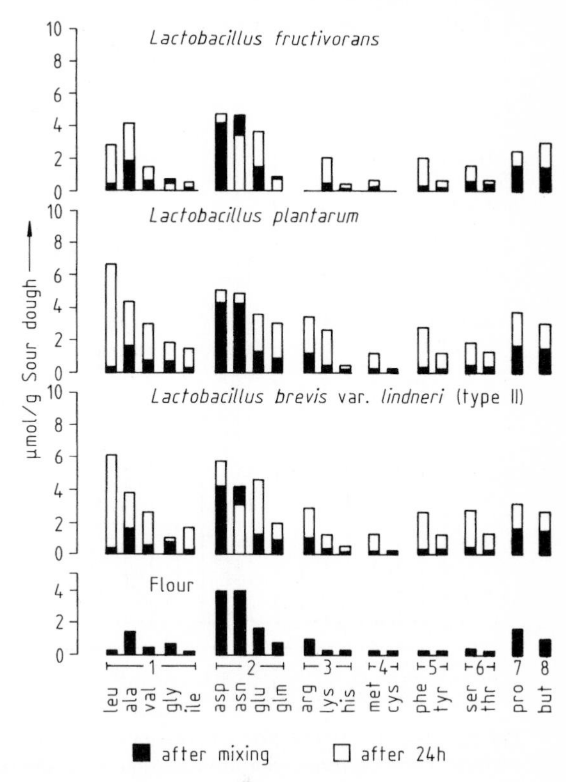

Fig. 21. Concentration of individual amino acids of a rye flour and doughs made with this flour with several species of *Lactobacillus.* (SPICHER and NIERLE, 1983).

Candida krusei. Of course, the total yeast count affects the uptake of amino acids (SPICHER and NIERLE, 1983).

6.2.11 Chemical Acidification

The required improvement of baked rye products by acidification may be carried out by the addition of acids such as lactic, tartaric or citric acids which are produced chemically or by fermentation. These acids may be used instead of the traditional sour dough methods. This direct acidification has been recommended because of the ready availability of these acids. However, the absence of prolonged hydration, the shorter shelf life, and the absence of aroma formation by sour dough organisms lowers the quality of bread

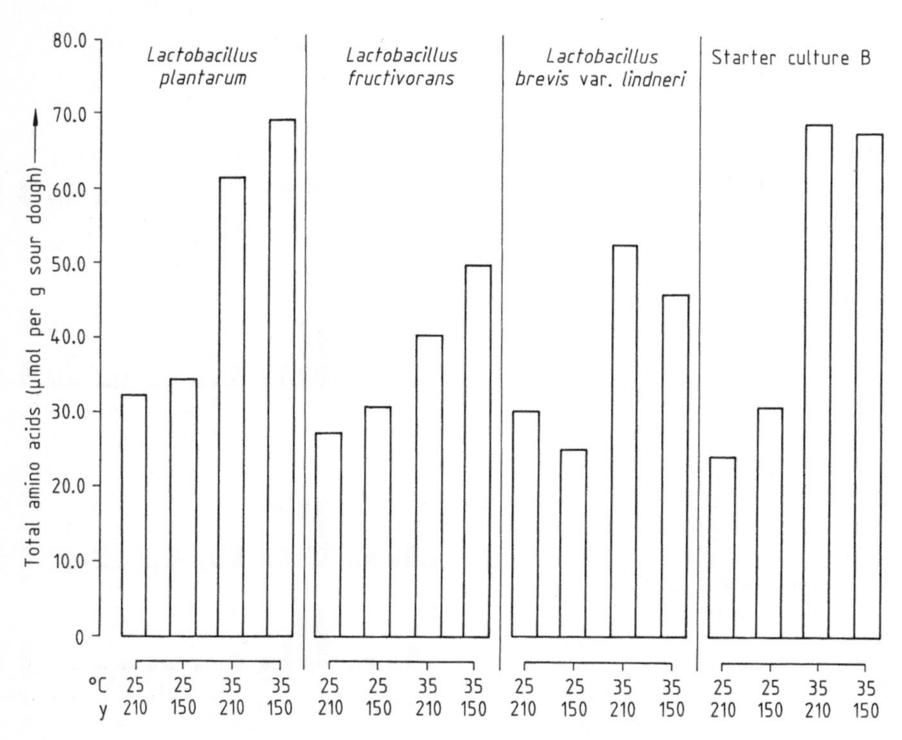

Fig. 22. Effect of temperature and dough yield on the concentration of total amino acids in sour doughs for various starter cultures. (SPICHER and NIERLE, 1983). – Y dough yield.

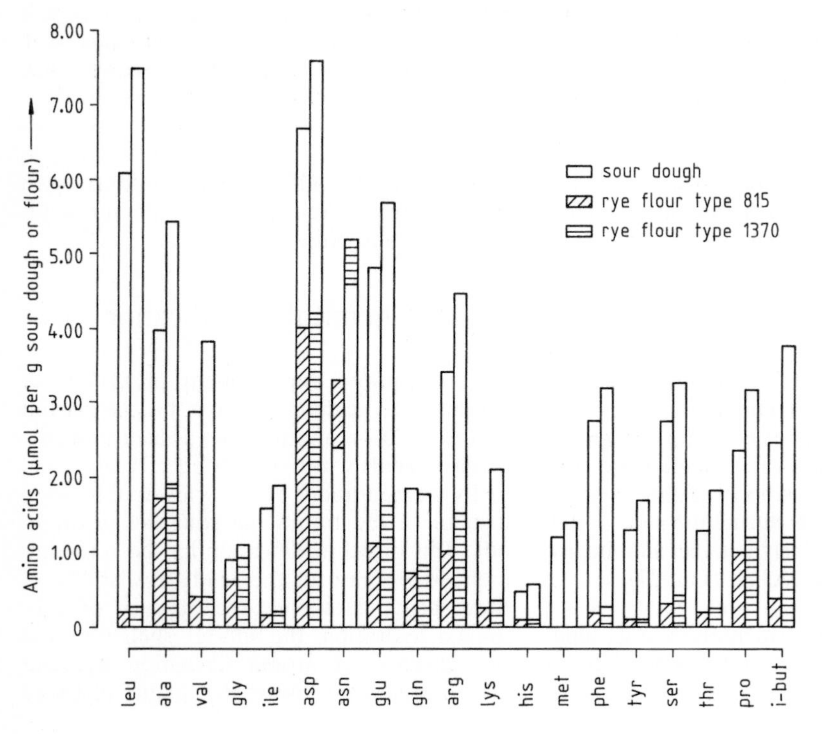

Fig. 23. Concentration of individual amino acids in two types of rye flour and in sour doughs made with these flours (*Lactobacillus brevis* spp. *lindneri* as starter culture). (SPICHER and NIERLE, 1983).

made with higher percentages of rye acidified with such acids (BRÜMMER and STEPHAN, 1979). These results have stimulated a systematic re-evaluation of the complex sour dough methods, not only for baked rye products but also for wheat products (BRÜMMER and LORENZ, 1991).

7 Processing of Fermented Doughs

Mixing of the dough is followed by the fermentation. The fully fermented dough is divided into dough pieces. These are rounded and permitted a rest period, the intermediate proof. The dough is then given its final form in the molder. The equipment used to carry out these steps is called the make-up equipment. Biological transformations continue during the make-up period and during the final fermentation period of the dough pieces, the proof. The membranes of the dough retain a major portion of the carbon dioxide and the dough rises. After the final proof the dough piece is baked.

7.1 Fermentation

After mixing the dough is often somewhat moist and sticky. During the fermentation the dough becomes somewhat drier and more plastic which facilitates the following make-up. Swelling of flour components continues with additional binding of water. Yeast activity increases and the dough rises. Enzymatic hydrolysis of the starch continues. The fermentation is completed when both the formation of a colloidal dough structure and the intensity of the fermenting yeast have reached an optimum. If the fermentation is prolonged beyond this point, then the gluten absorbs too much water and the dough becomes bucky, that is, it loses its extensibility and tears easily.

7.2 Dough Make-Up

Following the fermentation period the dough is divided into individual dough pieces. The weight of the dough pieces must be sufficiently heavy to allow for a 10–12% loss of weight during baking. Automatic dividers for bread production can produce 420–2000 dough pieces per hour with weights from 250 to 4000 g (weight deviation ±10 g). Dividers for rolls or buns can produce 8000 to 10000 pieces per hour.

Dividing is followed by rounding of the dough pieces. This eliminates the larger bubbles formed during the fermentation and leads to an even distribution of smaller bubbles throughout the dough. During the final proof the number of bubbles does not increase but the bubbles expand greatly. Rye doughs and mixed rye doughs may be molded immediately after rounding. Wheat doughs and mixed wheat doughs require an intermediate proof time between rounding and molding. During this period the dough relaxes and may be formed in the molder without tearing. The intermediate proof also improves the quality of the bread with an improved grain and better bread volume. For wheat breads the time for the intermediate proof is 5–30 min. For doughs containing rye flour it is shorter the higher the percentage of rye flour in the formula (Tab. 43). A shorter intermediate proof time is also desirable for doughs from relatively soft flours, for soft doughs, and for warm doughs.

Tab. 43. Intermediate Proof Times for Doughs with Various Percentages of Rye Flour

	Average Intermediate Proof Time (min)
Wheat dough	5–7
Wheat mixed grain dough	4–6
Mixed grain dough (50:50)	2–3
Rye mixed grain dough	1–1.5
Rye dough	no intermediate proof

(DOOSE, 1967)

The simplest way of providing time for intermediate proof is a long conveyor belt between the rounder and the molder. These belts can be open or closed. There are also special cabinets with pocket proofers in which the rounded dough pieces can be transported at variable speeds. The temperature and relative humidity in such cabinets can be regulated automatically.

Various devices are used for the final mechanical shaping of the dough. These may consist in a simple device for flattening the dough piece and then rolling in into a cylindrical shape or the more complex cross grain molders. The shaped or molded pieces are either deposited in pans or on trays and transported to the proof cabinet for the final rise.

7.3 Proofing

The final leavening is carried out in a so-called proof box or proof cabinet for periods ranging from 30–60 min depending largely on the size of the dough pieces and the activity of the yeast. The temperature in the proof box may vary between 30 and 40 °C. The relative humidity of the proof box is of greatest importance. It must be sufficiently high so that the surface of the dough piece retains its original elasticity. The highest relative humidities (65–86%) are required for dough pieces whose upper surface will also be the upper surface of the baked product, such as pan bread. The proof time for bread doughs is usually from 50–60 min and that for small dough pieces 25–30 min. The proof time can be regulated within limits by the amount of yeast used in the formula, by the temperature of the proof box, and the relative humidity. A dry proof box causes formation of an undesirable dry skin on the surface of the dough piece and prolongs the proof time. Firm doughs also require a somewhat longer proof. When the dough piece has reached the desired volume it should have also good tolerance, that is, it should be able to retain its volume for a sufficient time during transport into the oven.

Fermentation losses between the time the dough is formed and the entry into the oven depend largely on processing conditions. For straight dough processes with wheat flour they range from 1.0–1.5%. For multistage rye doughs they range from 2.0–2.5% (based on the total weight of flour).

8 Baking

Baking is the most energy intensive stage of the production of baked goods. It converts the dough into the final baked product by firming (stabilization of the structure) and by the formation of the characteristic aroma substances. The increased production of gas and its expansion causes a 40% increase in the volume of the dough piece and a 10% increase of its surface area (KRIEMS, 1970). This can be controlled within certain limits by the temperature of baking, the type of baking oven, the relative humidity, and the time of baking.

8.1 Baking Ovens

The oldest form of a baking oven is the peel oven, a domed structure of stone with a single loading door and a simple baking hearth. It is pre-heated by burning wood or coal on the hearth and the ashes must then be removed. The baking temperature is initially between 350–450 °C and decreases during baking to about 150 °C. The heating chamber also serves as the baking chamber (direct heating). Since the middle of the last century these ovens have been replaced by indirectly heated ovens in which the fire box was separated from the baking chamber. Such ovens provide more even heating and permit constant operation.

Today retail bakers use generally multideck ovens heated by gas, oil, or electricity. These are usually hot air ovens in which the baking chamber is heated from the outside or in which pure air is heated in a closed system and then introduced into the oven for heating by forced convection. More modern rack or tray ovens are the rotary hearth oven, the reel oven, or the pan rack ovens developed in Scandinavia (WENZ, 1971).

Wholesale bakeries started to use automatic ovens at the beginning of this century. This permits continuous baking. In such ovens the baking surface travels through the oven. The dough pieces enter through a loading door, traverse several temperature zones and usually leave the oven at the opposite end. Unloading is done by hand or with a conveyor belt. The different types of automatic ovens are mainly distinguished by the devices used to convey the dough through the oven. They may be conveyed by individual trays (swing tray travelling oven) or on continuous belts or wire mesh screens (tunnel oven). Such ovens have the advantage of lower and lighter construction, the ability to regulate the temperature for individual zones within the oven, the ability to apply heat from below and from the top, and easy regulation of the speed of the conveying belt. The dwell time in the oven may be varied from 20–120 min. Such wire mesh screen ovens may bake from 20000–30000 rolls or buns or 2000 1-kg loaves of bread.

It is not possible to reduce the baking period further by increasing the temperature. But baking times can be reduced drastically by the use of microwave ovens. These ovens are built as tunnel ovens in which the doughs pass through a high frequency field. Heating is uniform over the entire cross section of the dough. There is no crust formation since heat is not applied from the outside of the dough piece. Therefore, microwave ovens are usually equipped with infrared heating devices which apply heat by radiation to the outside of the doughs.

Cracked rye grain bread and pumpernickel are usually baked in a steam chamber at 100°C. The doughs are placed into covered pans and the baking chamber is heated for several hours with low pressure steam. This type of baking produces a very satisfactory crumb but does not lead to the formation of a real crust.

8.2 Processing Conditions During Baking

Energy is used to heat the dry matter of the dough to a temperature between 100 and 170°C, to heat the water to a temperature of 100°C, and to evaporate excess water. The requirements for a given temperature or for changes in temperature during baking depend on the type of baked goods. The baking temperature is normally between 200 and 250°C (Tab. 44). Wheat breads usually require a uniform temperature during baking. Rye breads and mixed grain rye breads require a higher temperature at the beginning of the baking process with a slowly decreasing temperature thereafter.

In directly heated ovens the temperature at the beginning is very high (350–450°C), followed by a slow decrease to 150°C. This permits the development of a strong bread aro-

Tab. 44. Baking Times and Temperatures for Various Baked Goods

	Time	Temp. (°C)
Waffles	2– 9 min	200–400
Cookies	8 min	350
Crisp (flat) bread, Knäckebrot	8 min	340
Wheat bread and wheat mixed grain bread (1 kg)	35–40 min	220–230
Wheat bread (2 kg)	50 min	220–230
Rye bread (1 kg)	40–60 min	220–230
Rye bread (2 kg)	75 min	250–270
Rye bread (3 kg)	90 min	250–270
Rhenish black bread	2– 4 h	210–230
Hamburg black bread	4– 5 h	200–210
Westphalian black bread	8–10 h	180–200
Pumpernickel	16–35 h	100–180

ma. In indirectly heated ovens the temperature is more uniform. The effect of direct heating may be imitated by baking breads containing more than 75% rye flour in a separate oven at 400 °C for 1–5 min. The bread is then fully baked in another oven at a uniform, lower temperature. This has a drying effect on the outer layers of the dough piece; it increases the thickness of the crust and leads to more browning and a higher intensity of the taste and aroma of the bread.

The dough is heated by conduction through the hearth or the baking pan, through convection by hot air, and through radiation from the roof of the baking chamber. Heat is transmitted from the outside of the loaf to the inside. The temperature of the inside of the loaf rises only slowly and rarely exceeds 100 °C (VASSILEVA and SEILER, 1981; VASSILEVA et al., 1981).

The minimum time of baking is that which permits full gelatinization of the starch, denaturation of the protein, and the formation of aroma substances. The upper limit of the baking time is determined by a loss of vitamins, excess evaporation of water, and also by economic considerations. The optimal baking time is that at which the interior of the crumb has reached a temperature of 98 °C, and the bread is then left in the oven for another 10 min (SCHNEEWEISS, 1965).

The relative humidity of the atmosphere in the oven is also of great importance. The firming of the crust must be delayed to permit a satisfactory oven spring and an optimal bread volume. For this purpose low pressure steam is directed into the oven at the beginning of the baking process. A part of this steam condenses on the surface of the dough piece and keeps it moist and elastic for a certain time.

The yield of baked goods from 100 kg of flour (bread yield) is largely determined by the baking loss, and to a lesser extent by the fermentation loss and some loss during the cooling of the baked loaf. The baking loss depends on the weight and size of the dough piece and the time and temperature of baking. For bread it varies from 8–14% of the weight of the dough piece, and for smaller pieces from 17–23%. Hearth breads show higher baking losses (10–14%) than pan breads (7–11%). A short baking time results in a minimum baking loss, but the quality of the loaf is diminished because of insufficient browning and lesser elasticity and taste of the crumb. About 95% of the baking loss is due to evaporation of water. The remaining 5% are due to vaporization of alcohol, carbon dioxide, volatile acids, and esters. These losses occur mainly in the crust and the outer layers of the crumb. The interior of the crumb has practically the same moisture content as the unbaked dough. The moisture content of the whole loaf varies from 34–41% depending on the thickness of the crust, with crumb moistures between 42–47% (DREWS and STEPHAN, 1963). Rye breads generally have a higher moisture content than wheat breads. The fermentation loss occurs during the bulk dough fermentation, the proof and the first minutes in the oven and amounts to 1–3% based on the weight of flour solids. The loss is almost exclusively a loss in starch and sugars. The loss after baking occurs mostly during the first 30 min of the cooling period. This is a loss in moisture. Such losses continue during the shelf life of the bread but at a greatly decreasing rate.

8.3 Physico-Chemical Changes During Baking

One can distinguish several stages during the change from a dough to a baked bread: (1) an enzyme active zone (from 30 to 60 or 70 °C); (2) a stage of starch gelatinization (from 55–60 °C) to no higher than 90 °C; (3) a stage of water evaporation; and (4) a stage of browning and aroma formation. These changes differ in the outer portions of the dough (crust) from those in the interior of the crumb. During the first stage and at temperatures not exceeding 45 °C the fermentation by yeasts and lactic acid bacteria continues with the formation of CO_2 at an accelerated rate. The activity of enzymes is also increased with increasing temperatures which affects the fermentation positively. This and the expansion of CO_2 with increasing temperature causes additional leavening of the dough, the so-called oven spring. At temperatures higher than 50 °C the activity of microorganisms diminishes and the microbes are finally killed. However, the amylases of the flour continue to act and reach their optimum activity at 60–70 °C.

Starch gelatinization begins at 55 °C for rye starch and at 60 °C for wheat starch and reaches its optimum at 60–88 °C and 80–90 °C, respectively. Swelling (mainly of amylopectin) increases the volume of the starch granules by 25–50%. At higher temperatures gelatinization of the starch continues, the crumb becomes stabilized, and ultimately it becomes dry and can be sliced. Gelatinized starch is hydrolyzed by enzymes more easily than raw starch, and since the activity of amylases is greatest at 60–70 °C the concentration of soluble carbohydrates (dextrins, maltose, glucose) increases. The effect is greater the longer it takes to traverse this temperature zone. Therefore, pumpernickel may sometimes contain as much as 20% sugar. Such bread types also have a dark crumb because of the formation of melanoidins within this temperature range. Sour dough breads show a greater formation of sugars and dextrins than yeast raised breads.

The dough must contain sufficient water to permit sufficient gelatinization of the starch. This water becomes available from the water absorbed mainly by gluten (for wheat flour) and gums (for rye flour). During the dough phase the swelled gluten or gums provide the structure required for gas retention and dough viscosity. During baking and at about 70 °C the gluten is denatured and water is released again. A portion of this water is used for starch gelatinization. The gluten loses its tough and elastic state and becomes stiff and brittle. As a result of the gluten structure which has been formed during mixing of the dough the gluten stiffens the starch structure in wheat bread so that a firm and elastic crumb is formed. This also accounts for the larger volume of wheat breads in comparison with bread from other grains. Rye and barley bread has only a starch structure and no coherent gluten network.

In the crust the temperature ultimately exceeds 100 °C with increasing loss of moisture. Starch is degraded non-enzymatically to dextrins at 110–140 °C. Caramelized products are formed at 140–-150 °C and roasted flavors at 150–200 °C. Proteins are also degraded at these temperatures and their nitrogen containing split products (such as amino acids) react with sugars to form dark brown melanoidins (Maillard reaction). These compounds are responsible for the color and the taste of the crust. Crust color formation depends largely on baking temperature, and the thickness of the crust on baking time.

8.4 Formation of Aroma Substances

The bread aroma is due to the combined effect of many substances (ROTHE and THOMAS, 1959). So far more than 150 compounds have been identified although only a few make an important contribution to bread flavor (Fig. 24). As far as they could be identified these substances are (1) organic acids, such as lactic and acetic acids, and traces of pyruvic acid, propionic acid, *n*- and isobutyric acid, *n*- and isovaleric acid, *n*- and isocaproic acid, and 40 other compounds. These acids are mainly produced by fermentation. Their role in bread aroma has not been clearly identified; (2) ethyl esters of some organic acids such as formic acid, acetic acid, lactic acid, pyruvic acid, levulinic acid, succinic acid, itaconic acid, and benzoic aeid; (3) alcohols such as ethanol, *n*-propanol, isobutanol and isopentanol and others. These alcohols are also

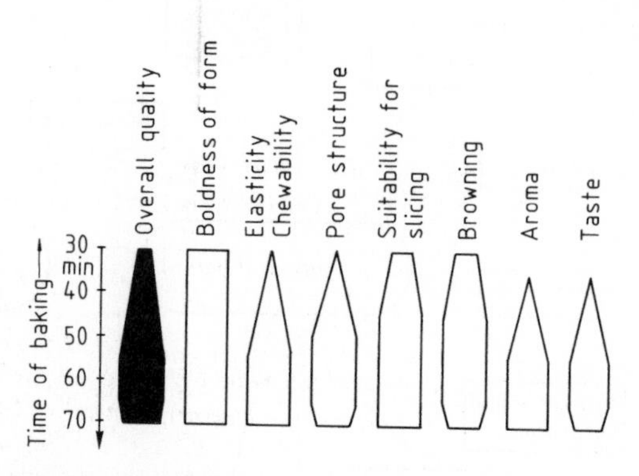

Fig. 24. Development of bread quality and the components of bread quality as a function of the time of baking (schematic). – The width of the column is a measure of quality; for 1.5 kg loaves of mixed grain rye bread – heart baked. (ROTHE et al., 1972).

formed by fermentation. They are volatile and only small concentrations remain in the baked products; (4) aldehydes such as furfural, hydroxymethylfurfural, and several isoaldehydes; (5) ketones; (6) sulfur compounds such as mercaptan, methional, and others. Finally, (7) there is a group of compounds such as maltol, isomaltol, and melanoidin type compounds produced by heating.

Newer investigations on wheat and rye bread have confirmed the multiplicity of flavor compounds as well as a differential composition of flavor compounds (Tab. 45). It was shown that the concentration of the following compounds, acetaldehyde, furfural, isovaleraldehyde, methylglyoxal, isobutyraldehyde, acetone, acetoin, and diacetyl, increase in the order: white bread, rye mixed grain bread, rye cracked grain bread, and pumpernickel; that is in the same order in which intensity of the flavor increases (ROTHE, 1966). The flavor and aroma compounds of bread are derived from the raw material components, the fermentation, and the baking (Fig. 25). The biochemical formation of aroma compounds or precursors dur-

Tab. 45. Concentration of Aroma Substances in Various Breads (ROTHE, 1974)

	Wheat Bread		Rye Mixed Grain Bread		Whole Grain Bread		Pumpernickel
	Crumb	Crust	Crumb	Crust	Crumb	Crust	
Ethanol	3900[a]	1800	3400	1100	2300	1000	1600
5-Hydroxymethyl-furfural	9	40	12	300	20	400	70
Acetaldehyde	4.3	12.8	4.7	22.6	4.6	26.2	7.1
Isopentanal	1.2	4.7	2.7	15.2	1.9	19.0	4.6
Furfural	0.3	5.5	1.5	12.4	2.3	28.7	27.4
Methylglyoxal	0.7	0.8	1.5	8.9	1.9	13.5	4.3
Isobutanal	0.3	2.6	0.9	6.0	0.8	12.9	1.8
Acetone	0.7	4.5	1.4	5.6	2.0	6.5	1.9
Acetoin	0.9	1.0	0.2	1.1	0.3	0.7	5.0
Diacetyl	0.2	0.9	0.2	1.3	0.2	1.3	0.7

[a] Average values in ppm based on fresh weight

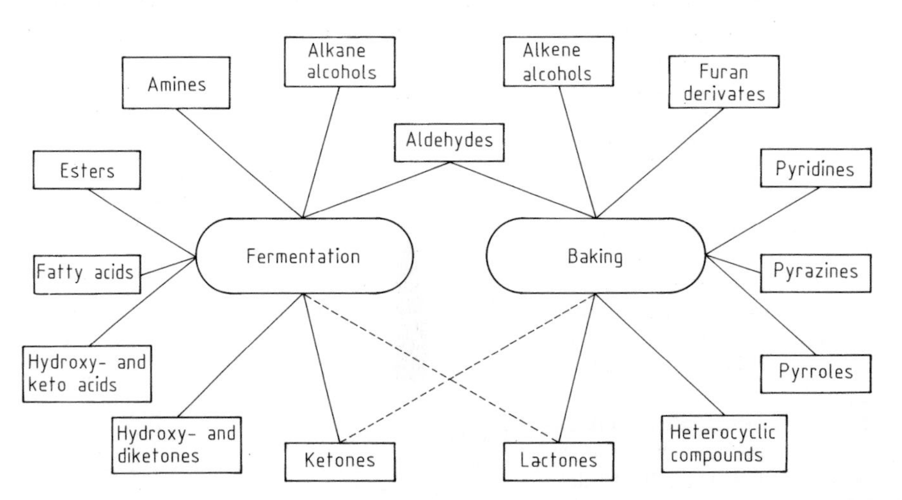

Fig. 25. Relationship of fermentation and baking to the major classes of aroma compounds. (ROTHE, 1980).

ing the fermentation and the baking process are mainly involved. However, 90% of the aroma substances are produced in the crust during baking (Fig. 26).

Raw materials. Raw materials and additives contribute only a minor portion of the aroma substances. Flour provides precursors such as acids, ethanol, and traces of aldehydes, ketones and esters. Starch and protein are important only as the medium in which the aromatic substances are dispersed.

Fermentation. In fermenting doughs one finds sugars, organic acids, amino acids (alanine, valine, leucine, methionine, glycine, phenylalanine), and alcohols. The more volatile substances escape during baking, while the less volatile substances are available for chemical reactions which may lead to aromatic substances. Sour dough breads are largely characterized by the presence of lactic and acetic acid. Their aromatic effect is slight, yet it is certain that these acids, as well as ethanol, affect the perception of aromatic compounds. In this way they contribute to the overall quality of the aroma (ROTHE, 1974). They may also contribute by retaining otherwise volatile compounds in the dough (HUNTER et al., 1961). Acetic acid is indispensable for the development of an excellent aroma. In such bread acetic acid amounts to 20–30% of the total acid (SCHULZ, 1942). The acidity of bread is perceived as pleasant if the molar ratio between lactic and acetic acid is between 1.5 and 4.0.

Baking. The largest amounts of aroma substances as well as the compounds with the most intense aroma are formed during baking, that is, at temperatures above 200°C. This is particularly true for rye breads in which aromatic substances produced during baking predominate. The higher acidity of sour dough breads and the longer baking are responsible for this predominance (ROTHE, 1974).

The principal reaction which takes place mainly in the outer (crust) portion of the dough piece is the non-enzymatic browning according to Maillard. In the course of this reaction the following may occur: Condensation of reducing sugars with free amino acids to form *N*-glycosides, the Amadori rearrangement, degrading of sugars to carbonyl compounds and of amino acids to aldehydes, the aldehyde amine condensation, the alcohol condensation of aldehydes, the formation of anhydro sugars, furfurol, and reductones (THOMAS and ROTHE, 1956). To a lesser extent amino acids are degraded according to the Strecker reaction to aromatic carbonyl compounds (in the presence of excess amino acids). This is an oxidative deamination and decarboxylation of amino acids to an aldehyde or ketone: leucine to 2-methylbutanal; isoleucine to 3-methylbutanal; methionine to methional; phenylalanine to phenylethanal. The resulting enaminols can either polymerize directly to melanoidins or split into acetaldehyde and aminoacetone. Basic amino acids

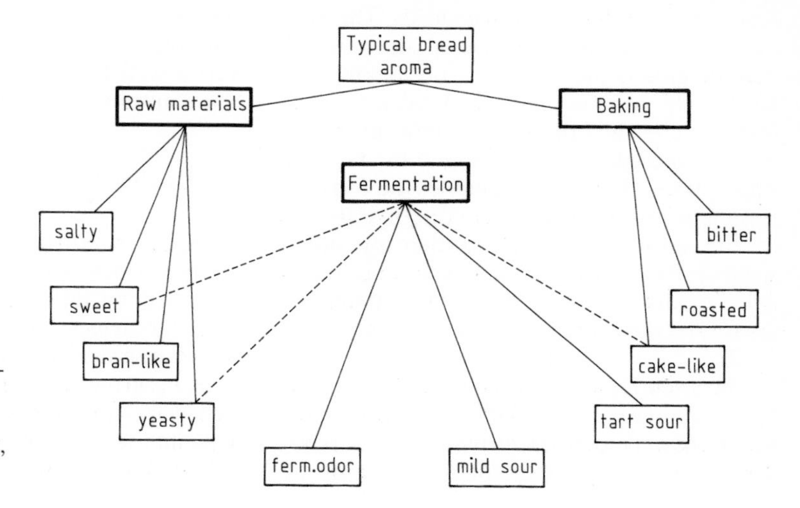

Fig. 26. Effect of raw material, fermentation and baking on the taste and aroma of bread. (ROTHE, 1980).

do not form aroma compounds but condense to melanoidin pigments and cause browning (JOHNSON, 1967).

Sugars caramelize when the outer layers of the dough piece have reached temperatures from 150–400 °C. Many compounds are formed by a complex series of reactions. Some of these compounds are aromatic: simple aldehydes, ketones, diketones, maltol, furan, cyclopentane derivatives as well as bitter substances of high molecular weight and unknown structure.

The total amount of aroma substances formed during baking depends very much on the time of baking. Non-enzymatic browning does not start until the temperature exceeds 100 °C, that is, not before the bread has been in the oven for 50–70 min. The stronger aroma of rye bread is due to the fact that rye breads of similar volume as wheat breads have a greater weight and require a longer period (about ⅓ longer) for evaporation of water from the crust region (ROTHE, 1970). But with increasing time of baking the ratio of less volatile aroma compounds to more volatile aroma compounds increases. This is readily apparent in the different taste of Knäckebrot (8 min baked), the crust of wheat bread (35 min), and the crust of rye bread (90 min).

However, it is not possible to prolong the baking time much more because of nutrient losses. The more pronounced aroma of rye or rye mixed grain bread is often attributed to the higher concentrations of pentosans which react more quickly with amino acids. The aroma of rye bread but not that of wheat bread may be greatly increased by "pre-baking" at 400 °C. The resulting rye bread has a thicker, browner crust and there are significant differences in the odor of the crust and the taste of bread slices. "Prebaking" doubles the concentration of furfural and hydroxymethylfurfural in the crust. The concentration of other strongly flavored substances such as isobutanal, 2-methylbutanal, and 3-methylbutanal increases on the average by 42% (ROTHE et al., 1972).

The crumb is not exposed to high temperatures. Therefore, the Maillard reaction and the aroma compounds formed by this reaction are not very important for the taste of the crumb. The flavor of the crumb is due mainly to aroma substances formed during the fermentation. However, the aroma compounds of the crust diffuse into the crumb during baking and particularly during cooling of the baked goods and affect its flavor.

9 Bread Quality and Its Preservation

The quality of bread is determined by its taste, digestibility and by appearance and volume of the loaf and the properties of the crust and crumb. Bread should have an even form that is not too heavily domed. The crust should be uniformly firm and free from tears. It should be neither hard and brittle nor soft, tough and leathery, and it should have a certain elasticity. The crust should be light brown without dark, red brown spots which are often seen in milk bread. In wheat bread the crumb should have a soft and elastic structure and a fine, even grain. Wheat mixed grain bread should also have good crumb elasticity and an even grain. A smeary or gummy crumb and the presence of streaks and cores can worsen the taste, the chewability or digestibility of the bread. In extreme cases the bread can become inedible. A crumb that is too dry and crumbly is also undesirable. It also lowers digestibility. The volume of the bread is also important since a high volume indicates good leavening. However, a well leavened bread dries out faster than a more dense loaf.

9.1 Bread Defects

It is often difficult to determine the cause of a bread defect since the same defect may be caused by different factors. But since a particular defect rarely occurs by itself one can often draw conclusions from the simultaneous occurrence of several defects. The following are the most common defects. Poor shape and low volume may be caused by any of the following: Over- or underfermentation, poor flour (sprouted grain), doughs are too firm or too soft, doughs are too highly acidified or too poorly acidified, the oven temperature is too high or too low, the concentration or the activity of the yeast is too low, or poor adjustment of the rounder and molder. Excessive browning is caused by use of flours from sprouted grain, insufficient acidification, high concentrations of malt preparations or of dairy or soy adjuncts, a high oven temperature, or an excessive baking time. A poor break and shred does not only detract from the appearance of a loaf but hinders proper leavening of the crumb by lowering the gas retention. It may be caused by too short a time of baking, by excessive gas pressure in the interior of the loaf, or by low relative humidity in the oven. An overly dense crumb may be due to insufficient leavening because the yeast concentration and/or the yeast activity are too low, the salt concentration is too high or the dough is too firm. A very short fermentation time or proof time or excessive working of the dough may also cause a dense crumb. A large, open grain is caused by the use of too much yeast, by doughs which are too cool or which are overfermented, or which have been undermixed. An irregular grain with sharp cornered pores may be caused by use of too much dusting flour at the make-up of the doughs. Larger openings in the crumb are due to soft doughs and insufficient dough development. Tears in the crumb can be distinguished by the following causes. A vertical tear is caused by a very firm dough. A lateral tear is due to insufficient swelling of the flour. A horizontal tear in the upper portion of the crumb occurs when the dough is too soft, when the dough is not sufficiently acidified or when the flour has been milled from sprouted grain. A horizontal tear in the lower portion of the crumb is due to excessive temperatures at the start of baking. Circular dense portions in the crumb (cores) or dense streaks sometimes occur in otherwise well leavened bread. They may be caused by poor adjustment of make-up equipment or with doughs that have been exposed to excessive temperatures. A crumb that lacks elasticity may result from the use of poor flour that does not swell well and that contains too much damaged starch or from insufficient acidification. A very soft dough, the use of too little salt or insufficient baking may cause the same defect. A lack of elasticity of the crumb is often accompanied by a moist crumbliness. In contrast dry crumbling may be caused by very firm doughs, by short fermentations or fermentations at high temperatures, or by excessive baking times. Lack of aroma and an insipid taste may be caused by insufficient acidification, a very cool or too short a fermentation, a very firm dough, use of too little salt, or a short baking time at high temperatures. A very sour taste results from doughs that are too cool or to firm, from the use of too much full sour, or excessive times for the fermentation and the proof or addition of too much acidification compounds. A straw taste occurs in wheat breads and wheat mixed grain breads if the dough is overmixed or if the fermentation is too long or carried out at too high a temperature. A bitter taste may be caused by contaminants such as rope, by rancid fats in high extraction flours, or the addition of flours from corn, oats, or barley.

9.2 Bread Staling

The quality of baked goods begins to deteriorate after baking. In some baked goods this is due to the uptake of moisture from the air (biscuits, waffles, other low moisture baked goods). In bread, rolls, and buns, on the contrary, this may be due to drying out. In such baked goods, in which the partial pressure of water vapor is high, moisture is given off to the air, and there is a more or less noticeable weight loss, the shelf loss. The loss is greatest during the first 30 min after baking. Bread loses on the average 2% of its water content during the first 24 hours if it is not wrapped. In succeeding 24 h intervals the loss is about 0.5% until the moisture content of the crust has been reduced to 15%. Small pieces may lose 2% of their water content within the first 2 hours. The loss depends on temperature of storage and relative humidity. Higher moisture losses are usually associated with the following: Hearth baking, a pronounced break (in the crust), sliced bread, high extraction of flour, and a higher moisture content of the baked loaf. However, a certain shelf loss is desired since it contributes to the optimum quality of the crust (crispness) and the crumb (elastic, slightly sticky, yet well chewable).

After prolonged storage one observes changes in the crust and the crumb which are called "staling". The crust takes up water from the air or from the crumb and loses its shean and crispness. It becomes leathery and wrinkled. The crumb loses its softness and ability to swell. It becomes unelastic, dry, firm, with a tendency to crumble, and the aroma of the bread becomes stale. The increase in the firmness of the crumb occurs also if there is no loss of moisture at all. Baked goods with a moisture content of 16–37% are most inclined to stale fast. Low moisture baked goods such as Knäckebrot, biscuits, crackers, and Zwieback do not stale in this manner. Also, breads with a high degree of starch hydrolysis such as pumpernickel do not show staling. Staling is greatly influenced by temperature. It is fastest at 0°C. Wheat bread stored at 4°C stales more in 1 day than stored for 3 days at 70°C. There is no staling below −7°C (WASSERMANN, 1972). The size of the pieces is also important. Rolls begin to stale within a few hours, bread only after 24 h. Wheat breads may be stale within 1–2 days, while rye breads, which normally have a firmer crumb, stale more slowly (rye cracked grain bread after 8–12 days) (Fig. 27). Breads with higher concentrations of fats, sugars, and skimmilk powder also stale more slowly. Economic losses due to staling are high in countries with wheat bread consumption and may amount to 3–10% of total bread production.

Staling is caused mainly by changes in the colloidal structure of starch polysaccharides, such as changes in water absorption and regeneration of a crystalline structure. During baking starch granules undergo limited swelling because of the low concentration of free water. A part of the linear amylose diffuses out of the starch granules and forms a concentrated solution. During cooling of the bread the amylose molecules form a gel in which the hydrated starch granules are imbedded. The firming of the crumb is due to the heat reversible association of the side chains of branched amylopectin within the

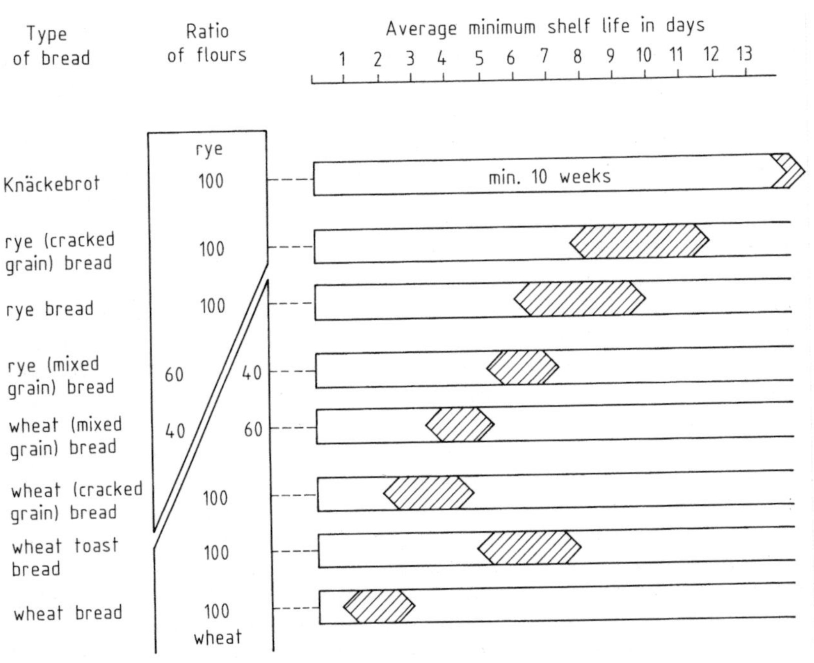

Fig. 27. Average minimum shelf life of various types of bread (packaged bread stored at ambient temperatures). (BRÜMMER and SEIBEL, 1979).

starch granules and to its retrogradation (SCHOCH, 1965; SCHOCH and FRENCH, 1947). It has been assumed that 70–80% of the amylose, 10% of the amylopectin, and 10% of the protein of the bread are involved in the staling process (AUGUSTAT et al., 1967). The association of starch molecules is temperature dependent. The water given off in the course of starch retrogradation (about 2%) is probably taken up by the gluten and the pentosans (HAMPEL, 1967).

Bread with a high protein content usually keeps better. The course of staling is affected by the mutual influence of the gluten/starch complex and the pentosan/starch association on each other. These associations do not only determine the structure and firmness of the crumb of freshly baked bread, but affect the course of subsequent staling (HAMPEL, 1968). Within the first 12 hours after baking the gluten becomes less extensible. Therefore, it may be assumed that gluten is responsible for staling at this early stage after baking, and that after 12 h the specific changes in starch granules cause staling (BANECKI, 1970).

The particular taste and odor which develops as the result of staling is in part due to the formation of melanoidin compounds, which are, however, not identical with those formed during baking (MAC MASTERS and BAIRD, 1954). On the other hand, the crust loses aroma compounds which diffuse increasingly into the crumb. Later on the concentration of aroma compounds in the crumb decreases. In wheat bread the concentration of formaldehyde, acetone, acetaldehyde, and isovaleraldehyde decreases drastically in the crumb as a result of staling (LINKO et al., 1962). In rye breads a decrease in the concentration of methylglyoxal, furfurol, acetaldehyde, and isovaleraldehyde was observed (ROTHE, 1966). The stale taste is not just due to the disappearance of aroma components. Aldehydes are easily oxidized in the presence of air to the corresponding acids. Some of these such as isobutyric acid and isovaleric acid have a disagreeable taste. Traces of such acids may already contribute to the stale taste.

9.3 Microbial Spoilage

Flour contains a microbial flora which is highly variable as to type and total numbers. However, this does not materially affect the quality of the baked goods since these organisms do not cause problems during processing of doughs and are inactivated during baking. Spoilage of baked goods is caused by bacteria or molds which may contaminate bread after baking and cooling. Contaminants may develop if the bread is stored under conditions of high relative humidity (above 70%) and between 20 and 30 °C. Such spoilage can lead to the formation of compounds with disagreeable odors or taste and possibly to compounds which make the product inedible.

Rope principally occurs in yeast leavened goods such as wheat bread, wheat mixed grain bread that has not been acidified, bread with higher concentrations of sugar or fat, fruit bread, etc. In rare cases rope can be seen in poorly acidified rye bread. The first manifestation about 12 h after baking is a slightly bitter taste of the crumb and a more or less fruity odor. With increasing spoilage and when the crumb is already discolored and decomposed a repulsive odor develops. This happens no sooner than 24 h after baking and usually later. The crumb assumes a weak yellow discoloration which darkens later on and finally becomes brown, dirty brown, or red brown. It loses elasticity and becomes sticky and smeary or plastic. Finally it may be pulled into fine, silky threads. Such bread is inedible and may cause diarrhea and vomiting.

Rope results from mixed infection of baked goods with *Bacillus* species. The organisms are mostly *Bacillus subtilis* but *B. licheniformis, B. megaterium,* and *B. cereus* cause similar spoilage (STREULI, 1955). The spores of the organisms reach the flour from the grain. Suitable conditions for their growth are a temperature of 30–35 °C, a relative humidity exceeding 85%, and a pH of 6.5. Therefore, rope is usually encountered in summer.

The contamination of baked goods with fungi occurs mainly during transport within the bakery, during cooling and storage, and during slicing and packaging. It may occur through spores carried in the air or by direct

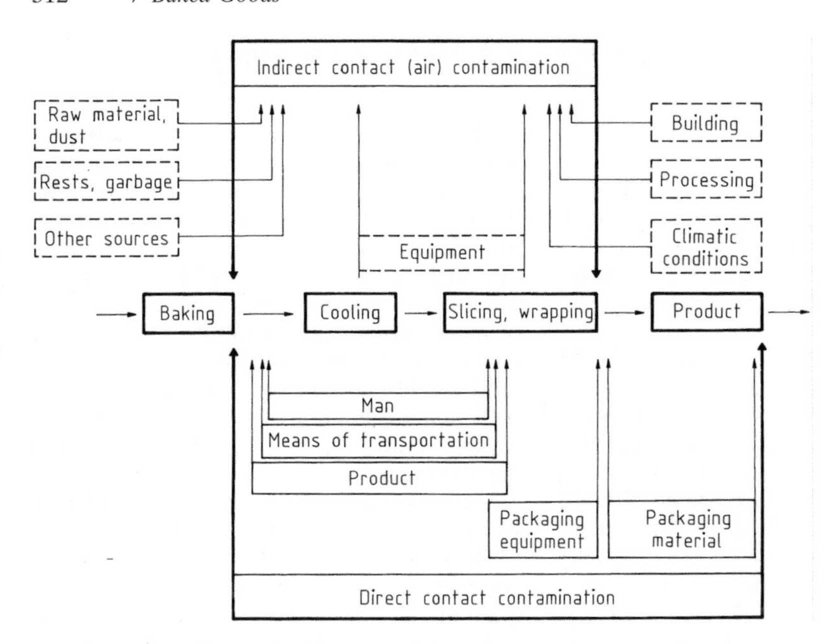

Fig. 28. Sources of mold contamination of bread in the course or production (SPICHER, 1980).

contact with the surface of equipment (Fig. 28).

After baking the products traverse areas of the bakery in which a m^3 of air may contain 30000 and up to 90000 mold spores. One can assume a contamination rate of 10–400 mold spores per hour for a surface of 100 cm^2. Under suitable conditions, that is at 27 °C and a relative humidity of more than 88%, the contamination can take hold within 3 days (SPICHER, 1980a). Breads are more susceptible to mold growth the greater the content of rye flour, that is, the higher the moisture content. Mold development is also favored by tight storage of baked goods. Sliced bread is particularly susceptible to contamination, while bread with a coherent crust free from breaks is much less susceptible. Mold develops rapidly in bread which has not been baked sufficiently (Fig. 29).

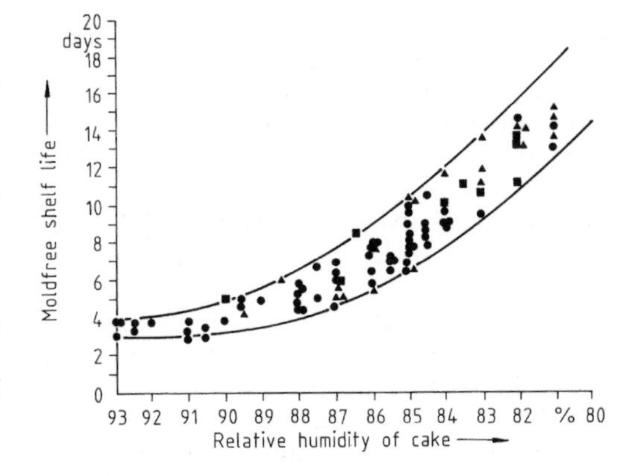

Fig. 29. Effect of relative humidity and extent of microbial contamination on the moldfree shelf life of baked goods (at 27 °C). (SEILER, 1964). ● Laboratory prepared plain cake; ▲ laboratory prepared fruit cake; □ commercially prepared plain cake.

Blue-green mycelia usually indicate contamination with aspergilli (*Aspergillus candidus, A. flavus, A. fumigatus, A. glaucus, A. nidulans, A. oryzae*) or penicillii (*Peniciliium crustaceum, P. glaucum, P. olivaceum*). Whitish gray to dark often cottony mycelia indicate contamination with *Mucor mucedo, M. plumbeus,* or *Rhizopus nigricans. Aspergillus niger* is recognized by its black color. *Trichosporon variabile* and *Oospora lactis* are recognized by white, chalky plaques. The red bread mold, *Monilia sitophila,* covers bread with a dusty pink to red mycelium.

Moldy baked products look unattractive and often smell and taste bad. There is also the danger of the production of mycotoxins (SPICHER, 1970).

9.4 Retention of Bread Quality

9.4.1 Storage

Storage conditions are chosen to prevent or minimize the loss of moisture from baked goods, the growth of microorganisms, and the action of enzymes, for instance those that cause fat hydrolysis. For bread the suggested storage conditions are between 15 and 20 °C at 65–75% rel. humidity; for Zwieback, waffles, crackers, etc. they are between 5 and 18 °C at 50–60% rel. humidity. Under these conditions even sliced bread may be kept without loss for 4 days (simplest packaging, no preservatives). Other storage conditions may have to be used for baked goods which are specially processed, for instance, bread that has been sliced and/or packaged. In such instances the conditions of storage between baking and packaging determine the quality of the product and a structure of crust and crumb most suitable for slicing.

Two different storage conditions have been shown to be suitable for preservation between the time of baking and the delivery of products to the market, preservation at elevated temperatures and by freezing. Preservation at elevated temperatures is best carried out between 45 and 65 °C and at rel. humidities of 75–80%. However, at this temperature some chemical reactions in the crust and the crumb are accelerated. Breads with a light colored crust show a darkening of the color. Baked goods lose some of their aroma, and on subsequent storage firming of the crust is accelerated. Therefore, storage under high temperature conditions is only practiced with darker breads such as rye and rye mixed grain breads. In general high temperature storage is limited to 12–15 h.

Preservation by freezing is mainly practiced with sweet goods and rolls and buns. Freezing of bread is done only for reasons of economy or scheduling of bakery production, for instance, with sliced and packaged wheat and toast bread. For optimum results the temperature of baked goods must be lowered to −18 °C, although complete inhibition of staling does not take place at temperatures above −25 °C. As has been mentioned staling is rapid between temperatures of −7 to +20 °C. Therefore, it is important to traverse this temperature zone rapidly. Freezing is done at −30 to −35 °C in air moving at 1.5 to 4 m per sec. Wheat bread is frozen directly as it leaves the oven hot. Wheat and rye mixed grain breads are frozen after a cooling period of 1–2 h following baking. Rolls are frozen 15–20 min after baking when the internal temperature has reached 50 °C.

9.4.2 Packaging

Baked goods are packaged for reasons of hygiene, to prevent loss of moisture (i.e., to preserve freshness), to prevent re-contamination with mold spores, and for promotional reasons. The growth of molds must be prevented by additional measures such as the use of preservatives, sterilization, or elimination of oxygen from the atmosphere. This may result in somewhat contradictory requirements for the properties of the packaging material (HARTMANN, 1980).

With some exceptions hermetic packaging is not desirable. While loss of moisture contributes to staling the optimal preservation of quality is obtained when the loss of moisture is 2% over a 4 week period. Mold spores may germinate if the packaging material is impervious to water vapor. This danger is acute if water condenses on the inner surface of the packaging material and the internal atmosphere is saturated with water vapor. Besides, defects in taste which develop during staling are more pronounced in a hermetic package. Sliced bread which has been baked for a long time should not be packaged in dense packaging material since the slices may stick to each other.

Wax paper is sufficient for a simple package, particularly for bread that is sold and consumed quickly. Protection against drying out is minimal. There is a loss of 16% by weight in 4 weeks at 35 °C and 90% rel. humidity. Re-contamination is not completely prevented since molds may grow through the packaging material. Aluminum foil has been used for packaging bread for quite some time. It must be coated to prevent attack by acetic acid and other volatile acids. It facilitates sterilization of bread because of its good heat conductance. It has very good gas retention, and the loss of weight of the bread is 0.4–2% after 4 weeks. Polyethylene is stable to temperatures of 95–120 °C (depending on type). It retains water vapor well. Weight losses of the bread are 2% in 4 weeks. The material is extensible and flexible. It is often used in the form of bags for packaging sliced bread.

Polypropylene sheets have a better tear resistance. If they are coated with polyvinylidene chloride emulsion the permeability for gases can be adjusted to suit different requirements. Compared with polyethylene the polypropylene sheets are somewhat stiffer and have better temperature resistance (to 155–160 °C). They may be used as an overwrap with aluminum foil, mainly for the packaging of sliced bread that will be heat-sterilized.

Shrink wraps have been introduced for the packaging of unsliced bread, and also for sliced bread. The packaging materials are polyethylene shrink wrap sheets and copolymers with polyvinyl chloride.

Cellophane is the oldest transparent packaging material. It is quite permeable to water vapor showing a weight loss of the bread of 10–19% in 4 weeks. But permeability can be reduced in controlled fashion by coating with nitrocellulose or polyvinylidene chloride copolymers.

Metal cans have also been used for packaging of baked goods. They can serve as baking pans and may be closed after filling with dough or after baking. Otherwise, the bread may be baked first and filled into the metal cans after baking and slicing. The cans are then sterilized after they have been closed.

9.4.3 Preservation

With current production methods it is not possible to prevent microbial contamination during storage (or slicing). Therefore, the shelf life data in Fig. 27 apply only if additional measures are taken to prevent growth of microorganisms, particularly with sliced bread. This can be done by heat sterilization of baked goods, by the addition of preservatives or packaging under an inert atmosphere. Sometimes heat sterilization is combined with use of a chemical preservative. In this case the chemical preservative is added to protect the product after the package has been opened by the consumer. German experience with such methods has been described by BRÜMMER (1988).

9.4.3.1 Physical Processes

Heating is the most reliable method of preventing mold contamination. It is used mainly for sliced, dark breads. In light breads the additional heating causes darkening of the crust and undesirable changes in flavor. A temperature of 90 °C must be applied for a minimum of 30 min to kill fungi. Hot air or saturated steam are generally used. The suggested conditions are a temperature of 150 °C for 15 min or a temperature of 98 °C for 12 hours. For intact loaves a temperature of 90 °C for 30–40 min (of the product) is sufficient. For sliced bread one to several hours are required at the same temperature. Packaging in polyethylene sheets limits the sterilization temperature to 95–100 °C, polypropylene sheets to 140–150 °C, and aluminum foil to 200 °C.

The simplest process is the return of the packaged sliced bread into the baking oven. The packaged bread can be placed into open baking pans or on trays. It can be exposed to a temperature between 120 and 130 °C for about 45 min. Sterilization chambers or steam chambers may also be used (BRÜMMER and MORGENSTERN, 1991).

Microwave sterilization and infrared sterilization provide certainty of results and a short sterilization time. Microwave ovens have the disadvantage of high investment costs and high energy demands. The sliced and packaged bread can pass through the microwave oven with a speed of 1–10 m/min (4 m length; 2450 megahertz). The energy consumption is about 0.2 kW per kg bread. The choice of the packaging material is critical. Metal foil is not suitable. Rye cracked grain bread, rye bread and pumpernickel may be sterilized by this method (BURG, 1968). For infrared sterilization an intermediate wavelength in the region of 250 nm is used. Surface sterilization of cakes and unsliced mixed grain bread can be obtained in 8–9 min (ZBORALSKI, 1973; BRÜMMER and MORGENSTERN, 1985b).

9.4.3.2 Chemical Processes

Chemical preservatives must be safe; they must be necessary; and they must not be used in a deceptive manner (LÜCK, 1977). In almost all countries regulations specify the preservatives which may be used, permissable concentrations, and the type of baked goods in which they may be used. Preservatives are generally added to the dough.

Sodium propionate, calcium propionate and propionic acid are very effective against molds and rope (*Bacillus mesentericus*). At the same time inhibition of yeast is not too great (LÜCK, 1969/1973). However, relatively high concentrations of propionates are required if baked goods are to be preserved for longer than a few days (BRÜMMER and STEPHAN, 1980). Usage levels in Germany are 0.03–0.06% calcium propionate and in the U.S. 0.15–0.3%. For cakes 0.2–0.3% sodium propionate are used.

Sorbic acid and its sodium, potassium and calcium salts are highly effective against molds and some yeasts (BRÜMMER and STEPHAN, 1980). They exert their effects without affecting taste or flavor significantly (LÜCK, 1969/1973). Commonly used levels are 0.1 to 0.2% as sorbic acid or 0.15 to 0.3% as sorbic acid salts. Both propionic acid and sorbic acid preservatives are only active at pH values below 5.0 (SEILER, 1964; BRÜMMER et al., 1988a, b). Both types of compounds inhibit yeasts and hence the fermentation but remarkably less so than propionic acid or propionates which may increase proof times by 30%. In contrast use of sorbic acid may require only a slight increase in the level of yeast or an extension of the fermentation time.

The development of rope may be prevented with acetic acid or preferably with calcium acetate. The latter may be added to unacidified wheat and wheat mixed grain breads in concentrations of 0.3–0.4% (based on weight of flour) (BRÜMMER and STEPHAN, 1980). Its principal effect is not only a lowering of the pH. Acetic acid lowers the heat resistance of bacteria but not that of yeasts and molds. Therefore, bacteria are killed more efficiently during baking. Sodium diacetate can be used as a fungicide at concentrations of 0.4–0.6%. However, it inhibits the fermentation and at higher concentrations affects the taste (BRÜMMER and STEPHAN, 1980; BRÜMMER et al., 1982; BRÜMMER and MORGENSTERN, 1985a).

Recently carbon dioxide has been used to replace air in packages of sliced bread and cakes. Mold growth can only be prevented if the residual oxygen concentration is below 0.2% (CERNY, 1976, 1979). It is difficult to reduce oxygen concentrations to that level in highly porous baked goods (BRÜMMER et al., 1980a; BRÜMMER, 1986).

Various baked goods and particularly bread may be preserved by treating with a mixture of ethanol and water followed by packaging under vacuum (KUCHEN, 1963). The preservative solution consists of 10 volumes of water and 1 to 5 (preferably 1 to 2) volumes of potable ethanol. This solution can be applied by dipping or spraying (GEIGES and GUNDLACH, 1981).

10 References

ABD-EL-MALEK, Y., EL-LEITHY, M. A., AWAD, Y. N. 1974), *Chem. Mikrobiol. Technol. Lebensm.* **3,** 148–153.

ANGERMANN, A., SPICHER, G. (1964), *Brot Gebäck* **18,** 162–169.

ANONYMOUS (1958), *Baker's Dig.* **32,** 49–52.

AUGUSTAT, S., SCHIERBAUM, F., RICHTER, M. (1967), *Ber. 3. Tagung Int. Probleme der modernen Getreideverarbeitung und Getreidechemie* Teil 2, pp. 122–128.

AXFORD, D. E., CHAMBERLAIN, N., COLLINS, T. H., ELTON, G. A. H. (1963), *Cereal Sci. Today* **8,** 265–268, 270.

AZAR, M., TER-SARKISSIAN, N., GHAVIFEK, H., FERGUSON, T., GHASSEMI, H. (1977), *J. Food Sci. Technol.* (Mysore) **14,** 251–254.

BACHMANN, B., KOSIEK, E., WLODARCZYK, C. (1973), *Mitt. Versuchsanst. Gärungsgewerbe* **27** (3), 45–46.

BADE, V. (1974), *Getreide Mehl Brot* **28,** 296–299.

BAKER, J. C. (1954), *Proc. Am. Soc. Bakery Eng.,* 65–71.

BAKER, J. C. (1960), *U.S. Patent* 2953460, Sept. 20.

BANECKI, H. (1970), *Ber. 5. Welt-Getreide- und Brotkongress* **5,** 179–184.

BECKER, E. (1966), in: *Brot in unserer Zeit* (SCHÄFER, W., Ed.), pp. 217–224, Detmold: Verlag Moritz Schäfer.

BLÜMEL, F., BOOG, W. (1977), in: *5000 Jahre Backofen,* Ulm: Deutsches Brotmuseum e. V.

BOND, E. E. (1967), *Brot Gebäck* **21,** 173, 176–180.

BRÜMMER, J.-M. (1977), in *Jahresbericht der Bundesforschungsanstalt für Getreide- und Kartoffelverarbeitung,* D21–D22, Detmold.

BRÜMMER, J.-M. (1979), *Getreide Mehl Brot* **33,** 147–153.

BRÜMMER, J.-M. (1985), *Brot Backwaren* **33,** 298–300.

BRÜMMER, J.-M. (1986), West German experience with controlled carbon dioxide packaging for bakery products, in: *Proc. Second Int. Conference and Exhibition on Controlled Atmosphere Packaging,* CAP '86, pp. 361–380.

BRÜMMER, J.-M. (1988), *Brot Backwaren* **36,** 25–32.

BRÜMMER, J.-M. (1989), *Getreide Mehl Brot* **43,** 19–21.

BRÜMMER, J.-M. (1990), *Brot Backwaren* **38,** 298–300.

BRÜMMER, J.-M. (1991), Modern equipment for sour dough production, *Cereal Foods World* **36,** 305–308.

BRÜMMER, J.-M. (1993), in: *Handbuch der Gärsteuerung,* Hamburg: Behr's Verlag.

BRÜMMER, J.-M., ELSNER, G. (1982), *Getreide Mehl Brot* **36,** 322–326.

BRÜMMER, J.-M., MORGENSTERN, G. (1975), *Getreide Mehl Brot* **29,** 233–239.

BRÜMMER, J.-M., MORGENSTERN, G. (1985a), *Getreide Mehl Brot* **39,** 198–200.

BRÜMMER, J.-M., MORGENSTERN, G. (1985b), *Brot Backwaren* **33,** 26, 28–29.

BRÜMMER, J.-M., MORGENSTERN, G. (1988), *Dtsch. Bäcker Ztg.* **75,** 1518–1520.

BRÜMMER, J.-M., MORGENSTERN, G. (1991), *Dtsch. Bäcker Ztg.* **78,** 936–939.

BRÜMMER, J.-M., SEIBEL, W. (1979), *Getreide Mehl Brot* **33,** 135–137.

BRÜMMER, J.-M., STEPHAN, H. 1979), *Back J.* **3,** 16–18, 43.

BRÜMMER, J.-M., STEPHAN, H. (1980), *Getreide Mehl Brot* **34,** 159–163.

BRÜMMER, J.-M., STEPHAN, H. (1986), *Getreide Mehl Brot* **40,** 51–58.

BRÜMMER, J.-M., STEPHAN, H., MORGENSTERN, G. (1980a), *Verpackungs-Rundschau* **31,** 21–27.

BRÜMMER, J.-M., SEIBEL, W., STEPHAN, H. (1980b), *Getreide Mehl Brot* **34,** 173–178.

BRÜMMER, J.-M., STEPHAN, H., MORGENSTERN, G. (1982), *Getreide Mehl Brot* **36,** 237–239.

BRÜMMER, J.-M., BRACK, G., SEIBEL, W. (1988a), *Getreide Mehl Brot* **42,** 17–21.

BRÜMMER, J.-M., MORGENSTERN, G., NEUMANN, H. (1988b), *Getreide Mehl Brot* **42,** 153–158.

BRUINSMA, B. L., FINNAY, K. F. (1981), *Cereal Chem.* **58,** 477–480.

BURG, F. (1968), *Brot Gebäck* **22,** 58–60.

CALVEL, R. (1972), *Getreide Mehl Brot* **26,** 75–78.

CAVALIER, G. (1963), *Baker's Dig.* **37,** 76–78, 85.

CERNY, G. (1976), *Chem. Mikrobiol. Technol. Lebensm.* **5,** 20–26.

CERNY, G. (1979), *Chem. Mikrobiol. Technol. Lebensm.* **6,** 8–10.

CHAMBERLAIN, N., COLLINS, T. H., ELTON, G. A. H. (1962), *Baker's Dig.* **36,** 52 ff.

CHAMBERLAIN, N., COLLINS, T. H., ELTON, G. A. H. (1965), *Baker's Dig.* **39,** 412–414, 457.

CHOI, R. C. (1955), *Proc. Am. Soc. Bakery Eng.,* 44–52.

CHOI, R. C., KONCUS, A. F., and staff (1954), in: *ADMI Stable Ferment Process,* Chicago: American Dry Milk Institute.

COLE, M. S. (1973), *Baker's Dig.* **47,** 21–23, 64.

COLE, E. W., HALE, W. S., PENCE, J. W. (1962), *Cereal Chem.* **39,** 114–122.

COLLYER, D. M. (1966), *J. Sci. Food Agric.* **17,** 440–445.

COPPOCK, J. B. M., COOKSON, M. A., LANEY, D. H., AXFORD, D. W. E. (1954), *J. Sci. Food Agric.* **5,** 8–26.

CREMER, H.-D., ACKER, L. (1960), in: *Die ernährungsphysiologische Bedeutung des Brotes,* Vereinigung Getreidewirtschaftliche Marktforschung e. V. (Ed.), Bonn.

CREMER, H.-D., SCHIELE, K., WIRTHS, W., MENGER, A. (1969), *Fortschr. Med.* **87,** 1257–1260.

DOOSE, O. (1967), *Brot Gebäck* **21,** 29–32.

DOOSE, O. (1969), in: *Arbeitskunde für Bäcker,* 4. Ed., Alfeld/Leine: Gilde-Verlag H. G. Dobler.

DREWS, E. (1958), *Bort Gebäck* **12,** 261–264.

DREWS, E. (1961), *Brot Gebäck* **15,** 33–35, 41–44, 105–113.

DREWS, E. (1971), *Brot Gebäck* **25,** 1–6.

DREWS, E., STEPHAN, H. (1956), *Brot Gebäck* **10,** 1–4.

DREWS, E., STEPHAN, H. (1963), *Brot Gebäck* **17,** 173–177.

ELTON, G. A. H. (1965), *Baker's Dig.* **39,** 38 ff.

ELTON, G. A. H., CHAMBERLAIN, N., COLLINS, T. H. (1966), *Ber. 4. Int. Getreide- und Brotkongress.*

ESCHENBECHER, F. (1968), *Brauwissenschaft* **21,** 424–437, 464–471.

ESCHENBECHER, F. (1969), *Brauwissenschaft* **22,** 14–28.

FAO (1978), in: *Production Yearbook 1977,* FAO Rome, Basic Data Unit, *Statist. Div.* **31,** 91.

FINK, H. (1968), *Industriebackmeister* **17,** 22–24.
FORTMANN, K. L., GERRITY, A. B., SNELL, P. E. (1966), *Cereal Sci. Today* **11,** 394ff.
FRETZDORFF, B., BRÜMMER, J.-M. (1992), Reduction of phytic acid during breadmaking of wholemeal breads, *Cereal Chem.* **69,** 266–270.
GASZTONYI, K. (1970), *Ber. 5. Welt-Getreide- und Brotkongress* **5,** 45–53.
GEIGES, O., GUNDLACH, J. (1981), *Getreide Mehl Brot* **35,** 306–309.
GROSSKREUTZ, J. C. (1960), *Biochim. Biophys. Acta* **38,** 400–409.
GROSS, A., REDFERN, S., BELL, R. L., FISCHER, F. (1968), *Cereal Sci. Today* **13,** 346–348, 358.
HAMPEL, G. (1967), *Ber. 3. Tagung Int. Probleme der modernen Getreideverarbeitung und Getreidechemie* Teil 2, pp. 136–147.
HAMPEL, G. (1968), *Bericht 19. Tagung für Getreidechemie,* pp. 147–160.
HARTMANN, G. (1980), *Getreide Mehl Brot* **34,** 137–140.
HEGAZI, F. Z., ABO-ELNAGA, J. G. (1980), *Zentralbl. Bakteriol. Parasitenkd. Infektionskr. Hyg.* Abt. 2, **135,** 212–222.
HENNEBERG, W. (1903), *Z. Spiritusind.* **26,** 226–227, 243–244, 255, 257, 270, 277–279, 288–289, 291, 302, 315–318, 329–332, 341, 343–344.
HERRMANN, K. (1974), *Z. Lebensm. Unters. Forsch.* **155,** 220–233.
HUBER, H. (1965), *Brot Gebäck* **19,** 205–216.
HUBER, H. (1966), *Brot Gebäck* **20,** 217–226.
HUBER, H. (1970), *Brot Gebäck* **24,** 46–52.
HUBER, H. (1972), *Getreide Mehl Brot* **26,** 62–67.
HUBER, H., BANSBACH, J. (1971), *Brot Gebäck* **25,** 75–79.
HÜTTINGER, R. (1972), *Goldschmidt informiert* **1,** 18, 8–13.
HUNTER, J. R., NG, H., PENCE, J. W. (1961), *J. Food Sci.* **26,** 578–580.
JOHNSON, J. A. (1967), *Brot Gebäck* **21,** 72–75.
JOHNSON, J. A., MILLER, B. S., CURNUTTE, B. (1961), *Agric. Food Chem.* **6,** 384–387.
KILBORN, R. H., TIPPLES, K. H. (1968), *Cereal Sci. Today* **13,** 25ff.
KLINE, L., SUGIHARA, T. F. (1971), *Appl. Microbiol.* **21,** 459–465.
KNIGHTLY, W. H. (1968), in: *Surface-active Lipids in Foods,* Soc. Chem. Ind. Monograph No. 32, London.
KNIGHTLY, W. H. (1973), *Baker's Dig.* **47,** 64–65, 70, 72–75.
KNUDSEN, S. (1924), *Den. Kgl. Veterinaer- og Landbohøiskoles,* Arsskrift.
KRAUT, H. (1963), in: *Brot und sein Nährwert.* Wissensch. Veröffentlichung der DGE Nr. 10, Frankfurt.
KRIEMS, P. (1970), *Ber. 5. Welt-Getreide- und Brotkongress* **5,** 125–141.
KUCHEN, W. 1963), *Patentschrift* Nr. 369087, Eidgen. Amt für geistiges Eigentum, Bern.
KUNKEL, O. (1966), *Brot Gebäck* **20,** 245–252.
LANGEJAN, A. (1972), *Ferment. Technol. Today,* 669–671.
LANGEJAN, A. (1974), *U.S. Patent* 3843800, Oct. 22.
LEMMERZAHL, J. (1937), *Mehl Brot* **35,** 1–4.
LINKO, Y., JOHNSON, J. A., MILLER, B. S. (1962), *Cereal Chem.* **39,** 468–476.
LUBIG, R. (1949), *Das Schaumsauerverfahren,* Köln: Greven Verlag.
LÜCK, E. (1969–1973), *Sorbinsäure, Chemie-Biochemie-Mikrobiologie-Technologie-Recht,* Hamburg: B. Behr's Verlag.
LÜCK, E. (1977), *Chemische Lebensmittelkonservierung,* Berlin-Heidelberg-New York: Springer-Verlag.
LUDEWIG, H. G. (1975), *Getreide Brot Mehl* **29,** 76–78.
MAC MASTERS, M. M., BAIRD, P. D. (1954), *Baker's Dig.* **28,** 32–35.
MARNETT, L. F. (1977), *Getreide Mehl Brot* **31,** 244–247.
MARSTON P. E. (1966), *Cereal Sci. Today* **11,** 530–532, 542.
MENDEN, E., ELMADJA, J., HORCHLER, V. (1975), *Getreide Mehl Brot* **29,** 253–257.
MENGER, A., BRETSCHNEIDER, F. (1972), *Getreide Mehl Brot* **26,** 120.
MILLER, B. S., JOHNSON, J. A. (1958), *Wallerstein Lab. Commun.* **21,** 115–137.
NEUKOM, H. (1972), *Getreide Mehl Brot* **26,** 299–303.
OPUSZYNSKA, H. (1965), *Kongressbericht Intern. Roggenkonferenz,* Poznan, May 10–15, XXI/1-XXI/22.
OPUSZYNSKA, H., KOWALCZUK, R. (1967), *Brot Gebäck* **21,** 7–13.
OSZLANYI, A. G. (1980), *Baker's Dig.* **54,** 16, 18–19.
PANZER, W. (1956). *Ernährungsforschung* **1,** 679–683.
PELSHENKE, P. F. (1941/42), *Mühlenlaboratorium* **11,** 106–110.
PELSHENKE, P. F. (1949), in: *Gebäck aus deutschen Landen,* 1st Ed., Alfeld/Leine: Gilde-Verlag H.-G. Dobler.
PELSHENKE, P. F. (1950), *Gertreide Mehl Brot* **4,** 257–258.
PIRRIE, P. (1954), *Baker's Weekly* **164** (2), 26–29.
PIRRIE, P., GLABAU, C. A. (1954), *Baker's Weekly* **163** (5/6/7), 25–28, (9), 25–28, (10), 29–31.
PONTE, J. G. Jr. (1971), in: *Wheat Chemistry and*

Technology (POMERANZ, Y., Ed.), pp. 675–735, Am. Assoc. Cereal Chem., St. Paul, Minnesota.
PONTE, J. G. Jr., Glass, R. L., Geddes, W. F. (1960), *Cereal Chem.* **37,** 263–279.
PYLER, E. J. (1970), *Baker's Dig.* **44,** 34ff.
RABE, E. (1981), *Getreide Mehl Brot* **35,** 129–135.
RABE, E., SEIBEL, W. (1981), *Getreide Mehl Brot* **35,** 129–135.
REDFERN, S., BRACHFELD, B. A., MASELLI, J. A. (1964), *Cereal Sci. Today* **9,** 190–191.
REDFERN, S., GROSS, H., BELL, R., FISCHER, F. (1968), *Cereal Sci. Today* **13,** 324ff.
REED, G. (1965), *Baker's Dig.* **39,** 32–36; 79.
REED, G., NAGODAWITHANA, T. (1991), *Yeast Technology,* 3rd Ed., New York: Van Nostrand Reinhold.
REIMELT, W. (1970), *Brot Gebäck* **24,** 36–40.
RITTER, K. (1966), in: *Brot in unserer Zeit,* 1st Ed. (SCHÄFER, W., Ed.). pp. 169–176, Detmold: Verlag Moritz Schäfer.
ROBINSON, R. J., LORD, T. H., JOHNSON, J. A., MILLER, B. S. (1958), *Cereal Chem.* **35,** 295–305.
RÖCKEN, W. (1976), *Branntweinwirtschaft* **116,** 1–11.
ROHRLICH, M. (1973), *Getreide Mehl Brot* **27,** 337–342.
ROHRLICH, M. (1976), *Brotindustrie* **1,** 9–14.
ROHRLICH, M., ESSNER, W. (1951), *Brot Gebäck* **5,** 85–91.
ROHRLICH, M., HERTEL, W. (1966), *Brot Gebäck* **20,** 109–113.
ROHRLICH, M., TÖDT, F., ZIEHMANN, G. (1959), *Zentralbl. Bakteriol. Parasitenkd. Infektionskr. Hyg.* Abt. 2 **112,** 351–358.
ROJTER, I. M., BERZINA, N. I., TIVOMENKO, G. P. (1967), *Chlebopek. Konditersk. Prom.* **11,** 10–12.
ROTHE, M. (1960), *Ernährungsforschung* **5,** 131–142.
ROTHE, M. (1966), *Brot, Gebäck* **20,** 189–193.
ROTHE, M. (1970), *Ber. 5. Welt-Getreide- und Brotkongress* **5,** 203–209.
ROTHE, M. (1974), in: *Handbuch der Aromaforschung – Aroma von Brot,* Berlin: Akademie-Verlag.
ROTHE, M. (1980), *Nahrung* **24,** 185–195.
ROTHE, M., BETHKE, E., REHFELD, G. (1972), *Nahrung* **16,** 517–524.
ROTHE, M., THOMAS, B. (1959), *Nahrung* **3,** 1–17.
ROTSCH, A. (1966), *Brot Gebäck* **20,** 213–217.
ROTSCH, A., SCHULZ, A. (1958), in: *Taschenbuch für die Bäckerei und Dauerbackwarenherstellung,* 1st Ed., Stuttgart: Wissensch. Verlags-Gesellschaft mbH.
ROTSCH, A., STEPHAN, H. (1966), *Brot Gebäck* **20,** 95–96.
DE RUITER, D. (1968), *Baker's Dig.* **4,** 24–33.
RUSSELL-EGGITT, P. W., COPPOCK, J. B. M. (1965), *Cereal Sci. Today* **10,** 406.
SCARBOROUGH, C. (1955), *Proc. Am. Soc. Bakery Eng.,* 52–62.
SCHÄFER, W., and co-workers (1966), in: *Brot in unserer Zeit* (Ergebnisse und Probleme der Getreideforschung), 1st Ed., Detmold: Verlag Moritz Schäfer.
SCHÄFER, W. (1972), *Getreide Mehl Brot* **26,** 350–352.
SCHAUZ, H. (1968), *Ber. 19. Tagung für Bäckerei-Technologie,* Detmold.
SCHAUZ, H. (1969), *Brot Gebäck* **23,** 54–57.
SCHNEEWEISS, R. (1965), *Ber. 2. Tagung Int. Probleme der modernen Getreideverarbeitung und Getreidechemie,* pp. 260–278.
SCHOCH, T. J. (1965), *Baker's Dig.* **39,** 48–57.
SCHOCH, T. J., FRENCH, D. (1947), *Cereal Chem.* **24,** 231–249.
SCHORMÜLLER, J., BRANDENBURG, W., LANGNER, H. (1961), *Z. Lebensm. Unters. Forsch.* **115,** 226.
SCHÜTZ, M., RADLER, F. (1973), *Arch. Mikrobiol.* **91,** 183–202.
SCHÜTZ, M., RADLER, F. (1974), *Arch. Mikrobiol.* **96,** 329–339.
SCHULZ, A. (1942), *Z. Gesamte Getreidewes.* **29,** 42–44.
SCHULZ, A. (1947), *Ber. 2. Jahrestagung der AG Getreideforschung in Detmold,* pp. 45–50.
SCHULZ, A. (1959), *Brot Gebäck* **13,** 141–144.
SCHULZ, A. (1968), *Brot Gebäck* **22,** 218–223.
SCHULZ, A. (1972), *Brot Gebäck* **24,** 161–163.
SCHULZ, A., UHLIG, H. (1972), *Brot Gebäck* **26,** 215–221.
SEIBEL, W. (1970), *Ber. 5. Welt-Getreide- und Brotkongress,* 15–20.
SEIBEL, W. (1979), *Ernährungs-Umschau* **26,** 107–112.
SEIBEL, W. BRÜMMER, J.-M., MENGER, A., LUDEWIG, H.-G. (1977), in: *Brot und Feine Backwaren,* Vol. 152, Frankfurt/Main: DLG-Verlag.
SEIBEL, W., BRÜMMER, J.-M., STEPHAN, H. (1978), *Getreide Mehl Brot* **32,** 301–310.
SEIBEL, W., MENGER, A., HAMPEL, G., STEPHAN, H. (1969), *Industriebackmeister* **17,** 106, 108, 110, 112.
SEIBEL, W., ZENTGRAF, H. (1981), in: *Brot in unserer Ernährung,* pp. 12–18, Vereinigung Getreide-, Markt- und Ernährungsforschung e. V. (Ed.), Bonn.
SEILER, D. A. L. (1964), in: *Microbial Inhibitors in Food* (ALMQVIST and WIKSELL, Eds.), pp. 211–220. 4th Int. Symp. on Food Microbiol., Göteborg–Stockholm.
SEILING, S. (1969), *Baker's Dig.* **43,** 54ff.

SNELL, P. E., TRAUBEL, I., GERRITY, A. B., FORTMANN, K. L. (1965), *Cereal Sci. Today* **10,** 434–436, 457.

SNYDER, E. (1963), *Baker's Dig.* **37,** 50.

SPICHER, G. (1961), *Brot Gebäck* **15,** 113–119.

SPICHER, G. (1968), *Brot Gebäck* **22,** 61–66, 127–132, 146–151.

SPICHER, G. (1970), *Zentralbl. Bakteriol. Parasitenkd. Infektionskr. Hyg.* Abt. 2 **124,** 697–706.

SPICHER, G. (1980a), *Zentralbl. Bakteriol. Parasitenkd. Infektionskr. Hyg.* Abt. 1 Orig. Reihe **B,** 508–528.

SPICHER, G. (1980b), *Getreide Mehl Brot* **34,** 128–137.

SPICHER, G. (1982), *Getreide Mehl Brot* **36,** 12–16.

SPICHER, G. (1983), *Z. Lebensm. Unters. Forsch.* **176,** 190–195.

SPICHER, G., NIERLE, W. (1983), *Getreide Mehl, Brot* **37,** 305–310.

SPICHER, G., RABE, E. (1980), *Z. Lebensm. Unters. Forsch.* **171,** 437–442.

SPICHER, G., RABE, E. (1981), *Z. Lebensm. Unters. Forsch.* **172,** 20–25.

SPICHER, G., RABE, E. (1983), *Z. Lebensm. Unters. Forsch.* **176,** 190–195.

SPICHER, G., SCHRÖDER, R. (1978), *Z. Lebensm. Unters. Forsch.* **167,** 343–354.

SPICHER, G., SCHRÖDER, R. (1979), *Z. Lebensm. Unters. Forsch.* **168,** 188–192.

SPICHER, G., SCHRÖDER, R. (1979), *Z. Lebensm. Unters. Forsch.* **168,** 397–401.

SPICHER, G., SCHRÖDER, R. (1980), *Z. Lebensm. Unters. Forsch.* **170,** 119–123, 262–266.

SPICHER, G., STEPHAN, H. (1982), in: *Handbuch Sauerteig – Biologie, Biochemie, Technologie,* Hamburg: BBV Wirtschaftsinformationen GmbH.

SPICHER, G., SCHRÖDER, R., SCHÖLLHAMMER, K. (1979), *Z. Lebensm. Unters. Forsch.* **169,** 77–81.

SPICHER, G., SCHRÖDER, R., STEPHAN, H. (1980), *Z. Lebensm. Unters. Forsch.* **171,** 119–124.

SPICHER, G., RABE, E., SOMMER R., STEPHAN, H. (1981), *Z. Lebensm. Unters. Forsch.* **173,** 21–25.

SPICHER, G., RABE, E., SOMMER, R, STEPHAN, H. (1982), *Z. Lebensm. Unters. Forsch.* **174,** 222–227.

SPICHER, G., RÖCKEN, W., BRÜMMER, J.-M. (1990), *Getreide Mehl Brot* **44,** 274–279.

SPRÖSSLER, B., UHLIG, H. (1972), *Getreide Mehl Brot* **26,** 210–215.

STEIN, E., VOM (1952), *Brot Gebäck* **6,** 165–168.

STEIN, E., VOM (1958), *Brot Gebäck* **12,** 92–97.

STEIN, E., VOM (1971), *Brot Gebäck* **25,** 130–133.

STEPHAN, H. (1956), *Weckruf* **43,** 703–704.

STEPHAN, H. (1957), *Weckruf* **44,** 556–558.

STEPHAN, H. (1959), *Brot Gebäck* **13,** 192–194.

STEPHAN, H. (1967), *Brot Gebäck* **21,** 235–238.

STEUB, G. (1970), *Brot Gebäck* **24,** 53–55.

STEWART, B. A. (1972), *Getreide Mehl Brot* **26,** 176–178.

STOKAR, W., VON (1951), in: *Die Urgeschichte des Brotes,* Leipzig: I. A. Barth.

STREULI, J. H. (1955), *Inaugural Dissertation* Zürich, p. 67.

SUGIHARA, T. F., KLINE, L., MILLER, M. W. (1971), *Appl. Microbiol.* **3,** 456–458.

SWANSON, C. O., WORKING, E. B. (1926), *Cereal Chem.* **3,** 65–83.

SWORTFIGUER, M. J. (1955), *Baker's Dig.* **29,** 65–69, 113–114.

TÄUFEL, K., BEHNKE, M. (1957), *Z. Lebensm. Unters. Forsch.* **105,** 274–283.

THOMAS, B., ROTHE, M. (1956), *Brot Gebäck* **10,** 157–162.

THOMAS, B., TUNGER, L. (1966), *Nahrung* **10,** 85–92.

THOMPSON, D. R. (1980), *Baker's Dig.* **54,** 28–37.

TIPPLES, K. H. (1967), *Baker's Dig.* **41,** 18–27, 75.

TRUM, G. W. (1964), *Cereal Sci. Today* **9,** 248 ff.

TSCHIRPE, W. (1967), *Kakao Zucker* **19,** 448–452.

TUNGER, L. (1971), *Nahrung* **15,** 439–463.

UHRICH, M. G. (1975), *Baker's Dig.* **49,** 43–45.

VASSILEVA, R., SEIBEL, W., STEPHAN, H. (1981), *Getreide Mehl Brot* **35,** 259–263.

VASSILEVA, R., SEILER, K. (1981), *Getreide Mehl Brot* **35,** 184–187.

WASSERMANN, L. (1972), *Getreide Mehl Brot* **26,** 34–40.

WASSERMANN, L. (1981), *Mühle Mischfuttertechnik* **118** (27/28), 395–397.

WEBB, R. B., INGRAHAM, J. L. (1960), *Am. J. Enol. Viticult.* **11,** 59–63.

WEILAND, P. J. (1976), *Getreide Mehl Brot* **30,** 82.

WEIPERT, D. (1976), *Getreide Mehl Brot* **30,** 219–222.

WENZ, E. M. (1971), *Brot Gebäck* **25,** 14–16.

WUTZEL, H. (1967), *Brot Gebäck* **21,** 45–48.

YUDKIN, J., RODDY, J. (1964), *Lancet* **2,** 6–8.

ZBORALSKI, U. (1973), *Getreide Mehl Brot* **27,** 213–216.

8 Commercial Production of Baker's Yeast and Wine Yeast

CLIFFORD CARON

Montreal, Canada

1 Introduction

Yeast, the tool, has had a profound relationship with man and civilization. Porridges made from wild grasses predate history, and once man started planting grain crops (BAILEY, 1975), spontaneous fermentations were a certainty. Historians tell us that middle eastern cultures were making bread and brewing wine and beer before 4000 BC. As a mysterious catalyst it transformed grain into bread, malt into beer and grape juice into wine. These transformations improved the palatability of the starting materials while making them more digestible. This process known as fermentation also improved the biological stability of the new foodstuffs, increasing their immunity to the ravages of spoilage organisms. Yet it was not until the investigation of wine by PASTEUR and HANSEN during the nineteenth century that the catalyst was revealed and the idea of pure culture technique was introduced. Prior to this, fermentations were at the mercy of indigenous microorganisms. Early bakers knew the importance of maintaining the "mother" fermentation, and this technique has continued into the present in the form of sourdoughs and natural leavenings.

Bakers during the early part of the nineteenth century relied on spent brewing and distilling yeasts to insure the success of their bread fermentation. Some attention had been given to "panary" yeasts late in the previous century, but it was not until PASTEUR discovered the positive impact aeration made on biomass yield that specific yeasts became popular. French bakers late in the nineteenth century confirmed the advantage of baking yeasts over spent brewing yeast (BOUTROUX, 1897). BOUTROUX also indicated the presence during that period of German yeast companies specializing in baker's yeasts. By 1860 more attention was paid to the benefit of aeration on the grain mash which resulted in a more active product due to the increase of cell production with reduced alcohol levels. Before this, even baking yeasts contained alcohol levels in excess of 5% which rendered them very unstable. Attempts were made by PASTEUR to dry yeast and extend its shelf-life by mixing pressed yeast with plaster (BOUTROUX, 1897). The next decade brought quantum leaps in the development of yeast production for baking. A chronology of the major sequences that ushered in the era of modern yeast production includes:

1. *Pure culture* techniques introduced by the research of PASTEUR and HANSEN.
2. *Aeration* and more precisely the "Vienna" process that stimulated by biomass yield cell activity and product stability.
3. *Incremental feeding* ("Batch-fed" or "Zulauf") introduced around 1910. This process, which is still the foundation of modern techniques, allowed the production of high biomass yields with little or no alcohol production.
4. *Centrifugation* of the yeast from the fermentation mash permitted the yeast producer at the turn of the century to sell a concentrated yeast product.
5. *Molasses* as a fermentation medium to replace more costly grains, was introduced by the German producers during World War I.
6. *Active Dry Yeast*, developed during the late 1930s, opened new horizons for baker's yeast because of its extended stability without the need for refrigerated storage.
7. *Accelerated Single Strain Baker's Yeast* packaged in bags, introduced in the late 1960s in the USA, greatly altered the industry fostering the growth of large wholesale baking plants.
8. *Instant Dry Yeast* introduced in Europe in the early 1970s revolutionized dry yeast for the baking industry.
9. *Bulk Liquid Yeast* introduced by the late 1970s allowed yeast to be handled as a bulk ingredient in the bakery.

The history of wine yeast as a commercial organism is much shorter. Wine fermentations until the introduction of commercial strains in the early 1970s relied heavily on spontaneous fermentations or the use of pure cultures. While the first scenario depended on the haphazard occurrence of "good" yeasts infecting the must, the second placed tremendous pressure on the winery to handle pure cultures at their busiest time of the year. Implantation of the right strain was determined by the winery environment rather than the vineyard (ROSSINI, 1984), and if a poor quality strain became dominant, it was very difficult to supplant it with another.

Wine yeast production has been very much influenced by the advances made in active dry yeast and instant dry yeast technology. Peculiar to the active dry wine yeast industry has been the proliferation of specialized strains and the importance of killer yeasts.

Much of the focus over the past decade in the commercial yeast community has shifted to automated, computer-driven fermentation processes. Equal attention has been paid to the selection of specific wine and baking yeast strains, genetic alteration of commercial strains, strain identification techniques, alternate feed sources and mixed culture products. The efforts of the industry are concerned with efficiency and productivity and at the same time search for some of the lost attributes of less clean fermentations.

2 Strain Management

Strain management within the concept of commercial yeast production is probably the single most important factor in the success of failure of a yeast producer. Other fermentation industries sell either an inactive biomass, a hydrolysate of the biomass, a primary fermentation product of the biomass or a fermentation by-product of the biomass. The producer of baker's yeast and wine yeast, however, is selling an active biomass. This active biomass must function properly in secondary fermentations, must have acceptable purity to the end user and must maintain its strain integrity.

In the yeast plant, the research and the quality assurance laboratories operate in unity to select, maintain and modify where necessary the organism grown. In the case of baker's yeast usually between one and four strains may be used in the production of commercial baker's yeast both fresh and dry. The number of wine yeast strains used in commercial production can be as few as two and can be in excess of one hundred. Yeast production facilities have spread out considerably over the past two decades to supply national and regional markets as is seen in Tab. 1.

Tab. 1. World Production of Baker's Yeast

Region	Number of Plants	Tons of Fresh Yeast[a]
North & Central America	30	363000
South America	14	83000
Asia	19	145000
Western Europe[b]	31	487000
Eastern Europe[c]	nd[d]	433000
Africa	7	52000

Source: PEPPLER (1983), in part from Universal Foods Inc., PERLMAN (1977) and OURA et al. (1982)

[a] 28 to 32% solids

[b] Includes Turkey

[c] Includes Soviet Union countries

[d] Not determined

2.1 Strain Selection

Selection criteria of yeast strains for commercial production differ greatly between baker's and wine yeast. Selection of baking strains has been largely effected by the yeast producer, while the selection of strains for wine-making has, in almost all cases, been accomplished at the user level either by wine producers or research institutes working in the oenology field.

The selection of bakery strains is influenced by its functionality within the parameters of the modern baking industry. Differences between major markets such as Europe and North America steer the selection of yeasts for end users. Large contrasts exist between these two markets (BAUR, 1991). The yeast producer must also give equal attention to the acceptability of the strain to processing (i.e., fermentation/post fermentation treatment) as well as to yields of the yeast. Generally two bakery strains are selected by the producer, one for fresh yeast and one for dry yeast.

Desirable qualities indicated for a successful baker's yeast include: performance, in terms of fermentation activity as well as consistency; stability of the yeasts over a 3 to 4 week storage period; flexibility of the yeasts within a range of different dough conditions from lean dough without sugar to rich dough with sugar levels over 15% by flour weight; freeze resistance of the yeast with the ability to remain viable and active for prolonged storage periods and finally purity of the product. Some of the necessary criteria in determining quality have been reported (ODA and OUCHI, 1989a).

The present baking industry has three distinct categories; retail bakeries, including in-store shops, wholesale baking plants and frozen dough producers. Ideally a unique strain for each application would be developed. But in reality the consequences of cost factors dictate that the same strains must function in all three applications. While selection of specific strains for bakery applications is continued, particularly for frozen dough use (HAHN and KAWAI, 1990; BAGUENA et al., 1991), the producer must rely on accurate screening of finished yeast to satisfy the requirements for each segment of the market (WOLT and D'APPOLONIA, 1984; GELINAS et al., 1993; NEYRENEUF, 1988). The yeast producer must assure that the strain he is selling exhibits the desired activity for well controlled bakery fermentation.

Selection of a dry yeast strain must take into account the above characteristics as well as the strain's resistance to the stresses associated with desiccation in the drying process.

Many researchers have indicated novel or modified strains for specific bakery applications (UNO, 1986; HINO et al., 1987; TAKANO et al., 1990; TRIVEDI and REED, 1988; NAKATOMI et al., 1986; UNO and ODA, 1990) including frozen dough, but the reality is that baker's yeast is a commodity with little room at present for the selection, development and commercialization of specific value added strains. As the baking industry innovates the need for specialized yeasts, the yeast industry will meet those needs.

Strain selection in the wine industry has followed a different route than baker's yeast. While the initial introduction of a commercial wine yeast was motivated by logistics in application, selection of new wine yeast strains over the past decade has focused on the suitability of the strain with respect to the wine process involved (fermentation conditions encountered) or the grape varieties used. The original strains found in the marketplace in the 1970s were often selected by the yeast producer for their ease of production and for the wide spectrum of applications.

The roles in strain selection shifted in the early 1980s from the yeast producer to the wine community. Yeast levels used in the wine making process were quite low, 1 to $2 \cdot 10^6$ cells per mL of must, compared with dough levels of around 200 to $300 \cdot 10^6$ cells per g, but the impact of the yeast in wine was much greater than that of the yeast in bread. Innovators in the wine industry realized the importance yeast strain could make on the finished wine and accepted the principle of value added yeast products.

The criteria used in the isolation and selection of commercial wine yeast strains must suit the needs of the oenologist (DAUMAS et al., 1990; FEUILLAT, 1986; BARRE, 1980; SCHMITT et al., 1984; CUINIER and LACOSTE, 1980; BENDA, 1986; POULARD et al., 1988) as well as the yeast producer. Researchers have attempted to define wine yeast quality in the following terms:

1. ferment to dryness over a wide temperature range and resist sticking at higher alcohol levels (>10%),
2. rapid onset of fermentation,
3. compete effectively with the other microorganisms encountered in the must (i.e., killer factor),
4. resist normal sulfite levels (20–100 ppm) and normal fungicides encountered,
5. settle rapidly after fermentation is finished,
6. produce little or no hydrogen sulfide and volatile acids,
7. ferment well using nutritionally poor musts,
8. ferment well under pressure to 6–7 bar (i.e., champagne),
9. reach a good sugar/alcohol yield (<17 g per ^{0} alcohol),
10. enhance varietal characteristics.

Wine making technology was influenced by both national and regional parameters, with regard to grape varieties, fermentation practices, etc., and this in turn affected the selection of wine yeast strains. Interesting strains are selected by the wine researcher who gives them to the yeast producer for evaluation as potential dry wine yeasts. Two major challenges facing any wine yeast strain's suitability for commercialization are the ability of the strain to grow well within the constraints of the industrial operation and the capacity to withstand the desiccation of the drying process. Fewer strains are capable of being commercialized than are received by the yeast producer. In the mid 1980s an extensive program to select wine yeast strains for the needs of a particular region took into account many of the criteria mentioned before (DAUMAS et al., 1990; CUINIER and LACOSTE, 1980).

Work done in the 1970s (BARRE, 1980) indicated the significant presence of indigenous killer yeasts in many of the wineries in Southern France. Subsequent studies (BENDA, 1986) continued to support the wide spread incidence of naturally occurring killers in musts and winery equipment. Some studies report the implantation of these killers as coming from equipment rather than the grape skin (ROSSINI, 1984; DELTEIL, 1988).

More recently selection of new strains is bound closely to the grape variety and wine quality desired. Specific strains for specific uses may be the prime criterium used in the future (FEUILLAT, 1986).

2.2 Strain Maintenance

Yeast strains are maintained using several basic methods within a culture collection in the company and comprise both commercial and non-commercial strains. The yeast culture bank is normally maintained within the function of the research department of the company. It is within their function to isolate/purify/and type the strains. Strains are normally kept on agar slants under oil, freeze-dried, or kept in liquid nitrogen. Using these methods, the strain can be maintained to be used as needed.

Proper culture maintenance and storage should maintain the genetic integrity of the strain by arresting metabolism and preventing the possibility of spontaneous mutations. Working slants are produced from the culture bank and supplied on agar slants to the production department. Even when the same personnel maintain and regenerate the strains as well as prepare starter fermentations for production, it is preferable to use separate working slants for production use. Working slants are normally used once to avoid possible contamination of the culture. Used slants are returned to the research department to confirm strain purity and integrity.

2.3 Strain Modification

Strain modification can be categorized into intentional alteration of the strain by classical or novel methods – or modification of the strain through fermentation changes. For our purposes we will consider strain modification within the section as intentional mutation of the strain and will look at fermentation changes under production.

Traditionally yeasts have been modified by using selective breeding techniques. Diploid cells were sporulated and their descendent haploid cells fused with haploid cells of another strain. Each strain was chosen with the intent of endowing the descendent diploid cell with the positive traits of the two parent cells. This form of breeding is rarely used today for modification of commercial strains, as over a period of domesticity most of the strains are either aneuploidal or polyploidal and resist classical methods employing sporulation (ODA and OUCHI, 1989b; MORRIS, 1983; CHAPMAN, 1991).

The procedures used today include selective mutation using specific mitochondrial inhibitors (DELTEIL, 1990; VALADE, 1990), developing hybrids resulting from spheroblast fusion (LILJESTROM-SUOMINEN et al., 1988; CARRAU et al., 1982) and recombinant DNA technology (DELOUVENCOURT et al., 1988; TRIVEDI et al., 1989; BUSSEY et al., 1988; STRASSER et al., 1990; SELIGY and JAMES, 1977).

Spheroblast fusion has had some limited success in the development of a commercial wine yeast (*Saccharomyces cerevisiae*) with *Schizosaccharomyces pombe.* The fusion product, a *Saccharomyces* strain, was able to degrade malic acid. This technique while easily applied requires extensive screening work and the daughter cells of fusions are not selective in their genetic make-up.

Two commercial wine yeast strains have been modified by the use of genetic markers, specifically resistant to two specific mitochondrial inhibitors. The technique was developed (VEZINHET and LACROIX, 1984; VEZINHET, 1985) as a simple tool for detecting the implantation rate of a pure yeast in a mixed culture fermentation. Culture yeasts are inoculated at a rate of approximately $2 \cdot 10^6$ cells per mL of must while indigenous yeast flora varies between 10^2 and 10^6 cells per mL, the higher levels usually found toward the end of the harvest. The marked strains were employed extensively over a wide range of conditions in two different regions and the success of implantation measured. The reasons for using this form of genetic marking were twofold. First, the mutation was mitochondrial without alteration of the nuclear genetic attributes. Second, the screening methods using replicate plating techniques were simple to apply.

While the use of recombinant DNA technology has been effectively used in modification of yeast for novel applications (insulin production), it has found little commercial application yet in the production of bakery and wine strains. Genetically engineered yeasts for the baking and wine industries have been successfully created (BUSSEY et al., 1988; CHAPMAN, 1991), yet they have either not been cleared for commercial use or, if licensed, have run into severe consumer and government opposition. It does not seem likely that governments will be quick to accept the rewards of genetic engineering in the near future, and in many cases they are implementing stricter controls on the release of genetically altered organisms. The principal obstacle appears to be the concern about unknown by-products that might cause allergenic reactions. A disciplined approach to development and evaluation of novel organisms (TIEDJE et al., 1989) appears to be the only solution, and while recombinant DNA technology promises new horizons for the biotechnology industries, its acceptance by consumers and government is not yet clear.

3 Production Trends

The production of both baker's yeasts and wine yeasts, while similar in many aspects, is radically different in others. A strain having been selected and characterized, must now be pushed through the system as it were. A healthy yield of active biomass with predictable characteristics is the goal. The reality is a maze of obstacles and pitfalls, physical, technological, and economical. The target of strain integrity, purity and functionality must be met within the real constraints of yeast production.

Basic fermentation concepts in the production of commercial yeast have changed little in 50 years. Much of the effort during this time has dealt with the subtleties of the craft with advances made in pushing the efficiency of the organism as well as the equipment and the process. It is true that the technology perfected in the production of baker's yeast, and particularly dry baker's yeast, has played a significant role in the success of the development of the active dry wine yeast process. It is equally true that the experience with active dry wine yeast and handling a multitude of strains with unique characteristics has allowed the yeast producer to better manipulate his baker's strain.

3.1 Media Selection and Preparation

Molasses whether cane, beet or a mixture of the two is still the substrate of choice for commercial yeast production. While other growth media have been tried either as replacements or supplements to molasses, none have ever succeeded in supplanting it. Molasses remains an inexpensive raw material

Tab. 2. Composition of Different Types of Molasses[a] Compared with Alternate Sugar Sources[b]

Constituent	Beet	Blackstrap Cane	Concentrate of Beet Juice	Glucose Syrup
Total reducing sugar after inversion (% w/w)	47–58	50–65	64	66
Total nitrogen (% w/w)	0.2–2.8	0.4–1.5	nd[c]	–
α-Amino nitrogen (N % w/w)	0.36	0.05	nd	–
Phosphorus (P_2O_5 % w/w)	0.02–0.07	0.2–2.0	nd	–
Calcium (CaO % w/w)	0.15–0.7	0.1–1.3	0.01–0.15	–
Magnesium (MgO % w/w)	0.01–0.10	0.3–1.0	<0.01	–
Potassium (K_2O % w/w)	2.2–5.0	2.6–5.0	0.5	nd
Zinc ($\mu g\ g^{-1}$)	30–50	10–20	1–15	nd
Total ash (% w/w)	4–11	7–11	1	0.14
Sulfur (SO_3 % w/w)	0.3–0.4	–	–	–
Biotin ($\mu g\ g^{-1}$)	0.01–0.13	0.6–3.2	0.01	–
Calcium pantothenate ($\mu g\ g^{-1}$)	40–100	20–120	15	–
Inositol ($\mu g\ g^{-1}$)	5000–8000	6000	1000	–
Thiamine ($\mu g\ g^{-1}$)	1–4	1.4–8.3	0.4	–
Pyridoxine ($\mu g\ g^{-1}$)	2.3–5.6	6–7	nd	–
Riboflavin ($\mu g\ g^{-1}$)	0–0.75	2.5	nd	–
Nicotinamide ($\mu g\ g^{-1}$)	37–51	20–25	nd	–
Folic acid ($\mu g\ g^{-1}$)	0.21	0.04	nd	–

[a] Based on BURROWS (1970)
[b] From BRONN (1985)
[c] Not determined

Tab. 3. Yeast Requirements for Mineral Elements[a] and Vitamins

Substance	Requirement (ppm)
Potassium	5000
Magnesium	500
Calcium	500
Sodium	250
Zinc	50
Iron	25
Manganese	5
Copper	2
Cobalt	1
Biotin	0.3
Pantothenic acid	40
m-Inositol	600
Thiamine[b]	11

From BRONN (1985)
[a] Excluding ammonia, phosphate and sulfate ions
[b] Not essential but beneficial

for yeast growth providing a good carbohydrate source along with other minor elements beneficial to the yeast (see Tabs. 2 and 3).

European production relies on sugar beet molasses as a raw material for media production due to the European Communities' beet sugar industry. The warmer countries tend to grow yeast using cane molasses only. In North America blends of both cane and beet molasses have been traditionally used, where the ratio has been decided based on either the economy or the cleanliness and suitability of the molasses. Cane molasses is richer in vitamins and marginally higher in fermentable sugars with a fairly equal balance between reducing sugars and sucrose. On the negative side, it is a dirtier product with a fairly high level of insoluble sludge which must be removed prior to fermentation. Beet molasses is cleaner and contains a higher ratio of sucrose than cane. It is, however, more expensive and somewhat poorer in minor constituents such as vitamins, in particular biotin. Biotin is essential for healthy yeast growth, and the minimum level of 25 μg/g of molasses can usually be found in cane molasses (WHITE, 1954). When using high levels of beet, biotin must be supplemented for good yields, and simple tests for monitoring biotin levels in the medium (DANIELSEN and ERIKSEN, 1968) can help in the optimal dosing of biotin.

Driven by effluent concerns, product availability, or rising molasses costs, yeast producers are continually evaluating alternate feed carbohydrate sources for yeast production. Whey can be used as the sole carbohydrate source for growth of *Kluyveromyces fragilis,* but its poor baking properties render it unsuitable for commercial application. Baker's yeast can be grown on whey permeate utilizing lactic acid which has been produced by inoculated bacteria as a carbohydrate source (CASSIO et al., 1987; CHAMPAGNE et al., 1989). Similarly inoculated whey has been successfully used to supplement the molasses at a 20% level in a commercial operation, but was discontinued because of variations arising from the lactic acid fermentation. A joint venture in the USA successfully operated a yeast plant using immobilized lactase enzymes to supply the baker's yeast with fermentable sugars (PEPPLER, 1983). The plant has since closed. More recently attention has been paid to corn steep liquor, corn molasses (hygrol), and other agriculturally based media as suitable substrates for yeast production. Interest in these substrates has been motivated from an environmental point of view as well as a concern for yield, cost, and yeast quality (MOEBUS and KIESBYE, 1982; BRONN, 1985). Attention may be directed in the future to the use of regional by-products as raw material for commercial yeast fermentation rather than importing by-products from another region. These agricultural by-products contribute BOD to the ecosystem. Logically our aim will be to reduce the overall BOD load on the environment by keeping the raw materials from the local region. Whatever the source used as fermentable substrate, attention must be paid to its chemical as well as its microbiological composition. The commercial yeast industry cannot afford the sterile environment of the pharmaceutical or even the enzyme industry and must control contaminants from entering the system whenever they can.

Cane and beet molasses are stored separately so that the ratio of the final blend can be altered at will by the producer. Often the yeast producer will use a higher ratio of beet molasses for seed and dry yeast production and a higher cane ratio for the commercial stages of fresh yeast production, the reason being the cleaner composition of the beet molasses as well as the ease in pasteurization. Additionally the lower level of invert sugars in beet molasses contributes to a whiter color which is beneficial in dry yeast.

In the preparation of molasses media for fermentation, the initial molasses is first diluted to a workable concentration – normally around 38–42 ^{0}brix (see Fig. 1). Control of the final dilution is effected by the use of a constant brix instrument. A constant dilution is critical for the efficiency of the fermentation process. Fermentation feed rates are based on inoculum level then calculated for the molasses dilution figure. Higher sugar levels in the medium can trigger the Crabtree effect, while reduced levels can create a starvation situation. The diluted molasses blend (wort) is acidified to pH 5.0–5.5 for better pasteurization efficiency. The wort is normally preheated to accelerate flocculation of insolubles, then clarified by centrifugation. The clarified wort is injected with wet steam and passed through a static mixer to achieve optimum gas/liquid dispersion. The wort/steam mixture is normally held at 120–125 °C for between 15 and 30 seconds. Increased bacterial destruction can be realized by dropping the pressurized slurry into an atmospheric vessel and "flashing" the wort. The pasteurized wort is then passed through the heat exchanger to cool down while preheating the raw wort. Two critical points in this step are pH control and the separation of the pasteurized and raw wort. Over-acidification of the wort can increase the caramelization of the sugars affecting both the yield and the finished yeast appearance. Contamination of the pasteurized wort by the raw wort can introduce bacteria and wild yeasts into the process.

Cross-contamination of the pasteurized wort is the more frequently found cause of wort-related contamination of the fermentation. The resulting microorganisms normally associated with this situation are wild yeasts, as well as mesophilic and thermophilic bacteria including *Leuconostoc* sp., *Lactobacillus* sp. and *Streptococcus* sp. Insufficient heat treatment of the wort usually results in con-

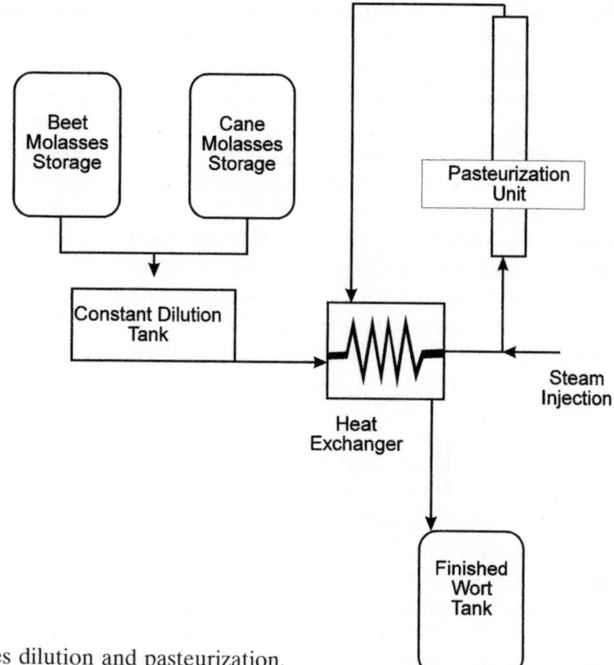

Fig. 1. Schematic flow diagram of molasses dilution and pasteurization.

tamination of the fermentation by the vegetative cells of spore-forming bacteria. It is rare to find both types of contamination in one process.

After cooling, the wort is stored in separate tanks which are used to feed the fermentations. Since sterility of the wort cannot be guaranteed, maximum storage times of between 4 to 8 hours are used to prevent any microbial growth in the storage tanks.

3.2 Seed Yeast Propagation

The importance of the seed yeast stages in commercial yeast production is threefold. The first concern is production of biomass in the shortest time possible. The second concern is maintaining the purity of the culture yeast, and the third concern is developing synchrony of the yeast population toward the end of the seed yeast process.

Between six and seven well defined fermentation stages are normally encountered in commercial yeast production. In the rare instance, the process can be contained in as few as four stages, but this is only practiced when unusual circumstances arise such as exist in some developing countries. Except for the last stages all other stages are considered seed yeast fermentation stages. Initial laboratory stages (1 and 2) are run using batch fermentation systems with sterility of the fermentation systems being a prime concern. Specific media are developed for these stages which are sterilized prior to yeast inoculation. The first stage is normally a flask fermentation of between 5 and 20 liters. This flask stage is inoculated with a production slant and fermented for 24 to 48 hours with mild agitation. Normal biomass yields in the initial flask fermentation are about 20 grams of dry matter. The flask is then used to inoculate the second lab stage which will run between 12 and 36 hours. The fermentation is aerated using sterile air for the entirety of the cycle. The efficiency of this stage is very low, probably less than 10% dry solid yield, but it is the rapid development of biomass along with strain purity that are the target. The size of the second stage fermentation will vary between 500 and 2000 liters of working volume and the final biomass produced will be approximately 50 kilos of dry matter. During the first two stages, no restrictions are placed on the yeast growth, nor has any emphasis on synchrony of the cells been addressed.

During the third and fourth seed stages, the prime concern is still biomass production and strain purity, but more attention is now paid to the synchrony of the cell population and the physical attributes of the biomass. The entire contents of the second stage are used to seed the third stage which will ferment between 10 and 18 hours. The volume of the third stage can be anywhere from 10000 liters to 30000 liters. For the first time, control of yeast growth is implemented by restriction of the sugar feed (batch-fed system). In addition to limiting the fermentable substrate, the mash is highly aerated. The final biomass yield from the third stage varies from 300 to 700 kilograms of dry matter. The entire third stage is used to seed the fourth stage. In essence the fourth stage is a larger replica of the third stage, and in many operations both stages are contained within one fermenter. The fourth stage will usually run from 12 to 16 hours with a final biomass production of around 2500 kilograms dry matter. By the end of the fourth stage, the cell population has begun to develop a suggestion of synchrony. The normal density at the termination of the fourth stage is between 50 and 70 grams of yeast dry matter per liter.

The yeast cells are separated from the fermentation medium for the first time. The need for separation is twofold. Fermentation capicity becomes a physical limitation, and it is less costly to separate and store seed cream yeast than to build enormous fermentors. Also, to achieve the synchrony in the following stages that is necessary for proper yeast functionality, seed inoculum for the final seed stage and the commercial stages must be precisely controlled. The fifth and last seed stage is seeded at 4–7% yeast solids based on final molasses. The fermentation duration lasts from 12 to 16 hours with normal working volumes of around 60000 liters. The growth of the cells is well controlled by sugar restriction. Due to the nature of the process, air becomes the limiting factor in this and all subsequent stages. Cell synchrony in the fifth stage approaches that of the final commercial stage. Synchrony is a critical factor in commercial yeast production not only in regulating the growth, but more importantly in the stability of the final yeast product. After the termination of the fifth stage, the yeast is again separated from the spent fermentation liquid. In both the fourth and fifth stages, the separated yeast is held in refrigerated storage tanks (4°C). During the seed fermentation stages, little attention is paid to the fermentation activity of the yeast. Protein and phosphate levels run normally higher than in the commercial stage. The final biomass yield from the fifth stage is around 3000 kilos of dry matter. The commercial yeast process is a branching effect. The fourth seed stage can usually feed 4 or 5 fifth seed stages, and each fifth stage can feed 2 to 4 commercial stages depending on the duration of the commercial stage and how many cycles are desired.

3.3 Commercial Yeast Propagation

During the commercial stage of yeast propagation emphasis is placed on controlled growth. The uniqueness of both baker's and wine yeast production is the emphasis on functionality of the finished biomass. What the yeast producer is selling, is a well defined activity within the parameters of the end use for the yeast. Not only must the functionality be well defined and quantifiable, it must also be stable over accepted periods of time and be reproducible from production lot to lot as well as from one season to the next. The commercial yeast producer must also be concerned with strain purity, integrity and the final cost of the product. There is always a compromise between quality and economy. Even though raw material costs, labor costs and processing costs have continued to increase, selling prices of baker's yeast have dropped necessitating the implementation of more efficient production methods. The baker's yeast industry has seen many recent changes, and some industry giants of the past no longer exist. The successful yeast producers are those who have reduced their production costs per kilogram of yeast.

As stated earlier, molasses blends constitute the main feed source for yeast growth.

As well as supplying carbohydrates for the cell, molasses is a source for some minor constituents – mainly vitamins. Cane molasses normally contains adequate biotin for the yeast growth, while beet molasses is deficient. The sugar industry has made many technological advances in its process resulting in poorer quality molasses to the yeast producer. As sugar levels decline and the insoluble content (sludge) increases, vitamin levels drop as a result of denaturing during the evaporation of the sugar syrups. With this constant decline in molasses quality, the producer must rigorously analyze new sources for fermentable sugars, inhibiting factors, and vitamin levels.

For protein synthesis in the yeast cell, nitrogen is furnished in many forms. The most common nitrogen source is aqua ammonia (NH_4 – 28%). Nitrogen can also be added as an ammonium salt (SO_4 or PO_4), and this has the added benefit of buffering the pH level during the fermentation where excess free ammonia can cause the pH to rise. In areas where aqueous ammonia is not available or where the cost is prohibitive, urea is used to supplement the ammonium salt. Except for a few fastidious brewing and wine yeast strains, inorganic nitrogen in any of the above mentioned forms is sufficient. In the case of these fastidious strains, free amino nitrogen can be added as either a yeast extract or a malt extract. In some special cases distinct amino acids can be used to supplement the nitrogen source (HO and MILLER, 1978).

Phosphorus is added as phosphoric acid (75%) or ammonium phosphate. While the phosphoric acid has cost benefit, the ammonium phosphate supplies both nitrogen and phosphorus with a buffering effect. Excessive pH shift during early stages of the fermentation can affect the yeast metabolism. Changes in yeast growth at this stage can affect carbohydrate uptake as well as nitrogen and phosphorus assimilation. Once an imbalance is created in the fermentation, the proper growth cannot be achieved, and excesses of residual sugar create catabolic repression, and excesses of residual phosphorus or nitrogen can inhibit (burn) the yeast growth.

Many trace elements are added to the fermentation to supplement deficiencies or prevent possible deficiencies in the molasses. Minerals such as magnesium and zinc are added as well as vitamins such as thiamine, pantothenic acid and biotin. Magnesium helps in optimizing yield, while zinc is added because it is a coenzyme factor in the synthesis of the protein content of the yeast. Creation of accelerated strains with high protein levels (greater than 50%) cannot be efficiently produced without the use of zinc. Copper is essential at very low levels (WHITE, 1954) and is usually present in sufficient quantities in the molasses. Both thiamine and pantothenic acid are essential vitamins for yeast growth and play an important role in the functionality of active dry wine yeast. A deficiency in thiamine promotes hydrogen sulfide production by the yeast cell; a deficiency in pantothenic acid promotes the production of volatile acids. It benefits the yeast producer to supplement his growth media with these vitamins. Biotin is an essential growth element for yeast, and a level of 25 micrograms per gram of yeast solids is a minimum requirement. If the molasses is deficient in biotin, it can be supplemented during the fermentation.

During the commercial stage, economics plays a major role. For this reason the size as well as the energy efficiency of the fermenter is a major concern. It is rare in today's industry to find commercial fermenters of less than 100000 liters in size. While the scale of economy dictates an increased size to reduce operating costs, cooling capacity limits the ultimately attainable size (see Fig. 2). During the course of the process, as many as six stages can be used or as few as four (see Fig. 3).

Commercial fermenter design has evolved since the 1960s. Higher energy costs in Europe drove engineers to high efficiency fermenters while North American design stressed increased capacities. When energy costs rose dramatically towards the end of the 1970s, efficiency became more of a concern as did emission standards for fermenters. Commercial fermenters can be broadly classified into one of the following:

1. Traditional tubular fermenter design with branched air sparging system. Fermenter height is increased to improve solubility of oxygen.

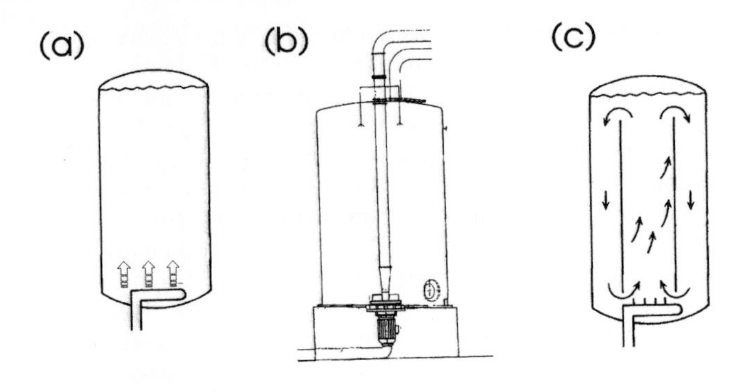

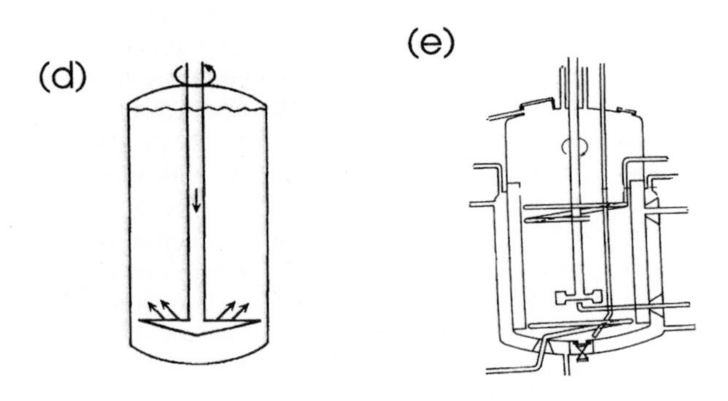

Fig. 2. Schematic diagrams of different fermenter designs including a simple sparge system (a), a high efficiency design with very low energy consumption (b), a simplified airlift design (c), a turbine generator design (d), and a stirred vessel design with internal baffles and exterior cooling jacket plus internal helical coils (e). From ATKINSON and MAVITUNA (1991), design (b) is from H. Frings Co.

2. Tubular design with mechanical agitation and side baffles that increase turbulence improving dissolution of oxygen by creating smaller bubble size. Air spargers can either be traditional branched systems or doughnut-shaped.
3. An airlift design which is fairly energy-efficient and achieves good oxygen dissolution by creating strong flow within the fermenter by use of a draught tube. The turbulence improves oxygen contact and prevents coalescence.
4. High energy-efficient systems such as the Vogelbusch or Frings design use high-speed rotating blades to disperse and dissolve the oxygen. Oxygen removal rates are reported greater than 15%, compared to less than 5% for simple sparger systems (BURROWS, 1970).

Many attempts have been made with continuous fermentation systems for the production of commercial yeasts, but have failed because of infection problems or reduced stability of the finished product.

The normal limiting factor in commercial stages of yeast propagation is oxygen and, it is always the focal point in fermenter design. Only the dissolved oxygen can be used by the yeast, and as the solubility of oxygen is very low in water (1.16 mmol/L) (FINN, 1967), emphasis is placed on increasing the dispersion of the oxygen. Oxygen is normally supplied as compressed air using either turbines or blowers. Saturation is the goal, but is usually impossible after the first few hours, as the biomass increases and the yeast consumes the oxygen faster than it can be supplied. The producer tries to correct this problem by improving the efficiency of the oxygen dispersion or by enriching the air with pure oxygen. The former approach tries to improve the efficiency of oxygen dispersion by:

1. Increasing the fermenter height, thereby increasing the pressure at the air inlet and increasing the solubility of the

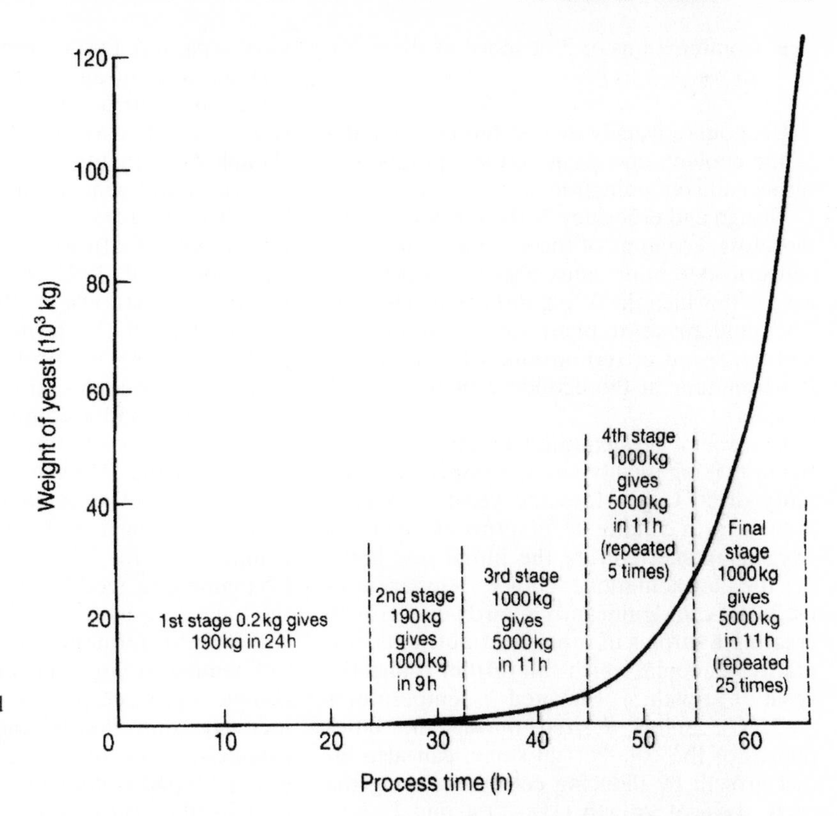

Fig. 3. The time related process showing the initial four seed yeast stages and the final commercial stage. From BURROWS (1970).

oxygen.

2. Improving dispersion by the use of mechanical agitation and increasing the turbulence in the fermenter by the use of baffles.
3. Increasing the shearing effect, thereby reducing the bubble size with the use of high-efficiency sparger systems such as the Frings generator.
4. Using air lift systems to decrease bubble size and improve oxygen dispersion.

Cooling of the fermenter is the second most challenging parameter the producer must face. The rate of evolution of heat is a function of growth rate and biomass concentration (REED and PEPPLER, 1973) and is proportional to the oxygen consumed and the CO_2 evolved (COONEY et al., 1969). It has been reported as 4.4 kcal (18.5 kJ) evolved per gram of dry matter. The production of baker's yeast implies maximum biomass yield per unit weight of substrate. Using total aerobic conditions, the yield attainable is reported as 54 kg of yeast dry matter per 100 kg of glucose (REED and NAGODAWITHANA, 1991) which can be loosely transposed as 90% yield of yeast at 30% solids based on molasses weight (assuming 50% fermentable sugars). In reality, yields above 70% are considered excellent. To achieve maximum yield the carbon source must be the limiting factor, yet in practical situations, the efficiency of heat removal often dictates the final fermentation parameters.

A notable amount of energy is liberated from the exothermic reactions into the medium and must be removed to maintain the fermentation temperature. Control of heat is achieved by one of the following practices:

- External jackets
- Internal helical coils
- External helical coils
- External heat exchanger
- External water cascade

- Combinations of 2 or more of the above.

Economics usually dictate the use of water as the coolant, and create critical points with respect to contamination.

Design and efficiency of the fermenter and, therefore, economy of the system is the major concern. Attention must also be paid to the ease with which the fermenter can be cleaned. The modern yeast plant runs continuously, and the speed of turn-around of a fermenter is paramount in the economy of the overall process.

Control of fermentation kinetics is essential in making quality yeast. Temperature stability of 30°C provides the yeast with optimum growth conditions, in terms of sugar and oxygen uptake. During the initial few hours of the fermentation, cooler temperatures (<28°C) can significantly retard yeast growth creating a surplus of other feed nutrients, primarily ammonia, which can further retard the yeast growth. Elevated temperatures (>32°C), unless desired during the latter phases of the commercial stage, can also affect growth by inducing cell lysis during the early stage of growth (HAGLER and LEWIS, 1974).

pH control is the second most important growth criterium. During the seed stage fermentations and also during the early part of the commercial stage, the pH can be intentionally maintained between 4.5 and 5.0 to suppress bacterial growth. During the latter half of the commercial stage, however, the pH is allowed to rise to 6–7. The increase is normal and attributable to the increasing concentration of fermentation metabolites and substances in the wort. Increased biomass concentration can be attained by continuous removal of metabolites (FUKUDA et al., 1981). pH control of the commercial stage is normally used to reduce bacteria levels or to impart an improved resistance to organic acids.

The dissolved oxygen (DO) is monitored for historical reasons, but can be used to reduce molasses and nitrogen feed if deemed necessary. We can assume that at 30°C saturation will be 7 ppm. During the course of a commerical fermentation, the DO will drop to 0.4 ppm or less than 5% of saturation (STROHM and DALE, 1961). Simple sparger systems without agitation may not solubilize more than 20% of the incoming oxygen (REED and NAGODAWITHANA, 1991). Even though the benefits of aeration to both baker's yeast and wine yeast quality are well established (CALCOTT and ROSE, 1982; PARK et al., 1985; KISHORE and KARANTH, 1986; ZIKMANIS et al., 1982; MAURICIO et al., 1991; CASEY and INGELDEW, 1986), excess oxygen at the start of the fermentation can create problems (TYAGI, 1984). Commercial fermentation schedules are precisely designed formulations and account for normal growth under normal conditions. Sudden unpredicted changes in the DO values should be treated as symptoms of an imbalance within the system, and managers should not try to overcompensate for this unexpected change. Changing one feed to suit a sudden shift in the balance usually produces a finished yeast with inferior fermentation qualities.

Foaming is an ever present problem in commerical yeast production and can be effectively controlled by the use of anti-foam products. Foam in a yeast fermentation is a gas in a liquid colloidal system protected by a protein film (BICKERMAN, 1973) and can be attributed to (1) an unusually high degree of cell lysis, liberating proteinaceous material, (2) a wild-yeast infection where the production of pseudomycelium on the surface of the fermenting liquid traps the CO_2, and (3) unusually high levels of molasses debris. Foaming can be controlled by the use of anti-foam agents, but large amounts of these products can reduce the solubility of the oxygen (SOLOMONS and PERKINS, 1958).

Anti-foam agents are added either by time-released prescribed doses or by sensor-activated solenoid switches. One of the problems associated with solenoid-activated systems is that the switch can become contaminated with dry foam film which can create "false" positive indications which activate the valve, dosing large quantities of anti-foam into the mash. Timer switches are normally used with a solenoid system that shuts the valve after several seconds and sounds an alarm.

Both the nitrogen feed rate and the phosphate feed rate are predetermined to follow the schedule of yeast growth. Fermentation operators conduct formol nitrogen analyses on the fermenting mash to monitor the ammonia assimilation by the yeast. Formol tests are a good monitoring device for gathering data, but should not be used as criteria for changing the nitrogen feed or altering the growth parameters. Changing the fermentation schedule because of a rise in formol nitrogen is treating the symptom and not identifying the cause.

Some of the data used to assess growth are:

1. Direct measurement of the DO in the mash,
2. Stack gas measurement of both O_2 and CO_2,
3. Energy (heat) production by the fermenting mash,
4. Head space alcohol.

Alcohol production during the commercial stage is virtually zero. Some individuals promote alcohol production during the first hour of the commercial stage with the intention of improving yeast growth (RICKARD and HOGAN, 1978) as well as sanitizing the starting medium (REED and NAGODAWITHANA, 1991).

Before the last decade, most yeast fermentations had the nutrient feed controlled manually or by cam driven systems. Today the microprocessor has replaced most of the traditional monitoring systems. Researchers have developed mathematical models as analogs of the fermentation process (PERINGER et al., 1974), and the first generation of computer-controlled fermentations utilize the PC to manage the various feed rates as well as pH and temperature control. Second generation systems will use direct and indirect measurement of key indicators to control feed rates to maintain RQ values and maximize yield (WHAITE et al., 1978; WANG et al., 1977).

3.3.1 Baker's Yeast

Until the mid 1960s most North American yeast producers used different strains for block yeast and bag yeast. Bag yeast used a single accelerated strain, while block yeast was a mixture of the accelerated and a dry yeast strain. The aim was to supply the wholesale bakery with a very active fermenting yeast while giving the retail baker a slightly slower yeast with better stability and keeping characteristics. Improvement of the strains being used has changed this practice so that today the same strain is used for both block and bag as well as for bulk liquid yeast. A separate strain is used for yeast destined for drying.

Fresh baker's yeast is used in three different strata within the commercial baking market: the retail operation which includes in-store bakeries, the wholesale baking plant, and the frozen dough producer. The division of these groups differs greatly between Europe and North America (BAUR, 1991), and the yeast industry in each market has changed to meet the needs of its market. Each segment of the market has common as well as individual requirements for successful operation. While separate strains can be used for specific market segments or indeed dough type (HAHN and KAWAI, 1990; TRIVEDI et al., 1989; NAKATOMI et al., 1986; BAGUENA et al., 1991; STRASSER et al., 1990; CHAPMAN, 1991), the economics of commercial yeast production inhibit this approach. Fresh yeast in the North American market is treated like a commodity, and while the cost of production has increased over the past half century, the price, modified to reflect inflation, has actually dropped. Scale of economy, automation and higher efficiency have been the target for the successful yeast producer. Multiplicity of strains is accompanied by higher costs to the producer which cannot be passed on to the end user.

The fresh baker's yeast should function well independently of the dough system used as well as over a varied range of formulae, both lean and sweet. The retail segments normally use either a traditional or no-time straight dough. They need acceptable activity from the yeast, but more important, they need good stability. They need stability in terms of moderate fermentation activity with a minimum of proteolytic activity, and stability in terms of consistency over the expected shelf-life of the yeast. Wholesale bakeries lean towards sponge and dough, flour or water brew and less frequently no-time straight dough systems. Their operations demand a yeast that performs very consistently and maintains a steady performance rate for 7 to 14 days. Consumer buying patterns for bread compel the baker to use preservatives, usually calcium propionate, and this requires a yeast resistant to normal preservative levels. The growth of frozen unbaked dough in the bake-off market has been meteorical since the early 1980s both in Europa (CALVEL, 1988) and in North America. This situation has stimulated selection and production of yeasts suitable for freezing. Storage periods for frozen dough can be as long as six months before thaw and bake-off, and while much of the structural weakness of frozen dough comes from damaged gluten structure (AUTIO and SINDA, 1992), the yeast can suffer from ice crystal formation causing cell rupture and reduced activity.

Most block production is destined for retail operations, so the solids can be reduced slightly to improve stability in the retail bakery. There is less surface area in the block yeast than in the bag yeast, and this benefits its storage stability.

For the wholesale operation speed, reproducibility, and versatility are the key criteria for the success of the yeast. The modern bakery can produce bread at a rate of 200 loaves per minute with proof times set at 58 min at 35 °C, 75 % relative humidity. While the yeast must have good activity, it should also demonstrate low proteolytic activity. Excessive fermentation activity in the sponge or flour brew can negatively affect the rheology of the final dough causing „ageing" and subsequent poor volume and open internal grain. To accomplish this, the producer maintains a fairly tight control on the enzyme activity of the yeast by controlling the protein level, as well as the zinc addition to the commercial fermentation. Both the amount of zinc as well as the sequence of addition play a role in optimizing protein functionality. Phosphate level or more precisely phosphate to protein ratio also influences the functionality of the yeast. Normal protein levels for fresh yeast are 54 to 58 % (N × 6.2) with phosphate levels of 2.8 to 3.3 % (expressed as P_2O_5). Yeast in the European Community tends to be less active, but more stable because of the structure of the baking industry. The solids of bag yeast are higher than of the block yeast and range between 29 and 33 %.

The other important factor for yeast sold to the wholesale baker is resistance to preservatives. Bakers normally use calcium propionate, potassium sorbate, vinegar, cultured whey or cultured flour or mixtures of two or more products to increase the mold-free shelf-life of the baked goods. Levels in bread products are 0.25 % (flour basis) for propionate, 0.1 % for sorbate with no limits for vinegar or cultured products. In specialized products such as sourdough breads and English muffins, the level of acids can be as high as 1 to 2 %. Pressure from the baker has forced the yeast producer to improve the yeast's resistance to organic acids. Commercial yeast propagation stages can be modified to improve the yeast's resistance. Techniques used include growth with reduced pH, addition of controlled levels of organic acids (WARTH, 1989), modifying the temperature during the latter stages of the fermentation (BURROWS, 1970) and fluctuating the oxygen to carbohydrate ratio during the latter part of the yeast cycle (WILLIAMSON and SCOPES, 1962). These techniques improve the cell's resistance to organic acid by modifying the membrane so that it retains its permeability under stress conditions.

For the frozen dough producer, the yeast must perform well over a range of product types. More important, it must perform well after being frozen and thawed even after four months of storage at 0 °C. The frozen dough producer is sensitive to yeast malfunction, as it often shows up only after several weeks of storage by which time the frozen product is in distribution. The yeast producer can achieve excellent results in frozen dough by (1) selecting a specific strain with improved attributes for freezing, (2) modifying the existing strain to improve its freeze tolerance or (3) use cryo-protectants to improve the yeast's resistance to the freezing stress as well as the storage period. While novel strains can be effectively used to extend frozen storage, the yeast industry relies on the last two methods for commercial production.

Yeast trehalose content has been reported as beneficial in producing yeasts with improved resistance to freezing, desiccation and high osmotic pressure (PANEK, 1985; HSU et al., 1979; GODKIN and CATHCART, 1949; RUDOLPH and CROWE, 1985; COUTINHO et al., 1988; HINO, 1990; HOTTINGER et al., 1987). Other information confirms the benefit of cell membrane lipid content (KRUUV et al., 1978; CALCOTT and ROSE, 1982; GELINAS et al., 1991a, b), moderate protein levels (DUNAS and WENDT, 1988), and increased initial lag phase (NEYRENEUF, 1990) to improved frozen dough performance. Fermentation of yeast for frozen dough can be modified to promote increased trehalose levels (PANEK, 1975; GRBA et al., 1975; KELLER et al., 1982; LONDESBOROUGH and VARIMO, 1984), or improve the membrane integrity (ZIKMANIS et al., 1982; MALANEY and TANNER, 1988; REED et al., 1987).

In the production of dry yeast, the concerns in producing a quality fresh yeast are increased by the need for the yeast cell to survive the desiccation process and to perform acceptably after rehydration and reactivation. Normally the protein is kept at a lower level (45 to 50%) as well as the phosphate level (2.5 to 2.9%). Maturity (ripeness) of the biomass is essential in achieving proper results. The fermentation is drastically altered to ripen the cells and to increase the carbohydrate content by combined feed/starvation. Trehalose levels above 14% are often encountered. The ripening process is also achieved by restricting phosphorus early in the commercial stage and by limiting nitrogen. Warmer temperatures are used for the last third of the cycle (ZIKMANIS et al., 1982). Salt can also be used at the beginning to increase the osmotic pressure in the fermentation mash and to increase the glycerol content of the cells. The resultant cells are always all mature with no budding cells present. The accentuated maturity also induces a lag phase to the yeast.

3.3.2 Dry Wine Yeast

The production problems associated with active dry wine yeast are similar to those encountered in the production of baker's yeast both fresh and dry. The principal difference between the two types of yeast lies in the number of strains and the complications that arise when handling a large number of strains within a yeast plant. While most baker's yeast operations have to contend with two strains, the same plant will have to manage as few as four strains and as many as one hundred wine yeast strains.

While world production of baker's yeast can be estimated at around 1600000 tons fresh yeast (see Tab. 1), the production of active dry wine yeast is probably less than 1000 tons. Unlike the baker's yeast operations, the scale of economy does not play an important role in the success of wine yeast production. Most probably no single producer sells more than 100 tons of the same strain in one year, and in the sphere of speciality wine strains, annual production can be as little as one ton. The main reason for the number of strains lies in the uniqueness of the commercial wine industry. The success of spontaneous fermentation lay in the winery itself playing host to a beneficial yeast strain. The early efforts in active dry wine yeast focused on their ease of use for the large winery. While this was, and still is, an important economic factor, the real growth of the wine yeast industry started with the introduction of speciality strains (SPONHOLZ, 1983).

Our concept of speciality wine strains is based on the assumption that the dry wine yeast is sound, both in genetic terms and in physical terms. Once this truth is accepted, the yeast ceases to be a commodity. The use of a specific wine yeast strain is tailored to the region, the grape type, the wine-making style and the fermentation parameters. The wine yeast is not seen simply as a convertor of sugar to alcohol, but as an insurance policy for the winery. The tank of grape must have a real value, but it also has a less intrinsic implied value as wine. The oenologist uses yeast in a more sophisticated manner than the baker. He wants an efficient fermentation, but he also wants to control as many aspects of the final outcome as possible.

The characteristics sought by the wine-making industry start with the strain. Unlike the baking strains, which have been the result of yeast research, wine yeast strains are the offspring of wine-related research. These wine yeast strains are like thoroughbreds with their lineage well documented. Almost all of the strains can be traced to a specific winery, vineyard or region.

Yeast strains that have survived until the present are still used because of consistent results. The yeast producer can only have an esoteric interest in the actual functionality of the yeast; his concern must remain with the quantifiable parameters of the product. Functionality of the baker's yeast in bread applications is the most important criterium for a commercial baker's yeast. The functionality of the wine yeast strains is best left to the oenologist. Because of the number of grape varieties, wine styles, and vinification techniques, it would be impossible for the yeast producer to run each batch in every possible situation. Much of the work done on the negative effect of fatty acids on yeast fermentation and the benefit of adding yeast hulls (ghosts) (LAFON-LAFOURCADE et al., 1984) and the correlation of aerated growth on the formation of sterols and sterol precursors (MAURICIO et al., 1991) has generated much interest in the functionality of dry wine yeasts.

The yeast producer is selling a pedigree. If the strain integrity can be guaranteed and the viable cell count controlled, then the functionality should follow. The wine yeast producer is selling a controlled fermentation where the sum of all the fermentation parts makes the wine. The baker's yeast producer is selling a defined activity.

Wine yeasts can be loosely divided into the *Saccharomyces cerevisiae* and *S. bayanus* strains with the former representing about 60% of the strains produced. More unusual strains including for Sherry, *S. rosei*, *S. uvarum* and *Schizosaccharomyces pombe* and *maldevorans* have been produced commercially, but remain on the fringe. Their fields of application are very narrow, and they are very problematic either from a fermentation point of view, requiring complex nitrogen sources and vitamins, or from a downstream handling perspective. Many wine strains because of their morphology or characteristics do not lend themselves to the physical demands during the separation stages or dewatering stages. This is particularly true of non-*Saccharomyces* strains.

With few exceptions, the majority of the wine yeast strains grow more slowly than do baker's yeast strains. They have poor yields and have to be almost starved during the commercial stage to prevent any alcohol accumulation in the cell. Even low levels of ethanol will reduce the viability of a dried product by over 20%. A four-hour doubling period during the non-restricted growth of the commercial stage for baker's yeast is normal, these times can easily climb up to eight hours for some wine yeast strains.

Active dry wine yeasts were designed to be seeded into the must at a level of 10 grams per hectoliter. The culture yeast must compete in a very hostile environment, and to be successful it must supplant any wild yeasts (FLEET et al., 1988) in the must und must stabilize the fermentation before bacteria can proliferate. Aerobic growth and promotion of high trehalose levels along with membrane integrity are critical for producing a dry wine yeast that will function well (MUNOZ and INGELDEW, 1989; HO and MILLER, 1978; CASEY et al., 1986; MAURICIO et al., 1991). With a viable population of 20×10^9 cells per gram inoculating with 10 grams per h would yield an initial cell count of 2×10^6 culture yeast cells per cm^3 of must, which increases to 10^8 million cells per cm^3 before true alcoholic fermentation takes over.

Bacteria counts of 10^7 per gram of yeast and wild yeast counts of 10^4 per gram are not common but can be tolerated in baking. They are a potential problem in wine making, especially when we find that most wineries do not use the dry yeast only once, but pitch it several times. The Office International du Vin has attempted to establish guidelines with regard to dry wine yeast in an effort to standardize the wine yeast industry (RADLER, 1984). Many wine yeast strains do not go past four stages prior to drying – rather than the six associated with baker's yeast.

During the separation dewatering stage, some wine yeast strains fare poorly. Whether it is the pressure of centrifugation or the physical compression of the cells during filtration and extrusion is not known, but some strains leak cell material blocking equipment. Particularly difficult to handle are flocculant yeasts which can easily block the centrifuge nozzles or the extruder plate (see Fig. 8). Elevated compression in the extruder can develop excessive heat which reduces viability.

3.4 Biomass Recovery

Normal yeast solids by the end of the commercial stage of yeast production rarely exceed 5 or 6%. The biomass is concentrated and "washed" by the use of centrifuges. By varying the nozzle size, and therefore the pressure, the solids concentration is increased to between 16 and 20%.

Separation is also used for the latter seed stages. An added part of the commercial stage separation is referred to as the "washing". This term encompasses one or two repasses of the commercial yeast diluted with water to wash away unused non-fermentables and color. Bakers have become accustomed to a white product and associate any dark tones with inferior quality. Washing can be achieved with only water or with added product to reduce the bacterial count of the finished yeast.

The most common practice is acidification of the first separation cream yeast to pH 2.5 – 3.0 for varying periods. The acidified yeast is then either re-separated with water or is neutralized back to pH 4 – 5 with a mild caustic and re-separated. Acidification is widely used to control bacterial counts in the seed cream yeasts and to varying degrees in the commercial cream. The acidification of the commercial cream must be judiciously controlled to avoid negatively affecting the activity of the finished product. Prolonged acidification and especially poorly dispersed acid in the cream creates stability problems. Acids normally used are sulfuric, hydrochloric and less frequently sulfamic acid. Phosphoric acid is never used, as it can be assimilated by the cell and initiate mitosis conditions which reduce yeast stability.

Some recent work has indicated the benefits of certain bacteriocins, primarily Nissin, as an effective control of Gram-positive bacteria without reducing activity of the yeast (OGDEN, 1987). While the concept works, it is not widely used due to cost.

The industry uses disc, nozzle type centrifuges to recover and concentrate the biomass. During the separation phase careful attention is paid to the flow and pressure of the operation. Maximum pressure, achieved by nozzle size, is important in removing the spent wort. Excessive pressure, however, can lead to biomass loss or grittiness in the final yeast. To control this the spent wort and wash water is checked for biomass content, either visually or by using turbidity meters.

Following separation the concentrated cream yeast is stored in refrigerated tanks often referred to as "receivers". To allow quality assurance time to confirm the suitability of the lot for further processing, it is a common practice to blend several lots of cream yeast as a means of equalizing the quality of the yeast prior to further concentration and packaging.

3.5 Biomass Concentration

The yeast is further concentrated in order to produce compressed yeast or dry yeast. At this stage the cream yeast can be routed to bakeries using bulk liquid yeast (VAN HORN, 1989).

During this concentration or "dewatering" stage the solids content of the cream yeast is raised to 28–34% depending on whether the yeast will be packaged into blocks or extruded to produce dry yeast. Dewatering is achieved with either a rotary vacuum filter (RVF) or a plate and frame filter. Both systems perform well, and each offers advantages over the other while also having equal disadvantages.

The most common method today is the RVF. Cream yeast is mixed with a saturated salt brine at a 1 to 2% level (see Fig. 4). The osmotic pressure of the extracellular water is increased causing the cell to shrink in an effort to balance the pressure. The shrunken cells in suspension are picked up or sprayed onto the rotating drum. A precoat of potato starch or sago flour is used to prevent the cells from being sucked into the vacuum tubes. A partial vacuum around 26 mm is pulled on the surface of the filter cover which

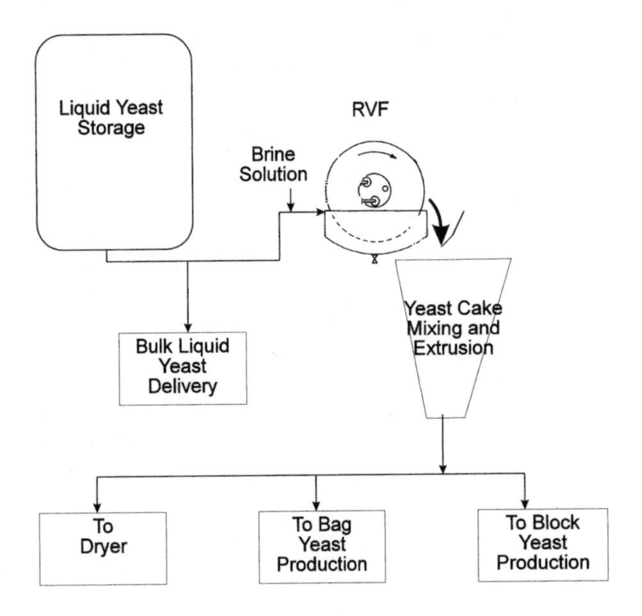

Fig. 4. Schematic flow diagram of the dewatering stage prior to packaging or drying.

draws the salted water. Water is sprayed onto the yeast coat in order to remove any residual salt from the yeast layer. The yeast layer is cut from the surface of the precoat along with a minute amount of starch and falls into a mixer which breaks up the yeast cake. The shrunken cells draw back much of the extracellular water, and the consistency of the yeast cake goes from that of putty to a more brittle solid. The main advantage of the RVF is its efficiency and economy. The disadvantages are the minute residues of salt, usually less than 0.04% and starch, usually less than one half percent. There is also the additional problem of the salt in the RVF effluent.

With the plate frame filter, the yeast cream is pumped into the frame chambers which use a canvas or woven plastic filter cloth. As pressure is exerted on the yeast cream, the extracellular water is forced through the filter cloths. Once the desired solids level is achieved, the frames are separated and the yeast cake removed. The advantage of this system is the reduced stress on the cell and the independence from using starch and salt. Economics are the only disadvantage. Whereas the RVF is a continuous system, the plate frame system is a slower batch system.

In most operations the solids level for fresh baker's yeast is maintained around 31 to 33% and the solids then reduced slightly as needed by the addition of water prior to packaging. For dry yeast the yeast is often brought out drier (32 to 34% solids) to improve the extrusion process.

3.6 Drying

Yeast can be effectively dried using several methods. The most common systems being used in the industry are fluidized-bed dryers, tunnel dryers and rotary dryers. Fluidized-bed dryers represent more than three quarters of the systems in daily operation (see Fig. 5).

Yeast cake from the filter is broken up and mixed with controlled amounts of processing aids or additives including antioxidants and emulsifiers (Hill, 1987). The most commonly used products include BHA, BHT and ascorbic acid as antioxidants along with emulsifiers such as diacetyltartaric acid esters of mono- and di-glycerides (DATA) and sorbitan polystearate (SPAN). Several patents describe the use of novel processes and protectants in the production of stable dry yeasts (Tent et al., 1986; Burrows, 1985; Tonge, 1983; Clément and Rossi, 1982). The blended yeast cake is extruded in an elongated noodle form

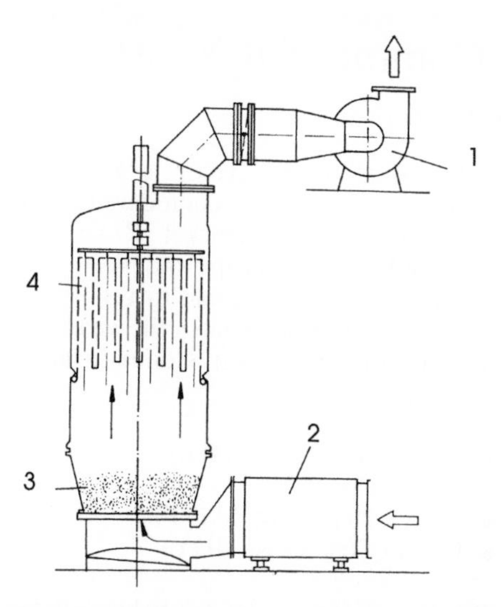

Fig. 5. Schematic diagram of a fluidized-bed drier showing the exhaust fan (1), the air inlet filter (2), the yeast bed (3), and the dust filter (4). Courtesy Aermatic AG.

through a perforated stainless-steel plate into the dryer. The size of the yeast noodle, the inlet temperature and the type of dryer will determine the drying time.

Most of the dried baker's yeast being sold today is of the "instant" type. The diameter of the noodle is approximately 0.5 to 0.8 mm. At the start of drying the inlet temperature is high (>70 °C). It gradually decreases as the moisture level of the yeast decreases. Because of high initial evaporation rates, the temperature differential between the inlet temperature and the yeast temperature reaches 35 °C. By the end of the drying this differential drops to almost zero. In most operations the air is dehumidified to improve the water transfer from the yeast.

In the tunnel dryer extruded yeast noodles (1–2 mm in diameter) are dropped onto a perforated conveyor which passes through several drying chambers with alternating air current. The air temperature in each chamber is programmed to lower with each stage. This allows maintenance of the correct temperature to guard against the yeast overheating. Drying times average about 2 to 4 hours.

Rotary dryers, also called Rotolouver dryers, rotate horizontally and have large internal baffles. The yeast is dried as it tumbles inside the dryer while hot air (60 °C) is forced along the horizontal axis of the dryer. These dryers were never used for instant yeast, as the drying times were long (4 to 12 hours), and this type of drying formed a hard shell of dead yeast material.

The most common form of drying employed today in commercial yeast operations is the fluidized-bed system. Fluidized-bed drying systems can be either batch- or continuous systems. The yeast cake is extruded into the dryer where air, forced through a perforated plate in the bottom of the dryer, raises the yeast particles in such a way as to fluidize them. The ratio of open area to total area in the perforated plate is critical for fluidization. Being fluid the contents of the dryer maintains all of the dynamics of a fluid system and can be poured from one section of the dryer to a lower section. The ability to handle the drying yeast as a fluid permits the operation to be handled as a continuous stream.

Drying times vary between 30 minutes and 2 hours, rarely longer. As in other systems the air inlet temperature starts off much higher than the yeast bed temperature and gradually decreases, as the evaporation rate of the yeast bed decreases.

In a continuous fluidized bed dryer, the dryer itself is rectangular with between four and six compartments. Each subsequent compartment is lower than its predecessor. The yeast bed flows via internal baffles from one stage to the next. The inlet temperature drops with each stage until the last stage where the temperature differentials between the air inlet and the yeast bed are identical.

3.7 Packaging

Fresh baker's yeast is packaged in a compressed block form or in a crumbled bulk bag form. The choice of packaging is designed to address specific needs in the baking field. Block yeast can be found in many weights with the most common being single blocks ranging from 1 pound (454 g) to 5 kg blocks or in multiple block packs such as 5×1 pound. Yeast cake is mixed with minor ingredients such as emulsifiers, plasticizers or lecithin along with enough water to control the solids contents, then extruded through a teflon mold into a continuous block. A blade or wire cuts the extruded block into the desired length and the cut block is wrapped in either wax paper or plastic wrap. The wrapped blocks are packaged in a carton and sealed. In some countries the wrapped blocks are overwrapped with a second wrap, often polyethylene to act as a vapor barrier. Block yeast is favored by the retail baker.

Bag yeast, also referred to as crumbled yeast, is sold in bags (22.5 – 25.0 kg) and remained the choice for the wholesale baker until the introduction of bulk liquid. The yeast cake is broken up into a medium coarse texture and packaged as it is or with the addition of a plasticizer that emulsifies the extracellular water creating a more uniformly colored yeast whiter in appearance. When requested by the baker, small levels of antifoam material are added (<0.1%) to the bag yeast to control foaming in brew systems. Bags are either the multi-wall kraft bag with inner poly lining or more recently the all poly bag.

Active dry yeast is normally packaged in consumer packs (8 to 20 g), tins (150 g to 10 kg) or bulk poly bags (20 to 40 kg). The consumer packs and tins are either packaged under ambient conditions or with nitrogen. Instant dry yeast is packaged in multi-laminate foil packs (500 g to 20 kg). Because of the porosity of the product and its sensitivity to oxidation, the yeast is always packaged without air either in a vacuum or by using CO_2 gas which the yeast absorbs pulling its own vacuum.

4 Commercial Yeast Forms

Commercial baker's yeast is available in four different forms to accomodate the bakery application for the market. Broad differences exist between Europe and North America in the ratio of wholesale to retail bakery (see Fig. 6), and this dictates the yeast forms best suited to that market. Commercial yeast forms include:

- Bulk Liquid Yeast (BY)
- Fesh Yeast (FY)
- Active Dry Yeast (ADY)
- Instant Dry Yeast (IDY)

The choice of yeast for the baker is decided upon by one or more of the following criteria, either (1) the baking application of the user, (2) the yeast volumes used, (3) the delivery logistics, (4) the cost restraints. General specifications for different yeast forms are found in Tab. 4.

Active dry wine yeast (WADY) is the only practical choice for commercial wine-making. This is a result not only of the seasonal nature of the industry where all of the annual yeast requirements fall within a four week period, but also because of the multitude of different strains produced and the actual size of the wine yeast market. A large wholesale bakery will consume the equivalent of 10 tons of dry yeast per week, while a large winery might use 2 tons of dry wine yeast annually split among four different strains.

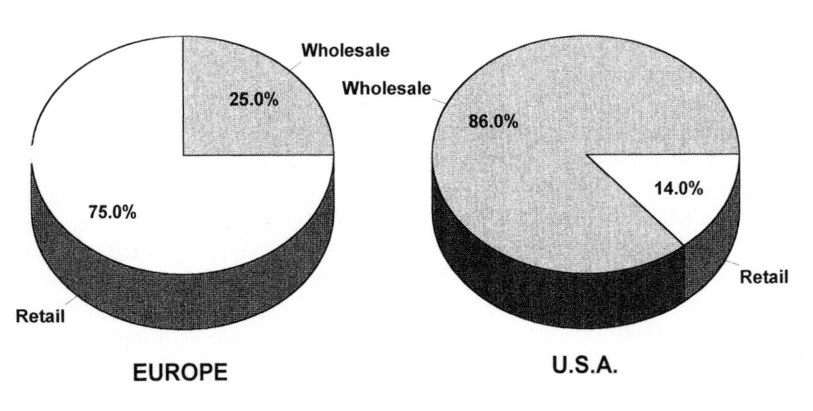

Fig. 6. Comparison of baking industry categorization between Europe and the U.S.A. From BAUR (1991). (The author grouped in-store bakeries within the wholesale category.)

Tab. 4. Specifications for Commercial Yeast Types

Parameter	Liquid Yeast	Compressed Yeast	Instant Dry Yeast	Active Dry Yeast
Solids (% w/w)	16–18	28–32	94–96	91–93
Nitrogen (% w/w dry basis)	7.5–9.0	7.5–9.0	7.0–8.0	6.5–7.5
Phosphorus (P_2O_5 % w/w)	2.8–3.3	2.8–3.3	2.5–2.7	2.0–2.4

4.1 Bulk Liquid Yeast

The application of bulk liquid yeast or cream yeast as a commercial product was at best sporadic until the late 1970s. As wholesale bakeries became larger and more automated, relying more on bulk handling of ingredients, the use of bulk liquid yeast became a commercially viable alternative, responding to the need for a way of handling yeast in the modern bakery that was less stressed, more uniform in performance and more stable during storage.

Bulk liquid yeast employs the principle that the baker buys biomass with consistent activity. The yeast producer attempts to standardize the activity by adjusting the cell population through controlling solids. In crumbled and compressed yeast, this is possible, but in a limited way. Bulk liquid yeast is ideally suited to this refinement of quality control. Given that the raw performance data of the cream yeast fall within certain parameters, they can be corrected to supply yeast with a predetermined activity.

Yeasts from the cold storage tanks are screened for fermentation activity and can then be allocated to further treatment prior to packaging or can be directed into the delivery channels for bulk liquid. Most of the bulk liquid yeast in the North American market is sold at either a 1.5:1 or 1.7:1 (v/w) ratio. Delivery from the yeast plant to the bakery is done by dairy tank trucks. The yeast farm tank system at the bakery usually consists of twin tanks, each tank holding a full truck load (see Fig. 7).

A computer control system monitors all aspects of the yeast handling system including (1) inventory, (2) storage conditions, (3) process data, (4) delivery, (5) batching to the mixer and (6) C.I.P. schedules of delivery lines, tanks and process lines. Batching of yeast to the dough mixers can be accomplished volumetrically or by weight.

The stability of bulk liquid yeast is marginally better than crumbled or compressed, due in part to the cell being less stressed and also because there is significantly less contact of the yeast cell with oxygen. The negative side of bulk yeast systems is the installation cost which limits their economic feasibility.

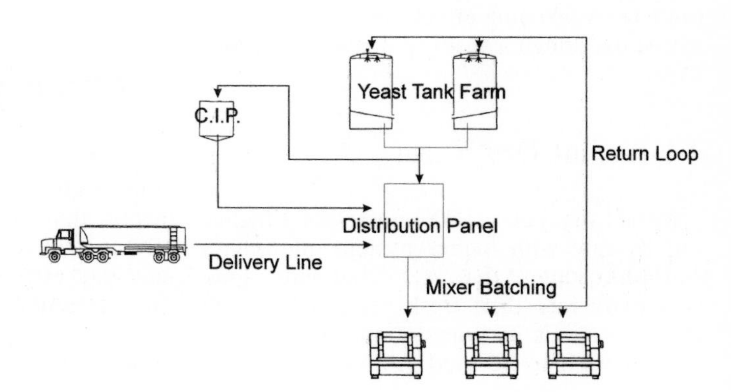

Fig. 7. Schematic flow diagram of a bulk liquid yeast system.

4.2 Compressed Yeast

Compressed yeast, fresh yeast, crumbled yeast are all synonyms for yeast which is concentrated to around 30% dry matter and sold either in bags or as blocks. Compressed yeast still represents more than 70% of the North American market. The ratio of bags to block yeast sold is roughly 3 to 1 in North America. Historically solids contents were higher in North America than Europe.

Compressed yeast can maintain 90% of its initial activity for up to 14 days at refrigerated conditions, even though deliveries to large bakeries are done on a bi-weekly basis. While compressed yeast is the leading form in North America, it is slowly being replaced by liquid systems.

4.3 Active Dry Yeast

Active dry yeast (ADY) had all but disappeared from commercial use following the success of instant dry yeast, but is now making a minor comeback. Produced by long drying schedules the ADY granule is surrounded by a hard shell comprised of dead yeast matter. This shell endows the active dry yeast with excellent stability, even when packaged under atmospheric conditions and stored at ambient temperatures. This protective shell requires the ADY to be rehydrated in warm liquid prior to incorporation in the dough. It is recommended to rehydrate the yeast in 5 to 10 times its weight in water at 40 to 45 °C for 10 to 20 min. Addition of the ADY to cool fluids can cause "cold shock" that can release reducing substances, usually glutathione, that can have weakening effects on the gluten matrix of the dough, causing gas retention problems.

4.4 Instant Dry Yeast

Instant dry yeast (IDY) possesses a higher activity rate with higher protein, phosphate and solids content than ADY, but with a lower activity rate than fresh yeast. Its unique characteristic is its porosity and particle size that allows it to be used in a dough without prior rehydration. It is the porosity and particle density that require it to be packaged under inert gas or vacuum.

Much of the dry wine yeast produced today is of the instant variety. While some winemakers add the dry yeast directly onto the must, rehydration and reactivation of the yeast is preferable (MONK, 1986; KRAUS et al., 1981) due to the negative effect of free SO_2 or residual fungicides that may be in the must. The use of dry wine yeast immobilized in alginate gels for use in the bottle fermentation of sparkling wines has been described (COULON et al., 1984; FUMI et al., 1988) and is usually re-activated and conditioned to alcohol prior to immobilization.

A process for producing a partially dried yeast product containing around 22 to 25% moisture has been described (OSZLANYI, 1989) and patented (GOUX and CLÉMENT, 1987). The benefit of this technology is inhibition of any pre-freezing activity and reduced ice crystal formation in the cell. Normal instant dry yeast has been reported to be applicable for frozen dough production (BRUINSMA and GIESENSCHLAG, 1984) but has never seen acceptance by the frozen dough producers because of poor storage stability.

4.5 Protected Active Dry Yeast

Protected active dry yeast (PADY) is a coated IDY that is designed for use in a blended bread mix (SANDERSON et al., 1983; TRIVEDI et al., 1989). The PADY is normally stable in the bread mix for several months.

5 Quality Assurance

Quality assurance for the yeast producer must address two distinct yet interdependent needs: that of the yeast producer as a measurement of the success/failure of the process and that of the end user. The quality criteria that determine the success or failure of the product and process can be separated into four groups:

1. Microbiological purity
2. Functionality
3. Physical and chemical analysis
4. Strain integrity.

5.1 Microbiological Quality Assurance

The commercial yeast producer takes a small amount of pure yeast culture and grows it by a log factor of 10^8. Given that a broad spectrum of different microorganisms can flourish well within the commercial system, contamination of the pure strain by other yeasts and bacteria is a very real concern. Because of the economic constraints in yeast production, sterility of media and air cannot be totally achieved. It becomes the role of Quality Assurance to inspect all raw materials used in intermediate and final stages of the process to ascertain the soundness of the process.

Valid programs of process control including HACCP are important tools in assuring the quality of the yeast and the process. Microbiological analyses start with the raw materials of which molasses and water are the most important.

The yeast itself and the production stages are sampled at appropriate points, usually critical transfer steps, to confirm the purity of the yeast. Finally the intermediate seed yeasts and the commercial yeast are evaluated with respect to problematic organisms that affect functionality, biomass yield, sanitation indicator organisms and pathogens.

While most analysis involves selective plating techniques, some visual microscopic evaluations are done with regard to wild yeast, lactobacilli chains, yeast budding and dead cells. Newer techniques using immunofluorescent techniques have been reported (Bouix, 1990). The accounting of yeast viability by means of methyleneblue staining is a quick test for the producer and is supported by a more extensive program of viability counts using plating technique to control the activity potential of wine yeasts.

5.2 Functionality of Baker's Yeast

Baker's yeast is used in both large and small operations, in lean dough and sweet dough, as well as in a myriad of dough systems. Functionality of the yeast is paramount to the baker. All of the intrinsic qualities of yeast mean little to the baker, if the yeast does not raise the dough efficiently.

While the yeast producer cannot compensate for all possible applications, he can selectively test the product for functionality in the more critical systems. In essence the yeast should function well in both lean ($<1\%$ sugar) and sweet ($>15\%$ sugar) doughs. Additionally, the yeast should reactivate with a minimum of lag phase in doughs with or without preservatives, and it should maintain this activity with minimal loss for up to 14 days from the date of delivery. Yeast functionality can be expressed in two ways: either as the time required to raise a dough to a specific volume, or the amount of CO_2 released within a preset period. Measurements for the latter are reported as cm^3 of CO_2 per period, or as cm^3 of CO_2 per gram of yeast solids per minute. In either case, controlled preparation and handling of the dough as well as constant temperature is necessary to yield reliable results. The two instruments most used to measure fermentation activity are the SJA Fermentograph or the RDesign Risograph.

Bake tests normally encountered include:

1. Straight dough test; where the yeast must produce the gas within a preset time. Bread dough formulae are normally used (sugar levels 3 to 10%).
2. Straight dough test with calcium propionate.
3. Sweet dough test (15 to 20% sugar), in addition to the osmotic stress of the sugar, there is also the stress from the salt and the non-functional weight of high levels of shortening (5–15%).
4. Sweet dough stability test; yeast is incubated at an elevated temperature (25–32°C) for 18 to 24 hours, then retested in the sweet dough formula. This test coupled with the sweet dough is a good indicator of the functional potential of the yeast in frozen systems.

5. Water brew test; the yeast ferments a water brew (30 to 120 min), then it is incorporated into a bread dough formula, with or without propionate and the gas evolution measured.
6. Flour brew test; the yeast is allowed to ferment a thin sponge (40–50% flour) for a preset period whereupon it is incorporated into a dough formula and the gas evolution measured.
7. Sponge and dough test; the yeast ferments a normal sponge for 3 to 4 hours, then it is remixed with the remaining ingredients. The gas evolved in the dough is measured.

In addition to these or similar tests, active and instant dry yeast are subjected to "cold shock" tests, where the dry yeast is rehydrated in colder than recommended water. This can leach some of the cell constituents including reducing substances which have deleterious effects on the dough. Cold shock evaluation is particularly important in the export market, where dry yeasts are often used with weak flours that react poorly to excessive glutathione levels.

Yeast functionality in frozen dough applications while researched extensively cannot easily be measured within a normal quality assurance program. The problems associated with yeast in frozen dough are usually due to reduced gas retention within the dough matrix and occur only after several weeks sometimes months. Evolved gas that is lost because of a damaged gluten matrix will be measured by the normal instruments. More appropriate instruments such as the Chopin Rheofermentometer (NEYRENEUF, 1988) are used to measure the differential between gas produced and gas lost. The more serious problem with assessing yeast in frozen dough is the time parameter. Results of poor gas retention show up only after weeks, and by then the information has only historical value.

The baking industry in North America by necessity must insure a mold-free shelf-life of its bread for periods of a week to ten days and relies to this end on the use of preservatives, principally calcium propionate. For this reason the yeast must perform well in the presence of normal preservative levels, and standard dough tests are constructed to include these preservatives. The differential between doughs with and without calcium propionate (0.25%) can be as wide as a 30% reduction in activity to no reduction in activity.

5.3 Functionality of Wine Yeast

The ability of the commercial yeast producer to assess the functionality of wine yeast subjected to differing conditions is limited by the large number of potential variables that could exist, and the multitude of strains produced. The yeast producer is marketing functionality, but as this functionality is measured differently from one region to another, the producer must simplify functionality testing and focus on strain integrity. The success of wine yeast strains and wine yeast producers in the international wine industry has been the result of collaboration between the yeast producer, the wine researcher and the wine-maker.

The evaluation of commercial wine yeasts is generally simplified to include the measurement of CO_2 production at normal and high sugar levels with and without the presence of SO_2 and at abnormal and cold temperatures. Some newer strains selected for sparkling wine production by bottle fermentation are extremely agglomerant which is beneficial to the wine-maker, but problematic for the yeast producer (see Fig. 8). Also of interest to the producer is the extent of the lag phase for the dry yeast and the activity stability after storing the dry yeast at warm temperatures for extended periods of time. By this method the yeast producer attempts to predict the long-term shelf-life of the packaged product.

5.4 Physical and Chemical Parameters

Quality assurance programs must address the physical and chemical qualities of the raw materials, the process itself, intermediate yeast stages and the finished yeast product.

Fig. 8. Photograph showing the flocculant behavior of an agglomerating wine yeast strain used for bottle fermentation of sparkling wine. This flocculant functionality eliminates the necessity of riddling. Courtesy of Lallemand S.A.

Analyses of raw materials must ascertain the purity, concentration and microbiological stability. In relation to quality and economics, molasses is the most critical raw material. The yeast producer must evaluate the molasses for total sugars, fermentable sugars, contamination by foreign materials and, under certain circumstances, for vitamin content. Chemical analysis of the process is normally restricted to pH, temperature, formol nitrogen and sugar concentration. The intermediate yeast stages are analyzed to confirm protein and phosphate levels. The finished product is analyzed for solids content, protein level, P_2O_5 level, residual salt and starch. Since the yeast producer is selling functionality, the chemical and physical testing of the yeast and process, while invaluable to the producer, is less important than functionality to the end user.

5.5 Strain Integrity

Considering the number of commercialized wine yeast strains available and with some commercial yeast producers handling in excess of 100 strains, the strain integrity of wine yeast has become very important. Oenologists exert a great deal of effort and research in matching yeast strain characteristics to grape strains and specific production methods. The ideal yeast strains are selected based on this research, and getting the same strain back in a dry form is vital. Traditional taxonomic methods, while adequate in the past, can no longer guarantee strain integrity. Today's yeast producer combines classic taxonomy with modern techniques for strain identification. Modern technologies to control strain integrity have been reported (THOMAS, 1990; LAVALLEE, 1990; BISSON, 1990; HALLET, 1990; DUBOURDIEU and FREZIER, 1990; DEGRÉ et al., 1989; VAN VUUREN and VAN DER MEER, 1987; KUNZE et al., 1993) and include:

- Yeast morphology
- Classical biochemical methods
- Killer typing
- Protein electrophoresis
- Fatty acid profiles
- Chromosomal electrophoresis
- DNA fingerprinting
- Pulsed-field gel electrophoresis
- Mitochondrial DNA analysis
- Chromosomal karyotyping.

A new proficiency in the guaranty of strain integrity by the use of this methodology has become invaluable to the yeast producer for controlling strains within the production as well as certifying the strain to the end user. The marriage of DNA fingerprinting to computer image analysis (PEDERSEN, 1990; LAVALLEE, 1990) enables the producer to map specific chromosomal sections of an unknown strain and compare them to the same segments of all existing strains in a data bank. This method allows the producer to guarantee the strain to the customer and protect him against fraudulent misuse of the yeast.

6 Recommended Reading

For more information on the production and application of baker's yeast and wine yeast, the following books or book chapters are recommended: WHITE (1954), BURROWS (1970), REED and PEPPLER (1973), PYLER (1988), REED and NAGODAWITHANA (1991), ROSE and VIJAYALAKSHMI (1993).

7 References

ATKINSON, B., MAVITUNA, F. (1991), *Biochemical Engineering and Biotechnology Handbook*, 2nd Ed., New York: Stockton Press.

AUTIO, K., SINDA, E. (1992), Frozen dough: rheological changes and yeast viability, *Cereal Chem.* **69** (4), 409–413.

BAGUENA, R., SORIANO, M. D., MARTINEZ-ANAYA, M. A., BENEDITO DE BARBER, C. (1991), Viability and performance of pure yeast strains in frozen wheat dough, *J. Food Sci.* **56** (6), 1690–1694

BAILEY, A. (1975), *The Blessing of Bread*, New York: Paddington Press Ltd.

BARRE, P. (1980), *O.I.V. Bull.* **53**, 560.

BAUR, J. (1991), Bread baking in Europe, *Ind. Cereales* Sept./Oct., 39–48.

BENDA, I. (1986), Über den Einfluß von Killerhefen auf die Struktur von Hefepopulationen bei der Mostgärung, *Weinwirtschaft* **41**, 345–355.

BICKERMAN, J. J. (1973), *Foams*, New York: Springer-Verlag.

BISSON, L. (1990), Yeast DNA analysis by chromosomal karyotyping, in: *Deuxième Rencontre Lalvin*, Bordeaux.

BOUIX, M. (1990), Application de l'immunofluorescence à la differentiation fine des souches de levures, in: *Deuxième Rencontre Lalvin*, Bordeaux.

BOUTROUX, L. (1897), *Le Pain et la Panification*, Paris: J. B. Bailliere & Fils.

BRONN, W. K. (1985), Possibilities of substituting alternative raw materials for molasses in producing baker's yeast (In German), *Branntweinwirtschaft*, July.

BRUINSMA, B. I., GIESENSCHLAG, J. (1984), *Baker's Digest*, Nov., 6–11.

BURROWS, S. (1970), Baker's yeast, in: *The Yeasts*, Vol. 3, pp. 349–413, London: Academic Press.

BURROWS, S. (1985), *European Patent* C12N118.

BUSSEY, H., VERNET, T., SDICU, A. M. (1988), Mutual antagonism among killer yeasts: competition between K1 and K2 killers and a novel cDNA based K1-K2 killer strain of *Saccharomyces cerevisiae*, *Can. J. Microbiol.* **34**, 38–44.

CALCOTT, P. H., ROSE, A. H. (1982), Freeze-thaw and cold shock resistance of *Saccharomyces cerevisiae* as affected by plasma membrane lipid composition, *J. Gen. Microbiol.* **128,** 549–555.

CALVEL, R. (1988), La surgelation en panification ou la medaille et son revers, *Le Boulanger Patissier*, Nov., 15–18.

CARRAU, J. L., LUCIO DE AZEVEDO, J., SUDBERY, P., CAMPBELL, D. (1982), Methods for recovering fusion products among oenological strains of *Saccharomyces cerevisiae* and *Schizosaccharomyces pombe*, *Rev. Bras. Genet.* **1**, 221–226.

CASEY, G. P., INGELDEW, W. M. (1986), Ethanol tolerance in yeasts, *CRC Crit. Rev. Microbiol.* **13** (3), 219–280.

CASSIO, F., LEAO, C., VAN UDEN, N. (1987), Transport of lactate and other short chain monocarboxylates in the yeast *Saccharomyces cerevisiae*, *Appl. Environ. Microbiol.*, March, 509–513.

CHAMPAGNE, C. P., GOULET, J., LACHANCE, R. A. (1989), Fermentative activity of baker's yeast cultivated on cheese whey permeate, *J. Food Sci.* **54** (5), 1238–1254.

CHAPMAN, J. W. (1991), The development and use of novel yeast strains for food and drink, *Trends Food Sci. Technol.*, July, 176–180.

CLÉMENT, P., ROSSI, J. P. (1982), *U.S. Patent* 4328250.

COONEY, C. L., WANG, D. I. C., MATELES, R. I. (1969), Measurement of heat evolution and correlation with oxygen consumption during microbial growth, *Biotechnol. Bioeng.* **11**, 269.

COULON, P., DUTEUTRE, B., CHARPENTIER, M., PARENTHOEN, A., BADOUR, C., MOULIN, J. P., VALADE, M., LAURENT, M., LEMENAGER, Y. (1984), Nouvelles perspectives dans la méthode Champenois: utilisation de levures incluses lors du tirage, *Le Vigneron Champenois*, 516–531.

COUTINHO, C., BERNARDES, E., FELIX, D., PANEK, A. D. (1988), Trehalose as cryoprotectant for preservation of yeast strains, *J. Biotechnol.* **7,** 23–32.

CUINIER, C., LACOSTE, J. (1980), Essai d'utilisation de levures sèches actives en Tourraine, controle de l'efficacité du levurage, *Conaissance Vigne Vin* **1**, 53–64.

DANIELSEN, S., ERIKSEN, B. W. (1968), A rapid method for determining biotin activity in raw materials for fermentation, *J. Inst. Brew.* **174** (6), 540–544.

DAUMAS, F., SEGUIN, S., PUISAIS, J., CUINIER, C.

(1990), Etude écologique de la microflore des Cores du Rhône et implantation de la souche L-2056, in: *Deuxième Rencontre Lalvin*, Bordeaux.

DEGRÉ, R., THOMAS, D. Y., ASH, J., MAILHIOT, K., MORIN, A., DUBORD, C. (1989), Wine yeasts strain identification, *Am. J. Enol. Vitic.* **40** (4), 309–315.

DELOUVENCOURT, L., FUKUHARA, H., HESLOT, H., WESOLOWSKI, M. (1988), *Eur. Patent* 0095986.

DELTEIL, D. (1988), *Personal communication.*

DELTEIL, D. (1990), Contrôle de l'efficacité du levurage par suivi de l'implantation d'une souche oenologique marquée génétiquement, in: *Deuxième Rencontre Lalvin*, Bordeaux.

DUBOURDIEU, D., FREZIER, V. (1990), Application de l'electrophorèse en champs pulses à l'etudé de l'ecologie des levures en fermentation, in: *Deuxième Rencontre Lalvin*, Bordeaux.

DUNAS, F., WENDT, G. (1988), Effect of nitrogen content of a fed batch cultured strain of baker's yeast on the stability in frozen dough, *7th Int. Symp. Yeasts, Industrial Uses of Yeasts* (II), Abstr. p. 27.

FEUILLAT, M. (1986), Influence des techniques de vinification sur composition des vins de Bourgogne, *Ind. Bevande*, Aug., 272–284.

FINN, R. K. (1967), Agitation and aeration, *Biochem. Biol. Eng. Sci.* **1**, 69–99.

FLEET, G. H., HEARD, G. M., GAO, G. (1988), The effect of temperature on the growth and ethanol tolerance of yeasts during wine fermentation, *7th Int. Symp. Yeasts, Industrial Uses of Yeasts* (IV), Abstr. p. 31.

FUKUDA, H., TAKESHI, S., WATARU, O. (1981) *U. S. Patent* 4284724.

FUMI, M. D., TRIOLI, G., COLOMBI, M. G., COLAGRANDE, O. (1988), Immobilization of *Saccharomyces cerevisiae* in calcium alginate gel and its application to bottle fermented sparkling wine production, *Am. J. Enol. Vitic.* **39** (4), 267–272.

GELINAS, P., TOUPIN, C. J., GOULET, J. (1991a), Cell water permeability and cryotolerance of *Saccharomyces cerevisiae*, *Lett. Appl. Microbiol.* **12**, 236–240.

GELINAS, P., FISET, G., WILLEMOT, C., GOULET, J. (1991b), Lipid content and cryotolerance of baker's yeast in frozen dough, *Appl. Environ. Microbiol.* **57** (2), 463–468.

GELINAS, P., LAGIMONIERE, M., DUBORD, C. (1993), Baker's yeast sampling and frozen dough stability, *Cereal Chem.* **70** (2), 219–225.

GODKIN, W. J., CATHCART, W. H. (1949), Fermentation activity and survival of yeast in frozen fermented and unfermented dough, *Food Technol.*, April, 139–146.

GOUX, J., CLÉMENT, P. (1987), *European Patent* 0237427.

GRBA, S., OURA, E., SUOMALAINEN, H. (1975), On the formation of glycogen and trehalose in baker's yeast, *Eur. J. Appl. Microbiol.* **2**, 29–37.

HAGLER, A. N., LEWIS, M. J. (1974), Effect of glucose on thermal injury of yeast may define the maximum temperature of growth, *J. Gen. Microbiol.* **80**, 101–109.

HAHN, Y. S., KAWAI, H. (1990) Isolation and characterization of freeze-tolerant yeasts from nature available for the frozen dough method, *Agric. Biol. Chem.* **54** (3), 829–831.

HALLET, J. N. (1990), Differentiation des souches de levures par l'analyse de profile de restriction de l'ADN mitochondrial, in: *Deuxième Rencontre Lalvin*, Bordeaux.

HILL, F. F. (1987), Dry living microorganisms – Products for the food industry, in: *Biochemical Engineering*, Stuttgart: Gustav Fischer Verlag.

HINO, A. (1990), Trehalose levels and survival ratio of freeze-tolerant versus freeze-sensitive yeasts, *Appl. Environ. Microbiol.* **64** (4), 269–275.

HINO, A., TAKANO, H., TANAKA, Y. (1987), New freeze-tolerant yeast for frozen dough preparations, *Cereal Chem.* **64**, 269.

HO, K. H., MILLER, J. J. (1978), Free proline content and sensitivity to desiccation and heat during yeast sporulation and spore germination, *Can. J. Microbiol.* **24**, 312–320.

HOTTIGER, T., BOLLER, T., WIEMKEN, A. (1987), Rapid changes of heat and desiccation tolerance correlated with changes of trehalose content in *Saccharomyces cerevisiae* cells subjected to temperature shift, *FEBS Lett.* **220** (1), 113–115.

HSU, K. H., HOSENEY, R. C., SEIB, P. A. (1979), Frozen dough: I. Factors affecting stability of frozen unfermented dough, *Cereal Chem.* **56** (5), 419–424.

KELLER, F., SCHELLENBERG, W., WIEMKEN, A. (1982), Localization of trehalose in vacuoles and of trehalose in the cystol of yeast, *Arch. Microbiol.* **131**, 298–301.

KISHORE, P. V., KARANTH, N. G. (1986), Critical influence of dissolved oxygen on glycerol synthesis by an osmophilic yeast *Pichia farinosa*, *Process Biochem.*, Oct., 160–162.

KRAUS, J. K., SCOPP, R., CHEN, S. L. (1981), Effect of rehydration on dry wine yeast activity, *Am. J. Enol. Vitic.* **32** (2), 132–134.

KRUUV, J., LEPOCK, J. R., KEITH, A. D. (1978), The effect of fluidity of membrane lipids on freeze-thaw survival of yeast, *Cryobiology* **15**, 73–79.

KUNZE, G., KUNZE, I., BARNER, A., SCHULZ, R. (1993), Genetical and biochemical characterization of *Saccharomyces cerevisiae* industrial strains, *Fresenius Z. Anal. Chem.* **346**, 868–871.

LAFON-LAFOURCADE, S., GENEIX, C., RIBEAU-GAYON, L. (1984), *Appl. Environ. Microbiol.*, June, 1246–1249.

LAVALLEE, F. (1990), Identification informatisée de souches de levure de vin par analyse de l'ADN a l'aide de sondes moleculaires specifiques, in: *Deuxième Rencontre Lalvin*, Bordeaux.

LILJESTROM-SUOMINEN, P. L., JOUTSJOKI, V., KORHOLA, M. (1988), Construction of a stable α-galactosidase producing baker's yeast strain, *Appl. Environ. Microbiol.,* Jan., 245–249.

LONDESBOROUGH, J., VARIMO, K. (1984), Characterization of two trehalases in baker's yeast, *Biochem. J.* **219**.

MALANEY, G. W., TANNER, R. D. (1988), The effect of sodium chloride on the urea cycle amino acids within baker's yeast during aerated fermentation of glucose, *Biotechnol. Appl. Biochem.* **10**, 42–48.

MAURICIO, J. C., GUIJO, S., ORTEGA, J. M. (1991), Relationship between phospholipid and sterol contents in *Saccharomyces cerevisiae* and *Torulopsis delbrueckii* and their fermentation activity in grape musts, *Am. J. Enol. Vitic.* **42** (4), 301–308.

MOEBUS, O., KIESBYE, P. (1982), *U.S. Patent* 4327179.

MONK, P. R. (1986), Rehydration and propagation of active dry wine yeasts, *Aust. Wine Ind. J.*, May, 3–5.

MORRIS, C. E. (1983), Genetically engineered yeasts, *Food Eng.*, May, 114–118.

MUNOZ, E., INGELDEW, W. M. (1989), Effect of yeast hulls on stuck and sluggish fermentations: importance of the lipid component, *Appl. Environ. Microbiol.*, June, 1560–1564.

NAKATOMI, Y., HARA, K., UMEDA, F., KONO, T., NIIMOTO, H. (1986), *Eur. Patent* 0197497.

NEYRENEUF, O. (1988), Efficacité d'une levure en surgelation de patons crus: analyse des principaux paramètres qui avant congelation influencent directement les performances, *Ind. Cereales* **55**, 45–51.

NEYRENEUF, O. (1990), Surgelation de patons ensemences a la levure: du fermenteur au consommateur. Quelles exigences pour la filière panification?, *Ind. Cereales*, March/Apr., 5–13.

ODA, Y., OUCHI, K. (1989a), Principal component analysis of the characteristics desirable in baker's yeast, *Appl. Environ. Microbiol.*, June, 1495–1499.

ODA, Y., OUCHI, K. (1989b), Genetic analysis of haploids from industrial strains of baker's yeast, *Appl. Environ. Microbiol.*, July, 1742–1747.

OGDEN, K. (1987), Cleansing contaminated pitching yeast with nisin, *J. Inst. Brew.* **93**, 302.

OSZLANYI, A. G. (1989), Stable yeast for frozen dough, *J. Am. Soc. Brew. Eng.*, 133–144.

OURA, E., SUOMALAINEN, H., VISKARI, R. (1982), in: *Economic Microbiology* (ROSE, A. H., Ed.), Vol. 7, p. 87, London: Academic Press.

PANEK, A. D. (1975), Trehalose synthesis during starvation of baker's yeast, *Eur. J. Appl. Microbiol.* **2**, 39–46.

PANEK, A. D. (1985), Trehalose metabolism and its role in *Saccharomyces cerevisiae*, *J. Biotechnol.* **3**, 121–130.

PARK, D. H., BAKER, D. S., BROWN, K. G., TANNER, R. D., MALANEY, G. W. (1985), The effect of carbon dioxide, aeration rate and sodium chloride on the secretion of proteins from growing baker's yeast, *J. Biotechnol.* **2**, 337–346.

PEDERSEN, M. B. (1990), Brewing yeast RFLP studies, in: *Deuxième Rencontre Lalvin*, Bordeaux.

PEPPLER, H. J. (1983), Ventures in yeast utilization, *Annu. Rep. Ferment. Processes* **6**, 237–251.

PERINGER, P., BLACHERE, H., CORRIEU, G., LANE, A. G. (1974), A generalized mathematical model for the growth kinetics of *Saccharomyces cerevisiae* with experimental determination of parameters, *Biotechnol. Bioeng.* **16**, 431–454.

PERLMAN, D. (1977), *ASM News* **43** (2), 82.

POULARD, A., COUIENCEAU, D., ETIENNE, F., REBERTEAU, R., DECHAIRE, P., BACABARA, I. (1988), Identification and yeast selection for the improvement of Ajou sweet wines, *7th Int. Symp. Yeasts*, *Yeast Ecology*, Poster Session (1), Abstr., p. 159.

PYLER, E. J. (1988), *Baking Science and Technology*, 2nd Ed., Merriam, KS: Sosland Publishing.

RADLER, F. (1984), Microbiological testing of active dry yeast preparations for wine-making, *Unpublished manuscript.*

REED, G., PEPPLER, H. J. (1973), *Yeast Technology*, Westport, CT: AVI Publishing Co.

REED, G., NAGODAWITHANA, T. W. (1991), *Yeast Technology*, 2nd Ed., New York: Van Nostrand Reinhold.

REED, R. H., CHUDEK, J. A., FOSTER, R., GADD, G. M. (1987), Osmotic significance of glycerol accumulation in exponentially growing yeasts, *Appl. Environ. Microbiol.*, 2119–2123.

RICKARD, P. A. D., HOGAN, C. B. J. (1978), Effect of glyoxylic acid cycle on the respiratory quotient of *Saccharomyces cerevisiae*, *Biotechnol. Bioeng.* **20**, 1111–1115.

ROSE, A. H., VIJAYALAKSHMI, G. (1993), Baker's yeast, in: *The Yeasts*, Vol. 5, 2nd Ed., pp. 357–392, London: Academic Press.

ROSSINI, G. (1984), Assessment of dominance of added yeast in wine fermentation and origin of *Saccharomyces cerevisiae* in wine-making, *J. Gen. Appl. Microbiol.* **30**, 249–256.

RUDOLPH, A. S., CROWE, J. H. (1985), Membrane stabilization during freezing: the role of two natural cryoprotectants, trehalose and proline, *Cryobiology* **22**, 367–377.

SANDERSON, G. W., REED, G., BRUISMA, B., COOPER, E. J. (1983), Yeast fermentation in bread baking, *A.I.B. Bull.* **5**, 12.

SCHMITT, A., CURSCHMANN, K., MILTENBERGER, R., KOHLER, H. J., WAGNER, K., KREUTZER, P. (1984), Trockenreinzuchthefen im mehrjährigen Vergleich, *Dtsch. Weinbau* **25/26**, 1126–1137.

SELIGY, V. L., JAMES, A. P. (1977), Multiplicity and distribution of rDNA cistrons among chromosome I and II aneuploids of *Saccharomyces cerevisiae*, *Exp. Cell Res.* **105**, 63–72.

SOLOMONS, G. L., PERKINS, M. D. (1958), The measurement and mechanism of oxygen transfer in submerged culture, *J. Appl. Chem.* **8**, 251.

SPONHOLZ, W. R. (1983), Trends in German winemaking practice, in: *5th Australian Wine Industry Conference*, pp. 339–351.

STRASSER, A. W., JANOWICZ, Z. A., DOHMEN, R. J., ROGGENKAMP, R. O., HOLLENBERG, C. P. (1990), Prospects of yeast in biotechnology, *Agro-Industry Hi-Tech.*, 21–24.

STROHM, J. A., DALE, R. F. (1961), Dissolved oxygen measurement in yeast propagation, *Ind. Eng. Chem.* **53**, 760–764.

TAKANO, H., HINO, A., ENDO, H., NAKAGAWA, N., SATO, A. (1990), *Eur. Patent* 0388262.

TENT, W., FEISTLE, L., BOSCHEINEN, H. (1986), *Eur. Patent* 0167643.

THOMAS, D. Y. (1990), Use of oligonucleotide probes for yeast DNA characterization, in: *Deuxième Rencontre Lalvin*, Bordeaux.

TIEDJE, J. M., COLWELL, R. K., GROSSMAN, Y. L., HODSON, R. E., LENSKI, R. E., MACK, R. N., REGAL, P. J. (1989), The planned introduction of genetically engineered organisms: ecological considerations and recommendations, *Ecology* **70** (2), 298–315.

TONGE, G. M. (1983), *U.K. Patent* 2108150A.

TRIVEDI, N. B., REED, G. (1988), *Eur. Patent* 0268012.

TRIVEDI, N., HAUSER, J., NAGODAWITHANA, T., REED, G. (1989), Update on baker's yeast, *A.I.B. Bull.* **11**, 2.

TYAGI, R. D. (1984), Participation of oxygen in ethanolic fermentation, *Process Biochem.* **19**, 136–141.

UNO, K. (1986), Freeze resistant dough and novel microorganism for use therein, *Appl. Environ. Microbiol.* **53**, Oct., 941–943.

UNO, K., ODA, Y. (1990), *Eur. Patent* 0196233.

VALADE, M. (1990), Implantation de souches selectionnées dans les mouts de Champagne, in: *Deuxième Rencontre Lalvin*, Bordeaux.

VAN HORN, D. R. (1989), Cream yeast, *J. Am. Soc. Brew. Eng.*, 144–153.

VAN VUUREN, H. J. J., VAN DER MEER, L. (1987), Fingerprinting of yeasts by protein electrophoresis, *Am. J. Enol. Vitic.* **38** (1), 49–53.

VEZINHET, F. (1985), Le marquage genetique de souches de levures oenologiques, *Rev. Fr. Oenol.* **97**, 47–51.

VEZINHET, F., LACROIX, S. (1984), Marquage génétiquement de levures: outil de contrôle des fermentations en souches pures, *Bull. O.I.V.*, 43–644; 759–777.

WANG, H. Y., COONEY, C. L., WANG, D. I. C. (1977), Computer aided baker's yeast fermentations, *Biotechnol. Bioeng.* **19**, 69–86.

WARTH, A. D. (1989), Relationship between the resistance of yeasts to acetic, propanoic and benzoic acids and to methyl paraben and pH, *Int. J. Food Microbiol.* **8**, 343–349.

WHAITE, P., ABORHEY, S., HONG, E., ROGERS, P. L. (1978), Microprocessor control of respiratory quotient, *Biotechnol. Bioeng.* **20**, 1459–1463.

WHITE, J. (1954), *Yeast Technology*, London: Chapman & Hall.

WILLIAMSON, D. H., SCOPES, A. W. (1962), A rapid method for synchronizing division in the yeast *Saccharomyces cerevisiae*, *Nature* **193**, 256–257.

WOLT, M. J., D'APPOLONIA, B. (1984), Factors involved in the stability of frozen dough. II. The effect of yeast type, flour type and dough additives on frozen dough stability, *Cereal Chem.* **61** (3), 213–221.

ZIKMANIS, P. B., AUZINA, L. P., AUZANE, S. I., BEKER, M. J. (1982), Relationship between the fatty acid composition of lipids and the viability of dried yeast *Saccharomyces cerevisiae*, *J. Appl. Microbiol. Biotechnol.* **15**, 100–103.

9 Cheese

NORMAN F. OLSON

Madison, WI 53706, USA

1 Introduction

The initial application of biotechnology to cheese manufacturing undoubtedly occurred during the first accidental souring and clotting of milk to form a rudimentary curd. All subsequent research and development efforts have characterized and refined that prehistoric use of microbial metabolism, enzymology and process engineering. Historical descriptions of cheese manufacturing are sketchy, but drawings in a Ramesid tomb (100 BC) show goats being led to pasture and skin bags hanging from poles (SCOTT, 1986). Contamination of milk with acid-producing bacteria undoubtedly led to curdling and the subsequent motion in the bags produced curds and whey, both of which were consumed out of necessity and preference. Cheese, whey and fermented milks offered a logical alternative protein source to meat which would require slaughtering of an essential animal. The subsequent evolution of usage, characterization and development of lactic acid bacteria was reviewed by TEUBER (1993a).

Although not documented, it seems reasonable that use of milk-clotting enzymes originated from an observation of clotted milk in the stomach of suckling animals. Our perceptive prehistoric ancestors could have related that transformation of milk to a substance in the stomach with subsequent evolution to practices of dipping stomach linings into milk to cause clotting. The stomachs of hares and kids served as sources of milk-clotting enzymes, but extracts of plant materials such as thistle flowers, fig tree, and saffron seeds also were used as clotting agents. The successful use of plant extracts probably relates to the prevalence of ewe's and goat's milk cheese which would not become bitter as would that from cow's milk.

Early records indicate that foods such as cheese and bread were staples as early as 6000 to 7000 BC in the Fertile Crescent located in present-day Iraq (SCOTT, 1986). Cheese was a favored food of ancient royalty; 13 of the 500 cooks serving the Persian king Darius were experts in cheesemaking and cheese is thought to be amongst the remains in the tomb of Pharaoh Horus. Although goats and sheep were the preferred animals, a Sumerian frieze dating between 3500 and 3000 BC shows cows being milked and the subsequent curdling of milk (HARRIS, 1984). The spread of cattle husbandry and the concomitant cheesemaking was fostered by the migrant Vikings.

Advances in cheesemaking were stagnant during the Dark Ages except for Scandinavia and the isolated West coast of Ireland (HARRIS, 1984). Subsequent evolution of cheese production took place through individuals, farmer cooperatives and monasteries with the Po Valley in Northern Italy becoming one of the principal commercial cheese exchanges in Europe. Interest was rekindled in agricultural technology in the sixteenth century. One of the first treatises in that era by an Italian, Agostino Gallo, indicated that cow's milk had replaced that from sheep and goats in the production of cheese. The Age of Enlightenment fostered technological developments and the consolidation of cheesemaking into commercial enterprises. This trend accelerated during the nineteenth century with VON LIEBIG, PASTEUR, METCHNIKOV and TYNDALL establishing scientific bases for cheese fermentation, microbiology and pasteurization. Several developments were especially important in the rationalization of cheese manufacturing: use of heat to destroy microorganisms by PASTEUR in 1857 evolved into specific processes (including pasteurization) and equipment to heat milk before cheese manufacturing, the introduction of pure cultures of lactic acid bacteria by STORCH in 1890 and ORLA-JENSEN in 1919, refinement of extraction of rennet from calf vells and standardization of the extract by HANSEN in 1870 and the development of the acidimeter by LLOYD in 1899 to objectively measure acid production by lactic acid bacteria during cheese manufacturing (SCOTT, 1986).

The advent of international trade, development of railway systems, the Industrial Revolution and urbanization of the population fostered improved processes and facilities, and the consolidation of cheesemaking operations and marketing systems. Cheese manufacturing plants that purchased milk from farmers arose throughout Europe and the United States during the nineteenth century. This specialization naturally led to systematic control of the biology, chemistry and composition during cheesemaking and automated systems of handling the ingredients and the resulting cheese (OLSON, 1970, 1975). Presently, highly automated manufacturing plants are capable of converting millions of liters of milk per day into cheeses that constitute major varieties on the world market; Cheddar, Gouda, Mozzarella, Swiss, Camembert and Brie are some examples. However, the cheese industry is still heterogeneous and includes some varieties being made by family units with techniques not greatly different from those used in the early origins of cheesemaking.

2 Cheese Types

It is not surprising, with the development of the cheese industry, when travel and communications were relatively limited, that a large number of cheese varieties would evolve. The exact number of cheese varieties would be impossible to determine and probably meaningless to ascertain. It is estimated that 2000 different varieties have been developed; 400 varieties have been described (WALTER and HARGROVE, 1972). This diversity of products hampers defining cheese. The Food and Agricultural Organization devised a Code of Principles which included the following definitions of cheeses. Cheese is the fresh or matured product obtained by the drainage (of liquid) after the coagulation of milk, cream, skimmed or partly skimmed milk, butter milk or a combination thereof. A second definition was added for cheeses made from the liquid whey obtained during manufacturing of cheese. Whey cheese is the product obtained by concentration or coagulation of whey with or without the addition of milk or milk fat. Newer cheese manufacturing techniques deviate slightly from the details of these definitions but not from the general concepts.

The diversity of cheeses prompted the need for classification to more effectively describe and compare cheeses from different regions. Several approaches exist depending upon the need for the classification (SCOTT, 1986). Marketers of cheese often classify by country of origin, which is logical to create a merchandizing image but creates confusion and overlap of many cheese varieties. More systematic classifications use composition, firmness and maturation agents as criteria as shown in Tab. 1. Alternatively, cheeses are categorized in Tab. 2 as natural cheeses meaning that they are manufactured by acid or enzymatic clotting of milk or of milk fractions or as processed cheeses that are manufactured from natural cheeses.

Categorization by composition obviously groups cheeses of greatly different flavor characteristics into a single class. This approach is useful for regulatory purposes and for comparing physical properties of cheese types. The term, water in fat free substance, is relevant since it is effectively a ratio of water content to the protein (caseins) content; the latter being the structural matrix of cheeses. Firmness of cheeses is closely related to that ratio but is also influenced by the percentage

Tab. 1. Classification of Cheeses According to Composition, Firmness and Maturation Agents (VEDAMUTHA and WASHAM, 1983)

I. Soft Cheese (50% to 80% moisture)

Unripened – low fat
Cottage
Quark
Baker's

Unripened – high fat
Cream
Neufchâtel

Unripened stretched curd or pasta filata cheese
Mozzarella
Scamorze

Ripened by external mold growth
Camembert
Brie

Ripened by bacterial fermentation
Kochkäse
Handkäse
Caciotta (ewe or goat)

Salt-cured or pickled
Feta – Greek
Domiati – Egyptian

Surface-ripend
Liederkranz

Tab. 1. (Continued)

II. Semi-soft Cheese (39%–50% moisture)
Ripened by internal mold growth
Blue
Gorgonzola
Roquefort (sheep's milk)

Surface-ripened by bacteria and yeast (surface smear)
Limburger
Brick
Trappist
Port du Salut, St. Paulin
Oka

Ripened primarily by internal bacterial fermentation but may also have some surface growth
Münster
Bel Paese
Tilsiter

Ripened internally by bacterial fermentation
Pasta Filata
Provolone
Low-moisture Mozzarella

III. Hard Cheese (maximum 39% moisture)
Internally ripened by bacterial fermentation
Cheddar
Colby
Caciocavallo

Internally ripened by bacterial fermentation plus CO_2 production resulting in holes or "eyes"
Swiss (Emmental)
Gruyere
Gouda
Edam
Samsoe

Internally ripened by mold growth
Stilton

IV. Very Hard Cheese (maximum 34% moisture)
Asiago Old
Parmesan, Parmigiano, Grana
Romano
Sardo

V. Whey Cheese
Heat and acid denaturation of whey protein
Ricotta (60% moisture)

Condensing of whey by heat and water evaporation
Gjetost (goat milk whey; 13% moisture)
Myost, Primost (13–18% moisture)

Tab. 1. (Continued)

VI. Spiced Cheese
Caraway – caraway seeds
Noekkelost – cumin, cloves
Kuminost – cumin, caraway seeds
Pepper – peppers
Sapsago – hard grating, clover

Tab. 2. Classification of Cheese by Manufacturing and Maturation Processes (OLSON, 1979)

I. Natural Cheeses
A. Cheese varieties in which milk is clotted by acid:
1. Cottage cheese
2. Baker's cheese
3. Cream cheese
4. Neufchâtel cheese

B. Cheese varieties is which milk is clotted by proteases:
1. Cheddar cheese
2. Colby and stirred curd (granular) cheese
3. Surface-ripened cheeses – Brick cheese, Limburger cheese, Port du Salut, Bel Paese, Tilsit cheeses
4. Other semi-soft cheeses – Edam, Gouda, Monterey, Münster cheeses
5. Cheeses with eyes – Swiss, Gruyère, Samsoe
6. Italian type
   a) Very hard (grating) – Parmesan, Romano
   b) Other hard – Asiago, Fontina
   c) Pasta Filata – Provolone, Mozzarella
7. Mold-ripened
   a) Blue, Roquefort
   b) Cheese with surface mold – Camembert, Brie, Coulommiers

II. Process Cheese
1. Processed Swiss, processed Cheddar, etc.
2. Cold-pack cheese

of fat in dry matter of cheese which is a ratio of the fat content to fat + protein + mineral contents. Classifying cheese by firmness and maturation processes, as in Tab. 2, characterizes types more definitively. However, varieties such as Cheddar and Provolone cheeses that differ greatly in characteristics are placed in the same group illustrating the difficulties in defining a complex food group such as cheese.

3 Cheese Manufacturing Overview

In spite of the heterogeneity of cheese varieties, there are common ingredients and processes that apply to all cheeses as illustrated in Fig. 1. The diagram indicates that cheese manufacturing is continuous through virtually all of the process since it is biologically driven. Cheese is a dynamic, viable organism from the point at which enzymes and/or microorganisms are added or activated until it is digested by consumers and/or converted into more stable process cheese by heat processing. Cheese is probably one of the more complex fermented foods to manufacture, since biological actions (fermentation of lactose to acids) affect chemical changes (expulsion of water and the sugar, lactose, and solubilization and expulsion of calcium phosphate). These, in turn, influence biological actions and their impacts by altering lactose availability and buffering capacity of cheese. Both of these influence physical properties of cheese (firmness and brittleness). The one over-riding principle of cheese manufacturing is control of rate and timing of acid production. This coincides with control of expulsion of serum (whey) that contains the substrate and buffering constituents which regulate the amount and impact of acid production. Numerous profiles of acid production are possible during cheese manufacturing if whey expulsion is coordinated with acid production profiles. The ultimate requisites are a cheese with the correct moisture content and pH. The two factors form a substantial basis for differentiating cheese varieties shown in Tab. 1 and their physical properties as discussed later.

3.1 Milk Analysis and Quality Control

The first step in cheese manufacturing as shown in Fig. 1 is analysis and quality control of milk, since these factors greatly influence the economics of cheese manufacturing, composition of cheeses and their sensory qualities. The amount of milkfat has traditionally served as the basis of payment for milk but value is now determined by levels of fat, protein and quality factors. Since casein and fat constitute about 90% of the solids in most cheese varieties, it is essential to measure concentrations of these accurately in milk because they, along with water, dictate the yield of cheese from milk (EMMONS et al., 1990). Casein is now estimated from protein concentrations until a feasible measurement can be developed.

3.2 Milk Pretreatments

Treatments of milk before cheese manufacturing vary with types of cheese and are reviewed in detail by VEDAMUTHU and WASHAM (1983), SCOTT (1986) and JOHNSON (1988). Some of the common treatments of milk are (1) heating, including pasteurization, to reduce bacterial populations and heat-labile enzymes, and (2) adjustment of milk composition by removing milk fat by centrifugal separation and by adding nonfat solids or cream. Conditions for heating vary with the type of cheese, the intended use for the cheese and legal requirements. In the United States, pasteurization at 71.7 °C for 15 s is required for cheese varieties that are traditionally consumed fresh and for any cheese that is not stored for at least 60 days at 1.67 °C or higher. The holding requirement recognizes the typical reduction in numbers of pathogens in the environment of a cheese with a pH of 5.4 or lower. Higher than normal heat treatments adversely affect the clotting properties of milk and the physical characteristics of the cheese.

Adjustment of milk composition is dictated by the traditional composition of a cheese variety. The choice of removing milk fat or adding nonfat solids usually is determined by economics. Concentrated skimmilk or nonfat dry milk are commonly used sources of solids. These must be of high microbiological quality and should not have been heat-treated excessively. The amount of nonfat solids added is limited to a few percent of the milk weight; excess levels will unduly increase lactose levels in cheese and can also impair the physical characteristics of the cheese.

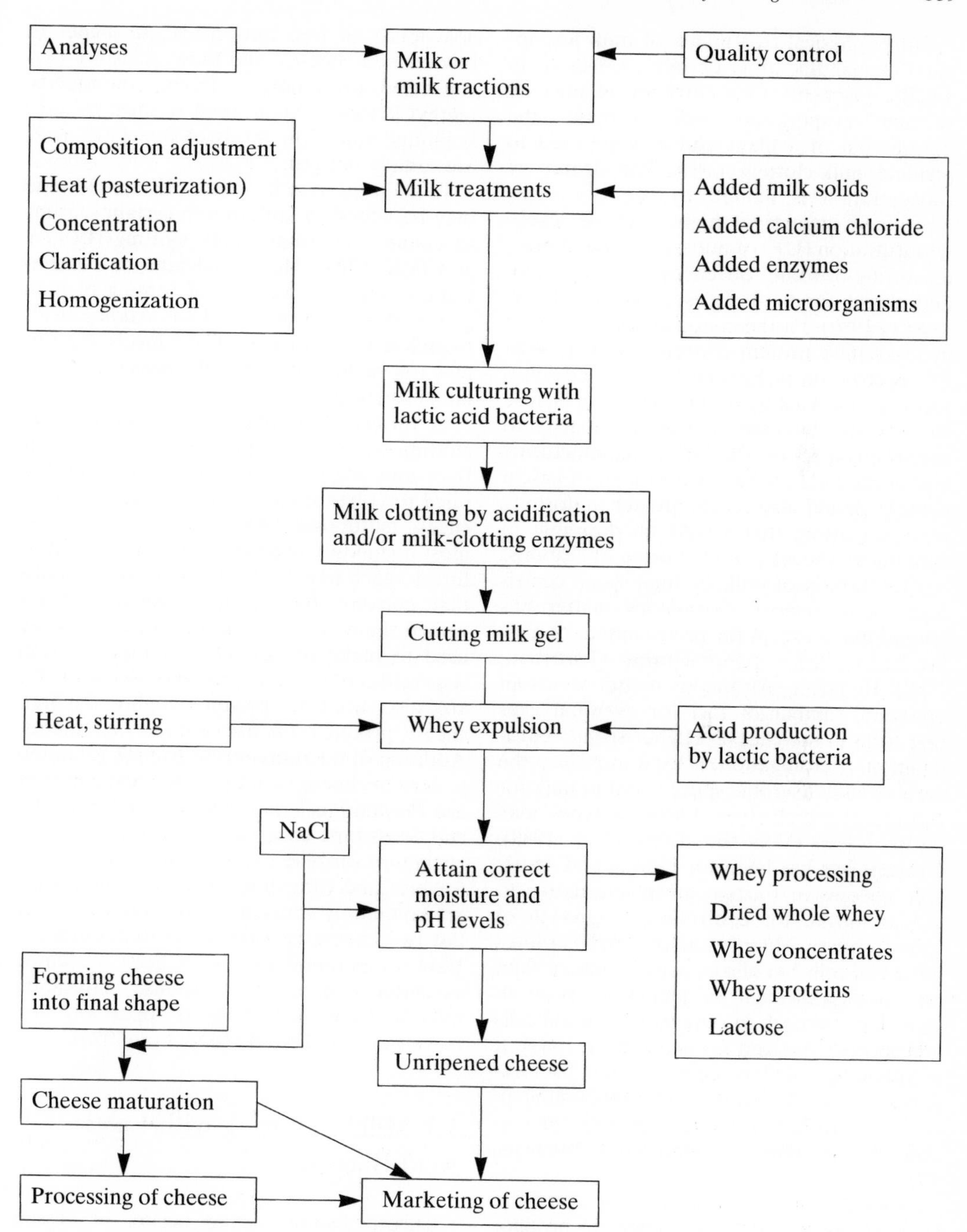

Fig. 1. Flow chart of the cheese manufacturing process showing treatments of milk, cheese curd and finished cheese and ingredients used in the process.

Other physical treatments of milk are applied in manufacturing certain cheeses or in specific processes. Concentration of milk by vacuum evaporation will increase the through-put of a plant and may be used to regulate milk-clotting rates. The degree of concentration is limited by the previously mentioned impacts of higher lactose levels. Ultrafiltration (UF) of milk is applied in manufacturing of some soft cheeses and to a limited extent for hard cheese varieties (LAWRENCE, 1989). Partial concentration by UF to increase milk protein concentrations to 4 to 5% is common in Europe for soft and semisoft cheeses. This treatment regulates lactose content and increases buffering capacity to enhance control of pH during manufacturing and in the final cheese. Adjustment of casein content should also create greater uniformity of milk-clotting that would be desirable in continuous cheese manufacturing operations.

Clarification of milk by high-speed centrifugation to remove extraneous matter has limited usage except for pretreatment of milk for Swiss cheese manufacturing (JOHNSON, 1988). Reducing extraneous matter yields an optimum number of foci for eventual eyes that form in the cheese. Higher-speed centrifugation, bactofugation, is used to reduce the aerobic and anaerobic spore count in milk for cheeses in which these bacterial types may create defects (VAN DEN BERG et al., 1989). Greatest use has been for Gouda and Swiss-type cheeses in Europe as an alternative to use of nitrate for controlling outgrowth of spore-formers. Microfiltration which captures microbial cells but allows constituents of skim milk to pass through the membrane is an alternative approach to removing bacterial cells and spores (MALMBERG and HOLM, 1988). It is technologically less appealing than bactofugation at present, but may offer interesting alternatives to heat treatment of milk for certain cheeses that are matured (MAUBOIS, 1991).

Homogenization imposes high-pressure shear to disrupt milk fat globules to produce substantially smaller globules that are recoated with milk proteins (VEDAMUTHU and WASHAM, 1983). Principal uses are to enhance lipolysis of blue-veined cheeses and the physical properties of cream cheese. Desirable levels of free fatty acids are higher in blue-veined cheeses, and these acids are also converted to important flavor compounds, methyl ketones. Application to most cheeses is limited since low levels of free fatty acids are usually desired.

In addition to milk solids, calcium chloride may be added to milk during certain seasons to enhance enzymatic milk clotting (LUCEY and FOX, 1993). Higher calcium ion concentration increases the rate of firming of milk gels as described in Sect. 4.3.1. Adding acids to milk will also increase Ca^{2+} levels, but this may not be permitted for all cheeses by regulatory agencies.

A variety of enzymes, in addition to milk-clotting enzymes, are permitted as food additives and are presently used or may by applied to enhance cheese flavor and/or rate of cheese maturation (ANONYMOUS, 1990). The most commonly used are lipases from oral or forestomach tissues of calf, kid goats or lambs that enhance flavors of Italian-type, blue-veined, Feta cheeses, and of Cheddar cheese used to make process cheese. Lipases from *Aspergillus niger*, *A. oryzae* and *Mucor miehei* are also used to produce highly flavored cheese products for use as food ingredients. Addition of microorganisms to milk is limited to certain cheese varieties. The most common are *Propionibacterium* species for Emmental and Swiss-type cheeses, spores of *Penicillium roqueforti* for blue cheese, and *P. camemberti* for brie and camembert cheeses. The spores are commonly sprayed on the surfaces of the last two cheese varieties prior to maturation. Various microbial species are being evaluated as cultures to accelerate cheese maturation rates, but commercial use is apparently limited or not publicized (EL SODA, 1993).

3.3 Milk Culturing with Lactic Acid Bacteria

Cheese manufacturing occurs in vessels (vats) that vary widely in capacity, in cheese manufacturing plants that differ greatly in size and with a substantial diversity in mechanization and automation of the processes (SCOTT, 1986). There appears to be a trend towards a dichotomy in which commodity cheeses, i.e., Cheddar and Mozzarella cheeses, are manufactured in large, highly mechanized plants, and specialty cheeses, i.e., trappist and blue, in smaller, less mechanized plants. However, specialty cheeses are also made in highly mechanized plants. Modern facilities utilize covered vats in which initial stages of cheese manufacturing are automated (WALSTRA, 1987). Subsequent handling of curd after removal of whey usually is mechanized but procedures vary widely for different cheese varieties.

The first step in cheese manufacturing carried out in the above equipment is addition of lactic acid bacteria. Acid-producing activity and metabolism of lactic starter cultures are the most important factors to control in cheese manufacturing, since they greatly influence cheese manufacturing efficiency and the composition, quality and safety of the finished cheese. Characteristics, functions and propagation procedures for these bacteria are discussed in Sect. 4.2. Facilities and technologies for preparing cultures and inoculating milk for cheese manufacturing vary between manufacturers. Modern operations will grow cultures under conditions approaching asepsis and will have procedures for accurately measuring cultures added to cheese milk. Strains of lactic acid bacteria will differ in acid-producing activity which necessitates adding different quantities to attain the same rate of acid production during manufacture of all lots of cheese.

3.4 Milk Clotting

In virtually all cheeses that are matured to develop desired flavor, the milk is clotted with selected enzymes which are described in Sect. 4.3. Uniformity of clotting and strength of the milk gel is critical for maximum retention of milk proteins (caseins) and milk fat in cheese and to minimize variations in cheese moisture levels. Milk-clotting enzymes are handled to avoid exposure to high temperatures and pH environments and to oxidizing agents such as hypochlorites. The enzymes usually are diluted in cold water and added uniformly to milk in the vat; inadequate distribution in milk will create variability in gel strength throughout the vat with the previously mentioned consequences.

Each type of cheese will require an optimum gel firmness at the point at which the gel is cut into smaller pieces. The choice of firmness level was developed subjectively, but a firmer gel will generally expel whey slower than a softer gel after cutting. The mechanisms regulating these effects are described in Sect. 4.3.1. Other factors such as size of curd pieces, temperature, pH, stirring of curd in whey and fat content influence syneresis of the curd (WALSTRA et al., 1987a).

3.5 Whey Expulsion

Whey is expelled rapidly from curd after cutting. This process is aided by raising the temperature of the curd–whey slurry which is being stirred in the vat. Most of the lactic acid bacteria are trapped in the curd and ferment lactose to lactic acid which diffuses from the curd. This is a dynamic system, since the substrate lactose is also being removed from the curd with the expelled whey. The relationship between the rate of moisture (and lactose) removal versus rate of lactic acid production by the lactic acid bacteria, to lower the curd pH, has profound effects on the characteristics of the final cheese as shown in Fig. 2 (LAWRENCE et al., 1984; LUCEY and FOX, 1993). These impacts result from the rate and extent of solubilization of calcium phosphate from the protein (casein) matrix of the curd. Calcium phosphate has a substantial effect on the physical proteins of the casein aggregates as described in Sects. 4.1.1 and 4.1.4. Rapid and extensive acid production will remove more calcium and phosphate, albeit less phosphate relative to calcium, to produce a brittle cheese with a lower mineral content. Several varieties of cheese illustrate the range of these interrelationships. In manufacturing Emmental cheese, acid production is slow when most of the whey is expelled from the curd. This solubilizes less calcium phosphate and yields a cheese that is more pliable. Acid production is more rapid and extensive during whey expulsion in manufacturing Cheshire and blue cheeses which are more brittle

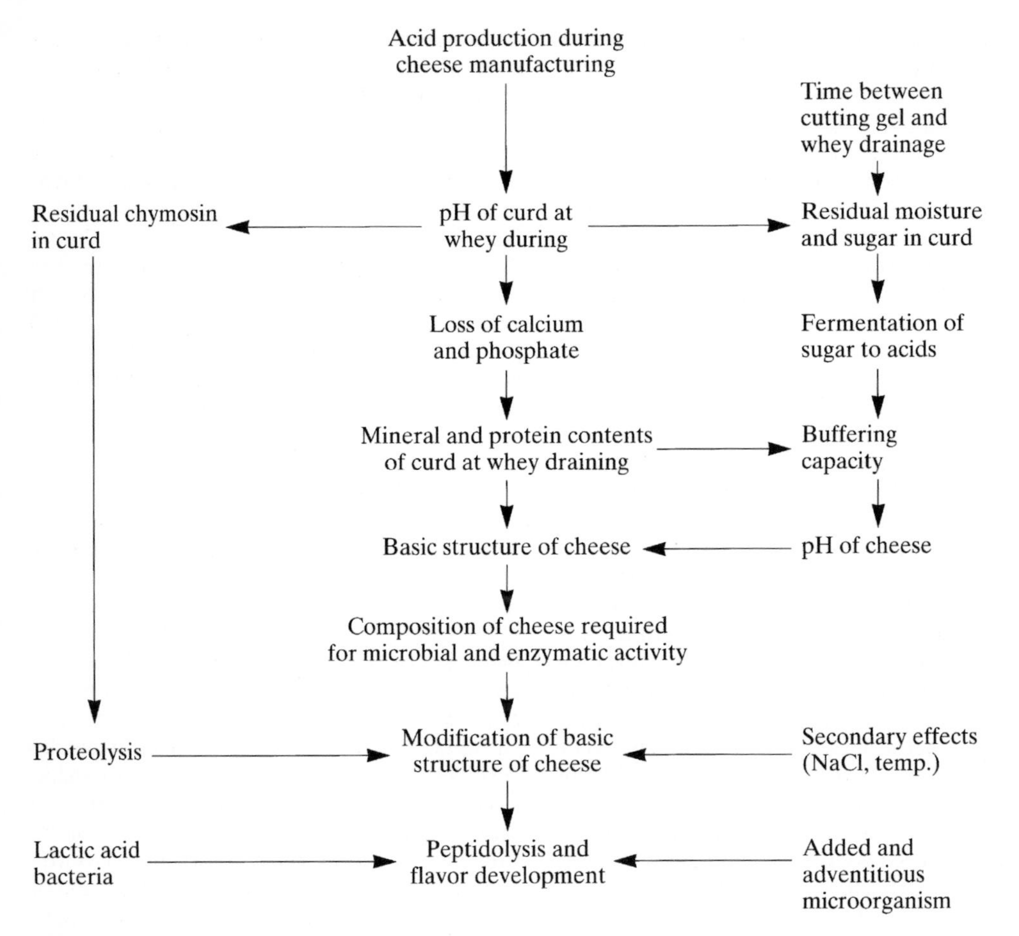

Fig. 2. Interrelationships between lactose fermentation, acid production and whey expulsion and the calcium retention, structure of cheese and proteolysis during maturation.

and less firm. Other varieties can be positioned between these extremes.

3.6 Moisture and pH Control

Physical properties of cheese are also influenced by the pH of the cheese which dictates the state of the calcium–phosphate–casein structure. The minimum pH of cheeses is usually reached within the first few days of maturation. It is regulated by the amount of lactose fermented to lactic acid and the buffering capacity of the curd during manufacturing of the cheese. Buffering capacity is determined by concentrations of undissolved calcium phosphate, caseins and lactate remaining in the cheese (LUCEY and FOX, 1993). Acid produced during early stages of cheese manufacturing will not be buffered as extensively because of higher moisture content of the curd. Acid produced later during manufacturing will be buffered to a greater extent by the higher concentration of buffering constituents. The pH of curd during whey expulsion also affects the degree of retention of the milk-clotting enzyme, chymosin, as shown in Fig. 2; lower pH values cause greater retention. This will accentuate the impact of low pH on depletion of calcium phosphate from the cheese matrix through enhanced proteolysis of that structure by the enzyme to create a weaker and more brittle cheese such as Cheshire. The opposite effects occur in cheeses such as Swiss and grana-type Italian cheeses in which pH values are higher and higher temperatures are used in manufacturing which will partially inactivate chymosin. Other milk-clotting enzymes described in Sect. 4.3.2 do not exhibit this effect of pH on retention.

3.7 Curd Handling

When the appropriate moisture and pH levels have been attained in a particular type of cheese curd, the curd particles and whey are partitioned. The technologies used include settling of curd and removing whey, transferring the curd–whey slurry into perforated forms from which the whey drains, fusing the curd particles under the whey while removing the whey or dipping the curd particles from the whey and transferring them to forms for curd fusion. Physical properties and pH of the curd at this stage affect curd fusion and appearance of the finished cheese. Fusion of Cheddar cheese curd does not occur until a pH of 5.8 is reached (LUCEY and FOX, 1993). Presumably, this applies to other varieties unless higher temperatures and higher moisture contents permit fusion at slightly higher pH values. Removal of whey from curd before fusion yields cheese with numerous openings; fusion in the presence of whey produces a dense-bodied cheese. A dramatic decrease in moisture content of the curd occurs with any curd fusion system because of the external pressure applied to the curd (WALSTRA et al., 1987a).

Sodium chloride may be applied in the crystalline form to curd after whey drainage or as a brine to cheese after manufacturing. Applying sodium chloride during the manufacturing, as is done for Cheddar cheese, controls acid production and influences final cheese quality (LAWRENCE et al., 1984). Concentrations of sodium chloride in the moisture phase of cheese exceeding 6% will inhibit acid production by *Lactococcus lactis* ssp. *cremoris* strains. Sodium chloride influences the subsequent flavor development and flavor perception, since cheese without sodium chloride is virtually tasteless.

The final stages of cheese manufacturing involve the fusing of the curd particles into a final shape which can vary greatly in dimensions and in size from <1 kg to >300 kg. Fusion can be done under the pressure of only the cheese, high external pressures or a combination of pressure combined with a vacuum treatment to induce fusion (VEDAMUTHU and WASHAM, 1983). The fused cheeses are usually protected with a coating such as wax, special emulsions or plastic films to minimize growth of molds or other microorganisms on the cheese surface. Varieties such as Camembert are not wrapped immediately since initial maturation involves growth of microorganisms on the cheese surface. Maturation of cheese is discussed in Sect. 6.

3.8 Process Cheese Products

A major alternative to marketing cheese in the form produced from milk (this is termed natural cheese) is to convert it to process cheese products by the combined action of mixing, heat and emulsifying salts (primarily phosphates and citrates). A detailed description of types, processing procedures, ingredients, characteristics and pertinent regulations for these products has been made by ZEHREN and NUSBAUM (1992) and CARIC and KALAB (1987). A diversity of process cheese products are made to satisfy consumer and food ingredient usage. These range in composition that closely simulates the moisture and fat contents of the natural cheese variety to products that contain virtually no fat, higher moisture contents and added ingredients to impart desired functionalities. The first step in processing cheese is selecting a blend of natural cheeses to impart the desired composition, flavor and physical properties. Emulsifying salts are selected for their ability to disperse and increase hydration of the cheese proteins which creates smoothness and fat emulsification. The effects on proteins result from the calcium sequestering ability of emulsifying salts and an increase in pH. Other ingredients such as milk solids, gums, sweetening agents, mold inhibitors and selected foods such as vegetables and meats are permitted in certain process cheese products. Process cheeses are popular because of their uniform properties and their stability achieved by heat processing. Cold-pack cheeses are also made from natural cheeses, but this process involves comminuting the cheese into a smooth paste without the aid of heat. Additional ingredients such as milk solids, gums and foods may be added as flavoring agents. These products must be stored under refrigeration.

4 Ingredients for Cheese Manufacturing

The definition previously given for cheese implies a degree of heterogeneity for ingredients that are used in the manufacturing of cheese. However, there are core components that are used for all major types of cheese: milk or milk components; clotting agents which may be lactic acid-producing bacteria, acids, milk-clotting enzymes, heat or combinations of these agents; and a seasoning agent such as sodium chloride. Ingredients used in some cheeses are pigments, enzymes and microorganisms to enhance flavor or alter physical properties and condiments to impart flavors.

4.1 Milk and Milk Components

Most cheese varieties are made from the milk of cows, water buffaloes, goats or sheep. It is used as secreted or processed to control biological and compositional properties. Fractions of milk such as whey and buttermilk are the principal ingredients for some cheeses. Milk consists of water, carbohydrates, proteins, lipids, minerals, and numerous minor constituents such as enzymes and vitamins; milk may contain as many as 10^5 different kinds of molecules (JENNESS, 1988). The constituents are dissolved, colloidally dispersed and emulsified in the water. The impact of many constituents on cheese manufacturing and cheese characteristics has been characterized but, undoubtedly, the effects of many minor constituents have not been elucidated. Milk proteins, particularly caseins, form the structural matrix of most cheeses. Fat exists as spherical globules in milk and is entrapped within the protein matrix, curd, during cheese manufacturing. The principal carbohydrate, lactose, is fermented to lactic acid and other products during cheese manufacturing but most is expelled with the whey; that remaining in cheese is generally metabolized during maturation. Calcium phosphate exists in various states in cheese and contributes to its physical properties.

4.1.1 Proteins

The proteins in milk have been very well characterized, and their role and alteration during cheese manufacturing and maturation have been studied extensively. Milk proteins have historically been divided into two general classes, caseins and whey proteins; the former being retained almost entirely in cheese and the latter being expelled with the whey because of their water solubility. Other proteins such as the milk fat globule membrane proteins, minor proteins and enzymes impact to varying degrees on cheese characteristics.

The principal milk proteins, caseins, were classified as phosphoproteins that precipitate from raw milk when acidified to pH 4.6 at 20°C (WHITNEY, 1988). Improved isolation techniques using electrophoresis and chromatography and determination of primary structure by chemical sequences or by cDNA or genomic DNA sequences indicated a great diversity within the caseins (SWAISGOOD, 1993). The known caseins, listed in Tab. 3, differ in composition between families (α_{s1}-casein(CN), α_{s2}-CN, β-CN and κ-CN), in amino acid substitutions between genetic variants and through post-translational modification by phosphorylation or glycosylation or partial proteolysis. For example, β-casein(CN) B-1P (f29-209) is in the β-family of caseins, is genetic variant B, underwent one post-translational phosphorylation and is a proteolytic

Tab. 3. Casein Fractions Found in Bovine Milk as Differentiated by Major Groups (Families), Genetic Variants and Post-Translational Modification by Phosphorylation and Proteolysis (WHITNEY, 1988)

Caseins (24–28 g L^{-1})	Genetic Variants
α_{s1}-Caseins (12–15 g L^{-1})	
α_{s1}-Casein-8P	A, B, C, D-9P, E
α_{s1}-Casein-9P	A, B, C, D-10P, E
α_{s1}-Casein fragments	
α_{s2}-Caseins (3–4 g L^{-1})	
α_{s2}-Casein-10P	A, B, C-9P, D-7P
α_{s2}-Casein-11P	A, B, C-10P, D-8P
α_{s2}-Casein-12P	A, B, C-11P, D-9P
α_{s2}-Casein-13P	A, B, C-12P, D-10P
β-Caseins (9–11 g L^{-1})	
β-Casein-5P	A^1, A^2, A^3, B, C-4P, D-4P, E
β-Casein-1P (f29-209)	A^1, A^2, A^3, B
β-Casein (f106-209)	A^2, A^3, B
β-Casein (f108-209)	A, B
β-Casein-4P (f1-28)	[a]
β-Casein-5P (f1-105)	[a]
β-Casein-5P (f1-107)	[a]
β-Casein-1P (f29-105)	[a]
β-Casein-1P (f29-107)	[a]
κ-Caseins (2–4 g L^{-1})	
κ-Caseins-1P	A, B
Minor κ-caseins -1, -2, -3, etc.	A, B

[a] Genetic variants not identified

fragment encompassing amino acid residues 29 through 209 of β-CN B.

Caseins tend to associate through electrostatic and hydrophobic interactions because of their high hydrophobicity and unique charge distribution (ROLLEMA, 1992). These driving forces in the presence of calcium phosphate produce large aggregates, micelles, during milk synthesis. Most of the bovine casein micelles approximate 15 nm in diameter but these comprise a small mass fraction. The micelle size distribution ranges up to 600 nm. The micelles exist as a colloidal suspension in milk and impart the unique characteristics to milk clotting and cheese structure. The major caseins would be unstable and precipitate in the presence of the calcium ion concentration in milk; stability is imparted by κ-CN. The stabilizing power of κ-CN arises from a low ester phosphate content, which imparts solubility in the presence of calcium, and its amphophilic nature. The C-terminal portion of the molecule has a negative charge and is less hydrophobic than the positively charged N-terminal domain. The latter property is important in enzymatic clotting of milk as described in Sect. 4.3.1.

Electron photomicrographs prepared by different techniques show the micelles as spheres exhibiting an inhomogeneous structure. The non-homogeneity purportedly indicates that the micelle is composed of submicelles which are held together by calcium phosphate binding and hydrophobic interactions (ROLLEMA, 1992). The submicelle structure has been criticized because of disparities between micelle size estimates from electron photomicrography and voluminosity measurements. It was proposed that the micelle was comprised of compact protein domains and regions of lower protein density. Casein micelles can be dissociated by removal of calcium phosphate under appropriate conditions. However, solubilization of a major portion of the calcium phosphate by acidification, especially slow, does not substantially dissociate the caseins from the micellar form. The conformation of casein micelles appears similar in acid gels, at pH 4.4 to 4.6, to that in milk. Acidification affects voluminosity and hydrodynamic diameter of micelles which manifests itself in the properties of cheeses at different pH values.

Consensus has not been reached on the location of the caseins and calcium phosphate in micelles and the function of the latter. This has resulted in a number of models of casein micelles being proposed; one is shown in Fig. 3. This model incorporates structural features satisfying properties of micelles and illustrates characteristics important in cheese chemistry. Localization of κ-CN primarily on the micelle surface imparts colloidal stability, since the protruding C-terminal domain increases hydrophilicity and the negative charge on the surface. The accessible C-terminal region of κ-CN is cleaved by milk-clotting enzymes and is released. This lowers the negative charge and hydrophilicity of the micelle surface leading to micelle aggregation

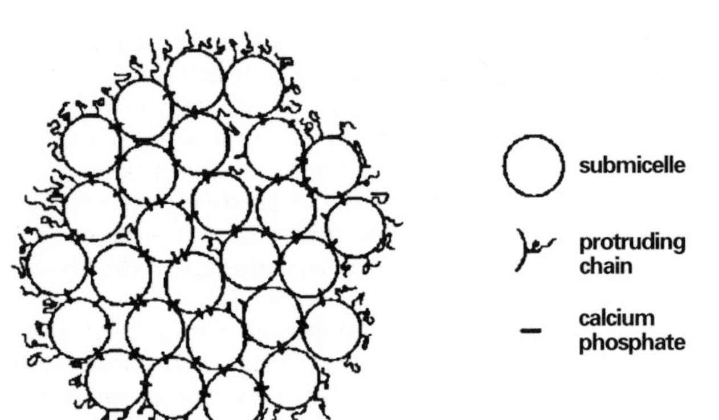

Fig. 3. Model of casein micelles as proposed by WALSTRA showing submicelles and protruding threads of C-terminal regions of κ-CN (ROLLEMA, 1992).

and clotting of milk. The other major caseins, α_s-CN and β-CN, are the building blocks of the internal micelle structure, but are also present on the micelle surface. There is some indication that α_s-CN is the more important structural component, since cleavage of one peptide bond between the hydrophobic N-terminal domain and the polar domain was related to softening of Cheddar cheese during maturation (CREAMER and OLSON, 1982; CREAMER et al., 1982). β-CN undoubtedly contributes to the physical properties of cheese, but its dissociation at low temperatures, similar to those for cheese maturation, and at low ionic strengths raises questions about its structural role.

Genetic polymorphism of the caseins affects the clotting of milk and the ability to capture the caseins in the cheese matrix (JAKOB and PUHAN, 1992). The rate of firming of gels formed by chymosin was faster with milks from cows of κ-CN BB phenotype than κ-CN AA, with the AB phenotype being intermediate. Part of the differences in firming rates could be explained by the higher casein concentrations in the BB and AB phenotypes. However, the superiority of κ-CN B was demonstrated in one study when milks of equivalent casein concentrations were compared. The differences in renneting time between κ-CN BB and AA can be minimized by adding calcium chloride and/or slight acidification of milk. Contradictory results have been published on the influence of α_{s1}-CN variants on clotting time of renneted milk. Firmness of chymosin-generated gels was greater for milk containing α_{s1}-CN C than milk containing the BB variant; extremely soft gels were obtained with milk containing α_{s1}-CN A. JAKOB and PUHAN (1992) concluded from the literature that clotting time and gel firmness were affected almost as significantly by β-CN variants as by κ-CN variants. Milk containing β-CN B variant, especially β-CN BB, exhibited shorter clotting time and formed firmer gels when treated with chymosin. β-CN C milk clotted faster than milk containing β-CN A. Clotting time and gel firmness data for milks containing the latter variant were inconsistent.

However, caution must be taken in applying the above effects, since alleles for single caseins are not inherited independently which necessitates consideration of composite phenotypes or haplotypes. Markedly shorter clotting time and greater gel firmness were observed with milk containing the B and C alleles of α_{s1}-CN combined with B alleles of both β- and κ-CN. Milks of the Norman breed containing combinations of α_{s1}-CN C/β-CN A^2/κ-CN B or α_{s1}-CN B/β-CN B/κ-CN B were found to be superior in their cheesemaking ability to milk containing α_{s1}-CN B/β-CN A^2/κ-CN A.

Similarly, the effect of genetic variants on the yield of cheese per unit weight of milk must be interpreted cautiously. Yields must be adjusted for differences in cheese moisture content and casein and fat concentrations in milk. Retention of casein and fat in cheese depends upon optimum strength and integrity of the casein gel formed by chymosin action. Gel strength is greatly influenced by casein concentration of milk and cheese manufacturing practices. Hence, retention of casein and fat in cheese may not be influenced directly by the inherent properties of the genetic variants, but indirectly by their effects on casein concentration and clotting properties of milk. In assessing previous research, JAKOB and PUHAN (1992) found the greatest impact was reported for variants of κ-CN. However, quantitative effects varied between studies. Some reported that yield of cheese from milk with κ-CN BB was 10% higher than that from milk with κ-CN AA, but no differences were found in other studies. Differences were likely to have resulted from lower retention of fat and casein in the weaker gel formed in the κ-CN AA milk. Milk containing β-lactoglobulin (β-Lg) BB formed a firmer chymosin-generated gel and yielded more cheese than other β-Lg variants. Undoubtedly these effects resulted from higher casein concentrations in the BB milk, since less than 10% of β-Lg would be retained in cheese. Strategies to use this information in breeding programs is under consideration. Technological advantages of higher casein concentrations, better milk clotting and higher cheese yields create economic incentives. Further substantiation of these incentives would seem prudent before widespread adoption of breeding practices to tailor-make properties of milk.

Goat's milk consists of α_{s1}-CN, α_{s2}-CN, β-CN and κ-CN; all are polymorphic (PIRISI et al., 1994). Polymorphism of α_{s1}-CN is linked to variations in synthesis of that fraction and total casein. Milks containing variants linked with greater casein synthesis exhibited the expected firmer milk gel formed by milk-clotting enzymes, higher recovery of milk solids in cheese, lower cheese moisture content, and greater serum retention during maturation of the cheese.

4.1.2 Lipids

Over 96% of the lipids in milk exist in the form of globules, ranging in diameter from less than 0.2 to 20 μm, in a colloidal suspension (KEENAN et al., 1988). Small globules, less than 1 μm, comprise 80% of the total number of globules. Intermediate sized globules, 1 to 8 μm, constitute over 90% of the total lipid weight. The lipid globules are enrobed in an emulsion-stabilizing membrane during milk synthesis and secretion (KEENAN et al., 1983). The membrane has a typical bilayer structure with a protein layer at the lipid interface. Proteins and lipids constitute over 90% of the membrane with neutral lipids and phospholipids comprising almost all of the lipid fraction. The membrane is essential in maintaining the physical integrity of the fat globule during handling of milk and manufacturing of cheese. Disruption causes loss of fat retention in cheese and free fat in cheese which imparts an oily mouthfeel and undesirable appearance of foods, i. e., pizza, that contain such cheese.

Triacylglycerols account for 97–98% of the total milk lipids (JENSEN and CLAR, 1988). Phospholipids are present at levels less than 1%, and other fractions such as diacylglycerols, monoacylglycerols, free fatty acids and free sterols range in concentration from 0.1 to 0.6%. Milkfat contains more than 400 different fatty acids but only 20 are present in more than trace quantities. The fatty acids in bovine milk fat, in order of their concentrations on a molar basis, are 18:1, 16:0 > 18:0, 4:0, 14:0 > 6:0, 12:0, 18:2. Concentrations will vary with feed and season. The presence of short-chain fatty acids, 4:0 and 6:0, contributes to flavor in several cheese varieties in which limited lipolysis occurs. These two fatty acids are located primarily, if not exclusively, on the sn-3 position of the triacylglycerol; many of the lipases and esterases that contribute to maturation of certain cheeses exhibit a preference for the sn-1 and sn-3 positions (HA and LINDSAY, 1993; JENSEN et al., 1990). In other cheese varieties, perceptible concentrations are considered undesirable.

Bovine, ovine and caprine milk fats differ in relative concentrations on *n*-chain fatty acids and their positions in glycerides (HA and LINDSAY, 1993). Of special interest are branched-chain fatty acids that impart the goaty and muttony flavors to meats and cheeses from goats and sheep (HA and LINDSAY, 1991). The goaty aroma of goats' milk cheese was associated with 4-ethyloctanoic acid which also appeared to be an important flavor note in Roquefort cheese which is made from sheeps's milk. The distinctive flavor of Romano cheese made from sheep's milk was thought to be influenced by 4-methyloctanoic and 4-ethyloctanoic acids combined with selected phenolic compounds (HA and LINDSAY, 1991).

4.1.3 Carbohydrates

The major carbohydrate in milk is lactose; milks of cows, sheep and water buffalo contain about 4.8%, while levels in goats' milk approximate 4.1% (HOLSINGER, 1988). Small amounts of other carbohydrates are in milk in the free form or bound to proteins, lipids or phosphate. Glycosylation of κ-CN facilitates its ability to stabilize casein micelles. Most of the lactose is carried out with the whey during cheese manufacturing with the levels remaining in cheese depending upon the moisture content of cheese, fermentation of sugars during cheese maturation and whether lactose may be incorporated as part of added ingredients in some cheese types. Virtually all of the lactose disappears from cheese during the first few months of maturation by hydrolysis to glucose and galactose and by fermentation of the lactose or the two monosaccharides. Higher concentrations of the sugars are present in fresh cheeses and may be present in cheeses containing higher levels of moisture that are not matured for more than a few weeks or months. Depletion of sugars in cheese to be matured is essential, since they can serve as energy sources for undesirable bacteria. Residual sugars can also participate in the Maillard browning reaction and create an unacceptable color in foods containing the heated cheese.

4.1.4 Minerals

The major minerals in milk are cationic K, Na, Ca and smaller amounts of Mg plus anionic phosphate, Cl and smaller amounts of citrate (HOLT, 1985). The soluble ions, especially K, Na and Cl, influence the ionic strength of milk, but are largely removed with the whey during cheese manufacturing. The multivalent ions' partition between the cheese curd and whey depends upon the pH of cheese during manufacturing (LUCEY and FOX, 1993). Approximately 60 and 55% of calcium and phosphate in milk are retained in Cheddar cheese; only 40 and 45% of these two salts are retained in the more acidic Cheshire cheese.

In milk which has a pH value of about 6.7, approximately two-thirds of the Ca, one-third of the Mg and one-half of the P_i are in a colloidal form. All colloidal P_i and citrate are presumed to be in the micellar calcium phosphate, whereas colloidal Ca and Mg are partially incorporated into the micellar calcium phosphate and some is bound directly to the caseins. Small amounts of these cations are bound to whey proteins. Concentrations of Ca^{2+} and colloidal calcium phosphate influence clotting rates of milk and the strength of the resulting gels formed by milk-clotting enzymes as discussed in Sect. 4.3. Calcium phosphate provides part of the buffering capacity of cheese that is important in pH control and contributes to the physical properties of the casein matrix of cheese (LUCEY and FOX, 1993). The latter function is highly dependent upon the pH of cheese which influences the proportion of the undissolved form of the salt.

4.1.5 Minor Constituents

Although enzymes are present in low levels (w/w) in milk, some significantly influence the properties of milk and cheese. About 60 enzymes have been identified in milk which originate from the mammary gland tissue cells, blood plasma and blood leukocytes (KITCHEN, 1985). Only a few of these enzymes have been demonstrated to have substantial effects on cheese. Alkaline phosphatase is used as an indicator of adequate heat treatment (pasteurization) of milk. Proteinases and lipases can attack substrates in milk and cheese to produce desirable and undesirable effects.

The alkaline milk proteinase, plasmin, is a serine proteinase exhibiting trypsin-like activity. It can attack all the major caseins, but its activity is much faster on β-CN. The peptides released from β-CN have been characterized; the major peptides were designated as γ-CN but now are renamed as fragments of β-CN (see Tab. 3). Both the active enzyme and the inactive precursor, plasminogen, exist in milk, but the latter at five times the level of the acitve form. Constituents of somatic cells are able to activate plasminogen which can attack the caseins and reduce the yield of cheese and alter protein functionality and milk quality (VERDI and BARBANO, 1991). In contrast, plasmin addition has been reported to enhance the quality of Cheddar cheese, and plasminogen activation by heat is a normal occurrence in manufacturing of some cheeses such as Swiss and Mozzarella cheeses (DELACROIX-BUCHET and TROSSAT, 1992; FARKYE and FOX, 1990, 1992; FARKYE et al., 1991). Higher plasmin levels by plasminogen activation, higher pH values and lower NaCl concentrations resulted in more extensive hydrolysis of β-CN during maturation of cheeses in which temperatures of 50–60 °C are reached during manufacturing. An acid proteinase purified from milk has properties that suggest that it is cathepsin D. It should be active under the ionic conditions in cheeses during maturation, but its role may be masked by chymosin because of similar specificities (KAMINOGAWA et al., 1980).

Lipolytic activity in raw bovine milk is due to one enzyme that can act as a lipoprotein lipase or a true lipase (KITCHEN, 1985). Approximately 80% of the activity is associated with the casein micelles. The lipase action is rather non-specific with the rate of hydrolysis of tributyrin being about the same as that of long-chain triacylglycerols. However, inhibitors such as 1 M NaCl have little effect on the tributyrin reaction. The milk lipase is destroyed by high-temperature short-time (HTST) pasteurization which is a minimum treatment of 71.6 °C for 15 s. This limits effects of the lipase to cheese made from raw or moderately heated milks. Homogenization, a physical process of disrupting the milk fat globules as described later, is a strong activator of lipolysis of raw milk (DEETH and FITZGERALD, 1983). This process is used to accelerate lipolysis in some blue-veined cheeses that depend upon free fatty acids and their metabolites for typical flavor. In many other cheeses, this level of lipolysis would be undesirable which obviates the use of milk homogenization.

4.2 Lactic Acid Bacteria

4.2.1 Characteristics and Functions

Lactic acid bacteria are added in the manufacturing of all cheeses except those in which curdling of milk results from acidification by adding acids and/or heat treatments. A few cheese types use a combination of acid addition and milk-clotting enzymes to form the curd. The taxonomy, cultivation, metabolism and genetics of lactic acid bacteria have been reviewed by TEUBER (1993a) in Volume 1 of this series. These bacteria can be classified into three general groups, as shown in Tab. 4, relative to their ability to produce acid at different temperatures during cheese manufacturing and their ability to produce carbon dioxide and diacetyl. The latter attribute contributes to eye formation in Gouda cheese and flavor in fresh cheeses such as Cottage cheese. Lactic acid bacteria can be grouped into those that are used in cheeses in which temperatures may reach 50–55 °C or those for cheeses in which temperatures may reach about 40 °C during manufacturing (Tab. 4). These groups evolved from the selective conditions and environments in cheeses as they were manufactured by early artisans. Subsequent isolation and strain improvement have yielded the presently used strains of the streptobacteria, *Lactococcus lactis* ssp. *lactis* and *Lc. lactis* ssp. *cremoris*, that are typical species for Cheddar cheese. Thermobacteria, *Lactobacillus helveticus* and *Streptococcus thermophilus* are species commonly used for cheeses such as Emmental, Parmesan and Provolone that receive higher heat treatments during manufacturing.

Lactic acid bacteria serve a number of functions during manufacturing and maturing cheese as illustrated in Tab. 5. Fermentation of lactose to lactic acid is the primary function, since this impacts upon the manufacturing process, gross composition of cheese, microbiological flora, enzymatic activities, and the texture, appearance and flavor of matured cheese. Sugar transport and metabolism differ between the various lactic acid bacteria used in the different types of cheese (POOLMAN, 1993). Species such as *Lactococcus lactis* ssp. *cremoris* utilize the phosphoenolpyruvate

Tab. 4. Typical Species of Lactic Acid Bacteria and Typical Cheeses Manufactured with Them

Species of Lactic Acid Bacteria	Typical Cheese Manufactured with the Species
Mesophilic	
Homofermentative	
Lactococcus lactis ssp. *cremoris* *Lactococcus lactis* ssp. *lactis*	Cheddar, Gouda, blue, Limburger, cottage, cream
Heterofermentative	
Leuconostoc mesenteroides ssp. *cremoris* *Lactococcus lactis* ssp. *lactis* biovar. *diacetylactis*	Gouda, cottage, cream, blue
Thermophilic	
Streptococcus thermophilus *Lactobacillus helveticus* *Lactobacillus delbrueckii* ssp. *bulgaricus*	Emmental/Swiss, Italian grana-types, Italian pasta types (Provolone, Mozzarella)

Tab. 5. Functions of Lactic Acid Bacteria During Manufacturing and Maturation of Cheeses

Fermentation and depletion of sugars in cheese to control growth and metabolism of adventitious bacteria
Fermentation of citrate to form flavor compounds and carbon dioxide for eye formation desired in some cheeses
Creating low oxidation–reduction potential in cheese during early maturation to influence microbial metabolism and flavor compound synthesis and stability
Establishing selective competition and synergism with adventitious bacteria during cheese maturation
Participating in proteolysis and peptidolysis of cheese during maturation
Synthesis of flavor compounds

(PEP):galactoside phosphotransferase system (PTS) for assimilation of sugars from milk and the serum phase of cheese. The high binding affinity of the PTS allows efficient transport of nanomolar concentrations of sugars that create a favorably low concentration of these energy sources in the maturing cheese. Other species such as *Streptococcus thermophilus* and *Lactobacillus delbrueckii* ssp. *bulgaricus* possess lactose transporters driven by an electrochemical proton gradient.

The subsequent metabolism of lactose by the streptobacteria and thermobacteria differs and can impact on characteristics of cheese. Following the PTS transport, β-phosphogalactosidase cleaves the phosphorylated lactose. The glucose moiety is phosphorylated and is metabolized via the EMP pathway. Galactose-phosphate proceeds through the tagatose 6-phosphate pathway and enters the EMP pathway after cleavage by an aldolase. Lactose transported by a proton gradient is cleaved by β-galactosidase, and the glucose moiety follows the glycolytic pathway. Galactose enters the Leloir pathway via phosphorylation by galactokinase. Most "wild type" strains of *S. thermophilus* and *Lb. delbrueckii* ssp. *bulgaricus* cannot metabolize galactose which has been attributed to low galactokinase activity (HUTKINS et al., 1985). The unmetabolized galactose is expelled back into the cheese and creates brown discoloration from the non-enzymatic browning reaction when the cheese is heated in food preparation or drying (JOHNSON and OLSON, 1985). Galactose metabolism in Emmental cheese also impacts upon the properties of this cheese, especially eye formation. Subsequent fermentation of galactose by *Lb. helveticus*, after lactose metabolism by *S. thermophilus*, effectively controls pH and eye formation which is not attained when non-galactose metabolizing lactobacilli are used (TURNER et al., 1983).

Heterofermentative lactococci are used in varieties such as Gouda and cottage cheeses to enhance flavor and to produce small eyes. Strains of *Leuconostoc* species and of *Lc. lactis* ssp. *lactis* biovar. *diacetylactis* ferment milk sugars and citrate to a variety of compounds including diacetyl. It is proposed that diacetyl is formed by chemical decarboxylation of α-acetolactate produced by metabolism of the heterolactics under conditions that greatly enhance intracellular accumulation of pyruvate (HUGENHOLTZ, 1993). Others feel that diacetyl is produced by direct metabolism of citrate in the presence of lactose, but factors regulating diacetyl and acetoin production are still not elucidated (COGAN, 1984). Accumulation of pyruvate and its diversion to diacetyl could result from a decrease in V_{max} and K_m of lactate dehydrogenase when the internal pH of *Leuconostoc* species decreases during fermentation (FITZGERALD et al., 1992).

Other characteristics of lactic acid bacteria shown in Tab. 5 will be discussed in subsequent sections except for their effect on the oxidation–reduction potential of cheese and production of metabolites that are inhibitory to other microorganisms. It is well-recognized that the redox potential of cheese decreases during early maturation (OLSON, 1990). This results from the metabolism of the lactic acid bacteria in cheese made from pasteurized milk; other species like adventitious lactobacilli contribute to a decrease at later stages. The utilization of oxygen and production of reducing groups contributes to desirable development and stabilization of flavor and aids in fusion of curd in concert with a decrease in curd pH during the first few days of maturation.

Lactic acid bacteria produce a variety of metabolites such as organic acids, diacetyl and hydrogen peroxide that are broad-spectrum antagonists plus bacteriocins which tend to inhibit species that are closely related to the strain producing the bacteriocin (BAREFOOT and NETTLES, 1993). The organic acids, lactic, acetic and propionic, are effective microbial inhibitors. Since they have higher pK_as, a greater proportion of the molecules exists in the undissociated, lipophilic form. Propionic and acetic acids would be more effective at the pHs of most cheeses. It is unlikely that diacetyl would be effective in most cheese, because inhibitory levels are higher than acceptable concentrations in cheeses. Hydrogen peroxide accumulates in cultures of *Lactobacillus, Leuconostoc* and *Pediococcus* species, but probably has minimal effect in most cheeses. However, a hydrogen peroxide super-producing mutant of a *Lc. lactis* ssp. *lactis* strain increased the shelf-life of cottage cheese by 17%. There is substantial interest in bacteriocins as inhibitors of spoilage microorganisms and pathogens in cheese. Nisin was approved as an antibotulinal additive in pasteurized process cheese spread in the US in 1988. Numerous other bacteriocins are produced by strains of *Lc. lactis* ssp. *cremoris, Lc. lactis* ssp. *lactis, Lc. lactis* ssp. *lactis* biovar. *diacetylactis, Leuconostoc* species and lactobacilli. Considerable research is still required to genetically and biochemically characterize bacteriocins. Similarly, more information is needed to permit synthesis of these compounds with appropriate heat stability and inhibitory properties. They should also be nontoxic, not affect sensory properties of foods, be simple to use and be cost-effective.

4.2.2 Culture Propagation

Culture propagation has evolved from reuse of whey from previous cheese manufacturing, to propagating semi-purified cultures in cheese plants, to the present day practice of purchasing highly characterized cultures routinely from suppliers (LODICS and STEENSON, 1993). Considerable development and commercial application of modern technologies for lactic starter cultures has occurred in the last 15 years (SANDERS et al., 1991). Cultures composed of numerous undefined strains have been replaced by cultures containing single or several defined and characterized strains. These cultures are developed and maintained by commercial suppliers and are sold in a variety of forms. The type of products and the method of distribution differ between regions of the world and type of cheese, but contemporary products include concentrates of cells in metal cans, frozen pellets of cell concentrates and freeze-dried cells (ROBINSON, 1991). The cans of concentrate are initially frozen in liquid nitrogen, maintained in liquid nitrogen by the culture supplier and stored at −40 to −60 °C.

Volumes of lactic starter cultures required in a cheese manufacturing plant can be substantial, ranging up to 10^4 liters per day. Culture preparation takes place in specialized, isolated rooms and in equipment designed to virtually eliminate contamination by other microorganisms and bacteriophages. The rooms are typically under positive pressure by air filtered through an effective system. One recommendation indicated use of a prefilter to remove 30% of 5.0 μm particles followed by an electrostatic precipitator to remove 95% of the 0.3 μm particles. Domed vats that are capable of being sealed are used to propagate cultures for cheese manufacturing. These may also be fitted for filtered air infusion to impose a positive pressure and minimize contamination. Specially designed inoculation ports that can be pre-sterilized before the inoculation step are becoming more common.

In spite of the above precautions, bacteriophage contamination of lactic starter cultures has not been eliminated. Total elimination for the cheese manufacturing environment is probably impossible because lactococci are lysogenic, raw milk contains bacteriophages that survive pasteurization and spray-drying temperatures, and complete asepsis is not attainable (LODICS and STEENSON, 1993). Environmental control is essential and beneficial when combined with other measures. Bacteriophage-inhibitory media, which function by chelation of calcium that is needed for bacteriophage proliferation, have been used for the past 40 years in the U.S. These were combined with rotation of strains with supposed unrelated bacteriophage sensitivity. Recently, fewer defined strains are used without rotation, but with constant monitoring for environmental bacteriophages to permit replacement of sensitive bacterial strains with unrelated strains. In Europe, propagation and storage technologies have been developed to maintain mixed strain cultures that have developed bacteriophage resistance mechanisms by constant exposure to bacteriophages. These starters have served as sources of resistant strains, but the resistance usually is lost during propagation under sterile conditions. Substantial research is underway to understand genetic determinants of bacteriophage resistance in lactic acid bacteria. One commercial lactococcal starter, transformed with plasmid pTR2030 that encodes for restriction-modification and abortive infection, was used commercially for seven years before a new bacteriophage evolved to attack the strain. Strains are being developed with multiple defense mechanisms that may lengthen the usage time limit in commercial cheese factories.

The damaging effects of low pH values during later stages of propagation and storage under acidic conditions on the activity of lactic acid bacteria is well known. However, the mechanisms by which these bacteria maintain a higher intracellular pH during active growth and the implications of that regulation have been recently defined (HUTKINS and NANNEN, 1993). All lactic acid bacteria are capable of maintaining a pH differential of one to two units between the outside medium and cell cytoplasm during growth even to pH values of 4.0. The pH differential collapses at cessation of growth. Extrusion of protons by lactococci is thought to occur by an electrogenic H^+-lactate symporter and by a proton-translocating ATPase. The latter is also thought to be operative in lactobacilli.

The need for increasingly active lactic starter cultures prompted development of systems to minimize the detrimental effect of low pH after cessation of growth (HUTKINS and NANNEN, 1993; WHITEHEAD et al., 1993). Two approaches are being used: one in which the medium is neutralized with a base such as NH_4OH, or another in which the medium is buffered. External neutralization can be continuous to maintain a desired pH value which will vary between different species, or a single neutralization can be done after culture growth to a given pH. After the single neutralization, the culture is allowed to grow and lower the pH of the medium to a level that will not harm the culture (KHOSRAVI et al., 1991). Internal buffering of the growth medium is attained by using encapsulated neutralizers for gradual release or by a sufficient reservoir of insoluble neutralizer.

Lactic starter media are also formulated to maximize acid-producing capability and shorten the lag phase during cheese manufacturing (WHITEHEAD et al., 1993). Nitrogen sources such as milk proteins, caseins, whey proteins and their hydrolysates provide a ready source of assimilable nitrogenous compounds and minimize loss of cell proteinase activity. Vitamins and minerals are supplied by yeast extract and corn steep liquor. Ascorbic acid and ferrous sulfate are added as antioxidants. Phosphates and citrates chelate calcium for bacteriophage control, and carbonates, hydroxides, oxides and phosphates serve as neutralizers. Commercial formulations have yielded lactic starter cultures that are substantially more active, which has reduced the volume required as inoculum for cheese manufacturing and has provided greater consistency in acid production. The changes in culture performance also have altered the cheese manufacturing schedule, especially for Cheddar cheese. The pH values reached during cheese manufacturing, to attain the desired final pH in the cheese, are higher as compared to those attained with non-neutralized cultures. The shift in acid production regime has resulted in higher lactate concentrations in cheese, because more lactic acid is produced after whey drainage. The implications of this are not fully known, but a higher lactate concentration in cheese is one factor associated with calcium lactate crystals on surfaces of Cheddar cheese (JOHNSON et al., 1990).

4.3 Milk-Clotting Enzymes

4.3.1 Functions

The principal function of milk-clotting enzymes, as the name implies, is clotting milks at or slightly below their normal pH. Secondarily, they play an important role in the initial proteolysis during cheese maturation. Enzymatic clotting of milk results from two processes, a primary, specific hydrolysis of κ-CN which reduces its casein micelle stabilizing capacity and a subsequent aggregation of the destabilized micelles (DALGLEISH, 1992). Hydrolysis occurs at one specific bond, Phe_{105}-Met_{106}, which releases the hydrophilic N-terminal region of the molecule which had projected out from the micelle surface into the solvent and imparted the micelle stability. The specific cleavage of this bond is probably related to the amino acid sequence surrounding the bond and the conformation of κ-CN in the casein micelle. Chymosin will not cleave the dipeptide Phe-Met, but was active against a pentapeptide simulating the region around the sensitive bond. Increasing the length of the peptide to mimic the Pro_{101}-Lys_{111} region of κ-CN substantially increased hydrolysis rates. It has been proposed that the sensitivity of the bond results from its position in a region of the protein that forms a projecting β-structure or is situated on a β-turn. Molecular modeling proposed that Pro-His-Pro and Pro-Pro residues adjacent to Phe-Met form a kink that projects this normally hydrophobic region into the solvent for easy access by the enzyme (FARRELL et al., 1993). Proteolytic activity exhibits the expected effects of pH, temperature and ionic strength. The pH optimum of chymosin in milk is approximately 6.0; activity also shows a maximum as ionic strength is increased. Heating milk reduces the rate and extent of κ-CN cleavage by chymosin, although the effect is much greater on the subsequent clotting of the milk. The mechanism of inhibition of proteolysis is unclear, but may be related to complex formation between κ-CN and β-lactoglobulin, thereby preventing approach of the enzyme to the active site on a portion of the surface κ-CN (DALGLEISH, 1992).

Aggregation of casein micelles does not occur until 60 to 80% of their κ-CN is hydrolyzed. The percentage conversion of κ-CN necessary to induce micelle aggregation and coagulation decreases from 90% at 15°C to about 60% at 30°C (CARLSON et al., 1986). Loss of micelle stability probably results from additive factors, but loss of steric stabilization after release of the hydrophilic portion of κ-CN that projected from the micelle is a major factor. DALGLEISH (1992) concluded that the rate of coagulation is enhanced by neutralization of charge within the casein micelles, lowering surface charge repulsion and allowing hydrophobic interactions to occur after sufficient steric stabilization imparted by κ-CN has been removed. Concentrations of chymosin have virtually no influence on the rate of aggregation but shorten the lag time before aggregation is initiated. The "molecular weight" of micelle clusters increases at a linear rate subsequent to the lag phase. Clotting rates are increased by reduction of micelle charge with cations such as Ca^{2+}, by increased temperatures and by lowering pH. The latter may impact primarily on the proteolytic stage.

Further aggregation leads to visible flocs of micelles and gelation under quiescent conditions. Gel strength increases at a rate that can be expressed by the following three parameter models (DEJMEK, 1987).

$$G = G_{\infty} e^{-T/(t-t_0)}$$

where G is the gel strength modulus, G_{∞} the asymptotic gel strength, T the time constant of constant gel build-up, and t_0 the time constant of incipient gelling. A mechanistic kinetic model to express gel strength development that combines the primary proteolytic and secondary gel development stages and incorporates effects of enzyme deactivation has been proposed by CARLSON et al. (1987). Their results indicate that enzymatic milk-clotting should be considered as a continuous phenomenon.

A variety of methods have been developed to monitor gel firmness including light scattering measurement, ultrasound, mechanical measurement of gel strength and heat flow as influenced by gel formation (BROWN and ERNSTROM, 1988). Most have been used in laboratory and pilot plant settings; a few have been used under commercial conditions to indicate the desired point of cutting the milk gel. Choosing the optimum gel firmness for cutting is important in cheese manufacturing, since it affects the retention of milk fat and casein in cheese and the moisture loss during cheese manufacturing. Gels that are too firm or soft at cutting can cause increased losses of milk solids. A firm gel at cutting, if all other factors are equal, will reduce the rate of whey expulsion (WALSTRA et al., 1987a). The mechanism for this effect relates to the endogenous syneresis pressure that develops in a milk gel after adding the milk-clotting enzyme and the permeability of the gel (VAN VLIET et al., 1991). The pressure depends upon the reactivity, bending stiffness and yielding of the casein strands in the gel. Yielding and reforming of the strands produces a dynamic system and increases the permeability of the gel. The endogenous pressure reaches a maximum and then stabilizes, when casein strand aggregation slows and breaking and reforming of the strands is minimized. Cutting the gel into smaller pieces during the aggregation phase should enhance syneresis, whereas cutting the stabilized gel should establish slower rates of whey expulsion.

4.3.2 Types

Although a variety of plant, bacterial, fungal and animal sources have been used as enzyme preparations for clotting milk, only the latter two have been major sources (FOLTMANN, 1987). The nomenclature and sources of the fungal and animal derived milk-clotting enzymes are shown in Tab. 6. The gastric proteinases are excreted as inactive precursors which are activated under acidic conditions. Chymosin is the principal proteinase in a semi-purified extract (rennet) of stomachs from suckling calves. The proportion of chymosin to bovine pepsin decreases as the animal ages. The percentage of total milk-clotting activity in calf rennet extracts attributable to chymosin may vary between 55 and 95%, whereas extracts from adult bovines may contain only 5 to 45% chymosin (GUINEE and WILKINSON, 1992).

Shortages of calf stomachs prompted the search for alternatives. One of the first was porcine pepsin which was used in combination with rennet extract. The instability and lower clotting activity of porcine pepsin at higher pH values made it unsuitable as the sole enzyme for many types of cheese. Bovine pepsin is less pH sensitive than porcine pepsin but more sensitive than chymosin. The microbial enzymes listed in Tab. 6 were used widely as substitutes for rennet, even though their proteolytic activity was higher. Rate of action of the microbial enzymes on casein is in the following order: *Endothia parasitica* > *Mucor miehei* > *M. pusillus*. The higher proteolytic activity has been related to slightly

Tab. 6. Nomenclature and Source of Animal and Microbial Derived Milk-Clotting Enzymes Used in Cheese Manufacturing (ANONYMOUS, 1990)

Commercial Name	Systematic Name IUB	Source
Calf rennet	Chymosin	Calf stomach
Bovine rennet	Pepsins A and B	Adult bovine stomach
Porcine pepsin	Pepsin A	Pig stomach
Microbial coagulant	Microbial carboxyl proteinase	*Mucor miehei*
Microbial coagulant	Microbial carboxyl proteinase	*Mucor pusillus* Lindt
Microbial coagulant	Microbial carboxyl proteinase	*Endothia parasitica*
Fermentation-derived rennets	Chymosin	*Escherichia coli* K12
		Kluyveromyces lactis
		Aspergillus niger

IUB, International Union of Biochemistry

lower cheese yields and bitterness in cheese, especially for the *E. parasitica* enzyme. However, the temperature sensitivity of this enzyme has permitted its commercial use for Swiss cheese which is subjected to higher temperatures during manufacturing. More heat-labile forms of the *Mucor miehei* proteinase are more similar to chymosin in general proteolytic activities.

The major development in milk-clotting enzymes in the last decade has been the successful cloning and expression of the chymosin gene in host microorganisms and its synthesis during fermentation (TEUBER, 1993b). Commercial fermentation-produced chymosin is derived from genetically modified *Aspergillus niger, Kluyveromyces lactis* and *Escherichia coli* K12. All available evidence indicates that these recombinant chymosin preparations have identical properties to calf chymosin. Nineteen varieties of cheese have been manufactured with these enzymes on an experimental or pilot-plant scale. Use of these products has grown substantially; it is estimated that over 50% of the US Cheddar cheese is manufactured with the fermentation-produced products. No significant differences were observed between cheeses produced with the fermentation-produced enzymes and those with calf chymosin in terms of yield, textural properties, sensory properties and maturation patterns of cheese. The products appear to be safe based on the inability to recover recombinant cells or vector cDNA from cheese.

5 Classes of Cheese

5.1 Major Distinctions Between Classes

The discussion in Sects. 3 and 4 described commonalities of microbiology, chemistry and physical properties of most cheese varieties plus the usual ingredients to manufacture cheese. The great diversity of cheeses precludes thorough discussion of the technologies and characteristics of even the major types of cheeses. This section will focus on principal attributes that distinguish the major types of cheese and provide an understanding of what causes those unique attributes. The reader is referred to OLSON (1970), SCOTT (1986) and FOX (1987) for detailed description of the technology, microbiology chemistry and physicochemistry of various cheeses. The following will supplement those references.

5.2 Cheddar Types

Cheddar cheese is one of the major varieties in the world, but most production is concentrated in the former British colonies. A substantial body of information has been published on its manufacture and characteristics (LAWRENCE and GILLES, 1987; SCOTT, 1986; WILSON and REINBOLD, 1965). The manufacturing procedures follow the general outline in Sect. 3, but handling curd after whey drainage is unique. The curd particles are allowed to fuse into a contiguous mass as the whey is drained. Traditionally, the mass of fused curd was then cut into smaller blocks which were turned and piled manually in a process called cheddaring. The blocks were cut into smaller pieces, by a process called milling, when the desired curd pH was attained. Sodium chloride was applied, and the cheese pieces were pressed into shape in forms called hoops.

Current technology for Cheddar cheese manufacturing is highly mechanized with equipment to drain the whey, fuse the curd, convey the curd during cheddaring, apply salt, fuse the salted curd in towers, remove blocks of fused cheese and package the cheese. Computer-controlled systems require very little labor input. An alternative, efficient, less capital-intensive system called stirred-curd Cheddar is popular in the United States. The procedure is similar to the traditional method except the pieces of curd are stirred mechanically and not allowed to fuse during whey drainage. The rest of the process is essentially the same as the traditional method except the fused blocks of cheese are subjected to vacuum to eliminate openings in the cheese.

Several variants of Cheddar, such as Colby and Monterey cheeses, are made similarly to the stirred-curd procedure. These cheeses typically contain more moisture than Cheddar cheese, so the curd is washed with cool water after whey drainage. This minimizes syneresis of the curd (VAN VLIET et al., 1991) to create higher cheese moisture contents, and removes some lactose and lactic acid to control the pH of the cheese.

5.3 Cheeses with Eyes

Two major varieties typify this category; Swiss-type (Emmental) and Gouda cheeses. Swiss-type cheeses are one of the best illustrations of the tightly-linked interactions of biological processes, chemical properties and physical characteristics that are required to manufacture a desirable cheese (REINBOLD, 1972; STEFFEN et al., 1987). Lactic acid bacteria used as cultures are predominantly *Streptococcus thermophilus* with small amounts of *Lactobacillus* species and possibly *Lactococcus lactis* ssp. *cremoris*, if more rapid acid production is needed. Sugar fermentation is sequential with *S. thermophilus* metabolizing all of the lactose, but it is not able to utilize the galactose moiety (TURNER et al., 1983). *Lactobacillus* species ferment the remaining galactose in the cheese with *Lb. helveticus* exhibiting more efficient utilization.

After curd fusion and after the proper pH has been attained, the blocks or wheels of cheese are salted in a sodium chloride brine and then stored for several days to allow more extensive curd fusion. The cheeses are placed in a room at 22 °C to allow growth and metabolism of lactic acid and residual sugars by *Propionibacterium* species. Metabolic products are acetic and propionic acids that impart flavor, and CO_2 that accumulates at weak points in the cheese to form eyes. Formation of uniform, round eyes without cracking depends upon the proper rate of CO_2 production and a cheese structure that has adequate plasticity and cohesiveness. These physical properties are exhibited in a curd having a pH between 5.2 and 5.4. Casein gels (and by inference, the casein matrix in cheese) undergo a marked transition at about pH 5.2 and exhibit minimum elasticity (G') at that pH value (ROEFS et al., 1990). This may relate to a transition from electrostatic interactions involving micellar calcium phosphate at pH values above 5.4 to plus–minus interactions involving protein functional groups below pH 5.1. Apparently, the intermediate range around pH 5.2 yields a casein matrix with the desired physical properties. This effect is critical also in manufacturing Italian pasta filata type cheeses as described in Sect. 5.6. Successful eye formation depends upon many other factors such as pH control through expulsion of whey (lactose) before lactic acid fermentation occurs. Also, the higher temperature for cooking the curd inactivates most of the chymosin and thereby maintains the integrity of the casein matrix which is essential for eye formation. The higher cooking temperature activates the alkaline milk proteinase, plasmin, which becomes more dominant in the proteolysis of Swiss-type cheeses.

Gouda cheese must possess many of the same characteristics as Swiss-type cheeses since small eyes typically form in this variety (WALSTRA et al., 1987b). This variety is manufactured with *Lactococcus lactis* ssp. *cremoris* and *Lc. lactis* ssp. *lactis* plus *Leuconostoc mesenteroides* ssp. *cremoris*. The latter species is responsible for eye formation in this cheese. The higher moisture content of this variety necessitates partial dilution of whey with water to control lactose levels and regulate the pH of cheese close to the desirable value of 5.2 during early stages of maturation when eyes form. Several cheese varieties such as Jarlsberg and Maasdam merge the Swiss-type and Gouda cheese technologies by using lactococcal cultures and propionibacteria (REINBOLD, 1972). Control of cheese pH through water dilution of whey is typical for these varieties.

5.4 Mold-Ripened Types

(GRIPON, 1987; MORRIS, 1981)

Cheeses in this category have blue pigmented mold in the interior (Roquefort, Gorgonzola, Stilton, blue) or white mold on the surface (Camembert, Brie). Lactococcal starter cultures are used for both types of cheese; *Leuconostoc* species may be added to produce more open texture for cheeses containing the interior mold. The pH of these cheeses reaches lower levels than other varieties, which creates a brittle body during early stages of maturation because the micellar calcium phosphate is solubilized. Subsequent mold metabolism will raise the pH and create a soft, smooth cheese texture. Variants of Camembert cheese, as discussed later, avoid the low pH to create a more stable cheese during storage.

The most important attribute to attain during manufacturing of blue-veined cheeses is an open texture to allow growth of *Penicillium roqueforti* in a uniform pattern throughout the cheese interior. Openness results from cutting the milk gel when it is firm into large cubes and maintaining the integrity of the cubes during subsequent handling the curd, until it is transferred to hoops for desired fusion. The wheels of fused curd are pierced with needles to allow diffusion of O_2 into and CO_2 out of the cheese to stimulate growth of *P. roqueforti* which will tolerate lower O_2 and higher CO_2 tensions than other molds. Maturation of these varieties depends heavily on the proteolytic and lipolytic activities of the mold. Conversion of free fatty acids to methyl ketones by the mold is essential for typical flavor development. The mold also raises the pH by metabolism of lactic acid and forms proteolytic by-products which favors methyl ketone production and improves physical properties of the cheese.

Growth of the white mold, *Penicillium camemberti*, is the distinguishing feature of numerous French cheeses. These varieties are typically formed into small pieces which are inoculated with the mold and stored under conditions to create the white mycelia mat over the entire cheese surface. These cheeses ripen from the surface inward as evidenced by the progressive inward softening. This phenomenon apparently results from metabolism of lactic acid and production of NH_3 by the mold and other microbial flora on the cheese surface. This produces a pH gradient from the cheese center to surface. Higher pH values increase water sorption of the casein matrix which had been depleted of micellar calcium phosphate. This softening effect plus the action of chymosin creates the typical ripening pattern. Knowledge of this phenomenon permitted the development of surface mold-ripened cheeses that are more stable during storage. In this process, the pH of the curd is not allowed to drop below 5.0 to 5.2 by washing the curd or using cultures, such as *Streptococcus thermophilus*, that will not lower the pH as drastically. Maintaining a higher pH will retain more of the micellar calcium phosphate in the casein matrix which will not exhibit the dramatic softening. Methyl ketones and secondary alcohols are abundant in Camembert cheese with oct-1-en-3-ol being especially important.

5.5 Surface-Ripened Types

(OLSON, 1969; REPS, 1987)

Numerous varieties including Tilsiter, Munster, brick, Romadour, Limburger, Saint

Paulin and even the eye-cheese, Gruyère, fall in this category. All are characterized by a progressive growth of microorganisms on the cheese surface to impart the typical cheese flavor. Manufacturing procedures for these varieties typically involve transferring the curd–whey slurry to perforated forms and fusing the curd particles as the whey is being expelled. The curd may be washed or the whey diluted with water to regulate the pH of these higher-moisture varieties.

The blocks of cheese usually are salted in sodium chloride brine and then placed in a humid room at 12 to 20°C, depending upon the variety, to facilitate microbial growth on the cheese surface. Although there are some exceptions, the progression of microorganisms are yeasts and possibly molds followed by micrococci and culminating with coryneform bacteria. The stages are not segmented, but overlap of types is typical. Yeasts and molds form the initial population because of their salt tolerance and ability to metabolize lactic acid. Yeasts raise the pH of the cheese surface, synthesize vitamins and hydrolyze cheese proteins to produce amino acids. All these activities stimulate succeeding growth of micrococci and coryneform bacteria. The impact of molds and micrococci on flavor has not been elucidated but both are proteolytic. Coryneform bacteria form the typical orange-red surface and are associated with the intensive flavor of surface-ripened cheeses. The dominant species, *Brevibacterium linens*, is highly proteolytic and produces lipases.

5.6 Italian Cheeses

(FOX and GUINEE, 1987; REINBOLD, 1963)

5.6.1 Pasta-Filata Types

The term pasta-filata, filamentous dough, accurately describes the physical structure of varieties in this class such as Mozzarella and Provolone cheeses. Manufacturing of these varieties is similar to traditional or stirred-curd Cheddar cheese, but thermophilic lactic bacteria are used as the starter culture. Typically, a 1:1 ratio of *Streptococcus thermophilus* and *Lactobacillus* species is used. Pregastric lipases are added to impart the typical flavor to matured Provolone cheese. The unique filamentous structure is obtained by heating, mixing and molding the cheese curd at a pH approximating 5.2. The pieces of cheese are cooled and immersed in a sodium chloride brine. Recent technologies for Mozzarella cheese allow salting before or during curd heating and mixing.

Mozzarella cheese is used within a few days to weeks after manufacturing, primarily as a food ingredient. Its principal use as a topping on pizza necessitates that it can be shredded, does not undergo excessive browning when heated, exhibits desired meltability and stringing qualities and has the correct "chewing" characteristics (KINDSTEDT, 1991). Control of these physical properties is attained by regulating composition, sodium chloride concentration, degree of proteolysis and pH during manufacturing and maturation of the cheese. Non-enzymatic browning is caused by excess sugars in cheese resulting from milk solids addition, inadequate removal of lactose during manufacturing or inadequate fermentation of galactose as described in Sect. 4.2.1. These factors also apply to Provolone cheese.

5.6.2 Grana Types

Parmesan and Romano cheeses are the major varieties in this class of hard, grating cheeses. Both are made with the thermophilic cultures used for pasta-filata cheeses. Pregastric lipases are usually added to Romano cheese and may be used in Parmesan cheese, if more flavor potency is desired. The manufacturing process simulates Emmental cheese except the curd is firmer at pressing and the structure of the cheese is more brittle and granular. These varieties are matured for extended periods, ranging up to 2 years. Non-enzymatic browning of these varieties is observed when galactose metabolism is lacking, but may also be associated with the reaction of α-dicarbonyls with amino groups. Pink discoloration during maturation has been associated with tyrosine metabolism by certain *Lactobacillus* strains (SHANNON et al., 1977).

5.7 Lowfat Types

Consumer demand for cheese of lower fat contents has prompted research and development to create cheeses with desirable flavor and physical characteristics but with fat contents that approach that of skim-milk. This has been difficult because fat serves a number of vital functions in cheese (OLSON and JOHNSON, 1990). These include enhancement of physical properties, source of flavorful fatty acids and other lipophilic compounds and as a solvent reservoir for flavor compounds formed during maturation. Cheese in which the fat content is reduced by 25 to 33% can be manufactured to possess physical properties fairly similar to its full-fat counterpart if the ratio of cheese moisture to non-fat portion (MNFP) is similar or slightly higher than the full-fat cheese (BANKS et al., 1989). This is essentially attaining the same ratio of water to intact casein in cheese which has been shown to be a dominant influence on physical properties of cheese (VISSER, 1991).

Flavor characteristics of Cheddar-type cheese and Swedish semi-hard, round-eyed cheese were preferred, if the fat contents were reduced by 33% rather than 50% (ARDO, 1993; JOHNSON, M., University of Wisconsin, Madison, personal communication). Important factors in producing good-flavored reduced-fat Cheddar-type cheese are: (1) choosing a lactococcal culture that does not exhibit high acid-producing and proteolytic activities and (2) slower rates of acid production during cheese manufacturing (CHEN et al., 1992). Bacterial cultures have been used to enhance flavor of reduced-fat cheeses with variable results. Some improvements in flavor also yielded a flavor profile that was pleasing but not typical for the full-fat counterpart. Such treatments might be suitable for developing new reduced-fat cheese varieties, but this would require substantial marketing efforts. A major hurdle is minimizing the growth of adventitious bacteria which produce undesirable flavors in these higher-moisture cheeses.

5.8 Soft Unripened Types
(SCOTT, 1986)

It is not surprising that there are soft unripened cheeses manufactured throughout the world where milk is available, since this probably was the first type of cheese to be made and manufacturing processes can be rudimentary. The most popular variety in this class in the United States is Cottage cheese (EMMONS und TUCKEY, 1967). It is manufactured from skim-milk using mesophilic lactococci as starter cultures. Clotting occurs when the pH of the milk approaches the isoelectric point of milk and the casein micelles interact to form a gel. Small amounts of milk-clotting enzymes are added to enhance the firmness of the milk gel when it is cut into cubes at a pH of about 4.8 at 30 to 32°C. The curd is more fragile than that formed by milk-clotting enzymes, so it is stirred very carefully while the curd–whey slurry is heated to 50 to 55°C. The lactic starter culture continues to produce acid until the temperature reaches about 40°C; the minimum pH should be about 4.5.

The curd is allowed to settle, and whey is drained off when the proper curd firmness is reached. The drained curd is resuspended twice to three times in batches of water that are progressively colder to cool the curd and leach some of the acid to raise the pH of the curd. Cream of differing fat contents is mixed with the curd to yield a mixture that should have a pH no higher than 5.0. Control of pH during manufacturing is essential to produce curd with the correct physical properties that will absorb the added cream. Diacetyl levels in the cream may be increased by fermentation processes using *Leuconostoc mesenteroides* ssp. *cremoris.*

Cream cheese is also manufactured by acid-induced clotting. The gel is not cut but is disrupted by stirring, and whey is removed by centrifugation. Most cream cheese is made by the "hot-pack" method in which the curd or mixtures of curd and cream are heated to about 70°C, homogenized and packaged hot. This produces a more stable product. Quark and similar cheeses have a smooth texture like cream cheese but typically the curd is not heat-treated before packaging (SCOTT, 1986).

The fat contents of these products can range from <1% to over 15%.

6 Cheese Maturation

Except for the cheese varieties discussed in Sect. 5.8, all varieties are held under controlled storage to develop the physical and flavor attributes that are characteristic for that variety. Each type of cheese has its own profile of microbial and enzymatic transformations which obviates a brief discussion of cheese maturation. Some of the maturation processes were described for the major cheese types in Sect. 5. All of these processes evolved from accidental imposition of environmental conditions on curd. The particular microbial flora that dominantes in a given cheese is able to thrive at the pH, sodium chloride concentration, oxidation-reduction potential for that cheese or for a specific region of the cheese such as the surface. The substantial amount of research on cheese maturation has attempted to elucidate and rationalize these natural processes. FOX and STEPANIAK (1993) have reviewed the importance and impact of enzymes on cheese maturation.

6.1 Glycolysis and Lipolysis

Metabolism of carbohydrates dominates the manufacturing and early maturation of all cheeses. It has been described in detail by FOX et al. (1990). The L-isomer of lactic acid is the principal product of glycolysis except for small amounts of D-lactic acid in cheeses made with thermophilic lactic bacteria. Metabolism of the lactate occurs in eye-cheeses, mold-ripened cheeses and smear-ripened cheeses, but not in most other varieties. Citrate fermentation by lactobacilli and heterofermentative lactococci and *Leuconostoc* species produces acetate, diacetyl, acetoin and other compounds. Lipolysis occurs during maturation of all cheeses; the relative importance in various cheeses was discussed in Sects. 5 and 4.1.2 as was fatty acid metabolism.

6.2 Proteolysis

Hydrolysis of caseins in cheese has received more attention than any other aspect of cheese maturation. It has been reviewed by FOX (1988), FOX and LAW (1991) and GRAPPIN et al. (1985). It is generally agreed that initial hydrolysis of caseins is carried out by the milk-clotting enzymes (VAN DEN BERG and EXTERKATE, 1993). One specific bond of α_{s1}-CN is cleaved initially, and this correlates with softening of cheeses, such as Cheddar and Gouda, during the first few months of maturation. These enzymes continue their action, primarily on α_{s1}-CN to produce polypeptides.

The polypeptides are attacked by microbial peptidases, initially those from the lactic starter culture predominate (CROW et al., 1993). Presumably, proteinases of lactococci are involved also in the degradation of proteins and polypeptides in cheese. Lactococcal proteinases are plasmid-encoded and, therefore, activity can be lost. The lactococcal proteinases are anchored to the cell membrane and extend out through the cell wall. They are released when Ca^{2+} is depleted from the medium, but remain active possibly with slightly different specificity. Peptidase systems of lactococci have been thoroughly characterized (PRITCHARD and COOLBEAR, 1993). These include aminopeptidases, pyrrolidone carboxylyl peptidase, peptidases capable of cleaving bonds adjacent to proline residues, dipeptidases, tripeptidases and endopeptidases. No carboxypeptidase activity has been observed. This complement of peptidases can facilitate virtually complete hydrolysis of caseins or polypeptides generated from caseins. This capability is important for growth of these bacteria in milk and for their action during cheese maturation, since caseins are distinguished by a high proline content. Most, if not all, peptidase activity of lactococci appears to be intracellular. This location is not reconciled readily with the need for substantial peptidolytic activity outside the cell to supply essential amino acids. Active research is underway on peptide transport to resolve this contradiction. Peptidases could be released by cell lysis in cheese, which is another area of active research (EL SODA, 1993). Preliminary results indicate that using lactococci, which were more susceptible to lysis, correlated with higher levels of amino nitrogen and faster maturation of cheese.

The dominant secondary flora of Cheddar cheese and varieties with similar internal environments are adventitious lactobacilli (PETERSON and MARSHALL, 1990). Their numbers can exceed 10^6 g^{-1} within a few months of maturation. Metabolic activities, especially proteolytic, peptidolytic and lipolytic, of this group have been partially characterized. Various species and strains have been added to cheese and have generally increased proteolysis and lipolysis, but have not had consistent impact on flavor enhancement. Fairly consistent reports of decreased bitterness indicates the beneficial action of peptidases on bitter peptides. It has also been suggested that suppression of the adventitious flora which can create flavor defects is a significant attribute of the successful strains (MARTLEY and CROW, 1993).

6.3 Cheese Flavor

Many cheese varieties have a dominant flavor note that has been partially characterized (URBACH, 1993). Examples are heptan-2-one resembling blue-vein cheese flavor and oct-1-en-3-ol creating a flavor note reminiscent of Camembert. Short-chain free fatty acids are important in many Italian-type cheeses, and methanethiol is characteristic of many surface-ripened cheeses. Cheeses like Cheddar and Gouda that have a more subtle flavor are more difficult to characterize. Researchers in Australia were able to demonstrate a linear relationship between perceived flavor and "fitted" flavor based upon the logarithms of concentrations of H_2S, heptan-2-one, butanone, δ-decalactone and propan-2-ol. Although the adjusted coefficient of determination (R^2) was 0.84, the researchers stated that it is not clear why the relationship holds and that it may be a spurious product of statistical analysis. They concluded that sulfur compounds were essential for Cheddar cheese flavor, but no sulfur compound was detected whose concentration correlated with flavor.

7 References

ANONYMOUS (1990), Use of enzymes in cheesemaking, *Int. Dairy Fed. Bull.* **247**, pp. 24–38, Brussels: International Dairy Federation.

ARDO, Y. (1993), Characterizing ripening in low-fat, semi-hard round-eyed cheese made with undefined mesophilic DL-starter, *Int. Dairy J.* **3**, 343–357.

BANKS, J. M., BRECHANY, E. Y., CHRISTIE, W. W. (1989), The production low fat Cheddar-type cheese, *J. Soc. Dairy Technol.* **42**, 6–9.

BAREFOOT, S. F., NETTLES, C. G. (1993), Antibiosis revisited: bacteriocins produced by dairy starter cultures, *J. Dairy Sci.* **76**, 2366–2379.

BROWN, R. J., ERNSTROM, C. A. (1988), Milk-clotting enzymes and cheese chemistry, Part 1: Milk-clotting enzymes, in: *Fundamentals of Dairy Chemistry* (WONG, N. P., Ed.), 3rd. Ed., pp. 609–633, New York: Van Nostrand Reinhold Co.

CARIC, M., KALAB, M. (1987), Processed cheese products, in: *Cheese: Chemistry, Physics and Microbiology – Major Cheese Groups* (FOX, P. F., Ed.), Vol. 2, pp. 339–383, London: Elsevier Applied Science Publishers, Ltd.

CARLSON, A., HILL, C. G., OLSON, N. F. (1986), The coagulation of milk with immobilized enzymes: a critical review, *Enzyme Microbiol. Technol.* **8**, 642–650.

CARLSON, A., HILL, C. G. JR., OLSON, N. F. (1987), The kinetics of milk coagulation: IV. The kinetics of the gel-firming process, *Biotechnol. Bioeng.* **29**, 612–624.

CHEN, C. M., JOHNSON, M. E., OLSON, N. F. (1992), Optimizing manufacturing parameters in 33 % reduced-fat Cheddar cheese, *J. Dairy Sci.* **75**, Suppl. 1, 104.

COGAN, T. M. (1984), Mesophilic lactic cultures, in: *Fermented Milks, IDF Document 179,* pp. 77–88, Brussels: International Dairy Federation.

CREAMER, L. K., OLSON, N. F. (1982), Rheological evaluation of maturing Cheddar cheese, *J. Food Sci.* **47**, 631–636.

CREAMER, L. K., ZOERB, H. F., OLSON, N. F., RICHARDSON, T. (1982), Surface hydrophobicity of α_sI, α_{s1}-casein A and B and its implication in cheese structure, *J. Dairy Sci.* **65**, 902–906.

CROW, V. L., COOLBEAR, T., HOLLAND, R., PRITCHARD, G. G., MARTLEY, F. G. (1993), Starters as finishers: starter properties relevant to cheese ripening, *Int. Dairy J.* **3**, 423–460.

DALGLEISH, D. G. (1992), The enzymatic coagulation of milk, in: *Advances in Dairy Chemistry –1: Proteins* (FOX, P. F., Ed.), pp. 579–619, London: Elsevier Applied Science Publishers Ltd.

DEETH, H. C., FITZ-GERALD, C. H. (1983), Lipolytic enzymes and hydrolytic rancidity in milk and milk products, in: *Developments in Dairy Chemistry – 2: Lipids* (FOX, P. F., Ed.), pp. 195–239, London: Elsevier Applied Science Publishers Ltd.

DELACROIX-BUCHET, A., TROSSAT, P. (1992), Proteolysis and texture of scalded-curd hard cheeses. I. Influence of water activity, *Lait* **71**, 299–311.

DEJMEK, P. (1987), Dynamic rheology of rennet curd, *J. Dairy Sci.* **70**, 1325–1330.

EL SODA, M. A. (1993), The role of lactic acid bacteria in accelerated cheese ripening, *FEMS Microbiol. Rev.* **12**, 239–252.

EMMONS, D. B., TUCKEY, S. L. (1967), *Cottage Cheese and Other Cultured Milk Products, Pfizer Cheese Monographs,* Vol. 3, New York: Pfizer Inc.

EMMONS, D. B., ERNSTROM, C. A., LACROIX, C., VERRET, P. (1990), Predictive formulas for yield of cheese from composition of milk: a review. *J. Dairy Sci.* **73**, 1365–1394.

FARKYE, N. Y., FOX, P. F. (1990), Observations on plasmin activity in cheese, *J. Dairy Res.* **57**, 413–418.

FARKYE, N. Y., FOX, P. F. (1992), Contribution of plasmin to Cheddar cheese ripening: effect of added plasmin, *J. Dairy Res.* **59**, 209–216.

FARKYE, N. Y., KIELY, L. J., ALLSHOUSE, R. D., KINDSTEDT, P. S. (1991), Proteolysis in Mozzarella cheese during refrigerated storage, *J. Dairy Sci.* **74**, 1433–1438.

FARRELL, H. M. JR., BROWN, E. M., KUMOSINSKI, T. F. (1993), Three-dimensional molecular modeling of bovine caseins, *Food Struct.* **12**, 235–250.

FITZGERALD, R. J., DOONAN, S., MCKAY, L. L., COGAN, T. M. (1992), Intracellular pH and the role of D-lactate dehydrogenase in the production of metabolic end products by *Leuconostoc lactis, J. Dairy Res.* **59**, 359–367.

FOLTMANN, B. (1987), General and molecular aspects of rennets, in: *Cheese: Chemistry, Physics and Microbiology – General Aspects* (FOX, P. F., Ed.), Vol. 1, pp. 33–61, London: Elsevier Applied Science Publishers, Ltd.

FOX, P. F. (1987), *Cheese: Chemistry, Physics and Microbiology – Major Cheese Groups,* Vol. 2, London: Elsevier Applied Science Publishers, Ltd.

FOX, P. F. (1988), Review: rennets and their action in cheese manufacturing and ripening, *Biotechnol. Appl. Biochem.* **10**, 522–535.

FOX, P. F., GUINEE, T. P. (1987), Italian cheeses, in: *Cheese: Chemistry, Physics and Microbiology – Major Cheese Groups* (FOX, P. F., Ed.), Vol. 2, pp. 221–255, London: Elsevier Applied Science Publishers, Ltd.

FOX, P. F., LAW, J. (1991), Enzymology of cheese ripening, *Food Biotechnol.* **5**, 239–262.

FOX, P. F., STEPANIAK, L. (1993), Enzymes in cheese technology, *Int. Dairy J.* **3**, 509–530.

FOX, P. F., LUCEY, J. A., COGAN, T. M. (1990), Glycolysis and related reactions during cheese manufacture and ripening, *Crit. Rev. Food Sci. Nutr.* **29**, 237–253.

GRAPPIN, R., RANK, T. C., OLSON, N. F. (1985), Primary proteolysis of cheese proteins during ripening, *J. Dairy Sci.* **68**, 531–540.

GRIPON, J. C. (1987), Mould-ripened cheeses, in: *Cheese: Chemistry, Physics and Microbiology – Major Cheese Groups* (FOX, P. F., Ed.), Vol. 2, pp. 121–149, London: Elsevier Applied Science Publishers, Ltd.

GUINEE, T. P., WILKINSON, M. G. (1992), Rennet coagulation and coagulants in cheese manufacture, *J. Soc. Dairy Technol.* **45** (4), 94–104.

HA, J. K., LINDSAY, R. C. (1991), Contributions of cow, sheep and goat milks to characterizing branched-chain fatty acid and phenolic flavors in varietal cheeses, *J. Dairy Sci.* **74**, 3267–3274.

HA, J. K., LINDSAY, R. C. (1993), Release of volatile branched-chain and other fatty acids from ruminant milk fats by various lipases, *J. Dairy Sci.* **76**, 677–690.

HARRIS, S. (1984), *Glynn Christian's World Guide to Cheese,* London: Ebury Press.

HOLSINGER, V. H. (1988), Lactose, in: *Fundamentals of Dairy Chemistry* (WONG, N. P., Ed.), 3rd Ed., pp. 279–342, New York: Van Nostrand Reinhold Co.

HOLT, C. (1985), The milk salts: their secretion, concentrations and physical chemistry, in: *Developments in Dairy Chemistry – 3, Lactose and Minor Constituents* (FOX, P. F., Ed.), pp. 143–181, London: Elsevier Applied Science Publishers, Ltd.

HUGENHOLTZ, J. (1993), Citrate metabolism in lactic acid bacteria, *FEMS Microbiol. Rev.* **12**, 165–178.

HUTKINS, R. W., NANNEN, N. L. (1993), pH homeostasis in lactic acid bacteria, *J. Dairy Sci.* **76**, 2354–2365.

HUTKINS, R. W., MORRIS, H. A., MCKAY, L. L. (1985), Galactokinase activity in *Streptococcus thermophilus, Appl. Environ. Microbiol.* **50**, 777–780.

JAKOB, E., PUHAN, Z. (1992), Technological properties of milk as influenced by genetic polymorphism of milk proteins – a review. *Int. Dairy J.* **2**, 157–178.

JENNESS, R. (1988), Composition of milk, in: *Fundamentals of Dairy Chemistry* (WONG, N. P., Ed.), 3rd Ed., pp. 1–38, New York: Van Nostrand Reinhold Co.

JENSEN, R. G., CLARK, R. W. (1988), Lipid composition and properties, in: *Fundamentals of Dairy Chemistry* (WONG, N. P., Ed.), 3rd Ed., pp. 171–213, New York: Van Nostrand Reinhold Co.

JENSEN, R. G., GALLUZZO, D. R., BUSH, V. J. (1990), Selectivity is an important characteristic of lipases (acylglycerol hydrolases), *Biocatalysis* **3**, 307.

JOHNSON, M. E. (1988), Part II – Cheese chemistry, in: *Fundamentals of Dairy Chemistry* (WONG, N. P., Ed.), 3rd Ed., pp. 634–654, New York: Van Nostrand Reinhold Co.

JOHNSON, M. E., OLSON, N. F. (1985), Nonenzymatic browning of Mozzarella cheese, *J. Dairy Sci.* **68**, 3143–3147.

JOHNSON, M. E., RIESTERER, B. A., OLSON, N. F. (1990), Influence of non-starter bacteria on calcium lactate crystallization on the surface of Cheddar cheese, *J. Dairy Sci.* **73**, 1145–1149.

KAMINOGAWA, S., YAMAUCHI, K., MIYAZAWA, S., KOGA, Y. (1980), Degradation of casein components by acid protease of bovine milk, *J. Dairy Sci.* **63**, 701–704.

KEENAN, T. W., DYLEWSKI, D. P., WOODFORD, T. A. (1983), Origin of milk fat globules and the nature of the milk fat globule membrane, in: *Developments in Dairy Chemistry – 2, Lipids* (FOX, P. F., Ed.), pp. 83–118, London: Elsevier Applied Science Publishers, Ltd.

KEENAN, T. W., MATHER, I. H., DYLEWSKI, D. P. (1988), Physical equilibria: lipid phase, in: *Fundamentals of Dairy Chemistry* (WONG, N. P., Ed.), 3rd Ed., pp. 511–582, New York: Van Nostrand Reinhold Co.

KINDSTEDT, P. S. (1991), Functional properties of mozzarella cheese on pizza: a review, *Cult. Dairy Prod. J.* **26**, 3, 27–31.

KITCHEN, B. J. (1985), Indigenous milk enzymes, in: *Developments in Dairy Chemistry – 3, Lactose and Minor Constituents* (FOX, P. F., Ed.), pp. 239–279, London: Elsevier Applied Science Publishers, Ltd.

KHOSRAVI, L., SANDINE, W. E., AYRES, J. W. (1991), Evaluation of a newly-formulated bacteriophage inhibitory bulk starter medium for the cultivation of thermophilic lactic acid bacteria, *Cult. Dairy Prod. J.* **26** (2), 4–9.

LAWRENCE, R. C. (1989), The use of ultrafiltration technology in cheesemaking, *Int. Dairy Fed. Bull.* **240**, Brussels: International Dairy Federation.

LAWRENCE, R. C., GILLES J. (1987), Cheddar cheese and related dry-salted varieties, in: *Cheese: Chemistry, Physics and Microbiology – Major Cheese Groups* (FOX, P. F., Ed.), Vol. 2, pp. 1–44, London: Elsevier Applied Science Publishers, Ltd.

LAWRENCE, R. C., HEAP, H. A. GILLES, J. (1984), A controlled approach to cheese technology, *J. Dairy Sci.* **67**, 1632–1645.

LODICS, T. A., STEENSON, L. R. (1993), Phage-host interactions in commercial mixed-strain dairy starter cultures: practical significance – a review, *J. Dairy Sci.* **76**, 2380–2391.

LUCEY, J. A., FOX, P. F. (1993), Importance of calcium and phosphate in cheese manufacture: a review. *J. Dairy Sci.* **76**, 1714–1724.

MALMBERG, R., HOLM, S. (1988), Producing low-bacteria milk by microfiltration, *North Eur. Food Dairy J.* **54**, 30–32.

MARTLEY, F. G., CROW, V. L. (1993), Interactions between non-starter microorganisms during cheese manufacture and ripening, *Int. Dairy J.* **3**, 461–483.

MAUBOIS, J.-L. (1991), New applications membrane technology in the dairy industry, *Aust. J. Dairy Technol.* **46**, 91–95.

MORRIS, H. A. (1981), *Blue-veined Cheeses, Pfizer Cheese Monographs,* Vol. VII, New York: Pfizer Inc.

OLSON, N. F. (1969), *Ripened Semisoft Cheeses, Pfizer Cheese Monographs,* Vol. IV, New York: Pfizer Inc.

OLSON, N. F. (1970), Automation in the cheese industry: a review, *J. Dairy Sci.* **53**, 1144–1150.

OLSON, N. F. (1975), Mechanized and continuous cheesemaking processes for Cheddar and other ripened cheese, *J. Dairy Sci.* **58**, 1015–1021.

OLSON, N. F. (1979), Cheese, in: *Microbial Technology II* (PEPPLER, H. J., PERLMAN, D., Eds.), 2nd Ed., pp. 39–77, New York: Academic Press.

OLSON, N. F. (1990), The impact of lactic acid bacteria on cheese flavor, *FEMS Microbiol. Rev.* **87**, 131–148.

OLSON, N. F., JOHNSON, M. E. (1990), Light cheese products: characteristics and economics, *Food Technol.* **44**, 10, 93–96.

PETERSON, S. D., MARSHALL, R. T. (1990), Non-starter lactobacilli in Cheddar cheese. A review, *J. Dairy Sci.* **73**, 1395–1410.

PIRISI, A., COLIN, O., LAURENT, F., SCHER, J., PARMENTIER, M. (1994), Comparison of milk composition, cheesemaking properties and textural characteristics of the cheese from two groups of goats with a high or low rate of α_{s1}-casein synthesis, *Int. Dairy J.* **4**, 329–345.

POOLMAN, B. (1993), Biochemistry and molecular biology of galactoside transport and metabolism in lactic acid bacteria, *Lait* **73**, 87–96.

PRITCHARD, G. G., COOLBEAR, T. (1993), The physiology and biochemistry of the proteolytic system in lactic acid bacteria, *FEMS Microbiol. Rev.* **12**, 179–206.

REINBOLD, G. W. (1963), *Italian Cheese Varieties,* New York: Pfizer Inc.

REINBOLD, B. W. (1972), *Swiss Cheese Varieties,* New York: Pfizer Inc.

REPS, A. (1987), Bacterial surface-ripened cheeses, in: *Cheese: Chemistry, Physics and Microbiology – Major Cheese Groups* (FOX, P. F., Ed.), Vol. 2, pp. 151–184, London: Elsevier Applied Science Publishers, Ltd.

ROBINSON, R. K. (1991), Starter cultures for milk and meat processing, in: *Biotechnology – a Comprehensive Treatise* (REHM, H.-J., REED, G., Eds.), 1st Ed., Vol. 3, pp. 191–202, Weinheim: Verlag Chemie.

ROEFS, S. P. F. M., VAN VLIET, T., BIJGAART, H. J. C. M., DE GROOT-MOSTERT, A. E. A., WALSTRA, P. (1990), Structure of casein gels made by combined acidification and rennet action, *Neth. Milk Dairy J.* **44**, 159–188.

ROLLEMA, H. S. (1992), Casein association and micelle formation, in: *Advanced Dairy Chemistry – 1: Proteins* (FOX, P. F., Ed.), pp. 111–140, London: Elsevier Applied Science Publishers, Ltd.

SANDERS, M. E., KONDO, J. K., WILLRETT, D. L. (1991), Application of lactic acid bacteria, in: *Biotechnology and Food Ingredients* (GOLDBERG, I., WILLIAMS, R., Eds.), pp. 433–459, New York: Van Nostrand Reinhold Company.

SCOTT, R. (1986), *Cheesemaking Practice,* 2nd Ed., London: Elsevier Applied Science Publishers, Ltd.

SHANNON, E. L., OLSON, N. F., DEIBEL, R. H. (1977), Oxidative metabolism of lactic acid bacteria associated with pink discoloration in Italian cheese, *J. Dairy Sci.* **60**, 1693–1697.

STEFFEN, C., FLUECKIGER, E., BOSSET, J. O., RUEGG, M. (1987), Swiss-type varieties, in: *Cheese: Chemistry, Physics and Microbiology – Major Cheese Varieties* (FOX, P. F., Ed.), Vol. 2, pp. 93–120, London: Elsevier Applied Science Publishers, Ltd.

SWAISGOOD, H. E. (1993), Review and update of casein chemistry, *J. Dairy Sci.* **76**, 3054–3061.

TEUBER, M. (1993a), Lactic acid bacteria, in: *Biotechnology* (REHM, H.-J., REED, G., Eds.), 2nd Ed., Vol. 1, pp. 325–366, Weinheim: VCH mbH.

TEUBER, M. (1993b), Genetic engineering techniques in food microbiology and enzymology, *Food Rev. Int.* **93**, 389–409.

TURNER, K. W., MORRIS, H. A., MARTLEY, F. G. (1983), Swiss-type cheese. II. The role of thermophilic lactobacilli in sugar fermentation, *N.Z. J. Dairy Sci. Technol.* **18**, 117–123.

URBACH, G. (1993), Relations between cheese flavour and chemical composition, *Int. Dairy J.* **3**, 389–422.

VAN DEN BERG, G., EXTERKATE, F. A. (1993), Technological parameters involved in cheese ripening, *Int. Dairy J.* **3**, 485–507.

VAN DEN BERG, G., DAAMEN, C. B. G., STADHOUDERS, J. (1989), Bactofugation of cheese milk, *North Eur. Food Dairy J.* **55**, 63–68.

VAN VLIET, T., VAN DIJK, H. J. M., ZOON, P., WALSTRA, P. (1991), Relation between syneresis and rheological properties of particle gels, *Colloid Polymer Sci.* **269**, 620–627.

VEDAMUTHU, E. R., WASHAM, C. (1983), Cheese, in: *Biotechnology – A Comprehensive Treatise* (REHM, H.-J., REED, G., Eds.), Vol. 5, pp. 231–313, Weinheim: Verlag Chemie.

VERDI, R. J., BARBANO, D. M. (1991), Effect of coagulants, somatic cell enzymes, and extracellular bacterial enzymes on plasminogen activation, *J. Dairy Sci.* **74**, 772–782.

VISSER, J. (1991), Factors affecting the rheological and fracture properties of hard and semi-hard cheese, *Bulletin 261,* Brussels: International Dairy Federation.

WALSTRA, P. (1987), Progress in cheese technology: from curd to cheese, in: *Milk The Vital Force* (Org. Comm. 22nd Int. Dairy Congr., Ed.), pp. 109–155, Dordrecht: D. Reidel Publ. Co.

WALSTRA, P., VAN DIJK, H. J. M., GEURTS, T. J. (1987a), The syneresis of curd, in: *Cheese: Chemistry, Physics and Microbiology – General Aspects* (FOX, P. F., Ed.), Vol. 2, pp. 135–177, London: Elsevier Applied Science Publishers, Ltd.

WALSTRA, P., NOOMEN, A., GEURTS, T. J. (1987b), Dutch-type varieties, in: *Chemistry, Physics and Microbiology – Major Cheese Groups* (FOX, P. F., Ed.), Vol. 1, pp. 45–92, London: Elsevier Applied Science Publishers, Ltd.

WALTER, H. E., HARGROVE, R. C. (1972), *Cheeses of the World,* New York: Dover Publications, Inc.

WHITEHEAD, W. E., AYRES, J. W., SANDINE, W. E. (1993), A review of starter media for cheese making, *J. Dairy Sci.* **76**, 2344–2353.

WHITNEY, R. MCL. (1988), Proteins of milk, in: *Fundamentals of Dairy Chemistry* (WONG, N. P., Ed.), 3rd Ed., pp. 81–169, New York: Van Nostrand Reinhold Co.

WILSON, H. L., REINBOLD, G. W. (1965), *American Cheese Varieties, Pfizer Cheese Monographs,* Vol. II, New York: Pfizer Inc.

ZEHREN, V. L., NUSBAUM, D. D. (1992), *Process Cheese,* Madison: Cheese Reporter Publ. Co., Inc.

10 Other Fermented Dairy Products

RAMESH C. CHANDAN

New Brighton, MN 55112, USA

KHEM M. SHAHANI

Lincoln, NE 68583-0919, USA

1 Introduction

The 1987 annual per capita consumption of various fermented fluid milks in various countries has been reported to range from 1.7 to 37.1 kg (IDF, 1989). Fermented dairy foods have historically constituted a vital part of the human diet. The motivation of fermentation processes relative to milk of various domesticated mammals may be conjectured to be preservation of the milk nutrients essential from the standpoint of nutrition and well-being of the human populations. In addition, novel flavors, textures and functional properties created as a result of intense metabolic activity of fermenting cultures provided portability, variety and versatility to the use of milk in human nutrition.

Modern processes for cultured dairy foods are based on reasonable scientific understanding of the microorganisms involved as opposed to spontaneous souring of milk practiced historically. Accordingly, modern cultured dairy products are expected by the consumer to deliver reliable organoleptic quality and safety from disease-causing pathogenic organisms. In addition, yogurt and certain fermented foods are perceived to contain a significant level of live and active cultures to fulfill the associated health benefits.

This chapter discusses technical aspects of the manufacture of cultured dairy products. More emphasis has been placed on practical considerations for the industrial manufacture of yogurt, frozen yogurt, cultured milk (buttermilk), and sour cream. Other cultured milks are not considered here in any depth. For more extensive treatment of various aspects of fermented dairy products, the reader is referred to several books and monographs on the subject, e.g., CHANDAN (1982), CHANDAN and SHAHANI (1993), DEVUYST and VANDAMME (1994), FERNANDEZ et al. (1992), GOLDIN and GORBACH (1992), IDF (1988a), NAKAZAWA and HOSONO (1992), ROBINSON (1991), SALMINEN and VON WRIGHT (1993).

2 Trends

Trends in the consumption of fermented milks presented in Tab. 1 reveal that, in general, the total fermented milk consumption is on the increase. The U.S. per capita sales of significant fermented milks are shown in Fig. 1. Cultured buttermilk is on the decline, while sour cream and yogurt are registering a significant growth. Fermented milks consumed in various regions of the world are summarized in Tab. 2.

Fermented milks may be consumed in original form, or they may be mixed with fruits, grains and nuts to yield delicious beverages, snacks, desserts, breakfast foods, or a light lunch. A variety of textures and flavors are generated as a result of selection of lactic acid bacteria. A combination of lactic acid bacteria and their strains allows an interesting array of products to suit different occasions of consumption. Tab. 3 shows the microorganisms used in the manufacture of different major fermented milks.

Fermented milk foods of desirable characteristics of flavor, texture and probiotic profiles can be created by chemical composition of the milk substrate mix, judicious selection of lactic acid bacteria, fermentation conditions, and blending with fruit juices, purees, preserves, along with compatible cereals, spices and flavorings. The lactic acid bacteria commonly employed (*Lactobacillus acidophilus*, *Streptococcus thermophilus*, *Lactococcus lactis* subsp. *lactis*, *Lactococcus lactis* subsp. *cremoris*, *Lactobacillus delbrueckii* subsp. *bulgaricus*) are responsible for the acidic taste arising from lactic acid. In some fermented foods *Lactococcus lactis*, *Lactobacillus helveticus*, *Leuconostoc dextranicum*, *Streptococcus durans* and *Streptococcus faecalis* are employed for acid and distinct flavor development while *Propionibacterium shermanii* secretes propionate, a natural shelf-life extender. Furthermore, it is possible to deliver a health-promoting microflora to the consumer of the food. In this regard, for yogurt culture, *Lactobacillus acidophilus*, *Lactobacillus casei*, *Bifidobacterium* species are notable examples.

Tab. 1. Annual Per Capita Consumption of Fermented Milks in Some Countries

Country	Total Consumption (kg)					1987 Consumption (kg)	
	1977	1980	1983	1986	1987	Yogurt	Other
Australia	1.4	1.8	2.2	2.8	2.8	2.8	
Austria	6.2	7.8	8.6	9.1	9.5	7.0	2.5
Belgium	4.2	4.9	4.9	7.0	7.7	7.0	0.7
Canada	1.2	1.7	2.0	2.9	3.2	3.2	
Chile	0.7	1.4	1.8	2.6	3.7	3.7	
Czechoslovakia	3.9	4.2	4.7	5.8	6.1	2.9	3.2
Denmark	15.6	16.0	17.1	15.7	15.4	8.3	7.1
Finland	34.5	36.9	37.9	37.4	37.1	10.8	26.3
France	8.0	9.4	11.8	13.0	13.6		
Germany (West)	7.1	7.9	8.6	9.9	10.6	9.7	0.9
Greece				5.6	6.0	5.5	0.5
Hungary	1.6	1.9	2.0	2.7	2.8	1.2	1.6
Iceland	20.5	20.3	21.3	21.2	22.1	7.9	14.2
India		3.7	4.0	4.1	4.2	4.2	
Israel	15.8	14.3	16.2	16.8			
Italy	1.2	1.5	2.3	3.2	3.4	2.1	1.3
Japan	6.9	6.5	6.8	6.9	7.2	3.3	3.9
Luxembourg	4.4	5.1	5.7	7.6	7.8		
Netherlands	15.0	16.9	17.7	19.1	19.1		
Norway	14.7	15.1	15.4	14.6	14.7	4.2	10.5
Poland	2.5	0.8	0.7	1.3	1.7	0.4	1.3
Ireland	1.5	2.3	3.4	3.5	3.3	3.3	
South Africa			3.8	3.8	3.7	1.5	2.2
Spain	4.3	6.0	6.8	6.9	7.3	7.3	
Sweden	21.0	23.9	26.1	27.3	27.2	5.8	21.4
Switzerland	12.6	13.8	15.0	16.5	16.5	16.5	
UK	1.8	2.1	2.3	3.5	3.6	3.6	
USA	1.1	1.2	1.4	1.9	2.1	2.1	
Former USSR	6.8	6.2	7.0	7.4	7.5		7.5

Adapted from IDF (1989)

3 Definitions

Standards of identity established by regulatory authority in each country assure the consumer a defined product. In the U.S., the following essential parameters are prescribed by the Food and Drug Administration regarding composition and processing of yogurt, cultured milk and sour cream. Tab. 4 lists the standards.

3.1 Yogurt

Yogurt is a semi-solid fermented product made from a heat-treated standardized milk mix by the activity of a symbiotic blend of *Streptococcus salivarius* subsp. *thermophilus* (ST) and *Lactobacillus delbrueckii* subsp. *bulgaricus* (LB) cultures. In certain countries, the nomenclature yogurt is restricted to the product made exclusively from LB and ST cultures, whereas in other countries, it is possible to label the product "yogurt" made with cultures additional to LB and ST. The adjunct cultures are primarily *Lactobacillus acidophilus* and *Bifidobacterium* spp.

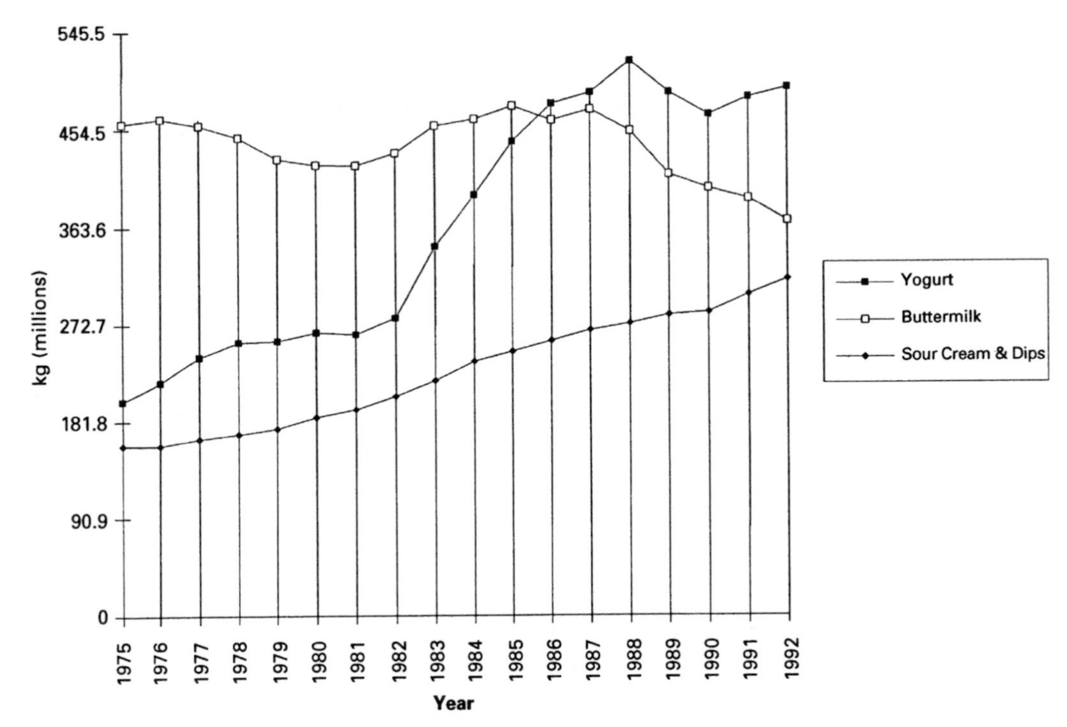

Fig. 1. Trends in cultured dairy products consumption in the United States. Source: MILK INDUSTRY FOUNDATION (1993).

Yogurt is produced from milk of various animals (cow, water-buffalo, goat, sheep, yak, etc.) in various parts of the world. Cow's milk is the predominant starting material in industrial manufacturing operations. In order to achieve custard-like, semi-solid consistency, the cow's milk is fortified with dried or condensed milk.

Vitamin addition at a level of 2000 IU of vitamin A and 400 IU of vitamin D per quart (946 mL) is allowed. Permissible dairy ingredients are cream, milk, partially skimmed milk, skim milk, alone or in combination. Other optional ingredients include:

(1) Concentrated skim milk, nonfat dry milk, buttermilk, whey, lactose, lactalbumins, lactoglobulins, or whey modified by partial or complete removal of lactose and/or minerals, to increase the nonfat solids content of the food, provided that the ratio of protein to total nonfat solids of the food and the protein efficiency ratio of all protein present shall not be decreased as a result of adding such ingredients.
(2) Nutritive carbohydrate sweeteners. Sugar (sucrose), beet or cane; invert sugar (in paste or syrup form); brown sugar, refiner's syrup; molasses (other than blackstrap); high fructose corn syrup; fructose; fructose syrup; maltose; maltose syrup, dried maltose syrup; malt extract, dried malt extract; malt syrup, dried malt syrup, honey, maple sugar, except table syrup.
(3) Flavoring ingredients
(4) Color additives
(5) Stabilizers.

3.2 Frozen Yogurt

The U.S. Food and Drug Administration has proposed tentative standards for frozen yogurt. The following is a summary of the proposal.

Tab. 2. Major Fermented Dairy Foods Consumed in Different Regions of the World

Product Name	Major Country/Region	Kind of Milk Used
Sour cream or cultured cream, Smetana	United States, Russia, CIS, Central Europe	Cow
Cultured half and half	United States	Cow
Cultured buttermilk	United States	Cow
Ymer	Denmark	Cow
Taettmelk	Norway	Cow
Filmjolk	Sweden	Cow
"Long" milk	Scandinavia	Cow
Pitkapiima	Finland	Cow
Viili	Finland	Cow
Lactofil	Sweden	Cow
Acidophilus milk	United States, Russia, CIS[a]	Cow
Yakult	Japan	Cow
Yogurt, yoghurt, yoghaurt, yoghourt, yahourth, yaaurt, yourt, jugart, yaert, yaoert	United States, Europe, Asia	Cow, goat, sheep, buffalo
Dough or abdoogh	Afghanistan and Iran	Cow, buffalo
Ayran	Turkey	Cow
Leben Rayeb/Matared/ Laban Zabaidi	Egypt, Lebanon, Syria, Jordan	Cow, buffalo
Dahi	Indian subcontinent	Cow, buffalo
Mazurn	Armenia	Cow
Kisselo maleko	Balkans	Cow
Ergo	Ethiopia	Cow
Gioddu	Sardinia	Cow
Kefir	Russia/CIS, Central Asia[a]	Cow, goat, sheep
Koumiss/Kumys	Russia/CIS, Central Asia[a]	Mare
Kurunga	Western Asia	Cow
Chal	Turkmenistan	Camel
Quark	Germany	Cow
Shrikhand	India	Cow, buffalo
Cream cheese	United States, Europe	Cow
Cottage cheese	United States	Cow
Tvorog	Russia/CIS[a]	Cow

Sources: IDF (1988a), CHANDAN (1982), KOSIKOWSKI (1982)
[a] refers to former USSR

Frozen yogurt is the food produced by freezing, while stirring, a mix containing safe and suitable ingredients including, but not limited to, dairy ingredients. The mix may be homogenized, and all of the dairy ingredients shall be pasteurized or ultrapasteurized. All or a portion of the dairy ingredients shall be cultured with a characterizing live bacterial culture that shall contain the lactic acid-producing bacteria *Lactobacillus bulgaricus* and *Streptococcus thermophilus* and may contain other lactic acid-producing bacteria. After culturing, the unflavored frozen yogurt mix shall have a titratable acidity of not less than 0.3%, calculated as lactic acid. Where the titratable acidity of the frozen yogurt mix is $<0.3\%$, the manufacturer may establish compliance with this section by disclosing to the Federal Food and Drug Administration (FDA) quality control records that demonstrate that as a result of bacterial culture fermentation, there has been at least a 0.15% increase in the titratable acidity, calculated as lactic acid, of the product above the apparent titratable acidity of the uncultured dairy ingredients in the frozen yogurt mix. The direct

Tab. 3. Starter Cultures Used in Manufacture of Commercial Fermented Milk Products

Product	Primary Microorganism(s)	Secondary/Optional Microorganism(s)	Incubation Temperature and Time	Major Function of Culture
Yogurt	*Lactobacillus delbrueckii* subsp. *bulgaricus* *Streptococcus salivarius* subsp. *thermophilus*	*Lactobacillus acidophilus* *Bifidobacterium longum/bifidum/infantis* *Lactobacillus casei/lactis/jugurti/helveticus*	43–45 °C/ 2.5 hours	Acidity, texture, aroma, flavor, probiotic
Cultured buttermilk and sour cream	*Lactococcus lactis* subsp. *lactis* *Lactococcus lactis* subsp. *cremoris* *Lactococcus lactis* subsp. *lactis* var. *diacetylactis*	*Leuconostoc lactis* *Leuconostoc mesenteroides* subsp. *cremoris*	22 °C/ 12–14 hours	Acidity, flavor, aroma
Fermented milk	*Streptococcus salivarius* subsp. *thermophilus* *Lactobacillus acidophilus* *Bifidobacterium longum/bifidus*	*Lactococcus lactis* subsp. *lactis/cremoris*	22–37 °C/ 8–14 hours	Acidity, flavor, probiotic
Acidophilus milk	*Lactobacillus acidophilus*		37–40 °C/ 16–18 hours	Acidity, probiotic
Bulgarian buttermilk	*Lactobacillus delbrueckii* subsp. *bulgaricus*		37–40 °C/ 8–12 hours	Acidity, probiotic
Kefir	*Lactococcus lactis* subsp. *lactis/cremoris* *Lactobacillus delbrueckii* subsp. *bulgaricus* *Lactobacillus delbrueckii* subsp. *lactis* *Lactobacillus casei/helveticus/brevis/kefir* *Leuconostoc mesenteroides/dextranicum* Yeasts: *Kluyveromyces marxianus* subsp. *marxianus* *Torulaspora delbrueckii* *Saccharomyces cerevisiae* *Candida kefir* Acetic acid bacteria: *Acetobacter aceti*		15–22 °C/ 24–36 hours	Acidity, aroma, flavor, gas (CO_2), alcohol, probiotic
Koumiss	*Lactobacillus delbrueckii* subsp. *bulgaricus* *Lactobacillus kefir/lactis*			

Tab. 3. Starter Cultures Used in Manufacture of Commercial Fermented Milk Products (Continued)

Product	Primary Microorganism(s)	Secondary/Optional Microorganism(s)	Incubation Temperature and Time	Major Function of Culture
Koumiss	Yeasts: *Saccharomyces lactis* *Saccharomyces cartilaginosus* *Mycoderma* spp. Acetic acid bacteria: *Acetobacter aceti*		20–25 °C 12–24 hours	Acidity, alcohol, flavor, gas (CO_2)
Yakult	*Lactobacillus casei*		30–37 °C/ 16–18 hours	Acidity, probiotic
Cheeses	*Lactococcus lactis* subsp. *lactis/cremoris* *Lactobacillus delbrueckii* subsp. *bulgaricus/lactis* *Streptococcus salivarius* subsp. *thermophilus* Fungi: *Penicillium roqueforti/candidum*	*Lactobacillus casei/helveticus* *Leuconostoc* spp. *Enterococcus* sp. *Propionibacterium shermanii* *Brevibacterium linens*	4–30 °C/ variable time	Acidity, enzymes for ripening

Sources: Koroleva (1988), De Vuyst and Vandamme (1994)

addition of food-grade acids or other acidogens for the purpose of raising the titratable acidity of the frozen yogurt mix to comply with the prescribed minimum is not permitted, and no chemical preservation treatment or other preservation process, other than refrigeration, may be utilized that results in reduction of the live culture bacteria. Sweeteners, flavorings, color additives, and other characterizing food ingredients, unless otherwise provided in the regulations of the FDA, may be added to the mix before or after pasteurization or ultrapasteurization, provided that any ingredient addition after pasteurization or ultrapasteurization is done in accordance with current good manufacturing practice. Any dairy ingredients added after pasteurization or ultrapasteurization shall have been pasteurized.

Frozen yogurt may be sweetened with any sweetener that has been affirmed as generally regarded as safe (GRAS) or approved as a food additive for this use by FDA and may or may not be characterized by the addition of flavoring ingredients. Frozen yogurt, before the addition of bulky characterizing ingredients or sweeteners, shall contain not less than 3.25% milkfat and 8.25% milk-solids-non-fat. Frozen yogurt shall contain not less than 156 g/L of total solids (1.3 lb/gal) and shall weigh not less than 480 g/L (4.0 lb/gal).

The nomenclature for frozen yogurt is based on the same line as refrigerated yogurt. Frozen yogurt contains a minimum of 3.25% milkfat; frozen lowfat yogurt contains a minimum of 0.5% milkfat and a maximum of 2.0% milkfat; frozen nonfat yogurt contains $<0.5\%$ milkfat.

Criteria for live and active yogurt have been established by the industry with a view to maintain the integrity of refrigerated and frozen yogurt. The National Yogurt Association's criteria are summarized below.

Live and active culture yogurt (refrigerated

Tab. 4. Essential Standards for Composition of Certain Fermented Milks in the U.S.A.

Product	Milkfat (%)	Milk Solids-Non-Fat (%)	Titratable Acidity (%) Expressed as Lactic Acid
Yogurt	Not less than 3.25	Not less than 8.25	Not less than 0.9
Lowfat yogurt	Not less than 0.5 and not more than 2.0	Not less than 8.25	Not less than 0.9
Non-fat yogurt	Not more than 0.5	Not less than 8.25	Not less than 0.9
Cultured milk	Not less than 3.25	Not less than 8.25	Not less than 0.5
Cultured lowfat milk	Not less than 0.5 and not more than 2.0	Not less than 8.25	Not less than 0.5
Cultured non-fat milk (buttermilk)	Not more than 0.5	Not less than 8.25	Not less than 0.5
Cultured (sour) cream	Not less than 18.0 Not less than 14.4 after the addition of bulky flavors, etc.[a]	No standard	Not less than 0.5
Cultured half & half	Not less than 10.5 and less than 18.0 Not less than 8.4 after addition of bulky flavors, etc.[a]	No standard	Not less than 0.5

Source: FDA (1993)
[a] Bulky flavors' refers to sugar preserved fruits, nuts, etc.

cup and frozen yogurt) is the food produced by culturing permitted dairy ingredients with a characterizing bacterial culture in accordance with the FDA standards of identity for yogurt. In addition to the use of the bacterial cultures required by the referenced Federal standards of identity and by these National Yogurt Association criteria, live and active culture yogurt may contain other safe and suitable food-grade bacterial cultures. Declaration of the presence of cultures on the label of live and active culture yogurt is optional.

Heat treatment of live and active yogurt is inconsistent with the maintenance of live and active cultures in the product; accordingly, heat treatment that is intended to kill the live and active organisms shall not be undertaken after fermentation. Likewise, manufacturers of live and active culture yogurt should undertake their best efforts to ensure that distribution practices, code dates, and handling instructions are conducive to the maintenance of living and active cultures.

In order to meet these criteria, live and active culture yogurt must satisfy each of these requirements:

1. The product must be fermented with both *L. delbrueckii* subsp. *bulgaricus* and *S. thermophilus*.
2. The cultures must be active at the end of the stated shelf life as determined by the activity test described in item 3. Compliance with this requirement shall be determined by conducting an activity test on a representative sample of

yogurt that has been stored at temperatures between 0 and 7 °C (32 and 45 °F) for refrigerated cup yogurt and at temperatures of −18 °C (0 °F) or colder for frozen yogurt for the entire stated shelf life of the product.

3. The activity test is carried out by pasteurizing 12% solids nonfat dry milk (NFDMS) at 92 °C (198 °F) for 7 min, cooling to about 43 °C (110 °F), adding 3% inoculum of the material under test, and fermenting at 43 °C (110 °F) for 4 h. The total organisms are to be enumerated in the test material both before and after fermentation by the International Dairy Federation (1988b) methodology. The activity test is met if there is an increase of 1 log or more during fermentation.
4. a) In the case of refrigerated cup yogurt, the total population of organisms in live and active culture yogurt must be at least 10^8 per gram at the time of manufacture.
   b) In the case of frozen yogurt, the total population of organisms in live and active culture yogurt must be at least 10^7 at the time of manufacture. (It is anticipated that if proper distribution practices and handling instructions are followed, the total organisms in both refrigerated cup and frozen live and active culture yogurt at the time of consumption will be at least 10^7.)
5. The product shall have a total titratable acidity expressed as lactic acid at least 0.3% at all times. At least 0.15% of total acidity must be obtained by fermentation. This is confirmed by demonstrating the presence of both D-(−) and L-(+) forms of lactic acid.

3.3 Cultured Milk

The U.S. Food and Drug Administration defines cultured milk and its reduced fat analogs. A summary of the published standards (FDA, 1993) is given below.

Cultured milk is the food produced by culturing one or more of the optional dairy ingredients specified below with characterizing microbial organisms. One or more of the other optional ingredients may also be added. When one or more of the dairy-derived ingredients are used, they shall be included in the culturing process. All ingredients used are safe and suitable. Cultured milk contains not less than 3.25% milkfat and not less than 8.25% milk solids nonfat and has a titratable acidity of not less than 0.5%, expressed as lactic acid. The food may be homogenized and shall be pasteurized or ultrapasteurized prior to the addition to the microbial culture, and when applicable, the addition of flakes or granules of butterfat or milkfat.

Vitamin addition (optional). If added, vitamin A shall be present in such quantity that each 946 milliliters (quart) of the food contains not less than 2000 International Units thereof, within limits of good manufacturing practice.

If added, vitamin D shall be present in such quantity that each 946 milliliters (quart) of the food contains 400 International Units thereof, within limits of good manufacturing practice.

Optional dairy ingredients include cream, milk, partially skimmed milk, or skim milk, used alone or in combination. Other optional ingredients are concentrated skim milk, nonfat dry milk, buttermilk, whey, lactose, lactalbumins, lactoglobulins, or whey modified by partial or complete removal of lactose and/or minerals, to increase the nonfat solids content of the food: Provided that the ratio of protein to total nonfat solids of the food, and the protein efficiency ratio of all protein present, shall not be decreased as a result of adding such ingredients.

Permitted nutritive carbohydrate sweeteners are sugar (sucrose), beet or cane; invert sugar (in paste or syrup form); brown sugar; refiner's syrup; molasses (other than blackstrap); high fructose corn syrup; fructose; fructose syrup; maltose; maltose syrup, dried maltose syrup; malt extract, dried malt extract; malt syrup, dried malt syrup; honey; maple sugar; except table syrup.

In addition, the following food materials are allowed: flavoring ingredients, color additives that do not impart a color simulating that of milkfat or butterfat, stabilizers, butterfat or milkfat, which may or may not contain color additives, in the form of flakes or granules, aroma- and flavor-producing microbial culture, salt, citric acid, in a maximum amount of 0.15% by weight of the milk used, or an equivalent amount of sodium citrate, as a flavor precursor.

Depending upon the culture used in fermentation, the label "cultured milk" may indicate the type of culture used. Cultured buttermilk is a traditional name for milk cultured with specific organisms, and the label should reflect that.

Reduced fat cultured milk may be labeled cultured lowfat milk if it contains not less than 0.5% milkfat nor more than 2% milkfat. Cultured skim milk contains less than 0.5% milk fat. All other requirements are identical to cultured milk.

3.4 Sour Cream/Cultured Sour Cream

Sour cream results from fermenting cream with specified lactic culture. Sour cream contains not less than 18% milkfat; except that when the food is characterized by the addition of nutritive sweeteners or bulky flavoring ingredients, the weight of the milkfat is not less than 18% of the remainder obtained by subtracting the weight of such optional ingredients from the weight of the food; but in no case does the food contain less than 14.4% milkfat. Sour cream has a titratable acidity of not less than 0.5%, calculated as lactic acid.

Optional ingredients include safe and suitable ingredients that improve texture, prevent syneresis, or extend the shelf life of the product. Also, sodium citrate in an amount of not more than 0.1% may be added prior to culturing as a flavor precursor. Rennet; safe and suitable nutritive sweeteners; salt; flavoring ingredients, with or without safe and suitable coloring, may be used: Fruit and fruit juice (including concentrated fruit and fruit juice) and safe and suitable natural and artificial food flavoring are permitted.

Reduced fat sour cream analogs include sour half and half containing not less than 10.5% milkfat but less than 18% milkfat. After the addition of bulky sweeteners or flavors, the fat content cannot be less than 8.4% in order to qualify for the label "sour half and half". Other standards are similar to those for sour cream.

3.5 Kefir and Koumiss

Kefir and koumiss are relatively popular fermented milks in Russia, Eastern Europe and certain Asian countries. In addition to bacterial fermentation, these products employ yeast fermentation as well. Thus, a perceptible yeast aroma, fizziness and alcohol content characterize these products. Kefir utilizes natural fermentation of cow's milk with a mixture of lactic organisms and several yeasts contained in kefir grains. Koumiss is obtained from mare's milk or cow's milk, using a more defined culture containing *Lactobacillus delbrueckii* subsp. *bulgaricus*, *L. acidophilus* and yeasts. These products have perceived health benefits and are recommended for all consumers, especially with gastrointestinal problems, allergy, hypertension and ischaemic heart diseases (KOROLEVA, 1988).

4 Starter Cultures

The natural habitat of a number of lactic acid bacteria is milk. These microfloras gain entry into milk during and after milking via cow, feed, air, utensils, and milking equipment. Consequently under appropriate growth conditions, viz. warm temperature, milk undergoes spontaneous souring, yielding somewhat uncontrolled flavor and textural attributes. Advances in microbial fermentation technology and biotechnology have resulted in fairly predictable product quality. Also, there is an opportunity to incorporate desirable characteristics at the cellular level to enhance the utility of lactic acid bacteria in terms of natural preservative and bacteriocin production, flavor generation and stability, and boosted probiotic properties. The industrial processes utilize culture concentrates or starter cultures for the production of fermented milks.

4.1 Taxonomy

A starter is made up of one or more strains of food-grade microorganisms. Culturing milk base with the starter produces a fermented milk of predictable consumer attributes. For the composition of various starters in relation to fermented milks see Tab. 3.

Individual microorganisms utilized as single culture (single or multiple strains) or in combination with other cultures exhibit characteristics impacting the technology of manufacture of fermented milks.

4.2 Metabolism and Flavor Metabolites

To assure the ecological compatibility, survival and viability of diverse cultures and their various strains, it is important to understand their microbiological, physiological and biochemical attributes. Tabs. 5 and 6 present certain characteristics of mesophilic and thermophilic cultures utilized in major commercial fermented milks. An interesting range of acidity, flavor, texture, ethanol content, gas production (fizziness) and prophylactic effects are achievable by combining various genera and species of appropriate strains of lactic acid bacteria, acetic acid producing bacteria and yeasts.

4.3 Growth and Inhibition

During repeated growth cycles, the starter organisms may remain active and preserve their characteristics for some time. However, they may lose their activity rapidly depending on the compatibility of the species and strains. Also, activity is lost or changed due to the physical environment. In any case, change from the normal fermentation pattern is considered a defect. The common defects are:

Insufficient acid development. This is one of the common defects in lactic cultures. When 1 mL culture, inoculated into 10 mL of antimetabolite-free, heat-treated milk, produces less than 0.7% titratable acidity in 4 h at 35 °C, it is considered a slow starter. Factors contributing to a slow starter are:

(a) Composition of milk

Certain raw milks exert an inhibitory effect on many lactic starters, and this is attributed to various natural inhibitors including lactenins, lactoperoxidase, agglutinins, and lysozyme. These inherent inhibitors are present in all milk and show considerable variations with breed and season. All these factors are heat-labile, and their inhibitory property is arrested progressively on heating. When milk is pasteurized at 72 °C for 16 s or autoclaved for 15 min, the natural inhibitors are completely destroyed. Further, the growth of starter cultures is stimulated in heated or autoclaved milk due to partial hydrolysis of casein, liberation of sulfhydryl groups, and formation of formate from lactose. Rapid acid production by lactic acid bacteria is observed in milk heated at 90 °C for 1 h, or 116 °C for 15 min, or 121 °C for 10 min. Autoclaving treatments are generally avoided for intermediate and bulk starter preparation because of the introduction of undesirable caramelized color and flavor in milk. However, flavor producing strains of *Leuconostoc cremoris* grow better in milk sterilized at 121 °C for 15 min.

Recent trends in ultrapasteurization or ultra-high temperature treatment (UHT) of milk as a means of extending shelf life appear to have interesting implications for the cultured dairy product industry. It appears that UHT milk is a better medium for culture growth than milk processed by batch or short-time pasteurization procedures.

Milk from mastitis-infected animals generally does not support the growth of lactic cultures. This effect is ascribed to the infection-induced changes in chemical composition of milk. For example, mastitis milk contains lower concentrations of lactose and unhydrolyzed protein, a higher chloride content and a higher pH than normal milk. Furthermore, a high leukocyte count in mastitis milk inhibits bacterial growth by phagocytic action. Heat treatment restores the culture growth in mastitis milk.

Colostrum and late lactation milk contain nonspecific agglutinins which clump and precipitate sensitive strains of the starter. The agglutinins may possibly retard the rate of acid production by interfering with the transport of lactose and other nutrients.

Tab. 5. Microbiological and Biochemical Attributes of Typical Mesophilic Lactic Acid Bacteria Used in Cultured Milks

Characteristic	*Lactococcus lactis* subsp. *lactis*	*Lactococcus lactis* subsp. *cremoris*	*Lactococcus lactis* subsp. *lactis* biovar *diacetylactis*	*Leuconostoc mesenteroides* subsp. *cremoris*	*Leuconostoc mesenteroides* subsp. *dextranicum*
Cell shape and configuration	Cocci, pairs, short chains	Cocci, pairs, short/long chains	Cocci, pairs, short chains	Cocci, pairs, short/long chains	Cocci, pairs, chains
Catalase reaction	—	—	—	—	—
Growth temperature (°C)					
Optimum	28–31	22	28	20–25	20–25
Minimum	8–10	8–10	8–10	4–10	4–10
Maximum	40	37–39	40	37	37
Incubation temperature (°C)	21–30	22–30	22–28	22	22
Heat tolerance (60°C/30 min)	±	±	±	—	—
Lactic acid isomers	L (+)	L (+)	L (+)	D (−)	D (−)
Lactic acid produced in milk (%)	0.8–1.0	0.8–1.0	0.8–1.0	0.1–0.3	0.1–0.3
Acetic acid production (%)	−	−	−	0.2–0.4	0.2–0.4
Gas (CO_2) production	−	−	+	±	±
Proteolytic activity	+	+	+	±	±
Lipolytic activity	±	±	±	±	±
Citrate fermentation	−	−	+	+	+
Flavor/aroma compound	+	+	+++	+++	+++
Mucopolysaccharide production	±	±	±	No dextran from sucrose	Dextran from sucrose
Hydrogen peroxide production	+	+	+	±	±
Alcohol production	±	±	±	±	±
Salt tolerance (% max)	4–6.5	4.0	4–6.5	6.5	6.5

Sources: IDF (1992), CHANDAN (1982)

Tab. 6. Some Characteristics of Thermophilic Starters

Characteristic	*Streptococcus thermophilus*	*Lactobacillus delbrueckii* subsp. *bulgaricus*	*Lactobacillus delbrueckii* subsp. *lactis*	*Lactobacillus acidophilus*	*Lactobacillus casei* subsp. *casei*
Cell shape and configuration	Spherical to ovoid, pairs to long chairs	Rods with round ends, single, short chains, metachromatic granules	Rods with round ends, metachromatic granules	Rods with round ends, single, pairs, short chains, no metachromatic granules	Rods with square ends, short/long chains
Catalase reaction	–	–	–	–	–
Growth temperature (°C)					
Optimum	40–45	40–45	40–45	37	37
Minimum	20	22	22	20–22	15–20
Maximum	50	52	52	45–48	40–45
Incubation temperature (°C)	40–45	42	40–45	37	37
Heat tolerance (60°C/30 min)	++	+	+	–	–
Lactic acid isomers	L (+)	D (–)	D (–)	DL	L (+)
Lactic acid produced in milk (%)	0.7–0.8	1.5–4.0	1.5–3.0	0.3–2.0	1.2–1.5
Acetic acid (%)	Trace	Trace	Trace	+	+
Gas (CO_2) production	–	–	–	–	–
Proteolytic activity	±	+	+	±	±
Lipolytic activity	±	±	±	±	±
Citrate fermentation	–	–	–	–	–
Flavor/aroma compounds	++	++	+	+	±
Mucopolysaccharide production	±	++	–	–	±
Hydrogen peroxide production	±	+	+	+	+
Alcohol production	–	Trace	Trace	Trace	Trace
Salt tolerance (% max)	2.0	2.0	2.0	6.5	2.0

Seasonal variation of the solids-not-fat fraction of milk affects the growth and the balance of strains in culture. Generally, a higher solids-not-fat level in milk favors the growth of lactic cultures.

(b) Contaminating microorganisms

Prior degradation of milk constituents by contaminants affects the growth of lactic organisms. Careful screening of milk for psychrotrophs is necessary for quality flavor production by lactic cultures.

(c) Antibiotics and chemicals

Various antibiotics gain entry into milk during antibiotic treatment of mastitis, and these inhibit acid production by bacteriostatic action, depending on the type of starter and the kind and amount of antibiotic involved.

Concentrations as low as 0.005–0.05 International Units of antibiotics per mL milk, used in mastitis therapy, are high enough to impact partial or full inhibition of the culture. Accordingly, it is imperative that the residual antibiotic level in milk be monitored routinely to keep the milk supply suitable for cultured milk manufacture.

Many sanitizing chemicals like quaternary ammonium compounds, iodine and chlorine compounds retard acid development by starter cultures. One to 5 parts per million of these sanitizing compounds are bactericidal to lactic cultures. Consequently, it is important to exert care and control in the use of sanitizers in the plant. Fatty acids (C-10 to C-16) also inhibit starters. These fatty acids may be present due to partial hydrolysis of milk by lipases, or they may be produced by lipolytic organisms. The fatty acids, particularly lauric, caprylic, and capric, lower the surface tension of milk to less than 40 dyn/cm. The inhibition of lactic cultures by free fatty acids is apparently related to the surface activity of the growth medium.

Avoiding the use of rancid milk is important not only from the standpoint of culture growth, but more significantly because it would impart an objectionable flavor to the starter and the cultured dairy products derived therefrom.

(d) Change in fermentation behavior

After continuous use, the starters may change their fermentation activity and consequently produce lower amounts of lactic acid. This is attributed to genetic changes brought about by environmental factors.

Strain dominance in mixed cultures, causing changes in the behavior of the composite cultures, is well documented. Results show that *Lactococcus lactis* subsp. *diacetylactis* tends to dominate *Lactococcus lactis* subsp. *lactis* or *Lactococcus lactis* subsp. *cremoris*. Different strains of *Lactococcus lactis* subsp. *lactis* display major differences in domination during associative growth with *Lactococcus lactis* subsp. *cremoris*.

(e) Phage action

Attack by bacteriophages is an important cause of slow acid production by lactic cultures. When the phage has reached a maximum level, all sensitive bacterial cells are infected and lysed within 30–40 min. When lysis occurs, acid production by the affected culture stops unless some resistant bacteria are present to carry on fermentation. Phages, which are strain-specific viruses, consist of a head (70 nm wide) and a tail (200 nm × 30 nm). The phage attacks lactococci as well as lactobacilli by attachment of the tail to the bacterial cell wall, followed by injection of DNA from the phage head into the cell. This is followed by synthesis of new phage particles to carry on further attack and lysis of the bacteria. If at least 50% of the fast acid-producing bacteria are phage-resistant, a phage attack may not be discernible.

Phage control is effected in cultured milk plants by using 200–300 ppm chlorine on processing equipment and by fogging the culture rooms with 500–1000 ppm of chlorine. Heat treatment of milk (75 °C for 30 min or 80 °C for 20 s) is considered adequate to inactivate various phages which attack lactic acid bacteria. By using combinations of proper procedures, such as sanitation, culture selection, and culture rotation, the probability of an infection with phage can be minimized. Use of phage-resistant and multiple-strain cultures is generally preferred.

In certain instances, acidity may be too high in relation to product standards. This problem is encountered in yogurt production. High acidity is usually associated with high incubation temperature, long incubation period, or excessive inoculum.

Insufficient or abnormal flavor development. Adequate production of lactic acid is essential for lowering the pH to a level where acetaldehyde, diacetyl and other compounds are formed in sufficient quantity. For good flavor, any factor interfering with proper acid development will retard or prevent adequate flavor development.

The culture may be incapable of producing adequate amounts of flavor due to a change in fermentation pattern induced by oxygen tension or due to a change in the balance of various bacterial cultures.

Ropiness, gassiness, and bitterness. These defects are due to limited proteolytic activity of some starter strains, and are commonly observed with Cheddar cheese cultures. Also, they can be attributed to the presence of proteolytic bacteria in the starter culture. Some sporeformers which survive normal heat treatment of the milk may also be involved.

4.4 Genetic Engineering

Natural selection and mutation techniques have been historically used to improve the performance of lactic cultures. The new strains were, however, not stable in their selected attributes. The instability of the desirable factors was attributed to plasmid loss. More recently, work on the plasmids of mesophilic lactic acid bacteria has significantly elucidated basic principles relative to acid production, sugar utilization, proteolytic activity, citrate metabolism, bacteriophage resistance and bacteriocin production. The plasmids are known to code for several significant functions. Technology for transfer of plasmids has been developed by commercial culture manufacturers to introduce new strains with enhanced bacteriophage resistance, boosted health attributes, ability to accelerate cheese ripening, stability in flavor and texture production, production of antimicrobial compounds and natural preservatives.

Also, yogurt cultures which produce negligible acid under refrigerated storage have been commercialized. Future research in this area is considered to have great potential in alleviating associated problems and improving performance of lactic starters.

4.5 Production of Starters

Dairy plants purchase cultures on a regular basis from commercial suppliers. Dairy cultures are available in various freeze-dried, liquid or frozen forms. Culture concentrates may require limited propagation to the stage of bulk starter prior to use in fermentation of milk substrates. Direct-to-the-vat set concentrates are designed for direct inoculation into milk substrates pumped into fermentation tanks. Fig. 2 illustrates various steps needed to prepare starters for use in the production of fermented milks.

The advent and industry-wide acceptance of frozen and freeze-dried culture concentrates has simplified the management of cultures in most cultured dairy plants in the United States. However, working knowledge and employee training in lactic cultures are still advantageous in handling starters at the plant level. In countries where frozen culture concentrates are not yet fully developed or in case proprietary strains of cultures are preferred, it is essential to develop and maintain appropriate microbiological expertise in the propagation, maintenance, and control of lactic starter cultures.

The starter is the most crucial component in the production of high-quality fermented milks. Culture propagation should be conducted in a specified, secluded area of the plant where access of personnel is restricted. An effective sanitation program coupled with filtered air and positive pressure in the culture area, and preferably all the manufacturing areas including the packaging room, should significantly reduce the airborne contamination. Consequently, culture failure due to phages may be controlled, and extended shelf life of the product may be attained.

The media for culture propagation are generally composed of liquid skim milk or blends of nonfat dry milk dispersed in water so as to

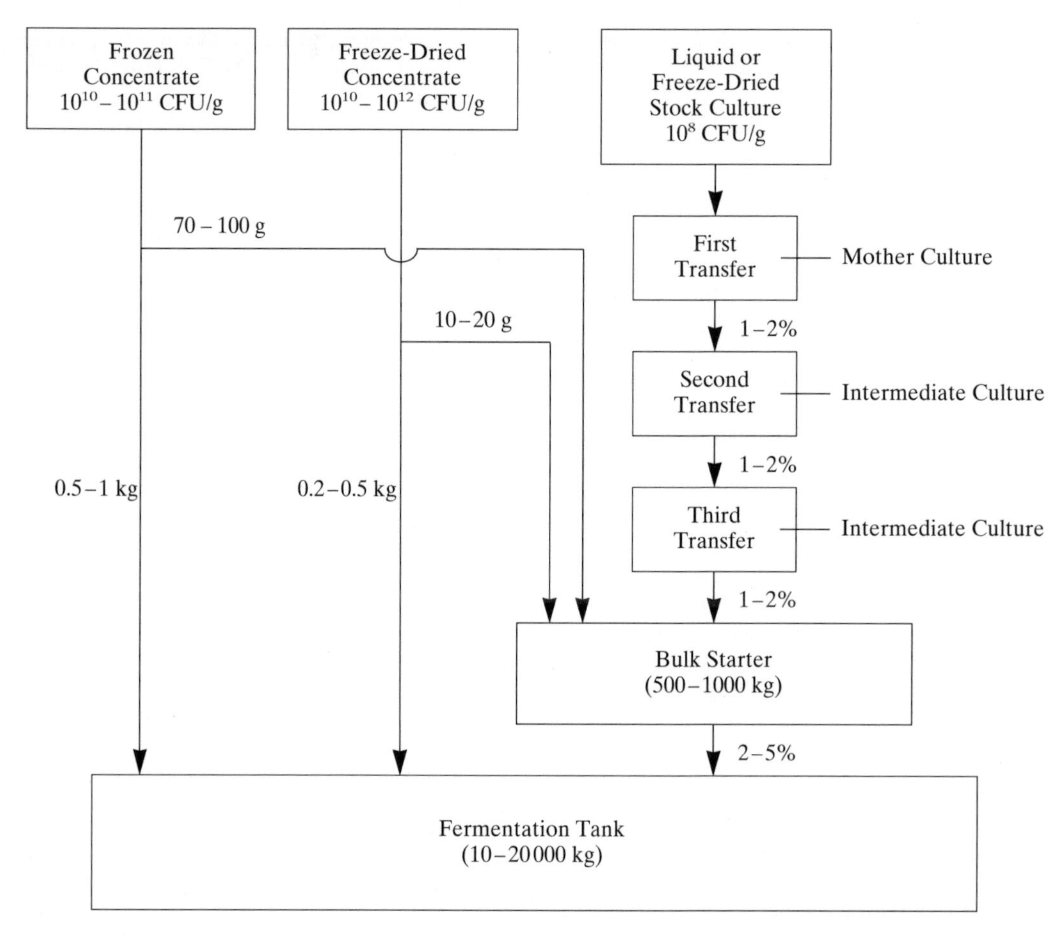

Fig. 2. Propagation and use of various starter types at dairy manufacturing plants. Sources: CHANDAN (1982), MAYRA-MAKINEN and BIGRET (1993).

contain 9–12% milk solids. Water, nonfat dry milk, and other media ingredients must be free from substances inhibitory to the growth of the starter. Such inhibitory substances include sanitizing chemicals such as chlorine, iodine, and quaternary ammonium compounds as well as antibiotics and phages. Special media for optimum culture activity and phage resistance are available from commercial culture companies. The media generally contain demineralized whey, nonfat dry milk, phosphate, citrate, and growth factors present in yeast extracts. Phosphate acts as a sequestrant of Ca^{2+} and thereby inhibits Ca^{2+}-dependent phage growth. Citrate provides a substrate for production of diacetyl and, along with phosphate, contributes to the buffering capacity. The powdered media are generally dispersed in water as such or blended with an equal weight of nonfat dry milk to attain 10–12% solids. Alternately, the media may be dispersed in liquid skim milk. The special media are particularly useful in cheese starter production.

Heat treatment is necessary to destroy the contaminants in the medium and to alleviate unnecessary competition to the growth of a desirable lactic culture. In addition, heating the medium produces desirable nutrients by heat-induced reactions in milk constituents.

Cultures purchased as frozen concentrates are shipped in dry ice. For extended storage, the culture concentrate cans must be stored in liquid nitrogen. However, for relatively short storage periods of 4–6 weeks, the cultures may be stored in special freezers (at −40°C). The use of freezers offers an economical alternative if the turnover of cultures at the plant level is high and proper care in culture can rotation is taken. The culture concentrates may be designed for bulk starter preparation or for direct inoculation into product mixes. The use of frozen culture concentrates eliminates the preparation of mother cultures and intermediate cultures. The procedure for cultivation of cultures at plant level involves the use of reconstituted nonfat dry milk and in cheese production a special medium developed by several manufacturers for producing bulk starters.

Control of acid and flavor development in lactic starter cultures may be achieved by understanding their growth characteristics. By modifying the inoculation rate, incubation temperature and time, it is possible to direct the fermentation, in a limited way, to fit the plant schedules. Care should be taken to preserve the balance of strains and organisms in the culture so that a symbiotic relationship is maintained, which is vital in the yogurt industry.

Yogurt culture consists of two lactose-fermenting organisms, *Lactobacillus delbrueckii* subsp. *bulgaricus* (rod) and *Streptococcus thermophilus* (coccus). Culturing the two organisms together results in a symbiotic relationship since the growth rate and acid production by each organism are greater when grown together than in a single culture. Optimum growth temperatures for the rod and coccus are 45°C and 40°C, respectively. Depending upon incubation temperature, a differential in the ratio of rod to coccus would be evident. Most commercial yogurt exhibits a ratio of 1:3 in favor of coccus. Upon repeated transfers, this ratio tends to change, depending upon the incubation temperature. A reasonable success in rectifying the balance has been achieved by varying the rate of inoculum, incubation time and temperature, acidity level in milk, and heat treatment of milk. Using a 2% inoculum and incubation at 44°C for 2.5 hours, a proper balance of rod and coccus can be maintained in a yogurt culture. If the ratio is not in the desirable range, the streptococci may be increased by lowering incubation time or temperature. Conversely, the lactobacilli population may be encouraged by higher incubation time or temperature.

For the preparation of bulk starter, the most common substrate is reconstituted nonfat dry milk (9–12% w/v). The substrate is heat-treated to 95°C and holding for 30 minutes. After cooling to an appropriate incubation temperature (22–30°C for mesophilic starters and 37–43°C for thermophilic cultures), the medium is inoculated with a frozen culture concentrate or, in some cases, with 2% of intermediate culture. The medium is subsequently held at an appropriate incubation temperature for 3 to 18 hours until a titratable acidity of 0.8% or a pH of 4.4–4.5 is attained. Thermophilic cultures, especially yogurt, grow rapidly in approximately 3 hours to the required acid level, whereas mesophilic cultures take 16–18 hours.

Typical steps including incubation conditions for bulk starter production in a dairy plant are shown in Fig. 3.

Mother cultures are propagated daily for a period of about 2–3 weeks after which a fresh lyophilized culture is activated. Different strains of frozen culture concentrates are rotated on a daily basis. General microscopic examination of the culture is recommended to ensure integrity of the culture.

5 Manufacturing Principles

5.1 Ingredients

Dairy. Yogurt is generally made from a mix standardized from whole, partially defatted milk, condensed skim milk, cream and nonfat dry milk. In rare practice, milk may be partly concentrated by removal of 15 to 20% water in a vacuum pan. Supplementation with nonfat dry milk is the preferred industrial procedure. Cultured milk is made from fat-stand-

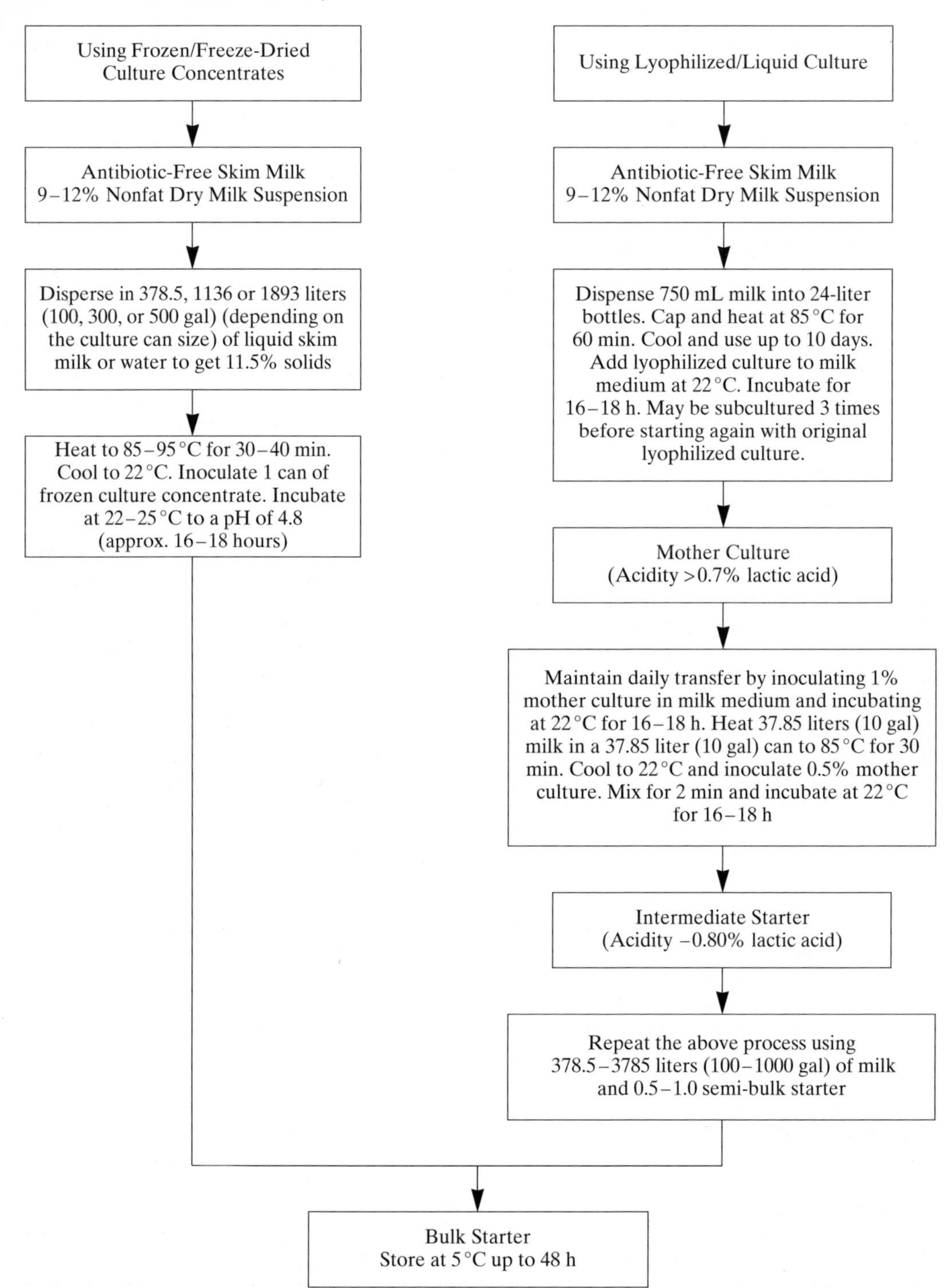

Fig. 3. Typical steps for bulk starter production in a dairy plant for manufacture of cultured buttermilk, sour cream and yogurt.

ardized milk. Cultured sour cream is made from 18% fat cream supplemented with some nonfat milk solids for textural improvement.

All dairy raw materials should be selected for high bacteriological quality. The procurement of all ingredients should be based on specifications and standards that are checked and maintained with a systematic sampling and testing program by the quality control laboratory. It is extremely important to standardize and control the day-to-day product in order to meet consumer expectations and regulatory obligations associated with a certain brand or label.

Sweeteners. Nutritive carbohydrates used in yogurt manufacture are similar to the sweeteners used in ice cream and other frozen desserts. Sucrose is the major sweetener used in yogurt production. Sometimes corn sweeteners may also be used, especially in frozen yogurt mixes. The level of sucrose in yogurt mix appears to affect the production of lactic acid and flavor by yogurt culture. A decrease in characteristic flavor compound (acetaldehyde) production has been reported at 8% or higher concentrations of sucrose. Sucrose may be added in a dry, granutated, free-flowing, crystalline form or as a liquid sugar containing 67% sucrose. Liquid sugar is preferred for its handling convenience in large operations. However, storage capability in sugar tanks along with heaters, pumps, strainers, and meters is required. The corn sweeteners, primarily glucose, usually enter yogurt via the processed fruit flavor in which they are extensively used for their flavor enhancing characteristics. Up to 6% corn syrup solids are used in frozen yogurt. High-intensity sweeteners (e.g., aspartame) have been used to produce a “light” product containing about 60% of the calories of normal sweetened yogurt.

Stabilizers. The primary purpose of using a stabilizer in yogurt and sour cream is to produce smoothness in body and texture, impart gel structure, and reduce wheying off or syneresis. The stabilizer increases shelf life and provides a reasonable degree of uniformity of the product. Stabilizers function through their ability to form gel structures in water, thereby leaving less free water for syneresis. In addition, some stabilizers complex with casein. A good stabilizer should not impart any flavor, should be effective at low pH values, and should be easily dispersed at the normal working temperatures in a dairy plant. The stabilizers generally used are gelatin; vegetable gums such as carboxymethyl cellulose, locust bean, and guar; and seaweed gums such as alginates and carrageenans.

The stabilizer system used in yogurt and sour cream mix preparations is generally a combination of various vegetable stabilizers to which gelatin may or may not be added. Their ratios as well as the final concentration (generally 0.5 to 0.7%) in the product are carefully controlled to get desirable effects. More recently, whey protein concentrate is being used as a stabilizer, exploiting the water binding property of denatured whey proteins.

Fruit preparations for flavoring yogurt. The fruit preparations for blending in yogurt are specially designed to meet the marketing requirements for different types of yogurt. They are generally present at levels of 10 to 20% in the final product.

Flavors and certified colors are usually added to the fruit-for-yogurt preparations for improved eye appeal and better flavor profile. The fruit base should meet the following requirements. It should (1) exhibit true color and flavor of the fruit when blended with yogurt, and (2) be easily dispersible in yogurt without causing texture defects, phase separation, or syneresis. The pH of the fruit base should be compatible with yogurt pH. The fruit should have zero yeast and mold populations in order to prevent spoilage and to extend shelf life. Fruit preserves do not necessarily meet all these requirements, especially of flavor, sugar level, consistency, and pH. Accordingly, special fruit bases of compatible composition are designed for use in stirred yogurt.

Calcium chloride and certain food-grade phosphates are also used in several fruit preparations. The soluble solids range from 60 to 65% and viscosity is standardized to 5 ± 1.5 Bostwick units (cm), 30 s reading at 24 °C. Standard plate counts on the fruit bases are generally <500/g. Coliform count, yeast, and mold counts of nonaseptic fruit preparations are <10/g. In general, more popular fruits are strawberry, raspberry, blueberry, peach, cherry, orange, lemons, purple plum, boysenberry, spiced apple, apricot, and pineapple. Blends of these fruits are also popular. Fruits used in yogurt base manufacture may be frozen, canned, dried, or combinations thereof. Among the frozen fruits are strawberry, raspberry, blueberry, apple, peach, orange, lemon, cherry, purple plum, blackberry, and cranberry. Canned fruits are pineapple, peach, mandarin orange, lemon, purple plum, and maraschino cherry. The dried fruit category includes apricot, apple, and prune. Fruit juices and syrups are also incorporated in the bases. Sugar in the fruit base functions in protecting fruit flavor against loss by volatilization and oxidation. It also balances the fruit and the yogurt flavor. The pH control of the base is important for fruit color retention. The base should be stored under refrigeration to obtain optimum flavor and extend shelf life. The current trend is to use aseptically packaged sterilized fruit preparations.

In *Fruit-on-the-bottom style yogurt*, typically, 59 mL (2 oz) of fruit preserves or special fruit preparations are layered at the bottom followed by 177 mL (6 oz) of inoculated yogurt mix on the top. The top layer may consist of yogurt mix containing stabilizers, sweeteners, and the flavor and color indicative of the fruit on the bottom. After lids are placed on the cups, incubation and setting of the yogurt takes place in the cups. When a desirable pH of 4.2 to 4.4 is attained, the cups are placed in refrigerated rooms for rapid cooling. For consumption, the fruit and yogurt layers are mixed by the consumer. If used, fruit preserves have a standard of identity. A fruit preserve consists of 55% sugar and a minimum of 45% fruit which is cooked until the final soluble solids content is 68% or higher (65% in the case of certain fruits). Frozen fruits and juices are the usual raw materials. Commercial pectin, 150 grade, is normally utilized at a level of 0.5% in preserves, and the pH is adjusted to 3.0 to 3.5 with a food-grade acid such as citric during manufacturing of the preserves.

Stirred style yogurt is also known as Continental, French, and Swiss yogurt. The fruit preparation is thoroughly blended in yogurt after culturing. Stabilizers are commonly used in this form of yogurt unless milk-solids-not-fat levels are relatively high (14 to 16%). In this style, cups are filled with a blended mixture of yogurt and fruit. On refrigerated storage for 48 hours the clot is reformed to exhibit a fine body and texture. Overstabilized yogurt possesses a solid-like consistency and lacks a refreshing character. Spoonable yogurt should not have the consistency of a drink. It should melt in the mouth without chewing.

5.2 Equipment

Yogurt and other cultured dairy products are produced in various parts of the world from the milk of several species of mammals. The animals include cow (*Bos taurus*), water buffalo (*Bubalus bubalis*), goat (*Capra hircus*), sheep (*Ocis aries*), mare (*Equus cabalus*), and sow (*Sus scrofa*). The composition of these milks is summarized in Tab. 7. Because the total solids in milk of various species range from 11.2 to 19.3%, the cultured products derived from them vary in consistency from a fluid to a custard-like gel. The range in casein content also contributes to the gel formation because on souring this class of proteins coagulates at its isoelectric point of pH 4.6. The whey proteins are considerably denatured and insolubilized by heat treatments prior to culturing. The denatured whey proteins are also precipitated along with caseins to exert an effect on the water binding capacity of the gel.

In the United States, bovine milk is practically the only milk employed in the industrial manufacture of cultured dairy products (see Tab. 2).

A cultured milk plant requires a special design to minimize contamination of the products with phage and spoilage organisms. Filtered air is useful in this regard. The plant is generally equipped with a receiving room to receive, meter or weigh, and store milk and other raw materials. In addition, a culture propagation room along with a control laboratory, a dry storage area, a refrigerated storage area, a mix processing room, a fermentation room, and a packaging room form the backbone of the plant. The mix processing room contains equipment for standardizing

Tab. 7. Composition of Milks Used in the Preparation of Cultured Dairy Foods in Various Parts of the World

Mammal	Fat (%)	Caseins (%)	Whey Proteins (%)	Lactose (%)	Ash (%)	Total Solids (%)
Cow	3.7	2.8	0.6	4.8	0.7	12.7
Water buffalo	7.4	3.2	0.6	4.8	0.8	17.2
Goat	4.5	2.5	0.4	4.1	0.8	13.2
Sheep	7.4	4.6	0.9	4.8	1.0	19.3
Mare	1.9	1.3	1.2	6.2	0.5	11.2
Sow	6.8	2.8	2.0	5.5	—	18.8

Source: CHANDAN (1982)

and separating milk, pasteurizing and heating, and homogenizing along with the necessary pipelines, fittings, pumps, valves, and controls. The fermentation room housing fermentation tanks is isolated from the rest of the plant. Filtered air under positive pressure is supplied to the room to generate clean room conditions. A control laboratory is generally set aside where culture preparation, process control, product composition, and shelf life tests may be carried out to ensure adherence to regulatory and company standards. Also, a quality control program is established by laboratory personnel. A utility room is required for maintenance and engineering services needed by the plant. The refrigerated storage area is used for holding fruit, finished products, and other heat-labile materials. A dry storage area at ambient temperature is primarily utilized for temperature-stable raw materials and packaging supplies.

The sequence of stages of processing in a yogurt plant is given in Tab. 8.

5.3 Mix Preparation

Milk is commonly stored in silos which are large vertical tanks with a capacity up to 100000 L. A silo consists of an inner tank made of stainless steel containing 18% chromium, 8% nickel, and <0.07% carbon. Acid and salt resistance in the steel is attained by incorporating 3% molybdenum. To minimize corrosion, this construction material is used for the storage of acidic products. The stainless steel tank is usually covered with 50 to 100 mm of insulation material which in turn is surrounded by an outer shell of stainless or painted mild steel or aluminum. The silo tanks generally have an agitation system (60 to 80 rpm), spray balls mounted in the center for cleaning in place (CIP), an air vent, and a manhole. The air vent must be kept open during cleaning with hot cleaning solutions. This precaution is necessary to prevent a sudden development of vacuum in the tank and consequent collapse of the inner tank upon rinsing with cold water.

For reconstitution of dry powders, such as nonfat dry milk, sweeteners, and stabilizers, the use of a powder funnel and recirculation loop, or a special blender is convenient.

5.4 Heat Treatment

The common pasteurization equipment consists of vat, plate, triple-tube, scraped, or swept surface heat exchangers. In case of milk, vat pasteurization is conducted at 63 °C with a minimum holding time of 30 min. This temperature is raised to 66 °C in the presence of sweeteners in the mix. For a high temperature-short time (HTST) system, the equivalent temperature–time combination is 73 °C for 15 s, or 75 °C for 15 s in the presence of sweeteners. An ultra-high temperature (UHT) system employs temperatures >90 °C and as high as 148 °C for 2 s. Alternatively, the culinary stream may be used directly by injection or infusion to raise the temperature to 77 to 94 °C, but allowance must be made for an increase in water content of the mix

Tab. 8. Sequence of Processing Stages in the Manufacturing of Yogurt

Step	Salient Feature
1. Milk procurement	Sanitary production of grade A milk from healthy cows is necessary. For microbiological control, refrigerated bulk milk tanks should cool to 10 °C in 1 h and <5 °C in 2 h. Avoid unnecessary agitation to prevent lipolytic deterioration of milk flavor. Milk pickup is in insulated tanks at 48 h intervals.
2. Milk reception and storage in manufacturing plant	Temperature of raw milk at this stage should not exceed 10 °C. Insulated or refrigerated storage up to 72 h helps in raw material and process flow management. Quality of milk is checked and controlled.
3. Centrifugal clarification and separation	Leucocytes and sediment are removed. Milk is separated into cream and skim milk or standardized to desired fat level at 5 °C or 32 °C.
4. Mix preparation	Various ingredients to secure desired formulation are blended together at 50 °C in a mix tank equipped with powder funnel and agitation system.
5. Heat treatment	Using plate heat exchangers with regeneration systems, milk is heated to temperatures of 85–95 °C for 10–40 min, well above pasteurization treatment. Heating of milk kills contaminating and competitive microorganisms, produces growth factors by breakdown of milk proteins, generates microaerophilic conditions for growth of lactic organisms, and creates desirable body and texture in the cultured dairy products.
6. Homogenization	Mix is passed through an extremely small orifice at a pressure of 2000–2500 psi, causing extensive physicochemical changes in the colloidal characteristics of milk. Consequently, creaming during incubation and storage of yogurt is prevented. The stabilizers and other components of a mix are thoroughly dispersed for optimum textural effects.
7. Inoculation and incubation	The homogenized mix is cooled to an optimum growth temperature. Inoculation is generally at the rate of 0.5–5%, and the optimum temperature is maintained throughout the incubation period to achieve a desired titratable acidity. Quiescent incubation is necessary for product texture and body development.
8. Cooling, fruit incorporation and packaging	The coagulated product is cooled down to 5–22 °C, depending upon the product. Using fruit feeder or flavor tank, the desired level of fruit and flavor is incorporated. The blended product is then packaged.
9. Storage and distribution	Storage at 5 °C for 24–48 h imparts in several yogurt products desirable body and texture. Low temperatures ensure desirable shelf life by slowing down physical, chemical, and microbiologial degradation.

due to steam condensation in this process. In some plants, steam volatiles are continuously removed by vacuum evaporation to remove certain undesirable odors (feed, onion, garlic) associated with milk.

In yogurt processing, the mix is subjected to much more severe heat treatment than conventional pasteurization temperature–time combinations. Heat treatment at 85 °C for 30 min or 95 °C for 10 min is an important step in manufacture. The heat treatment (1) produces a relatively sterile medium for the exclusive growth of the starter, (2) removes air from the medium to produce a more conducive medium for microaerophilic lactic cultures to grow; (3) effects thermal breakdown of milk constituents, especially proteins, releasing peptones and sulfhydryl groups which provide nutrition and anaerobic conditions for the starter; and (4) denatures and coagulates milk albumins and globulins which enhance the viscosity and produce a custard-like consistency in the product.

5.5 Homogenization

The homogenizer is a high-pressure pump forcing the mix through extremely small orifices. It includes a bypass for safety of operation. The process is usually conducted by applying pressure in two stages. The first stage pressure, of the order of 2000 psi, reduces the average milkfat globule diameter size from approx. 4 μm (range 0.1 to 16 μm) to <1 μm. The second stage uses 500 psi and is designed to break the clusters of fat globules apart with the objective of inhibiting creaming in milk. Homogenization aids in texture development, and additionally it alleviates the surface creaming and wheying off problems. Ionic salt balance in milk is also involved in the wheying off problem.

5.6 Fermentation

Fermentation tanks for the production of cultured dairy products are generally designed with a cone bottom to facilitate draining of relatively viscous fluids after incubation.

For temperature maintenance during the incubation period, the fermentation vat is provided with a jacket for circulating hot or cold water or steam located adjacent to the inner vat containing the mix. This jacket is usually insulated and covered with an outermost surface made of stainless steel. The vat is equipped with a heavy-duty, multispeed agitation system, a manhole containing a sight glass, and appropriate spray balls for CIP. The agitator is often of swept surface type for optimum agitation of relatively viscous cultured dairy products. For efficient cooling after culturing, plate or triple-tube heat exchangers are used.

The fermentation vat is designed only for temperature maintenance. Therefore, efficient use of energy requires that the mix not be heat-treated in the culturing vat.

5.7 Manufacturing Procedures

Plain yogurt. Plain yogurt is an integral component of the manufacture of frozen yogurt. The steps involved in the manufacturing of set-type and stirred-type plain yogurts are shown in Fig. 4. Plain yogurt normally contains no added sugar or flavors in order to offer the consumer natural yogurt flavor for consumption as such or an option of flavoring with other food materials of the consumer's choice. In addition, it may be used for cooking or for salad preparation with fresh fruits or grated vegetables. In most recipes, plain yogurt is a substitute for sour cream, providing lower calories and less fat. The fat content may be standardized to the levels preferred by the market. Also, the size of the package may be geared to the market demand. Plastic cups and lids are the chief packaging materials used in the industry.

Fruit-flavored yogurt. A general manufacturing outline for both set-style and stirred-style yogurts is presented in Fig. 5. Several variations of this procedure exist in the industry. Fruit is conveniently added by the use of a fruit feeder at a 10 to 20% level. Prior to packaging, the stirred-yogurt texture can be made smoother by pumping it through a valve or a stainless steel screen.

The incubation times and temperatures are coordinated with the plant schedules. Incubation temperatures lower than 40 °C in general tend to impart a slimy or sticky appearance to yogurt.

Postculturing heat treatment. The shelf life of yogurt may be extended by heating yogurt after culturing to inactivate the culture and the constituent enzymes. Heating to 60 to 65 °C stabilizes the product so the yogurt shelf life will be 8 to 12 weeks at 12 °C. However, this treatment destroys the "live" nature of yogurt, which may be a desirable consumer attribute to retain. Federal Standards of Identity for refrigerated yogurt permit the thermal destruction of viable organisms with the objective of shelf-life extension, but the parenthetical phrase "heat treated after culturing" must show on the package following the yogurt labeling. The postripening heat treatment may be designed to (1) ensure destruction of starter bacteria, contaminating organisms, and enzymes; and (2) redevelop the texture and body of the yogurt by appropriate stabilizer and homogenization processes.

Frozen yogurt. Both soft-serve and hard-frozen yogurts have gained immense popularity in recent years. Market value in frozen yogurt has exceeded that of refrigerated yogurt. Consumer popularity for frozen yogurt has been propelled by its low-fat and nonfat attribute. The recently developed frozen yogurt is a very low acid product resembling ice cream

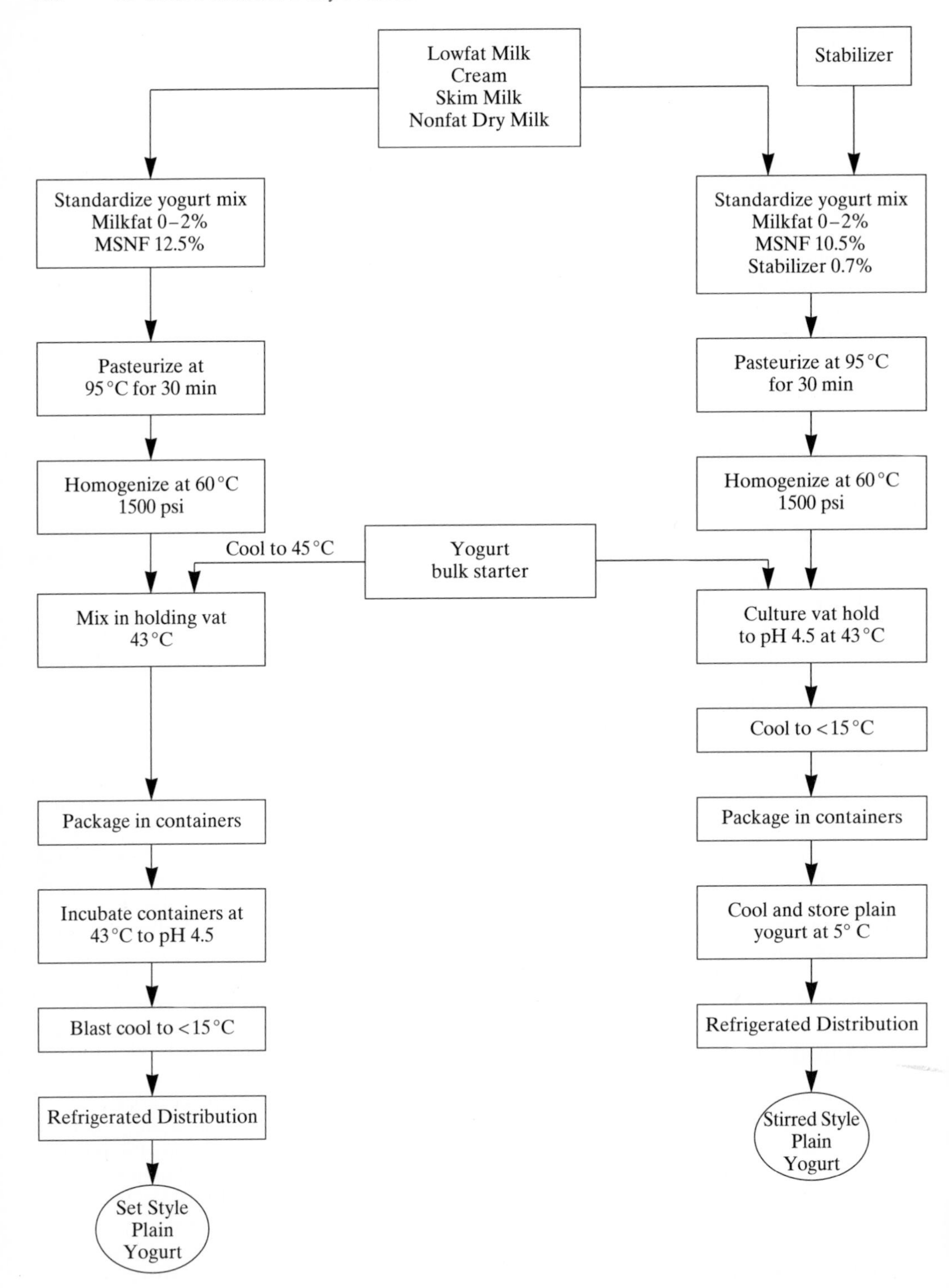

Fig. 4. Flow sheet outline for the manufacture of plain yogurt.

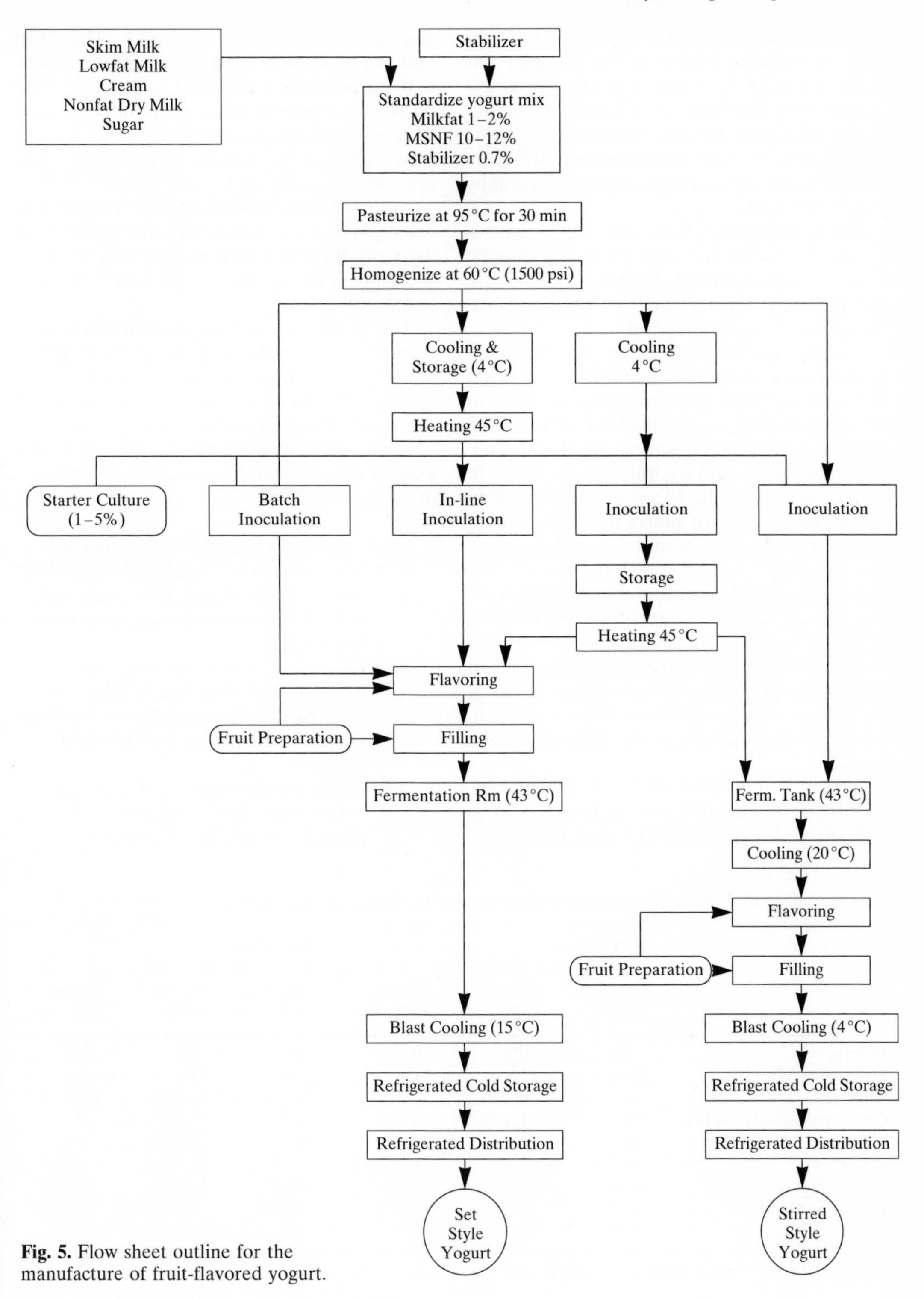

Fig. 5. Flow sheet outline for the manufacture of fruit-flavored yogurt.

or ice milk in flavor and texture. A significant shift in reduced acidity in the product has been observed in relation to the products available 10 years before. Essentially, the industry standards require minimum titratable acidity of 0.3%, with a minimum contribution of 0.15% as a consequence of fermentation by yogurt bacteria.

The frozen yogurt base mix may be manufactured in a cultured dairy plant and shipped to a soft-serve operator or an ice cream plant. Alternatively, the mix may be prepared and frozen in an ice cream plant.

Technology for production of frozen yogurt involves limited fermentation in a single mix and arresting further acid development by rapid cooling, or a standardization of titratable acidity to a desirable level by blending plain yogurt with ice milk mix (Tab. 9). In certain instances, the blend is pasteurized to ensure destruction of newly emerging pathogens, including *Listeria* and *Campylobacter* in the resulting low-acid food. To provide live and active yogurt culture in the finished product, frozen culture concentrate is blended with the pasteurized product. Alternatively, some processes are boosting the yogurt culture count by adding frozen culture concentrates to the fermented base. Figs. 6 and 7 illustrate typical processes for making frozen yogurt.

Cultured milk (buttermilk). Cultured buttermilk is obtained from pasteurized skim milk or part skim milk cultured with lactic and aroma-producing organisms. The term buttermilk is also used for a phospholipid-rich fluid fraction obtained as a by-product during the churning of cream in butter manufacture. However, cultured buttermilk is a viscous, cultured, fluid milk, containing a characteristic pleasing aroma and flavor.

Cultured buttermilk is usually produced in dairy plants processing milk and other fluid dairy products by a process similar to the flow sheet given in Fig. 8. It is packaged in traditional milk cartons.

The processes used in the manufacture of cultured buttermilk include pasteurization, homogenization, and culturing systems.

Under refrigeration, the keeping quality of cultured buttermilk is extended to 3–4 weeks. Wheying off may occur but can be avoided by using a suitable stabilizer and proper processing conditions.

Buttermilk defects and possible causes.

(1) *Flat, "Green".* – Acidity too low, low setting temperature, low solids milk, lack of flavor organisms, short incubation period.

(2) *High Acid, Sour.* – Setting temperature too high, incubation time too long, lack of fat, inoculation too heavy, poor refrigeration of finished product.

(3) *Biting on the Tongue. "Carbonated".* – Too many flavor organisms, overripening, contaminated culture.

Tab. 9. Typical Composition of Nonfat Soft-Serve and Hard-Pack Frozen Yogurt

Component	Soft Serve			Hard Pack		
	Stream 1 (20%)	Stream 2 (80%)	Blended Final Mix	Stream 1 (20%)	Stream 2 (80%)	Blended Final Mix
Milkfat	0%	0%	0%	0%	0%	0%
Milk solids not fat	11%	11%	11%	13%	13%	13%
Sucrose	0	16.25%	12%	0	16.25%	13%
Corn syrup solids, 36 DE	0	7.50%	6%	0	7.5%	6%
Maltodextrin, 10 DE	0	2.5%	2%	0	2.5%	2%
Stabilizer	0	1.5%	1.2%	0	1.5%	1.2%
Total solids	11%	38.75%	32.2%	13%	40.75%	35.20%
Titratable acidity	1.15	0.15	0.35%	1.15%	0.16%	0.35%
pH	4.4	6.7	5.5	4.4	6.7	5.5

Source: GERMANTOWN MANUFACTURING CO. (1991)

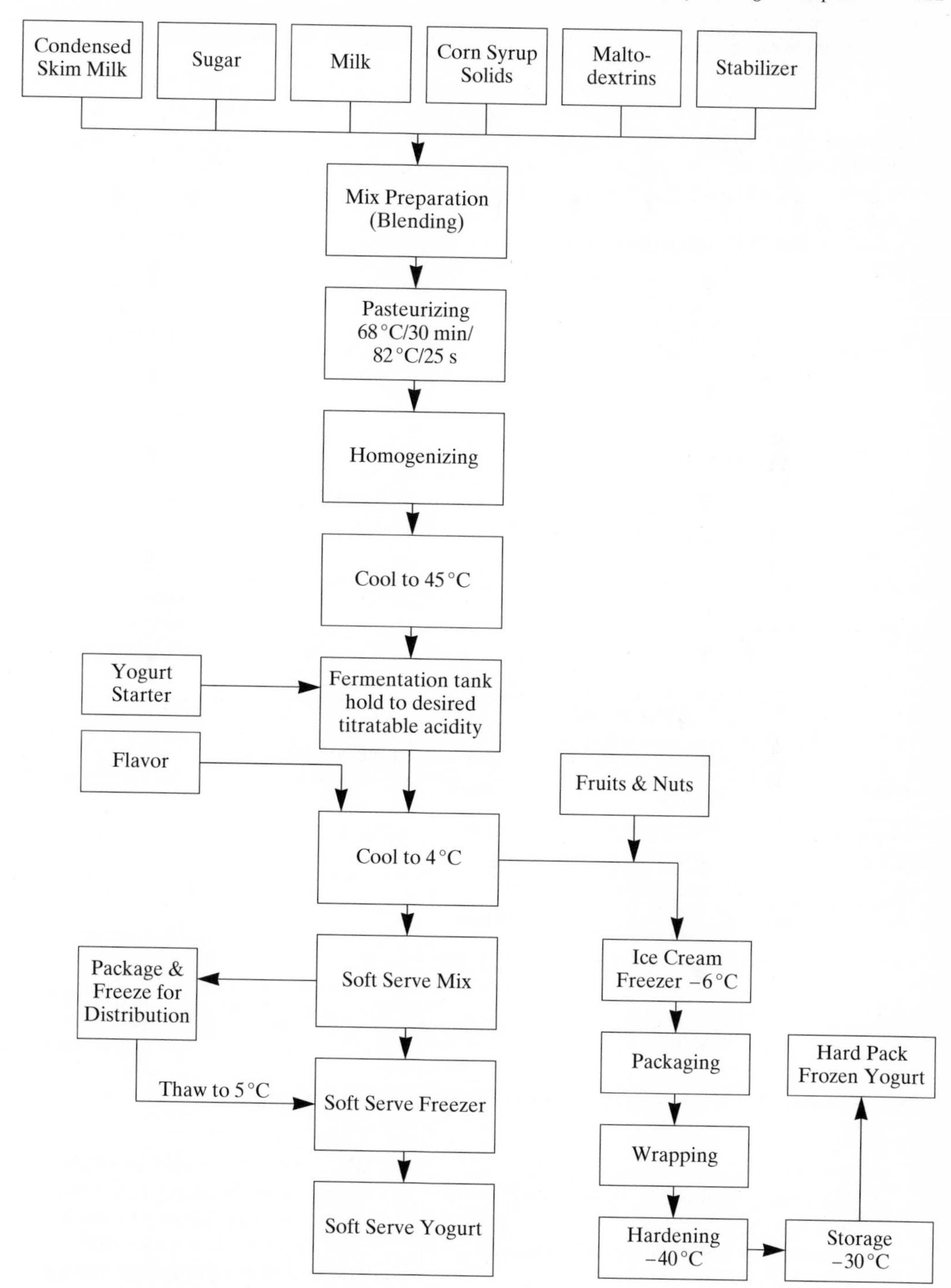

Fig. 6. Flow chart for frozen yogurt (single-stream process).

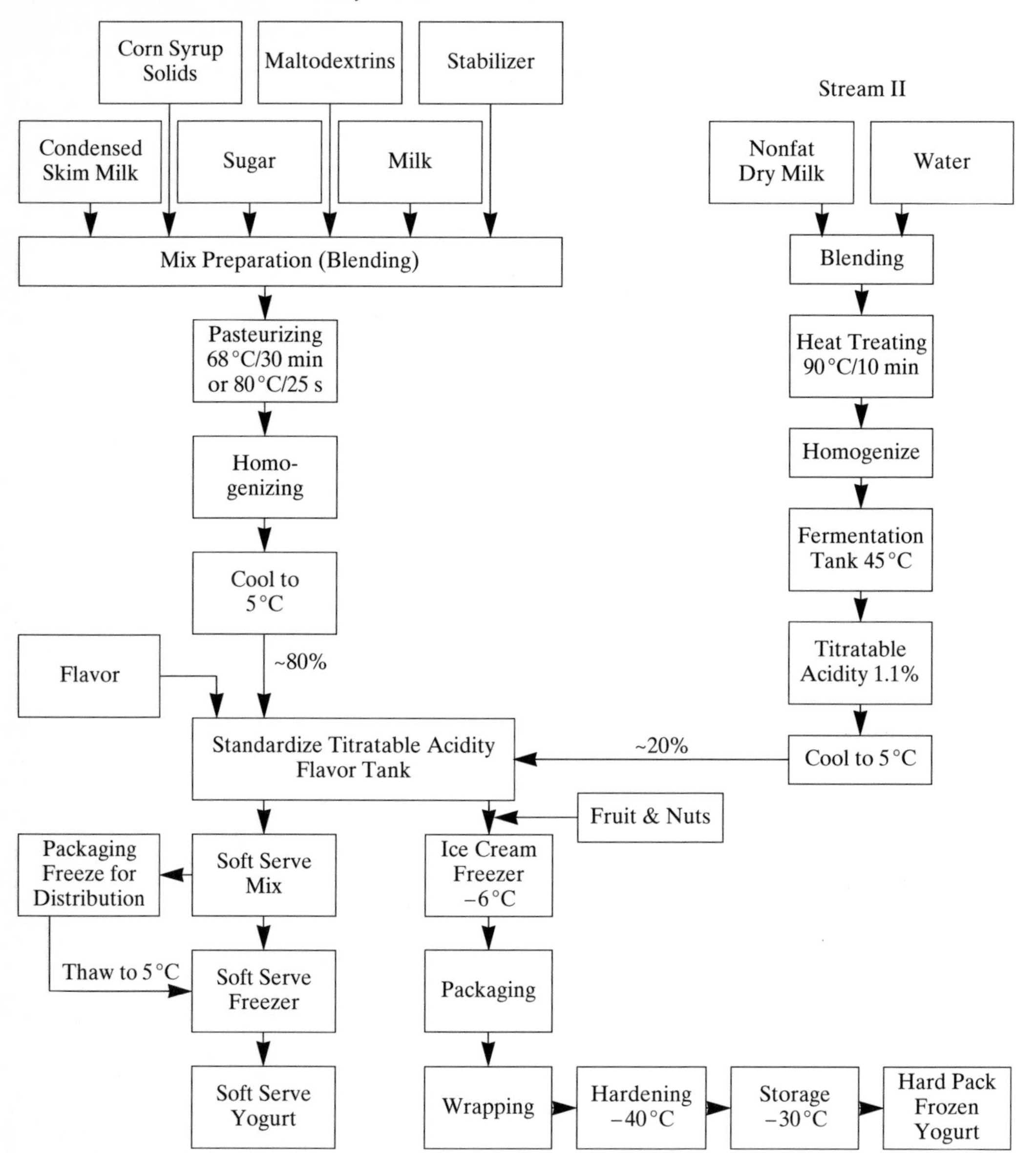

Fig. 7. Flow chart for frozen yogurt (two-stream blending process).

(4) *Metallic.* – Contact with copper or iron, rusty equipment, high acidity and age.
(5) *Oxidized.* – Metal contamination, use of oxidized dairy products.
(6) *Rancid.* – Mixing homogenized milk with raw dairy products, use of rancid milk or cream.
(7) *Whey Off.* – Excessive agitation when warm, breaking at too low acidity, adding skimmed milk to reduce viscosity, low solids milk, high storage temperature, high setting temperature, too low or too high treatment of skim milk.
(8) *Lumpiness.* – High acidity, high set-

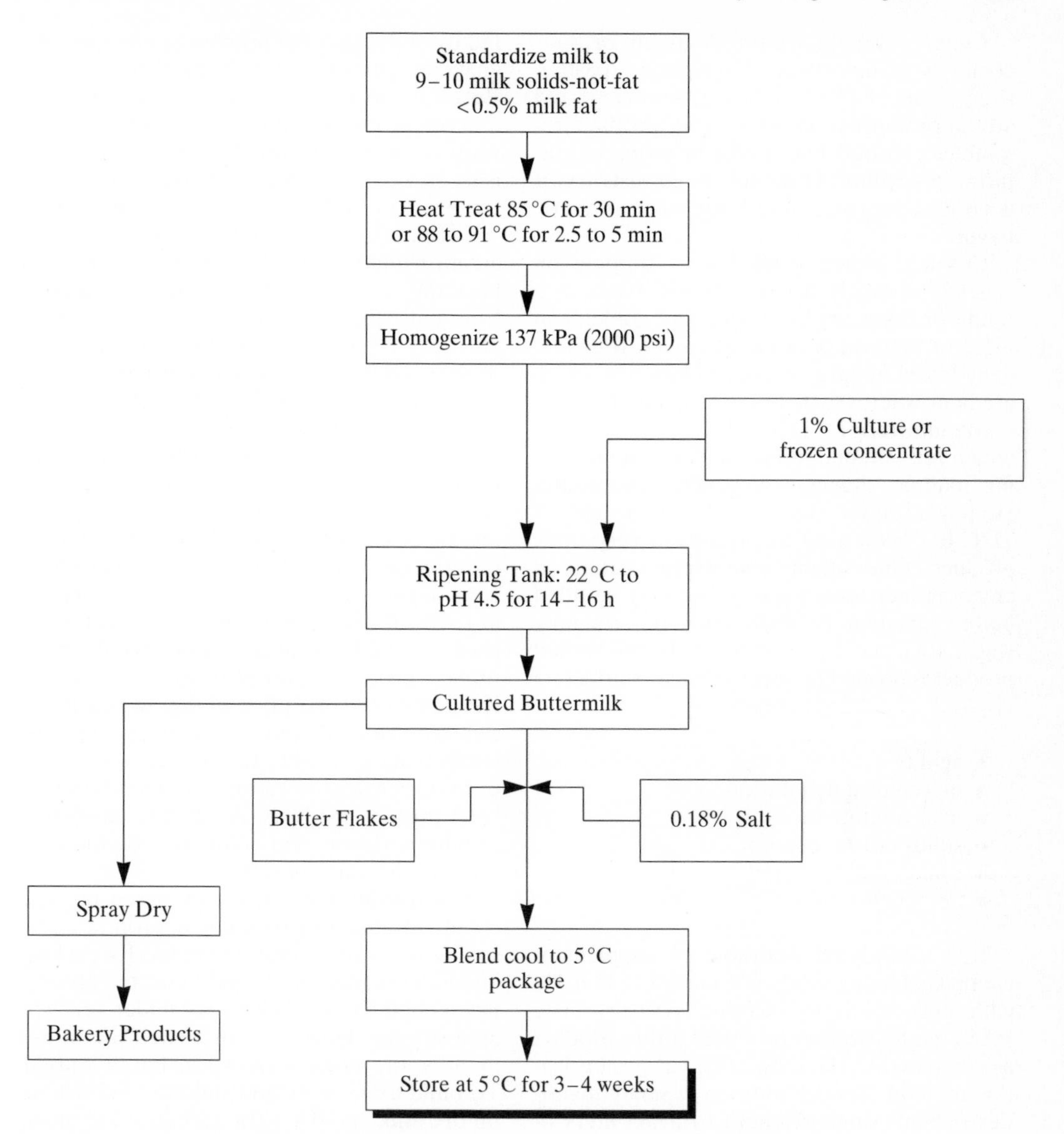

Fig. 8. Flow sheet diagram for the manufacture of cultured nonfat milk (buttermilk).

ting temperature, poor agitation during breaking, low heat treatment of milk.

(9) *Thick, Heavy Body.* – High acidity, ropy culture, solids too high, high incubation temperature, long incubation time, high heat treatment of milk, homogenization of high fat milk.

(10) *Thin, Light Body.* – Inactive culture, low setting temperature, low heat treatment of milk, low acidity, low solids, too much agitation during breaking.

(11) *Incorporated Air, Foamy.* – Agitation too vigorous, pumping with centrifugal pump, leaky pumps, or valves.

Cultured cream. Cultured cream or sour cream is manufactured by ripening pasteurized cream of 18% fat content with lactic and aroma-producing bacteria. This product resembles cultured buttermilk in terms of culturing procedure. However, in consistency, it is an acid gel containing butter-like aromatic flavor.

Cultured cream is used as a topping on vegetables, salads, fish, meats, and fruits, as a filling in cakes, and in soups and cookery in place of buttermilk or sweet cream. It can be dehydrated by spray-drying and used as an ingredient wherever its flavor is needed.

Manufacturing principles for cultured cream are outlined in Fig. 9. The manufacturing method shown is a general procedure. Homogenization twice at 17.2 kPa and at 71 °C has been used to produce a very thick product. One might manufacture cultured cream in individual packs which may be filled before ripening or soon after the ripening stage, followed by cooling. A heavy-bodied product is formed on setting. Factors affecting viscosity of cultured cream are:

- acidity,
- mechanical agitatation,
- heat treatment,
- solids-not-fat content,
- rennet addition, and
- homogenization.

It is considered desirable to supplement cream containing solids-not-fat less than 6.8% with milk solids to increase viscosity. The HTST pasteurization produces a thin product as compared to the long hold, vat pasteurization method. Rennet addition in small quantities (0.5 mL single-strength to 37.85 liters of cream) aids in thickening the product. As in buttermilk, flavor may be improved by incorporating into cream 0.15–0.20% sodium citrate, which is metabolized by *Lactococcus lactis* subsp. *diacetylactis* and *Leuconostoc cremoris* to produce more aroma compounds (diacetyl and volatile acids). A lactic culture enhances the smoothness and to some extent the viscosity of cultured cream. The body of the product appears to be independent of the culture used.

The hot-pack process ensures long shelf life by destroying the microorganisms and the enzymes present in the finished product. Packaging in a plastic or metal container with a hermetically sealed lid further ensures prevention of recontamination by microorganisms as well as protection from oxidative deterioration of milk fat in the finished product.

Cream used in the manufacture of cultured cream should be fresh with a relatively low bacterial count. During cream separation from milk, the bacteria tend to concentrate in the lighter phase, cream, thereby enhancing its vulnerability to spoilage. Pasteurization at 74 °C for 30 min or at 85 °C for 1 min is satisfactory from a bacteriological standpoint.

Artificial cultured cream. This product may be defined as cultured cream in which part or all of the milk fat has been replaced by other oils or fats. Such a product is sold commercially in the United States. It appears to have advantages over conventional cultured cream in terms of price and caloric value. In this regard, a suitable product may be manufactured using a process identical to that for cultured cream with the exception of the starting material. An artificial cream may be prepared by emulsifying a suitable fat in either skim milk or in suspensions of casein compounds or soybean protein products. A suitable emulsifier, stabilizer, flavor, and color may be incorporated in the starting mix.

Sour cream dip. Party dips based on sour cream are made by blending appropriate seasoning bases into cultured cream. By packaging under refrigerated conditions, the product has a shelf life of 2–3 weeks under refrigerated storage. However, for a shelf life of 3–4 months, the process shown in Fig. 9 is used. To build extra body and stability, 1–2% nonfat dry milk and 0.8–1.0% stabilizer are incorporated at 80 °C. The mixture is pasteurized by holding for 10 min, and homogenized at 17.2 kPa to resuspend and smooth the product. The seasonings are blended at this stage while the mix is still at 80 °C, followed by hot-packing in sealed containers. Upon cooling and storage at 5 °C, partial vacuum inside the container assists in the prevention of oxidative deterioration to yield an extended shelf life of 3–4 months.

Sour half and half. To satisfy the consumer demand for relatively low-fat and less expen-

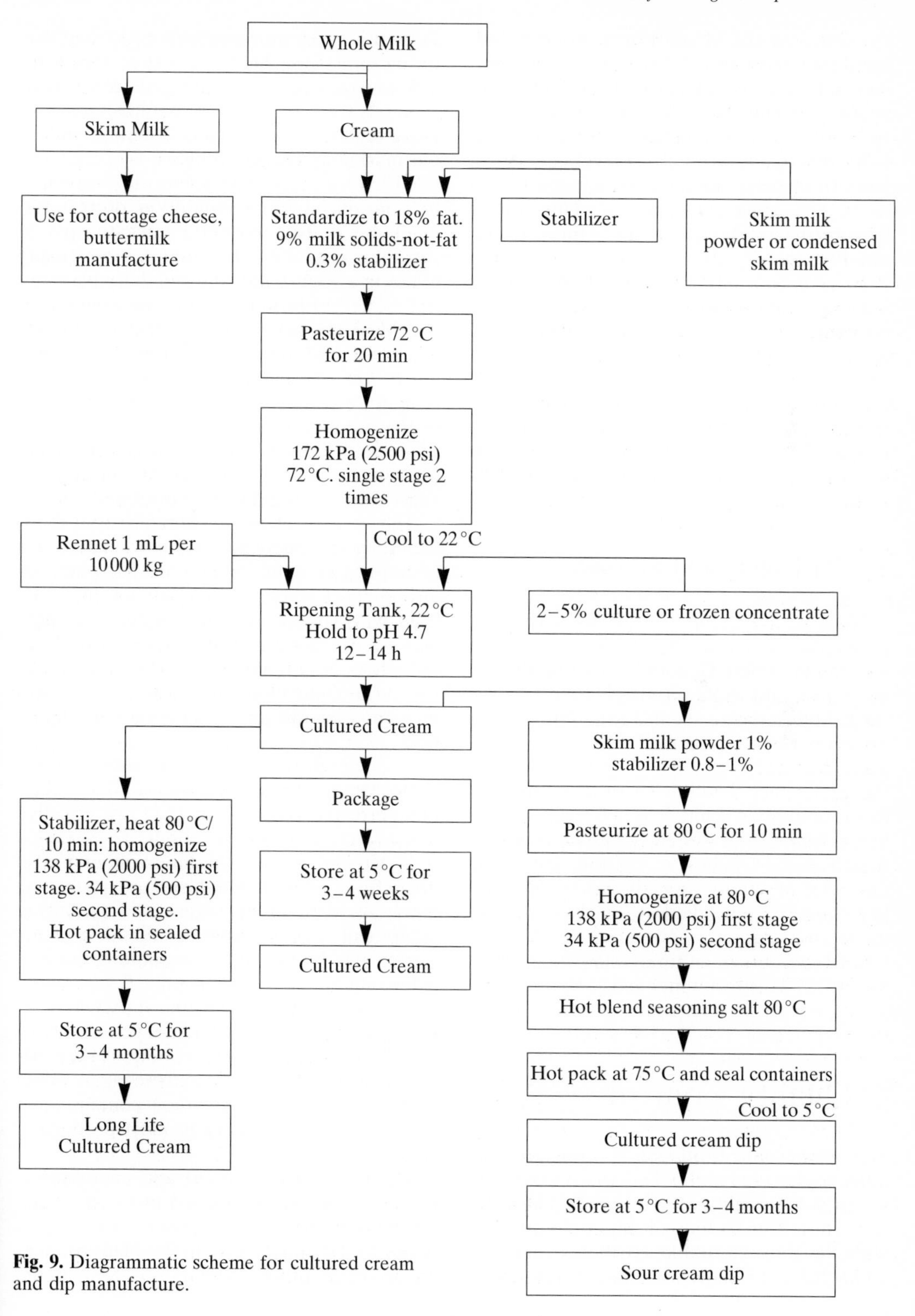

Fig. 9. Diagrammatic scheme for cultured cream and dip manufacture.

sive cultured cream substitutes, sour or cultured half and half has been developed commercially. It is manufactured from a mix containing a minimum of 10.5 and a maximum of 18% milk fat. To compensate for reduction in solids due to a lower milk fat level, it is customary to increase the milk solids-not-fat level to 10–12% milk solids-not-fat. This mix is then processed along the same lines as cultured cream (Fig. 9).

Salad dressing. Using sour half and half as a base, appropriate flavor bases may be blended to produce a distinctive creamy salad dressing. This dressing contains about 50–75% fewer calories than conventional salad dressings, but has a comparable flavor and texture. The reduction in calories is primarily due to the lower fat level of 10.5% in sour half and half dressing as compared to 30–80% oil in regular dressing.

5.8 Packaging and Storage

Most plants attempt to synchronize the packaging lines with the termination of the incubation period. Generally, textural defects in yogurt and cultured dairy products are caused by excessive shear during pumping or agitation. Therefore, positive drive pumps are preferred over centrifugal pumps for moving the product after culturing or ripening. For incorporation of fruit, it is advantageous to use a fruit feeder system adapted from the frozen dessert industry. Various packaging machines of suitable speeds (up to 400 cups per minute) are available to package various kinds and sizes of yogurt products. For extended shelf life of cultured dairy foods, storage at 1–4 °C is recommended.

6 Quality Control

A well-planned quality control program must be executed in the dairy plant to maximize shelf-life quality of a cultured milk product. To deliver to the consumer the product with the most desirable attributes of flavor and texture, it is imperative to enforce a strict sanitation program along with good manufacturing practices. Shelf-life expectations for commercial yogurt vary but generally exceed a month from the date of manufacture, provided temperature during distribution and retail marketing channels does not exceed 7 °C (45 °F). Lactic acid and some other metabolites produced by fermentation protect fermented milk from most Gram negative psychrotrophic organisms. In general, most quality issues in a yogurt and cultured dairy plant are not related to proliferation of spoilage bacteria. Most spoilage flora in fermented milks are yeasts and molds, which are highly tolerant to low pH and can grow at refrigeration temperatures. Yeast growth during shelf life of the product constitutes more of a problem than mold growth. The fungal growth manifests itself within 2 weeks of manufacture, if yeast contamination is not controlled.

The control of yeast contamination is effected by aggressive sanitation procedures related to equipment, ingredients, and plant environment. CIP chemical solutions should be used with special attention to their strength and proper temperature. Hypochlorites and iodophors are effective sanitizing compounds for fungal control on the contact surfaces and in combating the environmental contamination.

Hypochlorites at high concentrations are corrosive. Iodophors are preferred for their non-corrosive property as they are effective at relatively low concentrations.

Yeast and mold contamination may also arise from contaminated starter, packaging materials, fruit preparations, and packaging equipment. Organoleptic examination of the starter may be helpful in eliminating the fungal contamination therefrom. If warranted, direct microscopic view of the starter may reveal the presence of budding yeast cells or mold mycelium filaments. Plating of the starter on acidified potato dextrose agar would confirm the results. Avoiding contaminated starter for cultured dairy product production is necessary.

Efficiency of equipment and environmental sanitation can be verified by enumeration techniques involving exposure of poured plates to the atmosphere in the plant or making a smear of the contact surfaces of the equipment, followed by plating. Filters on the air circulation system should be changed frequently. Walls and floors should be cleaned and sanitized frequently and regularly.

The packaging materials should be stored under dust-free and humidity-free conditions. The filling room should be fogged with chlorine or iodine regularly.

Quality control checks on fruit preparations and flavorings should be performed (spot checking) to minimize yeast and mold entry into fruit-flavored yogurt. Refrigerated storage of the fruit flavorings is recommended.

Quality control programs include control of product viscosity, flavor, body and texture, color, fermentation process, and composition. Daily chemical, physical, microbiological, and organoleptic tests constitute the core of quality assurance. The flavor defects are generally described as too intense (acid), too weak (fruit flavor), or unnatural. The sweetness level may be excessive, weak, or may exhibit corn syrup flavor. The ingredients used may impart undesirable flavors such as stale, metallic, old ingredients, oxidized, rancid, or unclean. Lack of control in processing procedures may cause overcooked, caramelized, or excessively sour flavor notes in the product. Proper control of processing parameters and

Tab. 10. Typical Quality Problems in Buttermilk and Sour Cream

Defect	Probable Cause	Remedy
Insufficient flavor	Low citrate level in milk	– Add 0.02–0.05% (up to 0.1%) sodium citrate prior to the mix
	Diacetyl destroyed	– Cool rapidly after culturing – Agitate gently alter breaking to incorporate oxygen in the product
Green/yogurt flavor	Acetaldehyde accumulation	– Avoid *Lactococcus diacetylactis* use
Oxidized (cardboard) flavor	Copper contamination and/or exposure to fluorescent light or sunlight	– Avoid exposure to copper utensils – Protect product from direct sunlight/UV light exposure
Yeast-like/ cheesy	Contaminating yeast growth	– Sanitation check – Avoid return milk
Rancid flavor	Lipolytic activity	– Do not mix pasteurized and raw dairy ingredients prior to homogenization
Weak body	Heat treatment of the mix is insufficient Milk solids-not-fat too low Agitation too severe after fermentation	– Heat treatment should not be less than 85 °C/30 min for buttermilk and 74 °C/30 min for sour cream – Fortify with 0.5–1% nonfat dry milk for buttermilk and 2–3% for sour cream – Use appropriate stabilizers – thickeners – Use rennet in sour cream mix
Grainy texture	Acidity too high Nonfat dry milk or salt not dispersed properly	– Exercise rigid acidity control – Dispersion equipment check – Use in-line screen
Chalky/ powdery texture	Too much of nonfat dry milk	– Check the quality and quantity of dry milk

ingredient quality ensure good flavor. Product standards of fats, solids, viscosity, pH (or titratable acidity), and organoleptic characteristics should be strictly adhered to. Wheying off or appearance of a water layer on the surface is undesirable and can be controlled by judicious selection of effective stabilizers and by following proper processing conditions.

In hard-pack frozen yogurt, a coarse and icy texture may be caused by formation of ice crystals due to fluctuations in storage temperatures. Sandiness may be due to lactose crystals resulting from too high levels of milk solids. A soggy or gummy defect is caused by too high a milk-solids-not-fat level or too high sugar content. A weak body results from too high overrun and insufficient total solids.

Color defects may be caused by the lack of intensity or authenticity of hue and shade. Proper blending of fruit purees and yogurt mix is necessary for uniformity of color. The compositional control tests are fat, moisture, pH, and overrun, and microscopic examination of yogurt culture to ensure a desirable ratio of constituent culture organisms in frozen yogurt. It is evident that good microbiological quality of all ingredients is necessary for fine organoleptic and shelf-life quality of the product.

In general, quality control considerations applicable to refrigerated yogurt are relevant to cultured dairy products industry. Tab. 10 summarizes typical quality problems and steps needed to rectify the problem. Sour half and half manufacture requires standardization of mix to 11% milkfat and 12% solids-not-fat.

7 References

CHANDAN, R. C. (1982), Other fermented dairy products, in: *Prescott and Dunn's Industrial Microbiology* (REED, G., Ed.), 4th Ed., pp. 113–184, Westport, CT: AVI.

CHANDAN, R. C., SHAHANI, K. M. (1993), Yogurt, in: *Dairy Science and Technology Handbook*, Vol. 2 (HUI, Y. H., Ed.), pp. 1–56, New York: VCH Publishers, Inc.

DE VUYST, L., VANDAMME, E. J. (1994), *Bacteriocins of Lactic Acid Bacteria: Microbiology, Genetics and Applications*, New York: Blackie Academic & Professional.

FDA (US Food and Drug Administration) (1993), *Code of Federal Regulations*, Title 21, Sections 131.112, 131.138, 131.146, 131.160, 131.185, 131.200, 131.203, 131.206, pp. 243–276, Washington, DC: US Government Printing Office.

FERNANDEZ, C. F., CHANDAN, R. C., SHAHANI, K. M. (1992), Fermented dairy products and health, in: *The Lactic Acid Bacteria*, Vol. 1: *The Lactic Acid Bacteria in Health and Diseases* (WOOD, J. B., Ed.), pp. 297–339, New York: Elsevier Applied Science.

GERMANTOWN MANUFACTURING CO. (1991), *Pioneer Stabilizer/Emulsifier in Non-Fat Frozen Yogurt*, Product Bulletin G-813, Broomall, PA.

GOLDIN, B. R., GORBACH, S. L. (1992), Probiotics for humans, in: *Probiotics – The Scientific Basic* (FULLER, R., Ed.), pp. 355–376, New York: Chapman & Hall.

IDF (International Dairy Federation) (1988a), *Fermented Milks: Science and Technology*, Bulletin 277/1988, Brussels.

IDF (International Dairy Federation) (1988b), *Yogurt: Enumeration of Characteristic Organisms – Colony Count Technique at 37°C*, IDF Standard No. 117A:1988, Brussels.

IDF (International Dairy Federation) (1989), *Per Caput Consumption Statistics for Milk and Milk Products*, 1977–87, Bulletin IDF **227,** 1–32, Brussels.

IDF (International Dairy Federation) (1992), *New Technologies for Fermented Milks*, Bulletin **277**/1992, Brussels.

KOROLEVA, N. S. (1988), Starters for fermented milk, Section 4: Kefir and kumys starters, in: *Bulletin of the IDF* **227,** 35–40, Brussels.

KOSIKOWSKI, F. V. (1982), *Cheese and Fermented Milk Foods*, 2nd Ed., p. 474, Brooktondale, NY: F. V. Kosikowski & Assoc.

MAYRA-MAKINEN, A., BIGRET, M. (1993), Industrial use and production of lactic acid bacteria, in: *Lactic Acid Bacteria* (SALMINEN, S., VON WRIGHT, A., Eds.), pp. 65–95, New York: Marcel Dekker, Inc.

Milk Industry Foundation (1993), *Milk Facts*, Washington DC: The Milk Industry Foundation.

NAKAZAWA, Y., HOSONO, A. (1992), *Functions of Fermented Milk – Challenges for the Health Services*, pp. 1–78, New York: Elsevier Applied Science.

ROBINSON, R. K. (Ed.) (1992), *Therapeutic Properties of Fermented Milks*, New York: Elsevier Applied Science.

SALMINEN, S., VON WRIGHT, A. (1993), *Lactic Acid Bacteria*, pp. 1–107, New York: Marcel Dekker, Inc.

11 Brewing

INGEBORG RUSSELL
GRAHAM G. STEWART

London, Ontario
Canada

1 The Brewing Process – Introduction and Historical View

Biotechnology has been defined as creating products from raw materials using living organisms. By this definition the brewing of beer is one of the oldest biotechnology industries. The origin of brewing reaches far back into the prehistory of man. Artifacts taken from the mines of ancient cities such as Mesopotamia give evidence that the formal practice of brewing existed over 5000 years ago. Writings and drawings recovered from ancient Egypt give details of commercial brewing and distribution of beers. There is evidence that the Chinese produced a kind of beer called 'Kiu' over 4000 years ago. These beers were produced from barley, wheat, millet and rice. Monasteries and other establishments which brewed beer during medieval times produced it much as we know it today, using malted barley, water, hops and yeast.

Brewing was one of the earliest processes to be undertaken on a commercial scale, and of necessity it became one of the first processes to be developed from an art into a technology. JOULE, who established the calculations of electrical energy and the physics of heat, was a practicing brewer. LOUIS PASTEUR, LIEBIG and WÖHLER debated the nature of fermentation, and in 1876 PASTEUR published his paper "Etudes sur la Bière" describing how fermentation was carried out by yeast. Further research by PASTEUR identified that souring of beer and wine was caused by undesirable microorganisms. By gently heating the beer mash and then inoculating it with yeast, PASTEUR showed that a more desirable fermentation could be achieved. This heating technique to destroy microorganisms came to be known as 'pasteurization'. Danish scientist EMIL HANSEN showed that 'wild' yeast could contaminate brewer's yeast and destroy the flavor of the product and as a result, the technique of pure yeast culture was developed. SÖRENSEN's concept of the pH scale, developed in the Carlsberg laboratory in Denmark, has been adopted throughout science and technology, as has the Kjeldahl technique of measuring nitrogen. In 1897 the BUCHNER brothers discovered that sucrose could be fermented by yeast juice (cell-free extracts) and this finding led to a series of studies by a large number of scientists to elucidate the nature of the steps in the fermentation of sugars to ethanol and carbon dioxide. A thorough review of the history of brewing chemistry in the British Isles and highlights of the history of international brewing science have been published by ANDERSON (1992, 1993). These articles are enjoyable to read and highly recommended for those interested in brewing history.

Beer production is divided into four distinct processes:

(1) malting (germination of the barley or other cereal and drying of the germinated cereal)
(2) mashing (the extraction of the ground malted barley with water)
(3) wort boiling
(4) fermentation.

The production of beer is a relatively simple process. Yeast cells are added to a nutrient medium (the wort) and the cells take up the nutrients and utilize them so as to increase the yeast population. The cells excrete ethanol and carbon dioxide into the medium together with a host of minor metabolites, many of which contribute to beer flavor. The fermented medium, after the yeast is removed, is the 'green' beer. The beer is then aged, clarified, carbonated and packaged. Fig. 1 is a schematic diagram of the brewing process.

2 Raw Materials

The basic raw materials employed in brewing are barley, malt, hops, water and yeast. In addition, non-malt cereals such as corn and rice are often used as adjuncts with barley malt. Non-malted cereals are employed both for economic reasons and to produce a lighter product.

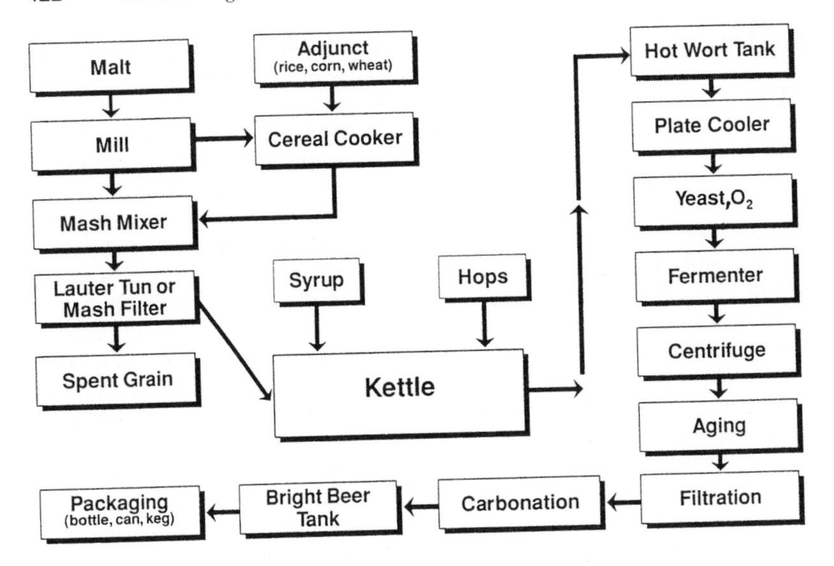

Fig. 1. The brewing process.

2.1 Brewing Water

About 92–95% of the weight of beer is water. Breweries have traditionally been located where water quality was consistent and often drawn from underground sources where its composition remained relatively uniform and where the geological strata protected it from pollution. Certain centers have become renowned for a certain type of beer, and these locations were often associated with water supplies of a special composition. Pilsen is known for its well water which is naturally soft. Burton-on-Trent became famous for its strong pale ale. It has deep wells that produce water low in bicarbonate, yet rich in sulfate. The Burton water is characterized by a high permanent hardness, whereas water with less calcium sulfate and more calcium carbonate was used to produce the sweeter darker beers of Munich and Dublin. The term 'Burtonizing' the water refers to the addition of calcium sulfate and sometimes other salts to the water so that its composition is similar to that of Burton well water.

If the water is not acceptable, it can be treated in a number of ways, such as the use of activated carbon for removal of phenolic compounds (chlorophenols) and trihalomethanes (THMs). Problems encountered with a bacterial build-up on the carbon surface must be treated by steaming at regular intervals. Salts can be added (e.g., calcium sulfate) or ions removed (e.g., bicarbonate). The latter is labor-intensive and the sludge produced carries expensive disposal costs. Ion exchange can be used to de-alkalize. The water is passed over special food grade resins, and calcium is replaced by sodium in the exchanger and bicarbonate is converted to carbon dioxide which is blown out. The acidic ion exchangers are regenerated with sulfuric acid. Reverse osmosis has been used by a number of breweries to desalt brackish water and can reduce the mineral content to about 5%. This technology also removes all suspended solids and most organic contaminants (ORMROD, 1986).

The basic requirements of good brewing water are as follows (HOUGH et al., 1971; BERNSTEIN and WILLOX, 1977): (1) meets the standards for potable water (chlorination is the most common method of sterilizing water supplies); (2) is clear, colorless, odorless and free of any objectionable taste and if surface water, free of organic matter; (3) alkalinity should be 50 mg/L or less and preferably less than 25 mg/L (BRIGGS et al., 1981); and (4) the brewing water should have a calcium concentration of 50 mg/L and since over 50% of the calcium is lost during the mashing process, a further 50 mg/L is usually added to the kettle. A final calcium content of 60–80 mg/L in the beer is desirable. Calcium is critical for enzyme activity in mashing and in wort has a number of functions including: (1) pH control; (2) yeast growth and flocculation; (3) oxalate removal (haze/gushing); (4) decrease in wort color; and (5) also limits extraction of certain colored and astringent substances and also silicous materials. The wort chloride level (as NaCl) can be adjusted in the kettle if desired. NaCl at 75–100 mg/L gives beer a mellow palate and fullness.

The cost of purchasing water has increased dramatically over the years while the cost of treating brewery effluent has increased even more. In the USA, water purchase costs have doubled and brewery effluent costs have increased eight-fold over the past 10 years. It is important for the 90s brewery to be very cognizant of water conservation (BLAND, 1993).

2.2 Hops

2.2.1 Introduction

The hops used for brewing are the dried blossoms (cones) of the female hop plant (*Humulus lupulus*). This plant is native to many North temperate areas (Northern Europe, West Central Asia, Japan and North America), and their successful growth is limited by the photoperiodic requirements of the plant. Today over 60% of the commercial crop is processed into powder, pellets or extracts. Hops have been used since ancient times, but their regular use as a beer additive is attributed to the German monks in the 12th century. It is believed that the use of hops became popular due to their bacteriostatic effect when present at high hopping ratios (ca. 400 g/hL). The description of the characteristic flavors of hops in beer is a subject that is widely debated, but what is clear is that hop flavor is an integral part of the total organoleptic impact of beer. Unhopped beer is reminiscent of sweet, malty acidic lemonade with an alcohol taste. The hop bitter compounds transform the taste and are critical to balancing it. Hops also contribute to aroma, flavor stability and foam retention.

2.2.2 Hop Chemistry – Historical

The first important chemical report on hops was published in 1888 when HAYDUCK described the separation of hop bitter compounds into a lead precipitable α-acid fraction and another fraction then named the β-acid fraction. WIELAND in 1925 and WOLLMER in 1916 described the general structure of the α- and β-acids, and in 1952 RIGBY showed that the α- and β-acids are mixtures of homologs and analogs and named the three major α-acids cohumulone, humulone and adhumulone and the β-acids colupulone, lupulone and adlupulone (VERZELE, 1986).

The next major historical step was the recognition that the α-acids are the major bittering compounds of hops and that they are chemically transformed during the kettle boil into the more soluble *iso*-α-acids, which account for most of the bitterness of beer (HOWARD, 1956; VERZELE, 1979). This understanding led to changes on how the industry approached hop breeding, growing, processing and usage in the brewery. The bitterness contribution of hops is relatively well understood, however, the flavor contribution is nowhere as clear, and currently the evaluation of hops to beer flavor and aroma, as opposed to bitterness, is still primarily a sensory evaluation. This presents problems when new varieties require evaluation for aroma and flavor.

2.2.3 Botany of Hops

There are only three recognized species of *Humulus*: *Humulus lupulus*, the species used for brewing; *H. japonicus*, an annual ornamental climbing plant from Japan which is devoid of resin and thus has no brewing value; and *H. yunnanensis*, grown in the Yunnan province of China of which there are very few herbarium specimens and no plants in cultivation. The genus *Humulus* is included in the family Cannabinaceae which includes the species *Cannabis sativa* which is more commonly known as Indian hemp, marijuana or hashish. Although there are similarities between the two plants, the resins of the species are distinct, with those of the hop plant providing the bitter principles of beer, while the *Cannabis* plant includes the psychotomimetic principles of the drug (HOUGH et al., 1982). *Cannabis* and *Humulus* spp. have been grafted onto each other, and the characteristic resins did not cross the grafts (CROMBIE and CROMBIE, 1975).

2.2.4 Harvesting of Hops

The hop is a hardy climbing herbaceous perennial plant. The root stock stays in the ground from year to year and the vines are trained onto strings which are supported on a wire trellis. In the USA, the hop fields are plowed in March, the vines trained onto the strings in May and harvest is from August to September. The first resins are usually detected in early August, and synthesis of α- and β-acids is almost complete by the end of the month. Varieties mature at typical but different rates, and some late varieties are usually grown to spread harvesting over three to four weeks in September. Hops should be picked within ten days of ripening. Overripe hop cones tend to open and are more fragile and easily shattered during harvesting.

The harvested hop cones are separated from leaf and stem waste. The freshly picked cones contain 80% (w/w) moisture, and they must quickly be dried in a kiln to a moisture level of 6–12% to maintain quality. The hops are cooled, packed into bales, and stored cold, to slow deterioration. The baled and kilned hops are not very stable even when stored cold, and cold storage entails significant costs. Dried and pressed hop cones are a bulky product which contains only 5–15% of active principles. Thus any concentration process will be beneficial in reducing handling, transportation and storage costs. Modified hop products are now used extensively, and such modification is designed to convert the hops to a packable, storable, more easily transportable material that requires less storage space and is more concentrated and stable with respect to the resin content. Ease of handling in brewing and improved utilization are also important advantages.

2.2.5 Hop Chemistry

Hop varieties have been divided into two classes, bittering hops and aroma hops. Traditionally hops were boiled with wort for 1–2 hours. The hop resins would go into solution and become isomerized to produce the so-called *iso*-α-acids, which are more soluble in wort and are much more bitter than other hop bitter acids. The majority of the essential oil constituents were lost during the wort boiling process. To increase the hop aroma of the beer, brewers have traditionally added a selection of aroma hops late in the boil or have added hops to the beer during conditioning in the tank or cask – a process called dry hopping. Beer typically contains around 25 mg/L of *iso*-α-acids (Tab. 1).

Tab. 1. Chemical Composition of Hops (Adapted from VERZELE, 1986; STEWART and RUSSELL, 1985)

Component	Percentage	Relative Importance
α-Acids	2–12	×××
β-Acids	1–10	××
Essential oils	0.5–1.5	××
Polyphenols	2– 5	××
Oil and fatty acids	Traces to 25%	×
Wax and steroids	Trace	×
Protein	15	
Cellulose	40–50	
Water	8–12	
Chlorophyll	Trace	
Pectins	2	
Ash (salts)	10	

The α- and β-acids, essential oils and polyphenols, have been shown to exert a significant effect on beer bitterness and flavor. Because the α-acid compounds yield the most bitter and abundant of the hop derived bitter compounds, the hop industry tries to maximize the α-acid content of the hops, and the brewing industry uses the α-acid content as a criterion for purchase. In addition to chemical analysis, hops are also evaluated on the basis of smell and feel. Chemically the desirable characteristics are a high content of α-acids and a low content of β-acids, a low level of the oxidation products of the α- and β-acids, oils (oxidation products can lead to off-flavors), seeds, and essential oils (a high oil content is indicative of an overripe hop).

2.2.6 Hop Products

The main reasons for using hop products as a replacement for kettle hops are: (1) more consistent bitterness between successive brews; (2) improved hop utilization; (3) improved stability on long-term storage; and (4) reduced transport, storage and handling costs. The improvement in bitterness consistency of successive brews is usually most marked when brewing on the small or pilot scale.

Significant losses of potential bittering substances normally occur when hops are boiled with wort in the kettle and also during fermentation, but only small losses occur during aging and subsequent packaging. As a result, the utilization of the bitter substances rarely exceeds 40% in commercial breweries and is often as low as 25%. In contrast, when isomerized extracts are added to beer after fermentation the utilization of the bitter substances is frequently in excess of 80%. Substantial savings can be made by employing isomerized extracts to bitter beer, and the North American, Australian and British brewing industries all use significant amounts of such extracts (NEVE, 1991).

The following definitions are taken from a review of hop products by CLARKE (1986):
(1) Hop Powder/Hop Pellets – A type of hop product consisting of dried, hammer-milled hop cones packaged as a powder or more generally, converted into pellet form prior to packaging.
(2) Enriched Hop Powder or Enriched Hop Pellets – A type of hop product consisting of dried, hammer-milled hop cones that have subsequently been concentrated by mechanical sieving at temperatures of $-20\,°C$ or less. The resultant powder contains 45% of the original hop weight and 90–95% of the original α-acids and can be packaged in this form or pelletized before packaging.
(3) Specialty Hop Powders and Hop Pellets – A type of product normally packaged and used in the pelleted form. Prior to pelleting, materials such as hop extract or inorganic salts are blended into the powder.
(4) Hop Extract – A type of product that is essentially a single solvent or mixed solvent extract of hops, the totality of the extract depending upon the solvent or solvents used for extraction. The extraction is designed to be effected with minimal or no change to the extracted components. The effect of hop deterioration on extraction efficiency and extract quality has been described by DAOUD and KUSINSKI (1992).
(5) Specialty Hop Extracts – A type of hop product, as in the case of Specialty Hop Pellets, where a hop extract has incorporated into it inorganic materials or increased amounts of hop oil (the latter by selective extraction).
(6) Hop Oil – A steam distilled oil of hops, either in concentrated form or as an emulsion in water.
(7) Isomerized Hop Extract – A sophisticated form of hop product in which a specific group of compounds, the α-acids, has been isolated from the hops, converted to the isomeric bitter form, with or without chemical reduction, and formulated into a concentrated liquid or solid product in which the active component consists of the salts of the isomers or reduced isomers of the α-acids.

The off-flavor compound in beer most easily recognized by the customer and not produced by yeast during fermentation is "lightstruck" flavor, better known as "skunky" aroma. This results when hop *iso*-acids react with sulfur compounds (amino acids or proteins) and flavins or polyphenols, in the presence of light, to yield 3-methyl-2-butenethiol, or "skunky thiol". The customer can identify this at levels as low as 8 parts per trillion (IRWIN et al., 1993). Cans and brown glass offer light protection to the beer. Imported beer in green or clear glass bottles is particularly susceptible to this chemical reaction. The new hop specialty products, tetrahydro *iso*-α-acids and hexahydro *iso*-α-acids, are not subject to the reactions which affect unreduced *iso*-α-acids (i.e., the *iso*-α-acids produced by the isomerization of natural α-acids during the boiling of hopped wort) in the presence of light and cause lightstruck off-flavor in beer

Tab. 2. Comparison of Hop-Derived Bitter Substances (Adapted from GARDNER, 1993)

Product	Relative Bitterness	Addition for Equivalent Bitterness	Relative Foam Enhancement at Equivalent Bitterness
Iso-α-acids	1.00	25	××
Tetrahydro-*iso*-α-acids[a]	2.03	12	×××
Hexahydro-*iso*-α-acids[a]	1.15	22	××××

[a] Light stability is achieved only in the complete absence of *iso*-α-acids.

(GUZINSKI and STEGINK, 1993). These products are intended to be added to beer at a late stage in the production process instead of the conventional hopping of the boiling wort, with a view to increasing the consistency of bitterness and hoppy flavor, prolonging the shelf life of bottled beers and enhancing foam stability (Tab. 2).

More than 60% of the hop crop is now processed, and the bulk of the crop is used for its bittering value (NEVE, 1986, 1991). Some of the crop is more highly valued for the fine quality of its aroma, and these hops demand a premium on the market. The chemical composition of hops will vary with variety, climate, geographical growing area, time of harvest, method of processing, storage conditions, etc.

2.3 Malt

2.3.1 Introduction

It is a truism that "good barley makes good malt".

The quality of the barley is affected by many factors, but of most importance are barley variety, growing conditions (weather and soil), farming practices and diseases. When barley is selected for malting, it is selected for: variety, physical appearance (e.g., broken kernels, mold, staining, weather damage), physical analysis (sizing, foreign seeds, germination and pearling), and lastly chemical analysis (protein and moisture). When malt is selected by the brewer, it is judged on barley variety, appearance, physical and chemical analysis.

The art of malting originated in prehistoric times, and it is reasonable to assume that by processes of trial and error, the early maltsters developed their craft to a modest level of efficiency. Over the last 25 years, there has been an accelerated understanding of the physiology and biochemistry of cereal seed germination, and much of this knowledge has been translated into malting practice.

2.3.2 Barley

Barley heads on the cereal may have either six rows or two rows. In six-row barley, there are three kernels at each node on alternate sides of the head, resulting in six rows of kernels. In two-row barley, only one kernel develops at each node on alternate sides of the head and results in two rows of barley. Efforts to improve barley by classical breeding and selection programs have resulted in new varieties with:

- much greater yields per acre
- shorter, stiffer straw that is resistant to lodging even after the application of nitrogenous fertilizers
- early ripening
- disease resistance
- generally greater uniformity.

Although other cereal grains can be employed for beer production, only barley has withstood the test of time and experience. The tough outer shell, or husk, of the barley kernel protects it during harvesting and during malting, the husk is left substantially intact and becomes a filter bed in the lauter tun. Prior to mashing, the malt is ground in such a way that the husk is left intact and the endosperm forms a coarse flour suitable for extraction.

Using the mutagen ethyl methanesulfonate (EMS), barley mutants have been produced that have very low levels of polyphenolic compounds called anthocyanogens. Anthocyanogens are largely responsible for the formation of chill-haze in beer. Beers brewed with anthocyanogen-free malt showed significantly improved chill-haze stability (VON WETTSTEIN et al., 1980). However, the barley mutants produced to date have poor agronomic characteristics (reduced yield per acre and decreased disease resistance).

2.3.3 The Malting Process

Only a very small portion of a barley kernel can be extracted and fermented by yeast, consequently the barley is malted to render the kernel more soluble and readily fermentable by yeast. The malting process involves: (1) the collection of stocks of suitable barley; (2) the storage of cereal until it is required; (3) steeping the grain in water; (4) germination of the grain; and finally (5) drying and curing in a kiln. The complete malting process usually takes 6–9 days.

Malting can be divided into three stages, i.e., steeping, germination and kilning. Malt cereal usually is allowed to germinate (grow) for a limited period of time (4–6 days) under carefully controlled conditions of temperature and humidity, and then growth of the embryo is terminated by drying at controlled temperatures (71–92 °C).

The goal in malting is to provide the correct environment to cause the kernels to render their food reserves such that they yield a fermentable extract upon mashing. These food reserves are found mainly in the endosperm which consists of numerous cells bounded by walls and containing starch granules set in a protein matrix.

For successful malting there are three basic requirements, namely grain viability, water and oxygen. Grain viability means that the embryo must be healthy and capable of strong and vigorous growth. If the embryo is dead then the kernel cannot germinate. Water is essential as dry seeds cannot germinate, as is oxygen, for most viable seeds will not germinate in the absence of oxygen.

During malting the following changes take place: (1) destruction of the endosperm; (2) breakdown of the protein matrix; (3) starch granules made available for subsequent attack by enzymes; and (4) formation of hydrolytic enzymes to carry out these changes.

The purpose of malting is to provide the correct environment for the grain to synthesize hydrolytic enzymes. The controlled action of these enzymes hydrolyzes the cell wall and reserve proteins of the endosperm. β-Amylase is present in the barley grain but *endo*-β-glucanases, α-amylase and peptidases are synthesized *de novo* from reserve material in the endosperm by the aleurone cells. There are three major categories of polymer in the barley endosperm which undergo modification (i.e., alteration in structure) during malting: cell wall polysaccharides (β-glucans and pentosans); reserve protein hordein; and starch granules. The germination is triggered by giberellins (plant growth hormones) which are released by the embryo as germination proceeds or these may be applied in sprays or in steep water to the barley that has been mechanically abraded by the maltster.

The moisture content of malt is reduced from ca. 45% to ca. 5% during kilning, and at these high temperatures the malt loses its gherkin-like taste and obtains the typical malt aroma. The degree of color and aroma can be controlled by adjusting the kilning conditions in regards to humidity, time and temperature (RUNKEL, 1975). Fig. 2 is a schematic diagram of the malting process.

Kilning the barley arrests biological activity. As the moisture content of the malt decreases, enzymatic activity slows down and eventually stops. Kilned malt at a moisture content of 4–5% is a stable product that is not subject to bacterial spoilage. It is crisp, readily cracks open during milling, and through chemical reactions the kilning process develops color, aroma and the flavor in the malt.

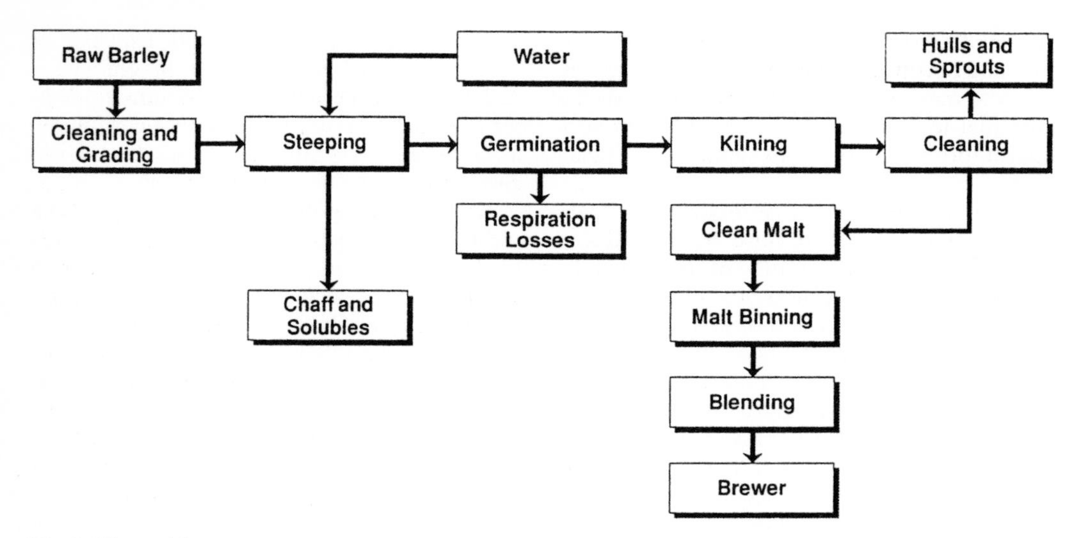

Fig. 2. The malting process.

2.3.4 Gushing

Gushing frequently occurs after malting barley has been harvested during periods of rainy weather, and it is probable that this barley has been contaminated by a large number of fungal species. Gushing is the violent uncontrolled ejection of beer when the package is first opened and involves the loss of a significant portion of the contents. Gushing is completely unacceptable to the consumer, and if the problem is not corrected this condition could prove catastrophic to the brewing company concerned. Although fungal contamination of the malting barley is not the only cause of gushing, it is certainly one of the primary factors (HAIKARA, 1980).

2.4 Adjuncts

2.4.1 Introduction

Many brewers today employ some form of brewing adjunct to supplement malt. Adjuncts are defined as non-malted carbohydrate materials of suitable composition and properties that beneficially complement or supplement barley malt. The Germany Purity Law defines an adjunct (or secondary brewing agent) as "... anything that is not malt, yeast, hops and water" (NARZISS, 1984). The United Kingdom's Foods Standards Committee defines a brewing adjunct as "... any carbohydrate source other than malted barley which contributes sugars to the wort" (COLLIER, 1986). Adjuncts are usually considered to be non-malt sources of fermentable sugars. These adjuncts vary considerably in their carbohydrate, nitrogenous, lipid and mineral composition and are used firstly for the influence they have on regulating the composition of the resulting wort and secondly as a cost-saving initiative due to the fact that most non-malted carbohydrate sources are less expensive than barley malt.

Although adjuncts are employed primarily because they provide extract at a lower cost than that available from malt, other advantages can be achieved. Beers produced with adjuncts are lighter in color with a less satiating thinner taste, greater brilliancy, enhanced physical stability and superior chill-proof qualities. In addition, the use of certain adjuncts makes it possible to increase production in cases of limited brewhouse capacity. This is especially true with the use of syrups and sugar adjuncts that can be added directly to the kettle and thus by-pass the mashing operation.

A wide range of materials can be used as unmalted brewer's adjuncts: yellow corn grits, refined corn starch, rice, sorghum, barley, wheat, wheat starch, cane and beet sugar (sucrose), rye, oats, potatoes, tapioca (cassava) and triticale. In addition, processed adjuncts include corn, wheat and barley syrups, torrified cereals, cereal flakes and micronized cereals (CANALES and SIERRA, 1976).

Adjuncts can contribute specific characteristics to the finished beer. For example, corn tends to give a fuller flavor to beer than wheat. Wheat imparts a certain dryness, while barley gives a stronger harsh flavor. Rice will also give a very characteristic flavor to beer. Both wheat and barley adjuncts can significantly improve head retention.

In the United States, non-malt adjuncts are estimated to total 38% of total brewing materials employed, excluding hops (MARCHBANKS, 1987). The most commonly used adjunct materials are corn (maize) (46% of total adjunct), rice (31%), barley (1%), and sugars and syrups (22%).

2.4.2 Corn Grits

Corn grits, produced by dry-milling yellow corn, are the most widely used adjunct in the United States and Canada. The milling process removes the hull and outer layers of the endosperm along with the oil-rich germ, leaving behind almost pure endosperm fragments. These fragments are further milled and classified according to brewer's specifications. Corn grits produce a slightly lower extract than other unprocessed adjuncts and contain higher levels of protein and fat. The gelatinization temperature range for corn grits (62–74 °C) is slightly lower than that of rice grits (64–78 °C).

MEILGAARD (1976) has reviewed the composition of worts and beers prepared with a variety of adjuncts. He has shown that a carbohydrate profile similar to an all malt wort can be attained with either 20% rice or 20% corn grits, although levels of sucrose and fructose decline as the adjunct level increases. Corn grits at the 30% level produce volatile aroma compounds similar to that of an all malt beer.

2.4.3 Rice

Rice is currently the second most widely used adjunct material in the United States (COORS, 1976) and on an extract basis, it is approximately 25% more expensive than corn grits. Brewer's rice is a by-product of the edible rice milling industry. Hulls are removed from brown paddy rice, and then it is dry milled to remove the bran, aleurone layers and germ. The objective of rice milling is to completely remove these fractions with minimal damage to the starchy endosperm, yielding whole kernels for domestic consumption. Up to 30% of the kernels can be fractured in the milling process, and these broken pieces are considered esthetically undesirable for domestic use and are sold to brewers at a lower price than the whole kernel rice price. Rice is preferred by some brewers because of its lower oil content compared to corn grits and its neutral aroma and flavor. When converted properly in the brewhouse, rice yields a light, clean tasting beer (VINH et al., 1993).

2.4.4 Barley

Unmalted barley can be used as an adjunct for brewing. However, the raw grain is abrasive and difficult to mill, and yields a high percentage of fine material which is problematic during lautering. These difficulties can be avoided if the grain is conditioned to 18–20% moisture prior to milling, but this process has not been widely employed in brewing.

In the past, barley has normally been partially gelatinized before use either by mild pressure-cooking, or by steaming at atmospheric pressure, followed by passage of the hot grits through rollers held at approx. 85 °C. Finally, the moisture content of the flakes is reduced to 8–10%. This process of pre-gelatinization can also be applied to corn. Pre-gelatinization of barley affects the ease of extraction of β-glucan during mashing and, hence, the β-glucan content of the wort. Prolonged steaming prior to rolling the barley results in a product which produces higher viscosity sweet worts. Barley starch is more readily hydrolyzed than corn or rice starch. Barley may be dehusked before use, to increase extract yields, but this may lead to run-off difficulties because the husk provides material for filter bed formation. Fine grinding improves extraction efficiency but also leads to slow run-off.

In barley brewing it is possible to approximate the starch hydrolysis profile and the degree of fermentability of 100% malt worts. This is possible by substituting malt with barley at levels of 50% (extract basis) and by controlling the main mash schedule (enzyme concentration, time and temperature). Most breweries can employ their normal fermentation and aging technology with barley brewing. In general, no significant difference in organoleptic properties between barley beers and 100% malt beers have been observed. A harshness of barley beers can be avoided by lowering the pH of the wort to 4.9 prior to boiling.

2.4.5 Sorghum (Millet)

Sorghum is the fifth most widely grown cereal crop in the world; only wheat, corn, rice and barley are produced in greater quantities (TAYLOR, 1992). Africa is a major source of sorghum as is Central America. But it is only in the past two decades that real interest has been shown in this cereal as a brewing adjunct or as a malted product. As well as its use as an adjunct, sorghum can be employed as a malt.

Sorghum's advantage in agronomic terms is its ability to survive under extreme water stress conditions and, hence, the cereal is ideally suited for cultivation in tropical countries such as Africa and Central America.

Sorghum is the traditional raw material (TAYLOR, 1992) in Africa for the production of local top-fermenting beers which are known by various names (e.g., "Bantu beer" in South Africa, "dolo" in Burkina Fase and "billi billi" in Chad). These beers, mainly found in the rural regions, are produced without hops, are slightly sour in taste and are drunk unfiltered. Sorghum offers the lowest cost source of available fermentable sugar (PALMER, 1989).

The chemical composition of sorghum grain is very similar to corn. Both grains contain starch consisting of 75% amylopectin and 25% amylose. Starch granules are similar in range, shape and size. Sorghum starch has a higher gelatinization temperature (68–76 °C) than corn starch (62–68 °C). In the brewhouse, sorghum brewer's grits perform within acceptable limits. No special handling or cooking techniques are required. Five percent malt in the cooking mash is sufficient. Conversion of starch occurs within the mashing time allowed. The beers produced are fully equivalent in chemical analysis, flavor and stability, to beers produced with other adjuncts.

2.4.6 Refined Corn Starch

Refined corn starch, a product of the wet milling industry, is by far the purest starch available to the brewer (COORS, 1976). It has not found widespread use, because the price is higher relative to corn grits and brewer's rice. An obvious drawback for refined starch usage is handling. The powder is extremely fine, and must be contained in well grounded lines and tanks to prevent explosions resulting from static electricity sparks produced during conveying. The starch bridges easily and is nearly impossible to flow from tanks unless they have special fluidizing bottoms.

Refined corn starch can be utilized as the total adjunct or can be mixed with rice or corn grits at the option of the brewer. Brewhouse yield can be increased 1–2% by the use of refined starch in place of rice. There are no run-off problems, fermentations often attenuate better, and colloidal stability is unaffected. Beer flavor is not affected, except the beer is considered slightly thinner, because of higher attenuation limits.

2.4.7 Wheat Starch

Refined wheat starch is currently not economically attractive in the United States because of its high price compared with other more readily available adjuncts. It has been employed, in the past, in Canada where it allowed the usage of surplus wheat grown in Canada. Chemically, wheat starch is very similar to refined corn starch, but is somewhat higher in β-glucans. The beer is quite comparable to beer brewed with corn grits in analysis and flavor.

2.4.8 Torrified Cereals

Torrification is a process by which cereal grains, usually barley or wheat, are subjected to heat at 260 °C and rapidly expanded or popped. This process renders the starch pregelatinized and thereby eliminates the cooking step in the brewhouse. It also denatures a major portion of the protein in the kernel such that the wort soluble protein is only 10% of the total. The flavor of beer produced with torrified adjuncts is reported to be unchanged, and if torrified cereals became economically competitive with other adjuncts they could be employed as an alternate adjunct source.

2.4.9 Liquid Adjuncts

The major liquid adjuncts used in brewing are glucose syrups, cane sugar syrups and invert sugar syrups. Glucose syrups used in brewing are in fact solutions of a large range of sugars and contain, in varying proportions depending upon the method of manufacture, dextrose, maltose, maltotriose, maltotetraose and larger dextrins.

Cane sugar syrups contain sucrose derived from sugar cane and sometimes, depending upon the grade, small quantities of invert sugar. Invert syrups, as the name suggests, are solutions of invert sugar – a mixture of glucose and fructose. Invert sugar is produced, in nature and commercially, by the hydrolysis of sucrose which, together with glucose and fructose, occurs abundantly in nature. Commercially, sucrose is extracted from sugar cane or beet, and glucose syrups are usually manufactured from starch derived from corn or wheat.

Glucose syrups have been available since the mid-1950s, and they were originally produced by an acid conversion of starch. In the mid-60s, new developments in enzyme technology, in addition to the poor quality of straight acid converted syrups, led to the process of acid conversion followed by enzyme conversion followed by activated carbon filtration.

As the use of liquid adjuncts continued in the 1980s, it became apparent that they had shortfalls, and the high level of glucose became a concern (CHANTLER, 1990). Conversion of starch with the aid of an acid produces predominately glucose as the hydrolysis product. When brewer's yeast is exposed to high concentrations of glucose, a phenomenon referred to as the "glucose effect" may be experienced which can result in sluggish and 'hung' fermentations.

Although acid/enzyme carbon refined syrups possess disadvantages, they were acceptable and popular until the late 1980s. Many changes have recently occurred in the brewing and wet milling industries that have led to the development of a new liquid adjunct. The two most important factors have been the increase in adjunct levels in North American beer (up to 50% of the wort content) and high gravity brewing, the latter being employed to ease capacity constraints and improve brewhouse efficiencies. The need for capital investment has been alleviated by increasing output as much as 50% through the practice of fermenting wort at 16 °P (Plato scale is similar to the Brix scale) or higher.

With the advent of new technology in enzyme liquefaction and downline multistage enzyme hydrolysis, production of corn syrups of virtually any carbohydrate profile is now possible. This new generation of high maltose corn syrups (enzyme/enzyme) now permits the brewer to introduce liquid adjunct at any level without changing the carbohydrate profile of the wort. Brewing with these syrups is now routine, and no difficulties in either the brewhouse or fermentation cellar have been reported. Sensory evaluations of beer produced from third generation syrups have revealed no significant differences from beers produced with acid/enzyme syrups. The sodium levels decreased by as much as 60% in comparison to beers produced from earlier generation syrups.

2.4.10 Malt from Cereals Other than Barley

Although the principal cereal employed as the raw material for malt is barley, a number of other cereals can be used and these include: wheat, oats, rye and sorghum.

2.4.10.1 Wheat Malt

Wheat malt is used in the production of some specialty beers, such as Berlin Weissbier, where it may constitute up to 75% of the grist. Wheat malt gives beer an outstanding head retention. The limited use of wheat malt in ordinary beers is due mainly to the difficulty experienced in malting the naked grain without damaging the exposed acrospire. As a result, much of the wheat malt made has been under-modified. The absence of husk however tends to result in a high extract.

2.4.10.2 Oat and Rye Malt

Malted oats are used to a limited extent, usually blended with barley malt, to produce stout beers. Today, malted rye is seldom used, although 50 years ago it was often used in specialty beers. Unmalted rye is sometimes used for vinegar brewing and also for certain distilled beverages such as Canadian Rye Whiskey.

2.4.10.3 Sorghum

The development of sorghum for use as malt in conventional beer (ales, lagers and stouts) production has been accelerated by the large foreign debt crisis in developing tropical countries which has made it increasingly difficult for them to import either barley or barley malt for their local breweries. For example, the Government of Nigeria has prohibited the importation of barley malt since 1988. Consequently, considerable research into local raw materials has occurred (especially sorghum) for use not only as an adjunct, but as a complete replacement for barley malt. As a result of these research efforts, as much as 30% of the sorghum harvest in Africa is being used for malting and brewing.

2.4.11 Conclusions

The use of unmalted carbohydrates or adjuncts in brewing is widespread with the exception of those countries that adhere to the German Purity Law. In most countries there are one or two dominant adjuncts and these are usually the most suitable and inexpensive carbon source. In the USA and Canada, corn grits (and corn syrup) and rice grits are predominantly used; in France, Belgium and Italy corn grits are extensively used. In the cane sugar producing countries such as Brazil, Australia and the West Indies, cane sugar syrups are widely used together with corn grits. Corn remains the most popular adjunct in Africa, however, sorghum use both as an adjunct and as malt is increasing. In the United Kingdom a wide range of cereals and sugars processed by a number of methods are used.

Although developments in the use of brewing adjuncts have been relatively stable for a number of years, the advent of "new generation" syrups produced principally, but not exclusively from corn, has currently a great impact on some parts of the brewing industry. Biotechnological advances such as wet milling, immobilized thermotolerant enzyme systems and ion exchange downstream processing techniques permit the production of syrups with virtually any carbohydrate profile. Currently, syrups are available that allow the brewer to introduce them at any level without changing the carbohydrate profile of the wort. The future will see the commercial ability to separate and isolate individual sugars according to their molecular weight and, subsequently, produce a blended syrup of any specific sugar profile.

3 Wort Production

3.1 Mashing

The process of mashing involves: (1) dissolving the substances in malt that are immediately soluble in warm water containing specified salts and a specific pH; and (2) rendering the substances which are insoluble in their natural state soluble through enzymatic action. The most important reaction in mashing is the conversion of starch to smaller dextrins and maltose by the action of barley α-amylase and β-amylase, which produces wort that is approximately 70–80% fermentable carbohydrates. The term 'fermentable carbohydrate' will henceforth refer to low molecular carbohydrates of wort including glucose, maltose, fructose, sucrose and maltotriose. There are many methods of mashing practiced in the world today, but the most popular procedures in current use can be classed as either infusion systems, decoction systems or a combination thereof (DOUGHERTY, 1977; MCFARLANE, 1993; NARZISS, 1994). The most widely used system in North America is the double mash upward infusion system. This system utilizes a cereal cooker in which adjuncts (non-malted carbohydrate material) are prepared by boiling and a mash mixer (mash tun) in which the malt mash is prepared and in which the two mashes are ultimately combined. This system permits the use of corn grits and rice as adjuncts, and the beer thus produced is lighter in taste.

The final wort will vary depending on: (1) the enzyme complement contributed to the mash by the malt; (2) the mashing temperature; and (3) the duration of the mashing. The finished wort contains simple sugars, dextrins, more complex polysaccharides, amino acids, peptides, proteins, other nitrogenous materials, vitamins, organic and inorganic phosphates, mineral salts, polyphenols, tannin precursors and tannins, small quantities of lipids and numerous other components, many of which have not been identified.

3.2 Wort Filtration

There are two types of equipment used to separate the wort from the mash solids, the lauter tun (or tub) and the mash filter. Which one is used is a matter of choice and often tradition, although it is not unusual to see the different methods employed in plants of the same brewing company.

The separation of the wort from the mash solids is usually by lauter tun (or tub) or by mash filter. Lautering is the name given to the process of separating the liquid part of the mash (the wort) from the undissolved part (the spent grains). Lautering in German means to clarify, purify or filter. The lauter tun is a vertical cylinder with a large diameter to depth ratio and fitted into the bottom of the tun is a wort collecting system. A mash filter consists of a series of alternating plates and roller frames in which a filter of polyethylene or polypropylene fiber is suspended. The merits of mash filter versus lauter tun is a controversial subject not yet resolved.

3.2.1 Mash Filter 2001

Novel technology in mash filtration was introduced in 1988 with the introduction of the mash filter 2001. This fully automatic filtration process was divided into six phases: filling of the filter, filtration of the mash, precompression, sparging, compression and cake discharge (EYBEN et al., 1989).

The objectives with the new filter were the following:

- produce a clear wort (low fatty acid content) with high yield
- use a small volume of sparging water
- produce dry spent grains
- obtain high efficiency (12 brews in 24 hours)
- ensure that there is low oxygen uptake
- the wort composition and beer quality must be at least equivalent to that obtained from a traditional lauter tun
- flexibility in load and composition must be good and lifetime of cloths and membranes must be acceptable.

To reach these objectives, the following principles were employed: thin layer filtration, mashing with fine flour (hammermill, filtration under constant pressure), uniform repartition of sparging liquor and pneumatic compression of spent grains.

The 2001 filter not only met the original objectives, but also provided:

- simplified milling technology
- no heat radiation
- no waste water
- a shorter mashing time.

Analytically, the beers showed no significant difference from traditional beers except for a higher ester content. The wort composition showed a lower fatty acid content compared to the classical mash filter thus explaining the higher ester content in the finished beer (MELIS and EYBEN, 1992).

3.3 Wort Boiling

The objectives of wort boiling are summarized as follows (MIEDANER, 1986):

- inactivation of enzymes to fix wort composition
- sterilization of wort, particularly if liquid adjuncts are added directly to the wort kettle
- coagulation of proteins (hot break formation)
- hop extraction (isomerization)
- evaporation of water
- formation of flavor compounds (Maillard reaction)
- evaporation of undesired volatiles.

Evaporation is required to remove unwanted volatile substances, and agitation ensures the best conditions for coagulation of excess protein. Traditional wort boiling systems involve open top vessels operating at atmospheric pressure which exploit thermosyphon circulation to keep the wort well agitated. Circulation of the hot wort in the kettle can be via heating jackets or internal heating elements. Both types are designed to produce a 'rolling boil' to ensure agitation of the wort. Generally, a 1.5–2 hour boil time is used with an evaporation rate of 4–8% per hour (ORMROD et al., 1991).

After mashing is complete, the filtered sweet wort is transferred to the kettle (or copper) and boiled for up to 2 hours. This is a very energy-intensive operation. The commonly quoted reasons for wort boiling are: (1) to extract bitter and aroma substances from hops and promote necessary chemical changes (i.e., isomerize hop α-acids to form soluble bitter *iso*-α-acids); (2) to precipitate unwanted nitrogenous material; (3) to terminate enzymatic action; (4) to remove undesirable volatile compounds; (5) to sterilize the wort; (6) to complete ionic interactions conducive to a drop in pH; (7) to denature and precipitate proteins and tannins ('hot break'); (8) to dissolve any sugar adjuncts added to the wort; and (9) to concentrate the wort by evaporating excess water. It is at the wort boil stage that various additions may be made to the wort such as sugars or syrups, materials which encourage precipitation of proteins and tannins, and hops. The objectives achieved by wort boiling are:

- stabilization
- flavor development (SEATON et al., 1981) and hop isomerization
- concentration
- sterilization.

Several ways of reducing the energy consumption of the brewhouse have been proposed in the past ten years. They include the reduction of the total evaporation rate, improved utilization of the energy of the kettle vapor, and reduction of boiling time by means of higher temperatures and pressures.

3.4 Wort Cooling

After boiling, the spent hops, precipitated proteins and other insoluble material, referred to as 'trub', are separated from the wort which is cooled, usually in a plate heat exchanger. During cooling, proteins and tannins are precipitated as a fine coagulum referred to as the 'cold break'. The purpose of the 'hot and cold breaks' is to reduce the material present that could later precipitate in the finished beer as a haze. During wort cooling, aeration/oxygenation of the wort is carried out to permit the yeast to ferment efficiently.

4 Yeast

4.1 Taxonomy of Yeast

It is at the strain level that interests in brewing yeast centers, and there are at least 1000 separate strains of *Saccharomyces cerevisiae.* These strains include brewing, baking, wine, distilling and laboratory cultures. There is a problem classifying such strains in the brewing context; the minor differences between strains that the taxonomist dismisses, are of great technical importance to the brewer.

Taxonomically, the two species *Saccharomyces uvarum (carlsbergensis)* and *Saccharomyces cerevisiae* have been distinguished on the basis of their ability to ferment the disaccharide melibiose. Strains of *S. uvarum (carlsbergensis)* possess the *MEL* gene(s). They produce the extracellular enzyme α-galactosidase (melibiase) and are able to utilize melibiose, whereas strains of *S. cerevisiae* do not produce α-galactosidase and, therefore, are unable to utilize melibiose. Also, ale strains can grow at 37 °C, lager strains cannot (max. 34 °C). Recently, yeast taxonomists have consolidated *S. uvarum (carlsbergensis)* and *S. cerevisiae* into one species, *S. cerevisiae* (BARNETT et al., 1983; KREGER-VAN RIJ, 1984). The number of species within the genus *Saccharomyces* was reduced from 41, as described by LODDER (1970), to ten (BARNETT et al., 1990). Some workers have not agreed with the current amalgamation of so many species, and there is still much dissent over the terminology.

The two main types of beer, lager and ale, are fermented with strains of *Saccharomyces uvarum (carlsbergensis)* and *S. cerevisiae*, respectively. Traditionally, lager is produced by bottom-fermenting yeasts at fermentation temperatures between 7 and 15 °C, and at the end of fermentation, these yeasts flocculate and collect at the bottom of the fermenter. Top-fermenting yeasts, used for the production of ale at fermentation temperatures between 18 and 22 °C, tend to be somewhat less flocculent, and loose clumps of cells, adsorbed to carbon dioxide bubbles, are carried to the surface of the wort. Consequently, top yeasts are collected for reuse from the surface of the fermenting wort (a process called skimming), whereas bottom yeasts are collected (or cropped) from the fermenter bottom. The differentiation of lagers and ales on the basis of bottom and top cropping has become less distinct with the advent of vertical conical bottom fermenters and centrifuges.

4.2 Classical Techniques of Strain Improvement

Classical approaches to strain improvement include mutation and selection, screening and selection, and cross-breeding. Mutation is any change that alters the sequence of bases along the DNA molecule, thus modifying the genetic material. Chemical and physical treatments such as ultraviolet light, ethyl methane sulfonate (EMS) and *N*-methyl-*N*-nitrosoguanidine (NTG) are used to induce mutation frequencies to detectable levels. Problems are often encountered with the use of mutagens for brewing strains, since mutagenesis is a destructive process and can cause a gross re-arrangement of the genome. The mutagenized strains often no longer exhibit many of the desirable properties of the parent strain and in addition may exhibit a slow growth rate and produce a number of undesirable taste and aroma compounds during fermentation. Mutagenesis is seldom employed with industrial strains due to their polyploid/aneuploid nature, since after mutagenic treatment, the mutations do not reveal themselves due to the presence of non-mutated alleles. CASEY (1990) cautions on the use of mutagens as these hidden undesirable mutations can become expressed after a time lag by such events as chromosome loss, mitotic recombination, or mutation of other wild-type alleles. He suggests an analogy to computer viruses in computer software that can surface months later. In a similar way, hidden mutations can later spoil the efforts of a protracted strain improvement program.

Screening of cultures to obtain spontaneous mutants or variants has proved to be a more successful technique as this avoids the use of destructive mutagens. To select for brewery yeast with improved maltose utilization rates, 2-deoxy-glucose, a glucose analog, was employed, and spontaneous mutants selected which were resistant to 2-deoxyglucose (STEWART et al., 1985). These isolates were also found to be derepressed for glucose repression of maltose uptake, thus allowing the cells to take up maltose without first requiring a 50–60% drop in wort glucose levels. This resulted in faster fermentation rates and no alternation in the final flavor of the product (STEWART et al., 1985; JONES et al., 1986; NOVAK et al., 1991).

4.2.1 Hybridization

The study of yeast genetics was pioneered by WINGE and his co-workers at the Carlsberg laboratory in Denmark, and in 1935 they established the haploid-diploid life cycle (WINGE, 1935). *Saccharomyces* species can alternate between the haploid (a single set of chromosomes) and diploid (two sets of chromosomes) state. Yeast can display two mating types, designated *a* and α, which are manifested by the extracellular production of an *a* or an α mating pheromone. When *a* haploids are mixed with α haploids, mating takes place and diploid zygotes are formed. Under conditions of nutritional deprivation, diploids undergo reduction division by meiosis and differentiate into tetranucleate asci, containing four uninucleate haploid ascospores, two of which are *a* mating type and two of which are α mating type. Ascus walls can be removed by the use of glucanase preparations such as snail gut enzyme. The four spores from each ascus can be isolated by use of a micromanipulator, induced to germinate, tested for their fermentation ability, and subsequently employed for further hybridization work. Both haploid and diploid organisms can exist stably and undergo cell division via mitosis and budding.

Although the technique of hybridization fell into disfavor for a number of years, when recombinant DNA was thought to be the solution to all future gene manipulation requirements, it has gradually again come to be accepted as a very valuable technique. There are three prerequisites to the production of a hybrid yeast by this technique. First, it must be possible to induce sporulation in the parent strains; secondly, single spore isolates must be viable; and, thirdly, hybridization (mating) must take place when the spores of single-spore cultures are placed in contact with one another. The great stability of industrial strains has been attributed to the following characteristics of industrial strains:

- little or no mating ability
- poor sporulation
- low spore viability.

Research in recent years has shown that it is possible to increase sporulation ability in many of these industrial strains thus making them much more amenable to hybridization (BILINSKI et al., 1986).

Employing the classical techniques of spore dissection and cell mating, with the aid of a micromanipulator, it has been possible to produce diploid strains with multiple genes for carbohydrate utilization. These strains can then be employed for specific purposes or further improved by fusing them to industrial polyploids with additional desirable characteristics (JONES et al., 1987).

GJERMANSEN and SIGSGAARD (1981) carried out extensive cross breeding studies with a *Saccharomyces uvarum (carlsbergensis)* strain and produced hybrids that were tested at the 575 hL scale and which produced beer of acceptable quality. SIGSGAARD and HJORTSHOJ (1992) reported the construction of a yeast using traditional genetic techniques such as sporulation, mutation, mating and selection that produced a beer with 10% of the normal diacetyl at the end of primary fermentation. BILINSKI et al. (1987) reported similar results with crosses between ale and lager meiotic segregants. These hybrids exhibited faster attenuation rates and produced beers of good palate which lacked the sulfury character of the lager but retained the estery aroma of the ale. The aforementioned examples demonstrate that although hybridization is an old and 'classical' technique, it is by no means outdated. One of the major advantages to cross-breeding is that this technique carries none of the burden of ethical questions and fears that can sometimes accompany the use of recombinant DNA technology.

4.2.2 Rare Mating

Rare mating, also called forced mating, is a technique that disregards ploidy and mating type and thus is ideal for the manipulation of polyploid/aneuploid strains where normal hybridization procedures cannot be utilized. When non-mating strains are mixed at a high cell density, a few hybrids with fused nuclei form and these can usually be isolated using appropriate selection markers. A possible disadvantage to this method is that while incorporating the nuclear genes from the brewing strain, the rare mating product can also inherit undesirable properties from the other partner, which is often a non-brewing strain. A good example of this is the work of TUBB and colleagues (TUBB et al., 1981; GOODEY and TUBB, 1982). They constructed dextrin-fermenting brewing strains, but introduced the *POF* gene (*P*henolic-*O*ff-*F*lavor) which imparts the ability to decarboxylate wort ferulic acid to 4-vinyl guaiacol, giving beer a phenolic or clove-like off-flavor. This made the hybrid product unsuitable for commercial use from a taste perspective but acceptable from a dextrin utilization standpoint.

4.2.3 Cytoduction

Cytoduction is a specialized form of rare mating in which only the cytoplasmic components of the donor strain are transferred into the brewing strain. The process of cytoduction requires the presence of a specific nuclear gene mutation designated *Kar,* for karyogamy defective. This mutation impairs nuclear fusion (CONDE and FINK, 1976). Cytoduction can be used in three ways: substitution of the mitochondrial genome; introduction of DNA plasmids; or transfer of double-stranded RNA species. When used in the substitution of the mitochondrial genome, it is possible to study the effects of these genetic elements on various cell functions. Mitochondrial substitution has been demonstrated to bring about variations in respiratory functions, cell surface activities (WILKIE and NUDD, 1981) and various other strain characteristics (CONDE ZURITA and MASCORT SUÁREZ, 1981; RUSSELL and STEWART, 1985). In addition rare mating, used to introduce DNA plasmids, has been successful in the introduction of specific genetic elements constructed by gene-cloning experiments (KIELLAND-BRANDT et al., 1979). Lastly, rare mating has also been used to transfer 'zymocidal' or 'killer' factor from laboratory haploid strains to brewing yeast strains without altering the primary fermentation characteristics of the brewer's yeast strain (YOUNG, 1983; HAMMOND and ECKERSLEY, 1984; RUSSELL and STEWART, 1985).

4.2.3.1 Killer Yeast

The 'killer' character of *Saccharomyces* spp. is determined by the presence of two species of cytoplasmically located dsRNA plasmids. The M-dsRNA 'killer' plasmid is 'killer' strain specific and codes for 'killer' toxin (an extracellular protein) and also for a protein or proteins that make the host immune to the toxin. The L-dsRNA, which is also present in many 'non-killer' yeast strains, codes for the production of a protein that encapsulates both forms of dsRNA, thereby yielding virus-like particles. These virus-like particles are not naturally transmitted from cell to cell by any infection process. The 'killer' plasmid behaves as a true cytoplasmic element, showing dominant non-Mendelian segregation. It depends, however, on a number of chromosomal genes for its maintenance in the cell and for expression (for review see MITCHEL and BEAVAN, 1987; YOUNG, 1987a).

Brewing strains can be modified such that they are both resistant to killing by a zymocidal yeast and so that they themselves have zymocidal activity, thereby eliminating contaminating yeasts (HAMMOND and ECKERSLEY, 1984; YOUNG, 1987b). Rare mating has been successfully employed to produce brewing 'killer' yeast strains by crossing a brewing lager yeast with a *Kar* 'killer' strain (RUSSELL and STEWART, 1985). Wort fermentations (40 L) were conducted with this strain, and the beer was bottled and evaluated by a trained taste panel. The beer brewed with the 'killer' strain was acceptable but contained an ester note that was not present in the control.

This suggested that the cytoplasm of the 'killer' strain appeared to exert more influence on the brewing strain than originally predicted. HAMMOND and ECKERSLEY (1984) recommended the use of a two-step procedure that retains more brewing strain mitochondria to produce 'killer' brewing strains. A question often asked is whether the toxin is still active in the finished beer. The toxin is extremely heat-sensitive, and a brewery pasteurization cycle of eight pasteurization units was shown to completely inactivate it.

The introduction of a killer strain into a brewery where several yeasts are employed for the production of different beers can present logistical problems. An error on an operator's part in keeping lines and yeast tanks separate could have serious consequences, since accidental mixing would prove fatal for the normal brewer's yeast. In a brewery with only one yeast strain, this would not be a cause for concern. It is worthy to note that a number of commercially available wine yeasts contain the killer characteristic, the purpose being to eliminate some of the yeasts that occur in the must that originates from the natural flora of the grapes.

4.2.4 Mutation

Yeast mutations are a common occurrence throughout the growth and fermentation cycle, but they are usually recessive mutations, due to loss of function of a single gene. Since industrial strains are usually at least diploid, the dominant gene will function adequately in the strain and it will be phenotypically normal. Only if the mutation takes place in both alleles will the recessive character be expressed. If the mutation weakens the yeast, the mutated strain will not be able to compete and soon be outgrown by the non-mutated yeast population.

Only three characteristics are routinely encountered resulting from yeast mutation that can be harmful to a fermentation. These are: (1) the tendency of yeast strains to mutate from flocculent to non-flocculent; (2) the loss of ability to ferment maltotriose; and (3) the presence of respiratory deficient mutants.

The respiratory deficient (RD) or 'petite' mutation is the most frequently identified spontaneous mutant found in brewing yeast strains. The mutant arises spontaneously when a segment of the wild-type mitochondrial genome, excised by an illegitimate site-specific recombination, is amplified to form a defective mitochondrial genome. The mitochondria are then unable to synthesize certain proteins. This mutation shows non-Mendelian segregation when crosses of the wild type with the 'petite' mutant are carried out. The respiratory deficient mutation normally occurs at frequencies of between 0.5% and 5% of the yeast population but in some strains, figures as high as 50% have been reported. The mutant is characterized by deficiencies in mitochondrial function resulting in a diminished ability to function aerobically, and as a result these yeasts are unable to metabolize non-fermentable carbon sources such as lactate, glycerol, or ethanol. Many phenotypic effects occur due to this mutation and these include alterations in sugar uptake, metabolic by-product formation, and tolerance to stress factors such as ethanol and temperature. Flocculation, cell wall and plasma membrane structure, and cellular morphology are affected by this mutation.

Beer produced with a yeast that is respiratory deficient or that produces a high number of respiratory deficient mutants, is more likely to have flavor defects and fermentation problems. ERNANDES et al. (1993) have reported that beer produced from these mutants contained elevated levels of diacetyl and some higher alcohols. Wort fermentation rates were slower, higher dead cell counts were observed, and biomass production and flocculation ability were reduced.

A significant reduction in diacetyl production has been achieved by the selection of spontaneous mutants from brewer's yeast cultures using resistance to the herbicide sulfometuron methyl (FALCO and DUMAS, 1985). The sulfometuron methyl resistant strains produce 50% less diacetyl than the parent strain due to partial inactivation of the acetohydroxy acid synthetase. Workers at the Carlsberg Research Institute in 1988 described the isolation of low diacetyl producing strains and successful plant trials at the 4500 hL scale (KIELLAND-BRANDT et al., 1988).

4.3 Novel Methods for Genetic Manipulation

4.3.1 Spheroplast Fusion

Spheroplast (protoplast) fusion, first described for yeast by VAN SOLINGEN and VAN DER PLAAT (1977), is a technique that can be employed in the genetic manipulation of industrial strains. The method does not depend on ploidy and mating type and consequently has great applicability to brewer's yeast strains because of their polyploid nature and absence of mating type characteristics. The yeast cell wall is removed with lytic enzymes such as extracts of snail gut or enzymes from various microorganisms. Removal of yeast cell walls results in osmotically fragile spheroplasts, which must be maintained in an osmotically stabilized medium such as 1 M sorbitol. The spheroplasting enzyme is removed by thorough washing, and the spheroplasts are then mixed and suspended in a fusion agent consisting of polyethylene glycol (PEG) and calcium ions in buffer. Subsequently, the fused spheroplasts must be induced to regenerate their cell walls and recommence division. This is achieved in solid media containing 3% agar and sorbitol. The action of PEG as a fusing agent is not fully understood, but it is believed to act as a polycation, inducing the formation of small aggregates of spheroplasts.

Some examples of fusions with commercial brewing strains are: (1) the construction of a brewing yeast with amylolytic activity by the fusion of *Saccharomyces cerevisiae* with *Saccharomyces diastaticus* (FREEMAN, 1981); (2) a polyploid capable of high ethanol production by fusion of a flocculent strain with Saké yeasts (SEKI et al., 1983); and (3) construction of industrial strains with improved osmotolerance by fusion of *S. diastaticus* with *S. rouxii* (SPENCER et al., 1985; CRUMPLEN et al., 1990). Although spheroplast fusion is an extremely efficient technique, it relies mainly on trial and error and is not specific enough to modify strains in a predictable manner. The fusion product is nearly always very different from both original fusion partners because the genome of both donors becomes integrated. Consequently, it is difficult to selectively introduce a single trait such as flocculation into a strain using this technique (STEWART and RUSSELL, 1981). Protoplast fusion has been found to be a viable technique when flavor of the final product is not critical. STEWART et al. (STEWART, 1981; STEWART et al., 1983; RUSSELL and STEWART, 1985) described a number of strains created by fusing brewing lager yeast with *Saccharomyces diastaticus* strains. The resultant hybrids had unsatisfactory beer flavor/taste profiles but could survive higher osmotic pressure, higher temperatures and produced higher ethanol yields and thus could be of use in the industrial alcohol industry (WHITNEY et al., 1985; STEWART et al., 1986).

4.3.2 Recombinant DNA

Although the techniques of hybridization, rare mating and spheroplast fusion have met with success, they have their limitations, the principal one being the lack of specificity in genetic exchange. It is only since 1978 that a DNA transformation system for yeast has been available (HINNEN et al., 1978), and great strides have been made in the past 15 years. It is now possible to modify the genetic composition of a brewer's yeast strain without disrupting the many other desirable traits of the strain and it is also possible to introduce genes from other sources.

A comprehensive review on the use of recombinant DNA technology for improving brewing yeast has recently been published by CASEY (1990), and some examples cited are: (1) glucoamylase activity from the fungus *Aspergillus niger*; (2) β-glucanase activity from the bacterium *Bacillus subtilis,* the fungus *Trichoderma reesii,* or barley; and (3) α-acetolactate decarboxylase activity from the bacterium *Enterobacter aerogenes*.

In a recent publication ABBOTT et al. (1993) reviewed the use of genetic engineering to construct yeasts capable of super attenuating wort, secreting proteases to chillproof the beer, secreting β-glucanases to degrade β-glucans in wort and thus enhance filterability as well as to construct strains that produce less diacetyl and with altered flocculation characteristics.

What are the future prospects for use of recombinant DNA in the brewing industry? At this time this is a difficult question to answer. It is quite surprising that there are not a number of recombinant brewer's yeasts commercially in use today. Permission has already been granted in the UK for the use of a baker's yeast strain that is genetically manipulated to enhance baking properties (GIST-BROCADES et al., 1988) and for a brewing strain that secretes glucoamylase to produce low-calorie beer (ADVISORY COMMITTEE ON NOVEL FOODS AND PROCESSES, 1994; MCLAIN, 1994). Perhaps the availability of alternative inexpensive traditional solutions for many of the problems that, it was hoped, a cloned yeast could solve, such as inexpensive sources of β-glucanase and gluco- and α-amylase, has retarded implementation. Also in many ways recombinant DNA technology is ahead of the knowledge base in yeast biochemistry. We can clone a gene but, in many cases, we lack the understanding of the biochemical pathway that is coded by the gene and the complex interactions of the various pathways. There is also still concern over consumer acceptance. Although this is a difficult hurdle, it is thought that as people become accustomed to pharmaceuticals produced by recombinant DNA, and more plants with improved characteristics for farming/food gain regulatory approval and consumer acceptance, the current reluctance to use this technology in the brewing industry will slowly disappear.

5 Yeast Performance

During the brewing process overall yeast performance is controlled by a plethora of factors. These factors include:

- the yeast strains employed
- the concentration and category of assimilable nitrogen
- the concentration of ions
- the fermentation temperature
- the pitching rate
- the tolerance of yeast cells to ethanol
- the wort gravity
- the wort oxygen level at yeast pitching
- the wort sugar spectrum.

These factors influence yeast performance either individually or in combination with others (D'AMORE, 1992; ZHENG et al., 1994a).

5.1 Yeast Nutritional Requirements

5.1.1 Oxygen

Wort fermentation in beer production is largely anaerobic, but when the yeast is first pitched into the wort, some oxygen must be made available to the yeast. Brewing yeasts, in the absence of molecular oxygen, are unable to synthesize sterols and unsaturated fatty acids which are essential membrane components. Sterols and unsaturated fatty acids are normally present in the wort in suboptimal quantities. They are abundant in malt, but normal manufacturing procedures prevent them from passing into the wort, and consequently oxygen must be added to wort to allow their synthesis by yeast. When ergosterol and an unsaturated fatty acid, such as oleic acid, are added to wort, the requirement for oxygen is diminished. When a brewer's yeast is grown aerobically, it accumulates sterols and unsaturated lipids within the cell in excess of the yeast's minimum requirements and the lipids can be 'diluted' to a degree by subsequent growth without negative effects. Thus cells prepared aerobically can, to some extent, grow anaerobically. However, if yeast is harvested at the end of fermentation and used to inoculate a second batch of wort, then oxygen is required, as the new inoculum contains no reserve of the necessary lipids.

5.1.2 Lipid Metabolism

The addition of lipids, especially ergosterol and unsaturated long-chain fatty acids, has a pronounced effect on the growth and metabolism of yeast. These compounds are an integral part of the plasma membrane where they regulate the movement of compounds in and out of the cell, the activities of membrane bound enzymes and enhancement of the yeast's ability to resist high ethanol concentrations. The wort oxygen content at pitching is important with regard to lipid metabolism, yeast performance and beer flavor. Underaeration leads to sub-optimal synthesis of essential membrane lipids and results in limited yeast growth, low fermentation rates and concomitant flavor problems. Over-aeration results in the production of unnecessary yeast biomass thus lowering fermentation efficiency and also produces beer with flavor defects such as high levels of diacetyl due to excessive α-acetolactate production.

5.1.3 Uptake and Metabolism of Wort Carbohydrates

When yeast is pitched into wort, it is introduced into an extremely complex environment due to the fact that the wort is a medium consisting of simple sugars, dextrins, amino acids, peptides, proteins, vitamins, ions, nucleic acids and other constituents too numerous to mention. One of the major advances in brewing science during the past 25 years has been the elucidation of the mechanisms by which the yeast cell, under normal circumstances, utilizes, in a very orderly manner, the plethora of wort nutrients.

Wort contains the sugars sucrose, fructose, glucose, maltose and maltotriose together with dextrin material. In the normal situation, brewing yeast strains [i.e., *Saccharomyces cerevisiae* and *Saccharomyces uvarum (carlsbergensis)*] are capable of utilizing sucrose, glucose, fructose, maltose and maltotriose in this approximate sequence, although some degree of overlap does occur. The majority of brewing strains leave the maltotetraose and other dextrins unfermented, but *Saccharomyces diastaticus* is able to utilize dextrin material. The initial step in the utilization of any sugar by yeast is usually either its passage intact across the cell membrane or its hydrolysis outside the cell membrane followed by entry into the cell by some or all of the hydrolysis products. Maltose and maltotriose are examples of sugars that pass intact across the cell membrane whereas sucrose (and dextrin with *S. diastaticus*) is hydrolyzed by an extracellular enzyme, and the hydrolysis products are taken up into the cell.

Maltose and maltotriose are the major sugars in brewer's wort and as a consequence, a brewer's yeast's ability to use these two sugars is vital and depends upon the correct genetic complement. It is probable that brewer's yeast possesses independent uptake mechanisms (maltose and maltotriose permease), to transport the two sugars across the cell membrane into the cell (ZHENG et al., 1994b). Once inside the cell, both sugars are hydrolyzed to glucose units by the α-glucosidase system. The transport, hydrolysis and fermentation of maltose is particularly important in brewing, since maltose is the major sugar component of brewing wort.

Maltose fermentation in *Saccharomyces* yeasts requires at least one of five unlinked *MAL* loci consisting of three genes encoding the structural gene for α-glucosidase (maltase) (*MAL S*), maltose permease (*MAL T*) and a *trans*-acting activator (*MAL R*) whose product coordinately regulates the expression of the α-glucosidase and permease genes. The expression of *MAL S* and *MAL T* is regulated by induction by maltose and repression by glucose. Work by SHIBANO et al. (1993) suggests that the constitutive expression of the *MAL T* gene is critical and that maltose fermentation ability of brewing yeasts depends mostly on maltose permease activity.

It has been reported that maltose transport is achieved by two separate components – one with a high affinity for maltose and the second with a low affinity (BUSTURIA and LAGUNAS, 1985). The low affinity system is effective only at high maltose concentrations and is not well studied. However, the permease or transport protein has been characterized both genetically and biochemically for the high affinity system. The gene coding for the permease is closely linked to the maltase (α-glucosidase) gene, and it is believed that brewing strains probably contain several copies of each of the two genes scattered as pairs around several different chromosomes.

Strains constructed with multiple *MAL* genes increase the rate at which sugar can be taken up from wort by yeast (STEWART et al., 1983).

The uptake and hydrolysis of maltose and maltotriose from the wort is also dependent on the glucose concentration. When glucose concentrations are high [greater than 1% (w/v)] the *MAL* genes are repressed, and only when 40–50% of the glucose has been taken up from the wort will the uptake of maltose and maltotriose commence. Thus, the presence of glucose in the fermenting wort exerts a major repressing influence on wort fermentation rate (ERNANDES et al., 1993). Using the glucose analog 2-deoxy-glucose, which is non-metabolizable by *Saccharomyces* yeast strains, spontaneous mutants of brewing strains have been selected in which the maltose uptake is not repressed by glucose, and as a consequence these strains have significantly increased fermentation rates (RUSSELL et al., 1987).

5.1.4 Uptake and Metabolism of Wort Nitrogen

Active yeast growth involves the uptake of nitrogen, mainly in the form of amino acids, for the synthesis of protein and other nitrogenous components of the cells. Later in the fermentation as yeast multiplication stops, nitrogen uptake slows or ceases. In wort, the main nitrogen source for synthesis of protein, nucleic acids and other nitrogenous cell components is the variety of amino acids formed from the proteolysis of barley protein. Brewer's wort contains 19 amino acids, and the yeast takes them up in an orderly manner, different amino acids being removed at various points in the fermentation cycle (JONES and PIERCE, 1967). Short chains of amino acids in the form of di- or tripeptides can also be taken up by the yeast cell. These peptides, do not permeate freely into the cell by simple diffusion, but rather their uptake is facilitated by a limited number of transport enzymes. At the start of fermentation the following eight amino acids are absorbed rapidly: arginine, aspartic acid, asparagine, glutamic acid, glutamine, lysine, serine and threonine. The other amino acids are absorbed only slowly, or not at all until later in the fermentation. Under strictly anaerobic conditions such as those encountered late in a brewery fermentation, proline, the most plentiful amino acid in wort, has scarcely been assimilated by the end of the fermentation, whereas over 95% of the other amino acids have disappeared. Proline is usually still present in the finished product at 200–300 mg/L, however, under aerobic laboratory conditions proline is assimilated after exhaustion of the other amino acids.

5.2 Yeast Excretion Products

Although ethanol is the major excretion product produced by yeast during wort fermentation, this primary alcohol has little impact on the flavor of the final beer. It is the type and concentration of the many other yeast excretion products produced during wort fermentation that primarily determine the flavor of the product. The formation of these excretion products depends on the overall metabolic balance of the yeast culture, and there are many factors that can alter this balance and consequently the flavor of the product. Yeast strain, fermentation temperature, adjunct type and level, wort pH, buffering capacity, wort gravity, etc. are all influencing factors.

Some volatiles are of great importance and contribute significantly to beer flavor, whereas others are important in building the background flavor of the product. The composition and concentration of beer volatiles depends upon the raw materials used, brewery procedures in mashing, fermentation, etc., and the yeast strain employed. The following groups of substances are found in beer: alcohols, esters, carbonyls, organic acids, sulfur compounds, amines, phenols and a number of miscellaneous compounds.

5.2.1 Alcohols

In addition to ethanol there are a great number of other alcohols found in beer, and these higher alcohols or fusel oils constitute an important part of the by-products formed during wort fermentation. Their formation is linked to yeast protein synthesis and they are formed from β-acids, which in turn may be formed by transamination and deamination of the amino acids in wort, or synthesized from wort carbohydrates. The particular yeast strain employed is of great significance in determining the level of higher alcohols in the beer. In addition, formation of higher alcohols is very dependent upon the fermentation temperature, with an increase in temperature resulting in increased concentrations of higher alcohols in beer (ENGAN, 1981).

5.2.2 Esters

Many factors in addition to the yeast strain employed have been found to influence the amount of esters formed during a wort fermentation. These include: fermentation temperature, where an increase in temperature from 10 to 25 °C has been found to increase the concentration of ethyl acetate from 12.5 to 21.5 mg/L; fermentation method, where continuous fermentation results in higher levels of ester formation than conventional batch fermentation; pitching (inoculation) rate, where higher rates have been reported to result in the formation of lower amounts of ethyl acetate; and wort aeration, where low levels of oxygen appear to enhance ester formation.

5.2.3 Sulfur Compounds

Sulfur compounds are a significant contributor to the flavor of the beer (VAN DEN EYNDE, 1991; DERCKSEN et al., 1992; LEEMANS et al., 1993). Although small amounts of sulfur compounds can be acceptable or even desirable in beer, in excess they give rise to unpleasant off-flavors, and special measures such as purging with CO_2 or prolonged maturation times are necessary to remove them. Although volatile sulfur compounds such as hydrogen sulfide, dimethyl sulfide, sulfur dioxide and thiols are contributed to the wort and beer by hops, adjuncts and malt, a significant proportion of those present in finished beer are formed during or after fermentation. During fermentation, yeasts usually excrete significant amounts of hydrogen sulfide and sulfur dioxide. Sulfur dioxide is usually present at concentrations below taste threshold, and normal beers, when free of infection, contain low levels of free hydrogen sulfide.

5.2.4 Carbonyl Compounds

Carbonyl compounds exert a significant influence on the flavor stability of beer. Excessive concentrations of carbonyl compounds are known to cause a stale flavor in beer. The effects of aldehydes on beer flavor are reported as grassy notes (propanol, 2-methyl butanol, pentanal) and give a papery taste (*trans*-2-nonenal, furfural) (GRÖNQVIST et al., 1993).

The carbonyl found in highest concentration in beer is acetaldehyde. It is formed during fermentation and is a metabolic branch point in the pathway leading from carbohydrate to ethanol. The acetaldehyde formed may either be reduced to ethanol, or oxidized to acetic acid, and in the final step of the alcoholic fermentation, acetaldehyde is reduced to ethanol by an enzymatic reaction. The concentration of acetaldehyde varies during fermentation and aging/conditioning and reaches a maximum during the main fermentation and then decreases. Excessive quantities of acetaldehyde in beer can also be the result of bacterial contamination. Acetaldehyde levels in bottled beer have been observed to increase during pasteurization and storage, especially if there is a high air content in the bottle head-space.

5.2.5 Diacetyl and Pentane-2,3-dione

Diacetyl and pentane-2,3-dione impart a characteristic aroma and taste to the beer; often described as 'buttery', 'honey- or toffee-like' or as 'butterscotch'. The taste threshold concentration for diacetyl in lagers is 0.1–0.14 mg/L with somewhat higher levels

(WAINRIGHT, 1973; HARDWICK et al., 1976) in ales.

Diacetyl and pentane-2,3-dione are formed outside the yeast cell, by the oxidative decarboxylation of α-acetolactate and α-acetohydroxybutyrate, respectively. These α-acetohydroxy acids are intermediates in the biosynthesis of leucine and valine (acetolactate) and isoleucine (acetobutyrate) and are leaked into the medium by yeast during fermentation. Once diacetyl and pentane-2,3-dione have been formed in the fermenting wort, they are normally converted to acetoin or pentane-2,3-diol, respectively, by the action of yeast reductases. Thus, the final concentration of diacetyl in beer is the net result of three separate steps: (1) synthesis and excretion of α-acetohydroxy acids by yeast; (2) oxidative decarboxylation of α-acetohydroxy acids to their respective diketones; and (3) reduction of diacetyl and pentane-2,3-dione by yeast.

The presence of diacetyl in beer at above threshold levels occurs when α-acetolactate has decomposed to give diacetyl at a time when the yeast cells are either absent or have lost their ability to reduce diacetyl to acetoin. Commonly the fault arises because α-acetolactate breakdown has been curtailed by the use of temperatures conducive to yeast settling when the potential to produce diacetyl remains. When the beer becomes warm, which is usually when it is packaged and pasteurized, but may not be until the beer is disposed at the point of sale, diacetyl is produced, and in the absence of yeast this diacetyl is not converted to acetoin and therefore accumulates. Diacetyl levels can thus be controlled by ensuring that there is sufficient active yeast in contact with the beer at the end of the fermentation to reduce diacetyl to acetoin. Diacetyl formation from α-acetolactate has been shown to depend upon pH, the concentration of α-acetolactate, temperature, the presence of oxygen, the vigor of the fermentation, and certain metal ions. Vigorous fermentations produce more acetohydroxy acids, but the decomposition of acetohydroxy acids to vicinal diketones is also more rapid. In addition, since diacetyl is formed earlier in the fermentation, there is more time for diacetyl removal by the yeast. Excessive levels of diacetyl can also be the result of beer spoilage by certain strains of bacteria such as *Pediococcus* and *Lactobacillus*.

5.2.6 Dimethylsulfide (DMS)

Dimethylsulfide (DMS) in beer originates from two sources, from the hydrolysis of malt *S*-methylmethionine (SMM) during mashing and from the reduction of dimethylsulfoxide (DMSO) by the yeast. LEEMANS et al. (1993) have recently shown that in lager beer brewed on the pilot plant scale, the majority of the DMS is produced by yeast and 80% of the DMS comes from DMSO. The DMS evaporation ratio can vary between 0 and 65% throughout the formation of this compound during fermentation, the total amount of DMS produced including that which had been evaporated ranged from 70 to more than 120 mg/L. When the influence of wort DMSO concentration on the production of DMS during fermentation was studied, it was observed that there is a direct proportional relationship between the concentrations of these compounds at the end of fermentation and at every stage of fermentation as well. They concluded that the variety of the malt has a direct influence on the DMSO quantity and, therefore, an indirect influence on the level of DMS in beer. When the concentration of DMSO in the wort at pitching is high, then the concentration of DMS in the beer will also be high.

5.3 Yeast Flocculation

Flocculation can be defined as the phenomenon wherein yeast cells adhere in clumps and either sediment rapidly from the medium in which they are suspended or rise to the medium's surface. This definition excludes other forms of aggregation, such as chain formation where daughter cells do not separate from mother cells.

Individual strains of yeast differ considerably in their flocculating power. At one extreme are the very flocculent strains referred to as 'gravel-like' and at the other extreme are totally non-flocculent strains sometimes referred to as 'powdery'. The strongly flocculating yeasts can sediment out of the fermentation broth prematurely giving rise to sweeter and less fermented beers, whereas the weakly flocculating strains can remain in the beer during aging and cause yeasty flavor and filtration difficulties.

To produce a high quality beer, it is axiomatic that not only must the yeast culture be effective in removing the required nutrients from the wort, be able to tolerate the prevailing environmental conditions (e.g., high ethanol levels), impart the desired flavor to the beer, but the yeasts themselves must be effectively removed from the wort by flocculation, centrifugation and/or filtration after they have fulfilled their metabolic role. Consequently, the flocculation properties, or conversely lack of flocculation, of a particular brewing yeast culture is very important when considering factors affecting wort fermentation.

In addition to flocculation there is the phenomenon of co-flocculation. Co-flocculation is defined as the phenomenon where two strains are non-flocculent alone but flocculent when mixed together. To date co-flocculation has only been observed with ale strains, and there are no reports of co-flocculation between two lager strains of yeast. There is a third flocculation reaction which has been described, where the yeast strain has the ability to aggregate and co-sediment with contaminating bacteria in the culture. Again this phenomenon is confined to ale yeast, and co-sedimentation of lager yeast with bacteria has not been observed (WHITE and KIDNEY, 1979; ZARATTINI et al., 1993).

The genetic control of yeast flocculation is far from being understood. Research has been complicated by the polyploid or aneuploid nature of most brewing strains and the large number of different strains in use. Dominant and recessive flocculation genes and flocculation suppressor genes have been described, and much still remains to be discovered regarding the genes that regulate flocculation. It is generally agreed that there are at least two dominant flocculation genes present in brewing yeast. The existence of multiple gene copies, as well as the possibility of more than one operating flocculation mechanism, underscores the difficulty researchers encounter when studying the genetics of brewing yeast flocculation.

Extensive research has been undertaken on the biochemical aspects of yeast flocculation but controversy as to the mechanism of flocculation still exists (STRATFORD and WILSON, 1990; SPEERS et al., 1992a, b; GUINARD and LEWIS, 1993). It is believed that flocculation is due in part to mannose specific lectin-like adhesion that is modulated by electrostatic and hydrophobic interactions. Yeast flocculation is mediated by specific interactions between cell wall proteins and carbohydrates from neighboring cells (MIKI et al., 1982). The yeast cell wall consists of an outer layer of *o*-mannosylated proteins which overlays a layer of glucan to which hyperglycosylated proteins (mostly *N*-) are covalently attached. Recent work by STRATFORD and CARTER (1993) describes a specific lectin–receptor interaction and VAN DER AAR et al. (1993) describe a correlation between cell hydrophobicity and flocculence. The FLO 1 gene product is believed to be a hydrophobic cell wall protein located on fimbriae-like structures which are absent in non-flocculent cells (DAY et al., 1975; STEWART and RUSSELL, 1981; MIKI et al., 1982). STRAVER et al. (1993) suggest that cell surface hydrophobicity is a major determinant for yeast cells to become flocculent during growth in wort. Increased hydrophobicity of the cells may facilitate cell contact leading to calcium-dependent lectin–sugar binding. They also suggest that oxygen may be an indirect growth-limiting factor and that shortage of sterols and unsaturated fatty acids precedes flocculence under brewing conditions.

6 Yeast Handling Techniques

6.1 Introduction – Pure Yeast Cultures

Yeasts belonging to the genus *Saccharomyces* are often referred to as the oldest plants cultivated by man. The practice of using a pure yeast culture for brewing was started by EMIL C. HANSEN in the Carlsberg laboratories over 100 years ago. Using dilution techniques, he was able to isolate single cells of brewing yeast, test them individually and then select the specific yeast strains that gave the desired brewing properties. The first pure yeast culture was introduced into a Carlsberg brewery on a production scale in 1883, and the benefits of using a pure culture quickly became clear. Soon twenty-three countries had installed HANSEN's pure culture plant, and in North America, Pabst, Schlitz, Anheuser Busch and 50 smaller breweries were using pure lager yeast cultures by 1892 (VON WETTSTEIN, 1983; ANDERSON, 1993).

It is normal practice in many breweries, to propagate fresh yeast every 8–10 generations (fermentation cycles) or earlier if contamination or a fermentation problem is identified. The systematic use of clean, pure and highly viable cells ensures that bacteria, wild yeast or yeast mutations, such as respiratory deficiency, do not lead to inconsistent fermentations and off-flavor development (HOUGH, 1973; SMITH, 1991).

Lager yeast normally consists of one pure culture, whereas ale yeast cultures have often in the past been a mix of strains. The pure culture practice is invaluable in ensuring that any wild yeasts are quickly detected and not allowed to proliferate into a significant problem for the brewer. As well, undesirable mutations of the parent strain, which often occur over long usage, are kept to a minimum.

6.2 Propagation and Scale-Up

The first yeast propagation plant was developed by HANSEN and KUHLE and consisted of a steam-sterilizable wort receiver and propagation vessel equipped with a supply of sterile air and an impeller. The basic principles of propagation devised in 1890 have changed little. Propagation can be batch or semi-continuous and usually consists of three stainless steel vessels of increasing size equipped with attemperation control sight glasses and non-contaminating venting systems. They are equipped with a CIP system and often have in-place heat sterilizing and cooling systems for both the equipment and the wort. The yeast propagation system is ideally located in a separate room from the fermenting area with positive air pressure, humidity control, an air sterilizing system, disinfectant mats in doorways and limited access by brewing staff.

During yeast propagation the aim is to obtain a maximum yield of yeast but also to keep the flavor of the beer similar to a normal fermentation so that it can be blended into the production stream. As a result, the propagation is often carried out at only a slightly higher temperature and with intermittent aeration to stimulate yeast growth. The propagation of the master culture to the plant fermentation scale is a progression of fermentations of increasing size (typically 4–10X) until enough yeast is grown to pitch (inoculate) a half size or full commercial size brew.

Wort sterility is normally achieved by boiling for 30 minutes or the wort can be 'pasteurized' using a plate heat exchanger, passed into a sterile vessel and then cooled. Wort gravities typically range from 10 °P to 16 °P. Depending on the yeast strain, zinc or a commercial yeast food can be added. Aeration is important for yeast growth, and the wort is aerated using oxygen or sterile air, and antifoam may be added depending on the yeast. Agitation is not normally necessary as the aeration process and CO_2 evolved during active fermentation are sufficient to keep the yeast in suspension. Scale-up steps are kept small at the early stages to ensure good growth.

6.3 Contamination of Cultures

6.3.1 Bacteria

Bacteria are a common spoilage agent of beer, and the most troublesome Gram-positive bacteria are the lactic acid bacteria belonging to the genera *Lactobacillus* and *Pediococcus*. At least 10 species of lactobacilli can cause beer spoilage. When viewed under a light microscope, lactobacilli are very pleomorphic in appearance and can range in shape from long slender rods to short cocco-bacilli. Brewing lactobacilli are heterofermentative and homofermentative, and produce lactic and acetic acid, carbon dioxide, ethanol and glycerol as end products with some strains also producing diacetyl. They are acid-tolerant and have complex nutritional requirements. Some species such as *Lactobacillus brevis* and *L. plantarum* can grow quickly during fermenting, aging or yeast storage, whereas others such as *L. lindneri* grow relatively slowly. Lactobacilli spoilage is most problematic during conditioning of beer and after packaging, where spoilage gives rise to a 'silky turbidity' (PRIEST, 1987).

Pediococci are homofermentative cocci that occur in pairs and tetrads. Six species of pediococci have been identified, but the species predominantly found in beer is *Pediococcus damnosus* (LAWRENCE and PRIEST, 1981). *Pediococcus* infection in the beer is characterized by lactic acid and diacetyl formation. They may also cause ropiness in beer due to the production of polysaccharide capsules.

Many Gram-positive bacteria are inhibited by hop bittering substances, when these compounds are present in very high concentrations, but Gram-negative bacteria are usually unaffected. Unfortunately, lactobacilli and pediococci, although Gram-positive, are generally insensitive to hops. Some members of the family Micrococcaceae can survive in beer, grow and cause spoilage as can some aerobic spore forming bacteria belonging to the genus *Bacillus* under certain conditions. Generally these two genera are inhibited by hop components, prefer an aerobic environment, and, therefore, are not as serious a threat.

Important Gram-negative beer spoilage bacteria include acetic acid bacteria (*Acetobacter, Gluconobacter*), certain members of the family Enterobacteriaceae (*Escherichia, Aerobacter, Klebsiella, Citrobacter, Obesumbacterium*) as well as *Zymomonas, Pectinatus*, and *Megasphaera* (VAN VUUREN, 1987). Acetic acid bacteria can convert ethanol to acetic acid, producing a vinegar flavor in the beer and tend to produce a ropy slime. This type of spoilage is most often observed in draft beer. The bacteria are airborne and prefer an aerobic environment but can survive under microaerophilic conditions and infect the keg as a result of air entering or beer standing too long on tap in a partly filled keg. The Enterobacteriaceae are aerobes or facultative aerobes and do not tolerate high ethanol levels. They are usually found early in the fermentation and can produce celery-like, cooked cabbage, cooked vegetable and rotten egg aromas, especially if pitching of the wort is delayed.

6.3.2 Wild Yeast

A wild yeast is any yeast, other than the culture yeast that was intentionally inoculated. With breweries producing different types of beer, each with its own yeast or mixture of yeasts, it is important that cross contamination does not occur. Wild yeast can originate from very diverse sources as shown by a study by BACK (1987) where 120 wild yeast strains from beer, brewing yeast and cleaned empty bottles were isolated. In addition to various *Saccharomyces* strains, species of the genera *Brettanomyces, Candida, Debaromyces, Hansenula, Kloeckera, Pichia, Rhodotorula, Torulaspora* and *Zygosaccharomyces* were isolated. The potential of a wild yeast to cause adverse effects varies with the specific contaminant. If the contaminating wild yeast is another culture yeast, the primary concern is with rate of fermentation, final attenuation, flocculation, and taste implications. If the contaminating yeast is a non-brewing strain and can compete with the culture yeast for the wort constituents, inevitably problems will arise as these yeasts can produce a variety of off-flavors and aromas often similar to those produced by contaminating bacteria. Some wild yeasts can utilize wort dextrins resulting in an over-attenuated beer that lacks body. These yeasts are found as both contaminants of fermentation and as post fermentation contaminants. In addition, wild yeasts often produce a phenolic off-flavor due to the presence of the *POF* gene (RUSSELL et al., 1983). However, under controlled conditions, such as in the production of a German wheat beer or 'weiss bier', this phenolic clove-like aroma, produced when the yeast decarboxylates wort ferulic acid to 4-vinyl guaiacol, can be a positive attribute of the beer.

6.4 Yeast Washing

Some breweries incorporate a yeast wash into their process as a routine part of the operation, especially if there are concerns over eliminating bacteria responsible for the production of apparent total *N*-nitroso compounds (ATNC), whereas others only wash the yeast when there is evidence of a bacterial infection. There has been considerable controversy over the practice of yeast washing and its effect on subsequent fermentations. Studies carried out at the Brewing Research Foundation International in the United Kingdom (SIMPSON, 1987; SIMPSON and HAMMOND, 1989) suggest that the problems often ascribed to yeast washing, i. e., reduced cell viability, vitality, reduced rate of fermentation, changes in flocculation, fining, yeast crop size and excretion of cell components, are only problems if yeast washing is carried out incorrectly.

There are three commonly used procedures for washing yeast: (1) sterile water wash, (2) acid wash, and (3) acid/ammonium persulfate wash.

(1) Sterile Water Wash: With the water wash, cold sterile water is mixed with the yeast slurry, the yeast is allowed to settle and the supernatant water is discarded. Bacteria and broken cells are removed through this process. This can be repeated a number of times.

(2) Acid Wash: There are a number of acids that can be used. Most common are phosphoric, citric, tartaric or sulfuric acids. The yeast slurry is acidified with diluted acid to a pH of 2.0, and it is important that agitation is continuous through the acid addition period. The yeast is usually allowed to stand for a maximum period of two hours.

(3) Acid/Ammonium Persulfate: An acidified ammonium persulfate treatment has also been found to be effective and can yield material cost savings. It is recommended that 0.75% (w/v) ammonium persulfate is added to a diluted yeast slurry (2 parts water:1 part yeast) and then the slurry is acidified with phosphoric acid to pH 2.8 (BRENNER, 1965; SIMPSON, 1987). If a pH of 2.0 is employed, a one hour contact time is maximum.

Many brewers have a strong preference for a certain regime of yeast washing, and a number of factors must be taken into account when choosing the method, such as food-grade quality of the acid, hazards involved in using the acid and cost. Phosphoric and citric acids offer the advantage of being weak acids and yeast pH is more easily controlled, whereas with strong acids, such as sulfuric acid, there are special handling procedures required for the operators, and a slight overdose will yield excessively low pH values.

SIMPSON and HAMMOND (1989) have listed a set of criteria which, if followed, should alleviate many of the problems that are associated with washing yeast. They are as follows: (1) use a food-grade acid – phosphoric or citric acid are good choices; (2) wash the yeast as a beer or water slurry; (3) chill both the yeast slurry and the acid to less than 5 °C; (4) stir constantly and slowly while adding the acid to the yeast; (5) if possible stir throughout the wash; (6) never let the temperature exceed 5 °C during the wash; (7) check the pH of the yeast slurry; (8) do not wash for more than two hours; (9) pitch yeast immediately after washing; and (10) do not wash unhealthy yeast or yeast from fermentations with greater than 8% ethanol present (if a wash is unavoidable, use a higher pH and/or a shorter contact time).

6.5 Yeast Pitching and Cell Viability

Microscopic examination of brewery pitching yeast is a rapid way to ensure that there is not a major contaminant or viability problem with the yeast. When a sample of pitching yeast in either water, wort or beer is examined under the microscope, it can be difficult to distinguish a small number of bacteria from the trub and other extraneous non-living material. The trub material, however, is irregular in size and outline, and dissolves readily in dilute alkali.

A trained microbiologist becomes familiar with the typical appearance of the yeast cytoplasm and shape of the yeast cells, whether the cells are normally chain formers, or in clumps, etc., and thus one can sometimes identify the presence of a wild yeast due to cells with an unusual shape or differences in budding or flocculating behavior.

The use of a viability stain such as methylene blue gives a good indication of the health of the cells. Although there are a number of other good stains and techniques available, in experienced hands, methylene blue will still quickly identify a problem if there is a known history of what the typical viability of the yeast strain is, before the yeast is pitched.

Yeast pitching is governed by a number of factors such as wort gravity, wort constituents, temperature, degree of wort aeration and previous history of the yeast. Ideally, one wants a minimum lag in order to obtain a rapid start to fermentation, which then results in a fast pH drop, and ultimately assists in the suppression of bacterial growth. Pitching rates employed vary from 5–20 million cells/mL, but 10 million cells/mL is considered an optimum level by many and results in a lager yeast reproducing four to five times. Increasing the pitching rate results in fewer doublings, since yeast cells under given conditions multiply only to a certain level of cells/unit volume, regardless of the original pitching rate. The pitching rate can be determined by a number of methods such as dry weight, turbidimetric sensors (REISS, 1986; BOULTON and BESFORD, 1992), haemocytometer and electronic cell counting. More recently, use has been made of commercially available in-line biomass sensors which utilize the passive dielectrical properties of microbial cells and can discriminate between viable and non-viable cells and trub (BOULTON et al., 1989). The amount of yeast grown is limited by a number of factors including oxygen supply, nutrient exhaustion and accumulation of inhibitory metabolic products.

6.6 Yeast Collection

Yeast collection techniques vary depending on whether one is dealing with a traditional ale top fermentation system, a traditional lager bottom fermentation system, a non-flocculent culture where the yeast is cropped with a centrifuge, or a cylindroconical fermentation system. With the traditional ale top fermentation system, although there are many variations on this system, a single, dual or multi-strain yeast system can be employed and the timing of the skimming can be critical to maintain the flocculation characteristics of the strains. Traditionally, the first skim or 'dirt skim' with the trub present is discarded, as is the final skim in most cases. The middle skim is normally kept for repitching. With the traditional lager bottom fermentation system, the yeast is deposited on the floor of the vessel at the end of fermentation. Yeast cropping is non-selective and the yeast contains entrained trub. With the cylindroconical fermentation system (now widely adopted for both ale and lager fermentations), the angle at the bottom of the tank allows for effective yeast plug removal.

The use of centrifuges for the removal of yeast and the collection of pitching yeast is now commonplace. There are a number of advantages such as shorter process time, cost reduction, increased productivity and reduced shrinkage. Care must be taken to ensure that high temperatures (i.e., >20 °C) are not generated during centrifugation and that the design ensures low dissolved oxygen pickup and a high throughput. This is usually accomplished by use of a self-desludging and low heat induction unit. Timing control of the desludge cycle is important and allows for a more frequent cycle for yeast for the pitching tank and resultant lower solids and a longer frequency for yeast being sent to waste with the higher solids and resulting reduced product shrink.

6.7 Yeast Storage

Ideally the yeast is stored in a room that is easily sanitized, contains a plentiful supply of sterile water and a separate filtered air supply with positive pressure to prevent the entry of contaminants, and a temperature of 0 °C. Alternatively, insulated tanks in a dehumidified room are employed. When open vessels were commonly used, great care had to be taken to ensure that sources of contamination were eliminated. Reduction of moisture levels to retard mold growth and elimination of difficult to clean surfaces and unnecessary equipment and tools from the room are useful.

Yeast is most commonly stored under six inches of beer, or under a water or 2% potassium dihydrogen phosphate solution. When high gravity brewing is used, it is important to remember that the ethanol levels are significantly higher and that this can affect the viability of the stored yeast. As more sophisticated systems have become available, storage tanks with external cooling (0–4 °C) and equipped with low shear stirring devices have become popular. Reduction of available oxygen is important during storage, and minimal yeast surface areas exposed to air are desirable. Low dead cell counts and minimum storage times are sought with the yeast being cropped "just-in-time" for repitching if possible. In this context, when cylindroconical fermenters are employed, often the yeast collected in the cone of one vessel is pitched directly into another fermenter, without use of a yeast storage system (HENSON and REID, 1988).

7 Flavor Maturation

'Aging', 'storage' and 'lagering' are all terms used to describe the process of holding beer in a tank at refrigerated temperatures for a period of time (up to four months) following fermentation. A storage period is not essential, but is desirable, for beer production. In the United Kingdom, ale beers are often produced and put in kegs without a storage period. However, a storage period (usually at low temperatures, 2–6 °C) is normal for lagers (and for North American ales) with the following process functions: (1) flavor maturation, (2) chillproofing and stabilization, (3) clarification and filtration, and (4) carbonation. Storage periods today usually range from one to four weeks.

An important function in a traditional brewing process is flavor maturation, particularly a reduction in the levels of hydrogen sulfide, acetaldehyde and diacetyl. These three reactions only occur in the presence of residual amounts of yeast fermentation, and minimal amounts of diacetyl, acetaldehyde and other undesirable flavors always remain at the end of a fermentation. Since it is known that at elevated temperatures, some lager yeast strains are susceptible to autolysis, a normal procedure following fermentation entails cooling the beer to 0 °C, and separating the yeast with a centrifuge. At this point, flavor maturation is virtually complete. The beer is then stored at 0–2 °C for a few more days to enhance physical stability, usually with the aid of an adsorbent such as silica gel or polyvinylpolypyrrolidone, and the beer is then carbonated and final polish filtered. This process requires less capital equipment than traditional processes, i.e., less tankage, but operating costs can be higher.

The stability of beer (physical, flavor and foam) has assumed increasing importance in the past 10–20 years, and measures to increase stability are usually taken during aging. This is particularly true of physical stability (haze). Several techniques are available such as the addition of proteolytic enzymes, adsorbents and tannic acid to remove the constituents that contribute to physical instability of the beer. However, some of these techniques, particularly proteolytic enzymes, can have a negative effect on foam stability. In addition, initiatives can be taken at this point to remove molecular oxygen from beer, which if allowed to remain can cause flavor instability. These activities usually involve the use of antioxidants such as sodium/potassium metabisulfite or ascorbic acid.

After fermentation, beer can be cloudy due to the presence of significant amounts of yeast still in suspension and protein/tannin materials which precipitate out of solution due to the cold temperatures, lower pH (approximately 4.2) and lower insolubility in alcohol solutions. This turbidity must be significantly reduced in order to render the beer marketable and with the advent of centrifuges, it is possible to have little or no yeast present during aging.

A modification of flavor maturation is the krausen storage process. The term "krausen" is a term applied to the most active stage of fermentation at which time foaming is most prevalent. Krausen storage is a process in which wort (usually 10 to 20% of the total liquid volume) is added to nearly or completely fermented wort. The krausen creates a secondary fermentation which produces a beer with a characteristic estery flavor. Krausening can also be employed to reduce the diacetyl level of a fermentation if it is unacceptably high.

Since flavor maturation is a prime reason for storage, many brewers have adopted novel technologies to reduce processing time. In some brewing processes, the beer is fermented beginning at 10–15 °C and allowed to warm up by the heat of fermentation to 16–18 °C and is held at that temperature until it is brilliant enough to market. There are a variety of processing techniques employed to clarify beer, and these include gravity sedimentation, fining and centrifugation.

During storage temperatures of 0 to 5 °C, the majority of the suspended yeast and turbidity will settle to the bottom of the storage vessel if fermentation has ceased and if the tank is under counter-pressure. At least a tenfold reduction in turbidity can be expected by this approach. Although this is a simple and traditional technique for reducing turbidity and is employed by many brewers, it has some drawbacks. The precipitated material on the tank bottom can be self-insulating and warm up permitting yeast autolysis to occur, and this can impart an unclean sulfury aroma to the beer. Also cleaning of the tank can be a problem. Fining agents can be added at the onset of aging in an attempt to accelerate clarification and to effect more complete sedimentation. Some fining agents employed are bentonite, tannic acid, isinglass, Irish moss (carrageenan) and silica gel. Obviously the use of one or more of these agents will add to overall operating costs.

7.1 Beer Stability

The flavor stability of a beer primarily depends on the oxygen content of the bottled beer. During wort production the following steps are critical (NARZISS, 1986):

- preservation of reducing substances by avoidance of oxygen pick-up during mashing, lautering and wort boiling
- elimination of substances which are prone to react with flavor active compounds like carbonyls by good mash and wort separation procedures
- prevention of the pick-up of ions such as iron and copper (IRWIN et al., 1991)
- avoidance of an excessive exposure of the wort to heat, to limit the formation of Maillard reaction products and related substances.

In many foods such as milk, butter, vegetables, vegetable oils and beverages, staling is caused by the appearance of various unsaturated carbonyl compounds, and it is now becoming increasingly clear that the same is true of beer staling. Packaged beer has a limited shelf-life. The phenomenon of beer staling has been intensively investigated by the brewing industry with a view to understanding and controlling it. Despite these studies, the mechanism of staling is still not fully understood. The actual compounds responsible for stale flavor vary during prolonged storage as evidenced by changes in the flavor profile of beer. Although the compounds causing the sweetish, leathery character of very old beers have not been identified, there is evidence that the papery, cardboard character of 2–4 month old beer is due to unsaturated aldehydes. The most flavor-active aldehyde which has been conclusively proven to rise beyond threshold levels is *trans*-2-nonenal. Others such as nonadienal, decadienal and undecadienal may also exceed threshold levels.

The adverse effects of oxidation on the flavor of finished beer have been known for a considerable time, and some brewers add bisulfites or other antioxidants, such as ascorbic acid, to beer prior to packaging to provide protection against oxygen pick-up. This can improve flavor stability. The effectiveness of bisulfite, besides its antioxidant properties, is also its ability to bind carbonyl compounds into flavor neutral complexes (COLLIN et al., 1991; IRWIN et al., 1991). Its addition to fresh beer reduces increases in free aldehyde concentration during aging. In addition, when added to stale beer, bisulfite lowers the concentration of free aldehydes and effects the removal of the cardboard flavor. However, over time, the bisulfite will be oxidized to sulfate thus increasing the concentration of free aldehydes again.

When beer is sold, the stability of the foam in a glass of beer is considered by some consumers to reflect the quality of the product. The increasing use of adjuncts and the associated decrease in malt being employed today has a negative effect on foam values in many beers. Many researchers have endeavored to identify the ingredients or process conditions that contribute to a good head retention value, and it is known that the proteinaceous material in the beer is the major foaming agent.

8 New Developments

8.1 High Gravity Brewing

High gravity brewing is a procedure which employs wort of higher than normal concentration and consequently requires dilution with water at a later stage in processing. Today, more beer is produced in North America according to this production method than by conventional means. By reducing the amount of water, increasing production demands can be met without expanding existing brewing, fermenting and storage facilities (PFISTERER and STEWART, 1976; MURRAY and STEWART, 1991). Generally, the lower the hopping levels and the higher the adjunct level, the more suited the beer will be to higher gravity without significant flavor change.

There are a number of advantages and disadvantages to this process. The advantages can be summarized as follows:

(1) increased brewing capacity, more efficient use of existing plant facilities
(2) reduced energy (heating, refrigeration, etc.), labor, cleaning and effluent costs
(3) improved physical and flavor stability
(4) more alcohol per unit of fermentable extract – reduced yeast growth
(5) high gravity worts may contain higher adjunct rates
(6) beers produced from high gravity worts are often rated smoother in taste
(7) high gravity brewing offers greater flexibility in product type.

However, there are a number of disadvantages:

(1) due to the more concentrated mash (increased ratio of carbohydrate to water), there is a decreased brewhouse material efficiency and reduced hop utilization
(2) decreased foam stability (head retention)
(3) beers produced from high gravity worts are sometimes difficult to flavor match with an existing normal gravity product
(4) high gravity worts can influence yeast performance with effects usually apparent upon both fermentation and flocculation characteristics (for example, tolerance to elevated osmotic pressure and ethanol).

The dilution step in the production process is the major innovation in the procedure, and it may be carried out before or during fermentation, or at some point during the aging and final filtration process. Each system offers advantages over the regular brewing procedure. Dilution in the fermenter improves fermentation cellar capacity as less head space is required. Further, when water is added during fermentation rather than immediately prior to final filtration, the oxygen in the dilution water will be removed by the yeast, and the requirement for expensive oxygen deaeration equipment is circumvented. However, the longer the concentrated beer is maintained undiluted, the greater is the capacity efficiency, and investigations have indicated that beers are more stable (physical and flavor), when they are processed in the undiluted stage. Consequently most breweries add the water to the concentrated beer immediately after the final polishing filter but prior to the trap filter. The water for dilution at this point in the process requires special treatment, in order to ensure the quality of the finished beer. Such treatment is to secure biological purity and chemical consistency and encompasses filtration, pH adjustment and, occasionally, ozonization, UV treatment or pasteurization. In addition to these measures, the dissolved oxygen content of the water must be reduced to a level of approximately 0.1 mg/L. This can be achieved by vacuum deaeration using either a hot or cold process. The hot system flashes water at 77 °C, and the cold system flashes water at a temperature of 3–24 °C through the vacuum deaeration.

The blending of the deaerated water with the concentrated beer is usually based on the concept of the primary flow of beer. The dilution rate is set according to the alcohol level required, and dilution may reach as high as 40% using conventional brewing techniques with lauter tuns. Indeed, wort concentrations above 16 °P may be achieved by employing alternative procedures such as using syrups in the kettle or applying mash filters to separate the spent grains from the extractable matter. Beer produced under high gravity conditions has a marked improvement in both flavor and physical stability. Recent studies at the 16 hL pilot plant level with 25 °P wort were successfully performed by altering such parameters as the fermentation temperature, wort oxygen level and yeast pitching rate (D'AMORE et al., 1991). Differences in the production of acetaldehyde, esters, higher alcohols and yeast intracellular trehalose were observed. The physical stability was improved, but there was a marked decrease in foam stability. Nevertheless, the resulting beers were found to be comparable to commercial products.

8.2 Immobilized Cell Technology

The first reported industrial application of immobilized cell technology was in China in the mid 1980s (HU, 1988), but pilot plant and full industrial scale processes encountered difficulties that were not easily solved. Engineering problems linked to the choice of carrier and reactor design were complicated by the effects of immobilization on the flavor profile of the beer products. Incomplete knowledge of the effects of immobilization on the physiology of brewer's yeast led to the incomplete and partly empirical use of the immobilized cell technology for brewing (NAKANISHI et al., 1985; PAJUNEN et al., 1987, 1989; LOMMI, 1990; AIVASIDIS et al., 1991).

Immobilized yeast systems can be and are used for the production of alcohol-free or low alcohol beer. The suppression of alcohol formation by arrested batch fermentation is widely accepted as a basic principle for the production of such beers (NARZISS et al., 1992). The technology is attractive in terms of low capital costs and operating simplicity but often fails when rated on product quality and flavor consistency. The beers produced are frequently characterized by an undesirable residual wort taste and aroma. Controlled ethanol production for low and non-alcoholic beers has been successfully achieved by partial fermentation using DEAE cellulose immobilized yeast columns. An industrial-scale packed-bed reactor is operating at the Bavaria brewery in the Netherlands for the production of alcohol-free beers (LOMMI, 1990).

An alternative approach has been developed at the Brussels Center for Microbial and Food Engineering (CERIA), in association with Schelde Delta, to produce beer with a final alcohol content below 0.5% in volume (VAN DE WINKEL et al., 1991). Immobilization was by passive colonization of sintered silicon carbide matrices.

A process, which has proven to be advantageous for the production of alcohol-free beer, is continuous fermentation through a fluidized-bed reactor using yeast immobilized on 1–2 mm diameter sintered macroporous glass particles having pore volumes up to 60% (60–300 μm pores). The system has been developed by Becks in association with the Institute for Biotechnology in Jülich, Germany (AIVASIDIS et al., 1991).

Immobilized yeast cell technology (cells attached to a solid matrix) is an area of current interest in the brewing industry (RUSSELL and STEWART, 1992; NORTON and D'AMORE, 1994). Although the technology was actively researched in the early 1970s, the concept fell into disfavor due to technical problems and flavor problems with the finished product, particularly due to high diacetyl levels. As previously discussed, the vicinal diketones, diacetyl and 2,3-pentanedione, are by-products of amino acid biosynthesis, specifically the isoleucine-valine pathway. They give a sweet butterscotch flavor to the beer. A low level of diacetyl is characteristic of some ales, but in most lagers it is undesirable and is considered a defect. At the end of fermentation, an active yeast is required to convert the diacetyl into the less flavor-active compounds acetoin and butanediol. In the past, diacetyl removal was one of the major problems encountered with immobilized cell systems. There are, currently, a number of processes that are at, or are approaching, the commercialization stage that address the problems associated with high beer diacetyl levels.

8.2.1 Multi-State Fermentation

A Japanese brewing company has a pilot plant which employs a two-stage fermentation system. The wort is fermented quickly in a fermenter using free cells and a stirrer. The young beer is then centrifuged and the resultant clear beer is passed through an immobilized yeast column (diacetyl is removed at this stage). The system is rapid (process time is two–three days) and can operate for more than three months before column regeneration (NAKANISHI et al., 1985). Studies are continuing with a commercially operating mini brewery employing the technology to further improve the process.

8.2.2 Main Fermentation Employing a Genetically Modified Yeast

The VTT Biotechnology Laboratory in Helsinki, Finland, has approached the problem of eliminating diacetyl in a different way. The main fermentation is conducted with a conventional brewer's yeast that has been immobilized, but it was found that high levels of diacetyl remained in the beer (LINKO and KRONLÖF, 1991). Their solution to the problem is to employ a genetically modified yeast cloned with the gene that encodes for an α-acetolactate decarboxylase (ALDC). This enzyme decarboxylates α-acetolactate, thereby preventing the formation of diacetyl. It is derived from the bacterium *Klebsiella terrigena*. The ultimate goal of this process is to conduct the entire fermentation and aging process in one continuously operated bioreactor with a residence time of one day, as compared to 7–10 days for fermentation and 10–30 days for aging. The problem of obtaining approval from the appropriate authorities to use a genetically modified yeast is a separate issue and could prove to be a major obstacle.

8.2.2.1 Continuous Secondary Fermentation

This system is currently in operation at the recently commissioned Sinebrychoff Kerava Brewery in Finland (PAJUNEN and JÄÄSKELÄINEN, 1993) and employs a continuous secondary fermentation to shorten beer aging time. Following a standard primary brewing fermentation, the green beer is passed at a flow rate of 40 hL/h through an immobilized yeast reactor system consisting of 6000 L and two 1000 L yeast reactors. Initial problems were encountered because of the slow rate of conversion of α-acetolactate to diacetyl. In order to overcome this problem, a rapid heat treatment of the beer (90 °C for 7 minutes) was employed before the immobilized yeast column to effect the conversion of α-acetolactate to diacetyl. The Sinebrychoff maturation system did not demonstrate any detectable differences in beer flavor to beer matured using traditional technology, and the beer quality falls into the acceptable range. Total lagering time was reduced from 10–14 days to 2–3 hours. The major economic benefits of this process are savings in capital cost and residence time. Currently 30% of the brewery's product is produced in this manner and then blended into a beer stream produced in the traditional way.

8.2.3 Summary

In the last five years, important breakthroughs have been achieved in the understanding of the effect of immobilization on yeast, resulting in a growing interest by the brewing and other fermentation-based industries in this technology. The main problems to solve are still linked to the flavor of beer. It is clear that a better understanding of yeast physiology, especially with regard to amino acid utilization and fatty acid type and content, will improve our capacity to design proper fermentation processes. Further research aimed at the design and optimization of carriers and process parameters such as fluidization, controlled aeration and temperature programming is still required.

8.3 Beer Categories

Light (Lite) beers are defined differently in Canada and the United States. In Canada, the beer category is defined by alcohol concentration, and light beers contain between 2.6 and 4.0% (v/v) alcohol. In the United States, light beers are defined by calorie content and must have at least a 20% reduction in calories when compared to regular beers.

Non-alcohol beers in Canada and the United States must have an alcohol content of 0.5% (v/v) or less. Alcohol-free beers, for importation into a number of Islamic countries, must contain less than 0.05% (v/v) alcohol. There are two main groups of processes in use to produce low-alcohol beer. One group consists of manipulation of the fermentation process and the second involves alcohol separation after fermentation (MARCHBANKS, 1987; GONZALES DEL CUETO, 1992).

The first group can involve: (1) the use of low gravity wort; (2) premature termination of fermentation (interrupted fermentation); (3) immobilized yeast processes; and (4) reduced ethanol production by use of special yeasts with restricted fermentation capabilities (LIEBERMAN, 1984). These beers often have a worty taste due to the presence of appreciable amounts of non-fermented extract and are reported to be missing some of the flavor compounds produced during a normal fermentation (STEIN, 1993).

The second group depends on different methods of alcohol separation. They include: (1) evaporation of alcohol (centrithermal evaporation, falling film); (2) use of membranes to remove alcohol (dialysis, reverse osmosis, pervaporation); (3) adsorption of alcohol; and (4) solvent extraction of alcohol (super-critical CO_2). It should be remembered that pure alcohol, although largely tasteless, works as a taste intensifier on other flavors and as a result, a dealcoholized beer will probably never taste the same as the original product (STEIN, 1993).

To overcome the flavor problems encountered in the two aforementioned processes, a number of innovative techniques have been used including the brewing of special flavor-rich beers. For example, use can be made of specialty malts (e.g., Crystal) to produce a strongly flavored beer which after dealcoholization still retains sufficient volatile constituents to produce a flavor similar to the regular beers. Work has been carried out on the recovery and restoration to the beer of flavor volatiles by BARRELL's patented process (BARRELL, 1990), where flavor volatiles are removed from a high gravity beer by carbon dioxide stripping and added to low gravity beer. Other methods include production of a fuller flavored low alcohol-free beer with a beer of normal alcohol content, and the use of flavoring additives in countries where permitted (KULANDAI et al., 1994).

Dry beers had their genesis in Japan. Consumer studies by a major Japanese brewer (Asahi) revealed that although the overall consumption of Saké in Japan had fallen throughout the 1970s and early mid 80s, the consumption of low sugar (dry) Saké had increased. This led to the development of dry beers. A dry alcohol beverage is not very sweet and is described as smooth, refreshing and not sticky. To produce this the brewers increased the original extract content in their dry beers above that of their regular beer while lowering the bitterness factor (USUBA, 1989). The beers are fermented to a greater degree than regular beers and consequently have a higher alcohol content and lower apparent and real extract.

Genuine draft beers have a variety of definitions depending largely on the country of production and sale. In Canada (where bulk draft beer is usually not pasteurized) the Federal government has agreed with the Brewer's Association of Canada on the following definition: "cans and bottles containing the word 'draught' or 'draft' signifies that the contents have not been pasteurized". In Germany, Britain and a number of other countries, where bulk draft beer is routinely pasteurized, this definition would not apply.

Kirin Ichiban is a Japanese beer produced from the fermentation of first wort. First wort in this context means the early wort from the lauter tun with little sparging of the grain bed, no last runnings and no post run-off re-cycling. It is claimed that such a beer is smoother and more drinkable due to the absence of polyphenols and tannins which tend to induce astringency. This concept had its genesis in tea-making where tea leaves that are only steeped in hot water for a short while produce a beverage that lacks astringency and is very drinkable.

8.3.1 Ice Beer™

A novel brewing process, termed "Ice Brewing™" was patented by a Canadian brewer in 1994 (MURRAY and VAN DER MEER, 1994). The "Ice Brewing™" process begins with the transfer of fermented beer to scraped surface heat exchangers (SSHE) at a temperature of about −1 °C to −2 °C. The SSHEs are fed by positive displacement pumps which fix the rate of flow through the heat exchangers. The SSHEs further cool the beer to about its freezing point, i.e., about −3.5 °C to −4.5 °C (product dependent), when small ice crystals form in the beer. The "Ice Brewing™" process utilizes a heat exchanger which is not plugged by the small ice crystals. This is unlike conventional beer refrigeration processes which generally do not use temperatures as cold as −4 °C because of risk of ice crystal formation blocking and damaging the heat exchanger. Conventional processes therefore tend to control temperatures to avoid approaching the freezing point of the beer. In the "Ice Brewing™" process, the mixture of beer and small ice crystals is then mixed with a slurry of larger ice crystals in beer maintained at the same beer freezing temperature. This latter slurry is usually contained in a "Recrystallizer", a vessel originally developed by Netherlands-based NIRO Process Technologies B.V. for use in its freeze concentration process. The vessel is equipped with a mixer which keeps the ice and beer slurry completely mixed and homogeneous. A series of rotating screens retain the "bank" of ice crystals (automatically controlled at a level of 10% to 25% of the total beer volume) within the "Recrystallizer" while allowing ice crystal-free beer to pass from the vessel into the aging tank. At the completion of a cycle, the beer in the system is pushed out of the "Recrystallizer" using CO_2. In this process it should be noted that the beer is not concentrated, although for convenience, the bank of ice in the "Recrystallizer" may be formed initially from the first portion of beer being treated in the process.

Young beer when maintained cooled over time creates a "chill" haze that appears in the liquid. The aforementioned process provides optimum temperature conditions for the early formation of such haze in a consistent manner. When formed, this haze is fine-filtered from the product, resulting in a well-balanced beer that remains brilliant. The process provides a simple to operate continuous system to produce a balanced beer which is less harsh, more mellow and has greatly increased shelf life due to increased physical stability compared with regular beers. This latter characteristic provides significant economic benefit in greatly reducing the time required for regular aging.

This "Ice Brewing™" process may be contrasted with that used to produce the specialty extra strong bock beers of Germany, termed "Eisbock" beers, in which the beer is partially frozen to form ice, which is then removed, to concentrate the brew to 8% alcohol by weight or more. A number of other brewers, in North America and elsewhere, now produce their own so-called "ice" beers, details of the processes used by these brewers are not available.

9 The Future

Compared to other industries, the brewing process has undergone little change over the past ten years. The adoption of high gravity brewing, use of high maltose syrups, the introduction of a new type of mash filter and the use of specialty hop products are the major areas of change. Greater care is also being taken by the breweries with regard to oxygen reduction throughout the process to prevent oxidation or staling of the product. The use of genetically manipulated yeast to date has not found general acceptance. The most novel technology on the horizon is the use of immobilized cell technology. There are still problems to be addressed with this technology, but if these can be solved and an acceptable product produced, major changes in the process will result.

In the future, the brewing industry will be very concerned with productivity and efficiency at minimum cost but without detriment to beer quality. At the same time the marketplace will require a constant stream of novel products. Productivity will concern both capital and operating costs, consequently shorter process times will be critical and this will stimulate further developments in high gravity brewing (with yeast strains that possess enhanced stress tolerance) and further decreases in fermentation and aging times. Efficiency will require increased process automation, on-line analysis and control, and consistent high-quality raw materials but at low cost. The consumer will expect quality beer with consistency in all areas – flavor, physical, microbiological, foam and light stability.

10 References

ABBOTT, M. S., PUGH, T. A., PRINGLE, A. T. (1993), Biotechnological advances in brewing; in: *Beer and Wine Production* (GUMP, B. H., Ed.), pp. 150–180, Washington: American Chemical Society.

ADVISORY COMMITTEE ON NOVEL FOODS AND PROCESSES (1994), *Food Safety Clearance for New Brewing Yeast* (press release), ACNFT 2/94.

AIVASIDIS, A., WANDREY, CH., EILS, H.-G., KATZKE, M. (1991), Continuous fermentation of alcohol-free beer with immobilized yeast in fluidized bed reactor, in: *Proc. 23rd Eur. Brew. Conv. Congr.*, Lisbon, pp. 569–576, Oxford: IRL Press.

ANDERSON, R. G. (1992), The pattern of brewing research: a personal view of the history of brewing chemistry in the British Isles, *J. Inst. Brew.* **98**, 85–109.

ANDERSON, R. G. (1993), Highlights in the history of international brewing science, *Ferment* **6**, 191–198.

BACK, W. (1987), Detection and identification of wild yeasts in the brewery, *Brauwelt Int.* **II**, 174–177.

BARNETT, J. A., PAYNE, R. W., YARROW, D. (1983), *Yeasts: Characteristics and Identification*, Cambridge: Cambridge University Press.

BARNETT, J. A., PAYNE, R. W., YARROW, D. (1990), *Yeasts: Characteristics and Identification, 2nd Ed.*, Cambridge: Cambridge University Press.

BARRELL, G. W. (1990), Process for the production of fermented beverages, *Brit. Patent* 2033:424.

BERNSTEIN, L., WILLOX, I. C. (1977), Water, in: *The Practical Brewer* (BRODERICK, H. M., Ed.), pp. 13–20. Madison: MBAA.

BILINSKI, C. A., RUSSELL, I., STEWART, G. G. (1986), Analysis of sporulation in brewer's yeast: Induction of tetrad formation, *J. Inst. Brew.* **92**, 594–598.

BILINSKI, C. A., RUSSELL, I., STEWART, G. G. (1987), Crossbreeding of *Saccharomyces cerevisiae* and *Saccharomyces uvarum (carlsbergensis)* by mating of meiotic segregants: Isolation and characterization of species hybrids, in: *Proc. 21st Eur. Brew. Conv. Congr.*, Madrid, pp. 499–504, Oxford: IRL Press.

BLAND, J. (1993), Water reuse and energy consideration in the 90's brewery ... practical considerations, advantages and limitations, *MBAA Tech. Quart.* **30**, 86–89.

BOULTON, C. A., BESFORD, R. P. (1992), A critical comparison of NIR turbidometry and radiofrequency permittivity for in-line measurement of yeast concentration, in: *Eur. Brew. Conv. Monogr.* **XX**, Symposium on Instrumentation and Measurement, Copenhagen, pp. 81–94, Oxford: IRL Press.

BOULTON, C. A., MARYAN, P. S., LOVERIDGE, D., KELL, D. B. (1989), The application of a novel biomass sensor to the control of yeast pitching rate, in: *Proc. 22nd Eur. Brew. Conv. Congr.*, Zurich pp. 653–661, Oxford, IRL Press.

BRENNER, M. W. (1965), Disinfection of brewing yeast with acidified ammonium persulfate (communication to the editor), *J. Inst. Brew.* **71**, 290.

BRIGGS, D. E., HOUGH, J. S., STEVENS, R. (1981), Water economy in maltings and breweries, in: *Malting and Brewing Science,* 2nd Ed., pp. 209–211, London: Chapman & Hall.

BUSTURIA, A., LAGUNAS, R. (1985), Identification

of two forms of the maltose transport system in *Saccharomyces cerevisiae* and their regulation by catabolite inactivation, *Biochim. Biophys. Acta* **82**, 324–326.

CANALES, A. M., SIERRA, J. A. (1976), Use of sorghum, *MBAA Tech. Quart.* **13**, 114–116.

CASEY, G. P. (1990), Yeast selection in brewing, in: *Yeast Strain Selection* (PANCHAL, C. J., Ed.), Vol. 7, pp. 65–111, New York: Marcel Dekker.

CHANTLER, J. (1990), Third generation brewers adjunct and beyond, *MBAA Tech. Quart.* **27**, 78–82.

CLARKE, B. J. (1986), Centenary review: Hop products, *J. Inst. Brew.* **92**, 123–130.

COLLIER, J. (1986), Trends in U.K. usage of brewing adjuncts, *Brew. Distill. Int.* **16**, 15–17.

COLLIN, S., MONTESINOS, M., MEERSMAN, E., SWINKELS, W., DUFOUR, J.-P. (1991), Yeast dehydrogenase activities in relation to carbonyl compounds removal from wort and beer, in: *Proc. 23rd Brew. Conv. Congr.*, Lisbon, pp. 409–416, Oxford: IRL Press.

CONDE, J., FINK, G. R. (1976), A mutant of *Saccharomyces cerevisiae* defective for nuclear fusion, *Proc. Natl. Acad. Sci. USA* **73**, 3651–3655.

CONDE ZURITA, J., MASCORT SUÁREZ, J. L. (1981), Effect of the mitochondrial genome on the fermentation behavior of brewing yeast, in: *Proc. 18th Eur. Brew. Conv. Congr.,* Copenhagen, pp. 177–186, Oxford: IRL Press.

COORS, J. (1976), Practical experience with different adjuncts, *MBAA Tech. Quart.* **13**, 117–123.

CROMBIE, L., CROMBIE, W. M. L. (1975), Cannabinoid formation in *Cannabis sativa* grafted interracially, and with two *Humulus* species, *Phytochemistry* **14**, 409–412.

CRUMPLEN, R. M., D'AMORE, T., RUSSELL, I., STEWART, G. G. (1990), The use of spheroplast fusion to improve yeast osmotolerance, *J. Am. Soc. Brew. Chem.* **48**, 58–61.

D'AMORE, T. (1992), Cambridge Prize Lecture – Improving yeast fermentation performance, *J. Inst. Brew.* **98**, 375–382.

D'AMORE, T., CELOTTO, G., STEWART, G. G. (1991), Advances in the fermentation of high gravity wort, in: *Proc. 23rd Eur. Brew. Conv. Congr.*, Lisbon, pp. 337–344, Oxford: IRL Press.

DAOUD, I. S., KUSINSKI, S. (1992), Liquid CO_2 and ethanol extraction of hops. Part I: Effect of hop deterioration on extraction efficiency and extract quality, *J. Inst. Brew.* **98**, 37–41.

DAY, A. W., POON, N. H., STEWART, G. G. (1975), Fungal fimbriae III. The effect on flocculation in *Saccharomyces cerevisiae, Can. J. Microbiol.* **21**, 558–564.

DERCKSEN, A. W., MEIJERING, I., AXCELL, B. (1992), Rapid quantification of flavor-active sulfur compounds in beer, *J. Am. Soc. Brew. Chem.* **50**, 93–101.

DOUGHERTY, J. J. (1977), Wort production, in: *The Practical Brewer* (BRODERICK, H. M., Ed.), pp. 62–98, Madison: MBAA.

ENGAN, S. (1981), Beer composition: Volatile substances, in: *Brewing Science* (POLLOCK, J. R. A., Ed.), Vol. 2, pp. 93–165, London: Academic Press Inc.

ERNANDES, J. R., WILLIAMS, J. W., RUSSELL, I., STEWART, G. G. (1993), Respiratory deficiency in brewing yeast strains – effects on fermentation, flocculation, and beer flavor components, *J. Am. Soc. Brew. Chem.* **51**, 16–20.

EYBEN, D., HERMIA, J., MEURENIS, J., RAHIER, G., TIGEL, R. (1989), Industrial results using the new mash filter 2001, in: *Proc. 22nd Eur. Brew. Conv. Congr.,* Zurich, pp. 275–282, Oxford: IRL Press.

FALCO, S. C., DUMAS, K. S. (1985) Genetic analysis of mutants of *Saccharomyces cerevisiae* resistant to the herbicide sulfometuron methyl, *Genetics* **109**, 21–37.

FREEMAN, R. J. (1981), Construction of brewing yeasts for production of low carbohydrate beer, in: *Proc. 18th Eur. Brew. Conv. Congr.*, Copenhagen, pp. 497–504, Oxford: IRL Press.

GARDNER, D. J. S. (1993), Hop products: Their use in brewing, *Ferment* **6**, 279–282.

GIST-BROCADES, N. V., OSINGA, K. A., BEUDEKER, R. F., VAN DER PLAAT, J. B., DE HOLLANDER, J. A. (1988), New yeast strains providing for an enhanced rate of the fermentation of sugars, a process to obtain such yeasts and the use of these yeasts, *Eur. Patent* EP 0306107 A2.

GJERMANSEN, C., SIGSGAARD, P. (1981), Construction of a hybrid brewing strain of *Saccharomyces carlsbergensis* by mating of meiotic segregants, *Carlsberg Res. Commun.* **46**, 1–11.

GONZALES DEL CUETO, A. (1992), A review of the U.S. patents related to the preparation of low caloric and low alcohol content beer, *Brew. Dig.* **67**, 16–19.

GOODEY, A. R., TUBB, R. S. (1982), Genetic and biochemical analysis of the ability of *Saccharomyces cerevisiae* to decarboxylate cinnamic acids, *J. Gen. Microbiol.* **128**, 2615–2620.

GRÖNQVIST, A., SIIRILÄ, J., VIRTANEN, H., HOME, S., PAJUNEN, E. (1993), Carbonyl compounds during beer production and in beer, in: *Proc. 24th Eur. Brew. Conv. Congr.,* Oslo, pp. 421–428, Oxford: IRL Press.

GUINARD, J.-X., LEWIS, M. J. (1993), Study of the phenomenon of agglomeration in the yeast *Saccharomyces cerevisiae, J. Inst. Brew.* **99**, 487–503.

GUZINSKI, J. A., STEGINK, L. J. (1993), Stable aqueous solutions of tetrahydro and hexahydro *iso-alpha* acids. *U.S. Patent* 5,200,227.

HAIKARA, A. (1980), Gushing induced by fungi, in: *Eur. Brew. Conv. Monogr.* **VI**, Symposium on the Relationship between Malt and Beer, pp. 251–259, Nürnberg: Verlag Hans Carl.

HAMMOND, J. R. M., ECKERSLEY, K. W. (1984), Fermentation properties of brewing yeast with killer character, *J. Inst. Brew.* **90**, 167–177.

HARDWICK, B. C., DONLEY, J. R., BISHOP, G. (1976), The interconversion of vicinal diketones and related compounds by brewers' yeast enzymes, *J. Am. Soc. Brew. Chem.* **34**, 65–67.

HENSON, M. G., REID, D. M. (1988), Practical management of lager yeast, *Brewer* **74**, 3–8.

HINNEN, A., HICKS, J. B., FINK, G. R. (1978), Transformation of yeast, *Proc. Natl. Acad. Sci. USA* **75**, 1929–1933.

HOUGH, J. S. (1973), Some aspects of scientific control within the brewery – a review on yeast quality control, *Brewer* **59**, 69–73.

HOUGH, J. S., BRIGGS, D. E., STEVENS, R. (1971), Brewing water, in: *Malting and Brewing Science,* pp. 170–195, London: Chapman & Hall.

HOUGH, J. S., BRIGGS, D. E., STEVENS, R., YOUNG, T. W. (1982), *Malting and Brewing Science,* Vol. II, *Hopped Wort and Beer.* 2nd Ed., New York: Chapman & Hall.

HOWARD, G. (1956), The configuration of Humulones (communication to the editor), *Chem. Ind.* **75**, 1504.

HU, J. (1988), Enzyme technology in China, *Int. Ind. Biotechnol.* **8**, 16–17.

IRWIN, A. J., BARKER, R. L., PIPASTS, P. (1991), The role of copper, oxygen and polyphenols in beer flavor instability, *J. Am. Soc. Brew. Chem.* **49**, 140–149.

IRWIN, A. J., BORDELEAU, L., BARKER, R. L. (1993), Model studies and flavor threshold determination of 3-methyl-2-butene-1-thiol in beer, *J. Am. Soc. Brew. Chem.* **51**, 1–3.

JONES, M., PIERCE, J. S. (1967), The role of proline in the amino acid metabolism of germinating barley, *J. Inst. Brew.* **73**, 577–583.

JONES, R. M., RUSSELL, I., STEWART, G. G. (1986), The use of catabolite derepression as a means of improving the fermentation rate of brewing yeast strains, *J. Am. Soc. Brew. Chem.* **44**, 161–166.

JONES, R. M., RUSSELL, I., STEWART, G. G. (1987), Classical genetic and protoplast fusion techniques in yeast, in: *Yeast Biotechnology* (BERRY, D. R., RUSSELL, I., STEWART, G. G., Eds.), pp. 55–79, London: Allen and Unwin.

KIELLAND-BRANDT, M. C., NILSSON-TILLGREN, T., HOLMBERG, S., PETERSEN, J. G. L., SVENNINGSEN, B. A. (1979), Transformation of yeast without the use of foreign DNA, *Carlsberg Res. Commun.* **44**, 77–87.

KIELLAND-BRANDT, M. C., GJERMANSEN, C., NILSSON-TILLGREN, T., HOLMBERG, S., PETERSEN, J. G. L. (1988), Diacetyl and brewers' yeast, in: *Yeast* (SUIHKO, M.-L., KNOWLES, J., Eds.), Vol. 4 (Spec. Iss.), p. S470, Chichester: Wiley.

KREGER-VAN RIJ, N. J. W. (1984), *The Yeasts: A Taxonomic Study,* 3rd Ed., Amsterdam: Elsevier Science Publishers.

KULANDAI, J., HAWTHORNE, D. B., KAVANAGH, T. E. (1994), Low alcohol beers – the continuing development, *Food Aust.*, April, 175–178.

LAWRENCE, D. R., PRIEST, F. G. (1981), Identification of brewery cocci, in: *Proc. 18th Eur. Brew. Conv. Congr.*, Copenhagen, pp. 217–228, Oxford: IRL Press.

LEEMANS, C., DUPIRE, S., MACRON, J.-Y. (1993), Relation between wort DMSO and DMS concentration in beer, in: *Proc. 24th Eur. Brew. Conv. Congr.*, Oslo, pp. 709–716, Oxford: IRL Press.

LIEBERMAN, C. E. (1984), Low alcohol beers, *Brew. Dig.* **59**, 30–31.

LINKO, M., KRONLÖF, J. (1991), Main fermentation with immobilized yeast, in: *Proc. 23rd Eur. Brew. Conv. Congr.*, Lisbon, pp. 353–360, Oxford: IRL Press.

LODDER, J. (1970), *The Yeasts: A Taxonomic Study*, Amsterdam: North-Holland.

LOMMI, H. (1990), Immobilized yeast for maturation and alcohol-free beer, *Brew. Distill. Int.* **21**, 22–23.

MARCHBANKS, C. (1987), A review of carbohydrate sources for brewing, *Brew. Distill. Int.* **17**, 16–18.

MCFARLANE, I. K. (1993), Mash conversion and preparation, *Ferment* **6**, 177–183.

MCLAIN, L. (1994), Genetically engineered yeast 'breakthrough' for brewers, *Brew. Distill. Int.* **25**(5), 16–17.

MEILGAARD, M. C. (1976), Wort composition: With special reference to the use of adjuncts, *MBAA Tech. Quart.* **13**, 78–90.

MELIS, M., EYBEN, D. (1992), The mash filter 2001: Latest industrial results, *MBAA Tech. Quart.* **29**, 18–19.

MIEDANER, H. (1986), Centenary Review: Wort boiling today – old and new aspects, *J. Inst. Brew.* **92**, 330–335.

MIKI, B. L. A., POON, N. H., JAMES, A. P., SELIGY, V. L. (1982), Possible mechanism for flocculation interactions governed by gene *FLO1* in *Saccharomyces cerevisiae, J. Bacteriol.* **150**, 878–889.

MITCHEL, D. J., BEAVAN, E. A. (1987) dsRNA killer systems in yeast, in: *Yeast Biotechnology* (BERRY, D. R., RUSSELL, I., STEWART, G. G., Eds.), pp. 104–156, London: Allen and Unwin.

MURRAY, C. R., STEWART, G. G. (1991), Experience with high gravity lager brewing, *Birra Malto.* **44**, 52–64.

MURRAY, C. R., VAN DER MEER, W. (1994), Improvements in production of fermented malt beverages, *U.S. Patent* 5,304,384.

NAKANISHI, K., ONAKA, T., INOUE, T., KUBO, S. (1985), A new immobilized yeast reactor system for rapid production of beer, in: *Proc. 20th Eur. Brew. Conv. Congr.,* Helsinki, pp. 331–338, Oxford: IRL Press.

NARZISS, L. (1984), The German beer law, *J. Inst. Brew.* **90**, 351–358.

NARZISS, L. (1986), Centenary Review: Technological factors of flavour stability, *J. Inst. Brew.* **92**, 346–353.

NARZISS, L. (1994), Beer quality 1993, *Brauwelt Int.* **I**, 20–28.

NARZISS, L., MIEDANER, H., KERN, E., LEIBHARD, M. (1992), Technology and composition of non-alcoholic beers, *Brauwelt Int.* **IV**, 396–410.

NEVE, R. A. (1986), Hop breeding worldwide – its aims and achievements, *J. Inst. Brew.* **92**, 21–24.

NEVE, R. A. (1991), Botany, in: *Hops,* pp. 1–23, London: Chapman & Hall.

NORTON, S., D'AMORE, T. (1994), Physiological effects of yeast immobilization: applications for brewing, *Enzyme Microb. Technol.* **16**, 365–375.

NOVAK, S., D'AMORE, T., RUSSELL, I., STEWART, G. G. (1991), Sugar uptake in a 2-deoxy-D-glucose resistant mutant of *Saccharomyces cerevisiae, J. Ind. Microbiol.* **7**, 35–40.

ORMROD, I. H. L. (1986), Centenary Review: Modern brewhouse design and its impact on wort production, *J. Inst. Brew.* **92**, 131–136.

ORMROD, I. H. L., LALOR, E. F., SHARPE, F. R. (1991), The release of yeast proteolytic enzymes into beer, *J. Inst. Brew.* **97**, 441–443.

PAJUNEN, E., JÄÄSKELÄINEN, K. (1993), Sinebrychoff Kerava – a brewery for the 90's, in: *Proc. 24th Eur. Brew. Conv. Congr.*, Oslo, pp. 559–567, Oxford: IRL Press.

PAJUNEN, E., MÄKINEN, V., GISLER, R. (1987), Secondary fermentation with immobilized yeast, in: *Proc. 21st Eur. Brew. Conv. Congr.*, Madrid, pp. 441–448, Oxford: IRL Press.

PAJUNEN, E., GRÖNQVIST, A., LOMMI, H. (1989), Continuous secondary fermentation and maturation of beer in an immobilized yeast reactor, *MBAA Tech. Quart.* **26**, 147–151.

PALMER, G. H. (1989), *Cereal Science and Technology*, Aberdeen, Scotland: Aberdeen University Press.

PFISTERER, E. A., STEWART, G. G. (1976), High gravity brewing, *Brew. Dig.* **51**, 34–42.

PRIEST, F. G. (1987), Gram-positive brewery bacteria, in: *Brewing Microbiology* (PRIEST, F. G., CAMPBELL, I., Eds.), pp. 121–154, London: Elsevier Applied Science.

REISS, S. (1986), Automatic control of the addition of pitching yeast, *MBAA Tech. Quart.* **23**, 32–35.

RUNKEL, U.-D. (1975), Malt kilning and its influence on malt and beer quality, in: *Eur. Brew. Conv. Monogr.* **II**, Barley and Malting Symposium, pp. 223–235, Nürnberg: Verlag Hans Carl.

RUSSELL, I., STEWART, G. G. (1985), Valuable techniques in the genetic manipulation of industrial yeast strains, *J. Am. Soc. Brew. Chem.* **43**, 84–90.

RUSSELL, I., STEWART, G. G. (1992), Contribution of yeast and immobilization technology to flavor development in fermented beverages, *Food Technol.* **46**, 146–150.

RUSSELL, I., HANCOCK, I. F., STEWART, G. G. (1983), Construction of dextrin fermentative yeast strains that do not produce phenolic off-flavors in beer, *J. Am. Soc. Brew. Chem.* **41**, 45–51.

RUSSELL, I., JONES, R. M., BERRY, D. R., STEWART, G. G. (1987), Isolation and characterization of derepressed mutants of *Saccharomyces cerevisiae* and *Saccharomyces diastaticus*, in: *CRC Biological Research on Industrial Yeasts* (STEWART, G. G., RUSSELL, I., KLEIN, R. D., HIEBSCH, R. R., Eds.), Vol. II, pp. 57–65, Boca Raton: CRC Press.

SEATON, J. C., FORREST, I. S., SUGGETT, A. (1981), High temperature wort boiling – consequence for beer flavour, in: *Proc. 18th Eur. Brew. Conv. Congr.*, Copenhagen, pp. 161–168, Oxford: IRL Press.

SEKI, T., MYOGA, S., LIMTONG, S., UEDONO, S., KAMNUANTA, J., TAGUCHI, J. (1983), Genetic construction of yeast strains for high ethanol production, *Biotechnol. Lett.* **5**, 351–356.

SHIBANO, Y., KODAMA, Y., FUKUI, N., NAKATANI, K. (1993), Improvement of maltose fermentation efficiency in brewing yeasts by molecular breeding. Poster presented in 1993 at the *Am. Soc. Brew. Chem. Ann. Meetg.,* Tucson, Arizona (Publication date 1994 *J. ASBC*).

SIGSGAARD, P., HJORTSHOJ, B. (1992), The yeast cell for tomorrow's brewing, in: *Proc. 5th Int. Brew. Conf.*, Harrogate, pp. 119–127, Richmond, Surrey: Brewing Technology Services Ltd.

SIMPSON, W. J. (1987), Kinetic studies of the decontamination of yeast slurries with phosphoric acid and acidified ammonium persulphate and a method for the detection of surviving bacteria involving solid medium repair in the presence of catalase, *J. Inst. Brew.* **93**, 313–318.

SIMPSON, W. J., HAMMOND, J. R. M. (1989), The response of brewing yeasts to acid washing, *J. Inst. Brew.* **95**, 347–354.

SMITH, I. B. (1991), Yeast fermentation systems – a review, *Brew. Guard.* **120**, 22–26.

SPEERS, R. A., TUNG, M. A., DURANCE, T. D., STEWART, G. G. (1992a), Biochemical aspects of yeast flocculation and its measurement: A review, *J. Inst. Brew.* **98**, 292–300.

SPEERS, R. A., TUNG, M. A., DURANCE, T. D., STEWART, G. G. (1992b), Colloidal aspects of yeast flocculation: A review, *J. Inst. Brew.* **98**, 525–531.

SPENCER, J. F. T., BIZEAU, C., REYNOLDS, N., SPENCER, D. M. (1985), The use of mitochondrial mutants in hybridization of industrial yeast strains. VI. Characterization of the hybrid, *Saccharomyces diastaticus* and *Saccharomyces rouxii*, obtained by protoplast fusion, and its behavior in simulated dough-raising tests, *Curr. Genet.* **9**, 649–652.

STEIN, W. (1993), Dealcoholization of beer, *MBAA Tech. Quart.* **30**, 54–57.

STEWART, G. G. (1981), The genetic manipulation of industrial yeast strains, *Can. J. Microbiol.* **27**, 973–990.

STEWART, G. G., RUSSELL, I. (1981), Yeast flocculation, in: *Brewing Science* (POLLOCK, J. R. A., Ed.), Vol. 2, pp. 61–92, London: Academic Press Inc.

STEWART, G. G., RUSSELL, I. (1985), Modern brewing biotechnology, in: *Comprehensive Biotechnology: The Principles, Applications, Regulations of Biotechnology in Industry, Agriculture, and Medicine* (BLANCH, H. W., DREW, S., WAND, D. I. C., Eds.), Vol. 3, pp. 335–381, Toronto: Pergamon Press.

STEWART, G. G., PANCHAL, C. J., RUSSELL, I. (1983), Current developments in the genetic manipulation of brewing yeast strains – a review, *J. Inst. Brew.* **89**, 170–188.

STEWART, G. G., JONES, R. M., RUSSELL, I. (1985), The use of derepressed yeast mutants in the fermentation of brewery wort, in: *Proc. 20th Eur. Brew. Conv. Congr.*, Helsinki, pp. 243–250, Oxford: IRL Press.

STEWART, G. G., RUSSELL, I., PANCHAL, C. J. (1986), Genetically stable allopolyploid somatic fusion product useful for the production of fuel alcohols, *Canadian Patent* 1199593.

STRATFORD, M., CARTER, A. T. (1993), Yeast flocculation: lectin synthesis and activity, *Yeast* **9**, 371–378.

STRATFORD, M., WILSON, P. D. G. (1990), A review: Agitation effects on microbial cell interactions, *Lett. Appl. Microbiol.* **11**, 1–6.

STRAVER, M. H., KIJNE, J. W., VAN DER AAR, P. C., SMIT, G. (1993), Determinants of flocculence of brewer's yeast during fermentation in wort, *Yeast* **9**, 527–532.

TAYLOR, J. R. N. (1992), Mashing with malted grain sorghum, *J. Am. Soc. Brew. Chem.* **50**, 13–18.

TUBB, R. S., SEARLE, B. A., GOODEY, A. R., BROWN, A. J. P. (1981), Rare mating and transformation for construction of novel brewing yeast, in: *Proc. 18th Eur. Brew. Conv. Congr.*, Copenhagen, pp. 487–496, Oxford: IRL Press.

USUBA, H. (1989), Asahi Super Dry Beer – Its identity and history, *MBAA Tech. Quart.* **26**, 85–88.

VAN DE WINKEL, L., VAN BEREREN, P. C., MASSCHELEIN, C. A. (1991), The application of an immobilized yeast loop reactor to the continuous production of alcohol-free beer, in: *Proc. 23rd Eur. Brew. Conv. Congr.*, Lisbon, pp. 577–584, Oxford: IRL Press.

VAN DEN EYNDE, E. (1991), The DMS story. From malt to beer, *Cerevisia Biotechnol.* **16**, 45–49.

VAN DER AAR, P. C., STRAVER, M. H., TEUNISSEN, A. W. R. H. (1993), Flocculation of brewers' lager yeast, in: *Proc. 24th Eur. Brew. Conv. Congr.*, Oslo, pp. 259–266, Oxford: IRL Press.

VAN SOLINGEN, D., VAN DER PLAAT, J. B. (1977), Fusion of yeast spheroplasts. *J. Bacteriol.* **130**, 946–947.

VAN VUUREN, H. J. J. (1987), Gram-negative spoilage bacteria, in: *Brewing Microbiology* (PRIEST, F. G., CAMPBELL, I., Eds.), pp. 155–186, London: Elsevier Applied Science.

VERZELE, M. (1979), The chemistry of hops, in: *Brewing Science* (POLLOCK, J. R. A., Ed.), Vol. 1, pp. 279–323, London: Academic Press.

VERZELE, M. (1986), Centenary Review: 100 years of hop chemistry and its relevance to brewing, *J. Inst. Brew.* **92**, 32–48.

VINH, N. T. T., VIET, N. V., MAI, N. T. (1993), The use of high percentage of rice as an adjunct in beer brewing, *MBAA Tech. Quart.* **30**, 42–44.

VON WETTSTEIN, D. (1983), "Emil Christian Hansen Centennial Lecture": From pure yeast culture to genetic engineering of brewer's yeast, in: *Proc. 19th Eur. Brew. Conv. Congr.*, London, pp. 97–120, Oxford: IRL Press.

VON WETTSTEIN, D., JENDE-STRID, B., AHRENST-LARSEN, B., ERDAL, K. (1980), Proanthocyanidin-free barley prevents the formation of beer haze, *MBAA Tech. Quart.* **17**, 16–23.

WAINRIGHT, T. (1973), Diacetyl – a review. Part I. Analytical and biochemical considerations, Part II. Brewing experience, *J. Inst. Brew.* **79**, 451–470.

WHITE, F. H., KIDNEY, E. (1979), The influence of yeast strain on beer spoilage bacteria, in: *Proc. 17th Eur. Brew. Conv. Congr.*, Berlin, pp. 801–815, Oxford: IRL Press.

WHITNEY, G. K., MURRAY, C. R., RUSSELL, I., STEWART, G. G. (1985), Potential cost savings for fuel ethanol production by employing a novel hybrid yeast strain, *Biotechnol. Lett.* **7**, 349–354.

WILKIE, D., NUDD, R. C. (1981), Aspects of mitochondrial control of cell surface characteristics in *Saccharomyces cerevisiae,* in: *Current Developments in Yeast Research* (STEWART, G. G., RUSSELL, I., Eds.), pp. 345–349, Toronto: Pergamon Press.

WINGE, O. (1935), On haplophase and diplophase in some Saccharomycetes, *Série de Physiologie* **21**, 7–111.

YOUNG, T. W. (1983), Brewing yeast with anti-contaminant properties, in: *Proc. 19th Eur. Brew. Conv. Congr.*, London, pp. 129–136, Oxford: IRL Press.

YOUNG, T. W. (1987a), Killer yeasts, in: *The Yeasts* (ROSE, A. H., HARRISON, J. S., Eds.), Vol. 2, pp. 131–161, San Diego: Academic Press.

YOUNG, T. W. (1987b), The biochemistry and physiology of yeast growth, in: *Brewing Microbiology* (PRIEST, F. G., CAMPBELL, I., Eds.), pp. 15–48, London: Elsevier Applied Science.

ZARATTINI, R. DE A., WILLIAMS, J. W., ERNANDES, J. R., STEWART, G. G. (1993), Bacterial-induced flocculation in selected brewing strains of *Saccharomyces, Cerevisia Biotechnol.* **18**, 65–70.

ZHENG, X., D'AMORE, T., RUSSELL, I., STEWART, G. G. (1994a), Factors influencing maltotriose utilization during brewery wort fermentations, *J. Am. Soc. Brew. Chem.* **52**, 41–47.

ZHENG, X., D'AMORE, T., RUSSELL, I., STEWART, G. G. (1994b), Transport kinetics of maltotriose in strains of *Saccharomyces, J. Ind. Microbiol.* **13**, 159–166.

12 Wine and Brandy

HELMUT H. DITTRICH

Geisenheim, Federal Republic of Germany

Introduction

Wine is the product obtained exclusively by the complete or partial fermentation of fresh, crushed grapes or grape must. *White wine* is produced by fermentation of the must obtained draining the juice from crushed grapes or by pressing of the grapes. The characteristic color of *Red wine* is extracted from the skins of crushed, red grapes by the alcohol formed during the fermentation. The requirement for the use of "fresh" grapes eliminates the use of dried grape berries (raisins).

Champagne is a foaming wine which contains dissolved CO_2. The CO_2 must be formed by fermentation, and the gas pressure at 20°C must be at least 3 bar. *Sparkling wine* may be produced by carbonation, that is, the injection of CO_2 into the wine.

Fruit wines are made from pomes, pit fruits, or berries. They resemble grape wines. Wines may also be produced from the juice of rhubarb, (sugar) maple, palms, agaves, or from honey or malt. *Liqueur wines* are sweet and have a higher alcohol content. They are produced from fruit (also from grape) wines. The concept of "wine" also includes the *wines for distillation*. These are wines to which distillates of wines have been added to increase the alcohol content to 18–24 vol.-%. Fortifying brandy is distilled from wine or lees (so-called distilling material) at 85 vol.-% ethanol. Distillates of fermented fruit juices or mashes are "fruit brandies".

These definitions which can be found in the regulations of wine-producing countries indicate the basics of the production technologies of the various alcoholic beverages.

White wine: The harvested grapes are destemmed and crushed. Most of the insoluble material of the must is removed, and – if required – the must may be fortified with sucrose. Fermentation starts either spontaneously or by the addition of starter cultures. The wine is separated from the yeast and other insoluble materials, SO_2 is added, and other legal treatments may be applied. After a shorter or longer storage period additional SO_2 may be added. The wine is then filtered through a filter which retains yeasts and bacteria and is then bottled. Some wines may be sweetened with grape juice or grape juice concentrate.

Red wine: The characteristic color is obtained by extraction from the grape skins of dark colored grapes by the alcohol produced in the early stages of the fermentation. The alcohol extracts the anthocyanins located in the grape skins. The mash is pressed after completion of the fermentation. The young wine is treated generally in the same way as white wine. However, storage in oak casks and the malo-lactic fermentation occur more frequently.

Champagne is made from white or red wine to which sucrose (25 g/L) and yeast have been added. Fermentation of this added sugar which produces the required CO_2 is carried out in sealed bottles or tanks. After lengthy storage the yeast is removed in such a way that the formed fermentation CO_2 remains in the champagne. Varying amounts of sugar solutions may be added, and the champagne is filtered and bottled.

Brandy is the distillate of wine diluted to an alcohol concentration of 40–45 vol.-%. EU regulations (1576/89) distinguish between "Branntwein aus Wein" (37.5 vol.-% ethanol) and Weinbrand (36 vol.-% ethanol). Particularly in France the distillation is often carried out before the fermentation yeast has been removed. Wine distillates are stored for varying lengths of time in oak barrels. Color and flavor are extracted from the wood into the distillate. After dilution to the desired alcohol concentration the brandy may be sweetened. It is then filtered and bottled. *Fruit brandies:* Ripe, undamaged fruit is macerated. It is then pressed and the juice is fermented or the entire mash is fermented after acidification (if required) and/or the addition of "nutrient" salts (see Sect. 4.6). In this case the entire mash is distilled.

These and other alcoholic beverages have been described by AMERINE et al. (1982) and ROSE (1977); sparkling wines by TROOST and HAUSHOFER (1980); fruit wines by SCHANDERL et al. (1981). Appropiate chapters can be found in WÜRDIG and WOLLER (1989). The technology of wine making has been described by TROOST (1988) and the distillation of fruit wines by PIEPER et al. (1993) as well as by TANNER and BRUNNER (1987).

The most important ingredient of all wines, all fermented mashes and all alcoholic beverages is ethanol. It must be produced by the fermentation of sugar. The producing organism, the yeast *Saccharomyces cerevisiae* (see Fig. 1a, b) is a microscopic fungus. It is the "mother" of wine and all alcoholic beverages.

This yeast has been described by TUITE and OLIVER (1991). It and others yeasts have been treated by ROSE and HARRISON (1987–1991) and KOCKOVÁ-KRATOCHVÍLOVÁ (1990). Yeast and other microorganisms which are important for the production and quality of wine and other alcoholic beverages have been described by RIBÉREAU-GAYON et al. (1975), LAFON-LAFOURCADE (1983), DITTRICH (1987, 1989a), BENDA (1982), FLEET (1993) and BACK (1994).

History

Wine is as old as humanity. The oldest civilized societies used wine in their culture and in their religious services. Wine was also the object of the first regulation of a food: In 1498 the Assembly at Freiburg/Br. (Germany) issued a regulation on "sulfiting". The scientific basis of the production of wine has only been developed during the past 200 years. In 1810 GAY-LUSSAC established the fermentation formula:

$$C_6H_{12}O_6 = 2\,C_2H_5OH + 2\,CO_2$$

In 1818 ERXLEBEN characterized yeast as a living organism, and a few years later the name of the genus, *Saccharomyces* (sugar mold) was coined. The glucophily of yeast and the formation of succinic acid were discovered in 1847. In 1858 PASTEUR described glycerol as a fermentation product, and in 1860 he established the equation for the more important products of fermentative metabolism. His book "Études sur le vin" was published in 1866. There he also described spoilage of wines and methods of "pasteurization".

The use of pure yeast culture starters, common in the brewing industry, was applied to wine production by SEIFERT and MÜLLER-THURGAU. The first pure wine yeast culture facility was founded by WORTMANN in Geisenheim (Germany) in 1894. This started a fruitful epoch in the investigation of wine microorganisms: The occurrence of wine yeasts, their physiology and their behavior during the production of wine were described, as well as the formation of by-products. Many "wild" yeasts and their significance in wine production were described. The microbial reduction of malic acid by yeasts and lactic acid bacteria was characterized as a microbial process. This clarifies the significance of another group of microorganisms after MÜLLER-THURGAU recognized the concept of the noble rot of grapes and the quality improvement of wines by *Botrytis* in 1888.

Beginning in 1915 wine production was revolutionized by introduction of the sterile filtration before bottling. This permitted the production of sugar-containing wines, juices and other sweet beverages. At this time one also explored the effect of SO_2 on wine yeast and the fermentation by the recognition of the binding of SO_2 by acetaldehyde. The role of preservatives was also considered.

The value of the "steered" fermentation was recognized, the "cold fermentation" was generally introduced, and the quality of wines was improved. In most wine-producing countries the quality of the wine was insured by permission to use sorbic acid as a preservative. In the US the use of dimethyldicarbonate was permitted.

Investigations into the metabolism of microorganisms which participate in wine fermentations have produced important results during the past decades. The investigations were required because of significant changes in technology. Fifty years ago a 5000 liter fermenter was considered to be very large. Today 10 to 20 times larger fermentation vessels are not rare. In addition, today's consumer is more critical. He demands information on wine components and their significance for his health. In all wine-producing countries strict regulations protect the consuming public. Consequently the wine of today is one of the best investigated beverages.

1 The Yeasts

Saccharomyces cerevisiae is the most important yeast for the fermentation of the juice of the grape, generally called "must". It is the typical "wine yeast" (Fig. 1a, b) because it can ferment considerable quantities of sugar in musts and mashes quickly and almost quantitatively. Champagne and distiller's yeasts belong to the same species and have similar positive properties.

In addition to this yeast there are numerous other species of importance in wine making. These have been described by KREGER-VAN RIJ (1984) and BARNETT et al. (1990).

1.1 Yeasts on Grape Berries, in Musts and Mashes

Yeasts occur often in fruits. After maturation the skin of grapes shows small fissures which permit the yeasts to enter the must and to multiply greatly. If the grapes are damaged or squashed, the possibilities for infection and multiplication increase greatly.

Normally there is a sufficient number of well fermenting yeasts on the grape skins. Hence the must obtained by pressing or the mash obtained by maceration carry with them the yeasts required for their fermentation.

Yeasts are spread from one vine to another by insects, for instance, by fruit flies (*Drosophila*) which are attracted by the odor of the incipient fermentation or by biting insects such as wasps which transmit the yeasts with their legs and mandibles.

Besides suitable strains of *S. cerevisiae*, which are essential for the production of alcoholic beverages, there are other yeasts which are not required for the fermentation or whose effect is detrimental. These are called "wild" yeasts.

Besides the true wine yeasts there are also weakly fermenting yeasts (Tab. 1). These are generally smaller than *Saccharomyces*. Many of them are not elliptic in shape but lemon-shaped, the so-called apiculate yeasts. A typical species is *Hanseniaspora uvarum* (Fig. 2). Particularly fruits and berries contain almost no other species. Apiculate yeasts account for 90% of the yeasts in freshly pressed grape must. These yeasts and some *Candida* species start a spontaneous fermentation of the must unless starter cultures of *S. cerevisiae* are added to accelerate the start of the fermentation.

The berries also harbor some oxygen-requiring yeasts. These do not ferment or ferment only poorly. They can grow on the surface of wines with a low alcohol concentration and on fruit juices. These film-forming yeasts belong mostly to the genera *Candida*, *Metsch-*

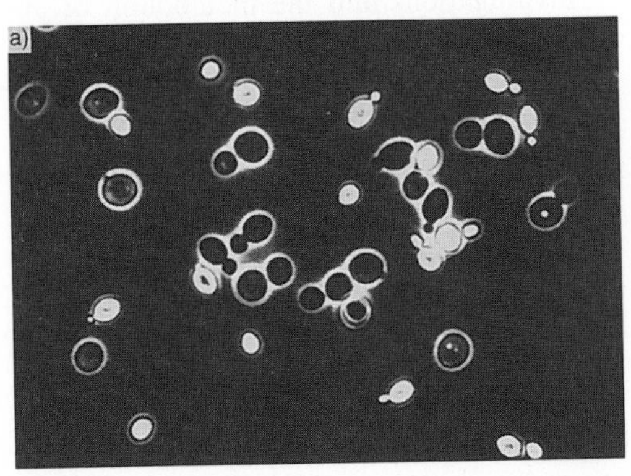

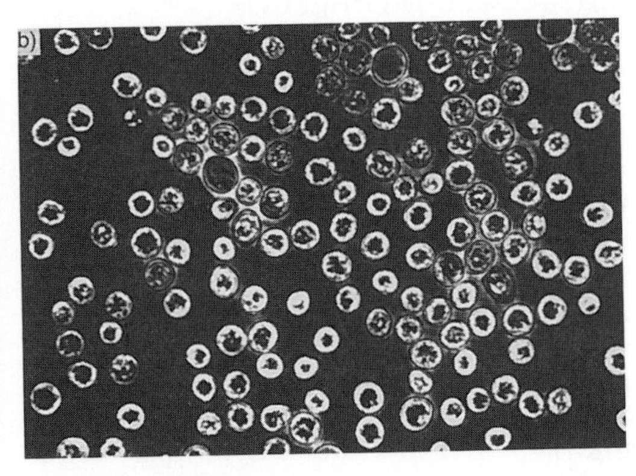

Fig. 1. The wine yeast *Saccharomyces cerevisiae.* Magnification 800-fold (DITTRICH, 1987).
a) Three day old budding pure culture in grape must. The transparent cell content and the presence of vacuoles are typical.
b) The stationary stage in young wines. The granular cell content is typical for these old cells.

Tab. 1. Changes in the Presence of Various Yeast Species and Their Strains During the Fermentation of Must (in %) (SCHÜTZ and GAFNER, 1993)

	Before the Fermentation	Middle of the Fermentation	End of the Fermentation
Hanseniaspora uvarum			
strain 1	8		
strain 2	47		
strain 3	22		
Metschnikowia pulcherrima	22		
Saccharomyces cerevisiae			
strain 1	—	89	100
strain 2	—	11	—

(Grape variety: Blauer Spätburgunder; Marugg, Flaesch, Switzerland)

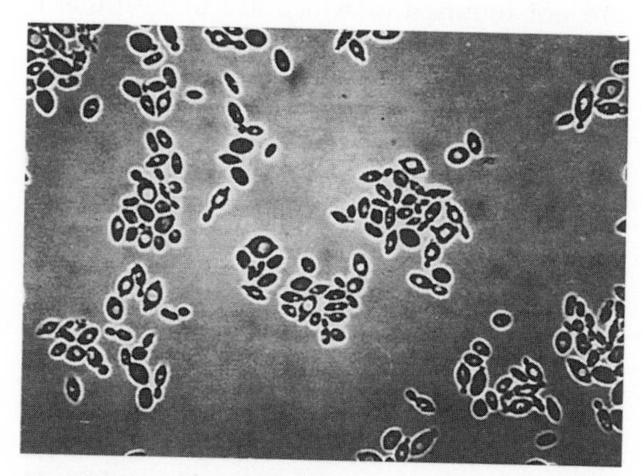

Fig. 2. The typical apiculate yeast, *Hanseniaspora uvarum*. Young budding cells of a pure culture. Magnification 800-fold.

nikowia and *Pichia*. They and other undesirable yeasts are suppressed during professional wine production.

The number of yeast cells is quite variable. Aseptically drawn grape juice contains 10 to 100 cells/mL. In the free run juice this number has increased to 1000 to 10000/mL. The number of yeast cells increases greatly through contact with presses, crusher/destemmers, hoses, pipes and other cellar equipment which has been in contact with must. At this point the musts begin to carry many yeasts. Subsequent sedimentation of the must or centrifuging decreases the number of cells again.

In Switzerland 22 to 808×10^6 cells have been found on undamaged berries per 100 g; in Austria 2.7 to 124×10^6. In Bordeaux freshly pressed must contained 100000 live yeast cells per mL. This corresponds to the presence of 100000 cells on each berry, since the individual berries weighed about 1 g each.

Numbers of yeast cells in the pressed must of a Japanese grape variety varied between 10 and 100000 cells per mL. Forty to 72% of these were apiculate yeasts, 0 to 18% were *Saccharomyces,* 15% *Candida* species, 3 to 22% were film-forming yeasts, and 1 to 4% were red and other yeasts.

The relative distribution of yeast species seems to be quite similar in all grape-growing areas. But a generally applicable minimum number or average number per berry or per mL of pressed must cannot be given.

Grape musts always contain – apart from the yeasts – acetic acid bacteria and lactic acid bacteria. Particularly in humid and cooler vine growing areas the number of yeast and bacterial cells depends on the infection with *Botrytis.* Musts from healthy berries are likely to have lower cell numbers (left column). Musts from moldy grapes show higher cell numbers (right column):

S. cerevisiae	10000 –	20000/mL
apiculate yeasts	50000 –	1000000/mL
red yeasts	5000 –	100000/mL
other yeasts	10000 –	100000/mL
acetic acid bacteria	10000 –	100000/mL
lactic acid bacteria	1000 –	1000/mL

In mashes for the production of distilled beverages the number of cells is highly variable depending on damage to the raw material and hygienic conditions. The number of contaminating microbes is particularly high in raw materials containing soil particles, or dropped or moldy fruits. In such mashes the risk of spoilage is very high. Even undamaged fruit should be acidified if necessary and inoculated with a yeast starter culture, since such raw materials rarely contain *S. cerevisiae* cells.

For fruits the total number of microbial cells is between 10^2 to 10^8 (KARDOS 1966). Each apple contained 10^2 to 10^6 cells (MARSHALL and WALKLEY, 1951; TOROK and KING, 1991). The composition by species and the numbers of microbes varies for different fruits or other raw material depending on the methods of harvesting, transport and storage.

Semi-solid residues after pressing (pomace) are highly contaminated. They must be fermented rapidly and distilled promptly. The spent yeast from wine fermentations must also be disposed of rapidly.

1.2 Changes of the Species Composition During Fermentation

Musts contain many yeasts. Therefore, they can start to ferment spontaneously. During the fermentation the spectrum of yeast species changes dramatically. The percentage of weakly fermenting yeasts diminishes quickly. But the strongly fermenting *S. cerevisiae* (wild strains) multiply greatly. At the end of the fermentation there remain generally no more than one to two strains of *S. cerevisiae* in the young wine (see Tab. 1).

During the fermentation "wild" yeasts die off more or less quickly. This holds also for *S. cerevisiae*.

The "yeast" which sediments during and after the fermentation consists of the mass of yeast cells and almost an equal weight of grape berry constituents, K-bitartrate, protein and polysaccharides. Usually these lees are transported to the vineyards. In large wine-producing plants the alcohol of the lees (distilling material) is recovered by distillation. Sometimes tartrates can be recovered.

2 Spontaneous Fermentation with Wild Strains or with Starter Cultures

Yeasts contained in the must multiply and start to ferment. The rate of fermentation is greatest when the number of yeast cells is the highest.

During the spontaneous fermentation the number of various strains of *S. cerevisiae* is relatively small, and that of other yeasts which are not required or which are undesirable is relatively large. Besides the true wine yeasts, these wild yeasts affect the composition and the taste of the wine.

During the "pure yeast" fermentation – or shortly "pure" fermentation – large numbers of cells of a physiologically suitable strain are added as a starter culture. This suppresses all wild yeasts, and the chosen starter strain provides greater numbers of yeast cells. Hence it alone characterizes the fermented beverage, and the wines have purer tones and are more typical of the grape variety.

In older, traditional wine-growing areas spontaneous fermentations predominate. In larger wine-producing plants, especially in South Africa, Australia, New Zealand and North America, the use of pure culture starters is generally practiced. Yet 80% of the worldwide wine production is still produced with spontaneous fermentations.

2.1 Selection and Dominance in the Spectrum of Naturally Occurring Species

The composition of the must and the conditions under which the must affects various yeast species lead to considerable variations in the multiplication of the yeasts. Depending on their genetic composition some species or strains have certain selective advantages. Their better chance of multiplication eventually leads to the dominance of such strains.

The reasons for the selection or the dominance of a particular species or strain are complex. For *S. cerevisiae* it is the polyploidy of the cells and their strong fermentative ability. Most "wild" yeasts, which occur in substrates with lesser sugar concentrations, have a larger requirement for oxygen. The generally low temperature of the must may contribute to this effect. For many species the alcohol concentration is a factor for the selection. In a mash which contains many more microorganisms than the must, *S. cerevisiae* strains do not attain dominance as easily.

The substrate and the fermentation conditions exert an indirect effect on the selection of yeast strains. In addition, there is a direct effect of yeast strains on each other. It may be due to exhaustion of a required nutrient, such as a vitamin, or the production of an inhibitor. Another possibility is the formation of a killer toxin. The toxins of *S. cerevisiae* seem to act only on other *S. cerevisiae* strains (RADLER and SCHMITT, 1987; SHIMIZU, 1993). Therefore, the killer properties of *S. cerevisiae* strains have no significance for wine production with *S. cerevisiae* starter cultures.

Wild yeasts can also inhibit fermentation by *S. cerevisiae*. Some strains of *H. uvarum* produce a killer toxin (RADLER and KNOLL, 1988). However, the practical inhibition of *S. cerevisiae* by *H. uvarum* does not depend on the killer properties but only on the total number of cells of *H. uvarum* (SPONHOLZ et al., 1990a).

2.2 Differences in the Quality of Wines Produced by Spontaneous Fermentation or with Starter Cultures

The use of starter cultures greatly reduces the numbers of all other yeast cells. Therefore, the metabolic processes during the fermentation depend practically on the starter culture. Pure culture yeasts are added in larger numbers. Therefore, the fermentation starts sooner and is completed sooner. The wine is "ready" sooner, residual sugar concentrations are lower and the alcohol concentration is higher.

This reduces the likelihood of spoilage of the wine. In addition, the requirement for SO_2 is reduced; it could be reduced by 40%.

The wines also have purer tones and the sensory qualities of the various varieties of grapes are better maintained. The formation of undesirable products, such as the increased production of acetic acid or ethyl acetate by apiculate yeasts or of musty odors by other yeasts, is inhibited, as well as the production of excess SO_2 by other yeast species. The use of starter cultures makes the fermentation of the must more reliable.

Cooling is definitely required if larger volumes of must are fermented, particularly if larger numbers of yeast cells are added. Otherwise, the fermentation may be completed too quickly with undesirable consequences (see Sect. 4.1).

Spontaneously fermented musts can also produce good wines with a satisfactory character of the grape variety. Some trained tasters prefer such wines. The differences, generally only recognized by trained tasters, are due to higher residual sugar values after the slower fermentations. This residual sugar consists mostly of fructose which is twice as sweet as glucose. The greater body of such wines results, at least in part, from the higher concentrations of glycerol in the wine. Finally, wild yeasts usually form more ethyl acetate and acetic acid. Spontaneously fermented musts have up to an 8-fold higher concentration of 2-phenyl ethanol. This compound is believed to increase the character of the wine. Other alcohols are also increased.

Spontaneously fermented musts which often do not ferment to complete dryness contain more yeast metabolites which bind SO_2. Therefore, they may require more SO_2. The differences between spontaneously fermented musts and musts with pure culture yeasts are not always significant. If the must has been pasteurized, then yeast starter cultures must be added. This is also true if the wine is made from diluted juice concentrate, as is usually the case with apple cider.

Often the fermentation of a must is started by the addition of a certain volume of a strongly fermenting must. In that case one can assume that the *S. cerevisiae* yeast has already inhibited the other yeasts, and one can consider this a "relatively" pure culture fermentation.

2.3 Properties and Use of Yeast Starter Cultures

Pure culture yeasts are yeasts specifically selected for use in wine making. Once their suitability has been established, they are propagated with retention of their desirable properties. Usually these are clones of top-fermenting yeast strains of *S. cerevisiae.* Most of these clones have been selected – for obvious reasons – from fermenting musts or mashes. Many have been named after the wine-producing area or their origin: Bordeaux, Geisenheim, Epernay, Steinberg, Ay, etc. It must be emphasized that the origin of the yeasts does not determine their quality for the fermentation or their suitability for wine making. Each strain must, therefore, be tested for its suitability and must always be selected for maintenance of its desirable characteristics.

The most important property of a strain is its ability to ferment. It must readily ferment musts of more than 20% sugar (Spätlese quality) under normal conditions of wine making to residual sugar concentrations of 0.2–2 g/L. A certain tolerance to cold, osmotic pressure and alcohol is also assumed. Under the given conditions the fermentation rate presumes a certain multiplication of yeast cells in the must.

Beyond these requirements, the product formation of the tested strain of *S. cerevisiae* should improve the quality of the wine, not lower it. The formation of glycerol is desired, while the formation of acetic acid, ethyl acetate, and higher alcohols should be slight, as well as the formation of H_2S, SO_2, and SO_2-binding compounds (LEMPERLE and KERNER, 1990).

The physical properties must also be satisfactory. The yeast should not foam excessively and should sediment rapidly after the fermentation. Yeast strains for champagne production must be sufficiently tolerant to ethanol and CO_2; when used on bottle fermentations they should be easy to riddle.

Starter cultures should be used for all musts and mashes, in order to assure a risk-free production. In the following instances their use is mandatory:

1. for the fermentation of high sugar musts such as those from frozen berries (Eiswein), very ripe berries (Spätlese) or dry berries (Beerenauslese, Trockenbeerenauslese),
2. for the fermentation of *Botrytis*-infected grapes which have many undesirable microorganisms,
3. if some inhibition of the fermentation is expected,
4. for the production of fruit and berry wines and dessert wines,
5. for the fermentation of pasteurized musts and diluted concentrates,
6. for the fermentation of "stuck" wines and the additional fermentation of wines low in alcohol concentration,
7. for the secondary champagne fermentation,
8. for the fermentation of mashes for the production of distilled beverages.

Items 1 and 2 concern similar musts. They contain many wild yeasts and acetic acid bacteria, and the mold infection has depleted nutrients required by the yeast.

Inhibitions of the fermentation may also occur if fungicides are used illegally until shortly before the harvest. Often the yeast starter culture is only added after inhibition can be observed. This is incorrect. In any case it is best to add the starter culture as soon as possible, that is, for white wines after the separation of the must and for red wines after crushing and mash formation.

Fruit and berry juices or mashes must be fermented with starter cultures, since the fruits contain few if any *Saccharomyces* cells. If the raw material has been damaged one must expect the presence of many undesirable microbes.

For dessert wines the use of starter cultures is required because of the higher alcohol concentrations which have to be reached. The fruit juices which serve as raw material contain relatively little sugar and lack highly fermentative yeasts. Both reasons require the use of starter cultures. The same is true for the fermentation of diluted fruit juice concentrates.

Wines which have not been fermented completely may be spoiled by lactic acid bacteria because of their residual sugar content (see Sect. 7.1). They also show high concentrations of SO_2 binding compounds (see Sects. 3.3.2, 8.1). Therefore, the remaining sugar must be fermented. The wines should be promptly separated from the old yeast and fortified with thiamine (see Sect. 3.3.2), yeast nutrients and possibly with yeast cell wall preparations (see Sect. 4.4.3). Finally new yeast starters should be added to complete the fermentation. The yeast may also be stirred up a few times; but care should be taken because of the liberation of CO_2.

During efforts to increase the alcohol concentration of wines with a low alcohol content by addition of sucrose, it must be assumed that the original yeast has used up the yeast nutrients, as one would have found in stuck fermentations. One should, therefore, proceed as described above and, of course, the calculated amount of sugar must be added. The same is true of the secondary champagne fermentation.

Pure yeast starter cultures may be applied in two forms: (1) as liquid yeast, and (2) as active dry yeast.

Liquid pure culture yeast is obtained by inoculation of a suitable strain into sterile must. The yeast multiplies in this medium to 80 to 100×10^9 cells per L. This yeast suspension is given to the plant management in quantities of 0.1 to 1 L. The yeast can be kept for a short time under refrigeration. It must be again propagated before use.

Active dry yeast is available commercially in the form of granules. A suitable strain has been produced aerobically on a large scale, and the yeast is then dried. The moisture content of these preparations is about 8%. They do not have to be propagated again and can be added directly to the must or the mash.

The multiplication of the liquid yeast can be carried out in must or commercial grape juice which may be diluted. For fruit and berry wines one uses juices of the appropriate fruits. For the secondary champagne fermentation one uses a wine from which the alcohol has been removed by boiling and to which 30 g sucrose per liter has been added.

The purchased *liquid yeast* is propagated in 2 to 10 liters of boiled must in a clean glass vessel at room temperature. The yeast has multiplied after 2 to 3 days; the must ferments. Two liters of this rapidly fermenting must are sufficient to inoculate 100 to 200 liters of a fresh must. For the fermentation of still larger volumes the yeast inoculum has to be propagated again by adding a two- to fivefold amount of fresh must. This addition can only be carried out if the inoculum is fermenting rapidly. Therefore, one has to start the propagation cycle early if one wants to ferment very large volumes of must.

Under normal circumstances inoculation of musts with 2% of the propagated yeast is sufficient. If difficulties are expected one should use 4%.

The amount of *active dry yeast* for normal musts is 3 g/L. Larger amounts are only required if the fermentation temperature is lower than 12 °C, when one expects the presence of undesirable microorganisms or when the fermentation is inhibited or a stuck wine has to be restarted.

The number of colony-forming units (CFUs) decreases during storage of the wine depending on temperature and time. Unopened packages of active dry yeasts should be stored under refrigeration for no more than 6 months. One gram of active dry yeast contains between 12 and 27×10^9 living cells. Up to 0.1×10^6 yeasts other than *S. cerevisiae* can be expected and between 9×10^3 and 7.6×10^6 bacteria, mostly lactic acid bacteria. The wild yeasts and the bacteria die during the fermentation (RADLER and LOTZ, 1990).

Baker's yeast (compressed yeast) is suitable for the fermentation of fruit-, cereal- and potato mashes. It is inexpensive and everywhere available. It requires a starting temperature of at least 20 °C. Mashes of stone fruit require 100 to 250 g/hL. Mashes that are difficult to ferment may require 500 g/hL. The use of pure yeast cultures for the production of wines has been treated by REED and NAGODAWITHANA (1988).

3 Fermentation and Formation of Metabolic Products

The metabolysis of grape musts or any other sugar-containing raw material takes place through the alcoholic fermentation by *S. cerevisiae*. The fermentable sugars of musts are glucose and fructose. They occur in a ratio of about 48:52. Pentoses which occur in fruit juices (about 1 g/L in grape juice) are not fermented.

Sucrose (saccharose), a disaccharide, is permitted in many countries for fortification of musts with low sugar concentrations. In the US many states permit the addition, California does not. The β-fructofuranosidase (invertase, saccharase) of *S. cerevisiae* hydrolyzes the disaccharide to glucose and fructose. Typical apiculate yeasts do not contain invertase, and, therefore, cannot ferment sucrose.

Saccharomyces cerevisiae is glucophilic: Glucose is fermented more quickly than fructose (Fig. 3). After completion of the fermentation the residual sugar consists of some fructose and much less glucose. The ratio of glucose/fructose in fermenting musts can be calculated (PRIOR et al., 1992).

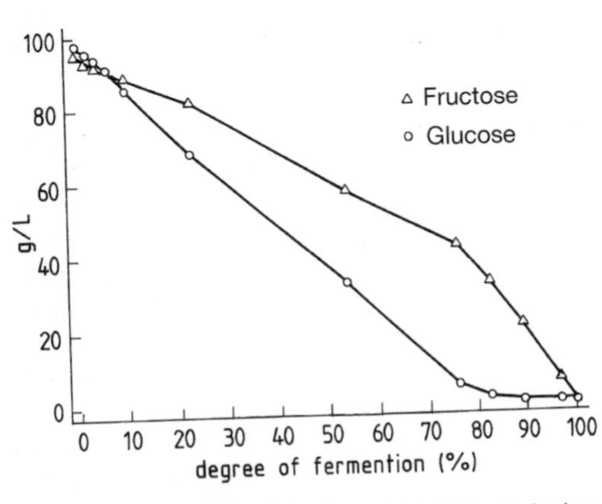

Fig. 3. The decrease of fructose and glucose during the fermentation of a grape must with *Saccharomyces cerevisiae* (Must Morio-Muscat; Rheinhessen quality wine; active dry yeast "Oenoferm"; vat 2000 liters.

Cereal and potato mashes do not contain carbohydrates which can be directly fermented, but starch, which must be split by amylase preparations to glucose and maltose (KREIPE 1981). The fermentation of Jerusalem artichoke mashes require inulase-producing strains of *Kluyveromyces* (SCHWARZ and HAMMES, 1991).

The ability of yeasts to split glucosides is important for the color of red wines. The pigments of red grapes are β-glucosides. The loss of color is greatest at the beginning of the fermentation. The extent of loss of color depends on the yeast strain. It may be as much as 70%. Cyanidin shows the greatest color loss (about 95%). Therefore, the grape varieties of the Trollinger group show significant losses of color (WENZEL, 1990).

The β-glucosidases of yeasts can also split terpene glucosides. The liberated monoterpenes may influence the bouquet of the wine (DARRIET et al., 1988).

The sugar uptake into yeast cells of glucose and fructose takes place by means of the same carrier (BISSON, 1993). This is the rate-determining step of the fermentation.

The biochemistry of the fermentation, its metabolites and enzymes have been treated extensively in this series. Therefore, only data particularly important for wines and mashes will be mentioned here.

In this view, a very important reaction of the fermentation is the decarboxylation of pyruvate: This liberation of CO_2 makes the beginning of the fermentation, its approximate rate, and its end directly visible. CO_2 has also technological importance in other areas of the wine industry.

3.1 Major Metabolic Products

Two molecules of ethanol and two molecules of CO_2 are formed from each molecule of glucose or fructose. This leads to the following quantitative formula:

$$180.15 \text{ g hexoses} \rightarrow 2 \cdot 46.05 \text{ g ethanol} + 2 \cdot 44 \text{ g } CO_2$$

theoretical yield: 51.1% and 48.9%

Practically alcohol yields vary between 45 and 48%. Usually one assumes a yield of 47%. The considerable difference between the theoretical and the practical yield is mainly due to the formation of fermentation by-products such as glycerol (see Sect. 3.2.1). Variations in the practical yield figures are generally due to variations in the conditions of the fermentation. The most important factor is the volume of must. Losses of ethanol are smallest for smaller must volumes, since the temperatures of the fermentation do not increase as rapidly. At higher must temperatures more ethanol is distilled off with the CO_2: at 35 °C, for instance, 1.2 vol.-%; at 20 °C 0.65 vol.-%; at 5 °C only 0.17 vol.-%. At temperatures of 20 °C the loss of ethanol is still modest. At 22 and 27 °C one can lose 0.3 to 0.8 vol.-% of the formed ethanol (HAUSHOFER and MAIER, 1979; ENKELMANN, 1979).

Very large fermentation tanks should not be used in order to avoid losses of ethanol. Large tanks must be cooled promptly and efficiently; and most of the insoluble particles should be removed, since they accelerate the fermentation and increase the temperature (see Sect. 4.3). Mashes ferment particularly fast. WILLIAMS and BOULTON (1983) have developed a model which permits an estimate of the alcohol losses.

For fermentations in large containers the difference between the actual weight of the must and the calculated weight was up to 8 °Oe. (Grad Oechsle is a measure of the specific weight in Germany and Switzerland, which indicates the approximate weight of the sugar in the must. English speaking countries calculate usually in Brix = Balling.) This could have undesirable consequences for the legal standing of the wine. Alcohol formation by yeasts also depends on the formation of by-products. Yeasts which form more glycerol produce less alcohol than those which form less glycerol. Quality wines must contain at least 7 vol.-% ethanol, wines from selected overripe berries (Beerenauslese) and from dehydrated berries (Trockenbeerenauslese) 5.5 vol.-%. (German quality wines are designated by names which have a specific meaning. Since this is lost in translation it is shown in Tab. 2.)

The volume of CO_2 produced during the fermentation is 40 to 50 times as large as the volume of the fermenting must. One can estimate the amount of ethanol from the volume of CO_2 as follows:

Tab. 2. Designations and Legal Requirements for German "Qualitätswein[a] mit Prädikat" as Shown for the Grape Variety Riesling of the Rheingau District

Designation	Minimum Must Weight in °Oe	Legal Requirements
Kabinett	73	Ripe berries; official control of harvest. Control number obtained after official analysis and taste test. No addition of sugar
Spätlese (late harvest)	85	As for "Kabinett" but late harvest of fully ripe grapes
Auslese (selection)	95	As for "Spätlese" but only fully ripe berries may be pressed after removal of spoiled or unripe berries
Beerenauslese (berry selection)	125	As for "Auslese" but berries which show noble rot (or are at least overripe) must be used
Trockenbeerenauslese (dry berry selection)	150	As with "Beerenauslese" but only shrunken, overripe berries with noble rot may be used
Eiswein (ice wine)	125	The berries must be frozen during harvest and pressing

[a] Produced from true wine grapes of designated viticultural areas requiring a minimum must specific gravity depending on the grape variety. Alcohol minimum concentration of 56 g/L; must have a control number.

$E = 1.85 \cdot V_c + 2.7$

where E is ethanol in g/L and V_c is the volume of CO_2 produced in liters per liter of fermenting substrate (EL HALOUI et al., 1988).

The formation of large volumes of CO_2 presents a hazard. A candle will be exstinguished at a 10% CO_2 content of the atmosphere. So this is a reliable warning sign.

The difference between the formed ethanol and the ethanol in the young wine is not known in each case because of the unknown losses of ethanol by evaporation during the fermentation.

The ethanol concentrations vary greatly depending on the sugar content of the must (50–140 g/L). More than 144 g/L ethanol cannot be expected (about 18.2 vol.-%). Even strongly fermenting yeasts cannot produce more ethanol under normal circumstances.

3.2 Primary and Secondary Metabolic Products

The general fermentation formula shows only ethanol and CO_2 as end products. Actually many other substances are formed, the so-called fermentation by-products. These can be classified into:

a) Primary by-products. These are metabolites of the fermentation which have been formed by simple reactions as well as the metabolites of the citric acid cycle.
b) Secondary by-products. These are products resulting from complex reactions, such as the higher alcohols and/or products derived from them such as esters or other important compounds. The profound chemical reactions which take place during the fermentation of a must are shown in Fig. 4.

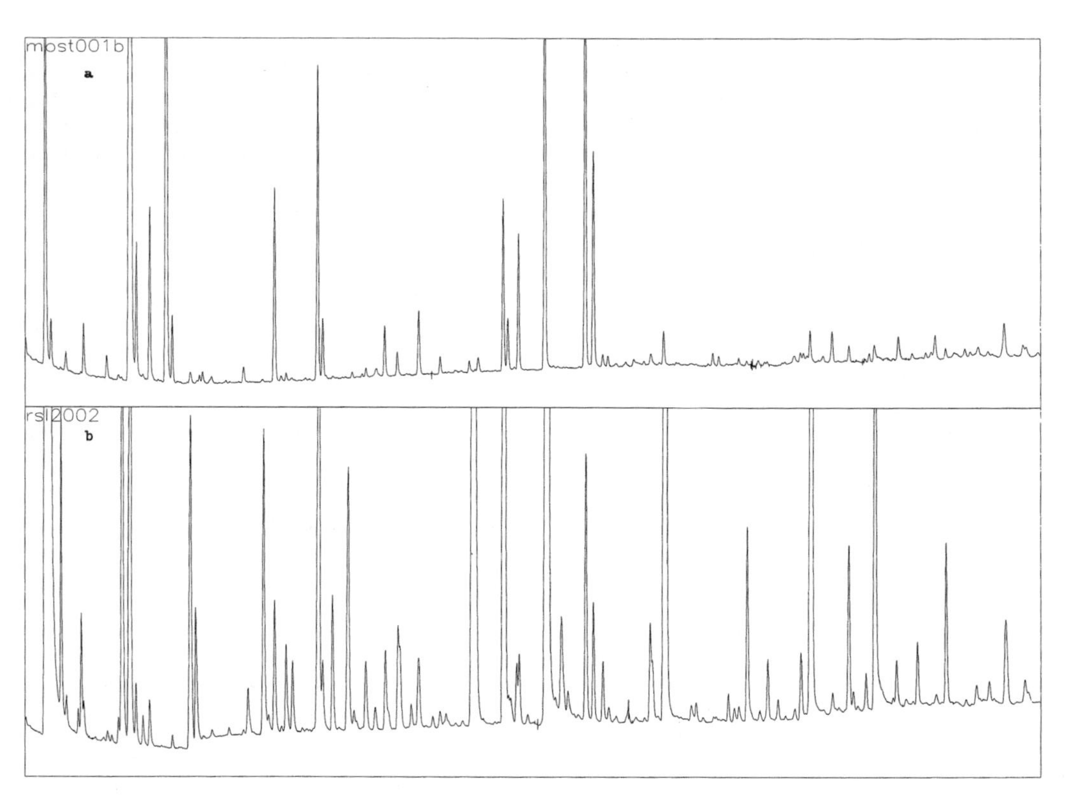

Fig. 4. Gas chromatograms of volatile constituents (Courtesy Dipl. Oenol. THOMAS ESS).
a) Must of the variety Rheinriesling, Johannisberg/Germany, 1993,
b) Wine produced from this must.

Primary By-Products

Glycerol

During the fermentation H is transferred from glyceraldehyde-3-phosphoric acid (GAP) largely to acetaldehyde and partially to DHAP (dihydroxyacetone phosphate) and other H acceptors. In these reactions alcohol dehydrogenase and glycerolphosphate dehydrogenase compete for NADH and its hydrogen.

Glycerol is a viscous liquid, which is almost as sweet as glucose. It determines largely the "body" of a wine, and it rounds off its mouthfeel. Normal wines contain from 5 to 9 g/L. Up to 30 g/L of glycerol – and sometimes more – can be found in wines made from selected grapes (Beerenauslese). After ethanol it is the quantitatively most important compound in wine.

In spontaneous fermentations glycerol amounts to up to 9% of the formed ethanol. In fermentations with pure culture yeasts the concentration of glycerol is usually lower. For very sweet musts (more than 30% sugar) glycerol formation may be even higher. For normal wines higher glycerol concentrations than 10% of the ethanol suggest the illegal addition of the compound: Much glycerol can be found in wines from selected grapes whose musts contain already some "must glycerol". This glycerol is formed by *Botrytis cinerea* and adds to the glycerol formed by the yeast fermentation (DITTRICH, 1989b). Higher fermentation temperatures also accelerate glycerol formation.

During the fermentation yeasts take up *m*-inositol and form small amounts of other polyols. These polyols become part of the "extract" of a wine and contribute to its sensory fullness (SPONHOLZ, 1988).

Acetaldehyde, Pyruvate, and 2-Oxoglutarate and Their Significance for the SO_2 Requirement of a Wine

Acetaldehyde, pyruvate and 2-oxoglutarate are by-products which occur only in quantities of several mg/L in wine (Tab. 3). They are significant because of their reactivity with SO_2. Acetaldehyde binds SO_2 the strongest. After the wine has been sulfited these compounds are bound to SO_2. They are not found free.

Wines are mainly sulfited to remove the undesirable odor of acetaldehyde and to prevent browning of the wine by oxidation. The prophylactic inhibition of potential spoilage microorganisms is only secondary.

The production of wines low in SO_2 is most desirable. Therefore, the formation of SO_2-binding metabolites should be minimized. Such metabolites occur in higher concentrations in very slow fermentations. Such wines require more SO_2 than rapidly fermented wines.

Acetaldehyde accounts for most of the SO_2-binding compounds. With improved must quality pyruvate and 2-oxoglutarate concentrations increase. The formation of these metabolites depends on the yeast strain and the conditions of the fermentation, on the degree of *Botrytis* infection, and also on the malo-lactic fermentation.

At the start of the fermentation acetaldehyde and pyruvate are excreted by the yeast in larger quantities. Then the concentration of these compounds decreases. The yeast re-

Tab. 3. Average Concentrations (in mg/L) of SO_2-Binding Metabolites in Wines of German Viticultural Areas (DITTRICH and BARTH, 1984)

	300 Quality and Kabinett Wines	83 Spätlese Wines	27 Auslese Wines
Acetaldehyde	50	60	78
Pyruvate	27	32	64
2-Oxoglutarate	48	46	50

sorbs these metabolites and converts them to ethanol. This resorption is prevented if the yeast sediments too fast, for instance, if the temperature is too low or if it is so high that the yeast is killed. In such cases acetaldehyde and pyruvate remain in the young wine in higher concentrations.

The production of wines low in SO_2 presumes a fast and complete fermentation. Addition of SO_2 to a fermentation which has not been completed increases total SO_2 because the compounds bound by SO_2 cannot be resorbed by the yeast and metabolized to ethanol. Grape must which is to be used to sweeten a wine should never have started fermenting.

The addition of up to 0.6 mg/L of thiamine to must is permitted to lower the SO_2 requirement. The diphosphate is the co-factor of pyruvate and 2-oxoglutarate decarboxylating enzymes. Thiamine hydrochloride may be used up to 0.76 mg/L.

Fermentation in the presence of thiamine lowers pyruvate formation greatly; that of 2-oxoglutarate less so. Yeast does synthesize thiamine but the addition of thiamine accelerates its metabolism which may reduce the SO_2 requirement by 50% (DITTRICH, 1983).

Spontaneous fermentations produce wines with more acetaldehyde than fermentations with selected yeast starter cultures. Therefore, they require more SO_2. It is best to ferment with pure yeast starters and to add thiamine.

The addition of thiamine is also required for musts from *B. cinerea*-infected grapes. The mold may reduce the thiamine content of the must to one tenth. In this case the metabolism of pyruvate by the yeast is inhibited, and the compound is partially excreted into the young wine. The addition of thiamine removes the inhibition and accelerates the fermentation.

A completed fermentation with musts with very high sugar concentrations is neither intended nor possible. Such wines may contain higher concentrations of SO_2-binding compounds and, therefore, they have a higher SO_2 requirement.

The thiamine content is also reduced in musts which have been partially fermented. The second generation of yeasts which should ferment the must to completion is, therefore, deprived of thiamine. Such young wines contain more pyruvate. The addition of thiamine prior to the inoculation with yeast can restore pyruvate formation to normal levels.

During the secondary fermentation of sparkling wines concentrations of acetaldehyde and pyruvate increase, e.g., from 30 to 58 mg/L and from 9 to 24 mg/L, respectively.

Acetic Acid

Higher concentrations of this acid lower the sensory quality of wines and distillates. Concentrations above 0.6 g/L indicate spoilage and formation by bacteria. Therefore, the concentration of "volatile acidity" which consists mostly of acetic acid is limited by regulations.

Suitable strains of *S. cerevisiae* form 0.2 to 0.5 g acetate per liter. Wild yeasts, such as, for instance, *Hanseniaspora uvarum* may form up to 1 g/L.

S. cerevisiae forms acetic acid by the oxidation of acetaldehyde by a NADP-specific acetaldehyde dehydrogenase. With increasing sugar contents of musts acetate formation increases. At very high sugar concentrations ethanol formation decreases again, while the maximum concentration of acetate is reached later on (Fig. 5).

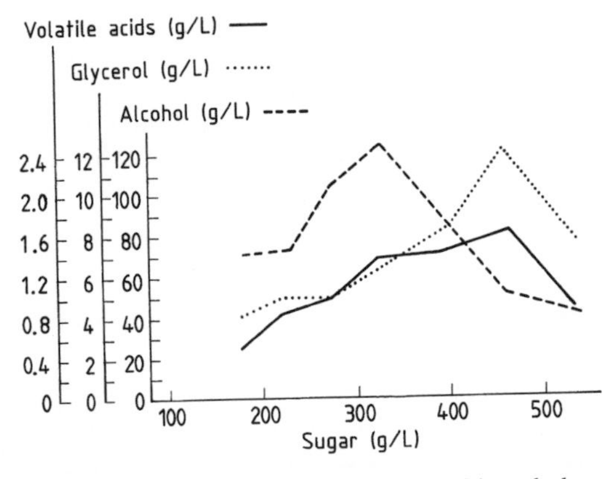

Fig. 5. Production of ethanol, acetic acid, and glycerol as a function of the sugar concentration of the must (22°C, minimal inoculation with *Saccharomyces cerevisiae*) (DITTRICH, 1989a).

In addition to acetic acid, octanoic and hexanoic acids are formed in significant quantities. Formic acid which occurs in must up to 60 mg/L decreases during the fermentation (SPONHOLZ and DITTRICH, 1986). Acetate and formate increase with increasing quality of the wine; while other fatty acids decrease.

Succinic Acid and Lactic Acid

German white wines contain from 200 to 700 mg of succinic acid per liter, occasionally more. The quantity does not seem to depend on the must (SPONHOLZ and DITTRICH, 1977). Succinic acid formation occurs principally in the early stages of the fermentation. It increases between 10 and 30 °C. It also increases with increasing pH (SHIMAZU and WATANABE, 1981).

Normal strains of *S. cerevisiae* form from 100 to 200 mg of lactic acid per liter. In the absence of the malo-lactic fermentation there is on the average 191 mg/L D-lactate and 91 mg/L L-lactate (DITTRICH and BARTH, 1984). Larger concentrations are formed during the malo-lactic fermentation, and in that case L-lactate predominates.

Secondary By-Products

These products are formed besides the primary by-products. In the narrower sense they are not fermentation products but compounds formed synthetically from sugar metabolites. Butanediol and the higher alcohols, are, for instance, by-products of yeast growth. The resulting products are esters which participate in the wine bouquet. Finally such by-products are compounds formed by other chemical reactions during the fermentation.

2,3-Butanediol and Other Diols

2,3-Butanediol occurs in wines in concentrations of 400 to 800 mg/L. In wines from selected berries (Beerenauslese and Trockenbeerenauslese) 1100 to 1800 mg/L have been found (SPONHOLZ et al., 1994). Its formation increases with increasing temperature and if the must has been aerated. Its synthesis occurs mainly by acetoin synthase and reduction of the formed acetoin (Fig. 6).

Ethanediol, 1,2- and 1,3-propanediol as well as 2,3-pentanediol are present only in concentrations of a few mg/L each (SPONHOLZ et al., 1994).

Higher Alcohols

The so-called "higher" alcohols are not true fermentation by-products, but result from excess accumulation during the synthesis of valine, leucine and isoleucine. Energy formation during the fermentation is slight, and, therefore, the resulting oxo-acids cannot be aminated to the corresponding amino acids. Hence, they are decarboxylated, and the resulting aldehydes are hydrated to the corresponding alcohols (Fig. 6).

Concentrations of the higher alcohols are quite variable. The major component is 3-methyl-1-butanol (isopentyl alcohol) which is present in normal wines at concentrations of 60–150 mg/L. 2-Methyl-1-propanol (isobutanol) follows with 20–80 mg/L and 2-methyl-1-butanol (optically active pentyl alcohol) with 10–30 mg/L. The ratio of 3- to 2-methyl-1-butanol is approximately 100:20. Together with isobutanol these alcohols account for 70% of the "fusel oil" fraction of a wine. Only 3-methyl-1-butanol may have an impact on the sensory quality of a wine if it is present in larger concentrations.

These alcohols are formed from sugars. Hence their concentrations are higher with a higher sugar concentration of the musts. But musts from selected berries (Beerenauslese) and from dry berries show much lower concentrations because of their lower ethanol content.

Addition of ammonium salts to the must lowers fusel oil formation; and addition of thiamine lowers it greatly. Isobutanol formation is slightly greater in musts from *Botrytis*-infected grapes (DITTRICH, 1983).

1-Propanol is probably formed by degradation of threonine via 2-oxobutyrate. Normal wines contain about 10–40 mg/L.

During spontaneous fermentations more 2-phenylethanol is formed (25–28 mg/L) than with pure culture yeast fermentations (6–

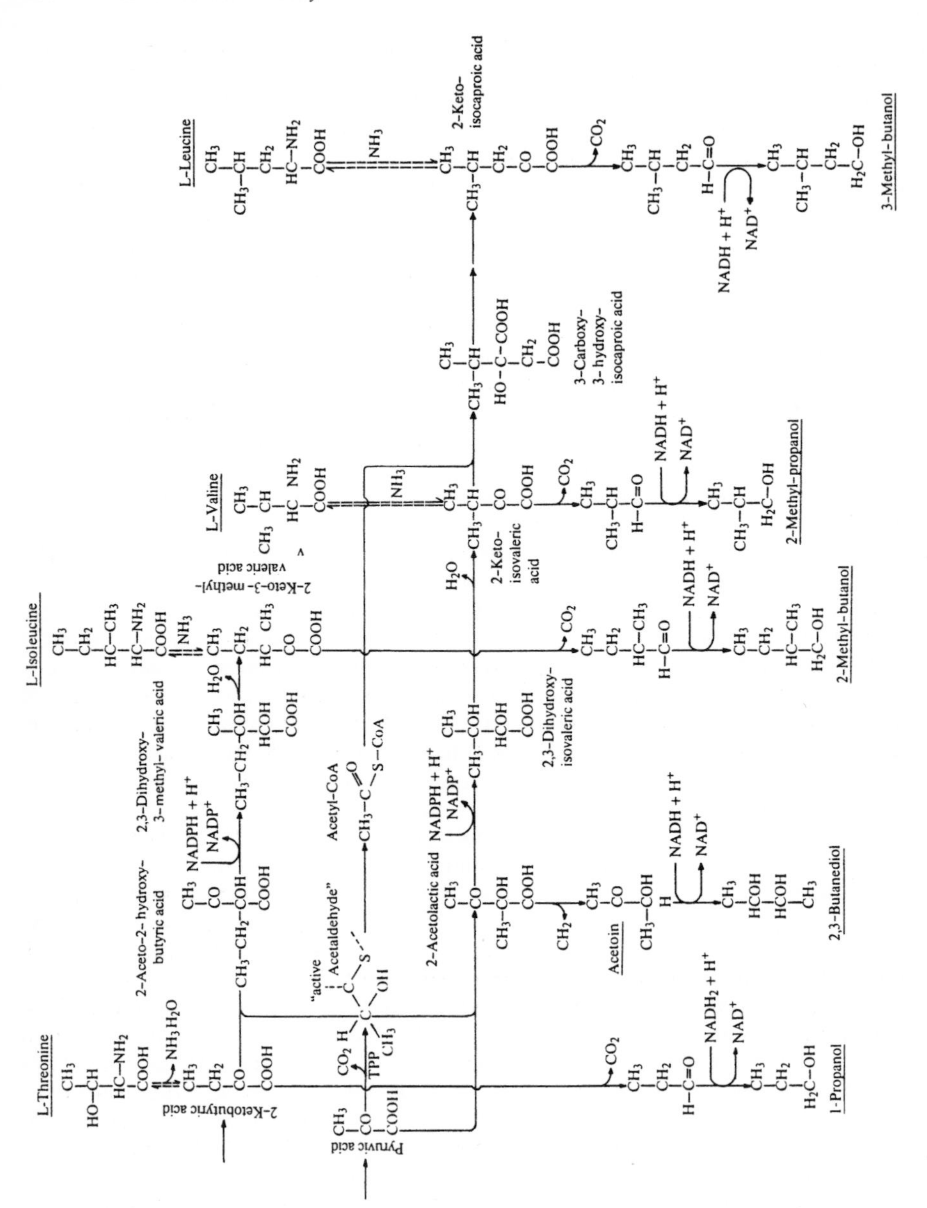

Fig. 6. Formation of 1-propanol, 2,3-butanediol, 2-methyl-1-propanol (isobutanol), 2-methyl-1-butanol (sec. butylcarbinol), and 3-methyl-1-butanol (isopentylalcohol) (DITTRICH, 1987).

29 mg/L) (SPONHOLZ and DITTRICH, 1974). The presence of tyrosol and tryptophol has also been determined as well as that of methyl mercaptopropanol, a sulfur-containing alcohol. German white wines contain 145–410 μg/L.

1-Hexanol (1–5 mg/L) is formed from the unsaturated linoleic and linolenic acids of the grape pits and from yeasts. It has a grassy odor and taste.

Methanol is principally formed from pectins by the action of pectin methylesterases of the grapes after the grapes have been crushed. Therefore, red wines contain more methanol than white wines. They should contain no more than 300 mg/L or 150 mg/L, respectively. These concentrations raise no health concerns.

Esters and Other Fermentation Flavorants

Esters are odorous compounds. They are important for the sensory quality of a wine. Obviously, acetic acid and ethanol are the most important compounds which form esters.

The formation of acetic acid and ethanol esters increases with increasing concentrations of sugars, with a faster rate of fermentation, and with increasing amounts of yeast. A rise in temperature from 11 to 15 °C increased the concentration of ethyl acetate threefold, and doubled that of other esters (HOUTMAN et al., 1980). Between 16 and 21 °C the formation of ethyl acetate, propyl acetate, and 3-methylbutyl acetate was maximal. The formation of higher alcohols and esters depended greatly on the yeast strain used, and also on the fermentation substrate (CABRERA et al., 1988).

Ester formation is preceded by the "activation" of acids to their coenzyme A derivatives. Ethyl acetate is produced by alcohol acetyltransferase.

The principal ester is ethyl acetate (10–50 mg/L). The presence of more than 60 mg/L indicates the presence of wild yeasts: *Pichia anomala*, *Candida krusei*, *Metschnikowia pulcherrima* and *Hanseniaspora uvarum*. In that case the "solvent odor" may lower the quality of the wine. Isoamyl acetate (3-methylbutyl acetate) has also a strong odor if it is present at more than 1 mg/L. The formation and importance of esters has been described by NYKÄNEN (1986).

The ethyl esters of amino acids are also present, e.g., ethyl methionine (~150 μg/L) and ethyl proline (~3600 μg/L). Ethyl carbamate (urethane) is not formed by *S. cerevisiae*, nor by "wild" yeasts, nor by *Botrytis* or acetic acid bacteria. Only a few very old wines may contain more than 30 μg/L. Wines which have undergone the malolactic fermentation always have larger concentrations than those which have not (SPONHOLZ et al., 1991). More than 6 mg/L could be found in brandy from stone fruit. There, the precursors are ethanol and cyanide. Avoidance of urethane formation or its elimination have been described by CHRISTOPH et al. (1988). In fruit brandy up to 0.8 mg/L can be tolerated. At such concentrations urethane is undesirable but poses no health problem.

Aldehydes and ketones have also considerable odor and taste. At the end of the fermentation only acetaldehyde is present in higher concentrations (see Tab. 2). The sulfiting of the young wine produces compounds which affect the bouquet and flavor of the wine to a lesser extent. The "fermentation bouquet" disappears.

Sulfur-containing compounds also have sensory properties. Methylthioacetic acid occurs in German wines at 7–11 μg/L.

The alcohols are rather volatile with the exception of phenyl ethanol. The ethyl esters of caprylic and higher fatty acid esters are less volatile, and, therefore, they are present in distillates in lower concentrations.

Galacturonic Acid and Other Acids

Galacturonic acid is formed through hydrolysis of pectins by polygalacturonidases. These enzymes split unesterified pectic acid in various ways. Galacturonic acid is present in white wines from 150 to 500 mg/L and in red wines from 500 to 1100 mg/L. In red wines the concentrations are higher because of the fermentation of the entire mash. In selected wines they are higher because of the *Botrytis* infection (300 to 1000 mg/L), and in selected berry wines they are 300 to 1000 mg/L. Up to 1537 mg/L have been found in dry berry wines (DITTRICH, 1989b).

2-Hydroxyglutarate is formed by hydrogenation of oxo-glutarate. Depending on the extent of *Botrytis* infection the wines may contain 30 to 160 mg/L (SPONHOLZ et al., 1981). 2-Methyl malic acid (citramalate) and 2-methyl-2,3-dihydroxybutyric acid (dimethyl glycerate) occur in similar concentrations.

Sulfurous Acid (SO_2) and Hydrogen Sulfide (H_2S)

Both compounds are produced by reduction of the sulfate of the must, and both compounds are important in wine making, SO_2 because its addition must be limited, and H_2S because of its undesirable odor.

SO_2 is formed by *S. cerevisiae* during the fermentation to a concentration of about 3 mg/L. But strains could be isolated from wines which contained up to 130 mg/L SO_2. These SO_2-forming yeasts are characterized by their high activity of sulfite-forming enzymes and the low activity of sulfite-degrading enzymes. Therefore, such strains hardly produce any H_2S. They ferment weakly and do not occur frequently. Use of starter cultures suppresses their growth.

H_2S is formed by the action of sulfite reductase. The presence of particulate matter in must increases H_2S formation. The protein of the particulate matter is degraded by the yeast to use the nitrogen for growth. Addition of nitrogenous nutrients has the opposite effect. Fast rates of fermentation also increase H_2S formation, and for this reason particulate matter additionally increases H_2S concentrations.

The sulfur used on grapes for the suppression of mildew is the primary source of H_2S. The sulfur is reduced to H_2S during the fermentation. H_2S is carried out with evolving CO_2 during a rapid fermentation. As the fermentation slows down, the H_2S content of the young wine increases again. The concentrations of H_2S can be decreased by centrifuging of the must.

H_2S participates in the odor of the young wine. During subsequent processing it can be eliminated by sulfiting. High concentrations of H_2S which lower the quality of the wine can be removed by precipitation with $CuSO_4$ (WENZEL et al., 1980).

Dimethyl sulfide has been found in wines in concentrations up to 400 μg/L. RAUHUT and KÜRBEL (1994) have dealt with S-containing compounds in wine.

Yeasts

A small portion of the sugar of the must is used by the yeast for growth. Only about one third of the lees consists of yeast cells. The larger part consists of grape substances and potassium bitartrate. In normal wooden barrels one finds about 3–5% of viscous lees, in large tanks of 60000 to 150000 L about 3%. For wines from centrifuged musts the lees amount only to about 1.5 to 2%. Wines from selected berries and secondary champagne fermentations produce only about 1 to 2‰ lees. Under such circumstances yeasts hardly multiply.

Yeast multiplication is required for the fermentation, and this growth occurs largely before the fermentation starts. A particular Riesling must contained 38.5×10^6 cells/liter. Yeast multiplication had almost ceased when the fermentation started. At that time the cell count was 175×10^9 cells/liter. A similar number of cell counts is reached, although the cell counts in the must may be quite different. An increasing number of cells dies already during the fermentation (Fig. 7). During a slow fermentation yeast may sediment rapidly during the final phase of the fermentation. Yeast multiplication can be greatly accelerated by the addition of thiamine and ammonium salts (DITTRICH, 1983).

The "yeast press wine" made from the lees is subject to bacterial infections. It contains compounds which diffuse out of the dead yeast cells, and the percentage of the "sugar free extract" is increased by the nitrogenous and phosphate compounds (STOISSER et al., 1988). The diffusing compounds and the higher pH favor bacterial multiplication. Malate is metabolized. The "yeast press wines" can spoil rapidly if they are not sulfited. They are often used as "distilling material".

Wines with larger concentrations of yeast compounds (distilling material) result in distillates with higher concentrations of ethyl esters of octanoic, decanoic and dodecanoic acids. The use of distilling material does not greatly affect the concentrations of other esters, carbonyl compounds, alcohols and terpenes (POSTEL and ADAM, 1984). The variable ratios of the esters of the higher fatty acids is suitable for judging the quality and

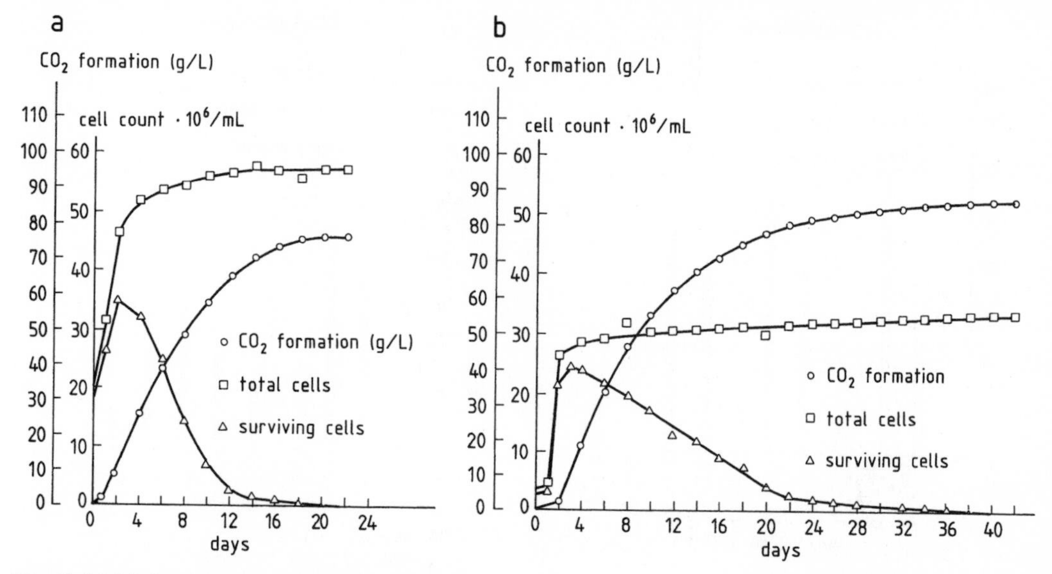

Fig. 7. Multiplication, survival and fermentation by *Saccharomyces cerevisiae* (Geisenheim strain 74) at normal and high sugar concentrations of the must.
a) 75°Oe, about 185 g/L fermentable sugar,
b) 200°Oe, about 420 g/L fermentable sugar (SPONHOLZ et al., 1990).

authenticity of German brandies and French cognacs and armagnacs (Fig. 8). The concentrations of individual volatile compounds are not as variable with cognacs and armagnacs as with German brandies, depending on the raw material and process of distillation (POSTEL and ADAM, 1980). The ratios of higher alcohols and the ratios of octanoate to decanoate permit conclusions as to the particular process of distillation used (ADAM and POSTEL, 1992).

4 Factors Affecting the Fermentation

The conditions under which the must is prepared and the conditions of the fermentation affect the metabolism of the yeast. The rate of fermentation is slow if the yeast cannot multiply rapidly. If it does multiply fast one can expect a fast fermentation unless inhibitors interfere with the metabolism of the cells.

4.1 Temperature

The optimal temperature for yeast growth is close to 30°C. Therefore, the fermentation of the must is faster at higher temperatures because of the presence of larger numbers of yeast cells, and also because of the higher activity of the enzymes of the fermentation pathway.

The temperature of the must depends on the temperature of the grapes which in turn depends on the weather at the time of harvest. But the increase of the temperature during the fermentation is more important.

During the fermentation of a must of 76°Oe (about 1 mol hexose, 180 g) about 23.5 kcal (99 kJ) are produced under practical conditions. Musts with a lesser sugar concentration produce fewer calories. For instance, a must of 66°Oe with about 156 g sugar/liter produces only

$$\frac{156 \cdot 23.5}{180} = 20.4 \text{ kcal (86 kJ)}$$

About 20% of the developed heat is car-

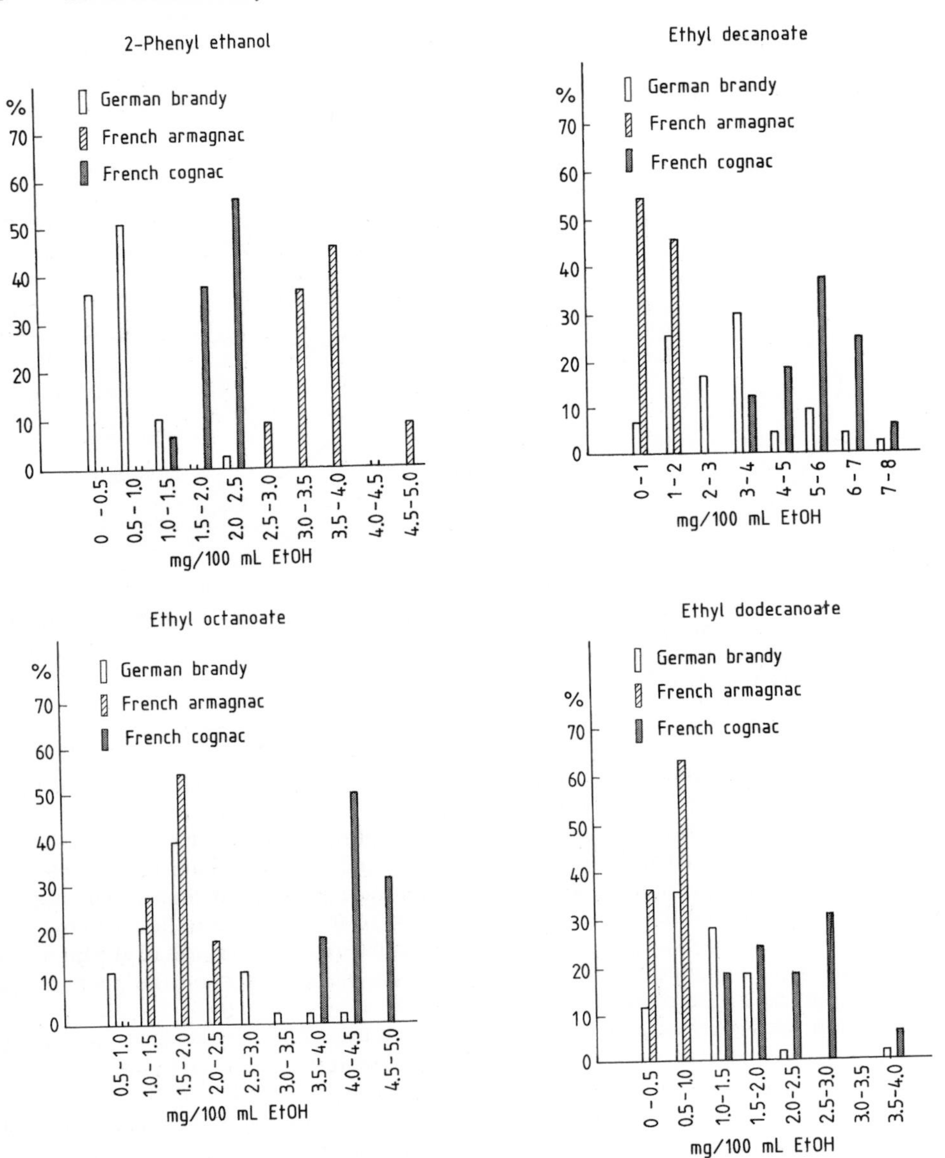

Fig. 8. Concentrations and frequency distributions of 2-phenyl ethanol, ethyl octanoate, ethyl decanoate, and ethyl dodecanoate in German brandy, French cognac and French armagnac (POSTEL and ADAM, 1980).

ried out with the CO_2. Heat develops throughout the fermentation, but there is a temperature maximum when the fermentation is half completed. Large volumes of must develop considerable heat and must promptly be cooled. The fermentation rate increases with increasing temperature up to a certain limit. At must temperatures above 30°C the fermentation starts faster. But then the fermentation rate diminishes, and the residual sugar values increase. This is caused by the toxicity of the warm alcohol whose concentration increases. With increasing must temperature the inhibiting activity of the ethanol increases. In a must with 300 g sugar/liter the fermentation stopped completely at the following concentrations of ethanol (EtOH in g/L): 9°C (139 g/L), 18°C (122 g/L); 27°C (98 g/L); and 36°C (71 g/L).

The fungicidal effect of the warm ethanol is used during the warm bottling of the wine. At an ethanol concentration of 10 vol.-% a holding time of 5 min at 55°C is sufficient. One has to consider the loss of heat through the glass of the bottle.

If the temperature rises too high the fermentation ceases, although the must still contains large amounts of sugar. This overheating may be one cause of a stuck fermentation. The fermenting yeast has been killed by its "overheated" metabolism.

Such musts and mashes are threatened by lactic acid bacteria. These bacteria tolerate a higher temperature; indeed they grow faster. Therefore, such wines are subject to malate degradation and spoilage. These musts or young wines must be separated from the yeast, possibly filtered, and fermented with active dry yeast after the addition of thiamine.

The temperature of musts leaving the presses is usually 15–18°C. They must be cooled so that the temperature does not exceed 20–22°C (at most 25°C) if they are fermented in large tanks. The rise in temperature depends on the amount of inoculated yeast and also on the volume of the fermenting must. The larger the volume, the faster the rise in temperature.

The rise in must temperature affects the composition of the wine. The escaping CO_2 carries with it not only heat but also ethanol and other volatile substances. Losses in ethanol can be larger than one volume percent. If one tries to calculate the original sugar content of the must on the basis of the ethanol concentration of the wine, one arrives at a value that is too low (HAUSHOFER and MEIER, 1979; DITTRICH, 1985).

With increasing must volume the following increases have been observed for wine components: butanediol: 0.15–0.40 g/L; glycerol: 1–2 g/L; succinic acid: 0.1–0.3 g/L; volatile acidity: up to 0.3 g/L (ENKELMANN, 1979).

Musts which have been inoculated with pure culture yeast starters and whose temperature does not exceed 25°C guarantee a quick start and a complete fermentation. Their sugar is completely fermented, and a lowering of the quality of the wine by bacteria need not be feared. The SO_2 requirements are low because few SO_2-binding compounds will have been formed.

Cold fermentations at temperatures below 10°C are rarely practiced today. The wines usually have somewhat higher residual sugar concentrations and a greater requirement for SO_2. The advantages of such cold fermentations are: Prevention of the malo-lactic fermentation, increased precipitation of potassium bitartrate, and smaller losses of ethanol and "bouquet" compounds.

Fruit mash fermentations are also usually set at 15–20°C; 27°C should never be exceeded to avoid losses of the aroma. The fermentation is usually completed in 10–20 days.

4.2 Ethanol and Carbon Dioxide

Yeast growth is already slightly inhibited when the alcohol concentration of the fermenting must reaches 2 vol.-%. The inhibition is stronger in a fermenting must than in an unfermented must to which 2% ethanol has been added, because the ethanol formed in the cell increases the internal ethanol pool and acts directly on yeast metabolism. In contrast added ethanol is kept back by the yeast cell membrane. In addition, toxic metabolites such as acetaldehyde and acetate are formed within the cell.

The alcohol tolerance of a yeast depends not only on its tolerance to the internal ethanol pool, but also on the composition of the plasma membrane which in turn depends on the properties of the substrate. Yeast growth is inhibited at lower ethanol concentrations than the fermentative ability of the cells. During the fermentation of musts under usual conditions no more than 15 to 16 vol.-% of ethanol can be reached.

CO_2 is also inhibitory. The growth of yeast does not take place above 15 g CO_2/L. The fermentation is inhibited at a concentration of 30 g/L CO_2. That means, that a large number of yeast cells may be present when the CO_2 concentration reaches 15 g/L. But fermentation may continue for quite some time. This effect is used for the storage of grape juices in the Böhi process. Such CO_2 concentrations do not inhibit lactic acid bacteria. That means that musts or mashes can spoil under these conditions.

The inhibiting effect of CO_2 is used to control fermentation rates in a pressure-controlled fermentation (TROOST, 1988, pp. 171–173). The inhibition by CO_2 is also important for champagne and to a lesser extent for sparkling wines. Such wines need not be sterile filtered before bottling.

4.3 Insoluble Particles of Must

The fermentation of mashes proceeds quickly and to completion. Musts which still contain insoluble particles after pressing also ferment quickly, while centrifuged (clarified) musts start to ferment later and ferment more slowly. Juices completely free from insoluble particles usually ferment slower even after the addition of yeast starter cultures. The fermentation will be incomplete, if they are also cold. The addition of some insoluble materials such as filter aids or wheat flour (apart from the yeast) accelerates the fermentation and reduces the residual sugar concentration.

This shows that the effect which accelerates the fermentation is based on the mechanical properties of the insoluble materials. The CO_2 formed by the yeasts is constantly liberated at the surfaces of the insoluble materials of the must. Their natural mobility leads to the formation of CO_2 bubbles at every location of the fermenting liquid. This causes a turbulence in the must. The particles of the must are the vehicles which are distributed vigorously in the must by fermentation CO_2. Yeast cells are also insoluble particles and act in the same manner. They are driven into all parts of the fermenting liquid which constantly supplies sugar for the fermentation. In this manner the continuous supply of sugar is coupled to the fast removal of the inhibiting products of the fermentation, the CO_2 and the ethanol.

The yeast sediments rapidly in clarified musts or in musts which contain no insoluble particles at all, as well as in diluted concentrates, for instance, for the production of apple wine. The sedimented yeast uses only the sugar available in its vicinity and excretes the inhibiting CO_2 and ethanol into this zone. As a result the yeasts are often not distributed throughout the entire fermenter volume. A simple stirring of the fermenter liquid removes this inhibition. Caution has to be used because of the rapid evolution of CO_2 gas. For secondary champagne fermentations in tanks frequent stirring is essential.

The fermenting yeast cells are also insoluble particles. At the start of the main fermentation their presence has made the must cloudy. One liter may contain 100×10^9 cells or more. The total surface area of the cells has been estimated as 18 m^2 per liter or 1800 m^2 per hectoliter of wine. This tremendous surface area also explains the rapid fermentation of the grape sugars.

Musts which have been clarified by separation of the bulk of the insoluble matter ferment more slowly with a lesser increase in temperature. The composition of such wines is, therefore, characterized by this lower temperature. Cloudy musts produce lower alcohol concentrations, higher concentrations of glycerol and lower concentrations of residual sugar and usually also less total acidity and total SO_2 (TROOST, 1988, p. 134).

Removal of particulate matter also removes most of the adhering microorganisms. This removal of wild yeasts, acetic acid bacteria and lactic acid bacteria and fermentation with pure yeast starters yields wines with a purer nose because of the lesser temperature of the fermentation. This is particularly true for musts from *Botrytis*-infected berries.

Clarification also removes residues of pesticides which adhere to particles of the grape skins and which can inhibit the fermentation and lower the sensory quality of the wine (RAUHUT and DITTRICH, 1991). In the EU the addition of 40 g of yeast cell wall preparations is permitted per hectoliter. These cell wall particles act by releasing CO_2. They can also contribute nutrients for the growth of yeasts (SPONHOLZ et al., 1990c).

4.4 Sugar Content of Fermentation Substrates

Musts with lower sugar concentrations start to ferment sooner and the sugar is completely fermented. In contrast, musts with a high sugar content (selected berries or dried selected berries) ferment slowly and incompletely. With increasing sugar concentrations of musts the alcohol concentration increases at first, but then it declines and residual sugar values increase: for musts with the following amounts of sugar (ethanol amounts) 110 g/L (59 g/L); 160 g/L (82 g/L); 210 g/L (107 g/L); 265 g/L (120 g/L); 300 g/L (110 g/L); 400 g/L (87 g/L); 540 g/L (12 g/L).

High sugar concentrations inhibit the fermentation by their high osmotic pressure which draws water from the yeast cells. The reduction of the water activity inhibits yeast cell metabolism. This is seen primarily in a reduced rate of yeast cell multiplication. Musts with sugar concentrations between 15 and 18% (63 to 76°Oe) ferment fastest. Within these limits and under normal conditions their fermentation rates are similar.

But sugar concentration is important for musts from selected berries. Some of these musts will still ferment to completion but others will not. With musts from specially selected berries or from partly dried berries (>125°Oe or >150°Oe, respectively) the multiplication of yeasts is definitely inhibited. The smaller number of cells leads to a slower fermentation and to high concentrations of residual sugar (see Fig. 7). Yeast starter cultures should be used to accelerate the fermentation. Thiamine and possibly ammonium salts should be added since these are musts from *Botrytis*-infected grapes. The yeast strains used must be able to tolerate high sugar concentrations, i.e., they must be osmotolerant.

With increasing osmotic stress the yeast cells shrink. The formation of glycerol and acetate still increases when the maximal rate of ethanol formation has been passed (see Fig. 5).

The starting materials for fruit and dessert wines have low sugar concentrations. However, it is desirable to add sugar to reach ethanol concentrations of 13–18 vol.-%. One can avoid the inhibiting effect of the sugar by adding it in small amounts at staggered intervals. In this manner the yeast is never exposed to high sugar concentrations, and the desired ethanol levels can easily be reached.

4.5 Sulfurous Acid (SO_2)

In aqueous solutions such as musts and wines the compound is present as SO_2 (associated with H_2O), HSO_3^- and SO_3^{2-}. The content is expressed as mg/L of free and bound SO_2.

The fermentation is delayed for a considerable time if larger amounts of SO_2 are added; i.e., if the must is sulfited. But the rate of the fermentation remains the same and so does its completeness. The delaying action of SO_2 on the start of the fermentation is based on its inhibition of yeast multiplication due to a variety of effects (HINZE and HOLZER, 1986).

Undissociated SO_2 penetrates the cell membrane more easily, and therefore, it alone is responsible for the inhibiting reaction. Depending on the pH its percentage of total SO_2 is between 1 and 10%. Its dissociation and, therefore, its effect depend on the pH. The dissociation declines with lower pH values, and a more sour wine requires less SO_2. For the same reason the pH affects the time span before start of the fermentation. With higher pH values and at the same SO_2 content the fermentation starts sooner.

SO_2 tolerance is quite variable between yeast species and strains. Normal *S. cerevisiae* strains can tolerate up to 4 mg/L of undissociated SO_2. SO_2-tolerant yeasts can survive much higher concentrations. Therefore, the concentrations of SO_2 permitted in wine are not sufficient to guarantee a quantitative kill of yeast cells. In musts yeasts tolerate more SO_2 than in wine, and musts require several fold higher concentrations of SO_2 for their conservation.

Free SO_2 added to the must is *bound* during the fermentation. Bound SO_2 has hardly any microbicidal effect. Therefore, the inhibition is only effective as long as free SO_2 is available. At the time of the start of the fermentation there is no free SO_2 available, since it has reacted quantitatively with SO_2-binding yeast metabolites.

Apiculate yeasts are very sensitive to SO_2. Occasionally yeasts can be found in bottled wines which multiply in the presence of normal levels of free SO_2 (50 mg/L). *Saccharomycodes ludwigii* is one of these yeasts. It consists of large cells formed like the soles of shoes or like sausages. It ferments only weakly, and forms little sediment in the bottle. Another such yeast is *Zygosaccharomyces bailii.* It forms branched masses of cells of the size of a pin head. It ferments faster than *S. ludwigii* but not as fast as *S. cerevisiae*. It is osmotolerant and tolerates sorbic acid well. Therefore, it can multiply in concentrates. In countries in which wines are sweetened with concentrates the yeast can enter the bottles if the sterile filtration is not carried out satisfactorily.

Lactic acid bacteria and acetic acid bacteria are strongly inhibited by SO_2. Therefore, sulfiting is a preventive measure which excludes bacterial growth and its consequences.

The human body oxidizes the SO_2 ingested with the wine to sulfate, and it is excreted in this form. There is no indication that the SO_2 of the wine is harmful in any way (CREMER and HÖTZEL, 1970).

4.6 Nitrogen Assimilable by Yeasts

Yeasts can synthesize all their N containing components from NH_4^+. But the presence of nitrogenous materials in must aids in this process and lowers the requirements for this synthesis, and most of these compounds can be taken up directly from the must. During yeast multiplication the N content of the must consequently decreases.

Most of the N requirements of yeasts are provided by the uptake of amino acids. Musts of normal berries contain about 3 g of total amino acids per liter, and wines about 2 g per liter. The reduction of amino acids in the must has already taken place at the end of yeast multiplication, that is, at the start of the fermentation.

During the fermentation and particularly after the fermentation yeast cells die. Therefore, the amino acid concentration in the lees is again higher, and the yeast press wine contains rather high concentrations of amino acids. For instance, a wine contained 1199 mg/L, while the press wine made from the lees after 3 months had 1989 mg amino acids per liter (RAPP, 1989).

Musts of rather poor quality also contain enough N for the entire fermentation. But musts of the best quality made from selected berries or partly dried berries are poor in nitrogenous compounds. The infecting *Botrytis* mycelium has already assimilated most of the soluble nitrogenous substances. Therefore, yeast multiplication in such musts is also reduced. In hot viticultural areas the N content of musts may be smaller than in areas of moderate climate.

The following nutritional additives are permitted to prevent a possible deficiency of nitrogen and consequent difficulties of the fermentation and to stimulate yeast multiplication: $(NH_4)_2HPO_4$ or $(NH_4)_2SO_4$ up to 0.3 g/L (DITTRICH, 1983).

For some fruit and berry mashes the nitrogen content is insufficient. The addition of up to 0.4 g/L of ammonium phosphate, sulfate or chloride is permitted, particularly for the fermentation of dessert wines.

The nitrogen content is usually sufficient for the secondary champagne fermentation. The yeast takes up half to one eighth of the nitrogen. During extensive storage of the champagne on the yeast the content of amino acids in solution increases again. It is then about as high as before the secondary fermentation (KOENIG and DIETRICH, 1991). Several aromatic compounds are formed from the amino acids (FEUILLAT, 1980).

4.7 Volatile Acidity (Acetic Acid)

Musts from *Botrytis*-infected grapes or from damaged grapes often contain more than 0.6 g/L of volatile acid. The fermentation is inhibited at concentrations which are still higher. DITTRICH (1989b) found 0.7 g/L of acetate and 90 mg/L formate in musts of partly dried berries. Under such conditions of osmotic stress the yeast also produces more volatile acid.

Acetic acid inhibits the multiplication of yeast cells. This effect is greater in young vinegary wines which are to be re-fermented than in musts because of the lesser concentration of yeast nutrients. The alcohol concentration and in secondary champagne fermentations the CO_2 act additionally as inhibitors.

4.8 Metal Content and Pesticides

Acid fruit juices such as grape juice can dissolve and take up Fe, Cu, Zn and Al by corrosion of presses and other cellar equipment. Copper can get into the must during pressing if the vines have been treated with Cu preparations against the "false" mildew (*Plasmopara viticola*). The Cu concentrations can then reach 1–5 mg/L.

Such small concentrations of Cu are today considered advantageous. During the fermentation H_2S and other sulfur compounds are formed which may lead to musty odors in the wine. They react with Cu to form CuS and other poorly soluble Cu compounds.

The fermentation is not affected by normal concentrations of trace metals. The effect of higher concentrations will not cause any inhibition except maybe during the re-fermentation of a stuck wine or the secondary champagne fermentation. Even higher concentrations of heavy trace metals are almost completely removed during the fermentation. Partly these metals are taken up by the yeast. The rest is precipitated as the poorly soluble sulfides and also removed from the wine (MOHR, 1979).

The fermentation may be inhibited if wines are treated with pesticides, which are toxic for yeasts, too late in the season or at too high a concentration. The actual amount of the pesticide present at the time of harvest which dissolves in the must during pressing is critical. Separated musts are, therefore, less contaminated.

The inhibition can be avoided by using yeast starter cultures. Currently fungicides against *Botrytis*, Metalaxyl (Ridomil) used against false mildew, and Triadimefon (Bayleton) used against true mildew, cause no inhibition. Insecticides with phosphoric acid esters are toxic to yeasts.

5 Preservatives

By far the greatest number of wines do not contain any preservatives. Until bottling the wine is stored in such a way that the few microorganisms which may be present will not multiply greatly. During filling the wine is sterile filtered which removes all or almost all microorganisms. Microbial spoilage occurs, therefore, only in exceptional circumstances. The composition of the wine and the storage conditions also combine to inhibit or prevent the multiplication of microorganisms.

The preservatives used are only those which inhibit organisms capable of multiplication in wine; namely yeasts and lactic acid bacteria. They can spoil the wine by the degradation of residual sugar or possibly by malate degradation. Sugar is, therefore, required for this multiplication and subsequent spoilage. Wines which have been sweetened with grape must are, therefore, susceptible to infection by yeasts. The wine must be sterile filtered if it contains more than 1 g sugar per liter, or the organisms must be killed by filling the warmed wine. *S. cerevisiae* is the most frequent contaminant because large numbers remain in the wine after the wine has been racked. The poorly fermenting, SO_2-tolerant yeasts, *Zygosaccharomyces bailii* and *Saccharomycodes ludwigii* frequently occur in wines with residual sugar.

Sometimes yeasts pass the filter during rapid filling operations in spite of good manufacturing practices. Such yeasts cloud the wine and ferment the residual sugar. Even isolated instances are undesirable if the bottles are sold directly to the consumer. The defect is usually not seen at the winery but becomes obvious to the consumer.

Such defects can be avoided by the addition of legal preservatives. The preservative action of some natural compounds will be discussed before added preservatives are mentioned.

5.1 Yeast Metabolites as Preservatives

For the production of sweet dessert wines the preservative effect of alcohol is used. The fermentation is interrupted or completely stopped by fortification with alcohol to a concentration of 17–20 vol.-% ethanol (GOSWELL and KUNKEE, 1977; AMERINE et al., 1982). Under normal conditions yeasts will not ferment residual sugar if the DU (Delle units) are at least 80. Delle units are calculated as follows:

$$DU = a + 4.5 \cdot c$$

where a is the weight percentage of sugar and c the vol.-% of ethanol. The equation is based on the fact that fermentation is completely inhibited by either 18 vol.-% ethanol or by 80% sugar, and that ethanol is 4.5 times as inhibitory as sugar (AMERINE and KUNKEE, 1965).

Port is fortified with 76 vol.-% EtOH to a final concentration of 18% (EGGENBERGER, 1974). Madeira is also fortified with ethanol before the fermentation is completed. A part of the must is sulfited to 100 mg/L SO_2 and fortified with 96% ethanol to 17–20 vol.-%. This prevents any further yeast multiplication. The "Mistelle" is used to sweeten the wine.

Liqueurs rarely show infections because of their alcohol concentration of at least 15 vol.-% (14 vol.-% for egg liqueurs). However, growing yeasts have been found in liqueurs made with the berries of the mountain ash; and a brandy with 38 vol.-% of EtOH showed deposits of *Bacillus megaterium.* Growth of the bacillus was explained by the high pH of 4.9. The same authors found the bacillus also in a "French brandy" (MURELL and RANKINE, 1979). WEGER (1984) found spore-forming bacteria in a pear distillate. The infection had spread through the water used for diluting the spirits.

Creams used in the production of cream liqueurs must be immediately pasteurized. The total number of live cells must not exceed 100 per g. Products containing 17 vol.-% EtOH show a gradual decline of vegetative cells during storage. Egg yolk was experimentally infected and used in the production of egg liqueur. After a 24 h storage at 20°C neither *Salmonella* nor any other coliform organisms could be found. Of course egg yolk is pasteurized routinely for use in liqueurs.

Many species which do not normally survive in alcoholic beverages still may have strains which can multiply in such beverages. A particular strain may become resistant if it is not recognized and, hence, not eliminated.

Lactobacillus homohiochii and *L. fructivorans* have caused cloudiness in dessert wines high in alcohol concentration and in Japanese sake (RADLER and HARTEL, 1984).

The CO_2 content protects champagne against the yeast cells which normally occur in the bottled beverage. In spite of the relatively high sugar concentration yeast cells do not multiply. Even in sparkling (efferverscent) wines with only 4–5 g/L CO_2 a sterile filtration is not necessary. Still lower CO_2 concentrations of 1.8 g/L protected a wine with 100 yeast cells per liter and at a low ethanol concentration (HAUBS et al., 1974).

Pathogenic bacteria quickly die in wine. *E. coli* died in a white wine in 24–45 min and in a red wine in 60 min. *Vibrio cholerae* survived only 0.5–5 min (BENDA, 1984).

5.2 Legally Permitted Preservatives

Sorbic Acid (2,4-Hexadienoic Acid)

Addition of this acid is permitted up to 200 mg/L in the EU and up to 300 mg/L in the US. The K salt is used because of the low solubility of the acid (200 mg sorbic acid correspond to 265 mg potassium sorbate).

Sorbic acid is quite effective against yeasts and molds, but hardly against lactic and acetic acid bacteria (LÜCK, 1980). It inhibits substrate uptake and sugar metabolism (REINHARD and RADLER, 1981; BURLINI et al., 1993).

Sorbate – just as SO_2 – is only effective as the undissociated molecule. Therefore, its effectiveness depends on the pH. The effect of sorbate is increased by SO_2 and higher ethanol and sugar concentrations, but the number of colony forming units (CFUs) should not be too high.

Wines which contain more than 2 g sugar per liter must be prefiltered, if no more than 200 mg of sorbate are used. This concentration inhibits yeast growth but not the malolactic fermentation or bacterial spoilage. *Zygosaccharomyces bailii* and *Saccharomycodes ludwigii* are only inhibited by sorbate concentrations above the legal limit.

In wines which have been preserved with sorbate lactic acid bacteria can cause the so-called "geranium" tone. The principal reaction is the reduction of sorbic acid to sorbinol (2,4-hexadiene-1-ol). H^+ ions isomerize the compound to 3,5-hexadiene-2-ol. This compound forms an ether with ethanol and yields 2-ethoxyhexa-3,5-diene which is responsible for the "geranium" odor (CROWELL and GUYMON, 1975; VON RYMON LIPINSKI et al., 1975). The ethyl ester of sorbic acid also occurs in wines preserved with sorbic acid.

Allylisothiocyanate is only permitted as a preservative in Italy.

Dimethyldicarbonate (DMDC)

DMDC (trade name Velcorin) is permitted as a preservative in the EU for de-alcoholized wines, but not for wines. In the US 200 mg/L may be used in wine. In wine DMDC acts at concentrations above 60 mg/L against yeasts and spoilage bacteria. In beverages it disintegrates quickly and dependent on the temperature to $2\,CH_3OH + 2\,CO_2$ or with ethanol to CH_3OH and methylethylcarbonate. The products of this disintegration are not harmful, but the preservative effect is lost. Because of its rapid disintegration DMDC must be used during filling of the wine bottles and immediately before closure (for details refer to: Bayer Information, Velcorin).

Sulfurous Acid (SO_2), Peracetic Acid, Ozone

The concentrations of SO_2 which are added to wine do not ordinarily inhibit yeasts, – at most some bacteria. The larger part reacts with wine compounds and becomes bound SO_2. This does not act on microorganisms. The presence of more than 50 mg/L of free SO_2 would produce an objectionable odor.

In higher concentrations SO_2 is used for the preservation of must which is used for sweetening wine. It is also used for the preservation of empty wine barrels with SO_2-containing water, for the sterilization of wine bottles prior to bottling, and eventually for the sterilization of equipment, such as the filler.

Since SO_2 has raised some health concerns, oxidants are now used for the sterilization of bottles. They have a good sterilizing effect and lower the BOD of the waste water. Combinations of peracetic acid and hydrogen peroxide (products: Stellanal, Divin steril) are also effective against all microorganisms at low temperature (2–10 °C). Ozone (O_3) is also quite effective. Only 28 yeast cells out of 2 million live yeast cells per bottle survived treatment with 2.7 mg/L ozone (BAUER et al., 1981).

6 Microbial Degradation of Acids

The acid content of grape musts is higher in years with poor climatic conditions. The total acid content consists principally of tartaric and malic acids. Tartaric acid is not normally degraded by microorganisms. It crystallizes in part during and after the fermentation as the acidic potassium salt. But malic acid can be degraded by many microorganisms, the wine is then less sour, its pH is somewhat higher, and under favorable conditions its taste is improved.

6.1 Malic Acid Degradation by Yeasts

During the fermentation *S. cerevisiae* degrades some of the naturally occurring L-malate, – in German viticultural areas about 10–32% (23% average) (WENZEL et al., 1982). These differences as well as differences in the formation of glycerol and butanediol account for differences in the sugar-free extract of the wine depending on the yeast strain.

Malate is metabolized by yeasts to ethanol and CO_2:

$$HOOC{-}CHOH{-}CH_2{-}COOH \xrightarrow[\text{malic enzyme (E.C. 1.1.1.40)}]{NAD^+ \quad NADH+H^+} HOOC{-}CO{-}CH_3 + CO_2$$

$$HOOC{-}CO{-}CH_3 \xrightarrow[\text{pyruvate decarboxylase}]{} CO_2 + OCH{-}CH_3$$

$$OCH{-}CH_3 \xrightarrow[\text{alcohol dehydrogenase}]{NADH+H^+ \quad NAD^+} \underset{\text{ethanol}}{HOCH_2{-}CH_3}$$

Schizosaccharomyces pombe in pure culture can degrade malate completely, but the species cannot be used in practice.

About 50% of the citrate is degraded by *S. cerevisiae* during the fermentation. Wines contain 0.1 to 0.5 g/L; wines from specially selected grapes often contain higher concentrations.

6.2 Malate Degradation by Lactic Acid Bacteria

In English speaking countries this is known as the malolactic fermentation and in French speaking countries as fermentation malolactique. This expresses the formation of the monobasic lactic acid from the dibasic malic acid. The liberated CO_2 can be mistaken for the fermentation of sugars.

The rate of the malolactic fermentation is quite variable. It is not always welcome. It would be most welcome in sour musts, but these require supportive measures. In warm viticultural areas and with grapes with low acid concentrations the malolactic fermentation often proceeds already during the alcoholic fermentation, particularly in large vats. The acid concentration in the wine may then be too low. The fact that the malolactic fermentation has taken place, as shown by a minimal malate concentration and an increased L-lactate concentration, does not prove that the malolactic fermentation was intended. It occurs more frequently in red wines. But a partial malate degradation is not infrequent in white wines.

Opinions on the desirability of the malolactic fermentations are not uniform. With red wines it is often desired or in exceptional cases tolerated. In white wines it is desired in some viticultural areas, but in most areas it is considered objectionable. The acid content of the wine is, however, only one aspect of its evaluation. Sensory substances are also formed during the malolactic fermentation which often change the typical character of a wine. Therefore, the malolactic fermentation is often objectionable in wines with a well defined varietal character. On the other hand, these compounds may improve the flavor of wines which lack the strong varietal character, and which often taste too thin or too neutral. Newer reviews have been published by DAVIS and WIBOWO, 1988; DITTRICH (1987) and HENICK-KLING (1993).

Malate-Degrading Bacteria and Their Multiplication in Wine

Plant juices and wine are not natural habitats for these organisms. In juices, mashes and wines one usually encounters species of *Lactobacillus* (*L. casei*, *L. plantarum*, obligately and facultatively heterofermentative), *Leuconostoc (L. oenos*, obligately heterofermentative), and *Pediococcus* (*P. damnosus*, *P. pentosaceus*, homofermentative). The typical organism in wine is *Leuconostoc oenos* (Fig. 9A).

Cocci are usually more frequent than rods; homofermentative organisms more frequent than heterofermentatives. Almost all strains degrade malate; hardly any strain degrades tartrate. Citrate is degraded by 17–50% of the strains; arabinose by 30–50%.

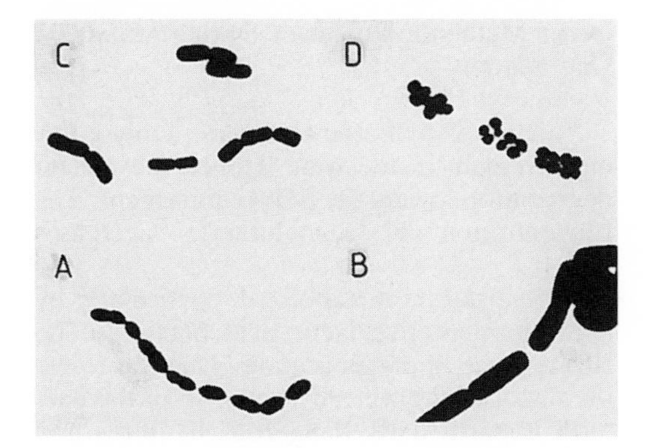

Fig. 9. Lactic acid bacteria capable of growing in wine.

A. *Leuconostoc oenos.* The most important malate-degrading species (500-fold magnification, photo WEILER). At the start of the malate degradation the diplococci occur singly. At the end of the malate degradation and in wines with high alcohol concentrations they form shorter or longer chains. The individual diplococci can be well recognized in the figure.

B. *Lactobacillus casei.*

C. *Lactobacillus brevis.*

D. *Pediococcus damnosus.*

Wine is a poor substrate for lactic acid bacteria. Therefore, the occurrence of the malolactic fermentation is uncertain. Often it occurs too late or not at all. One of the causes is the high nutritional requirement. Therefore, the bacteria grow better in juices than in wines. The nutritional requirements differ from strain to strain. The presence of malate is not required. Sugar concentrations in wines are usually sufficient for growth. They decrease only by about 0.4 to 0.8 g/L. This corresponds to a small mass of bacteria, and the wines are only slightly cloudy. The cell number is about 10×10^6 cells/mL.

The pH value of musts, mashes, and wines is very important. The lower limit is a pH of 3.2. Below this value malate degradation takes place only if all other conditions are optimal. *Pediococcus* which is considered an undesirable microorganism for the malolactic fermentation hardly grows below a pH of 3.5. The preferred *L. oenos* grows at lower pH values.

Lactic acid bacteria do not multiply at wine temperatures below 15°C. Usually temperatures above 20°C are required. The temperature of the fermenting must or mash is already of importance. Smaller concentrations of acids can only be retained if the musts are fermented cold. Large volumes of must which get warmer during the fermentation lose some acid during the fermentation. Mashes for the production of distilled beverages must usually be acidified.

The strongest inhibitor in wine is free SO_2. Already 10 mg/L inhibit growth; higher concentrations stop growth completely. Bound SO_2 inhibits only at much higher concentrations. Therefore, sulfiting does not always prevent malate degradation during the alcoholic fermentation or in the young wine.

Ethanol hardly inhibits the malolactic fermentation although most lactic acid bacteria are sensitive to ethanol. In red wines with more than 12 vol.-% ethanol the malolactic fermentation takes place without difficulties unless there are other negative conditions. The higher concentrations of polyphenols and tannic acid compounds do not inhibit the malolactic fermentation appreciably.

Yeasts stimulate the growth of lactic acid bacteria through nutrient compounds which are provided by dead cells. Following sedimentation the few bacterial cells are surrounded by the densely packed yeasts. The bacteria can then multiply rapidly in the sediment. If the yeast is stirred up, the bacteria are also distributed and can grow rapidly. This does not occur if the yeast is separated early from the young wine. Such early separation tends to preserve the malic acid.

The presence of bacteriophages which attack *Leuconostoc oenos* and lactobacilli is another factor which may explain the uncertainty of the occurrence of the malolactic fermentation (HENICK-KLING et al., 1986; SORRI and MIGNOT, 1988).

In the EU starter cultures of *Leuconostoc*, *Lactobacillus* or *Pediococcus* species may be added to the must or wine. However, much higher cell numbers must be added than yeast cells to initiate the malolactic fermentation; namely, 10^{10} bacterial cells per liter (KRIEGER and HAMMES, 1988). Current practice tends toward the use of liquid cultures of *L. oenos*, which can be added directly to the young wine without further activation. The use of currently available preparations is not always successful. Wild strains of lactic acid bacteria can multiply independently of the inoculation and simulate the malolactic fermentation with the starter culture.

Biochemistry of the Bacterial Degradation of Malate

Only naturally occurring L-malate is degraded to L-lactate. This decarboxylation is carried out by the "malolactic enzyme" (so far without enzyme number) (CASPRITZ and RADLER, 1983).

$$\begin{array}{c} COOH \\ | \\ HOCH \\ | \\ CH_2 \\ | \\ COOH \end{array} \xrightarrow[NAD^+,\ Mn^{2+}]{\text{malolactic enzyme}} \begin{array}{c} COOH \\ | \\ HOCH + CO_2 \\ | \\ CH_3 \end{array}$$

L-malate L-lactate

The dicarbonic acid is degraded to a monocarbonic acid. This increases the pH. The total weight of lactic acid is lower than that of degraded malic acid. The wine is less sour.

The malate degradation causes a lowering of the extract value; 134 g of malate yield theoretically 90 g of lactate and 44 g of free CO_2; i.e., 1 g malate yields 0.67 g lactate. In the usual calculation as tartaric acid the loss of 2 g lactate corresponds to a loss of 1 g of total acid (calculated as tartaric acid).

The actual loss of extract is greater. Apart from malate, there are losses of citrate, pyruvate, amino acids, sugar and eventually glycerol. The total amount of L-lactate is greater than the 67% which can be expected from the degradation of malate since, depending on the bacterial species, L-lactate (but also D-lactate) are produced from sugar.

The lactate concentration of a wine shows whether the wine has undergone the malolactic fermentation, and approximately what the original malate concentration has been. Lactate concentrations are usually between 1.5 and 3.5 g/L. In rare instances 4 g/L are exceeded (DITTRICH and BARTH, 1984).

Other Metabolic Processes During Malate Degradation

Pyruvate is hydrated to lactate. Only a few mg/L remain in the wine. Hence the malate degradation lowers the SO_2 requirement. The concentration of ketoglutarate decreases also.

Citrate can be metabolized, particularly by heterofermentative lactic acid bacteria. The citrate lyase liberates acetate. This can lower the flavor of the wine. The growth of the bacteria is particularly dangerous in juices and fruit wines which contain much citrate.

Gluconic acid occurs in wines from specially selected grapes. In principle it could be metabolized by lactic acid bacteria. But even in wines spoiled by bacteria one cannot find a significant loss of gluconic acid (BANDION et al., 1980).

Acetaldehyde can be partly hydrated to ethanol. But overall the SO_2 sparing effect of malate degradation is small.

N containing substances can also be metabolized. The protein concentration is only slightly lower. The concentration of ammonia increases. A few species can metabolize arginine that is present in relatively large concentrations (about 1000 mg/L). This produces ornithine and urea. The latter disintegrates to NH_3 and CO_2.

7 Decrease of Wine Quality by Microbial Action

The malate degradation by lactic acid bacteria is a useful metabolic process, if its occurrence is desired. But it is only one of several processes and not always the most important one. The bacteria primarily metabolize sugar. If they multiply too much and if enough sugar or other substrates are present, they can metabolize substrates other than malate. The lactic acid bacteria first considered useful can then become objectionable. They can decrease the sensory quality of the wine. The same is true for mashes. Besides, lactic acid bacteria can reduce the ethanol yield by their sugar metabolism.

The following substrates are most subject to attack by lactic acid bacteria: juices, wines, and mashes with low acid concentrations, as well as wines following malate degradation or de-acidification with $CaCO_3$, particularly if their pH is greater than 3.5. They are more in danger of spoilage if they contain sugar. Therefore, mashes with low acid concentrations must be acidified.

Acetic acid bacteria are only potentially spoilage organisms. "Wild yeasts" are also of importance. Some fungi can lower the quality of wines by producing musty and bitter substances. But lactic acid bacteria are of the greatest significance in this respect. The various types of spoilage caused by them are often difficult to distinguish.

The best remedy is sulfiting of the wine. Free SO_2 should not exceed 25 mg/L. The wine should be quickly centrifuged or filtered if the total bacterial count is too high. It should be strongly sulfited and stored cool. Juices of stone fruits may also be sulfited if they are used for the production of distilled beverages.

7.1 Bacterial Spoilage

Vinegary Spoilage

This is the most frequent and most objectionable defect. In spoiled wines the odor of volatile acids is disagreeable. It consists mainly of acetic acid (SPONHOLZ et al., 1982).

In the EU the legal limit for concentrations of volatile acidity is 1.08 g/L for white and rosé wines; 1.20 g/L for red wines; 1.80 g/L for ice wines, wines from selected berries and from partly dried berries; and 1.50 g/L for some French and Italian wines; always calculated as acetic acid. But in normal wines 0.7–0.8 g/L of volatile acidity are already organoleptically unacceptable. Austrian wines show vinegary spoilage if they contain more than 0.8 g/L of volatile acidity in addition to more than 90 mg/L of ethyl acetate; or if they contain less volatile acidity but more than 200 mg/L of ethyl acetate (BANDION and VALENTA, 1977a). Wines with vinegary spoilage may not be sold. They must be used for the production of vinegar.

For the production of fruit brandies it is important to start with undamaged fruit. Damaged fruit often shows vinegary spoilage. Such mashes may be neutralized with $CaCO_3$ and then immediately distilled. A slight amount of vinegary spoilage may be corrected after dilution of the distillate by addition of 300–500 g/hL of magnesium oxide or basic magnesium carbonate (TANNER and BRUNNER, 1987, p. 91).

The odor due to increased concentrations of ethyl acetate at normal concentrations of volatile acidity is called "ester tone". It is usually caused by "wild" yeasts. Such distillates can be treated with alkali. Following neutralization of the liberated acetic acid with $CaCO_3$ the material is redistilled (TANNER and BRUNNER, 1987).

Vinegary Spoilage by Lactic Acid Bacteria

This is usually caused by heterofermentative bacteria. They form the acetate anaerobically from sugar. Therefore, juices, mashes and incompletely fermented wines are particularly subject to this spoilage.

The formation of D-lactate from sugars and of mannitol from fructose occurs simultaneously with the formation of acetate. The presence of more than 1 g/L of D-lactate is a sign of spoilage (BANDION and VALENTA, 1977b).

Vinegary Spoilage by Acetic Acid Bacteria

Acetic acid can only be formed with adequate availability of oxygen, that is, on fruits, in juices, and in mashes. Wines are usually kept in the absence of oxygen, and vinegary spoilage is rare. There is a positive correlation between acetic acid and ethyl acetate if spoilage does occur.

The formation of acetic acid on grapes or in mashes in the presence of oxygen may be followed by the anaerobic formation of acetic acid by lactic acid bacteria from sugar. Such wines contain increased concentrations of D-lactate in addition to acetic acid and ethyl acetate (SPONHOLZ et al., 1982).

Grapes which show vinegary spoilage must be discarded during the selection.

The Lactic Acid Tone and Lactic Acid Spoilage

These designations only relate to lactic acid because quality-reducing metabolites may be formed during the formation of lactic acid. Even with a normal malate degradation the wines show modifications of taste and odor. With stronger changes one refers to the "degradation tone" or the "whey" or "lactic acid" tone. The cause is mainly the formation of diacetyl (2,3-butanedione). Acceptable wines contain 0.2 to 0.3 mg/L while wines with "lactic acid" tone had 0.9 mg/L or more (DITTRICH and KERNER, 1964). Pediococci seem to leave a larger residue of diacetyl after the malolactic fermentation than *Leuconostoc.* This particular fault can be corrected with fermenting yeast by hydration of the diacetyl to 2,3-butanediol (DITTRICH and KERNER, 1964). The formation of ethyl lactate also contributes to the lowered quality after the malolactic fermentation. The total ethyl lactate concentration may be 60–200 mg/L of which L-ethyl lactate may account for 65–200 mg and D-ethyl lactate for 25–50 mg (or more in undesirable wines).

With progressive activity of lactic acid bacteria the "lactic acid tone" turns into the "lactic acid spoilage". Spoilage in this context means the noticeable contribution of acetic acid. Such wines have an undesirable odor, and a sweet/sour scratchy taste. The spoiled wines also contain diacetyl and mannitol, and they are often viscous. The concentrations of 1-propanol and 2-butanol are increased. The diacetyl odor can be removed (see above), but acetic acid cannot be eliminated. Such wines are irrevocably spoiled.

The risk of such spoilage is great with stone fruit and pomace wines. Therefore, acidification of fruit mashes is permitted, for instance, the addition of 50 mL H_2SO_4 (95–98% purity) per 100 kg of Williams pears. Apricots do not require acidification. Mashes for the production of distilled spirits which have been stored for more than three months should be acidified after the fermentation (PIEPER et al., 1993). Phosphoric acid and lactic acid may also be used (TANNER and BRUNNER, 1987). National legislation should be observed.

Increased Viscosity

Grape and fruit wines with low acid concentrations can become viscous. They are designated "slimy" or "ropy" wines, and they have a stale, somewhat disagreeable taste because of the loss of acid and the formation of diacetyl. Such viscous wines can also exhibit vinegary spoilage.

In the presence of bacteria capable of raising the viscosity the process can start during the malolactic fermentation. It often ends at the end of the alcoholic fermentation if the malate degradation takes place at the same time. Therefore, such wines can already be viscous after the alcoholic fermentation.

The viscosity is increased by polysaccharides which are produced by many bacterial species. But not all malate-degrading bacteria can synthesize polysaccharides. The most important producer of polysaccharides is *Pediococcus damnosus*. In the presence of 10^6 to 5×10^6 cocci a wine can become slightly to moderately viscous (MAYER, 1974). *Leuconostoc mesenteroides* and *L. dextranicum* as well as some strains of *L. oenos* are also known to produce polysaccharides.

P. damnosus strains produce a 1,3:1,2-β-D-glucan from traces of glucose. Every second glucose molecule of the chain has a glucose molecule attached to its side (CANAL-LLAUBERES et al., 1989). The presence of 100 mg sugar per liter suffices to produce a slimy wine. The increase in viscosity begins after the alcoholic fermentation in the yeast sediment because this is the best substrate for bacterial growth. A slight increase in viscosity can often be observed after the alcoholic fermentation. This disappears again during normal cellar practice.

Mannitol Spoilage

Heterofermentative bacteria such as *Lactobacillus brevis* and *Leuconostoc* species produce mannitol besides lactic acid, acetic acid and ethanol from fructose but not from glucose. The mannitol dehydrogenases reduce fructose (E.C. 1.1.1.67) and fructose-6-phosphate (E.C. 1.1.1.17). Mannitol formation is coupled with the formation of acetate. Concentrations of D-lactate, 1-propanol and 2-butanol are increased. Such wines may also become viscous. Mannitol spoilage cannot occur if the sugar has been completely fermented by yeasts, since not enough fructose is available for mannitol formation.

High Acid Formation by Lactic Acid Bacteria

Sometimes wines have a disagreeably high concentration of acids although the musts have a normal acid content. Analytically they show a complete degradation of malate but abnormally high concentrations of L- and D-lactate (e.g., 5.4 g/L and 4.5 g/L, respectively). Volatile acidity is hardly increased. Homofermentative lactic acid bacteria are probably the cause of the acid formation.

Lactate formation and other metabolic processes using sugar as the substrate and which do not lead to ethanol formation (for instance, mannitol formation) lower the yield of alcohol in mashes for the production of distilled beverages.

Mousiness

A typical odor can be discerned if a few drops of a mousy wine are rubbed between the palms. The wines have a disagreeable, long lasting odor and sometimes a spoiled aftertaste. This condition is not rare in wines which have not been sulfited or in fruit wines.

The cause are usually lactic acid bacteria and in warmer climates also *Brettanomyces* yeasts. *Lactobacillus* and *Brettanomyces* species form 2-acetyltetrahydropyridine from lysine and ethanol. The tautomers of this compound are in equilibrium (HERESZTYN, 1986).

A slight degree of mousiness can usually be eliminated by sulfiting or by fermentation with fresh must. Strongly mousy wines cannot be used for distillation nor for the production of vinegar.

Glycerol Degradation and Acrolein Spoilage

In wines the degradation of glycerol is almost without significance. But it is common in mashes with low acid concentrations. Its degradation product, acrolein, lowers the quality and is harmful to one's health (WESENBERG and LAUBE, 1990). In Germany the government brandy monopoly has set lower prices for acrolein-containing raw spirits (WELTER, 1991). In Switzerland the limits for acrolein concentration have been set at 0.2 to 0.4 mg/100 mL pure ethanol.

Only 4 out of 42 strains of lactic acid bacteria could degrade glycerol (SCHÜTZ and RADLER, 1984). *Lactobacillus coryneformis*, an acrolein producer, has been isolated from potato wash water. Acrolein is frequently found in distilleries which use fresh potatoes as raw material. Clostridia may also be the cause of acrolein formation if the potatoes still contain soil or if dropped fruits are used. *Citrobacter freundii* is another producer of acrolein (BUTZKE et al., 1990).

In the presence of sugar glycerol is reduced via 3-hydroxypropanal largely to 1,3-propanediol (SCHÜTZ and RADLER, 1984; Fig. 10). 2,3-Butanediol can also form 2-butanol in the presence of glucose. The unreacted 3-hydroxypropionaldehyde forms acrolein non-enzymatically by loss of one molecule of water. This reaction takes largely place during the distillation of the fermented mash.

In alcoholic solutions acrolein is associated with ethanol and water. On re-distillation the same percentage of acrolein is found again. In spite of its low boiling point (53 °C) it can only partly be separated with the most volatile fraction (TANNER and BRUNNER, 1987, pp. 94–95).

Acrolein can be reduced to allyl alcohol. Its presence is a sign of bacterial spoilage. Acrolein can react non-enzymatically with polyphenols to form bitter compounds (Fig. 10). This can account for the occasional bitterness of red wines.

Butyric acid spoilage is rare in wines. It may occur in mashes from starchy substrates

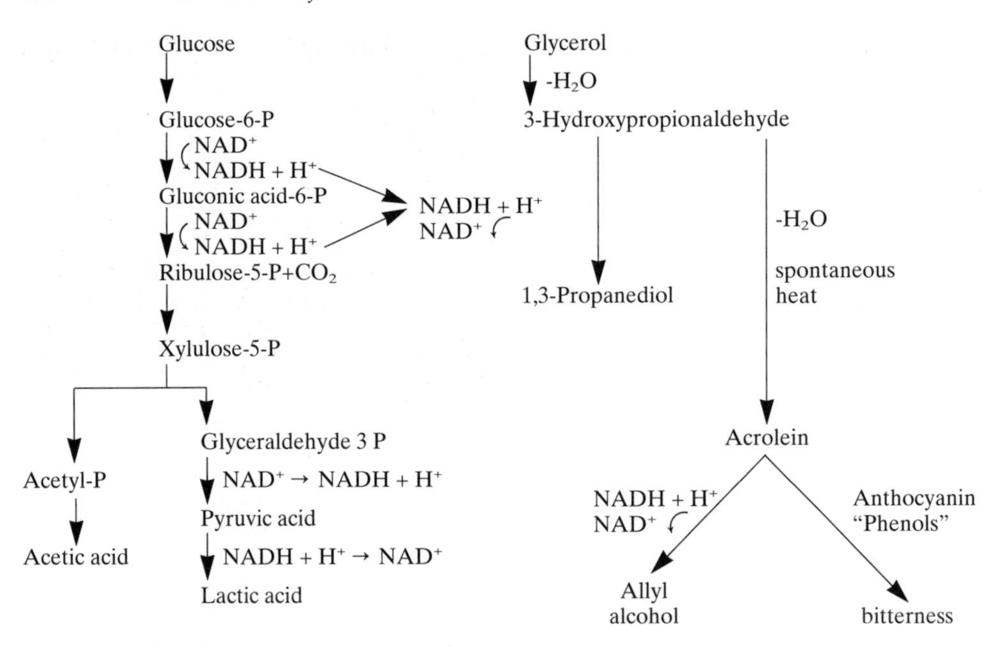

Fig. 10. Glucose and glycerol fermentation by the heterofermentative *Lactobacillus brevis* to 1,3-propanediol (SCHÜTZ and RADLER, 1984), and the formation of acrolein, allyl alcohol and bitter substances as products of the glycerol fermentation.

which are often heavily contaminated with butyric acid bacteria and have low acid concentrations. The distillates have the odor of rancid butter. The butyric acid is often associated with propionic acid, 1-butanol and acetone. Such distillates can be improved by addition of 500 g/hL of calcium hydroxide (TANNER and BRUNNER, 1987, p. 94).

Degradation of tartaric acid is very rare. Only 5 strains of lactic acid bacteria out of 78 strains could metabolize tartaric acid. The degradation can follow two separate paths (RADLER and YANISSIS, 1972), however, only after the complete degradation of malate, and usually only after complete spoilage of the wine.

7.2 Quality Defects Due to Yeasts

Yeasts of the *S. cerevisiae* group can multiply in wine if they enter the bottles through faulty filling procedures. *Zygosaccharomyces bailii* and *Saccharomycodes ludwigii* can also multiply under these conditions.

Some yeast strains which enter the must from the grapes can also form more H_2S, SO_2, or volatile acidity than normal. To avoid such occurrences the musts should be centrifuged, possibly pasteurized, and inoculated with pure yeast starter cultures.

Ethyl acetate has an objectionable odor. Wines containing more than 200 mg/L (BANDION and VALENTA, 1977a) are undesirable because of their "ester" or "solvent" odor (SPONHOLZ et al., 1982). Ethyl acetate is mainly produced by yeast species of the following genera: *Hanseniaspora*, *Kloeckera*, *Candida*, *Metschnikovia*, and *Hansenula* (SPONHOLZ and DITTRICH, 1974).

Production of the mousy odor by *Brettanomyces* species has already been mentioned. Quality defects caused by fungi will be treated in Sect. 8 below. Detailed descriptions of bacterial spoilage of wines have been published by DITTRICH (1987) and SPONHOLZ (1993).

8 Effect of Fungi on Wine Quality

Fungi which infect grapes change the composition of the must already on the vine. They change some compounds, eliminate others, and produce compounds not normal in grape musts. The site of the infection also permits entry of other microorganisms into the grapes. Therefore, it is difficult to separate the effects of the fungi from those of the secondary bacterial invaders.

Fig. 11. Conidia of *Botrytis cinerea* (photo BLAICH).

8.1 *Botrytis cinerea* (Noble Rot)

The gray mold (Fig. 11) permits the production of wines from selected grapes if the weather in fall is suitable. The mold changes the composition of the must (Tab. 4) and increases the quality of the premium wines which can be produced from these infected berries (RINEREAU-GAYON et al., 1980; DITTRICH, 1989b; DONÈCHE, 1993). These "Auslese" (select), "Beerenauslese" (grape select) and "Trockenbeerenauslese" (partly dried grapes) wines, as well as Tokays and Haut Sauterne wines are world famous. More recently they are also produced with selected grapes in California, South Africa and Japan.

Tab. 4. Compounds in Musts of Increasing Quality from *Botrytis*-Infected Berries of Increasing Dryness (Variety Ruländer) (SPONHOLZ, DITTRICH and LINSSEN, 1986; unpublished)

Quality of Grapes	Kabinett[a]	Spätlese	Auslese	Beeren-auslese	Trockenbeeren-auslese
°Oechsle	82	91	97	128	231
Weight of 100 berries, g	209	175	143	85	36
Sugar, G and F, g/L	182	204	210	295	500
G and F ratio	0.98	0.94	0.86	0.80	0.72
Total acids, g/L	11.8	11.8	12.8	15.2	20.8
Tartaric acid, g/L	7.3	6.5	4.2	2.6	2.4
Malic acid, g/L	4.2	5.7	6.3	8.0	10.1
Glycerol, g/L	0.1	0.8	3.2	8.0	20.7
Gluconic acid, g/L	0.1	0.17	0.56	1.46	2.17
Galacturonic acid, g/L	0.1	0.64	0.65	0.61	1.12
Mucic acid, g/L	0.1	0.49	0.61	1.01	1.19
Citric acid, mg/L	104	182	195	204	237
Acetic acid, mg/L	0	46	202	450	129
L-Lactic acid, mg/L	8	13	38	105	176
Mannitol, mg/L	12	75	253	516	2132
Arabitol, mg/L	0	10	37	463	818
Inositol, mg/L	148	171	218	335	634
Sorbitol, mg/L	30	191	317	371	362
Ethanol, mg/L	122	1038	1170	618	254

[a] For an explanation of terms see Tab. 2
G, glucose; F, fructose

Changes in Must Components

Grapes infected by *Botrytis* lose water in dry fall weather. The berries shrink more or less and ultimately become "dried berries". The evaporation of water may lead to an almost seven-fold increase in concentration; from 74–78°Oe to 208–231°Oe (see Tab. 4). The specific gravity and the sugar concentration increase.

The mold requires sugar for its growth besides other nutrients. The increase in sugar concentration is, therefore, only relative. The total amount of sugar in the berries decreases greatly.

The mold degrades the cell walls of the grapes. Therefore, the concentration of some sugars, which are of no importance in normal musts and wines, increases. For instance, the sum of galactose and arabinose in "quality" wines is 0.1 g/L; in wines from selected berries 0.75 g/L, and in wines from partly dried berries 1.95 g/L (DITTRICH and BARTH, 1992).

In infected grapes the total amount of acids increases. The amount of tartaric acid decreases while the amount of malic acid strongly increases. The degradation of the pectic substances of the grape causes an increase in galacturonic acid and of gluconic acid, the product of glucose oxidation.

The increase in acetic acid shows the participation of acetic acid bacteria. These bacteria most likely produce the larger portion of gluconate and its oxidation products 5- and 2-oxogluconic acid (DITTRICH, 1989b). An oxidation product of fructose, 5-oxofructose, probably occurs in *Botrytis*-infected musts.

Galactaric acid, the product of glucuronic acid oxidation, is a typical constituent of musts from *Botrytis*-infected grapes. Musts of selected berries contain 0.2 to 0.5 g/L and of partly dried berries 2.0 g/L. The calcium salt of the acid which is poorly soluble is of importance. It precipitates only after the fermentation in the form of irregular, slimy clumps which may be 3 mm in length. Even if the wines are cooled the precipitation usually occurs in the bottle. It is a sign of quality.

Glycerol is an important product of the metabolism of the mold. Its greatly increased concentration is the cause of the high extract values of wines from selected grapes and partly dried grapes (DITTRICH, 1989b, analytical section). Fermenting yeasts produce more glycerol in such musts in addition to the high glycerol concentration produced by the mold.

Polyols also occur in such musts and wines in increased concentrations. Sorbitol concentration may be almost 1 g/L. Mannitol which is thought to be produced by lactic acid bacteria may occur in concentrations up to 13 g/L (SPONHOLZ, 1988).

The polysaccharides formed by *Botrytis* in berries cannot be fermented. Musts of infected berries (90–118°Oe; 21.0–27.2% sugar) contained 1.5–2.0 g/L more polysaccharides than sound berries (59–80°Oe; 14.0–18.2% sugar) (DITTRICH, 1964). Besides mannan, the most important polysaccharide seems to be glucan. One of the glucans, 1,3:1,6-β-D-glucan seems to have practical importance (DUBOURDIEU et al., 1981). It causes problems in the clarification and filtration of wines from *Botrytis*-infected grapes. It forms molecular aggregates which plug up the filters. A concentration of 2–3 mg/L of this glucan slows the filtration. Up to 50 mg/L and more have been found. It is practically impossible to filter such wines (WUCHERPFENNIG and DIETRICH, 1983) (Fig. 12). The polysaccharides surround the hyphae of the mold. The more the berries are mechanically crushed, the more glucan can enter the must. Therefore, the concentrations of glucans in such musts and wines are quite variable. A glucanase preparation from *Trichoderma viride* is

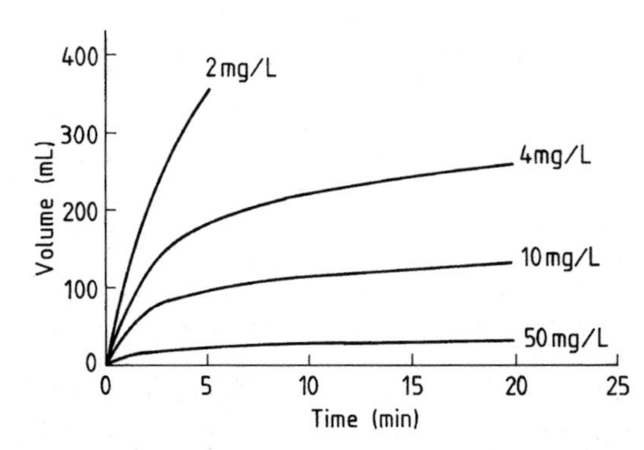

Fig. 12. Rate of filtration of ultrafiltered wine as a function of added amounts of *Botrytis* glucan (WUCHERPFENNIG and DIETRICH, 1983).

suitable for the hydrolysis of this glucan (WUCHERPFENNIG and DIETRICH, 1982).

The mold hydrolyzes various glycosides with a glucosidase; among others the anthocyanins of red wines. Therefore, red wines from *Botrytis*-infected musts have a lighter color. Red or blue grapes are rarely used for the production of wines from selected berries.

In the presence of oxygen polyphenols are oxidized by laccase, a special polyphenol oxidase, of the mold. The soluble enzyme is quite stable in acid musts. It oxidizes more compounds than the native tyrosinase of the grape. Therefore, musts and wines of *Botrytis*-infected grapes are subject to browning. This can be prevented by a sufficient concentration of free SO_2. Laccase activity decreases during the fermentation. Subsequent sulfiting inactivates the enzymes, so that relatively light colored wines can be produced from *Botrytis*-infected grapes.

Botrytis reduces the differences in taste and odor of wines from different grape varieties. Wines of such *Botrytis*-infected grapes are indeed "enobled" but have largely lost their varietal character. The difficulties to define "noble rot" character and the "selected berry" character are typical for such wines. This can possibly be explained by the decrease in monoterpenes (BOIDRON, 1978). The monoterpene disaccharides are hydrolyzed by glucosidases. The liberated terpenes can evaporate. Besides, the mold transforms, for instance, linalool into other terpenes (SHIMIZU, 1982).

Botrytized wines, just as flor sherries, old sake (rice wine) and molasses, contain the Maillard product 4,5-dimethyl-3-hydroxy-2(5)-furanon (Sotolon). Its taste is sweet, sugar- and caramel-like (SPONHOLZ and HÜHN, 1994). Another compound believed to participate in the bouquet is ethyl-9-hydroxynonanoate.

Botrytis requires for its growth nitrogen. Its uptake of amino acids reduces the nitrogen available for growth of yeast. A must from sound berries contained about 2500 mg N per liter, while infected berries contained about 1450 mg per liter (DITTRICH, 1989b).

Other must compounds are also required for growth of the mold. Thiamine was reduced from 318 ng/L to 35 ng/L in comparable musts from *Botrytis*-infected grapes. Pyridoxal was reduced to one half of its concentration. Minerals (ash) were also reduced, and the content of potassium and magnesium increased greatly (WAGNER and KREUTZER, 1977).

Consequences of the Changed Composition of Musts of *Botrytis*-Infected Grapes for the Alcoholic Fermentation

The must of *Botrytis*-infected grapes with their high sugar concentrations is difficult to ferment. Residual (unfermented) sugar concentrations of 100–150 g/L remain in the wine. Therefore, the alcohol concentration of wines from selected berries or dried berries is often below 10 vol.-%. The legal minimum concentration of alcohol is 43.4 g/L which is 5.5 vol.-%.

The high sugar concentrations also draw water osmotically from the yeast cells. This is the decisive factor for the inhibition of the fermentation of wines from *Botrytis*-infected musts (DITTRICH, 1964). The growth of yeasts is also inhibited. For example, the fermentation of 100 mL of the must of sound berries leads to the formation of 230 mg yeast dry substance, but in *Botrytis*-infected grapes only to 155 mg. The addition of yeast cell wall preparations and of ammonium salts can improve the fermentation.

The reduction in thiamine concentration is also of considerable importance. The remaining thiamine is not sufficient for the growth requirements of the yeast, and the yeast cannot synthesize the required thiamine fast enough. The deficit inhibits the fermentation. More pyruvate and oxoglutarate are excreted by the yeast since thiamine diphosphate is the coenzyme of the pyruvate decarboxylase. This increases the SO_2 requirement still further.

The legally authorized addition of 0.6 mg thiamine per liter improves the fermentation. It leads to a normal excretion of oxo-acids during the fermentation of such musts. This contributes significantly to a reduction of the SO_2 requirements. The legal limits for SO_2 additions are 350 mg/L for wines from selected berries and dried berries in German and Austrian viticultural areas and to 400 mg/L for wines of comparable quality from certain French areas.

The relatively high requirements for SO_2 are caused by the higher concentrations of SO_2-binding metabolites such as acetaldehyde, pyruvate and oxoglutarate. To this one has to add the galacturonic acid from the must, and possibly the 2,5-dioxogluconic acid. Finally, the high residual sugar concentrations also bind SO_2.

The inhibition of the alcoholic fermentation also affects the formation of higher alcohols. They are sugar metabolites, and, therefore, formed in lesser concentrations than in wines from sound berries.

8.2 Harmful Fungi

Besides *Botrytis* grapes are often infected with *Penicillium.* Of 222 fungal strains (excepting *Botrytis*) 133 were Penicillia (RADLER and THEIS, 1972); the most common species is *P. expansum.* In particular, damaged grapes are attacked by this "green rot". Unripe berries are completely destroyed by the strong degradation of pectins. Wines of *Penicillium*-infected berries have high sugar concentrations, and they clarify quickly. The color of red wines may be deeper. The concentrations of sugar, total acidity, and gluconic acid increase in *Penicillium*-infected berries; that of nitrogen decreases (ALTMAYER, 1983).

Penicillium-infected grapes cannot be used for the production of grape juice because of the formation of the mycotoxin patulin. This has no significance for the production of wine since patulin cannot be found after the fermentation and sulfiting (SCOTT, 1977). The formation of bitter substances lowers the quality of the wine.

Penicillium expansum, P. roquefortii and other species can produce volatile, disagreeable moldy odors of varying identity, which may also be the cause for the "cork odor". The quality of the wine can be lowered without the direct effect of the fungus. The molds cannot grow in the wine, and the spores are killed by the alcohol. It is sufficient if a bottle is closed with a cork which is moldy with *P. expansum.* The detrimental substances with strong taste and strong odor can then dissolve in the wine.

Penicillium species, like other fungi, produce formic acid (DIZER, 1980).

Eleven *Aspergillus* strains could be found on moldy berries besides *Botrytis* and *Penicillium.* Ten of these were strains of the *A. fumigatus* species (RADLER and THEIS, 1972). These fungi have little significance.

A. niger is used industrially for the production of pectic enzymes. The enzymes accelerate the pressing operation and improve color extraction of mashes from red berries.

Aflotoxin-forming fungi do not occur on grapes. The mycotoxins could not be found in wines made from moldy grapes.

The "pink rot" of grapes by *Trichothecium roseum* is rare. The mold forms the bitter tasting trichothecins besides a musty odor. The antibiotic has been found in some isolated cases in moldy grapes (FLESCH and VOGT-SCHEUERMANN, 1993). Infected berries should be discarded because of the strong bitterness.

Berries are sometimes infected by *Mucor* species. Hyphae extending into the must liberate round cells. Those of *Mucor racemosus* can produce 4–5 vol.-% ethanol. They form glycerol, succinate, oxalate and lactate from sugars. Mucoraceae also produce formic acid. They rarely interfere with the fermentation. Both hyphae and the round cells are killed by more than 5 vol.-% alcohol.

Rhizopus stolonifer attacks grapes in South Africa during wet fall weather. The berries rot within a few days. This causes great losses in the harvest since the juices run out.

Very different fungi can produce musty or moldy odors which lower the quality of the wine. Such fungi are Mucoraceae, Ascomycetes and their imperfect forms, but also the Basidiomycete *Armillaria mellea.* These fungi can infect cork, and the volatile odors produced by them are a cause of the "cork odor" of wines (DAVIS et al., 1981; LEE and SIMPSON, 1993; SPONHOLZ and MUNO, 1994).

The cause of this objectionable odor is the formation of 2,4,6-trichloroanisole by molds. It is caused mainly by hypochlorite bleaching of corks. Apparently phenols are liberated from the cork and chlorinated during this operation. During storage of the corks infecting molds can methylate the formed 2,4,6-trichlorophenol to the corresponding anisole. The 2,4,6-trimethylanisole enters the wine from the cork. As little as 10–50 ppt can give a perceptible odor.

2,3,4,6-Tetrachloroanisole also has a musty odor. At some time pentachlorophenol was used as a bleach in barrels. The material contained about 10% tetrachlorophenol. This compound could be methylated by fungi to tetrachloroanisole. Molds can also synthesize chloroanisoles without acting on the chlorophenols liberated from corks.

The highly odorous chloroanisoles can be adsorbed from the cellar atmosphere on bentonite or active carbon, just as other odors can be so adsorbed. They can thus get into the wine through use of these well known additives.

9 References

ADAM, L., POSTEL, W. (1992), Behaviour of brandy volatiles in dependence on the alcohol strength of the distillate by means of continuous still, *Chem. Microbiol. Technol. Lebensm.* **14,** 95–103.

ALTMAYER, B. (1983), Beeinflussung der Most- u. Weinqualität durch den Pilzbefall reifer Trauben, *Dtsch. Weinbau* **38,** 1702–1704.

AMERINE, M. A., KUNKEE, R. E. (1965), Yeast stability tests on dessert wines, *Vitis* **5,** 187–194.

AMERINE, M. A., BERG, H. W., KUNKEE, R. E., OUGH, C. S., WEBB, A. D. (1982), *The Technology of Wine Making,* 4. Ed., Westport, CT: AVI Publ. Comp.

BACK, W. (1994), *Farbatlas und Handbuch der Getränkebiologie* I, Nürnberg: Verlag H. Carl.

BANDION, F., VALENTA, M. (1977a), A contribution to the proof of acescence in wines and fruit wines in Austria, *Mitt. Klosterneuburg* **27,** 18–22.

BANDION, F., VALENTA, M. (1977b), A contribution to the evaluation of the D(−) and L(+) lactic acid contents in wines, *Mitt. Klosterneuburg* **27,** 4–10.

BANDION, F., ROTH, I., MAYR, E., VALENTA, M. (1980), Assessment of gluconic acid contents in wine in connection with possible changes during storage, *Mitt. Klosterneuburg* **30,** 32–36.

BARNETT, J. A., PAYNE, R. W., YARROW, D. (1990), *Yeasts. Characteristics and Identification,* 2. Ed., Cambridge: University Press.

BAUER, H., ZÜRN, F., GIES, P., KARASAIT, T. (1981), Peressigsäure und Ozon, *Weinwirtschaft* **117,** 436–439.

BENDA, I. (1982), Wine and brandy, in: *Prescott & Dunn's Industrial Microbiology* (REED, G., Ed.), 4th Ed. pp. 293–402, Westport, CT: AVI Publ. Comp.

BENDA, I. (1984), The occurrence of coli- and coliform bacteria in wine, *Mitt. Klosterneuburg* **34,** 249–251.

BISSON, L. F. (1993), Yeasts – metabolism of sugars, in: *Wine Microbiology and Biotechnology* (FLEET, G. H., Ed.), pp. 55–75, New York: Harwood Academic Publishers.

BOIDRON, J. N. (1978), Relation entre les substances terpéniques et la qualité du raisin (Role du *Botrytis cinerea*), *Ann. Technol. Agric.* **27,** 141–145.

BURLINI, N., PELLEGRINI, R., FACHERIS, P., TORTORA, P., GUERRITORE, A. (1993), Metabolic effects of benzoate and sorbate in the yeast *Saccharomyces cerevisiae* at neutral pH, *Arch. Microbiol.* **159,** 220–224.

BUTZKE, Ch. E., BOSSMEYER, M., SCHEIDE, K., MISSELHORN, K. (1990), Anmerkungen zur Acrolein-Problematik in der Alkoholindustrie, *Branntweinwirtschaft* **130,** 286–289.

CABRERA, M. J., MORENO, J., ORTEGA, J. M., MEDINA, M. (1988), Formation of ethanol, higher alcohols, esters, and terpenes by five yeast strains in musts from Pedro Ximenes grapes, *Am. J. Enol. Vitic.* **39,** 283–287.

CANAL-LLAUBERES, R. M., DUBOURDIEU, D., RICHARD, B., LONVAUD-FUNEL, A. (1989), Structure moleculaire du β-D-glucane exocellulaire de *Pediococcus* sp., *Connaiss. Vigne Vin* **23,** 49–52.

CASPRITZ, G., RADLER, F. (1983), Malolactic enzyme of *Lactob. plantarum*, *J. Biol. Chem.* **258,** 4907–4910.

CHRISTOPH, N., SCHMITT, A., HULDENBRAND, K. (1988), Ethylcarbamat in Steinobstbränden I, *Kleinbrennerei* **40,** 154–158.

CREMER, H. D., HÖTZEL, D. (1970), Thiaminmangel und Unbedenklichkeit von Sulfit für den Menschen, 4. *Int. Z. Vitaminforschg.* **40,** 52–57.

CROWELL, E. A., GUYMON, J. F. (1975), Wine constituents arising from sorbic acid addition and identification of 2-ethoxyhexa-3,5-diene as source of geranium-like off-odor, *Am. J. Enol. Vitic.* **26,** 97–102.

DARRIET, P., BOIDRON, J. N., DUBOURDIEU, D. (1988), L'hydrolyse des hétérosides terpéniques du Muscat a petits grains par les enzymes périplasmiques de *Saccharomyces cerevisiae*, *Connaiss. Vigne Vin* **22,** 185–195.

DAVIS, C. R., FLEET, G., LEE, T. (1981), The microflora of wine corks, *Aus. Grape-grower Winemaker* **18,** 42–44.

DAVIS, C. R., WIBOWO, D. (1988), Properties of wine lactic acid bacteria, *Am. J. Enol. Vitic.* **39,** 137–142.

DITTRICH, H. H. (1964), Zur Vergärung edelfauler u. hochkonzentrierter Moste, *Wein-Wissenschaft* **19,** 169–182.

DITTRICH, H. H. (1983), Einfluß von Thiamin und Ammoniumsalzen auf die Weinqualität, *Dtsch. Weinbau* **38,** 1366–1372.

DITTRICH, H. H. (1985), Die Bedeutung der Gärungswärme für die Weinqualität, *Dtsch. Weinbau* **40,** 1029–1035.

DITTRICH, H. H. (1987), *Mikrobiologie des Weines.* 2. Ed., Stuttgart: Ulmer.

DITTRICH, H. H. (1989a), Die Gärung, in: *Chemie des Weines* (WÜRDIG, G., WOLLER, R., Eds.), pp. 184–221, Stuttgart: Ulmer.

DITTRICH, H. H. (1989b), Influence of *Botrytis cinerea* on berries compounds of worth and on wine quality, *Vitic. Enol. Sci.* **44,** 105–131.

DITTRICH, H. H., BARTH, A. (1984), The SO_2-content, SO_2-binding substances and malo-lactic fermentation in German wines, *Wein-Wissenschaft* **39,** 184–200.

DITTRICH, H. H., BARTH, A. (1992), Galactose and arabinose contents in musts and wines of the Auslese group, *Vitic. Enol. Sci.* **47,** 129–131.

DITTRICH, H. H., KERNER, E. (1964), Diacetyl als Weinfehler. Ursache und Beseitigung des "Milchsäuretones", *Wein-Wissenschaft* **19,** 528–535.

DIZER, H. (1980), Die Ameisensäurebildung durch Schimmelpilze in Nährlösungen, Fruchtsäften u. Früchten, *Dissertation*, Universität Gießen.

DONÈCHE, B. J. (1993), Botrytized wines, in: *Wine Microbiology and Biotechnology* (FLEET, G. H., Ed.), pp. 327–351, New York: Harwood.

DUBOURDIEU, D., RIBÉREAU-GAYON, P., FOURNET, B. (1981), Structure of the extracellular β-D-glucan from *Botrytis cinerea*, *Carbohydr. Res.* **93,** 294–299.

EGGENBERGER, W. (1974), Die Portweine und ihr Produktionsgebiet, *Schweiz. Z. Obst Weinb.* **110,** 166–169.

EL HALOUI, N., PIQUE, D., CORRIEU, G. (1988), Alcoholic fermentation in wine-making: On-line measurement of density and CO_2 evolution, *J. Food Eng.* **8,** 17–30.

ENKELMANN, R. (1979), Untersuchung zur Klärung der Alkoholausbeute bei Weinen des Jahrgangs 1978, *Badischer Winzer*, 322–326.

FEUILLAT, M. (1980), Vieillissement du vin de Champagne sur levures: Phenomènes d'autolyse, *Rev. Franc. Oenol.* **16,** 35–46.

FLEET, G. H. (Ed.) (1993), *Wine Microbiology and Biotechnology*, New York: Harwood Academic Publishers.

FLESCH, P., VOGT-SCHEUERMANN, I. (1993), Isolation and identification of iso-trichothecin of the fungus *Trichothecium roseum*, *Vitic. Enol. Sci.* **48,** 15–19.

GOSWELL, R. W., KUNKEE, R. E. (1977), Fortified wines, in: *Alcoholic Beverages* (ROSE, A. H., Ed.), London–New York: Academic Press.

HAUBS, H., MÜLLER-SPÄTH, H., LOESCHER, T. (1974), Über den Einfluß von CO_2 auf den Wein, *Dtsch. Weinbau* **29,** 930–934.

HAUSHOFER, H., MEIER, W. (1979), Die Alkoholausbeute bei der Vergärung von Traubenmosten als Funktion der Gärmasse und der Temperatur, *Weinwirtschaft* **115,** 1247–1254.

HENICK-KLING, Th. (1993), Malolactic fermentation, in: *Wine Microbiology and Biotechnology* (FLEET, G. H., Ed.), pp. 289–326, New York: Harwood Acad. Publ.

HENICK-KLING, Th., LEE, T. H., NICHOLAS, D. J. (1986), Inhibition of bacterial growth and malolactic fermentation in wine by bacteriophage of *Leucon. oenos*, *J. Appl. Bacteriol.* **61,** 287–293.

HERESZTYN, T. (1986), Formation of substituted tetrahydropyridines by species of *Brettanomyces* and *Lactobacillus* isolated from mousy wines, *Am. J. Enol. Vitic.* **37,** 127–132.

HINZE, H., HOLZER, H. (1986), Analysis of energy metabolism after incubation of *Saccharomyces cerevisiae* with sulfite or nitrite, *Arch. Microbiol.* **145,** 27–31.

HOUTMAN, A. C., MARAIS, J., DU PLESSIS, C. S. (1980), The possibilities of applying present-day knowledge of wine aroma components influence of several juice factors on ester production, *S.A. Enol. Vitic.* **1,** 27–33.

KARDOS, E. (1966), *Obst- und Gemüsesäfte*, Leipzig: VEB Fachbuchverlag.

KOCKOVÁ-KRATOCHVÍLOVÁ, A. (1990), *Yeasts and Yeast-Like Organisms*, Weinheim: VCH.

KÖNIG, P., DIETRICH, H. (1991), influence of the yeast contact and storage time on the chemical composition and the colloids of sparkling wine, *Vitic. Enol. Sci.* **46,** 85–92.

KREGER-VAN RIJ, N. J. (1984), *The Yeasts, a Taxonomic Study*, 3. Ed., Amsterdam: Elsevier Sci. Publ.

KREIPE, H. (1981), *Getreide- und Kartoffelbrennerei.* 3. Ed., Stuttgart: Ulmer.

KRIEGER, S., HAMMES, W. P. (1988), Biologischer Säureabbau im Wein unter Einsatz von Starterkulturen, *Dtsch. Weinbau* **43,** 1152–1154.

LAFON-LAFOURCADE, S. (1983), Wine and brandy, in: *Biotechnology* (REHM, H.-J., REED, G., Eds.), 1st. Ed., Vol. 5, pp. 81–163, Weinheim: VCH.

LEE, T. H., SIMPSON, R. F. (1993), Microbiology and chemistry of cork taints in wine, in: *Wine Microbiology and Biotechnology* (FLEET, G. H., Ed.), pp. 353–372, New York: Harwood.

LEMPERLE, E., KERNER, E. (1990), Trockenhefen auf dem Prüfstand, *Weinwirtsch. Tech.*, 15–20.

LÜCK, E. (1980), *Antimicrobial Food Additives*, Berlin–Heidelberg–New York: Springer Verlag.

MARSHALL, C. R., WALKLEY, V. T. (1951), Some aspects of microbiology applied to commercial apple juice production I., *Food Res.* **16**, 448–456.

MAYER, K. (1974), Nachteilige Auswirkungen auf die Weinqualität bei ungünstig verlaufendem biologischem Säureabbau, *Schweiz. Z. Obst Weinb.* **110**, 385–391.

MOHR, H. D. (1979), Untersuchungen zum Verbleib von Schwermetallen, die Traubenmost zugesetzt wurden, *Weinberg Keller* **26**, 277–288.

MURELL, W. G., RANKINE, B. C. (1979), Isolation and identification of a sporing *Bacillus* from bottled brandy, *Am. J. Enol. Vitic.* **30**, 247–249.

NYKÄNEN, L. (1986), Formation and occurrence of flavor compounds in wine and alcoholic beverages, *Am. J. Enol. Vitic.* **37**, 84–96.

PIEPER, H. J., BRUCHMANN, E. E., KOLB, E. (1993), *Technologie der Obstbrennerei*, 2. Ed., Stuttgart: Ulmer.

POSTEL, W., ADAM, L. (1980), Gaschromatographische Charakterisierung von Weinbrand, Cognac und Armagnac II, *Branntweinwirtschaft* **120**, 154–164.

POSTEL, W., ADAM, L. (1984), Einfluß des Hefeanteils in Wein auf die Gehalte an flüchtigen Verbindungen in Weindestillaten, *Dtsch. Lebensm. Rundsch.* **80**, 268–273.

PRIOR, B., KIRCHNER-NESS, R., DITTRICH, H. H. (1992), Mathematical description of the glucose/fructose ratio in fermenting and totally fermented musts, *Vitic. Enol. Sci.* **47**, 145–152.

RADLER, F., HARTEL, S. (1984), *Lactobacillus trichodes*, an alcohol dependent lactic acid bacterium, *Wein-Wissenschaft* **39**, 106–112.

RADLER, F., KNOLL, C. (1988), Formation of killer toxin by apiculate yeasts and interference with fermentation, *Vitis* **27**, 111–132.

RADLER, F., LOTZ, B. (1990), The microflora of active dry yeast and the quantitative changes during fermentation, *Vitic. Enol. Sci.* **45**, 114–122.

RADLER, F., SCHMITT, M. (1987), Killer toxins of yeasts, *J. Food Protect.* **50**, 234–238.

RADLER, F., THEIS, W. (1972), Über das Vorkommen von *Aspergillus*-Arten auf Weinbeeren, *Vitis* **10**, 314–317.

RADLER, F., YANISSIS, C. (1972), Decomposition of tartrate by lactobacilli, *Arch. Microbiol.* **82**, 219–238.

RAPP, A. (1989), Stickstoffverbindungen, in: *Chemie des Weines* (WÜRDIG, G., WOLLER, R., Eds.), pp. 540–550, Stuttgart: Ulmer.

RAUHUT, D., DITTRICH, H. H. (1991), Pflanzenschutzmittel und Weinqualität, *Weinwirtsch.-Tech.* **1**, 18–23.

RAUHUT, D., KÜRBEL, H. (1994), The production of H_2S from elemental sulfur residues during fermentation and its influence on the formation of sulfur metabolites causing off-flavors in wine, *Vitic. Enol. Sci.* **49**, 27–36.

REED, G., NAGODAWITHANA, T. W. (1988), Technology of yeast usage in wine making, *Am. J. Enol. Vitic.* **39**, 83–90.

REINHARD, L., RADLER, F. (1981), The action of sorbic acid on *Saccharomyces cerevisiae*, *Z. Lebensm. Unters. Forschg.* **154**, 279–284.

RIBÉREAU-GAYON, J., PEYNAUD, E., RIBÉREAU-GAYON, P., SUDRAUD, P. (1975), *Sciences et Techniques du Vin*, Tome 2, Paris: Dunod.

RIBÉREAU-GAYON, J., RIBÉREAU-GAYON P., SEGUIN, G. (1980), *Botrytis cinerea* in enology, in: *The Biology of Botrytis* (COLEY-SMITH, J. R., VERHOEF, K., JARVIS, W. R., Eds.), pp. 251–274, London: Academic Press.

ROSE, A. H. (Ed.), *Alcoholic Beverages*, London: Academic Press.

ROSE, A. H., HARRISON, J. St. (1987–1991), *The Yeasts*, 2. Ed., Vols. 1–4, London: Academic Press.

RYMON LIPINSKI, G. W., VON, LÜCK, E., OESER, H., LÖMKER, F. (1975), Entstehung und Ursachen des "Geranientons", *Mitt. Klosterneuburg* **25**, 387–394.

SCHANDERL, H., KOCH, J., KOLB, E. (1981), *Fruchtweine*, Stuttgart: Ulmer.

SCHÜTZ, H., RADLER, F. (1984), Anaerobic reduction of glycerol to propanediol-1,3 by *Lactobacillus brevis* and *Lactobacillus buchneri*, *System. Appl. Microbiol.* **5**, 169–178.

SCHÜTZ, M., GAFNER, J. (1993), Analysis of yeast diversity during spontaneous and induced alcoholic fermentations, *J. Appl. Bacteriol.* **75**, 551–558.

SCHWARZ, E., HAMMES, W. P. (1991), The application of inulase producing yeasts for production of brandy from jerusalem artichokes, *Chem. Mikrobiol. Technol. Lebensm.* **13**, 70–75.

SCOTT, P. M. (1977), Patulin content of juice and wine produced from moldy grapes, *J. Agric. Food Chem.* **25**, 434–437.

SHIMAZU, Y., WATANABE, M. (1981), Effects of yeast strains and environmental conditions on forming of organic acids in must during fermentation, *J. Ferment. Technol.* **59**, 27–32.

SHIMIZU, J. M. (1982), Transformation of terpenoid in grape and must by *Botrytis cinerea*, *Agric. Biol. Chem.* **46**, 1339–1344.

SHIMIZU, K. (1993), Killer yeasts, in: *Wine Micro-*

biology and Biotechnology (FLEET, G. H., Ed.), pp. 243–264, New York: Harwood.

SORRI, T., MIGNOT, O. (1988), Les bactériophages en oenologie, *Bull. O.I.V.* **61,** 705–716.

SPONHOLZ, W. R. (1988), Alcohols derived from sugars and other sources and full bodiedness of wines, in: *Modern Methods of Plant Analysis* (LINSKENS, H. F., JACKSON, J. F., Eds.), Vol. 6, pp. 147–172, Berlin: Springer Verlag.

SPONHOLZ, W. R. (1993), Wine spoilage by microorganisms, in: *Wine Microbiology and Biotechnology* (FLEET, G. H., Ed.), pp. 395–420, New York: Harwood.

SPONHOLZ, W. R., DITTRICH, H. H. (1974), Bildung von SO_2-bindenden Gärungsnebenprodukten, höheren Alkoholen und Estern bei einigen Reinzuchthefestämmen und einigen "wilden" Hefen, *Wein-Wissenschaft* **29,** 301–313.

SPONHOLZ, W. R., DITTRICH, H. H. (1977), Enzymatische Bestimmung von Bernsteinsäure in Mosten u. Weinen, *Wein-Wissenschaft* **32,** 38–47.

SPONHOLZ, W. R., DITTRICH, H. H. (1986), Volatile fatty acids in wines of differing qualities, *Z. Lebensm. Unters. Forschg.* **183,** 344–347.

SPONHOLZ, W. R., HÜHN, T. (1994), 4,5-Dimethyl-3-hydroxy-2(5)-furanon (Scotolon), an indicator for *Botrytis* infection? *Vitic. Enol. Sci.* **49,** 37–39.

SPONHOLZ, W. R., MUNO, H. (1994), Corkiness – a microbiological problem? *Vitic. Enol. Sci.* **49,** 17–22.

SPONHOLZ, W. R., WÜNSCH, B., DITTRICH, H. H. (1981), Enzymatic determination of (*R*)-2-hydroxyglutaric acid in musts, wines and other alcoholic beverages, *Z. Lebensm. Unters. Forschg.* **172,** 264–268.

SPONHOLZ, W. R., DITTRICH, H. H., BARTH, A. (1982), Über die Zusammensetzung essigstichiger Weine, *Dtsch. Lebensm. Rundsch.* **78,** 423–428.

SPONHOLZ, W. R., DITTRICH, H. H., HAN, K. (1990a), The influence on fermentation and acetic acid ethylester formation by *Hanseniaspora uvarum*, *Vitic. Enol. Sci.* **45,** 65–72.

SPONHOLZ, W. R., HEUER, C., DITTRICH, H. H. (1990b), Reproduction, survival and metabolism of yeasts in musts of increasing sugar concentrations, *Vitic. Enol. Sci.* **45,** 1–7.

SPONHOLZ, W. R., MILLIES, K. D., AMBROSI, A. (1990c), Action of yeast hulls on fermentation, *Vitic. Enol. Sci.* **45,** 50–57.

SPONHOLZ, W. R., DITTRICH, H. H., MUNO, H. (1994), Diols in wine, *Vitic. Enol. Sci.* **49**, 23–26.

SPONHOLZ, W. R., KÜRBEL, H., DITTRICH, H. H. (1991), Contributions to the formation of ethyl carbamate in wine, *Vitic. Enol. Sci.* **46,** 11–17.

STOISSER, B., BANDION, F., WURZINGER, A., CARDA, E. (1988), Detection of lees wine in grape wine, *Mitt. Klosterneuburg* **38,** 235–239.

TANNER, H., BRUNNER, H. R. (1987), *Obstbrennerei heute*, 3. Ed., Schwäbisch Hall: Heller Chemie- u. Verwaltungsgesellschaft.

TOROK, T., KING, A. D. (1991), Comparative study of the identification of foodborne yeasts, *Appl. Environ. Microbiol.* **57,** 1207–1212.

TROOST, G. (1988), *Technologie des Weines*, 6. Ed., Stuttgart: Ulmer.

TROOST, G., HAUSHOFER, H. (1980), *Sekt, Schaum- und Perlwein*, Stuttgart: Ulmer.

TUITE, M. F., OLIVER, St. O. (1991), *Saccharomyces*, New York–London: Plenum Press.

EWG Verordnung 1576/89 vom 29. 5. 1989 zur Festlegung der allgemeinen Regeln für die Begriffsbestimmung, Bezeichnung und Aufmachung von Spirituosen.

WAGNER, K., KREUTZER, P. (1977), Zusammensetzung u. Beurteilung von Auslesen, Beeren- u. Trockenbeerenauslesen, *Wein-Wirtschaft* **113,** 272–275.

WEGER, B. (1984), Are there coliform bacteria resistent to alcohol? *Mitt. Klosterneuburg* **34,** 13.

WELTER, K. (1991), Acrolein bei der Bundesmonopolverwaltung für Branntwein, *Alkohol-Industrie* **7,** 139–141.

WENZEL, K. (1989), The selection of a yeast mutant to reduce colour losses during red wine fermentation, *Vitis* **28,** 111–120.

WENZEL, K., DITTRICH, H. H., SEYFARDT, H. P., BOHNERT, J. (1980), Schwefelrückstände auf Trauben und im Most und ihr Einfluß auf die H_2S-Bildung, *Wein-Wissenschaft* **35,** 414–420.

WENZEL, K., DITTRICH, H. H., PIETZONKA, B. (1982), Untersuchungen zur Beteiligung von Hefen am Äpfelsäureabbau bei der Weinbereitung, *Wein-Wissenschaft* **37,** 133–138.

WESENBERG, J., LAUBE, K. (1990), Acrolein-Bildung und Eigenschaften einer bei der ethanolischen Gärung unerwünschten Verbindung, *Lebensm. Ind.* **37,** 156–159.

WILLIAMS, L. A., BOULTON, R. (1983), Modeling and prediction of evaporative ethanol loss during wine fermentation, *Am. J. Enol. Vitic.* **34,** 234–242.

WUCHERPFENNIG, K., DIETRICH, H. (1982), Verbesserung der Filtrierfähigkeit von Weinen durch enzymatischen Abbau von kohlenhydrathaltigen Kolloiden, *Weinwirtschaft* **118,** 598–603.

WUCHERPFENNIG, K., DIETRICH, H. (1983), Bestimmung des Kolloidgehaltes von Weinen, *Lebensmitteltechnik* **15,** 246–253.

WÜRDIG, G., WOLLER, R. (Eds.) (1989), *Chemie des Weines*, Stuttgart: Ulmer.

13 Indigenous Fermented Foods

LARRY R. BEUCHAT

Griffin, GA 30223-1797, USA

1 Introduction

Fermented foods, whether from plant or animal origin, are an intricate part of the diet of people in all parts of the world. It is the diversity of raw materials used as substrates, methods of preparation and sensory qualities of finished products that are so astounding as one begins to learn more about the eating habits of various cultures. The preparation of many indigenous or "traditional" fermented foods and beverages remains today as a household art. The preparation of others, e.g., soy sauce, has evolved to a biotechnological state and is carried out on a large commercial scale.

It will not be the objective of this chapter to review in detail or even to introduce the reader to the many hundreds of indigenous fermented foods eaten daily. Space does not permit a detailed account and, besides, we know very little or nothing about the biochemistry and microbiology of many of these foods. Fermented vegetables, dairy products and beverages will be covered in considerable detail in other chapters in this volume. Tab. 1 lists some of the more common indigenous fermented foods consumed in various parts of the world. Many of these are discussed in the following text which, out of need for some degree of organized approach, has been divided according to the areas of the world in which they are most likely to be prepared and consumed.

Several books (HESSELTINE and WANG, 1986; REDDY et al., 1986; STEINKRAUS, 1983; WOOD, 1985) and reviews (BEUCHAT, 1987; CAMPBELL-PLATT and COOK, 1989; CHAVAN and KADAM, 1989; NOUT and ROMBOUTS, 1990; SANNI, 1993) have been published on the subject of indigenous fermented foods. A book describing applications of biotechnology to traditional fermented foods was published by the U.S. National Research Council (RUSKIN, 1992). A dictionary and guide to fermented foods of the world (CAMPBELL-PLATT, 1987) and a glossary of indigenous fermented foods (WANG and HESSELTINE, 1986) provide excellent descriptions of known biochemical and microbiological processes associated with indigenous food fermentations. The reader is encouraged to consult these and other publications cited in the following text for more detailed descriptions of fermented foods.

2 Fermented Foods of the Orient

2.1 Soy Sauce

The written records of the Chinese show that they have been using soy sauce for over three thousand years (YONG and WOOD, 1974). Production of soy sauce in Japan probably was a result of the introduction of Buddhism from China and the consequent change to a vegetable diet in 552 A.D. (HESSELTINE, 1965). SMITH (1961) published a report on various methods of using soybeans as foods, including soy sauce, in China, Japan, and Korea. YOKOTSUKA (1960), YONG and WOOD (1974) and HESSELTINE (1983) have subsequently reviewed soy sauce fermentation in considerable detail. The technology of soy sauce preparation was at one time a closely guarded family art passed on from one generation to the next. While there are still unique formulae used on a domestic level, the major steps involved in the manufacture of soy sauce are no longer a secret. There is, however, much to be learned about the biochemical changes which occur during fermentation and lead to desirable as well as undesirable sensory qualities in the finished product.

Two distinct basic processes can be used to prepare soy sauce (BEUCHAT, 1984). The first involves fermentation with microorganisms and the second, i.e., chemical method, involves the use of acids to promote hydrolysis of ingredient constituents. The latter method will not be discussed here mainly because it cannot be considered as traditional or indigenous, but also because there are many who consider the end product to be inferior and not in a class deserving of recognition as a substitute for the fermented product. Further-

Tab. 1. Indigenous Fermented Foods[a]

Product	Geography	Substrate	Microorganism(s)	Nature of Product	Product Use
Ang-kak (anka, red rice)	China, Southeast Asia, Syria	Rice	*Monascus purpureus*	Dry red powder	Colorant
Bagoong	Philippines	Fish	Unknown	Paste	Seasoning agent
Bagni	Caucasus	Millet	Unknown	Liquid	Drink
Banku	Ghana	Maize, cassava	Lactic acid bacteria, yeasts	Dough	Staple
Bonkrek	Central Java (Indonesia)	Coconut press cake	*Rhizopus oligosporus*	Solid	Roasted or fried in oil, used as a meat substitute
Bouza	Egypt	Wheat	Unknown	Liquid	Thick acidic
Braga	Romania	Millet	Unknown	Liquid	Drink
Burukutu	Savannah regions of Nigeria	Sorghum and cassava	Lactic acid bacteria, *Candida* spp., *Saccharomyces cerevisiae*	Liquid	Creamy drink with suspended solids
Busa	Tartars of Krim, Turkestan, Egypt	Rice or millet, sugar	*Lactobacillus* and *Saccharomyces*	Liquid	Drink
Chee-fan	China	Soybean wheat curd	*Mucor* sp., *Aspergillus glaucus*	Solid	Eaten fresh, cheese-like
Chicha	Peru	Maize	*Aspergillus, Penicillium* spp., yeasts, bacteria	Spongy	Eaten with vegetables
Chichwangue	Congo	Cassava roots	Bacteria	Paste	Staple
Chinese yeast	China	Soybeans	Mucoraceous molds and yeasts	Solid	Eaten fresh or canned, used as a side dish with rice
Darassum	Mongolia	Millet	Unknown	Liquid	Drink
Dawadawa (daddowa, uri, kpalugu, kinda)	West Africa, Nigeria	African locust bean	Lactic acid bacteria, yeasts	Solid, sun-dried	Eaten fresh, supplement to soups, stews
Dhokla	India	Bengal gram and wheat	Unknown	Spongy	Condiment
Dosai (doza)	India	Black gram and rice	Yeasts, *Leuconostoc mesenteroides*	Spongy, pancake-like	Breakfast food
Fish sauce (nuoc-mam, patis, mam-pla, ngam-pya-ye)	Southeast Asia	Fish	Bacteria	Liquid	Seasoning agent

Gari	West Africa	Cassava root	*Corynebacterium manihot, Geotrichum candidum*	Wet paste	Eaten fresh as staple with stews, vegetables
Hamanatto	Japan	Whole soybeans, wheat flour	*Aspergillus oryzae, Streptococcus, Pediococcus*	Beans retain individual form, raisin-like, soft	Flavoring agent for meat and fish, eaten as snack
Idli	Southern India	Rice and black gram	Lactic bacteria (*Leuconostoc mesenteroides*), *Torulopsis candida* and *Trichosporon pullulans*	Spongy, moist	Bread substitute
Injera	Ethiopia	Teff, or maize wheat, barley, sorghum	*Candida guilliermondii*	Bread-like, moist	Bread substitute
Jalebies	India, Nepal, Pakistan	Wheat flour	*Saccharomyces bayanus*	Pretzel-like, syrup filled	Confection
Jamin-bang	Brazil	Maize	Yeasts and bacteria	Bread or cake-like	Bread substitute
Kaanga-kopuwai	New Zealand	Maize	Bacteria and yeasts	Soft, slimy	Eaten as vegetable
Kanji	India	Rice and carrots	*Hansenula anomala*	Liquid	Sour, added to vegetables
Katsuobushi	Japan	Whole fish	*Aspergillus glaucus*	Solid, dry	Seasoning agent
Kecap	Indonesia and vicinity	Soybeans, wheat	*Aspergillus oryzae, Lactobacillus, Hansenula, Saccharomyces*	Liquid	Condiment, seasoning agent
Kenim	Nepal, Sikkim, Darjeeling district of India	Soybeans	Unknown	Solid	Snack
Kenkey	Ghana	Maize	Unknown	Mush	Steamed, eaten with vegetables
Ketjap	Indonesia	Black soybeans	*Aspergillus oryzae*	Syrup	Seasoning agent
Khaman	India	Bengal gram	Unknown	Solid, cake-like	Breakfast food
Kimchi (kim-chee)	Korea	Vegetables, sometimes seafoods, nuts	Lactic acid bacteria	Solid and liquid	Condiment
Kishk (kushuk, kushik)	Egypt, Syria, Arab world	Wheat, milk	Lactic acid bacteria, *Bacillus* spp.	Solid	Dried balls dispersed rapidly in water
Lafun	West Africa, Nigeria	Cassava root	Bacteria	Paste	Staple food
Lao-chao	China, Indonesia	Rice	*Rhizopus oryzae, R. chinensis, Chlamydomucor oryzae, Saccharomycopsis* sp.	Soft, juicy, glutinous	Eaten as such as dessert or combined with eggs, seafood

Tab. 1. Indigenous Fermented Foods[a] (Continued)

Product	Geography	Substrate	Microorganism(s)	Nature of Product	Product Use
Mahewu (Magou)	South Africa	Maize	Lactic acid bacteria (*Lactobacillus delbrueckii*)	Liquid	Drink, sour and non-alcoholic
Meitauza	China, Taiwan	Soybean cake	*Actinomucor elegans*	Solid	Fried in oil or cooked with vegetables
Meju	Korea	Soybeans	*Aspergillus oryzae, Rhizopus* spp.	Paste	Seasoning agent
Merissa	Sudan	Sorghum	*Saccharomyces* sp.	Liquid	Drink
Minchin	China	Wheat gluten	*Paecilomyces, Aspergillus, Cladosporium, Fusarium, Syncephalastum, Penicillium, Tricothecium* spp.	Solid	Condiment
Miso (chiang, jang, doenjang, tauco, tao chieo)	Japan, China	Rice and soybeans or rice and other cereals such as barley	*Aspergillus oryzae, Torulopsis etchellsii, Lactobacillus*	Paste	Soup base, seasoning
Munkoyo	Africa	Millet, maize or kaffir corn plus roots of munkoyo	Unknown	Liquid	Drink
Nan (khab-z)	India, Pakistan, Afghanistan, Iran	Unbleached wheat flour	Unknown	Solid	Snack
Natto	Northern Japan	Soybeans	*Bacillus natto*	Solid	Cake, as a meat substitute
Ogi	Nigeria, West Africa	Maize	Lactic bacteria (*Cephalosporium, Fusarium, Aspergillus, Penicillium* spp., *Saccharomyces cerevisiae, Candida mycoderma* (*C. valida* or *C. vini*)	Paste	Staple, eaten for breakfast, weaning babies
Oncom (ontjom, lontjom)	Indonesia	Peanut press cake	*Neurospora intermedia*, less often *Rhizopus oligosporus*	Solid	Roasted or fried in oil, used as meat substitute
Papadam	India	Black gram	*Saccharomyces* spp.	Solid, crisp	Condiment
Peujeum	Java	Banana, plantain	Unknown	Solid	Eaten fresh or fried

Pito	Nigeria	Guineacorn or maize or both	Unknown	Liquid	Drink
Poi	Hawaii	Taro corms	*Lactobacillus* bacteria, *Candida vini* (*Mycoderma vini*), *Geotrichum candidum*	Semi-solid	Side dish with fish, meat
Pozol	Southeastern Mexico	Maize	Molds, yeasts, bacteria	Dough, spongy	Diluted with water, drunk as basic food
Prahoc	Cambodia	Fish	Unknown	Paste	Seasoning agent
Puto	Philippines	Rice	Lactic acid bacteria, *Saccharomyces cerevisiae*	Solid	Snack
Rabdi	India	Maize and buttermilk	Unknown	Semi-solid	Mush, eaten with vegetables
Sierra rice	Ecuador	Unhusked rice	*Aspergillus flavus, A. candidus, Bacillus subtilis*	Solid	Brownish-yellow, seasoning
Sorghum beer (Ibantu beer, kaffir beer, leting, joala, utshivala, mqomboti, igwelel)	South Africa	Sorghum, maize	Lactic acid bacteria, yeasts	Liquid	Drink, acidic and weakly alcoholic
Soybean milk	China, Japan	Soybeans	Lactic acid bacteria	Liquid	Drink
Soy sauce (Chaing-yu, shoyu, toyo, kanjang, kecap, seeieu)	Japan, China, Philippines, other parts of Orient	Soybeans and wheat	*Aspergillus oryzae* or *A. soyae, Lactobacillus* bacteria, *Zygosaccharomyces rouxii*	Liquid	Seasoning for meat, fish, cereals, vegetables
Sufu (tahur, taokaoan, tao-hu-yi)	China, Taiwan	Soybean whey curd	*Actinomucor elegans, Mucor hiemalis, M. silvaticus, M. subtilissimus*	Solid	Soybean cheese, condiment
Tao-si	Philippines	Soybeans plus wheat flour	*Aspergillus oryzae*	Semi-solid	Seasoning agent
Taotjo	East Indies	Soybeans plus roasted wheat meal or glutinous rice	*Aspergillus oryzae*	Semi-solid	Condiment

Tab. 1. Indigenous Fermented Foods[a] (Continued)

Product	Geography	Substrate	Microorganism(s)	Nature of Product	Product Use
Tapé	Indonesia and vicinity	Cassava or rice	*Saccharomyces cerevisae, Hansenula anomala, Rhizopus oryzae, Chlamydomucor oryzae, Mucor* sp., *Endomycopsis fibuliger* (*Saccharomycopsis* sp.)	Soft solid	Eaten fresh as staple
Tarhana	Turkey	Parboiled wheat meal and yoghurt (2:1)	Lactic acid bacteria	Solid powder	Dried seasoning for soups
Tauco	West Java (Indonesia)	Soybeans, cereals	*Rhizopus oligosporus, Aspergillus oryzae*	Liquid	Drink
Tempeh (tempe kedeke)	Indonesia and vicinity, Surinam	Soybeans	*Rhizopus* spp., principally *R. oligosporus*	Solid	Fried in oil, roasted, or used as meat substitute in soup
Thumba (bojah)	West Bengal	Millet	*Endomycopsis fibuliger*	Liquid	Drink, mildly alcoholic
Torani	India	Rice	*Hansenula anomala, Candida guilliermondii, C. tropicalis, Geotrichum candidum*	Liquid	Seasoning for vegetables
Waries	India	Black gram flour	*Candida* spp., *Saccharomyces* spp.	Spongy	Spicy condiment eaten with vegetables, legumes, rice

[a] Compiled from Beuchat (1983, 1987), Campbell-Platt (1987), Hesseltine (1979), Hesseltine and Wang (1980, 1986), Reddy et al. (1986) and Stanton and Wallbridge (1969)

more, since less technical information is available in Western literature on procedures for making Chinese type soy sauce, the information to follow pertains largely to soy sauce fermentation in Japan.

According to FUKUSHIMA (1979), there are five main types of soy sauce recognized by the Japanese government (Tab. 2). The Koikuchi soy sauce represents the largest amount prepared in Japan; approximately 90% of the production is of this type. It is an all-purpose seasoning characterized by a strong aroma and dark reddish-brown color. A second type of soy sauce, amounting to less than 10% of the production, is called Usukuchi. It is lighter in color, milder in flavor, and is used mainly for cooking when preservation of the original flavor and color of the foodstuffs is desired. Tamari style soy sauce, produced in China traditionally, represents about 2% of the production. This soy sauce is characterized by a strong flavor and dark brown color. Produced in lesser amounts are Saishikomi soy sauce, containing only a trace of alcohol, and Shiro soy sauce which is characterized by a high level of reducing sugars and a yellowish-tan color. All of these types of soy sauce contain relatively high levels of salt, in the range of 17 to 19%, and all are used as seasoning agents to enhance the flavor of meats, seafoods and vegetables. Known as shoyu in Japan, soy sauce is called chiang-yu in China, kecap in Indonesia, kanjang in Korea, toyo in the Philippines and see-ieu in Thailand (WANG and HESSELTINE, 1982).

2.1.1 Preparation of Soybeans

A flow sheet for manufacturing soy sauce is shown in Fig. 1. Two processes, viz., soaking and cooking soybeans, and roasting and crushing (cracking) wheat, are separate, yet simultaneous, in the early stages of production. Whole soybeans or defatted soybean meal or flakes can be used in the production of soy sauce. If whole beans are used, oil must eventually be removed from the fermented mash; otherwise an inferior product will result. This oil is used for making a low-grade soap or as a source of linoleic acid (YONG and WOOD, 1974). Pressed or solvent-extracted soybean meal is used. Cost is lower, utilization of nitrogen is higher, and fermentation time is shorter with defatted beans than with whole soybeans. This may be due to a lower surface:volume ratio in whole beans versus meal, hence a more pronounced physical restraint in whole beans with regard to the access of enzymes and organisms to soybean components during fermentation. Oil from whole soybeans which rises to the surface of the mash may also eventually retard microbiological activity during aging.

Whole beans or meal are soaked for 12 to 15 h at ambient temperature or, preferably, at about 30 °C until a 2.1 times increase in weight results. Soaking is done either by running water over the beans or by changing still water every 2 or 3 h. If water is not changed, spore-forming *Bacillus* may proliferate to levels eventually deleterious to end product quality. Also, depending upon the depth of soybeans in the water, those in the bottom layer of tanks may heat if water is not

Tab. 2. Types and Composition of Soy Sauces Recognized by the Japanese Government[a]

Type of Soy Sauce	Color	pH	Bé	NaCl (g/100 mL)	Total Nitrogen (g/100 mL)	Formol Nitrogen (g/100 mL)	Reducing Sugar (g/100 mL)	Alcohol (Vol/100 mL)
Koikuchi	Deep brown	4.7	22.5	17.6	1.55	0.88	3.8	2.2
Usukuchi	Light brown	4.8	22.8	19.2	1.17	0.70	5.5	0.6
Tamari	Dark brown	4.8	29.9	19.0	2.55	1.05	5.3	0.1
Saishikomi	Dark brown	4.8	26.9	18.6	2.39	1.11	7.5	Trace
Shiro	Yellow to tan	4.6	26.9	19.0	0.50	0.24	20.2	Trace

[a] Modified from FUKUSHIMA (1979)

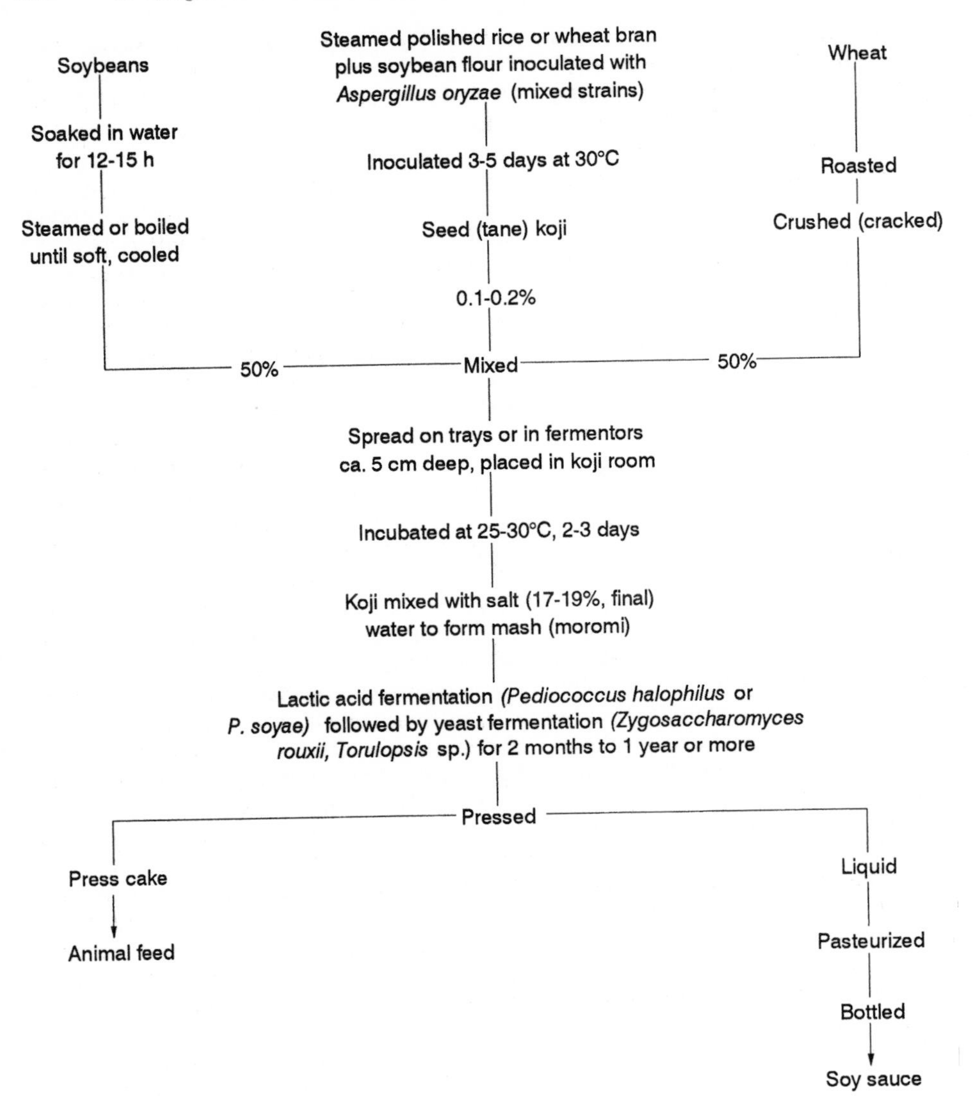

Fig. 1. Flow diagram for manufacture of soy sauce (Koikuchi type).

changed or circulated around them during soaking. The swollen beans or meal are then drained, covered with water again and steamed to achieve further softening and pasteurization. If pressure is used, the beans can be sterilized during the cooking process. On a smaller scale, beans are boiled in an open pan until soft enough to easily press flat between the thumb and finger. The conditions for cooking soybeans not only play an important role in influencing digestibility by enzymes during fermentation, but also may affect the turbidity of the final product when it is diluted or heated for home use (OGAWA and FUJITA, 1980). This turbidity is caused by insolubilization of undenatured protein when dispersed in concentrated salt solution and deflates the commercial value of soy sauce. With increased moisture or pressure during steaming, the soy protein tends to denature more readily; on the other hand, excessively denatured soy protein has reduced accessibility for enzyme reaction. Consequently, the yield of soluble nitrogen and other soluble compounds will be reduced. Thus, the procedure used for cooking soybeans is critical to fermentation patterns and end product quality.

Rapid cooling on an industrial scale is done by spreading the beans in about a 30-cm layer on tray-like platforms and forcing air through them (YONG and WOOD, 1974). It is important to reduce the temperature to less than 40 °C within a few hours. Otherwise, proliferation of microorganisms may ensue and thus spoil the beans before controlled fermentation can be initiated.

2.1.2 Preparation of Wheat

Concurrent with the preparation of soybeans is the roasting and crushing (cracking) of wheat. Wheat flour or wheat bran may be used in place of whole wheat kernels. The roasting of wheat contributes to aroma and flavor of soy sauce. Characteristic breakdown and conversion products produced by cooking wheat include the guaiacyl series compounds, such as vanillin, vanillic acid, ferulic acid, and 4-ethylguaiacol (ASAO and YOKOTSUKA, 1958). The free phenolic compound content increases with heating due to degradation of lignin and glycosides. Roasting causes the formation of an array of brown-colored reaction products which contribute to desired visual properties in the finished product.

2.1.3 Koji Process

The word koji, meaning "bloom of mold", refers to the enzyme preparation produced on cereals or sometimes pulses that is used as a seed or starter for larger batches of plant seed substrates when making several kinds of traditional fermented foods. In the case of soy sauce, seed (tane) koji is produced by culturing a number of mixed strains of *Aspergillus oryzae* or *A. sojae* on either steamed, polished rice (usual practice in Japan) or a mixture of wheat bran and soybean flour (China) (YONG and WOOD, 1974). It is added to a soybean/wheat mixture at a level of 0.1 to 0.2% to produce what is then simply called koji.

The strains of molds used as a starter culture must have high proteolytic and amylolytic activities and should contribute to the characteristic aroma and flavor of soy sauce. Lipase (YONG and WOOD, 1977) and cellulase (GOEL and WOOD, 1978) production may also occur. Several acid, neutral, and alkaline proteases as well as peptidases have been shown to be produced by *A. oryzae* and *A. sojae.* On an industrial level, the koji substrate, e.g. a 1:1 soybean:wheat mixture, is spread in 5-cm layers in trays made from bamboo strips or in stainless steel trays, inoculated with the seed koji and stacked in such a way as to allow good air circulation at the surface. The control of temperature (25 to 35 °C) and moisture is also important to the development of good koji; a temperature of 30 °C for 2 to 3 days is preferable. A rather high moisture content is required at the beginning of the approximate 3-day incubation period when mycelial growth occurs followed by a lower moisture content in later stages when spores are being formed (YONG and WOOD, 1974). A moisture content of 27 to 37% is necessary to maintain good enzyme activity (HARADA, 1951). The incubation period should be long enough to promote adequate production and accumulation of enzymes, but not so long as to encourage excess sporulation which may impart undesirable flavors to the finished product. Mature koji is clear yellow to yellowish-green in color.

Properties of enzymes produced by *A. oryzae* that contributes to soybean fermentation have been reported in a series of publications by NAKADAI et al. (1972a–d; 1973a–f). The effects of diisopropyl-phosphorofluoridate, SH reagents such as *p*-chloromercuribenzoate and monoiodoacetate, and metal chelating agents such as ethylenediaminetetraacetate, α,α'-dipyridyl and ρ-phenanthroline as well as carboxypeptidases, alkaline and neutral proteinases and leucine aminopeptidases have been studied.

IMPOOLSUP et al. (1981) characterized alkaline and neutral proteases produced by *As-*

pergillus flavus var. *columnaris*, a koji mold recommended for use in the production of Thai soy sauce. The mold was reported to not produce aflatoxin.

Koji infected with *Rhizopus* or *Mucor* species indicates that atmospheric moisture may have increased to a level which caused water droplets to be formed on the surface at some point during incubation. Excessively contaminated koji should be discarded, because of the undesirable flavor and aroma it will impart to soy sauce.

2.1.4 Mash (Moromi) Stage

When the koji is mature, it is ready for brining. The koji is mixed with an equal amount or more (up to 120% by volume) of saline to form the mash, or moromi. The sodium chloride content of the mash should range from 17 to 19%. Concentrations less than 16% salt may enable growth of undesirable putrefactive bacteria during subsequent fermentation and aging. On the other hand, concentrations in excess of 23% may retard the growth of desirable osmophilic yeasts and halophilic bacteria. The mycelium of the koji mold is killed during the very early stage of mash preparation.

If fermentation is allowed to proceed naturally without controlling temperature as would be the situation in the family home, a period of 12 to 14 months is necessary for the fermentation and aging process. If the mash is kept in large wooden or concrete containers such as those used by commercial manufacturers, the temperature is usually maintained at 35 to 40 °C, thus reducing the fermentation and aging period to 2 to 4 months. Regardless of the storage temperature, it is important to stir the mash intermittently. This is done with a wooden stick on a small scale or with compressed air in modern commercial facilities.

During the early stages of fermentation, koji enzymes hydrolyze proteins to yield peptides and free amino acids. Starch is converted to simple sugars which in turn are fermented by microorganisms to yield lactic, glutamic and other acids as well as alcohols and carbon dioxide. As a consequence, the pH of the mash drops from near neutrality to 4.5 to 4.8. Stirring must be correlated to a certain extent with the rate of carbon dioxide production. Elevated levels of carbon dioxide will enhance the growth of certain anaerobic microorganisms which may impart undesirable flavor and aroma to the finished product. Excessive aeration, on the other hand, will hinder proper fermentation.

The microbiology of mash is not clearly understood; however, it is known that various groups of bacteria and yeasts predominate in sequence during the fermentation and aging process (YONG and WOOD, 1976). The halophilic bacterium, *Pediococcus halophilus*, grows readily in the first stage of fermentation, converting simple sugars to lactic acid and causing a drop in pH. Later, *Zygosaccharomyces rouxii, Torulopsis* and other yeasts dominate. Older literature reports the presence of several *Zygosaccharomyces* species in mash, but some of these are now not recognized as distinct species and have been combined under the *Zygosaccharomyces rouxii* umbrella. While *A. oryzae, A. sojae, Monilia, Penicillium* and *Rhizopus* may appear on the surface of mash, these molds are believed to have no relation to proper fermentation or aging (YOKOTSUKA, 1960). The production of lactic acid by lactic acid bacteria enhances the inhibitory effect of acetic acid on yeasts (NODA et al., 1982).

Like most traditional fermented foods, soy sauce owes its pleasant aroma or flavor largely to the enzymatic activities of microorganisms. A partial list of flavor components identified in soy sauce is shown in Tab. 3. *Pediococcus halophilus* and, perhaps, *Lactobacillus* species produce lactic and other organic acids which, in themselves, contribute to aroma and flavor. However, it is the yeasts that probably make the greatest contribution to characteristic sensory qualities of soy sauce. By-products of fermentation such as 4-ethylguaiacol, 4-ethylphenol and 2-phenylethanol (YOKOTSUKA et al., 1967), furfuryl alcohol (MORIMOTO and MATSUTANI, 1969), pyrazines, furanones (NUNOMURA et al., 1976a, b, 1979, 1980) and ethyl acetate (YONG et al., 1981) are among the main flavor-contributing compounds.

The optimal aging period can be determined in part by analyzing for free glutamic acid content. At this point, liquid (sauce) is

Tab. 3. Some Flavor Components in Soy Sauce[a]

Acetaldehyde	Benzoic acid
Acetone	2-Acetyl furan
Propanal	Benzaldehyde
2-Methylpropanal	Furfuryl acetate
3-Methylbutanal	2-Methyl propanoic acid
Ethyl acetate	Bornyl acetate
2-Hexanone	4-Pentanolide
2,3-Hexanedione	Butanoic acid
Ethanol	Phenyl acetaldehyde
2-Propanol	Furfuryl alcohol
2-Methyl-1-propanol	Ethyl benzoate
3-Methylbutyl acetate	3-Methylbutanoic acid
1-Butanol	Diethyl succinate
3-Methyl-1-butanol	Borneol
3-Methyl-3-tetrahydrofuranone	3-Methylthio-1-propanol (methional)
2-Methylpyrazine	Ethylphenyl acetate
3-Hydroxy-2-butanone (acetoin)	2-Phenylethyl acetate
2,6-Dimethylpyrazine	2-Methoxyphenol (guaiacol)
2,3-Dimethylpyrazine	Benzylalcohol
Ethyl-2-hydroxypropanoate (ethyl lactate)	2-Phenylethanol
2-Ethyl-6-methylpyrazine	3-Hydroxy-2-methyl-4-pyrone (maltol)
Acetic acid	2-Acetylpyrrole
3-Ethyl-2,5-dimethylpyrazine	2-Methoxy-4-ethylphenol (4-ethylguaiacol)
Furfural	4-Ethylphenol
4-Hydroxy-2-ethyl-5-methyl-3(2H)-furanone	2,6-Dimethoxyphenol
4-Hydroxy-5-ethyl-2-methyl-3(2H)-furanone	Ethyl myristate
4-Hydroxy-5-methyl-3(2H)-furanone	

[a] Adapted from NUNOMURA et al. (1976a, b)

removed from the mash with a simple mechanical press on a domestic level or with a hydraulic press in commercial operations. Liquid may also be removed by siphoning off the top of the mash. Fresh salt water is then sometimes added to the residue, and a second fermentation is allowed to proceed for 1 or 2 months before a second drawing is made. The second drawing is of lower quality than the first. Oil is removed from the filtrate by decantation.

2.1.5 Pasteurization

The raw soy sauce is pasteurized at 70 to 80 °C, thus killing the vegetative cells of microorganisms. Enzymes are also denatured, and proteins are coagulated. Alum or kaolin may be added to enhance clarification. The sauce is then filtered, bottled and marketed. In Japan, preservatives may be added to prevent growth of yeasts during storage (YONG and WOOD, 1974). Butyl-*p*-hydroxybenzoate or, alternatively, sodium benzoate are most widely used. Preparation of soy sauce in the home does not usually include pasteurization and certainly does not involve the use of chemical preservatives other than salt present as an integral part of the product.

Advances in fermentation technology have enabled manufacturers to produce soy sauce with consistent quality. The use of pure culture inocula at all stages of production has reduced the risk of carrying unwanted contaminants from one batch to the next. Also, because of the predominance of groups of microorganisms at various stages of fermentation, the proper balance of microorganisms in a mature moromi is not likely to be present to give optimum initial levels when used as an inoculum in a fresh batch. This problem has been minimized by using pure culture inocula. The addition of enzymes, aside from those in koji, to the mash offers some promise relative to more rapid and controlled breakdown of soybean and wheat components. Needless to say, these practices will not be immediately or widely instituted in domestic or cottage industry schemes. Traditional methods for preparing soy sauce will hopefully persist for many years.

2.2 Miso

Fermented soybean pastes, known by various names, are prepared in the Orient using a variety of procedures. The prototype of these pastes is believed to have been introduced in China in 600 A.D. or before (EBINE, 1971). Today, fermented soybean pastes are known as chiang in China, jang or doenjang in Korea, miso in Japan, tauco in Indonesia, tao chieo in Thailand and taosi in the Philippines. In addition to soybeans and salt, most of these products also contain cereals, such as rice or barley. Most chiang in China is prepared at home, just as people in Western countries make their own jams, jellies or pickles (FUKUSHIMA, 1979). In Japan, however, miso is now manufactured commercially in modernized factories on a large scale.

Fermented soybean pastes are consumed in various ways. Chiang is used as a base for sauce served with meat, poultry, seafood and vegetable dishes. In Japan, miso is mainly used as a base for soups. From 80 to 85% of the miso used in Japan is consumed in the preparation of miso soup and the balance used as seasoning for other foods (FUKUSHIMA, 1979).

Methods for manufacturing miso differ somewhat, but the basic process is the same as that shown for rice miso in Fig. 2. Rice miso is made from rice, soybeans and salt; barley miso is made from barley, soybeans and salt; and soybean miso is made from soybeans and salt (Tab. 4). The three major types of miso are further classified on the basis of degree of sweetness and saltiness. The procedure for making miso consists of four major steps, two of which are carried out concurrently. These consist of preparation of the koji and the soybeans (simultaneous processes), brining or fermentation and finally aging. Each step for the preparation of rice miso will be considered separately here in order to illustrate the general procedure for making miso.

2.2.1 Preparation of Koji

Polished rice is used to prepare koji, the source of enzymes to hydrolyze soybean components later in the fermentation process; brown rice is not suitable for growth of the koji mold (*A. oryzae*) because the surface texture is hard and contains waxes which inhibit

Tab. 4. Types and Composition of Major Types of Miso Made in Japan[a]

Raw Material	Taste	Color	Fermentation/ Aging Time	Composition (%) Protein	Fat	Carbo-hydrate	Ash	Mois-ture
Rice	Sweet	Yellowish white	5–20 days	11.1	4.0	35.9	7.0	42.0
		Reddish brown	5–20 days	12.7	5.1	31.7	7.5	43.0
	Semisweet	Bright light yellow	5–20 days	13.0	5.4	29.1	8.5	44.0
		Reddish brown	3–6 months	11.2	4.4	27.9	14.5	42.0
	Salty	Bright light yellow	2–6 months	13.5	5.9	19.6	14.0	47.0
		Reddish brown	3–12 months	13.5	5.9	19.1	14.5	47.0
Barley	Semisweet	Yellowish/reddish brown	1–3 months	11.1	4.1	29.8	13.0	42.0
	Salty	Reddish brown	3–12 months	12.8	5.2	21.1	15.1	46.0
Soybeans	Salty	Dark reddish brown	5–20 months	19.4	9.4	13.2	13.0	45.0

[a] Adapted from FUKUSHIMA (1979)

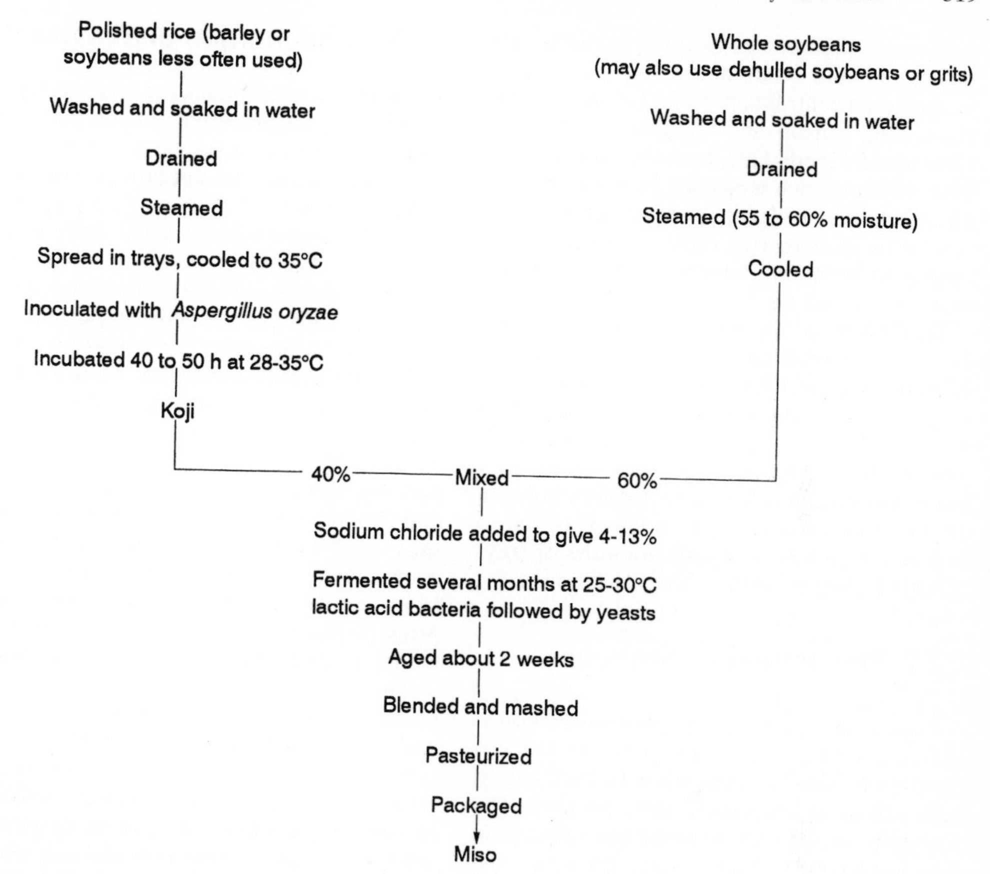

Fig. 2. Flow diagram for manufacture of miso.

penetration of the kernel by mycelium (SHIBASAKI and HESSELTINE, 1962). After washing, the rice is soaked in water overnight at about 15 °C to bring the moisture content to about 35%. Excess water is removed and the rice is steamed at atmospheric pressure for 40 min to 1 h. This may be done in a batch process or by using a continuous cooker. The cooked rice is spread out on large trays or platforms to cool to about 35 °C. Seed koji, prepared as described for use in soy sauce manufacturing, is added at a rate of 1 g per kg of rice. This application rate is based on a viable spore count of 10^9 per g of seed koji (EBINE, 1971). The use of koji rooms in which trays of inoculated rice are stacked has been replaced largely by modern rotary fermentors. Rice is put into a large drum of the fermentor in which the temperature, air circulation and atmospheric relative humidity is controlled within critical limits. The drum is rotated to prevent the rice from agglomerating during fermentation. The temperature is maintained between 30 and 35 °C to promote growth of *A. oryzae* and maximum production of saccharolytic and proteolytic enzymes. Overheating is usually caused by rapid growth of undesirable bacteria which may be present on the uncooked rice and survive the steaming process. A source of undesirable microorganisms may also be found on the surface of cooling trays and in the air surrounding the cooked rice and fermenting koji. Ventilation should be adequate to supply sufficient oxygen and to dissipate carbon dioxide, and humidity should be such that neither drying nor sticking of the rice occurs. Fermentation is complete after 40 to 50 h.

If rotary fermentors are not available, the inoculated and well mixed mass of rice is heaped on a central bed in the koji room, covered with canvas and incubated overnight (SHIBASAKI and HESSELTINE, 1962). The next morning, rice is spread in wooden trays and stacked. During an approximate 50-h fermentation period, rice must be stirred thoroughly to maintain uniform aeration, temperature and moisture.

Regardless of the system used to prepare koji, when fermentation is adequate, the rice is well covered with white mycelium of *A. oryzae.* Harvesting is done before sporulation with consequent pigmentation development. The product at this point is characterized by a sweet aroma and flavor; musty odors indicate an inferior quality koji. The addition of salt to koji as it is removed from fermentors or trays retards further growth of the mold.

2.2.2 Preparation of Soybeans

Concurrent with the preparation of koji is the preparation of whole soybeans for fermentation. Ideally, soybeans with thin, glossy, light yellow or white seed coats are used. The soybeans should also be large and uniform in size and have an ability to absorb water and cook very rapidly. After extraneous materials are removed by mechanical equipment or by hand, soybeans are washed and soaked in water for 18 to 22 h. Water should be changed during the soaking period, especially in the summer months when temperatures are elevated, to control the proliferation of bacteria. At the end of the soaking period, beans have increased in volume by about 240% and the weight by 220 to 260%, depending upon the ratio of protein and carbohydrate.

After being drained, the soybeans are cooked in water or steamed at 115 °C for about 20 min in a closed cooker, until they are sufficiently soft to be easily pressed flat between the thumb and finger. Flavor and color development can be achieved by varying the heating time and temperature.

2.2.3 Fermentation and Aging

Cooked, cooled beans are then mixed with salted koji and inoculum. In the past, a portion of miso from a previous batch was used as the inoculum, but modern technology involves the use of starters containing pure cultures of osmophilic yeasts and bacteria. Strains of *Zygosaccharomyces rouxii, Torulopsis, Pediococcus halophilus* and *Streptococcus faecalis* are the most important yeasts and bacteria in miso fermentation (WANG and HESSELTINE, 1982). The mixture, known as green miso, is packed into vats or tanks to undergo anaerobic fermentation and aging at 25 to 30 °C for various periods of time, depending upon desired characteristics of the end product. White miso takes about 1 week, salty miso 1 to 3 months, and soybean miso over 1 year. The green miso is transferred from one vat to another at least twice to improve fermentation. The aged miso is then blended, mashed, pasteurized and packaged.

The type of miso produced is largely dependent upon the proportions of koji, salt and cooked soybeans used, but not on the type of inoculum. Thus, white miso contains 50% or less soybeans, whereas yellow or brown miso contains 50% or more soybeans with the balance being rice and salt (SHIBASAKI and HESSELTINE, 1962). White miso contains 4 to 8% salt which permits rapid fermentation, and yellow or brown miso contains 11 to 13% salt. Moisture content ranges from 44 to 52%, protein from 8 to 19%, carbohydrate from 6 to 30%, and fat from 2 to 10%, depending on the levels of soybean, rice and barley used as ingredients. The proximate composition of various types of miso is shown in Tab. 4.

A large part of the soybean protein is digested by proteases produced by *A. oryzae* in the koji. Amino acids and their salts, particularly sodium glutamate, contribute to flavor. The addition of commercial enzyme preparations to enhance fermentation has met with some success. TAKEUCHI et al. (1972) reported that the pattern of amino acids liberated from soybean flour by a commercial protease was essentially the same as that liberated by koji enzymes. EBINE (1971) observed that when an enzyme preparation from a mold was added to cooked soybeans, mixed with brine and allowed to react, an improvement in miso flavor resulted. The time of fermentation was reduced by 33% and yield was increased by 8%.

The relative amount of carbohydrates in miso is a reflection of the amount of rice in the product. A large portion of starch is saccharified by koji amylases to yield glucose and maltose, some of which are in turn utilized as sources of energy by the microorganisms responsible for fermentation. Miso contains 0.6 to 1.5% acids, mainly lactic, succinic and acetic. Some esters are formed with ethyl and higher alcohols which, together with fatty acid esters derived from fatty acids of soybean lipid, are important in giving miso its characteristic aroma (SHIBASAKI and HESSELTINE, 1962).

Changes in lipid components, fatty acid composition and total tocopherol content during miso making were studied by YOSHIDA and KAJIMOTO (1972). Total tocopherol content was decreased in the cooking process but was unaffected during aging. Substantial hydrolysis of triglycerides was detected during the early stages of fermentation. EBINE (1971) noted that miso has strong antioxidative activity. This activity was attributed in part to the existence of isoflavones, tocopherols, lecithin, compounds of amino-carbonyl reactions and living microbial cells which tend to have a reducing activity.

The accumulation of unsightly white tyrosine crystals can result from proteolysis of soybean in the preparation of tao chieo, a Thai fermented soybean paste with some similarities to traditional Japanese miso (FLEGEL et al., 1981). The addition of polyethylene glycol at the beginning of the moromi fermentation reduces the severity of crystal formation.

There are only low levels of vitamins in miso. Cooking and steaming reduce thiamin and riboflavin levels. However, some commercial manufacturers may fortify miso by adding vitamin A, thiamin and riboflavin. According to SHIBASAKI and HESSELTINE (1962), these vitamins are stable during fermentation and aging as well as cooking of miso soup.

2.2.4 Modified Indigenous Procedures

There has been interest in developing miso with reduced salt content. This is due largely to a concern about the relationship between salt intake and its adverse effects on individuals suffering from high blood pressure. OKADA et al. (1975) studied the hydrolytic activities of traditional koji and a commercial enzyme preparation from *Aspergillus oryzae* in low-salt miso. Total and free amino acid contents and color intensity increased in accordance with a reduction of salt concentration. However, no marked differences were observed in the patterns of free and bound amino acids between conventional soybean miso and low-salt miso. Sensory evaluation revealed that a bitter taste was detected in low-salt miso, but there was no significant difference in the taste of miso soups prepared from low-salt and regular miso when the former preparation was supplemented with salt. This indicates that the bitter taste of soybean miso is suppressed by salt.

SHIEH and BEUCHAT (1982) and SHIEH et al. (1982) also investigated miso products containing reduced levels of salt. The use of *Rhizopus oligosporus* instead of *A. oryzae* as a koji mold and peanuts in place of soybeans was also studied. The total microbial population of miso during the fermentation period was altered by the salt level in miso containing *R. oligosporus* koji but not in miso containing *A. oryzae*. Significantly higher populations of *Z. rouxii* were noted in low-salt (6%) miso. Changes in color occurred earlier during fermentation in low-salt miso than in high-salt (12%) miso, but salt level had no effect on viscosity and soluble solids. Salt level had no apparent effect on fatty acid profiles or free fatty acids, but low-salt miso had higher soluble nitrogen and free amino acid content compared to high-salt miso. It was concluded that good quality low-salt miso can be prepared using *R. oligosporus* as a koji mold and peanuts as the oilseed ingredient.

RAO and RAO (1972) made a nutritional evaluation of miso-like products and studied the effect of their supplementation in rice diets. Miso-like products, blends of peanut flour and Bengal gram dhal along with either rice or corn, were evaluated for their protein efficiency ratio (PER) values. Results indicated that the PER of rice diets could be improved by supplementation with predigested products. The replacement of rice by corn in the preparation of koji had little effect on fermentation.

Others have examined various plant materials for their suitability as substrates for making koji and miso. ILANY-FEIGENBAUM et al. (1969) used defatted soybean flakes instead of whole soybeans to prepare miso. Koji prepared by growing *A. oryzae* on corn, wheat, barley, millet, oats, potatoes, sugar beets or bananas was suitable for use in preparing miso-like products. ROBINSON and KAO (1977) and KAO and ROBINSON (1978) investigated the use of chickpea and horsebean grits as replacements for soybeans in traditional miso. The color of chickpea miso was darker than soybean miso, whereas horsebean miso was lighter. An increase in concentration of reducing sugars, soluble protein and water-soluble vitamins (thiamin, riboflavin, pyridoxine, cobalamine, niacin, pantothenic acid, and ascorbic acid) was noted upon fermentation of chickpeas and horsebeans. It was concluded that horsebeans are more suitable than chickpeas for making miso.

2.3 Fermented Whole Soybeans (Natto Products)

Natto is a Japanese name given to fermented whole soybeans. Similar products are known as tou-shih by the Chinese, tao-tjo by the East Indians and tao-si by the Filipinos (WANG and HESSELTINE, 1982; REDDY et al., 1986). The color, aroma, and flavor of these products vary, depending upon the microorganisms used to ferment the soybeans; however, natto products are generally dark in color, having a pungent but pleasant aroma and often a harsh flavor due to the relatively high level of free fatty acids. Fermented whole soybeans are eaten with boiled rice or they can be used as a seasoning agent with cooked meats, seafoods and vegetables. They are thus served as a condiment much in the same fashion as one would serve soy sauce or mustard.

There are three major types of fermented whole soybeans prepared in Japan (KIUCHI et al., 1976; REDDY et al., 1986) (Fig. 3). Itohiki-natto is produced in large quantities in Eastern Japan and is referred to simply as "natto". Washed soybeans are soaked overnight or until they are approximately doubled in weight due to uptake of water. The beans are then steamed for about 15 min, a process which serves to inactivate a large portion of the natural microbial load, and inoculated with *Bacillus natto,* a variant strain of *Bacillus subtilis.* The beans are packaged in approximately 150-g quantities and allowed to ferment at 40 to 45 °C for 18 to 20 h (HESSELTINE, 1965). During fermentation, *B. natto* produces polymers of glutamic acid which cause the surface of the final product to have a viscous appearance and texture. High-quality natto characteristically is covered with a large amount of very viscous polymers.

A second type of fermented whole soybeans produced in Japan is known as Yukiwari-natto. This product is made by mixing Itohiki-natto with salt and rice koji, and then aging at 25 to 30 °C for about 2 weeks. As with soy sauce and miso, the mold of choice for preparing the rice koji is *A. oryzae.*

Hama-natto is the third major type of fermented whole soybean product in Japan, where it is limited to an area in the vicinity of Hamanatsu. Soybeans are soaked in water for about 4 h and then steamed without pressure for 1 h (REDDY et al., 1986). The cooled soybeans are inoculated with a koji prepared from roasted wheat and barley and fermented for about 20 h or until covered with the green mycelium of *A. oryzae.* After drying in the sun or by forced warm air to lower the moisture to about 12%, the beans are submerged in a salt brine along with strips of ginger and allowed to age under pressure for 6 to 12 months. Extensive breakdown of proteins, carbohydrates and lipids occurs during the aging process which contributes to desirable sensory qualities of Hama-natto. In addition to enzymes originating from *A. oryzae* in the koji and which are known to hydrolyze soybean components, enzymes produced by bacteria such as *Micrococcus, Streptococcus* and

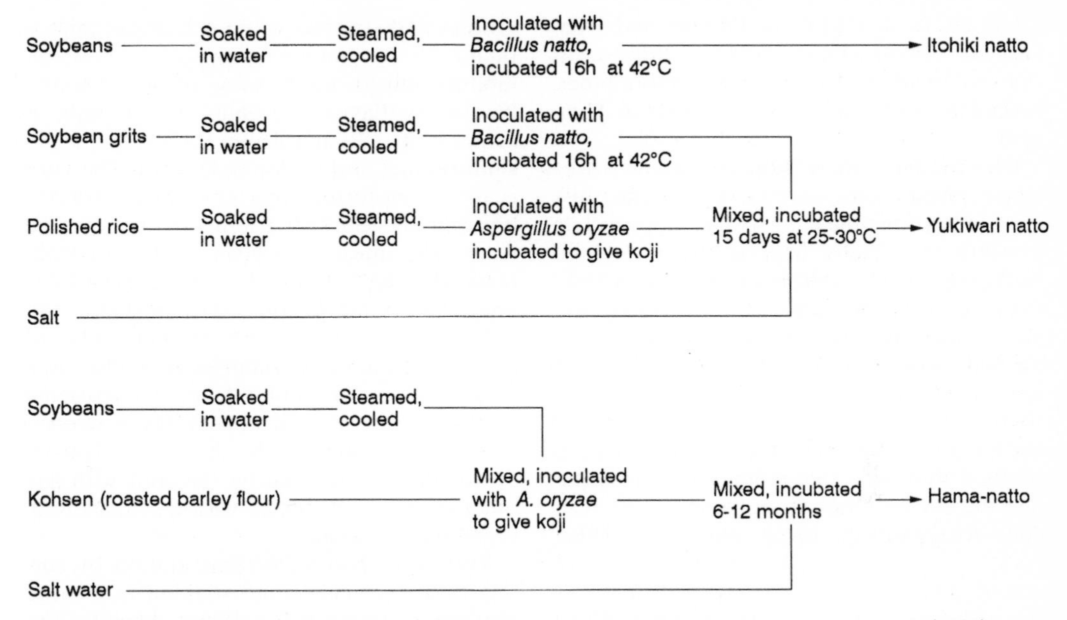

Fig. 3. Flow diagram for manufacture of various types of natto. Adapted from KIUCHI et al. (1976).

Pediococcus which are reported to be widely distributed on the surface and on the inner part of Hama-natto may also contribute to hydrolysis of soybean components.

KIUCHI et al. (1976) studied the lipid content and compositions of Itohiki-, Yuki-wari- and Hama-natto. The lipid contents were 2.8, 10.9 and 6.4%, respectively. The pattern of lipids of Itohiki-natto was similar to that of soybean; however, Yukiwari-natto contained 5 to 18% free fatty acids, and Hama-natto contained 78% free fatty acids in total lipids. The patterns of fatty acid composition of Itohiki- and Hama-natto were similar to those of raw soybeans, but that of Yukiwari-natto was observed to contain lauric and myristic acids in addition. Data indicated that *B. natto* did not produce lipase but the koji mold did. The free fatty acid content of Hama-natto amounted to 12% of the wet weight, suggesting that the harsh taste caused by free fatty acids should have been very strong. However, the sweet taste of sugars originally present in soybeans and resulting from microbial enzyme activity, as well as free amino acids and peptides produced by degradation of protein, tended to have a mellowing affect on this harsh taste. Nevertheless, free fatty acids play an important role in flavor development of fermented whole soybeans.

The proximate composition of fermented whole soybeans varies greatly, but the following ranges (as percentage of wet weight) are given by REDDY et al. (1986): water, 55.0 to 60.8; protein, 16.7 to 22.7; fat, 0.7 to 8.5; carbohydrate, 5.4 to 6.6; and ash, 2.1 to 3.0; a low level of fat in some products results from using defatted soybeans. Thiamin and riboflavin content of natto is reported to be about threefold higher than that of unfermented soybeans and the vitamin B_{12} content is increased nearly five-fold (SANO, 1961).

2.4 Sufu

Sufu is a mold-fermented soybean curd product consumed in the Orient, particularly among the Chinese, who also refer to it as fu-ju or tou-fu-ju (WANG and HESSELTINE, 1982). Because of the difficulties of phonetic translation from Chinese to English, many synonyms for sufu have appeared in the literature. These include tosu-fu, fu-su, toe-fu-ru, tou-fu-ru, fu-ru, and fu-yu (WANG and HESSELTINE, 1970). The product is known as chao in Vietnam, ta-huri in The Philippines, taokaoan in Indonesia and tao-hu-yi in Thailand.

The preparation of sufu consists of three major phases, viz., making a soybean milk curd, fermenting the curd with an appropriate mold(s) and finally brining the fermented curd (Fig. 4). The curd is prepared in a fashion essentially the same as that used to prepare tofu (WANG, 1967a). Soybeans are washed, soaked overnight and ground with water; a water to dry soybean ratio of about 10:1 is commonly used. After boiling or steaming for 20 to 30 min, the ground mass is strained through a fine metal screen to separate the soybean milk from the insoluble residue. Alternatively, beans may be soaked, ground and strained without heating; milk is then heated to boiling to inactivate trypsin inhibitors and to reduce some of the undesirable beany flavor. Curdling of the milk is achieved by adding calcium sulfate or magnesium sulfate, and occasionally acid. The curd is then transferred to a cloth-lined wooden box and pressed with a weight to remove the whey. The finished soybean curd (tofu) contains 80 to 85% water, 10% protein and 4% lipid. The Asian people attach a similar degree of importance to soybean milk and curd as people from dairy countries attach to cow's milk. Tofu is important from a nutritional standpoint, because it constitutes a needed source of calcium for the diet. It is bland in flavor and as such can be flavored with soy sauce or miso, or cooked with meat, seafood, vegetables or soup.

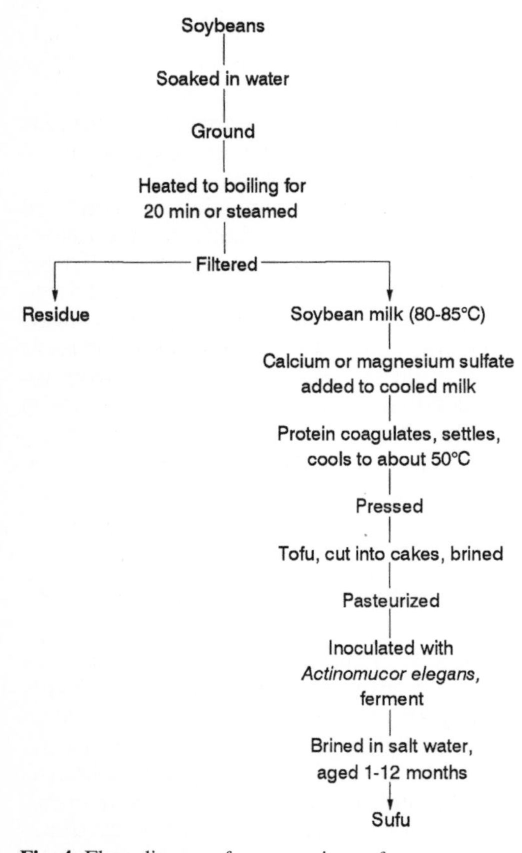

Fig. 4. Flow diagram for preparing sufu.

Tofu is prepared for fermentation by cutting into 3-cm cubes and soaking in a brine containing about 6% sodium chloride and 2.5% citric acid for 1 h. This treatment retards or prevents the growth of bacterial contaminants but has little effect on growth of desired molds during subsequent stages of sufu preparation. The brined cubes are heated at 100 °C for about 15 min, cooled and placed on a perforated tray in such a way that they are not touching. The cubes are surface-inoculated with a selected mold and incubated at 12 to 25 °C for 2 to 7 days, depending upon the type and rate of growth of the mold and the desired flavor characteristics of the final product. The freshly molded cubes are covered with white to yellowish-white mycelium, known as pehtze; they contain 74% water, 12% protein and 4.3% lipid (WAI, 1968).

Various molds belonging to the family Mucoraceae have been isolated from sufu during its preparation. These include *Mucor corticolus, M. hiemalis (dispersus), M. praini, M. racemosus, M. silvaticus, M. subtilissimus, Actinomucor elegans* and *Rhizopus chinensis* (FUKUSHIMA, 1979; HESSELTINE, 1965, 1983). These molds secrete proteases which hydrolyze the soybean protein to yield peptides and amino acids, thus contributing to flavor development. Lipid is probably used as a source of carbon, thus necessitating the production of lipase by the fermenting mold.

The last stage of making sufu involves brining and aging. Depending upon the desired flavor and color, pehtzes may be submerged in salted, fermented rice or soybean mash, fermented soybean paste or a solution containing 5 to 12% sodium chloride in rice containing 10% ethanol. Red rice and soybean mash impart a red color to sufu. Brines containing high levels of ethanol result in a product having a marked alcoholic bouquet. This product is known as tsui-fang or tsue-fan, which means drunk sufu (WANG and HESSELTINE, 1982). Flavor may be modified by adding hot pepper or rose essence to the aging brine. In addition to imparting a salty taste, sodium chloride also causes the release of mycelial enzymes which penetrate the molded cubes and hydrolyze soybean components. The aging period ranges from 1 to 12 months at which time the sufu is consumed as a condiment or used in cooking vegetables or meat. On a commercial scale, sufu is bottled in the same brine and heat-sterilized before marketing.

According to WANG and HESSELTINE (1982), because sufu has the texture of cream cheese, it would be suitable to use in Western countries as a spread for crackers or as an ingredient for dips and dressings. They suggest that the salt content of traditional sufu may be too high for the Westerner's palate and it may be necessary to reduce the sodium chloride content of the aging brine to not less than 3%, the minimal concentration to release mycelium-bound proteins (WANG, 1967b).

2.5 Meitauza

The solid waste collected from ground, steeped, strained soybeans in the preparation of tofu and sufu is sometimes fermented to give a product known as meitauza. Soybean cakes approximately 10 to 14 cm in diameter and 2 to 3 cm thick are allowed to ferment for 10 to 15 days with moderate aeration (HESSELTINE, 1965). The best meitauza is prepared during the cooler months of the year, because high temperatures may result in growth of undesirable bacteria. During the fermentation period, cakes become covered with white mycelium of *Mucor meitauza*, which is a synonym of *Actinomucor elegans*, the principal mold involved in sufu fermentation. At the end of fermentation, cakes are partially sun-dried and sold on the market. Meitauza is cooked in vegetable oil or with vegetables as a flavoring agent.

2.6 Lao-chao

Lao-chao, also known as chiu-niang or tien-chiu-niang by the Chinese, is a fermented rice product. Glutinous rice is first steamed and cooked, then mixed with a small amount of commercial starter known by the Chinese as chiu-yueh or peh-yueh (WANG and HESSELTINE, 1970).

The mass is incubated at ambient temperature for 2 to 3 days during which yeasts hydrolyze the starch, rendering the product soft, juicy, sweet, fruity and slightly alcoholic (1 to 2%). Filamentous fungi may also play a role in lao-chao preparation. According to WANG and HESSELTINE (1970), members of the mucoraceous fungi, including *Rhizopus oryzae*, *R. chinensis* and *Chlamydomucor oryzae* can be consistently isolated from lao-chao. *Endomycopsis*, one of the few yeasts capable of producing amylases and utilizing starch, is also an integral part of the necessary microflora. Lao-chao is consumed as such or it may be cooked with eggs and served as a dessert. Believed to help them gain strength, lao-chao has a unique place in the diet of new mothers.

2.7 Ang-kak

Red rice (ang-kak, ankak, anka, ang-quac, beni-koji, aga-koji) has been used in the fermentation industry for preparation of red rice wine (Shao-Hsing wine) and foods such as sufu, fish sauce, fish paste and red soybean curd, a cheese-like product used as a spice for many centuries in China. Pigments produced by *Monascus purpureus* and *Monascus anka* on a rice substrate have been used as household and industrial food colorants in many Oriental countries. Ang-kak may be marketed for these uses in the form of dried red rice or as its powder.

The traditional manufacture of ang-kak is known only in certain localities in China. However, PALO et al. (1960) and HESSELTINE (1965, 1983) have described procedures for preparing ang-kak on a laboratory scale. Rice is first washed, soaked in water for about 1 day and drained thoroughly. The moist rice is then transferred to a glass beaker or suitable container to allow plenty of air space above it and autoclaved for 30 min at 121 °C. Upon cooling, the rice is inoculated with a sterile water suspension of ascospores removed from a 25-day-old culture of *M. purpureus* grown on Sabouraud agar. At the time of inoculation, the rice should appear rather dry. A wet or mushy substrate is undesirable. The inoculated rice is thoroughly mixed and then incubated at 25 to 32 °C for about 3 days. By this time the rice will have taken on a red color and should be stirred and shaken to redistribute the moisture and kernels with respect to depth from the surface of the fermenting mass. It may be necessary to add some sterile water to replenish moisture lost during incubation. Within about 3 weeks, the rice should take on a deep purplish red color and kernels should not stick together. After drying at 40 °C, the kernels are easily crumbled by slight force and may be reduced to a powder before using in foods or beverages.

Corn can be used as a substrate for producing red pigment (PALO et al., 1960). However, non-glutinous varieties of rice are most suitable for preparing ang-kak, since kernels of glutinous varieties tend to stick together and thus reduce the surface to volume ratio of solid material which is so critical to pigment production. Large quantities of hydrolytic enzymes such as α-amylase, β-amylase, glucoamylase, protease, and lipase are produced by *Monascus* species which break down the rice constituents during growth and penetration of mycelium into the kernels.

The main pigments produced by *Monascus* species are monascorubrin, rubropunctatin, and monascorubramine (WONG, 1982) (Fig. 5). Monascorubrin (red) and monascin (yellow) pigments produced by *M. purpureus* have probably been studied most extensively. Several researchers have investigated the conditions necessary for good pigment production. The optimum cultural conditions for the production of pigment by a *Monascus* species isolated from the solid koji of Kaoliang liquor

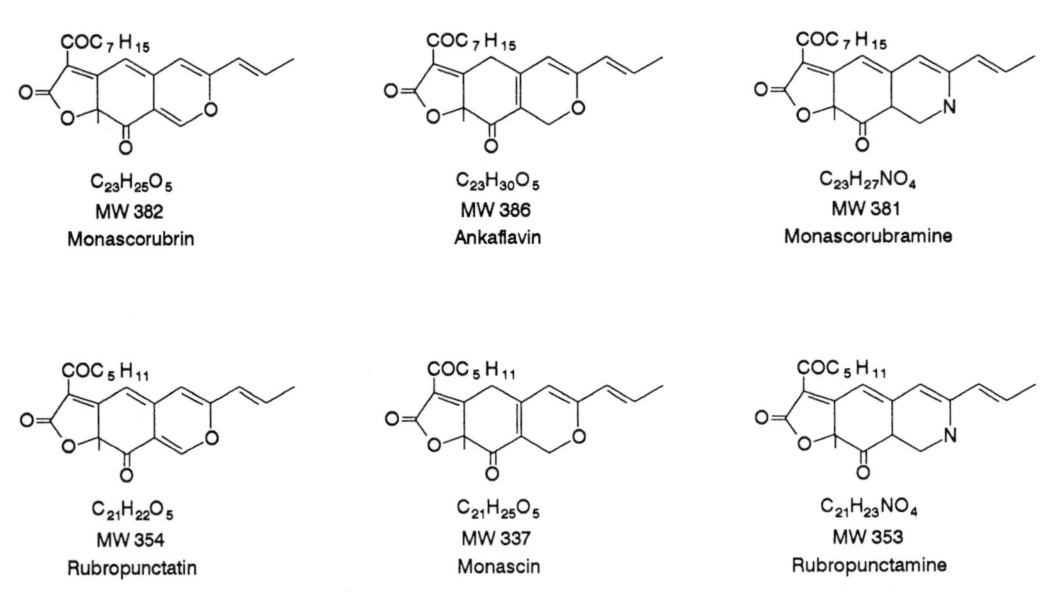

Fig. 5. Molecular structures of pigments produced by *Monascus purpureus* in ang-kak. From WONG (1982).

are reported to be pH 6.0 for a 3-day incubation at 32 °C (LIN, 1973). Among the carbon sources tested, starch, maltose and galactose were found to be suitable for pigment production; a starch content of 3.5% (5% rice powder) and sodium nitrate or potassium nitrate content of 0.5% gave maximum yield of pigment in laboratory media. In a later study (LIN and SUEN, 1973), it was shown that mutation by treatment with *N*-methyl-*N'*-nitro-*N*-nitrosoguanidine and successive isolation could be combined to improve the yield of pigment. BAU and WONG (1979) reported that zinc may act as a growth inhibitor of *M. purpureus* and concomitantly as a stimulant for glucose uptake and the synthesis of secondary metabolites such as pigments.

In addition to its value as a colorant, ang-kak may also possess therapeutic properties. According to WONG and KOEHLER (1981), red rice was first mentioned in LI SHIH-CHUN's "Pen Chaw Kang Mu" (a monograph of Chinese medicine). Ailments and diseases purportedly cured by ang-kak included indigestion, bruise of muscle, dysentery and anthrax. Indeed, *M. purpureus* was demonstrated by WONG and BAU (1977) to have considerable antibacterial activity. Several species of *Bacillus, Streptococcus aureus, Pseudomonas eisenbergii* and *P. fluorescens* were inhibited by a pale yellow compound given the name monascidin by its discoverers. The dose response of *B. subtilis* to the purified antibiotic was determined by WONG and KOEHLER (1981). The minimum effective dose was about 1.5 μg per 6-mm paper assay disc. Production of the antibiotic was usually accompanied by increased pigment production.

2.8 Puto

A fermented rice cake commonly called puto in The Philippines is usually prepared from one-year-old rice that is ground with sufficient water to allow fermentation before steaming (SANCHEZ, 1975) (Fig. 6). The product is similar to tofu prepared from rice in Thailand and to idli prepared from rice and

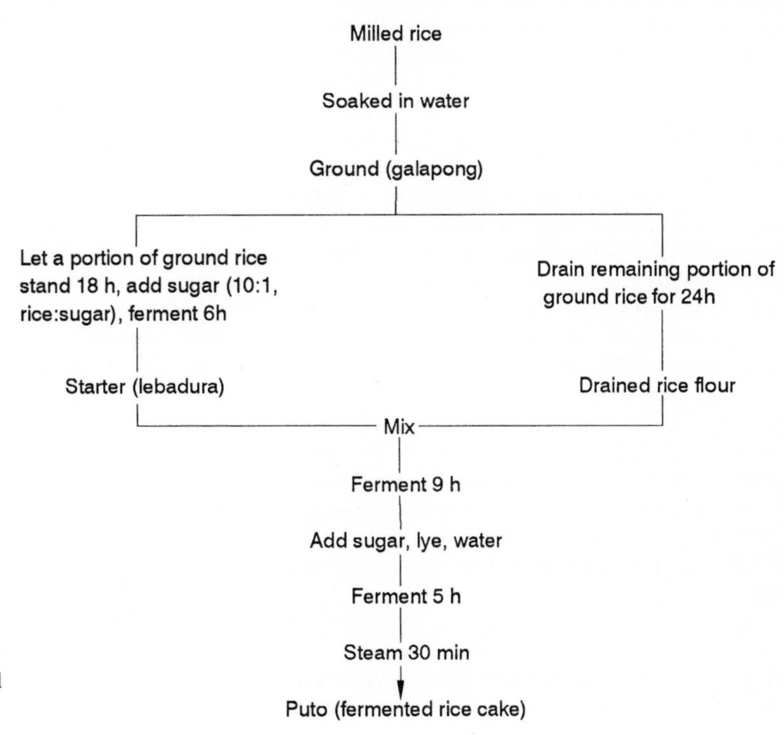

Fig. 6. Flow diagram for preparing puto. Adapted from SANCHEZ (1975).

black gram mungo (*Phaseolus mungo*) in India. The quality of puto is dependent upon the microflora present in the milled rice as well as the variety of rice used as a starting material. In an attempt to determine the influence of rice composition on puto-making qualities, SANCHEZ (1975) evaluated the organoleptic characteristics, volume expansion and fermentation activities of twelve rice varieties. There was a high degree of correlation between amylose content (within limits of 20 to 27%) and general acceptability. For texture and flavor, there was no correlation with amylose content; however, good correlation was observed between high amylose content and satisfactory volume expansion. Varieties having very fine textural characteristics tended to be dry.

A method for shortening the approximately 42 h required for preparing puto by the traditional process was reported by SANCHEZ (1977). This was accomplished with the use of a starter containing *Leuconostoc mesenteroides, Streptococcus faecalis* and *Saccharomyces cerevisiae.* Sensory evaluation showed that there was no significant difference in general acceptability, texture, flavor and volume expansion of puto produced by the traditional and shortened (21 h) methods.

Tab. 5. Mycological Profile of Various Types of Ragi Produced in Indonesia[a]

Type of Ragi	Fungus
Roti	*Saccharomyces cerevisiae*
	Candida solani
Tempeh	*Rhizopus oryzae*
	R. arrhizus
	R. oligosporus
	R. stolonifer
	Mucor rouxii
	M. javanicus
	Trichosporon pullulans
Tapé	*Candida parapsilosis*
	C. melinii
	C. lactosa
	Hansenula subpelliculosa
	H. anomala
	H. malanga
	Chlamydomucor oryzae
	Aspergillus oryzae
Ketjap	*Rhizopus microsporus*
	R. arrhizus
	Aspergillus oryzae

[a] Adapted from DWIDJOSEPUTRO and WOLF (1970)

2.9 Ragi

The Indonesian word "ragi" is roughly equivalent to the English word yeast, but is somewhat wider in scope, since it may include filamentous fungi as well, and connotes the starter or inoculum used to initiate various kinds of fermentations (DWIDJOSEPUTRO and WOLF, 1970). Thus, for example, Indonesian bread (roti) is made with the use of a baker's yeast preparation (ragiroti), fermented glutinous rice or cassava products called tapé-ketan and tapé-ketella, respectively, are prepared using ragi-tempeh. Each type of ragi has its own mycological profile (Tab. 5). Some fungi are common to more than one type of ragi, while others are not.

Ragi is a source of enzymes necessary for the breakdown of carbohydrates and proteins in grains, legumes and roots used as main fermentation substrates. Of particular importance is amylase produced by *Endomycopsis fibuliger* in ragi-tapé. KATO et al. (1976) isolated a strain of this yeast from ragi-tapé and found that it produced α-D-1,4-gluconoglucohydrolase (EC 3.2.1.3). It releases the β-form of glucose by hydrolysis and has high specific activities toward maltodextrins with four degrees of polymerization, amylose, amylopectin and glycogen, but little or no activity toward α-methyl- or ρ-nitrophenyl-α-glucoside. Glucamylase of *E. fibuliger* may be one of the principal enzymes involved in saccharification of rice and cassava starch in the fermentation of tapé.

2.10 Tapé

Indonesian tapé ketan is a fermented, partially liquefied, sweet-sour, mildly alcoholic rice paste (CRONK et al., 1977; ARDHANA and FLEET, 1989). In the traditional process, fermentation is initiated by the addition of powdered ragi made from rice flour containing the desired fungi. In a practical sense, yeasts and molds naturally present in the environment and on equipment used to manufacture tapé serve as inocula for preparing ragi.

Tapé ketan (tepej) is prepared by fermenting glutinous rice (*Oryza sativa glutinosa*) whereas tapé ketella (Indonesian), tapé télo (Javanese) or peujeum (Sudanese) are prepared by fermenting cassava roots (*Manihot utilissima*) (DJIEN, 1972). Rice or peeled, chopped cassava is steamed or cooked until soft, spread in thin layers in bamboo trays, inoculated with powdered ragi, covered with a banana or other suitable leaf and allowed to ferment for 1 to 2 days. At this time, the product will take on a white appearance, soft texture and pleasant, sweet alcoholic aroma and flavor. Both products may be consumed as such or, in the case of tapé ketella, fried in oil before consumption (HESSELTINE, 1965).

Several researchers have studied the microorganisms essential for tapé fermentation (WENT and GEERLINGS, 1895; BOEDIJN, 1958; DWIDJOSEPUTRO and WOLF, 1970; SAONO et al., 1974). The presence of amylolytic molds and alcohol-producing yeasts appears to be necessary for preparing good tapé. CRONK et al. (1977) investigated the basic biochemical changes that occur during a typical tapé-ketan fermentation under pure culture conditions. *Amylomyces rouxii* (formerly *Chlamydomucor oryzae*) in combination with eight yeasts that had been isolated from ragi-tapé, with particular emphasis on *Endomycopsis burtonii* (syn. *E. chodati*), were evaluated. *Amylomyces rouxii* used about 30% of the total rice solids, resulting in a crude protein content of 12% in 96 h, whereas a combination of *A. rouxii* and *E. burtonii* reduced total solids by 50% in 192 h, causing crude protein to increase to 16.5%. The mold alone reduced the starch content of rice from 78 to 10% in 48 h and to less than 2% in 144 h; the mold plus yeast reduced the starch content to about 18% in 48 h. The thiamin content of the rice increased nearly threefold as a result of fermentation by *A. rouxii* in combination with *E. burtonii*. The mold and at least one species of yeast were required to develop the rich aroma and flavor of typical tapé-ketan.

DJIEN (1972) investigated the use of pure culture starters for tapé fermentation on an industrial scale. A mixture of *Chlamydomucor oryzae* and *E. fibuliger* originally isolated from Indonesian ragi had good fermentation characteristics. Prepared starters, produced by growing the fungi on rice and then dehydrating them, were as active as cultures grown on a synthetic agar medium. Little change in activity of the *C. oryzae* starter was observed after 5 months of storage at 20 °C. Thus, technologies have been developed for a modernized industrial process.

Brem and arak are prepared in a fashion similar to that for tapé, but the fermentation period is longer, resulting in a greater liquefaction of rice. The liquid portion is sun-dried to form a sweet solid product known as brem or can be rehydrated and consumed as a mild alcoholic beverage known as arak.

2.11 Tempeh

A popular fermented soybean product in Indonesia, New Guinea and Surinam is tempeh (tempé). Kedelee or kedele, meaning soybean, is used to differentiate tempeh made using soybeans from tempeh bongkrek, a product prepared from coconut press cake (copra). Other beans, peas and cereals are also used to make tempeh (NOUT and ROMBOUTS, 1990). Tempeh kedelee is preferred to the less costly tempeh bongkrek. For purposes of simplification, the word tempeh as used in the following text will be synonymous with tempeh kedelee.

2.11.1 Preparation of Soybeans and Fermentation

Tempeh is made by fermenting dehulled soybeans with various *Rhizopus* species (Fig. 7). The traditional process is simple and rapid, although considerable variations exist. Soybeans are soaked in water at ambient temperature overnight or until hulls (testae) can be easily removed by hand. Others prefer to parboil the beans before soaking in water. NOUT et al. (1987b) reported that species of

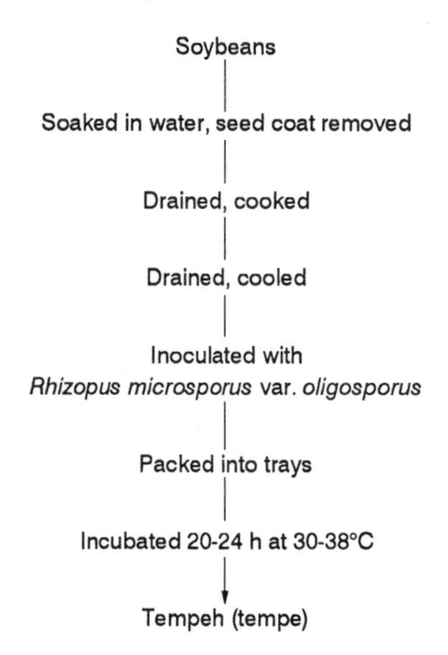

Fig. 7. Flow diagram for preparing tempeh.

lactic acid bacteria, Enterobacteriaceae and yeasts are predominant in water in which soybeans have been soaked, suggesting their involvement in the fermentation. A comprehensive, quantitative study of the ecology of soybean soaking for tempeh fermentation was done by MULYOWIDARSO et al. (1989). *Lactobacillus casei, Streptococcus faecium, Staphylococcus epidermidis* and *Streptococcus dysgalactiae* dominated the fermentation, but significant contributions were made by several other bacteria, and the yeasts *Pichia burtonii, Candida diddensiae* and *Rhodotorula rubra. L. casei, S. faecium* and *S. epidermidis* are the main species responsible for reduction of pH of soak water.

After the hulls are removed from soaked soybeans, cotyledons may be pressed slightly to remove more water and then mixed with small pieces of tempeh from a previous batch or ragi tempeh, a commercial starter. The inoculated beans are then spread onto bamboo frames, wrapped in banana leaves and allowed to ferment at ambient temperature for 1 to 2 days. At this point, the soybeans are covered with white mycelium and bound together as a cake (HESSELTINE et al., 1963).

A somewhat more elaborate method for preparing tempeh involves first washing and boiling soybeans (an operation taking 2 h), transferring to cold water and soaking for about 1 day (ILJAS et al., 1973). The hulls are removed and the cotyledons are boiled again and then steamed. Meanwhile the tempeh mold is prepared by wrapping a small portion of tempeh in a teak leaf and allowing it to incubate and dry for 2 days. The inoculum is then cut into small pieces and sprinkled over the soybeans which are in turn wrapped in banana leaves and left to ferment for about 1 day. MARTINELLI and HESSELTINE (1964) devised methods to make tempeh rapidly in large amounts by pure-culture fermentations in shallow wooden and metal trays with perforated bottoms and covers. Excellent tempeh was also made in perforated plastic bags and tubes. The authors considered the latter containers to be especially good, because soybeans could be fermented in the same package ultimately used for distribution and sale. These methods for preparing tempeh can be carried out within a 24-h period at 31 °C.

Certain modifications of traditional processes for making Indonesian tempeh have been suggested by researchers in the United States (STEINKRAUS et al., 1965a). They developed a pilot plant process for manufacturing tempeh in which the labor-intensive step of removal of hulls by rubbing the soaked soybeans between the hands is eliminated. The process involves size-grading of the dry soybeans, heating the beans in circulating hot (104 °C) air to shrivel the cotyledons, passing them through a burr mill to loosen the hulls and finally passing the mixture of hulls and cotyledons over a gravity separator to remove the hulls. According to STEINKRAUS (1978), village level manufacturers of tempeh in Indonesia are ready and willing to accept modification which improves production of their product.

The method of STEINKRAUS et al. (1960) involves the use of lactic acid as an acidifier in the soak water. After dehulling, cleaned soybeans are returned to the acidified soak water and boiled for 90 min, drained, cooled to about 37 °C and inoculated with *Rhizopus.* Acidification of the soak water to about pH 5.0 was suggested as a procedure for inhibiting the growth of some microorganisms which can cause spoilage.

Some degree of success has been achieved in the development of tempeh-like products containing ingredients other than soybean cotyledons. Tempeh can be made from soybean grits (HESSELTINE and WANG, 1967) and cereals such as wheat, oats, rye, barley, rice and combinations with or without soybeans (HESSELTINE et al., 1967; WANG and HESSELTINE, 1966; WANG et al., 1968), horsebean and chickpea grits (ROBINSON and KAO, 1977) and wingbeans (GANDJAR, 1978). The latter product is called tempeh-kecipir. Corn, sorghum, and peanuts offer less promise as soybean substitutes for making tempeh-like products (HESSELTINE et al., 1967).

As with other indigenous fermented foods, the temperature and moisture content of the fermenting substrate are critical if a good-quality tempeh is to be obtained. The most desirable temperature is between 30 and 38 °C at which fermentation is complete within 1 to 2 days, depending upon perceptions of optimal sensory qualities by the manufacturer and local consumers. The fermenting beans should be kept covered to retard the rate of loss of moisture. However, slow diffusion of air to and release of gas from the product is essential to promote proper growth and metabolic activities of the mold. Production of black sporangia and sporangiospores by *Rhizopus* is undesirable and usually indicates inadequate environmental conditions during fermentation or a product which has been kept beyond its normal shelf-life. Tempeh is extremely perishable and is usually eaten within a day of its preparation. Fermented, yeasty off-odors are produced during storage (WINARNO and REDDY, 1986). The production of ammonia as a result of enzymatic breakdown of mycelia and soybean protein causes the product to be inedible within a very short period of time. Properly prepared tempeh is either sliced, dipped in salt solution and deep-fat fried in coconut oil or cut into pieces and used in soups. Salt or soy sauce may also be added to tempeh after it is fried.

The shelf-life of tempeh can be prolonged by various methods. In Indonesia, freshly prepared tempeh is sliced and then sun-dried. STEINKRAUS et al. (1965b) reported on a pilot plant process they developed for dehydrating tempeh by a hot-air dryer at 93 °C for 90 to 120 min, and ILJAS (1969) described the acceptability and stability of tempeh preserved in sealed cans for 10 weeks. Acceptability was not significantly changed when the can was stored at −29 °C immediately after sealing or when the tempeh was packed in water, steam-vacuum sealed, heat-processed at 115 °C for 20 min and stored at room temperature. The acceptability of tempeh tended to decrease during storage if tempeh had been air-dried at 60 °C for 10 h prior to sealing in a can and storing at room temperature.

The presence of *Staphylococcus aureus, Clostridium botulinum, Salmonella, Yersinia enterocolitica* and *Bacillus* species, including *B. cereus,* in tempeh has raised concern about the microbiological safety of tempeh (NOUT et al., 1987b; SAMSON et al., 1987). These researchers emphasized the importance of acidification of soybeans prior to fungal fermentation for controlling the growth of these pathogens.

2.11.2 Biochemical Changes

Although other genera of molds are occasionally found in tempeh, none of them in pure culture, except species of *Rhizopus,* can produce tempeh (HESSELTINE, 1965). *Rhizopus* spp. are known to produce carbohydrases, lipases, proteases, phytases and other enzymes (NOUT and ROMBOUTS, 1990). These include *R. microsporus* var. *oligosporus, R. stolonifer, R. arrhizus, R. oryzae, R. formosaensis* and *R. achlamydosporus.* The principal species used in Indonesia is *R. microsporus,* a strain designated as NRRL-2710 having received considerable research attention by workers at the United States Department of Agriculture (HESSELTINE et al., 1963; HESSELTINE, 1965). Utilization of carbon and nitrogen compounds by this strain was studied by SORENSON and HESSELTINE (1966). They found that glucose, fructose, galactose and maltose supported excellent growth but that xylose, sucrose, stachyose and raffinose could not be utilized as the sole source of carbon. Among the carbohydrases produced by *R. microsporus* are endocellulase, xylanase and arabanase (NOUT and ROMBOUTS, 1990).

Since a substantial amount of soluble carbohydrate is lost due to soaking and cooking, carbohydrase activity by *R. microsporus* is minimal during fermentation. STAHEL (1946) stated that 7% of the dry matter in soybeans is lost at the first boiling and an additional 11% is lost during subsequent soaking and boiling. This dry matter consists mostly of water-soluble carbohydrates. VAN VEEN and SCHAEFER (1950) reported that no starch or dextrin and only 0.9% soluble carbohydrates were present in cooked soybeans, whereas a trace of starch, 2.5% dextrin and 7.8% soluble carbohydrates were present in raw soybeans. The carbohydrate content of cooked soybeans was observed by SHALLENBERGER et al. (1966) to be about half of that in raw beans, the decrease in content being attributed mainly to loss of sucrose.

Although *R. microsporus* cannot utilize flatulence-causing oligosaccharides because it does not produce α-galactosidase (WORTHINGTON and BEUCHAT, 1974), compared with unfermented soybean grits, tempeh made from grits has been reported to have a delayed effect on gas formation in humans (CALLOWAY et al., 1971). Delay was attributed to possible inhibition of intestinal microflora by an antibiotic substance produced by *Rhizopus*.

The hemicellulose content (as glucose) is reduced from 2.8% in raw soybeans to 2.0% as the beans are cleaned and cooked, and to 1.1% after fermentation (VAN VEEN and SCHAEFER, 1950). The fiber content, however, may actually increase upon fermentation due to the production of mycelium by the mold (MURATA et al., 1967). WANG et al. (1968), on the other hand, observed a decrease in fiber content of soybeans as a result of fermentation. In any event, the crude fiber content ranges from about 1.5 to 4.3% in tempeh at the end of fermentation.

Protease production by *R. microsporus* is substantial, and perhaps plays the most important role in developing good-quality tempeh. Two proteolytic enzyme systems were studied by WANG and HESSELTINE (1965). One has an optimum pH at 3.0 and the other at 5.5; both systems have maximum activities at 50 to 55 °C and are fairly stable at pH 3.0 to 6.0, but are rapidly denatured at pH below 2 or above 7. The pH 5.5 system dominates in tempeh fermentation. While the total crude protein content of soybeans may increase (MURATA et al., 1967; ILJAS, 1969) or decrease (VAN BUREN et al., 1972) somewhat during fermentation, total nitrogen content of cooked soybeans remains about the same, i.e., about 7.5%, but the soluble nitrogen increases from 0.5% to 2.0% (HESSELTINE, 1965; STEINKRAUS et al., 1960). Amino acid profiles of soybeans are not changed significantly upon fermentation; however, free amino acids may be increased as much as 85-fold (MURATA et al., 1967). Prolonged fermentation may result in losses of lysine (WINARNO and REDDY, 1986). Deep-fat frying causes some amino acids to decrease, whereas steaming of tempeh has no effect on the amino acids (STILLINGS and HACKLER, 1965). Increases in glucosamine content during fermentation are a reflection of increased levels of *R. microsporus* mycelium.

The lipid content of soybeans is reduced (on a dry weight basis) during the soaking process. The temperature of the soak water may influence the rate of release of lipid from soybeans. There may be a slight decrease in lipids as a result of fermentation (MURATA et al., 1971; WANG et al., 1968).

Rhizopus microsporus possesses strong lipase activity (BEUCHAT and WORTHINGTON, 1974) and has been shown to hydrolyze over one-third of the neutral fat in soybeans during a 3-day fermentation period (WAGENKNECHT et al., 1961). Palmitic, stearic, oleic, linoleic and linolenic acids were liberated in roughly the same proportions found in soybeans. During the most active growth phase, however, proportionately higher levels of free palmitic acid were found, and the level of linoleic acid was somewhat lower. Except for the depletion of some 40% of the linolenic acid in the late stages of fermentation, there apparently was no preferential utilization of any fatty acid. Bacteria (*Bacillus* species) have also been identified as a part of the microflora of tempeh produced in the laboratory (SUDARMADJI and MARKAKIS, 1978). These authors detected as much as 10 g of free fatty acids per 100 g of tempeh. Upon frying in coconut oil, tempeh underwent a sharp reduction in free fatty acid content with a concomitant increase in the free fatty acid content of the frying oil.

Lipid in tempeh has been found to be more resistant to autoxidation than that in raw soybeans, indicating the presence of an antioxidant as a result of fermentation (CHEN et al., 1972). The peroxide value of lyophilized tempeh stored at 37 °C for 5 months increased from 6 to 12 compared with 6 to 246 in nonfermented soybeans (IKEHATA et al., 1968). PACKETT et al. (1971) demonstrated that corn oil containing 50% tempeh oil showed higher antioxidant potential than oil containing 25% tempeh or 0.03% α-tocopherol, and GYÖRGY et al. (1975) reported that soybean, cottonseed, corn and safflower oils and lard could be protected from autoxidation by the addition of tempeh oil. GYÖRGY et al. (1964) had isolated earlier the antioxidants jenistin, deadzein and 6,7,4′-trihydroxyisoflavone in tempeh, which are assumed to be in a bound inactive form in raw soybeans. Hydroxylated aglycones of even higher activity are formed by *R. microsporus* (MURATA, 1988; JHA, 1985).

The phytic acid content of soybeans is reduced by about one-third as a result of fermentation with *R. microsporus* (SUDARMADJI and MARKAKIS, 1977), while an equivalent amount of phosphate is released in the tempeh. The pH optimum of phytase produced by the mold was 5.6.

2.11.3 Nutritional Characteristics

VAN VEEN and SCHAEFER (1950) observed the beneficial effects of tempeh on patients suffering with dysentery in the prison camps of World War II. They suggested that tempeh was much easier to digest than nonfermented soybeans; however, animal feeding studies have not confirmed this conclusion (HACKLER et al., 1964; SMITH et al., 1964; MURATA et al., 1967). The PER of tempeh is not changed substantially from that of raw soybeans but is reduced upon frying in oil; steaming for 2 h has no effect (HACKLER et al., 1964). ZAMORA and VEUM (1979), on the other hand, reported that diets containing soybeans fermented with *Rhizopus microsporus* had a greater apparent net protein utilization and apparent biological value than nonfermented soybeans. Tempeh-like products containing peanuts (BAI et al., 1975) and wheat (WANG et al., 1968) in place of part of the soybeans have PER values as good as or better than 100% soybean tempeh.

Niacin, riboflavin, pantothenic acid, and vitamin B_6 contents of tempeh are reported by some researchers to be higher than levels in raw soybeans (ROELOFSEN and TALENS, 1964; MURATA et al., 1967). Riboflavin is three to five times higher and niacin increased by a factor of 3.4. WINARNO (1979) stated that tempeh made with a pure culture of *R. microsporus* contains very little vitamin B_{12} compared with tempeh made by traditional methods. Apparently a bacterium present in commercial samples of tempeh produces vitamin B_{12} during fermentation. Indonesian tempeh may contain 30 μg/g of B_{12} (WINARNO, 1979).

WANG et al. (1969) found that *R. microsporus* produces an antibacterial agent which is especially active against some Gram-positive bacteria, including both microaerophilic and anaerobic types, e.g., *Streptococcus cremoris, B. subtilis, Staphylococcus aureus, Clostridium perfringens* and *Clostridium sporogenes.* The compound contains polypeptides having a high carbohydrate content; its activity is not affected by pepsin but is slightly decreased by trypsin and peptidase. The production of this agent by *R. microsporus* may represent a significant contribution to the health of Indonesians who consume tempeh on a regular basis.

2.12 Oncom

Oncom (onchom, ontjom, lontjom) is a fermented peanut press cake product prepared and consumed largely in Indonesia (BEUCHAT, 1976, 1978, 1982, 1987). It is consumed as a snack or for breakfast. According to HESSELTINE (1965), the flavor of fermented peanut press cake is fruit-like and somewhat alcoholic, but takes on a mincemeat or almond character if the product is deep-fried. Oncom may also be roasted, covered with boiling water and seasoned with salt or sugar before it is eaten. In another fashion, the roasted product is cut into pieces and covered with a ginger-flavored sauce before eating.

2.12.1 Preparation of Peanuts and Fermentation

The general scheme for preparing oncom is illustrated in Fig. 8. After oil has been extracted from peanuts, the press cake, called boongkil, is broken up and soaked in water for about 24 h (HESSELTINE and WANG, 1967). Technical-grade press cake is low in residual oil content, while village products contain considerable amounts (VAN VEEN et al., 1968). Oil that rises to the surface of the water during the soaking period is removed, and the press cake is then steamed for 1 to 2 h and pressed into a layer about 3 cm deep in a bamboo frame. The mass is inoculated with either *Neurospora intermedia* (formerly *N. sitophila*) or *R. microsporus,* the same mold used to prepare tempeh. *Neurospora* is a common mold in woody material under tropical rain forest conditions whenever sterilization or pasteurization has occurred, such as in the burning of forest land (STANTON, 1971) and presumably would be available to the Indonesian villager for use in inoculating a new batch of oncom. More often, as is commonly done in preparing traditional fermented foods, a small portion of a previous batch of oncom can be used as an inoculum.

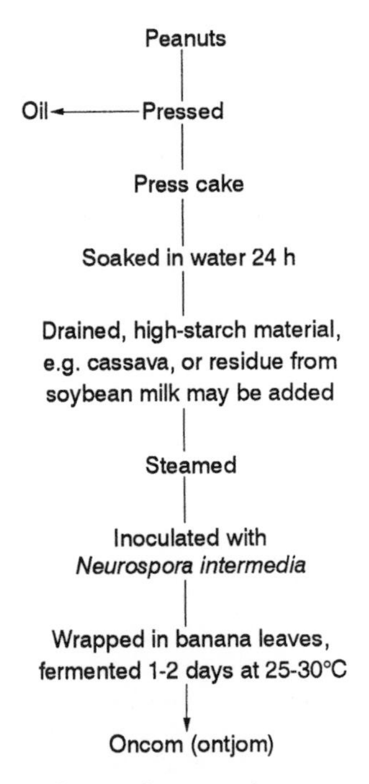

Fig. 8. Flow diagram for preparing oncom.

The inoculated press cake is covered with banana leaves and allowed to ferment for 1 to 2 days, at which time the internal portion of the mass has been invaded by mycelia. Constant aeration is important in the production of oncom, as are temperature, moisture content and degree of press cake granulation. Although covered with banana leaves, the oncom is exposed to the air during fermentation. This practice aids in sporulation of the mold inoculant, thus resulting in an orange to apricot-colored product if *N. intermedia* is used, or a gray to black product if *R. oligosporus* is used. Only the surface of the fermented press cake is covered with colored conidia or spores. A temperature range of 25 to 30 °C is suitable for producing the best oncom.

A high-carbohydrate material such as tapioca (cassava) or potato peels may be added to the press cake prior to inoculation to enhance fermentation. VAN VEEN et al. (1968) observed that culture of *Neurospora* isolated from oncom grew well on a substrate with a pH lower than 6.0. They concluded that extraction of peanut press cake with hot water, followed by the addition of 1% tapioca, pasteurization and adjustment of pH to 4.5 resulted in the best substrate for growth, sporulation, color and flavor development. The addition of tapioca to peanuts has been shown to promote the growth of *Rhizopus* sp. (HESSELTINE et al., 1967) and the addition of citric acid (1.25% by weight), tapioca (1%) and sodium chloride (0.63%) to defatted peanut flour has been demonstrated to enhance the growth of *N. intermedia, R. microsporus, R. delemar* and *Aspergillus* species (BEUCHAT and WORTHINGTON, 1974). Salt may exert a beneficial osmotic or ionic effect on fungal mycelium, whereby extracellular enzymes are readily freed to act upon substrate constituents. *Mucor dispersus,* a mold used in sufu fermentation, has been reported to have increased proteinase activity, when sodium chloride is added to a soybean substrate (WANG, 1967b).

2.12.2 Biochemical Changes

Both *Neurospora intermedia* and *Rhizopus microsporus* produce considerable amounts of hydrolytic enzymes that act upon peanut constituents during fermentation. The bulk of research data relating to changes in protein after fermentation of peanut press cake indicates that protein content is little altered, but solubility is greatly increased. VAN VEEN et al. (1968) reported that crude protein (Kjeldahl nitrogen × 6.25) was elevated slightly while true protein decreased from 94% of the crude protein in peanut press cake to 74% of the crude protein in oncom. QUINN et al. (1975) reported that the crude protein content of peanut flour increased upon fermentation with the oncom molds as well as with *Mucor hiemalis, Actinomucor elegans* and *Aspergillus oryzae.* It was suggested that there is not an actual increase in protein weight as a result of fermentation but rather a loss of non-protein volatiles during the fermentation process, thus accounting for proportionate increases in unaltered peanut constituents.

Qualitative changes in proteins resulting from hydrolytic activities of *N. intermedia* and *R. microsporus* and seven other molds used to ferment foods were observed by BEUCHAT et al. (1975). Large molecular weight globulins were hydrolyzed to smaller components, and the percentages of specific amino acids and proportions of specific amino acids within the free amino acid fraction varied greatly among the ferments as well as between fermented and unfermented peanut substrates. The percentages of total amino acids as free acids in peanuts fermented for 98 h at 28 °C were reported as 8.67 and 2.99 for *N. intermedia* and *R. microsporus,* respectively. In a related study, QUINN and BEUCHAT (1975) observed that the nitrogen solubility of peanuts increased to 24 and 19% when fermented with *N. intermedia* and *R. oligosporus,* respectively. The nitrogen solubility of unfermented peanuts was less than 5%. Using a peanut protein substrate, BEUCHAT and BASHA (1976) reported that maximal activity of protease extracted from *N. intermedia* occurred at pH 6.5. The greatest protease activity and mycelium production occurred during the first day of a 4-day test period.

WORTHINGTON and BEUCHAT (1974) investigated the ability of *N. intermedia* and *R. microsporus* to utilize flatulence-causing oligosaccharides in peanuts. *N. intermedia* showed strong α-galactosidase activity, utilizing raffinose and stachyose within a 21-h incubation period. However, *R. microsporus* did not utilize these sugars or utilized them only slowly during a 98-h test period. The increased digestibility of oncom compared with unfermented peanuts may be attributable in part to the decrease in levels of these sugars.

Lipid is hydrolyzed by the oncom molds as a consequence of fermentation of peanuts (BEUCHAT and WORTHINGTON, 1974). At 28 °C, as much as 9.4 and 31.5% of the lipid can be in the form of free fatty acids as a result of fermentation with *N. intermedia* and *R. microsporus*, respectively. The free fatty acid fraction contains a significantly higher level of saturated fatty acids, particularly palmitic and stearic acids, and lower levels of linoleic acid than does the total lipid of oncom. Differences in free fatty acid distribution are those which would be expected from the action of 1,3-lipases, since saturated acids are located primarily in the 1,3-position and linoleic acid is in the 2-position of peanut triglycerides. It is well known that flavor development in fermented dairy products is significantly influenced by lipolytic activities of various microorganisms employed. Undoubtedly, lipid hydrolysis by *N. intermedia* and *R. microsporus* also enhances the flavor and aroma of oncom.

The phytase activity of *N. intermedia* and *R. microsporus* was studied by FARDIAZ and MARKAKIS (1981). The phytic acid content of unfermented peanut press cake was reduced from a level of 1.36% on a dry weight basis to 0.70% by *Neurospora* and 0.05% by *R. microsporus* after 72 h of fermentation. Penta-, tetra-, tri-, di- and monophosphates of inositol, as well as inorganic phosphate and inositol, were found in the fermented cake. Phytic acid is nutritionally important because of its ability to form insoluble complexes with zinc and thereby reduce its bioavailability. Thus the degradation of phytic acid by oncom molds may contribute to an overall improvement in nutritional quality of raw peanut press cake.

The protein efficiency ratio of fermented peanuts apparently is not increased over that of properly heat-treated raw ingredients (VAN VEEN et al., 1968; VAN VEEN and STEINKRAUS, 1970; QUINN et al., 1975). However, the nutritive quality of the free amino acid fraction of fermented peanuts is improved over unfermented substrate (BEUCHAT et al., 1975).

2.13 Fish Products

Fermented fish sauce and paste are popular condiments prepared and consumed in the Orient. Whole small fish, with or without entrails, or shrimp are heavily salted (up to 30% sodium chloride), packed into containers and allowed to undergo fermentation for periods ranging from a few days to over a year (WANG and HESSELTINE, 1982). Roasted cereals, glutinous or red rice flour and bran may be added in varying amounts to prepare fish pastes. The process for making fish sauce and paste is essentially anaerobic and involves bacterial and autolytic breakdown of proteins and lipids in fish tissue to result in highly flavored products which complement an otherwise bland rice diet.

Because of their high salt content, consumption of large quantities of fish sauce and paste is limited. These products do, however, represent an important source of calcium for a population whose diet may be somewhat low in this mineral.

Fish sauce and paste are known as ngampya-ye and ngapi, respectively, in Burma, nuoc-mam and mams in Cambodia and Vietnam, patis and bagoong in The Philippines (MABESA et al., 1992) and mampla and kapi in Thailand (VAN VEEN, 1953; WANG and HESSELTINE, 1982). While traditional methods for preparing sauce or paste are similar in different countries, variations exist to result in desired appearance, aroma, texture and flavor characteristics in the final products. For exemplary purposes, a limited number of products will be discussed here.

2.13.1 Nuoc-mam

Nuoc-mam, a clear brown liquid, is produced on a large scale commercially as well as in cottage industries. Total production was estimated at 40 million liters per year in the early 1970s, and daily per capita consumption ranged from 15 to 60 mL (WANG and HESSELTINE, 1982). It may be produced from small marine or freshwater fishes. On a small scale, fish are kneaded and pressed by hand, salted, tightly packed into earthenware pots, sealed and allowed to ferment. The supernatant fluid that accumulates is known as nuoc-mam. On a commercial scale, fish are brined in large, wooden cylindrical vats. Six parts of uncleaned fish are mixed with four to five parts salt and piled above the top of the vat (VAN VEEN, 1953). After 3 days, the liquid (nuoc-boi), which is turbid and bloody in appearance, is collected. At this point the fish are tightly packed and a portion of the nuoc-boi (now clearer in appearance due to chemical changes and a settling of solid materials) is added back to cover the mass by a depth of about 10 cm. Aging then proceeds for a few months if small fish are being fermented, or as long as 18 months for large fish.

First-quality nuoc-mam is drawn off from the matured ferment using a tap; a product having lower quality can be obtained by extracting the residual mass with fresh brine solution. The total and amino acid nitrogen content of first-quality nuoc-mam is about double that of ordinary quality product (SUBBA RAO, 1967). First-quality nuoc-mam contains no less than 16 g of total nitrogen per liter; the extent of fermentation is based on formol titratable nitrogen, but no less than 50% total nitrogen; and the amount of ammonia nitrogen, an index of nutritive value, cannot exceed 50% of the formol nitrogen. One part of fish gives from two to six parts nuoc-mam, depending upon the quality of the final product.

Little is known about the contribution of microorganisms to nuoc-mam fermentation.

The high level of salt greatly inhibits proteolytic activity; however, bacteria and perhaps yeasts present on and within the fish as they are packed into vats, undoubtedly play some role in desired fermentation patterns and flavor development. The addition of proteolytic enzymes from *Aspergillus oryzae* to the fermenting fish has been shown to reduce the fermentation time and increase the yield of nuoc-mam (BAENS-ARCEGA et al., 1969). Fresh pineapple and papaya have also been used to increase the rate of proteolysis.

2.13.2 Bagoong

Bagoong is a fish paste prepared in The Philippines from sea fish, anchovies, ambassids or shrimp (VAN VEEN, 1953). Salt is mixed with three parts fish, placed in clay vats and left undisturbed for 3 months. The resulting paste-like product is eaten raw or cooked.

2.13.3 Prahoc

Two stages are involved in preparing prahoc, a fermented fish paste of Cambodia. The first stage takes place at the fishing place. Fish are beheaded, eviscerated, packed in wicker baskets and tamped by foot to loosen and remove the scales. The fish are then washed in water, drained, covered with banana leaves and weighted with stones or some other heavy object for about 24 h. Salt is added to the fish at a ratio of 1:10, and the fish are then spread out to sun-dry for an additional day. The salted, semi-dried fish are again tightly packed into wicker baskets.

For the second stage of prahoc preparation, the fish are moved to homes of villagers where they are pounded into a paste, packed into earthen jars, and placed in the sun for about a month. Fluid which accumulates on the surface of the fermenting paste is removed daily and used as nuoc-mam. The prahoc is then ready for consumption, principally in the preparation of soups. Three parts of fish will produce about one part of prahoc.

2.13.4 Phaak

Phaak (or mamchao) is a fermented paste product produced in parts of Indochina (Cambodia) containing eviscerated, salted fish with heads and glutinous rice pretreated with yeasts (VAN VEEN, 1953). Roasted rice, with or without sugar, and ginger may also be added to obtain a special flavor in the final product. In other parts of Southeast Asia, other flavoring agents such as peppers as well as colorants may be added to fish paste to satisfy local tastes.

2.13.5 Katsuobushi

A fermented fish product of Japan which is neither a sauce nor a paste but rather a hard, dried product is known as katsuobushi. The product is made from skipjack tuna or bonito which is dried and allowed to mold over a period of several weeks. According to GRAIKOSKI (1973), ripening of katsuobushi can be hastened by artificially inoculating the fish with species of the *Aspergillus glaucus* group of molds.

2.14 Kimchi

Kimchi (kim-chee) is a class designation for salted and fermented products prepared in Korea. Winter kimchi made in the fall contains cabbage (Chinese and Korean), large-rooted radishes, leeks, onions, garlic, red pepper, ginger and salt (CHOI, 1991). The mixture is allowed to undergo fermentation with lactic acid bacteria for several weeks or months before being used (WILLIAMS et al., 1959). Summer kimchi is made from young radishes and radish leaves, cucumbers, cabbage, lettuce and some of the above minor ingredients, and is fermented only 2 or 3 days before serving. Small, salted shrimp, ground raw fish or nuts may also be a part of the kimchi formula.

Kimchi serves as a supply of preserved vegetables, particularly during the winter months when fresh green vegetables are less abundant. Kimchi is regarded as an important source of ascorbic acid and may represent a vehicle to supply vitamin B_{12} if the bacteria responsible for fermentation can be selected and controlled (Ro et al., 1979). Anaerobic bacteria in kimchi include *Lactobacillus plantarum, L. brevis, Leuconostoc mesenteroides* and *Pediococcus cerevisiae.* The presence of *Propionibacterium freudenreichii* has been shown to increase the vitamin B_{12} content of kimchi twofold (Ro et al., 1979). Kimchi is treated in more detail in Chapter 17.

3 Fermented Foods of India

3.1 Idli

Idli is a steamed fermented dough made from various proportions of rice and black gram flours. It is typically eaten for breakfast and is especially popular in South India, although consumed throughout the country. Proportions of rice to black gram cotyledons used to prepare idle range from 1:4 to 4:1 (Reddy et al., 1986), depending upon taste preference and availability of ingredients. Idli containing higher amounts of rice is characterized by a more predominant starchy flavor. Other ingredients such as cashew nuts, ghee, chili peppers, ginger, fried cumin seeds or curry leaves may be added to the dough in small quantities to impart additional flavor.

A flow diagram for preparing traditional idli is shown in Fig. 9. Dehulled black gram and rice are washed and soaked in water separately for 5 to 10 h at ambient temperature. The amount of soaking water can vary from 1.5 to 2.2 times the dry weight of the black gram or rice (Hesseltine, 1979). Reddy and Salunkhe (1980) reported that 1.5 times water over dry ingredients was optimum for fermentation and for idli preparation. After soaking, the black gram is ground with water to give a coarse paste, whereas the rice is ground to give a smooth gelatinous paste. Salt (about 0.8%) is added to the combined mixture of pastes, and fermentation is allowed to proceed for 15 to 24 h. Upon steaming, the soft, spongy final product resembling a sour bread or pancake is consumed while still hot. The spongy open texture of idli is attributed

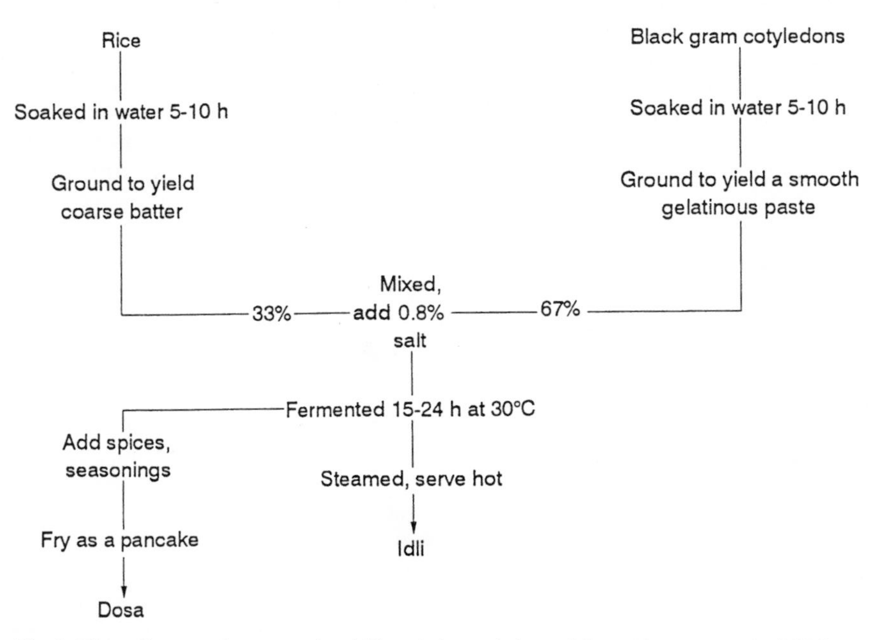

Fig. 9. Flow diagram for preparing idli and dosa. Adapted from Reddy et al. (1986).

to the protein (globulin) and polysaccharide (arabino-galactan) in black gram (SUSHEELAMMA and RAO, 1979).

Acidification and leavening are the most important processes which occur during fermentation. According to STEINKRAUS et al. (1967), idli batter volume increases 1.6 to 3.1 times, and the pH falls from an initial 6.0 to 4.3 during fermentation. Thus, the sour taste of idli is a necessary and desirable characteristic.

While several reports have been made on the microbiology of idli fermentation, no comprehensive studies on the microorganisms involved and the changes brought about by them have been made. Bacteria identified as part of the microflora responsible for production of good idli include *Leuconostoc mesenteroides, Lactobacillus delbrueckii, L. fermenti, L. lactis, Streptococcus faecalis* and *Pediococcus cerevisiae* (LEWIS and JOHAR, 1953; MUKHERJEE et al., 1965; RAJALAKSHMI and VANAJA, 1967; BATRA and MILLNER, 1974; RAMAKRISHNAN, 1976). Yeasts implicated in idli fermentation include *Oidium lactis* (*Geotrichum candidum*), *Torulopsis holmii, T. candida,* and *Trichosporon pullulans* (LEWIS and JOHAR, 1953; DESIKACHAR et al., 1960; BATRA and MILLNER, 1974). Soured buttermilk or yeast are sometimes added to the dough to reduce the fermentation time and, consequently, influence the microbial profile of the total fermentation process.

The lactic acid bacteria are obviously responsible for pH reduction in idli. They may also contribute significantly to improvement of the nutritional value of unfermented black gram and rice. LAKSHMI (1978) investigated the effects of *Leuconostoc mesenteroides, Lactobacillus fermenti, L. delbrueckii* and *Bacillus* species on changes in amino nitrogen, free sugar, thiamin, riboflavin and inorganic phosphate content as well as sensory qualities of traditional idli, rice-soybean idli (black gram replaced by soybeans), dhokla and khaman (Tab. 6). The increase in thiamin and riboflavin contents as a result of fermentation is particularly notable in light of inadequacies of these vitamins in the diets of some Indian children. Bacteria may also play a role in the breakdown of phytate present in black gram. RAO (1978) reported that a strain of *L. mesenteroides* isolated from soybean idli secretes β-*N*-acetylglucosaminidase and α-D-mannosidase which are involved in the hydrolysis of hemagglutinin.

There is some debate as to the effect of fermentation of idli on the PER. RAJALAKSHMI and VANAJA (1967) and RAO (1961) reported that the PER of idli was improved as a result of fermentation, whereas VAN VEEN et al. (1967) concluded that fermentation apparently did not improve PER or the digestibility of raw materials. Several researchers have observed an increase in methionine content resulting from fermentation (RAO, 1961; STEINKRAUS et al., 1967; PADHYE and SALUNKHE, 1978). A loss of 77.8% of the inorganic sulfur during black gram and rice blend fermentation was reported by REDDY and SALUNKHE (1980). Using *in vitro* digestion techniques, they also showed that there was an improve-

Tab. 6. Changes in Physical and Chemical Composition of Some Indian Fermented Foods[a]

Fermented Food	Increase in Vol. (%)	pH after Fermentation	Percentage of Unfermented Values					
			Amino Nitrogen	Free Sugar	Thiamin	Riboflavin	Niacin	Inorganic P
Rice-black gram idli	80	4.0	177	284	109	171	148	284
Rice-soybean idli	56	4.5	158	308	183	349	167	292
Dhokla	51	5.1	129	408	158	213	173	288
Khaman	55	5.4	219	676	136	300	190	548

[a] Experimental fermentations; from LAKSHMI (1978)

ment in the availability of essential amino acids during the fermentation.

Studies on the substitution of other *Phaseolus* species for black gram have shown that products with acceptability comparable to that of traditional idli can be prepared. RAMAKRISHNAN et al. (1976) made a rather extensive evaluation of soybean idli prepared using several types of bacterial inoculants. It was concluded that chemical, physical, and organoleptic characteristics of the soybean products are not unlike that of traditional idli. SATHE and SALUNKHE (1981) evaluated idli containing Great Northern beans (*Phaseolus vulgaris* L.) in place of black gram. Idli containing Great Northern beans had somewhat different flavor and a sticky top surface. The authors suggested that the incorporation of Great Northern beans into idli would offer the advantage of increased roughage.

3.2 Waries

Waries (Panjabi waries) are a spicy condiment shaped in the form of a ball about 3 to 8 cm in diameter which are used in cooking with vegetables, grain legumes or rice in India (BATRA and MILLNER, 1974). Dehulled grain legumes are soaked in water, ground into a coarse paste, and mixed with spices and a small amount of paste from a previous fermented batch. Some spices used include asafoetida, caraway, cardamon, cloves, fenugreek, ginger, red pepper and salt. After fermenting as a mass for 4 to 10 days at ambient temperature, the paste is formed into balls and air-dried in the sun. The surface of the balls becomes sealed with a mucilaginous coating during the drying process, thus entrapping gases produced by yeasts present inside (REDDY et al., 1986). Yeasts identified as contributing to fermentation of waries are *Candida* spp. and *Saccharomyces cerevisiae.* Waries can also be prepared from Bengal gram and mung bean flours.

3.3 Papadam

Papadam is similar to waries but does not contain fenugreek or ginger. These circular tortilla-like wafers are prepared from a mixture of black gram paste and spices which have been fermented for 4 to 6 h (BATRA and MILLNER, 1974). Yeasts responsible for fermentation are the same as those found in waries. Papadams are served roasted or deep-fat fried and consumed as a relish.

3.4 Dhokla

Dhokla is a steamed fermented food prepared from a mixture of wheat semolina and Bengal gram (2:1 ratio) flours (REDDY et al., 1986). The flours plus about 3% salt are made into a thick batter by adding water and then allowing to ferment for about 14 h. Chopped fenugreek leaves are added to the dough before steaming for 20 min in a pan with oil. The product is then cut into pieces and may be seasoned with cracked mustard seeds before eating as a condiment with other breakfast foods in India.

3.5 Khaman

Khaman is a steamed fermented condiment prepared from Bengal gram flour in India (REDDY et al., 1986). Dehulled seeds are washed, soaked for 4 h, ground with water (2:3, Bengal gram:water) and seasoned with salt before fermenting for 12 h. Like dhokla, the dough is then steamed and seasoned with mustard seeds before eating. RAJALAKSHMI and VANAJA (1967) reported that thiamin and riboflavin increased significantly, while RAMAKRISHNAN et al. (1976) did not find any variations due to fermentation.

3.6 Kenima

Kenima is a fermented soybean product resembling Indonesian tempeh but prepared in Nepal, Sikkim and the Darjeeling districts of India (REDDY et al., 1986). Soybeans are soaked overnight, dehulled and cooked in water for 2 to 3 h. The cooled soybeans are inoculated by the addition of a portion from a fermented batch, wrapped in leaves and fermented at 22 to 23 °C for 24 to 48 h. The microorganisms responsible for the mucilaginous end product known as kenima have not been identified. When deep-fat fried and salted, kenima has a nutlike flavor much like that of tempeh.

3.7 Jalebies

Jalebies (Fig. 10) are a spiral-shaped, deep-fat fried confectionery product made from fermented wheat flour (BATRA and MILLNER, 1974). Chickpea flour may be added to the fermented wheat flour before the paste is fried (CHAVAN and KADAM, 1989). The fresh fried jalebies are dipped in sugar syrup before serving. They are consumed throughout India, Nepal, and Pakistan.

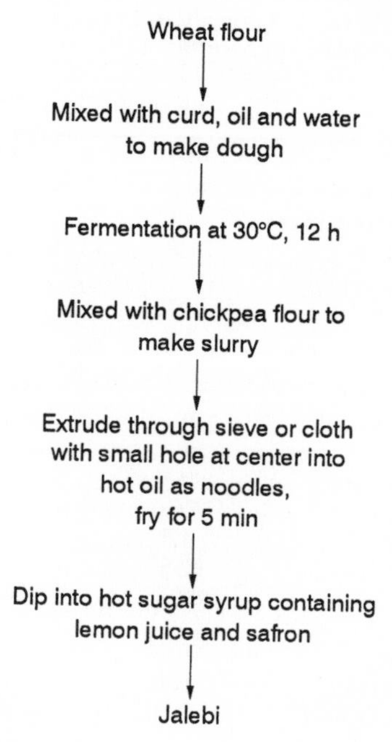

Fig. 10. Flow diagram for preparing jalebies.

3.8 Kurdi

Kurdi is a deep-fat fried, salty, crisp product prepared from fermented wheat (CHAVAN and KADAM, 1989) (Fig. 11). Wheat grains are soaked for about 5 days at 30 °C

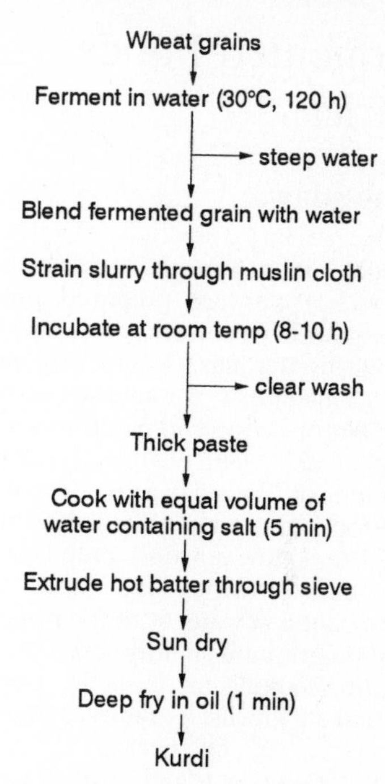

Fig. 11. Flow diagram for preparing kurdi.

before the slurry is fried to make the snack, which is consumed in Central India. Little is known about the fermentation process, although yeasts and lactic acid bacteria are undoubtedly involved.

3.9 Kanji

Kanji is a fermented beer-like beverage common in households and marketplaces in India (BATRA and MILLNER, 1974). North Indian kanji is prepared from purple or occasionally orange cultivars of carrots, beets, spices, a portion from a previous batch of kanji, and water. South Indian kanji is prepared in two steps, the first involving the preparation of torani (a fermented rice liquor) and the second involving the fermentation of torani and a mixture of vegetables, spices and water (HESSELTINE, 1965). *Hansenula anomala* var. *anomala* has been isolated from kanji collected in North India (Delhi) whereas *H. anomala, Candida guilliermondii, C. tropicalis* and *Geotrichum candidum* have been isolated from South Indian kanji.

4 Fermented Foods of Africa

4.1 Dawadawa

Dawadawa is a fermented locust bean (*Parkia filicoidea*) product prepared and consumed largely in West Africa. It is also known as kpalugu by the Kusasis and Dagombas of Northern Ghana, iru or daddowa in Nigeria, kinda in Sierra Leone and neteton in Gambia (REDDY et al., 1986; ODUNFA, 1988). The preparation of dawadawa is described by these authors and EKA (1980) essentially as follows. The yellow powdery pulp is removed from the dark brown to black seeds, and the seeds are boiled in water with the possible addition of potash until slightly soft. Gas boiling or pressure cooking increase the amount of water uptake by locust beans over that of firewood cooking, mainly because of sustained heat input (OYEWOLE and ODUNFA, 1990b). The beans are then stored overnight in earthenware, metallic pots or baskets to further soften the seed coat. On the following day, the black seed coats are removed by hand rubbing the seeds, rubbing them against the walls of the storage vessel, or by gently pounding them with a wooden pestle and mortar. Sand or wood ash may be used as abrasive agents to aid in the removal. The swollen cotyledons are then washed in water, boiled for 30 min and deposited at a depth of about 10 cm in a tray, pot or basket. Alternatively, the cotyledons may be put into a hole in the ground for fermentation. Also, the yellow portion of the first pulp and wood ashes may be sprinkled over the cotyledons at this time.

The preparation is covered with leaves or sheets of polyethylene and left to ferment for 2 to 3 days at ambient temperature. The microorganisms responsible for fermentation have not been determined, but undoubtedly include sporeforming bacilli, lactic acid bacteria and probably yeasts (ODUNFA, 1985). Metabolic activities of the microflora result in a mucilaginous, strongly proteolytic, ammoniacal smelling substance which covers and binds the individual beans (REDDY et al., 1986). During fermentation, the beans change in color from light to dark brown and become greatly softened. Moisture is partially removed from the fermented mass by sun-drying before pounding into flattened cakes. The cakes may be further dried to prolong shelf-life; darkening during sun-drying is due to polyphenol oxidation (CAMPBELL-PLATT, 1980).

The composition of unfermented locust beans and daddawa is shown in Tab. 7 (EKA, 1980). The amount of protein is increased by fermentation. While protein in daddawa remains low in tryptophan and cystine content, it is clear that fermentation generally improves the nutritional quality of locust beans.

Tab. 7. Composition of Unfermented Locus Beans and Daddawa, a Fermented Locus Bean Cake Product Prepared in Nigeria[a]

Component	Unfermented	Fermented
Moisture[b]	12.7	13.8
Crude protein[c]	30.6	38.5
Total carbohydrate[c]	49.1	23.6
Crude fiber[c]	7.8	6.2
Ether extractable material[c]	15.2	31.2
Ash[c]	5.1	6.8
Minerals[d]		
Potassium	250	550
Sodium	240	250
Zinc	15	18
Magnesium	80.5	83.3
Calcium	330	360
Copper	1.5	2.0
Iron	22.5	28.0
Phosphorus	280	320
Vitamins[d]		
Thiamin	0.65	1.35
Riboflavin	0.45	1.30
Ascorbic acid	7.50	5.20
Toxic substances[d]		
Oxalate	0.21	0.12
Hydrocyanic acid	0.0026	0.0012
Total phytic acid P	51.0	31.0
Phytic acid P as % of total P	15.0	7.5

[a] Excerpted from EKA (1980)
[b] g/100 g wet weight
[c] g/100 dry weight
[d] mg/100 g dry weight

Most notable are increased levels of thiamin and riboflavin. The composition of fermented locust beans may vary somewhat depending on the traditional method of preparation in diverse West African cultures. However, like soybeans, the digestibility of locust beans is probably vastly improved upon fermentation. There is a real need for research to determine the microbiological, toxicological, nutritional and technological characteristics of fermented locust bean products.

4.2 Gari

A stable food prepared by fermenting the root of the cassava plant (also known as manioc, mandioca, apiun, yuca, cassada or tapioca in various parts of the world) is known as gari in the rain forest belt of West Africa. OLAYIDE et al. (1972) estimated that about 70% of the cassava grown in Nigeria is used for gari manufacturing. As described by COLLARD and LEVI (1959), the traditional preparation of gari consists of the following stages as illustrated in Fig. 12. First, the corky outer peel and the thick cortex are removed, and the inner portion of the root is grated by hand on homemade raspers. The grated pulp is then packed into jute bags and weights are applied to express some of the juice. Fermentation takes place over a 3- to 4-day period at which time the cassava is sieved to remove coarse lumps and heated while constantly turning over a hot steel pan or in an oven. This process has been termed "garifying" (OGUNSUA, 1980). The moisture content of the fermented cassava is reduced to about 10 to 15% to yield a final product known as gari. In addition to water, gari contains 80 to 85% starch, 0.1% fat, 1 to 1.5% crude protein and 1.5 to 2.5% crude fiber. Palm oil may be added to the product just before or after drying to give it color. OKE (1966) identified trace amounts of sodium, manganese, iron, copper, boron, zinc, molybdenum and aluminum in gari.

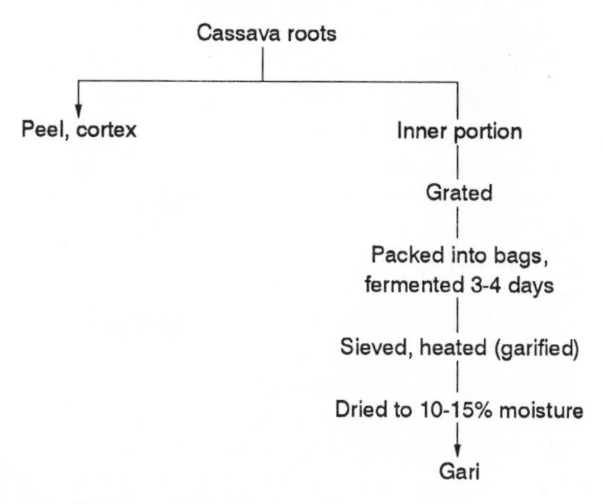

Fig. 12. Flow diagram for preparing gari.

While it has been recognized for some time that fresh cassava roots contain cyanogenic glucosides, it is also known that these glucosides decompose during traditional procedures for preparing gari with the liberation of gaseous hydrocyanic acid (COLLARD and LEVI, 1959). These researchers studied the microbiology of gari during its preparation and consistently isolated a bacterium (*Corynebacterium manihot*) and a fungus (*Geotrichum candidum*) from grated cassava during the holding stage. They suggested that the process of detoxification of cassava root should be regarded as a two-stage fermentation. In the first stage, *C. manihot* hydrolyzes the starch and produces various organic acids.

Lactobacillus plantarum and other lactic bacteria also contribute significantly to decreasing the pH (OYEWOLE and ODUNFA, 1990a), which causes spontaneous hydrolysis of cyanogenic glucosides with the liberation of gaseous hydrocyanic acid. The acid condition favors the growth of *G. candidum* which produces aldehydes and esters which gives gari its characteristic aroma and flavor. NGABA and LEE (1979) attributed flavor development in gari to lactic acid bacteria, particularly *Lactobacillus* species and, to a lesser degree, *Streptococcus* spp. SANNI (1989) studied the mycoflora of commercially processed gari and observed that several genera of molds were present.

Fermentation of gari also involves yeasts. EKUNSANMI and ODUNFA (1990) investigated the ethanol tolerance and invertase activities of yeasts isolated from steep water of fermented cassava tubers. Tolerance to as high as 10% ethanol and 25% glucose was observed. Yeasts were also observed to produce invertase.

AKINRELE (1964) attempted to elucidate the nature of biochemical reactions taking place during fermentation and to establish the optimum conditions which could be used in a modern industry based on this process. The fermentation was found to be self-sterilizing, exothermic and anaerobic, and to proceed in two stages at an optimum temperature of about 35 °C. Lactic and formic acids are produced with a trace of gallic acid. Frequent mixing of the mash and exposure to light appeared to accelerate fermentation. A satisfactory gari was produced within a 15-h period of fermentation, and a continuous system may possibly give a better product than a batch process.

While the traditional methods for preparing gari are known to result in a reduction in linamarin and lotaustralin, the glycosides present in fresh cassava roots, to yield hydrocyanic acid, the levels of this acid possibly present in final products were of some concern (KETIKU et al., 1978). Consequently, these researchers designed experiments to determine the effectiveness of traditional methods in eliminating or reducing to innocuous levels the hydrocyanic acid content of cassava roots processed into gari and lafun (flour). The acid was significantly reduced from an initial concentration of 90.1 mg/kg and 165.5 mg/kg of fresh grated pulp to 25.8 mg/kg and 19.6 mg/kg in gari and lafun, respectively. The detoxification process was most effective during the holding (fermentation) phase.

The effects of processing grated cassava roots by screw press and by traditional fermentation methods on the cyanide content of gari have been studied (MADUAGWU and OBEN, 1981). The relative amounts of hydrocyanic acid which disappear from gari prepared using the quick (1-day) screw press method and the slow (3-day) traditional one were comparable (92 to 100%). The screw press method resulted in gari which retained some of the bound cyanide, whereas the traditional method did not. The authors concluded that the hydrocyanic acid content of gari did not appear to be correlated with varietal (sweet versus bitter) differences.

In addition to changes in levels of cyanogenic glycosides during gari preparation, OGUNSUA (1980) measured concentrations of other constituents. As observed by others, the pH decreased to about 4 during 3 days of fermentation; the level of reducing sugars was doubled after 1 day, but decreased on subsequent days of fermentation and the titratable acidity (as % lactic acid) increased about threefold after 3 days. He concluded that if processed properly by fermenting for 4 days or longer, the final product is free of cyanide or any cyanide-yielding glycosides.

Recognizing the low protein content of cassava roots, researchers at the Tropical Products Institute in London mounted a substantial effort in the 1960s to upgrade the nutritional status of the West African staple via controlled fermentation with fungi. One of the more interesting reports evolving from these studies was made by HARRIS (1970), who demonstrated the appearance of γ-linolenic acid and changes in other fatty acids in cassava flour fermented for 4 days at 30 °C with *Rhizopus arrhizus*. γ-Linolenic acid is rarely found in higher plant tissues, so its production by *R. arrhizus* is noteworthy since, like linoleic acid, it is considered to be an essential fatty acid to animals. The total lipid content of cassava fermented with *R. arrhizus* was reduced to about half of the level present in raw material. Plant glycolipids disappeared and were replaced by mold phospholipids, while sterol glycosides present in the original cassava were not metabolized and became a major component of the final product.

AHONKHAI and KOLEOSO (1979) reported the presence of mannitol in gari. The percentage of mannitol present in gari ranged from 1.4 to 3.1%, depending upon whether the product was manufactured by traditional or mechanized processes, respectively.

4.3 Banku

A fermented starch-based Ghanaian food known as banku is prepared exclusively from maize or from a mixture of maize and cassava (OWUSU-ANSAH and VAN DE VOORT, 1980). The process involves steeping the raw material in water for 1 day, wet-milling and fermenting for 3 days. The dough is then mixed with water (5:2, maize dough/water or 4:1:2, maize dough/cassava dough/water). Considerable stirring and kneading of the fermented dough is required to attain the correct consistency during subsequent cooking. The product is then put into a mold and served. Lactic acid bacteria and perhaps yeasts are responsible for fermentation.

4.4 Ogi

Maize is eaten in West Africa principally in the form of porridge known as ogi (Nigeria) or kenkey (Ghana); the Bantu (South African) equivalent to ogi is called mahewu. To prepare ogi, kernels of maize are soaked in warm water for 1 to 3 days, after which they are wet-milled and sieved with water through a screen mesh to remove fiber, hulls and much of the germ (AKINRELE, 1970) (Fig. 13). The filtrate is fermented to yield a sour, white, starchy sediment known as ogi which is marketed as a wet cake wrapped in leaves. It may be diluted in water to 8 to 10% solids and boiled into a pap or cooked and turned into a stiff gel (eke) before eating. Ogi is an important traditional food for weaning babies and a major breakfast cereal for adults.

The fermentation of maize to prepare ogi proceeds naturally without the addition of inoculants or enzymes. AKINRELE (1970) investigated the microorganisms responsible for fermentation, including their contribution to nutritive value. Lactic acid bacteria were found to be mainly *Lactobacillus plantarum*; aerobic bacteria included *Corynebacterium* and *Aerobacter*; yeasts identified were *Candida mycoderma, Saccharomyces cerevisiae* and *Rhodotorula*; and molds consisted of *Cephalosporium, Fusarium, Aspergillus* and *Penicillium* species. *L. plantarum* was responsible for production of lactic acid, the main flavor base of ogi, whereas an increase in riboflavin and niacin content was attributed to *Aerobacter cloacae. S. cerevisiae* and *C. mycoderma* are also thought to contribute to flavor development in ogi. In contrast, *L. plantarum* is not a predominant lactic acid bacterium in mawè, a fermented maize dough made in Benin (HOANHOUIGAN et al., 1993).

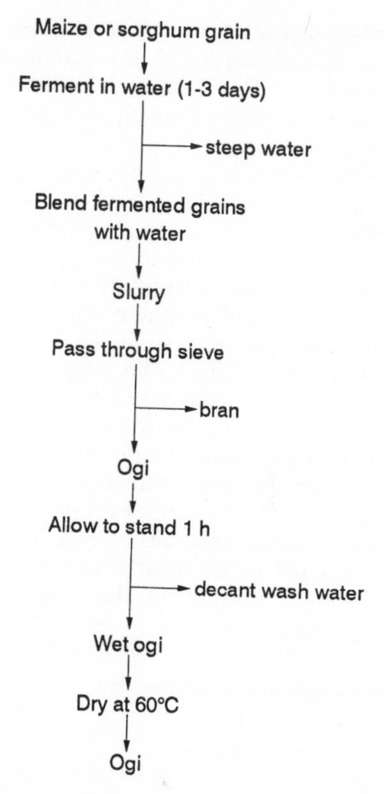

Fig. 13. Flow diagram for ogi. Adapted from CHAVAN and KADAM (1989).

BANIGO et al. (1974) evaluated high-lysine maize for its suitability as a substrate for producing ogi. It was concluded from studies using controlled inocula and fermentation conditions that this nutritionally improved maize variety could be substituted for traditional varieties without sacrificing final product sensory qualities. AKINRELE et al. (1969) reported that supplementation of maize with soybean meal not only enhances the rate of fermentation of ogi, but also may improve the utilization of cereal protein.

MUGULA (1992) evaluated the nutritive value of maize-soybean tempeh as a potential weaning food in Tanzania. Fermentation did not significantly change proximate composition, but dietary fiber was increased. Phytate content decreased by 67% and iron absorption increased 2.5-fold. Niacin, riboflavin and thiamin increased 2–2.5- and 0.5-fold, respectively. The digestibility coefficient, PER and net protein ratio improved to be comparable to skimmed milk. These researchers promoted maize-soybean tempeh as a potential weaning food.

4.5 Injera

Injera is an Amharic word for an Ethiopian bread made from teff sorghum, wheat, barley, corn or a mixture of these grains (STEWART and GETACHEW, 1962). To prepare injera (Fig. 14), teff flour and water are combined with irsho, a fermented yellow fluid saved from a previous batch. The resultant thin, watery paste is generally incubated for 1 to 3 days. A portion of the fermented paste is then mixed with three parts water and boiled to give a product called absit which is in turn mixed with a portion of the original fermented flour to yield a clean-looking, thin injera. Thick injera (aflegna), popular in rural Ethiopia, is teff paste which has undergone only minimal fermentation (12 to 24 h) and is characterized by a sweet flavor and a reddish color. A third type of injera (komtata) is made from over-fermented paste and, consequently, has a sour taste, probably due to proliferation of lactic acid bacteria. While the microflora responsible for fermentation of the sweeter types of injera have not been fully determined, yeasts probably play the greatest role. *Candida guilliermondii* apparently is a primary yeast in this process.

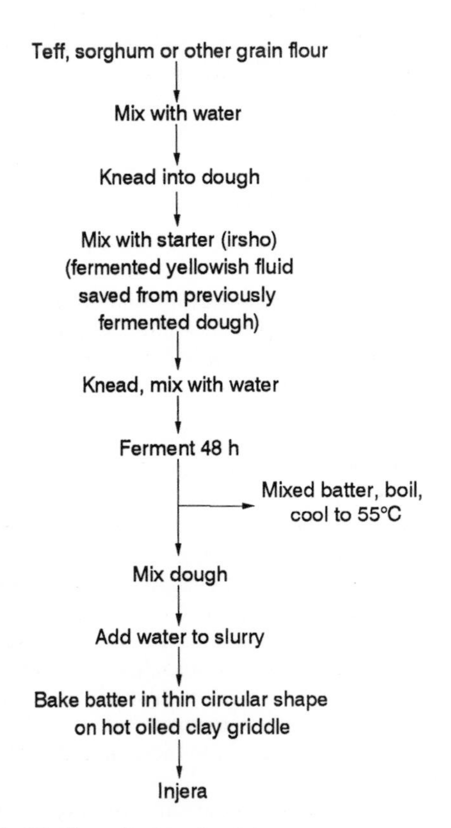

Fig. 14. Flow diagram for injera.

Regardless of the method used to prepare injera, the paste is baked or grilled to result in a bread-like product similar in appearance to pancakes in the United States. Injera is said to keep well in a mesob (cylindric container with a flat bottom and made of grass stems), although it is prepared daily in many Ethiopian households (STEWART and GETACHEW, 1962).

4.6 Kaffir Beer

Strictly speaking, kaffir beer is an African beverage made from kaffir corn, i.e., sorghum, a principal grain in some parts of the continent. However, similar beverages are prepared from other grains and starchy plant materials such as plantains and cassava (PLATT, 1964).

According to WILLIAMSON (1955), cereal grain is pounded using a wooden mortar and pestle, or by rubbing it between stones. Malt is prepared by soaking whole grain in water for 1 or 2 days after which it is sun-dried and allowed to mature before grinding. The pounded grain is made into a thin gruel, boiled, mixed with a small portion of uncooked malt and allowed to set overnight. On the second day, the mixture is boiled, cooled and allowed to stand until the fourth day when more pounded, boiled malted grain is added to the brew. On the fifth day the brew is strained through coarse baskets to remove some of the husks; the filtrate is now ready for consumption.

During the early stage of fermentation saccharification of starch occurs due to the growth of *Aspergillus flavus* and *Mucor rouxii* (HESSELTINE, 1965). The malt actually supplies little diastase to break down starch during the fermentation process.

4.7 Merissa

The process for preparing merissa beer has been described by DIRAR (1978). Sorghum grains are soaked in water for 1 day before spreading on the ground and covering with wet sacks or plant leaves. Germination proceeds for 2 days at which time the grains are sun-dried and milled into coarse flour. This represents the malting phase. Meanwhile, ungerminated sorghum is milled into flour and moistened with water. This dough undergoes fermentation for about 36 h and is then known as ajeen. The ajeen is cooked without further addition of water until it takes on a dark brown color. The intermediate moisture product (soorij) is extremely sour but has a pleasant caramelized flavor. An equal amount of water and about 5% of malt flour and 5% of good merissa are added to the soorij, and the mixture is allowed to ferment for 4 to 5 h. At this stage, the product is called deboba.

Meanwhile, an amount of sorghum flour twice the size used for preparing the ajeen is divided into two equal lots. One lot is cooked to yield a grayish-brown paste, and the other is cooked longer to produce a brown paste. The two are mixed and cooled to form a gelatinized material called futtara. Malt flour (5%) is mixed with the futtara which in turn is placed underneath the deboba surface as one solid mass. Fermentation (and liquefaction) proceeds for 8 to 10 h before the mass is strained to yield merissa (filtrate) and a residue (mushuk), which is used as animal feed. The principal microorganisms in merissa fermentation are lactic acid bacteria and a *Saccharomyces* species of yeast (DIRAR, 1978). Molds are not considered to play a role.

5 Other Fermented Products

5.1 Milk/Grain Products

Several indigenous fermented foods contain a mixture of animal milk and grain. PLATT (1964) gave an account of two such products, Iraqi kushik and Turkish tarhana.

5.1.1 Kushik

To prepare kushik, dried parboiled whole wheat meal and yoghurt (1:1) are combined and allowed to ferment for 1 week. Curd from an equivalent volume of milk is added, and the mass is fermented for an additional 4 to 5 days. The product is then sun-dried, ground into a powder and stored for later consumption. It is reconstituted to form a porridge to be eaten with pulses or bread, or cooked with chickpeas or green vegetables such as beet leaves.

5.1.2 Tarhana

Tarhana is also made from parboiled wheat meal and yoghurt, but in proportions of 2:1 (PLATT, 1964). Tomatoes, peppers, onions, garlic, salt and spices are added and, after fermentation for several days, the product is dried. It is used largely for soup making.

5.1.3 Kishk

Kishk is a fermented product prepared from parboiled wheat and milk, and consumed in much of the Arab world (MORCOS et al., 1973). In Egypt, more milk than can be consumed is available at certain times of the year, so the surplus is stored in earthenware containers. Wheat is slowly boiled until it is soft, washed with water and dried. It is then coarsely ground and sieved to remove the seed coats.

In the meantime, laban zeer is prepared by concentrating salted sour milk. The powdered parboiled wheat is moistened with slightly salted boiling water and mixed laban zeer to obtain a homogeneous paste called hamma. After 1 day of fermentation, the hamma is kneaded; twice the volume of laban zeer added before it is diluted with water or milk is added to the hamma and left for another day. Spices such as pepper or cumin may also be added with the laban zeer. At this point, the mass is thoroughly mixed, formed into small balls and dried. The product is known as kishk. Lactic acid bacteria responsible for fermentation include *Lactobacillus plantarum, L. casei* and *L. brevis*; yeasts also contribute to the process.

5.2 Kaanga-kopuwai

The process of fermenting maize in water before eating is carried out in parts of New Zealand where the final product is termed kaanga-kopuwai (maize soaked in water) in Northland and kaanga-pirau (rotten corn) and kaanga-wai (water corn) among the Maoris of Central North Island and the Bay of Plenty (YEN, 1959). The Maori process, as described by YEN, is briefly summarized: mature whole maize cobs are placed unhusked in a jute sack and submerged in water, usually in streams, stagnant ponds or run-offs from pastures. The time required for proper fermentation is about 3 months, but the corn is said to stay fit for eating for an indefinite period if left in the water. If the sack begins to disintegrate, bundles of ferns are sometimes tied around it for preservation.

After 12 weeks of fermentation, the husks are yellow and soft and the kernels are full, but very yellow and soft and often slimy on the surface. Preparation for eating consists of scraping the kernels off the cob, removing the pericarp (optional), mincing, pulverizing and boiling with water to form a sort of gruel which is often eaten hot with sugar and milk or cream. The kernels may also be fried with salt and animal fat and eaten in this fashion.

Since the kaanga-kopuwai has been described as "rotten" maize by Westerners, it is safe to assume that considerable proteolysis occurs as a result of microbial activity during the fermentation process. An account of the microflora responsible for fermentation, however, has not been given.

5.3 Poi

Taro (*Colocasia esculenta*) corms are the principal material used to prepare poi in Hawaii and islands in the South Pacific (ALLEN and ALLEN, 1933). Corms are first cooked, peeled and ground or pounded to a fine consistency. The addition of water at this point results in fresh poi. Based on consistency, poi is designated as "one finger", "two finger" or "three finger", depending upon the number of fingers required for a satisfying mouthful. The second phase of poi preparation involves fermentation at ambient temperature for 1 to 3 days or longer. As fermentation progresses, texture changes from a sticky mass to one having a more watery and fluffy consistency. Poi is marketed commercially in glass jars and plastic bags.

Lactic acid bacteria are the predominant microflora during early stages of fermentation. *Lactobacillus delbrueckii, L. pastorianus, L. pentacetius, Lactococcus lactis* and *Streptococcus kefir* produce large amounts of lactic acid and moderate amounts of acetic, propionic, succinic, and formic acids (ALLEN and ALLEN, 1933). *Candida vini* and *Geotrichum candidum* are prevalent in later stages of fermentation. These fungi are thought to be responsible for imparting the pleasant fruity aroma and flavor to older poi.

5.4 Chicha

The word chicha is mainly applied to fermented maize chicha in Peru; however, it may also be prepared from fruits, cassava or mesquite (NICHOLSON, 1960). Over the centuries chicha has played a predominant role in religious and primitive fertility rites. The derivation of the word chicha describes the principal way in which it was made in the past, i.e., using saliva to convert starches to sugars to facilitate fermentation and increase the alcohol content.

Methods for preparing chicha in various parts of Peru differ somewhat, but NICHOL-

SON (1960) described a procedure followed in the Pacific lowlands as illustrative of the process. Good quality Alaza maize is shelled and kernels are immersed in water in earthenware pots and placed in the ground. After 12 to 18 h, the swollen maize is removed from the pot and spread in layers 5 to 7 cm thick and kept wet over a period of about 3 days to promote germination. The optimum stage of germination is reached when the plumule is about 1 cm long and tastes sweet. The germinated maize is then piled into heaps and covered for a period of 2 days during which it is said to be "humeando" or smoking because of the increased temperature due to biochemical activity. At this point the parched maize is sun-dried for 2 to 5 days, and broken pieces of roots and seedling shoots, collectively known as jora, are separated from the maize kernels. Jora is milled into pachucho which is mixed with water and boiled for 3 to 4 h, gradually cooled, boiled again for about 4 h, cooled, strained through a suitable cloth or wire screen into a special pot reserved for chicha-making and allowed to ferment for at least 1 day. Presumably this pot would carry an inoculum of lactic acid bacteria and yeasts necessary for the desired fermentation. So long as the brew remains sweet, it is not ready for drinking, but as soon as it turns slightly acidic it is said to possess the correct flavor.

5.5 Pozol

Pozol is a fermented maize dough that, diluted in water, is consumed raw as a beverage by the Indian and Mestizo populations in Southeastern Mexico (ULLOA-SOSA, 1974). Their method for making pozol does not differ much from that used by Mayan ancestors. White maize kernels (1.5 kg) are boiled for about 1 h in 2 liters of water to which a handful of lime powder has been added. When the kernels are swollen and their pericarps are easily peeled, they are cooled, rinsed with water and drained to get what is called nixtamal. The nixtamal is ground to a coarse dough, shaped into rolls about 6 cm wide and 12 cm long, wrapped in banana leaves and allowed to ferment at ambient temperature for 5 to 8 days.

It is during the processing of nixtamal that inoculation of maize dough occurs, since vegetative cells and most spores are destroyed during the time kernels are boiled in lime water. Sanitary measures are minimal by those who prepare pozol, so the water and utensils used, hands and air all constitute sources of microorganisms. Despite an expected variation in the microflora of pozol due to the uncontrolled manner of preparation, there are several species of yeasts and molds which are commonly found in pozol prepared at different places and at different times (ULLOA-SOSA, 1974). These include *Geotrichum candidum, Trichosporon cutaneum* and various species of *Candida.* Molds such as *Cladosporium cladosporioides* or *C. herbarum, Monilia* (*Neurospora*) *sitophila* and *Mucor rouxianus* or *M. racemosus* are also common in pozol. Bacteria are predominant during early stages of fermentation. It is interesting that *Agrobacterium azotophilum*, a nitrogen-fixer, has been identified as contributing to an increased nitrogen content of pozol compared to unfermented maize (ULLOA and HERRERA, 1972).

5.6 Legume-Based Milk Products

Soybean milk is an aqueous extract of soybeans which is commercially marketed as a beverage in several Asian countries. One can also prepare peanut milk from peanuts. Both products are characterized by a distinct beany or green flavor which is judged undesirable by some individuals, but can be fermented with lactic acid bacteria to yield variously flavored and textured foods which have high sensory acceptability.

The use of bacteria to ferment soybean milk beverages has been studied most extensively. WANG et al. (1974) evaluated growth rates of eight *Lactobacillus acidophilus* strains and four *Lactobacillus bulgaricus* strains in soybean milk enriched with glucose, lactose and sucrose. Almost all of the cultures could adapt themselves to the growth media tested. Taste panel evaluation of a chilled soybean beverage made using *L. acidophilus* revealed that the refreshing sour drink with a raisin flavor was highly acceptable to panel members. The authors were not able to suggest the origin of the raisin flavor and odor, nor was it concluded that the beany flavor of soybeans was merely masked by other flavors derived from fermentation or modified somehow.

A comparison of procedures for extracting milk from soybeans which would then serve as substrates for lactic acid bacterial fermentations was made by MITAL and STEINKRAUS (1976). Their study suggests that an aqueous extract of dehulled, pulverized, defatted soybeans which was fortified with 2% sucrose and 2.5% fat before fermenting with *Lactococcus* species served as a good medium for preparing a product with flavor acceptability comparable to that of fermented cow's milk. Another report by the same authors (MITAL and STEINKRAUS, 1974) indicates that lactic acid bacteria possessing the ability to utilize sucrose can be successfully employed to manufacture products from soybean milks. Some of these organisms have also been shown to possess α-galactosidase, an enzyme necessary to split α-galactosidic bonds present in raffinose and stachyose (MITAL et al., 1973), thus being capable of reducing or eliminating the gas-producing carbohydrate fraction in legumes and legume milks. They found α-galactosidase to be constitutive in *Lactobacillus buchneri, L. brevis, L. cellobiosis, L. fermentum* and *L. salivarius* subsp. *salivarius* and present in the soluble fraction of the cell.

The use of lysine-excreting mutants of *L. acidophilus* and *L. bulgaricus* to increase the lysine content of fermented soybean milk was demonstrated by SANDS and HANKIN (1976). The lysine content of soybean milk was increased by as much as 270% in a yoghurt-like product prepared by fermenting soybean milk with the mutants.

Fermentation of soybean milk with *L. acidophilus* and subsequent flavoring, notably with lemon flavor, has been reported to yield an acceptable yoghurt-type product (KANDA et al., 1976). To achieve this success, the soybean milk was supplemented with 5% sucrose and 2% Cheddar cheese whey solids. A higher amount of whey solids caused the milk to coagulate during sterilization. To prevent whey separation, 0.5 to 1.5% gelatin was added to the yoghurt product; the optimum level of gelatin depended upon the level of protein in the product. The authors reported that soybean yoghurt can be kept at 5 °C for about 19 days without any significant change in acidity, pH and viable cell count.

In a series of experiments, PINTHONG et al. (1980a, b, c) evaluated methods of production as well as sensory and chemical properties of soybean-based yoghurts. The organoleptic quality of fermented products was directly related to levels of *n*-pentanal and *n*-hexanal, the former being produced by bacteria and the latter being naturally present in soybean milk. They suggested that although *n*-hexanal is important in terms of adverse flavor, other aldehydes, including *n*-pentanal, must be considered in fermented soybean products. Reduction in oligosaccharide content is dependent in part on the method of starter preparation.

A review of fermentation of soybean milk by lactic acid bacteria was made by MITAL and STEINKRAUS (1979). Concern was expressed about the lack of information available on the fate of oligosaccharides and the importance of proteolytic and lipolytic activities of lactic cultures in fermented products.

Fermented soybean and peanut milk products having yoghurt-like consistencies were prepared and evaluated by SCHMIDT and BATES (1976). Products were compared with fermented 25% toned milks containing cow's milk extended with commercial soybean protein concentrate or peanut flour. Although more acceptable yoghurt was manufactured from toned milks than from oilseed milks, homogenization of control and soybean toned milk significantly improved acceptance of yoghurt-like products.

BEUCHAT and NAIL (1978) examined several extraction procedures for their suitability to yield desirable peanut milks for fermentation by four lactic acid bacteria. Studies revealed that a procedure in which peanuts were soaked in 1.0% sodium bicarbonate for 16 to 18 h, drained, washed with tap water, ground, steeped for 4 to 5 h in tap water, and filtered resulted in milk most desirable for fermentation (Fig. 15). The addition of lactose (2%) to pasteurized peanut milk before fermenting with *Lactobacillus bulgaricus* and *L. acidophilus* for 3 days at 37 °C resulted in a

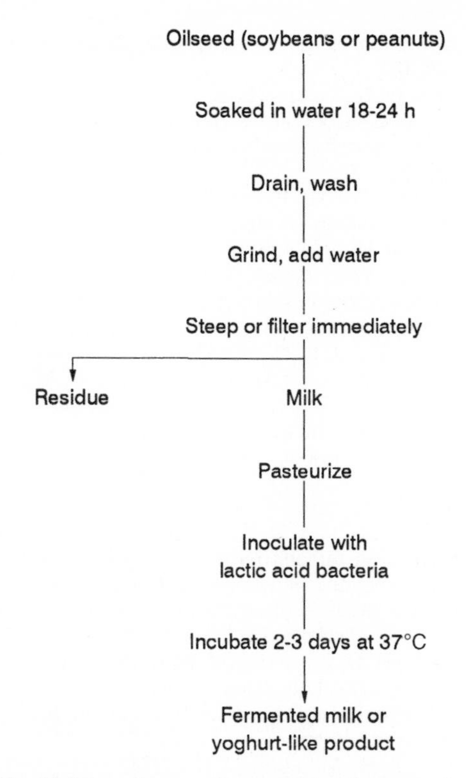

Fig. 15. Flow diagram for preparing fermented oilseed milk products.

custard-like product having 0.38 to 0.53% titratable acidity at pH 4.76 to 4.43, respectively. Sensory panel evaluations of blended, fermented peanut milks containing added sucrose (2%) and fruit flavorings showed that the products were acceptable and competed favorably with flavored buttermilk. Fermented peanut milk substituted for buttermilk in a corn muffin recipe resulted in products with organoleptic characteristics not significantly different from those of the control. The authors expressed the need for further investigations to determine the optimum conditions for preparing fermented peanut milk products.

6 Nutritional and Public Health Aspects

As noted in preceding discussions, many indigenous fermented foods are staples in the diets of vast populations of people who would otherwise have less than minimum intakes of protein and/or calories. Beyond this, the use of fermented foods as seasoning agents makes bland and unpalatable foods more interesting and attractive. While the quality or quantity of proteins in vegetable-based fermented foods generally are not increased substantially over raw substrates, the digestibility may be improved. Degradation of cellulose during fermentation is also thought to improve digestibility. And there is clear evidence that certain vitamins in plant seeds, e.g., vitamin B_{12}, are increased in concentration as a result of fermentation. Thus, the positive contribution of indigenous fermented foods to the nutritional well-being of those who consume them on a regular basis is fully recognized, although the impact perhaps has not been critically assessed.

An immediate and obvious point of interest to food microbiologists and sanitarians concerns the possibility of microorganisms producing toxic substances during fermentation or storage of finished products. In 1970, MATSUURA et al. reported the results of their investigation of aflatoxin-forming ability of 238 strains of *Aspergillus* used in the Japanese food industry. Fifty-two strains produced fluorescent compounds, but none produced aflatoxin. They also examined 46 domestic rice, 11 imported rice, 108 miso and 28 rice-koji samples and found none to contain aflatoxin. MURAKAMI et al. (1968) found another 214 strains of industrial koji mold to be non-aflatoxigenic. By-products of fungal growth such as aspergillic acid, koji acid, β-nitro-propionic acid, oxalic acid and formic acid have been demonstrated to be present in koji at levels not considered to represent a public health hazard (YOKOTSUKA, 1971). KINOSITA et al. (1968) isolated 37 strains of fungi from 24 samples of miso, katsuobushi and tane koji in Japan; many strains produced koji acid and β-nitropropionic acid, but none produced aflatoxin. Laboratory experiments have shown that some *Rhizopus* spp., including *R. oryzae,* may actually degrade aflatoxin to a less toxic form (COLE and KIRKSEY, 1972).

At least one report (SHANK et al., 1972) documents the presence of aflatoxin in soy sauce collected in Hong Kong. Whether the toxin was a contaminant of the raw ingredients or produced during fermentation is not known. EL-HAG and MORSE (1976) claimed that *Aspergillus oryzae* NRRL-1988 produced aflatoxin. However, there is overwhelming evidence to conclude that this strain does not in fact produce aflatoxin (FENNELL, 1976). We have worked with this same strain in our laboratory and did not detect any toxicity associated with fermented peanuts fed to rats (QUINN et al., 1975).

The per capita consumption of indigenous fermented foods seems destined to increase in the future, particularly in light of increasingly significant cost advantages in producing food proteins of plant origin versus animal origin. Populations in the Western world, too, are becoming more aware and receptive to fermented foods previously known only in other parts of the world. There is much to be learned about the biochemistry and microbiology of indigenous fermented foods, and it is hoped that the information provided here will assist those already involved in this research area as well as provide a starting point for those bold and adventuresome enough to begin.

7 References

AHONKHAIL, S. I., KOLEOSO, O. A. (1979), Isolation of L-mannitol from gari – a processed food from cassava (*Manihot esculenta* Crantz) tubers, *J. Sci. Food Agric.* **30**, 849–852.

AKINRELE, I. A. (1964), Fermentation of cassava, *J. Sci. Food Agric.* **15**, 589–594.

AKINRELE, I. A. (1970), Fermentation studies on maize during the preparation of a traditional African starch-cake food, *J. Sci. Food Agric.* **21**, 619–625.

AKINRELE, I. A., MAKANJU, A., EDWARDS, C. C. A. (1969), Effect of soya flour on the lactic fermentation of milled corn, *Appl. Microbiol.* **17**, 186–187.

ALLEN, O. N., ALLEN, E. K. (1933), The manufacture of poi from taro in Hawaii: With special emphasis upon its fermentation, *Hawaii Agric. Exp. Stn. Bull.* **70**.

ARDHANA, M. M., FLEET, G. H. (1989), The microbial ecology of tape ketan fermentation, *Int. J. Food Microbiol.* **9**, 157–165.

ASAO, Y., YOKOTSUKA, T. (1958), Studies on flavorous substances in soy sauce. Part XVI. Flavorous substances in raw soy sauce, *J. Agric. Chem. Soc. Jpn.* **32**, 617–623.

BAENS-ARCEGA, L., MARANON, J., PALO, M. A., ANDRADA, L., ANGLO, G., ARGULLES, L. (1969), New and quick process of manufacturing patis and bagoong, *Philippine Patent* 4113, Jan. 7.

BAI, R. G., PRABHA, T. N., RAO, T. N., RAO, T. N. R., SREEDHARA, V. P., SREEDHARA, N. (1975), Studies on tempeh fermented soybean food, I. Processing and nutritional evaluation of tempeh from a mixture of soybean and groundnut, *J. Food Sci. Technol.* **12**, 135–138.

BANIGO, E. O. I., DEMAN, J. M., DUITSCHAEVER, C. L. (1974), Utilization of high-lysine corn for the manufacture of ogi using a new, improved processing system, *Cereal Chem.* **51**, 559–572.

BATRA, L. R., MILLNER, P. D. (1974), Some Asian fermented foods and beverages and associated fungi, *Mycologia* **66**, 942–950.

BAU, Y.-S., WONG, H.-C. (1979), Zinc effects on growth, pigmentation and antibacterial activity of *Monascus purpureus* fungi, *Physiol. Plant.* **46**, 63–67.

BEUCHAT, L. R. (1976), Fungal fermentation of peanut presscake, *Econ. Bot.* **30**, 227–234.

BEUCHAT, L. R. (1978), Microbial alterations of grains, legumes and oilseeds, *Food Technol.* **32**(5), 193–198.

BEUCHAT, L. R. (1982), Flavor chemistry of fermented peanuts, *Ind. Eng. Chem. Res. Dev.* **21**, 533–536.

BEUCHAT, L. R. (1983), Indigenous fermented foods, in: *Biotechnology*, 1st Ed. (REHM, H.-J., REED, G., Eds.), Vol. 5, pp. 477–528, Weinheim–Deerfield Beach, FL–Basel: Verlag Chemie.

BEUCHAT, L. R. (1984), Fermented soybean foods, *Food Technol.* **36**(6), 64–70.

BEUCHAT, L. R. (1987), Traditional fermented food products, in: *Food and Beverage Mycology* (BEUCHAT, L. R., Ed.), 2nd Ed., pp. 269–306, New York: Van Nostrand Reinhold.

BEUCHAT, L. R., BASHA, S. M. M. (1976), Protease production by the ontjom fungus, *Neurospora sitophila, Eur. J. Appl. Microbiol.* **2**, 195–203.

BEUCHAT, L. R., NAIL, B. J. (1978), Fermentation of peanut milk with *Lactobacillus bulgaricus* and *L. acidophilus, J. Food Sci.* **43**, 1109–1112.

BEUCHAT, L. R., WORTHINGTON, R. E. (1974), Changes in lipid content of fermented peanuts, *J. Agric. Food Chem.* **22**, 509–512.

BEUCHAT, L. R., YOUNG, C. T., CHERRY, J. P. (1975), Electrophoretic patterns and free amino acid composition of peanut meal fermented with fungi, *Can. Inst. Food Sci. Technol. J.* **8**, 40–45.

BOEDIJN, K. B. (1958), Note on the Mucorales in Indonesia, *Sydowia Ann. Mycol.* **12**, 321–362.

CALLOWAY, D. H., HICKEY, C. A., MURPHY, E. L. (1971), Reduction of intestinal gas-forming properties of legumes by traditional and experimental food processing methods, *J. Food Sci.* **36**, 251–255.

CAMPBELL-PLATT, G. (1980), African locus bean (*Parkia* species) and its West African fermented food product, dawadawa, *Ecol. Food Nutr.* **9**, 123–132.

CAMPBELL-PLATT, G. (1987), *Fermented Foods of the World: A Dictionary and Guide*, London: Butterworth.

CAMPBELL-PLATT, G., COOK, P. E. (1989), Fungi in the production of foods and ingredients, *J. Bacteriol. Symp. Suppl. Ser.* **18**, 67, 117S–131S.

CHAVAN, J. K., KADAM, S. S. (1989), Nutritional improvement of cereals by fermentation, *Crit. Rev. Food Sci. Nutr.* **28**, 349–400.

CHEN, L. H., PACKETT, L. V., YUN, I. S. (1972), Tissue antioxidant effect of ocean hake fish and fermented soybean (tempeh) as protein sources in rats, *J. Nutr.* **102**, 181–186.

CHOI, S. Y. (1991), *Science and Technology of Kimchi,* 2nd Ed., Seoul: Korea Food Research Institute.

COLE, R. J., KIRKSEY, J. W. (1972), Conversion of aflatoxin B_1 to isomeric hydroxy compounds by *Rhizopus* spp., *J. Agric. Food Chem.* **20**, 1100–1106.

COLLARD, P., LEVI, S. (1959), A two-stage fermentation of cassava, *Nature* **183**, 620–621.

CRONK, T. C., STEINKRAUS, K. H., HACKLER, L. R., MATTICK, L. R. (1977), Indonesian tapé ketan fermentation, *Appl. Environ. Microbiol.* **33**, 1067–1073.

DESIKACHAR, H. S. R., RADHAKRISHNAMURTHY, R., RAMARAO, G., KADKOL, S. B., SRINIVASAN, M., SUBRAHMANYAN, V. (1960), Studies on idli fermentation. I. Some accompanying changes in the batter, *J. Sci. Ind. Res.* **19**, 168–172.

DIRAR, H. A. (1978), A microbiological study of Sudanese merissa brewing, *J. Food Sci.* **43**, 1683–1686.

DJIEN, K. S. (1972), Tapé fermentation, *Appl. Microbiol.* **23**, 976.

DWIDJOSEPUTRO, D., WOLF, F. T. (1970), Microbiological studies of Indonesian fermented foodstuffs, *Mycopathol. Mycol. Appl.* **41**, 211–222.

EBINE, H. (1971), Miso, in: *Conversion and Manufacture of Foodstuffs by Microorganisms* (KAWABATA, T., FUJIMAKI, M., MITSUDA, H., Eds.), p. 127, Tokyo: Saikon Publishing Co.

EKA, O. U. (1980), Effect of fermentation on the nutrient status of locust beans, *Food Chem.* **5**, 303–308.

EKUNSANMI, T. J., ODUNFA, S. A. (1990), Ethanol tolerance, sugar tolerance and invertase activities of some yeast strains isolated from steep water of fermenting cassava tubers, *J. Appl. Bacteriol.* **69**, 672–675.

EL-HAG, N., MORSE, R. E. (1976), Aflatoxin production by a variant of *Aspergillus oryzae* (NRRL strain 1988) on cowpeas (*Vigna sinensis*), *Science* **192**, 1345–1346.

FARDIAZ, D., MARKAKIS, P. (1981), Degradation of phytic acid in oncom (fermented peanut press cake), *J. Food Sci.* **46**, 523–525.

FENNELL, D. I. (1976), *Aspergillus oryzae* (NRRL strain 1988): A clarification, *Science* **194**, 1188.

FLEGEL, T. W., BHUMIRATANA, A., SRISUTIPRUTI, A. (1981), Problematic occurrence of tyrosine crystals in the Thai soybean paste tao chieo, *Appl. Environ. Microbiol.* **41**, 746–751.

FUKUSHIMA, D. (1979), Fermented vegetable (soybean) protein and related foods of Japan and China, *J. Am. Oil Chem. Soc.* **56**, 357–362.

GANDJAR, I. (1978), Fermentation of winged bean seeds (tempeh kecipir), *Workshop/Seminar on Development of the Potential of the Winged Bean, Los Banõs, Philippines*, University of the Philippines, p. 130.

GOEL, S. K., WOOD, B. J. B. (1978), Cellulose and exo-amylase in experimental soy sauce fermentations, *J. Food Technol.* **13**, 243–248.

GRAIKOSKI, J. T. (1973), Microbiology of cured and fermented fish, in: *Microbial Safety of Fishery Products* (CHICHESTER, C. O., Ed.), pp. 97–112, New York: Academic Press.

GYÖRGY, P., MURATA, K., IKEHATA, H. (1964), Antioxidants isolated from fermented soybeans, *Nature* **203**, 870.

GYÖRGY, P., MURATA, K., SUGIMOTO, Y. (1975), Studies on antioxidant activity of tempeh oil, *J. Am. Oil Chem. Soc.* **51**, 377–379.

HACKLER, L. R., STEINKRAUS, K. H., VAN BUREN, J. P., HAND, D. B. (1964), Studies on the utilization of tempeh protein by weaning rats, *J. Nutr.* **82**, 452–456.

HARADA, Y. (1951), Process for soy sauce fermentation, *Rep. Tatsuno Inst. Soy Sauce* **2**, 51.

HARRIS, R. V. (1970), Effect of *Rhizopus* fermentation on the lipid composition of cassava flour, *J. Sci. Food Agric.* **21**, 626–627.

HESSELTINE, C. W. (1965), A millennium of fungi, food, and fermentation, *Mycologia* **57**, 149–197.

HESSELTINE, C. W. (1979), Some important fermented foods of Mid-Asia, the Middle East, and Africa, *J. Am. Oil Chem. Soc.* **56**, 367–374.

HESSELTINE, C. W. (1983), Microbiology of oriental fermented foods, *Annu. Rev. Microbiol.* **37**, 575–601.

HESSELTINE, C. W., WANG, H. L. (1967), Traditional fermented foods, *Biotechnol. Bioeng.* **9**, 275–288.

HESSELTINE, C. W., WANG, H. L. (1980), The importance of traditional fermented foods, *BioScience* **30**, 402–404.

HESSELTINE, C. W., WANG, H. L. (1986), *Indigenous Fermented Food of Non-Western Origin, Mycologia Memoir No. 11*, Berlin: Cramer.

HESSELTINE, C. W., SMITH, M., BRADLE, B., DJIEN, K. S. (1963), Investigation of tempeh, an Indonesian food, *Dev. Ind. Microbiol.* **4**, 275–287.

HESSELTINE, C. W., SMITH, M., WANG, H. L. (1967), New fermented cereal products, *Dev. Ind. Microbiol.* **8**, 179–186.

HOUNHOUIGAN, D. J., NOUT, M. J. R., NAGO, C. M., HOUBEN, J. H., ROMBOUTS, F. M. (1993), Characterization and frequency distribution of species of lactic acid bacteria involved in processing of mawè, a fermented maize dough from Benin, *Int. J. Food Microbiol.* **18**, 279–287.

JHA, H. C. (1985), Novel isoflavonoides and its derivatives, new antioxidants derived from fermented soybeans (tempe). *Proc. Asian Symp. Non-salted Soybean Fermentation, Tsukuba, Japan*, July, 1985, pp. 199–202, Tsukuba Science City: National Food Research Institute.

IKEHATA, H., WAKAIZUMI, M., MURATA, K. (1968), Antioxidant and antihemolytic activity of a new isoflavone, "factor 2" isolated from tempeh, *Agric. Biol. Chem.* **32**, 740–746.

ILANY-FEIGENBAUM, J., DIAMANT, J., LAXER, S. H., PINSKY, A. (1969), Japanese miso-type products prepared by using defatted soybean flakes and various carbohydrate-containing foods, *Food Technol.* **23**, 554–556.

ILJAS, N. (1969), Preservation and Shelf-Life Studies of Tempeh, *M.S. Thesis*, Ohio State University, Columbus.

ILJAS, N., PENG, A. C., GOULD, W. A. (1973), Tempeh – An Indonesian fermented soybean food, *Hort. Ser.* **394**, Ohio Agricultural Research and Development Center, Wooster, OH, 36 pp.

IMPOOLSUP, A., BHUMIRATANA, A., FLEGEL, T. W. (1981), Isolation of alkaline and neutral proteases from *Aspergillus flavus* var. *columnaris*, a soy sauce koji mold, *Appl. Environ. Microbiol.* **42**, 619–628.

KANDA, H., WANG, H. L., HESSELTINE, C. W., WARNER, K. (1976), Yoghurt production by *Lactobacillus* fermentation of soybean milk, *Process Biochem.* **11**(4), 23–25, 46.

KAO, C., ROBINSON, R. J. (1978), Nutritional aspects of fermented foods from chickpea, horsebean, and soybeans, *Cereal Chem.* **55**, 512–517.

KATO, K., KUSWANTO, K., BANNO, L., HARADA, T. (1976), Identification of *Endomycopsis fibuligera* isolated from ragi in Indonesia and properties of its crystalline glucoamylase, *J. Ferment. Technol.* **54**, 831–837.

KETIKU, A. O., AKINYELE, I. O., KESHINRO, O. O., AKINNAWO, O. O. (1978), Changes in the hydrocyanic acid concentration during traditional processing of cassava into gari and lafun, *Food Chem.* **3**, 221–226.

KINOSITA, R., ISHIKO, T., SUGIYAMA, S., SETO, T., IGARASI, S., GOETZ, I. E. (1968), Mycotoxins in fermented food, *Cancer Res.* **28**, 2296–2311.

KIUCHI, K., OHTA, T., ITOH, H., TAKABAYASHI, T., EBINE, H. (1976), Studies on lipids of natto, *J. Agric. Food Chem.* **24**, 404–407.

LAKSHMI, I. (1978), Studies on Fermented Foods, *M.S. Project*, University of Baroda, India.

LEWIS, Y. S., JOHAR, D. S. (1953), Microorganisms in fermenting grain mashes used for food preparations, *Cent. Food Technol. Res. Inst.* **2**, 228.

LIN, C.-F. (1973), Isolation and cultural conditions of *Monascus* sp. for the production of pigment in a submerged culture, *J. Ferment. Technol.* **51**, 407–414.

LIN, C.-F., SUEN, S. J.-T. (1973), Isolation of hyperpigment-productive mutants of *Monascus* sp. F-2, *J. Ferment. Technol.* **51**, 757–759.

MABESA, R. C., CARPIO, E. V., MABESA, L. B. (1992), An accelerated process for fish sauce (patis) production, in: *Applications of Biotechnology to Traditional Fermented Foods* (RUSKIN, F. R., Ed.), pp. 146–149, Washington, DC: National Academy Press.

MADUAGWU, E. N., OBEN, D. H. E. (1981), Effects of processing grated cassava roots by the screw press and traditional fermentation methods on cyanide content of gari, *J. Food Technol.* **16**, 299–302.

MARTINELLI, A. F., HESSELTINE, C. W. (1964), Tempeh fermentation: packaging and tray fermentations, *Food Technol.* **18**, 761.

MATSUURA, S., MANABE, M., SATO, T. (1970), Surveillance for aflatoxins of rice and fermented-rice products in Japan, in: *Proc. First U.S.-Japan Conference on Toxic Microorganisms, Mycotoxins, Botulism* (HERZBERG, M.,

Ed.), pp. 48–55, Washington, DC: U.S. Government Printing Office.

MITAL, B. K., STEINKRAUS, K. H. (1974), Growth of lactic acid bacteria in soy milks, *J. Food Sci.* **39**, 1018–1022.

MITAL, B. K., STEINKRAUS, K. H. (1976), Flavor acceptability of unfermented and lactic fermented soy milks, *J. Milk Food Technol.* **39**, 342–344.

MITAL, B. K., STEINKRAUS, K. H. (1979), Fermentation of soy milk by lactic acid bacteria: A review, *J. Food Sci.* **42**, 895–899.

MITAL, B. K., SHALLENBERGER, R. S., STEINKRAUS, K. H. (1973), α-Galactosidase activity of lactobacilli, *Appl. Microbiol.* **26**, 783–788.

MORCOS, S. R., HEGAZI, S. M., EL-DAMHOUGHY, S. T. (1973), Fermented foods of common use in Egypt. II. The chemical composition of bouza and its ingredients, *J. Sci. Food Agric.* **24**, 1153–1161.

MORIMOTO, S., MATSUTANI, N. (1969), Studies on the flavor components of soy sauce; isolation of furfuryl alcohol and the formation of furfuryl alcohol by yeasts and molds, *J. Ferment. Technol.* **47**, 518–525.

MUGULA, J. K. (1992), Evaluation of the nutritive value of maize-soybean tempe as a potential weaning food in Tanzania, *Int. J. Food Sci. Nutr.* **43**, 113–119.

MUKHERJEE, S. K., ALBURY, M. N., PEDERSON, C. S., VAN VEEN, A. G., STEINKRAUS, K. H. (1965), Role of *Leuconostoc mesenteroides* in leavening the batter of idli, a fermented food of India, *Appl. Microbiol.* **13**, 227–231.

MULYOWIDARSO, R. K., FLEET, G. H., BUCKLE, K. A. (1989), The microbial ecology of soybean soaking for tempe production, *Int. J. Food Microbiol.* **8**, 35–46.

MURAKAMI, H. S., TAKASE, S., ISHII, T. (1968), Production of fluorescent substances in rice koji, and their identification by absorption spectrum, *J. Gen. Appl. Microbiol.* **14**, 97–110.

MURATA, K. (1988), Antioxidative stability of tempeh, *J. Am. Oil Chem. Soc.* **65**, 799–800.

MURATA, K., IKEHATA, H., MIYAMOTO, T. (1967), Studies on the nutritional value of tempeh, *J. Food Sci.* **32**, 580–584.

MURATA, K., IKEHATA, H., EDANI, Y., KOYANAGI, K. (1971), Studies on the nutritional value of tempeh. II. Rat feeding test with tempeh, unfermented soybeans, and tempeh supplemented with amino acids, *J. Agric. Biol. Chem.* **35**, 233–241.

NAKADAI, T., NASUNO, S., IGUCHI, N. (1972a), The action of peptidases from *Aspergillus sojae* on soybean proteins, *Agric. Biol. Chem.* **36**, 1239–1243.

NAKADAI, T., NASUNO, S., IGUCHI, N. (1972b), Purification and properties of acid carboxypeptidase I from *Aspergillus oryzae, Agric. Biol. Chem.* **36**, 1343–1352.

NAKADAI, T., NASUNO, S., IGUCHI, N. (1972c), Purification and properties of acid carboxypeptidase II from *Aspergillus oryzae, Agric. Biol. Chem.* **36**, 1473–1480.

NAKADAI, T., NASUNO, S., IGUCHI, N. (1972d), Purification and properties of acid carboxypeptidase III from *Aspergillus oryzae, Agric. Biol. Chem.* **36**, 1481–1488.

NAKADAI, T., NASUNO, S., IGUCHI, N. (1973a), Purification and properties of leucine aminopeptidase I from *Aspergillus oryzae, Agric. Biol. Chem.* **37**, 757–765.

NAKADAI, T., NASUNO, S., IGUCHI, N. (1973b), Purification and properties of leucine aminopeptidase II from *Aspergillus oryzae, Agric. Biol. Chem.* **37**, 767–774.

NAKADAI, T., NASUNO, S., IGUCHI, N. (1973c), Purification and properties of leucine aminopeptidase III from *Aspergillus oryzae, Agric. Biol. Chem.* **37**, 775–782.

NAKADAI, T., NASUNO, S., IGUCHI, N. (1973d), Purification and properties of alkaline proteinase from *Aspergillus oryzae, Agric. Biol. Chem.* **37**, 2685–2694.

NAKADAI, T., NASUNO, S., IGUCHI, N. (1973e), Purification and properties of neutral proteinase from *Aspergillus oryzae, Agric. Biol. Chem.* **37**, 2695–2701.

NAKADAI, T., NASUNO, S., IGUCHI, N. (1973f), Purification and properties of neutral proteinase II from *Aspergillus oryzae, Agric. Biol. Chem.* **37**, 2703–2708.

NGABA, P. R., LEE, J. S. (1979), Fermentation of cassava (*Manihot esculenta* Crantz), *J. Food Sci.* **44**, 1570–1571.

NICHOLSON, G. E. (1960), Chicha maize types and chicha manufacture in Peru, *Econ. Bot.* **14**, 290–299.

NODA, F., HAYASHI, K., MIZUNUMA, T. (1982), Influence of pH on inhibitory activity of acetic acid on osmophilic yeasts used in brine fermentation of soy sauce, *Appl. Environ. Microbiol.* **43**, 245–246.

NOUT, M. J. R., ROMBOUTS, F. M. (1990), Recent developments in tempe research, *J. Appl. Bacteriol.* **69**, 609–633.

NOUT, M. J. R., BEERINK, G., BONANTS-VAN LAARHOVEN, T. M. G. (1987a), Growth of *Bacillus cereus* in soybean tempeh, *Int. J. Food Microbiol.* **4**, 293–301.

NOUT, M. J. R., DE DREU, M. A., ZUURBIER, A. M., BONANTS-VAN LAARHOVEN, T. M. G.

(1987b), Ecology of controlled soyabean acidification for tempeh manufacture, *Food Microbiol.* **4**, 165–172.

NUNOMURA, N., SASAKI, M., ASAO, Y., YOKOTSUKA, T. (1976a), Identification of volatile components in shoyu (soy sauce) by gas chromatography, *Agric. Biol. Chem.* **40**, 485–491.

NUNOMURA, N., SASAKI, M., ASAO, Y., YOKOTSUKA, T. (1976b), Isolation and identification of 4-hydroxy-2(or 5)ethyl-5(or 2)methyl-3(2H)-furanone as a flavor component in shoyu (soy sauce), *Agric. Biol. Chem.* **40**, 491–496.

NUNOMURA, N., SASAKI, M., YOKOTSUKA, T. (1979), Isolation of 4-hydroxy-5-methyl-3(2H)furanone, a flavor component in shoyu (soy sauce), *Agric. Biol. Chem.* **43**, 1361–1367.

NUNOMURA, N., SASAKI, M., YOKOTSUKA, T. (1980), Shoyu (soy sauce) flavor components: acidic fractions and the characteristic flavor component, *Agric. Biol. Chem.* **44**, 339–345.

ODUNFA, S. A. (1985), African fermented foods, in: *Microbiology of Fermented Foods* (WOOD, B. J. B., Ed.), pp. 155–191, London: Elsevier.

ODUNFA, S. A. (1988), Review: African fermented foods. From art to science, *Mircen J.* **4**, 259–273.

OGAWA, G., FUJITA, A. (1980), Recent progress in soy sauce production in Japan, in: *Recent Progress in Cereal Chemistry* (INGLETT, G. E., MUNCK, L., Eds.), p. 381, New York: Academic Press.

OGUNSUA, A. O. (1980), Changes in some chemical constituents during the fermentation of cassava tubers (*Manihot esculenta*, Crantz), *Food Chem.* **5**, 249–255.

OKADA, Y., YOKOO, Y., TAKEUCHI, T. (1975), Studies on the reduction of salt concentration in fermented foods. Part II. Trial making of non- and low-salted soybean miso, *J. Jpn. Food Ind.* **22**, 379–386.

OKE, O. L. (1966), Chemical studies on some Nigerian foodstuffs–gari, *Nature* **212**, 1055–1056.

OLAYIDE, S. O., OLATUNBOSUN, D., IDUSOGIE, E. O., ABIAGOM, J. D. (1972), A quantitative analysis of food requirements, supplies and demands in Nigeria 1968–1985, Lagos, Nigeria: Federal Department of Agriculture.

OWUSU-ANSAH, J., VAN DE VOORT, F. R. (1980), Banku: Its degree of gelatinization and development of a quick cooking product, *Can. Inst. Food Sci. Technol. J.* **13**, 131–134.

OYEWOLE, O. B., ODUNFA, S. A. (1990a), Characterization and distribution of lactic acid bacteria in cassava fermentation during fufu production, *J. Appl. Bacteriol.* **68**, 145–152.

OYEWOLE, O. B., ODUNFA, S. A. (1990b), Effect of cooking method on water absorption and ease of dehulling in preparation of African locust beans for Iru, I. *J. Food Sci. Technol.* **25**, 461–463.

PACKETT, L. V., CHEN, L. H., LIU, J. Y. (1971), Antioxidant potential of tempeh as compared to tocopherol, *J. Food Sci.* **36**, 798–799.

PADHYE, V. W., SALUNKHE, D. K. (1978), Biochemical studies on black gram (*Phaseolus mungo* L.). III. Fermentation of the black gram and rice blend and its influence on the *in vitro* digestibility of proteins, *J. Food Biochem.* **2**, 327–347.

PALO, M. A., VIDAL-ADEVA, L., MACEDA, L. M. (1960), A study on ang-kak and its production, *Philippine J. Sci.* **89**, 1–22.

PINTHONG, R., MACRAE, R., ROTHWELL, J. (1980a), The development of soya-based yogurt. I. Acid production by lactic acid bacteria, *J. Food Technol.* **15**, 647–652.

PINTHONG, R., MACRAE, R., ROTHWELL, J. (1980b), The development of soya-based yogurt. II. Sensory evaluation and analysis of volatiles, *J. Food Technol.* **15**, 653–659.

PINTHONG, R., MACRAE, R., DICK, J. (1980c), The development of a soya-based yogurt. III. Analysis of oligosaccharides, *J. Food Technol.* **15**, 661–667.

PLATT, B. S. (1964), Biological ennoblement: improvement of the nutritive value of foods and dietary regimes by biological agents, *Food Technol.* **18**, 68–76.

QUINN, M. R., BEUCHAT, L. R. (1975), Functional property changes resulting from fungal fermentation of peanut flour, *J. Food Sci.* **40**, 475–478.

QUINN, M. R., BEUCHAT, L. R., MILLER, J., YOUNG, C. T., WORTHINGTON, R. E. (1975), Fungal fermentation of peanut flour: Effects on chemical composition and nutritive value, *J. Food Sci.* **40**, 470–474.

RAJALAKSHMI, R., VANAJA, K. (1967), Chemical and biological evaluation of the effects of fermentation on the nutritive value of foods prepared from rice and grams, *Br. J. Nutr.* **21**, 467–473.

RAMAKRISHNAN, C. V. (1976), Preschool child malnutrition. Pattern, prevalence and prevention, *Baroda J. Nutr.* **3**, 1–39.

RAMAKRISHNAN, C. V., PAREKH, L. J., AKROLKAR, P. N., RAO, G. S., BHANDARE, S. D. (1976), Studies on soya idli fermentation, *Plant Foods for Man* **2**, 15–33.

RAO, G. S. (1978), Studies on Fermented Foods with Reference to Hemagglutin in Hydrolyzing Bacteria Isolated from Rice-Soy Idli Batter, *Ph.D. Thesis,* University of Baroda, India.

RAO, M. V. R. (1961), Some observations on fermented foods, in: *Meeting the Protein Needs of Infants and Children,* Publ. No. **843**, pp. 291–293.

Washington, DC: National Academy of Sciences, National Research Council.

RAO, N. N., RAO, T. N. R. (1972), Nutritional evaluation of predigested protein-rich foods (miso-like products) and the effect on their supplementation to poor rice diets. III. *J. Food Sci. Technol. (India)* **9**, 127–128.

REDDY, N. R., SALUNKHE, D. K. (1980), Effects of fermentation on phytate phosphorus and mineral content in black gram, rice, and black gram and rice blends, *J. Food Sci.* **45**, 1708.

REDDY, N. R., PIERSON, M. D., SALUNKHE, D. K. (1986), *Legume-based Fermented Foods*, Boca Raton: CRC Press, Inc.

RO, S. L., WOODBURN, M., SANDINE, W. E. (1979), Vitamin B_{12} and ascorbic acid in kimchi inoculated with *Propionibacterium freudenreichii* ssp. *shermanii, J. Food Sci.* **44**, 873–877.

ROBINSON, R. J., KAO, C. (1977), Tempeh and miso from chickpea, horsebean, and soybean, *Cereal Chem.* **54**, 1192–1197.

ROELOFSEN, P. A., TALENS, A. (1964), Changes in some B vitamins during molding of soybeans by *Rhizopus oryzae* in the production of tempeh, kedelee, *J. Food Sci.* **29**, 224–229.

RUSKIN, F. R. (1992), *Applications of Biotechnology to Traditional Fermented Foods*, Washington, DC: National Academy Press.

SAMSON, R. A., VAN KOOIJ, J. A., DEBOER, E. (1987), Microbiological quality of commercial tempeh in the Netherlands, *J. Food Prot.* **50**, 92–94.

SANCHEZ, P. C. (1975), Varietal influence on the quality of Philippine rice cake (puto), *Philipp. Agric.* **58**, 376–382.

SANCHEZ, P. C. (1977), Shortened fermentation process for the Philippine rice cake (puto), *Philipp. Agric.* **61**, 134–140.

SANDS, D. C., HANKIN, L. (1976), Fortification of foods by fermentation with lysine-excreting mutants of lactobacilli, *J. Agric. Food Chem.* **24**, 1104–1106.

SANNI, A. I. (1993), The need for process optimization of African fermented foods and beverages, *Int. J. Food Microbiol.* **18**, 85–95.

SANNI, M. O. (1989), The mycoflora of gari, *J. Appl. Bacteriol.* **67**, 239–242.

SANO, T. (1961), Feeding studies with fermented soy products, in: *Meeting the Protein Needs of Infants and Children,* Publ. **843**, p. 274, Washington, DC: National Academy of Sciences, National Research Council.

SAONO, S., GANDJAR, T., BASUKI, T., KARSONO, H. (1974), Mycoflora of "ragi" and some other traditional fermented foods in Indonesia, *Annales Borgorienses* **5**, 187–204.

SATHE, S. K., SALUNKHE, D. K. (1981), Fermentation of the great northern bean (*Phaseolus vulgaris* L.) and rice blends, *J. Food Sci.* **46**, 1374–1378, 1393.

SCHMIDT, R. H., BATES, R. P. (1976), Sensory acceptability of fruit flavored oilseed milk formulations, *Proc. Florida State Hortic. Soc.* **89**, 217–220.

SHALLENBERGER, R. S., HAND, D. B., STEINKRAUS, K. H. (1966), Changes in sucrose, raffinose, and stachyose during tempeh formation, *Report Eighth Dry Bean Research Conference*, 11–13 August, 1966, ARS-74-71, pp. 68–74.

SHANK, R. C., WOGAN, G. N., GIBSON, J. B., NODASUTA, A. (1972), Dietary aflatoxins in human liver cancer. I. Aflatoxins in market foods and foodstuffs of Thailand and Hong Kong. *Food Cosmet. Toxicol.* **10**, 61–69.

SHIBASAKI, K., HESSELTINE, C. W. (1962), Miso fermentation, *Econ. Bot.* **16**, 180–195.

SHIEH, Y.-S. C., BEUCHAT, L. R. (1982), Microbial changes in fermented peanut and soybean paste containing kojis prepared from *Aspergillus oryzae* and *Rhizopus oligosporus, J. Food Sci.* **47**, 518–522.

SHIEH, Y.-S. C., BEUCHAT, L. R., WORTHINGTON, R. E., PHILLIPS, R. D. (1982), Physical and chemical changes in fermented peanut and soybean pastes containing kojis prepared using *Aspergillus oryzae* and *Rhizopus oligosporus, J. Food Sci.* **47**, 523–529.

SMITH, A. K. (1961), *Oriental Methods of Using Soybeans as Food*, Washington, DC: U.S. Department of Agriculture, ARS Circ. 71-17.

SMITH, A. K., RACKIS, J. J., HESSELTINE, C. W., SMITH, M., ROBBINS, D. J., BOOTH, A. N. (1964), Tempeh: Nutritive value in relation to processing, *Cereal Chem.* **41**, 173–181.

SORENSON, W. G., HESSELTINE, C. W. (1966), Carbon and nitrogen utilization by *Rhizopus oligosporus, Mycologia* **63**, 681–689.

STAHEL, G. (1946), Foods from fermented soybeans as prepared in The Netherlands, Indies. II. Tempe, a tropical staple, *J. New York Bot. Gard.* **47**, 285–296.

STANTON, W. R. (1971), Microbiologically produced foods in the tropics, in: *Proc. Int. Symp. Conversion and Manufacture of Foodstuffs by Microorganisms, Kyoto,* pp. 133–139, Tokyo: Saikon Publishing Co. Ltd.

STANTON, W. R., WALLBRIDGE, A. (1969), Fermented food processes, *Process Biochem.* **4**(4), 45–51.

STEINKRAUS, K. H. (1978), Tempeh – An Asian example of appropriate/intermediate food technology, *Food Technol.* **32**(4), 79–80.

STEINKRAUS, K. H. (1983), *Handbook of Indige-*

nous Fermented Foods, New York: Marcel Dekker.

STEINKRAUS, K. H., YAP, B. H., VAN BUREN, J. P., PROVIDENTI, M. I., HAND, D. B. (1960), Studies on tempeh – An Indonesian fermented soybean food, *Food Res.* **25**, 777–788.

STEINKRAUS, K. H., LEE, C. Y., BUCK, P. A. (1965a), Soybean fermentation by the ontjom mold *Neurospora, Food Technol.* **19**, 1301–1302.

STEINKRAUS, K. H., VAN BUREN, J. P., HACKLER, L. R., HAND, D. B. (1965b), A pilot plant process for the production of dehydrated tempeh, *Food Technol.* **19**, 63–68.

STEINKRAUS, K. H., VAN VEEN, A. G., THIEBEAU, D. B. (1967), Studies on idli – An Indian fermented black gram-rice food, *Food Technol.* **21**, 110–113.

STEWART, R. B., GETACHEW, A. (1962), Investigations of the nature of injera, *Econ. Bot.* **16**, 127–130.

STILLINGS, R. B., HACKLER, L. R. (1965), Amino acid studies on the effect of fermentation time and heat-processing of tempeh, *J. Food Sci.* **30**, 1043–1048.

SUBBA RAO, G. N. (1967), Fish processing in the Indo-Pacific area, Bangkok, *Indo-Pacific Fisheries Council Regional Studies* **4**, 75–76, 81.

SUDARMADJI, S., MARKAKIS, P. (1977), The phytate and phytase of soybean tempeh, *J. Sci. Food Agric.* **28**, 381–383.

SUDARMADJI, S., MARKAKIS, P. (1978), Lipid and other changes occurring during the fermentation and frying of tempeh, *Food Chem.* **3**, 165–170.

SUSHEELAMMA, N. S., RAO, M. V. L. (1979), Functional role of the arabinogalactan of black gram (*Phaseolus mungo*) in the texture of leavened foods (steamed puddings), *J. Food Sci.* **44**, 1309–1312, 1316.

TAKEUCHI, T., HOSOKAWA, N., YOSHIDA, M. (1972), Studies on the characteristics of proteolysate in soybean miso manufactured with enzyme preparation, *J. Ferment. Technol.* **50**, 21–27.

ULLOA, M., HERRERA, T. (1972), Descripcion de dos especies nuevas de bacterias aislalas del pozol: *Agrobacterium azotophilum* y *Achromobacter pozolis, Rev. Lat.-Am. Microbiol.* **14**, 15–24.

ULLOA-SOSA, M. (1974), Mycofloral succession in pozol from Tabasco, Mexico, *Biol. Soc. Mex. Microbiol.* **8**, 17–48.

VAN BUREN, J. P., HACKLER, L. R., STEINKRAUS, K. H. (1972), Solubilization of soybean tempeh constituents during fermentation, *Cereal Chem.* **49**, 208–211.

VAN VEEN, A. G. (1953), Fish preservation in Southeast Asia, *Adv. Food Res.* **4**, 209–231.

VAN VEEN, A. G., SCHAEFER, G. (1950), The influence of the tempeh fungus on the soya bean, *Doc. Neerl. Indones. Morbis Trop.* **2**, 270–281.

VAN VEEN, A. G., STEINKRAUS, K. H. (1970), Nutritive value and wholesomeness of fermented foods, *J. Agric. Food Chem.* **18**, 576–578.

VAN VEEN, A. G., HACKLER, L. R., STEINKRAUS, K. H., MUKHERJEE, S. K. (1967), Nutritive quality of idli, a fermented food of India, *J. Food Sci.* **32**, 339–341.

VAN VEEN, A. G., GRAHAM, D. C. W., STEINKRAUS, K. H. (1968), Fermented peanut press cake, *Cereal Sci. Today* **13**, 96–98.

WAGENKNECHT, A. C., MATTICK, L. R., LEWIN, L. M., HAND, D. B., STEINKRAUS, K. H. (1961), Changes in soybean lipids during tempeh fermentation, *J. Food Res.* **26**, 373–376.

WAI, N. S. (1968), Investigation of the various processes used in preparing Chinese cheese by the fermentation of soybean curd with *Mucor* and other fungi. *Technical Report, USDA, Public Law* **480** Proj. UR-A6-(40)-1, p. 89.

WANG, H. L. (1967a), Products from soybeans, *Food Technol.* **21**(5), 115–116.

WANG, H. L. (1967b), Release of proteinase from mycelium of *Mucor hiemalis, J. Bacteriol.* **93**, 1794–1799.

WANG, H. L., HESSELTINE, C. W. (1965), Studies on the extracellular proteolytic enzymes of *Rhizopus oligosporus, Can. J. Microbiol.* **11**, 727–732.

WANG, H. L., HESSELTINE, C. W. (1966), Wheat tempeh, *Cereal Chem.* **43**, 563–570.

WANG, H. L., HESSELTINE, C. W. (1970), Sufu and lao-chao, *J. Agric. Food Chem.* **18**, 572–575.

WANG, H. L., HESSELTINE, C. W. (1982), Oriental fermented foods, in: *Prescott and Dunn's Industrial Microbiology* (REED, G., Ed.), 4th Ed., p. 492, Westport: AVI Publishing Co.

WANG, H. L., HESSELTINE, C. W. (1986), Glossary of indigenous fermented foods, in: *Indigenous Fermented Food of Non-Western Origin* (HESSELTINE, C. W., WANG, H. L., Eds.), pp. 317–344, Berlin: Cramer.

WANG, H. L., RUTTLE, D. I., HESSELTINE, C. W. (1968), Protein quality of wheat and soybeans after *Rhizopus oligosporus* fermentation, *J. Nutr.* **96**, 109–114.

WANG, H. L., RUTTLE, D. I., HESSELTINE, C. W. (1969), Antibacterial compound from a soybean product fermented by *Rhizopus oligosporus* (33930), *Proc. Soc. Exp. Biol. Med.* **131,** 579–583.

WANG, H. L., KRAIDEJ, L., HESSELTINE, C. W. (1974), Lactic acid fermentation of soybean milk, *J. Milk Food Technol.* **37**, 71–73.

WENT, F. A. F. C., GEERLINGS, H. C. P. (1895), Beobachtungen über die Hefearten und zuckerbildenden Pilze der Arrakfabrikation, *Verh. Koninklij. Akad. Wetenschappen* **II.** 4, 3–31.

Williams, R. R., Combs, G. F., McGarrity, W. J., Kertesz, Z. I. (1959), A nutritional survey of the armed forces of the Republic of Korea, *J. Nutr., Suppl.* **I** 68, 1–80.

Williamson, J. (1955), Govt. Printer, Zomba, Nyasaland, p. 133.

Winarno, F. G. (1979), Fermented vegetable protein and related foods of Southeast Asia with special reference to Indonesia, *J. Am. Oil Chem. Soc.* **56**, 363–366.

Winarno, F. G., Reddy, N. R. (1986), Tempe, in: *Legume-based Fermented Foods* (Reddy, N. R., Pierson, M. D., Salunkhe, D. K., Eds.), pp. 95–117, Boca Raton: CRC Press.

Wong, H.-C. (1982), Antibiotic and Pigment Production by *Monascus purpureus, Ph.D. Dissertation,* University of Georgia, Athens.

Wong, H., Bau, Y.-S. (1977), Pigmentation and antibacterial activity of fast-neutron and X-ray induced strains of *Monascus purpureus* Went. *Plant Physiol.* **60**, 578–581.

Wong, H.-C., Koehler, P. E. (1981), Production and isolation of an antibiotic from *Monascus purpureus* and its relationship to pigment production, *J. Food Sci.* **46**, 589–594.

Wood, B. J. B. (1965), *Microbiology of Fermented Foods*, 2 Vols., London: Elsevier.

Worthington, R. E., Beuchat, L. R. (1974), α-Galactosidase activity of fungi on intestinal gas-forming peanut oligosaccharides, *J. Agric. Food Chem.* **22**, 1063–1066.

Yen, D. E. (1959), The use of maize by the New Zealand Maoris, *Econ. Bot.* **13**, 319–327.

Yokotsuka, T. (1960), Aroma and flavor of Japanese soy sauce, *Adv. Food Res.* **10**, 75.

Yokotsuka, T. (1971), Shoyu, in: *Conversion and Manufacture of Foodstuffs by Microorganisms,* pp. 117–125, Tokyo: Saikon Publishing Co.

Yokotsuka, T., Sakasai, T., Asao, Y. (1967), Studies on flavorous substances in shoyu. Part 35. Flavorous compounds produced by yeast fermentation, *J. Agric. Chem. Soc. Jpn.* **41**, 428–433.

Yong, F. M., Wood, B. J. B. (1974), Microbiology and biochemistry of the soy sauce fermentation, *Adv. Appl. Microbiol.* **17**, 157–194.

Yong, F. M., Wood, B. J. B. (1976), Microbial succession in experimental soy sauce fermentation, *J. Food Technol.* **11**, 525–536.

Yong, F. M., Wood, B. J. B. (1977), Biochemical changes in experimental soy sauce koji, *J. Food Technol.* **12**, 163–175.

Yong, F. M., Lee, K. H., Wong, H. A. (1981), The production of ethyl acetate by soy yeast (*Saccharomyces rouxii* Y-1096), *J. Food Technol.* **16**, 177–185.

Yoshida, H., Kajimoto, G. (1972), Changes in lipid components during miso making. Studies on the lipids of fermented foodstuffs (Part I), *J. Jpn. Soc. Food Nutr.* **25**, 415–421.

Zamora, R. G., Veum, T. L. (1979), The nutritive value of dehulled soybeans fermented with *Aspergillus oryzae* or *Rhizopus oligosporus* as evaluated by rats, *J. Nutr.* **109**, 1333–1339.

14 Cocoa Fermentation

ALEX S. LOPEZ

Itabuna, Bahia 45600-000, Brazil
and University Park, PA 16802, USA

PAUL S. DIMICK

University Park, PA 16802, USA

1 Introduction

Cocoa beans of commerce, the starting material for cocoa and chocolate manufacture, are the seeds of the species *Theobroma cacao*. Prior to being sold on the market or exported to chocolate processors, the seeds (beans) from ripe fruits (pods) are subjected sequentially to a fermentation and drying process often referred to as "curing". This process is carried out exclusively on the farms, estates, or cooperatives in the producing countries immediately following the harvesting of ripe fruit. It was thought at one time that the fermentation process was only an aid in removing the mucilaginous pulp surrounding the seed to facilitate drying and storage prior to marketing. This, in fact, is only a secondary purpose, and the main reason for fermentation is to induce biochemical transformations within the seeds that lead to formation of precursors of chocolate aroma, flavor and color. Only material pre-treated in this way possesses the intrinsic qualities essential and desirable for the manufacture of chocolate. On the other hand, unfermented seeds do not develop chocolate flavor when processed into chocolate; the product is excessively bitter and astringent and lacks the characteristic flavor and color. After curing, however, the seeds, now called beans, lose most of the bitterness and astringency, and upon roasting, develop the rich color, aroma and flavor typical of chocolate. Cocoa is therefore classed into that category of foods whose characteristic flavor is developed through a fermentation process.

Cocoa is not consumed for its nutritional value. Rather, it owes its great demand and worldwide appeal exclusively to its unique flavor and aroma. The quality and strength of this flavor are governed by the genetic constitution of the seed. This genetic potential is developed by the fermentation process and further ameliorated by manipulation of processing parameters during the manufacturing process. The unprocessed material, therefore, sets an upper limit to what can be achieved, and while it is impossible to improve genetically inferior material by superior processing techniques, it would be quite easy on the other hand, to ruin good quality cocoa by careless or inadequate curing.

2 The Cocoa Fruit

The fruit or "pod" of *T. cacao* varies among varieties in size, shape, external color and appearance; characters which often have been used to classify cacao. However, as far as the flavor quality is concerned, the only really important morphological distinctions are those which differentiate the Criollos of Central and South America from the Forasteros of the Amazon. The former type, which was the source of the original "fine" cocoa is, however, rapidly disappearing because of its susceptibility to disease and lower productivity. It is being replaced by the hardier, more prolific Forastero and their crosses. The following discussion refers primarily to Forastero cacaos which account for over 95% of the world production.

Flowers emerge from cushions on the bark of the trunk and stems. Bud burst to flowering takes about 30 days, and the flowers abscise if not fertilized 48 h after opening. The fruits mature 170 days after pollination at which time they also attain their maximum weight. During the season, fruits in all stages of development are present on the tree at the same time. Pod ripeness is judged by the external color of the pod which changes from green or dark red-purple to yellow, orange or red, depending on the variety. During the final days of ripening important changes occur within the fruit which have a bearing on quality. There is a build-up of sugars and an increase in acidity of pulp manifested in the pH drop; cotyledon tannins and carbohydrates also increase and the free fatty acids of the lipid fraction of the seed are converted to triacylglycerols. Changes in the enzymes and proteins have also been observed and stress the importance to harvest only ripe pods.

The mature fruit pods are thick-walled and contain between 30–40 seeds, each enveloped in a sweet, white, mucilaginous pulp. The cotyledons of the Forastero have various shades of purple in contrast to Criollo seeds which are white or pale pink in color. An exception to this rule is *Theobroma leiocarpa*, the Catongo cacao of Brazil which is a Forastero mutant with a white cotyledon. The Forastero/Criollo hybrids (Trinitarios), have cotyledons whose colors vary between white and purple. An excellent detailed account of the anatomy of the seed has been given by ROELOFSEN (1958), but for the purpose of describing the fermentation process it is only necessary to consider the basic morphology of the seed. The seed comprises two main parts: (1) the testa (seed coat) together with the adhering mucilaginous pulp, which is essential to fermentation but constitutes a waste in chocolate and cocoa manufacture, and (2) the embryo and cotyledons contained within; the former is removed during the winnowing process, and the latter provide the "nib" used in the manufacture of chocolate.

The mucilaginous pulp is made up principally of large, finger-like, spongy parenchyma cells containing the juices whose composition is given in Fig. 1 (LOPEZ, 1986). It forms the substrate for the microorganisms during fermentation. The pulp is separated from the seed cotyledons by a fibrous testa. The two cotyledons enclosing the embryo or radicle are made up principally of two types of parenchyma storage cells; the protein-lipid cells and the polyphenolic cells. In the former, lipid vacuoles are abundant and closely packed, and the plasm forms a grid in between them. Suspended in this plasm, in the spaces between the fat vacuoles, are organelles such as nuclei, mitochondria, starch grains and pro-

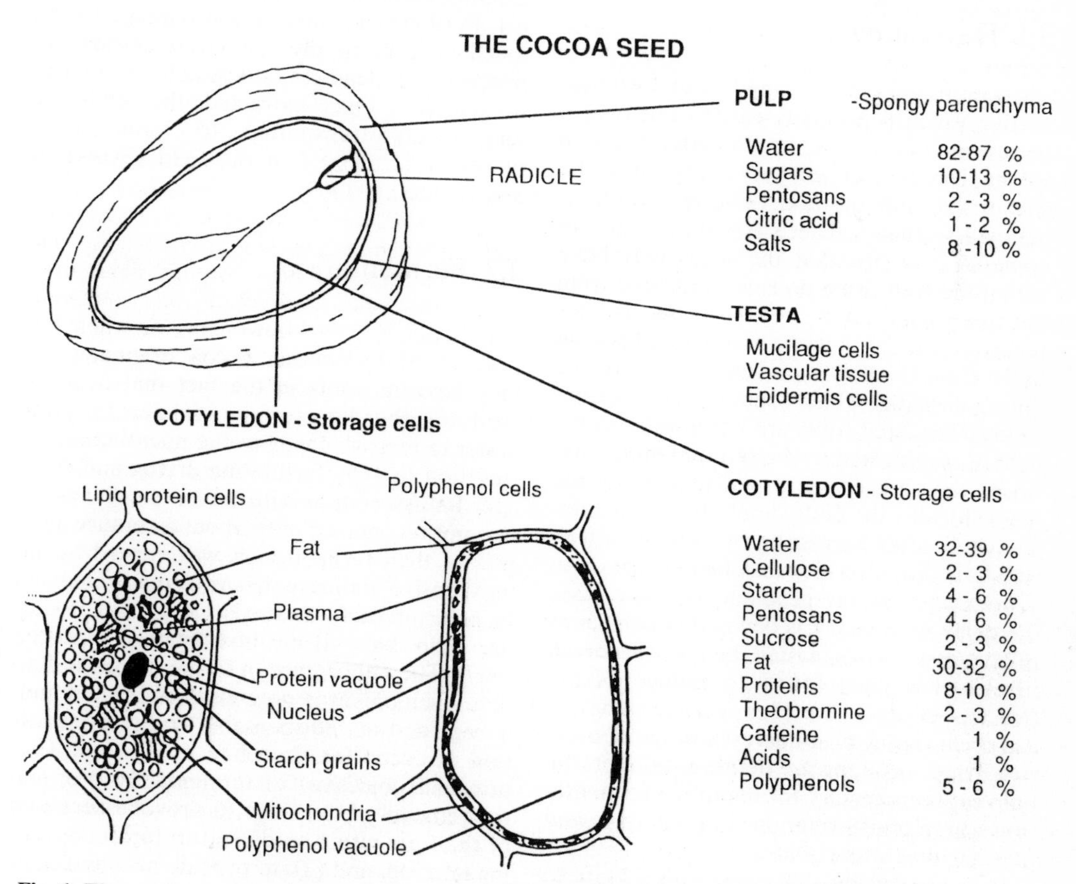

Fig. 1. The components and composition of the cocoa seed.

tein vacuoles (LOPEZ et al., 1987). Polyphenolic storage cells make up about 10% of the seed. They contain a single large storage vacuole which occupies most of the cellular space and contains polyphenols and alkaloids. The polyphenols in the Forastero varieties include anthocyanins which are responsible for the characteristic, deep purple color of the cotyledons. In the intact seed, enzymes and their substrates are separated by biological barriers which, during fermentation, break down allowing enzymes and substrates to freely mix to produce the flavor and aroma precursors of chocolate (ROELOFSEN, 1958).

3 Cocoa Processing

3.1 Harvesting

Cocoa primary processing begins with harvesting which is normally carried out over a period of 3 to 4 days, at intervals of 3 to 6 weeks. This practice will vary according to the size of the farm and pod yield. Harvesting is planned so that pods are not allowed to become so over-ripe that the seeds germinate within the fruit. Ripe pods are removed from the tree using knives, and those out of arms reach are harvested with special long-handled tools. Care is taken to avoid damage to the flower cushions which would reduce future yields. Harvested fruits are left in piles in the field or transported to the fermentation plant where they are broken open to remove the seeds. Ideally, the pods should be opened immediately after harvesting but this is not always practical, and they may be held for up to a week without any problem. In some cases the delay in breaking is deemed beneficial as in the case of Malaysian beans where the practice has been found to reduce acidity (MEYER et al., 1989). However, germination can occur during post-harvest storage of overripe fruits resulting in deterioration of the pulp sugars necessary for microbial fermentation and in transformations in seed lipids and proteins that affect quality.

Pods are usually split open with a knife to remove the seeds, although in many countries the use of a blunt instrument, such as a billet, is recommended to avoid cutting the seeds. The pod is broken across the middle, leaving the seeds adhering to the central placenta attached to the proximal end of the pod. Seeds are removed by hand and piled onto banana leaves, plastic sheets or gunny sacks to minimize contamination with earth and consequential discoloration of the beans. The placentas and diseased seeds should be rejected at this stage (ROHAN, 1963). The process is laborious, and mechanical pod breakers designed for use on large estates are employed on some farms in Mexico, West Africa and Brazil. However, they have not gained wide acceptance because they do not efficiently extract nor cleanly separate the seeds from the shell fragments produced by the machine. These must be hand-picked, and hence the machines have not yet proved to be economical. Furthermore, mechanical damage to the mucilage cells in the seed testa during the process often leads to undesirable "slimy fermentations". Once extracted, the seeds are immediately transported to fermentation plants or fermented in the field (FORSYTH and QUESNEL, 1963).

3.2 Fermentation

It is not exactly known how or when the practice of fermenting cocoa originated. It may have its roots in the fact that heaping seeds together for a few days renders the pulp easier to remove owing to the microbial deterioration thereby facilitating drying and storage. Its association with flavor development was not recognized until about a century ago. Before then fermentation was viewed as an inexpensive and easy means of pulp removal. In fact, alternate methods for getting rid of the pulp have been investigated (KNAPP, 1937). The importance of fermentation to flavor development is now well established and is regarded as indispensable in the preparation of cocoa for chocolate production. The principal objectives of fermentation are: (1) the removal of mucilage to provoke aeration of the fermenting seeds and to facilitate drying later on, and (2), to provide heat and acetic acid necessary for killing (preventing germination) and curing the seeds. Forastero varieties require a fermentation period of 5 to 8 days for the development of flavor, whereas Criollo cocoas requires just 2 to 3 days (FORSYTH and QUESNEL, 1963).

Although generally used, the term "fermentation" is misleading, when applied to the curing of cocoa, because even though typical alcoholic, lactic and acetic fermentations occur in the external pulp enveloping the seed, the changes that take place within the tissue, which actually result in the formation of flavor precursors, are biochemical, enzymatic reactions induced by the physico-chemical products of the pulp fermentation. Attempts at artificially providing these conditions for curing result in a product with chocolate flavor but which does not compare favorably with naturally cured beans.

3.3 Methods of Fermentation

It is recommended that the curing process commence immediately following pod breaking. This is the norm but in exceptional cases, for example, to reduce acidity, seeds may be spread out for several hours before being placed in the fermentation boxes (BIEHL and MEYER, 1990); another method, for the same reason, recommends partial removal of the seed mucilage prior to fermentation (LOPEZ, 1986). The actual methods of fermentation vary considerably from country to country, and in many instances even adjacent farms differ in their processing practices. The establishment of governmental extension agencies in the cacao growing regions has done a great deal towards standardizing fermentation practices and today, many of the primitive methods, such as fermentation in holes in the ground, in canoes, and in makeshift banana and bamboo frames, are the exception rather than the rule (KNAPP, 1937). In general, the seeds are placed in some kind of receptacle and covered and weighed down. Here naturally occurring microorganisms that contaminated the external pulp during handling ferment the sugars contained therein. Most of the world's cocoa is fermented on drying platforms, in baskets, banana leaf covered heaps or in boxes.

Fermentation on drying platforms is practiced in Ecuador and parts of Central America where Criollo cacao was once grown. Wet cocoa seeds are spread directly on drying platforms where they receive a mild aerobic fermentation while drying and an anaerobic fermentation when the seeds are heaped into piles each night. This practice, which suffices for the development of flavor in Criollo seeds, is insufficient for Forastero varieties requiring longer fermentation times. Although Criollo cacao has been largely replaced by Forastero hybrids in these countries, the method of fermentation still persists in may areas (ROHAN, 1963).

Baskets fermentation is practiced principally in Nigeria and by producers on small holdings in the Amazon region, the Philippines and some parts of Ghana. Small lots of cacao are placed in baskets lined with plantain leaves. The surface is covered with plantain leaves and the mass is weighed down. The seeds are mixed from time to time until the end of fermentation which generally lasts from 4 to 6 days (ROHAN, 1963).

Fermentation in heaps covered with plantain leaves is a popular method among the majority of farmers in Ghana and judging by the quality, it is evident that it can produce good quality cocoa. It is also a common practice among farmers in some parts of Nigeria and the Ivory Coast and is observed infrequently in Brazil. Cocoa seeds varying in quantities from 25 to 1000 kg are heaped on a floor of plantain leaves, perforated for drainage. The heap is covered with plantain leaves which are weighed down by branches, stems or other material. The seeds are mixed (turned) on the spot or by making into another heap. This process is laborious and hence is generally performed once or twice during fermentation; small heaps may not be turned at all. The duration of fermentation is from 4 and 7 days (ROHAN, 1963).

Box fermentation which requires a relatively large, fixed volume of cocoa, is the method of choice on large estates. The containers vary in size from region to region, but are generally very similar in design and function. A wooden box is subdivided by either fixed or movable internal partitions into compartments measuring approximately

1×1×1 m which can contain between 600 and 700 kg of freshly harvested ("wet") cocoa seeds. The box is raised above ground level over a drain which carries away the pulp juices (sweatings) liberated during fermentation. The floor of the box may be solid or slatted and contain holes and/or spaces between wooden slats for drainage and aeration. "Sweat-boxes", as these compartments are called, vary considerably in size, the smallest having about the size of a fruit box (0.4×0.4×0.5 m); the largest 7×5×1 m are found in Malaysia. Large estates and cooperatives often have rows of 20 to 30 sweat-boxes arranged in tiers, one below the other on a slope. The arrangement is to take advantage of gravity to facilitate turning which is effected by simply removing a movable wall and shoveling the seeds into the box below. Recommendations with respect to the type of wood used in the construction of sweat-boxes differ in various countries, but as a general rule, any local hardwood, absent of strong odors, which will not warp or split can be used. Variations occur not only in the size of boxes but also in methods of drainage, aeration ("turning") and duration of fermentation. In many countries, primitive methods of fermentation practiced on small holdings are disappearing with the establishment of cooperative plants where cocoa can be cured in boxes according to a prescribed method (Lehrian and Patterson, 1983).

In the majority of cases the boxes are filled to 10 cm from the top, and the surface is covered with a padding of banana leaves or jute sacking to maintain the heat and prevent the surface seeds from drying. The box contents are mixed, at least, every 48 hours by "turning" the seeds into an adjacent empty box to assure uniformity. Fermentation generally requires 6 to 8 days depending on the season and the state of the harvested fruits, both of which influence the quality of the mucilaginous pulp and consequently the fermentation. The progress of fermentation is assessed by the odor and the external and internal color changes of the seeds. When the process is judged complete, the fermented seeds now called beans are dried in the sun, or by employing artificial means.

3.4 Drying

The fermented beans have a moisture content of about 50% which must be reduced to between 6 and 8% for safe storage. Fermentation and oxidative chemical changes, which began during fermentation, continue during the drying phase at rates that progressively decrease with moisture loss. Sun drying is the preferred method because it allows these changes to occur, but in regions where harvesting coincides with the wet weather some form of artificial drying is necessary and desirable. In general, sun drying is employed on small farms, whereas large estates employ both natural and artificial dryers.

In sun drying, the beans are exposed in thin layers not exceeding 5 to 7 cm in thickness, on wooden platforms, mats, polypropylene sheets or concrete floors. The tacky, moist beans are constantly mixed to promote uniform drying, to break agglomerates that form and discourage mold growth. Under sunny conditions, drying is achieved in about a week but under cloudy or rainy conditions, drying time may be prolonged to up to 3 or 4 weeks, thus increasing the probability of mold development and deterioration. To overcome the dependence on the weather, artificial dryers are employed. The various types and their mode of operation have been described by McDonald et al. (1981). Electrical energy is usually costly and, in general, wood or oil is used as a source of energy. Heating of the beans may be by direct contact with the flue gases or indirectly via heat exchangers employing convection and radiant heat. In either instance, where installations are improperly used or poorly maintained, there is danger of contamination with smoke from wood or oil. This results in the "smoky" or "hammy" off-flavors characteristic of beans from some origins. Platform, tray and rotary dryers of various designs, coupled to furnaces are in use. In every case the initial drying must be slow to allow acid loss and oxidative biochemical reactions to occur. Drying temperatures must not exceed 60°C, and times should not be less than 48 hours. Rapid drying at elevated temperatures tends to produce acid, bitter cocoa with brittle shells and cotyledons.

3.5 Storage

The cured beans usually undergo a period of storage in warehouses on the farms, at the buying agencies on the wharfs prior and post shipping and at the factories. The product will suffer alterations in quality depending on the storage conditions and duration. A relative humidity of 65–70% will generally maintain the moisture level in the beans at 6–8% and avoid insect and mold problems and the necessity of repeated fumigation. During the initial storage period slow oxidation and loss of volatile acids may improve flavor but prolonged periods will eventually lead to "staling".

4 The Biochemistry of Fermentation

The biochemical changes that affect curing may be conveniently divided into those that occur in the external pulp owing to microbial action and those initiated within the seed as a consequence of the former.

4.1 Pulp Fermentation

The beans inside the ripe pod are microbiologically sterile. As soon as they are removed, they become inoculated with a variety of microorganisms from the pod walls, the laborer's hands, the containers used for transporting the beans to the plant, the dried mucilage of the previous fermentation that coats the sweat-boxes, and by insects. Fresh cocoa pulp contains sugars and citric acid which makes it an excellent medium for the growth of microorganisms. A comprehensive listing of the microorganisms isolated during cocoa fermentation has been cited by LEHRIAN and PATTERSON (1983). Diagrams which summarize the changes occurring in the pulp and cotyledons are presented in Figs. 2 and 3.

During the initial phase of fermentation, the low pulp pH (3.4–4.0), high sugar content (8–24%) and the low oxygen tension are most suitable for the growth of anaerobic yeasts which initially dominate the fermentation during the first 24 to 36 hours. Yeasts cause a typical alcoholic fermentation, converting sugars to alcohol, CO_2 and metabolizing citric acid. Some strains of yeasts produce pectinases which break down the pulp cells so that the juices drain as "sweatings", carrying away flakes of pulp. The collapse of the pulp parenchyma cells results in the formation of spaces between the beans through which air can percolate. Loss of citric acid by drainage and through microbial metabolism causes a rise in pH, which, together with the increasing alcohol concentration and better aeration, favors the lactic acid bacteria. These may dominate the yeasts until after the first mixing which introduces air and makes conditions conducive to acetic acid and other aerophilic spore-forming bacteria. These provoke the exothermic oxidation of alcohol resulting in an increase in temperature to about 45 to 50 °C. The rise in temperature beyond 45 °C is unfavorable to acetic acid bacteria. By the fourth day the spore forming aerophilic bacteria dominate the fermentation and may constitute 80% of the microflora (LOPEZ, 1986).

Molds are found in small numbers throughout the fermentation period in the more aerated, cooler areas at the corners, along joints in the walls and in the superficial areas of the fermentation box. Wherever conditions favor their proliferation, growth is obvious from the whitish color of the mycelia and the gray/green or black fruiting bodies. Mycelial growth results in the agglomeration of beans which must be broken during mixing. Some strains of molds and bacteria developing under anaerobic conditions and low temperatures break down the cells of the testa liberating slime from the slime cells.

4.2 Bean Curing

If cocoa seeds are dried immediately after harvesting they lose their germinating potential, and virtually no other changes occur in the cotyledon to produce flavor precursors. When beans are put to ferment, germination and the associated chemical changes are prevented due to the presence of inhibitors. The

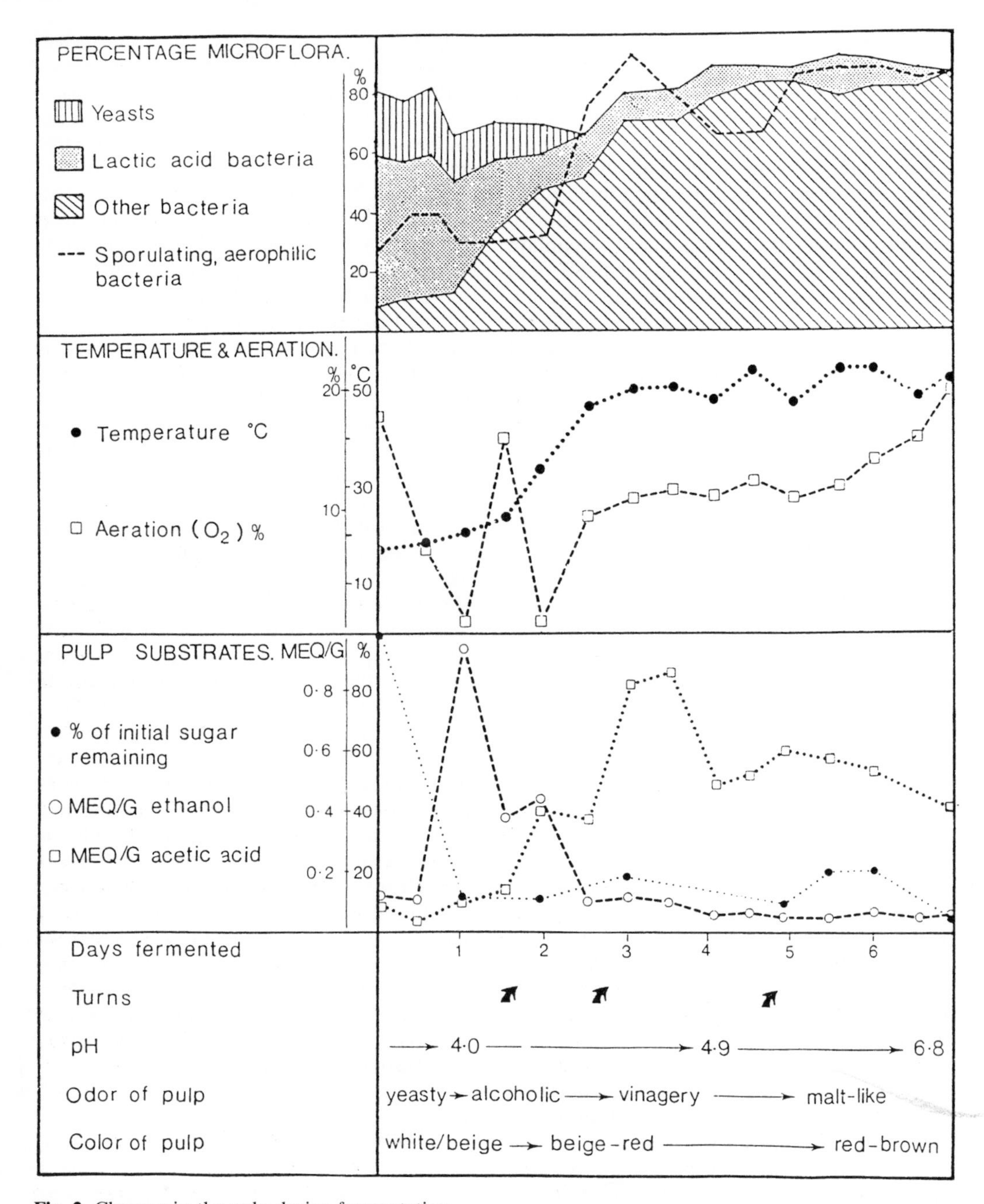

Fig. 2. Changes in the pulp during fermentation.

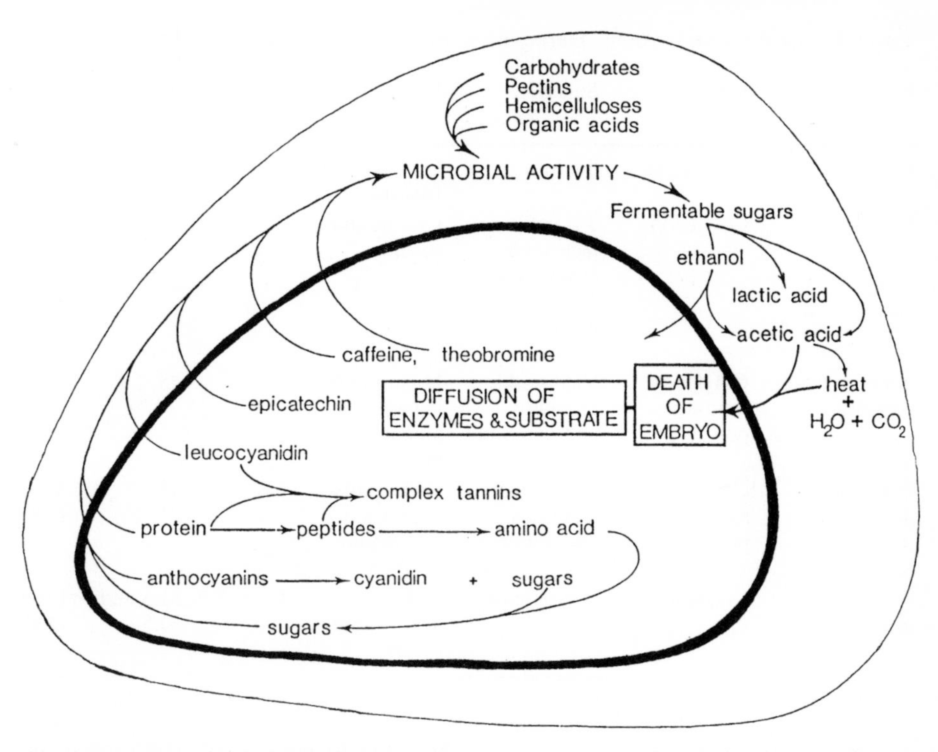

Fig. 3. Summary of the chemical changes in the pulp and cotyledon during fermentation.

degree of inhibition depends upon the state of maturity of the fruit. In the initial stages of fermentation, the inhibitors are destroyed and an incipient germination begins. The seeds absorb moisture and become turgid. The cellular components become hydrated. But the alcohol, acetic acid and the high temperature, produced in the pulp, are instrumental in killing the embryo. Prior to the death of the seed, few other changes occur within the cotyledon. Upon the death of the seed, the biological barriers that separate the enzymes and substrates break down and the cellular components freely mix. Alcohol, acetic acid and water diffusing into the seeds act as solvents for these cellular materials and transport them to sites of activity. A summary of the enzymes reported in the cocoa seed, their substrates, products, temperature and pH optimum are contained in Tab. 1 (LOPEZ, 1986).

Chemical changes in the cotyledon occur in two phases: an anaerobic hydrolytic phase, followed by an oxidative condensation phase. The activity of the cocoa seed enzymes once liberated, are governed by the oxygen tension, pH and temperature inside the seed tissue. At the start of pulp fermentation there is little or no oxygen in the fermenting mass, and the amount of dissolved oxygen diffusing into the seeds is negligible. During this anaerobic stage, hydrolytic enzymatic reactions occur manifested by the diffusion and bleaching of anthocyanin pigment from the storage cells following the absorption of water and acetic acid by the tissue. Browning by polyphenol oxidase at this stage is inhibited by citric and ascorbic acids, and by the lack of oxygen. At about the fourth day, oxygen begins to enter the seed and the concentration of organic acids no longer inhibits the oxidases. A series of enzymatic oxidative reactions are initiated that result in the browning of the cotyledons from the outside inwards. In actual fact, there is considerable overlap of both, hydrolytic and oxidative condensation reactions,

Tab. 1. Principal Enzymes Active During the Curing of Cocoa Beans

Enzyme	Location	Substrate	Products	pH	Temp. (°C)	Reference
Invertase	Bean testa	Sucrose	Glucose and fructose	4 and 5.25	37 and 52	LOPEZ and DIMICK (1991)
Glycosidase	Cotyledon	3-β-D-Galactosidyl cyanin and 3-α-L-arabinosidyl cyanin	Cyanin and sugars	3.8 and 4.5	45	FORSYTH and QUESNEL (1957)
Aspartic endo protease	Cotyledon acetone powders	Vicilin-like globulins and albumin	Hydrophobic peptides and amino acids	3.5	45	VOIGT et al. (1994)
Carboxy peptidase	Cotyledon acetone powders	Vicilin-like globulins and albumin	Hydrophilic peptides and amino acids	5.8	45	VOIGT et al. (1994)
Polyphenol oxidases	Cotyledon	Polyphenols	*o*-Quinones and *o*-diquinones	6	31.5 and 34.5	LOPEZ and DIMICK (1991)

because even though oxygen penetrates the surface of the seed cotyledons, parts of the interior are still anaerobic. The relative enzymic activity and the products formed are dependent on pH and temperature. The unfermented seed has a pH of between 6.3 and 6.8. The absorption of acetic acid can gradually lower this pH to about 4.2. The seed temperature, on the other hand, rises from a few degrees below ambient to between 45 and 55 °C in accordance with the fermentation (FORSYTH and QUESNEL, 1963).

4.3 Anaerobic Phase

The most obvious change results from the hydrolysis of the anthocyanin pigments early in fermentation. Cocoa glycosidases are activated as soon as the seeds die and the pigments 3-β-D-galactosidyl cyanin and 3-α-L-arabinosidyl cyanin, responsible for the purple color of Forastero cocoa are hydrolyzed to sugars and leucocyanidins, resulting in a bleaching effect of the color as fermentation progresses. The glycosidase enzymes have a pH optimum of 4 to 4.5 at 45 °C and become inactivated by oxidation products of polyphenolic complexes later in fermentation. Although the pigments themselves do not possess any marked taste or flavor potential, the hydrolysis of polyphenols is important in the fermentation since there is an inverse relationship between the flavor development and the purple color retained after fermentation. This suggests that the conditions required for breakdown of anthocyanins are also those necessary for the production of good flavor (Figs. 4 and 5). The polyphenols, as a group, are believed to be important indicators of overall seed chocolate flavor potential. A correlation has been shown between varieties with high polyphenol content and strength of cocoa flavor.

However, the most important precursor-forming reaction occurring during the anaerobic phase is the hydrolysis of storage protein to peptides and amino acids. Analysis of unfermented and fermented seeds shows that storage proteins disappear during fermentation, and more amino nitrogen is formed. In addition, a steady decrease in protein-bound nitrogen and a concomitant increase in amino acids and peptide nitrogen occurs. Protease in

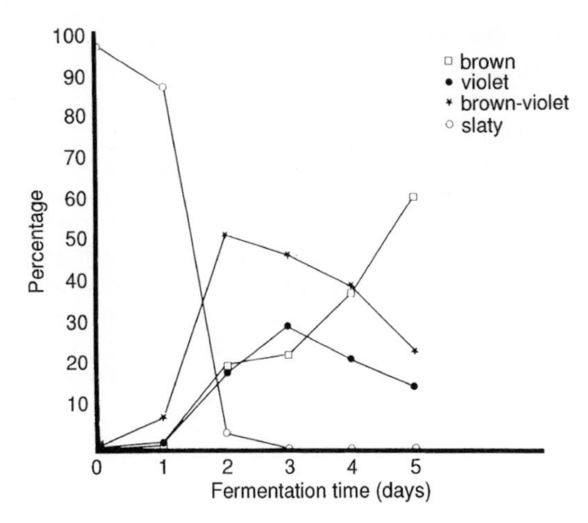

Fig. 4. Cocoa bean cotyledon color changes in a small fermenter as affected by fermentation time.

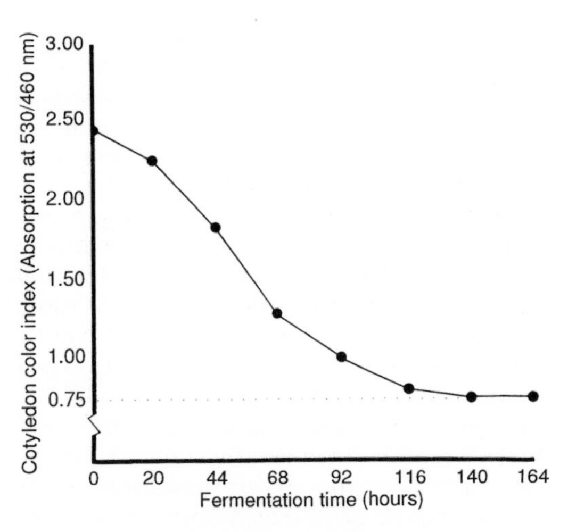

Fig. 5. Cocoa bean color index (absorption at 530/460 nm), as affected by fermentation time.

cocoa was believed to be a complex consisting of several enzymes since acetone preparations showed more than one pH and temperature optimum (QUESNEL, 1971). Further purification and separation have revealed an aspartic endoprotease with a pH optimum of 3.5 at 45 °C; a carboxypeptidase (optima at pH 5.8 and 45 °C), and a leucine-*p*-nitroanilide cleaving enzyme (pH 6.8, temperature 45 °C). However, optimum proteolysis in seed fragments, measured by the increase in amino acids, occurs at a temperature of 55 °C and a pH of 4.7. However, high accumulation of amino acids and peptides is not essential for good flavor. A correlation of the acidification, proteolysis, and the flavor potential during this anaerobic phase was demonstrated (BIEHL et al., 1985). Cocoa aroma specific precursors are obtained by the co-operation of aspartic endoprotease and carboxypeptidase on the cocoa protein. High acidification by acetic acid promotes proteolytic production of high concentrations of hydrophobic peptides by aspartic endoprotease but reduces the flavor potential. Conversion of these to hydrophilic peptides by carboxypeptidase generates aroma precursors (VOIGT et al., 1994).

The cocoa storage proteins comprise a vicilin-like globulin fraction which comprises principal polypeptide subunits of 47 kDa, 31 kDa, 15.5 kDa and 14.5 kDa and which generates the flavor precursors. The albumin fraction comprises mainly a 19 kDa polypeptide and contains most of the enzyme activity (SPENCER and HODGE, 1992; VOIGT et al., 1994). Although proteolysis generates amino groups, in general, there is a net decrease in total nitrogen due to the loss of soluble nitrogen compounds by exudation through the testa.

The dominant sugar in the unfermented seed is sucrose. During fermentation it is almost completely hydrolyzed to fructose and glucose by invertase present in the testa of the seed. Sorbose, mannitol and inositol appear, presumably from the diffusion of microbial metabolites from the pulp or as a result of the hydrolysis of polysaccharides. In addition to being produced from carbohydrate, glucose, arabinose and galactose in the fermenting seed, they may also arise by hydrolysis of the anthocyanins and polymeric phenolic glycosides. Although sugars are being formed during fermentation, the increase is not evident because of the losses in the exudate. There is no conclusive evidence to indicate a change in the starch content, however, some hydrolysis may occur. The apparent increase in the starch content during fermentation is attributed to the change in dry weight. This is also the case with the fats which remain unchanged (LOPEZ and DIMICK, 1991).

Apart from the above, there are other less understood reactions which result in a large number of complexes of polyphenols with proteins in various stages of degradation. These consequently result in a progressive decrease in the rate of enzymatic reactions. Other compounds that decrease are epicatechin and the purines which though chemically unchanged, are lost by exudation (HOLDEN, 1959).

4.4 Aerobic Phase

As oxygen diffuses into the seed tissue, oxidases become active, and the aerobic conditions and oxidation products of the polyphenols progressively inhibit the enzymes which were operative during the anaerobic phase. Cyanidins and protein-phenolic complexes, formed during the anaerobic phase, undergo oxidative reactions. Epicatechin is the preferred substrate of the polyphenol oxidase system, and its action is manifested in the brown color that appears in the inter-cotyledon liquid and on the surface of the cotyledons towards the end of fermentation. Proteins complex with polyphenols in reactions analogous to those in tanning which results in the progressive inactivation of enzymes as browning advances. During the subsequent drying period, the oxidative reactions that began during fermentation are accelerated by the increased exposure of the tissue to oxygen (FORSYTH and QUESNEL, 1963).

The first stage in the oxidation of polyphenols is the formation of *o*-quinones. These are active oxidizing agents and can react in a number of different ways with the vast variety of compounds that are formed during the preceding stages. They may also polymerize to form diphenols and diphenol-quinones. Quinones also take part in oxidation–reduction reactions with compounds generating active hydrogen. Thus, they can combine with amines, amino acids and thiols in reactions analogous to those occurring in tea fermentation.

Polyphenol oxidase has a pH optimum of 6 at 35.5 °C. Its activity decreases during drying because it is continuously destroyed by tanning as it oxidizes polyphenols. It is believed that the oxidative phase contributes by the formation of auxiliary flavors. However, the most important function of the enzyme is the reduction of astringency and bitterness by the oxidation of polyphenols which then complex with proteins and peptides. Consequently, soluble proteins are bound, with the result that the off-flavor attributed to the roasting of native protein is eliminated. Other aromatic compounds found in cocoa can also be produced during this stage by oxidative deamination of amino acids (LOPEZ and DIMICK 1991).

4.5 Drying Phase

Drying of cocoa, the second phase of the 'curing' process, appears to be as important as the fermentation. The oxidative enzyme reactions that begin during fermentation, continue during drying as long as there is sufficient moisture, however, eventually all biochemical reactions virtually stop. The seed tissue shrinks, and the folds of the cotyledon open resulting in increased aeration. Non-enzymatic oxidative reactions may continue. Rapid utilization of the surface acetic acid brings the testa pH to between 5 and 6, and if care is not exercised, there is a serious danger of mold and putrefactive organisms developing to the extent that off-flavors are produced. When drying is carried out under natural conditions, this problem is even more acute; but with proper care and continual moving of the cocoa beans on the drying platforms, mold growth can be discouraged until a safe moisture level is reached. The increasing use of artificial dryers reduces the risk of mold by the rapid removal of moisture. The system offers a method that is independent of climatic factors. However, when employed, the temperatures should be such that the inside of the bean remains below 60 °C so that enzymes are not deactivated before browning occurs, thus preventing harsh flavors. Rapid drying also tends to make the beans retain excessive amounts of acetic acid which is deleterious to flavor. When wood is used as a source of heating fuel, there is the added danger that the beans will be contaminated by smoke if the process is not well controlled. Such flavors appear in the manufactured chocolate as undesirable off-flavors. In general, slow sun dried beans possess a better flavor than rapidly artificially dried beans.

4.6 Factors Affecting Quality

The primary factor that influences the flavor of the bean is its genetic constitution, but the amount of flavor actually developed is dependent on fermentation. The two types of biochemical reactions that are responsible for the production of flavor precursors are the hydrolytic reactions in fermentation and the oxidative reactions during drying. In order to obtain a good product, these two phases must follow in the correct sequence.

Hydrolytic enzymes are inactivated by the products of polyphenol oxidation. If oxidation occurs in the cotyledon immediately after the death of the seed, the activity of the enzymes would be curtailed and the resulting product would be deficient in chocolate flavor. However, an extended anaerobic phase would result in the loss of excessive quantities of soluble materials that might be important as flavor precursors. The chocolate from such fermentations is characterized by having weak chocolate flavor. A short oxidative stage, on the other hand, will not allow sufficient tanning of the protein or oxidation of polyphenols thereby giving products, which although possessing chocolate flavor, are excessively astringent and bitter. This implies that in order to obtain the best flavor quality, sufficient time must be allowed for the formation of flavor precursors during fermentation before the oxidative stage occurs.

Furthermore, the role of pH on aroma precursors has been mentioned. Acid production is necessary for killing the seed, but excessive acid production and the resulting low cotyledon pH will result in the production of low precursor type peptides and amino acids. Weak chocolate flavors will result regardless of the fact that the other requirements have been fulfilled.

5 Measurement of Fermentation

Undoubtedly there is an optimal duration for fermentation. In the field this is judged empirically by evaluating a combination of factors such as appearance, odor and the internal and external bean color. Such appraisals rely heavily on the experience of farm workers and may lead to a good deal of variation in quality. Since fermentation is a critical step in the development of flavor, it is essential that manufacturers know the extent of the fermentation of beans so as to assess its quality prior to purchase.

5.1 The Cut Test

The simplest, and still the most widely used method is to assess the quality of a random sample of beans from a batch by visual evaluation of the cut beans. This 'cut test', however, is extremely subjective and at best is limited to the measurement of bean defects. It assumes that beans showing the least amount of the prescribed defects stand a good chance of producing good chocolate flavor. The defects most commonly looked for are: slaty beans – which occurs when beans are dried without undergoing fermentation; purple beans – a combination of the intensity and occurrence gives an indication of the degree of under fermentation; mold – which can produce off-flavors and rancidity as well as being an indicator of the quality of the product; insect infestation which also is an indicator of the sanitation condition; germinated beans; other physical defects such as flat beans, undersized beans and broken beans which relate more to the yield factor rather than to the flavor quality. The cut test is no guarantee of good flavor, and the development of more precise means of flavor/quality determinations has been attempted. The earliest was the correlation of the decrease in the anthocyanin pigment with flavor development (LOPEZ, 1984). The test assumes that the anthocyanin content is reasonably constant in all cocoas, which is probable in Forasteros, but it can be extremely variable in Trinitarios (Forastero/Criollo crosses). Modifications of the 'cut test', as described by SHAMSUDDIN and DIMICK (1986) and known as the 'cut-test score', have rendered the method quantitative. Thus, comparisons of the 'cut-test score' to other quantitative measurements of bean fermentation are also possible. The method is simple, and, therefore, suitable for use in the field and the laboratory.

Correlations between 'cut-test scores', fermentation indices, total polyphenols and (−)-epicatechin contents have been calculated, and a summary of their correlation coefficients (r) is given in Tab. 2. High correlations were obtained between 'cut-test scores' and fermentation indices ($r=0.92$) and between total polyphenols and (−)-epicatechin contents ($r=0.91$).

Tab. 2. Correlation Coefficients of Fermentation Measurement

Test	PPO[a]	EPI[b]	FI[c]	CTS[d]
Total polyphenols (a)	1.00	0.91	−0.75	−0.78
(−)-Epicatechin (b)	0.91	1.00	−0.77	−0.83
Fermentation index (c)	−0.75	−0.77	−1.00	0.92
Cut-test scored (d)	−0.78	−0.83	0.92	1.00

6 Development of Flavor

Flavor is developed during the roasting process in the manufacture of chocolate. The reducing sugars and amino compounds formed in the seed during fermentation undergo Strecker-type degradation reactions during which flavor volatiles are produced (ROHAN, 1969; MOHR et al., 1971). In roasting, two phases are recognized: an initial drying phase during which moisture is freely lost, followed by a roasting phase where browning reactions occur at low moisture content. The drying phase takes the moisture down to about 0.8–1%. While this is in progress, the bean temperature rises to between 100 and 120 °C. Although browning reactions begin at about 100 °C, very little aroma is developed at temperatures below 100 °C, the best flavor being produced between 120 and 140 °C (bean temperature). Above 140 °C an 'over-roast' index is reached. There is a difference in temperature between the hot air and the nib owing to a barrier created by the testa and the air space it encloses. The temperature boundary varies with the type of bean. The temperature difference between a 'best' roast and an 'over-roast' is quite small, and it is therefore advantageous to roast different varieties separately before blending. Over-roasts develop a burnt 'bready' flavor which is difficult to remove by conching, because the compounds responsible are not very volatile.

The body of evidence gained thus far suggests that chocolate aroma arises from Maillard-type, non-enzymic browning reactions between amino acids and reducing sugars during roasting (LOPEZ and QUESNEL, 1971, 1974). Indeed, it has been shown that the Strecker reaction is operative during the roasting of cocoa beans (HOSKIN and DIMICK, 1994). In contrast to the study of flavor precursors, there was very little investigation into the aroma constituents of cocoa until the middle of this century, chiefly because of the difficulties encountered in isolating and concentrating, by classical methods, the minute amounts of components present in the aroma complex. The development of new techniques and sophisticated instrumentation in analytical chemistry has greatly simplified this task and increased exponentially the number of flavor components identified over the last two decades.

Nearly 400 compounds (LOPEZ, 1974; HOSKIN and DIMICK, 1994) of the chocolate flavor complex have been isolated and identified by gas chromatography and supporting techniques, but all the reactions that are responsible for their formation have not yet been clearly elucidated. That such a large number of compounds arise despite the low roasting temperature can be attributed to the complexity of the chemical constituents of the bean after fermentation. Analyses of the bean constituents before and after roasting show that of the major constituents, free amino acids, amino peptides and reducing sugars, undergo the most change while the alkaloids, polymeric polyphenols and lipids are practically unaffected. Although pyrolysis does not occur to any extent, Maillard-type reactions and caramelization typical of non-enzymatic, non-oxidative browning occur.

The cocoas of commerce are classed into two broad categories; "bulk" and "fine". The "bulk" cocoas come from the Forastero–Amelonado variety which has a mild chocolate flavor and the "fine" or "quality" cocoas include the Criollos and the Trinitarios which possess a much stronger chocolate flavor often accompanied by "bouquets". The terms "fine" and "quality" refer to particular flavor characteristics rather than quality of the beans in respect of size, uniformity and other visual characteristics used in judging quality. Flavor cocoas are used mainly in mixing with bulk cocoas to produce a desired blend, and unlike the latter, the flavor is dependent, to a larger extent, on the fermentation and is subject to a great deal of fluctuation.

Although in commercial circles physical criteria such as 'color', 'plumpness' and 'break' are frequently employed as the basis of quality assessment of raw cocoa, the final criterion of quality is the flavor developed upon roasting. Not only is the intensity of the flavor of the bean important but also the auxiliary flavors which give the beans from different geographic regions their characteristic flavor properties. The desire of the manufacturer, in short, is for large, plump beans showing the strongest possible chocolate flavor together with those flavor characteristics which are traditionally associated with the particular region.

7 Conclusion

Cocoa curing proceeds in two stages: fermentation and drying. The course of fermentation is determined largely by the microorganisms involved and by the physical and chemical conditions which govern their proliferation and activity. These conditions are purely accidental and hence extremely variable; and although careful management of the process does considerably increase the likelihood of obtaining a good quality cocoa, even under good supervision, the variability between batches cannot be entirely eliminated. This situation underlies the need for more biochemical and technological research into the factors that determine chocolate flavor. Only after a thorough understanding is obtained of all the changes occurring during fermentation, the conditions that govern them and determine the quality of the processed bean, can proper measures be taken to ensure a uniformly high quality product. Besides the necessity for ensuring uniformity of flavor, it is also desirable to be able to develop the full flavor potential of the seed. There is reason to believe that the development of flavor precursors during processing is closely associated with the techniques employed, and stronger chocolate flavors may possibly be induced by suitable modifications in processing methods. There has been a great deal of research in the past on the biochemical mechanism of fermentation; however, our understanding of the curing process of cocoa is not yet complete. Many questions, particularly with respect to the formation of off-flavors and the effect of hybridization on flavor remain unanswered. Furthermore, it is quite clear that the drying process, following fermentation, is not a separate process designed solely to remove water from the bean, but is a continuation of curing. Many enzymatic reactions initiated during fermentation continue during drying and are important to the flavor quality of the product. In the past these reactions were guaranteed on the sun-drying floors; however, as artificial dryers replace natural drying, the tendency is to dry as fast as possible. Under these circumstances there is a likelihood of incomplete curing and flavor deterioration owing to enzyme inactivation by high temperatures or lack of moisture. In a complex system, as it has been described, in which there is a minimum control over the microflora and the biochemical and chemical parameters influencing the fermentation process, it would be surprising, indeed, to find no variation in the quality of the product. That this variation is small is even more surprising and, in general, it is found that the flavor quality of cocoa from any particular source is fairly constant. Differences in quality between sources, however, may be quite marked and serve as features for distinguishing between cocoa from different regions. Thus, for example, although Ghana, Malaysia and Brazil are producers of 'bulk' cocoa, the Brazilian and Malaysian products are characterized by high acidity. This difference cannot be entirely attributed to dissimilarities in processing methods in these countries. Small heap fermentations of the West African type with Brazilian beans do not reduce the acidity substantially. Nor does increased aeration by an increased number of turnings or other methods, unless these are extremely severe. This is probably because such procedures have only a small effect on the microbial pattern and population. For this reason, the problem of acid beans has more efficiently been resolved by influencing microbial activity through the mechanical or enzymic reduction of the substrate prior to fermentation, rather than through the manipulation of the aeration. The cost of the operation in such cases, is more than off-set by the sale of the pulp juice for the manufacture of by-products such as ice-cream and juices.

Flavor studies involving such a complex system are, to say the least, extremely difficult. This problem is compounded by the fact that the manufacturer's requirements are notoriously varied so that what is acceptable flavor quality to one may be unacceptable to another. Furthermore, flavor evaluation by subjective methods is difficult to interpret and the descriptive terminology employed can be extremely ambiguous. It is hoped that in the future these difficulties will be resolved by the development of standard methods of flavor analysis and a common language for its description. Such a system will be invaluable in the study of the fermentation process and the flavors that are developed by it.

8 References

BIEHL, B., MEYER, B. (1990), Bean spreading: a method for pulp preconditioning to impair strong nib acidification during cocoa fermentation in Malaysia, *J. Sci. Food Agric.* **51** (1), 35–45.

BIEHL, B., BRUNNER, E., PASSERN, D., QUESNEL, V. C., ADOMAKO, D. (1985), Acidification, proteolysis and flavor potential in fermenting cocoa beans, *J. Sci. Food Agric.* **36,** 583–598.

FORSYTH, W. G. C., QUESNEL, V. C. (1957), Cocoa glycosidase and colour changes during fermentation, *J. Sci. Food Agric.* **8**, 505–509.

FORSYTH, W. G. C., QUESNEL, V. C. (1963), Mechanisms of cocoa curing, *Adv. Enzymol.* **25,** 457–492.

HOLDEN, M. (1959), Processing of raw cacao III. Enzymic aspects of cacao fermentation, *J. Sci. Food Agric.* **10,** 691–698.

HOSKIN, J. C., DIMICK, P. S. (1994), Chemistry of flavor development in chocolate, in: *Industrial Chocolate Manufacture and Use* (BECKETT, S. T., Ed.), 2nd Ed., Glasgow: Blackie Academic of Professional.

KNAPP, A. W. (1937), *Cocoa Fermentation. A Critical Survey of its Scientific Aspects,* Curnow: John Bale.

LEHRIAN, D. W., PATTERSON, G. R. (1983), *Cocoa fermentation,* in: *Biotechnology* (REHM, H.-J., REED, G., Eds.), 1st Ed., Vol. 5, pp. 529–575, Weinheim: Verlag Chemie.

LOPEZ, A. S. (1974), Contribution of volatile compounds to the flavour of chocolate and their development during processing, *Ph.D. Thesis,* University of the West Indies, Trinidad.

LOPEZ, A. S. (1984), Limtação da "prova de corte" no controle de qualidade do cacau commercial, *Revista Theobroma* **14** (3), 199–207.

LOPEZ, A. S. (1986), Chemical changes occurring during the processing of cocoa, in: *Proceedings of Cocoa Biotechnology* (DIMICK, P. S., Ed.), pp. 19–54, Department of Food Science, Pennsylvania State University, University Park PA.

LOPEZ, A. S., DIMICK, P. S. (1991), Enzymes in cocoa production, in: *Food Enzymology* (FOX, F. F., Ed.), Vol. 2, pp. 211–236, London: Elsevier Applied Science.

LOPEZ, A. S., QUESNEL, V. C. (1971), An assessment of some claims relating to the production and composition of chocolate aroma, *Int. Chocolate Rev.* **26,** 19–24.

LOPEZ, A. S., QUESNEL, V. C. (1974), The contribution of sulphur compounds to chocolate aroma, in: *Erster Internationaler Kongress Kakao- und Schokoladeforschung, München,* pp. 91–104.

LOPEZ, A. S., DIMICK, P. S., WALSH, R. M. (1987), Scanning electron microscopy studies of the cellular changes in raw, fermented and dried cocoa beans, *Food Microstructure* **6,** 6–16.

MCDONALD, C. R., LASS, R. A., LOPEZ, A. S. (1981), Cocoa drying – a review, *Cocoa Growers' Bull.* **31,** 5–39.

MEYER, B., BIEHL, B., SAID, M. B., SAMARAKODDY, R. J. (1989), Post-harvest pod storage: A method for pulp preconditioning to impair strong nib acidification during cocoa fermentation in Malaysia, *J. Sci. Food Agric.* **48**, 285–304.

MOHR, W., ROHRLE, M., SEVERIN, T. (1971), Über die Bildung des Kakaoaromas aus seinen Vorstufen, *Fette, Seifen, Anstrichmittel* **73,** 515–521.

QUESNEL, V. C. (1971), Proteolysis during fermentation, *Cocoa Research Unit,* p. 49, University of the West Indies (Trinidad).

ROELOFSEN, P. A. (Ed.) (1958), Fermentation, drying and storage of cocoa beans, *Adv. Food Res.* **8.**

ROHAN, T. A. (1963), Processing of raw cocoa for the market, *Agricultural Studies* No. 60, p. 207. Rome: FAO.

ROHAN, T. A. (1969), The flavour of chocolate; its precursors and a study of their reactions, *Gordian* **69** (9–12), 443.

SHAMSUDDIN, S. B., DIMICK, P. S. (1986), Qualitative and quantitative measurements of cocoa bean fermentation, in: *Proceedings of Cocoa Biotechnology* (DIMICK, P. S., Ed.), pp. 55–78, Department of Food Science, The Pennsylvania State University, University Park, PA.

SPENCER, M. E., HODGE, R. (1992), Cloning sequence of a cDNA encoding the major storage proteins of *T. cacao:* Identification of the proteins as members of the vicilin class of storage proteins, *Planta* **186,** 567–576.

VOIGT, J., HEINRICHS, H., VOIGT, G., BIEHL, B. (1994), Cocoa-specific aroma precursors are generated by proteolytic digestion of the vicilin-like globulin of cocoa seeds, *Food Chem.* **50** (2), 177–184.

15 Vinegar

HEINRICH EBNER

Linz, Austria

HEINRICH FOLLMANN
SYLVIA SELLMER

Bonn, Federal Republic of Germany

1 Introduction

This chapter is the continuation of the chapter on Acetic Acid in Volume 6 of this series, where the microbial and technological aspects of the fermentation process and its economic significance have been discussed. The main emphasis here is on the processing of raw vinegar, on its use and characteristics, on the differentiation between the variety of vinegar types and on basic food law provisions.

1.1 Physical and Chemical Properties of Vinegar

Vinegar is a clear aqueous liquid, colorless or of the color of the raw material or colored by caramel with a prescribed content of acetic acid between 40 and 150 g/L. Further constituents have their origin in the raw material, in added nutrients and in the water used for dilution. In dependence on the contents of acetic acid and the raw material density, boiling point, freezing point, surface tension and viscosity differ more or less from the corresponding values of pure water. The pH value lies between 2.0 and 3.5.

1.2 Legal Definition in the European Union

The name "vinegar" is reserved for the product that is produced by the biological process of double fermentation, alcoholic and acetous, from liquids or other substances of agricultural origin. Some sorts of vinegar are: spirit vinegar, wine vinegar, fruit vinegar, cider vinegar, grain vinegar and malt vinegar.

1.3 Legal Definition in the USA

In the US the definition of 1.2 is extended to the point that synthetically produced alcohol may also be used for acetous fermentation.

1.4 Legal Definition in Japan

Brewed vinegar is a liquid condiment obtained by the acetous fermentation of mash prepared from grains or fruits where alcohol or saccharides may be added.

2 Treatment of Raw Vinegar

Vinegar produced by one of the processes mentioned above is called "raw vinegar". It contains less than 0.5% alcohol and depending on the raw material up to 20% acetic acid. It is more or less turbid as it contains acetic acid bacteria and deposits which originate from the raw material. Prior to consumption, it therefore needs further treatment.

2.1 Storage and Quality

In the course of acetous fermentation the pH value of the fermenting mash decreases. Vinegars obtained from natural raw material therefore show an instability with regard to the solubility of previously dissolved substances. This lability lasts the longer, the less the pH changes during fermentation. A freshly produced cider vinegar, for example, may need months to become stable. Alcohol vinegar does not show this instability. Vinegar for storage should preferably be undiluted, the way it is discharged from the fermenter. In storage the quality of vinegar improves. This is especially important for all types of vinegar from various wines. The aging of wine vinegar comprises many complicated reactions as described in detail by MECCA et al. (1979). Vinegar must have a pure aroma which imparts the flavor of the raw material.

2.2 Filtration

2.2.1 Traditional Methods

The traditional methods of vinegar clarification by fining and precoat filtration are still in use. Nowadays micro- and ultrafiltration are gaining ever increasing market shares in the large European and American factories in particular.

Filters with diatomaceous earth coating are well suited to aged or non-aged, fined or unfined vinegar. A filter layer of approx. 1 mm thickness of diatomaceous earth or cellulose is placed on an adequate coat of acid-resistant steel, wood, or nylon. The quality of this filter coat determines the flow rate of vinegar filtered through the layer, and the brilliance of the product. During filtration, diatomaceous earth is added continuously to keep the growing filter cake as permeable as possible. As soon as the filtration pressure has increased and the filter capacity has decreased too much, filtration has to be stopped and the filter has to be cleaned and prepared for a new filtration. Complete, largely automatic filters with both horizontal and with vertical filtering surfaces are available on the market.

Vats filled with special, pretraited wool, the so-called "vinifilter", may also be used for vinegar filtration, especially for wine vinegar. These vats are filled with liquid to be filtered. Each day, some 10% of the contents are drawn off at the bottom as clear vinegar, and are replaced by raw vinegar filled in at the top.

The vinegar can be discharged directly into the vinifilter. Some vinegar producers carry out the filtration step first, for example cross flow filtration, and afterwards the vinifilter step. The filtered vinegar still requires sterile filtration before bottling. Either deep bed filters with finely porous discs or cartridges or membrane filters of traditional design and construction are in use.

2.2.2 Continuous Ultrafiltration

The basic principle of the first long-term ultrafiltration process for vinegar, which was elaborated 1976 and performed in the Frings stream filter, has been described in detail by EBNER and FOLLMANN (1983) and FOLLMANN and EBNER (1985).

Vinegar is an inexpensive material, filter costs must, therefore, be low. Consequently a repeated and long-term use of the filter modules is important. The unavoidable formation of a film on the membrane reducing the filtration rate must be prevented as long as possible. Easy, repeated and mainly automatic cleaning of the membrane is also important. Today any cross-flow filter, partly fulfilling these requirements, may be more or less suitable for the filtration of vinegar.

2.3 Pasteurization

Vinegar made from wines of all kinds is frequently unstable due to enzymes and microorganisms which may later cause cloudiness in the bottles. To avoid this, vinegar is pasteurized prior to bottling or afterwards. Heating to 75–80 °C for 30–40 seconds, or to 50–55 °C for longer periods is a common procedure. The optimum solution depends on the type of vinegar and its constituents. At lower temperature, only the enzymes are inactivated, and at higher temperatures, the microorganisms as well. On the other hand, higher temperatures might influence the color, smell, and the taste of the vinegar, and again cause turbidities (MECCA et al., 1979).

2.4 Sulfiting

SO_2 is the traditional preserving agent and anti-oxidant which is allowed by law. Vinegar from natural raw material binds and very quickly oxidizes SO_2 added in gaseous form or as $K_2S_2O_5$. An addition immediately before bottling is therefore essential. Binding velocity depends on the raw material, especially on the free aldehyde and ketone groups, and on the pH and rH. In the bottles, free active SO_2 is mostly present in quantities of less than 10 mg/L.

2.5 Coloring and Decoloring

Spirit vinegar is of a slightly yellowish color due to the application of nutrients and due to substances liberated by the acetic acid bacteria. In a number of countries, spirit vinegar is sold completely colorless which requires a treatment with activated carbon. In other countries it is colored yellow with caramel or other colorants admitted for food. Vinegar from natural raw material sometimes also requires coloring. Red wine, for instance, is becoming lighter during vinegar fermentation so that, occasionally, oenocyanines are used for a deeper color. In some countries where the red color of the wine is undesired, e.g., in Italy, ultrafiltration is applied for decoloring.

2.6 Bottling and Shelf Life

Vinegar for use in households is filled into bottles of glass or polyethylene, sealed with plastic screw caps. Air tightness is essential to guarantee preservation. Filtered spirit vinegar does not present great difficulties, but all vinegars from wine do. Institutional users such as restaurants or hospitals get their vinegar frequently in plastic containers of about 25 L, industrial users in tank cars of stainless steel.

3 Use of Vinegar

NUNHEIMER and FABIAN (1940) found that acetic acid, compared with citric acid, lactic acid, malic acid and tartaric acid is at a higher pH a stronger growth inhibitor of microorganisms than the other acids. SZAKALL (1952) stated that vinegar has a specific, inhibitory effect on the growth of microorganisms, as compared to diluted acetic acid, whose effect is a function of acid concentration only. The reason of this phenomenon is yet unknown.

In the household, vinegar is mainly used for the acidification of salads. The food industry makes use of the long known capability of vinegar to preserve and season food at the same time. The most important preserved vegetables prepared with vinegar are gherkins, onions, red beets, peppers in the form of pickles and relishes. Vinegar is an important ingredient in mustard, tomato ketchup, Worcester sauce, mayonnaise and salad dressings. Fish and meat are also preserved with vinegar.

4 Constituents of Vinegar

Besides acetic acid and ethanol, vinegar contains secondary constituents which play an important role for its smell, taste and preserving qualities. These constituents have their origin in the raw material, in added nutrients, and in the water used for dilution. They are also formed by acetic acid bacteria, or they are a product of the interaction of produced components.

4.1 Spirit Vinegar

SCHANDERL and STAUDENMEYER (1956) were the first who discovered eight amino acids in a spirit vinegar produced by a submerged fermentation, where only glucose and inorganic salts had been added for nutrients. BERGNER and PETRI (1959, 1960a, b) increased the number of amino acids found in spirit vinegar to 18. According to PETRI (1958) these are mainly products of autolysis of the acetic acid bacteria. AURAND et al. (1966) and KAHN et al. (1966, 1972) identified in an ether–pentane extract of neutralized spirit vinegar 27 volatile compounds, among them ethanol, ethyl acetate, acetaldehyde, ethyl formate, higher alcohols, but also compounds whose origin is difficult to explain, such as bromoacetaldehyde diethylacetal.

4.2 Wine Vinegar

Wine vinegars contain the same spectrum of amino acids as spirit vinegar, but in larger amounts. GALOPPINI and ROTINI (1956) and ROTINI and GALOPPINI (1957) found that during acetous fermentation acetyl-methyl-carbinol develops in varying quantities. In an ether–pentane extract of wine vinegar KAHN et al. (1972) identified 42 compounds. Besides the substances which had been found in spirit vinegar, compounds derived from higher alcohols such as isopentyl acetate, isovaleryl aldehyde, or β-phenethyl acetate are of particular interest. GARCIA et al. (1973) analyzed 20 types of wine vinegar. Most of them contained acetoin and butylene glycol, but only seven contained diacetyl.

Polyphenolic compounds have been shown to be of great interest regarding the stability of wine vinegars. GALVEZ et al. (1994) identified polyphenolic compounds by HPLC separation. They stated that vinegars obtained from wines exhibit as a rule a greater number and content of polyphenolic compounds than vinegars from apples or honey. Those originating from Rioja wines exhibited a maximum, followed by those derived from sherries. In wine vinegars the following substances were identified: gallic acid, *p*-hydroxybenzaldehyde, caffeic acid, vanillic acid, syringic acid, *p*-coumaric acid, anisaldehyde, epicatechin, sinapic acid and salicylaldehyde. The phenolic compounds are generally contributed by the solid parts of the grapes. Therefore, in the case of wines kept in longer contact with the grapes (as in the case with Rioja wines) a larger amount of polyphenols will be extracted.

4.3 Cider Vinegar

KAHN et al. (1966, 1972) identified 33 compounds in the ether–pentane extract of cider vinegar, and only a few of them were found in the cider from which the cider vinegar was made. The compounds correspond approximately to those found in wine vinegar. The following phenolic compounds were identified in cider vinegar by HPLC separation: gallic acid, catechin, caffeic acid and *m*-coumaric acid (GALVEZ et al., 1994).

4.4 Malt Vinegar

JONES and GREENSHIELDS (1969, 1971) identified six esters, six alcohols, two acids, acetoin and acetaldehyde in malt vinegar. KAHN et al. (1972) found 54 compounds in malt vinegar, among which halogen compounds could be identified which obviously originated from the raw material.

4.5 Rice Vinegar

MAEKAWA and KODAMA (1964) identified five phenolic acids in rice vinegar. YANAGIDA et al. (1974) found that the percentage of amino acids in rice wine decreased during acetous fermentation. KOIZUMI et al. (1985a, b) compared the free amino acids, organic acids and aroma components in Chinese vinegars with those of Japanese rice vinegars. The Chinese vinegars were rich in alanine, glutamic acid, lysine and leucine. The amount of these amino acids was approximately ten times higher than that of Japanese rice vinegar. In Chinese vinegars lactic acid and pyroglutamic acid were the most dominant acids besides acetic acid. High levels of *n*-propyl alcohol, isobutyl alcohol, acetoin, diacetyl, *n*-butyl acetate, ethyl acetate and furfural were detected among the aroma components.

4.6 Whey Vinegar

BOURGEOIS (1957) analyzed the amino acids of whey vinegar, and HADORN and ZÜRCHER (1973) determined the specific compounds for its characterization.

5 Analysis of Vinegar

Vinegar is analyzed for two different purposes: for process control by general routine methods, and for a comprehensive knowledge of its chemical constituents by special methods. The latter is needed to distinguish between vinegar made by fermentation and diluted synthetic acid and between true wine or fruit vinegar and blends of such vinegars with spirit vinegar or acetic acid.

5.1 Detection of Acetic Acid in Vinegar

There are two possibilities of differentiation between vinegar and acetic acid. On one hand the specific impurities in the acetic acid which go back to its production process may serve as a criterion for differentiation. A survey on standard processes for synthetic acid has been given by STAEGE (1981). However, as acetic acid is produced with high purity these days, a secure identification of polluting substances has become difficult. Therefore, a characterization of the specific compounds in fermentation vinegars is growing in significance. A great variety of qualitative reactions and quantitative determinations are described in the literature.

5.1.1 Older Methods

^{14}C content

A critical presentation of this method can be found in the work of SCHMID et al. (1977). Acetic acid normally has no appreciable ^{14}C activity because of the gradual decay of the raw material radiation (mostly petroleum) during millions of years. Contrary to this, vinegar which ultimately has its origin in plants, has a relatively high ^{14}C radioactivity, as the plants, via their metabolism, are in balance with the CO_2 of the atmosphere and with its natural ^{14}C content.

However, acetic acid occasionally also shows a ^{14}C activity comparable to that of vinegar. A possible reason is that such acetic acid was produced by carbonization of wood. Wood is a recent material which shows a ^{14}C activity comparable to that of CO_2 in the atmosphere.

$^{13}C/^{12}C$ isotopic ratio

Based on the insight that the specific ^{14}C activity determination is a necessary, but not a sufficient criterion for the differentiation of vinegars, SCHMID et al. (1978a, b) performed a systematic determination of the $^{13}C/^{12}C$ isotopic ratio of acetic acid and of vinegar by mass spectroscopy. They found that acetic acid has a higher ^{13}C content than vinegar. Vinegar blends having a content of at least 15 to 20% of acetic acid can be identified as such.

Gas chromatography

KAHN et al. (1966) proved that there are clear characteristics in quality and quantity, especially with regard to esters, organic acids, alcohols, and other specific accompanying substances in different types of vinegar (see Sect. 5.3.1.2). LLAGUNO (1977) analyzed by spectrophotometry and by gas chromatography a number of vinegars to which differing quantities of acetic acid had been added. The author stressed the different contents of amino acids, polyphenols, and volatile acids. However, the great variety of secondary products of metabolism makes it difficult to verify additions of acetic acid which are lower than 20%.

5.1.2 SNIF–NMR Method

Meanwhile newly developed methods offer promising results. The authentication of the origin of vinegars is determined by modern isotopic methods, called SNIF–NMR (nuclear mass resonance spectroscopy) and MR (mass spectroscopy) by the Eurofins Laboratoires, Nantes (DUMOULIN, 1993). The SNIF–NMR method was originally developed at the University of Nantes. The first results have demonstrated the applicability to detect added synthetic acid to vinegar and to find the botanical origin of the raw materials of vinegar. This method is based on the quantitative determination of isotopes in alcoholic liquid (^{2}H-, ^{13}C- and ^{18}O-SNIF–NMR) and was first applied to detect adulterations of wines (CHRISTOPH, 1994). Because of the good correlation between the isotopic parameters of alcohol and those of the corresponding acetic acid, the collection of the alcoholic media was not necessary. To improve the precision in detecting adulterations, a database of all kinds of vinegars and synthetic acid was developed.

Now it is possible not only to determine whether the vinegar is made from wine, apples, malt, cane or beets, but also to determine the country and the region, where the raw material comes from.

The detection threshold of adulterations with synthetic acetic acid is:

- 5% of synthetic acid into a natural vinegar of known geographic origin,
- 10% of synthetic acid into a natural vinegar of unknown geographic origin.

The present procedure for the authentication of vinegars using ^{2}H-NMR and ^{13}C-MS on acetic acid and ^{2}H-MS on water is ready to be made official by including it in the European Code of Practice, which has been developed by the Vinegar Industry Associations CPIV.

5.2 Detection of Alcohol Vinegar in Wine Vinegar

A safe differentiation between true wine vinegars and their blends with spirit vinegar, or generally between vinegars rich in extract and their blends, is of particular importance. Generally, a good method is the identification of the specific fruit acids, such as tartaric acid in wine or malic acid in cider vinegar. A formol titration to identify amino acids may also be applied. On the other hand, fruit-specific acids and also amino acids can easily be added.

5.2.1 Older Methods

The following methods were in use for the differentiation of alcohol from spirit vinegar (for details see Ebner and Follmann, 1983):

- ratio: acidity/dry residue
- UV-absorption
- chromatography
- determination of potassium.

All these methods are not very specific or, as mentioned above, can easily be manipulated.

5.2.2 SNIF–NMR Method

For the detection of alcohol vinegar in wine vinegar, the SNIF–NMR method is the most specific and precise method (see Sect. 5.1.2).

The detection threshold of adulterations with alcohol vinegar is:

- 10% of alcohol vinegar into wine vinegar of known geographic origin,
- 20% of alcohol vinegar into wine vinegar of unknown geographic origin.

5.3 Methods

5.3.1 Routine Methods

5.3.1.1 Acetic Acid

The determination of acetic acid by titration with NaOH and phenolphthalein as indicator is the most frequently used and simplest method. Today this titration can be carried out fully automatically. In such general titrations, minor quantities of other acids normally present, such as tartaric acid, citric acid, lactic acid, and malic acids are calculated as acetic acid.

5.3.1.2 Alcohol

Determination of ethanol by gas chromatography or enzymatic methods is carried out in most laboratories. These are the most precise methods for the specific determination of ethanol. As a standard for the quantitative analysis by gas chromatography, for example, isobutanol can be used.

By the enzymatic method, alcohol is oxidized to acetaldehyde by means of a specific alcohol dehydrogenase (Bergmeyer, 1974), and the reduction of equimolecular quantities of formed $NADPH_2$ is measured at 340 nm.

In practice distillation with subsequent alcoholometric or refractometric determination is the usual method. Occasionally the boiling point of vinegar is determined by an ebulliometer. Results, however, are relatively inaccurate.

While a sufficient precision is obtained with these methods at lower alcohol concentrations (less than 0.5%), higher alcohol values (more than 3%) may lead to substantial distortions due to the presence of measurable quantities of other volatile compounds whose characteristics are similar to those of alcohol. More precise and more specific methods are therefore necessary (above-mentioned).

Occasionally, an analysis of the alcoholic constituents is required in order to decide whether distilled alcohol can be fermented or not. For the determination of the following constituents, gas chromatography is used most frequently: methanol, ethyl acetate, acetaldehyde, pentanol, butyric acid, isobutanol, and fusel oils. The quantity of these compounds greatly influences the quality of the alcohol.

5.3.2 Special Methods

Not all of the substances indicated below are frequently determined. In most cases, the analysis is restricted to total acid, alcohol and sulfurous acid. In wine vinegar and other vinegars rich in extract, sugar-free extract, tartaric acid, sorbite, methanol, glycerol, and acetyl-methyl carbinol are analyzed in addition. Further determinations are: citric acid in lemon vinegar, malic acid in cider vinegar, and phosphate in malt vinegar.

A modern total analysis of vinegar cannot be done without a total gas chromatogram for the identification and quantitative determination of all volatile components. A number of organic acids and sugars can be determined selectively and very accurately by the enzymatic method.

Determination of heavy metals can be realized very precisely with the atomic absorption spectrometer.

5.3.2.1 Total Acids, Volatile Acids and Non-Volatile Acids

Volatile acids consist mainly of acetic acid. Vinegar can be checked for the presence of other volatile acids, as there are tartaric acid, malic acid or citric acid. The determination of volatile acids is made by steam distillation with the addition of NaCl. The distillate is titrated with 0.5 N NaOH. The non-volatile acids can be titrated with 0.1 N NaOH in the distillation residue.

The kind of non-volatile acids is characteristic for the origin of the vinegar and whether it is a blend or not. Non-volatile acids are calculated from the difference between total acid and volatile acids. Calculation is made for acetic acid in g/L. For the determination of total acid see Sect. 5.3.1.1.

5.3.2.2 Sugars

The determination of sugars metabolizable by acetic acid bacteria is important for the assessment of raw material quality. For acetous fermentation the presence of sugar is indispensable. The concentration of sugar required depends on the fermentation process. Especially glucose is important, as this is the main carbon source for *Acetobacter*. One method of determination is the reduction of Fehling's solution with a titration of the cuprous oxide by 0.1 N potassium permanganate. The results obtained by this method are varying and not reproducible. Obviously, substances are formed during fermentation which have reducing effects and which apparently increase the quantity of sugar and give false results (MATHEIS et al., 1994). This is the reason why the enzymatic method (enzyme-test kits) is the recommended method for the determination of glucose and fructose.

5.3.2.3 Extract

Extract is defined as the quantity of soluble solids which the vinegar naturally contains. Determination of extract is only useful for wine and fruit vinegar, without addition of spices or other additives. Due to the addition of spices the results obtained by this method can be falsified. The density of the extract solution can be calculated from the density of the vinegar, the rediluted alcoholic distillate and the volatile acids with the formula of Tabarie. The quantity of extract can be determined with a table in g/L (TANNER and BRUNNER, 1987).

The direct method is the evaporation of filtered vinegar in a platinum dish almost to dryness. The residue is then dissolved in water, and the solution is thickened again. This procedure is repeated two times. The residue is dried for 3 hours at 103 °C. After cooling to room temperature in the desiccator the sample is weighed. It is difficult to obtain reproducible results from the latter method.

5.3.2.4 Ash, Ash Alkalinity

Determination of ash is only useful for wine and fruit vinegar without addition of NaCl or glutamate. The ash residue is determined after evaporation at approx. 500–525 °C and cooling to ambient temperature.

High values of ash alkalinity are an indication for the addition of mineral salts from organic acids such as sodium glutamate.

5.3.2.5 Sulfurous Acid

There are different methods for the determination of total sulfurous acid. Rapid analytical methods are common which are suited both for the raw material and for the finished vinegar. The sulfurous acid is titrated with 1/64 N iodine solution either for total sulfurous acid or for free sulfurous acid, using starch as indicator.

More accurate analyses are obtained by steam distillation. Vinegar is filled into a retort together with phosphoric acid and methanol. While nitrogen is introduced, the vinegar is heated to the boiling point, and sulfurous acid is oxidized to sulfuric acid in hydrogen peroxide. The amount of sulfuric acid is determined by titration of sodium hydroxide (TANNER, 1963).

Free sulfurous acid can be determined by direct iodometric tiration in vinegar using starch as indicator.

5.3.2.6 Density

The density of vinegar is calculated by dividing the mass of the liquid to be analyzed by the mass of the same volume of a reference liquid (water), weighing a pyknometer at 20 °C in an air-filled room.

5.3.2.7 Lactic Acid

For the determination of lactic acid the enzymatic analysis is as well a recommended and precise method. By a specific lactate dehydrogenase, lactic acid is oxidized to pyruvate, in the presence of nicotinamide adenine dinucleotide (NAD). The quantity of the NADH formed is equivalent to the quantity of lactic acid.

6 Food Law Provisions

Legally binding provisions do not yet exist within the member states of the European Union, although for more than ten years now, provisions on vinegar and on acetic acid have been available in some countries to local control committees as guidelines.

Today in four member states, products obtained from the dilution of synthetic acetic acid with water are found on the market. Two of them make no distinction as to the name of these products, and allow both (products obtained from double fermentation and diluted synthetic acetic acid) to be called "vinegar".

In ten EU member states the name "vinegar" is confined to the product obtained by the same process of double fermentation, that is, using alcohol and acetic acid, whatever the raw material used.

6.1 European Code of Practice for Vinegar (1990)

For Europe, the Permanent International Vinegar Committee (CPIV), has been trying for years to set up binding proposals for the harmonization of legal and administrative regulations pertaining to the production and marketing of vinegars in the EU member states. The latest version is that of the year

1990, based on the FAO/WHO Regional Standard for Vinegar from 1987 elaborated by the FAO/WHO Codex Alimentarius Commission. However, it was not adopted by the national legislatures of the EU member states. That is why the CPIV undertook to work for a "harmonization" of the different national provisions relating to vinegar.

The "European Code of Practice for Vinegar" contains the quality criteria as well as the descriptive regulations pertaining to this product group. Purpose of this European Code is to facilitate the free circulation of products within the Single Market, to guarantee product safety in order to protect consumers against fraud and misinformation, and to ensure fair competition (VERBAND DER DEUTSCHEN ESSIGINDUSTRIE, 1992).

There should as well exist a harmonization of methods of analysis within the EU members. For this reason methods have been ring-tested between members of the Technical Commission of the CPIV. The results of the ring-testing will be discussed in the near future.

The following articles are described in the Code of Practice for Vinegar:

I. Description
The name "vinegar" is reserved for the product that is produced exclusively by the biological process of double fermentation, alcoholic and acetous, from liquids or other substances of agricultural origin.

II. As raw materials are used:
1. Wine or grapes, fruit or berries, cider
2. Distilled alcohol of agricultural origin
3. Products of agricultural origin containing starch, sugars or starch and sugars including but not limited to: fruit, berries, cereal grains, malted barley, whey.

III. Processing aids
In order to feed the acetic bacteria organic substances may be used in the required quantities: e.g., malt preparation, liquid starch, glucose and inorganic substances such as phosphates and ammonium salts.

IV. The total acid content of vinegar is in 100 mL not less than 5 g, calculated as acetic acid free of water. The acid content of wine vinegar is not less than 6 g/100 mL, calculated as acetic acid free of water.

V. Admitted ingredients
The following substances may be added to vinegars provided that they are organoleptically discernible:
- Plants or parts of plants including spices and fruits
- Sugar up to 100 g per 1000 mL of the total volume of the product
- Salt up to 100 g per 1000 mL of the total volume of the product, but for malt vinegar up to 4 g per degree of acidity of the product
- Honey up to 100 g per 1000 mL of the total volume of the product, calculated as total sugar content
- Vinegars may have natural and/or concentrated fruit juices added. The addition must – with reference to natural fruit juices – amount to at least 100 g per 1000 mL and to no more than 150 g per 1000 mL of the total volume of the product diluted to 6 g/100 mL.

VI. Admitted additives
(maximum level)
1. 170 mg/kg sulfur dioxide and its salts (E220–E227), calculated as total sulfur dioxide for wine
2. Caramel color (E150), except for wine vinegar
3. L-Ascorbic acid (E300) as antioxidant
4. Flavor enhancers monosodium, monopotassium and calcium glutamate (E620, E621, E623)
5. Natural flavors and natural flavoring substances

VII. The use of the following substances is prohibited in the production of vinegars:
- artificial flavors of all types
- artificial and natural grape oils
- residues of distillation (washy wines of all kinds and derivates), residues of fermentation (dregs, cloudy wine elements) and their by-products

- substances extracted from marc of all types
- acids of all types with the exception of those naturally contained in the raw material used or any substances the addition of which is permitted
- colorings except that mentioned in VI.

VIII. Labeling

1. The name "vinegar" as such or in combination with other words is exclusively reserved for products defined in Article I.
2. Vinegars produced from a single raw material will be marketed as "vinegar" in connection with the indication of the raw material, e.g., "alcohol vinegar", "wine vinegar", "malt vinegar".
3. Vinegars produced from several raw materials are marketed as "vinegar" in conjunction with indication of all the raw materials.
4. Vinegars with additions of Article V (sugar, honey, fruit, juice) will be marketed under their designation completed by a mention like: "with … grams/liter of sugar" etc., provided these additions in the product exceed a quantity of 20 g per 1000 mL of the total volume of the product.
5. The name "vinegar" may be used in connection with an indication of origin (Sherry vinegar, Aceto de vino chianti) if either the raw material used for the product comes from the said region or the product itself was made in the said region or processed according to a special system.
6. The acetic acid content of vinegars, expressed as their total acidity, has to be mentioned on the label as "…% acidity".
7. The nominal volumes, expressed in milliliters or centiliters, have to be mentioned on the label.
8. The vinegars may not be labeled or made up in a way that can mislead consumers about their origin, especially when caramel color (E150) is added as mentioned in IV, 2.

7 References

AURAND, L. W., SINGLETON, J. A., BELL, T. A., ETCHELLS, J. L. (1966), Volatile components in the vapors of natural and distilled vinegars, *J. Food Sci.* **31,** 172–177.

BERGMEYER, H. U. (1974), *Methoden der enzymatischen Analyse*, Weinheim: Verlag Chemie.

BERGNER, K. G., PETRI, H. (1959), Aminosäuren des Branntweinessigs, *Angew. Chem.* **71,** 31–32.

BERGNER, K. G., PETRI, H. (1960a), Aminosäuren des Branntweinessigs. I. Nachweis der Aminosäuren im Branntweinessig und in Essigbakterien, *Z. Lebensm. Unters. Forsch.* **111,** 319–333.

BERGNER, K. G., PETRI, H. (1960b), Aminosäuren des Branntweinessigs. II. Quantitative Bestimmung der Aminosäuren; Unterscheidung von Gärungs- und Essenzessig, *Z. Lebensm. Unters. Forsch.* **111,** 494–504.

BOURGEOIS, J. (1957), Chromatographische und mikrobiologische Bestimmung von Aminosäuren in verschiedenen Essigarten, *Branntweinwirtschaft* **79,** 250–254.

CHRISTOPH, N. (1994), Einsatz der ^{2}H-, ^{13}C, ^{14}C- und ^{18}O-Isotopenanalytik in der Lebensmittel- und Weinüberwachung, *45. Arbeitstagung des Regionalverbandes Bayern* am 21. April 1994, Würzburg.

DUMOULIN, M. (1993), SNIF–NMR, *A New Analysis to Characterize the Plant Origin of Vinegar*, Nantes, France: Eurofins Laboratoires.

EBNER, H., FOLLMANN, H. (1983), Vinegar, in: *Biotechnology* (REHM, H. J., REED, G., Eds.), 1st Ed., Vol. 5, pp. 425–446, Weinheim: Verlag Chemie.

FOLLMANN, H., EBNER, H., (Heinrich Frings GmbH & Co KG) (1985), *US Patent* 4689153.

GALOPPINI, C., ROTINI, O. T. (1956), Ulteriori indagni nella formazione del acetilmetil-carbinolo della fermentazione acetica, *Ann. Fac. Agrar. Univ. Pisa* **17,** 99–111.

GALVEZ, M. C., BARROSO, G., PEREZ-BUSTAMANTE, J. A. (1994), Analysis of polyphenolic compounds of different vinegar samples, *Z. Lebensm. Unters. Forsch.* **199,** 29–31.

GARCIA, O. R., CASABALLIDO-ESTEVEZ, A., CASTANA-TORRES, M. (1973), Vinegars II, Content of diacetyl, acetoin, butylene glycol and alcohol, *Ann. Brom.* **25,** 121–145.

HADORN, H., ZÜRCHER, K. (1973), Herstellung, Analyse und Beurteilung von Molkenessig, *Mitt. Geb. Lebensmittelunters. Hyg.* **64,** 480–503.

JONES, D. D., GREENSHIELDS, R. N. (1969), Volatile constituents of vinegar, I. A survey of some commercially available malt vinegars, *J. Inst. Brew. London* **75,** 457–463.

Jones, D. D., Greenshields, R. N. (1971), Volatile constituents of vinegar, IV. Formation of volatiles in the Frings process and a continuous process of malt vinegar manufacture, *J. Inst. Brew. London* **77,** 160–163.

Kahn, H. J., Nickol, G. B., Conner, H. A. (1966), Vinegar compounds, Analysis of vinegar by gas–liquid chromatogaphy, *Agric. Food Chem.* **14,** 460–465.

Kahn, H. J., Nickol, G. B., Conner, H. A. (1972), Identification of volatile compounds in vinegars by gas chromatography–mass spectrometry, *Agric. Food Chem.* **20,** 214–218.

Koizumi, Y., Nakakoji, T., Yanagida, F. (1985a), General composition, sugar components and inorganic cations of vinegars made in China, *Nippon Shokuhin Kogyo Gakkaishi* **32,** 108–113.

Koizumi, Y., Nakakoji, T, Yanagida, F. (1985b), The free amino acids, organic acids and aroma components of Chinese vinegars, *Nippon Shokuhin Kogyo Gakkaishi* **32,** 288–294.

Llaguno, C. (1977), Quality of Spanish wine vinegars, *Processes Biochem.* **12,** 17–46.

Maekawa, K., Kodama, M. (1964), Isolation and identification of phenolic acids in rice vinegar, *Agric. Biol. Chem.* **28,** 436–442.

Matheis, W., Bourgeois, J., Caperos, J., Cerny, Z, Diller-Zulauf, A., Feusi, J., Helbling, J. (1994), *Gärungsessig, Schweizerisches Lebensmittelbuch.*

Mecca, F., Andreotti, R., Verdonelli, L. (1979), *L'Aceto*, pp 1–433, Brescia (Italy): Edizioni AEB S.p.A.S.

Nunheimer, T. D., Fabian, F. W. (1940), Influence of organic acids, sugars and sodium chloride on strains of food-poisoning staphylococci, *Am. J. Public Health*, 1040–1049.

Petri, H. (1958), Untersuchungen über die Extraktivstoffe des Spritessigs unter besonderer Berücksichtigung der Aminosäuren, *Doctoral Thesis*, Technische Hochschule Stuttgart, pp. 1–92.

Rotini, O. T., Galoppini, C. (1957), L'Acetilmetilcarbinolo, la fermentazione acetica e la genuinita degli aceti del commercio, *Ann. Sper. Agrar.* **11,** 1355–1372.

Schanderl, H., Staudenmayer, T. (1956), Unterscheidung von Gärungsessig und synthetischem Essig durch Aminosäuren, *Z. Lebensm. Unters. Forsch.* **163,** 121–122.

Schmid, E. R., Fogy, I., Kenndler, E. (1977), Beitrag zur Unterscheidung von Gärungsessig und synthetischem Säureessig durch die Bestimmung der spezifischen ^{14}C-Radioaktivität, *Z. Lebensm. Unters. Forsch.* **163,** 121–122.

Schmid, E. R., Fogy, I., Schwarz, P. (1978a), Beitrag zur Unterscheidung von Gärungsessig und synthetischem Säureessig durch die massenspektrometrische Bestimmung des ^{13}C/^{12}C Isotopenverhältnisses, *Z. Lebensm. Unters. Forsch.* **166,** 89–92.

Schmid, E. R., Fogy, I., Kenndler, E. (1978b), Beitrag zur Unterscheidung von Gärungsessig und synthetischem Säureessig durch die Bestimmung der spezifischen ^{3}H-Radioaktivität, *Z. Lebensm. Unters. Forsch.* **166,** 221–224.

Staege, H. (1981), Essigsäure aus Kohle unter Anwendung der Flugstromvergasung nach Koppers-Totzek, *Verfahrenstechnik* **15.**

Szakall, A. (1952), Untersuchungen über das biologisch wirksame Milieu im Gärungsessig, *Z. Lebensm. Unters. Forsch.* **95,** 100–107.

Tanner, H. (1963), Die Bestimmung der gesamten schwefligen Säure in Getränken, Konzentraten und in Essigen, *Mitt. Geb. Lebensmittelunters. Hyg.* **54,** 158–174.

Tanner, H., Brunner, H. R. (1987), *Getränke-Analytik*, Schwäbisch Hall: Verlag Heller Chemie- und Verwaltungsgesellschaft mbH.

Verband der Deutschen Essigindustrie e.V. (1992), Europäische Beurteilungsmerkmale für Essig, – Code of Practice –, *Dtsch. Lebensmittel-Rundsch.* **8,** 251–253.

Yanagida, F., Fukui, J., Kaneko, N., Yamamoto, Y., Koizumi, Y (1974), Acetic acid bacteria and their utilization: XI. Amino acids contained in the vinegar mash made by various material treatments, *Nippon Jozo Kyokai Zasshi* **69,** 759–764.

16 Olive Fermentations

ANTONIO GARRIDO FERNÁNDEZ
PEDRO GARCÍA GARCÍA
MANUEL BRENES BALBUENA

Sevilla, Spain

1 Definition

According to the Qualitative Standard Applying to Table Olives in International Trade prepared by the International Olive Oil Council (IOOC, 1991), "Table olives are the sound fruit of specific varieties of the cultivated olive tree (*Olea europaea sativa* Hoffm. et Link) harvested at the proper stage of ripeness and whose quality is such that, when they are suitably processed as specified in this standard, produce an edible product and ensure its good preservation as a marketable good. Such processing may include the addition of various products or spices of good table quality". In general, any processing method aims to remove the natural bitterness of this fruit, caused by the glucoside oleuropein, and improve its organoleptic characteristics. The elaborated fruit is used as a food or an appetizer.

2 Historical Evolution

The origin of the cultivation of the olive tree is known through legends and tradition. Olives grew wild in Syria and Asia Minor as early as 3500 B.C. When it was discovered that a flavorsome oil could be extracted from olives, the Assyrians cultivated the wild olive from a thorny bush into the tree we know today. Olive cultivation spread to Jerusalem some 2000 years before Christ. The first reference to the existence of the olive tree is found in the Old Testament where in Genesis the flight of the dove with the olive branch is described, announcing the end of the flood. Several Mediterranean people (Phoenicians, Hebrews, Carthaginians, Romans, Arabs, etc.) took turns through history to extend olive tree cultivation until it reached the Atlantic Ocean (Loussert and Brouse, 1980).

During the 16th century colonizers and Spanish Franciscan monks extended the planting of the tree to various points of the new continent (Fernández Díez, 1971). The first three trees were carried to Lima (Peru) in 1560, one of which was stolen and taken to Chile. From Peru the olive seeds and cuttings were carried to Mexico and California by Spanish monks at about 1775. In 1844 olives were taken to South Australia from Sicily (Hartmann, 1962). Olives are now also moderately distributed in Japan and South Africa, completing their current extension in both hemispheres within two bands distributed approximately between latitudes 25 to 45° North and 15 to 35° South (Fernández Díez, 1983). More detailed information on the history of the olive tree has been described by Garoglio (1950), Hartmann (1953), Patac et al. (1954), Foytik (1960), and Goor (1966).

Greeks and, later, Romans thought up many ways for preparing olives for consumption. The oldest written reference to the preparation of fruits for eating is that of Columela (54 B.C.). In his *Treatise on Agriculture* (Vol. II), he describes several recipes for treating table olives, according to their variety and degree of ripeness. The natural bitterness was removed by immersion in a suspension of alkaline ashes, vinegar or just water. Subsequently, fruits were kept in brine with fennel, mint and other aromatic herbs. Must, wine and grape syrup were also used as substitutes for brines. Some of these Greek and Roman preparations can still be found in rural Mediterranean areas.

Consumption of table olives was restricted to the producing regions until the 19th century, when new markets outside the Mediterranean area were found. In 1844, table olives from Seville were sent to Puerto Rico and Cuba, and later to New York. In 1900 it was discovered in California that olives could be canned like other fruits, and in 1916 the first factory was built. Since then, and especially during the last three decades, processing of table olives has expanded to practically all olive-growing zones.

During the 20th century, systematic basic and technological research has been carried out in several countries. Classic reports on the different table olive types were published by Cruess (1958), Borbolla y Alcalá and Gómez Herrera (1949) and Balatsouras (1966). Research has continued to the present day and plays an important role in supporting the well-established table olive industry throughout the world.

3 Table Olive Production

Table olive production has developed rapidly since 1970 when big fermentation vessels (15000 L capacity) were introduced to replace the old wooden barrels which were traditionally used. In 1969/70, world production was estimated at 461000 tons (GUERBAA, 1990). In 1991/92 world production had reached 893000 tons (ANONYMOUS, 1993), which represents an increase of approximately 94%, in the last twenty years. The increases in production and consumption have reflected increases in the population rather than real per capita increases in consumption. In any case, this industry is now an economically vital area that must be carefully studied by technologists to maintain and upgrade the quality demanded by the consumer (FERNÁNDEZ DÍEZ, 1991). The current main table olive producing countries, according to the IOOC estimation for the 1991/92 crop (ANONYMOUS, 1993), are listed in Tab. 1. The European Union is nowadays the greatest producer with 435000 tons, which represents about half of worldwide production. Three quarters of the processed olives (about 700000 tons per year) are still consumed in the producing countries and the rest (around 200000 tons per year) are exported. Only good quality olives are destined for exportation (mainly green olives in brine) the main exporters being: Spain (50%), Morocco (15%) and Greece (10%). The United States of America is the main importer (in addition to being a big producer), receiving about 40% of the international trade, followed by France (10%), Brazil (10%) and Canada (6%).

Tab. 1. Table Olive Major Producing Countries, 1991/92 Crop (ANONYMOUS, 1993)

Country	Production (·1000 tons)
Spain	227
Turkey	110
Italy	100
Morocco	85
Greece	80
United States of America	57
Syria	56
Argentina	30
Portugal	25
Israel	15
Tunisia	14
Jordan	13
Others	81
Total	893

The distribution of production among the main trade preparations is as follows: treated green olives in brine (Spanish style), 40%; natural black olives in brine, 30%; and treated black olives (ripe olives) in brine, 15%. The rest (about 15%) are produced by other methods and are mainly destined for local markets. There is a tendency towards an increase in the production of green and ripe table olives in brine, whereas, as a result of the higher processing costs and spoilage involved, production and processing of natural black olives in brine is fairly stable.

4 Characteristics and Composition of the Fresh Fruits

The olive is a meaty, intensely bitter, stoned fruit. Its shape is roundish and generally elongated. During ripening the color changes from green to purple or red. Its total weight may vary between 0.5 to 20 g, but it generally falls within the range of 0.5 to 10 g. The length of the fruit is usually between 2 and 3 cm. The specific weight is close to unity. The pulp accounts for 70 to 90% of the weight of the fruit and the endocarp accounts for the remaining 10 to 30%. The seed accounts for less than 10% of the weight of the stone. The physical and chemical characteristics of the fruits depend on many factors, including variety, ripeness at the time of harvesting, geographical location, type of cultivation, on irrigated land or on dry non-irrigated but arable land, etc. (FERNÁNDEZ DÍEZ, 1983).

A range of composition of the fruit and the fresh pulp is shown in Tab. 2. More details

Tab. 2. The Composition of the Fruit of the Olive Tree and of the Fresh Pulp (FERNÁNDEZ DÍEZ, 1983, 1991)

			Weight (%)
Fruit			
Pericarp	Epicarp		
	Mesocarp	Pulp	70–90
Endocarp		Stone	9–27
		Seed	1–3
Pulp			
Moisture			50–75
Lipids (oil)			6–30
Reducing sugars, soluble			2–6
Non-reducing sugars, soluble			0.1–0.3
Crude protein (N × 6.25)			1–3
Fiber			1–4
Phenolic compounds			1–3
Organic acids			0.5–1
Pectic substances			0.3–0.6
Ash			0.6–1
Others			3–7

may be found in the reviews by CRUESS et al. (1939), SANDRET (1957), VÁZQUEZ RONCERO (1963, 1964), FERNÁNDEZ DÍEZ et al. (1985), PETRUCCIOLI (1965) and BALATSOURAS et al. (1988). These give extensive information on the composition of Californian, Moroccan, Spanish, Italian and Greek varieties, respectively.

From the processing standpoint, the flesh/stone ratio should be as high as possible and always greater than 5. The epicarp must be thin, flexible and resistant to shock, although it must allow uniform penetration of sodium hydroxide and sodium chloride. The stone should be easily separated from the pulp or flesh. As a general rule, the weight and volume of the fruits, and the flesh to pit ratio increase during growth and maturation.

Oil, after water, is the major component of the olive pulp, but it plays hardly any role in table olive processing. Its content increases during growth and ripening. Table olive varieties must contain a certain amount of fermentable sugars, ideally higher than 4%. The major soluble sugars in order of decreasing importance are glucose (1–3%), fructose (0.1–1.1%) and, to a minor degree, sucrose (0.1–0.2%). Mannitol (0.2–1.5%) is also present, sometimes in high proportions (FERNÁNDEZ BOLAÑOS et al., 1982). The presence of xylan, galactose and mannose in overripe fruits has also been reported (SANDRET, 1958).

Organic acids (mainly citric, oxalic and malic acids) and their salts are present in the juice of the fruits in concentrations between 0.5 and 1% of the weight of pulp. These acids, as well as those produced during fermentation, are responsible for the buffering capacity of brines and fermented products (FERNÁNDEZ DÍEZ and GONZÁLEZ PELLISSÓ, 1956; VLAHOV et al., 1975). A ripening index based on the acid content (malic/citric ratio equal to 1), has been proposed to define the appropriate times for harvesting olives for processing as treated green olives in brine (VLAHOV, 1976).

The content in phenolic components, which are almost exclusively *ortho*-diphenols, is between 1 and 3% of the weight of the pulp, expressed as tannic acid. They play a fundamental role during fermentation and in the darkening process for preparing ripe olives in brine. Oleuropein is the phenolic compound chiefly responsible for the bitterness of olive fruits. Its structure was elucidated by PANIZZI et al. (1960). VÁZQUEZ RONCERO et al. (1974) and AMIOT et al. (1986) have studied in detail the phenols of olives, reporting the presence of glucosides, β-(3,4-dihydroxyphenyl)ethanol, cinnamic acids, flavonoids, verbascoside, rutin, and luteolin-7 glucoside, in addition to oleuropein. Their concentrations diminish as ripening progresses and drop to nearly half of their initial concentrations in ripe fruits. The lower the phenol content, the more appropriate is the variety for producing treated green olives or natural black olives in brine. It has been proposed that oleuropein and ligstroside form part of a multichemical mechanism defending bitter olives against insect and microbial attack (KUBO et al., 1985). Fresh fruits are also rich in catechol oxidase, which is responsible for the enzymatic browning of green olives, when they are damaged during picking. This enzyme has been studied in detail during ripening of the fruit by BEN SHALOM et al. (1977, 1978).

As ripening progresses, olives become pink to purple or black. This color change is due to the formation of anthocyanins. Vázquez Roncero et al. (1970) have identified six of them. In all cases, their structures were derived from the 3-cyanidin glucoside. A similar study with Italian varieties has been reported by Vlahov (1990) who found that of all the anthocyanins the monitoring of the formation of cyanidin-3-rutinoside gave the best measure of the progress of the ripening process. These pigments are responsible for the color of natural black olives.

The superficial color of fresh green olives is due to the presence of chlorophylls. Carotenes (β-carotene, lutein, luteoxanthin, violaxanthin and neoxanthin) are also chloroplastic pigments found in these fruits, although in lower concentrations. These pigments reach their highest levels during the first stage of growth. As the fruit ripens, their concentrations show a progressive decrease (Mínguez Mosquera and Garrido Fernández, 1989). Differences in the concentrations of these pigments between varieties may be important and may influence the quality of the final product (Mínguez Mosquera et al., 1990).

The olive flesh contains 1–3% of protein on a fresh basis. This level remains constant throughout ripening (Narasaki and Katakura, 1954). Non-protein nitrogen represents less than 1% of the total nitrogen. All the common amino acids are present in the free amino acid pool. Arginine, alanine, aspartic acid, glutamic acid, and glycine constitute approximately 60% of the free amino acids. Olive fruit protein contains all the common amino acids present in other plant proteins (Manoukas et al., 1973). Other researchers have found a rapid increase in protein content, when the superficial color of the fruit changes from green to yellow, maximum levels being reached during the subsequent ripening (Raya et al., 1977). Their rapid solubilization into the brine is important for an adequate lactic acid process, since lactobacilli need rich nutritive media for growth.

The pectic substances in olives, which are of great importance for the texture of the fruit, have been investigated by Chung et al. (1974) and Mínguez Mosquera (1982). They represent about 0.3–0.6% of the weight of the fresh pulp, this value remaining more or less stable throughout ripening, although qualitative changes may occur. Mínguez Mosquera et al. (1978) found the pectinolytic enzymes, pectin methyl esterase and polygalacturonase to be present, and Castillo Gómez et al. (1979) demonstrated the presence of a natural polygalacturonase inhibitor. Marsilio and Salinas (1990) have stressed the importance of the initial content in pectic substances and the behavior of these during processing, for the quality of the final product.

Fiber represents the fundamental support for the structure of the fruit. Its major components are: cellulose, hemicellulose and lignin, with cellulose accounting for more than 50%. When the degree of maturity advances, enzymatic degradation may produce changes in the proportions of the components. Thus, as ripening proceeds, the percentage of cellulose decreases from 40–45% to 23%, the percentage of lignin remains almost constant, ca. 33–38%, while hemicellulose increases from 22–23% to 41%, when fruits are completely ripe (Heredia Moreno and Fernández Bolaños, 1987). The presence of cellulolytic enzymes and glycosidases, which appear in completely mature fruits, and their role in the texture of the final products are still under investigation (Heredia Moreno et al., 1993).

The percentage of ash varies from 0.6 to 1.0%. The ash includes, in order of decreasing importance: K, Ca, P, Na, Mg, S and in lower amounts, Fe, Zn, Cu, and Mn. Their proportions remain unchanged during ripening (Fernández Díez, 1983).

5 Main Varieties Used for Table Olives

Of all the olive varieties that exist, only those having appropriate characteristics (see Sect. 3) are used for table olive processing, and even fewer varieties are used for industrial preparations and international trade. There are some varieties which are destined exclusively for industrial processing, although others may, and have been, used, especially during the last decades. The extent to which the other varieties are used depends on the availability of the main industrial varieties. Normal fluctuations in olive tree yields have been mitigated by this practice. In fact, dual purpose varieties play an important role in market stability.

The different types of table olive require specific quality attributes in the final product, and these are only obtained with certain varieties. Consequently, there is a close relationship between type of table olive and olive variety. Those varieties used to produce more than one type, do not achieve good standards in all of them. Varieties of olive trees in the different countries of the Mediterranean area were originally autochthonous, although a progressive dissemination of the most suitable varieties throughout this and other olive growing regions, is nowadays evident. Changes in climate, soil, and cultivation methods, however, may significantly modify the physico-chemical characteristics of the olive as well as induce differences in fermentability and final quality. Thus, importation of a good table olive variety is not always a guarantee of success. In many cases olive tree varieties are not yet well known, so national or local names rather than a botanical classification are used to identify them.

There are two Spanish varieties which are appreciated throughout the world for table olive production and have been introduced progressively in most of the other producing countries. These varieties also account for the highest percentages of international trade. Some of their characteristics follow:

Gordal (*Olea europaea regalis*, Clemente). This variety is also known as Gordal Sevillano, Sevillano or Sevillana and is mainly produced in lower Andalusia, particularly in the province of Seville. The fruit is large (an average of 100–120 olives/kg with a flesh to stone ratio of 7.5:1), ellipsoidal in shape and has an indentation at the stem that gives it a slightly heart-shaped form. The epidermis is thin and speckled with characteristic white spots; the flesh has an agreeable texture and has a fairly light-green color, which changes to purplish-black when ripe. The stone is regular in shape with deep striations. The oil content is low, about 15% of pulp weight. It ripens early and requires mild, warm climates. It is used mainly for green olives.

Manzanilla (*Olea europaea pomiformis*). It is also known as Manzanilla and is cultivated widely throughout Spain although it fruits best in a mild climate such as that found in the Guadalquivir valley and the province of Seville. Subvarieties are found in other areas: Fina (around Seville), Carrasqueña (in Extremadura) and Serrana (in mountainous zones). The fruit is medium-sized, yielding 200–280 olives/kg and has a flesh to stone ratio of 6:1. It is apple-shaped, thin-skinned and its flesh has an excellent texture. It is light green-colored and spotted with tiny whitish dots. It matures to a black shade with a hint of violet. More elongated than the fruit, the stone is relatively small and less ridgid than the Gordal stone. This variety reaches full ripeness later and its oil content is higher, accounting for about 20% of the pulp. Manzanilla crops are used almost entirely for processing as Spanish or Sevillian-style green olives. Along with the Gordal, it is one of the varieties that enjoys greatest international fame for the excellence of its flesh and its extraordinary organoleptic characteristics.

Tab. 3 shows the main table olive varieties in the major olive-producing countries. Exhaustive information on varieties from all table olive-producing countries can be found in FERNÁNDEZ DÍEZ et al. (1985), UEBA (1990), FERNÁNDEZ DÍEZ (1991), and MARSILIO (1993).

6 Trade Preparations

These have been well established by the International Olive Oil Council (1991) in the Unified Qualitative Standard Applying to Table Olives in International Trade. This section only gives a brief description of the three most important ones. A complete definition of all of them can be found in this standard.

The denomination of commercial preparations, or its equivalent trade preparations, is used for all products destined for consumers.

Tab. 3. Main Olive Varieties for Table Use in the Major Producing Countries (UEBA, 1990)

Country	Varieties
Algeria	Sigoise, Sevillana, Azeradj...
Argentina	Arauco, Empeltre, Sevillana, Manzanilla, Ascolano, Asignola...
Greece	Conservolea, Kalamata, Chalkidiki, Megaritici, Kothreici, Amigdalolea, Thrubolea, Karidolea...
Italy	Ascolana, Grossa di Spagna, Cucco, Itrana, Nocellana Etnea, Pasola, Sant' Agostino, Uovo di Piccioni...
Morocco	Moroccan Picholine or Zitoun, Soussia, Meslala, Gordal, Hojiblanca...
Portugal	Carrasquenha, Redondil, Cordovil, Bical, Negrinha, Verdial, Gordal, Manzanilla...
Spain	Gordal, Manzanilla, Morona, Hojiblanca, Cacereña, Verdial, Picual, Lechin, Rapazalla, Aloreña...
Syria	Jlot, Kaiamani-Kachali, Sourani...
Tunisia	Meski, Picholine, Zarasi, Barouni...
Turkey	Domat, Gomlik, Memecik, Edremit or Ayvalik, Izmir, Sofralik, Bugarree, Memeli, Uslu...
U.S.A.	Mission, Manzanilla, Sevillano, Ascolano, Barouni...
Israel	Manzanilla, Santa Catalina, Merharia, Kalamata, Souri, Nabali, Kadesh...
Chile	Azapa, Huasco, Manzanilla, Empeltre, Ascolano, Siguria, Santa Emiliana...
Australia	Verdial, Manzanilla, Gordal...
Cyprus	Ladoelia, Gordal, Manzanilla, Cucco, Kalamata...
Iraq	Ashasi...

They are diverse and include not only traditional methods of processing, but also those derived from them and improved by new technologies. Generally, the complete name includes information on: (a) the type of raw material; (b) the procedure used for eliminating the bitterness; and (c) the method of preserving the product. As mentioned above, classification of table olives is first made on the degree of ripeness (color) of the raw material (or of the final product if the color of this does not coincide with that of the original fruits). This establishes the so-called olive types. There are four: green, turning color, ripe olives (blackened by oxidation), and natural black olives.

In general, the aim of processing methods is mainly focused on decreasing or completely removing the natural bitterness of the olive. If this is achieved by alkaline hydrolysis – treating olives with an aqueous solution of sodium hydroxide before fermentation – the word "treated" is included in the denomination of the commercial preparation. Usually olives processed in this way become absolutely sweet. On the contrary, oleuropein may be slowly and partially eliminated during fruit preservation or fermentation in brine (or by any other aqueous solutions). In these cases, the word "untreated" appears in the denomination.

The third part of the definition refers to the preservation system used for the product. The most generally used medium for olive preservation is brine (NaCl solution). However, dry salt has also been traditionally used. Thus, expressions such as "in brine", "in dry salt", etc. may conclude commercial denominations. The main commercial preparations are described below.

Treated green olives in brine. These are treated in an alkaline lye and then packed in brine in which they undergo a complete or partial (spontaneous or controlled) lactic fermentation. Olives which have undergone complete fermentation are known as Sevillian or Spanish-style olives and need only appropriate physico-chemical conditions during final packing to ensure their preservation. Partially fermented olives must be subsequently preserved at a pH within the limits specified in the standards, by sterilization, pasteurization, addition of preservatives, refrigeration or by inert gas (without brine), this latter being a new method of commercial preservation which has little importance at present. These are generally known as green olives (in brine).

Untreated natural black olives in brine. These are placed directly in brine. They have a fruity flavor which is more marked than that of treated black olives, and they usually retain a slightly bitter taste. They are preserved by natural fermentation in brine alone, by sterilization or pasteurization or by the addition of a preservative. They are usually known as natural black olives (in brine).

Black (ripe) olives in brine (darkened by oxidation). These are obtained from fruits which, when not fully ripe, have been darkened by oxidation and whose bitterness has been removed by treatment with alkaline lye. They must be preserved by heat sterilization when packed under anaerobic conditions. These are known as ripe olives.

Other trade preparations are: Untreated black olives in dry salt; untreated naturally shrivelled black olives; untreated pierced black olives in dry salt; dehydrated black olives, etc. Several of these are also included in the so-called specialities which comprise any other preparation not explicitly defined.

Finally, according to the IOOC Standards (IOOC, 1991) olives may be presented in one of the following styles, depending on the type and trade preparation: whole, pitted, stuffed, halved, quartered, divided, sliced, chopped or minced, broken, olive paste, salad, as a mixture of olives and capers, etc. Products for stuffing olives are also diverse (peppers, pepper paste, anchovies, almonds, etc.).

7 Characteristics of the Major Fermentation Processes

The schemes for processing the three main commercial table olive preparations are shown in Fig. 1. Some operations are common to all of them, and these will be described together. They differ, however, in the degree of ripeness of the raw material. The optimum stage for picking is very different for each preparation. For the green type, fruits must have reached a green-straw yellow color, and the flesh should be easily freed from the pit by twisting, after the fruit has been cut around the transverse diameter. Natural black olives are picked when fruits are completely ripe. A good stage of maturity is considered to have been reached, if the red purple color extends at least half between the skin and stone. Ripe olives are harvested at almost the same time as green olives. Nevertheless, some producers prefer the olives to be harvested a little earlier and others a little later. Their preference depends on their experience and their markets.

7.1 Harvesting and Transportation

Picking costs nowadays are about 50% of the raw material price. Harvesting is still done by hand, although numerous attempts have been made to harvest mechanically. Tree shakers and other mechanical devices have not yet been accepted because of the damage they cause to the trees and, particularly, to the fruits. There is currently a great deal of research and experimentation underway to develop appropriate shakers and handling conditions that avoid fruit bruising. Immediately after picking, small fruits, leaves and other extraneous materials are removed.

Fresh fruits are generally transported to the factory in metallic bins with perforated plastic walls which allow access of air. The capacity of the bins is about 500 kg, and the fruit remains in them for periods varying from several hours to 3 to 4 days depending on the variety, trade preparation and processing method. Transportation in bulk, using 10 to 20 ton trucks, is increasingly used, although this seriously affects the fruit quality.

7.2 Prior Handling

Certain characteristics of shipments and fruits are recorded when the fruit reaches the factory to establish the origin of the raw material and ensure final product with appropriate characteristics. Among the details re-

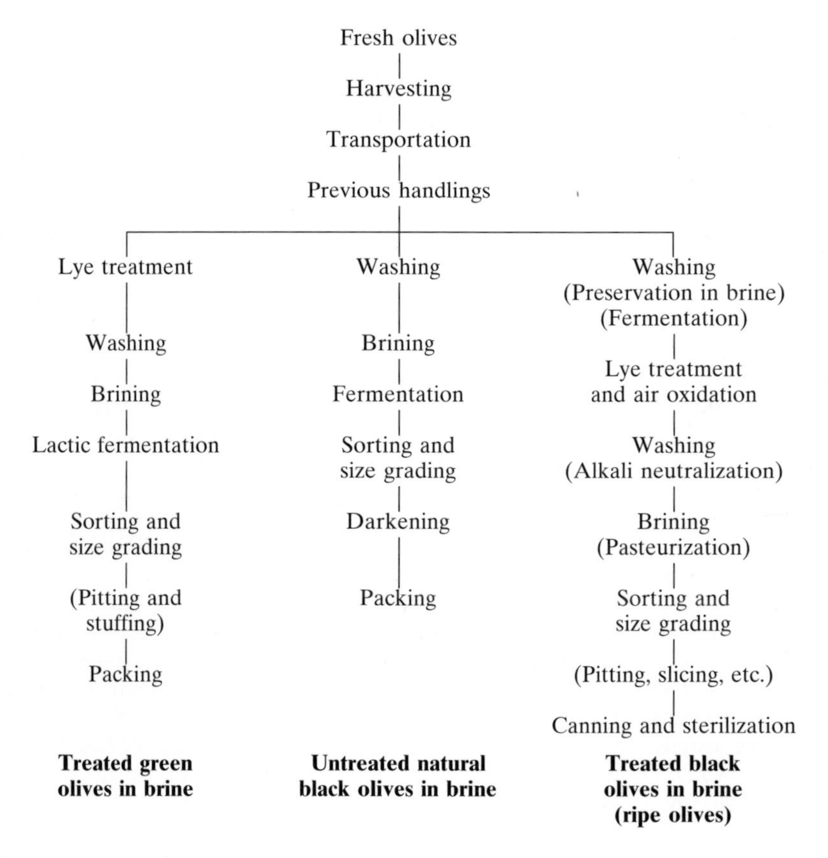

Fig. 1. Processing schemes for the main table olive trade preparations. Operations in parentheses are optional.

corded are: supplier, cultivation system, date of harvesting, date of arrival at the factory, variety, stage of maturity, net weight, and date of processing. Average size and size distribution as well as the percentage of defects, calculated from a suitable sample, are used for fixing the price.

The fresh fruits are sometimes sorted and graded according to size. This has its advantages (it gives more information on the available sizes, allows for more homogeneous lye treatment, etc.), but has the drawback of causing damage to fruits and, consequently, quality losses. For this reason, most companies prefer to sort and grade after fermentation.

The containers used for further processing of all commercial olive preparations, except darkening tanks, are nowadays cylindrical or spherical fermenters made of polyester and glass fiber. These can be closed completely to create anaerobic conditions and to exclude the growth of film yeasts. Such fermenters have a capacity of about 10 tons of fruit and 5000 L of brine. As most factories place these tanks underground, special pumps and other devices have been developed for good brine recirculation and fruit unloading.

7.3 Green Table Olives in Brine

According to the Unified Qualitative Standard Applying to Table Olives in International Trade (IOOC, 1991), these are defined as olives treated in an alkaline lye and then placed in brine in which they undergo a lactic fermentation.

7.3.1 Lye Treatment and Washing

Both treatments aim to facilitate the later lactic fermentation. The strength of the lye (NaOH solution), the time of immersion in this solution and the degree of penetration achieved may be varied according to a series of factors (BORBOLLA Y ALCALÁ, 1981b), the most important being variety, maturity of the fruit and temperature. Good control of these factors is essential for the regulation of the speed, the effectiveness and the yield of the fermentation. Furthermore, such control is critical in determining the organoleptic quality of the final product, principally its color and texture. The lye concentration to be used generally varies between 1.3 and 2.6% (w/v), although it may reach up to 3.5% (w/v) in the coldest zones. This percentage is fixed in such a way that the treatment takes a determined number of hours to give a suitable penetration – two-thirds to three-quarters the distance between the skin and the stone – for each variety. In general, an appropriate time for the majority of the varieties is 5–7 h, except for Gordal and Ascolano, which, because of their softer texture, require a slower treatment, around 9–10 h, with more diluted lyes. The equilibrium between lye concentration, lye penetration and treatment time constitutes the so-called strength of the treatment. Working around the optimum is essential, if a final product with an appropriate quality is to be achieved. Low NaOH concentrations (and, thus, long treatments) lead to olives with an only fairly acceptable color and to poor fermentations (even in the absence of lactobacilli). On the contrary, high lye concentrations cause texture deficiencies and excessive losses of fermentable substrates. Pilot experiments at the beginning of each season are recommended for establishing the most suitable lye concentrations every year.

The length and number of washes are equally decisive (BORBOLLA Y ALCALÁ and REJANO NAVARRO, 1978). A large part of the sodium hydroxide that has remained in the fruit as a result of the alkaline treatment, and a portion of the sugars in the pulp are removed by washing. An excessive number of washes can deplete the available substrates and these would have to be added later to achieve an adequate fermentation. Long washes can facilitate undesirable bacterial contamination of fruits and this, in turn, may prolong the first stage of the fermentation. Short washes lead to the retention of high concentrations of bound organic acids (residual lye) in the fruits and fermentation brines, which, as will be described later, prevents suitable final pH values to be achieved. A rapid rinse after the alkaline treatment followed by a first wash of 2–3 h and a second one of 10–20 h is adequate. This system has been changed recently. Now one of the washes has been replaced by a partial neutralization of the residual lye by food-grade HCl or another strong acid. The acid is added in two operations: half of the required amount is included in the fresh brine and the rest after fermentation. In this case, olives are given only one wash of 18–20 h. Results are similar to those obtained with the traditional procedure, but washing waters are reduced by 50%. This helps a great deal in cutting down the pollution caused by these solutions (REJANO NAVARRO et al., 1986).

7.3.2 Brining

The concentration of the initial brine is also very important (BORBOLLA Y ALCALÁ, 1979). If the concentration of salt is too low, the low osmotic pressure can lead to spoilage by sporulating microorganisms of the clostridial type during the first stage of the fermentation, and the pH may remain too high. If, on the contrary, the concentration of salt is too high, the olives may become wrinkled, sometimes irreversibly. Depending on the variety and the maturity of the fruits, the concentration of salt must be initially regulated between 10 and 12% (w/v).

After the initial decrease of the salt concentration, it should be increased gradually in order to maintain good fermentation conditions and, at the same time, protect the fruits from microbial spoilage. However, the concentration of salt must never be so high as to interfere with the growth of lactobacilli. As an approximation, the concentration of NaCl should be between 5 and 6% during the fermentation, rising to 7% at the end of it and to

8% or even higher during the period of storage, to avoid the growth of propionibacteria during this phase and the consequent depletion in the lactic acid content.

In spite of careful washing, the initial pH of brine may be higher than 7. Microbial growth rapidly initiates brine acidification, but to reduce the length of this period the initial brine may be acidified with food-grade HCl or acetic acid or the pH of the brine corrected by injecting CO_2 into the fermenters during the first week after brining (FERNÁNDEZ DÍEZ et al., 1985).

7.3.3 Characteristics of the Fermentation Media

When olives are brined, the aqueous solution becomes a good medium for growth of microorganisms. In addition to the salt content and pH conditions, which have already been described, osmotic exchange leads to leaching of substrates, vitamins and minerals into the brine, which is thus increasingly capable of supporting organisms with higher nutrient demands. A typical brine composition has been published by MONTAÑO et al. (1993) and is reproduced in Tab. 4. These authors have also shown that an equilibrium between olives and brine can be reached in 5–7 days. The main fermentable substrate is glucose, followed by fructose and mannitol. Sucrose and organic acids may be an important source of carbon during fermentation. Part of the organic acids remain as salts after the lye treatment, and these give the brine a strong buffer capacity. The organic acid concentration depends on the lye and washing operations. It is usually measured as moles per liter, and called "combined acidity" or "residual lye". It should be around 100 mmol/L. The comparatively high proportion of acetic acid found may be attributable to the transformation of sugars following alkaline treatment.

Tab. 4. Chemical Characteristics of Aseptic Olive Brine Seven Days After Brining (MONTAÑO et al., 1993)

pH	10.00
Sodium chloride (%)	5.60
Carbohydrates (g/L)	
Glucose	5.10
Fructose	3.10
Mannitol	1.40
Sucrose	0.50
Organic acids (g/L)	
Malic acid	0.80
Citric acid	0.40
Acetic acid	0.70
Volatile compounds (g/L)	
Methanol	0.40
Ethanol	0.08

Oleuropein is partially hydrolyzed during the alkaline treatment. Thus the phenolic compounds in brine are: a small proportion of oleuropein, hydroxytyrosol, tyrosol, and other derivatives in small amounts. However, when lye treatments are deficient, the oleuropein content is higher and that of the other phenolic compounds lower (BRENES BALBUENA et al., 1990). Total phenols, expressed as tannic acid, may reach over 1 g/L. They play an important role in olive fermentation, as will be discussed in Sect. 8.

The following vitamins have been found in green olive brines per mL: nicotinic acid (maximum amount 2.5 μg); biotin (40 ng); pantothenic acid (around 300 ng); vitamin B_6 (750 ng). Amino acids have also been shown to be present: tryptophan (less than 0.2 μg); methionine (0.6 μg); glutamic acid (5 μg) and phenylalanine (0.5 mg) (RUIZ BARBA, 1991).

7.3.4 Microbiological Growth During Fermentation

The fermentative process of green table olives in brine is still spontaneous in most cases. Growth of microorganisms begins as soon as olives are put into the brine. The physicochemical characteristics of the brines cause a natural selection of organisms as fermentation progresses. There is general agreement among authors that this fermentation can be considered in three stages (BORBOLLA Y ALCALÁ, 1981a; VAUGHN, 1985), although with the new technology that uses fermenters, a fourth stage is also evident.

7.3.4.1 First Stage

At the beginning of brining, a complex population of organisms coexists. According to BORBOLLA Y ALCALÁ and GONZÁLEZ CANCHO (1975), these organisms arise from contamination of fruits and fermenters. During the first few days of brining it is usual to find some molds. These are typical aerobic organisms that produce films, and their growth is non-viable under normal anaerobic conditions. However, it must be emphasized that their presence must be avoided throughout the fermentation process, since they may produce pectinolytic enzymes that cause fruit softening. Mold growth can be avoided by fitting fermenters with hermetic seals.

Organisms of the genus *Bacillus* are also common, as they are widespread in nature and have relatively simple nutritive requirements. Some of the species isolated (*B. subtilis, B. megaterium, B. pumilus, B. polymixa-macerans,* etc.) may also produce pectinolytic enzymes and cause softening in olives. Nevertheless, during the fermentation process these organisms are insignificant, except when there is much contamination, since the rapid drop in pH produced during the normal process prevents their growth.

The microorganisms that best characterize this stage are non-sporulating Gram-negative bacteria, which are present in abundance from the very start of brining, as a contaminant from the washing waters in which they develop progressively. They are capable of growing in simple, high pH media, such as waters and brines after lye treatment. They are always found during the first phase and part of the second and are responsible for the great volume of gas produced during the first days of fermentation. Tab. 5 shows the different species found. The most frequently isolated are: *Enterobacter cloacae, Citrobacter freundii, Klebsiella aerogenes, Flavobacterium diffusum, Aerochromobacter superficialis, Escherichia coli* and *Aeromonas* spp. The most important group is the *coli-aerogenes* including the genera: *Enterobacter, Citrobacter, Klebsiella* and *Escherichia.* These are characterized by their ability to grow anaerobically and their fermentation of sugars to give, depending on genus and species, different end

Tab. 5. The More Relevant Species of Non-Sporulating Gram-Negative Bacteria in the First Stage of Fermentation of Treated Green Table Olives in Brine (Spanish Style) (BORBOLLA Y ALCALÁ and GONZÁLEZ CANCHO, 1975)

Species	pH Range
Coli aerogenes	
Enterobacter cloacae	8.5–5.75
Citrobacter freundii	7.2–5.75
Citrobacter intermedius	8.2–5.75
Klebsiella aerogenes	8.40–5.75
Escherichia coli	5.75
Others	
Aeromonas spp.	8.50–6.00
Aeromonas hydrophila	5.75
Pseudomonas spp.	6.00
Achromobacter superficialis	8.35–6.20
Achromobacter spp.	6.40–5.75
Flavobacterium diffusum	8.60–8.20
Flavobacterium balustinum	8.35–8.30

products (CO_2, H_2, acetic acid, lactic acid, ethanol, etc.). In addition to *Aeromonas* spp., which also produce gas, they may produce overflows in fermenters, spoilage of fruits and depletion of the available sugars to the detriment of the main lactic fermentation. However, *Pseudomonas, Achromobacter* and *Flavobacterium* use glucose oxidatively and do not produce gas. The maximum growth of Gram-negative bacteria is reached around 48 h after brining, and from this moment their population diminishes gradually to disappear by the end of the second stage (12–15 days). Isolation of these organisms is normal at pH values ranging from 8.5 to 5.75. Below pH 5.75, they are hardly found, growth of these organisms being strongly inhibited when the pH is lower than 4.5. In fact, their decrease parallels the decrease in pH (BORBOLLA Y ALCALÁ and GONZÁLEZ CANCHO, 1975).

As fermentation progresses, Gram-positive lactic cocci also begin to grow. Some of the strains belong to the genus *Micrococcus.* These use glucose oxidatively and produce a small amount of acid. Their growth in brine is inhibited after a few days. However, strains of the genus *Pediococcus* are more frequently found in normal processes during the first stage and at the beginning of the second phase of fermentation. Members of this genus are relatively tolerant to acids, and together with the Gram-negative bacteria they contribute to the decrease in pH that facilitates growth of lactobacilli.

Improvement of the hygienic conditions and introduction of the 10 ton glass fiber fermenters, which are easier to clean and which suffer less contamination from the environment, has led to a considerable reduction in the undesirable initial probiota, especially Gram-negative bacteria, as well as to a marked increase of lactic cocci (GONZÁLEZ CANCHO and DURÁN QUINTANA, 1981).

VAUGHN (1985) has also described this stage in detail for Californian varieties, isolating Gram-negative bacteria of the genera *Aerobacter* and *Pseudomonas*, Gram-positive bacteria of the genera *Streptococcus* and *Leuconostoc* and some yeasts.

The length of this stage or phase is not constant and varies from 2 to 3 or more days, ending when the pH reaches 6.00 and the lactic acid bacteria appear. VAUGHN fixed it at approximately 7 days.

7.3.4.2 Second Stage

This period starts at the moment the pH reaches 6.00 and continues to the total disappearance of non-sporulating Gram-negative bacteria at a pH of around 4.5. It may last up to 10 to 15 days and is characterized by a rapid growth of lactobacilli and yeast and a decrease in the Gram-negative bacterial population. Sporulating Gram-positive bacteria disappear from the 4th to 5th day onwards. These changes are provoked by the active metabolism of lactobacilli which find an environmental pH very suitable for their growth as well as an abundance of nutrients, most of which are leached from the olives, although others are produced by the organisms present during the first and second stage. By this time the brine is able to supply the lactobacilli with all the complex nutrients they need (including some vitamins of the B group, amino acids and other growth factors, in addition to fermentable carbohydrates). The maximum population of lactic acid bacteria is reached on about the 7th or 10th day after brining. Subsequently the population diminishes very slowly up to 60–300 days of fermentation. In processes carried out in Spain the only two species of lactobacilli found have been *Lactobacillus plantarum* (90%) and *L. delbrueckii* (10%), which has a more limited duration.

As acid production increases the pH goes down, although there is no significant correlation between the amount of acid produced and the pH obtained, due to the buffer effect of the so-called "combined acidity" or "residual lye". This buffer effect has the advantage of facilitating complete exhaustion of the carbohydrates by maintaining the pH within comfortable limits for lactobacilli, but has the drawback of introducing some risks for the preservation of the fermented olives, if the final pH is not low enough (<4.0).

The use of new technologies and fermenters has also introduced some favorable changes in the traditional evolution of this phase. The relatively low Gram-negative populations normal nowadays (see Sect. 7.3.4.1) favor abundant growth of lactic acid-producing cocci. The genus *Pediococcus* (homofermentative) appears first with maximum growth between the 1st and 12th day of the fermentation, and disappears between the 12th and 20th day. *Pediococcus* accounts for 35% of the isolated species of cocci. The genus *Leuconostoc* (heterofermentative), which represents 60% of the isolated species, begins to develop between the 4th and 8th day in brine. It reaches its maximum development between the 10th and 24th day, and the organisms disappear after 23 to 50 days. The genus *Streptococcus* (homofermentative) represents only 4% of the original cocci and is not considered important. In this case, lactobacilli also start their development during the first 8–13 days, reach the maximum growth rate approximately between 19–30 days and remain viable practically up to the end of the fermentation.

In Californian processes, VAUGHN has demonstrated the presence of the *Aerobacter* Gram-positive bacteria already mentioned, during this second stage (VAUGHN's middle stage lasting until the 21st day in brine). VAUGHN (1985) also mentions that *Leuconostoc mesenteroides* and *Streptococcus* spp. were found in low salt fermentations of the

Sevillano variety, dominated the later part of the secondary stage of fermentation and disappeared from the population within 3 or 4 weeks. He also identified *Pediococcus cerevisiae* (now called *P. pentosaceus*) in some fermentations during the last part of the initial stage and the first phases of the intermediate stage, after which it declined rapidly. Finally, VAUGHN (1985) states, from his wide experience on table olive fermentation in California, that *Lactobacillus plantarum* always dominates the last part of the intermediate stage and the whole of the final stage of fermentation.

7.3.4.3 Third Stage

During the third stage which lasts until the fermentative substrates are exhausted, neither Gram-negative nor sporulating Gram-positive bacilli can be found. Only species of *Lactobacillus* abound, and these coexist with a yeast flora. The dominant species is *Lactobacillus plantarum* and, to a much lesser extent, *Lactobacillus delbrueckii* (BORBOLLA Y ALCALÁ and GONZÁLEZ CANCHO, 1975). VAUGHN (1985) reported the presence of the gas-forming species *Lactobacillus brevis* during the final stage of fermentation, but this never approached the population levels of *L. plantarum.* Furthermore, *L. brevis* was never found in the higher salt fermentations of the Manzanilla variety. In some fermentations of these fruits, the same author also isolated *Pediococcus cerevisiae.* However, he concludes that not all the Spanish-type olive brines contain *P. cerevisiae* and that nowadays all varieties of olives are fermented at salt concentrations high enough to prevent growth of *Lactobacillus brevis.* Consequently, the only species of lactic acid bacteria that are certain to be found in all fermentations of the Spanish type is *L. plantarum.*

Throughout this period a yeast flora is also present. Two groups can be differentiated: fermentative and oxidative yeasts. The first ones are characterized because they produce ethanol, ethylacetate, acetaldehyde, etc. Their presence is not considered detrimental to the fermentation, in spite of the fact that they use a certain proportion of the fermentable sugars, because their metabolites contribute to improving the organoleptic characteristics of the final product and many of them also excrete vitamins necessary for supporting *Lactobacillus plantarum* growth (RUIZ BARBA, 1991). The main species found are shown in Tab. 6. Other species isolated with a lower frequency are: *Candida catenulata, Candida guilliermondii, Candida pseudotropicalis, Candida solani, Candida tropicalis, Kluyveromyces veronae, Pichia fermentans, Saccharomyces cerevisiae, Saccharomyces italitus, Saccharomyces rosei, Torulopsis colliculosa, Torulopsis glabrata* and *Torulopsis holmii* (FERNÁNDEZ DÍEZ et al., 1985).

Tab. 6. The More Important Species of Fermentative Yeasts Isolated from Treated Green Table Olives Brines (Spanish Style) (FERNÁNDEZ DÍEZ, 1985)

Species	Time Interval (days)	pH Range
Hansenula anomala	1–365	7.45–3.70
Candida krusei	10–365	5.20–3.00
Saccharomyces chevalieri	1–365	8.90–3.75
Candida parasilopsis	0– 7	8.70–4.90
Hansenula subpelliculosa	2– 17	9.40–4.80

In contrast, the oxidative yeasts arise from superficial films and are harmful because they consume the lactic acid, lower the pH and lead to environmental conditions that favor later malodorous spoilage. There are also some facultative species, but their presence is also undesirable. BORBOLLA Y ALCALÁ and GONZÁLEZ CANCHO (1975) have described in detail the species found.

7.3.4.4 Storage of Fermented Olives

After fermentation, olives are kept in the same brine until they are sold. Usually, stabilization is done by increasing the salt content to 8% or higher. If this is not achieved in time there is an increase in volatile acids (acetic and propionic acids). This may occur during the storage stage, when the fermentable substrates have been exhausted and the microbial activity is supported almost exclusively at the expense of lactic acid. It leads to an increase in the pH and can produce undesirable changes in the fruits (REJANO NAVARRO et al., 1978). This phase is called the 4th stage of the fermentation, although its occurrence is not desirable. This biotransformation is due to species of *Propionibacterium* which are relatively tolerant to high salt concentrations. The effect of this phase is variable, producing in most cases a slight increase in pH (0.1–0.02). Greater pH increases (0.4 or more) may lead to eventual malodorous spoilage. Suppression of these organisms requires at least 8% salt at pH values between 3.7 and 4.0 at the end of the fermentation.

7.3.5 Biochemical Changes During Fermentation

The biochemical transformations produced by the growth of *Lactobacillus plantarum* have been followed by MONTAÑO et al. (1993). Carbohydrate consumption started about 2 days after inoculation. *Lactobacillus plantarum* metabolized glucose and fructose simultaneously but, at least initially, the consumption of glucose was faster than that of fructose. The levels of both of these carbohydrates reached values below 0.1% in about one week. Mannitol was also used, although more slowly. Hardly any sucrose degradation had occurred by 34 days, but after 70 days of fermentation 90% of it was degraded.

The main acids produced during fermentation were D- and L-lactic acid, which increased during exponential growth and reached maximum values during the stationary phase. Levels of D-lactic acid were always higher than those of L-lactic acid, in accordance with the results of BOBILLO and MARSHALL (1991). This indicates that in both studies D-lactate dehydrogenase was more active than L-lactate dehydrogenase.

The malic acid initially present disappeared rapidly, presumably being degraded to lactic acid and CO_2 via a malolactic pathway. Use of citric acid did not begin until near the end of the exponential phase of growth when the pH had fallen to about 4.5. It might have been degraded via the anaerobic pathway of citrate metabolism probably forming succinic acid and ethanol, serving as electron acceptor for anaerobic mannitol metabolism. Concentrations of final products are shown in Tab. 7.

Tab. 7. End Products Formed During the Fermentation of Green Table Olives by *Lactobacillus plantarum* (MONTAÑO et al., 1993)

Compound	Concentration (mmol/L)
Lactic acid	107.2
Acetic acid	11.5
Succinic acid	2.1
Ethanol	3.2

Biochemical changes in spontaneous fermentations, however, are considerably more complex due to the simultaneous activity of the different organisms present. A summary of these may be found in Tab. 8a (FERNÁNDEZ DÍEZ, 1983).

7.3.6 Fermentation Control

Appropriate lye treatment and washing are essential for a normal fermentation. The aim of these treatments, in addition to eliminating the bitterness of the olives, is to reduce the oleuropein and other polyphenols produced by its hydrolysis, to concentrations compatible with lactobacilli growth. The inhibitory effect of oleuropein and its derivatives was reported by JUVEN et al. (1968, 1971) and FLEMING et al. (1973). When their concentrations are high, lactic acid bacterial growth may be scarce or even non-existent. Partial substitution of brine with addition of fermentable substrate, and the use of a starter culture, if necessary, is recommended to restart the halted lactic processes.

In general, it is convenient to correct the initial pH to values around 6.0 or less so as to diminish the growth of Gram-negative bacte-

Tab. 8a. Processing and Compositional Changes of Spanish-Style Green Olives in Brine (FERNÁNDEZ DÍEZ, 1983)

Starting Phase Process	Changes in Composition	**Main Phase** Process	Changes in Composition	**Final Phase** Process	Changes in Composition	General Characteristics of the Final Product
Alkaline treatment and washing with water	Hydrolysis of oleuropein Loss of sugars and organic acids Formation of organic acids from sugars	Fermentation in brine (mainly lactic) Secondary action of other microbes	Lactic acid formation from sugars and other fermentable compounds Formation of organic acids and volatiles. Degradation of chlorophylls	Conservation in brine Bottling	None under normal conditions	Free acidity: about 1% as lactic acid pH approx. 3.6–4.2 NaCl: approx. 2–8% depending on commercial process

Tab. 8b. Processing and Compositional Changes of Natural Black Olives in Brine (FERNÁNDEZ DÍEZ, 1983)

Starting Phase Process	Changes in Composition	**Main Phase** Process	Changes in Composition	**Final Phase** Process	Changes in Composition	General Characteristics of the Final Product
None	None	Spontaneous fermentation in brine Yeasts predominate; lactic acid bacteria sometimes present	Slow loss of sugars, polyphenols and bitterness Formation of organic acids, ethanol, acetaldehyde and ethyl acetate	Conservation in brine Bottling	None under normal conditions	Free acidity: about 0.3–0.5% as lactic acid pH about 4.3–4.5 NaCl: 6–10% depending on commercial practice

Tab. 8c. Processing and Compositional Changes of Ripe Olives in Brine (Alkaline Oxidation)

Starting Phase Process	Changes in Composition	**Main Phase** Process	Changes in Composition	**Final Phase** Process	Changes in Composition	General Characteristics of the Final Product
Fermentation in brine by lactic acid bacteria and yeasts	Slow loss of sugars, polyphenols and bitterness	Alkaline treatment	Hydrolysis of oleuropein	None	None under normal conditions	pH about 5.8–8.0
		Washing				NaCl: 1.5% depending on commercial process
		Oxidation by air	Polymerization of polyphenols			
	Formation of organic acids, ethanol and other aromatic compounds	Brining	Loss of sugars and organic acids			
		Bottling				
		Heat sterilization	Organic acid formation from sugars			

ria at the beginning of the fermentation. This can be accomplished by bubbling CO_2 to saturation 10–12 h after brining, an operation which may be repeated 48 h later till the solution reaches a pH of 6.0. Other acids such as acetic, lactic, food-grade HCl, etc. may also be used. In any case, sharp drops in the pH must be avoided, because they may adversely affect the fermentation.

The spontaneous growth of lactobacilli is sufficient to produce an adequate population of these bacteria. However, the use of a starter culture (from commercial or laboratory cultures) may assure a faster lactic flora dominance, although the use of brines in a phase of active fermentation (pH around 4.0 and lactobacilli counts over 10^6 CFU/mL) is the most widespread practice. Strains of *Lactobacillus plantarum* with new properties such as the ability of producing bacteriocins are being investigated for the design of more appropriate starter cultures (JIMÉNEZ DÍAZ et al., 1993). If the microbial flora is satisfactory, a temperature between 23–30 °C for 15–20 days considerably improves the fermentation rate.

Apart from the microbiological aspects, periodical physico-chemical analyses are also required to adequately control the fermentation. The pH and free acidity are the variables which are most useful for monitoring the progress of fermentation. Salt concentration and combined acidity are more important for the final preservation. NaCl adjustment is achieved simply by addition of NaCl either in solid form or in saturated solution. Combined acidity is modified by replacing a predetermined volume of brine with a fresh solution of NaCl and lactic acid of similar concentrations; food-grade HCl is also used for this purpose. The best results, however, are achieved using a mixture of both.

Free acidity is improved by slight heating of the brine when the fermenting temperature is low, by addition of sugars if the microbial flora is adequate and there is a lack of nutrients, incorporation of a starter if the problem is the absence of lactobacilli or the addition of fermentable substrate and a starter culture if both are needed.

7.3.7 Other Operations and Packing

Before packing, fermented green olives are usually graded by size, sorted, pitted or pitted and stuffed with different products. If fermentation has been complete, these olives can be preserved using only appropriate physicochemical conditions, i.e., pH ≤ 3.5 and NaCl $\geq 5\%$. If fermentation is only partial, however, the olives need to be pasteurized. Heat resistance of lactobacilli and yeast was studied by FERNÁNDEZ DÍEZ and GONZÁLEZ CANCHO (1966), and the pasteurization conditions fixed by GONZÁLEZ PELLISSÓ and REJANO NAVARRO (1984). This treatment has, if properly applied, little influence on color and texture (SÁNCHEZ et al., 1991).

7.4 Natural Black Olives in Brine

Processing of natural black olives in brine has been traditionally practiced in Greece (BALATSOURAS, 1966) and still has great importance there and in other countries like Turkey, North African countries, etc. For preparing this type of olive the fruit should ideally be completely ripe but not overripe, i.e., when the skin color is between violet and black. There is a great difference between fruits picked at the beginning and at the end of the season. The olives harvested early have a good texture, but the color after processing is not good. Fruits picked at the end of the season retain an excellent color after processing but the texture is less firm. Harvesting and transportation are carried out as already described (see Sect. 7.1).

7.4.1 Traditional Anaerobic Fermentation

After an initial selection at the factory to separate damaged fruits, the olives are sometimes washed to remove superficial dirt. The fruits are then placed directly into brine with a salt concentration which is normally between 8 and 10% although in some cases lower concentrations (about 6% NaCl) are used.

In Greece, concrete tanks are generally used as fermentation vessels. These may be completely below ground level and hold 10 to 20 tons of fruits. They are lined with paraffin or epoxy-type resins to provide anaerobic conditions. In recent years, glass fiber fermenters similar to those used for processing green olives, have come into use. These are generally buried underground to avoid the high environmental temperatures in factories located in warmer zones. Avoiding high temperatures is important, because natural ripe fruits must be processed at a relatively low fermentation rate. In both types of containers, strict anaerobiosis is advisable to avoid the growth of film-forming yeasts and surface growth of molds which affect the texture as well as the flavor of the product.

This fermentation process takes a long time because diffusion of soluble components through the skin, when this has not been treated with alkali, is slow. A complex microbiological population develops during the fermentative process. From the very start of brining Gram-negative bacteria are present. Strains identified belong to the genera *Citrobacter* (64%), *Klebsiella* (20%), *Achromobacter* (9%), *Aeromonas* (4%) and *Escherichia* (2%) (GONZÁLEZ CANCHO et al., 1975). This population reaches its maximum during the first few days of brining and disappears after 7 to 15 days. In this spontaneous fermentation, however, the yeasts dominate. They start growing during the first few days of brining, reach their maximum at 10 to 25 days and are present throughout the whole time that olives are kept in brine. The species isolated from Spanish varieties are the following. Sporulating yeasts: *Saccharomyces oleaginosus* (35%), *Hansenula anomala* (27.3%), *Hansenula subpelliculosa* (<1%), *Debaryomyces hansenii* (19%), *Pichia membranaefaciens* (5%), *Pichia farinosa* (<1%), *Pichia fermentans* (<1%), *Kluyveromyces veronae* (<1%). Non-sporulating yeasts: *Torulopsis candida* (11%), *Torulopsis norvegica* (<1%), *Candida diddensii* (5%), *Candida boidinii* (<1%), *Candida krusei* (<1%), *Candida valida* (<1%), *Cryptococcus ater* (<1%). *S. oleaginosus* and *H. anomala* have been identified in fermentations of all varieties and were the most frequently found. They can, therefore, be considered as the species which best characterize this fermentation, followed by *T. candida, D. hansenii, C. diddensii* and *P. membranaefaciens*, although these are characterized by their predominantly oxidative metabolism and, consequently, have little ability for fermenting olive sugars. With certain varieties (Hojiblanca, Gordal, etc.) and conditions (salt concentration below 6–7%), Gram-positive lactic cocci and lactobacilli are also found. Strains isolated belong to the genera *Pediococcus* and *Leuconostoc*, and their presence is detected during the first days of fermentation, subsequently disappearing due to their relatively low acid tolerance, when *Lactobacillus* begins to grow abundantly. *Lactobacillus* may be viable during the whole fermentation period, if the salt level is not raised above 8%.

7.4.2 Fermentation Under Aerobic Conditions

To carry out fermentation under aerobic conditions, brines with the physico-chemical conditions described in Sect. 7.4.1 are used. The fermenter should be modified by introducing a central column through which air is bubbled. This air removes the CO_2 produced by fruit respiration and microbial metabolism. At the same time, the presence of air maintains a certain percentage of dissolved O_2, inducing the growth of facultative instead of fermentative yeasts. The rate of air injection depends on the technical design of the factory, but may range between 0.1 and 0.5 volumes per fermenter volume per hour.

From the microbial point of view, the new process is characterized by the growth of Gram-negative bacteria of the same type as those found in the traditional process. The species found are from the Enterobacteriaceae family. Yeasts are present during the whole fermentation process, their counts being higher than under anaerobic conditions. The most representative species identified have been: *Torulopsis candida* (20%), *Debaryomyces hansenii* (19%), *Hansenula anomala* (8%) and *Candida diddensii* (6%) which have a facultative metabolism. *Pichia membranaefaciens* (14%), *Hansenula mrakii* (11%) and *Candida bodinii* (7%) which have an oxidative metabolism have also been found. The proportions of these in relation to the facultative yeast flora depend on the air flux, increasing as the rate of air flow increases. However, a great volume of air is not advisable. Lactic acid bacteria do not grow, if the salt concentration is above 8%, but if this is lower, they appear about 6–7 days after brining and they are present up to 3 months later. At the beginning they are almost exclusively *Leuconostoc* and *Pediococcus*, but after 20 days there is a predominance of *Lactobacillus*. In this process, the level of CO_2 dissolved in brines and in the fruits is considerably lower than in the traditional fermentation. To maintain this low level, it is recommendable to inject air periodically even after the period of active fermentation. The amount of dissolved O_2 in the traditional process is about 30% of saturation. The aim of the new process is to increase this to about 40%, to prevent the presence of fermentative yeasts. The continuous circulation provoked by the air bubbling causes a rapid diffusion of sugars and a high fermentation rate, reducing the length of the process to about 3 months.

The main advantages of this new process when compared with anaerobic fermentation are: (a) a lower incidence of gas-pocket spoilage; (b) elimination of shrivelling in fruits; (c) more homogeneous brines and faster fermentations; (d) improved color, flavor and texture. When the air flux is excessive, the only drawbacks observed have been a predominance of oxidative yeasts, which raise the pH above 4.6, and an advance of the ripeness. Detailed information on this process may be found in García et al. (1985).

7.4.3 Biochemical Changes During Fermentation

Biochemical changes in natural black olives are relatively simple, due to the absence of lye treatment. The fermentable substrates present are the same as those already described for green olives, although their concentrations in the flesh are lower. They pass slowly and progressively into the brine, where they are used by the microorganisms. However, in industries where the temperatures are rather low (15–18°C), they may reach high levels in the brine (0.5–1.0%) during the winter period. These sugars are rapidly used when spring arrives and the temperatures rise.

Due to the slow diffusion in both directions, the olive flesh may be a more appropriate culture medium than the brine (lower NaCl concentration and higher level of fermentable sugars). It is common, therefore, to find an abundant population of yeasts (and lactic acid bacteria, if they are also in the brine) in the flesh of natural black olives in brine. This is not desirable, because these microorganisms can destroy the flesh, at least in part, produce gases and cause olive swelling. During this fermentation, sugars are transformed into organic acids (in a low proportion, except when lactic acid bacteria are present) and volatile compounds, mainly ethanol which accounts for 95–99% of them. Consequently, it can be considered to be a typical alcoholic fermentation. Other volatile compounds found are: acetaldehyde, acetone, ethyl-acetate, *n*-propanol, 2-methyl-propanol, 2-methyl-butanol, and other compounds in lower proportions (FERNÁNDEZ DÍEZ et al., 1985).

Elimination of the bitterness of olives is achieved only by solubilization of the oleuropein into the brine. Under anaerobic conditions, the equilibrium is reached in 6–12 months. However, when air is injected, the brine circulation facilitates all the osmotic exchange processes (NaCl introduction into the flesh and solubilization of the organic substrates into the brine), and olives become sweet after a much shorter period of time (around 3 months). Due to the acidic medium, oleuropein is hydrolyzed, producing similar compounds to those found in green olives (BRENES BALBUENA et al., 1992). Although olives elaborated by the aerobic system are more rapidly debittered and sweeter than those from anaerobic processes, there is still no scientific evidence to explain why. It is possible that some of the microorganisms that grow in the presence of air, can use oleuropein or its derivatives as a carbon source.

Anthocyanins are also partially solubilized and degraded, although there have been few studies on the changes undergone by these compounds. The superficial blackening observed in these olives when they are exposed to air is due to anthocyanin polymerization (VLAHOV, 1990). The nitrogen content of the brine fluctuates between 187 and 950 mg/L. No detailed studies on the amino acid composition have been published, but this level would seem sufficient to support normal microbial growth. High nitrogen contents have been associated with high NaCl concentrations (possibly because the higher osmotic pressure extracts more and the population of microorganisms is lower) and low nitrogen contents with the production of acids during fermentation; possibly due to nitrogen use by the lactic acid bacteria (BALATSOURAS, 1966).

Other characteristics of this fermentation are: low combined acidity (<50 mmol/L) due to the absence of lye treatment; low free acidity (<0.5%), except when there is growth of *Lactobacillus*; and volatile acidity ranges from 0.1 to 0.2% (mainly acetic acid). A complete revision of the physico-chemical changes and the microbiology of this process has been published by GONZÁLEZ CANCHO et al. (1975). A summary of these changes is shown in Tab. 8b (FERNÁNDEZ DÍEZ, 1983).

7.4.4 Fermentation Control

In general, fermentation is conditioned by the initial pH and NaCl concentration. In order to prevent excessive growth of Gram-negative bacteria, a certain proportion of acetic acid (0.2% v/v) is added to the brining solutions to reduce the pH to below 4.5. If this is not done, the population of Gram-negative bacteria is excessive and produces great volumes of gases which, in turn, cause overflow in fermenters and olive swelling, especially under anaerobic conditions. The NaCl content is also fixed according to the type of fermentation required. If the dominant organism required is *Lactobacillus*, an NaCl content of around 6% in the equilibrium is fixed, and the final product will have a low pH (about 4.0) and an acid concentration higher than 0.6%, expressed as lactic acid. The product will have a very different taste from the generally marketed product, and its color will be much lighter resembling that of treated black natural olives (treated with sodium hydroxide) described by BALATSOURAS and POLYMENAKOS (1964). If the dominant organisms required are yeasts, as is the case traditionally, the salt percentage is fixed at 8% or more. In this case the final product has pH values between 4.4 and 4.6 and the acid concentration varies between 0.1 and 0.6% expressed as lactic acid. The low acidity developed and the relatively high pH means that the salt concentration must be gradually increased to ensure adequate preservation of this product. At the end of the fermentation, the NaCl concentration should be 8–10%.

Temperature and available sugars also limit the fermentation rate. In general, the sugar content in brines is always low due to the relatively low sugar concentration of these fruits and to the slow diffusion through the skin. Consequently, a low fermentation rate is usually preferred for this product. This is achieved by maintaining temperatures between 13–18°C. At higher temperatures growth of organisms into the flesh is favored, and gas pocket and shrivelling spoilage increase. Fermenters are never heated in this process.

The concentration of phenols, mainly anthocyanins, also limits the microbial population. No scientific studies on this inhibition have been reported, but Gram-negative bacteria are never found after phenols reach a certain concentration and neither is *Lactobacillus* in the case of most varieties.

In aerobic conditions, the flux of air is either controlled by a flowmeter adapted to the air inlet of each fermenter, or it is fixed on the basis of past experience. Aeration periods, after the active fermentation process is ended, are determined by measuring the CO_2 content in the brine or the increase in the brine volume.

7.4.5 Other Operations and Packing

Once the fruits have been fermented, they are oxidized by exposure to air, a process which improves their skin color. This is performed especially in the case of the traditional anaerobic process. The olives are then selected and classified by size and finally packed in appropriately conditioned fresh brine. The containers are either wooden or plastic which hold about 130 to 150 kg of fruits, smaller tinplate containers holding 10 to 15 kg or plastic bags of the same capacity. Glass jars are rarely used for this product, although there is an increasing demand for them. The characteristics of the final product in glass jars and plastic containers have been studied by GARRIDO FERNÁNDEZ (1979). The most frequent values for commercial products are: pH of about 4; free acidity, expressed as lactic acid, between 0.5 and 0.6%; and a salt concentration of between 6 and 9%.

7.5 Black Olives in Brine (Ripe Olives)

For this commercial preparation, at the time of harvesting the superficial color of fruits varies considerably from a yellowish-green to a more or less pronounced purple, even though the pulp may still be light in color. The taste, texture and color of the final product may vary greatly depending on the maturity of the fruit and the variety. Objective measurements of the color to determine the best degree of ripeness at harvesting for each variety have been reported (FERNÁNDEZ DÍEZ and GARRIDO FERNÁNDEZ, 1971).

To produce ripe olives, fruits can pass directly to the oxidation process, without any fermentation. However, not all the fruits available at the time of harvest can be processed immediately, because the factories do not have the required capacity and because it is not desirable to keep large amounts of canned merchandise in stock. For this reason the greater portion of olives harvested at the correct stage of maturity are placed directly into brine or other solutions until market demand calls for their processing. This also allows the brined olives to be put to other uses, such as for processing turning color olives in brine. The time period for storage in brine generally varies from 2 to 12 months.

7.5.1 Fermentation of Stored Olives

The comments in this section are also applicable to untreated green or turning color olives brined directly without any lye treatment to eliminate bitterness. In Spain, these olives are kept in the same fermenters as treated green olives or natural black olives in brine. In America, open wooden fermentation tanks are still used in some cases. In general, this holding period does not favor an abundant development of acidity, and olives ferment slowly at the environmental temperatures of the brines, especially during the latter part of the harvest season. To prevent losses of acidity by superficial molds and yeasts, anaerobiosis is ensured in open tanks by covering the brine surface with plastic films. Physico-chemical conditions of the storing brines are similar to those used for natural black olives, although the salt levels are usually lower. An initial pH correction is advisable. The comments made in Sect. 7.4.4 are, therefore, also valid here.

7.5.1.1 Microbiological Growth

VAUGHN (1985) has described in detail the microbiology of this preservation stage. In summary, the initial population of extraneous organisms consisting mainly of coliform bacteria and bacilli disappears slowly in normal fermentations (they remain for about 10–14 days), but if the fermentation is abnormal, they may cause spoilage and have been known to do so. The comparatively high concentration of salt used restricts the sequence of lactic acid bacteria so that the only species still persisting is *Lactobacillus plantarum.*

In Spain this preservation phase has been studied in some varieties both for their use as speciality products and as ripe olives. Due to the advantages of the aerobic conditions, air injection can also be used with these olives. Experience until now has shown that the presence of oxygen has a favorable influence on the quality attributes of the final products. Depending on the phenol levels, and other circumstances related to the variety, a lactic process may or may not be induced. Some varieties can undergo lactic processes even with 6% NaCl (because of their low phenol content and high substrate concentrations) (DURÁN QUINTANA et al., 1991; GARCÍA GARCÍA et al., 1992). With other varieties, however, it has only been possible to achieve a lactic process using low salt concentrations. A detailed study of the microbial growth during this aerobic preservation phase and its dependence on the physico-chemical conditions of brines has been published by FERNÁNDEZ GONZÁLEZ et al. (1993). The yeast population begins to grow from the beginning of brining. Its growth rate is lower and its maximum reached later, as the salt level increases. The species found are: *Pichia membranaefaciens* (44%), *Pichia vinii* (11%), *Pichia fermentans* (9%), *Hansenula polymorpha* (9%), and others of less significance. These yeasts have little or no fermentative capacity and their presence is induced by air injection. When olives are placed initially in water containing no salt, a lactic acid bacterial population develops, but the sequence of species described in the case of green olives is not observed. *Lactobacillus plantarum* is the first species to appear, while *Pediococcus inopinatus* is detected later. The presence of lactic acid bacteria causes a marked decrease in the yeast counts.

In general, the lactic process can be prevented by the presence of inhibitory compounds (FLEMING et al., 1973) in brines or by insufficient amounts of available nutrients. Phenols pass into the brine relatively rapidly, the rate with which they pass depends on salt concentration, temperature and variety. Oleuropein reaches high concentrations in some varieties from the very beginning of fermentation and is slowly hydrolyzed during the fermentation process. In contrast, hydroxytyrosol levels increase as oleuropein is hydrolyzed. Concentrations of phenols may reach 1.4 mmol/L, and at these levels lactic acid bacteria are easily inhibited (BRENES BALBUENA et al., 1993). *Lactobacillus* growth may also be slowed down by the scarce diffusion of nutrients into the brine (sugars, vitamins, amino acids, etc.). However, research on *Lactobacillus plantarum* survival during the first days of ripe olive brining is still underway, to develop a lactic acid fermentative process during storage similar to that which occurs with treated green olives (Spanish style) (DURÁN et al., 1993). The use of aerobic conditions does not appear to have any deleterious effect on the inducement of a lactic acid fermentation.

7.5.1.2 Biochemical Changes

The biochemical changes occurring during the preservation of ripe olives from both anaerobic and aerobic processes using different fermentation conditions have been studied (GARRIDO et al., 1993). The concentrations of the main fermentable substrates in the flesh are shown in Tab. 9. The major reducing sugar is glucose, followed by fructose. The two account for about 80% of the fermentable compounds. Mannitol and malic acid are also present in significant amounts (around 20% of fermentable compounds), although normal monitoring of the fermentation does not involve the analysis of these substances. In general the total amount of fermentable substrates in these fruits is lower than in green olives. During ripening the levels of fructose decrease faster than those of glucose, thus increasing the glucose to fructose ratio. The concentration of glucose in the preservation brines is always lower than 5 mmol/L both under aerobic and anaerobic conditions, except when salt and acetic acids are present in concentrations higher than 6 and 0.5%, respectively. However, in this case when the sugars (and possibly other nutrients) reach a high enough level, they are rapidly utilized by yeasts and even lactobacilli in spite of the presence of polyphenols. This fact lends support to the hypothesis that inhibition of lactic acid bacteria depends on the equilibrium between inhibitors and growth factors. With air injection, the glucose levels in the flesh reach concentrations of about 5 mmol/kg after 3 months of brining. In an anaerobic medium, the decrease is much slower. Fructose follows a similar pattern, although its elimination from the flesh is faster, because its initial content is lower. Malic acid levels in the flesh diminish slowly, and mannitol diffusion is very slow. However, these compounds maintain relatively high concentrations in brines and show only a very slight decrease with time. This indicates that they are used during preservation to a very limited extent, although the degree to which they are used may depend on the fermentation conditions.

Tab. 9. Initial Content of Fermentable Substrate in the Flesh of Ripe Olives (Hojiblanca Variety) (GARRIDO et al., 1993)

	Concentration (mmol/kg)
Glucose	91.6
Fructose	11.8
Sucrose	1.6
Mannitol	14.3
Malic acid	9.9
Total	129.2

Among the end products formed during this storage period are, acetic acid, in anaerobic conditions where only yeasts were present and volatile compounds; ethanol (around 60 mmol/L) and methanol (10–15 mmol/L) being the major compounds. In general aerobiosis induces lower concentrations of all of the end products, and lactic acid fermentation leads to the most efficient utilization of carbohydrates and the most suitable conditions for olive preservation.

In some cases, fermentation is not desired, as is the case in America nowadays. In this case, strongly acidified solutions (lactic or acetic acids or a mixture of both) are used for keeping the olives during this period. This practice was developed by VAUGHN et al. (1969) and is widely used by the Californian industry.

7.5.2 Ripe Olive Blackening and Canning

Processing methods for American varieties have been described in detail by CRUESS (1958). In general, fruits are successively treated with sodium hydroxide solutions for varying periods of time to achieve a progressive penetration of the lye into the pulp. After each alkaline treatment, the olives are placed into water and oxidized by injecting air under pressure into the water. This oxidation of the phenolic compounds permits a complete blackening of the fruit skin and a uniform coloration of the pulp. Compounds responsible for this oxidation have been studied by BRENES BALBUENA et al. (1992). The promoters of such polymerization have been identified as hydroxytyrosol (3,4-dihydroxyphenyl acetic acid) and caffeic acid, the decrease of which in the flesh was strongly correlated with fruit darkening. Olives were darker and oxidation rates higher with higher pH values (GARCÍA et al., 1992).

The number of lye treatments is generally between 3 and 5. Penetration into the fruit is controlled so that the sodium hydroxide of the first treatment merely passes through the skin. Subsequent treatments are chosen so that they penetrate deeper into the pulp. The final lye treatment must reach the stone. The concentration of sodium hydroxide in the lye solution depends on the ripeness of the fruit, its variety, the environmental temperature and the desired penetration speed. It varies between 1 and 2% (w/v). The higher concentration is usually used for the first treatment. In general, oxidation rates can be improved by using high dissolved oxygen concentrations and moderately high temperatures (≤50°C) during rinsing with water (GARCÍA et al., 1991).

For treating with alkali and washing, aeration tanks are arranged in parallel. They may have different shapes and sizes and are made of concrete, stainless steel, polyester or glass fiber. In all cases they have the same network of pipes for the distribution of pressurized air so that the oxidation process is uniform. A detailed description of oxidation devices is given by FERNÁNDEZ DÍEZ et al. (1985).

The blackened olives are washed several times with water to remove most of the sodium hydroxide and lower the pH in the flesh to around 8. Generally, 0.1% (w/v) of iron gluconate is added to the last wash to stabilize the color achieved by oxidation. Nowadays, iron lactate can also be used for the same purpose (GARCÍA GARCÍA et al., 1986). The olives are then placed into brine containing 3% NaCl, and packed in cans varnished on the inside.

The final canned product has organoleptic properties very different from the fermented fruits obtained by other processes. The pH values are between 5.8 and 7.9, and the NaCl content is between 1 and 3%. The effect of processing on different quality attributes has been studied by GARRIDO FERNÁNDEZ (1980a, b). A summary of the changes in fruits during processing and the characteristics of the final product are shown in Tab. 8c (FERNÁNDEZ DÍEZ, 1983).

8 Role of Lactic Acid Bacteria in Table Olive Processing

Of all table olive processes, processing to produce treated green table olives in brine was the only one developed as a lactic acid fermentation, as is evident from the previous comments. Lactic acid bacteria are also present in other processes (natural black olives, ripe olives, etc.), and their presence may be advantageous in many of them. However, at present, all types of fermentative processes of table olives are still spontaneous. In general, control of the physico-chemical conditions is enough to get an adequate final product. The extent of growth of lactic acid bacteria, however, depends on the fermentation conditions and the restrictions imposed by the composition of the fruit itself.

The presence of inhibitory compounds in olives strongly affects the lactic acid bacteria in olive fermentations. ETCHELLS et al. (1966) found that olives brined directly did not support added pure cultures, suggesting the presence of an inhibitory compound. The activity of inhibitors was demonstrated by FLEMING et al. (1969), and the inhibitory effect was assigned to oleuropein, the bitter principle of olives, although the products of its hydrolysis, aglucone and elenolic acid, were more active (FLEMING et al., 1973; WALTER et al., 1973). RUIZ BARBA et al. (1990, 1991) have shown that oleuropein at 0.4% (w/v) has an immediate bactericidal effect on *Lactobacillus plantarum*, an effect that disappears after NaOH treatment but not after heat shock. Phenolic compounds alter the cell structure by adsorbing onto the cell wall and promoting the formation of mesosomes which cause rupture of the cell. Both effects are responsible for losses in viability. Thus, in general, a lactic acid process is not viable, if lye treatment is deficient (treated green olives) or if olives are brined directly (natural black, color turning, etc.) when the concentration of inhibitors is high. In addition to the effect of phenolic compounds, it seems that the level of nutrients is important in determining growth of lactic acid bacteria. The concentration of phenols in treated green olives is sometimes as high as in directly brined olives, but the former can support lactic acid bacteria and the latter cannot. The explanation is apparently that the level of nutrients in green olives is considerable and these contribute to overcoming the effect of the phenols. Growth of lactobacilli, in fact, in the case of some directly brined varieties can only be explained by their low phenol and high nutrient contents (GARCÍA GARCÍA et al., 1992). Further research is needed to elucidate these interactions completely.

Fortunately, lactic acid bacteria, and especially *Lactobacillus plantarum*, can grow and produce lactic acid in up to 7–8% (w/v) salt concentrations and resist low pH and acidity (MCDONALD et al., 1990). Tolerance to both of these conditions is exploited in table olive preparations to inhibit growth of Gram-negative bacteria and other anaerobic organisms that may cause softening or malodorous spoilage, at the beginning of brining, without producing any negative effect on the main fermentation by lactic acid bacteria. *L. plantarum* plays an important role by producing the acid needed for optimum storage of the product and, usually, it is this microorganism that terminates the fermentation, especially in treated green olives.

Lactic acid bacteria involved in olive fermentations are both homofermentative (e.g., *Pediococcus, Lactobacillus plantarum,* etc.) and heterofermentative (e.g., *Leuconostoc mesenteroides*). There is no definitive agreement on the best species or on the benefits or otherwise on using more than one, although homofermentative organisms are preferred as the amount of lactic acid produced should be as high as possible and formation of CO_2 in excess may contribute to gas-pocket spoilage. In fact, the metabolism of the *L. plantarum* strains that grow in olives might be classified as facultative homofermentative, especially under conditions of stress (high salt percentage, low sugar content, high polyphenol concentration, etc.). The ability of organisms to use organic acids as a carbon source is of limited significance in olives; although when the final product (especially green olives) is preserved only by its physico-chemical characteristics, it should be tested for the presence of such organisms. The decarboxylation of amino acids with production of significant amounts of biogenic amines has not been reported.

Vitamins and amino acids are not limiting factors for growth of lactic acid bacteria in the case of treated green table olives. After only 10 days of brining these nutrients were present in sufficient amounts to support development of these organisms. However, in directly brined olives, because these nutrients diffuse more slowly into the brine, their deficiency can cause a certain delay in bacterial growth. These deficiencies, in addition to the inhibitory action of phenols and the low availability of sugars, are responsible for the difficulties experienced by lactobacilli in growing in directly brined olives. Some of the yeasts usually present in the fermentative process have been demonstrated to produce vitamins. Thus, the presence of these, which has been described above, can contribute to better *L. plantarum* viability.

It is well known that optimum temperatures for *Lactobacillus plantarum* growth are between 25 and 30°C, and most processes are carried out when the environmental temperatures are within this range. This bacterium, however, can also support lower temperatures and is able to ferment olives properly at 18–20 °C, although at a slower fermentation rate. In any case fermenters can be heated to speed up processing of treated green olives in colder zones.

In general, strains of lactic acid bacteria isolated from olives have shown great genetic stability and resistance to bacteriophages (RUIZ BARBA, 1991).

In order to increase the microbiological control of the fermentation better starter cultures than those currently available commercially would be desirable. It would be particularly advantageous to increase the survival of these starters because, at present, the inoculated bacteria are replaced by wild strains in just 15–20 days. Research is being carried out to find bacteriocin-producing strains that could be effective against most of the other similar or undesirable organisms from the environment. There is one strain, LPCO-10, that excretes bacteriocins both in solid and liquid media (RUIZ BARBA, 1991). Further work is underway to study its application as a starter culture.

In spite of the essential role of lactic acid bacteria in olive fermentation, the use of only one species has failed to produce good quality table olives, especially in the case of green olives. A mixed population is necessary to achieve the required standards. The starter culture may be formed of lactic acid bacteria, but the presence of yeasts, which are present in all table olive fermentation processes, is also of great significance, and many of the compounds responsible for the flavor of the fermented product arise from the biological activity of these microorganisms. Thus, if the traditional flavor of table olives is to be reproduced, strains of yeasts should be included in any new starter culture. Finding the most appropriate strain, the best time for its addition, etc. still requires further research.

9 Microbiological Spoilage of Table Olives

Most spoilage that affects table olives is of microbial origin. The different types of spoilage will be described below.

9.1 "Alambrado", Gassy, "Floater", or Fish-Eye Spoilage

This deterioration is characterized by the development of blisters which result from accumulating gases causing the skin to separate from the flesh of the olives, and also by the formation of gas pockets which may extend to the pits of the fruit. In treated green table olives such spoilage is produced mainly during the first stage of the fermentation by Gram-negative species which produce great volumes of CO_2 and H_2: *Enterobacter aerogenes, Citrobacter* spp., *Citrobacter freundii, Aeromonas hydrophila, Aeromonas* spp. and *Enterobacter cloacae* (in decreasing order of importance) (FERNÁNDEZ DÍEZ et al., 1985). The problem is avoided by lowering the initial brine pH to 4.0–4.5, by bubbling CO_2 or adding acetic, hydrochloric or lactic acids.

VAUGHN (1985) also reports that species of *Bacillus polymyxa* and *B. macerans* as well as species of the anaerobic genus *Clostridium* also cause gassy deterioration in other types of olives. In addition, it has been found that in directly brined olives some yeasts of the genera *Saccharomyces* and *Hansenula* may produce typical gas blisters (VAUGHN, 1985) and fissures in natural black olives (DURÁN QUINTANA et al., 1979). *S. kluyveri* and *S. oleaginosus* also cause severe softening, but the cultures of *Hansenula* are not pectinolytic. Aerobic conditions during storage and pasteurization at the beginning of the washing periods during ripe olive darkening may prevent their development.

9.2 Malodorous Fermentations

There are four extremely malodorous bacterial fermentations which develop in olives. They are: the putrid and butyric acid fermentations, hydrogen sulfide fermentation, and "zapatera" spoilage. The odor of the putrid fermentation is similar to that produced during putrefaction of organic matter and is caused by certain species of *Clostridium* from contaminated water, devices, etc. The butyric acid fermentation is characterized by its butyric acid or rancid butter odor during the initial stages of the development of the fermentation, when sugars and other fermentable substrates are still abundant. All the cultures related to this abnormality were saccharolytic, non-proteolytic types of the genus *Clostridium*. Most of the cultures consisted of, or were very closely related to, *Clostridium butyricum* (VAUGHN, 1985). As with all malodorous fermentations, it begins at the bottom of the fermenters and is due to contamination from water, fermenters or other devices used during processing.

Hydrogen sulfide fermentations are characterized by the typical identifying odor of H_2S gas. The halophilic species *Desulfovibrio aestrearii* was associated by VAUGHN (1985) with this spoilage. Zapatera off-odor is first described as "cheesy" or "sagey" but as the spoilage progresses, the odor intensifies and develops into an unmistakably foul, fecal-like stench. It develops when the desirable lactic acid fermentation stops before the pH has dropped to around 4.5. There is a continuous loss of acidity as the spoilage progresses. VAUGHN (1985) identified some species of *Clostridium* as being responsible for this spoilage. Among the species found, *Clostridium bifermentans* and *C. sporogenes* predominated in association with species of *Propionibacterium* (*P. pentosaceum* and *P. zeae*), which produced propionic acid from lactate in both culture media and olive brine. Spanish researchers (FERNÁNDEZ DÍEZ et al., 1985) have also found the participation of these two genera of bacteria, and have identified *Clostridium sporogenes, C. lituseburense* and *C. limosum* as well as *Propionibacterium jansenii, P. acidipropionici* and other *Propionibacterium* spp.

All the malodorous fermentations can be prevented by working under good sanitary conditions and maintaining adequate control over the pH (≤ 4.6) and salt concentration ($\geq 5\%$).

10 Composition of the Final Products

Gross composition, moisture, oil, protein, residual soluble sugars, total carbohydrates, ash, fiber, organic acids produced during the fermentation and data on such vitamins as vitamins A, thiamine and riboflavin have been reported by CRUESS et al. (1939), USDA (1950), BORBOLLA Y ALCALÁ et al. (1956) and BALATSOURAS (1964). More detailed research on table olive composition was carried out by CASTRO RAMOS et al. (1979, 1980a, b), VÁZQUEZ LADRÓN et al. (1979), NOSTI VEGA et al. (1981, 1982), and VAMVOUCAS et al. (1980). Detailed information on table olive composition can also be found in FERNÁNDEZ DÍEZ et al. (1985). By way of illustration, Tab. 10, shows the concentration ranges of the main components and nutrients of the final product of treated green table olives in brine (FERNÁNDEZ DÍEZ, 1983).

In general, moisture and oil contents show an inverse relationship. Both components are quite variable, and their concentrations depend on variety and, to a lesser extent, on maturity. During processing the percentage of moisture slightly increases while that of oil remains unchanged. The oil fraction is of good quality because of its saturated fatty acid (12–19%) content and the presence of 5–8% of linoleic acid (NOSTI VEGA et al., 1982). The protein content is slightly reduced due to solubilization during lye treatment and brining. Fiber shows a similar trend. The fiber is well balanced and its composition suggests good digestibility; the ratio of lignin to cellulose being lower than 0.5 in almost all of the cases studied by HEREDIA MORENO and FERNÁNDEZ DÍEZ (1982), who reported the following values for its main constituents: cellulose 1–3%, lignin 0.5–0.9% and hemicellulose 0.3–0.9%.

The percentage of ash increases noticeably, mainly due to sodium adsorption with the lye treatment and brining. There are no significant differences in the protein, fiber, or ash composition between varieties. Practically all important amino acids are present in the final product. The concentration of leucine, and aspartic and glutamic acids are high. There are

Tab. 10. The Composition of Green Olives in Brine, Spanish Style, Concentration Ranges in the Pulp (FERNÁNDEZ DÍEZ, 1983)

Components	Concentration Range (weight %)
Moisture	61.00–80.56
Lipids	9.05–28.19
Protein	1.00–1.45
Fiber	1.40–2.06
Ash	4.19–5.46
Vitamins	Concentration Range
Carotene	0.02–0.23 mg/100 g
Vitamin C	1.44–2.87 mg/100 g
Thiamine	0.40–3.37 μg/100 g
Essential Amino Acids	Concentration Range (mg/100 g)
Valine	55–157
Isoleucine	43–121
Leucine	82–227
Threonine	5–64
Methionine	13–79
Phenylalanine	39–111
Lysine	5–31
Tryptophan	13–18
Minerals	Concentration Range (mg/100 g)
Phosphorus	7–21
Potassium	34–109
Calcium	35–86
Magnesium	6–40
Sodium	1313–1753
Sulfur	14–38
Iron	0.58–1.16
Manganese	0.06–0.12
Zinc	0.25–0.41
Copper	0.42–0.82
Caloric Value	Calories per 100 g
	102–280

appreciable quantities of provitamin A, vitamin C, thiamine and mineral elements (particularly calcium and magnesium).

In general, soluble sugars and oleuropein disappear during processing as do chlorophylls. In green olives the latter are degraded to their corresponding pheophytins and pheophorbides, and these are later oxidized during the subsequent conservation in brine (MÍNGUEZ MOSQUERA et al., 1993). Carotenoids remain unchanged. In treated green olives and natural black olives organic acids (mainly lactic acid) may reach high concentrations (0.5–1.5%). When lactic acid bacteria are absent the acidity is lower.

Finally, directly brined olives are characterized by their relatively high content of volatile components. Of special interest is the fermentation of natural black olives by yeasts, a typical alcoholic process in which ethanol accounts for 96–99% of the volatile compounds, but there are also other components (acetaldehyde, ethyl acetate, etc.) which contribute significantly to the highly appreciated organoleptic properties.

11 Waste Water Management

The volumes of waste waters produced during the processing of table olives are: treated green olives, 2 L/kg in lye treatment and washings, and about 7 L/kg in packing; natural black olives in brine, 1 L/kg (mainly in packing); treated black olives (ripe olives), between 7 and 15 L/kg, most of them in the blackening process. In general, lye solutions, washing waters used after these treatments and fermentation brines are the most polluting waste waters.

As an example, Tab. 11 shows the main characteristics of the different waste waters arising during processing of treated green olives (FERNÁNDEZ DÍEZ et al., 1985). The high content in polyphenols and sodium makes any treatment of these waste waters difficult. There is still no purifying procedure suitable for them. However, application of internal control measures to diminish the volume of liquid residues and appropriate management of the rest of them can considerably reduce the pollutant impact of these industries. In the case of treated green olives in brine, processing can be modified by re-using the lye during practically the whole season and by using only one washing water (GARRIDO FERNÁNDEZ et al., 1977). The remaining residual lye or combined acidity may be

Tab. 11. Some Characteristics of the Waste Waters from Processing-Treated Green Table Olives (FERNÁNDEZ DÍEZ et al., 1985)

Characteristics	Lye	1st Washing	2nd Washing	Fermentation Brine
pH	12.2	11.2	9.8	3.9
Free NaOH (g/L)	11.0	1.5		
NaCl (g/L)				97.0
Free acidity (g lactic acid/L)				6.0
Sugars (g glucose/L)	8.6	8.0	7.1	
Phenols (g tannic acid/L)	4.1	4.0	6.3	6.3
DQO (g O_2/L)	23.0	24.6	28.4	10.7
BOD_5 (g O_2/L)	15.0	12.3	15.6	9.5

reduced to the values normally found in the traditional processes by adding an appropriate amount of a strong acid to the initial brine and by further correcting the fermentation brines (removing a portion of them and replacing it with a fresh acidified one). No negative impact on the quality of the final product has been observed to date by applying these changes during processing. At present, re-used lyes and the washing waters are eliminated by emptying them into evaporating ponds.

When treated green olives must be packed, the fermentation brines are discarded, as are the washing waters of fruits and containers. The main contaminating agent is the fermentation brine. To cut down the contamination levels in the packing industry, re-using the brines as holding solutions for the final product has been suggested. Two procedures to regenerate brines are available: treatment by activated charcoal and ultrafiltration (BRENES BALBUENA and GARRIDO FERNÁNDEZ, 1988; BRENES BALBUENA et al., 1988). Experiments on an industrial scale have demonstrated that these measures are effective in eliminating suspended solids, color and microorganisms. The treated brine retains about 97% of the salt content and 90–95% of the lactic acid. Packing tests carried out re-using these brines in proportions up to 70% of the final holding solutions have not shown any negative influence on the quality attributes of the final product (GARRIDO FERNÁNDEZ et al., 1992). If fermentation brines are removed from the waste waters of the packing industry, their pollutant effect diminishes remarkably and a simple physicochemical treatment can be applied to bring their characteristics to levels acceptable by the authorities in charge of pollution control (GARCÍA GARCÍA et al., 1990).

In the case of untreated natural black olives, the only waste solutions are the fermentation brines. So far it has proved impossible to re-use them for successive fermentation processes (FERNÁNDEZ DÍEZ et al., 1985). However, their re-use for packing is attractive as they would only need to be filtered to eliminate suspended solids and microorganisms. A pore size of 0.45 nm would be sufficient.

The management of waste waters from treated black olives (ripe olives) is more complex. At present, storage brines cannot be reused. They are emptied into evaporating ponds, or a proportion of them is used for washing during the blackening process. Lye solutions are re-used as described for treated green olives; when excessively polluted they are emptied into evaporating ponds. Washing waters have been reduced to only one volume after each lye treatment. Instead of removing them, more frequently they are neutralized with strong acid (HCl or H_2SO_4) or CO_2. To date there is no treatment for these solutions, and they are also emptied into evaporating ponds. However, some assays using anaerobic digestion have given somewhat promising results, especially if the level of salt could be diminished (BORJA et al., 1993). Finally, gluconate or lactate solutions may be re-used for canning or the iron salts added directly in the final holding solution. A complete discussion on waste water problems associated with table olive processing can be found in FERNÁNDEZ DÍEZ et al. (1985).

As a final remark, the critical situation of this industry as a result of the heavily polluted waste waters that it generates and the lack of suitable procedures for their treatment should be emphasized. Research focused on these problems is urgently required.

12 Regulations

Table olives, as well as olive oil, constitute a basic food in the Mediterranean area and in all of the olive-producing countries. Furthermore, nowadays they are also used as an appetizer with many kinds of alcoholic and nonalcoholic beverages or as decorative or nutritional elements in various dishes. Because of their organoleptic characteristics, the healthy nature of the Mediterranean diet, their composition, and the promotion of olive oil consumption by the International Olive Oil Council, table olives are used worldwide. In fact, their production continues to increase steadily while that of other canned vegetables is decreasing. This situation is due, at least in part, to the contribution of the International Olive Oil Council in improving quality, facilitating technological innovation in less developed countries and in regulating international trade.

The International Olive Oil Council has been particularly important in promoting, through a Committee of Experts, the publication of the "Unified Qualitative Standard Applying to Table Olives in International Trade" (1991), which includes the table olives prepared by the different processes. It is revised periodically and used in all international transactions. Some countries (Spain, USA, European Union, etc.) also have their own national standards derived from this.

In the same manner, the JOINT PROGRAMME FAO/OMS (1974), with the cooperation of the International Olive Oil Council, has also evolved nutritional standards for table olives. Both standards are in close agreement and aid in both commerce and in providing protection for the consumer.

Acknowledgements

The authors express their thanks to CICYT (ALI-91-1166-CO3-O1) for the financial support of table olive research.

13 References

AMIOT, M. I., FLEURIET, A., MACHEIX, J. J. (1986), Importance and evolution of phenolic compounds in olive during growth and maturation, *J. Agric. Food Chem.* **34**, 823–826.

ANONYMOUS (1993), Situación del mercado internacional de la aceituna de mesa, *Olivae* **45**, 11–13.

BALATSOURAS, G. D. (1964), Composition chemique des olives noires de Grèce. Variation de quelques constituants en function de la région de production, *Inf. Oleic. Int.* **28**, 131–156.

BALATSOURAS, G. D. (1966), *Contribution to the Study of the Chemical Composition and the Microflora for the Stored in Brine Greek Black Olives,* Athens: National Printing Office.

BALATSOURAS, G. D., POLYMENAKOS, N. G. (1964), Resultats preliminaires sur la fermentation des olives noires par acide lactique, *Inf. Oleic. Int.* **27**, 153–167.

BALATSOURAS, G., PAPANTSIS, G., BALATSOURAS, V. (1988), Changes in olive fruit of "Conservolea" during development viewed from the standpoint of green and black pickling, *Olea* **19**, 43–55.

BEN SHALOM, N., KAHN, V., HAREL, E., MAYOR, A. M. (1977), Catechol oxidase from green olives: Properties and partial purification, *Phytochemistry* **16**, 1153–1158.

BEN SHALOM, N., HAREL, E., MAYOR, A. M. (1978), Enzymic browning in green olives and its prevention, *J. Sci. Food Agric.* **29**, 398–402.

BOBILLO, M., MARSHALL, V. (1991), Effect of salt and culture aeration on lactate and acetate production by *Lactobacillus plantarum, Food Microbiol.* **8**, 153–160.

BORBOLLA Y ALCALÁ, J. M. R. DE LA (1979), Preparación de aceitunas estilo sevillano, *Grasas Aceites* **30**, 31–36.

BORBOLLA Y ALCALÁ, J. M. R. DE LA (1981a), La preparación de aceitunas estilo sevillano. La fermentación. II, *Grasas Aceites* **32**, 103–113.

BORBOLLA Y ALCALÁ, J. M. R. DE LA (1981b), Sobre la preparación de la aceituna estilo sevillano. El tratamiento con lejía, *Grasas Aceites* **32**, 181–189.

Borbolla y Alcalá, J. M. R. de la, Gómez Herrera, C. (1949), La industria del aderezo de aceitunas verdes, *Rev. Cienc. Apl.* **3**, 120–132.

Borbolla y Alcalá, J. M. R. de la, González Cancho, F. (1975), Preparación de aceitunas en verde, in: *Proc. II. Seminario Oleicola Internacional*, Córdoba, Spain.

Borbolla y Alcalá, J. M. R. de la, Rejano Navarro, L. (1978), Sobre la preparación de aceitunas estilo sevillano. El lavado de los frutos tratados con lejia, *Grasas Aceites* **29**, 281–291.

Borbolla y Alcalá, J. M. R. de la, Gómez Herrera, C., González Cancho, F., González Díez, M. J., Gutiérrez González Quijano, R., Iquierdo Tamayo, A., González Pellissó, F., Vázquez Ladrón, R., Guzmán García, R. (1956), *El Aderezo de Aceitunas Verdes*, Madrid: Consejo Superior de Investigaciones Científicas.

Borja, R., Martin, A., Garrido, A. (1993), Aerobic digestion of black olive waste water, *Bioresour. Technol.* **45**, 27–32.

Brenes, M., García, P., Durán, M. C., Garrido, A. (1993), Concentration of phenolic compound changes in storage brine of ripe olives, *J. Food Sci.* **58**, 347–350.

Brenes Balbuena, M., Garrido Fernández, A. (1988), Regeneración de salmueras de aceitunas verdes estilo español con carbón activo y tierras decolorantes, *Grasas Aceites* **39**, 96–101.

Brenes Balbuena, M., García García, P., Garrido Fernández, A. (1988), Regeneration of Spanish style green table olive brines by ultrafiltration, *J. Food Sci.* **53**, 1733–1736.

Brenes Balbuena, M., Montaño, A., Garrido, A. (1990), Ultrafiltration of green table olive brines: Influence of some operating parameters and effect on polyphenol composition, *J. Food Sci.* **55**, 214–217.

Brenes Balbuena, M., García García, P., Garrido Fernández, A. (1992), Phenolic compounds related to the black color formed during the processing of ripe olives, *J. Agric. Food Chem.* **40**, 1192–1196.

Castillo Gómez, J., Mínguez Mosquera, M. I., Fernández Díez, M. J. (1979), Presencia de poligalactoronasa (PG) en la aceituna negra madura. Factores que influencian la actividad de dicha enzima, *Grasas Aceites* **30**, 333–338.

Castro Ramos, R., Nosti Vega, M., Vázquez Ladrón, R. (1979), Composición y valor nutritivo de algunas variedades españolas de aceitunas de mesa. I. Aceitunas verdes aderezadas al estilo sevillano, *Grasas Aceites* **30**, 83–91.

Castro Ramos, R., Nosti Vega, M., Vázquez Ladrón, R. (1980a), Composición y valor nutritivo de algunas variedades de aceitunas de mesa. IV. Efecto de la reutilización de la lejía de cocido y aguas de lavado, *Grasas Aceites* **31**, 91–95.

Castro Ramos, R., Nosti Vega, M., Vázquez Ladrón, R. (1980b), Estudio comparativo del valor nutritivo de las aceitunas de mesa, efecto de la variedad, del tipo de elaboración y del envasado, *Alimentaria* **XVII**, num. 115, 21–24.

Chung, J. I., Sakamura, S., Luh, B. S. (1974), Effect of harvest maturity on pectin and texture of canned black olives, *Confructa* **19**, 227–235.

Columela, L. J. M. (54 B.C.), *De re rustica*, Vol. II, Spanish Edition translated by Alvarez de Sotomayor and Rubio in 1979, Santander: Nestlé A.E.P.A.

Cruess, W. V. (1958), Pickling and canning of ripe olives, in: *Commercial Fruits and Vegetable Products,* 4th Ed., New York: McGraw Hill.

Cruess, W. V., El Saifi, A., Develter, E. (1939), Changes in olive composition during processing, *Ind. Eng. Chem. News Ed.* **31**, 1012–1014.

Durán, M. C., García, M., Brenes, M., Garrido, A. (1993), *Lactobacillus plantarum* survival during the first days of ripe olive brining, *Syst. Appl. Microbiol.* **16**, 153–158.

Durán Quintana, M. C., González Cancho, F., Garrido Fernández, A. (1979), Aceitunas negras al natural en salmuera. IX. Ensayos de producción de alambrado por incubación de diversos microorganismos aislados de salmueras de fermentación, *Grasas Aceites* **30**, 361–367.

Durán Quintana, M. C., Brenes Balbuena, M., García García, P., Fernández González, M. J., Garrido Fernández, A. (1991), Aceitunas tipo negras. Estudio comparativo de tres procedimientos para la conservación previa de frutos de la variedad Gordal (*Olea oleuropaea regalis*), *Grasas Aceites* **42**, 106–113.

Etchells, J. L., Borg, A. F., Kittel, I. D., Bell, T. A., Fleming, H. P. (1966), Pure culture fermentation of green olives, *Appl. Microbiol.* **14**, 1027–1032.

Fernández Bolaños, J., Fernández Díez, M. J., Rivas Moreno, M., Gil Serrano, A., Pérez Romero, T. (1982), Azúcares y polioles en aceitunas verdes. III. Determinación cuantitativa por cromatografía gas-líquido, *Grasas Aceites* **34**, 168–171.

Fernández Díez, M. J. (1971), The olive, in: *The Biochemistry of Fruits and Their Products*, Vol. 2 (Hulme, A. C., Ed.), pp. 255–279, London: Academic Press.

Fernández Díez, M. J. (1983), Olives, in: *Biotechnology*, 1st Ed. (Rehm, H. J., Reed, G., Eds.), Vol. 5, pp. 380–397, Weinheim, Germany: Verlag Chemie.

FERNÁNDEZ DÍEZ, M. J. (1991), Olives, in: *Encyclopedia of Food Science and Technology* (HUI, Y. H., Ed.), pp. 1910–1925, New York: John Wiley & Son.

FERNÁNDEZ DÍEZ, M. J., GARRIDO FERNÁNDEZ, A. (1971), Aceitunas negras por oxidación en medio alcalino. I. El color como criterio de madurez y calidad en el producto elaborado, *Grasas Aceites* **22**, 193–199.

FERNÁNDEZ DÍEZ, M. J., GONZÁLEZ CANCHO, F. (1966), Resistencia térmica de lactobacilos y levaduras. II, *Microbiol. Españ.* **19**, 119–129.

FERNÁNDEZ DÍEZ, M. J., GONZÁLEZ PELLISSÓ, F. (1956), Cambios en la composición de la aceituna durante su desarrollo. II. Acidez y pH del jugo. Determinación de los ácidos oxálico, cítrico y málico, *Grasas Aceites* **7**, 185–189.

FERNÁNDEZ DÍEZ, M. J., CASTRO RAMOS, R. DE, GARRIDO FERNÁNDEZ, A., HEREDÍA MORENO, A., GONZÁLEZ CANCHO, F., GONZÁLEZ PELLISSÓ, F., NOSTI VEGA, M., REJANO NAVARRO, L., DURÁN QUINTANA, M. C., MÍNGUEZ MOSQUERA, M. I., GARCÍA GARCÍA, P., CASTRO RAMOS, A. DE (1985), *Biotecnología de Aceitunas de Mesa,* Madrid: Consejo Superior de Investigaciones Científicas.

FERNÁNDEZ GONZÁLEZ, M. J., GARCÍA GARCÍA, P., GARRIDO FERNÁNDEZ, A., DURÁN QUINTANA, M. C. (1993), Microflora of the aerobic preservation of directly brined green olives from Hojiblanca cultivar, *J. Appl. Bacteriol.* **75**, 226–233.

FLEMING, H. P., WALTER JR., W. M., ETCHELLS, J. L. (1969), Isolation of a bacterial inhibitor in green olives, *Appl. Microbiol.* **18**, 850–860.

FLEMING, H. P., WALTER, W. M., ETCHELLS, J. L. (1973), Antimicrobial properties of oleuropein and products of its hydrolysis, *Appl. Microbiol.* **26**, 777–782.

FOYTIK, J. (1960), California olive industry, *Calif. Agric. Exp. Stn. Circ.* **493**.

GARCÍA, P., DURÁN, M. C., GARRIDO, A. (1985), Fermentación aeróbica de aceitunas negras maduras en salmuera, *Grasas Aceites* **36**, 14–20.

GARCÍA, P., BRENES, M., GARRIDO, A. (1991), Effect of oxygen and temperature on the oxidation rate during darkening of ripe olives, *J. Food Eng.* **13**, 259–271.

GARCÍA, P., BRENES, M., VATTAN, T., GARRIDO, A. (1992), Kinetic study at different pH values of the oxidation processes to produce ripe olives, *J. Sci. Food Agric.* **60**, 327–331.

GARCÍA GARCÍA, P., BRENES BALBUENA, M., GARRIDO FERNÁNDEZ, A. (1986), Uso de lactato ferroso en la elaboración de aceitunas tipo negras, *Grasas Aceites* **37**, 33–38.

GARCÍA GARCÍA, P., BRENES BALBUENA, M., VICENTE FERNÁNDEZ, J. DE, GARRIDO FERNÁNDEZ, A. (1990), Depuración de las aguas residuales de las plantas envasadoras de aceitunas verdes mediante tratamientos físico-químicos, *Grasas Aceites* **41**, 263–269.

GARCÍA GARCÍA, P., DURÁN QUINTANA, M. C., BRENES BALBUENA, M., GARRIDO FERNÁNDEZ, A. (1992), Lactic fermentation during the storage of "Aloreña" cultivar untreated green table olives, *J. Appl. Bacteriol.* **73**, 324–330.

GAROGLIO, P. J. (1950), *Tecnología de los Aceites Vegetables,* Vol. II: *El Aceite de Oliva y su Industria,* Mendoza (Argentina): Ministerio de Educación, University of Cuyo.

GARRIDO, A., GARCÍA, P., MONTAÑO, A., BRENES, M., DURÁN, M. C. (1993), Biochemical changes during the preservation stage of ripe olive processing, *Nahrung* **37**, 583–591.

GARRIDO FERNÁNDEZ, A. (1979), Aceitunas negras al natural en salmuera. VII. Estudio de algunos aspectos relacionados con su envasado en frascos de vidrio y bolsas de material plástico, *Grasas Aceites* **30**, 153–157.

GARRIDO FERNÁNDEZ, A. (1980a), Aceitunas negras por oxidación en medio alcalino. IV. Estudio de la influencia de algunos aditivos y formas de envasado sobre la textura, *Grasas Aceites* **31**, 323–329.

GARRIDO FERNÁNDEZ, A. (1980b), Aceitunas negras por oxidación en medio alcalino. V. Estudio de la influencia de algunos aditivos y formas de envasado sobre el color, *Grasas Aceites* **31**, 385–389.

GARRIDO FERNÁNDEZ, A., GONZÁLEZ PELLISSÓ, F., GONZÁLEZ CANCHO, F., SÁNCHEZ ROLDÁN, F., REJANO NAVARRO, L., CORDÓN CASANUEVA, J. L., FERNÁNDEZ DÍEZ, M. J. (1977), Modificaciones de los procesos de elaboración y envasado de aceitunas verdes de mesa en relación con la eliminación y reuso de vertidos, *Grasas Aceites* **28**, 267–285.

GARRIDO FERNÁNDEZ, A., GARCÍA GARCÍA, P., BRENES BALBUENA, M. (1992), The recycling of table olive brine using ultrafiltration and activated carbon adsorption, *J. Food Eng.* **17**, 291–305.

GONZÁLEZ CANCHO, F., DURÁN QUINTANA, M. C. (1981), Bacterias cocaceas del ácido láctico en el aderezo de aceitunas verdes, *Grasas Aceites* **32**, 373–379.

GONZÁLEZ CANCHO, F., NOSTI VEGA, M., DURÁN QUINTANA, M. C., GARRIDO FERNÁNDEZ, A., FERNÁNDEZ DÍEZ, M. J. (1975), El proceso de fermentación de las aceitunas negras maduras en salmuera, *Grasas Aceites* **26**, 297–309.

GONZÁLEZ PELLISSÓ, F., REJANO NAVARRO, L.

(1984), La pasteurización de aceitunas estilo sevillano. II, *Grasas Aceites* **35**, 235–239.

GOOR, A. (1966), The place of the olive in the Holy Land and its history through the ages, *Econ. Bot.* **20**, 233–243.

GUERBAA, H. (1990), Estudio retrospectivo del mercado mundial del aceite de oliva y las aceitunas de mesa, *Olivae* **32**, 26–32.

HARTMANN, H. T. (1953), Olive production in California, *Calif. Agric. Exp. Stn. Man.* **7**.

HARTMANN, H. T. (1962), Olive growing in Australia, *Econ. Bot.* **16**, 1–44.

HEREDIA MORENO, A., FERNÁNDEZ BOLAÑOS, J. (1987), Evolución de la fibra alimentaria durante el desarrollo y maduración del fruto del olivo (*Olea europaea arolonsis*), *Alimentaria*, April, 19–22.

HEREDIA MORENO, A., FERNÁNDEZ DÍEZ, M. J. (1982), Composición de la fibra. V. Hemicelulosas e indice de digestibilidad en aceitunas de mesa, *Grasas Aceites* **33**, 197–200.

HEREDIA MORENO, A., GUILLEN, R., JIMÉNEZ, A., FERNÁNDEZ BOLAÑOS, J. (1993), Activity of glycosidases during development and ripening of olive fruit, *Z. Lebensm. Unters. Forsch.* **196**, 147–151.

IOOC (International Olive Oil Council) (1991), *Unified Qualitative Standards Applying to Table Olives in International Trade*, Madrid: IOOC.

JIMÉNEZ DÍAZ, R. L., RIOS SÁNCHEZ, R. M., DESMAZEAUD, M., RUIZ BARBA, J. L., PIARD, J. C. (1993), Plantaricins S and T, two new bacteriocins produced by *Lactobacillus plantarum* LPC010 isolated from green olive fermentation, *Appl. Environ. Microbiol.* **59**, 1416–1424.

JOINT PROGRAMME FAO/OMS (1974), *Documento CAC/RS 6G,* 1974. Roma: Comission Codex Alimentarius.

JUVEN, B., SAMISH, Z., HENIS, Y. (1968), Identification of oleuropein as a natural inhibitor of lactic fermentations of green olives, *Israel. J. Agric. Res.* **18**, 137–138.

JUVEN, B., HENIS, Y., JACOBY, B. (1971), Studies on the mechanism of the antimicrobial action of oleuropein, *J. Appl. Bacteriol.* **35**, 559–567.

KUBO, J., MATSUMOTO, A., TOKASE, I. (1985), A multichemical defense mechanism of bitter olive *Olea europaea* (Oleaceae). Is oleuropein a phytoalexin precursor? *J. Chem. Ecol.* **11**, 251–263.

LOUSSERT, R., BROUSE, G. (1980), Origen y expansión del cultivo del olivo, in: *El Olivo,* pp. 23–30, Madrid: Editorial Mundi-Prensa.

MANOUKAS, A., MAZAMENOS, B., PATRINOU, M. A. (1973), Amino acid composition of three varieties of olive fruit, *J. Agric. Food Chem.* **21**, 215–217.

MARSILIO, V. (1993), Producción, elaboración y reglamentación de las aceitunas de mesa en Italia, *Olivae* **49**, 6–16.

MARSILIO, V., SOLINAS, M. (1990), Presenza e comportamento delle sostanze pectiche nelle olive da tavola. Influenza della varietà e della tecnologia, in: *Proc. Seminario Internazionale Olio d'Oliva e Olive da Tavola: Tecnologia e Qualità,* pp. 357–372, Pescara (Italy): Ist. Sper. Elaiotecnica.

MCDONALD, L. C., FLEMING, H. P., HASSAN, H. M. (1990), Acid tolerance of *Leuconostoc mesenteroides* and *Lactobacillus plantarum, Appl. Environ. Microbiol.* **56**, 2120–2124.

MÍNGUEZ MOSQUERA, M. I. (1982), Evolución de los constituyentes pécticos y de las enzimas pectinolíticas durante el proceso de maduracíon y almacenamiento de la aceituna Hojiblanca, *Grasas Aceites* **33**, 327–333.

MÍNGUEZ MOSQUERA, M. I., GARRIDO FERNÁNDEZ, J. (1989), Chlorophyll and carotenoid presence in olive fruit (*Olea europaea*), *J. Agric. Food Chem.* **37**, 1–7.

MÍNGUEZ MOSQUERA, M. I., CASTILLO GÓMEZ, J., FERNÁNDEZ DÍEZ, M. J. (1978), Presencia de pectinesterasa y su relación con el ablandamiento de algunos productos al alderezo, *Grasas Aceites* **29**, 29–36.

MÍNGUEZ MOSQUERA, M. I., GARRIDO FERNÁNDEZ, J., GANDUL ROJAS, B. (1990), Quantification of pigments in fermented Manzanilla and Hojiblanca olives, *J. Agric. Food Chem.* **38**, 1662–1666.

MÍNGUEZ MOSQUERA, M. I., GALLARDO GUERRERO, L., GANDUL ROJAS, B. (1993), Characterization and separation of oxidized derivatives of pheophorbide a and b by thin layer chromatography, *J. Chromatogr.* **633**, 295–299.

MONTAÑO, A., SÁNCHEZ, A. H., CASTRO, A. DE (1993), Controlled fermentation of Spanish type green olives, *J. Food Sci.* **58**, 842–844.

NARASAKI, T., KATAKURA, K. (1954), Fundamental studies on the utilization of olive fruits. II. Identification of the amino acids in the protein hydrolisate of ripe olive flesh by paper chromatography. *Tech. Bull. Kagawa Agric. Coll.* **6**, 194–198.

NOSTI VEGA, M., MATEOS NEVADO, B., MARTIN GARCÍA, E. (1981), Aspectos bromatológicos de las aceitunas de mesa. Determinación de sus amino ácidos por cromatografía en fase gaseosa, *Anal. Bromatol.* **XXVIII-1**, 35–45.

NOSTI VEGA, M., CASTRO RAMOS, R., VÁZQUEZ LADRÓN, R. (1982), Composición y valor nutritivo de las aceitunas de mesa. V. Efecto de la reutilización de la lejía y de la supresión de los lavados en las aceitunas verdes aderezadas, *Grasas Aceites* **33**, 5–8.

PANIZZI, L., SCARPATI, M. L., ORIENTE, G. (1960), Constituzione delle oleuropeina, glucoside amaro e ad azione ipotensiva dell' olivo, *Nota II, Gazz. Chem. Ital.* **90**, 1449–1485.

PATAC DE LAS TRAVIESAS, L., CADAHIA CICUENDEZ, P., CAMPO SÁNCHEZ, E. DEL (1954), *Tratado de Olivicultura*, Madrid: Sindicato Nacional del Olivo.

PETRUCCIOLI, G. (1965), Variazioni dei componenti delle olive durante il periodo di maturazione, *Olearia* **19**, 5–13.

RAYA, J., LÓPEZ-GEORGE, J., RECALDE, L. (1977), Etudes physiologiques et biochimiques de l'olive. I. Variation de la concentration de divers metabolites pendant son cycle évolutif, *Agrochimique* **XXI**, 311–321.

REJANO NAVARRO, L., GONZÁLEZ CANCHO, F., BORBOLLA Y ALCALÁ, J. M. R. DE LA (1978), La formación de ácido propiónico durante la conservación de aceitunas verdes de mesa. II, *Grasas Aceites* **29**, 203–210.

REJANO NAVARRO, L., CASTRO GÓMEZ-MILLÁN, A., GONZÁLEZ CANCHO, F., DURÁN QUINTANA, M. C., SÁNCHEZ GÓMEZ, A. H., MONTAÑO ASQUERINO, A., GARCÍA GARCÍA, P., SÁNCHEZ ROLDÁN, F., GARRIDO FERNÁNDEZ, A. (1986), Repercusión de diversas formas de tratamiento con ácido clorhídrico en la elaboración de aceitunas verdes estilo sevillano, *Grasas Aceites* **37**, 19–24.

RUIZ BARBA, J. L. (1991), Estudio de algunos factores que influyen en el desarrollo de una fermentación ácido-láctica apropiada de aceitunas de mesa, *PhD Thesis*, University of Sevilla.

RUIZ BARBA, J. L., RÍOS SÁNCHEZ, R. M., FEDRIANI IRISO, C., OLÍAS, J. M., RÍOS, J. L. (1990), Bactericidal effect of phenolic compounds from green olives on *Lactobacillus plantarum, Syst. Appl. Microbiol.* **13**, 199–205.

SÁNCHEZ, A. H., REJANO, L., MONTAÑO, A. (1991), Kinetics of the destruction by heat of colour and texture of pickled green olives, *J. Sci. Food Agric.* **54**, 379–385.

SANDRET, F. G. (1957), *Contribution à l'Etude de la Composition Chimique de la Pulpe d'Olive. Variations au cours de la Maturation,* Publication n° 2, Meknes (Morocco): Ecole Marocaine d'Agriculture.

SANDRET, F. G. (1958), La maturation de la pulpa d'olive. Les substances glucidiques solubles et leur evolution, *Oleagineaux* **13**, 459–464.

UEBA (Unidad Estructural de Biotecnología de Alimentos) (1990), *Table Olive Processing*, Madrid: International Olive Oil Council.

USDA (United States Department of Agriculture) (1950), *Composition of Food Raw, Processed, Prepared, Agricultural Handbook*, No. 8, Washington, DC: United States Department of Agriculture.

VAMVOUCAS, D., STEFANOUDAKIS-KATZOURAKIS, E., LOUPAKIS-ADROULAKIS, M., KIRITSAKIS, A. (1980), Results from chemical analysis and determinations on the main cultivar and styles of Greek table olives, in: *The Biological Value of Table Olives, IInd World Olive Year 1979/80*, Congress on the Biological Value of Olive Oil, Chania, Crete (Greece), *Agric. Res.* **IV**, 356.

VAUGHN, R. H. (1985), The microbiology of vegetable fermentations, in: *Microbiology of Fermented Foods*, Vol. 1 (BRIAN, WOOD, J. B., Eds.), pp. 49–109, Barking (England): Elsevier Applied Science Publishers.

VAUGHN, R. M., MARTIN, M. H., STERENSON, K. E., JOHNSON, M. G., CRAMPTON, V. M. (1969), Salt free storage of olives and other produce for future processing, *Food Technol.* **23**, 124–126.

VÁZQUEZ LADRÓN, R., CASTRO RAMOS, R., NOSTI VEGA, M. (1979), Composición y valor nutritivo de algunas variedades españolas de aceitunas de mesa. III. Aceitunas verdes aderezadas envasadas, *Grasas Aceites* **30**, 221–226.

VÁZQUEZ RONCERO, A. (1963), Química del olivo I. Los componentes orgánicos, *Grasas Aceites* **14**, 262–270.

VÁZQUEZ RONCERO, A. (1964), Química del olivo II. Los componentes orgánicos, *Grasas Aceites* **15**, 87–92.

VÁZQUEZ RONCERO, A., MAESTRO DURÁN, R., JANER DEL VALLE, M. L. (1970), Colorantes antocianicos de la aceituna madura II. Variaciones durante la maduración, *Grasas Aceites* **21**, 337–341.

VÁZQUEZ RONCERO, A., GRACIANI CONSTANTE, E., MAESTRO DURÁN, R. (1974), Componentes fenólicos de la aceituna. I. Polifenoles de la pulpa, *Grasas Aceites* **25**, 269–279.

VLAHOV, G. (1976), Gli acidi organici delle olive: Il rapporto malico/citrico quale "indice di maturazione", *Ann. Ist. Sper. Elaiotecnica* **35**, 93–112.

VLAHOV, G. (1990), La polimerizazione degli antociani delle olive nere da mesa. Ruolo de la tecnologia, in: *Proc. Seminario Internazionale Olio d'Oliva e Olive da Tavola: Tecnologia e Qualità*, pp. 357–372, Pescara (Italy): Ist. Sper. Elaiotecnica.

VLAHOV, G., CUCURACHI, A., BRIGHINNA, A. (1975), Non-volatile acids in table olives. Their evolution during ripening and working process, *II International Olive Seminar*, Córdoba, Spain.

WALTER JR., W. M., FLEMING, H. P., ETCHELLS, J. L. (1973), Preparation of antimicrobial compounds by hydrolysis of oleuropein from green olives, *Appl. Microbiol.* **26**, 773–776.

17 Vegetable Fermentations*

HENRY P. FLEMING

Raleigh, NC 27695-7624, USA

KYU H. KYUNG

Seoul, Korea

FRED BREIDT

Raleigh, NC 27695-7624, USA

* Mention of a trademark or proprietary product does not constitute a guarantee or warranty of the product by the U.S. Department of Agriculture or North Carolina Agricultural Research Service, nor does it imply approval to the exclusion of other products that may be suitable.

1 Introduction

The preservation of vegetables by fermentation is thought to have originated before recorded history and the technology developed by trial and error. PEDERSON (1979) presumed early man to observe that when vegetables were flavored with salt or brine and packed tightly in a vessel, they changed in character but remained appetizing and nutritious. He concluded that the Chinese were the first to preserve vegetables in this manner and assumed that fermentation in salt brines occurred first and that dry salting came later. The Chinese have been credited with introducing fermented vegetables into Europe.

Many, if not all, vegetables have been preserved by fermentation throughout the world. This chapter is devoted to summarizing the current scientific principles and technology involved in the commercial preservation of cucumbers (for pickles) and cabbage (for sauerkraut) by fermentation. The preservation of kimchi, a Korean fermented vegetable mixture of radishes, Chinese cabbage, cucumbers, and other components, is also summarized.

2 Cucumbers for Pickles

2.1 Raw Product

Various authors have attributed the origin of the cucumber (*Cucumis sativus* L.) to Africa, China, India, or the Near East (MILLER and WEHNER, 1989). Later domestication occurred throughout Europe, and cucumbers are now grown throughout the world using field or greenhouse culture, but with various characteristics, depending upon region.

Cucumbers are bred either for fresh market or processing (pickling). The fresh market varieties possess a relatively tough skin, which serves to extend their storage life as fresh produce. Pickling varieties, however, possess thin, relatively tender skin. Pickling cucumbers are harvested in a relatively immature stage, before the seeds mature and before the seed area becomes soft and starts to liquefy. The fruit contains an endo-polygalacturonase in the area surrounding the seeds which causes pectin hydrolysis and, thus, liquefaction in the seed area as the fruit matures (MCFEETERS et al., 1980). The value of pickling cucumbers to the processor varies inversely with fruit size, and growers are paid accordingly. Size grades are determined by fruit diameter, and grading devices sort the fruit by diameter. There are limited efforts to also grade the fruit by length, but it is a more common practice to cut overly long fruit to match jar sizes preferred by the processor/consumer.

Pickling varieties of cucumbers have been carefully bred to resist diseases and environmental stresses, grow well in the specific region for which they were developed, produce high yields, and possess the desired physical and chemical attributes (MILLER and WEHNER, 1989). Among the physical attributes desired are relatively small seed area, thin and tender skin, straight and uniform shape, a length to diameter ratio of approximately 3.0, firm texture, typical green color, and absence of internal defects. Comparatively little has been done to manipulate the chemical composition of cucumbers, although research has indicated several possibilities for consideration.

Cucurbitacins (ENSLIN et al., 1967) responsible for bitterness; sugar content (MCCOMBS et al., 1976; MCCREIGHT et al., 1978), important in fermentation; malic acid (MCFEETERS et al., 1982b), important in bloater formation during fermentation; and polygalacturonase activity (MCFEETERS et al., 1980), involved in softening as the fruit ripens, are some of the chemical constituents that have been considered for possible manipulation. Although most improvements in cucumber varieties to date have been accomplished by traditional breeding programs (MILLER and WEHNER, 1989), the tools of modern molecular genetics are in the early stages of application (STAUB and BACKER, 1995).

2.2 Processing

Cucumbers are harvested by hand or mechanical means, depending upon availability of labor, land size and conformation, and other factors. Cucumbers are a seasonal crop that is grown in various geographical regions and shipped to the processor. Great changes have occurred in the United States over the past 20 years as to origin of the fruit which a processor receives. While once a mainly regional and seasonal enterprise, some large processors now receive fresh cucumbers nearly the entire year. Figures of the production of cucumbers for pickles in the U.S. are shown in Tab. 1. The fruit are grown from Mexico to Canada and shipped fresh to processors according to their demands. Cucumbers grown near the processor may be processed within 24 hours. Cucumbers are shipped under refrigeration if grown at distant locations from the processor. The demand for a year-round supply of fresh cucumbers varies according to the types of products that the processors manufacture. Brined cucumbers, being more stable, are transported inter-continentally.

Pickling cucumbers are preserved by three basic methods, fermentation (40% of overall production), pasteurization (40%), and refrigeration (20%) (FLEMING and MOORE, 1983), as outlined in Fig. 1. Fermentation is the oldest method of preservation and was the only commercial method until about 1940. Pasteurization of fresh cucumber pickles was introduced into the United States industry in the 1940s and resulted in increased consumption of pickles because of their milder acid flavor and more uniform quality. The process involved heating properly acidified cucumbers to an internal temperature of 74°C and holding for 15 min (ETCHELLS and JONES, 1942; MONROE et al., 1969). Some processors deviate from this standard process, depending upon their products and experiences. Fermented cucumbers also may be pasteurized to increase shelf stability, but at lower temperatures and times (JONES et al., 1941). Refrigerated pickles were introduced on a national scale in the United States in the 1960s. Most of these products are preserved by addition of low concentrations of vinegar and a chemical preservative (e.g., sodium benzoate), in addition to refrigeration at 1–5°C. Microbial growth in these products is not desired. Non-acidified, refrigerated pickles, originally popular among certain ethnic groups, also are marketed in some metropolitan areas. These

Tab. 1. U.S. Production Statistics for Pickles and Sauerkraut for 1992

Crop/State	Harvested (1000 acres)	Yield per Acre (tons)	Total Production (1000 tons)	Total Value ($1000)
Cabbage for Sauerkraut				
New York	1.4	25.0	35.0	1365
Wisconsin	2.6	38.8	100.8	3348
Other states	1.5	23.6	35.9	1799
Total U.S.	5.5	31.1	171.7	6512
Cucumbers for Pickles				
North Carolina	20.1	4.0	80.4	17125
Michigan	21.5	5.2	111.8	17776
Other states	61.6	—	397.4	85912
Total U.S.	103.2	5.7	589.6	120813

Source: ANONYMOUS (1993a)

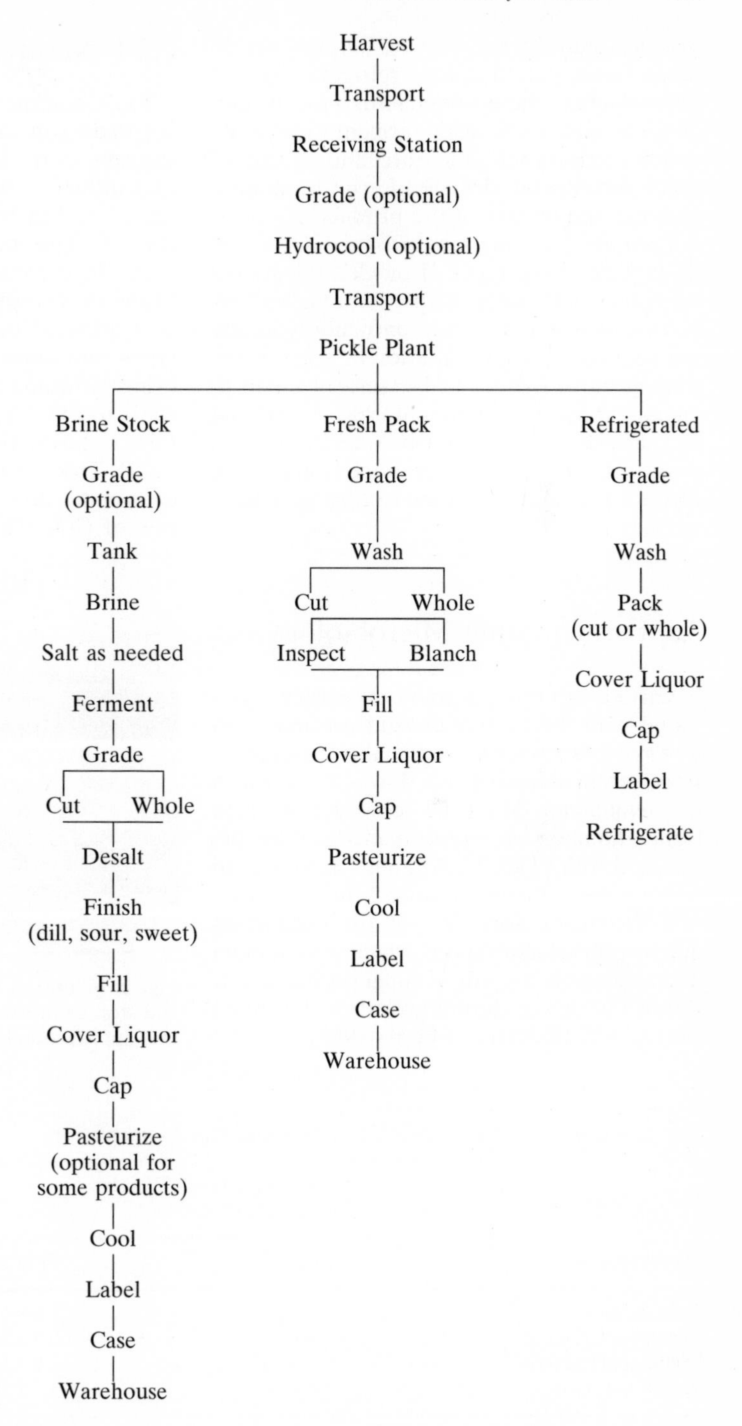

Fig. 1. Flow diagrams for three methods of cucumber processing. From FLEMING and MOORE (1983).

products may or may not be allowed to undergo fermentation before refrigeration. After packaging, these non-acidified pickles undergo a slow lactic acid fermentation while under refrigeration, the rate and extent of which dictates the storage life (up to about 3 months) and quality of the product.

Commercial firms may produce any or all of the three basic types of pickles, depending upon their size, area of distribution, and expertise. Waste generation, particularly excess salt generated by fermentation in brine, is becoming more significant because of environmental concerns about chloride in ground and surface waters of North America. In fact, problems with waste generation influence the types of products that some companies manufacture.

2.3 Fermentation Microbiology

Various groups of microorganisms associated with the cucumber fruit are indicated in Tab. 2. Some fruit, especially the smaller sizes, retain the withered flower from which they emanated. These flowers contain much higher numbers of microorganisms than the attached fruit (Tab. 2). Attempts are made to remove these flowers before brining to prevent enzymatic softening of the cucumbers due to polygalacturonases which may remain due to possible growth of fungi on the nectar within the flower during growth of the plant (BELL, 1951; ETCHELLS et al., 1958).

2.3.1 Natural Fermentations

Most commercial cucumber fermentations are the result of naturally occurring microorganisms and the environmental conditions that influence their growth. Fresh cucumbers are placed in brine contained in bulk tanks (Fig. 2). The microbial activities in the brine occur in various stages during fermentation (Tab. 3). The pH, temperature, and salt concentration of the brine greatly influence the types and rates of microbial activities. The brine typically is acidified (vinegar or acetic acid) to pH 4.5 or slightly below to facilitate CO_2 removal by purging (COSTILOW et al., 1977). This pH is well below the p*K* 6.1 of bicarbonate, thus resulting in a higher proportion of CO_2, which can be removed by purg-

Fig. 2. Fiberglass tanks used for the fermentation and storage of brined cucumbers. These tanks contain approximately 35000 L.

Tab. 2. Microorganisms on Raw Vegetables Used for Fermentation

	Number/g Fresh Weight Cucumber[a]		Cabbage[b]
Microorganisms	Fruit	Flower	
Total aerobes	1.6×10^4	1.8×10^7	1.3×10^5
Enterobacteriaceae	3.9×10^3	6.4×10^6	3.9×10^3
Lactic acid bacteria	5×10^0	2.6×10^4	4.2×10^1
Yeasts	1.6×10^0	3×10^3	<10

[a] From ETCHELLS et al. (1975)
[b] After trimming outer leaves
From FLEMING et al. (1988a)

Tab. 3. Stages of Microbial Activities During the Natural Fermentation of Vegetables

Stage	Prevalent Microorganisms (Conditions)
Initiation of fermentation	Various Gram-positive and Gram-negative bacteria
Primary fermentation	Lactic acid bacteria, yeasts (sufficient acid has been produced to inhibit most bacteria)
Secondary fermentation	Fermentative yeasts (when residual sugars remain and LAB have been inhibited by low pH)
	Spoilage bacteria (degradation of lactic acid when pH is too high and/or salt/acid concentration is too low, e.g., propionic acid bacteria, clostridia)
Post-fermentation	Open tanks: surface growth of oxidative yeasts, molds, and bacteria
	Anaerobic tanks: none (provided the pH is sufficiently low and salt or acid concentrations are sufficiently high)

Modified from FLEMING (1982)

ing. Acidification also influences the types of bacteria that grow during initiation of fermentation and suppresses growth of undesirable bacteria such as the Enterobacteriaceae (MCDONALD et al., 1991).

Fermentation by lactic acid bacteria (LAB) is preferred, and the rate and extent of growth by these bacteria is dictated by brine concentration (5–8% salt initially) and temperature (15–32 °C). Species of LAB involved in fermentation are summarized in Tab. 4. At about 5% NaCl and 21–27 °C, fermentation by LAB is relatively rapid, and fermentable sugars are converted mostly to lactic acid, with relatively little gas formation (ETCHELLS and JONES, 1943; JONES and ETCHELLS, 1943). At 10 to 15% salt, however, the rate and extent of acid production by LAB is re-

Tab. 4. Lactic Acid-Producing Bacteria Involved in Vegetable Fermentations

Genus and Species	Fermentation Type[a]	Main Product (molar ratio)	Configuration of Lactate
Streptococcus faecalis	Homofermentative	Lactate	L(+)
Streptococcus lactis	Homofermentative	Lactate	L(+)
Leuconostoc mesenteroides	Heterofermentative	Lactate:acetate:CO_2 (1:1:1)	D(−)
Pediococcus pentosaceus	Homofermentative	lactate	DL, L(+)
Lactobacillus brevis	Heterofermentative	Lactate:acetate:CO_2 (1:1:1)	DL
Lactobacillus plantarum	Homofermentative	Lactate	D(−), L(+), DL
	Heterofermentative[b]	Lactate:acetate (1:1)	D(−), L(+), DL
Lactobacillus bavaricus	Homofermentative	Lactate	L(+)

Adapted from KANDLER (1983)
[a] With respect to hexose fermentation
[b] Heterofermentative with respect to pentoses (facultatively heterofermentative)

duced, and gas production by yeasts is increased.

Various species of yeasts have been isolated from cucumber fermentations. Fermentative species include *Hansenula anomala, Hansenula subpelliculosa, Saccharomyces baillii, Saccharomyces delbrueckii, Saccharomyces rosei, Torulopsis holmii, Torulopsis lactis-condensii* (*Torulopsis caroliana*), and *Torulopsis versatilis* (*Brettanomyces versatilis*) (ETCHELLS et al., 1961). Oxidative species include *Candida krusei, Debaromyces hansenii* (*Debaromyces membranaefaciens* var. Holl.), *Pichia ohmeri* (*Endomycopsis ohmeri*), *Rhodotorula* sp., *Saccharomyces rouxii* (*Zygosaccharomyces halomembranis*) (ETCHELLS and BELL, 1950b).

During fermentation, the brine is purged with either nitrogen or air to prevent bloater formation. Nitrogen purging presents fewer problems with yeast and fungal growth, and with off-flavors and colors (FLEMING, 1979). However, air purging is used more commercially because of its lower expense. To offset the potential growth and softening spoilage by fungi, potassium sorbate is added (0.035%) to the brine to prevent their growth (GATES and COSTILOW, 1981). Typically, the fermentation of all sugars to acids and other products is completed within about 3 weeks, depending upon temperature and salt concentration.

2.3.2 Controlled Fermentation

PEDERSON and ALBURY (1961) found that *Lactobacillus plantarum* terminated cucumber fermentations, regardless of the species of LAB used for inoculation. Other LAB tested included *Streptococcus faecalis, Leuconostoc mesenteroides, Lactobacillus brevis,* and *Pediococcus cerevisiae.* Apparently, the greater acid tolerance of naturally occurring *L. plantarum* permitted this bacterium to grow after the added cultures had become inhibited by high levels of acidity. Later, ETCHELLS et al. (1973) used an acid-tolerant strain of *L. plantarum* as inoculum in a controlled fermentation procedure (Fig. 3). This procedure included addition of sodium acetate buffer to assure complete fermentation of sugars to lactic acid and, thereby, prevent residual carbohydrate for fermentation and gas production by yeasts. The *L. plantarum* strain used, however, was found to result in CO_2 production (FLEMING et al., 1973b) and bloater formation (FLEMING et al., 1973a). Thus, the procedure also included purging of CO_2 from the brine to prevent bloater formation. At the time, it was unclear why *L. plantarum* caused bloater formation, since it does not produce CO_2 from hexoses. Later it was found that cucumbers contain malic acid (MCFEETERS et al., 1982a), which is degraded to lactic acid and CO_2 (Fig. 4) and resulted in bloater formation in unpurged fermentations (MCFEETERS et al., 1982b).

Cucumbers	Cucumbers
↓	↓
Wash	Wash
↓	↓
Brine (salt)	Brine (salt, calcium acetate, culture)
↓	↓
Chlorination	N_2 Purge
↓	↓
Acidification	Ferment
↓	
Salt Addition	
↓	
Sodium Acetate Addition	
↓	
Culture Addition	
↓	
N_2 Purge	
↓	
Ferment	

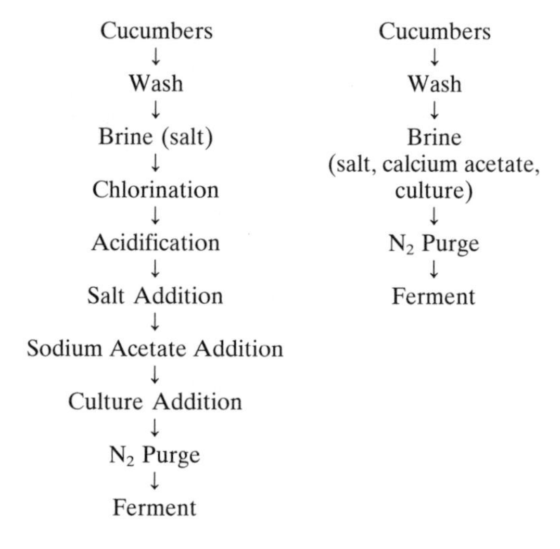

Fig. 3. Controlled fermentation procedures for brined cucumbers. From ETCHELLS et al. (1973), left, and FLEMING et al. (1988a), right.

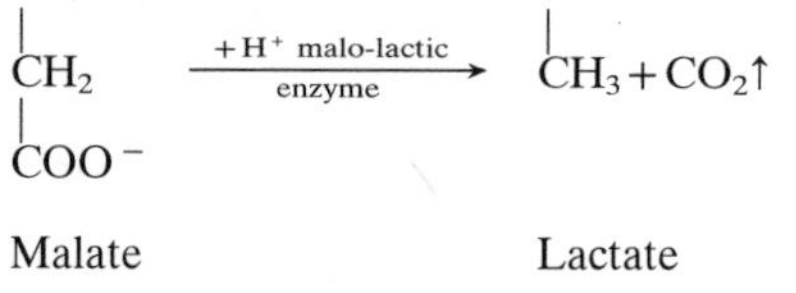

COO⁻
CHOH
CH2
COO⁻
+H⁺ malo-lactic
enzyme
COO⁻
CHOH
CH3 + CO2↑
Malate
Lactate

Fig. 4. Proposed malolactic reaction in cucumber fermentation (MCFEETERS et al., 1982b).

Most commercial firms now purge cucumber brines during fermentation (2–3 weeks), even for natural fermentations, as cited above. Nitrogen originally was recommended for purging because of its inertness (ETCHELLS et al., 1973; FLEMING, 1979). Air has been shown to encourage enzymatic softening by fungi and bleaching discoloration (FLEMING et al., 1975b). COSTILOW et al. (1980) confirmed that air results in softening of cucumbers by mold growth where aerated brine exits the side arm purger (Fig. 5). This is the location where dissolved oxygen in the brine contacting the cucumbers is presumed to be highest. GATES and COSTILOW (1981) determined, however, that addition of 0.035% potassium sorbate to the fermentation brine prevents mold and yeast growth and enzymatic softening of air-purged cucumbers. Also, it has been shown that addition of 0.16% acetic acid to the fermentation brine will prevent growth of fungi responsible for fruit softening (POTTS and FLEMING, 1982).

A procedure for fermentation of cucumbers in closed tanks has been proposed (FLEMING et al., 1983). Nitrogen is used for purging of the brine during fermentation, and an anaerobic headspace of nitrogen is maintained during storage. Calcium acetate addition to the fermentation brine was found to allow fermentation/storage of the cucumbers at a relatively low salt concentration (FLEMING et al., 1978). Addition of calcium acetate and *L. plantarum* to brines of cucumbers fermented in closed tanks resulted in a product of high quality (firm with desirable flavor and color) at relatively low salt concentrations (FLEMING et al., 1988a). See Fig. 3 for procedure. The closed tank concept has yet to be commercially adopted on a large scale probably because of greater expense of the tanks and the necessity for a different handling system. Removable, flexible covers have been proposed as a possible means of converting open to closed tanks, thereby achieving the advantages of open tanks for product handling and closed tanks for fermentation and storage (HUMPHRIES and FLEMING, 1991).

Fig. 5. A tank of fermenting cucumbers being purged to prevent bloater formation. Note the brine and foam that are exiting the side arm purger. The side arm is constructed of a polyethylene tube (10 cm diam.) that extends nearly to the bottom of the tank. The purging gas (air or nitrogen) is introduced through a gas diffuser into the brine as fine bubbles near the bottom of the tank and within the polyethylene tube. This causes the brine to rise in the tube and exit the side arm. Thus, the purging action serves to remove CO_2 from the brine and to circulate the brine within the tank. This tank is constructed of wood and contains approximately 23000 L.

2.3.3 Pure Culture Fermentation

Pure culture fermentation of cucumbers requires the inactivation of naturally occurring microorganisms. ETCHELLS et al. (1964) used hot water blanching (66 to 80 °C for 5 min) or gamma-radiation (0.83 to 1.0 Mrad) to accomplish this goal. They determined the rate and extent of fermentation of pasteurized cucumbers by *P. cerevisiae* (probably *Pediococcus pentosaceus*), *L. plantarum, L. brevis,* and six other LAB. The above three named species are common to cucumber fermentations, and their sequence of growth during fermentation of a three-species mixture as inoculum was similar to that thought to occur in natural fermentations. Although pure culture procedures can result in fermented cucumbers of high and consistent quality, such procedures have been considered impractical for commercial application. Cucumbers to be brined in bulk tanks during the hectic harvest season must be handled quickly and economically. Cucumbers for pasteurized or refrigerated products command first attention during this period.

2.4 Fermentation Chemistry

Glucose and fructose, the primary fermentable sugars of cucumbers, are converted to lactic and acetic acids, ethanol, mannitol, and CO_2 (see Tab. 4), depending upon the species that grow. *L. plantarum,* homofermentative to hexoses, normally predominates cucumber fermentations and converts hexoses primarily to lactic acid. In cucumber juice, 95% or more of the sugars are converted to lactic acid by *L. plantarum* (PASSOS et al., 1994). Malic acid, a natural constituent of cucumbers(MCFEETERS et al., 1982a), is degraded to lactic acid and CO_2 (MCFEETERS et al., 1982b) (see Fig. 4). Heterofermentative LAB, when active, produce mannitol, ethanol, acetic acid, and CO_2 from hexoses, in addition to lactic acid. Fermentative yeasts produce ethanol and CO_2.

Cucumber fruits typically constitute about 60% of the volume in bulk tanks, the remaining 40% being occupied by the cover brine. For cucumbers containing 2% fermentable sugars, about 1.2% sugar theoretically would be present if allowed to come to equilibrium (this does not occur because of fermentation). Thus, about 1.1% lactic acid is present after exclusive fermentation of hexoses by *L. plantarum.* Since malic acid (0.2–0.3% of cucumbers) is degraded to lactic acid, this additional amount must be considered in predicting total lactic acid formation. When less than 1.1% lactic acid is present after fermentation, growth by microorganisms other than homofermentative LAB or a leaking tank is implicated.

2.5 Sensory Properties

Textural properties are highly important in fermented, as well as pasteurized and refrigerated cucumber pickles. A firm, crisp texture is desired. In fermented products such as hamburger dill chips and sweet pickles, the tissue is translucent (referred to as a "cured" appearance) due to expulsion of gas during brine storage. This is normal for these products and, therefore, expected. This translucent appearance is not desired in pasteurized or refrigerated pickles since white, opaque appearance in these products is associated with freshness. Internal tissue voids are objectionable in all products. Cucumber products with small, immature seeds and a small seed area are preferred.

The flavor of cucumber pickles varies widely, depending upon the spices and flavorings that are added to the numerous products. Lactic acid is a natural component of fermented pickles, but its flavor is too tart and objectionable to some in certain products such as hamburger dill chips, where custom has resulted in a preference for an acetic acid (vinegar) flavor. Lactic acid is removed from the brined cucumbers by the desalting operation and is replaced with vinegar to acidify the final product. In other products such as "genuine" dill pickles, the lactic acid flavor is preferred, and these products can be sold in the brine in which they were fermented.

A clean flavor which does not detract from the added spices and flavorings is desired. Proper levels of acid and salt in the finished product are essential and vary among the many products that are made. However, pasteurized pickles typically contain 0.5 to 0.6% acetic acid and possess a pH of about 3.7. The salt concentration varies from 0% (in dietetic pickles) up to about 3%. Fermented pickles vary widely in types of products that are produced. In relish and salad cube products the acetic acid and sugar concentrations are varied to give products that are characterized by their degrees of sweetness and acidity. The ratios and concentrations of acetic acid and sugar can be varied to preserve sweet pickles without the need for pasteurization (BELL and ETCHELLS, 1952). Although the preservation of some mildly acidic and sweet, fermented pickles is assured by pasteurization today, before about 1940 pasteurization was not used.

2.6 Spoilage Problems

Gaseous spoilage (bloater formation) is due to the growth of gas-forming microorganisms such as yeasts (ETCHELLS and BELL, 1950a), heterofermentative LAB (ETCHELLS et al., 1968), and Enterobacteriaceae (ETCHELLS et al., 1945). *L. plantarum* also contributes CO_2 by decarboxylation of malic acid (McFEETERS et al., 1984), as mentioned earlier. CO_2 from the fermenting brine diffuses into the cucumber tissue faster than entrapped nitrogen can diffuse out, thus gas pressure greater than 1 atmosphere results within the tissue, which can result in bloater formation, depending upon resistance of the fruit to internal gas pressure, brine depth, and other factors (FLEMING and PHARR, 1980). Although bloated cucumbers can be used to make relish, the product is of lower value. As mentioned above, bloater formation can be prevented by purging CO_2 from solution during fermentation.

Softening spoilage of fermented cucumbers can arise from various sources. Polygalacturonase enzymes of fungal origin on the surface of the fruit, and especially within the flowers retained on small fruit, can cause softening of the fruit during brine storage (ETCHELLS et al., 1958). This type of softening is rather uniform throughout the tissue of individual fruit, and can occur within the entire tank. Softening of individual fruit in spots normally is due to fungal or bacterial growth on the fruit before brining. Softening of the interior of the fruit is normally associated with ripening of the fruit and the natural endo-polygalacturonase that develops in the fruit as it matures (McFEETERS et al., 1980). Softening can be reduced by removing flowers from the fruit before brining (ETCHELLS et al., 1958), by brining only disease-free fruit as soon after harvest as possible, and by avoiding over-mature fruit. High concentrations of salt (NaCl) will prevent softening by polygalacturonases (BELL and ETCHELLS, 1961), but can present a waste disposal problem. The addition of $CaCl_2$ to the brine (0.2–0.4%) has been shown to reduce the concentration of NaCl needed to prevent softening (BUESCHER et al., 1979, 1981; FLEMING et al., 1987; McFEETERS and FLEMING, 1990).

Bleaching of the green color from brined cucumbers can be due to exposure of the fruit to sunlight. Excessive concentrations of potassium sorbate (>0.035%) have been reported to cause bleached or gray-colored fruit.

Off-flavors and odors in fermented cucumbers result from the growth of undesirable microorganisms. Oxidative yeasts can grow on the surface of cucumber brines that are not exposed to sunlight, utilize lactic acid which causes the pH to rise, and allow spoilage bacteria to grow. Butyric and propionic acids have been shown to be produced, resulting in offensive odors, when the cucumbers are brined at very low (e.g., 2.3%) concentrations of NaCl (FLEMING et al., 1989).

3 Cabbage for Sauerkraut

3.1 Raw Product

The modern, hard-head cultivars of cabbage (*Brassica oleracea*) are reported to have descended from wild, non-heading brassicas originating in the eastern Mediterranean and in Asia Minor (DICKSON and WALLACE, 1986). Cabbage is grown for the fresh market and for the production of sauerkraut. For use in the production of sauerkraut it is desired that the cabbage heads be large (typically 8–12 lbs), compact (i.e., dense), have a minimum of green outer leaves, and possess desirable flavor, color, and textural properties when converted into sauerkraut. Cabbage varieties are bred for yield, disease and insect resistance, storage stability, and dry matter content (DICKSON, 1987; DICKSON and WALLACE, 1986). Although varieties have been developed for fresh market, as well as sauerkraut, when demand exceeds supply, varieties can serve multiple purposes.

Cabbage for sauerkraut is grown in cooler climates in the U.S., and primarily in the states of New York and Wisconsin (see Tab. 1), with production in lesser quantities in Ohio, Oregon and other states. Processing plants for sauerkraut production also are located mostly in these states.

3.2 Processing

Fresh cabbage for sauerkraut is harvested mechanically or by hand mostly in the months of August to November. The cabbage is transported to the processor, where it is graded, cored, trimmed, shredded, and salted. The waste from the coring and trimming operations typically is returned to the field, where it is plowed into the soil. This waste constitutes about 30% of the weight of the fresh cabbage.

After shredding (ca. 1 mm thick), the cabbage is conveyed by belt, where salt is added (Fig. 6A), to the fermentation tanks (Fig. 6B). The tanks typically hold 20–180 tons of shredded cabbage (STAMER, 1983). Most tanks today are constructed of reinforced concrete,

Fig. 6. Salting, conveying, tank filling, and covering sliced cabbage for fermentation into sauerkraut. (A) The sliced cabbage is conveyed by a belt, where it is salted. (B) The salted, sliced cabbage is conveyed into the tank. (C) The cabbage is heaped above the tank and loosely covered with plastic sheeting. (D) After about 24 h, the cabbage is allowed to be lowered by removal of brine from the tank bottom and is then leveled. (E) The cabbage is covered with one or more plastic sheets. A plastic tube may be inserted between the cabbage and the plastic sheet to allow for escape of gas during fermentation. The reinforced concrete tank shown contains approximately 80000 kg of cabbage.

but some wooden tanks remain. The tank is uniformly filled, heaped to extend slightly above the top of the tank, and loosely covered with plastic sheeting (Fig. 6C). After about 24 h, brine generated and located at the bottom of the tank is allowed to drain from the tank to allow the top of the cabbage to settle below the top of the tank. Then, the cabbage is manually distributed to create a slightly concave surface (Fig. 6D). The plastic sheeting is then placed on the surface and water is added on top to weight it down and to provide an anaerobic seal (Fig. 6E). Gas generated during fermentation escapes by forcing its way between the tank wall and the cabbage, or through a plastic tube placed between the cabbage and the cover (Fig. 6E). In cabbage with a low moisture content, excess brine may be insufficient to allow for settling of the salted cabbage from above the top of the tank. In this case the cabbage is not heaped above the tank at filling.

After filling and heading, the cabbage may present a heaving problem during the first few days. This is due to gas pockets being formed in the shredded cabbage from tissue respiration and microbial fermentation. In severe cases the cabbage may be lifted so high as to empty the water seal. In these instances more brine must be removed from the bottom of the tank to allow for settling of the cabbage, and then the tank reheaded with a plastic cover-water seal.

In the U.S. the cabbage is allowed to remain in the tanks until at least 1% lactic acid is formed (about 30 days minimum, depending upon the temperature), and is stored beyond this time and until such time as needed for processing. Thus, the sauerkraut tanks are used for fermentation as well as storage. Although extended periods of holding in the tanks can result in excess acid formation and waste generation due to the need to wash excess acid from the product before canning, this disadvantage is offset by the economic advantages of bulk storage. Also, most U.S. sauerkraut companies specialize in this product only, and desire the option of further processing of sauerkraut throughout the year so as to distribute labor and equipment needs. This U.S. procedure differs significantly from that of many European manufacturers who process the sauerkraut into finished products when it reaches the desired level of titratable acidity (calculated as lactic acid, typically 1%). Thus, European manufacturers have greater control over product uniformity, but lose the economic advantages of bulk storage. Also, the European method results in less waste generation since less excess acid is produced.

The sauerkraut is removed pneumatically, by mechanical fork, or by hand from the tanks as needed for processing during the year. The sauerkraut may be packaged in cans, glass, or plastic containers. When packaged in glass or plastic, the product is not heated. Rather, sodium benzoate (0.1%, w/w) and potassium metabisulfite are added as preservatives, and the product is held under refrigeration (5 °C) (STAMER and STOYLA, 1978).

Canned sauerkraut is preserved by pasteurization without the addition of preservatives. Heating is performed either by steam injection into a thermal screw, or by a sauerkraut juice immersion-type cooker. The product is heated to 74–82 °C for about 3 min and hot-filled into the cans (STAMER, 1983). After closure, the cans are immediately cooled to less than 32 °C. The shelf life of enamel-lined canned products is estimated at 18–30 months, and that of glass- and plastic-packaged products is 8–12 months (STAMER, 1983).

3.3 Fermentation Microbiology

3.3.1 Natural Fermentation

Cabbage, like many fresh vegetables, contains numerous species of microorganisms and relatively high numbers of total aerobic bacteria. LAB constitute a relatively small proportion of the total bacteria (see e.g., Tab. 2). When properly shredded and salted, the cabbage (at a proper temperature) undergoes fermentation by a sequence of LAB that results in the distinctive flavor of sauerkraut. The most comprehensive review of the sauerkraut fermentation is that by PEDERSON and ALBURY (1969), which should be consulted by serious students of the subject. These researchers listed five species of LAB as important in the sauerkraut fermentation in order of increasing total acid production: *Streptococcus faecalis* < *Leuconostoc mesenteroides* < *Lactobacillus brevis* < *Pediococcus cerevisiae* (*P. pentosaceus*) < *Lactobacillus plantarum*. Although the names of some of these species have been changed, it is believed that all of these, and perhaps others, are involved. However, it seems clear that *L. mesenteroides* is a major species in the early, heterofermentative stage of fermentation, and that *L. plantarum* is a major species involved in the late, homofermentative stage of fermentation. Salt concentration and temperature influence the relative extent of growth by these two species of LAB, and thus the quality of the sauerkraut. Low salt concentration and low temperature (e.g., 1%, 18°C) favor growth of heterofermentative LAB, while high salt concentration and high temperature (e.g., 3.5%, 32°C) favor growth of homofermentative LAB (PEDERSON and ALBURY, 1954).

The fact that sauerkraut first undergoes fermentation by heterofermentative and then homofermentative species is illustrated in Fig. 7. During the first 8 days of fermentation at 2% salt and 18.3°C, heterofermentative species were predominant. This stage of the fermentation resulted in much gas production. After this period, homofermentative species predominated and relatively little gas was formed.

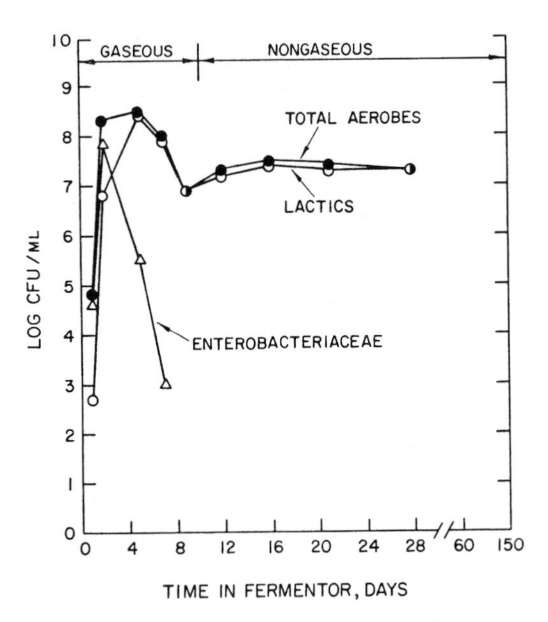

Fig. 7. Microbial growth in sauerkraut fermentation. From FLEMING et al. (1988b). Heterofermentative LAB predominated during the gaseous stage and homofermentative during the non-gaseous stage.

3.3.2 Controlled Fermentation

Proper salt concentration and temperature have been concluded to be primary means for control of the sauerkraut fermentation (PEDERSON and ALBURY, 1954). Some have concluded that addition of cultures is not needed for desirable sauerkraut if these conditions are appropriate (PEDERSON and ALBURY, 1969; STAMER, 1983). However, cultures could be desirable if they have sufficiently unique and valuable properties, or if the environmental conditions (salt, temperature) are undesirable for traditional sauerkraut manufacture. Culture development for vegetable fermentations is discussed later in this chapter.

Some European manufacturers inoculate sliced cabbage with brine from a previous fermentation, and this seems to result in a desirable heterolactic fermentation. The mild acidity (1%), relatively low salt concentration (e.g., 1%), rapid fermentation (e.g., 1 week), and immediate canning of German sauerkraut probably account for its high quality and uniformity. Since American sauerkraut may be held in bulk storage for several months, it is even more important that heterofermentation by *L. mesenteroides* attain its maximum potential. If temperature or salt is not ideal, homofermentation by *L. plantarum* can result in a harshly acidic product. Thus, culture selection and predominance could have a significant impact on quality and uniformity of American sauerkraut.

3.4 Fermentation Chemistry

The fermentable sugars of cabbage are primarily glucose and fructose, with a smaller

Tab. 5. Composition of Raw Cabbage Used in Fermentations

Compound	Concentration[a] in Leaves mM	SD[b]	Concentration[a] in Core mM	SD[b]
Sucrose	7.0	4.6	53.1	16.4
Glucose	132.5	9.9	75.7	15.9
Fructose	114.2	3.2	60.3	7.3
Malic acid	12.2	1.6	7.1	3.0

Source: FLEMING et al. (1988b)
[a] Averages of four replicates
[b] SD = standard deviation

concentration of sucrose (see, e.g., Tab. 5). In the cabbage used for the following discussion, the core accounted for 23% and the leaves for 77% of the cabbage weight. The core region contained a relatively high concentration of sucrose (53.1 mM) compared to the leaves (7.0 mM). Fermentation of this cabbage at 18.3°C in laboratory fermentors resulted in sugar depletion and product formation, as illustrated in Figs. 8 and 9 (FLEMING et al., 1988b). These chemical changes are consistent with the microbiological changes noted in Fig. 7, where the fermentation was characterized into gaseous and non-gaseous stages. After coring, shredding, salting, and packing the cabbage into the fermentors, the brine generated was analyzed periodically, as indicated in Figs. 8 and 9. Although fructose and glucose were in similar concentrations in the raw cabbage, the fructose concentration was 30 mM in the initial brine and declined thereafter. However, glucose concentration continued to increase in the brine up to 75 mM after 7 days, and then declined to near 0 after 60 days. Mannitol, acetic acid, and ethanol increased rapidly until about 7 days and plateaued. Lactic acid production continued until all of the fermentable sugars were de-

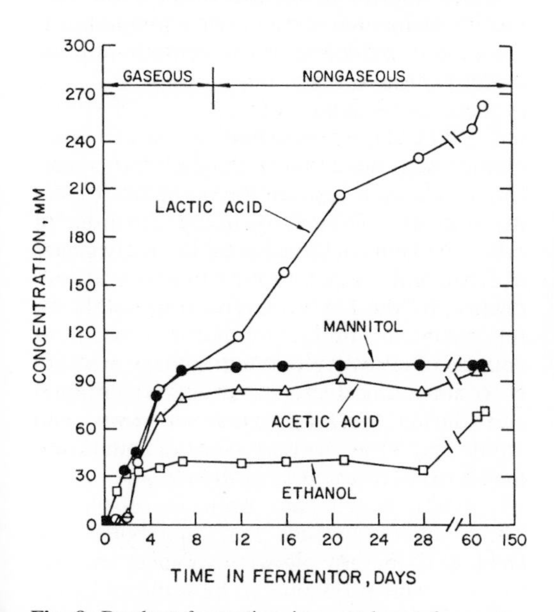

Fig. 8. Product formation in sauerkraut fermentation. From FLEMING et al. (1988b).

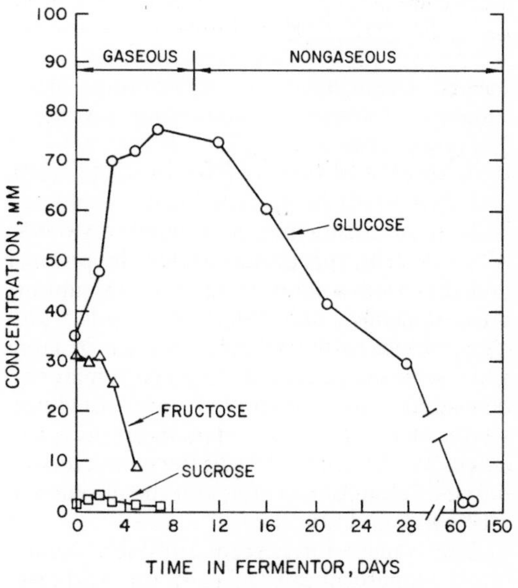

Fig. 9. Substrate depletion during sauerkraut fermentation. From FLEMING et al. (1988b).

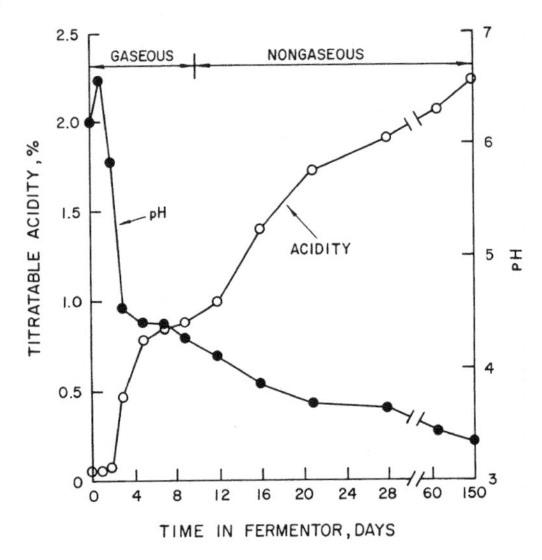

Fig. 10. Titratable acidity and pH changes during sauerkraut fermentation. From FLEMING et al. (1988b).

pleted. During fermentation, the pH was reduced from about 6 to about 3.4 and the titratable acidity, calculated as lactic acid, increased to 2.2% (Fig. 10).

3.5 Sensory Properties

U.S. Department of Agriculture grade standards for canned sauerkraut involve a 100-point scale with color (30 points), cut (10), absence of defects (20), character (10), and flavor (30) being individually evaluated and these factor scores totaled (ANONYMOUS, 1993b). It is preferred that the product be light to straw color. "Cut" refers to uniformity, thickness, and length of shreds. "Defects" refers to the relative absence of large and coarse pieces of leaves and core material, blemished pieces, spotted or discolored pieces, etc. "Character" refers to firmness and crispness. "Flavor" refers to the characteristic flavor of clean sauerkraut with the absence of off-flavors and off-odors.

Ideal ranges for percent titratable acidity (1.1–1.5), percent salt (1.7–2.4), salt/acid ratio (1.0–1.7), and lactic/acetic acid concentrations (3.0–5.0) were reported by PEDERSON (1940) for canned sauerkraut. More recently, a survey of U.S. commercial sauerkraut revealed a slightly lower range in salt concentration (1.4–2.0%) and a lower lactic/acetic ratio (1.4–4.0) (CORBET et al., 1995). This may be due to the industry's trend to use lower salt concentrations for fermentation than the 1.8–2.25% recommended earlier by PEDERSON and ALBURY (1969).

3.6 Spoilage Problems

Although sauerkraut is simple to make, the commercial manufacture of a product of consistently high quality is not simple. American manufacturers of sauerkraut are confronted with the problem of maintaining product quality during storage of the sauerkraut for many months. In contrast, many European manufacturers can their product when it reaches the desired level of acidity (about 1 week) and, thus, are not confronted with quality maintenance during bulk storage.

Inferior or off-flavor can be a serious defect, depending upon the extent, and can result from an improper fermentation or growth of spoilage microorganisms during bulk storage. An inferior flavor can result if the cabbage is fermented at too high a temperature, resulting in predominant fermentation by homofermentative LAB and an improper ratio of lactic:acetic acids (PEDERSON and ALBURY, 1954). The production of butyric, propionic, and other short-chained fatty acids can result in a serious flavor defect (VORBECK et al., 1961). Growth by clostridia or other spoilage bacteria during the early stages of fermentation, before growth and acid production by the LAB, may be responsible for this problem. Improper salting procedures and use of excessively soiled cabbage are likely contributing factors. After fermentation and during storage, oxidative yeasts and molds may grow at the surface of sauerkraut that is exposed to air (PEDERSON and ALBURY, 1969), which can result in off-flavors. These microorganisms may metabolize the lactic acid, thereby allowing spoilage bacteria to grow. This problem can be reduced by assuring a proper seal of the plastic covers of the tanks.

Discoloration of sauerkraut in the fermentation tank can result from growth of pigmented (pink) yeast species (perhaps *Rhodotorula* sp.) due to improper salting (PEDERSON and KELLY, 1938). A red color has been shown to be produced by the non-pigmented bacterium, *L. brevis* (STAMER et al., 1973). In this case, the color formation was catalyzed by aeration and inhibited by cysteine and ascorbic acid, natural reductants of cabbage. Darkening of sauerkraut during fermentation or bulk storage, or after canning seems to be due to oxidative changes, perhaps influenced by the content of natural constituents of the cabbage and exposure to air. It is considered important that the ascorbic acid concentration in the sauerkraut be sufficiently high, e.g., 20 mg/100 g sauerkraut or higher, to serve as an antioxidant and color preservative. Exclusion of air during fermentation, bulk storage, and in consumer containers is important for color retention.

Soft sauerkraut can result from cabbage stored at less than 1.8% salt, whereas greater than 2.5% salt results in a tough texture (PEDERSON, 1946). Mold growth can result in enzymatic softening, perhaps by polygalacturonases and other enzymes, thus reinforcing the need to exclude air from the sauerkraut.

4 Kimchi

Kimchi is a general term applied to a Korean product made by the lactic acid fermentation of salted vegetables (dry salted or brined) with or without secondary ingredients. Unless otherwise specified, however, kimchi generally means a product made from Chinese cabbage as the primary vegetable with secondary ingredients. Kimchis made from vegetables other than Chinese cabbage as the primary vegetable are specified with a qualifying word, e.g., radish kimchi, cucumber kimchi, and green onion kimchi. Both solids and liquids of kimchis are consumed.

Kimchi has been a major table condiment in Korea for about 200 years (LEE, 1975). Kimchi provided a spicy and flavorful adjunct to a rather narrow choice of foods for Koreans in difficult times, complementing the bland taste of cooked rice. Although kimchi still is a major dish in Korea today, its consumption has declined from 200–300 g per person per day in the 1950s (LEE et al., 1960) to 100 g today (CHO, 1993). Consumption is greater during the winter months (150–200 g/day) than in the summer months (50–150 g/day) (ANONYMOUS, 1974). Koreans now enjoy a more diversified diet thanks to industrialization of the country. Interestingly, however, demands for the product may be increasing in the United States and certain other countries where ethnic and regional foods seem to be gaining in popularity.

4.1 Raw Product

Raw materials used for kimchi preparation are divided into two basic groups, primary vegetables and secondary ingredients (Tab. 6). The division is not absolute and relative amounts used vary. Radishes (roots or greens), green onions, or Indian mustard greens, for example, can be primary vegetables in some kimchis but secondary ingredients in other kimchis, depending on the percentages comprising the kimchis. The two most frequently used primary vegetables are Chinese cabbage (*Brassica campestris* subsp. *pekinensis*) and radishes (*Raphanus sativus* L.). Other primary vegetables such as cucumbers, green onions, Chinese leeks, Indian mustard greens, turnips, sowthistle, and green peppers are used less frequently. Spinach, pumpkins, cabbages, and eggplants are used much less frequently.

More than 38 different varieties of kimchi are made and consumed in Korea, depending on vegetables and other ingredients used (SOHN, 1992). More kimchis are made in individual households than commercially, resulting in varying formulations to appeal to the preferences and traditions of each household.

4.2 Kimchi Processing

Winter kimchi has been most traditional and is made between mid-November and ear-

Tab. 6. Raw Materials Used for Kimchi Preparation

	Raw Materials
Primary vegetables	Chinese cabbage, radish (roots with or without leaves), cucumber, green onion, Chinese leek, Indian mustard greens, turnip, green pepper, sowthistle, spinach, pumpkin, cabbage, eggplant
Secondary ingredients	*Vegetables:* radish (root and/or greens), Indian mustard greens, water celery, carrot *Seafoods:* shrimp, oyester, Alaskan pollack, squid, flounder, yellow corvenia *Spices:* red pepper powder, green onion, garlic, ginger, Chinese leek, onion *Fruit:* pine nut, gingko nut, pear, apple, chestnut, jujube *Seasoning:* salt, salted fermented fish sauce (anchovy, shrimp, yellow corvenia), sugar, MSG *Cereals:* cooked rice, cooked rice flour, cooked wheat flour

Adapted from LEE and CHO (1990)

ly December, depending on the climate of the particular year. It is consumed until the following spring. The fresh ingredients are tightly filled into large earthen jars with earthen lids. The jars are partially buried (80–90% of container depth) under ground (Fig. 11) and are covered with bundles of rice straw for protection from direct sunlight and the climate. This procedure is still practiced in much of the rural areas and some households in urban areas. Households living in urban areas usually do not have a place to bury their kimchi jars in the close vicinity of their house and instead keep the jars covered with clothing and other insulating materials in the shaded places of apartments or in home cellars. The sub-surface temperature of the earth during the winter season is fit for slow but excellent fermentation of kimchi and also for the subsequent storage. Over-acidification and development of yeasty flavor are the two most important problems in the long-term storage of kimchi. However, as the temperature goes up late in the following spring, the quality of winter kimchi is reduced.

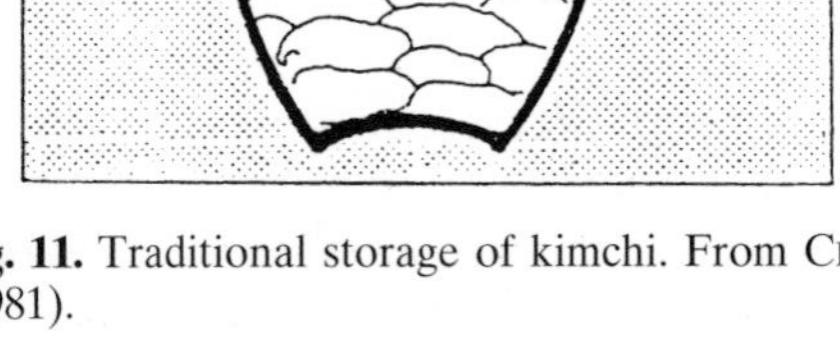

Fig. 11. Traditional storage of kimchi. From CHUN (1981).

Today, lesser amounts of winter kimchi are made due to more availability of food items, including meats, fish, and fresh vegetables, and due to year-round availability of vegetables for kimchi preparation. Koreans now enjoy freshly fermented kimchi anytime of the year due to both the year-round availability of fresh vegetables and the general availability of household refrigerators. It is common for Korean families to make small batches of kimchi as needed and to store it under refrigeration for short periods to extend the fresh quality.

Kimchis can be divided into two types simply depending on the way the primary vegetable is salted. Primary vegetables are salted (either by dry salting or brining) and then the liquid is drained off completely before blending with secondary ingredients in the preparation of general kimchis. However, sufficient brine is poured over the packed ingredients to cover whole cucumber fruits or radish roots. Pickles are usually prepared without secondary ingredients.

4.3 Fermentation Microbiology

The total aerobic population in raw vegetables used to make the kimchi mixture typically is in the range of 10^3–10^7 cells/g. Total LAB, including *Lactobacillus* and *Leuconostoc* species, are in the range of 10^4–10^6 cells/mL (KIM and CHUN, 1966; MHEEN and KWON, 1984; KIM and LEE, 1988; LEE et al., 1992; KIM et al., 1989; SHIM et al., 1990a, b). SHIM et al. (1990b) reported the population of *Lactobacillus plantarum* on individual kimchi ingredients (Tab. 7) as determined by the most probable number (MPN) method to be in the range of 0.36 to 240/g (mean 7.3) for mixed ingredients and 0.0 and 0.36/g (mean 0.0) for Chinese cabbage.

LAB found in raw kimchi mixture include *Leuconostoc mesenteroides, L. dextranicum, Lactobacillus leichmannii,* and *L. sake.* Other LAB found during the course of the kimchi fermentation are *Lactobacillus fermentum, Pediococcus pentosaceus, Streptococcus faecalis, Lactobacillus plantarum,* and *L. brevis* (SHIM et al., 1990b). When LAB isolated from raw kimchi mixture and fermenting kimchi were inoculated into filter-sterilized fresh cabbage juice, *L. mesenteroides, L. leichmannii,* and *L. fermentum* produced 0.5–1.0% acid as lactic. *P. pentosaceus* and *L. plantarum* produced more acid than other species and kept producing acid after cell growth ceased.

The types and numbers of initial microorganisms in kimchi mixture are heavily dependent on the quality of ingredients and the washing procedure. Red pepper powder, which is not washed before use, is a particular source of variation in microorganisms. Kimchi fermentation typically is initiated by *L. mesenteroides* and terminated by *L. plantarum* and/or *L. brevis.*

Good quality primary vegetables and proper combination of secondary ingredients are believed to be important prerequisites for the production of good-quality kimchi. Salt concentration and fermentation temperature are the two most important variables influencing bacterial growth and metabolism. The maximum number of microorganisms attainable and the time required to reach maximum numbers vary depending on the temperature and salt concentration (Fig. 12; MHEEN and KWON, 1984). As the salt concentration is increased, maximum microbial populations attained are lower and the time required to reach maximum numbers is longer. At all temperatures tested, *L. mesenteroides* was shown to reach the maximum population before any other LAB. *L. mesenteroides* declines fast after reaching the maximum populations and disappears from fermenting kim-

Tab. 7. Estimated Number of *Lactobacillus plantarum* in Individual Ingredients (cells/g) and Kimchi Mixture (cells/mL)

	Number of *Lactobacillus plantarum* Mean	Range
Kimchi	7.3	0.36–240
Chinese cabbage	0.0	0.0 –0.36
Radish	2.3	0.0 –9.3
Ginger	4.6	0.0 –30.0
Green onion	1.82	0.0 –4.6
Garlic (unpeeled)	0.0	0.0 –37.2
Red pepper powder	0.0	0
Red pepper (wet)	4.3	0.0 –240

From SHIM et al. (1990b)
These estimates were based on the most probable number (MPN) technique and the fact that only *L. plantarum* and *L. brevis,* of the LAB isolates from kimchi, were able to grow in MRS broth containing 7% ethanol. It was possible to distinguish between these two species by the fact that *L. brevis* produces CO_2 from hexoses, but *L. plantarum* does not.

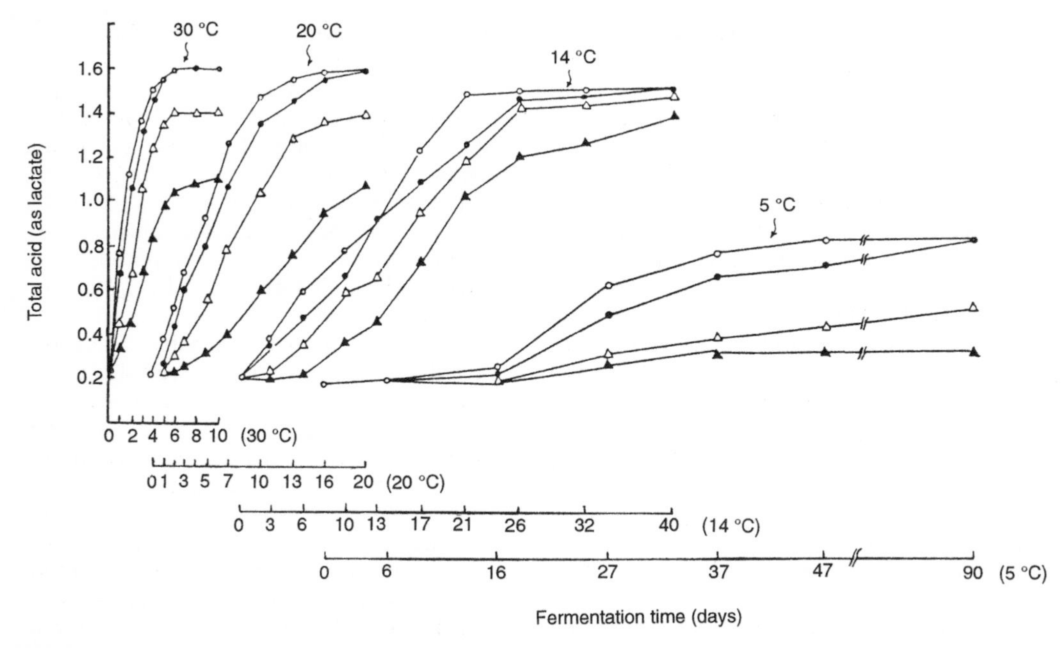

Fig. 12. Changes of total acid during kimchi fermentation at different temperatures and salt concentrations. NaCl: —○—, 2.25%; —●—, 3.5%; —△—, 5.0%; —▲—, 7.0%. From MHEEN and KWON (1984).

chis when other LAB, including *L. plantarum,* become predominant when the temperature is 14°C or above. At 5°C, however, *L. mesenteroides* is the dominant bacterium during the early phase of fermentation and does not decline as fast as at higher temperatures. High populations are maintained throughout fermentation. *L. plantarum* and *L. brevis,* which are blamed for over-acidification, do not appear in kimchi fermented at 5°C.

Various scientists (KIM and CHUN, 1966; MHEEN and KWON, 1984; SHIM et al., 1990a; LEE and YANG, 1975) agree that *L. mesenteroides* appears at the early stage of kimchi fermentation, followed by *L. plantarum,* as in the case with sauerkraut (PEDERSON and ALBURY, 1969).

4.4 Fermentation Chemistry

The pH and acidity of optimally fermented kimchis are 4.2–4.5 and 0.4–0.8% (as lactic acid), respectively (SONG et al., 1966; MHEEN and KWON, 1984; CHUN, 1981). Acid production, as well as microbial growth, is greatly influenced by salt concentration and fermentation temperature. More total acid is produced at lower salt concentrations and higher temperatures (Fig. 12). At the lower concentration of salt, maximum acidity is reached in a shorter period of time. Optimum acidity (0.6%) of kimchi is reached within 1 day at 30°C and at 2.25–3.5% salt, and the same level of acidity is reached in 2 and 4 days at 5.0 and 7.0% salt, respectively.

More volatile acid is produced when low concentrations of salt are used. The ratios of volatile to non-volatile acids are highest after only 2 days of fermentation, decline quickly, and are stabilized thereafter. Also, the ratios were higher at lower fermentation temperatures. Kimchis fermented at lower temperature typically are judged to be superior in quality than those fermented at higher temperatures. The time when the ratio of volatile to non-volatile acids reaches the maximum is considered to be the time when kimchi tastes best (MHEEN and KWON, 1984).

As the salt concentration is increased, the volatile to non-volatile acid ratio decreases, producing a less palatable product. *L. planta-*

rum is not as sensitive to salt as *L. mesenteroides,* which produces more volatile end products.

Organic acids other than lactic and acetic are found in kimchi and kimchi ingredients. These include citric, fumaric, oxalic, malonic, malic, and succinic acids as non-volatile acids and formic, propionic, valeric, butyric, caproic, and heptanoic acids as volatile organic acids. All these organic acids, except caproic and heptanoic, have been reported in small amounts in fresh raw ingredients such as Chinese cabbage and radishes (KIM and RHEE, 1975; RYU et al., 1984; TSUYUKI and ABE, 1979). Malic acid is in highest concentration among the natural organic acids found in fresh Chinese cabbage. Butyric, caproic, and heptanoic acids have been reported to contribute to the off-flavor of sauerkraut (VORBECK et al., 1961) and may be important in kimchi flavor.

The kinds of organic acids found in kimchi depend on the secondary ingredients used, as well as the fermentation. Concentrations of certain organic acids produced seem to be influenced by individual secondary ingredients when secondary ingredients are tested separately. More lactic acid is produced in kimchi made with red pepper powder, garlic, and green onion, and more acetic acid is produced in kimchi with garlic (RYU et al., 1984).

Mannitol has been shown to be produced during the fermentation of kimchi (HA et al., 1989). Fructose is said to be reduced by *L. mesenteroides* to mannitol, which is consumed by *L. plantarum* in the following stage of fermentation (FRAZIER and WESTHOFF, 1978).

4.5 Sensory Properties

Factors influencing kimchi flavor are sugar, amino acids, organic acids, salt, and volatile sulfur compounds. The sugar content of vegetables influences the taste of final products due to acidity formed or residual sweetness. Sweetness in the final product is desired. The soluble solids content (° Brix) varies for Chinese cabbage (1.2–6.6°) and radishes (2.6–5.1°) (KIM et al., 1989; SHIM et al., 1990a). The sugar composition of Chinese cabbage has been reported to be about 70% glucose, 17% mannose, and 10% fructose. No sucrose was reported (HA et al., 1989).

Free amino acids also influence the flavor of kimchi. Eighteen amino acids have been identified in raw and fermented kimchi, and reports on the amino acid changes during the fermentation are conflicting. CHO and RHEE (1979) reported that the content of free amino acids decreased as fermentation proceeded, while TAKAMA et al. (1986) and HAWER et al. (1988) reported that it doubled. TAKAMA et al. (1986) maintained that the content of free amino acids increased in kimchi because of the liberation of amino acids from plant proteins. Kimchi made with fermented fish (anchovy) sauce has a higher content of free amino acids and a richer flavor.

Acetic and lactic acids, generated by LAB activity, and several other acids contributed by the vegetable material, are important to kimchi flavor. Excess acidity is considered a spoilage problem, as will be discussed later. The optimum pH and total acidity for kimchi are 4.0–4.5 and 0.4–0.8%, respectively. Kimchis stored for an extended period of time may contain propionic, butyric, caproic, and heptanoic acids, which are believed to contribute to off-flavor (MHEEN and KWON, 1984).

The quality of kimchi fermented with 3% salt was reported to be superior to that with higher concentrations (MHEEN and KWON, 1984), and is the concentration used in commercial kimchi (Tab. 8). KIM and KIM (1990), who evaluated the sensory properties of low sodium kimchi, reported that they could replace more than 50% salt with potassium chloride without sacrificing kimchi flavor.

A firm, crisp texture is highly desired in the vegetable components of kimchi. Pectinmethylesterase exclusion (CHEONG et al., 1993) and heat treatments (55 °C, YOOK et al., 1985) in the presence of 0.05 M $CaCl_2$ helped retain kimchi firmness. However, pre-heated kimchi was inferior in sensory quality to unheated controls (SONG et al., 1967). Fermentation temperature and time influence the firmness of kimchi (LEE and RHEE, 1986; JUNG and RHEE, 1986). Kimchi fermented at lower temperatures (6–10 °C) was firmer than that fermented at higher temperatures (22–24 °C). A

Tab. 8. Ingredient Composition for Commercially Produced Chinese Cabbage Kimchi

	Composition (%) of Four Different Companies			
Ingredients	A	B	C	D
Cabbage	73	85	76	85
Radish (shredded)	18	0	8	0
Salt	3	3	3	3
Red pepper powder	2	4	3	2
Garlic	0.7	2	0.8	1.7
Ginger	0.2	0	0.4	0.7
Green onion	0.4	0	2	3.8
Fermented fish sauce	0	5	2.5	0
MSG	0	<0.1	<0.1	0
Sugar	0	0.3	0.9	0.9

Adapted from LEE and CHO (1990)

similar result was reported by THOMPSON et al. (1979) that brined cucumbers retained firmness, provided the cucumbers are washed to remove softening enzymes and the storage temperature is 15.5 °C or lower.

Freshly fermented, good quality kimchi should have rather distinct red and green colors contrasted against the white color of the cabbage stem. Kimchi made with red pepper powder of inferior quality may have a dull red color with a dark brownish tint, which is not desired. Kimchi loses its bright color and develops dullness in appearance when it is exposed to air after a long storage period, especially in the case of winter kimchi.

4.6 Spoilage Problems

Quality deterioration of kimchi during storage is mainly due to over-acidification, softening of vegetable tissues, the development of yeasty flavor, and a darkened appearance.

Over-acidification has been the problem most studied by Korean workers without much success. The growth and acid production by *L. plantarum* and *L. brevis* following the initial growth and activity of *L. mesenteroides* are regarded as undesirable in kimchi because they impart a harsh, strong acidic taste. Various methods have been tested to extend the storage period for good quality kimchi and to maintain pH 4.0–4.5 and acidity of 0.4–0.8%. These include pasteurization after canning (85 °C/25.2 min; LEE et al., 1968; GIL et al., 1984; LEE and CHUN, 1982) or in retort pouch (80–90 °C/>10 or 95 °C/>7 min; PYUN et al., 1983), gamma irradiation (KIM, 1962; CHA et al., 1989; LEE and LEE, 1965), addition of buffering agents (KIM and LEE, 1988; KIM, 1985), addition of antimicrobial agents (KWON and CHOI, 1967; SONG et al., 1966; CHOI et al., 1990), and others (KIM et al., 1991a; LIM et al., 1989; LEE et al., 1993; HONG and YOON, 1989; UM and KIM, 1990). Even though many workers claim they were successful, the effects were only marginal in most cases. Pasteurization seems to be more successful in preventing over-acidification, but the flavor defect caused by heating is undesirable.

Only a method which allows the growth and activity of *L. mesenteroides* but restricts *L. plantarum* and *L. brevis* will successfully prevent over-acidification of kimchi. Any method or chemicals used have to be microbially selective to be successful.

BREIDT et al. (1994a) inoculated nisin-resistant *L. mesenteroides* into a kimchi mixture with appropriate amounts of nisin. Nisin inhibited the growth of Gram-positive bacteria except for the nisin-resistant inoculum, which successfully initiated fermentation. Further research is needed to evaluate the addition of nisin to control kimchi fermentation, as was earlier proposed by CHOI et al. (1990).

Softening is another problem of stored kimchi but is not as serious a problem as over-acidification. It is caused by autolytic pectic enzymes of the vegetables. Pectinolytic enzymes of Chinese cabbage have been isolated and studied (BAEK et al., 1989). These scientists recommended that preheating cabbage at 50 °C for 1.5 h in 0.05 M $CaCl_2$ solution will ensure crispness and firmness of kimchi. YOOK et al. (1985) made kimchi with preheated radish root and obtained a maximum firmness at 55 °C for 2 h in 0.05 M $CaCl_2$ solution. This condition was mentioned to be optimal for pectinesterase activity, but inhibitory for polygalacturonase, a softening enzyme. Free carboxyl groups exposed due to pectinesterase activity are believed to form cross-linkages with Ca^{2+} to make tissues firmer. There are reports that firmness is increased in kimchi by adding sodium acetate (UM and KIM, 1990) and calcium acetate (KIM et al., 1991b).

Yeasty off-flavor occurs in winter kimchi due to air exposure after being stored for an extended period of time. Winter kimchi loses its bright red, green, and white colors and gains a dull dark color under these conditions. Very little research has been done to deal with these problems, but air exclusion seems important for their prevention.

5 Culture Development for Vegetable Fermentations

Currently, most vegetable fermentations rely on the natural microflora, although use of starter cultures has been suggested by various authors (ETCHELLS et al., 1973; DAESCHEL and FLEMING, 1984; HAMMES, 1990; LUCKE et al., 1990; BUCKENHUSKES, 1993). Various reasons have been proposed to explain the lack of commercial use of cultures, including economics, lack of sufficiently unique and valuable properties to justify their use, and the fact that vegetables will undergo a natural lactic fermentation under proper environmental conditions (FLEMING et al., 1985). However, recent research with cucumbers, sauerkraut, and olives indicates that use of special cultures for vegetable fermentations may find application in the near future. This section is focused upon methods that have been used for characterization and development of LAB strains for use in vegetable fermentations. Such methods include mutation and/or selection of unique cultures, marking of cultures for differential enumeration, and determination of growth kinetics.

5.1 Mutation and/or Selection of Cultures

In selecting cultures for food use, the first concern must be for impact of the culture on safety of the food for human consumption. Health concerns have been raised about the production of D(−)-lactic acid and biogenic amines by LAB during vegetable fermentations. Most LAB involved in vegetable fermentations produce DL-lactic acid, with the exception of *L. mesenteroides,* which produces D(−)-lactic acid (reviewed in FLEMING et al., 1985). Due to the difficulty of metabolism of D(−)-lactic acid, the WHO (1974) has suggested that infants should not consume D(−)- or DL-lactic acid. However, no limitation was placed on adult consumption of DL-products. Consumer demand for L(+)-lactate containing foods in Europe has led to the development of LAB starter cultures producing only this isomer for the fermentation of vegetables and vegetable juices (STETTER and STETTER, 1980; HAMMES, 1990; BUCKENHUSKES et al., 1990; BUCKENHUSKES, 1993). The production of biogenic amines (including histamine, putrescine, spermidine, and others) during the fermentation of vegetables has also been investigated (RICE et al., 1976; BRINK et al., 1990; HUIS IN 'T VELD et al., 1990; BUCKENHUSKES, 1993). Biogenic amines are formed by the decarboxylation of amino acids and are potentially hazardous to human health. Many LAB species are capable of decarboxylating one or more amino acids, producing biogenic amines. KUNSCH et al. (1989) demonstrated that a natural fermentation of sauerkraut produced an excess of 200 μg/mL of putrescine, along with other biogenic amines. They also showed that the production of biogenic amines could be substantially reduced using an *L. plantarum* starter culture.

Other consequences of LAB metabolism, such as the production of carbon dioxide by malolactic fermentation, can be considered in the development of bacterial starter cultures. McFEETERS et al. (1982b) found that the malolactic fermentation by LAB was a source of carbon dioxide in the fermentation of cucumbers. Prior to the introduction of purging technology for the pickle industry (ETCHELLS et al., 1973), product losses due to bloater damage were a major concern. While carbon dioxide is formed in cucumber fermentations from the respiration of cucumber tissue (FLEMING et al., 1973a), McFEETERS et al. (1984) showed that carbon dioxide formed from the malolactic reaction of *L. plantarum* was a direct cause of bloater damage. Toward this end, DAESCHEL et al. (1984) mutagenized *L. plantarum* and selected strains, using a differential medium, that do not produce CO_2 from malate. One mutant, MOP3-M6, has been further characterized and tested for use in vegetable fermentations (McDONALD et al., 1993; BREIDT and FLEMING, 1992). BREIDT and FLEMING (1992) developed a selective medium (MS agar medium) that only allows LAB strains that carry out a malolactic fermentation (MDC^+) to grow. Using MS medium, they found that the LAB population in cucumber fermentations is predominantly composed of MDC^+ strains. They found that the rapid degradation of malate in cucumber fermentations could be prevented by inoculation of the fermentations with 10^6 CFU/mL of the MOP3-M6 starter culture. The conditions that would ensure the predominance of the starter culture, however, and therefore prevent the fermentation of malate, remain unclear (McDONALD et al., 1993).

Many LAB strains have been shown to produce bacteriocins; these peptide antibiotics have been investigated for use in a wide variety of food products (RAY, 1992a, b; DAESCHEL, 1992). Intensive research has focused on the use of the bacteriocin, nisin (produced by *L. lactis*), and pediocins (produced by *Pediococcus* strains as food preservatives (reviewed by DE VUYST and VANDAMME, 1994; RAY, 1992a, respectively). Nisin has a broad spectrum of bacteriocidal activity against Gram-positive bacteria. Genes encoding nisin production, and immunity have been cloned and sequenced (reviewed in RAUCH et al., 1994), and the mechanism of action of nisin has been investigated (DE VUYST and VANDAMME, 1994). Bacteriocin-producing LAB have been investigated for use in vegetable fermentations. FLEMING et al. (1975a) reported that a *P. pentosaceus* strain, isolated from cucumber fermentation brines, produced a pediocin that was inhibitory to a variety of LAB species. HARRIS et al. (1992a) isolated a nisin-producing *Lactococcus* strain from fermenting cabbage and proposed using this bacteriocin-producing strain to control the fermentation of cabbage in the production of sauerkraut. In mixed culture broth fermentations, HARRIS et al. (1992b) demonstrated the production of nisin and the inhibition of *L. plantarum* (Fig. 13). BREIDT et al. (1994a), using this nisin-producing *Lactococcus,* demonstrated the effect of the nisin produced on the indigenous microflora in cabbage fermentations and isolated a nisin-resistant mutant of *L. mesenteroides* for use in cabbage fermentations (BREIDT et al., 1993). JIMENEZ-DIAZ et al. (1993) isolated a bacteriocin-producing *L. plantarum* strain from green olive fermentations. RUIZ-BARBA et al. (1994) demonstrated the use of this strain in controlling the microflora of olive fermentations. These experiments have shown that bacteriocin-producing LAB starter cultures can be effective in controlling the microflora in vegetable fermentations and will likely inspire the development of commercial starter cultures in the future.

The potential for bacteriophage problems with starter culture systems for vegetable fermentations should be considered (BUCKENHUSKES, 1993). The presence of bacteriophages in dairy fermentations has been recognized since 1935 (for review see LUNDSTED, 1983). It seems possible that any pure culture system may fall prey to bacteriophages. One reason that bacteriophages have not been observed in natural vegetable fermentations may be due to variation in the microflora and

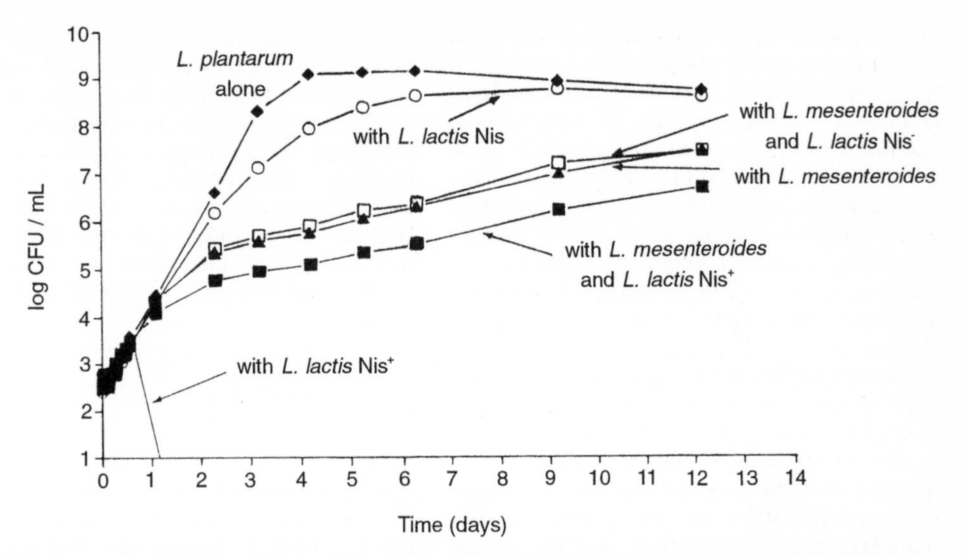

Fig. 13. Growth of *Lactobacillus plantarum* in a model sauerkraut fermentation. Growth of Nis^s *L. plantarum* ATCC 14917 in pure culture and mixed culture with Nis^r *Leuconostoc mesenteroides* NCK293, Nip$^+$ *Lactococcus lactis* subsp. *lactis* NCK401, or Nip$^-$ *L. lactis* subsp. *lactis* NCK402 in a model sauerkraut fermentation. Initial inoculum levels were 4×10^5 CFU/mL (*L. lactis* subsp. *lactis*), 2×10^3 CFU/mL (*L. mesenteroides*), and 3×10^2 CFU/mL (*L. plantarum*). These studies were done in cabbage juice broth. From HARRIS et al. (1992b).

a succession of LAB species during the course of the fermentation. RUIZ-BARBA et al. (1994) showed the rise and fall of naturally present *Lactobacillus* species between 20 and 30 days into an olive fermentation. They found 14 different *Lactobacillus* strains by following plasmid profiles. Little is known about the numbers of different strains of any given species in vegetable fermentations.

5.2 Enumeration of Starter Cultures in Vegetable Fermentations

Genetically marked cultures can be used to investigate the growth of starter cultures in food fermentations (RUIZ-BARBA et al., 1994; FOEGEDING et al., 1992; BREIDT and FLEMING, 1992; WINKOWSKI and MONTVILLE, 1992; FLEMING et al., 1988a). The availability of plasmid vectors for LAB with antibiotic resistance genes has greatly simplified the task of marking LAB cultures. A marked culture can be selectively enumerated from a mixed population by taking advantage of the antibiotic resistance phenotype. It should be noted that these genetically marked strains are restricted to laboratory studies and are not meant for human consumption. The technology for isolation of plasmids from LAB (ANDERSON and MCKAY, 1983; KLAENHAMMER, 1984), transformation of LAB by electroporation (LUCHANSKY et al., 1988; SUVOROV et al., 1988; DAVID et al., 1989), and rapid analysis of LAB strains for the presence of plasmids (ANDERSON and MCKAY, 1983; ORBERG and SANDINE, 1985) has opened LAB to the tools of molecular biology. Plasmid cloning vectors, such as pGK12 (KOK et al., 1984), have been generated by marking cryptic plasmids from various strains of LAB with antibiotic resistance genes. The chloramphenicol acetyltransferase gene and the *erm*C methyltransferase gene (encoding resistance to erythromycin and other macrolide antibiotics) originally from the staphylococcal plasmids pC194 (IORDANESCU and SURDEANU, 1980) and pE194

(IORDANESCU et al., 1987), respectively, have been widely used for this purpose. pGK12 has been used to enumerate LAB starter culture strains in a variety of foods (FOEGEDING et al., 1992; BREIDT and FLEMING, 1992; WINKOWSKI and MONTVILLE, 1992). An alternative strategy has been to develop streptomycin or rifampin-resistant LAB starter culture strains by sequential selection for these phenotypes on increasing concentrations of the antibiotics (FLEMING et al., 1988a; RUIZ-BARBA et al., 1994). The use of either method must be validated by determining the stability of the antibiotic resistance marker. Experiments using genetically marked LAB starter cultures can be limited by the interference of bacterial populations in the indigenous microflora that are resistant to the antibiotic(s) used for selection of the marked strain. Antibiotic-resistant lactic acid bacterial isolates, from a variety of sources, including meats, dairy, and other products, have been reported (ORBERG and SANDINE, 1985; VESCOVO et al., 1982; VIDAL and COLLINS-THOMPSON, 1987), as well as the spontaneous appearance of antibiotic-resistant LAB mutants (CURRAGH and COLLINS, 1992).

Experiments with marked starter cultures to investigate the change in the microflora have been carried out. The dominance of a *L. plantarum* starter culture inoculated at 10^6 CFU/mL and the effect of the starter culture on the indigenous microflora of a laboratory cucumber fermentation were demonstrated by BREIDT and FLEMING (1992). In these experiments the starter culture achieved a cell concentration of 1×10^9 CFU/mL, while the concentration of the natural microflora was shown to reach a maximum of 5×10^5 CFU/mL. The indigenous microflorae were enumerated with the use of a selective medium that prevented the growth of the starter culture but permitted the growth of the naturally present LAB. However, similar experiments with commercial-scale fermentations showed a starter culture inoculated at 10^6 CFU/mL did not predominate in an experimental, anaerobic tank, cucumber fermentation (FLEMING et al., 1988a). The controlled olive fermentations carried out by RUIZ-BARBA et al. (1994) showed that a bacteriocin-producing *L. plantarum* would dominate a brined live fermentation with an inoculum of 10^5, but a non-bacteriocin-producing derivative strain added to brined olives at the same concentration would not predominate, and was lost from the fermentation. Clearly, further experiments with marked starter cultures and methods to determine the effect of the starter cultures on the indigenous microflora will be needed to understand the factors affecting the ecology of controlled vegetable fermentations.

5.3 Growth Kinetics

Determining the growth kinetics of the starter culture strain can be an important test of the fitness of the strain for use in controlling vegetable fermentations. For laboratory analysis of growth kinetics, it is convenient to use broth cultures for optical density readings. Automated microtiter plate methods have been developed that allow convenient and rapid determination of bacterial growth kinetics (THOMAS et al., 1985; BREIDT et al., 1994b). Large amounts of data can be generated using these automated methods (Fig. 14). Multi-well microtiter plates can allow replicates of batch fermentations with a variety of growth media, salt concentrations, organic acids, pH, and other conditions to be tested simultaneously. A limitation of these methods is the need for the optical clarity of the growth medium. Vegetable juices prepared as a growth medium for kinetic tests can have inhibitory compounds or stimulatory compounds present. For example, inhibitors in cabbage juice have been identified, which are present in fresh cabbage juice, and may affect the growth of LAB species (KYUNG and FLEMING, 1994). The effect of temperature on the growth kinetics of LAB species can be investigated with a temperature gradient or Arrhenius block. To determine the "temperature character" of a bacterial strain, an Arrhenius block can be used to grow batch cultures over a range of temperatures (DROST-HANSEN, 1977; ROMICK, 1994). Temperature character is a measurement of the range of temperatures over which a culture can grow. Models for determining the temperature character of bacterial strains have been devel-

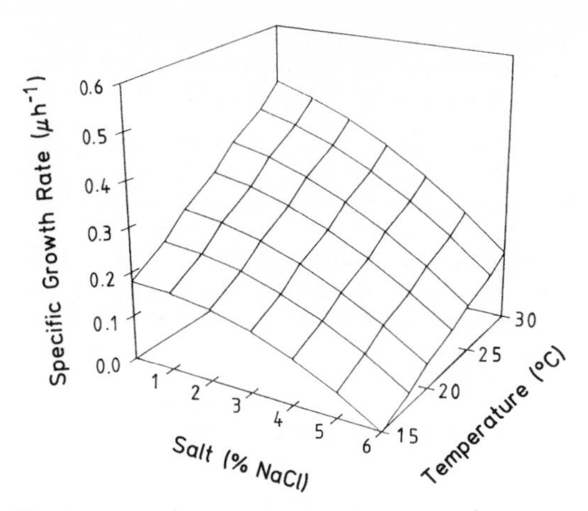

Fig. 14. The effects of temperature and NaCl on the growth kinetics of *Leuconostoc mesenteroides*. Response surface graph showing the surface determined from the predicted values for specific growth rates. From BREIDT et al. (1994b).

oped and characterized by a number of workers (RATKOWSKY et al., 1982; ADAIR et al., 1989; HANUS and MORITA, 1986).

Knowledge of the kinetics of growth, product formation, and substrate utilization by the fermenting bacterium is essential for modeling of vegetable fermentations. PASSOS et al. (1993) assessed the effects of key variables involved in cucumber fermentations and developed a model for predicting the growth of *L. plantarum*. Limiting conditions for growth in cucumber juice were pH 3.37 (lower limit), 69 mM undissociated lactic acid, 150 mM undissociated acetic acid, or 11.8% NaCl. A predictive equation, based on data collected, was developed and found to predict growth of *L. plantarum* reasonably well in batch cultures in cucumber juice. Later, a model was developed to describe substrate utilization and lactic acid production, in addition to growth by *L. plantarum* in cucumber juice (PASSOS et al., 1994).

6 References

ADAIR, C., KILSBY, D. C., WHITTALL, P. T. (1989), Comparison of the Schoolfield (non-linear Arrhenius) model and the square root model for predicting bacterial growth in foods, *Food Microbiol.* **6,** 7–18.

ANDERSON, D. G., MCKAY, L. L. (1983), Simple and rapid method for isolating large plasmid DNA form lactic streptococci, *Appl. Environ. Microbiol.* **46,** 549–552.

ANONYMOUS (1974), *Report on Nutritional Survey (1970–1974):* Ministry of Health and Social Affairs, Republic of Korea.

ANONYMOUS (1993a), *The Almanac of the Canning, Freezing, Preserving Industries,* Vol. 2, Westminster: E. E. Judge & Sons, Inc.

ANONYMOUS (1993b), United States Standards for grades of canned sauerkraut, in: *The Almanac of the Canning, Freezing, Processing Industries,* Westminster: E. E. Judge & Sons, Inc.

BAEK, H. H., LEE, C. H., WOO, D. H., PARK, K. H., PEK, U. H., LEE, K. S., NAM, S. B. (1989), Prevention of pectinolytic softening of kimchi tissue, *Korean J. Food Sci. Technol.* **21,** 149–153.

BELL, T. A. (1951), Pectolytic enzyme activity in various parts of the cucumber plant and fruit, *Bot. Gaz.* **113,** 216–221.

BELL, T. A., ETCHELLS, J. L. (1952), Sugar and acid tolerance of spoilage yeasts from sweet-cucumber pickles, *Food Technol.* **6,** 468–472.

BELL, T. A., ETCHELLS, J. L. (1961), Influence of salt (NaCl) on pectinolytic softening of cucumbers, *J. Food Sci.* **26,** 84–90.

BREIDT, F., FLEMING, H. P. (1992), Competitive growth of genetically marked malolactic-deficient *Lactobacillus plantarum* in cucumber fermentations, *Appl. Environ. Microbiol.* **58,** 3845–3849.

BREIDT, F., CROWLEY, K. A., FLEMING, H. P. (1993), Isolation and characterization of nisin-resistant *Leuconostoc mesenteroides* for use in cabbage fermentations, *Appl. Environ. Microbiol.* **59,** 3778–3783.

BREIDT, F., CROWLEY, K. A., FLEMING, H. P. (1994a), Controlling cabbage fermentations with nisin and nisin-resistant *Leuconostoc mesenteroides, Food Microbiol.,* submitted.

BREIDT, F., ROMICK, T. L., FLEMING, H. P. (1994b), A rapid method for the determination of bacterial growth kinetics, *J. Rapid Methods Microbiol.* **3,** 59–68.

BRINK, B., DAMINK, C., JOOSTEN, H. M. L. J., HUIS IN 'T VELD, J. H. J. (1990), Occurrence and formation of biologically active amines in foods, *Int. J. Food Microbiol.* **11,** 73-84.

BUCKENHUSKES, H. J. (1993), Selection criteria for lactic acid bacteria to be used as starter cultures for various food commodities, *FEMS Microbiol. Rev.* **12,** 253–272.

BUCKENHUSKES, H., HIEMANN, M., SCHWARZ, E., GIERSCHNER, K., HAMMES, W. P. (1990), Lactic acid fermentation of cauliflower by L(+)-lactate producing lactobacilli, in: *Processing and Quality of Foods* (ZEUTHEN, P., CHEFTEL, J. C., ERICKSSON, C., GORMLEY, T. R., LINKO, P., PAULUS, K., Eds.), pp. 2225–2230, London: Elsevier Applied Science.

BUESCHER, R. W., HUDSON, J. M., ADAMS, J. R. (1979), Inhibition of polygalacturonase softening of cucumber pickles by calcium chloride, *J. Food Sci.* **44,** 1786–1787.

BUESCHER, R. W., HUDSON, J. M., ADAMS, J. R. (1981), Utilization of calcium to reduce pectinolytic softening of cucumber pickles in low salt conditions, *Lebensm. Wiss. Technol.* **14,** 65–69.

CHA, B. S., KIM, W. J., BYUN, M. W., KWON, J. H., CHO, H. O. (1989), Evaluation of gamma irradiation for extending the shelf life of kimchi, *Korean J. Food Sci. Technol.* **21,** 109–119.

CHEONG, T. K., MOON, T. W., PARK, K. H. (1993), Thermostability of polygalacturonase from Chinese cabbage, *Korean J. Food Sci. Technol.* **25,** 576–581.

CHO, J. S. (1993), Personal communication.

CHO, Y., RHEE, H. S. (1979), A study on flavorous taste components in kimchis. On free amino acids, *Korean J. Food Sci. Technol* **11,** 26–31.

CHOI, S. Y., LEE, I. S., YOO, J. Y., CHUNG, K. S., KOO, Y. J. (1990), Inhibitory effect of nisin upon kimchi fermentation, *Korean J. Appl. Microbiol. Biotechnol.* **18,** 620–623.

CHUN, J. K. (1981), Chinese cabbage utilization in Korea: Kimchi processing technology, in: *Proc. First Int. Symp. on Chinese Cabbage* (TELEKAR, N. S., GRIGGS, T. D., Eds.), Tainan, Taiwan: Asian Vegetable Research and Development Center.

CORBET, A., FLEMING, H. P., YOUNG, C. T., MCFEETERS, R. F. (1995), Chemical and sensory characterization of commercial sauerkraut, *J. Food Qual.,* in press.

COSTILOW, R. N., BEDFORD, C. L., MINGUS, D., BLACK, D. (1977), Purging of natural salt stock pickle fermentations to reduce bloater damage, *J. Food Sci.* **42,** 234–240.

COSTILOW, R. N., GATES, K., LACY, M. L. (1980), Molds in brined cucumbers: cause of softening during air-purging of fermentations, *Appl. Environ. Microbiol.* **40,** 417–422.

CURRAGH, H. J., COLLINS, M. A. (1992), High levels of spontaneous drug resistance in *Lactobacillus, J. Appl. Bacteriol.* **73,** 31–36.

DAESCHEL, M. A. (1992), Bacteriocins of lactic acid bacteria, in: *Food Preservatives of Microbial Origin* (RAY, B., DAESCHEL, M. A., Eds.), pp. 323–345, Boca Raton: CRC Press.

DAESCHEL, M. A., FLEMING, H. P. (1984), Selection of lactic acid bacteria for use in vegetable fermentations, *Food Microbiol.* **1,** 303–313.

DAESCHEL, M. A., MCFEETERS, R. F., FLEMING, H. P., KLAENHAMMER, T. R., SANOZKY, R. B. (1984), Mutation and selection of *Lactobacillus plantarum* strains that do not produce carbon dioxide from malate, *Appl. Environ. Microbiol.* **47,** 419–420.

DAVID, S., SIMONS, G., DE VOS, W. M. (1989), Plasmid transformation by electroporation of *Leuconostoc paramesenteroides* and its use in molecular cloning, *Appl. Environ. Microbiol.* **55,** 1483–1489.

DE VUYST, L., VANDAMME, E. J. (1994), Nisin, a lantibiotic produced by *Lactococcus lactis* subsp. *lactis:* properties, biosynthesis, fermentation, and applications, in: *Bacteriocins of Lactic Acid Bacteria* (DE VUYST, L., VANDAMME, E. J., Eds.), pp. 151–221, London: Blackie Academic and Professional.

DICKSON, M. H. (1987), What is new in cabbage breeding, *N.Y. Agric. Exp. Stn. Geneva Spec. Rep.* **61.**

DICKSON, M. H., WALLACE, D. H. (1986), Cabbage breeding, in: *Breeding Vegetable Crops* (BASSETT, M. A., Ed.), pp. 395–432, Westport, CT: Avi Publishing Co.

DROST-HANSEN, W. (1977), Gradient device for the study of temperature effects on biological systems, *J. Wash. Acad. Sci.* **7,** 187–201.

ENSLIN, P. R., HOLZAPFEL, C. W., NORTON, K. B. (1967), Bitter principles of the Cucurbitaceae. XV. Cucurbitacins from a hybrid of *Lagenaria siceraria, J. Chem. Soc. Commun.* 964–972.

ETCHELLS, J. L., BELL, T. A. (1950a), Classification of yeasts from the fermentation of commercially brined cucumbers, *Farlowia* **4,** 87–112.

ETCHELLS, J. L., BELL, T. A. (1950b), Film yests on commercial cucumber brines, *Food Technol.* **4,** 77–83.

ETCHELLS, J. L., JONES, I. D. (1942), Pasteurization of pickle products, *Fruit Prod. J.* **21,** 330–332.

ETCHELLS, J. L., JONES, I. D. (1943), Bacteriological changes in cucumber fermentation, *Food Ind.* **15,** 54–46.

ETCHELLS, J. L., FABIAN, F. W., JONES, I. D. (1945), The *Aerobacter* fermentation of cucumbers during salting, *Mich. State Agric.* Exp. *Stn. Tech. Bull.* **200.**

ETCHELLS, J. L., BELL, T. A., MONROE, R. J., MASLEY, P. M., DEMAIN, A. L. (1958), Populations and softening enzyme activity of filamentous fungi on flowers, ovaries, and fruit of pickling cucumbers, *Appl. Microbiol.* **6,** 427–440.

ETCHELLS, J. L., BORG, A. F., BELL, T. A. (1961), Influence of sorbic acid on populations and species of yeasts occurring in cucumber fermentations, *Appl. Microbiol.* **9,** 139–144.

ETCHELLS, J. L., COSTILLOW, R. N., ANDERSON, T. E., BELL, T. A. (1964), Pure culture fermentation of brined cucumbers, *Appl. Microbiol.* **12,** 523–535.

ETCHELLS, J. L., BORG, A. F., BELL, T. A. (1968), Bloater formation by gas-forming lactic acid bacteria in cucumber fermentations, *Appl. Microbiol.* **16,** 1029–1035.

ETCHELLS, J. L., BELL, T. A., FLEMING, H. P., KELLING, R. E., THOMPSON, R. L. (1973), Suggested procedure for the controlled fermentation of commercially brined pickling cucumbers – the use of starter cultures and reduction of carbon dioxide accumulation, *Pickle Pak Sci.* **3,** 4–14.

ETCHELLS, J. L., FLEMING, H. P., BELL, T. A. (1975), Factors influencing the growth of lactic acid bacteria during the fermentation of brined cucumbers, in: *Lactic Acid Bacteria in Beverages and Food* (CARR, J. G., CUTTING, C. V., WHITING, G. C., Eds.), pp. 281–305, New York: Academic Press, Inc.

FLEMING, H. P. (1979), Purging carbon dioxide from cucumber brines to prevent blaoter damage – a review, *Pickle Pak Sci.* **6,** 8–22.

FLEMING, H. P. (1982), Fermented vegetables, in: *Economic Microbiology,* Vol. 7, *Fermented Foods* (ROSE, A. H., Ed.), pp. 227–258, New York: Academic Press, Inc.

FLEMING, H. P. (1991), Mixed cultures in vegetable fermentations, in: *Mixed Cultures in Biotechnology* (ZEIKUS, J. G., JOHNSON, E. A., Eds.), pp. 69–103, New York: McGraw-Hill, Inc.

FLEMING, H. P., MOORE, W. R. (1983), Pickling, in: *CRC Handbook of Processing and Utilization in Agriculture* (WOLFF, I. A., Ed.), Vol. II, pp. 397–463, Boca Raton: CRC Press, Inc.

FLEMING, H. P., PHARR, D. M. (1980), Mechanism for bloater formation in brined cucumbers, *J. Food Sci.* **45,** 1595–1600.

FLEMING, H. P., THOMPSON, R. L., ETCHELLS, J. L., KELLING, R. E., BELL, T. A. (1973a), Bloater formation in brined cucumbers fermented by *Lactobacillus plantarum, J. Food Sci.* **38,** 499–503.

FLEMING, H. P., THOMPSON, R. L., ETCHELLS, J. L., KELLING, R. E., BELL, T. A. (1973b), Carbon dioxide production in the fermentation of brined cucumbers, *J. Food Sci.* **38,** 504–506.

FLEMING, H. P., ETCHELLS, J. L., COSTILOW, R. N. (1975a), Microbial inhibition by an isolate of *Pediococcus* from cucumber brines, *Appl. Microbiol.* **30,** 1040–1042.

FLEMING, H. P., ETCHELLS, J. L., THOMPSON, R. L., BELL, T. A. (1975b), Purging of CO_2 from cucumber brines to reduce bloater damage, *J. Food Sci.* **40,** 1304–1310.

FLEMING, H. P., THOMPSON, R. L., BELL, T. A., HONTZ, L. H. (1978), Controlled fermentation of sliced cucumbers, *J. Food Sci.* **43,** 888–891.

FLEMING, H. P., HUMPHRIES, E. G., MACON, J. A. (1983), Progress on development of an anaerobic tank for brining of cucumbers, *Pickle Pak Sci.* **7,** 3–15.

FLEMING, H. P., MCFEETERS, R. F., DAESCHEL, M. A. (1985), The lactobacilli, pediococci, and leuconostocs: vegetable products, in: *Bacterial Starter Cultures for Foods* (GILLILAND, S. E., Ed.), pp. 97–118, Boca Raton: CRC Press, Inc.

FLEMING, H. P., MCFEETERS, R. F., THOMPSON, R. L. (1987), Effects of sodium chloride concentration on firmness retention of cucumbers fermented and stored with calcium chloride, *J. Food Sci.* **52,** 653–657.

FLEMING, H. P., MCFEETERS, R. F., DAESCHEL, M. A., HUMPHRIES, E. G., THOMPSON, R. L. (1988a), Fermentation of cucumbers in anaerobic tanks, *J. Food Sci.* **53,** 127–133.

FLEMING, H. P., MCFEETERS, R. F., HUMPHRIES, E. G. (1988b), A fermentor for study of sauerkraut fermentation, *Biotechnol. Bioeng.* **31,** 189–197.

FLEMING, H. P., DAESCHEL, M. A., MCFEETERS, R. F., PIERSON, M. D. (1989), Butyric acid spoilage of fermented cucumbers, *J. Food Sci.* **54,** 636–639.

FOEGEDING, P. M., THOMAS, A. B., PILKINGTON, D. H., KLAENHAMMER, T. R. (1992), Enhanced control of *Listeria monocytogenes* by an *in situ*-produced pediocin during dry fermented sausage production, *Appl. Environ. Microbiol.* **58,** 884–890.

FRAZIER, W. C., WESTHOFF, D. C. (1978), *Food Microbiology,* New York: McGraw-Hill Book Co.

GATES, K., COSTILOW, R. N. (1981), Factors influencing softening of salt-stock pickles in air-purged fermentations, *J. Food Sci.* **46,** 274–277, 282.

GIL, G. H., KIM, K. H., CHUN, J. K. (1984), Pasteurization of Chinese radish kimchi by a pilot scale continuous kimchi pasteurizer, *Korean J. Food Sci. Technol.* **16,** 95–98.

HA, J. H., HAWER, W. S., KIM, Y. J., NAM, Y. J. (1989), Changes of free sugars in kimchi during fermentation, *Korean J. Food Sci. Technol.* **21,** 633–638.

HAMMES, W. P. (1990), Bacterial starter cultures in food production, *Food Biotechnol.* **4,** 383–397.

HANUS, F. J., MORITA, R. Y. (1986), Significance of the temperature characteristic of growth, *J. Bacteriol.* **95,** 736-737.

HARRIS, L. J., FLEMING, H. P., KLAENHAMMER, T. R. (1992a), Characterization of two nisin-producing *Lactococcus lactis* subsp. *lactis* strains isolated from a commercial sauerkraut fermentation, *Appl. Environ. Microbiol.* **58,** 1477–1483.

HARRIS, L. J., FLEMING, H. P., KLAENHAMMER, T. R. (1992b), Novel paired starter culture system for sauerkraut, consisting of a nisin-resistant *Leuconostoc mesenteroides* strain and a nisin-producing *Lactococcus lactis* strain, *Appl. Environ. Microbiol.* **58,** 1484–1489.

HAWER, W. S., HA, J. H., SEOG, H. M., NAM, Y. J., SHIN, D. W. (1988), Changes in the taste and flavour compounds of kimchi during fermentation, *Korean J. Food Sci. Technol.* **20,** 511–517.

HONG, W. S., YOON, S. (1989), The effects of low temperature heating and mustard oil on the kimchi fermentation, *Korean J. Food Sci. Technol.* **21,** 331–337.

HUIS IN 'T VELD, J. H. J., HOSE, H., SCHAAFSMA, G. J., SILLA, H., SMITH, J. E. (1990), Health aspects of food biotechnology, in: *Processing and Quality of Foods* (ZEUTHEN, P., CHEFTEL, J. C., ERIKSSON, C., GORMLEY, T. R., LINKO, P., PAULUS, K., Eds.), pp. 2.73-2.97, London: Elsevier Applied Science.

HUMPHRIES, E. G., FLEMING, H. P. (1991), Flexible restraining covers for cucumber brining tanks, *Appl. Eng. Agric.* **7,** 582–586.

IORDANESCU, S., SURDEANU, M. (1980), New incompatibility groups for *Staphylococcus aureus* plasmids, *Plasmid* **4,** 256-260.

IORDANESCU, S., SURDEANU, M., LATTA, P. D., NOVICK, R. (1987), Incompatibility and molecular relationships between small staphylococcal plasmids carrying the same resistance marker, *Plasmid* **1,** 468–479.

JIMENEZ-DIAZ, R., RIOS-SANCHE, R. M., DESMAZEAUD, M., RUIZ-BARBA, J. L., PIARD, J. C. (1993), Plantaricins S and T, two new bacteriocins produced by *Lactobacillus plantarum* LPC010 isolated from a green olive fermentation, *Appl. Environ. Microbiol.* **59,** 1416–1424.

JONES, I. D., ETCHELLS, J. L. (1943), Physical and chemical changes in cucumber fermentations, *Food Ind.* **15,** 62–64.

JONES, I. D., ETCHELLS, J. L., VELDHUIS, M. K., VEERHOFF, O. (1941), Pasteurization of genuine dill pickles, *Fruit Prod. J.* **20,** 304–305, 316, 325.

JUNG, G. H., RHEE, H. S. (1986), Changes of texture in terms of the contents of cellulose, hemicellulose and pectic substances during fermentation of radish kimchi, *Korean J. Soc. Food Sci.* **2,** 68–75.

KANDLER, O. (1983), Carbohydrate metabolism in lactic acid bacteria, *Antonie van Leeuwenhoek* **49,** 209–224.

KIM, C. S. (1962), Preservation of Korean kimchi by the method of irradiation with gamma rays of ^{60}Co, *Kyungpook Univ. Res. Rep.* **5,** 21–27.

KIM, S. D. (1985), Effect of pH adjuster on the fermentation of kimchi, *J. Korean Soc. Food Nutr.* **14,** 259–264.

KIM, H. S., CHUN, J. K. (1966), Studies on the dynamic changes of bacteria during the kimchi fermentation, *Res. Rep. Atom. Energy Res. Inst.* **6,** 112–118.

KIM, I. H., KIM, K. O. (1990), Sensory characteristic of low sodium kakdugi, *Korean J. Food Sci. Technol.* **22,** 380–385.

KIM, S. D., LEE, S. H. (1988), Effect of sodium malate buffer as a pH adjuster on the fermentation of kimchi, *J. Korean Soc. Food Nutr.* **17,** 358–364.

KIM, H. O., RHEE, H. S. (1975), Studies on the nonvolatile organic acids in kimchis fermented at different temperatures, *Korean J. Food Sci. Technol.* **7,** 74–81.

KIM, W. J., KANG, K. O., KYUNG, K. H., SHIN, J. I. (1991a), Addition of salts and their mixtures for improvement of storage stability of kimchi, *Korean J. Food Sci. Technol.,* **23,** 188–191.

KIM, S. Y., UM, J. Y., KIM, K. O. (1991b), Effect of calcium acetate and potassium sorbate on characteristics of kakdugi, *Korean J. Food Sci. Technol.* **23,** 1–5.

KIM, K. J., KYUNG, K. H., MYUNG, W. K., SHIM, S. T., KIM, H. K. (1989), Selection scheme of radish varieties to improve storage stabilities of fermented pickled radish cubes with special reference to sugar content, *Korean J. Food Sci. Technol.* **21,** 100–108.

KLAENHAMMER, T. R. (1984), A general method for plasmid isolation in lactobacilli, *Curr. Microbiol.* **10,** 23–28.

KOK, J., VAN DER VOSSEN, J. M. B. M., VENEMA, G. (1984), Construction of plasmid cloning vectors for lactic streptococci which also replicate in *Bacillus subtilis* and *Escherichia coli, Appl. Environ. Microbiol.* **48,** 726–731.

KUNSCH, U., SCHARER, H., TEMPERLI, A. (1989), Biogene Amine als Qualitätsindikator von Sauerkraut, *XXIV. Tagung der Deutschen Gesellschaft für Qualitätsforschung,* Kiel.

KWON, S. P., CHOI, K. W. (1967), Prevention of overacidification of kimchi, *Korean Patent* 305.

KYUNG, K. H., FLEMING, H. P. (1994), Antibacterial activity of cabbage juice against lactic acid bacteria, *J. Food Sci.* **59,** 125–129.

LEE, S. W. (1975), Studies on movements and interchanges of kimchi in China, Korea, and Japan, *J. Korean Soc. Food Nutr.* **4,** 71–95.

LEE, C. Y., CHO, J. S. (1990), Reviews of history and researches on kimchi, *Res. Rep. Miwon Res. Inst. Korea Food Diet. Cult.* **1,** 193–256.

LEE, N. J., CHUN, J. K. (1982), Studies on the kimchi pasteurization. Part II. Effects of kimchi pasteurization conditions on the shelf-life of kimchi, *J. Korean Agric. Chem. Soc.* **25,** 197–200.

LEE, H. S., LEE, K. B. (1965), Food preservation by ionizing radiations. 1. Studies on the preservation of kimchi by ionizing radiations, *Res. Rep. Atom. Energy Res. Inst.* **5,** 64–69.

LEE, Y. H., RHEE, H. S. (1986), The changes of pectic substances during the fermentation of kimchis, *Korean J. Soc. Food Sci.* **2,** 54–58.

LEE, T. Y., KIM, J. S., CHUNG, D. H., KIM, H. S. (1960), Studies on the composition of kimchi. Part 2. Variation of vitamins during kimchi fermentation, *Kua Yean Whi Bo* **5,** 43–50.

LEE, C. Y., KIM, H. S., CHUN, J. K. (1968), Studies on the manufacture of canned kimchi, *J. Korean Agric. Chem. Soc.* **10,** 33–38.

LEE, C. W., KO, C. Y., HA, D. M. (1992), Microfloral changes of the lactic acid bacteria during kimchi fermentation and identification of the isolates, *Korean J. Appl. Microbiol. Biotechnol.* **20,** 102–109.

LEE, S. K., KIM, I. H., CHOI, S. Y., JEON, K. H. (1993), Effect of lysozyme, glycine and EDTA on kimchi fermentation, *J. Korean Soc. Food Nutr.* **22,** 58–61.

LEE, H. O., YANG, I. W. (1975), Studies on the packaging and preservation of kimchi, *J. Korean Agric. Chem. Soc.* **13,** 207–218.

LIM, B. S., KIM, Y. S., LEE, B. H., CHANG, K. W., LIM, H. B. (1989), Preparation of kimchi with extended storage stability, *Korean Patent* 89-4895.

LUCHANSKY, J. B., MURIANA, P. M., KLAENHAMMER, T. R. (1988), Application of electroporation for transfer of plasmid DNA to *Lactobacillus, Lactococcus, Leuconostoc, Listeria, Pediococcus, Bacillus, Staphylococcus, Enterococcus,* and *Propionibacterium, Mol. Microbiol.* **2,** 637–646.

LUCKE, F.-K., BRÜMMER, J.-M., BUCKENHUSKES, H., GARRIDO FERNANDEZ, A., RODRIGO, M., SMITH, J. E. (1990), *Starter Culture Development,* Vol. 2, *Processing and Quality of Foods,* London: Elsevier Applied Science.

LUNDSTEDT, E. (1983), Some reflections on the development of starters for the cultured dairy products industry, *Cult. Dairy Prod. J.* **18,** 10–15.

MCCOMBS, C. L., SOX, H. N., LOWER, R. L. (1976), Sugar and dry matter content of cucumber fruits, *Hort. Sci.* **11,** 245-247.

MCCREIGHT, J. D., LOWER, R. L., PHARR, D. M. (1978), Measurement and variation of sugar content of pickling cucumber, *J. Am. Soc. Hort. Sci.* **103,** 145–147.

MCDONALD, L. C., FLEMING, H. P., DAESCHEL, M. A. (1991), Acidification effects on microbial populations during initiation of cucumber fermentation, *J. Food Sci.* **56,** 1353–1356, 1359.

MCDONALD, L. C., SHIEH, D.-H., FLEMING, H. P., MCFEETERS, R. F., THOMPSON, R. L. (1993), Evaluation of malolactic-deficient strains of *Lactobacillus plantarum* for use in cucumber fermentations, *Food Microbiol.* **10,** 489–499.

MCFEETERS, R. F., FLEMING, H. P. (1990), Effect of calcium ions on the thermodynamics of cucumber tissue softening, *J. Food Sci.* **55,** 446–449.

MC FEETERS, R. F., BELL, T. A., FLEMING, H. P. (1980), An endo-polygalacturonase in cucumber fruit, *J. Food Biochem.* **4,** 1–16.

MCFEETERS, R. F., FLEMING, H. P., THOMPSON, R. L. (1982a), Malic and citric acids in pickling cucumbers, *J. Food Sci.* **47,** 1859–1861, 1865.

MCFEETERS, R. F., FLEMING, H. P., THOMPSON, R. L. (1982b), Malic acid as a source of carbon dioxide in cucumber juice fermentations, *J. Food Sci.* **47,** 1862–1865.

MCFEETERS, R. F., FLEMING, H. P., DAESCHEL, M. A. (1984), Malic acid degradation and brined cucumber bloating, *J. Food Sci.* **49,** 999–1002.

MHEEN, T. I., KWON, T. W. (1984), Effect of temperature and salt concentration on kimchi fermentation, *Korean J. Food Sci. Technol.* **16,** 442–450.

MILLER, C. H., WEHNER, T. C. (1989), Cucumbers, in: *Quality and Preservation of Vegetables* (ESKIN, N. A. M., Ed.), pp. 245–264, Boca Raton: CRC Press.

MONROE, R. J., ETCHELLS, J. L., PACILIO, J. C., BORG, A. F., WALLACE, D. H., ROGERS, M. P., TURNEY, L. J., SCHOENE, E. S. (1969), Influence of various acidities and pasteurizing temperatures on the keeping quality of fresh-pack dill pickles, *Food Technol.* **23,** 71–77.

ORBERG, P. K., SANDINE, W. E. (1985), Survey of antimicrobial resistance in lactic streptococci, *Appl. Environ. Microbiol.* **49,** 538–542.

PASSOS, F. V., FLEMING, H. P., OLLIS, D. F., HASSAN, H. M., FELDER, R. M. (1993), Modeling the specific growth rate of *Lactobacillus plantarum* in cucumber extract, *Appl. Microbiol. Biotechnol.* **40,** 143–150.

PASSOS, F. V., FLEMING, H. P., OLLIS, D. F., FELDER, R. M., MCFEETERS, R. F. (1994), Kinetics and modeling of lactic acid production by *Lactobacillus plantarum, Appl. Environ. Microbiol.* **60,** 2627–2636.

PEDERSON, C. S. (1940), The relation between quality and chemical composition of canned sauerkraut, *N.Y. State Agric. Exp. Stn. Geneva Tech. Bull.* **693.**

PEDERSON, C. S. (1946), Improving methods for salting sauerkraut, *Food Packer* **27,** 53–57.

PEDERSON, C. S. (1979), *Microbiology of Food Fermentations,* 2nd Ed., Westport, CT: AVI Publishing Co., Inc.

PEDERSON, C. S., ALBURY, M. N. (1954), The influence of salt and temperature on the microflora of sauerkraut fermentations, *Food Technol.* **8,** 1–5.

PEDERSON, C. S., ALBURY, M. N. (1961), The effect of pure culture inoculation on fermentation of cucumbers, *Food Technol.* **15,** 351–354.

PEDERSON, C. S., ALBURY, M. N. (1969), The sauerkraut fermentation, *N.Y. State Agric. Exp. Stn. Geneva Tech. Bull.* **824.**

PEDERSON, C. S., KELLY, C. D. (1938), Development of pink color in sauerkraut, *Food Res.* **3,** 583–588.

POTTS, E. A., FLEMING, H. P. (1982), Prevention of mold-induced softening in air-purged, brined cucumbers by acidification, *J. Food Sci.* **47,** 1723–1727.

PYUN, Y. R., SHIN, S. K., KIM, J. B., CHO, E. K. (1983), Studies on the heat penetration and pasteurization conditions of retort pouch kimchi, *Korean J. Food Sci. Technol.* **15,** 414–420.

RATKOWSKY, D. A., OLLEY, J., MCMEEKIN, T. A., BALL, A. (1982), Relationship between temperature and growth rate of bacterial cultures, *J. Bacteriol.* **149,** 1–5.

RAUCH, P. J. G., KUIPERS, O. P., SIEZEN, R. J., DE VOS, W. M. (1994), Genetics and protein engineering of nisin, in: *Bacteriocins of Lactic Acid Bacteria* (DE VUYST, L., VANDAMME, E. J., Eds.), pp. 223–249, London: Blackie Academic and Professional.

RAY, B. (1992a), Bacteriocins of starter culture bacteria as food biopreservatives: an overview, in: *Food Preservatives of Microbial Origin* (RAY, B., DAESCHEL, M. A., Eds.), pp. 177–205, Boca Raton: CRC Press.

RAY, B. (1992b), Pediocin(s) of *Pediococcus acidilactici* as a food biopreservative, in: *Food Preservatives of Microbial Origin* (RAY, B., DAESCHEL, M. A., Eds.), pp. 207–264, Boca Raton: CRC Press.

RICE, S. L., EITENMILLER, R. R., KOEHLER, P. E. (1976), Biologically active amines in food: a review, *J. Milk Food Technol.* **39,** 353–358.

ROMICK, T. L. (1994), Biocontrol of *Listeria monocytogenes,* a psychotropic pathogen model in low salt, non-acidified, refrigerated vegetable products, *Ph. D. Thesis,* North Carolina State University, Raleigh.

RUIZ-BARBA, J. L., CATHCART, D. P., WARNER, P. J., JIMENEZ-DIAZ, R. (1994), Use of *Lactobacillus plantarum* LPC010, a bacteriocin producer, as a starter culture in Spanish-style green olive fermentations, *Appl. Environ. Microbiol.* **60,** 2059–2064.

RYU, J. Y., LEE, H. S., RHEE, H. S. (1984), Changes of organic acids and volatile flavor compounds in kimchis fermented with different ingredients, *Korean J. Food Sci. Technol.* **16,** 169–174.

SHIM, S. T., KIM, K. J., KYUNG, K. H. (1990a), Effect of soluble-solids contents of Chinese cabbages on kimchi fermentation, *Korean J. Food Sci. Technol.* **22,** 278–284.

SHIM, S. T., KYUNG, K. H., YOO, Y. J. (1990b), Lactic acid bacteria isolated from fermenting kimchi and their fermentation of Chinese cabbage juice, *Korean J. Food Sci. Technol.* **22,** 373–379.

SOHN, K. H. (1992), Kimchi varieties and their utilization, *Kimchi Sci. Ind.* **1,** 68–85.

SONG, S. H., CHO, J. S., KIM, K. (1966), Studies on the preservation of kimchi. Part 1. Effects of preservatives on kimchi fermentation, *Rep. Army Res. Test. Lab (Korea)* **5,** 5–9.

SONG, S. H., CHO, J. S., PARK, K. C. (1967), Studies on the preservation of the kimchi. Part 2. On the control of enzyme action for overfermented kimchi, *Rep. Army Res. Test. Lab. (Korea)* **6,** 1–4.

STAMER, J. R. (1983), Lactic acid fermentation of cabbage and cucumbers, in: *Biotechnology* (REHM, H. J., REED, G., Eds.), Vol. 5, pp. 365–378, Weinheim: Verlag Chemie.

STAMER, J. R., STOYLA, B. O. (1978), Stability of sauerkraut packaged in plastic bags, *J. Food Prot.* **41,** 525–529.

STAMER, J. R., HRAZDINA, G., STOYLA, B. O. (1973), Induction of red color formation in cabbage juice by *Lactobacillus brevis* and its relationship to pink sauerkraut, *Appl. Microbiol.* **26,** 161–166.

STAUB, J. E., BACKER, J. (1995), Cucumbers as a processed vegetable, in: *Processing Fruits and Vegetables: Science and Technology,* Vol. III, Lancaster, PA: Technomic Publ. Co, Inc.

STETTER, H., STETTER, K. O. (1980), *Lactobacillus bavaricus* sp. nov., a new species of the subgenus *Streptobacterium, Zentralbl. Bakteriol. Parasi-*

tenkd. *Infektionskr. Hyg. Abt. 1 Org. Reihe* C, 70.

SUVOROV, A., KOK, J., VENEMA, G. (1988), Transformation of group A streptococci by electroporation, *FEMS Microbiol. Lett.* **56,** 95–100.

TAKAMA, F., ISHII, H., MARUKI, S. (1986), Quality changes of salted Chinese cabbages during storage and by freeze-drying, *Nip. Shok. Kogyo Gakk.* **33,** 701–707.

THOMAS, D. S., HENSCHKE, P. A., GARLAND, B., TUCKNOTT, O. G. (1985), A microprocessor-controlled photometer for monitoring microbial growth in multi-welled plates, *J. Appl. Bacteriol.* **59,** 337–346.

THOMPSON, R. L., FLEMING, H. P., MONROE, R. J. (1979), Effects of storage conditions on firmness of brined cucumbers, *J. Food Sci.* **44,** 843–846.

TSUYUKI, H., ABE, T. (1979), Studies on the free organic acids in kimchi, *Bull. Coll. Agric. Vet. Med. Nihon Univ.* **36,** 163–170.

UM, J. Y., KIM, K. O. (1990), Effect of sodium acetate and calcium chloride on characteristics of kakdugi, *Korean J. Food Sci. Technol.* **22,** 140–144.

VESCOVO, M., MORELLI, L., BOTTAZZI, V. (1982), Drug resistance plasmids in *Lactobacillus acidophilus* and *Lactobacillus reuteri, Appl. Environ. Microbiol.* **43,** 50–56.

VIDAL, C. A., COLLINS-THOMPSON, D. L. (1987), Resistance and sensitivity of meat lactic acid bacteria to antibiotics, *J. Food Prot.* **50,** 737–740.

VORBECK, M. C., MATTICK, L. R., LEE F. A., PEDERSON, C. S. (1961), Volatile flavor of sauerkraut: gas chromatographic identification of a volatile acidic flavor, *J. Food Sci.* **26,** 569–572.

WHO (World Health Organization) (1974), Toxicological evaluation of some food additives, including anticaking agents, antimicrobials, antioxidants, emulsifiers, and thickening agents, *WHO Food Addit. Ser.* **5,** 461–464.

WINKOWSKI, K., MONTVILLE, T. J. (1992), Use of meat isolate, *Lactobacillus bavaricus* MN, to inhibit *Listeria monocytogenes* growth in a model meat gravy systems, *J. Food Saf.* **13,** 19–31.

YOOK, C., CHANG, K., PARK, K. H., AHN, S. Y. (1985), Pre-heating treatment for prevention of tissue softening of radish root kimchi, *Korean J. Food Sci. Technol.* **17,** 447–453.

18 Enzymes in Food Processing

HANS SEJR OLSEN

Bagsvaerd, Denmark

1 Introduction

Industrial enzymology is an important branch of biotechnology. Enzymatic processes permit natural raw materials to be upgraded and finished. Enzymes offer alternative ways of making products previously made using conventional chemical processes.

Enzymes were used in ancient Greece for the production of cheese. Early references to this are found in Greek epic poems dating from about 800 BC. Fermentation processes for brewing, baking, and the production of alcohol have been known since prehistoric times.

The development of submerged fermentation technique resulted in tremendous progress in the field of industrial enzymology. This technique was used already in the early fifties for production of bacterial amylase for the textile industry. Industrially microbial enzymes for the starch industry were used earlier than in the detergent industry (the largest use of enzymes). Special types of syrups that could not be made by means of conventional chemical hydrolysis were the first compounds made entirely by enzyme processes.

The actual turning point for the food industry was reached in the early 1960s, when an enzyme preparation (amyloglucosidase) was launched for the first time. Now starch could be broken down completely into glucose. Soon thereafter almost all glucose production was changed to enzymic hydrolysis from acid hydrolysis; not least because of advantages that were obvious to the producer – namely greater yields, a higher degree of purity and facilitated crystallization. The most significant event was in 1973: The development of immobilized glucose isomerase, which made the industrial production of fructose syrup feasible. Since then the starch conversion industry has become the second largest enzyme-consuming industry, and due to the many different products made out of starch significant research efforts have been devoted to it. Modern biotechnology using the latest techniques of molecular biology and genetic engineering has been concentrated very much on this practical industry. Besides the enzyme development, a significant amount of application process research has been the necessary back-up to support the breakthroughs in microbiology.

Due to their efficiency, specific action, working under mild conditions, their high purification and standardization, enzymes are ideal catalysts for the food industry. Simple equipment can be used, and as moderate temperature and pH values are required for the reactions only few by-products influencing taste and color are formed. Furthermore, enzyme reactions are easily controlled and can be stopped when the desired degree of conversion is reached.

1.1 World Consumption of Industrial Enzymes in Food Processing

Worldwide consumption of industrial enzymes for the food industry amounted to approximately US$ 300 million in 1990, the main industries being starch and dairy. Brewing, wine and juice, and baking together accounted for less than the dairy industry, even though significant potential exists for increasing usages of industrial enzymes in these industries.

The total growth (including non-food applications) in volume of the enzyme business from 1974 to 1986 was 10–15% per year although no major new applications emerged in this period. From 1980 to 1990 the growth can be estimated as 5–10% per year. Lately new non-food applications have been developed within the textile industry. The major worldwide sales values of enzyme product types used by the food industry can be seen in Tab. 1.

1.2 Regulatory Considerations

Existing legislation and regulations are used by national authorities when dealing with enzymes in the food industry.

For general purpose the Joint FAO/WHO Expert Committee on Food Additives (JECFA) and the Food Chemicals Codex (FCC) have made guidelines available for the appli-

Tab. 1. Estimated World Sales Values of Food Enzymes by Product Type, 1990

Enzyme	Consumption, 10^6 US$
Rennet (animal and microbial)	75
Glucose isomerase	40
Glucoamylase	75
α-Amylase	50
Papain	8
Trypsin	8
Other food proteases	8
Invertase	8
Pectinase	7
Other (β-glucanase, cellulase, dextranase, glucose oxidase, hemicellulase, lactase, lipase, pullulanase)	20

cation of enzymes as food additives. The Association of Microbial Food Enzyme Producers (AMFEP) in Europe and the Enzyme Technical Association (ETA) in the United States, work nationally as well as internationally for harmonization of regulations. In the European Community, common guidelines are being developed for the evaluation of enzymes for food and feed uses.

Obviously a microbial source for a food enzyme must be non-pathogenic and non-toxicogenic. Manufacturers of microbial food enzymes have always selected their production microorganisms from the safe end of the spectrum of available sources. Consequently, a few species have acquired a record of safe use as sources of a wide variety of food enzymes. For enzymes from recombinant microorganisms, the primary regulatory status is determined by the host organism, the donor organism, and any vectors involved in the genetic transfer.

The majority of food enzymes are used as processing aids, and have no function in the final food. In this case they do not need to be declared on the label, as they will not be present in the final food in any significant amount. Some enzymes (a few) are used both as processing aids and as food additives. When used as additives, they must be declared on the food label using the appropriate class name, e.g., preservative or antioxidant, E-number and generic name such as lysozyme or glucose oxidase.

1.3 Quality Assurance of Industrial Enzymes

AMFEP has defined Good Manufacturing Practice (GMP) for microbial food enzymes. The most important element is to ensure a pure culture of the production organism. Product specifications for microbial food enzymes have been established by JECFA and FCC. They prescribe the absence of contaminants such as arsenic, heavy metals, lead, coliforms, e.g., *Escherichia coli* and *Salmonella.* Furthermore, they prescribe the absence of antibacterial activity and mycotoxins for fungal enzymes only.

Commercial enzyme products are usually formulated in aqueous solutions or processed to non-dusty dry products (granulated or as microgranulates). Both types of preparations must be formulated to be suitable for the final application.

Requirements mainly regarding the storage stability (e.g., enzyme activity stability, microbial stability, physical stability and formulations) as significant quality parameters of industrial enzyme products must be taken into consideration by both the producer and the user.

2 Enzymes for Starch Modification

Depending on the enzymes used and controlling the enzyme reactions, various valuable products may be produced to suit nearly any particular requirement of the food industry. Syrups and modified starches with different compositions and physical properties can be obtained. The syrups are used in a wide variety of foodstuffs: soft drinks, confectionery, meat products, baked goods, ice cream, sauces, baby food, canned fruit, preserves, etc.

The major steps in conversion of starch are liquefaction, saccharification and isomerization. In simple terms, the further a starch processor proceeds, the sweeter is the syrup that can be obtained.

A slurry of the starch is cooked in the presence of a heat-stable bacterial endo-α-amylase. The enzyme hydrolyzes the α-1,4 glycosidic bonds in pregelatinized starch, whereby the viscosity of the gel rapidly decreases and the so-called maltodextrins are produced. The process may be terminated at this point, the solution purified and dried, and the maltodextrins utilized as bland tasting functional ingredients in, for instance, dry soup mixes, infant foods, sauces, gravy mixes and so on. Further hydrolysis by means of amyloglucosidase leads to the formation of sweet-tasting, fermentable sugars. Sweet starch hydrolysates with special functional properties may be obtained by using fungal α-amylase either alone or in combination with amyloglucosidase. Alternatively, a vegetable β-amylase may be used to increase the yield of maltose.

Also many non-food products made by fermentation are based on the use of enzymically modified starch products. Enzyme-hydrolyzed starches are used, for instance, for the production of alcohol, ascorbic acid, enzymes, lysine and penicillin. The particular steps are described in the following.

2.1 Starch Liquefaction

As native starch is only slowly degradable by use of α-amylases, a gelatinization and liquefaction of a 30–40% dry matter suspension is needed to make the starch susceptible to enzymatic breakdown. The gelatinization temperature is different for different starches (LII and CHANG, 1991). Maize starch is the most common source followed by wheat, tapioca and potato. A good liquefaction system is necessary for successful saccharification and isomerization. For liquefaction an efficient temperature-stable α-amylase is added to the starch milk. The mechanical way of performing the liquefaction may be by use of stirred tank reactors, continuous stirred tank reactors (CSTR) or a jet cooker.

In most plants for sweetener production the starch liquefaction is made in a single-dose jet cooking process as shown in Fig. 1.

Cooking extruders have been studied for the liquefaction of starch, but due to the high temperature in the extruder inactivation of the enzyme activities they demand 5–10 times higher dosages than in a jet cooker (CHOUVEL et al., 1983).

2.2 Alpha-Amylases

During the liquefaction the α-1,4 linkages are hydrolyzed at random. This reduces the viscosity of the gelatinized starch and increases the dextrose equivalent (DE) – a measure of the degree of hydrolysis of the starch. The liquefaction is carried out in such a way as to give the required DE for the subsequent process. For saccharification to dextrose, a DE of 8–12 is commonly used. Higher DE values are often necessary for maltodex-

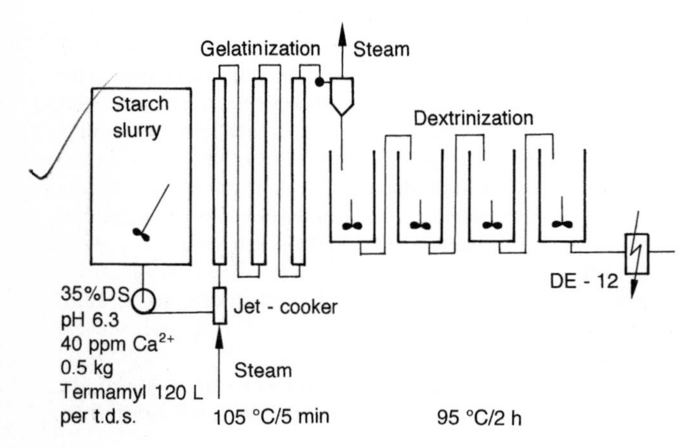

Fig. 1. Starch liquefaction process. Termamyl is a bacterial thermophilic α-amylase (NOVO NORDISK, 1990).

Tab. 2. Industrially Important Starch-Degrading Enzymes

Enzyme	Microorganism	Application Temperature Range (°C)	Application pH Range	Minimum Ca^{2+} Dosage (ppm)
Bacterial mesophilic α-amylase	*Bacillus subtilis*	80– 85	6–7	150
Bacterial thermophilic α-amylase	*Bacillus licheniformis*	95–105	6–7	20
Fungal α-amylase	*Aspergillus oryzae*	55– 70	4–5	50
Amyloglucosidase	*Aspergillus niger*	55– 65	3.5–5	0
Pullulanase	*Bacillus acidopollulyticus*	55– 65	35–5	0

trin production. The maximal DE obtainable is about 40.

Industrially important α-amylases are those made from *Bacillus licheniformis*, *Bacillus subtilis* and *Aspergillus oryzae*. In Tab. 2 some application conditions and a few comparative characteristics are shown for various enzymes available for starch processing.

2.3 Beta-Amylases

β-Amylases are exo-enzymes, which attack amylose chains resulting in efficient successive removal of maltose units from the non-reducing end. In the case of amylopectin the cleavage stops two to three glucose units from the α-1,6 branch points. β-Amylase is used for the production of maltose syrups and in breweries for adjunct processing. The most important commercial products are made from barley or soybeans.

2.4 Isoamylase and Pullulanase

Isoamylase (glycogen-6-glucanohydrolase) and pullulanase (pullulan-6-glucanohydrolase) hydrolyze α-1,6 glucosidic bonds of starch. When amylopectin is treated with pullulanase, linear amylose fragments are obtained. Using heat- and acid-stable pullulanase in combination with saccharification enzymes makes the starch conversion reactions more efficient (NORMAN, 1982).

2.5 Saccharification of Liquefied Starch

Maltodextrin (DE 15–25) produced from the liquefied starch is commercially valuable for its rheological properties. Maltodextrins are used in the food industry as fillers, stabilizers, thickeners, pastes, and glues. When saccharified by further hydrolysis using amyloglucosidase or fungal α-amylase, a variety of sweeteners can be produced having dextrose equivalents in the ranges 40–45 (maltose), 50–55 (high maltose), and 55–70 (high conversion syrup) (REICHELT, 1983). Applying a series of enzymes including β-amylase, glucoamylase and pullulanase as debranching enzymes, intermediate-level conversion syrups having maltose contents close to 80% can be produced (NORMAN, 1982).

A high content of 95–97% glucose may be produced from most starch raw materials (corn, wheat, potatoes, tapioca, barley, and rice). The action of amylases and debranching enzymes is shown in Fig. 2.

2.6 Production of High-Dextrose Syrups

The amyloglucosidase was introduced in the early 1960s, and today this saccharification enzyme is considered a commodity. The

α-1-4-link
Amylose
α-1-6-link
Amylopectin

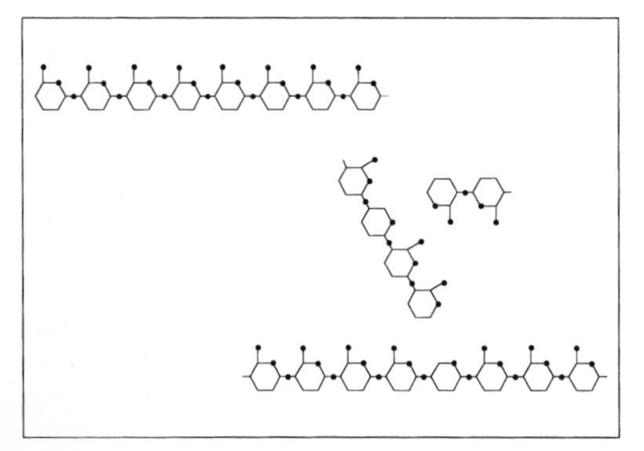

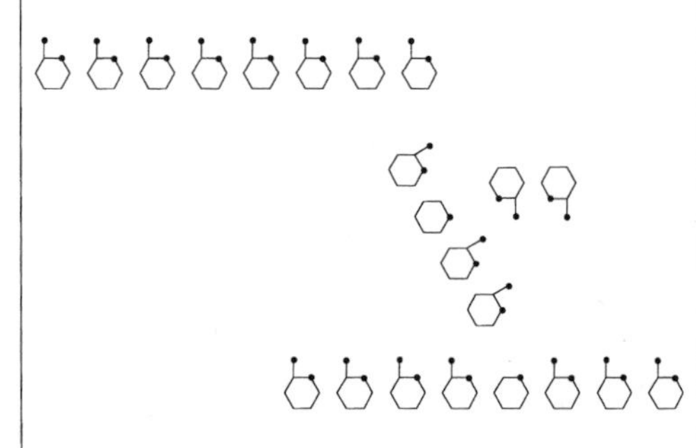

Fig. 2. Scheme of the action of starch-degrading enzymes. Top left: Structure of amylose and amylopectin. Top right: Effect of fungal α-amylase on amylose and amylopectin. Bottom left: Effect of a debranching enzyme on amylase and amylopectin. Bottom right: Effect of amyloglucosidase and debranching enzyme on amylose and amylopectin.

amyloglucosidase hydrolyzes the α-1,4 linkages rapidly, but during the saccharification process the α-1,6 linkages of the highly branched amylopectin are hydrolyzed much slower. Using a pullulanase, starch is debranched by rapid hydrolysis of the α-1,6 bonds. Using a pullulanase together with the amyloglucosidase at the start of the saccharification, the α-1,6 linkages of the branched dextrins are rapidly hydrolyzed. In consequence fewer branched oligosaccharides accumulate towards the end of the saccharification. The point at which reversion outbalances dextrose formation is thus shifted towards a higher DX level. This is illustrated in Fig. 3 using AMG® (glucoamylase) and a balanced mixture of amyloglucosidase and pullulanase, Dextrozyme®. The latter enzyme mixture produces higher DX values than can be obtained if amyloglucosidase is used alone. This is further illustrated in Fig. 4, which at the same time illustrates the influence of the dry substance content on the maximal obtainable DX-value – the yield of glucose.

The purification processes of saccharified starch may be adapted to each raw material and may be different from plant to plant. When the starch milk is liquefied in the jet cooker (see Fig. 1), the saccharification process is carried out at 55–65 °C, pH 4–4.5 for 24–72 hours. The additional steps consist of filtrations or centrifugations, ion exchange, isomerization (see below), treatment with activated carbon, and evaporation to a storage-stable product.

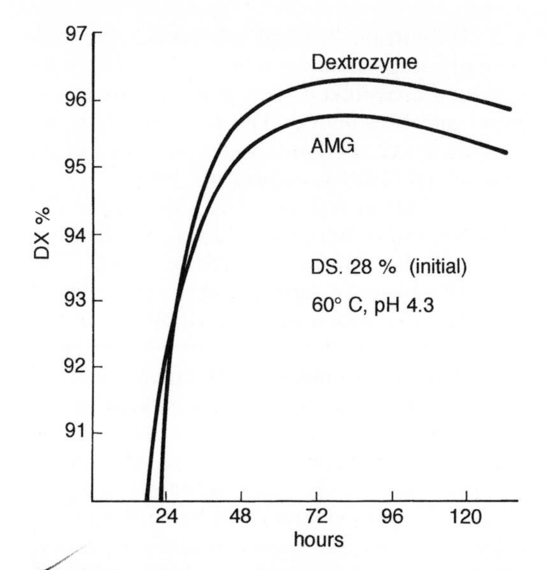

Fig. 3. The influence of Dextrozyme on saccharification.

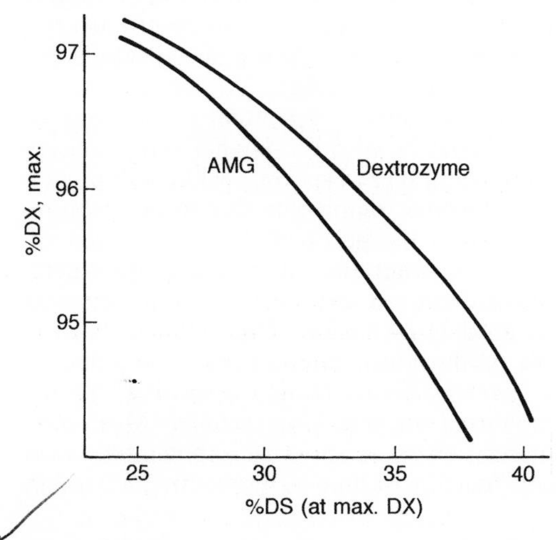

Fig. 4. The effect of dry substance on maximal obtainable % *DX*.

2.6.1 Saccharification in a Membrane Reactor

A continuous hydrolysis process for the saccharification of liquefied corn starch has been developed using a membrane reactor (SIMS and CHERYAN, 1992a). A residence time distribution confirmed that the membrane could be modelled as a simple continuous stirred tank reactor (CSTR). Kinetic studies indicated that the continuous reactor operated in the first-order region with respect to substrate concentration at substrate concentrations greater than 200 g/L. The productivities were claimed to be 10 to 20 times higher than those obtained in a batch reactor (SIMS and CHERYAN, 1992b). All saccharifications were carried out using the amyloglucosidase from *Aspergillus niger* (AMG 200 L) at 55 °C and pH 4.5 in a dosage of 0.5 vol./vol.% (16.6 liters per ton of starch). This is a dosage, which is about 15 times higher than normally recommended in batch processes using a saccharification time of 40–70 hours. At a residence time of 1 hour in the membrane reactor the apparent Michaelis constant (K_m) was 562 g/L, which is 2–7 times greater than that obtained in a batch reactor (SIMS and CHERYAN, 1992b). It seems evident that more work is needed in order to clear up all technical and economical aspects and possibilities of the use of membrane technology for the saccharification of starch.

2.6.2 Enzymes as Processing Aid in Purification of Saccharified Wheat Starch

Especially with wheat starch a minor content of impurities in the starch milk adheres to the starch granules. To facilitate the refining, e.g., filtration operations, cellulases, pentosanases, glucanases, proteases, and pectinases are used in some special cases. Wheat starch is known to form precipitates or hazes, which are difficult to filter. Arabinoxylan, pentosanes, and lysophospholipids are claimed to be responsible for this problem (KONIECZNY-JANDA and RICHTER, 1991).

2.7 Glucose Isomerization

Enzymatic isomerization of glucose to fructose was the process that provided a real alternative to white sugar (sucrose) from cane or beets. The commercial product obtained was high fructose corn syrup (HFCS). Two grades of the syrup have established themselves in the world market, HFCS-42 and HFCS-55. They contain 42% or 55% fructose based on dry substance. These products account for over a third of the caloric sweetener market in USA. Annually more than 8 million tons of HFCS are produced using glucose isomerase. This represents the largest commercial application of immobilized biocatalysts in industry.

2.8 The Isomerization Reaction

Glucose can reversibly be isomerized to fructose. The equilibrium conversion for glucose ⇌ fructose is 50% under industrial conditions, and the reaction is slightly endothermic (TEWARI and GOLDBERG, 1984). The reaction is usually carried out at 60°C, pH 7–8. To avoid excessive reaction time, the conversion is normally limited to about 45%.

The isomerization reaction can only be economic by using immobilized enzyme. The reaction parameters in this system have to be optimal in order to obtain a reasonable yield of fructose. The pH must be approx. 7.5 or higher in order to secure high activity and stability of the enzyme. Under these conditions glucose and fructose are rather unstable and decompose easily to organic acids and colored by-products. To overcome these problems the reaction time must be limited. This is done by using an immobilized isomerase in a fixed-bed reactor process in a column through which glucose flows continuously. The enzyme granules must be rigid enough to prevent compaction during the operation.

2.9 The Immobilized Enzyme System

The glucose isomerases used are immobilized and granulated to a particle size between 0.3 and 1.0 mm. As an example Sweetzyme® T from Novo Nordisk A/S is produced by a mutant of a selected *Streptomyces murinus* strain. The immobilization procedure for Sweetzyme® T consists of a disruption of a cell concentrate through a homogenizer with a single stage homogenizing valve. The cells are then crosslinked with glutaraldehyde, diluted and flocculated. The concentration aggregate is extruded and finally fluid-bed dried and sieved (JØRGENSEN et al., 1988).

Mg^{2+} acts as activator and stabilizer of the enzyme, and is therefore added to the feed syrups in the form of $MgSO_4 \cdot 7H_2O$. The amount necessary depends on the presence of calcium ions, which act as an inhibitor in the system by displacing the magnesium ion activator from the isomerase molecule. Therefore the calcium ion content should be kept as low as possible. At a calcium ion content in the feed syrup of 1 ppm or less, the addition of 45 ppm Mg^{2+} (e.g., approximately 0.6 g of $MgSO_4 \cdot 7H_2O$ per liter) will be sufficient. At higher calcium ion concentrations, a proportionate weight ratio between Mg^{2+} and Ca^{2+} ions should be provided.

The main criteria for selecting the feed syrup specifications are optimization of enzyme productivity and limitation of by-product formation. Typical feed syrup specifications are shown in Tab. 3.

The dry-substance content of the feed syrup should be 40–50%. Higher syrup concentration and higher viscosity will result in a reduced isomerization rate due to diffusion resistance in the pores of the immobilized enzyme. A deaeration step is desirable to remove dissolved oxygen that would increase by-product formation. The pH is adjusted to the productivity optimum of the enzyme.

Enzyme decay. During operation, the immobilized enzyme loses activity. Most commercial enzymes show an exponential decay as a function of time as shown for SwQ in Fig.

Tab. 3. Typical Feed Syrup Specifications

Temperature	55–60°C
pH	7.5–8.0
Dry-sbustance content	40–50 wt%
Glucose content	≥95%
SO_2	0–100 ppm
Calcium ion	≤1 ppm
$MgSO_4 \cdot 7H_2O$ (activator)	0.15–0.75 g/L
Conductivity	≤100 μS/cm
UV absorbance (280 nm)	≤0.5

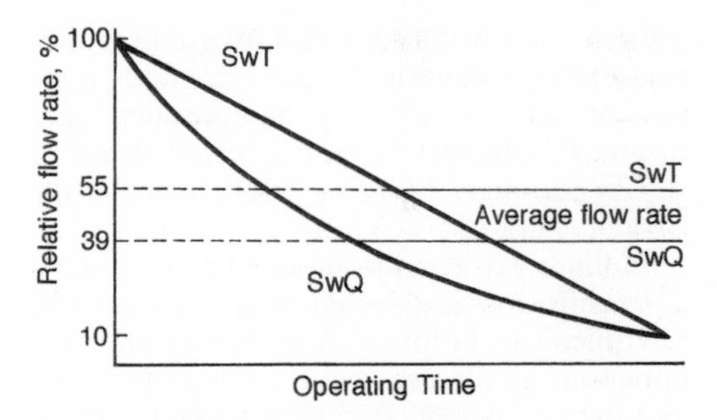

Fig. 5. Activity (syrup flow rate) versus operating time for typical immobilized isomerases.

5, but with SwT (Sweetzyme® T) a linear activity decay is found.

Typically, a reactor load of glucose isomerase is replaced after three half-lives, i.e., when the activity has dropped to around 12.5% of the initial value. The most stable commercial glucose isomerases have half-lives of around 200 days in industrial practice.

To maintain a constant fructose concentration in the product syrup, the feed flow rate is adjusted according to the actual activity of the enzyme. With only one isomerization reactor in operation, excessive variations in syrup production rate would be the result. To avoid this, several reactors containing enzyme of different age are operated in combination.

Reactor design for glucose isomerization in the United States has been described (BLANCHARD and GEIGER, 1984). Reactor diameters are normally between 0.6 and 1.5 m. Typical bed heights are 2–5 m. Minimum bed height–diameter ratio for one reactor is 3:1 to ensure good flow distribution. Plants producing more than 1000 t of HFCS (based on dry matter) per day use at least 20 individual reactors.

2.10 Economics of High Fructose Corn Syrup Production

25 kg (approximately one bushel) of corn yields in the wet milling process about 0.75 kg of corn oil, 6.3 kg corn gluten feed, 1.2 kg corn gluten meal and 14 kg HFCS-55 (on a dry basis). Based upon values given in BIO-

Tab. 4. Estimated Mass Balance of the Corn Wet-milling and HFCS Process

Products and Fractions Corn (90% dry solids)	Mass (kg) 100
Corn oil	3.0
Corn gluten feed	25.2
Corn gluten meal	4.8
HFCS (78% dry solids)	73.0

Tab. 5. Production Costs of HFCS 55 in US$ per 100 kg of HFCS (70–86% dry solids, average = 78%)

Process Step	Value in US$ per 100 kg of HFCS
Sales values	
Value of by-products	11
Value of HFCS 55	28–40 (1989)
Costs of HFCS production	
Raw material:	
Corn (137 kg)	(10.70)[a]
Net starch costs	8.80
Capital depreciation	2.30
Labor (2.5 men)	2.00
Milling and feed preparation	7.00
Refinery utilities	1.00
Liquefaction enzymes	0.17
Saccharification enzymes	0.35
Clarification filter aid	0.12
Carbon treatment	
– dextrose	0.30
– HFCS	0.20
Ion exchange	
– resins	0.23
– chemicals	0.35
Isomerization	
– enzyme	0.38
– chemicals	0.03
Fractionation	
– resin	0.12
– evaporation	0.16
Final evaporation	0.32
Total production cost	23.85

[a] not used for calculation

TOL (1991) and the mass balance shown in Tab. 4, the production cost has been calculated for HFCS-55 (70–86% dry solids) as shown in Tab. 5. These data vary dependent on a lot of parameters related to the actual production plant, country and production year, but they can be used as an illustration of the various costs of the different operation steps involved. A more detailed description of the developments of glucose syrup, its science and technology and its progression to a highly complex, biotechnologically based industry has been given by DZIEDZIC and KEARSLEY (1984).

2.11 Other Sweeteners

Later developments in enzymic modification of starch are among others concerned with production of cyclodextrins. The cyclodextrins are cyclic molecules consisting of 6–8 glucose units linked together in α-1,4 bonds. They can bind other compounds in their cavity and thereby stabilize, solubilize or precipitate compounds (PSZCZOLA, 1988). Cyclodextrins are produced during liquefaction of starch by use of the class of enzymes called cyclodextrin glycosyl transferase (CGT-ase) (NIELSEN, 1991).

Production of fructose from inulin is one use of enzymes for production of sweeteners, and the use of invertase is another alternative to change sweetness and functional properties of sweeteners.

2.11.1 Enzymatic Hydrolysis of Inulin in the Production of Fructose

Fructose is the sweetest naturally occurring nutritive sweetener. An alternative source of fructose is inulin, which occurs in the roots of plants such as chicory and Jerusalem artichoke. Inulin is a linear, β-2,1 linked fructose polymer initiated by a glucose unit. The fructose content varies depending on the inulin source, but is typically approximately 90%, corresponding to an average chain length of 10 units.

Inulin may be saccharified by use of a mixture of exo-inulinase (EC 3.2.1.80) and endo-inulinase (EC 3.2.1.7), e.g., the product Fructozyme™ obtained from a selected strain of *Aspergillus niger* (ZITTAN, 1981; NOVO NORDISK, 1993a).

Inulin is extracted from sliced plant material by diffusion at elevated temperatures. The raw juice can be purified by liming and carbonation, similar to sugar beet processing, and then be decolorized by activated carbon. A dry substance content in the raw juice of 15–25% is optimal for Fructozyme. A degree of hydrolysis above 98% is obtained within 12 to 48 hours depending on enzyme dosage. The fructose syrup produced can be used as is or for blending for the production of 55% fructose syrup. Alternatively, the purified syrup can be used for production of crystalline fructose (NOVO NORDISK, 1993a).

2.11.2 Enzymatic Hydrolysis of Sucrose

Invertase hydrolyzes sucrose to glucose and fructose. Invertase is used in the confectionery industry to convert the easily crystallized sucrose into the less easily crystallized glucose–fructose mixture. "After Eight Mints", for example, have invertase in their centers. Thereby the hard sucrose core (coated with chocolate) is turned into the soft center, which we finally eat (BAINS, 1993).

3 Use of Enzymes in the Baking Industry

Wheat flour contains enzymes, the most important of which are amylases and proteases. However, the quantities of these enzymes are not always ideal for baking purposes and supplementary enzymes often have to be added.

Addition of α- and β-amylases and proteases to wheat doughs has been practiced since the 1800s, when malt was added as the

Tab. 6. Potential Bread Improving Properties Using Microbial Enzymes

Enzyme	Purpose	Reference
Amylases	Maximize the fermentation process to obtain an even structure of the bread crumb and a high bread volume	MILLER et al., 1953
	Prevent staling	RUBENTHALER et al., 1965
Glucose oxidase/catalase	Oxidation of free sulfhydryl groups in gluten (like bromate). Strengthen weak doughs. More elastic doughs	BIOTOL, 1991
Lipases	Dough conditioning	SI and HANSEN, 1994
Lipoxygenases	Bleaching and strengthening effects on dough	HAAS and BOHN, 1934
Pentosanases	Easier dough-handling and improved crumb structure	KULP, 1968
Proteases	Provide the plastic properties required in a dough used for biscuits	CONN et al., 1950
Sulfhydryl oxidase	Strengthen weak doughs by -S-S- formation	SCOTT, 1989

source of enzyme to reduce dough viscosity and increase rate of fermentation. This results in bread of improved volume and texture. Already TAKAMINE discovered in 1844 that *Aspergillus oryzae* was a better source of α-amylase and proteases for bread-making than those from malt. In Tab. 6 many potential bread-improving properties using microbial enzymes are listed. The references are shown with year of origin in order to demonstrate that most of the ideas of using enzymes in bread-making have been demonstrated several years ago.

In recent years enzyme preparations without disturbing side activities and non-dusting granulated products have been developed which can be used with the highest possible safety in industrial baking or by consumers.

The main applications are for obtaining improvement of dough tolerance, loaf volume, crumb structure and shelf life. Today the enzyme products used are free-flowing microgranulates, which are easy to handle and mix with flour during the bread-making.

3.1 Amylases

To standardize the α-amylase content of flour a fungal α-amylase is used. β-Amylase is generally available in sufficient quantity in the grain, but α-amylase is often deficient and must be supplemented. The amount of additional enzyme must be sufficient for the desired CO_2 production by the yeast. If CO_2 production is too small, the bread becomes hard mainly due to the reduced volume. Amyloglucosidases are also used in bread-making to break down starch, oligosaccharides and dextrins into glucose. It is thus possible to develop crust coloring and, together with fungal α-amylase, to stabilize cool/freeze doughs. The fungal α-amylase also improves dough-handling, crumb structure and loaf volume in industrial baking plants and in home-baking.

Of course, a too slow production of CO_2 could be compensated for by addition of fermentable sugar for the yeast. However, as the CO_2 production rate should coincide with the ability of the dough to retain the produced gas, sugar addition leads to a too rapid fermentation before the dough is able to form a sufficiently rigid network. Fungal α-amylase has an advantage over sugar addition by providing a gradual formation of fermentable sugars by hydrolyzing damaged or mechanically modified starch.

The α-amylase breaks down the starch polymers to fragments we call dextrins. These compounds are further broken down to maltose by β-amylase (which is normally present in wheat flour in sufficient quantities).

An important phenomenon during baking is the gelatinization of starch. By gelatinization the starch granules swell by the uptake of water, lose their crystalline structure and exude the amylose. As a result of this process the concentrated starch suspension in the dough is transformed into a starch gel. The strength of this gel is determined by the amylose concentration. Recrystallization of amylose during baking and cooling results in the formation of a rigid 3-dimensional network and this together with the gluten network is responsible for the fixation of the crumb structure. A fast crumb setting, however, reduces "ovenspring" resulting in a small loaf volume. Degradation of the gel-forming starch by α-amylase maintains the flexibility of the dough for a longer period during baking, allows greater expansion and produces a larger loaf volume. The temperature of inactivation of fungal α-amylase (55–70 °C) is lower than the gelatinization temperature of wheat starch, and so only a small part of the starch is degraded during baking.

Excessive starch degradation is detrimental to bread quality, because it reduces gel formation and leads to an extremely soft, gummy and open crumb structure. As bacterial α-amylase is heat-stable (see Sect. 2.2), fungal α-amylase is preferred, as it is less heat-stable and does not as such require a narrow dosing range. Apart from higher loaf volume, addition of fungal α-amylase results in a softer crumb. This effect can be explained by (1) a lower crumb density as a result of higher loaf volume and a more uniform crumb structure; (2) the partial breakdown of starch leading to reduced firmness of the formed starch gel.

Practical use and characterization of a commercial fungal amylase

The example is given for the use of Fungamyl® BG, developed for use in baking as an ingredient in bread improvers and pre-mixes, or to be added to the flour at the flour mill.

Fungamyl is based on a purified preparation from a selected strain of *Aspergillus oryzae*. This enzyme hydrolyzes the α-1,4 glucosidic linkages in amylose and amylopectins forming dextrins and maltose. Addition of the enzyme to flour will ensure continuous production of dextrins and maltose during leavening, resulting in a good and even structure of the bread crumb and a larger bread volume. Fungamyl BG ("baking granulate") is inactivated in the oven before the starch gelatinizes. This excludes any risk of excessive dextrination, which otherwise could lead to sticky crumb.

Typical dosages of the product type Fungamyl 2500 BG are 0.2–0.4 g/100 kg flour and correspondingly higher in bread improvers and pre-mixes. The BG is a free-flowing, non-dusting, agglomerated powder with an average particle size of around 150 microns. No particles are larger than 200 microns, and less than 0.5% are smaller than 50 microns (NOVO NORDISK, 1993b).

The advantage of a BG granulate is that it has a particle size distribution much like flour – without the dust.

3.2 Enzymes for Antistaling

After a bread leaves the oven, various processes lead to deterioration of quality. The major problem is hardening (crumb firming), caused mainly by the recrystallization (retrogradation) of starch during storage. This bread "staling" phenomenon is a problem caused by the presence of more of the B-form than of the A-form of crystalline starch. The A-form binds less than 7% water, while the B-form binds over 25% water (WHITAKER, 1990). This can be avoided by use of shortening (fats) to produce a softer crumb structure or by limited degradation by α-amylase at low dosage (BIOTOL, 1991).

Bacterial maltogenic amylase preparations have unique qualities as antistaling agents. Such products cannot be overdosed, which means that there is no risk of sticky doughs and gummy crumb. An independent laboratory has carried out trials, which, compare the effect of the antistaling ingredient GMS (glycerol monostearate) and Novamyl™ in white pan bread. Novamyl™ is a purified maltogenic amylase produced by a genetically modified strain of *Bacillus subtilis* (host), which has received the gene for maltogenic amylase

Tab. 7. Effect of Novamyl and GMS on Bread Freshness for White Pan Bread

Ingredient (per 100 kg of flour)	Crumb Firmness[a] after			
	1 day	3 days	6 days	9 days
Control	172	295	417	465
750 g GMS	173	268	348	408
190 g GMS + 30 g Novamyl	126	179	206	236
30 g Novamyl	125	203	226	252
100 g Novamyl	129	165	196	209

[a] Measured by means of an Instron Texture Analyzer. The units are in gram of force required to compress a 1.5 inch thick slice of crumb with a 28 mm diameter flat disk by 4 mm at a rate of 50 mm per min.

Tab. 8. Effect of Novamyl and GMS on Bread Freshness for 60% Whole Wheat

Ingredient (per 100 kg of flour)	Crumb Firmess[a] after			
	1 day	3 days	6 days	9 days
Control	183	255	343	419
750 g GMS	158	225	278	333
190 g GMS + 30 g Novamyl	148	188	227	263
30 g Novamyl	149	194	228	288
100 g Novamyl	149	171	208	250

[a] see Footnote Tab. 7

from a strain of *Bacillus stearothermophilus* (donor) (MEI-HING et al., 1993). The flour was a "bread flour" with 11.3% protein. The breads were made according to a sponge-and-dough recipe, with the enzyme and GMS being added at the pre-mix stage. The bread volumes were measured, and then the breads were kept in plastic bags at 25 °C. Breads were evaluated for organoleptic properties and crumb firmness after 1, 3, 6 and 9 days. The important differences were in crumb firmness (the higher the figure for crumb firmness, the firmer the crumb). The crumb firmness was measured using a Texture Analyzer® probe, which compresses 5 mm of a 20 mm bread slice. The same equipment can be used for measuring elasticity. As shown in Tab. 7, Novamyl is significantly more effective as an antistaling agent than GMS. Tests also have shown that addition of Novamyl increases the crumb elasticity significantly. Crumb elasticity refers to a better mouth-feel as "freshness" (SI and SIMONSEN, 1994).

SI and SIMONSEN (1994) explain the mechanism of Novamyl's antistaling effect as being due to the significant amounts of small fragments of saccharides, mainly maltose, produced *in situ*. These fragments may prevent the interaction between starch and gluten.

The same trials were carried out with "60% whole wheat" bread. The sponge contained the whole wheat flour, and the dough a bread flour with high protein content (13.8%). The results are shown in Tab. 8. Owing to the high water-holding capacity of the fiber material, whole wheat breads do not stale as fast as white pan breads. Both GMS and Novamyl have a significant effect on the newly baked bread. The normal dosage of Novamyl (30 g/100 kg) is better at keeping the bread fresh than GMS, although the difference is much smaller than for the white pan bread. The high dosage of Novamyl is extremely effective, by keeping the 9 day old bread as soft as the 3 day control.

Tab. 9. Effect of Pentopan on Dough and Bread Quality for White Pan Bread

	Dosage (g/100 kg flour)	Dough Stability[a]	Bread Volume Index	Crumb Structure (0–10)[b]	Freshness[c] (48 hours %)
Blank	–	1	100	3	100
Pentopan 500 BG	4	2	109	6	209
Pentopan 500 BG	8	2	110	6.5	240
Pentopan 500 BG	12	2	115	7	281
Lecithin	600	2	103	4	297

[a] 1, low stability; 2, sufficient dough development, dry and soft dough
[b] scale 0, poor; 10, very good (optimal)
[c] based on compressibility

The conclusion of the trials, mentioned in Tabs. 7 and 8, is that the dosage recommendations for Novamyl 1500 MG, are 6–60 g/100 kg flour, depending on the type of process and raw materials, whether or not GMS is used, and on the desired degree of antistaling.

3.3 Pentosanases

The endosperm of wheat contains 2.5–3% pentosans or hemicellulosic material composed of arabinoxylans. About 2/3 of the pentosans are insoluble and about 1/3 are soluble. Pentosans have a water-binding capacity of about 6.5 times their weight of water, and probably contribute about 25% of the water-binding properties of doughs; 25% of the water is bound by the raw starch, 25% are bound by the damaged starch and the last 25% of water are bound to the protein (MEI-HING et al., 1993). Pentosans interact with the gluten and thereby contribute to the physical properties of the dough. Depending on the content of phenolic constituents such as ferulic acid and vanillic acid in the arabinoxylan fraction, oxidizing conditions may result in the formation of a structural network with proteins. Therefore, both proteases and pentosanases may cause this gel to solubilize.

By a specific hydrolysis of the pentosan fraction it will lose a great deal of its water-binding capacity. The water released will then primarily become available for the gluten to form a strengthened network. Secondarily the starch utilizes part of the water released. When starch has an optimal amount of water, it gelatinizes more easily during baking, giving a softer crumb structure and a prolonged freshness of the bread.

The pentosanase Pentopan® 500 MG works on the gluten/pentosan fraction of flour, resulting in easier dough-handling and improved crumb structure. Also this pentosanase is able to replace 50–100% of the amount of emulsifiers used, i.e., such additives can be replaced by a more natural ingredient. In Tab. 9 the results of baking trials with Pentopan and an emulsifier, lecithin, show that both increased dough stability and improved crumb structure and freshness are seen after 48 hours. Also it has been demonstrated that Pentopan gives a higher bread volume than lecithin, thus a softer bread.

The dosage of Pentopan 500 BG is 2–8 g per 100 kg flour, depending on the flour quality. If too high dosages are used, the water-binding capacity of the dough might be too low, resulting in a sticky dough. Therefore, overdosage should be avoided (MEI-HING et al., 1993).

Most commercially available pentosanases contain several other enzymes, often different hemicellulases and xylanases. Such activities, maybe also the presence of minor protease activities, interfere with the analysis of baking results, so that no correlation is found between the activities measured and the baking effect. This problem can be solved by preparation of pure arabinoxylanases using expression cloning techniques (SI et al., 1993).

Dynamic rheological studies have shown that the addition of mono-component xylanase results in a less rigid and more viscous dough complex, and at the same time the xylanase can increase gluten strength and elasticity. An improved baking effect in terms of larger loaf volume and better crumb structure was demonstrated (SI et al., 1993).

3.4 Proteases

The proteins of wheat flour form a network with disulfide bridges. The structure of gluten gets stronger, the more sulfur bridges are present. Oxidizing agents such as bromate or dehydroascorbic acid strengthen the dough in this manner. Proteases can break down this network by hydrolyzing the peptide bonds. Thus, for example, neutral bacterial endoproteases, which modify the gluten fraction of wheat will weaken a too strong flour or provide the plastic properties required in a dough used for biscuits. Fig. 6 shows a simplified model of the gluten network with disulfide bridges and the possibilities for chemical and enzymatic modification.

Owing to the content of ferulic acid and vanillic acid in the arabinoxylan chains a network with the gluten proteins may be formed. Using, for example, peroxidases or the system glucose oxidase and glucose, free radicals may be formed. Thereby linkages between the arabinoxylan part of the pentosans and the free gluten-SH groups, formed by oxidation, can be assumed to strengthen the gluten network. Sulfhydryl oxidase as discussed by SCOTT (1989) has not yet found use in the baking industry.

In biscuit manufacturing either sodium meta-bisulfite or neutral bacterial protease like Neutrase® are used to weaken the gluten, so that the biscuits, cookies or crackers keep their shape and size.

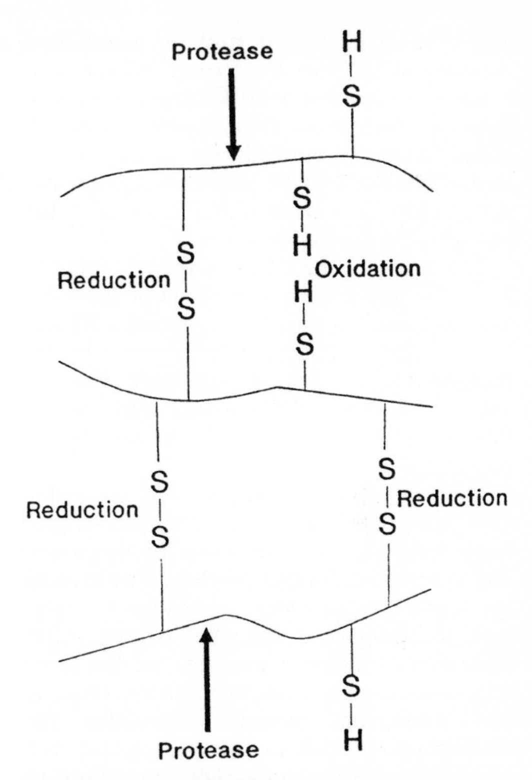

Fig. 6. Model of the gluten network and chemical and enzymatic modifications.

4 Dairy Products

Milk is processed into a variety of products. The application of enzymes in processing of milk is well-established. Even in ancient times rennets were used for coagulation during cheese production. Proteases of various kinds are used for acceleration of cheese ripening, for modification of the functional properties, and for modification of milk proteins to reduce allergenic properties of cow milk products. Lactase (β-galactosidase) is used to hydrolyze the lactose in order to increase digestion, or to improve solubility or sweetness in various dairy products. Lipases are used mainly in cheese ripening.

4.1 Production of Cheese

For industrial cheese production the milk must be standardized with respect to protein and fat content. This can normally be achieved by blending milk batches, skimming (by centrifugation) to remove fat, or fat addition. The standardized milk is pasteurized normally at 72°C for 15 seconds. Subsequently the milk is cooled to 30°C before it is transferred to the cheese tank. Starter culture is added to the pasteurized milk to initiate fermentation. Rennet (the milk-clotting enzyme) and calcium chloride is added to promote the milk protein-clotting reaction.

The milk-clotting effect of rennet is due to a specific and limited hydrolysis of the κ-casein, surrounding the protein micelles. Thereby the micelles become electrostatically discharged and are able to aggregate with the help of calcium and phosphate ions, and to form a network entrapping the fat micelles. A gel structure is thus formed.

After about 25 minutes at 30°C the gelled milk (the coagulum) is cut or stirred to promote syneresis, the exudation of the whey. The syneresis is further promoted by heating the curd–whey mixture. Specialized equipment for drainage of the whey is used. After drainage the cheese is pressed, and the lactic acid fermentation continues during pressing and subsequent storage. Hereafter the cheeses are steeped in brine. However, the fermentation of the lactose continues, until all lactose has been fermented. After removal of the brine the cheese is free of lactose. The cheese is then ripened during an appropriate period of storage.

4.2 Chymous Enzymes

Rennet, also called rennin, is a mixture of chymosin and pepsin. The rennet is extracted from the gastric mucosa of the abomasum of young mammals, e.g., calves and lambs. The pepsin-to-chymosin ratios differ in rennet preparations, because chymosin is present only in the stomachs of unweaned mammals, and is later replaced by pepsin. Thus the content of pure chymosin depends on the age and species of the animal. Chymosin is a highly specific endo-proteinase. It only splits κ-casein into a glycomacropeptide and *para*-κ-casein by selectively cleaving the 105–106 bond between phenylalanine and methionine. Pepsin and other proteolytic enzymes are less specific and can give rise to a number of degradation products, which tend to taste bitter (Fox, 1981).

4.3 Rennet Substitutes

Microbial rennets from a number of producers, e.g., Novo Nordisk, Gist Brocades, Chr. Hansen, and Miles have been available since the 1970s, and have proved satisfactory for production of different kinds of cheese. They are marketed under various trade names (Rennilase, Fromase, Marzyme, Hannilase). The enzyme cost for treating of the same volume of milk is considerably lower than using standard rennet (Burgess and Shaw, 1983).

The microbial rennet Rennilase®, from Novo Nordisk A/S, is produced by submerged fermentation of a selected strain of the fungus *Rhizomucor miehei.* Various modifications of this enzyme have been developed – the main differences being the thermolability of the enzyme itself. This helps cheese makers to develop the many different types of cheeses under local conditions (Budtz, 1989).

The properties of microbial rennets have proven very similar to those of chymosin. Only slight modifications of the cheese-making technique are made in practice (Fedrick and Fuller, 1988; van den Berg et al., 1987).

Recently Novo Nordisk has succeeded in expressing just one proteolytic enzyme from the fungus *Rhizomucor miehei* to the well-known production organism *Aspergillus oryzae.* This host organism is then able to produce the single protease that cleaves the κ-casein into a glycomacropeptide and *para*-κ-casein by hydrolyzing only at the 105–106 bond between phenylalanine and methionine. This "mono-component enzyme" product has the trade name Novoren® (*Biotimes*, 1994a).

The gene for calf chymosin has been cloned into selected bacteria, yeasts and molds. Chymosins made with recombinant DNA techniques by *Kluyveromyces lactis* (Gist Brocades), *E. coli* (Pfizer) and *Aspergillus nidulans* (Hansens) are now commercially available (Fox, 1993).

4.4 Lactase

Lactase (β-1,4-galactosidase) is used to manufacture milk products with reduced content of lactose by hydrolyzing it to glucose and galactose. Many people do not have sufficient lactase to digest the milk sugar.

Using lactase, lactose can be broken down, and a whole range of lactose-free milk products can be made. Manufacturers of ice cream, yoghurt and frozen desert use lactase to improve scoop and creaminess, sweetness and digestibility and to reduce sandiness of their products.

Lactases based on *Kluyveromyces fragilis*, *Aspergillus niger* or *A. oryzae* are strongly inhibited by galactose. This problem can be overcome by hydrolyzing at low concentrations, by use of immobilized enzyme systems, or by recovering the enzyme using ultrafiltration after batch hydrolysis as described by NORMAN et al. (1979). LÜTZEN and NORMAN (1979) have compared *Bacillus* and yeast lactases. The *Bacillus* sp. enzyme was found to be superior with respect to thermostability, pH operation range, product inhibition, and sensitivity against high substrate concentration. *Bacillus saccharomyces* or *B. kluyveromyces* are known as producers of β-1,4-galactosidase. Also *Bacillus stearothermophilus* and *B. subtilis* are known to be used for lactase production.

4.5 Other Enzyme Applications in Dairies

Modification of proteins with proteases (to be discussed in Sect. 6) to reduce allergenic properties of cow milk products for infants, and lipolysis of milk with lipases for development of lipolytic flavors in speciality cheeses, are other practical applications of enzyme technology used in dairies.

4.6 Ripening of Cheese

Fresh curd obtained by milk clotting and drainage is composed of caseins, fat, carbohydrates and minerals. Cheese ripening is defined as the enzymatic modification of these substrates, so that it will progressively obtain the texture and flavor of mature cheese. Enzymes, synthesized by curd microorganisms, play a major role in these biochemical modifications. The great varieties of cheeses are developed due to different physico-chemical conditions, and due to the utilization of various microfloras. The main role is played by the lactic acid bacteria which constitute the basic flora of all cheeses, but mesophilic streptococci, propionic acid bacteria, and the microflora, which colonize the surface of cheeses, are responsible for the flavor. These microbes include coryneform bacteria, micrococci and yeast. Surface-mold-ripened soft cheeses are ripened with *Penicillium camenberti*. *P. roqueforti* develops within blue-veins of the cheeses, and provides them with their typical appearance and flavor (Blue Cheese). An extended description of the changes of the curd components, the characteristics of microbial enzymes involved, and their role during ripening has been presented by GRIPON et al. (1991).

4.6.1 Accelerated Cheese Ripening

Originally cheese manufacturing was carried out to preserve the principal nutrients of milk. During storage of cheese, chemical and biochemical changes occur, but it was most important that stability could be maintained as long as possible. However, stability of the cheese during storage is no longer as important as in the old days. Since ripening of cheese needs space and controlled temperatures, this process is rather expensive. FOX (1993) mentions that a cost of approximately $ 1.5 per ton per day may justify an acceleration, especially of low-moisture, slow-ripening varieties, at least under certain circumstances, provided that the whole process can be maintained in balance.

Research on acceleration of cheese ripening has concentrated on proteolysis in Cheddar (FOX, 1993). However, with some exceptions, notably Blue and some Italian varieties, lipolysis in cheese is quite limited, and is usually not rate-determining in the ripening of most varieties. Lipases are used in Blue and Italian cheeses to develop the "piquant" flavor, which is due primarily to short-chain fatty acids. Originally this flavor was produced as a result of the action of lipases (of oral origin) in rennet pastes, traditionally added during the preparation of these cheeses. Rennet pastes were prepared by a special technique from the stomachs of the calves, lambs or kids slaughtered after suckling. Because of possible health risks to the public, these rennet pastes are now prohibited in some countries. Several attempts have been made to develop safe processes for oral lipase, but the most obvious and safe technique seems to be the use of lipases manufactured from microorganisms. The lipase from *Mucor miehei* has been shown to give satisfactory results in Italian cheese manufacture (PEPPLER et al., 1976).

4.6.2 Method of Enzyme Addition

One of the technical problems associated with adding of ripening enzymes to cheese is the method of addition itself. Essentially, three methods are available: addition of free enzyme to the cheese milk, or to the curd, or addition of encapsulated enzymes to the cheese milk. Each method has advantages and limitations. Since the diffusion of large molecules, like proteases, in cheese is very low, direct addition of enzymes to cheese curd is only applicable to Cheddar-type cheeses, which are salted as chips during the end of manufacture. Enzymes cannot be added to surface-salted cheese (FOX, 1993).

5 Enzymes in the Brewing Industry

The traditional enzyme source used for the conversions of cereals to beer has been a variety of malted grains. The malting of barley to produce enzymes is one of the central steps in the beer brewing process. If too little enzyme activity is present during malting or mashing, the extract yield will be low; the wort separation will take too long; the fermentation process will be too slow, and too little alcohol will be produced; the beer filtration rate will be reduced; and the beer flavor and stability become inferior. A schematic presentation of the beer brewing process is shown in Fig. 7.

Industrial enzymes are used to supplement malt enzymes to prevent these problems. Furthermore, industrial enzymes are used to give a better liquefaction of adjunct; to produce low-carbohydrate beer, also called "light beer"; to shorten the beer maturation time; and to produce beer from cheaper raw materials. The various steps of the brewing operations, where microbial enzymes are occasionally added, are shown in Tab. 10, where enzymes, enzymic action and their functions are summarized.

5.1 Brewing with Adjuncts

Malt is the traditional α-amylase source for liquefaction of adjuncts. The action of α-

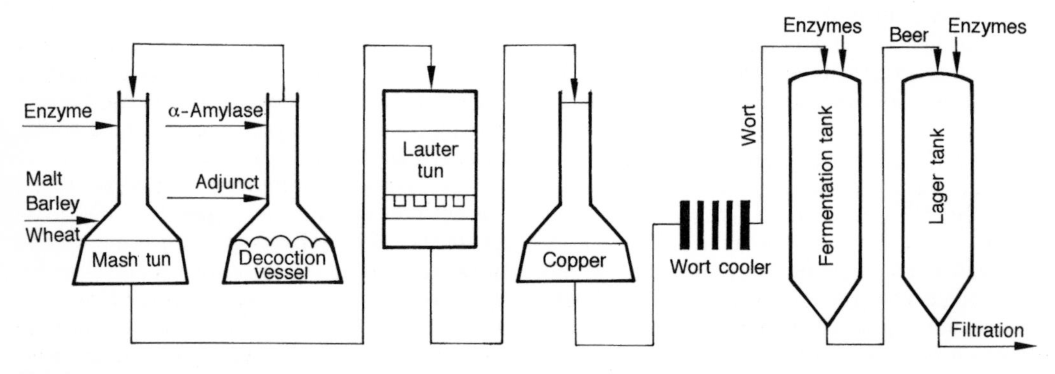

Fig. 7. Operational sequences of the beer brewing process.

Tab. 10. Steps of the Brewing Operations Where Microbial Enzymes are Occasionally Added (inspired by GODFREY, 1983a and by POWER, 1993)

Operation	Enzymes	Enzyme Action	Function
Decoction vessel (cereal cooker)	α-Amylase	Degrade starch	Adjunct liquefaction. Reduce viscosity
	β-Glucanase	Degrade glucans	Aid the filtration
Mashing	α-Amylase	Degrade starch	Malt improvement
	Protease	Increase soluble protein	Malt improvement
	β-Glucanase	Degrade glucans	Improve wort separation
	Pentosanase	Degrades pentosans of esp. wheat	Improves extraction
Fermentation	Debranching enzyme	Degrades α-1,6 branch points of starch	Secures maximum fermentability of the worts
	Fungal α-amylase	Increases maltose and glucose content	Increases % fermentable sugar in "light" beer
	Amyloglucosidase	Increases glucose content	Increases % fermentable sugar in "light" beer
	β-Glucanase	Hydrolyze glucans	Reduce viscosity and aid filtration
	α-Acetolactate decarboxylase	Converts α-acetolactate to acetoin directly	Decreases fermentation time by avoiding formation of diacetyl
Conditioning tank	Protease	Modifies protein-polyphenolic compounds	Reduces the chill haze formed in beer

amylase ensures a simpler liquefaction stage and a reduced process time. There is a strong trend to use heat-stable α-amylase preparations (e.g., Termamyl®, Sect. 2.2). This is first of all, because safety and simplicity of production are secured. Heat-stable α-amylases are much more stable than malt α-amylase. This gives a simpler liquefaction diagram and a reduced process time, and therefore an overall increased productivity.

Secondly, the malt enzymes are preserved for the saccharification process, where they may be fully needed. This safeguards the brewhouse operation, results in a better wort, and in the end a better beer.

Thirdly, eliminating the malt from the adjunct cooker gives a smaller adjunct mash and thus more freedom to balance out volumes and temperatures in the mashing program – a problem for many brewers who use a high adjunct ratio.

5.2 Brewing with Barley

Traditionally, the use of barley has been limited to 10–20% of the grist when using a good-quality malt. When going to higher levels of replacement, or when using a lower quality malt, processing becomes more difficult. In these cases the mash will need to be supplemented with extra enzyme activity in order to benefit from the advantages of using barley, and still maintain performance.

Either a malt-equivalent blend of α-amylase, β-glucanase, and proteinase is added at mashing-in, or the enzymes are added individually according to need.

Tab. 11. Some Effects of Addition of Enzymes on the Yield of Fermentable Sugar (POWER, 1993)

Operation	Enzymes Used	% Yield of Fermentable Sugar Based on Raw Material	% Yield of Fermentable Sugar Based on Wort Carbohydrates
Mashing	Malt only	65	
	Malt + α-amylases	75	
Fermentation	None		65–70
	Debranching enzyme		75
	Fungal α-amylase		85
	Amyloglucosidase + debranching enzyme		95

5.3 Filtration Problems – Overall

Wort separation and beer filtration are two common bottlenecks in the brewing process. Poor lautering not only causes a loss in production capacity, but can also lead to losses in extract yield. Furthermore, a slow lautering negatively affects the quality of the wort, which may give beer filtration problems, and problems with the flavor and stability of the beer.

A thorough breakdown of β-glucans and pentosans during mashing is essential to fast wort separation. Undegraded β-glucans and pentosans, carried over into the fermenter, reduce the beer filter capacity, and increase the consumption of diatomaceous earth.

A wide selection of β-glucanase/pentosanase preparations to be used during mashing or fermentation/maturation are available to solve these problems.

5.4 Enzyme Aid During Fermentation

Small adjustments in fermentability can be obtained by adding a debranching enzyme (Sect. 2.4) or a fungal α-amylase at mashing-in, or by adding a fungal α-amylase at the start of fermentation.

Beer types with a very high attenuation, "light beers", can be made with saccharifying enzymes. Fungal α-amylases produce mainly maltose and dextrins, and amyloglucosidase produces glucose, from both linear and branched dextrins.

The amount of alcohol in a beer is limited by the amount of solids, or extract, transferred from the raw materials to the wort, and by the percentage of the extract, which is fermentable sugar. The sugar content is controlled by the extent of starch degradation due to the effect of the amylases in the mash, and the saccharifying enzymes used during the fermentation. POWER (1993) mentions the approximate percentage yields of fermentable sugars shown in Tab. 11.

5.5 Nitrogen Control

If the yeast does not get enough free amino nitrogen, the fermentation will be poor, and the beer quality inferior. A neutral bacterial proteinase added at mashing-in can be used to raise the level of free amino nitrogen. This is useful, when working with poorly modified malt or with high adjunct ratios.

5.6 Diacetyl Control

Diacetyl is formed by a non-enzymatic oxidative decarboxylation of α-acetolactate, which is produced by the yeast during primary fermentation. The diacetyl is removed again by the yeast during the maturation stage by conversion to acetoin, which has a much higher flavor threshold value.

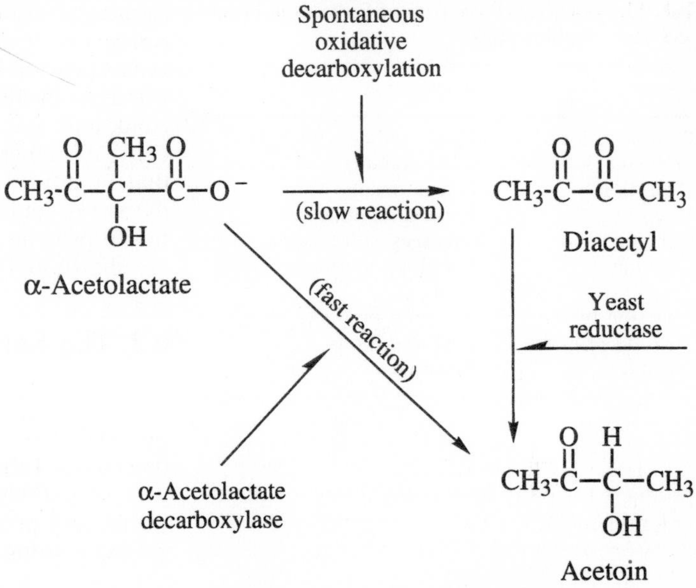

Spontaneous oxidative decarboxylation

$CH_3-C(=O)-C(CH_3)(OH)-C(=O)-O^-$

α-Acetolactate

(slow reaction)

(fast reaction)

$CH_3-C(=O)-C(=O)-CH_3$

Diacetyl

Yeast reductase

α-Acetolactate decarboxylase

$CH_3-C(=O)-CH(OH)-CH_3$

Acetoin

Fig. 8. Removal of *α*-acetolactate during fermentation.

By adding the enzyme *α*-acetolactate decarboxylase (e.g., Maturex™) at the beginning of the primary fermentation, it is possible to by-pass the diacetyl stage (see Fig. 8), and to go directly to acetoin already during the primary fermentation. Thereby it is possible to shorten or completely eliminate the maturation period (AUNSTRUP and OLSEN, 1986; ROSTGAARD JENSEN et al., 1987).

5.7 Chillproofing Enzymes

In spite of the final cold filtration of the beer before packaging cloudiness or haze may form during chilling. The amount of haze varies with the type of beer, and it increases with age, exposure to oxygen, and agitation during shipping. The haze is a result of the formation of complexes of peptides with polyphenolic procyanidins (HOUGH et al., 1982).

In a series of patents, LEO WALLERSTEIN (1911) proposed the use of proteases to prevent the formation of chill haze in beer. In fact he patented the addition to beer of "a proportion of proteolytic enzymes, active in slightly acid media, sufficient to modify the proteins contained in the beer in such a manner, that they will not be precipitated upon chilling subsequent to pasteurization, the beer being rendered chillproof in the sense, that it is capable of remaining brilliant, even when kept on ice for a considerable time". The original patents described the application of bromelain, papain and pepsin for chillproofing. Papain is far superior to other enzymes used for chillproofing (POWER, 1993).

6 Enzymes and Food Protein

The enzymic modification of proteins is often an attractive means of obtaining better functional and nutritional properties of food proteins. The conversion of milk to cheese is an effect of the action of the protease in a limited, controlled, and specific way. Enzymes are also applied to food proteins for manufacturing new and valuable products of vegetable origin or from animal proteins that appear in by-products, for example from slaughterhouses. Enzymatic hydrolysis of milk proteins (whey and caseins) is used to produce non- and low allergenic cow milk

Tab. 12. Functional Properties of Proteins in Foods and Their Applications

Property	Application
Emulsification	Meats, coffee whiteners, salad dressing
Hydration	Dough, meats
Viscosity	Beverages, doughs
Gelation	Sausages, gel desserts
Foaming	Toppings, meringues, angel food cakes
Cohesion binding	Textured products, dough
Textural properties	Textured foods
Solubility	Beverages

products for baby food (substitutes for mother's milk).

6.1 Protein Modification

The protein structure is modified to improve solubility, emulsification, and foaming properties. Chemical modification is not very desirable for food applications because of the harsh reaction conditions, non-specific chemical reagents, and the difficulties of removing residual reagents from the final product. Enzymes provide, on the other hand, several advantages including fast reaction rates mild conditions and, most importantly high specificity. Also there is a need for gentle methods that modify the food itself in order to limit the use of additives. Some important properties of proteins and their application in foods are shown in Tab. 12.

6.2 The Substrates

Over the years many different protein raw materials have been used and with different objectives. Tabs. 13a and b show some examples of extraction processes with enhanced yields, and processes for producing food ingredients using proteases.

6.3 Proteases in Use

Proteases are classified according to their source of origin (animal, plant, microbial), their catalytic action (endo-peptidase or exopeptidase) and the nature of the catalytic site.

Tab. 13a. Examples of Extraction Processes and Processes for Producing Food Ingredients Using Protease

Extration Processes
Soya Milk
Scrap meat recovery
Bone cleaning
Gelatin
Fish/meat stickwater
Rendering of fat
Deskinning of fish roe
Meat extracts
Yeast extracts

Tab. 13b. Examples of Extraction Processes and Processes for Producing Food Ingredients Using Protease

Functional Protein Ingredients
Cheese
Isoelectric soluble soy protein
Egg white substitute of soy protein
Emulsifier of soy protein
Soluble wheat gluten (E-HVP)[a]
Foaming wheat gluten
Blood cell hydrolysate
Whey protein hydrolysates
Casein hydrolysates
Soluble meat proteins (E-HAP)[b]
Gelatin hydrolysates
Meat tenderizing

[a] enzyme-hydrolyzed vegetable protein
[b] enzyme-hydrolyzed animal protein

Tab. 14. Some Commercial Proteolytic Enzymes

Product Name	Microorganism or Other Origin	State of Product	Activity Anson Units/g[a]	Practical Application Range for pH	Practical Application Range °C
Alcalase®	*Bacillus licheniformis*	Liquid	2.4	6–10	10–80
Esperase®	*Bacillus lentus*	Liquid	2.4	7–12	10–80
Neutrase®	*Bacillus amyloliquefaciens*	Liquid or granulate	0.5	6–8	10–65
Protamex™	*Bacillus* sp.	Microgranulate	1.5	6–8	10–65
Flavourzyme™	*Aspergillus oryzae*	Granulate	[b]	4–8	10–55
Rennilase®	*Rhizomucor miehei*	Liquid or granulate	[c]	3–6	10–50
Trypsin	Pancreas	Granulate	3.3	7–9	10–55

[a] One Anson Unit (AU) is the amount of enzyme, which under standard conditions digests hemoglobin at an initial rate, liberating per minute an amount of TCA soluble products, which give the same color with phenol reagent as one milliequivalent of tyrosine (ANSON, 1939)
[b] Standarized in leucine amino peptidase units per gram (LAPU/g)
[c] Standarized in milk clotting activities

In his latest review on proteases ADLER-NISSEN (1993) compiled the most important proteases for food protein hydrolysis and systematically classified them into types based on nature of the active site, source of origin, common names and trade names, typical pH range and preferential specificity. Based on a comparison of active sites, catalytic residues, and three-dimensional structures, four major protease families are recognized today: the serine, the thiol, the aspartic and the metalloproteases. The serine protease family contains two subgroups: the chymotrypsin-like and the subtilisin-like proteases. Many industrially important proteases are mixtures of the different types of proteases. This is especially the case for pancreatin, papain (crude), some proteases from *Bacillus amyloliquefaciens*, *Aspergillus oryzae*, *Streptomyces* and *Penicillium duponti*.

Some commercial enzyme products used for industrial conversion of food protein products have the characteristics shown in Tab. 14. In Tab. 14 the enzymes Alcalase, Esperase and trypsin are serine proteases, while the enzyme called Neutrase is a metalloprotease having Zn^{2+} in its active site. Furthermore, this enzyme is stabilized by Ca^{2+}. The milk-clotting enzyme product called Rennilase is an aspartic protease. Examples of thiol (or cysteine proteases) are plant proteases from papain latex (papain), from pineapple stem (bromelain) and from fig latex (ficin).

6.3.1 Protease Inhibitors

Some proteins are natural inhibitors of proteases. Trypsin and chymotrypsin inhibitors of legume and cereal grains have been reviewed and studied by BIRK (1976). Protease inhibitors for bacterial proteases, such as subtilisin, have been purified from legume seeds and characterized by CHAVAN and HEJGAARD (1981).

Tab. 15. Kjeldahl Conversion Factors and Content of Peptide Bonds for Various Food Proteins (NOVO NORDISK, 1989)

Protein	Kjeldahl Conversion Factor, f_N	h_{tot} eqv/kg ($N \times f_N$)
Casein	6.38	8.2
Whey protein isolate	6.38	8.8
Meat	6.25	7.6
Fish muscle	6.25	8.6
Egg white	6.25	approx. 8
Soy meal, concentrate and isolate	6.25	7.8
Cotton seed	6.25	7.8
Red blood cells	6.25	8.3
Wheat protein	5.7	8.3
Gelatin	5.55	11.1

The probably best known and most studied protease inhibitors are the trypsin inhibitors of soybeans. The nutritional significance of the antitryptic activity of soybeans and soybean foods (and feed materials) has been known for a long time. The way to inactivate these inhibitors has been an inherent part of most manufacturing processes for soybeans (LIENER, 1978).

6.4 Control of the Hydrolysis

The technical discipline of enzymic hydrolysis of food proteins has been developed over the last 20 years to become a significant area of modern food processing. The number of literature references has grown enormously. Fundamental principles, the methods of controlling the hydrolysis reaction, ways of carrying out industrial processing and characterization of functional properties of proteins, treated with proteases, has been intensively researched and developed at Novo Nordisk. A great deal of the work has been described by ADLER-NISSEN (1986), and examples and highlights of the continuing work will be described in the following.

The most important parameters of the hydrolysis reaction are S (% protein in the reaction mixture), E/S (enzyme–substrate ratio in activity units per kg protein), pH and the temperature. Together with the specificity and properties of the enzyme itself, these parameters are responsible for the course of reaction for a given protein raw material. The quantitative criterion of a proteolytic reaction is the degree of hydrolysis, calculated from a determination of the number of peptide bonds cleaved and the total number of peptide bonds in the intact protein.

As a quantitative measure for the hydrolytical reaction, the hydrolysis equivalent, h, is used. Its unit is equivalent peptide bonds cleaved per kg protein. h is used for kinetic investigations of the protein hydrolysis, but in other connections the derivative quantity DH (degree of hydrolysis) is preferred. DH is calculated from the following equation:

$$DH = \frac{h}{h_{tot}} \cdot 100\% \quad (1)$$

h_{tot} is estimated on the basis of the amino acid composition of the protein. For most food proteins, the average molecular weight of amino acids is about 125 g/mol, making h_{tot} about 8 g equivalents per kg protein (calculated as $6.25 \times N$). Tab. 15 gives some more exact figures for Kjeldahl conversion factors f_N and for h_{tot} for some common proteins.

The degree of hydrolysis of enzyme-treated proteins determines the properties of relevance to food applications. It is, therefore, of utmost importance that the degree of hydrolysis can be measured, while the reaction is going on. Only in this way is it possible to stop the reaction at a well defined stage, when the desired property of the product has been obtained.

Relatively simple analytical tools that can to some extent be applied directly during the reactions are often used. Some of them are the pH-stat technique, pH drops, osmometry, viscosimetry, and chemical determination of free amino groups using, for example, a measurement of liberated α-amino groups by reaction with trinitrobenzenesulfonic acid (TNBS), whereby a colored derivative of the peptide is formed. The TNBS method has been adapted to be usable for hydrolysates of food proteins, which usually are partly soluble, that is, the sample should be dispersed in SDS (sodium dodecyl sulfate) (ADLER-NISSEN, 1979). The pH-stat technique and the osmometry technique are quicker and less labor-intensive than the TNBS method.

6.4.1 Calculation of Degree of Hydrolysis of Proteins Using the pH-Stat Technique

Originally the pH-stat method was developed by JACOBSEN et al.(1957) at the Carlsberg Laboratory. It is based on the principle that pH is kept constant during hydrolysis by means of automatic titration with a base, when the hydrolyses are carried out under neutral to alkaline conditions. When the hydrolysis is carried out under acid conditions, as, for example, by use of pepsin at pH 3, the titration must be made with acid. The high buffer capacity of proteins means that at extreme pH values (i.e., pH > 11 or pH < 3) the pH-stat is inoperable (JACOBSEN et al., 1957).

The free carboxyl and free amino groups found after hydrolysis will be more or less ionized, depending on the pH of the hydrolysis reaction. Working in the region pH 6–9.5, the amino groups will be partially protonated. This means that the hydrolysis of protein in this pH region, is accompanied by a release of H^+ as already noted by SÖRENSEN (1908). Consequently at pH values above 7.5–7.8 (pK values at 25 °C), the amino group will be less than half protonated, but the carboxyl groups will be fully dissociated. This leads to a net release of 0.5–1 mol H^+ for each mole of peptide bonds cleaved. This is the principle behind the pH-stat technique for continuously following of the degree of hydrolysis (DH) during the reaction. DH is calculated on the basis of the titration equations as follows:

$$DH = \frac{h}{h_{tot}} \cdot 100\% \quad (1)$$

$$DH = B \cdot N_b \cdot \frac{1}{\alpha} \cdot \frac{1}{MP} \cdot \frac{1}{h_{tot}} \cdot 100\% \quad (2)$$

where B is the base consumption (in mL or L),
N_b the normality of base,
α the average of dissociation of the α-NH_2 groups (see below),
MP the mass of protein ($N \times f_N$) (in g or kg),
h the hydrolysis equivalents (in meqv/g protein or eqv/kg protein),
h_{tot} the total number of peptide bonds in the protein substrate (in meqv/g protein or eqv/kg). The degree of dissociation is

Tab. 16. Degree of Dissociation (α-values)

	T (°C)	40	50	60	70	75	80
	pK	7.3	7.1	6.9	6.7	6.6	6.5
pH							
6.5			0.20	0.29	0.39	0.44	0.50
7.0		0.33	0.44	0.55	0.67	0.71	0.76
7.5		0.61	0.71	0.80	0.86	0.89	0.91
8.0		0.83	0.89	0.93	0.95	0.96	0.97
8.5		0.94	0.96	0.97	0.98	0.99	0.99

$$\alpha = \frac{10^{\mathrm{pH\text{-}p}K}}{1 + 10^{\mathrm{pH\text{-}p}K}} \tag{3}$$

pK is the average pK value of the α-amino groups liberated during the hydrolysis, and it can be determined from a direct assay of these amino groups by use of the TNBS methods. pK also varies significantly with temperature because the ionization enthalpy of the amino group is considerable (ADLER-NISSEN, 1986). Inserting the ionization enthalpy (+45 kJ/mol) in the Gibbs–Helmholtz equation, it was found, that there is a pK change of about 0.23 pH units for a change of 10 °C in the hydrolysis temperature (ADLER-NISSEN, 1982). Tab. 16 gives results of calculations of the degree of dissociations α from Eq. (3) for various pH values as a function of the temperature using the above described method.

6.4.2 Calculation of Degree of Hydrolysis of Proteins Using the Osmometry Technique

This method can be used for calculation of the degree of hydrolysis of proteins, when no soluble components are added during the reaction, i.e., in ranges of the pH scale where the pH-stat principle is not functioning. A modern freezing-point osmometer can record the osmolality of a sample in a few minutes on a sample taken directly from the hydrolysis mixture. By drawing of samples as a function of time during a hydrolysis reaction, a good reaction curve can be constructed. The first description of the use of osmometers for determination of degree of hydrolysis of protease-treated substrates was published by ADLER-NISSEN (1984). With the osmometer ΔT is measured and converted to osmol through the simple proportionality: $\Delta T = K_f \cdot$ osmolality.

From the increase in osmolality, ΔC, DH is calculated by use of the following equation:

$$DH = \frac{\Delta C}{S\% \cdot f_{\mathrm{osm}}} \cdot \frac{1}{\omega} \cdot \frac{1}{h_{\mathrm{tot}}} \cdot 100\% \tag{4}$$

where ΔC is the increase in osmolality (in osmol/kg H_2O),
$S\%$ the protein substrate concentration ($N \times 6.25$) (in w/w %),
ω the osmotic coefficient ($\omega = 0.96$ for most actual concentrations of protein, ADLER-NISSEN, 1986),
h_{tot} the total number of peptide bonds in the protein substrate (in meqv/g protein or eqv/kg),
f_{osm} the factor to convert % to g per kg H_2O. This can be calculated using the following equation:

$$f_{\mathrm{osm}} = \frac{1000}{100 - D\%} \tag{5}$$

where $D\%$ is the % dry matter present in the reaction mixture.

6.4.3 Some Kinetic Aspects of Protein Hydrolysis

Proteases catalyze the hydrolytic degradation of the peptide chain, as shown for reactants and products in Fig. 9. In aqueous solutions and suspensions of protein the equilibrium lies so far to the right that degradation and not synthesis of larger molecules is thermodynamically favored.

When a protease acts on a protein substrate (see Fig. 10), the catalytic reaction consists of three consecutive reactions:

1. Formation of the Michaelis complex between the original peptide chain (the substrate) and the enzyme (referred to as ES).

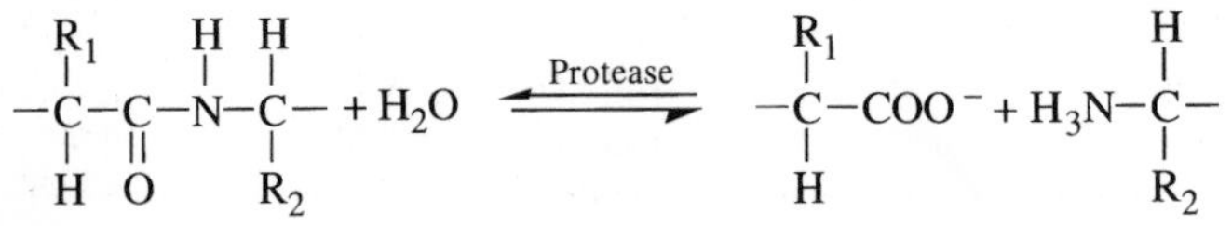

$$-\overset{\overset{R_1}{|}}{\underset{\underset{H}{|}}{C}}-\underset{\underset{O}{\|}}{\overset{\overset{H}{|}}{C}}-\overset{\overset{H}{|}}{N}-\underset{\underset{R_2}{|}}{\overset{\overset{H}{|}}{C}}- + H_2O \xrightarrow{\text{Protease}} -\overset{\overset{R_1}{|}}{\underset{\underset{H}{|}}{C}}-COO^- + H_3N-\underset{\underset{R_2}{|}}{\overset{\overset{H}{|}}{C}}-$$

Fig. 9. The hydrolysis reaction: enzymatic cleavage of a peptide bond.

$$E + S \underset{k_{-1}}{\overset{k_{+1}}{\rightleftharpoons}} ES \xrightarrow{k_{+2}} EP + H\text{–}P' \xrightarrow[+H_2O]{k_{+3}} E + P\text{–}OH + H\text{–}P'$$

Fig. 10. The catalytic mechanism of a protease. E, enzyme; S, substrate; P, P′, resulting peptide; k, reaction velocity constants.

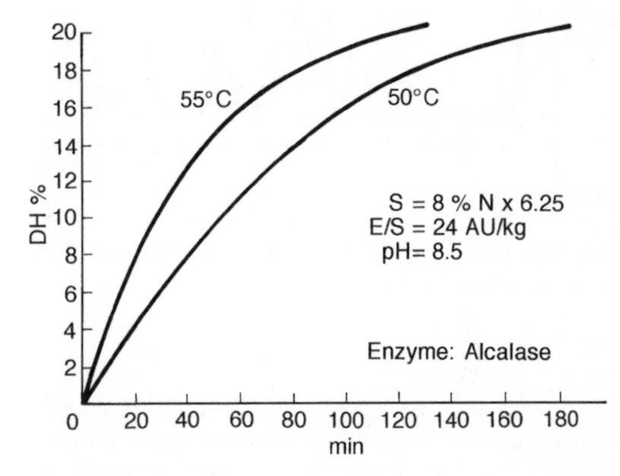

Fig. 11. Hydrolysis curves for a hemolyzed red blood cell fraction.

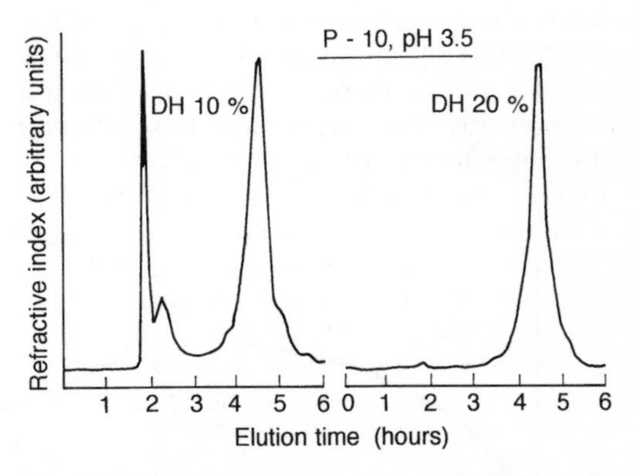

Fig. 12. Gel chromatogram of hemoglobin hydrolysate (ADLER-NISSEN, 1986).

2. Cleavage of the peptide bond to titrate one of the two resulting peptides.
3. A nucleophilic attack on the remains of the complex to split off the other peptide and to re-constitute the free enzyme.

This reaction mechanism is simplified by the scheme shown in Fig. 10. The rate-determining step is the acylation step characterized by the reaction velocity constant k_{+2}. The ratio k_{-1}/k_{+1} is equal to the Michaelis constant K_w for serine proteases. This is to say in other words, that classical Michaelis–Menten kinetics are valid for protein hydrolysis.

The pH-stat technique has been used at Novo Nordisk to study the proteolytic degradation of hemoglobin. Highlights from this work will be presented here with regard to the practical utilization of this technique and understanding of the reaction mechanism.

In the discussions of initial proteolysis, LINDERSTRØM-LANG (1952) mentioned the reaction of a protease on the native hemoglobin molecule "one-by-one", indicating that a particular protease molecule degraded one substrate molecule at a time. No appreciable amounts of intermediary products will be present. The reaction mixture will consist of native proteins and end products only.

The opposite case mentioned was, where the native protein molecules were rapidly converted into intermediary forms, which would be degraded to end products more slowly ("Zipper reaction").

As seen on the hydrolysis curves in Fig. 11, made by using Alcalase 2.4 L (see Tab. 14) at 50 and 55 °C in the pH-stat, the slight bending at low DH values of the blood cell fraction hydrolysis curves (i.e., hemoglobin) indicates a thorough initial degradation to small peptides (a zero-order reaction). This degradation is close to the ideal one-by-one reaction. The hydrolysis proceeds rapidly with little decrease in velocity until the stage, where the substrate concentration is so low that first-order kinetics takes over.

A gel chromatogram of the soluble part of the hemoglobin hydrolysate at DH 10% and DH 20% (Fig. 12), fully confirms the degradation pattern of the hemoglobin substrate. At DH 10% the hydrolysate consists of high-molecular-weight material and small peptides in accordance with the one-by-one mecha-

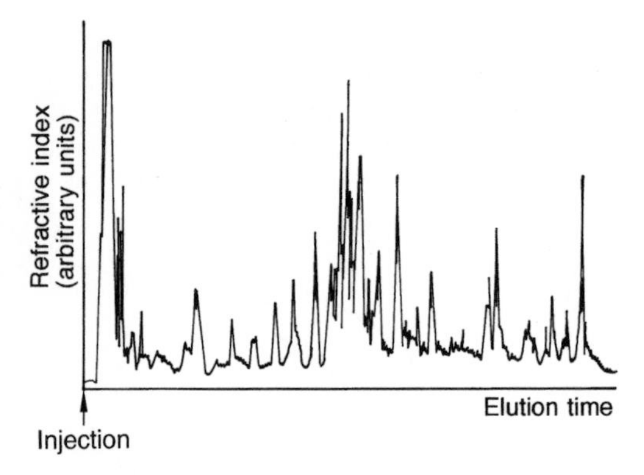

Fig. 13. HPLC chromatogram of BCH, *DH* 20% (ADLER-NISSEN, 1986).

nism. At *DH* 20% the high-molecular-weight material has completely disappeared, as predicted by LINDERSTRØM-LANG already in 1952 (ADLER-NISSEN, 1986).

This is an important factor behind the successful application of ultrafiltration in the industrial process for separation of the hydrolysate from the sludge (OLSEN, 1983). If the hydrolysis was not of the one-by-one type, large peptides would also have been present and consequently the yields and fluxes would have been much lower.

One should be careful not to draw too simple a conclusion about kinetic models on enzymatic hydrolysis of even a fairly simple and well-defined substrate such as hemoglobin. The HPLC spectrum of BCH is shown in Fig. 13. At least fifty different peptide compounds are present under the single peak in the gel chromatogram of *DH* 20%.

6.5 The Effect of Proteolysis

When the protein structure is modified by proteolysis, the various effects are considered either positive or negative, dependent on wether the effect is wanted or unwanted. The developments within enzymic hydrolysis has of course been within both the criteria. Wanted effects are studied to be utilized and found of importance like increased solubility, emulsification, and foaming properties. Unwanted effects like increased bitterness and other unwanted taste developments have been the key issues concerning problems to solve.

As a consequence of the enzymic degradation of proteins, the key indices, psi (protein solubility index), PCL (peptide chain length), and TCA index (protein solubility in 0.8 M TCA), are changed as shown in Fig. 14.

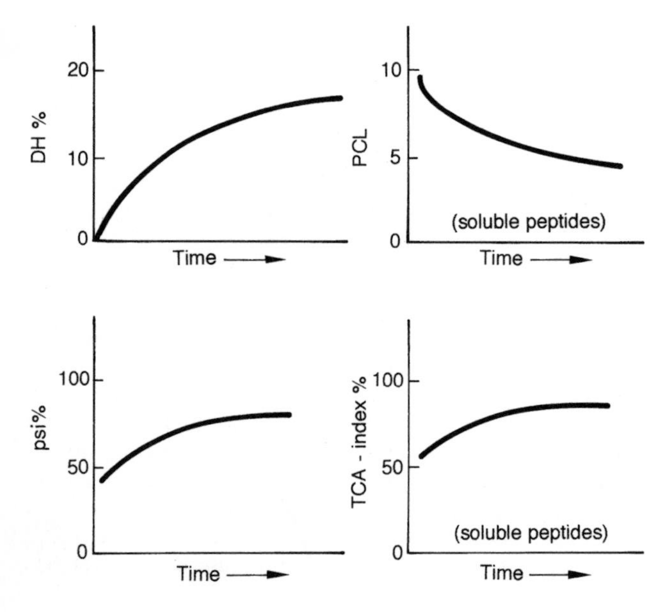

Fig. 14. Time dependence of key indices during enzymatic hydrolysis of protein (ADLER-NISSEN, 1986).

6.5.1 The Bitterness Problem

Unpleasant bitterness was early observed for many protein hydrolysates. Today this problem can be managed by proper selection of the reaction parameters and the enzymes used. Bitterness is a complex problem, which can be influenced by at least the following variables (ADLER-NISSEN, 1986).

1. Hydrophobicity of the substrate, since hydrolysates become bitter, if amino acid side chains containing hydrophobic groups become exposed due to hydrolysis of the protein (NEY, 1971).
2. High *DH*, and therefore the concentration of soluble hydrophobic peptides, and their chain length.
3. The specificity of the enzyme used plays an enormous role in the bitterness of protein hydrolysates.
4. Separation steps applied in the downstream processing; for example, centrifuging at isoelectric pH, and carbon treatment.
5. Masking effects from other components in the protein hydrolysate. Organic acids such as malic acid and citric acid exert a significant masking effect in soy protein hydrolysates (ADLER-NISSEN and OLSEN, 1982). This masking effect can be utilized in the final formulation of the food, where the hydrolysate is incorporated.

In addition to the above, recent recognition of the following effects points out other solutions to solve bitterness problems for production of protein hydrolysates with high *DH* values:

A debittering of protein hydrolysates can be performed by a selective extraction or removing of hydrophobic peptides into a 2-butanol phase (ADLER-NISSEN, 1986, 1988). This knowledge can be used to explain, why it "surprisingly" was possible to produce a *DH* 20% soy protein hydrolysate using Alcalase (Tab. 14), from full-fat soybean meal, which was fully acceptable from an organoleptic point of view (OLSEN, 1982). The oil was simultaneously recovered by centrifugation.

Reaction parameters such as pH and temperature which change the specificity of the proteases also play a very important role for managing of bitterness. Also combinations of enzymes can change the overall mode of degradation of a protein, so that bitterness is reduced. The application of exo-peptidases is a generally recognized way of removing bitterness of high-*DH* hydrolysates. The following case study illustrates these effects.

General hypotheses

A. Terminal hydrophobic amino acids result in less bitterness than non-terminal.
B. Bitterness is highest when the hydrophobic amino acids are non-terminal.

The specificity of, e.g., Alcalase and Neutrase is shown in Fig. 15 for the bonds attacked in the oxidized B chain of insulin (JOHANSEN et al., 1968).

As it can be seen from the specificity of Alcalase and Neutrase, on hydrolysis of the oxidized B-chain of insulin, more terminal hydrophobic amino acid residues may be formed by Neutrase under conditions, where all bonds possible to cleave are cleaved. This will result in less bitterness compared to a hydrolysate using Alcalase at the same degree of hydrolysis, where the hydrophobic groups are in the chain.

Therefore, it is obvious that mixing the two enzymes, or using the one enzyme before the other one, the bitterness will be reduced compared to the same degree of hydrolysis.

Conclusion of the case study

Combination of proteases is a method to reduce bitterness of protein hydrolysates.

Use of single proteases outside their optimum reaction parameters (pH, temperature) may reduce bitterness, due to the reduced number of hydrophobic side chains released.

6.5.1.1 Other Off-Flavor Problems in Relation to Bitterness

Animal raw materials considered useful for protein hydrolysis usually do not contain many *in situ* bitter compounds of significance. Bitterness can appear from rancid traces of fat or from extended degradation of proteins.

Alcalase:

NH_3–Phe–Val–Asn–Gln–|His–Leu–$CySO_3$–Gly–Ser–|His–Leu–|
1 5 10

Val–Glu–Ala–Leu–|Tyr–Leu–Val–$CySO_3$–Gly–Glu–Arg–Gly–
12 15 20

Phe–Phe–Tyr–|Thr–Pro–Lys–Ala
25 30

Hydrophobic groups: **Leu–Val**(11–12), **Leu–Tyr**(15–16), **Tyr**–Thr(26–27). 5 terminal hydrophobic amino acids.

Hydrophilic groups: **Gln–His**(4–5), **Tyr**-Thr(26–27). 3 terminal hydrophobic amino acids.

Neutrase:

NH_2–Phe–Val–Asn–Gln–His–|Leu–$CySO_3$–Gly–Ser–His–|Leu–
1 5 10

Val–Glu–Ala–|Leu–Tyr–|Leu–Val–$CySO_3$–Gly–Glu–Arg–Gly–|
12 15 20

Phe–|Phe–Tyr–Thr–Pro–Lys–Ala
25 30

Hydrophobic groups: His–**Leu**(5–6), His–**Leu**(10–11), Ala–**Leu**(14–15), **Tyr–Leu**(16–17), Gly–**Phe**(23–24), **Phe–Phe**(24–25). 8 terminal hydrophobic amino acids.

Hydrophilic groups: His–**Leu**(5–6), His–**Leu**(10–11), Ala–**Leu**(14–15), Gly–**Phe**(23–24). 4 terminal hydrophobic amino acids.

Fig. 15. Specific bonds of the oxidized B-chain of insulin hydrolyzed by two proteases.

Oilseeds contain a variety of phenolic compounds, in particular bitter-tasting phenolic acids (SOSULSKI, 1979). Some volatile compounds are hydrophobic and bind to the protein phase, in particular to denatured protein. Off-flavor may be released by enzymatic hydrolysis of such proteins. Some can be removed but usually not all of them. A solution to such problems has been described for industrial production of a soluble enzymatic hydrolysate of soya protein (OLSEN and ADLER-NISSEN, 1979). An isoelectric soluble, enzymatic hydrolysate of soya protein without bitterness and a bland taste was produced economically from soya white flakes. The white flakes were washed at pH 4.5 with water in a four-step extraction process to produce a concentrate with a low level of beany off-flavor. The concentrate was immediately hydrolyzed with Alcalase (Tab. 14) to a specific degree of hydrolysis. The hydrolysis reaction was terminated by lowering the pH, and the supernatant was recovered by centrifugation at the isoelectric point. The hydrolysate was carbon-treated, concentrated and used for protein fortified fruit juices and for extension of cured, whole meat products.

6.5.2 Functional Protein Hydrolysates

A limited hydrolysis of a protein product may improve the whipping and emulsifying capacities, as already investigated by PUSKI (1975) on soy protein isolates. They were hydrolyzed with *Aspergillus oryzae* protease at varying ratios of enzyme to substrate. The degree of hydrolysis (*DH*) was not directly measured, but from the data given in the publication the maximum *DH* value obtained was 5%.

ADLER-NISSEN and OLSEN (1979) used as the first the pH-stat technique for a controlled production of enzymatically hydrolyzed soy proteins. The enzymes used were Alcalase at pH 8 and Neutrase at pH 7. The isoelectric solubility, the foaming capacity and the emulsifying capacity were all increased in agreement with other authors. If the reactions were continued to higher *DH* values than 5%, the foaming capacity and the emulsifying capacity decreased again. Also a significant difference was observed in the measured functional capacities dependent on the two different proteases applied. Alcalase was found to have a much better performance. The reason for this effect could be due to the fact that foams and oil emulsions are stabilized better by peptides having a higher content of non-terminal hydrophobic amino acids, than by peptides having terminal hydrophobic amino acids. Fig. 16 shows the whipping expansion versus *DH* from the above mentioned early demonstration.

In the further research on functional ingredients based on enzymatic modification, improved products were developed with other soy raw materials or when physical treatments such as ultrafiltration were used during purification of the active components responsible for the specific effect. ADLER-NISSEN et

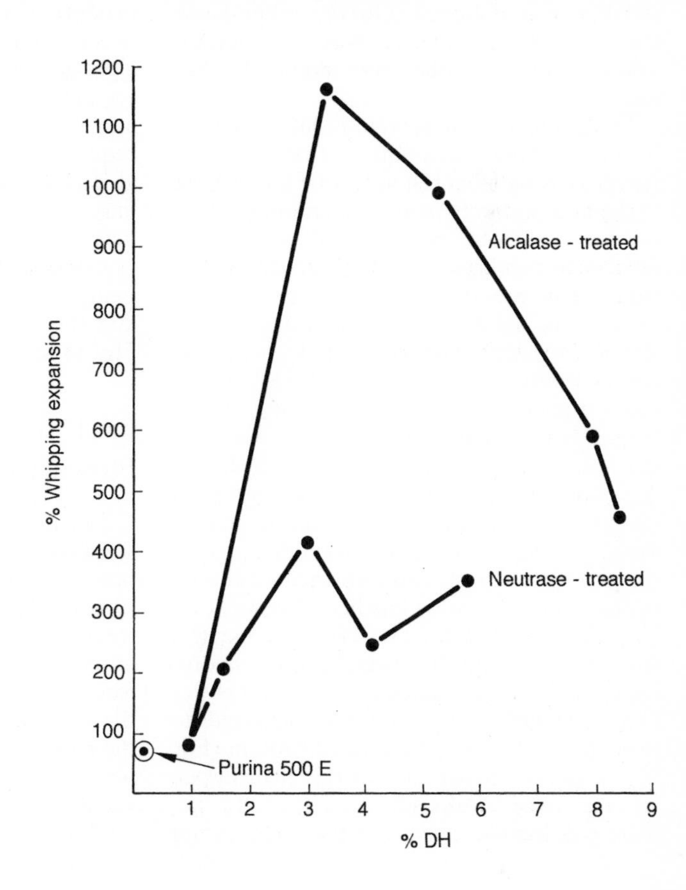

Fig. 16. Whipping expansion as a function of the degree of hydrolysis (*DH*) for soy protein hydrolysates (ADLER-NISSEN and OLSEN, 1979).

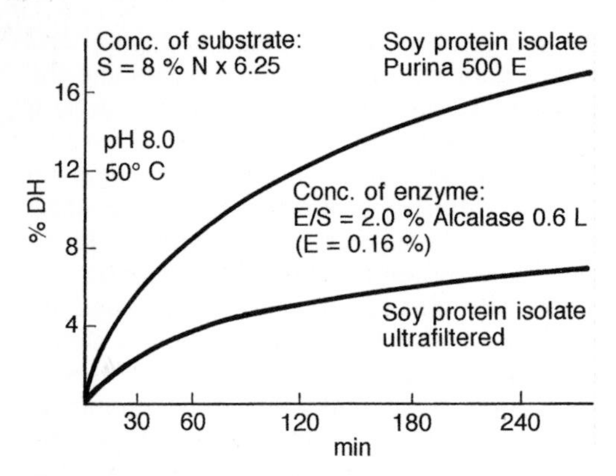

Fig. 17. Hydrolysis curves for soy protein isolates (OLSEN and ADLER-NISSEN, 1981).

al. (1983) found that when soya concentrate was hydrolyzed to *DH* 3 to 5% with Alcalase, a functional protein hydrolysate with high foam expansion, and remarkably good foam stability, was obtained. On the other hand, the emulsifying capacity was considerably lower than if soy isolate was modified in the same way.

Using a native soy protein isolate produced by ultrafiltration of an aqueous extract of defatted soybean meal (OLSEN, 1978), the shape of the hydrolysis curve was remarkably different from that obtained for acid precipitated protein, as can be seen in Fig. 17. The hydrolysis curves are shown for acid precipitated soy protein isolate (denatured and nearly insoluble in water) and ultrafiltered soy protein isolate (native and soluble in water). It appears from Fig. 17 that the ultrafiltered protein isolate is hydrolyzed more slowly than the acid precipitated protein. This is due to the compact molecular structure of the ultrafiltered protein which is still in its native stage. The degree of denaturation of a protein substrate has a profound influence on the kinetics of proteolysis as has been known for a long time (CHRISTENSEN, 1952). Using this soy protein isolate, functional protein hydrolysates with better foam expansion and better foam stability were obtained compared to those from the acid precipitated protein (Fig. 16). If the small peptides in this hydrolysate were removed by ultrafiltration, the foam stability was improved even further. The use of ultrafiltration before and/or after the enzymatic hydrolysis was investigated on a pilot plant scale in a number of process combinations (OLSEN and ADLER-NISSEN, 1981; OLSEN, 1984). The high-molecular-weight hydrolysates had excellent whipping properties and, furthermore, they were also able to heat-coagulate, so that they could substitute egg white 100% on a protein basis in a meringue batter.

6.6 Processing of Enzymatic Modifications

The following are examples of the use of enzymes for extraction processes and for modification processes with regard to production of new ingredients. Enzymatic processing methods have become an important and indispensable part to be evaluated for new processes used by modern food industry to produce a large and diversified range of products based on protein raw materials. The principal advantage offered by enzymatic processes are the enzymes' specific mode of attack and the mild temperature and pH conditions required. Gentle treatment of the raw material is achieved, and only few by-products that may affect taste and smell of the final products are produced.

Considering a typical enzyme application, many effects influence the enzyme reaction and the final product. The scheme shown in Fig. 18 may be used for this purpose.

6.6.1 Inactivation of Enzymes and Downstream Processing

The managing of degree of hydrolysis of a substrate may be made by the selection of the reaction parameters.

The degree of hydrolysis of enzyme-treated substrates determines the properties of relevance to food applications. It is, therefore, of utmost importance that the effects of the enzyme reaction can be measured while the reaction is going on. Only in this way is it possible to stop the reaction at a well defined stage, when the desired property of the product has been obtained.

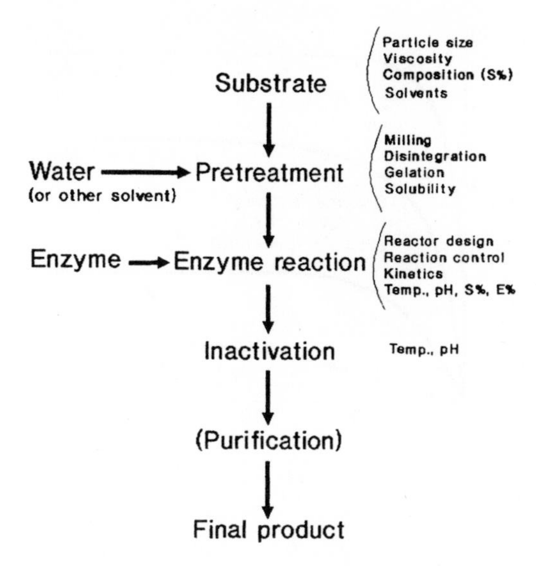

Fig. 18. Practical aspects of enzyme applications.

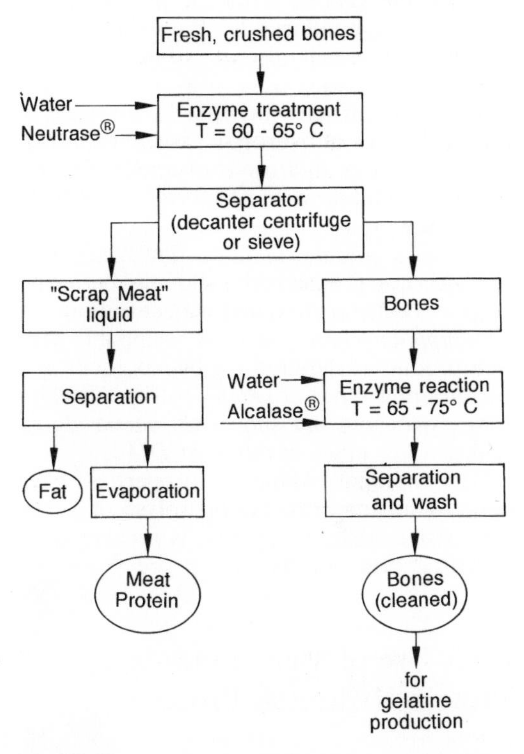

Fig. 19. Enzymatic scrap meat recovery and bone cleaning.

In most enzymatically treated food products active enzyme is not wanted due to changes of the product that could appear. Therefore, an efficient method for irreversible inactivation of the enzyme must be developed. An efficient heat treatment is usually used, but neutral and alkaline proteases are also effectively inactivated when the pH is adjusted to 4 or below. Engineering aspects with regard to the up-scaling of inactivation has to be taken into account when the enzyme treatment is going to be carried out on a production scale. Examples on inactivation procedures are shown in the following. Purification of the products made enzymatically may in some cases be required, when the enzymatic method does not produce the product *in situ.* The selection of unit operations and the understanding of the performance of downstream processes are therefore also an important part of the application technology.

6.6.2 Bone Cleaning

As an alternative to rendering, an enzymatic process can be used to upgrade the fresh bones to valuable products, e.g., cleaned bone material suitable for gelatin production and a meat protein hydrolysate for the food industry. The process (Fig. 19) can be performed as a two-step enzyme process (OLSEN, 1988a).

Scrap meat recovery (the first enzymatic extraction). The fresh bone material is crushed and mixed with water to a dry solid level of around 25%. Neutrase (Tab. 14) is added, and the slurry is agitated at 60–65 °C and at natural pH (6.5–7).

In the hydrolysate, the degree of hydrolysis, determined by the TNBS method (ADLER-NISSEN, 1979), should be in the range of 5–10 percent in order to ensure a high yield and to avoid formation of bitterness.

The inactivation of the enzyme is simultaneously achieved, when the reaction is carried out at 60–65 °C.

Centrifuging at 90 °C may be used to defat the hydrolysate. The fat is a valuable by-product that may be further refined for food use. The protein solution may be concentrated or dried, and be used as ingredient in the meat industry, or, possibly as a clear soluble meat extract for soups and seasonings.

Bone cleaning (second enzymatic extraction). The solid bone fraction from the first separation is mixed 1:1 with hot water (65–75 °C) and treated with the alkaline proteases, Alcalase or Esperase. After 1 hour reaction the bones are separated and washed with water. The cleaned bones make an excellent raw material for production of gelatin.

6.6.3 Enzymatic Tenderization of Meat

The use of the vegetable proteinases papain, ficin and bromelain for enzymatic tenderization of meat has been known for many years. These enzymes have relatively good potency on muscle tissue components such as collagen and elastin. Microbial proteinases also have shown the ability to tenderizing meat, but not to the extent as the vegetable ones.

The most successful meat tenderizing enzymes are those having the ability to hydrolyze the connective tissue proteins as well as the proteins of the muscle fibers. The enzymes are applied to the meat by sprinkling the enzyme powder on the thin slice of meat, by dipping the meat in an enzyme solution, or by injecting an enzyme solution into the meat. A method has been developed, in which the enzyme is introduced directly into the circulatory system of the animal shortly before slaughter (BERNHOLDT, 1975), or after stunning the animal to cause brain death (WARREN, 1992).

Whatever method is used, the difficulties in using external enzymes to obtain an even distribution of the enzyme throughout the tissues seem to be the controlling factor.

6.6.4 Modification of Wheat Gluten

The inactivation of the protease is conveniently carried out simultaneously with the hydrolysis, if the reaction is carried out at a temperature above the denaturation temperature of the enzyme (OLSEN, 1988a). Examples of hydrolysis curves drawn, when using a pH-stat for the monitoring of the progress of the reaction, are shown in Fig. 20.

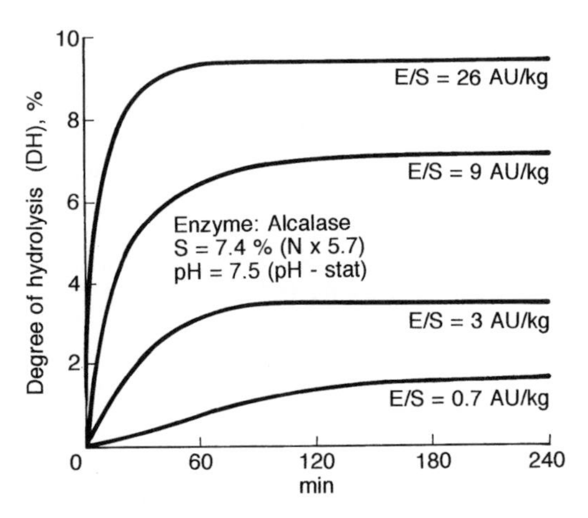

Fig. 20. Enzymatic hydrolysis of wheat gluten at 72.5 °C.

Soluble wheat gluten: A completely soluble wheat gluten hydrolysate is desirable for application as protein enrichment of food and soft drinks. Also such a protein hydrolysate is an excellent raw material for production of enzymatically produced HVP for flavoring. It may be produced by use of exo-peptidases or by acid.

High protein solubility may be obtained at higher degrees of hydrolysis, as shown in Fig. 21. A degree of hydrolysis of about 10% results in a solubilization of more than 90% of the gluten.

A 100% soluble, bland-tasting wheat gluten hydrolysate with high yield may be recovered by centrifugation and concentration.

Whipping wheat gluten: A whippable protein product is required for baked goods and for different types of candy. In Fig. 21 whipping expansions are shown for wheat gluten hydrolysates made at different *DH* %.

The optimal whipping properties have been found at a degree of hydrolysis of 2–3%. The active whipping protein is recovered by centrifugation and drying.

6.6.5 Use of Membranes in Protein Hydrolysis Processes

Membrane processes are used with many techniques of industrial enzymatic hydrolysis

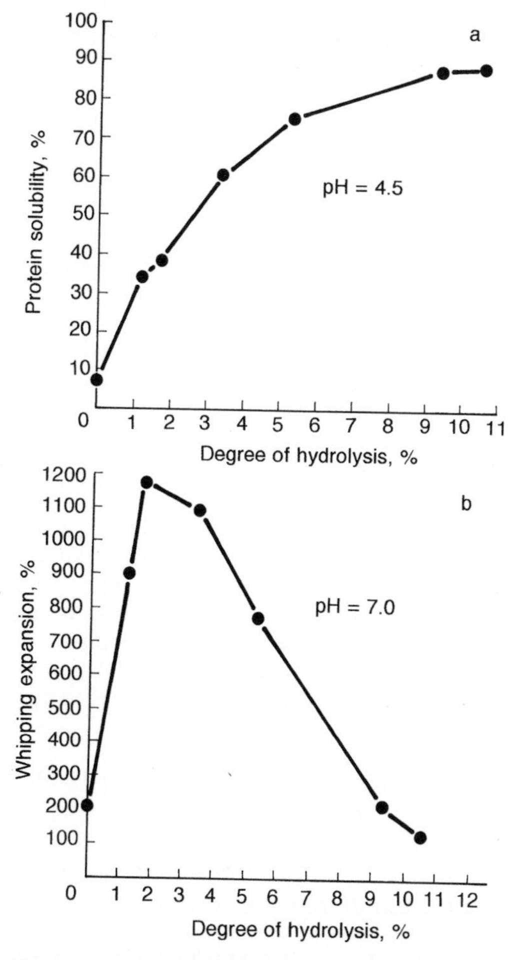

Fig. 21. Protein solubility (a) and whipping expansion (b) for high temperature enzyme hydrolyzed wheat gluten.

of proteins. Tab. 17 shows how membrane processes can be applied to pre-treatment of proteins, the reaction step, and the purification step.

Functional protein hydrolysates. Examples of the application of ultrafiltration processes in various process combinations for production of highly functional protein hydrolysates of soya protein are mentioned in Sect. 6.5.2. Principally the technique of removing small peptides from total hydrolysates may be used for many substrates other than soy. Excellent foaming and coagulating protein products may be isolated (purified). The first process step after inactivation is normally a centrifugation, which is carried out in order to remove insoluble material, which may reduce the whipping expansion and the foam stability. In order to perform the purification of the supernatant as effective as possible, the ultrafiltration can be carried out in 3 steps. In the first step the concentrate reaches a protein content of about 10%. Hereafter a diafiltration process is performed by addition of water in a flow similar to the permeate flux. During this step small low-functional peptides are washed out. The last and final ultrafiltration can bring the protein content up to approximately 20%. Finally this concentrate can be spray-dried. The efficiency of these processes and the overall yields are dependent on the actual ultrafiltration equipment used, the membrane cut-off value, and the character of the hydrolysate (enzyme used, *DH* %, pH, protein type) (OLSEN and ADLER-NISSEN, 1981; OLSEN, 1984).

Tab. 17. Application of Membrane Processes During Enzymatic Modification of Proteins (OLSEN and ALDER-NISSEN, 1981)

Types of Enzymatic Modified Protein	Pretreatment	Enzymatic Reaction Step	Posttreatment
Highly functional proteins	Production of native protein isolate by UF	–	Molecular separation by UF
Low molecular weight protein hydrolysates	–	–	Concentration and/or desalination by HF
	–	Membrane reactor for UF	Concentration and/or desalination by HF

UF, ultrafiltration
HF, hyperfiltration or nanofiltration

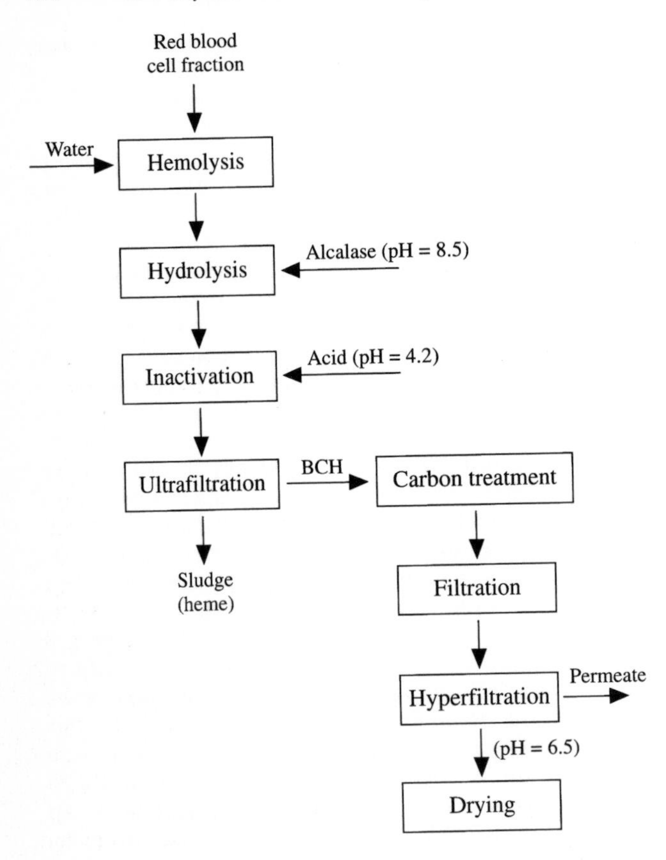

Fig. 22. Flow-chart for the production of blood cell hydrolysate.

Low-molecular-weight hydrolysates are produced from raw materials, which are hydrolyzed to a high degree. One example of such products are the mild-tasting decolorized blood cell hydrolysates developed from the hygienically collected red cell fraction of slaughterhouse blood (OLSEN, 1983). Another example is the isoelectric soluble soy protein hydrolysate (ISSPH) mentioned in Sect. 6.5.1.1 (OLSEN and ADLER-NISSEN, 1979). Flow-charts for these hydrolysates are shown in Figs. 22 (blood cell hydrolysate) and 23 (isolectric soluble hydrolysate of soy). These two examples demonstrate how the membrane processes can be used for different purposes.

In the blood cell hydrolysis process the heme product (insoluble at pH 4.2) may be removed by centrifugation or by ultrafiltration. Comparisons of the centrifugation process and the ultrafiltration process showed that the ultrafiltration process was more economical when new plants were installed (OLSEN, 1983; KRISTENSEN, 1985). Ultrafiltration equipment from various suppliers have been tried with technical success.

Application trials using blood cell hydrolysate (BCH). The earliest application trials were for pumping of meat cuts. Using up to 3% BCH on a final meat product basis, nutritional protein equivalence was obtained, yields were increased, costs were reduced, toughness of meat was reduced and the fat/protein ratio was reduced.

In a study using BCH in low concentrations in meat, it has been demonstrated that BCH has a polyphosphate-like character for binding of water in water-added meat systems. Also BCH has been compared to sodium tripolyphosphate in luncheon meat, heat-treated at 75°C and at 108°, respectively (OLSEN, 1991).

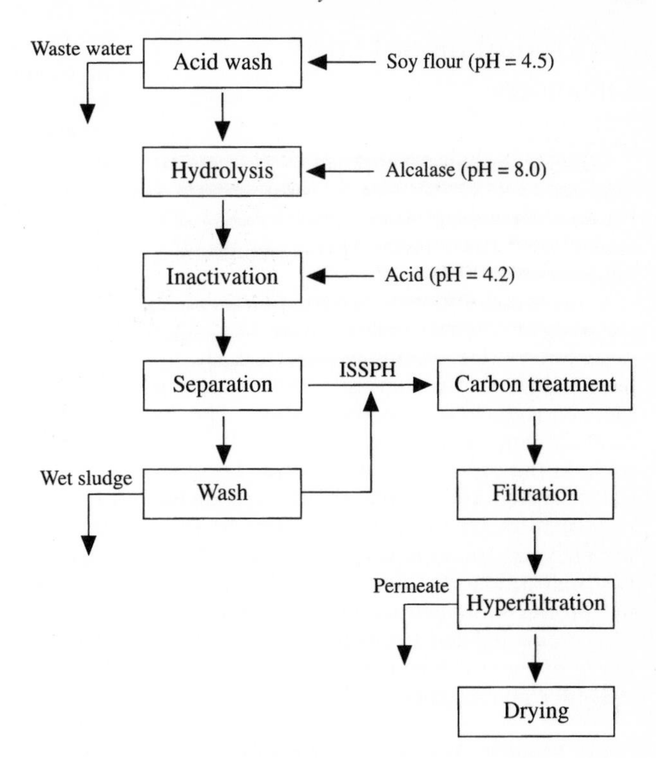

Fig. 23. Flow-chart for the production of isoelectric soluble soy protein hydrolysate.

During production of the soya protein hydrolysate (Fig. 23) the clear filtrate after carbon treatment is concentrated by hyperfiltration as an alternative to evaporation. A falling film evaporator can concentrate the hydrolysate to more than 50% dry matter. Hyperfiltration can only be used efficiently up to about 25% dry matter, but allows a simultaneous removal of salt, if nanofiltration membranes are used for concentration. Thus about 75% of the salt in ISSPH was removed by direct concentration, and 93% if a diafiltration was included in a pilot plant trial on a selected cellulose acetate membrane (OLSEN and ADLER-NISSEN, 1981).

The membrane reactor is an ultrafiltration system, in which a high concentration of hydrolytic enzyme is confined. High-molecular-weight substrate is fed continuously to the reactor, and the low-molecular-weight products are removed simultaneously as permeate through the membrane. Ideally, a steady state is reached, in which the degradation of substrate is carried out indefinitely with high efficiency and negligible loss of enzyme. A considerable number of publications on the use of membrane reactors for enzymatic hydrolysis of proteins appeared 10–15 years ago. However, the practical use of the system has not been widespread. One of the reasons may be due to the fact that the economical advantage of the membrane reactor depends critically on the need for purging of the reactor. A small part of the substrate is always non-degradable. Therefore, inert material will rapidly build up in the reactor causing mechanical problems. A considerable purge is therefore necessary, so that the concentration of this material will be kept at a reasonable low level. This has a drastic, negative influence on the yield of product. Also the enzyme loss during purging will be considerable. OLSEN and ADLER-NISSEN (1981) have presented kinetic investigations on protein hydrolysis in the membrane reactor and established a mathematical model. Based on these calculations, it has been found that the concomitant loss of enzyme in certain cases may lead to an overall enzyme consumption close to that of the batch hydrolysis process.

6.7 Production of Flavor Enhancers

Proteins in their natural state do not contribute much flavor to a food. The products of the hydrolysis of proteins, peptides and amino acids, are much more reactive, as are other breakdown products, generally.

Hydrolyzed protein, mainly produced by use of hydrochloric acid, is one of the largest contributors to flavoring produced from protein. Using various sources a wide variety of savory products are on the market. The hydrolysis with hydrochloric acid usually takes place in the presence of triglyceride fat material. It is well known that chloro- and dichloro-propanols are formed during acid hydrolysis. They are found in very small quantities in the final products. Concern for the safety of the products of hydrochloric acid hydrolysis of proteins has led to investigations into the safety of these chloro-compounds:

Under investigation are:

(a) Monols: Monochloropropanols, e.g., 1-, 2- and 3-chloropropanol. Dichloropropanols, e.g., 2,3-dichloropropane-1-ol, 1-3-dichloropropane-2-ol.

(b) Diols: Monochlorodipropanols, e.g., 2-chloro-1,3-propanediol, 3-chloro-1,2-propanediol.

Concern over these contaminants has led to the imposition of limits in Germany and Holland for various chloro-propanols. Other governments are considering their positions (WEIR, 1992).

Glutamic acid, as monosodium glutamate (MSG), is by far the most widely consumed single flavor enhancer of proteinogenic origin. About 270000 tons are produced per year (BIOTOL, 1991). Glutamates are known as the fifth basic taste, in addition to sweet, sour, salty, and bitter. This fifth basic taste is called umami by the Japanese, and savoriness by the Americans. MSG is used at concentrations of 0.2–0.8% in a variety of foods such as soups, broth, sauces, gravies, flavoring and spice blends, as well as in canned and frozen meats, poultry, vegetables and finished dishes. In order to be able to reduce the intake of sodium and to produce *in situ* flavor-enhancers in protein hydrolysates, peptidoglutaminases are of interest for industrial purposes (HAMADA, 1992, 1994; NIELSEN, 1994) – see Sect. 6.9.

Peptides, through their reactions with taste sites, may give flavors, which are bitter, sweet, salty or umami. Also sour taste and astringency have been attributed to peptides. These tastes have been obtained from peptides isolated from protein hydrolysates or by synthesis. The effect on flavor, as a result of enzymatic activity, has been the subject of many studies of dairy products, meat and fish products, and yeast extracts. When it comes to a direct production of flavor products, enzymes may be used alone or in combination with fermentation processes, or chemical, or heat treatments (for example, Maillard reactions). Meat flavors and similar condiments are produced from several sources, and their production often involves a natural microbial fermentation. This includes the oriental products based on soy protein. Wheat and maize glutens are also hydrolyzed to flavor intense peptides.

Soy and wheat hydrolysates are used for flavoring and flavor enhancement of soups and seasonings, meat hydrolysates for addition of flavor to meat products, soups bouillon and sauces, and milk/cheese hydrolysates for production of cheese flavors (NIELSEN, 1994).

6.7.1 Yeast Extracts

The demand for yeast extracts has increased as a result of concerns over the safety of acid hydrolyzed proteins. Yeast extract is a source of amino acids, peptides, sugars, nucleotides, lipids and B vitamins, all of which can act as meat flavor precursors. AMES and ELMORE (1992) have analyzed aroma components of yeast extracts with different aroma properties. Several hundreds of components have been identified. Most of the aroma components arose from sugar and/or amino acid interactions, or thiamin degradation. Variations between conditions used to manufacture the yeast extracts reflected the aroma.

Mixtures of a primary yeast extract and protein hydrolysate are used as both a savory flavor and as a flavor enhancer. To avoid the use of strong acids like hydrochloric acid for degradation of proteins and carbohydrates, most yeast extracts are produced by use of the following steps: plasmolysis, autolysis, pasteurization, clarification and concentration (REED and NAGODAWITHANA, 1991).

6.7.1.1 Plasmolysis

Plasmolysis is a simple and rapid method for initiation of cell disruption. During this process the yeast is killed, but its enzymes are still active to be used in the following steps. Salt or organic solvents such as ethyl acetate or isopropanol are used as plasmolysing agents. The conditions used may be several hours (5–8 h) treatment at 50–60 °C. The use of cell wall degrading enzymes for lysing of yeast cells has already been discussed by PHAFF (1977) and later by ANDREWS and ASENJO (1987). BALDWIN and ROBINSON (1994) have demonstrated an enhanced disruption of *Candida utilis* using a combination of partial enzymatic lysis with the enzyme Zymolyase-20T from Seikagaku Kogyo Co., Ltd. followed by mechanical disruption in a microfluidizer high-pressure homogenizer. A total disruption of 95% was obtained with four passes at a pressure of 95 MPa, as compared with only 65% disruption using only mechanical homogenization.

Zymolyase was tested for lysis activity using brewers' yeast. This is a most application adapted test; no further characterization of the activity was used.

HAMLYN et al. (1981) have tested several commercial polysaccharidases for their ability to liberate protoplasts from fungi. The lytic components of the enzymes used in the experiments were assessed by determining their hydrolytic activities against different polysaccharides known to be components of the walls of the fungi used. It was found that most of the enzyme preparations contained side activities capable of hydrolyzing the following polymers: Acid swollen chitin (chitinase); nigeran (α-glucanase); laminarin (β-glucanase); baker's yeast mannan (mannanase). High levels of β-glucanase and chitinase activities were found in the enzymes that gave the best protoplast yields. Mannanase was detectable only in very small amounts. The most effective of the enzymes tested was Novozyme 234. This enzyme also contained considerable proteolytic activity. By use of this preparation a protoplast yield of 98.9% was obtained from *Saccharomyces cerevisiae* by incubation of the lytic mixtures at 28 °C for up to 3 hours with gentle shaking. The efficiency of this enzyme preparation was demonstrated for cell wall removal without any mechanical treatment.

6.7.1.2 Autolysis

KNORR et al. (1979) demonstrated that addition of proteolytic enzymes during incubation of yeast cells with lytic enzymes caused the concurrent hydrolysis of the yeast proteins. This technique has been utilized by DE ROOIJ and HAKKAART (1992) in a process for the preparation of food flavor by inactivating yeast and then degrading it with an enzyme having proteolytic activity. Simultaneously or subsequently a fermentation was carried out with yeast or with lactic acid producing microorganisms. Furthermore, an RNA degrading enzyme was used in order to obtain 5′-ribonucleotides. After downstream processing, such as removal of insoluble material by centrifugation or filtration, concentration, pasteurization, and drying the yeast extract obtained, typically had the following composition:

Protein material	20–45%
peptides	10–30%
free amino acids	5–20%
Guanosine-5′-monophosphate	0.5– 6%
Lactic acid	8–20%

The yeast extract is claimed to be an excellent material to improve the flavor of soups, meat products, instant gravies, margarine, frying fat, drinks, bakery products, cheese and confectionary products.

HOBSON and ANDERSON (1991) have described the development of a co-hydrolytic process for the production of novel extracts from yeast and non-yeast proteins, which could be used with maize gluten, corn gluten, wheat gluten, soybean meal, whey solids, soup stock, dried red blood, oat bran and wheat bran. The autolysis and the hydrolysis processes may be carried out simultaneously. A preferred protease system includes papain and Neutrase®. The hydrolysis steps involved maintaining the reaction mixtures under the following conditions:

Step 1: 40–50 °C for 5–15 hours
Step 2: 55–65 °C for 1–5 hours
Step 3: Pasteurization and enzyme inactivation at 90 °C.

Usual downstream processing was carried out.

6.7.2 Hydrolyzed Vegetable Protein (e-HVP)

Enzymatic hydrolysis usually leads to products, which are light in color and have a much less pronounced meaty or savory flavor. Meat flavor and similar condiments are produced from many sources and their production often involves a natural microbial fermentation. Among these can be included the oriental products based on soya protein. Wheat and maize glutens are also hydrolyzed to flavor intense peptides.

High glutamic acid levels are characteristic of soya flavor products, and this has been enhanced by application of specific peptidoglutaminases. GODFREY (1983c) mentions that in many cases glutamic acid levels of 30% above control values have been obtained without enzyme addition.

The content of free glutamic acid is important for the umami flavor. HAMM (1992) has described a way of exhibiting flavor enhancement characteristics, by using mild acid hydrolysis of the protein followed by enzymatic hydrolysis. The preferred protein for producing savory flavors is wheat gluten. The mild acid hydrolysis is made in order to obtain a deamidation in the first place. This step was carried out for one hour at about 95 °C, at a hydrogen ion concentration of about 1.0 M. The enzymatic hydrolysis was preferably carried out by use of an *Aspergillus oryzae* protease (Prozyme 6 from Amano) having both endo- and exo-proteases. The overall degree of hydrolysis was 50–70%. The conditions during the acid treatment were mild enough to avoid the formation of substantial amounts of monochloropropanols.

Recently NIELSEN (1994) has announced Flavourzyme™ to be used in combination with Alcalase® as a possibility to obtain a degree of hydrolysis of 60–70% for hydrolysis of soy protein isolate. Except for the deamidation step the same degree of hydrolysis could thus be obtained without the acid treatment.

6.7.3 Hydrolyzed Animal Protein (e-HAP)

Products, which have a high meat extract flavor, may be used in soups, sauces and prepared meals. Proteinaceous material can be recovered with enzymes from coarse and fine scrap-bone residues from mechanical fleshing of beef, pig, turkey or chicken bones. The flavor intensity is dependent on the content of free amino acids and peptides and their reaction products. The reactions which develop flavor are, for example, reactions between reducing sugars and amino acids (Maillard), thermal degradations due to the Maillard second stage, deamination, decarboxylation or degradation of cysteine and glutathione. The latter reaction can give rise to a large number of volatile compounds of importance to aroma and taste (WEIR, 1992).

For the production of protein hydrolysates from meat the first step is an efficient solubilization of the product. For this purpose the bacterial serine proteases and metalloproteases are very efficient (see Sect. 6.3). However, the hydrolysates are usually bitter when the degree of hydrolysis is above 10%, which is needed for sufficient solubilization (O'MEARA and MONRO, 1984). PEDERSEN et al. (1994) have patented a method comprising a series of steps, using raw meat which is hydrolyzed with a specified combination of neutral and alkaline proteases. This meat hydrolysate exhibits excellent organoleptic properties and can be used as a meat flavored additive to soup concentrate. A degree of hydrolysis above 20% did not show any bitterness, when such specified combinations of enzymes were used. The reason for this effect may be due to the preferential specificity being favorable, when metalloprotease and serine protease are used simultaneously.

A method for further enhancement of the flavor of a natural beef juice concentrate is a hydrolysis using endo-proteases (metalloprotease, serine protease or a combination) followed by treatment with an exo-protease like Flavourzyme™ (see Tab. 14) or by incubation with a culture of a food-grade microorganism, which is capable of producing exoproteases (e.g., aminopeptidase) as described by KWON et al. (1992).

6.8 Transglutaminases

Transglutaminase catalyzes protein-modifying reactions such as amine incorporation, crosslinking and deamidation. IKURA et al. (1992) describe attempts to make practical use of transglutaminase in quality improvement and processing of food proteins. The mass production of transglutaminase has become possible from a system for production of recombinant animal transglutaminase, and by screening microorganisms producing the enzyme.

By use of the amine incorporation catalyzed by transglutaminase, desired amino acids were incorporated into food proteins to improve their nutritive and functional properties. Casein and soybean globulins were polymerized through intermolecular crosslinking catalyzed by the enzyme. Solutions of several food proteins at high concentrations can be gelatinized by transglutaminase.

6.9 Deamidation of Proteins

Transglutaminase, protease and peptidoglutaminase (PG-ase) are the only enzymes reported in the literature for deamidation.

Deamidation improves the solubility, emulsification, foaming, and other functional properties of the proteins. Deamidation of peptides or proteins is defined as the removal of ammonia from peptides or protein by the hydrolysis of the amide groups. For most proteins, deamidation refers to the conversion of the amide groups to carboxylic groups with concomitant release of ammonia. The reaction, in which glutamine is hydrolyzed to glutamic acid and ammonia, can take place both at the amide bond of free amino acids and at peptide- and protein-bound amino acid residues. This reaction is shown in the following scheme:

$$\underset{\displaystyle NH_2}{HOOC\underset{|}{C}H_2}-CH_2-CH_2-CO-NH_2+H_2O \longrightarrow \underset{\displaystyle NH_2}{HOOC\underset{|}{C}H_2}-CH_2-CH_2-COOH+NH_3$$

The taste of protein hydrolysates depends on the primary structure as discussed in Sect. 6.5.1. In addition to bitterness as a result of the content of hydrophobic amino acid side chains of peptides, other residues in the amino acids and peptides are responsible for sweet, sour, brothy, or beefy tastes. Glutamic acid residues in intact proteins are not a flavor enhancer, but when bound in peptides the glutamic acid has flavor enhancing properties. HAMADA (1994) mentions the possibility of using deamidated protein hydrolysates for flavor enhancement. In connection with possible peptidoglutaminases, proteases or a mixture of proteases, that are tailored to produce protein hydrolysates without bitterness, powerful meaty protein hydrolysates may be produced, when suitable glutaminases are developed for commercial use. Thus the deamidated protein hydrolysates may eventually substitute in part for the use of monosodium glutaminate.

Use of enzymes for protein modification is more desirable than chemical treatments because of their speed, mild reactions conditions, and their high specificity (HAMADA, 1994).

7 Extraction Processes of Vegetable Raw Material

Plant material is used to a large extent for production of valuable food products. Many ingredients used by the food and brewing industries are produced by extraction of vegetable raw material. Some examples are protein, starch, sugar, fruit juice, oil, flavor, color, coffee, and tea, all products found intracellularly in plant material (seeds, fruits, etc.).

The development of cell wall degrading enzymes for plants has improved the use of enzymes in liquefaction processes and for extraction of intracellular components of fruits and seeds. Thus a new potential outlook for the industrial use of enzymes has been introduced.

7.1 Plant Cell Walls and Specific Enzyme Activities

An important development is the degradation of very complex polysaccharides found in the cell walls of unlignified plant material. CARPITA and GIBEAUT (1993) have described a model of the type I primary cell wall, generalized for flowering plants. These cell walls are composed of cellulose fibers, to which strands of xyloglucan (hemicelluloses) are attached. The fibers are embedded in a matrix of pectic polysaccharides, polygalacturonic acid and rhamnogalacturonan, the latter being substituted with small polymeric side groups of araban, galactan and arabinogalactan. Some of these groups are linked to a structural protein as indicated in Fig. 24 (KEEGSTRA et al., 1973).

The proximate composition of cell wall polysaccharides varies according to the crop or fruit considered. In Tab. 18 some of the compositions are shown for the plant material considered in the following.

When enzymatic opening, liquefaction or degradation are considered, the total content of cell wall in fruits or seeds is of importance, as it will determine the actual enzyme dosage based on raw material. Tab. 19 is a simple list showing the total content of cell wall material for some fruit and seeds.

The optimal composition of an enzyme complex for cell wall degradation will always depend on the kind of material to be degraded and on the desired effect. This is qualitatively indicated in Tab. 20.

Enzyme preparations capable of attacking plant cell walls contain different specific enzymes belonging to the groups of pectinases, hemicellulases and cellulases. Commercial cell wall degrading enzymes are mainly from *Aspergillus* species. PILNIK and VORAGEN (1993) inform that commercial pectinases, regardless of their production by different companies, Gist-Brocades, Novo Nordisk, Röhm, Biocon etc., usually contain mixtures of pectin degrading activities, hemicellulases and endo-β-glucanases.

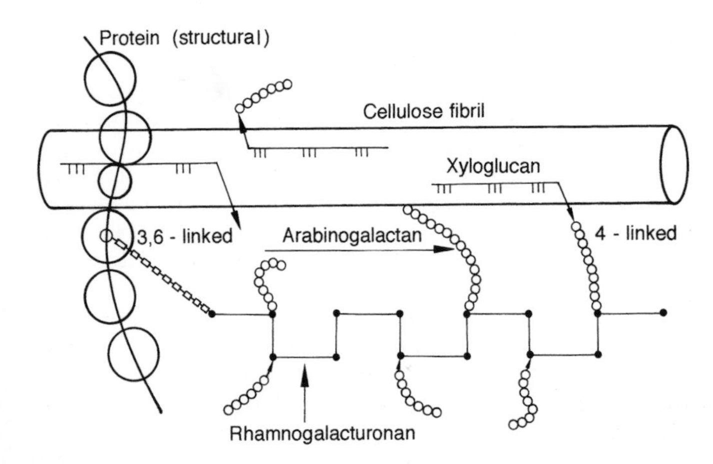

Fig. 24. General model of a primary plant cell wall (KEEGSTRA et al., 1973).

Tab. 18. Proximate Composition of Plant Cell Wall Polysaccharides of Some Plant Material (% of weight of cell wall dry matter)

Plant Material (Reference)	Pectins	Hemicellulosics					Proteins	α-Cellulose
		Xyloglucans	Arabinoxylans	β-Glucans	Mannans	Other		
Apples (BIOTOL, 1991)	31		20				9	40
Tomatoes (GROSS, 1984)	55		15					30
Carrot (SHEA et al., 1989)	37	16	11	–	16		n.d.	21
Banana (KOFFI et al., 1991)	20		40				n.d.	30
Cocconut kernel (SAITTAGAROON et al., 1983)	6–8[a]				61	21[b]		
Rape seed (THEANDER and ÅMAN, 1977)	39	29				8[c]	n.d.	22
Corn germ (OLSEN, 1988b)	<1		40			10	n.d.	39
Sunflower (DÜSTERHÖFT et al., 1992)	24	4.5			5	24[d]		42
Palm-kernel meal (DÜSTERHÖFT et al., 1992)		3	3		78			12

[a] Reference: DEL ROSARIO and GABUYA (1980),
[b] galactomannan, [c] arabinogalactan, [d] glucuronoxylan
n.d., not determined

In order to obtain high degradation of cell wall material from various plants, selected carbohydrases must be used in combination with pectinases or other specific preparations. This has been illustrated in several cases, for example, by DÜSTERHÖFT et al. (1993) for cell wall materials (CWM) of palm-kernel and sunflower meals, by MASSIOT et al. (1989) for CWM of carrots or by OLSEN (1988b) for CWM of coconut and corn germ.

7.1.1 Pectinases

Pectins are polymers of galacturonic acid and its methoxyl- or acetyl esters. Most pectin in cell walls is found in a complex form as a chain molecule with a rhamnogalacturonan backbone. This backbone consists of "smooth" regions, mainly a linear polymer of α-D-1,4 linked galacturonic acid and 1,2 linked α-L-rhamnose. The galacturonic acid is esterified to a various degree with methoxyl or acetyl groups. The rhamnogalacturonan part of the pectin has a so-called "hairy" character (ASPINALL, 1980). This is due to side chains, which can be attached to the backbone by a glycosidic linkage to carbon atom number 3 or 4 in rhamnose, and to carbon atom number 2 or 3 of galacturonic acid oligomers or polymers of neutral sugars such as galactose, arabinose and xylose.

Tab. 19. The Total Content of Cell wall Material in Some Plant Material (reference: various tables of dietary fibers)

Plant Material	Total Content of Plant Cell Wall Material (% w/w of fruit or dry seed)
Apple	1.7
Cherries	1.1
Pineapple	1.0
Mango	2.2
Orange	2.0
Pear	1.2
Tomatoes	1.9
Carrot	2.7
Banana	3.5
Coconut kernel	16.9
Rape seed	12
Corn germ	12
Soy bean	18

The enzymes involved in the degradation of pectins are pectin esterases (methoxyl- or acetyl esterases), and depolymerizing enzymes attacking the interior bonds of pectin chains (endo-enzymes) and exo-enzymes, which liberate galacturonic acid from the end of the pectin molecule. All types are found in commercial pectinases from *Aspergillus* species.

Pectin esterases

In fruit and vegetables that are used for juice production, more than 70% of the carboxyl groups are esterified with methanol. This may be the reason why pectin esterases are usually referred to as "pectin methylesterase". The pectin esterases de-esterify pectins to low methoxyl pectins or pectic acid by releasing methanol (Fig. 25). Acetyl groups are predominant in the cell walls of, e.g., sugar beet or soy beans (PILNIK and VORAGEN 1993).

Polygalacturonases, pectate lyases and pectin lyases

Polygalacturonases split the α-1,4 bonds of the pectin or rhamnogalacturonan chain by hydrolysis of the glycosidic linkages next to free carboxyl groups. Pectate lyases split the glycosidic linkages next to free carboxyl groups by β-elimination, see Fig. 25. Pectate and low-methoxyl pectins are, therefore, the preferred substrates for these enzymes. High-methoxyl pectins are degraded very slowly by these enzymes. They are degraded by pectin lyases, also referred to as endo-pectin transeliminase. The mode of action of these three pectinases is shown in Fig. 26.

Tab. 20. Desired Effect and Properties of Enzymes Needed for Cell Wall Degrading Enzymes

Desired Effect	Enzymes Needed
Liquefaction of fruits and oil extraction	Structures important for the integrity of the cell wall need to be degraded. These are not necessarily the major constituents of the cell wall
Maceration (e.g., baby food consistency)	The same type of enzymes as for liquefaction, but degradation of major wall constituents must be limited
Fiber extraction	As for liquefaction, but the specific fiber must not be attacked
Cloud-stable juice	Partial degradation of major wall components
Clarification	Extensive degradation of major cell wall components
Lowering viscosity	Endo-attack of specific polysaccharides

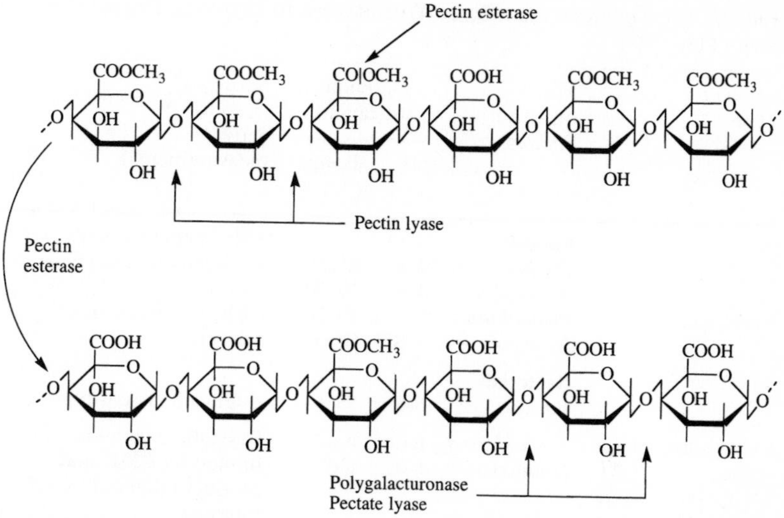

Fig. 25. Points of attack of pectin esterase, pectin lyase, pectate lyase and polygalacturonase.

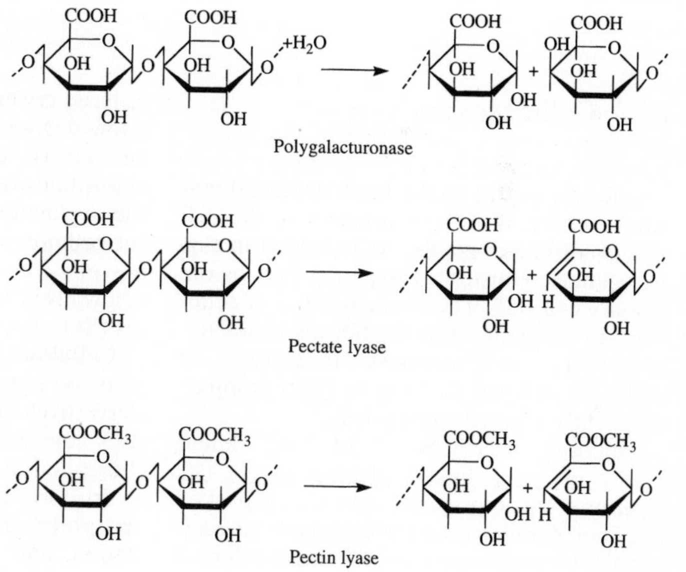

Fig. 26. Mode of action of pectinases.

7.1.2 Hemicellulases

Hemicelluloses are polysaccharides made up predominantly of heteropolymeric xylans, arabinans, arabinoxylans, mannans and galactomannans. Enzymes, which can degrade these polysaccharides, can be important for cell wall degradation, as indicated in Tabs. 18 and 20. Hemicellulases are mainly found in *Aspergillus* and *Trichoderma* fungi. Also from these microorganisms most commercial xylanases and pentosanases are produced. These products are often used in combination with pectinases.

Tab. 21. The Cellulases Considered With Regard to Cell Wall Degradation (after GODFREY 1983b and SCHÜLEIN, 1992)

Cellulase Type	Micro-organism Species	Optimum (or range)		Substrate (also for activity measurements)	Products Formed by the Reaction
		pH	Temp. (°C)		
Endoglucanases	**Fungal:**				
	Trichoderma	4–6	45–55	Filter paper or acid swollen cellulose (C_1-activity)	Glucans
	Aspergillus	4–6	40–50	CMC (decrease of viscosity)	Cellobiose
	Penicillium	5	40–50		Cellooligomers
	Humicola	5–8	50–60		
	Bacterial:				
	Bacillus	6–8	55–60	*β*-Glucan (viscosity)	Cellooligomers
Cellobiohydrolases	*Trichoderma*	n.d.	n.d.	Crystalline cellulose (inhibited by CMC and product inhibited by cellobiose)	Cellobiose
	Humicola	n.d.	n.d.		Cellooligomers
Cellobiase/ *β*-glucosidases	*Aspergillus*	4–5	60	Cellobiose	Glucose

n.d., not determined

7.1.3 Cellulases

Cellulose makes up the basic structural material and has a skeletal function of the cell wall of all higher plants. As indicated in Fig. 24, cellulose constitutes the rigid part of the primary cell wall. The characteristics of cellulose are its great strength, fibrous character, insolubility, and inertness. It consists of chains in ordered packing to form compact and tightly bonded aggregates.

Cellulose is composed of long, linear chains of *β*-1,4 linked glucose units. The structures can in principle have chains packed in either parallel (most widespread) or antiparallel bundles; where parallel means that all non-reducing or all reducing ends are at the same ends; antiparallel means that they are at different ends. The flattened sheets of the chains lie side by side and are joined by hydrogen bonds. These sheets are laid on top of each other in a way that staggers the chains, just like bricks are staggered to give strength and stability to a wall (REES et al., 1982). Cellulose chains pass repeatedly through highly ordered crystalline and amorphous regions of a low degree of order. The proportion of the amount of cellulose in crystalline and in amorphous regions may be of importance for the explanation of the different properties of the cellulose of different cells and to the varying physical and chemical properties of the cell walls of fruits, vegetables, seeds and cereals (SELVENDRAN, 1983).

Cellulases are enzyme complexes, which in stepwise reactions degrade native cellulose or derivatives of cellulose to glucose. The cellulases considered are grouped in Tab. 21 and illustrated in Fig. 27. The effect of enzymatic hydrolysis of cellulose depends very much on the pretreatment (temperature, alkaline oxidation, and mechanical degradation) of the substrate. The direct action of cellulases is hindered by the presence of the pectic matrix material and in addition xylans, *β*-glucans and lignin may be bound to cellulose fibrils. This is another reason why cellulose present in cell walls is difficult to degrade.

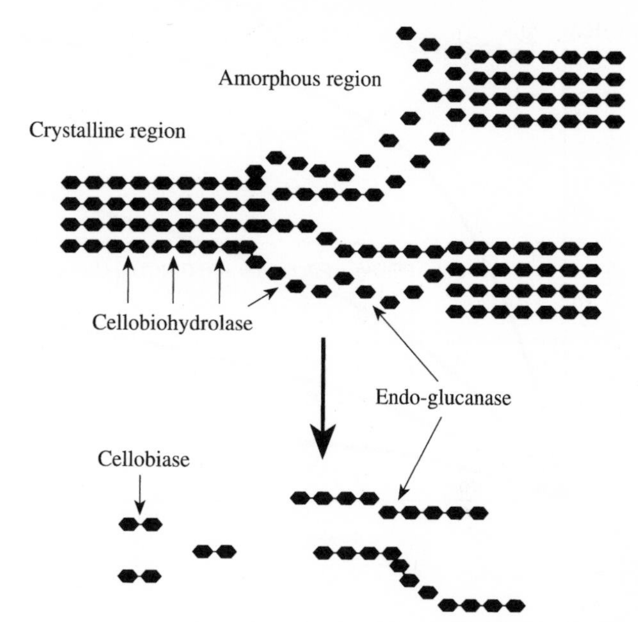

Fig. 27. Degradation of a cellulose chain (DALBØGE and HELDT-HANSEN, 1992).

7.1.4 SP-249 – A Multienzyme Complex

Several multienzyme complexes for cell wall degradation have been developed and are used for extraction, liquefaction and downstream processing in the food and feed industry (OLSEN and CHRISTENSEN, 1987; CHRISTENSEN, 1989). As shown in Tab. 18, cell wall polysaccharides of industrial plant materials differ in composition. Also as shown in Tab. 19, the total content of cell wall material differs. Owing to these differences the optimal composition of enzyme complexes for industrial use must differ according to the kind of material to be treated. As one of the first efficient multienzyme complexes Novo Nordisk developed a strain of *Aspergillus aculeatus,* which was able to produce 10–15 different enzyme activities. This offers the possibility of producing a variety of commercial enzyme products designed for various applications based on these activities. Besides the essential pectinase complex (see Sect. 7.1.1), a wide range of hemicellulases, endoglucanases and β-glucosidases are present. Conventional pectinases are unable to degrade the rhamnogalacturonan backbone of the pectic substances completely. This is illustrated by the following model experiment.

Isolation of cell wall

Cell wall was isolated from the endosperm of rape seed. The process used consisted of a dry milling, an aqueous wet milling and a wet sieving in order to remove the hulls from the endosperm. A thorough and radical hydrolysis of the proteins using a high dosage of Alcalase at pH = 8, T = 50 °C for 4 hours could secure an efficient release of the protein by solubilization and simultaneous release of oil in more or less emulsified form. After carefully washing by use of water and a repeated hydrolysis using Alcalase the product was freeze-dried. The residual amount of lipids was extracted with hexane.

Degradation trial using osmometer

4 × 100 mL suspensions containing 5% w/w dry matter of the rape seed cell wall material, prepared according to the above method, was adjusted to pH 4.5 using 4 N HCl. An Erlenmeyer flask was used. The flask was placed in a waterbath with magnetic stirring. The temperature was adjusted to 50 °C. Celluclast®, a *Trichoderma reesei* cellulase without pectinase and Pectinex®, an *Aspergillus niger* pectinase without cellulase activity were added to the

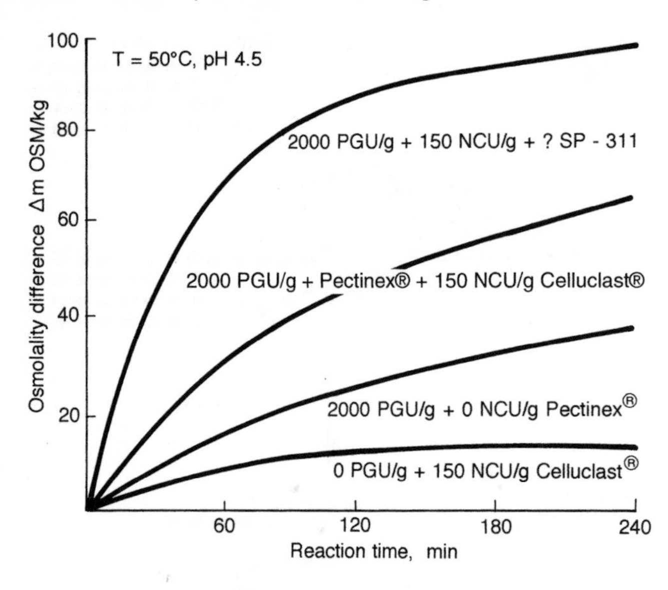

Fig. 28. Degradation of isolated rape seed cell wall (substrate concentration $S = 5\%$ dry matter).

flask in dosages based on their activities, so that the polygalacturonase activity units (PGU) and the cellulase activity units in Novo Cellulase Units (NCU) were exactly as for SP-311 (a variant of SP-249). The osmolality was measured versus time with an osmometer. Reaction curves could be drawn demonstrating the increase of low molecular weight components in the soluble phase.

As shown in Fig. 28, the enzyme SP-311 is more suitable than Pectinex and Celluclast for degradation of the cell walls of rape seed.

SP-311 was a special experimental version of the so-called SPS-ase preparation, which was based on a selected strain of *Aspergillus aculeatus* (ADLER-NISSEN et al., 1984). The enzyme complex, "SPS-ase", was named after its intended commercial use as a soya polysaccharide degrading enzyme. The SPS-ase activity is presumably connected to the rhamnogalacturonase activity, but this has still not been confirmed.

7.2 Fruit Juice Processes

Pectinases have been used for the production of fruit juice since 1930. Over the years enzymes have been used with almost all fruits.

The purpose of the use of enzymes is the following:

(1) to improve the yield of juice,
(2) to liquefy the entire fruit for maximal utilization of the raw material,
(3) to improve color and aroma,
(4) to clarify juice,
(5) to break down all insoluble carbohydrates such as pectins, hemicellulose and starch.

For clarification of juice a mixture of pectinases – pectin lyase, polygalacturonase (PG), and pectin methylesterase (PE) – have to be present. A treatment of apple mash with highly active mash enzymes or other dejuicing technologies leading to high juice yields may also dissolve araban (a high molecular weight polysaccharide) from the cell walls. Araban may cause haze in concentrates. Clarification enzymes should therefore also contain a substantial amount of arabanase activity besides the mentioned pectinase activities. The pH of apple juice is 3–3.8 and of citrus juice 2–2.5.

The use of cell wall degrading enzymes in wine making is new. However, all the same advantages as mentioned above will be valid for wineries, too. The product family Pectinex® has over the years been important for the juice and wine industry.

7.2.1 Apple and Pear Juices and Concentrates

Enzymes are currently used in apple and pear juice production in the two steps, (1) mash treatment and (2) juice treatment.

(1) Mash treatment

During enzymatic mash treatment a total or a partial liquefaction of the fruit flesh is obtained for production of fruit pulps or nectars (DÖRREICH, 1983). By this process the juice yield is increased and the extraction of important fruit components such as flavor and color is also ameliorated. The result is an overall better economy in juice production (JANDA, 1983). The press capacity and press performance can be improved wherever horizontal presses, belt presses or decanters are applied (DÖRREICH, 1986). The pomace produced will be drier than usual, no press aids are needed and less energy is used (JANDA and DÖRREICH, 1984). BEVERIDGE et al. (1992) have described a mash treatment using the enzyme Pectinex Ultra SP followed by a decanter centrifugation as an alternative method to pressing for juice extraction. The decanter has a number of advantages in relation to juice yields (80–90%) and fewer suspended solids.

VORAGEN et al. (1992) has described the effect of the mash treatment on a more molecular basis, and GRASSIN (1992) has described the apple juice processing using enzymes from Gist Brocades.

(2) Juice treatment

After pressing the aroma is usually recovered. This process consists of a short time heat treatment (98°C for 30 seconds) and a flashing. Due to the content of extracted pectin (if the mash treatment is not performed effectively enough) precipitate and haze may form during this flashing. During juice treatment, pectin which has not been completely degraded during mashing, can result in high viscosity. The application of effective pectinases reduces this viscosity and facilitates clarification. The amounts of fining aids such as gelatin, diatomaceous earth or bentonite may be reduced if depectinization completes. Furthermore, the filtrations will be better and the concentrated product clearer and more stable. When the fruit is not completely matured, it may still contain a small amount of starch. This starch becomes gelatinized during the processing of the juice and may cause problems during fining. Amyloglucosidase is therefore often added to avoid this problem.

Ultrafiltration of the juice

In most modern apple juice plants ultrafiltration is used as a substitute for chemical fining. The mashing enzymes used to avoid fouling of the membranes must be very efficient in the degradation of soluble high molecular weight polysaccharides. Flux rates and economy of ultrafiltration can thereby be considerably improved (STUTZ, 1993). Fig. 29 shows how commercial enzymes are used for the production of apple and pear juice concentrates.

Apple juice stabilization with laccase

GIOVANELLI and RAVASINI (1993) reported that prevention of physical-chemical damage is a major problem in fruit juice production. They evaluated the effect of enzymatic oxidation of polyphenols on "active filtration" and membrane filtration in terms of juice characteristics and stability. Laccase oxidation of polyphenols results in a significant decrease in the phenolic content of juices and is associated with a remarkable color increase. Active filtration with polyvinylpolypyrrolidone and active charcoal partly decolorizes the oxidized juices and stabilizes these products by removing a large quantity of phenolic compounds. Membrane filtration is an efficient technique of stabilization, but only if it is performed with low molecular weight cutoff membranes. The laccases used in these experiments were obtained from *Polyporus versicolor* or *Myceliophtora thermophyla.*

The possibility of using an immobilized laccase for enzymatic removal of phenolic compounds from must and wine was evaluated by BRENNA and BIANCHI (1994). The immobilized enzyme was packed in a column, and

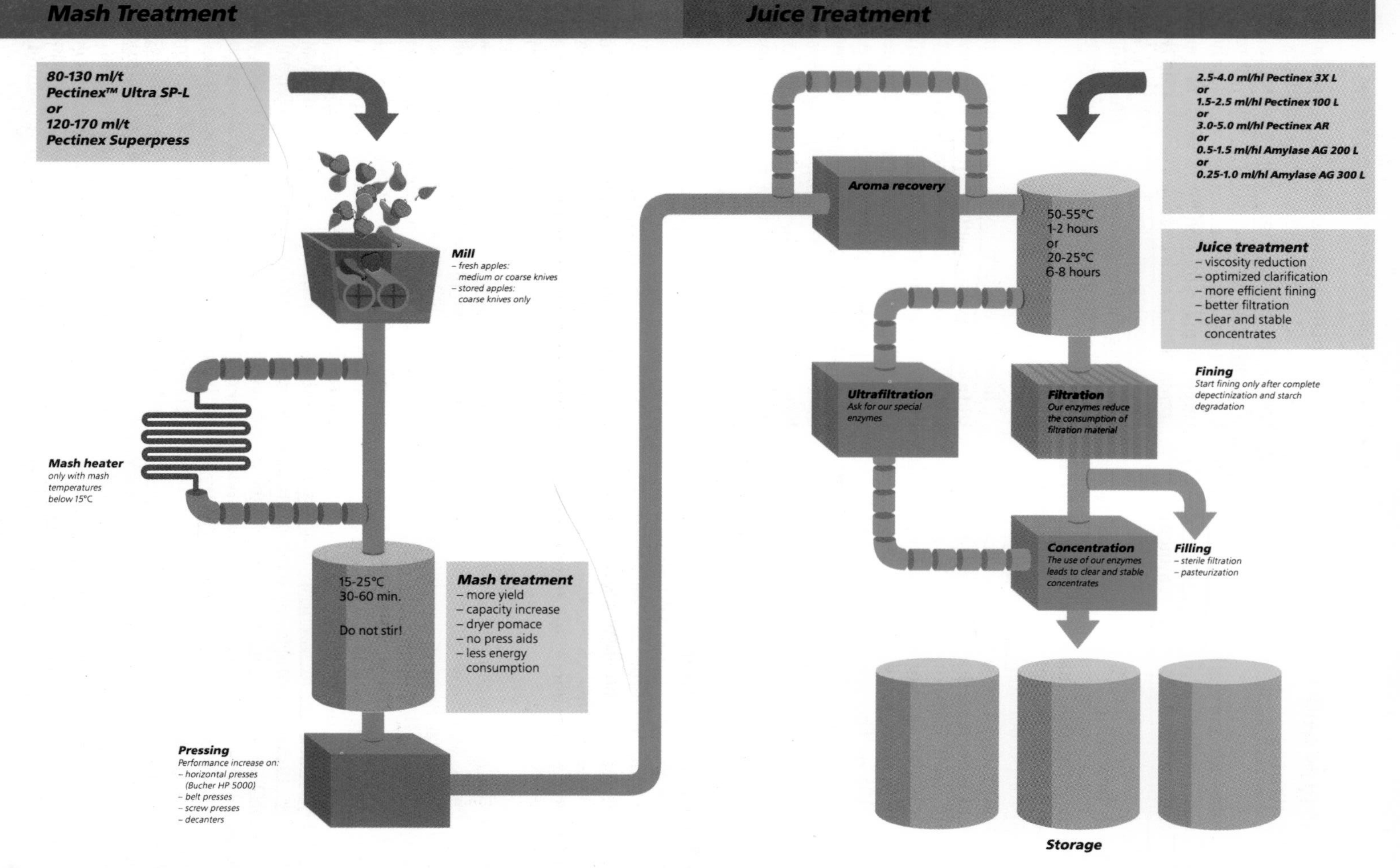

Fig. 29. Use of enzymes for the production of apple and pear juices and concentrates (Novo Nordisk Ferment Ltd.).

Peeling of fruit

Typical conditions:
- flavedo rasping
- enzyme reactor
- incubation:
  60-180 minutes
  temperature: 20-45°C
  700-1200 ppm Peelzym

Objectives of enzyme application:
- nicely peeled fruit or segments
- easy to remove the albedo and the peel
- low juice losses during storage
- no sliminess during storage

Fig. 30. Enzymatic peeling of fruit.

model solutions of chatechin or white grape must were passed through. Oxygen saturation, flow rate and immobilized enzyme concentration are the important parameters to evaluate in this case.

7.2.2 Citrus Juice

Citrus fruit processing is a mechanical process as it has been traditionally. The fruit is sorted, squeezed, milled, and juiced by machine. The efficiency of the machinery determines the yield and quality of the fruit. Numerous products are manufactured from citrus. They include juice, concentrates, peel juice, clarified juice, comminuted products, natural cloudy peel products, essential oils, aromatic essences, pectin, natural pigments, and feed materials.

Tailor-made industrial enzyme preparations offer the citrus processors the possibility to improve economy (reduction of processing costs, yield improvements, improvements of filtration rates and easier cleaning steps) and the flexibility to produce new and different products and by-products (DÖRREICH, 1993a, b). Furthermore, new enzyme preparations make it feasible to increase the efficiency in citrus oil recovery and to reduce waste water production. The enzyme series Citrozym™ is used for peel extraction, pulp wash, viscosity reduction, juice clarification, oil recovery and fruit peeling, see Fig. 30.

SOFFER and MANNHEIM (1994) have optimized an enzymatic peeling process for Valencia oranges and grapefruit. The procedure for Valencia oranges includes: scoring of peels; vacuum infusion of an enzyme solution which consisted of pectinases and cellulase; incubation (20–25 min) at 40 °C, and rinsing. The peeling process for grapefruit included: scalding; removing of the peel near the stem; scoring; vacuum infusion of enzyme solution; incubation at 40 °C for 50 min; and rinsing. The fruit obtained by enzymatic peeling was firm and had a very good shiny appearance. The enzyme solution could be used for two consecutive peeling cycles in the case of oranges, and in the case of grapefruit after restoring it to the initial activity.

Tab. 22. Enzyme Activities of Preparations Tested for Peeling of Fruits (SOFFER and MANNHEIM, 1994)

Enzyme Preparation	PG (PGU/mL)	Pectin lyase (PTEU/mL)	PE (PEU/mL)	Cellulase (CEU/mL)
Pectinex Ultra SP-L	26000	6000	600	<50
Rohapect D5S	25000	2500	400	50
Rohapect 9-90	42000	7000	115	50
Rohament CT	750	0	0	2500
Celluclast 1.5 L	0	0	0	1900

Tab. 23. Appearance of Valencia Oranges Peeled with Different Combinations of Enzyme Preparations (SOFFER and MANNHEIM, 1994)

Pectolytic Enzymes	Cellulolytic Enzymes	Incubation Time (min)	Appearance
Pectinex Ultra SP-L	Celluclast 1.5 L	60–80	Very good
Rohapect D5S	Rohament CT	95–120	Not good
Rohapect 9-90	Rohament CT	50–95	Good

Enzyme preparations from Novo Nordisk Ferment AG and Röhm, Germany have been characterized as to their content of pectin esterase (PE), polygalacturonase (PG), pectin lyase, and cellulase (CE) activities. Generally a set of standard analysis methods was used. The enzymatic activities of five different preparations tested for peeling of the fruits are shown in Tab. 22, and in Tab. 23 the qualitative appearance of Valencia oranges peeled with different combinations of enzyme preparations are shown.

The evaluation was based on the time required for peel removal and peeling quality. The data show that it is necessary to use effective pectolytic and cellulolytic enzyme preparations. The enzyme solutions used for immersion of the fruits contained 2 g of pectinase and 1 g of cellulase per kg of 0.02 mol/L sodium citrate/citric acid at 40°C, pH 4–5. The limiting activity with regard to incubation time was the cellulolytic. However, too high concentrations damaged the fruit. Lower and/or higher pectolytic enzyme concentrations did not affect peeling time but reduced appearance of the peeled oranges.

This example of an enzyme application within the technique of "enzyme infusion" as discussed by MCARDELE and CULVER (1994) is highlighted here as an illustration of adaptation of enzymes already marketed to be applied in a new process.

7.2.3 Carrot Juice

Carrot juice is characterized by being cloud stable. The polysaccharides of carrots have been described by MASSIOT et al. (1989). The carrot cell wall contains more than 80% of polysaccharides composed of high-branched pectins (45%) highly methylated and weakly acetylated, of cellulose (25%) and of hemicellulose. The methylation analysis of pectic side chains shows the presence of arabinogalactans (types I and II), linear galactans and branched arabinans. The hemicelluloses are xyloglucans, xylans and mannans.

The composition and structure of the polysaccharides solubilized by purified enzymes confirmed that pectins are highly branched, that cellulose is wrapped in pectins, and suggest possible associations between glucans and galacturonans, or between pectic side chains and xylans.

The two enzyme preparations, SP-249 which is rich in pectinase, and Celluclast, which is an endoglucanase, have been shown to act with synergy when used in combination, and lead to the liquefaction of 95% and depolymerization of 70% of the cell-wall polysaccharides.

7.2.4 Coconut Milk

Coconut milk is usually prepared by aqueous extraction of, or expression from comminuted coconut meat. The oil content differs markedly from that of cow's milk. While cow's milk has about equal amounts of fat and protein, coconut milk has about ten times as much oil (fat) as protein. The overall dry matter content of coconut milk is influenced by the amount of water used in its extraction process. Sometimes the dry matter content is so high that the product is a cream and not a milk. Most of the coconut milk is consumed fresh in the rural areas by squeezing the coconut meat. Markets for convenient, high quality products with long shelf life are well defined.

Alfa-Laval has developed aseptic coconut processes for production of the following products: coconut milk, coconut skimmilk, coconut water (LEUFSTEDT, 1990). Coconut products are extremely sensitive to heat treatment and other processing. Therefore, many attempts have been made to modify the processing or to add stabilizers (HAGENMAIER et al., 1974; HAGENMAIER, 1980).

CHRISTENSEN and OLSEN (1990) have patented a method for production of an upgraded coconut product comprising the following steps:

1. Treatment of an aqueous homogenized suspension of coconut meat by cell wall degrading enzyme and galactomannase. Dosage of enzyme in the preliminary pilot plant trials was 1% w/w Viscozyme 120 L + 1% Gammanase 1.5 L based on coconut dry matter (temperature 50 °C, pH = 4.5, t = 4 hours).
2. After heating to 90 °C separation was carried out by using a decanter, a solids ejecting centrifuge and a three-phase oil separator. Clear virgin coconut oil was recovered directly from the separator. The solids-free aqueous phase was evaporated and spray-dried and the solid phase was freeze-dried. The following approximate mass balance was obtained:

Raw material:	100 kg dry coconut meat (97% dry matter, 57% fat, 6% protein)
Virgin oil:	54 kg oil
Coconut skimmilk powder:	30 kg powder (30% protein, 55% carbohydrates, 6–9% ash, 2% fat)
Low fat meat:	14 kg (6% fat, 8.5% protein, 85% polysaccharides)

A process for production of enzymatically extracted and stabilized full fat coconut milk has also been developed. In principle, such product is made by a process as the above but without the separation of oil after the decanter. A homogenization secures the physical stability of the milk product.

7.2.5 Banana

In most banana producing countries large amounts of fruit are lost every year due to poor storage, transportation, and a large surplus. As an example of the use of surplus fruit, bananas are processed into clear juice or concentrate (KOFFI et al., 1991). Since bananas have a high sugar content and a recognizable, desirable flavor, high-value clarified juices from excess bananas could become a valuable product for drinking juice or for baby food. Among the problems associated with banana juice processing, a high viscosity and difficulties with juice extraction as well as browning are the most severe. A most efficient method seems to be usage of a multicomponent enzyme which includes pectinases, hemicellulases and cellulases followed by a decanter centrifugation. Ripe banana also contains starch, which affects the viscosity and the clarity of the juice. KOFFI et al. (1991) conducted experiments to determine the effects of commercial enzyme preparations on viscosity reduction and filterability of banana juice.

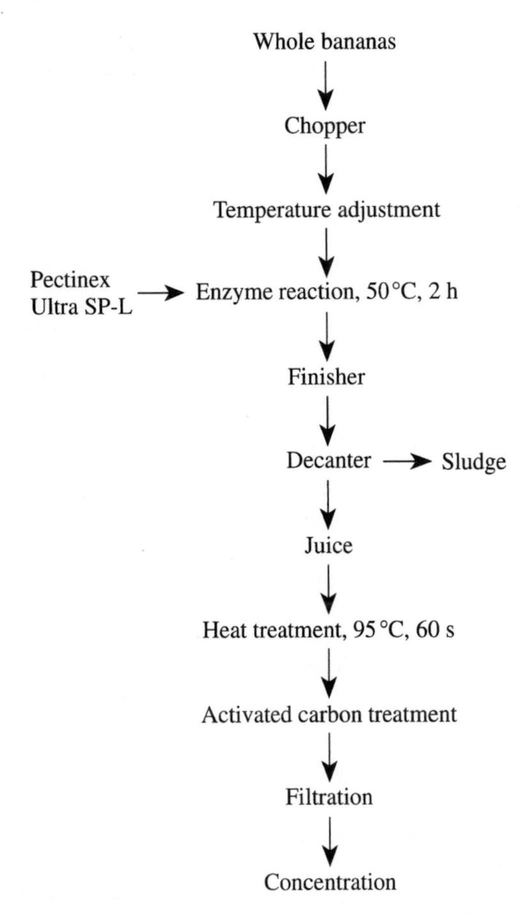

Fig. 31. Processing of whole bananas to clear banana juice.

Novo Nordisk Ferment (*Biotimes,* 1994b) has developed three banana juice processes with a yield over 80% after separation, based on whole bananas, peeled bananas or banana purée. With banana purée, it is important to gelatinize and completely degrade the starch. This is done by the use of fungal α-amylase and amyloglucosidase (see Tab. 1). A dosage of 200 ppm of the multicomponent cell-wall degrading enzyme Pectinex Ultra SP-L has been used for 2 hours in a stirred tank reactor at 50 °C.

A complete flow sheet for a process based on processing of whole bananas to clear banana juice is shown in Fig. 31.

7.3 *In situ* Gelation of Fruit

An undesirable gel formation in orange juice concentrate has been related to the action of pectin esterase (PE) on insoluble pectin (SCHWIMMER, 1981). A correlation between the action of PE on insoluble pectin and the degree of clarification has been observed. Cloudiness of orange juice is desired, but a PE-induced formation of the gel of Ca and Mg pectate can drag down occluded cloud particles.

In concentrated citrus and other fruit juices gels were formed due to the action of very little PE at favorable pH, high sugar concentrations and increased divalent cation concentrations. Thereby even lightly demethylated pectin was rendered easily susceptible to gelation as in jams and jellies (SCHWIMMER, 1981). The albedo fraction of the peel is known as a rich source of PE, which finds its way into the juice and causes the gelling.

YAMAOKA et al. (1983) have purified pectin methylesterase from mung bean (*Phaseolus aureus*) and demonstrated that it could cause gelation of apple pectin solution. Furthermore, KOMAE et al. (1989) isolated a pectin esterase that was responsible for spontaneous gel formation of the awkeotsang polygalacturonide, isolated from the red tepals attached to the pedicels of seeds of *Ficus awkeotsang* Makino. The methylesterase was partially purified by ion-exchange column chromatography. ISHII et al. (1979) purified a pectin esterase from the culture medium of *Aspergillus japonicus* completely free of pectin depolymerizing enzymes. The purified enzyme was able to convert high-methoxyl pectin into low-methoxyl pectin capable of forming strong gels with calcium ions.

Conventionally produced high-methoxyl pectins (HM-pectin) from citrus fruits require large amounts of sugar and low pH for gel formation. Low methoxyl pectins (LM-pectin), on the other hand, can form gels with or without sugar in the presence of divalent cations such as calcium. Conventional LM-pectins are prepared from HM-pectins by using either of the four demethylating agents: (1) acid, (2) alkali, (3) ammonia in alcohol and (4) enzyme.

DALBØGE and HELDT-HANSEN (1994) developed a cloning system, which is independent of prior knowledge of the amino acid sequence of the enzyme of interest. This was achieved by combining the ability of *S. cerevisiae* to express heterologous genes with the utilization of functional enzyme screening assays. A number of different enzymes including the monocomponent pectin methylesterase (PME) have successfully been cloned by use of this method. HOCKAUF et al. (1994) have characterized this cloned PME and found that it resembled in regard to kinetics those properties expected for a PME of fungal origin like the one produced by ISHII et al. (1979).

GRASSIN (1994) has disclosed the use of PME for treatment of fruits or vegetables for preparation of jams, jellies, compotes, sauces and soups by demethoxylation. When sufficient calcium is present or added, this method requires no pectin or gelling agent to prepare the final food. The disclosure also suggests a modification of the standard US apple sauce preparation process.

In situ gelation with PME has thus been demonstrated based on a development that started already with the first observations reported by SCHWIMMER (1981) on gelling of orange juice.

7.4 Oil Extraction

Oil from rape seed (canola), coconut, corn germ, sunflower seed, palm kernel and olives is traditionally produced by a combined process using expeller pressing followed by extraction with organic solvents. For the oil industry the unrecovered solvent from vegetable oil extraction processes is a problem. The solvent most commonly used in this process is hexane, which has been identified as a hazardous air pollutant by recent environmental regulations. In the early years of solvent extraction of vegetable oil, about 1 gallon of solvent was lost for every ton of soybean processed. In the 1970s about 0.5 gallons per ton of soybean were lost, and in the 1980s and 1990s 0.3 gallons of hexane were lost per ton of soybean (KEMPER, 1994).

Cell wall degrading enzymes may be used to extract vegetable oil in an aqueous process by liquefying the structural cell wall components of the oil-containing crop. This concept is already commercialized in connection with olive oil processing (CHRISTENSEN, 1991), and it has been investigated for rapeseed oil extraction (OLSEN, 1987; OLSEN and CHRISTENSEN, 1987). DOMÍNGUEZ et al. (1994) reviewed the area of enzymatic pretreatment to enhance oil extraction from fruits and oilseeds and concluded that there is a close relation between enzymatic action and the amount of oil released. Regardless of the type of enzyme, the quality of the oil is good, and its composition is not affected by enzymatic treatment as long as the enzyme system used is essentially free of lipase activity.

OLSEN (1988b) showed that the enzyme system used for opening up the cell wall must be selected and adapted to each kind of cultivar. To deal with a variety of cell walls, broad-activity, but nevertheless very different, plant cell wall degrading enzyme preparations had to be developed. As shown in Tab. 18 for rape seed and sunflower, a high activity of pectinase, xyloglucanase and cellulase is necessary, while for coconut kernel a high mannase and galactomannanase activity is necessary.

The impact of the two widely different extraction processes – the expeller process followed by hexane extraction of residual oil versus the aqueous enzymatic process – on the final product and its quality is significant. In the enzymatic process, all the components of the cell, i.e., protein, oil and polysaccharides, are transferred to the aqueous phase, which facilitates a clean separation of these components by centrifuge processes; the traditional expeller process gives rise only to two products: an oil and a press cake. This is illustrated for rape seed in Fig. 32.

7.4.1 Olive Oil Extraction

Production and use of olive oil in foods have been a tradition for millenia, particularly in the Mediterranean countries. Today the production of this oil takes place in a large number of highly diverse factories, ranging

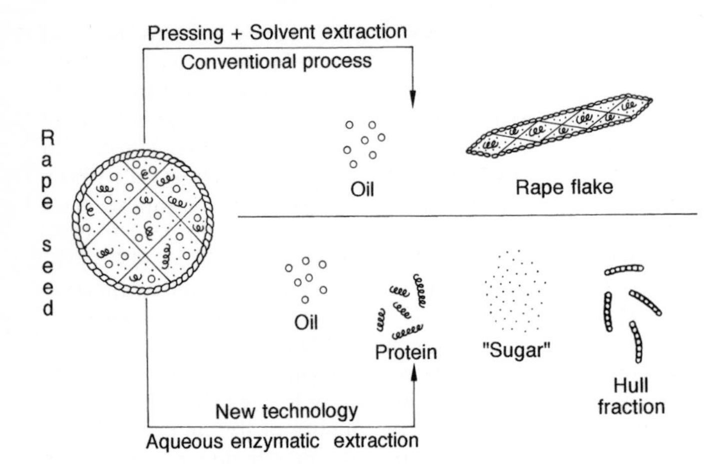

Fig. 32. Conventional process versus "new technology" aqueous enzymatic extraction.

from large, modern, stainless-steel plants to small traditional oil mills. The process is basically the same, although different equipment may be used in the various plants. The olives are ground in a so-called malaxer, then the oil/water is separated from the cake by a decanter or a press. The oil/water emulsion is subsequently separated in an oil centrifuge.

During recent years plant cell wall degrading enzyme preparations have begun to be used in olive oil processing. The enzymes are added to the malaxer when the olives are ground. Thereby the oil is released easier in the subsequent separation operations. The enzyme treatment thereby causes an increased yield of virgin olive oil. The yield increase depends on the type and ripeness of the olives, the enzyme used, temperature, pH and, of course, the dosage of enzyme applied.

Olivex®, an enzyme preparation derived from *Aspergillus aculeatus,* contains, besides the different pectolytic main activities, various side activities, hemicellulases and cellulases. A dosage of 200 mL/ton of olives is added to the mills or to the first compartment of the malaxing unit. Depending on the conditions, an additional 10–20 kg oil from one ton of olives can be obtained using Olivex. An example of a mass balance for a continuous line is shown in Fig. 33. Besides quality-preserving properties of Olivex during the process, Olivex-treated olives yield an oil with a better storage stability. The oil shows an increased content of polyphenols and vitamin E (tocopherol) which stabilizes the oil against rancidity. When olives with a higher degree of acidity are processed, the application of an enzyme like Olivex yields oil with a reduced amount of free fatty acids.

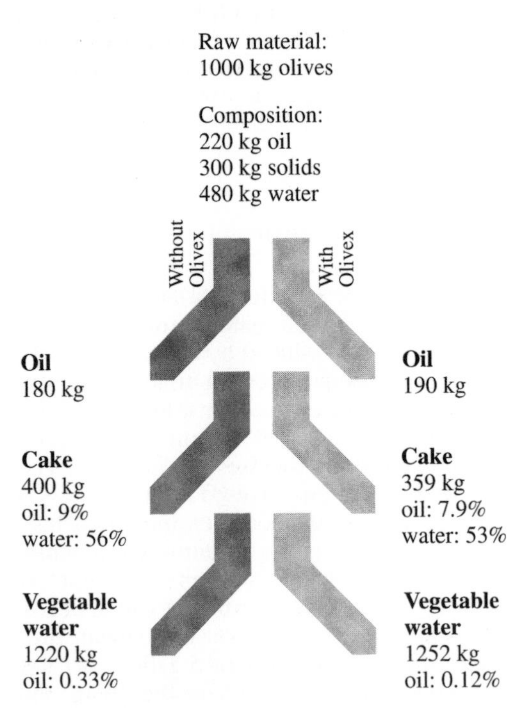

Fig. 33. Example of a mass balance (continuous line) olive oil extraction.

7.4.2 Canola/Rape Seed Oil

Aqueous enzymatic rape seed processing has been developed as a mild and environmentally safe technique for production of high-quality rape seed products. The efficient plant cell wall degrading enzyme, SP-311 (see Sect. 7.1.4), was used for the release of the oil, protein and hull fraction at pH 4.5, 50 °C for 4 hours. The fractions were separated and purified applying usual wet separation equipment like sieves, centrifuges, evaporators and dryers (OLSEN, 1988b).

Rape seed proteins have a well-balanced amino acid composition, which is reflected by a high biological value (JENSEN et al., 1990). However, too high a content of glucosinolates and their degradation products, aromatic choline esters and hulls in rape seed products can cause serious problems in feeds (BILLE et al., 1983). Too high levels of glucosinolates in the diets of monogastric animals decrease the protein utilization by lowering the biological value. The degradation products of the glucosinolates increase this effect (BILLE et al., 1983; BJERG et al., 1989). In other words, degraded glucosinolates are known to be more antinutritional than intact products. Therefore, e.g., Danish pig producers are advised to ensure that diets to pigs contain less than 1 μmol glucosinolates per g of diet (SØRENSEN, 1988).

Traditional rape seed processing (pressing and hexane extraction) often results in significant degradation of glucosinolates. The enzymatic oil recovering process has been shown to leave the glucosinolates intact. In this stage they are water soluble and can easily be washed out of the solid phases (JENSEN, 1990).

The principle, details of the technique, mass balances and economic aspects of this type of rape seed processing has been intensively studied in the pilot plant of Novo Nordisk, with oil processors and in the period 1992–1994 within the research project entitled "The Whole Crop Biorefinery Project" carried out under the ECLAIR (European Collaborative Linkage of Agriculture and Industry through Research) program. From this work the following can be summarized:

An engineering concept description for treatment of 100000 tons of rape seed per year shows that the process is compatible with conventional oil extraction technique if the protein meal finds its application in specialty feed mixtures for young animals. This means that the quality of the by-products, and thereby their prices, are more closely related to the whole process than in the conventional hexane extraction process. Economic feasibility within the Biorefining process has been demonstrated at a scale of 12000 tons rape seed/year providing, that wet feed products can be delivered to local farmers either as slurry or as fractionated slurry, i.e., as high-quality protein, hulls and syrup, and that the oil can be sold for non-food applications. If the oil should be refined for food use, at least the 100000 tons scale is required. This is due to the heavily economic governmental subsidizing of non-food production based on agricultural raw materials.

7.4.2.1 Processing Details

Twelve varieties of rape seed and canola have been selected and tested on a pilot scale. About 80% of the oil content could be extracted by the aqueous enzymatic extraction, depending on the rape seed variety and using economical dosages of enzymes.

The process proposed for large-scale operation is shown in Fig. 34.

Roller-milled rape seed has been shown to be the best substrate for the enzymatic hydrolysis. In a roller-mill two rolls revolve towards each other at the same or differential speed. The distance between the rolls is adjustable and the rolls can be either smooth, grooved or more or less toothed.

Myrosinases present in the milled seed material were initially inactivated by heat treatment of the milled seed in water at 85–90 °C for a few minutes to avoid hydrolysis of glucosinolates. The amount of water to be added should be adjusted to obtain a proper stirring during the following steps and to facilitate the wet milling operations. Fresh water is used for washing the protein fraction, thereby maintaining the water consumption at a minimum.

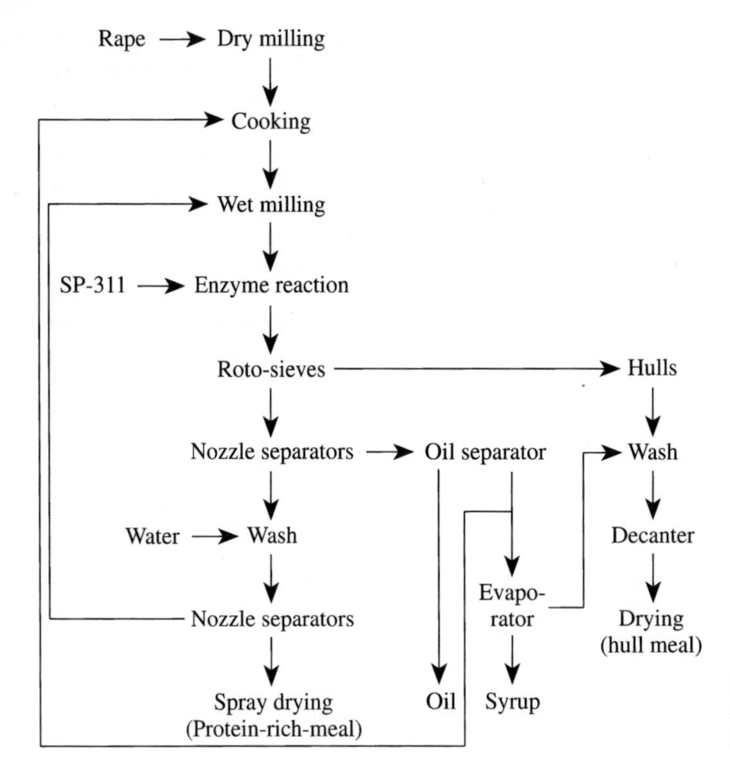

Fig. 34. Aqueous enzymatic oil extraction process. Layout proposal for large-scale operation.

Wet milling is performed to ensure a sufficient surface of the particles in order to render as many free cells as possible ready for the enzymatic attack.

In a pilot plant wet milling is performed in a rotary crusher (e.g., FRYMA MZ-130) at different settings of the clearance between the cone and the shell. If the seeds are milled leaving the cells intact, the lowest protein loss is seen, together with the maximum degradation of the cell wall material (OLSEN, 1988b). A water/solid ratio of 3.6 and three wet milling steps improved the possibilities for high oil yields in a simple batch reaction system. During preparation of the reaction mixture pH and temperature should be adjusted before adding the enzyme and for the tolerance of the substrate and the products. Optimum pH and temperature for the enzyme SP-311 are pH 4.0–5.0 and $T=50\,°C$.

The pilot plant trials have shown that a dosage of SP-311 based on rape seed of 0.50 w/w % can be used in a four hour enzyme reaction when recirculation of syrup is included as shown in Fig. 34. The effect of recirculation of syrup on the yield of oil is illustrated in Fig. 35. The enzyme dosage can be reduced to 0.25 w/w % by increasing the hydrolysis time to twelve hours.

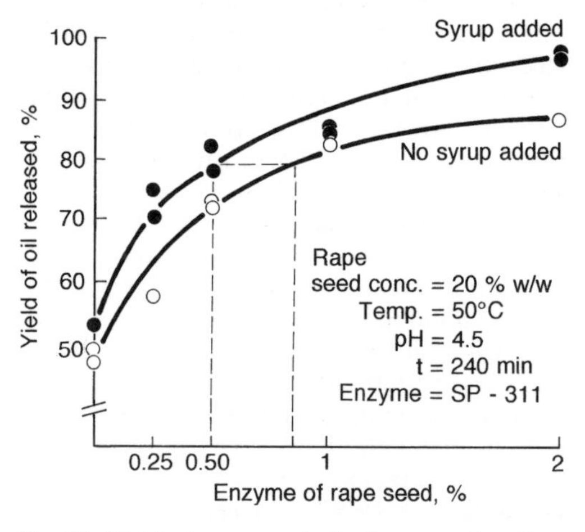

Fig. 35. Yield of rape seed oil after centrifugation.

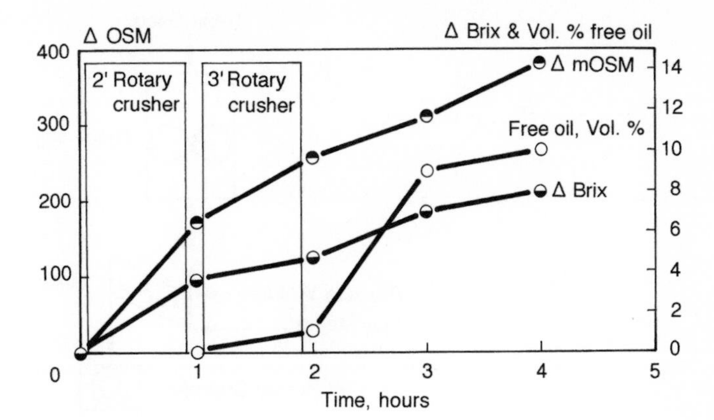

Fig. 36. Enzyme reaction: °Brix, osmolality and vol.-% of free oil.

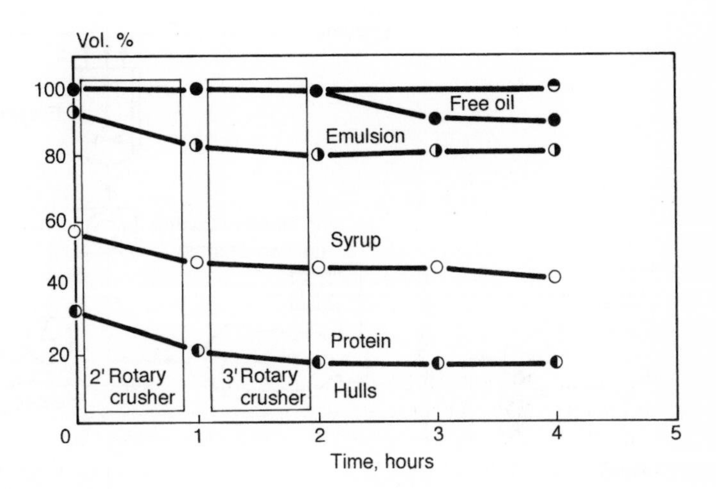

Fig. 37. Phase distribution in vol.-% versus time.

During the reaction simple analytical measurements are made in order to evaluate the effects of the reaction parameters. A spin test is made on the reaction mixture by centrifuging 10 mL in a graduated centrifuge tube at 4000 rpm (eqv. to $3000 \times g$) for 3 min in a table centrifuge. From this separation the following tests can be performed.

Refraction index on supernatant (°Brix): All soluble compounds, released from the cell content or hydrolyzed from the insoluble compounds contribute to the increase (Δ) of refraction index in relation to their concentration.

Osmolality (mOSM/kg water): Using an osmometer the decrease of the freezing point compared to the freezing point of a standard solution is quickly measured. As the freezing point is related to the concentration of "small molecules" in a solution, depending on the molecular weight, the osmolality is used as a measurement of the concentration of degraded cell wall components in the solution.

Phase distribution: From the centrifuged sample the percentage of phases are recorded on the graduation of the tube.

In Figs. 36 and 37 reaction curves from the "Biorefinery project" are reproduced which demonstrate the development of the above parameters in a single trial.

In the project a simplified solid decanter based separation process was developed by Westfalia Separator AG for fractionation of

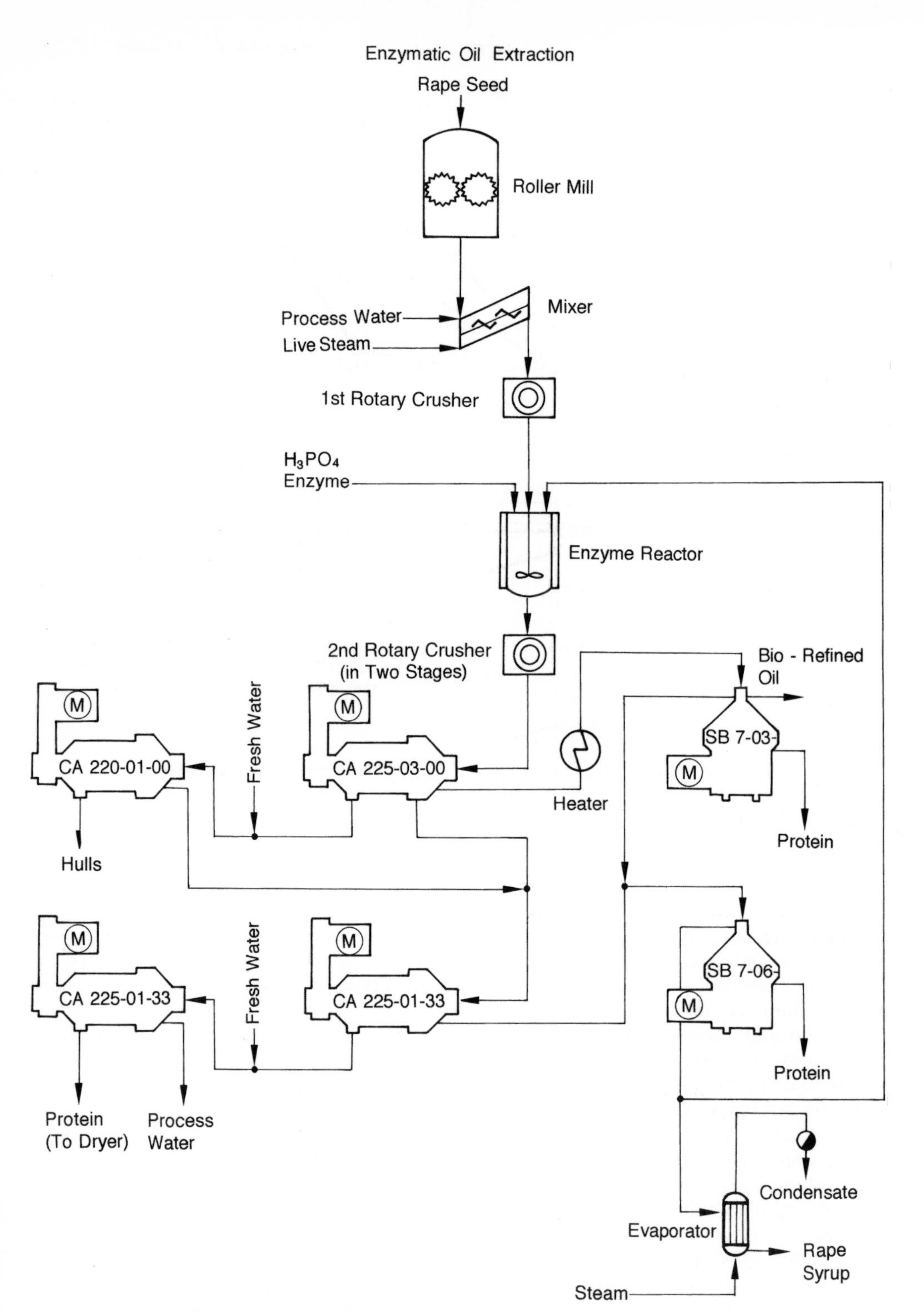

Fig. 38. Centrifuge plan for the "Biorefinery project". A: decanter centrifuges (Westfalia Separator), SB: solids ejecting centrifuges (Westfalia Separator).

rape seed into oil, protein-rich meal, hulls and syrup (Fig. 38).

7.4.2.2 Product Properties

The aqueous extracted oil

The following characteristics of aqueous enzymatic extracted oil have been identified: a content of phosphorus <10 ppm, non-hydratable phosphorus <5 ppm, total sulfur <7 ppm, a low process contribution to the oxidation state of the oil and a low content of free fatty acids (FFA of 0,43%). According to the results of the refining tests conducted by Karlshamn Oils and Fats the refining could consist of neutralization, bleaching with an approximately 30% clay reduction and deodorization to obtain a good-quality edible oil (BAGGER and OLSEN, 1995).

Aqueous extracted protein meals

Rape seed processing based on use of the cell wall degrading enzymes for twelve hours instead of four hours as suggested in earlier experiments (OLSEN, 1988a; JENSEN et al., 1990) resulted in a protein-rich meal of increased true digestibility TD (from 84 to 91%) and digestible energy DE (from 71 to 85%). While the nutritional value of protein-rich meal was slightly improved, the biological value (BV) of rape seed syrup was reduced.

Further improvement of the protein-rich meal could be obtained by a simple and easy sieving procedure of the defatted slurry suspension before separation into syrup and protein. Hereby, the fiber contents was reduced by approximately 50% compared with the contents found in protein-rich meal produced in the original process. This fiber reduction resulted in both TD and DE values of about 95%, which cannot be obtained for rape seed protein product using the centrifuge process only. Protein-rich meal containing between 50–62% protein, depending on the seed variety, and dry matter yields between 16–23 w/w% were obtained.

The four products, oil, protein-rich meal, syrup and hulls have been chemically and physiologically characterized (JENSEN, 1990). Feeding trials have been performed with minks, piglets and young calves during the weaning period. Based on these trials it was concluded that this technique results in a feed protein product of a quality comparable to, e.g., high quality fish meal, meat meal or skimmilk powder. The practical use of this process is dependent on the demand and supply of the conventional feed protein products mentioned. Further, the process is available as an alternative to the use of organic solvents in the oil milling industry.

Over the years many attempts to develop new rape seed products have always been concentrated on the possibilities of reducing the chemical and biological effects of the harmful compounds of the raw seeds. Glucosinolates and their transformation products are still of major concern in rape seed meal. Also aromatic cholin esters are undesirable. Dietary fibers (rape seed contains about 20% DF) in rape seed products have been recognized as being responsible for reduced digestibility of rape seed meal.

The aqueous enzymatic rape seed process as a means of increasing the quality of rape seed products can solve the above problems.

This technology leads to fractionation of rape seed into four products: oil, protein-rich meal, syrup and hulls. The majority of water soluble low molecular weight (LMW) rape seed constituents are extracted into the syrup fraction. The process does not involve use of organic solvents, and it can be performed at mild conditions thus limiting degradation of labile, water soluble compounds as glucosinolates. The problems related to the occurrence of degradation products of glucosinolates in the oil are hereby reduced.

The composition of the other rape seed fractions obtained, syrup, hulls and protein-rich meal, has been determined by use of various chemical-biochemical methods. The meal fraction contains very low concentrations of LMW-compounds, a high content of protein, and a relatively high content of fat and dietary fibers (DF). DF are the quantitatively dominating constituents of hulls, and DF in hulls are different with regard to several properties from the DF residues of the protein-rich meal.

7.4.3 Coconut Oil

An enzymatic oil extraction process was carried out in continuation of the coconut milk process mentioned in Sect. 7.2.4. A virgin coconut oil was produced using this technique. LEUFSTEDT (1990) discussed the need for such a product.

7.4.4 Other Enzymatic Oil Extraction Possibilities

Improvement of extractability of vegetable oils from fruits and seeds has already been demonstrated a long time ago (SOROA, 1967), although complete evaluations of the subject were scarce. Though mentioned in the "SPS-ase" patent (ADLER-NISSEN et al., 1984) – filed on 23 December 1981 (priority date) –, FULLBROOK (1983) applied enzymes to soybean and rape seed oil. DOMÍNGUEZ et al. (1993) have described the effects of variables affecting the process (mechanical treatment, moisture percentage, enzyme concentration and time of hydrolysis) on the optimum operation for soybeans and sunflower kernels as a function of the extractability of samples pretreated with various enzyme formulations. At 40–60% moisture the enzymes increased the extractability of the seeds. The data obtained were in agreement with the work carried out by SOSULSKI et al. (1988) for canola. Also the extractability of oil using hexane could be enhanced after treatment with cell wall degrading enzymes at about 30% seed moisture. Incubations were carried out for 12 hours using 0.12% w/w of enzyme (SPS-ase-like enzymes). Hereafter the mixture had to be dried to 4% moisture in order to demonstrate that the hexane extraction could be enhanced. SMITH et al. (1993) investigated the use of enzymatic hydrolysis as a pretreatment for mechanical expelling of soybeans and optimized the process parameters to yield enhanced oil recovery. The enzyme used was a crude product from *Aspergillus fumigatus*. The moisture content during hydrolysis was 23% and a dosage of 11% v/w of enzyme product was used. The incubation time used was 13 hours. The moisture content during pressing was 9%. It was shown that 64% of the total extractable oil was extrated. SOSULSKI and SOSULSKI (1993) studied the commercial enzymes SP-249 (Novo Nordisk) and Olease (Biocon) in the same process (pressing in a laboratory expeller) on canola. The seeds were flaked, autoclaved, moistened to 30% moisture and treated with 0.01–0.1% enzyme for 6 hours at 50°C. Prior to pressing in the expeller the samples were dried to 6% moisture. The treatment with enzymes improved the throughput of the expeller 30–50%. The recovery of oil was increased from 72% of the seed oil for control samples to 90–93% for enzyme treated samples. The oil quality was inferior to a cold-pressed control, but was better than has been reported for solvent extracted oil. With regard to cost evaluation only the energy consumption was considered. The enzyme pressing used 25% less energy but the extra drying costs and enzyme consumption have not been considered.

BUENROSTRO and LÓPEZ-MUNGIA (1986) succeeded in the extraction of avocado oil using α-amylase, cellulase and papain. α-Amylase was shown to be most efficient as single enzyme. Even though cellulase and papain also had effects on the oil yields, they were not able to boost the α-amylase when added in combinations. Avocado oil is produced in tropical and subtropical climates. Mexico is one of the biggest producers. Avocado oil is used in the food industry and in cosmetology due to its adsorption and penetration properties.

BADR and SITOHY (1992) extracted sunflower oil using substrate concentrations of 30 and 50%, 1–3 hours reaction and 0.5–2% enzyme based on sunflower seed. The yields of oil were improved by 30–50% over the control depending on conditions. These data were in agreement with those of various other authors.

SZAKÁCS-DOBOZI et al. (1988) have demonstrated that the yield of mustard oil could be enhanced by cellulolytic pretreatment. They used Celluclast® from *Trichoderma reesei* in a dosage of 0.2% of the brown mustard seeds. The yield of resulting oil from ground samples was approximately 50% higher than that obtained without cellulolytic pretreatment.

OLSEN (1988b) demonstrated the enzymatic opening of isolated corn germ cell wall and proposed the aqueous enzymatic extraction of corn oil. The most efficient degradation was obtained with a mixture of two cellulases as judged by osmolality measurements. The degrading effect was also demonstrated by the fact that the volume of hydrated cell wall material could be significantly reduced during a 4 h enzyme reaction. BOCEVSKA et al. (1993) showed that by hydrothermal pretreatment of corn germ it was possible to inactivate native enzymes present in the germ and to loosen its structure. The corn germ was then ground and treated with enzymes. After the oil had been released by the enzymatic reaction, it was separated by centrifugation. The oil quality had a lower content of colored substances and an extremely low content of phosphatides compared to expeller crude oil or degummed oil. The oil yield was shown to be 85% of the available oil by using Celluclast® (BOCEVSKA et al., 1992).

Selected grape varieties with aroma potential, e.g.:
- Gewürztraminer
- Muscat
- Riesling
- Morio Muscat
- Chenin Blanc

Temperature: Min. 14°C
Time: 30 days

Novoferm 12 L: 10 ml/hl

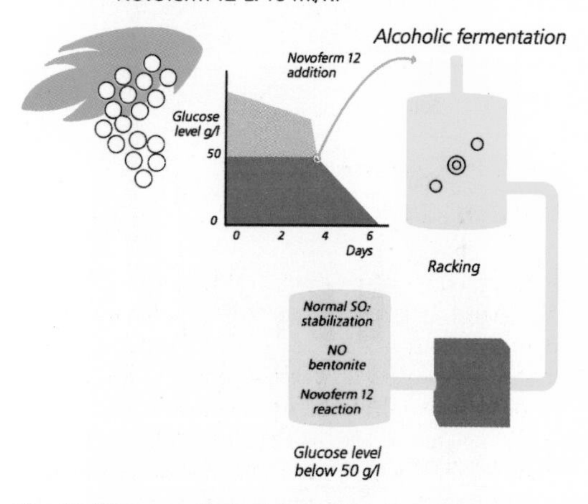

Fig. 39. The use of Novòferm 12 to enhancement of the varietal bouquet.

7.5 Winemaking

The grapes own enzyme activities are weak and often insufficient to break down pectic substances. The use of cell wall degrading enzymes in winemaking is new and offers the winemaker the advantages of improved quality and flexibility in the production of new types of wine. All the same advantages as mentioned for fruit juice processes will be valid for wineries, too. The main applications in the steps of wine and grape juice production are to improve maceration, skin contact, debourage, color extraction, clarification and filtration.

Vinozym® EC, is a blend of pectinases, arabanases and cellulases, which allow optimal release of the red color from skin. This enzyme is used for extraction and stabilization of color in the production of red wines and red grape juice. Furthermore, it has been observed that red wines produced with Vinozym EC have a pleasant fruity aroma and the clarification of the young wine is improved. Vinozym EC is dosed directly into the crusher or after the crusher into the mash tank line. Typical dosages will be 20–50 mL/ton mash (NOVO NORDISK FERMENT, 1991).

β-1,3-Glucans with very short β-1,6 branched side chains, are produced in wine made by grapes attacked by the fungus *Botrytis cinerea.* Such β-glucans when passed into the wine hinder clarification and rapidly clog filters. These troublesome β-glucans can easily be removed by adding a highly specific β-glucanase, which breaks down the glucans completely to glucose. Glucanex™ has been especially developed in order to improve the clarification and filtration of wines made from botrytized grapes. This improvement is the result of a selective enzymatic degradation of the glucans produced in the grapes by *Botrytis cinerea.*

To enhance the bouquet of wine, glycosidases have been developed which hydrolyze terpenyl glycosides, known as bound terpenes

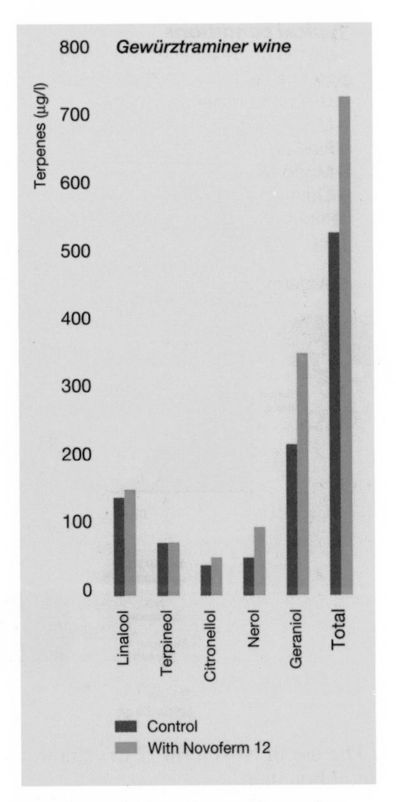

Fig. 40. Effect of Novoferm 12 on free terpenes after 30 days of fermentation.

(WILLIAMS et al., 1992; DUPIN et al., 1992; *Biotimes,* 1993). Terpenes are one of the important constituents of the bouquet, particularly for varieties with an intense floral aroma such as Muscat, Riesling and Gewürztraminer. By the application of such an enzyme as Novoferm 12 it is possible to liberate more aromatic compounds and improve the bouquet. The best time to add Novoferm 12 is towards the end or after the fermentation (see Fig. 39). The enzyme is completely inactivated by addition of fining agents such as bentonite, which is normally added after racking to stabilize the wine. Trials made in Germany have proven that the enzyme does not influence the wine's ageing properties. Fig. 40 shows that the total content of major aromatic terpenes in a Gewürztraminer wine could be increased significantly by the enzyme usage (*Biotimes,* 1993).

8 Enzymatic Modification of Lipids

In synthetic chemistry and lipid technology enzymes have been considered of potential value for many years. Due to the high selective mode of action and the ability of specific enzymes to catalyze reactions at organic–aqueous interfaces enzymes are found useful in some synthetic organic chemistry. To enhance stability and reaction rates immobilized enzymes are often used.

Concentrated efforts to introduce immobilized lipase technology into the fats and oils industry have been made since the mid-1980s (EIGTVED, 1992; GODTFREDSEN, 1993).

A number of specific lipases are used for ester synthesis, interesterification and hydrolysis reactions. The components involved in the reactions are oils (triglycerides), glycerol, free fatty acids, esters and alcohols (see Fig. 41). Lipases provide an opportunity for the fats and oils industry to produce new types of triglycerides, esters and fatty acids and to produce existing products of a higher quality than with conventional technology. A few examples of products are edible oil, which is nutritionally balanced with respect to saturated and unsaturated fatty acids, cocoa butter extenders, esters for lubricants and cosmetics, monoglycerides as emulsifiers, etc.

The lipase-catalyzed interestification of lipids for use in the fats and oils industry has been researched since the first publication of MACRAE (1983). The practical use for production of nutritionally improved fats has recently been initiated (QUINLAN and MOORE, 1993).

The laundry industry was the first user of industrial lipases. The first detergent lipase, Lipolase®, has been introduced in 1988 into this industry. Lipolase was originally isolated from the fungus *Humicola lanuginosa* with low levels of enzyme expression. Using recombinant DNA techniques the lipase is now expressed in acceptable yields in the host microorganism *Aspergillus oryzae* (BOEL and HUGE-JENSEN, 1989).

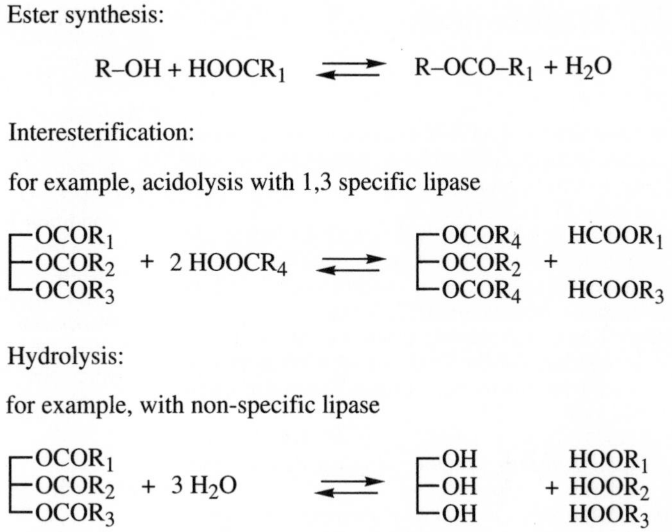

Ester synthesis:

$$R\text{–}OH + HOOCR_1 \rightleftharpoons R\text{–}OCO\text{–}R_1 + H_2O$$

Interesterification:

for example, acidolysis with 1,3 specific lipase

$$\left[\begin{array}{l}-OCOR_1\\-OCOR_2\\-OCOR_3\end{array}\right. + 2\,HOOCR_4 \rightleftharpoons \left[\begin{array}{l}-OCOR_4\\-OCOR_2\\-OCOR_4\end{array}\right. + \begin{array}{l}HCOOR_1\\ \\HCOOR_3\end{array}$$

Hydrolysis:

for example, with non-specific lipase

$$\left[\begin{array}{l}-OCOR_1\\-OCOR_2\\-OCOR_3\end{array}\right. + 3\,H_2O \rightleftharpoons \left[\begin{array}{l}-OH\\-OH\\-OH\end{array}\right. \begin{array}{l}HOOR_1\\+\,HOOR_2\\HOOR_3\end{array}$$

Fig. 41. Lipase reactions.

9 Conclusion

A number of enzyme applications have been mentioned. The emphasis of the sections of this chapter has been put on the subjects of the writer's own research over the years and on the industrially most significant applications, as seen from an enzyme producer's point of view. Some enzyme applications have not been mentioned in detail. This is the case, for example, with regard to the use of enzymes in fish processing, where enzymes are used for production of fish sauce, for salted matjes herrings, for deskinning of fish, for production of protein hydrolysate as a cryoprotective agent, and for roe caviar purification. For some of these applications reference is made to STEFANSSON (1993) who has reviewed the area recently.

Enzymatic synthetic reactions for food ingredients could cover synthesis of artificial sweeteners like aspartame. The area of fat replacers and soluble indigestible bulking agents is another area where enzyme processes are proposed, even these areas have not yet matured to industrial use. Also the area of "enzymes and food flavor" could deserve more emphasis, however, the review of CHRISTEN and LÓPEZ-MUNGUÍA (1994) covers the use of enzymes in flavor generation in great details. Important products derived from fats, proteins, nucleic acids and flavor precursors are discussed as one part. Enzyme reactions used for elimination of natural or process-induced off-flavors are discussed in a second part as well as the use of enzymes for direct synthesis of flavoring compounds.

Acknowledgements

I am grateful to the product managers and present project leaders at Novo Nordisk whom I have consulted for corrections of significant sections like starch modification (ERIK ANDERSEN and TOMMY REX CHRISTENSEN), baking (JOAN QI SI), dairy (PETER BUDTZ and JAN BOEG HANSEN), brewing (SØREN JEPSEN), wine and juice (Dr. KURT DÖRREICH), and protein (PER MUNK NIELSEN). I am grateful for their help and advice during the preparation of the manuscript.

Most of all I thank my wife (Ulla) and my two children (Dorte and Søren) for their patience with me during the week-ends and nights and early mornings, when I have been sitting by the desk with all the references behind the portable computer. Novo Nordisk is thanked for permission to undertake this work.

10 References

ADLER-NISSEN, J. (1979), Determination of the degree of hydrolysis of food protein hydrolysates by trinitrobenzenesulfonic acid, *J. Agric. Food Chem.* **27**, 1256–1262.

ADLER-NISSEN, J. (1982), Limited enzymatic degradation of proteins: A new approach in the industrial application of hydrolases, *J. Chem. Technol. Biotechnol.* **32**, 138–156.

ADLER-NISSEN, J. (1984), Control of the proteolytic reaction and the level of bitterness in protein hydrolysis processes, *J. Chem. Technol. Biotechnol.* **34B**, 215–222.

ADLER-NISSEN, J. (1986), *Enzymic Hydrolysis of Food Proteins,* London: Applied Science Publishers.

ADLER-NISSEN, J. (1988), Bitterness intensity of protein hydrolysates – Chemical and organoleptic characterization, in: *Frontiers of Flavor* (CHARALAMBOUS, G., Ed.), pp. 63–77, Amsterdam: Elsevier.

ADLER-NISSEN, J. (1993), Proteases, in: *Enzymes in Food Processing,* 3rd Ed. (NAGODAWITHANA, T. W., REED, G., Eds.), New York: Academic Press, Inc.

ADLER-NISSEN, J., OLSEN, H. S. (1979), The influence of peptide chain length on taste and functional properties of enzymatically modified soy protein, *Am. Chem. Soc. Symp. Ser.* **92**, 125–146.

ADLER-NISSEN, J., OLSEN, H. S. (1982), Taste and taste evaluation of soy protein hydrolysates, in: *Chemistry of Foods and Beverages – Recent Developments* (CHARALAMBOUS, G., INGLETT, G., Eds.), pp. 149–169, New York: Academic Press.

ADLER-NISSEN. J., ERIKSEN, S., OLSEN, H. S. (1983), Improvements of the functionality of vegetable proteins by controlled enzymatic hydrolysis, *Qual. Plant. Plant Foods Hum. Nutr.* **32**, 411–423.

ADLER-NISSEN, J., GÜRTLER, H., JENSEN, G. W., OLSEN, H. S., RIISGAARD, S., SCHÜLEIN, M. (1984), SPS, SPS-ase and method for producing SPS-ase, *US Patent* 4478939 (to Novo Industri A/S, Denmark).

AMES, J. M., ELMORE, J. S. (1992), Aroma components of yeast extracts, *Flavour Fragrance J.* **7**, 89–103.

ANDREWS, B. A., ASENJO, J. A. (1987), Enzymatic lysis and disruption of microbial cells, *Trends Biotechnol.* **5**, 273–277.

ANSON, M. L. (1939), The estimation of pepsin, trypsin, papain and cathepsin with haemoglobin, *J. Gen. Physiol.* **22**, 79–89.

ASPINALL, G. O. (1980), Chemistry of cell wall polysaccharides, in: *The Biochemistry of Plants* (PREISS, J., Ed.), Vol. 3, pp. 473–500, New York: Academic Press.

AUNSTRUP, K., OLSEN, F. (1986), Alpha-acetolactate decarboxylase enzyme and preparation thereof, *US Patent* 4617273.

BADR, F. H., SITOHY, M. Z. (1992), Optimizing condition for enzymatic extraction of sunflower oil, *Grasas Aceites* **43** (5), 281–283.

BAGGER, C. L., OLSEN, H. S. (1995), *The Whole Crop Biorefinery Project – Aqueous Enzymatic Extraction of Agricultural Crops.* Final Report from Task Nr. 5 for the ECLAIR Project Contract Nr. AGRE – CT91 – 0061 (in preparation).

BAINS, W. (1993), *Biotechnology from A to Z,* Oxford University Press.

BALDWIN, C. V., ROBINSON, C. W. (1994), Enhanced disruption of *Candida utilis* using enzymatic pretreatment and high-pressure homogenization, *Bioeng. Biotechnol.* **43**, 46–56.

BERNHOLDT, H. F. (1975), Meat and other proteinaceous foods, in: *Enzymes in Food Processing,* 2nd Ed. (REED, G., Ed.), New York: Academic Press, Inc.

BEVERIDGE, T., HARRISON, J. E., GAYTON, R. R. (1992), Decanter centrifugation of apple mash: effect on centrifuge parameters, apple variety and apple storage, *Food Res. Int.* **26**, 125–130.

BILLE, N., EGGUM, B. O., JACOBSEN, I., OLSEN, O., SØRENSEN, H. (1983), The effects of processing on antinutritional constituents and nutritive value of double low rapeseed meal, *Z. Tierphysiol. Tierernähr. Futtermittelkd.* **49**, 148–163.

Biotimes (1993), Novoferm™ 12 – the bouquet booster, Vol. VIII, No. 3, 12–13 (a quarterly magazine from Novo Nordisk A/S).

Biotimes (1994a), A mold can do what a calf does. Watch out chymosin. Here comes Novoren®, Vol. IX, No. 2, 2–4.

Biotimes (1994b), New markets await tropical juices, Vol. IX, No. 4, 8–9.

BIOTOL (1991), *Biotechnological Innovations in Food Processing,* Oxford: Butterworth-Heinemann.

BIRK, Y. (1976), Proteinase inhibitors from plant sources, *Methods Enzymol.* **45**, 695–739.

BJERG, B., EGGUM, B. O., JACOBSEN, I., OTTE, J., SØRENSEN, H. (1989), Antinutritional and toxic effect in rats of individual glucosinolates (±myrosinases) added to standard diet 2, *Z. Tierphysiol. Tierernähr. Futtermittelkd.* **61**, 227–44.

BLANCHARD, P. H, GEIGER, E. O. (1984), Production of high-fructose corn syrup in the USA, *Sugar Technol. Rev.* **11**, 1–94.

BOCEVSKA, M., KARLOVIĆ, Dj., TURKULOV, J. (1992), in: *Proc. World Conf. Oilseed Technology and Utilization* (APPLEWHITE, T. H., Ed.) September 13–18, Budapest, AOCS Press, Champaign, 1993, pp. 470–74.

BOCEVSKA, M., KARLOVIĆ, Dj., TURKULOV, J., PERICIN, D. (1993), Quality of corn germ oil obtained by aqueous enzymatic extraction, *J. Am. Oil Chem. Soc.* **70** (12), 1273–1277.

BOEL, E., HUGE-JENSEN, B. (1989), *Eur. Patent Appl.* 305216 (to Novo Nordisk).

BRENNA, O., BIANCHI, E. (1994), Immobilised laccase for phenolic removal in must and wine, *Biotechnol. Lett.* **16** (1), 35–40.

BUENROSTO, M., LÓPEZ-MUNGUÍA, C. A. (1986), Enzymatic extraction of avocado oil, *Biotechnol. Lett.* **8**, 505–506.

BUDTZ, P. (1989), Microbial rennets for cheese making, *Dairy Ind. Int.* **55** (5), 15–20.

BURGESS, K., SHAW, M. (1983), Dairy, in: *Industrial Enzymology* (GODFREY, T., REICHELT, J., Eds.), London: Macmillan Publishers Ltd.

CARPITA, N. C., GIBEAUT, D. M. (1993), Structural models of primary cell walls in flowering plants: Consistency of molecular structure with the physical properties of the walls during growth, *Plant J.* **3**, 1–30.

CHAVAN, J. K., HEIJGAARD, J. (1981), Detection and partial characterization of subtilisin inhibitors in legume seeds by isoelectric focusing, *J. Sci. Food Algric.* **32**, 857–862.

CHOUVEL, H., CHAY, P. B., CHEFTEL, J.-C. (1983), Enzymatic hydrolysis of starch and cereal flours at intermediate moisture contents in a continuous extrusion-reactor, *Lebensm. Wiss. Technol.* **16**, 346–353.

CHRISTEN, P., LÓPEZ-MUNGUÍA, A. (1994), Enzymes and food flavour – a review, *Food Biotechnol.* **8** (2&3), 167–190.

CHRISTENSEN, F. M. (1989), Enzyme technology versus engineering technology in the food industry, *Biotechnol. Appl. Biochem.* **11**, 249–265.

CHRISTENSEN, F. M. (1991), Extraction by aqueous enzymatic processes, *Inform* **2** (11), 984–987.

CHRISTENSEN, F. M., OLSEN, H. S. (1990), Method for production of an upgraded coconut product, *US Patent* 4904483 (to Novo Nordisk).

CHRISTENSEN, L. K. (1952), Denaturation and enzymatic hydrolysis of lactoglobulin, *C. R. Trav. Lab. Carlsberg Sér. Chim.* **28** (1), 39–169.

CONN, J. F., JOHNSON, J. A., MILLER, B. S. (1950), An investigation of commercial fungal and bacterial alpha amylase preparations in baking, *Cereal Chem.* **27**, 191.

DALBØGE, H., HELDT-HANSEN, H. P. (1992), *International Publication,* Novo Nordisk A/S.

DALBØGE, H., HELDT-HANSEN, H. P. (1994), A novel method for efficient expression cloning of fungal enzyme genes, *Mol. Gen. Genet.* **243**, 253–260.

DE ROOIJ, J. F. M., HAKKAART, M. J. J. (1992), A process for the preparation of food flavours, *Eur. Patent* 0191513B2 (new European patent specification to Unilever N. V.).

DEL ROSARIO, R. R., GABUYA, E. S. (1980), Preliminary studies on the polysaccharide composition of coconut and makapuno cell wall, *Philipp. J. Coconut Stud.* **5** (1), 16–21.

DOMÍNGUEZ, H., NÚÑEZ, M. J., LEMA, J. M. (1993), Oil extractability from enzymatically treated soybean and sunflower: range of operational variables, *Food Chem.* **46**, 277–284.

DOMÍNGUEZ, H., NÚÑEZ, M. J., LEMA, J. M. (1994), Enzymatic pretreatment to enhance oil extraction from fruits and oilseeds: a review, *Food Chem.* **49**, 271–286.

DÖRREICH, K. (1983), Totalverflüssigung von Äpfeln, *Fluess. Obst* **50** (7), 304–307.

DÖRREICH, K. (1986), Investigations on production of apple juice without the utilisation of presses. *Proc. IFU-Symp.*, Den Haag, The Netherlands, pp. 183–197.

DÖRREICH, K. (1993a), New fruit juice technologies with enzymes, *Proc. IFU-Symp. "New Trends in Juice Processing"*, Budapest, Hungary, pp. 51–62.

DÖRREICH, K. (1993b), Influence of technologies on juice characteristics and properties, *Shokuhin Kogyo (Food Industry)* **36** (18), 33–41.

DUPIN, I., GÜNATA, Z., SAPIS, J. C., BAYONOVE, C., M'BAIARAOUA, O., TAPIERO, C. (1992), Production of β-apiosidase by *Aspergillus niger:* partial purification, properties and effect on terpenyl apiosylglucosides from grape, *J. Agric. Food Chem.* **40** (10), 1886–1891.

DÜSTERHÖFT, E.-M., POSTHUMUS, M. A., VORAGEN, A. G. J. (1992), Non-starch polysaccharides from sunflower (*Helianthus annuus*) meal and palm-kernel (*Elaeis guineensis*) meal – Investigation of the structure of major polysaccharides, *J. Sci. Food Agric.* **52**, 151–160.

DÜSTERHÖFT, E.-M., BONTE, A. W., VORAGEN, A. G. (1993), Solubilization of non-starch polysaccharides from oil-seed meals by polysaccharide degrading enzymes, *J. Sci. Food Agric.* **53**, 211–220.

DZIEDZIC, S. Z., KEARSLEY, M. W. (1984), *Glucose Syrups: Science and Technology,* London–New York: Elsevier Applied Science Publishers Ltd.

EIGTVED, P. (1992), Enzymes and lipid modification, in: *Advances in Applied Lipid Research* (PADLEY, F. B., Ed.), pp. 1–64, London: Vol. 1, JAI Press Ltd.

FEDRICK, I. A., FULLER, S. C. (1988), Comparison of calf rennet and modified *Mucor miehei* coagulant in cheddar cheese, *Aust. J. Dairy Technol.* (May), 12–15.

FOX, P. F. (1981), Proteinases in dairy technology, *Neth. Milk Dairy J.* **35**, 233–253.

FOX, P. F. (1993), Exogenous enzymes in dairy technology – A review, *J. Food Biochem.* **17**, 173–199.

FULLBROOK, P. D. (1983), The use of enzymes in the processing of oilseeds, *J. Am. Oil Chem. Soc.* **60**, 476–478.

GIOVANELLI, G., RAVASINI, G. (1993), Apple juice stabilization by combined enzyme–membrane filtration process, *Lebensm. Wiss. Technol.* **26**, 1–7.

GODFREY, T. (1983a), Brewing, in: *Industrial Enzymology* (GODFREY, T., REICHELT, J. Eds.), London: Macmillan Publishers Ltd.

GODFREY, T. (1983b), Comparison of key characteristics of industrial enzymes by type and sources, in: *Industrial Enzymology* (GODFREY, T., REICHELT, J., Eds.), London: Macmillan Publishers Ltd.

GODFREY, T. (1983c), Flavouring and colouring, in: *Industrial Enzymology* (GODFREY, T., REICHELT, J., Eds.), London: Macmillan Press, Ltd.

GODTFREDSEN, S. E. (1993), Lipases, in: *Enzymes in Food Processing*, 3rd Ed. (NAGODAWITHANA, T. W., REED, G., Eds.), New York: Academic Press, Inc.

GRASSIN, C. (1992), Preßenzyme in der apfelverarbeitenden Industrie, *Flüss. Obst* **59** (7), 418–422.

GRASSIN, C. (1994), Use of pectinesterase in the treatment of fruit and vegetables. *PCT WO94/12055* (filed by Gist Brocades N. V. [NL/NL]).

GRIPON, J.-C., MONNET, V., LAMBERET, G., DESMAZEAUD, M. J. (1991), Microbial enzymes in cheese ripening, in: *Food Enzymology,* Vol. I (FOX, P. F., Ed.), London–New York: Elsevier Applied Science.

GROSS, K. C. (1984), Fractionation and partial characterization of cell walls from normal and non-ripening mutant tomato fruit, *Physiol. Plant.* **62** (1), 25–32.

HAAS, L. W., BOHN, R. M. (1934), *US Patent* 1957333.

HAGENMEIER, R. (1980), Coconut aqueous processing, *San Carlos Publications,* University of San Carlos, Cebu City, Philippines.

HAGENMAIER, R., MATTIL, K. F., CARTER, C. M. (1974), Dehydrated coconut skim milk as a food product: Composition and functionality, *J. Food. Sci.* **39**, 196–199.

HAMADA, J. S. (1992), Modification of proteins by enzymatic methods, in: *Biochemistry of Food Proteins* (HUDSON, B. J. F., Ed.), New York–London: Elsevier Applied Science.

HAMADA, J. S. (1994), Deamidation of food proteins to improve functionality, *Crit. Rev. Food Sci. Nutr.* **34** (3), 283–292.

HAMLYN, P. F., BRADSHAW, R. E., MELLON, F. M., SANTIAGO, C. M., WILSON, J. M., PEBERDY, J. F. (1981), Efficient protoplast isolation from fungi using commercial enzymes, *Enzyme Microb. Technol.* **3**, 321–325.

HAMM, D. J. (1992), A process for production of hydrolyzed proteins and the products thereof. *Eur. Patent Appl.* No. EP 0495390 AI (Applicant: CPC International).

HOBSON, J. C., ANDERSON, D. A. G. (1991), A cohydrolytic process for the production of novel extracts from yeast and non-yeast proteins, *PCT WO 91/16447* (filed by CPC International, Inc. [US/US]).

HOCKAUF, M., BUDOLFSEN, G., HELDT-HANSEN, H. P., DALBØGE, H. (1994), Monocomponent enzymes – a key for the modification of biopolymers, Poster presented at the *Conference: Biopolymer Mixtures* 19–21 September, UK.

HOUGH, I. S., BRIGGS, D. E., STEVENS, R., JOUNG, T. W. (1982), *Malting and Brewing Science*, Vol. II, pp. 826–828, London: Chapman and Hall.

IKURA, K., SASAKI, R., MOTOKI, M. (1992), Use of transglutaminase in quality-improvement and processing of food proteins, *Comments Agric. Food Chem.* **2** (6), 389–407.

ISHII, S., KIHO, K., SUGIYAMA, S., SUGIMOTO, H. (1979), Low-methoxyl pectin prepared by pectinesterase from *Aspergillus japonicus, J. Food Sci.* **44**, 611–614.

JACOBSEN, C. F., LÉONIS, J., LINDERSTRØM-LANG, K., OTTESEN, M. (1957), The pH-stat and its use in biochemistry, *Methods Biochem. Anal.* **4**, 171–210.

JANDA, W. (1983) Totalverflüssigung von Äpfeln, Technologie und Ökonomie, *Fluess. Obst* **50** (7), 308–311.

JANDA, W., DÖRREICH, K. (1984), Optimized enzymatic apple treatment, *Proc. IFU-Symp. Advances in the Fruit and Vegetable Industry*, Tel Aviv, Israel, pp. 205–217.

JENSEN, S. K. (1990), *Biochemical and Physiological Investigations of the Meal and Syrup Fractions from Aqueous Enzymatic Rapeseed Processing,* Novo Nordisk A/S and Chemistry Department, The Royal Veterinary and Agriculture University, Copenhagen, 130 pp.

JENSEN, S. K., OLSEN, H. S., SØRENSEN, H. (1990), Aqueous enzymatic processing of rapeseed for production of high quality products, in: *Rapeseed/Canola: Production, Chemistry, Nutrition and Process Technology* (SHAHIDI, F. Ed.), pp. 331–343, New York: Van Nostrand Reinhold.

JOHANSEN, J. T., OTTESEN, M., SVENDSEN, I., WYBRANDT, G. (1968), The degradation of the oxidized B-chain of insulin by two subtilisins and their succinylated and N-carbamylated derivatives, *C. R. Trav. Lab. Carlsberg* **36**, 365–384.

JØRGENSEN, O. B., KARLSEN, L. G., NIELSEN, N. B., PEDERSEN, S., RUGH, S. (1988), A new immobilized glucose isomerase with high productivity produced by a strain of *Streptomyces murinus*. *Starch/Stärke* **40** (8), 307–313.

KEEGSTRA, K., TALMADGE, K. W., BAUER, W. D., ALBERTSHEIM, P. (1973), The structure of plant cell walls, *Plant Physiol.* **51**, 188–196.

KEMPER, T. G. (1994), Minimizing solvent loss, *Inform,* **5** (8), 898–901.

KOFFI, E. K., SIMS, C. A., BATES, R. P. (1991), Viscosity reduction and prevention of browning in the preparation of clarified banana juice, *J. Food Qual.* **14**, 209–218.

KOMAE, K., SONE, Y., KAKUTA, M., MISAKI, A. (1989), Isolation of pectinesterase from *Ficus awkeotsang* seeds and its implication in gel-formation of Awkeotsang polygalacturonide, *Agric. Biol. Chem.* **53** (5), 1247–1254.

KONIECZNY-JANDA, G., RICHTER, G. (1991), Progress in the enzymatic saccharification of wheat starch, *Starch/Stärke* **43** (8), 308–315.

KNORR, D., SHETTY, K. L., HOOD, L. F., KINSELLA, J. E. (1979), An enzymatic method for yeast autolysis, *J. Food Sci.* **44**, 1362–1365.

KRISTENSEN, S. (1985) Use of UF for separation of soluble proteins from enzymatic blood cell hydrolysis reaction mixture, *A/S De Danske Sukkerfabrikker,* File 1972-GB-1182.

KULP, K. (1968), Enzymolysis of pentosans of wheat flour, *Cereal Chem.* **45** (4), 339–350.

KWON, S. S.-Y., MARSICO, M. A., VADEHRA, D. V. (1992), Beef flavour, *Eur. Patent Appl.* EP 0505733 A1 (filed by Societé des Produits Nestlé S.A.).

LEUFSTEDT, G. (1990), Opportunities for future diversification of the coconut industry. *Oleagineux* **45** (11), 505–510.

LIENER, I. E. (1978), Toxicological considerations in the utilization of new protein foods, in: *Proc. 11th FEBS Meeting,* Vol. 44: *Biochemical Aspects of New Protein Food* (ADLER-NISSEN, J., EGGUM, B. O., MUNCK, L., OLSEN, H. S., Eds.), Oxford: Pergamon Press.

LII, C.-Y., CHANG, Y.-H. (1991), Study of starch in Taiwan, *Food Rev. Int.* **7** (2), 185–203.

LINDERSTRØM-LANG, K. (1952), Proteins and enzymes. III. The initial stages in the breakdown of proteins by enzymes, *Lane Medical Lectures,* Vol. VI, pp. 53–72, Stanford, CA: Stanford University Press.

LÜTZEN, N. W., NORMAN, B. E. (1979), Comparison of β-galactosidases from *Kluyveromyces fragilis* and *Bacillus* sp. and their application in whey treatment, *Paper* presented at the *Food Process Engineering Congress,* Espoo, Finland, August 1979 (available as file A-5570a from Novo Nordisk A/S).

MACRAE, A. R. (1983), Lipase-catalyzed interesterification of oils and fats, *J. Am. Oil Chem. Soc.* **60** (2), 291–294.

MASSIOT, P., THIBAULT, J. F., ROUAU, X. (1989), Degradation of carrot (*Daucus carota*) fibres with cell wall polysaccharide-degrading enzymes, *J. Sci. Food Agric.* **49**, 45–57.

MCARDELE, R. N., CULVER, C. A. (1994), Enzyme infusion: a developing technology, *Food Technol.* (November), 85–89.

MEI-HING, S. L., SI, J. Q., JEPSEN, S. (1993), Enzymes for the baking industry, *Lecture* presented in Guangzhou, P. R. of China, Novo Nordisk A/S Publication File A-06275.

MILLER, B. S., JOHNSON, J. A., PALMER, D. L. (1953), A comparison of cereal, fungal and bacterial α-amylases as supplements for bread baking, *Food Technol.* **7**, 38–42.

NEY, K. H. (1971), Voraussage der Bitterkeit von Peptiden aus deren Aminosäurezusammensetzung, *Z. Lebensm. Untersuch. Forsch.* **147**, 64–71.

NIELSEN, H. K. (1991), Novel bacteriolytic enzymes and cyclodextrin glycosyl transferase for the food industry, *Food Technol.* **45** (1), 102–104.

NIELSEN, P. M. (1994), Enzyme technology for production of protein based flavours, *Lecture* presented at the *FIE Conference,* London, October 1994.

NORMAN, B. E. (1982), A novel debranching enzyme for application in the glucose syrup industry, *Starch/Stärke* **34** (10), 340–346.

NORMAN, B. E., SEVERINSEN, S. G., NIELSEN, T., WAGNER, J. (1979), Enzymatic treatment of whey permeate with recovery of enzyme by ultrafiltration, *The World Galaxy for the World Dairy Industry,* No. 7.

NOVO NORDISK Application Sheet (1989), *Enzymatic Modification of Proteins Using Novo Proteinases,* Bagsvaerd, Denmark: Enzyme Process Division of Novo Nordisk A/S.

NOVO NORDISK Application Sheet (1990), *Use of Termamyl® for Starch Liquefaction,* Bagsvaerd, Denmark: Enzyme Process Division of Novo Nordisk A/S.

NOVO NORDISK Application Sheet (1993a), *Use of Fructozyme™ in the Production of Fructose from Inulin,* Bagsvaerd, Denmark: Enzyme Process Division of Novo Nordisk A/S.

NOVO NORDISK Product Sheet B-697 (1993b), *Fungamyl® BG*, Bagsvaerd, Denmark: Enzyme Process Division of Novo Nordisk A/S.

NOVO NORDISK FERMENT Ltd. (1991), *Vinozym EC,* Product Sheet B-448.

OLSEN, H. S. (1978), Continuous pilot plant production of bean protein by extraction, centrifugation, ultrafiltration, and spray drying. *Lebensm. Wiss. Technol.* **11**, 57–64.

OLSEN, H. S. (1982), Method of producing soy protein hydrolysate from fat-containing soy material, and soy protein hydrolysate. *US Patent* 4324805 (to Novo Industri A/S, Denmark).

OLSEN, H. S. (1983), Herstellung neuer Proteinprodukte aus Schlachttierblut zur Verwendung in Lebensmitteln, *ZFL – Int. Z. Lebensm. Technol. Verfahrenstechn.* **34** (5).

OLSEN, H. S. (1984), Method of producing an egg white substitute material, *US Patent* 4431629 (to Novo Industri A/S, Denmark).

OLSEN, H. S. (1987), Aqueous enzymatic extraction of rape seed oil, *Lecture* given at the *Workshop on Agricultural Refineries – A Bridge from Farm to Industry,* September 16th to 18th, Bornholm. Internal Novo file Nr. A-06008a/HSO, Novo Nordisk A/S, Denmark, 10 pp.

OLSEN, H. S. (1988a), Enzymes and food proteins, *Food Technology International Europe,* pp. 245–250, Sterling Publications Limited.

OLSEN, H. S. (1988b), Aqueous enzymatic extraction of oil from seeds, *Proc. Food Conf.* '88, Bangkok, Thailand 22–24 Oct. (MANEEPUN, S., VARANGOON, P., PHITHAKPOL, B., Eds.), Kasetsart University, Bangkok. Novo file Nr. A-06041, Novo Nordisk A/S, Denmark, 11 pp.

OLSEN, H. S. (1991), Enzymatic processing of slaughterhouse blood, *Paper* presented at the *Meat Symposium* by Zalahuis, Zalaergerszeg, Hungary on 3rd April, 1991 (Novo Nordisk file Nr. A-06143).

OLSEN, H. S., ADLER-NISSEN, J. (1979), Industrial production and application of a soluble enzymatic hydrolysate of soya protein, *Process Biochem.* **14** (7), 6–11.

OLSEN H. S., ADLER-NISSEN, J. (1981), Application of ultra- and hyperfiltration during production of enzymatically modified proteins, *Am. Chem. Soc. Symp. Ser.* **154**, 133–169.

OLSEN, H. S., CHRISTENSEN, F. M. (1987), Novel uses of enzymes in food processing, in: *Proc. 7th World Congress of Food Science & Technology,* pp. 139–146, Singapore, Sept. 28th–Oct. 2nd, 1987.

O'MEARA, G. M., MONRO, P. A. (1984), Effects of reaction variables on the hydrolysis of lean beef tissue by Alcalase, *Meat Sci.* **11**, 227–238.

PEDERSEN, H. H., OLSEN, H. S., NIELSEN, P. M. (1994), Method for production of a meat hydrolyzate and a use of the meat hydrolyzate, *PCT WO94/01003* (filed by Novo Nordisk A/S [DK/DK]).

PEPPLER, H. J., DOOLEY, J. G., HUANG, H. T. (1976), Flavour development in Fontina and Romano cheese by fungal esterase, *J. Dairy Sci.* **59**, 859–862.

PHAFF, H. J. (1977), Enzymatic yeast cell degradation, in: *Food Proteins. Improvement Through Chemical and Enzymatic Modification* (FEENEY, R. E., WHITAKER, J. R., Eds.), *Adv. Chem. Ser.* **160**, American Chemical Society, Washington, DC.

PILNIK, W. VORAGEN, A. G. J. (1993), Pectic enzymes in juice manufacture, in: *Enzymes in Food Processing,* 3rd Ed. (NAGODAWITHANA, T. W., REED, G., Eds.), New York: Academic Press, Inc.

POWER, J. (1993), Enzymes in brewing, in: *Enzymes in Food Processing,* 3rd Ed. (NAGODAWITHANA, T. W., REED, G., Eds.), New York: Academic Press, Inc.

PSZCZOLA, D. E. (1988), Production and potential food application of cyclodextrins, *Food Technol.* **42** (1), 96.

PUSKI, G. (1975), Modification of functional properties of soy proteins by proteolytic enzyme treatment, *Cereal Chem.* **52**, 655–664.

QUINLAN, P., MOORE, S. (1993), Modification of triglycerides by lipases: Process technology and its application to the production of nutritionally improved fats, *Inform* **4** (5), 580–585.

REED, G., NAGODAWITHANA, T. W. (1991), *Yeast Technology,* 2nd Ed., an AVI Book, New York: Van Nostrand Reinhold.

REES, D. A., MORRIS, E. D., THOM, D., MADDEN, J. (1982), Stapes and interactions of carbohydrate chains, in: *The Polysaccharides,* Vol. 1 (ASPINALL, G. O., Ed.), New York: Academic Press Inc.

REICHELT, J. R. (1983), Starch, in: *Industrial Enzymology* (GODFREY, T., REICHELT, J. R., Eds.), London: Macmillan Publishers Ltd.

ROSTGAARD JENSEN, B., SVENDSEN, I., OTTESEN, M. (1987), Isolation and characterization of β-acetolactate decarboxylase useful for accelerated beer maturation, in: *Proc. 21st Congr. Eur. Brewery Convention,* Madrid; as cited in *Brewers' Guardian* **116**, 10–11.

RUBENTHALER, G., FINNEY, K. F., POMERANZ, Y. (1965), Effects on loaf volume and bread characteristics of alpha-amylases from cereal, fungal and bacterial sources, *Food Technol.* **19**, 239–241.

SAITTAGAROON, S., KAWAKISHI, S. NAMIKI, M. (1983), Characterization of polysaccharides of copra meal, *J. Sci. Food Agric.* **34**, 855–860.

SCHÜLEIN, M. (1992), Cellulases, *Internal presentation,* Novo Nordisk A/S.

SCHWIMMER, S. (1981), *Source Book of Food Enzymology,* pp. 541–544, Westport, CT: AVI Publishing Company, Inc.

SCOTT, D. (1989), in: *Biocatalysis in Agricultural Biotechnology* (WHITAKER, J. R., SONNET, P., Eds.), *ACS Symp. Ser.* **389**, 176–192.

SELVENDRAN, R. R. (1983), The chemistry of plant cell walls, in: *Dietary Fibre* (BIRCH, G. G., PARKER, K., Eds.), London–New York: Applied Science Publishers.

SHEA, E. M., GIBEAUT, D. M., CARPITA, N. C. (1989), Structural analysis of the cell walls regenerated by carrot protoplasts, *Planta* **179** (3), 293–308.

SI, J. Q., HANSEN, T. T. (1994), Effect of lipase on breadmaking in correlation with their effects on dough rheology and wheat lipids, *Proc. Int. Symp. of AACC/ICC/CCOA,* Beijing, November (available as file A-06352 from Novo Nordisk A/S).

SI, J. Q., SIMONSEN, R. (1994), Functional mechanism of some microbial amylases' anti-staling effect and correlation with their effect on wheat starch, in: *Proc. Int. Symp. of AACC/ICC/CCOA,* Beijing, November (available as file A-06353 from Novo Nordisk A/S).

SI, J. Q., KOFOD, L. V., GODDIK, I. (1993), Effect of microbial xylanases on water insoluble wheat pentosans and in correlation with their baking effect, *Paper* presented at the 1993 *AACC Annual Meeting* (available as file A-06279 from Novo Nordisk A/S).

SIMS, K. A., CHERYAN, M. (1992a), Hydrolysis of liquefied corn starch in a membrane reactor, *Biotechnol. Bioeng.* **39**, 960–967.

SIMS, K. A., CHERYAN, M. (1992b), Continuous saccharification of corn starch in a membrane reactor, *Starch/Stärke* **44** (9), 341–346.

SMITH, D. D., AGRAWAL, Y. C., SARKAR, B. C., SINGH, B. P. N. (1993), Enzymatic hydrolysis pretreatment for mechanical expelling of soybeans, *J. Am. Oil Chem. Soc.* **70** (9), 885–890.

SOFFER, T., MANNHEIM, C. H. (1994), Optimization of enzymatic peeling of oranges and pomelo, *Lebensm. Wiss. Technol.* **27**, 245–248.

SØRENSEN, H. (1988), Analysis of glucosinolates and acceptable concentrations of glucosinolates in oilseed rape and products thereof used as feed to different animals, GCIRC *Bull.* **4**, 17–19.

SÖRENSEN, S. P. L. (1908), Enzymstudien I. Über die quantitative Messung proteolytischer Spaltungen. "Die Formoltitrierung", *Biochem. Z.* **7**, 45–101.

SOROA, J. M. (1967), Extracción, mejora, empleos y subproductos del aceite de oliva, *Elayotécnica,* Chapter X, pp. 297–299.

SOSULSKI, F. (1979), Organoleptic and nutritional effects of phenolic compounds on oilseed protein products: A review, *J. Am. Oil Chem. Soc.* **56**, 711–715.

SOSULSKI, K., SOSULSKI, F. W. (1993), Enzyme-aided vs. two-stage processing of canola: Technology, product quality and cost evaluation, *J. Am. Oil Chem. Soc.* **70** (9), 825–829.

SOSULSKI, K., SOSULSKI, F. W., COXWORTH, E. (1988), Carbohydrase of canola to enhance oil extraction with hexane, *J. Am. Oil Chem. Soc.* **65** (3), 357–361.

STEFANSSON, G. (1993), Fish processing, in: *Enzymes in Food Processing,* 3rd Ed. (NAGODAWITHANA, T. W., REED, G., Eds.), New York: Academic Press, Inc.

STUTZ, C. (1993), The use of enzymes in ultrafiltration. *Fruit Processing* **7**, 366–369.

SZAKÁCS-DOBOZI, M., HALÁSZ, A., KOZMA-KOVÁCS, E., SZAKÁCS, G. (1988), Enhancement of mustard oil yield by cellulolytic pretreatment, *Appl. Microbiol. Biotechnol.* **29**, 39–43.

TEWARI, Y. B., GOLDBERG, R. N. (1984), Thermodynamics of the conversion of aqueous glucose to fructose, *J. Solution Chem.* **13** (8), 523–547.

THEANDER, O., ÅMAN, P. (1977), Fractionation and characterization of polysaccharides in rapeseed (*Brassica napus*) meal, *Swed. J. Agric. Res.* **7**, 69 ff.

VAN DEN BERG, G. et al. (1987), The use of Rennilase XL for the manufacture of Gouda cheese, *Nizo Report* R 126.

VORAGEN, A. G. J., SCHOLZ, H. A., BELDMAN, G. (1992), Maßgeschneiderte Enzyme in der Fruchtsaftherstellung, *Flüss. Obst* **59** (7), 404–410.

WALLERSTEIN, L. (1911), *US Patents* No. 995820; 995823; 995824; 995825; 995826.

WARREN, S. J. (1992), Method of tenderising meat before slaughtering, *Eur. Patent Appl.* EP 0471470A2.

WEIR, G. S. D. (1992), Proteins as a source of flavour, in: *Biochemistry of Food Proteins* (HUDSON, B. J. F., Ed.), London–New York: Elsevier Applied Science.

WHITAKER, J. R. (1990), New and future uses of enzymes in food processing, *Food Biotechnol.* **4** (2), 669–697.

WILLIAMS, P. J., SEFTON, M. A., FRANCIS, L.

(1992), Glycosidic precursors of varietal grape and wine flavor, in: *Flavor Precursors* (TERANISHI, R., TAKEOKA, G., GÜNTERT, M., Eds.), pp. 74–86, *ACS Symp. Ser.* **490**, Washington DC.

YAMAOKA, T., TSUKADA, K., TAKAHASHI, H., YAMAUCHI, N. (1983), Purification of a cell wall-bound pectin-gelatinizing factor and examination of its identity with pectin methyl-esterase, *Bot. Mag. Tokyo* **96**, 139–144.

ZITTAN, L. (1981), Enzymatic hydrolysis of insulin – an alternative way to fructose production, *Starch/Stärke* **33** (11), 373-377.

19 Carbohydrate-Based Sweeteners

RONALD E. HEBEDA

Summit-Argo, IL 60501-0345, USA

1 Introduction

This chapter covers the manufacture, properties and applications of carbohydrate-based sweeteners. For the purpose of this chapter, carbohydrate-based sweeteners are defined as those products that are produced from starch by enzymatic or acid conversion procedures and used in a variety of food and non-food applications. Products of interest are crystalline and liquid dextrose (D-glucose), high fructose syrup (HFS, isosyrup, isoglucose), crystalline fructose, regular syrup (glucose syrup, starch syrup), maltodextrin (hydrolyzed cereal solids) and cyclodextrin (cycloamylose, Schardinger dextrin).

1.1 Terminology

Although the title of this chapter is "Carbohydrate-Based Sweeteners", neither this term nor any other singular term adequately defines the group of products mentioned above. In certain cases, the term "sweetener" is appropriate, since some of these products are used to provide sweetness in food applications. For instance, a 55% fructose syrup and crystalline fructose are about 100 and 140% as sweet as sucrose, respectively, and are used primarily in applications where added sweetness is desired. In other applications such as dextrose for intravenous solutions, sweetness is obviously not a consideration. Furthermore, a product such as maltodextrin does not exhibit perceptible sweetness and is used in applications for other purposes.

To further confound the definition issue, the term "starch hydrolysis product" can only be correctly applied to maltodextrins, syrups and dextrose products, since these materials are produced by acid and/or enzymatic hydrolysis of starch. Fructose, however, is produced by enzymatic isomerization of dextrose and is not a starch hydrolysis product, *per se.*

Consequently, since no singular term encompasses all of the products described in this chapter, phrases such as "carbohydrate-based sweeteners", "starch-based sweeteners", and "starch hydrolysis products" will be used to refer to products of interest.

1.2 Product Definition

Starch hydrolysis products such as dextrose, syrups and maltodextrin are often described on the basis of dextrose equivalent (DE). DE provides an estimate of the degree of hydrolysis and is defined as the percent of reducing sugars present expressed as dextrose on a dry weight basis. This analysis has traditionally been conducted by copper-reducing methods but can also be calculated from liquid chromatography data (BERNETTI, 1992).

Starch exhibits a DE of essentially zero, since very few reducing groups are present relative to the total number of dextrose units. Complete hydrolysis of starch yields dextrose, and since every dextrose molecule is a reducing sugar, a DE of 100 is attained. An intermediate degree of hydrolysis yields a product exhibiting a DE between zero and 100. Dextrose is defined as any hydrolysis product of 99.5 DE or greater, corn syrups are products of 20–99.4 DE, and maltodextrins are products of less than 20 DE.

DE, however, does not always adequately characterize a starch hydrolysis product. This is especially true in the case of syrups that are of equivalent DE but vary widely in saccharide composition due to differences in methods of production. In this case, the product is best described in terms of saccharide distribution, i.e., concentration of dextrose (DP-1), maltose (DP-2), maltotriose (DP-3) and higher saccharides (DP-4+), where DP refers to the degree of polymerization.

Fructose-containing products are simply defined on the basis of fructose level. Standard syrup products contain 42, 55 or 90% fructose. Crystalline fructose is 99% + pure.

Cyclodextrins are closed ring forms of either 6, 7 or 8 glucose units and are referred to as alpha-, beta- and gamma-cyclodextrin, respectively.

1.3 History

The history of starch-based sweeteners can be traced back to at least the ninth century when syrup was produced in Japan from arrowroot starch by the action of the natural enzymes in malt (YOSHIZUMI et al., 1986). By the late 1700s (DEAN and GOTTFRIED, 1950) and early 1800s (KIRCHOFF, 1811), laboratory studies had shown that a sweet material was formed by the action of acid on starch. Shortly thereafter, it was determined that the action of acid on starch proceeded by hydrolysis rather than dehydration (DE SAUSSERE, 1815). These and other studies eventually led to the development of a starch hydrolysis industry in the United States and in Europe; by 1876, 47 factories were in operation in the U.S. alone.

The range of commercial products available during the first several decades of commercial operation was limited. Important discoveries in two areas during the early 1900s radically altered the industry. First in 1923, a commercial process for producing crystalline monohydrate dextrose was developed (NEWKIRK, 1923). Fifteen years later in 1938, the first commercial enzymatic process was developed for producing a corn syrup (DALE and LANGLOIS, 1940); acid-hydrolyzed starch was treated with a fungal amylase preparation to produce a 65 DE product that was sweeter and less bitter than a total acid syrup.

These two innovations resulted in rapid growth within the starch hydrolysis industry; monohydrate dextrose production increased from 50 to 320000 mt (metric tons) between 1923 and 1940 (SCHENCK and HEBEDA, 1992) and corn syrup consumption increased by 86% between 1940 and 1942 (JONES and THOMASON, 1951).

During the following 25 years, a number of important developments took place in the area of industrial enzymes that eventually led to the most recent major product from the starch industry – a high fructose syrup that would replace sucrose as a sweetener in various applications. In order to produce a fructose syrup, enzymatic processes had to be first developed for (1) liquefying starch, (2) converting liquefied starch to dextrose and (3) isomerizing dextrose to fructose. With the exception of the one enzymatic corn syrup process mentioned earlier, all other industrial starch hydrolysis processes utilized acid catalysis. It was recognized that acid hydrolysis had drawbacks such as (1) a limited range of saccharide compositions, (2) precipitation of long-chain starch polymers in syrups under 30 DE, and (3) production of color-forming degradation products in syrups over 55 DE. Consequently, the industry began searching for enzymes that could replace acid for starch hydrolysis.

Certain amylases (enzymes that hydrolyze starch) were already known and available at the time. For instance, commercial production of fungal alpha-amylase had been developed by the end of the 19th century (TAKAMINE, 1894) and the enzyme was used for starch removal in pectin (DOUGLAS, 1932). Bacterial alpha-amylases were developed in the early 1900s (BOIDIN and EFFRONT, 1917) and used in the brewing, textile and paper industries (WALLERSTEIN, 1939).

Studies in the early 1940s confirmed the presence of a dextrose-producing enzyme (glucoamylase) in fungal strains of *Aspergillus.* In 1951 glucoamylases from *Aspergillus niger* (KERR et al., 1951) and *Rhizopus delemar* (PHILLIPS and CALDWELL, 1951) were characterized. By 1960, the enzyme from *A. niger* was being used commercially for producing dextrose from acid liquefied starch. Later in the decade, a bacterial alpha-amylase from *Bacillus subtilis* replaced acid for starch liquefaction. As a result, a total enzyme process was implemented that successfully increased dextrose yield and reduced refining requirements.

The industrial utilization of enzymes for starch liquefaction and saccharification was soon followed by the development of an enzyme for isomerizing dextrose to fructose. Production of fructose by alkaline isomerization of dextrose was reported in 1895 (LOBRY, DEBRUYN and VAN ECKENSTEIN, 1895), but various attempts at commercializing such a process never materialized due to low fructose yield and formation of color, off-flavor and other by-products. In the 1950s, several enzymes were discovered that had the ability to convert three, four, and five carbon aldoses to the corresponding ketoses. In 1957, an enzyme from *Aerobacter cloacae* (originally reported as *Pseudomonas hydrophila*) was found to isomerize glucose to fructose (MARSHALL and KOOI, 1957). Due to low fructose yields, additional work in the U.S. was discontinued. However, investigations in Japan proceeded and enzymes were discovered that produced fructose at commercially viable yields. An industrial process for isomerization of dextrose to fructose was developed (TAKASAKI and TANABE, 1971) and commercialized in 1966. A joint effort between Japan and the U.S. was successful in initiating production in the U.S.; a 14–16% fructose syrup was first produced in 1967, and manufacture of 42% fructose syrup was initiated in 1968.

1.4 Production and Consumption of Starch Sweeteners

1.4.1 Manufacturers

Today, starch-based sweeteners are manufactured world-wide from a variety of starches. About 225 plants are in operation in 57 different countries; these include 78 plants in Asia, 54 in Europe, 39 in North America, 32 in South America/Cuba, 13 in Africa/Middle East and 9 in Australasia (SCHENCK and HEBEDA, 1992). Although corn (maize) starch is the primary raw material, starches from sorghum (milo), wheat, rice, potato, tapioca (yucca, cassava), arrowroot and sago are also used to varying degrees.

1.4.2 Production

In the U.S., corn starch is used as the raw material in essentially all production of starch hydrolysis products. The amount of corn utilized for this purpose has nearly tripled from the 207 MM bushels used in the 1975/76 crop year (CORN REFINERS ASSOCIATION, 1988) to the 620 MM bushels used in the 1992/1993 crop year (U.S. DEPARTMENT OF AGRICULTURE, 1992). From the 1989/90 to the 1992/93 crop years, 7.3–8.2% of the corn crop was used for sweetener production; about two-thirds of this corn was used for high fructose syrup (HFS) and one-third for dextrose and glucose syrups (Tab. 1).

Production capacity for corn sweeteners in the 27 U.S. wet-milling plants was about 10.4–11.8 MM mt (dry basis) in 1992 and is estimated to increase to 11.3–13.2 MM mt for 1995 (U.S. DEPARTMENT OF AGRICULTURE, 1992).

Actual 1992 U.S. production of corn sweeteners (dry basis) was about 6000 M mt high fructose syrup (59% HFS-55, 41% HFS-42), 2800 M mt glucose syrup, and 680 M mt dextrose (U.S. DEPARTMENT OF AGRICULTURE, 1992). About 91 M mt maltodextrin is sold per year (ALEXANDER, 1992), and the estimated annual world production of cyclodextrin is 0.85 M mt (TEAGUE and BRUMM, 1992).

1.4.3 Consumption

Corn sweetener consumption in the U.S. has increased markedly since the late 1970s when the soft drink industry began to substitute high fructose syrups for sucrose. In 1975, the consumption of corn sweeteners was less than one-third the consumption of sucrose (U.S. DEPARTMENT OF AGRICULTURE, 1992). By the mid 1980s, corn sweetener consumption had surpassed that of sucrose and between 1988 and 1992, consumption of both

Tab. 1. Corn Utilization for Sweetener Production in the U.S. (in million bushels)[a]

	1989/1990	1990/1991	1991/1992	1992/1993[b]
Total U.S. corn crop	7525	7934	7474	8770
High fructose syrup	368	379	400	404
Dextrose, glucose syrup	193	200	213	216
Total sweetener	561	579	613	620
Sweetener share of crop (%)	7.5	7.3	8.2	7.1

[a] Data from U.S. DEPARTMENT OF AGRICULTURE (1992)
[b] Forecast

Tab. 2. Sweetener Consumption in the U.S.[a] (in 1000 metric tons, dry basis)

	1988	1989	1990	1991[b]	1992[c]
Total sweeteners	15030	15318	15768	15949	16435
High fructose syrup	5401	5467	5565	5683	5883
Dextrose	477	488	507	519	522
Glucose syrup	2085	2154	2224	2279	2338
Total corn sweeteners	7963	8109	8295	8478	8742
% of Total	53.0	52.9	52.6	53.2	53.2
Sucrose	6903	7045	7309	7308	7529
% of Total	45.9	46.0	46.4	45.8	45.8

[a] Data from U.S. DEPARTMENT OF AGRICULTURE (1992)
[b] Preliminary
[c] Forecast

corn sweeteners and refined sucrose stabilized at about 53 and 46%, respectively, of total caloric sweeteners (U.S. DEPARTMENT OF AGRICULTURE, 1992) (Tab. 2). Total consumption of each sweetener increased by 9–10% between 1988 and 1992; corn sweetener and refined sucrose consumption was expected to reach 8.7 and 7.5 MM mt, respectively, in 1992 (U.S. DEPARTMENT OF AGRICULTURE, 1992).

2 Starch

The basic raw material used for the production of carbohydrate-based sweeteners is starch. Starch is composed of linear (amylose) and branched (amylopectin) polymers that contain anhydroglucose units connected by either alpha-1,4 or alpha-1,6 linkages. The amylose fraction contains nearly all alpha-1,4 linked glucose units. The amylopectin fraction contains about 5% alpha-1,6 linkages; therefore, a linear chain is interrupted on the average of every 20 glucose units by an alpha-1,6 linkage "branch point". The relative amount of each fraction varies, and amylose/amylopectin ratio ranges from 0/100 to 85/15 in different starches (ZOBEL, 1992). Regular corn starch contains about 27% amylose and 73% amylopectin, whereas high amylose corn starch, potato starch and waxy maize starch contain 50–70, 17–23 and <2% amylose, respectively (BEMILLER, 1992).

Starch is present in plants as small granules that range in size from 0.5 to 175 microns depending on the particular starch source. The granules may be present in various locations within the plant including the root, tuber, stem-pith, leaf, seed, fruit and pollen (ZOBEL, 1992).

When heated in an aqueous slurry, granules hydrate and swell (gelatinize) resulting in a loss in crystallinity. Depending on the type of starch, gelatinization generally begins between 50 and 68°C and is completed between 64 and 78°C (ZOBEL, 1992). Regular corn starch, for example, exhibits a gelatinization range of 62–70°C, waxy maize starch 63–72°C, potato starch 58–62°C, tapioca starch 52–64°C and 70% high-amylose corn starch in excess of 100°C (BEMILLER, 1992).

3 Enzymes

The primary enzymes used to produce starch-based sweeteners are hydrolases (amylases) and isomerases. The amylases, such as alpha-amylase, beta-amylase, glucoamylase and pullulanase, have the ability to catalyze the hydrolysis of alpha-1,4 and/or alpha-1,6 linkages in starch to produce lower molecular weight saccharides. Isomerases convert dextrose to fructose. Cyclodextrin-producing en-

Tab. 3. Commercially Available Enzymes Used in Starch Based Sweetener Production[a]

Enzyme Source	Number of Products	Number of Suppliers
Starch Liquefaction – alpha-Amylase		
Bacillus amyloliquefaciens	13	11
Bacillus licheniformis	9	6
Bacillus stearothermophilus	3	3
B. lichen./B. stear. blend	1	1
Dextrose Production – Glucoamylase		
Aspergillus niger	14	12
Rhizopus niveus	1	1
Rhizopus oryzae	1	1
Rhizopus sp.	3	3
Dextrose Production – Mixed Systems		
Glucoamylase in combination with other enzymes	4	4
Dextrose Production – Transferase		
Bacillus megaterium	1	1
Dextrose and Syrup Production – Debranching Enzymes		
Bacillus acidopullulyticus	1	1
Klebsiella aerogenes	1	1
Klebsiella planticola	1	1
Pseudomonas amyloderamosa	1	1
Syrup Production – Fungal alpha-Amylase		
Aspergillus niger	1	1
Aspergillus oryzae	9	9
Rhizopus oryzae	1	1
Syrup Production – Plant beta-Amylase		
Barley	6	5
Soybean	1	1
Wheat	1	1
Fructose Syrup Production – Glucose Isomerase		
Actinoplanes missouriensis	1	1
Bacillus coagulans	1	1
Microbacterium arborescens	1	1
Streptomyces olivochromogenes	3	3
Streptomyces griseofuscus	1	1
Streptomyces murinus	1	1
Streptomyces phaeochromogenes	1	1
Streptomyces rubiginosus	1	1
Cyclodextrin Production		
Bacillus stearothermophilus	1	1

[a] From TEAGUE and BRUMM (1992)

zymes act by first hydrolyzing starch and then forming ring structures via a transglycosylation reaction.

During the production of starch-based sweeteners, thermostable alpha-amylases from bacterial sources are used at high temperature to liquefy starch and produce soluble dextrins. Glucoamylases from fungal sources are used in the saccharification process to convert dextrins to dextrose. Other amylases are often used in conjunction with glucoamylase to provide an increase in dextrose yield. Alpha- and beta-amylases from bacterial, fungal and plant sources are used to saccharify dextrins to a wide range of syrups that exhibit varied saccharide compositions. Glucose isomerases from a number of different bacterial sources are used to isomerize dextrose to fructose.

A recent listing (TEAGUE and BRUMM, 1992) itemized 86 commercially available enzymes that are used in starch-based sweetener production. These enzymes are available from 19 different suppliers in the U.S., England, Denmark, Germany and Japan. A summary of this information is shown in Tab. 3. Many of these enzymes are similar in properties and are interchangeable for similar applications, differing only in supplier, concentration, form or source. Overall, the list includes 26 enzymes for starch liquefaction, 28 enzymes for dextrose production, 10 enzymes for glucose syrup production, 8 enzymes for fructose syrup production and 1 enzyme for cyclodextrin production.

3.1 Liquefaction Enzymes

Alpha-amylases (1,4-alpha-D-glucan glucanohydrolase, EC 3.2.1.1) derived from bacterial sources are used to liquefy (thin) a starch slurry to a low DE, soluble hydrolysate that can be processed to a maltodextrin or converted to a variety of sweeteners. The enzymes are specific for hydrolysis of alpha-1,4 linkages and are endoenzymes, i.e., they have the ability to hydrolyze internal bonds of the starch molecule on either side of alpha-1,6 linkages. In the presence of substrate and calcium, bacterial alpha-amylases exhibit good thermostability at near neutral pH and, depending on the particular source, can be used at temperatures between 90 and >100 °C.

The first enzymatic liquefaction process was developed in the 1960s and utilized a *Bacillus subtilis* (*B. amyloliquefaciens*) alpha-amylase at 85–90 °C to thin starch. In the following decades, alpha-amylases exhibiting even greater thermostability were commercialized; this was accomplished first in the 1970s with the introduction of a *B. licheniformis* alpha-amylase and later in the 1980s with a *B. stearothermophilus* amylase.

These highly thermoduric enzymes are sufficiently stable to be used at temperatures in excess of 100 °C under the proper conditions of starch concentration, pH and calcium level. At 102–105 °C and varied pH, substrate solids and calcium level, the *B. licheniformis* and *B. stearothermophilus* enzymes exhibit half-lives

Tab. 4. Stability of *Bacillus stearothermophilus* and *Bacillus licheniformis* alpha-Amylases under Process Conditions

Temp. (°C)	pH	Calcium (ppm) as is	Solids (% w/w)	Half-life (min) BS[a]	BL[b]
102	6	25	37.3	213	70
102	6	25	31.4	126	44
102	6	25	26.5	81	25
105	6.5	40	35	132	50
105	5.5	40	35	43	10
105	6.5	40	25	75	20
105	6.5	10	35	42	15

[a] *B. stearothermophilus* alpha-amylase (HENDERSON and TEAGUE, 1988)
[b] *B. licheniformis* alpha-amylase (ROSENDAL et al., 1979)
Reproduced with permission of Academic Press, Inc. (HEBEDA, 1993)

of several minutes to several hours (Tab. 4) (HEBEDA, 1993). Based on these data, the *B. stearothermophilus* enzyme is three to four times more stable than the *B. licheniformis* enzyme when used under similar conditions. A blend of the two thermostable amylases has been commercialized and reportedly provides an improved starch liquefaction enzyme system able to operate at low pH with an overall reduction in dosage (CARROLL and SWANSON, 1990).

B. stearothermophilus alpha-amylase has been cloned in other bacteria and petitions for generally recognized as safe (GRAS) status have been accepted for filing by the Food and Drug Administration (FDA). The enzyme was first cloned into a *B. subtilis* host (ZEMAN and MCCREA, 1985) and became the first genetically engineered food ingredient for which an affirmation of GRAS status petition was filed and accepted (CPC INTERNATIONAL, INC., 1986). A GRAS petition was also filed and accepted for a *B. stearothermophilus* alpha-amylase cloned into *B. licheniformis* (NOVO NORDISK BIOINDUSTRIAL, INC., 1991).

Future improvements in the area of liquefaction enzymes may be the development of "hyperthermostable" alpha-amylases of increased stability at higher temperature and lower pH. Enzymes of this type have been identified, evaluated and shown to operate efficiently at high temperature and low pH (4.2–4.8) without the need for added calcium to maintain stability. Examples are amylases from *Pyrococcus* archaebacteria (BROWN et al., 1990; ANTRANIKIAN et al., 1990) and from a genus of *Thermoanaerobacter* (STARNES, 1990).

3.2 Saccharification Enzymes – Dextrose Process

3.2.1 Glucoamylase

Glucoamylase (1,4-alpha-D-glucan glucohydrolase, amyloglucosidase, gamma-amylase, EC 3.2.1.3) is used in saccharification (conversion) to produce dextrose from liquefied starch hydrolysate. The major source of commercial glucoamylase is the fungus *Aspergillus niger*, although *Rhizopus* enzymes are also available. Several isozymes are present in glucoamylase preparations and differ by molecular weight, isoelectric point and degree of starch binding although enzymatic properties are similar. *A. niger* glucoamylase is stable up to about 60 °C, exhibits maximum activity between pH 4 and 5, and does not require metal ions for activity. At 60 °C, pH 4.3 and a substrate concentration of 30–35% solids, the enzyme has a half-life of one or more days.

Glucoamylase catalyzes the hydrolysis of both alpha-1,4 and 1,6 linkages. The enzyme is an exo-enzyme, i.e., hydrolysis occurs sequentially from the non-reducing end of a polysaccharide and dextrose is released in a stepwise manner. Hydrolysis rate of a specific linkage varies depending on the sequence of linkages and size of the substrate molecule (ABDULLAH et al., 1963). Alpha-1,4 linkages are hydrolyzed at a faster rate than alpha-1,6 linkages; for instance, the hydrolysis rate of maltose is 100 times faster than that of isomaltose. Increasing the size of the substrate from maltose (one linkage) to a linear pentasaccharide (four linkages) increases the rate of hydrolysis of the first glycosidic bond by a factor of 10. The hydrolysis rate of an alpha-1,4 linkage is slowed if the adjacent bond is alpha-1,6; however, the rate of hydrolysis of an alpha-1,6 bond is increased when adjacent to an alpha-1,4 bond. In no case is the hydrolysis rate of an alpha-1,6 bond any more than 3% that of the hydrolysis rate of an alpha-1,4 bond in an equal size saccharide (MEAGHER et al., 1989).

Glucoamylase also catalyzes the reverse reaction (reversion) in which maltose, isomaltose or higher saccharides are formed by the condensation of a beta-anomer of D-glucopyranose with either an alpha- or beta-D-glucose molecule. The level of reversion products formed is dependent on substrate solids level and reaction time, and is the major deterrent to attaining a 100% yield of dextrose in an industrial process.

Altering the properties of glucoamylase by site-directed mutagenesis has shown some reduction in extent of reversion; it has been reported that this technique has been successful in reducing the amount of isomaltose formed via glucoamylase catalyzed reversion by about 20% (REILLY, 1993).

Side activities, such as alpha-amylase or transglucosidase, may be present in varying degrees in commercial glucoamylase preparations. The presence of alpha-amylase is important in maximizing dextrose yield during the saccharification process; the enzyme hydrolyzes polysaccharides to smaller molecules and provides additional substrate for the action of glucoamylase (KOOI and ARMBRUSTER, 1967). Conversely, transglucosidase is not desired, since the enzyme lowers the dextrose yield by forming maltose and isomaltose via a transfer reaction. Strain improvement has essentially eliminated transglucosidase as a problem. If, however, transglucosidase is present, a number of procedures have been developed for removing the enzyme from fermentation broth (HEBEDA, 1983).

3.2.2 Debranching Enzymes

Debranching enzymes such as pullulanase (alpha-dextrin 6-glucanohydrolase, EC 3.2.1.41) and isoamylase (glycogen 6-glucanohydrolase, EC 3.2.1.68) specifically catalyze the hydrolysis of alpha-1,6 linkages and are used in combination with glucoamylase to increase dextrose yield during saccharification.

Pullulanase hydrolyzes alpha-1,6 linkages in amylopectin, glycogen and pullulan. Commercial sources for this enzyme are *Bacillus acidopullulyticus* and *Klebsiella planticola* (*K. aerogenes*) (see Tab. 3). Isoamylase is a debranching enzyme that has the ability to completely debranch glycogen but does not hydrolyze pullulan. The only commercial source is *Pseudomonas amyloderamosa* (Tab. 3).

The pullulanase from *K. planticola* and the isoamylase from *P. amyloderamosa* exhibit a maximum operational temperature of about 55°C. As a result, these enzymes are not often used with glucoamylase at the normal saccharification temperature of 60 °C. The pullulanase from *B. acidopullulyticus*, however, is sufficiently stable for use at 60°C and can be used with glucoamylase under standard saccharification conditions.

3.2.3 *Bacillus megaterium* Amylase

A commercialized *B. megaterium* alpha-amylase (1,4-alpha-D-glucan glucanohydrolase, EC 3.2.1.1) provides many of the same advantages as pullulanase in saccharification. The enzyme utilizes a unique hydrolysis/transferase mechanism that converts branched saccharides to smaller oligomers that are more readily hydrolyzable by glucoamylase (DAVID et al., 1987). The enzyme is sufficiently stable to be used at the normal saccharification temperature of 60°C.

B. megaterium amylase hydrolyzes alpha-1,4 linkages in starch, glycogen, cyclodextrin and pullulan (BRUMM et al., 1991a) and also catalyzes transfer reactions in which glucose or glucosides act as acceptor molecules (BRUMM et al. 1991b).

The *B. megaterium* amylase is produced commercially using a genetically engineered strain of *B. subtilis*, and a petition for affirmation of GRAS status has been accepted for filing by the FDA (ENZYME BIO-SYSTEMS LTD., 1988).

3.3 Saccharification Enzymes – Syrup Process

3.3.1 Maltose-Producing Enzymes

The fungal alpha-amylase produced by *Aspergillus oryzae* (1,4-alpha-D-glucan glucanohydrolase, EC 3.2.1.1) or the plant beta-amylases from barley malt or soybean (1,4-alpha-D-glucan maltohydrolase, EC 3.2.1.2) produce maltose by cleaving alternate alpha-1,4 linkages in starch substrates. They differ, however, in the mode of action.

The *A. oryzae* alpha-amylase is an endoamylase and has the ability to hydrolyze an alpha-1,4 linkage on either side of a branch point. Maltose units in the alpha-configuration are then released sequentially in a stepwise fashion. Maltotriose and a small amount of dextrose are produced from polymers containing an odd number of dextrose units. Small dextrins containing branched linkages remain unhydrolyzed. The enzyme exhibits maximum activity at 50°C and pH 5–6 and requires calcium ions for stability and activity (OIKAWA and MAEDA, 1957).

Plant beta-amylases are exo-enzymes and release maltose in the beta-configuration in a stepwise manner from the non-reducing end of the substrate molecule. The enzyme cannot by-pass an alpha-1,6 bond and, therefore, a high molecular weight branched fraction referred to as a beta-limit dextrin accumulates during saccharification. Calcium and a free sulfhydryl group are required for activity (THOMA et al., 1965; NITTA et al., 1983).

Maltose-producing enzymes from bacterial sources have also been identified. These include maltogenic enzymes from *Clostridium thermosulfurogenes* (ZEIKUS and HYUN, 1987), *B. stearothermophilus* (OUTTRUP, 1986), *B. cereus* (TAKASAKI, 1976), *Streptomyces praecox* (WAKO et al., 1978), *S. hygroscopicus* (HIDAKA et al., 1974), *B. megaterium* (ARMBRUSTER and JACAWAY, 1970) and *B. polymyxa* (ROBYT and FRENCH, 1964). None of these are important commercially.

3.3.2 Oligosaccharide-Producing Enzymes

Enzymes that produce high levels of DP-3, 4, 5 and 6 oligosaccharides have been discovered and studied (OKADA and NAKAKUKI, 1992). These include *S. griseus* and *B. subtilis* amylases that form maltotriose (WAKO et al., 1979; TAKASAKI, 1985), a *Pseudomonas stutzeri* amylase that forms maltotetraose (ROBYT and ACKERMAN, 1971), *B. licheniformis* and *B. stearothermophilus* amylases that form maltopentaose (HEBEDA, 1993) and a *Klebsiella pneumoniae* amylase that forms maltohexaose (KAINUMA et al., 1972). The maltotriose-forming enzyme from *B. subtilis* (exomaltotriohydrolase) and the maltotetraose-forming enzyme from *Pseudomonas stutzeri* (exo-maltotetraohydrolase, EC 3.2.1.60) are used commercially for the purpose of oligosaccharide production.

3.3.3 Glucoamylase

Glucoamylase, as described earlier (Sect. 2.4.1), is used in combination with maltogenic enzymes (alpha-amylases or beta-amylases) to produce dextrose in certain types of syrups.

3.3.4 Debranching Enzymes

Debranching enzymes, as described earlier (see Sect. 3.2.2), are used in combination with maltogenic enzymes to produce increased levels of maltose.

3.4 Fructose-Producing Enzymes

Glucose isomerase (D-xylose ketol-isomerase, EC 5.3.1.5) catalyzes the isomerization of D-glucose to D-fructose. The enzyme is actually a xylose isomerase that also isomerizes D-xylose to D-xylulose, as well as D-ribose to D-ribulose. The isomerase is specific for the alpha-anomeric form of the substrate.

Glucose isomerase is used in an immobilized form in continuous column operation to produce fructose syrup from a dextrose feed. Immobilization is conducted by one of two procedures; the intracellular enzyme is either immobilized within the bacterial cells to produce a whole-cell product or the enzyme is released from the cells, recovered and immobilized onto an inert carrier.

An example of the whole-cell process is one in which cells are disrupted by homogenization, cross-linked with glutaraldehyde, flocculated with a cationic flocculent and extruded (JORGENSEN et al., 1988). In a second example, a cell–gelatin mixture is cross-linked with glutaraldehyde (HUPKES, 1978).

When soluble isomerase is used for binding, the enzyme is first released from the cell and then recovered and concentrated. Examples of this type of immobilization include binding enzyme to a DEAE-cellulose/titanium dioxide/polystyrene carrier (ANTRIM and AUTERINEN, 1986) or absorbing enzyme onto alumina followed by cross-linking with glutaraldehyde (LEVY and FUSEE, 1979; ROHRBACH, 1981).

As shown in Tab. 3, commercial enzymes are produced from several different bacterial sources. All of the enzymes are similar in action, and exhibit a pH optimum in the range of at least 7–8; specific enzymes may have a slightly broader pH range. In some cases, the immobilized form of the enzyme has a higher pH optimum than the soluble form. Temperature optima also vary with enzyme source and all commercial enzymes are sufficiently stable for use at 60°C or higher. The enzyme is a metalloprotein and calcium ions inhibit activity; magnesium and cobalt ions enhance activity and stability.

3.5 Cyclodextrin-Producing Enzymes

Cyclodextrin enzymes (CGTase, EC 2.4.1.19) are produced from *Bacillus circulans*, *B. macerans* and *B. stearothermophilus* (TEAGUE and BRUMM, 1992). The enzyme has the ability to hydrolyze starch and form rings of alpha-1,4 linked glucose units by intramolecular transglycosylation (cyclization). The *B. macerans* and *B. stearothermophilus* enzymes produce primarily a ring containing six glucose units (alpha-cyclodextrin) while the *B. circulans* enzyme produces primarily a seven-unit ring (beta-cyclodextrin). An eight-unit ring (gamma-cyclodextrin) is also produced. The ratio of types of cyclodextrins formed is dependent on the particular enzyme used as well as a number of different reaction parameters.

4 Starch Liquefaction

The first step in the production of starch-based sweeteners is partial hydrolysis of starch (liquefaction) to achieve a soluble, low viscosity substrate that can be converted to a range of products.

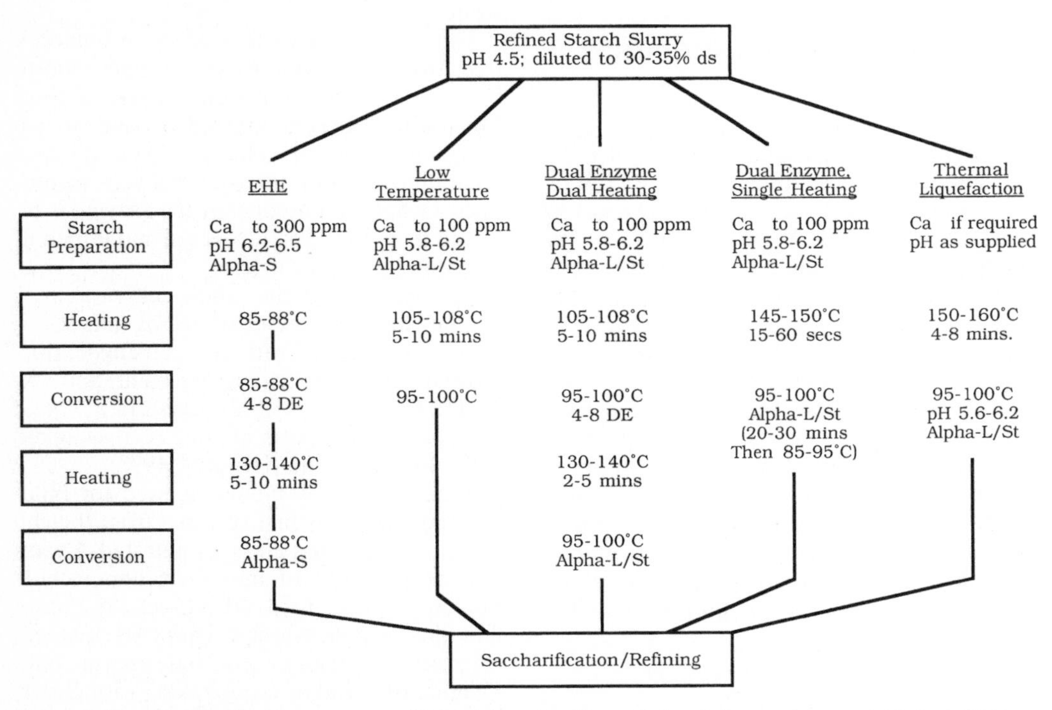

Fig. 1. Industrial liquefaction processes. Alpha-S is a *Bacillus subtilis* alpha-amylase and Alpha-L/ST is a *Bacillus licheniformis* or *Bacillus stearothermophilus* alpha-amylase (REEVE, 1992).

Liquefaction was conducted entirely by acid hydrolysis until the mid 1960s. When thermostable alpha-amylases became available, enzyme liquefaction systems were developed that are now used in most industrial processes as the first step for producing dextrose and fructose products as well as some syrups. Acid liquefaction is still used for certain maltodextrins and syrups.

4.1 Enzyme Processes

Several different enzyme liquefaction processes have been developed as described by REEVE (1992) and shown in Fig. 1. These processes are referred to as (1) enzyme-heat-enzyme (EHE), (2) low temperature, (3) dual enzyme/dual heating, (4) dual enzyme/single heating and (5) thermal liquefaction.

4.1.1 EHE Process

The "enzyme-heat-enzyme" process was the first industrial enzymatic liquefaction procedure developed and used a *B. subtilis* alpha-amylase for hydrolysis. This enzyme could be used at a maximum temperature of only about 90 °C before a significant loss in activity occurred; therefore, a high temperature heat treatment step was needed to solubilize residual starch present as a fatty acid/amylose complex. The heat treatment inactivated the alpha-amylase and, as a result, a second addition of enzyme was required to complete the reaction. Consequently, the process was termed "enzyme-heat-enzyme" (EHE).

In the EHE process (Fig. 1), a starch slurry at 30–40%, pH 6–6.5, and containing 200–400 ppm (dry basis) added calcium (as the chloride or hydroxide) is dosed with enzyme, passed through a stream injection heater at 85–90°C and held at temperature for about one hour. The resulting 4–8 DE hydrolysate is then subjected to a heat treatment at 120–140°C for 5–10 minutes in a holding tube, redosed with enzyme and reacted for an additional hour at 85°C until a DE level of 10–15 is reached.

4.1.2 Low Temperature Process

The EHE liquefaction process was simplified when *B. licheniformis* and *B. stearothermophilus* alpha-amylases became commercially available. Compared to the *B. subtilis* enzyme, these enzymes are more thermostable, more aciduric and require less calcium for stability. Consequently, a heat treatment step was not required and the maximum temperature was reduced substantially.

The "low temperature" process (Fig. 1) is conducted at pH 5.8–6.2 with an added calcium level of no more than 100 ppm (dry basis). Enzyme is added to the starch, the slurry is heated to 105–108°C and held at temperature for 5–10 minutes. The resulting 1–2 DE hydrolysate is flashed to atmospheric pressure and held at 95–100°C for 1–2 hours in a batch or continuous reactor. Since the enzyme is not significantly deactivated at the first stage temperature, a second enzyme addition is not needed. This process is used world-wide throughout the starch-based sweetener industry and has been judged the most efficient process for dextrose production.

4.1.3 Dual Enzyme Processes

In some cases (especially in syrup production in Europe), a liquefaction process incorporating a thermostable enzyme *and* a high temperature heat treatment step provide better filterability than an acid liquefaction process (REEVE, 1992). Consequently, dual enzyme processes were developed that utilized multiple additions of either *B. licheniformis* or *B. stearothermophilus* alpha-amylase and a heat treatment step (Fig. 1).

In these processes, the starch slurry is prepared in the same manner as in the low temperature process. In the dual enzyme/dual heating process, the low temperature process scheme is followed. However, after completion of the second stage reaction, a 2–5 minute heat treatment at 130–140°C is used followed by a second enzyme addition and another reaction step at 95–100°C.

In the dual enzyme/single heating process, the starch slurry is immediately heated to

145–150 °C for a minute or less. Although the enzyme is rapidly inactivated, sufficient hydrolysis takes place to provide a partially thinned hydrolysate that can be pumped to a second stage where additional enzyme is added and the reaction continued at 95–100 °C for 20–30 minutes. The temperature is then lowered to 85–95 °C for the remainder of the reaction.

4.1.4 Thermal Liquefaction Process

In the thermal liquefaction process (Fig. 1), a starch slurry containing no enzyme or added calcium is heated at 150–160°C for several minutes at as is pH. Since the slurry pH is slightly acidic, sufficient acid liquefaction is achieved to reduce viscosity. The hydrolysate (at essentially zero DE) is flash cooled to 95–100 °C, alpha-amylase is added, and the reaction is continued at pH 5.6–6.2 to completion.

4.2 Acid Process

Acid liquefaction is now only used for production of certain maltodextrins and syrups. The reaction is conducted in a continuous or batch mode using a 35–40 % starch slurry adjusted to 0.015–0.2 N hydrochloric acid and at a pressure at 140–160°C for 15–20 minutes. When the desired DE is reached, the hydrolysate is neutralized with sodium carbonate.

5 Dextrose

5.1 Manufacture

5.1.1 Saccharification Process

Until about 1960, dextrose was produced by a straight acid hydrolysis process. A dextrose yield of only about 88 % was achieved due to the formation of acid-catalyzed degradation products (KOOI and ARMBRUSTER, 1967). The availability of a commercial glucoamylase product resulted in the development of an acid–enzyme process that produced 92–94 % dextrose from a 10–20 DE acid liquefied hydrolysate. Dextrose yield was increased additionally to 95–97 % with the development of a total enzymatic process that utilized an alpha-amylase for liquefaction and glucoamylase for saccharification.

In a typical enzymatic process, liquefied hydrolysate at 30–35 % solids, pH 4.0–4.5 and about 60°C is dosed with glucoamylase. Saccharification is conducted either batchwise or in a continuous mode using large tanks that may contain 200000 to several million liters of hydrolysate. Depending on the glucoamylase dosage, reaction is carried out for 1–4 days until the maximum dextrose level is reached. Glucoamylase dosage is indirectly related to reaction time, i.e., doubling the dosage reduces time to reach maximum dextrose by half (Fig. 2).

A maximum dextrose level of about 96.1 % is attained at a typical industrial substrate solids level of 30 %. The 3.9 % non-dextrose saccharides are composed of about 2.7 % disaccharides (1.5 % isomaltose, 1.0 % maltose, 0.2 % maltulose), 0.3 % trisaccharide and 0.9 % higher molecular weight oligosaccharides (HEBEDA, 1993).

Maltose and isomaltose are formed by the reverse action of glucoamylase via condensation of two dextrose molecules; this reaction is termed reversion. As a result of reversion, maltose reaches an equilibrium level early in the saccharification and remains constant thereafter. Isomaltose formation, however, is slow and the disaccharide continues to accumulate during the reaction. Consequently, continuing saccharification beyond maximum dextrose results in a reduction in dextrose yield (Fig. 2) due to the slow continual formation of isomaltose by the reverse reaction. If the reaction was allowed to proceed to equilibrium, a level of 10–12 % isomaltose would be reached (HEBEDA, 1993).

Maltose and isomaltose levels are reduced as substrate water concentration is increased (HEBEDA, 1993). Consequently, dextrose yield is increased as solids level is lowered.

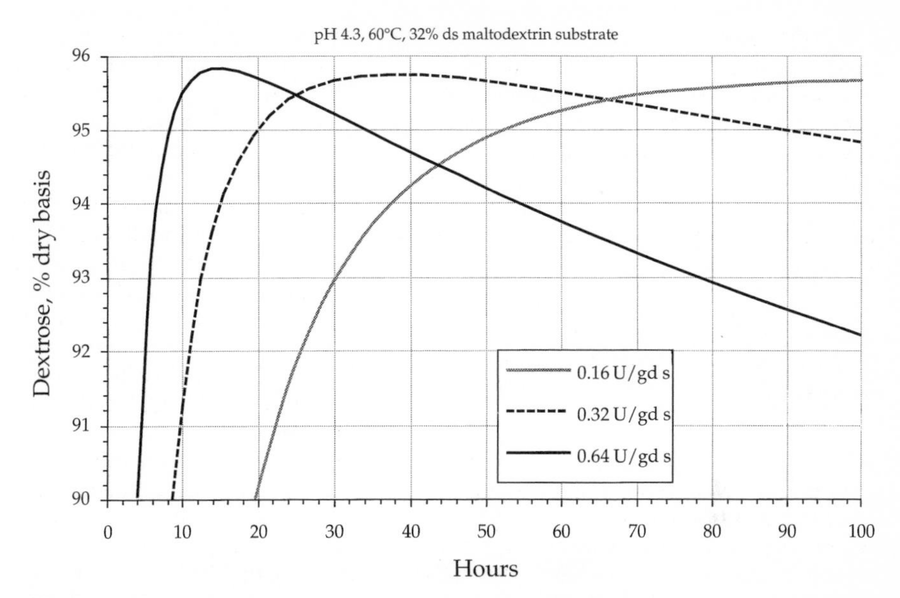

Fig. 2. Dextrose production during saccharification with glucoamylase. Figure provided by Enzyme Bio-Systems Ltd., Beloit, WI; ds, dry substance.

For instance, dextrose yields of 98.8, 98.2, 97.5 and 96.9% are achieved at solids levels of 10, 15, 20 and 25%, respectively (HEBEDA, 1987). However, problems associated with microbial contamination and cost of water removal preclude low solids operation.

Maltulose (4-*O*-alpha-D-glucanopyranosyl D-fructose) is present in hydrolysate as a result of high pH during liquefaction; a maltulose precursor is formed when the end dextrose unit of an oligomer is converted to a fructose unit by alkaline isomerization. The presence of the terminal fructose unit prevents hydrolysis of the last alpha-1,4 linkage by glucoamylase during saccharification resulting in the formation of maltulose. The maltulose level is minimized by maintaining liquefaction at the lowest pH possible, preferably at 6.0 or less. At higher pH levels, significant levels of maltulose can be formed with an equivalent reduction in dextrose yield.

Saccharides of DP-3 or higher that remain in dextrose hydrolysate are often generally branched oligomers that are only slowly hydrolyzed by glucoamylase.

Dextrose concentration can be increased beyond the normal maximum level by the use of either pullulanase or *B. megaterium* amylase in the saccharification step. Pullulanase provides two advantages; (1) branched oligosaccharides are more readily hydrolyzed resulting in a lower DP-4+ saccharide level and (2) less glucoamylase is needed resulting in a slower rate of reversion and less isomaltose formation. Consequently, dextrose yield is increased by 0.5–1.5% when pullulanase is used with glucoamylase for saccharification (JENSEN and NORMAN, 1984).

Dextrose yield is also increased by the use of *B. megaterium* amylase. During saccharification, highly branched oligosaccharides accumulate that are resistant to hydrolysis by glucoamylase. The *B. megaterium* amylase hydrolyzes these oligosaccharides to panosyl units that are then transferred to dextrose to form maltotetraose (DP-4 saccharides) (6^3-alpha-glucosylmaltotriose). The DP-4 saccharides are then readily hydrolyzed to dextrose by glucoamylase. As with pullulanase, the use of *B. megaterium* amylase (Megadex) results in lower isomaltose and DP-4+ levels, and dextrose is increased by 0.5–1.0% (Fig. 3). Other potential advantages are shorter reaction time, lower glucoamylase dosage or higher solids operation (HEBEDA et al., 1988).

Recommended pH ranges for saccharifica-

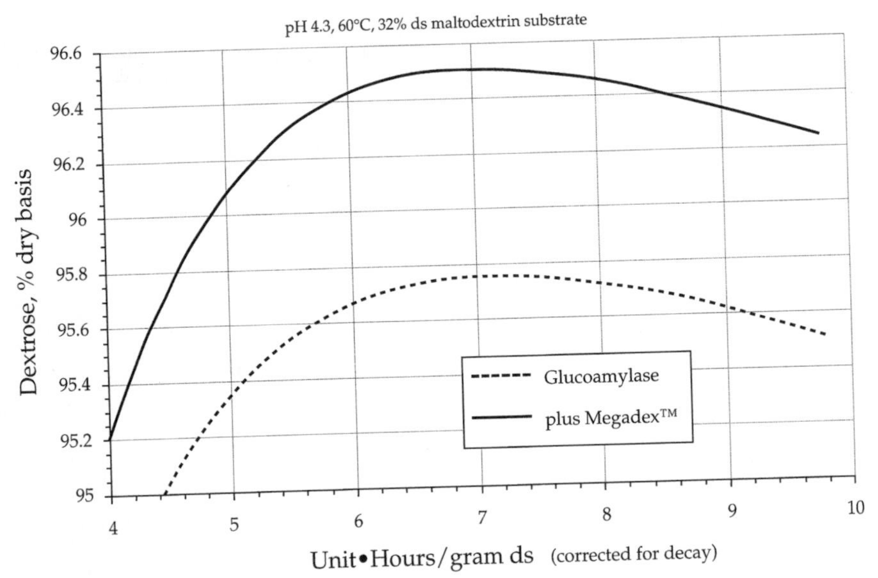

Fig. 3. Dextrose production saccharification with glucoamylase alone or in combination with *Bacillus megaterium* amylase (Megadex™). Figure provided by Enzyme Bio-Systems Ltd., Beloit, WI.

tions with pullulanase or *B. megaterium* amylase are 4.5–4.8 and 5.0–5.2, respectively (REEVE, 1992).

5.1.2 Refining

Saccharified dextrose hydrolysate is clarified, refined and processed to several different products including a high dextrose syrup, liquid dextrose or crystalline dextrose.

Clarification is generally conducted by vacuum filtration using a rotary filter precoated with a filter aid such as diatomaceous earth (diatomite, kieselgur, infusorial earth), perlite (silicate) or cellulosic fibers (wood flour) (BASSO, 1992). The precoated filter drum rotates within an agitated filter bowl containing the hydrolysate. Vacuum is applied and the hydrolysate is drawn through the precoat depositing a layer of insoluble fat, protein and residual starch on the surface. A continuously advancing knife blade removes the surface of the precoat to expose fresh filter aid. Important variables are precoat permeability, precoat concentration, precoat type, drum speed, drum submergence, vacuum, and knife blade advance.

Other possible types of separation processes include membrane filtration with tubular crossflow membranes, centrifugation and flotation (BASSO, 1992).

Subsequent prove-up filtration may be used at various points in the process to remove small insoluble particles that pass through the precoat filtration/refining steps or are precipitated during processing. For this purpose, vertical or horizontal pressure leaf filters are used with a fine filter aid.

The clarified hydrolysate is generally refined using a sequential carbon/ion-exchange treatment. In some cases, carbon has been eliminated in preference to total ion-exchange refining (SCHENCK and COTILLON, 1992).

Powdered or granular activated carbons are used to remove color, soluble proteins and other impurities. Treatment with powdered carbon is conducted by either passing hydrolysate through a carbon layer precoated on a pressure filter or by batch treatment where carbon is added directly to the clarified hydrolysate and, after treatment, removed by filtration. Granular carbon is used in columns employing a continuous hydrolysate flow in either fixed-bed or pulse-bed operation. Regardless of whether powdered or granular carbon is used, the refining step also removes any insoluble material that passed through the initial filtration step. Important carbon treatment factors include feed concentration, temperature, contact time, and pH.

Ion exchange treatment involves demineralization with a double pass cation–anion system; resins are macroporous strong-acid cation and medium-base anion (SCHENCK and COTILLON, 1992). Treatment with an adsorbent resin alone or in combination with a weak-acid cation resin is often also used in a final polishing step. Ion-exchange treatment removes salts, proteins, color and color precursors.

5.1.3 Commercial Products

Commercial dextrose products are produced in both solid and liquid form.

5.1.3.1 Crystalline Dextrose

Crystalline dextrose monohydrate is produced using a refined feed of 95–96% dextrose syrup at 75–78% solids. The syrup at about 46°C is mixed with as much as 20–25% of a previous batch of crystalline dextrose to act as a seed. Alternatively, a continuous precrystallizer may be used to form seed crystals (MULVIHILL, 1992). The mass is agitated and slowly cooled to 20–40°C over 3–5 days generating a crystalline yield of about 60%. The material is spun in perforated screen centrifuge baskets and sprayed with water to wash out solubles. The wet crystal mass at about 99.5% or higher purity and 14% moisture is dried in rotary driers to a final moisture of 8.5–8.9%. Crystals are screened to produce fine and coarse grades.

The mother liquor at 90% dextrose from the first crystallization is concentrated and crystallized in the above manner at a lower temperature and for a longer time to generate a second crop of crystals. The mother liquor from the second crystallization contains less than 80% dextrose and is evaporated to 71% solids and sold as hydrol.

Crystalline alpha-anhydrous dextrose is prepared in an evaporative crystallizer at 60–65°C in 6–8 hours, centrifuged, washed and dried to a moisture level of about 0.1%. Crystalline beta-anhydrous dextrose requires a higher solids of about 90% and a higher temperature above 100°C for crystallization.

5.1.3.2 Total Sugar

Total sugar products in which hydrolysate is dehydrated to a mix of crystals and amorphous glass material are also produced. One such product produced in the U.S. is a partially crystallized, spray-dried material used for tableting (MULVIHILL, 1992). Large amounts of total sugar products, however, are not produced in either the U.S. or Europe. Total sugar is more popular in Japan and Korea where it represents 40 and 50%, respectively, of total crystalline dextrose sold (SCHENCK, 1989).

5.1.3.3 Liquid Dextrose

A 95% dextrose syrup at 98 DE is produced by refining dextrose hydrolysate and evaporating to 71% solids. Concentration is accomplished with a falling-film evaporator using a multiple effect, thermal recompression or mechanical vapor recompression design (DEDERT, 1992). A liquid 99.5% dextrose product is produced by dissolving crystalline monohydrate dextrose in water to 71% solids. Chromatographic separation is also used to produce a liquid high dextrose product. In this case, a 95% dextrose refined hydrolysate at 60% solids is fed to a simulated moving bed column containing a cation-exchange resin in the sodium form. The separation process produces a product containing 99% dextrose and an oligosaccharide by-product syrup containing 60–80% dextrose (MULVIHILL, 1992).

5.1.3.4 Pharmaceutical Dextrose

Pharmaceutical-grade dextrose can be produced by dissolving crystalline monohydrate dextrose in water, carbon treating to remove color and 5-hydroxymethylfurfural (HMF), ultrafiltering to remove pyrogens and crystallizing an alpha-anhydrous fraction containing 99.9% dextrose (MULVIHILL, 1992). This material is then sold to the pharmaceutical industry. Mother liquor from this process contains 99.2% dextrose and is recycled to monohydrate crystallization.

5.2 Properties

During saccharification, dextrose is released as the beta-D-glucose anomer but mutarotates rapidly to an equilibrium level of 63% beta-D-glucose and 37% alpha-D-glucose. Both anomers exist in the pyranose form. A small amount of the open-chain aldehyde form is present and is responsible for reducing properties.

Crystallization yields alpha-D-glucose, alpha-D-glucose hydrate or beta-D-glucose depending on temperature. In aqueous solution at 25°C, each form will eventually reach an equilibrium solubility level of 51.2 g glucose/100 g solution. The initial solubility, however, depends on the particular crystalline form. Monohydrate dextrose dissolves rapidly to a level of 30.2% w/w. As the more soluble beta-anomer is formed, solubility increases to 51.2%. Anhydrous alpha- and beta-dextrose initially dissolve to levels of 62% and 72%, respectively. These levels are beyond the solubility of the hydrate form resulting in crystallization of the hydrate.

Dextrose undergoes a variety of reactions including alkaline isomerization to fructose and mannose, alkaline degradation to carboxylic acids, acid degradation to HMF, hydrogenation to sorbitol, alkaline degradative hydrogenation to glycol, 1,2-propanediol and glycerol, and oxidation to gluconic and glucaric acids (KIEBOOM and VAN BEKKUM, 1985).

The important functional properties of dextrose in food applications include sweetness, fermentability, osmotic pressure and flavor enhancement; in addition, dextrose exhibits browning, stability, bulking and tableting properties (CORN PRODUCTS, 1991).

As a sweetener, dextrose generally exhibits 60–80% of the sweetness of sucrose. In some formulations, however, a dextrose/sucrose blend is perceived as more sweet than sucrose alone. Sweetness perception depends on presence of other ingredients, temperature, pH, viscosity and solids level; for instance, dry and liquid dextrose are perceived as 76% and 65–70% as sweet as sucrose (HANOVER, 1982). Dextrose also shortens the perception of sweetness in some foods, acting to enhance flavor.

As a monosaccharide, dextrose is fermented rapidly by yeast and other organisms, often in preference to other substrates. Since dextrose is about half the molecular weight of sucrose, it produces a greater osmotic effect than sucrose on an equal weight basis. By reacting with nitrogenous compounds via the Maillard reaction, dextrose produces a variety of colored products depending on conditions such as pH, temperature, type of nitrogen and concentration of reactants.

5.3 Applications

The physical, chemical and nutritive properties of dextrose are utilized in a variety of food applications such as brewing, baking and confections as well as in non-food areas such as the chemical, drug and pharmaceutical industries. Total distribution of dextrose (monohydrate basis) in the U.S. was 604, 513 and 547 M mt in 1991, 1980 and 1970, respectively (HEBEDA, 1993). The primary industries and trades where dextrose is used are shown in Tab. 5. More than 75% of the dextrose is used in four market segments; these

Tab. 5. Distribution of Dextrose in the United States, 1992

Industry	% of Total
Baking and mixes	23.4
Chemicals, drugs, pharmaceuticals	21.4
Alcoholic beverages and brewing	16.6
Confections	14.0
Miscellaneous	24.6

are baking/mixes, chemicals/drugs/pharmaceuticals, alcoholic beverages/brewing and confections.

Dextrose is used in baked goods such as breads, buns and rolls to provide fermentables, supply strength in sidewalls for ease of slicing and handling, develop color and flavor, and optimize texture. In cakes, dextrose improves volume, texture and symmetry and in cake mixes, dextrose helps to reduce lumping.

In cookies, dextrose controls sweetness and flavor and spread during baking while providing uniform surface color and a tenderizing effect.

In icings, dextrose provides smooth texture and mild sweetness as well as excellent flow properties in icing mixes.

In brewing, dextrose is used as a completely fermentable adjunct for reducing residual carbohydrate and caloric level in low-caloric beer production. In wine, dextrose is used as an additive to increase the level of fermentables, to contribute to flavor, and to add sweetness.

In confections, dextrose provides sweetness, softness and crystallization control. Blending dextrose with sucrose enhances gum flavor while balancing sweetness, improving color and coating gloss, and generating a cool mouthfeel. In panned candy, dextrose is ideal for coating, reducing panning/drying time and providing strength/hardness and enhancing flavor/gloss. Dextrose assists in tableting by providing good flowability in tableting machines as well as good binding and release properties. Dextrose also adds to sweetness, tenderness and whippability in marshmallow and nougat products.

In dairy applications, dextrose is used in ice cream and frozen desserts for sweetness, crystallization control and smooth texture.

In beverages, dextrose is used for sweetness, body and osmotic pressure. In beverage powders, dextrose enhances flavor and reduces excessive sweetness while controlling flowability and improving shelf stability.

In canned products such as sauces, soups, gravies, fruits and juices, dextrose is used to provide sweetness, flavor, body, osmotic pressure, improved texture, and better aesthetic quality.

Dextrose is also used in seasoning formulations for sausages and for color in hams, in peanut butter for mouthfeel and chewability, in jams and jellies for controlling sweetness and providing stability by controlling osmotic pressure, as a substrate for vinegar fermentation and in prepared mixes for biscuits, pancakes, waffles, doughnuts and icings.

In the pharmaceutical industry, dextrose is used for intravenous solutions and in tableting. Dextrose is used as the raw material in many different fermentation processes to produce organic acids, vitamins, antibiotics, enzymes, amino acids and polysaccharides (DASINGER et al., 1985). A major industrial fermentation use of dextrose is for the production of fuel ethanol as an oxygenate or octane enhancer (VENKATASUBRAMANIAN and KEIM, 1985).

Dextrose is also hydrogenated to sorbitol for use in various food and non-food applications (VERWAERDE and SICARD, 1984). Acid catalyzed polymerization of dextrose in the presence of a polyol is used for the production of polydextrose, a cross-linked polymer used as a reduced-calorie bulking agent (TORRES and THOMAS, 1981). Other industrial applications of dextrose include use in wallboard as a humectant, in concrete as a setting retardant, in resin formulations as a plasticizer, in adhesives for flow control and for the production of methyl-glucoside.

6 Fructose

6.1 Manufacture

6.1.1 Isomerization

The immobilized enzyme is placed into packed-bed reactors that are generally about 6 m high and 1.5 m in diameter (WHITE, 1992) and contacted with a 40–50% solids refined 93–96% dextrose feed. Magnesium is added at a level of 0.5–5 mM to activate and stabilize the enzyme and to counteract the negative effect of residual calcium. The feed may also be deaerated or treated with sodium bisulfite at a level of 1–2 mM SO_2 to prevent oxidation of the enzyme and loss in activity. Temperature is controlled in the range of 55–61 °C to maintain maximum enzyme activity while minimizing the possibility of microbial contamination. The pH is controlled in the range of 7.5–8.2; the specific level is dependent on the particular enzyme. Flow through the column is maintained at a rate that produces an effluent containing 42% fructose and 53% dextrose.

Under these conditions, a typical column can be used efficiently for several months until the residual level of activity is reduced by 80–90% through several enzyme half-lives.

6.1.2 Commercial Products

6.1.2.1 Fructose Syrups

The 42% fructose isomerisate is refined in the same manner as previously described for dextrose products (see Sect. 5.1.2). A carbon treatment is used first to remove color and off-flavors; ion-exchange treatment with strong-acid cation resin in the hydrogen form followed by a weak-base anion resin in the free-base form is then used to remove salts and residual color. The product is evaporated to 71% solids at low temperature and stored at 30–32 °C to prevent crystallization of dextrose.

A 55% fructose syrup is produced by a combination of enrichment and blending. The 42% HFS is first enriched to a 90% HFS by chromatographic separation. Separation is achieved on a strong-acid, cation-exchange resin in the calcium form using a feed of refined 42% HFS at about 50% solids. Fructose is preferentially retained on the resin and eluted with water. A simulated moving-bed process is used in which feed is added and products are withdrawn continuously in a manner that maximizes separation. In a typical operation, a product containing 80–90% fructose and 7–19% dextrose is obtained. A low fructose fraction (raffinate) at about 20% solids and containing 80–90% dextrose and 5–10% fructose is a by-product of the separation and is recycled to the process at either the saccharification or isomerization stage.

The 90% HFS is then blended with 42% HFS to a fructose level of 55% and the product is evaporated to 77% solids for shipment. Crystallization is not a concern because of the low dextrose level.

The 90% HFS product containing 9% dextrose is evaporated to a solids level of 80%. High fructose products at 80% fructose/18% dextrose and 95% fructose/1% dextrose are also produced (WHITE, 1992).

6.1.2.2 Crystalline Fructose

A number of different aqueous and solvent processes have been proposed for the crystallization of fructose (HANOVER, 1992; HEBEDA, 1987). Aqueous processes have the disadvantages of high fructose solubility and possible formation of hemihydrate, dihydrate and dianhydride forms of fructose that inhibit crystallization of pure fructose. Fructose solubility is reduced by replacing at least a portion of the water with alcohol, however, removal and recovery of the solvent is costly. As a result, crystalline fructose production in the U.S. and elsewhere is based primarily on the aqueous system with the exception of some aqueous/ethanol processes in Europe and the Far East (HANOVER, 1992).

A composite crystallization process based on several sources and proposed by HANOVER (1992) involves the use of a 90% fructose feed in an aqueous, alcohol or combined solvent system at a temperature of 60–85 °C and a pH of 3.5–8. Crystallization is conducted either in a batch or continuous mode with a crystalline fructose seed. Temperature is lowered to 25–35 °C and/or pressure is reduced, and crystallization is accomplished during 2–180 hours with constant agitation. Crystals are separated by centrifugation or filtration, impurities are washed out, and the mass is dried and screened.

6.2 Properties

Pure fructose is a keto-hexose that crystallizes as beta-D-fructopyranose and mutarotates in a 36 °C aqueous solution to an equilibrium mixture of 57% beta-fructopyranose,

Tab. 6. Relative Sweetness of Sucrose, Fructose and Dextrose

Sweetener	Reference 1	Reference 2	Reference 3
Sucrose	100	100	100
Fructose	110	150–170	120–180
Dextrose	70	70–80	60–70

1 INGLETT (1981)
2 HOBBS (1986)
3 HANOVER (1992)

Tab. 7. Distribution of Fructose Syrup in the United States, 1992

Industry	% of Total
Beverage	66.2
Canning	7.6
Baking	5.7
Dairy and ice cream	3.8
Miscellaneous	16.7

31% beta-fructofuranose, 9% alpha-fructofuranose and 3% alpha-fructopyranose (DODDRELL and ALLERHAND, 1971). The concentration of beta-fructopyranose in solution varies from 77% at 0.5°C to 48% at 61°C (HANOVER, 1992).

Fructose exhibits the same reactions as other carbohydrates to form esters, ethers and acetals. Degradative reactions at high and low pH occur more readily than with dextrose.

Beta-fructopyranose is the sweetest tautomer and, in the crystalline state, has been reported to be about twice as sweet as sucrose (HYVONEN et al., 1977). However, in solution mutarotation of fructose to less sweet tautomers reduces the perception of sweetness. Other factors that affect sweetness include temperature, pH, concentration and presence of other additives. Depending on the particular test, fructose is generally reported to be 10–80% sweeter than sucrose and 50–200% sweeter than dextrose (Tab. 6).

Important properties of fructose syrups include sweetness, fermentability, solubility, osmotic pressure, humectancy, viscosity, color development and flavor development. Sweetness is the most important property; 55% fructose syrup exhibits about the same sweetness as sucrose, 42% fructose syrup is about 8% less sweet than sucrose and 90% fructose syrup is about 6% sweeter than sucrose (WHITE, 1992).

6.3 Applications

Distribution of fructose syrup (commercial basis) in the U.S. in 1991, 1980 and 1970 was 7748, 2659 and 99 M mt, respectively (HEBEDA, 1993). The primary application of fructose syrup is as a sweetener in soft drinks; about two-thirds of the HFS is used in the beverage industry with the remainder used in canning, baking, dairy/ice cream and other food applications (Tab. 7).

The significant increase in fructose syrup utilization over the last two decades is due to replacement of sucrose in soft drinks. HFS-42 was first used for this purpose in 1974, when partial replacement of sucrose was approved. Since HFS-55 exhibits the same level of sweetness as sucrose, it can be used as a direct replacement for sucrose in soft drinks, and as a result, HFS-55 was approved for 50% substitution in 1980 and for 100% substitution in 1984.

HFS is also used for properties other than sweetness. Generally 42% HFS is used for these applications, although 55% HFS can also be used. In baked goods, HFS is used as a source of fermentables, to retain moisture for maintaining shelf-life, provide flavor and aroma, improve crumb texture and softness. In ice cream, HFS helps to control crystallization and add body. In dairy applications, HFS provides body and improves mouthfeel and texture. In confectionery, HFS helps in grain control and humectancy. In canned goods, HFS is used for preservation and for adding sheen. HFS-90, because of its greater sweetness compared to other fructose syrups, can be used in reduced-calorie applications. Crystalline fructose is used primarily for added sweetness in low-calorie foods. Other areas of interest for crystalline fructose include use as a flavor enhancer, synergy with gel-forming ingredients, synergy with other sweeteners and moisture management (HANOVER, 1992).

Tab. 8. Composition of Acid Syrups[a]

Saccharide DP[b]	Syrup DE 30	35	42	55
1	10	14	19	31
2	9	12	14	18
3	10	10	11	13
4+	71	64	56	48

[a] From HOWLING (1992)
[b] DP, degree of polymerization

7 Syrups

7.1 Manufacture

Syrups are produced by either a one-step straight acid hydrolysis process or by a two-stage acid–enzyme (A–E) or enzyme–enzyme (E–E) process.

7.1.1 Acid Converted Syrups

Straight acid converted syrups are manufactured by the process described in Sect. 4.2. The standard industrial acid syrup is typically about 42 DE; higher and lower DE products are also manufactured, although the range of product DE is limited to about 30–55. At lower DE, insufficient hydrolysis may cause starch retrogradation and a haze in the final product; at higher DE, poor color stability and off-flavors are produced due to acid-catalyzed side reactions (HOWLING, 1992). A typical commercial acid converted syrup at 41–47 DE contains 16–24% dextrose, 12–18% disaccharides, 8–17% trisaccharides and 42–61% higher saccharides (CORN PRODUCTS, 1993a). Composition of 30, 35, 42 and 55 DE acid syrups is shown in Tab. 8.

7.1.2 High Conversion Syrups

Acid–enzyme syrups are manufactured by saccharification of acid liquefied hydrolysate with one or more enzymes. The most common product of this type is referred to as a high conversion syrup and is prepared using a combination of glucoamylase and maltogenic enzyme for saccharification. The maltogenic enzyme is generally a fungal *Aspergillus oryzae* alpha-amylase, although beta-amylase from malted barley can also be used. Relative amounts of the two enzymes and the total dosage is dependent on the exact syrup composition desired and the reaction time used. Saccharification is generally conducted at pH 4.8–5.2 and 55–60°C for anywhere from several hours to 1 or 2 days until a DE of about 63 is reached. A typical commercial product at 61–67 DE contains 22–42% dextrose, 21–35% disaccharide, 5–15% trisaccharide and 16–30% higher saccharides (CORN PRODUCTS, 1993b).

7.1.3 Maltose Syrups

High maltose syrups are produced using only a maltogenic enzyme for saccharification. An enzyme–enzyme process is generally used for manufacture of maltose syrups to minimize the dextrose level, although an acid-enzyme process can also be used. In either case, substrate DE is maintained at 10–20 or lower to maximize maltose production. As substrate DE is increased, more oligomers containing an odd number of glucose units are present resulting in the production of more maltotriose and less maltose.

In a typical process, a 35–45% solids, 10–20 DE substrate is hydrolyzed with a maltogenic enzyme at about pH 5 and 50–55 °C for 24–48 hours (FULLBROOK, 1984) or until a maltose level of 40–60% is reached. Higher maltose levels up to about 80% can be achieved by using a debranching enzyme during saccharification to hydrolyze alpha-1,6 linkages and provide additional substrate for the action of the maltogenic enzyme. Typical syrup compositions are shown in Tab. 9.

7.1.4 Other Oligosaccharide Syrups

Syrups containing 50 and 60% maltotriose are made from high maltose syrup by chromatographic separation or by using a maltotriose-forming enzyme in combination with a debranching enzyme (OKADA and NAKAKUKI, 1992).

Tab. 9. Composition of High Maltose Syrups[a]

Process	Saccharification Enzyme(s)	Dextrose	Maltose	DP-3	DP-4+
A-E	beta-Amylase	8	40	15	37
A-E	beta-Amylase	9	52	15	24
E-E	alpha-Amylase	4	52	23	21
E-E	beta-Amylase	2	54	19	25
E-E	beta-Amylase + pullulanase	1	76	16	7

[a] A-E data from CORN REFINERS ASSOCIATION, INC. (1979); E-E data from HEADY and ARMBRUSTER (1971)

A syrup containing 50% or higher maltotetraose can be prepared with a *Pseudomonas stutzeri* DP-4-forming enzyme (ABDULLAH, 1972). Commercial production of 50 and 72% maltotetraose syrups is conducted by a continuous process using an immobilized enzyme system (KIMURA and NAKAKUKI, 1990).

7.1.5 Refining

Syrups are refined by the same carbon- and ion-exchange techniques described earlier and evaporated to 75–86% solids. In some cases, sulfur dioxide is added prior to evaporation to minimize color development. Corn syrup solids are prepared by spray-drying to a moisture level of 5%.

7.2 Properties

Functional properties of corn syrups are generally dependent on the relative levels of high and low molecular weight saccharides present, i.e., the average molecular weight. DE, which is a measure of degree of hydrolysis and average molecular weight, is usually a good indicator of specific properties. As an example, syrup viscosity increases as average molecular weight increases (i.e., as DE decreases) assuming other factors such as solids and temperature are constant. Examples of other properties that increase as DE decreases include body development, cohesiveness, crystallization control, emulsion, foam stabilization, sheen production and thickening.

Degree of sweetness, on the other hand, is dependent on the concentration of low molecular weight saccharides, and, therefore, increases as average molecular weight decreases (i.e., as DE increases). Other properties that increase as DE increases include browning and color formation, fermentability, flavor enhancement, freezing point depression, hygroscopicity, nutritive value, osmotic pressure and preservation.

The relationship between DE and functional properties is only true when saccharide distribution changes gradually as DE changes. It is possible, however, to produce equivalent DE syrups that exhibit varied saccharide distribution depending on the process used (Tab. 10). Since the levels of high and low molecular weight saccharides differ, properties vary as the average molecular weight changes. For instance, as shown in Tab. 10, the acid syrup (with the highest level of higher molecular weight saccharides) exhibits the greatest viscosity, browning and hygroscopicity; the E–E syrup (with the highest level of

Tab. 10. Composition of 42 DE Syrups Produced by Different Processes[a]

Process	DE	Dextrose	Maltose	DP-3+
Acid	42	19	14	67
A-E	42	6	45	49
E-E	42	2	56	42

[a] From HOWLING (1992)

Tab. 11. Distribution of Corn Syrups in the United States, 1992

Industry	% of Total
Confections	20.1
Food, miscellaneous	13.4
Brewing/alcoholic beverages	12.8
Drugs and chemicals	11.3
Baking	6.8
Formulated dairy	6.7
Canning	6.0
Ice cream/frozen desserts	5.6
Miscellaneous	17.3

low molecular weight saccharides) provides the most sweetness, flavor enhancement, and fermentability (HOWLING, 1992).

7.3 Applications

Syrups are used in a wide range of food and non-food applications. Total distribution of syrups (commercial basis) in the U.S. was 2845, 2201 and 1449 M mt in 1991, 1980 and 1970, respectively (HEBEDA, 1993). The primary industries and trades where syrups are used are shown in Tab. 11. Three major segments represent 44.2% of the total; these are confections, brewing/alcoholic beverages and drugs/chemicals.

In the confectionery industry, straight acid syrups, high conversion syrups and maltose syrups are used in a variety of products including hard candies, toffees, caramels, fudge, gums, fondants, chewing gum and marshmallows. Syrups are used to control sweetness, crystallization, hygroscopicity, viscosity, texture, mouthfeel, grain and flavor.

In beverages, syrups are used primarily in the brewing industry as adjuncts to supplement malted barley. High conversion and maltose syrups provide fermentable sugars, enhance flavor and provide body.

In the area of dairy products (ice cream and other frozen desserts), syrups impart body and smoothness, provide texture, control crystallization, adjust flavor and sweetness, and improve shelf-life and melt-down properties.

In baking, high conversion or maltose syrups are often used as a source of fermentables in yeast-raised baked goods. Syrups also provide sweetness, aid in color development, increase shelf-life by increasing humectancy, retard crystallization, and enhance tenderness.

Syrups are used in a variety of icings, toppings, glazes and fillings to control crystallization, and provide sheen, smoother texture and body.

In canned goods, syrups contribute viscosity and body, provide sheen, impart desirable mouthfeel, prevent crystallization, accentuate fruit flavors and improve color and texture.

Syrups are also used in jams and jellies to add sweetness, control texture, body and brilliance, inhibit crystallization and improve shelf stability.

In breakfast cereals, syrups are used to coat individual pieces to improve shelf-life, enhance flavor, develop color and provide resistance to breaking. In ketchup, syrups are used to control sweetness, body and texture. In meats, syrups enhance flavor and color.

Syrup is used in pharmaceuticals such as medicated confections that contain active ingredients and as a carrier for liquid cough mixtures and medicines.

In non-food applications, syrups are used in adhesives to improve stability, in concrete as a setting retardant, in air fresheners and tobacco as a humectant, in colognes and perfumes to control evaporation and in dyes, inks, explosives, metal plating, shoe polish, leather tanning, paper and textiles.

Hydrogenation of syrups produces products containing varied levels of sorbitol, maltitol and higher molecular weight polyols (VERWAERDE and SICARD, 1984).

8 Maltodextrins

8.1 Manufacture

Maltodextrins are manufactured by a variety of starch liquefaction processes. The primary processes in use today are referred to as "single-stage" or "dual-stage" as described by

ALEXANDER (1992). In the single-stage process, acid is used to liquefy regular starch at a temperature above 105 °C; alternatively, a bacterial alpha-amylase is used to liquefy waxy starch at 82–105 °C. Other reaction conditions are similar to those described earlier for starch liquefaction (see Sect. 4). Following the initial reaction, hydrolysate is held batchwise to continue hydrolysis until the desired DE is attained. In the dual-stage process, a starch slurry is first liquefied to less than 3 DE with acid or bacterial alpha-amylase at a temperature above 105 °C. A high temperature heat treatment is then used to ensure complete starch gelatinization followed by the addition of bacterial alpha-amylase and reaction at 82–105 °C until the desired DE is reached.

In either process, the reaction is stopped by inactivating the enzyme with heat or by pH adjustment, and the hydrolysate is clarified by vacuum filtration and refined with carbon. Final product is generally available in a spray-dried form at 3–5% moisture although liquid products at about 75% solids are also produced.

A recent survey of commerical maltodextrins (ALEXANDER, 1992) indicated that there were 43 products (14 conventional, 29 specialty) available from six U.S. producers and 24 products (23 conventional , 1 specialty) available from five non-U.S. producers (Australia, Belgium, France, Holland, Mexico). Products of 1 to 19.5 DE are made primarily from dent and waxy corn starch, although, tapioca and potato starches are also used.

8.2 Properties

Maltodextrins are generally non-hygroscopic and bland tasting; lack of perceived sweetness is due to the minimal level of low molecular weight saccharides. The low level of reducing sugars is also important in minimizing the browning reaction. Other properties of importance include viscosity, cohesiveness, freezing point control, crystallization prevention, and osmolality. These properties increase or decrease in degree relative to DE in the same manner as for corn syrups.

8.3 Applications

Maltodextrins are often used in applications where there is a need for a non-sweet, non-hygroscopic, water holding additive; for instance, as a spray-drying aid in flavor encapsulation or as a bulking agent in puddings, soups, frozen desserts and dry mixes. In the medical area, maltodextrin is used as an excipient in tableting and as a carbohydrate source in nutritional fluids such as infant formulas and enteral products. A rapidly growing application of maltodextrin is as a fat replacer to reduce caloric content in a variety of foods.

Approximately 91000 mt of maltodextrin are sold per year; application areas include spray-drying aids (25% of total), bulking agents (23%), medical and nutritional (17%), fat replacers (7%), frozen desserts (6%) and other uses in the areas of non-dairy coffee whiteners, salad dressings, confections, coatings and drilling fluids (5% or less each) (ALEXANDER, 1992).

9 Cyclodextrins

9.1 Manufacture

Cyclodextrins are produced by the action of cyclodextrin enzyme on starch hydrolysate in either a solvent (ARMBRUSTER, 1988) or non-solvent process (HORIKOSHI et al., 1982). Solvents direct the formation of specific cyclodextrins by forming complexes; for example, alpha-, beta-, and gamma-cyclodextrins are formed in the presence of decanol, toluene and naphthol/methylethyl ketone, respectively (HEDGES, 1992). In the absence of solvent, a mixture of products is obtained.

In the solvent process, the complex is removed by filtration or centrifugation, refined and crystallized, In the non-solvent process, the product is refined, dewatered by evaporation and crystallized.

Cyclodextrins are shipped as fine powders in 25 and 50 kg containers as well as in bulk.

9.2 Properties

Alpha-, beta- and gamma-cyclodextrins are soluble in water at 25 °C to the extent of 12.8, 1.8 and 25.6 g/100 mL, respectively (HEDGES, 1992). Increasing temperature from 25 to 45 or 60 °C increases solubility by about 2.5 and 5 times, respectively, for each form. The degree of solubility in organic solvents varies depending on the particular solvent.

Cyclodextrins are resistant to heat up to about 300 °C and are not hydrolyzed by bases even at high temperatures; hydrolysis with acid and oxidation with oxidizing agents does occur (HEDGES, 1992). Exo-enzymes, such as glucoamylase and beta-amylase, do not hydrolyze cyclodextrins due to the lack of a reducing end on the molecule. Some endo-enzymes, such as alpha-amylases, have the ability to hydrolyze cyclodextrins to dextrose and oligosaccharides.

9.3 Applications

Cyclodextrins have the ability to complex with a "guest" molecule. This property is utilized in a number of applications. For instance, in foods cyclodextrins can be used to solubilize, disperse or mask flavors, decrease volatility of odors, and remove unwanted materials such as bitter flavors in citrus juice, pigments in fruit juices and caffeine in tea. In pharmaceuticals, cyclodextrins can increase solubility of drugs and allow an overall lower dose, as well as control drug release over a prolonged time. In analytical and diagnostic assays, cyclodextrins serve to solubilize dyes, indicators and substrate.

10 References

ABDULLAH, M. (1972), *U.S. Patent* 3654082.

ABDULLAH, M., FLEMING, I. D., TAYLOR, M. WHELAN, W. J. (1963), Substrate specificity of the amyloglucosidase of *Aspergillus niger, Biochem. J.* **89** (1), 35–36.

ALEXANDER, R. J. (1992), Maltodextrins: production, properties, and applications, in: *Starch Hydrolysis Products* (SCHENCK, F. W., HEBEDA, R. E., Eds.), pp. 233–275, New York: VCH Publishers, Inc.

ANTRANIKIAN, G., KOCH, R., SPREINAT, A. (1990), *International Patent Application* PCT/DK90/00074.

ANTRIM, R. L., AUTERINEN, A. L. (1986), A new regenerable immobilized glucose isomerase, *Starch/Stärke* **38** (4), 132–137.

ARMBRUSTER, F. C. (1988), Use of cyclohexane in the production of pure alpha and beta cyclodextrins, in: *Proceedings of the Fourth International Symposium on Cyclodextrins* (HUBER, O., SZEJTLI, J., Eds.), pp. 33–39, Dordrecht: Kluwer Academic Publishers.

ARMBRUSTER, F. C., JACAWAY, W. A. (1970), *U.S. Patent* 3549496.

BASSO, A. J. (1992), Hydrolysate clarification processes and equipment, in: *Starch Hydrolysis Products* (SCHENCK, F. W., HEBEDA, R. E., Eds.) pp. 483–504, New York: VCH Publishers, Inc.

BEMILLER, J. N. (1992), Starch, in: *Encyclopedia of Food Science and Technology* (HUI, Y. H., Ed.), Vol. 4, pp. 2418–2424, New York: John Wiley & Sons, Inc.

BERNETTI, R. (1992), Quality assurance and analytical methods, in: *Starch Hydrolysis Products* (SCHENCK, F. W., HEBEDA, R. E., Eds.), pp. 367–394, New York: VCH Publishers, Inc.

BOIDIN, A., EFFRONT, J. (1917), *U.S. Patent* 1227374.

BROWN, S. H., COSTANTINO, H. R., KELLY, R. M. (1990), Characterization of amylolytic enzyme activities associated with the hyperthermophilic archaebacterium *Pyrococcus furiosus, Appl. Environ. Microbiol.* **56,** 1985–1991.

BRUMM, P. J., HEBEDA, R. E., TEAGUE, W. M. (1991a), Purification and characterization of the commercialized, cloned *Bacillus megaterium* alpha-amylase, Part 1: Purification and hydrolytic properties, *Starch/Stärke* **43** (8), 315–319.

BRUMM, P. J., HEBEDA, R. E., TEAGUE, W. M. (1991b), Purification and characterization of the commercialized, cloned *Bacillus megaterium* alpha-amylase, Part 2: Transferase properties, *Starch/Stärke* **43** (8), 319–323.

CARROLL, J. O., SWANSON, T. R. (1990), *U.S. Patent* 4933279.

CORN PRODUCTS (1991), *The Dextrose Advantage,* Technical Data Sheet.

CORN PRODUCTS (1993a), *Globe® Corn Syrup 1132,* Technical Data Sheet.

CORN PRODUCTS (1993b), *Globe® Corn Syrup 1632,* Technical Data Sheet.

CORN REFINERS ASSOCIATION, INC. (1979), *Nutritive Sweeteners from Corn,* Washington, D.C.

CORN REFINERS ASSOCIATION, INC. (1988), *Corn Annual,* Washington, D.C.

CPC INTERNATIONAL, INC. (1986), Filing of a petition for affirmation of GRAS status (Petition No. GRASP 4G0293, *Fed. Reg.* **51,** 10571.

DALE, J. K., LANGLOIS, D. P. (1940), *U.S. Patent* 2201609.

DASINGER, B. L., FENTON, D. M., NELSON, R. P., ROBERTS, F. F., TRUESDELL, S. J. (1985), Enzymic and microbial processes in the conversion of carbohydrate derived from starch, in: *Starch Conversion Technology* (VAN BEYNUM, G. M. A., ROELS, J. A., Eds.), pp. 237–262, New York: Marcel Dekker, Inc.

DAVID, M.-H., GÜNTHER, H., RÖPER, H. (1987), Catalytic properties of *Bacillus megaterium* amylase, *Starch/Stärke* **39** (12), 436–440.

DEAN, G. R., GOTTFRIED, J. B. (1950), The commercial production of crystalline dextrose, in: *Advances in Carbohydrate Chemistry* (HUDSON, C. S., CANTOR, S. M., Eds.), Vol. 5, pp. 127–143, New York: Academic Press.

DEDERT, T. W. (1992), Hydrolysate concentration processes and equipment, in: *Starch Hydrolysis Products* (SCHENCK, F. W., Hebeda, R. E., Eds.), pp. 505–529, New York: VCH Publishers, Inc.

DE SAUSSERE, T. (1815), *Ann. Phys.* **49,** 129.

DODDRELL, D., ALLERHAND, A. (1971), Study of anomeric equilibrium of ketoses in water by natural-abundance carbon-13 Fourier transform nuclear magnetic resonance. D-Fructose and D-furanose, *J. Am. Chem. Soc.* **93,** 2779–2781.

DOUGLAS, R. (1932), *U.S. Patent* 1858820.

ENZYME BIO-SYSTEMS LTD. (1988), Filing of a petition for affirmation of GRAS status (Petition No. GRASP 7G0328, *Fed. Reg.* **53,** 16191.

FULLBROOK, P. D. (1984), The enzymic production of glucose syrups, in: *Glucose Syrups: Science and Technology* (DZIEDZIC, S. Z., KEARSLEY, M. W., Eds.), pp. 65–115, New York: Elsevier Applied Science Publishers.

HANOVER, L. M. (1982), Functionality of corn-derived sweeteners in formulated foods, in: *Chemistry of Foods and Beverages: Recent Developments* (CHARALAMBOUS, G., INGLETT, G., Eds.), pp. 211–233, New York: Academic Press.

HANOVER, L. M. (1992), Crystalline fructose: Production, properties, and applications, in: *Starch Hydrolysis Products* (SCHENCK, F. W., HEBEDA, R. E., Eds.), pp. 201–231, New York: VCH Publishers, Inc.

HEADY, R. E., ARMBRUSTER, F. C. (1971), *U.S. Patent* 3565765.

HEBEDA, R. E. (1983), Syrups, in: *Kirk-Othmer Encyclopedia of Chemical Technology*, Third Ed., Vol. 22, pp. 499–522, New York: John Wiley & Sons, Inc.

HEBEDA, R. E. (1987), Corn sweeteners, in: *Corn: Chemistry and Technology* (WATSON, S. A., RAMSTAD, P. E., Eds.), pp. 501–534, St. Paul, MN: American Association of Cereal Chemists, Inc.

HEBEDA, R. E. (1993), Starches, sugars and syrups, in: *Enzymes in Food Processing* (NAGODAWITHANA, T., REED, G., Eds.), 3rd Ed., pp. 321–346, San Diego: Academic Press, Inc.

HEBEDA, R. E., STYRLUND, C. R., TEAGUE, W. M. (1988), Benefits of *Bacillus megaterium* amylase in dextrose production, *Starch/Stärke* **40** (1), 33–36.

HEDGES, A. R. (1992), Cyclodextrin: Production properties, and applications, in: *Starch Hydrolysis Products* (SCHENCK, F. W., HEBEDA, R. E., Eds.), pp. 319–333, New York: VCH Publishers, Inc.

HENDERSON, W. E., TEAGUE, W. M. (1988), A kinetic model of *Bacillus stearothermophilus* alpha-amylase under process conditions, *Starch/Stärke* **40** (11), 412–418.

HIDAKA, H., KOAZE, Y., YOSHIDA, K., NIWA, T., SHOMURA, T., NIIDA, T. (1974), Isolation and some properties of amylase from *Streptomyces hygroscopicus* SF-1084, *Starch/Stärke* **26** (12), 413–416.

HOBBS, L. (1986), Corn syrups, *Cereal Foods World* **31** (12), 852–857.

HORIKOSHI, K., NAKAMURA, N., MATSUZAWA, N., YAMAMOTO, M. (1982), Industrial production of cyclodextrins, in: *Proceedings of the First International Symposium on Cyclodextrins* (SZEJTLI, J., Ed.), pp. 25–39, Budapest: Akademiai Kiado.

HOWLING, D. (1992), Glucose syrup: Production properties, and applications, in: *Starch Hydrolysis Products* (SCHENCK, F. W., HEBEDA, R. E., Eds.), pp. 277–317, New York: VCH Publishers, Inc.

HUPKES, J. V. (1978), Practical process conditions for the use of immobilized glucose isomerase, *Starch/Stärke* **30** (1), 24–28.

HYVONEN, L., VARO, P., KOIVISTOINEN, P. (1977), Tautomeric equilibria of D-glucose and D-fructose: Polarimetric measurements, *J. Food Sci.* **42** (3), 652–653.

INGLETT, G. E. (1981), Sweeteners – A review, *Food Technol.*, March, 37–41.

JENSEN, B. F., NORMAN, B. E. (1984), *Bacillus acidopullulyticus* pullulanase: Application and regulatory aspects for use in the food industry, *Process Biochem.,* Aug., 129–134.

JONES, P. E., THOMASON, F. G. (1951), *U.S. Department of Agriculture, Bulletin 48,* Production and Marketing Administration, Washington, D.C.

JORGENSEN, O. B., KARLSEN, L. G., NIELSEN, N. B., PEDERSEN, S., RUGH, S. (1988), A new immobilized glucose isomerase with high productivity produced by a strain of *Streptomyces murinus, Starch/Stärke* **40** (8), 307–313.

KAINUMA, K., KOBAYASHI, S., ITO, T., SUZUKI, S. (1972), Isolation and action pattern of maltohexaose-producing amylase from *Aerobacter [Enterobacter] aerogenes, FEBS Lett.* **26**, 281–285.

KERR, R. W., CLEVELAND, F. C., KATZBECK, W. J. (1951), The action of amylo-glucosidase on amylose and amylopectin, *J. Am. Chem. Soc.* **73**, 3916–3921.

KIEBOOM, A. P. G., VAN BEKKUM, H. (1985), Chemical conversion of starch-based glucose syrups, in: *Starch Conversion Technology* (VAN BEYNUM, G. M. A., ROELS, J. A., Eds.), pp. 263–334, New York: Marcel Dekker, Inc.

KIMURA, T., NAKAKUKI, T. (1990), Maltotetraose, a new saccharide of tertiary property, *Starch/Stärke* **42** (4), 151–157.

KIRCHOFF, G. S. C. (1811), Memoires, *Acad. Imp. Sci. St. Petersbourg* **4**, 27.

KOOI, E. R., ARMBRUSTER, F. C. (1967), Production and use of dextrose, in: *Starch: Chemistry and Technology* (WHISTLER, R. L., PASCHALL, E. F., Eds.), Vol. 2, pp. 553–568, New York: Academic Press, Inc.

LEVY, J., FUSEE, M. C. (1979), *U.S. Patent* 4141857.

LOBRY DEBRUYN, C. A., VAN ECKENSTEIN, W. A. (1895), Action of alkalis on sugars. Reciprocal transformation of glucose and fructose, *Rev. Trav. Chim.* Pays-Bas **14**, 201–216.

MARSHALL, R. O., KOOI, E. R. (1957), Ezymatic conversion of D-glucose to D-fructose, *Science* **125**, 468–469.

MEAGHER, M. M., NIKOLOV, Z. L., REILLY, P. J. (1989), Subsite mapping of *Aspergillus niger* glucoamylases I and II with malto- and isomalto-oligosaccharides, *Biotechnol. Bioeng.* **34** (5), 681–688.

MULVIHILL, P. J. (1992), Crystalline and liquid dextrose products: Production, properties and applications, in: *Starch Hydrolysis Products* (SCHENCK, F. W., HEBEDA, R. E., Eds.), pp. 121–176, New York: VCH Publishers, Inc.

NEWKIRK, W. B. (1923), *U.S. Patent* 1471347.

NITTA, Y., KUNIKATA, T., WATANABE, T. (1983), Difference spectroscopic study of the interaction between soybean beta-amylase and substrate or substrate analogs, *J. Biochem.* (Tokyo) **93**, 1195–1201.

NOVO NORDISK BIOINDUSTRIAL, INC. (1991), Filing of a petition for affirmation of GRAS status (Petition No. GRASP 0G0363, *Fed. Reg.* **56**, 32435.

OIKAWA, A., MAEDA, A. (1957), Role of calcium in Taka-amylase A, *J. Biochem.* **44**, 745–752.

OKADA, M., NAKAKUKI, T. (1992), Oligosaccharides: Production, properties, and applications, in: *Starch Hydrolysis Products* (SCHENCK, F. W., HEBEDA, R. E., Eds.), pp. 335–366, New York: VCH Publishers, Inc.

OUTTRUP, H. (1986), *U.S. Patent* 4604355.

PHILLIPS, L. L., CALDWELL, M. L. (1951), Study of the purification and properties of a glucose-forming amylase from *Rhizopus delemar,* gluc amylase, *J. Am. Chem. Soc.* **73**, 3559–3563.

REEVE, A. (1992), Starch hydrolysis processes and equipment, in: *Starch Hydrolysis Products* (SCHENCK, F. W., HEBEDA, R. E., Eds.), pp. 79–120, New York: VCH Publishers, Inc.

REILLY, P. J. (1993), Economics of developing an improved glucoamylase: Interaction of science and the marketplace, Presented at the *Bernard Wolnak and Associates Conference: Enzymes, A Billion Dollar Business by 2000,* Chicago, June.

ROBYT, J. F., ACKERMAN, R. J. (1971), Isolation, purification and characterization of a maltotetraose-producing amylase from *Pseudomonas stutzeri, Arch. Biochem. Biophys.* **145**, 105–114.

ROBYT, J. F., FRENCH, D. (1964), Purification and action pattern of an amylase from *Bacillus polymyxa, Arch. Biochem. Biophys.* **104**, 338–345.

ROHRBACH, R. P. (1981), *U.S. Patent* 4268419.

ROSENDAL, P., NIELSEN, B. H., LANGE, N. K. (1979), Stability of bacterial alpha-amylase in the starch liquefaction process, *Starch/Stärke* **31**, 368.

SCHENCK, F. W. (1989), Glucose and glucose-containing syrups, in: *Ullmann's Encyclopedia of Industrial Chemistry,* Vol. A 12, pp. 457–476, New York: VCH Publishers, Inc.

SCHENCK, F. W., COTILLON, M. (1992), Refining: Carbon treatment and reactive precoats – ion exchange, in: *Starch Hydrolysis Products* (SCHENCK, F. W., HEBEDA, R. E., Eds.), pp. 531–554, New York: VCH Publishers, Inc.

SCHENCK, F. W., HEBEDA, R. E. (1992), Starch hydrolysis products: An introduction and history, in: *Starch Hydrolysis Products* (SCHENCK, F. W., HEBEDA, R. E., Eds.), pp. 1–21, New York: VCH Publishers, Inc.

STARNES, R. L. (1990), Industrial potential of cyclodextrin glycosyl transferases, *Cereal Foods World* **35** (11), 1094–1099.

TAKAMINE, J. (1894), *U.S. Patent* 525823.

TAKASAKI, Y. (1976), Studies on amylases from *Bacillus* effective for production of maltose, *Agric. Biol. Chem.* **40** (8), 1515–1522.

TAKASAKI, Y. (1985), An amylase producing maltotriose from *Bacillus subtilis, Agric. Biol. Chem.* **49** (4), 1091–1097.

TAKASAKI, Y., TANABE, O. (1971), *U.S. Patent* 3616221.

TEAGUE, W. M., BRUMM, P. J. (1992), Commercial enzymes for starch hydrolysis products, in: *Starch Hydrolysis Products* (SCHENCK, F. W., HEBEDA, R. E., Eds.), pp. 45–77, New York: VCH Publishers, Inc.

THOMA, J. A., KOSHLAND JR., D. E., SHINKE, R., RUSCICA, J. (1965), Influence of sulfhydryl groups on the activity of sweet potato beta-amylase, *Biochemistry* **4**, 714–722.

TORRES, A., THOMAS, D. (1981), Polydextrose and its application in foods, *Food Technol.* **35** (7), 44–49.

U.S. DEPARTMENT OF AGRICULTURE (1992), *Sugar and Sweetener Situation and Outlook Report,* Economic Research Service, September.

VENKATASUBRAMANIAN, K., KEIM, C. R. (1985), Starch and energy: Technology and economics of fuel alcohol production, in: *Starch Conversion Technology* (VAN BEYNUM, G. M. A., ROELS, J. A., Eds.), pp. 143–173, New York: Marcel Dekker, Inc.

VERWAERDE, F., SICARD, P. J. (1984), Production of hydrogenated glucose syrups and some aspects of their use, in: *Glucose Syrups: Science and Technology* (DZIEDZIC, S. Z., KEARSLEY, M. W., Eds.), pp. 117–135, New York: Elsevier Applied Science Publishers.

WAKO, K., TAKAHASHI, C., HASHIMOTO, S., KANAEDA, J. (1978), Studies on maltotriose- and maltose-forming amylases from *Streptomyces, J. Jpn. Soc. Starch Sci.* **25** (2), 155–161.

WAKO, K., HASHIMOTO, S., KUBOMURA, S., YOKOTA, K., AIKAWA, K., KANAEDA, J. (1979), Purification and some properties of a maltotriose-producing amylase, *J. Jpn. Soc. Starch Sci.* **26** (3), 175–181.

WALLERSTEIN, L. (1939), Enzyme preparations from microorganisms, *Ind. Eng. Chem.,* October, 1218–1224.

WHITE, J. S. (1992), Fructose syrup: Production, properties, and applications, in: *Starch Hydrolysis Products* (SCHENCK, F. W., HEBEDA, R. E., Eds.), pp. 177–199, New York: VCH Publishers, Inc.

YOSHIZUMI, S., ITOH, H., KOKUBU, T. (1986), Amami No Keifu To Sono Kagaku (*Geneology and Science of Sweeteners*), translated by S. ENOKIZONO, pp. 24–25, Tokyo: Korin Publishing.

ZEIKUS, J. G., HYUN, H.-H. (1987), *U.S. Patent* 4647538.

ZEMAN, N. W., MCCREA, J. M. (1985), Alpha-amylase production using a recombinant DNA organism, *Cereal Foods World* **30** (11), 777–780.

ZOBEL, H. F. (1992), Starch: Sources, production, and properties, in: *Starch Hydrolysis Products* (SCHENCK, F. W., HEBEDA, R. E., Eds.), pp. 23–44, New York: VCH Publishers, Inc.

IV. Fermented Feeds

20 Fermented Feeds and Feed Products

RANDY D. SHAVER
KESHAB K. BATAJOO

Madison, WI 53706, USA

1 Introduction

Fermented feeds and feed products are of major importance to the livestock feed industry. Fermentation by-product feeds from the brewing and distilling industries have long been utilized as protein and energy supplements for livestock, particularly ruminants. Numerous by-products of other fermentation industries are also used as sources of nutrients in compounded feeds. Ensiled forages form the cornerstone of feeding programs for ruminants, particularly dairy cattle. Silage additives, bacterial inoculants and enzymes, have been developed to aid ensiling and reduce storage losses. These additives are used widely in the preservation of silage. Bacterial and fungal treatments have been used to enhance the nutritive value of crop residue for livestock feeding. This practice is of major importance in developing countries where high quality feedstuffs are lacking. A myriad of feed products have resulted from fermentation techniques. These are used in feed industry formulations to enhance animal health and productive efficiency, and include antibiotics and ionophores, yeast and fungal additives, bacterial direct-fed microbials, and enzymes.

2 Fermentation By-Product Feeds

2.1 Brewers By-Products

Major feed by-products of the brewing industry are malt sprouts resulting from the malting process and brewers wet and dried grains resulting from the brewing process. Malting, or the controlled germination of barley, was a part of the brewing process in traditional breweries, but with modern breweries malting is a separate operation (STENGEL, 1991). Barley is malted prior to being used in brewing to develop the enzyme systems required for the breakdown of starch to sugars during the mashing process. Residual animal feed products of the malting process are malted barley, barley screenings, and malt sprouts. Malt sprouts are a common component of protein and energy supplements for ruminants. Rice provides a source of starch for the mashing process. Residual animal feed products of rice milling operations are rice bran, rice hulls, and rice millfeed (combination of bran and hulls). Residual animal feed products of the brewing process are brewers yeast, brewers condensed solubles, and brewers grains. Residual brewers yeast may be spray or drum dried and sold to the feed industry for use in pet foods or calf mixes (STENGEL, 1991). Brewers condensed solubles is a liquid brewing residual resulting from the evaporation, to a 50% solids level, of several of the dilute liquid streams generated during the brewing process (STENGEL, 1991). The solids in brewers condensed solubles are high in digestible carbohydrates. Brewers condensed solubles are used in poultry rations, alcohol fermentation, bacterial fermentation, and ruminant feeds to a limited degree (STENGEL, 1991). The major by-product of the brewing process for use in livestock rations is brewers grains. After their respective stages in the brewing process, the spent grains and hops are blended together into what is referred to as wet brewers grains (STENGEL, 1991). Without further processing this product contains approximately 80% moisture and 25% crude protein (CP) on a dry matter (DM) basis (NRC, 1989). When dried to approximately 10% moisture, this product is commonly referred to as brewers dried grains. The primary market for wet brewers grains is dairy farms and beef cattle feedlots located in relatively close proximity to the brewery. Wet brewers grains are incorporated directly into rations at the farm as a medium-protein supplement. Brewers dried grains are commonly used by the feed industry as a component of protein supplements for ruminants. The higher fraction of ruminally undegradable protein relative to soybean meal makes wet or dry brewers grains particularly attractive in diets for lactating dairy cows.

2.1.1 Brewers Dried Grains

Brewers dried grains (International Feed Number (IFN) 5-00-516) is defined (FEED INDUSTRY RED BOOK, 1994) as the dried extracted residue of barley malt alone or in mixture with other cereal grains or grain products resulting from the manufacture of wort or beer and may contain pulverized dried spent hops in an amount not to exceed 3%, evenly distributed. Typical nutrient analyses (DM basis) are 25.4% CP, 24% acid detergent fiber (ADF), 46% neutral detergent fiber (NDF), and 6.5% ether extract (EE) (NRC, 1989). Approximately 50% of the protein is ruminally undegradable compared with 35% in soybean meal (NRC, 1989). The formula feed industry generally limits brewers dried grains to less than 50% of protein supplements and 25% of complete feeds for dairy cattle.

2.1.2 Brewers Wet Grains

Brewers wet grains (IFN 5-00-517) is defined (FEED INDUSTRY RED BOOK, 1994) as the extracted residue from the manufacture of wort from barley malt alone or in mixture with other cereal grains or grain products. The guaranteed analysis shall include the maximum moisture. Typical nutrient analyses are similar to brewers dried grains (NRC, 1989), except for moisture content which may range from 70 to 80%. The high moisture content of brewers wet grains limits its use to livestock operations near the point of production or within 200 miles of major breweries (STENGEL, 1991). Feeding levels are generally limited to less than 15 and 5 kg/head/day (as fed basis) for dairy and beef cattle, respectively.

2.1.3 Malt Sprouts

Malt sprouts (FEED INDUSTRY RED BOOK, 1994) is obtained from malted barley (IFN 5-00-545) by the removal of the rootlets and sprouts, which may include some of the malt hulls, other parts of malt, and foreign material unavoidably present. It must contain no less than 24% CP. The term "malt sprouts" when applied to the corresponding portion of other malted cereals must be used in qualified form, as, for example: "rye malt sprouts" (IFN 5-04-048) and "wheat malt sprouts" (IFN 5-29-796). Typical nutrient analyses (DM basis) are 28.1% CP, 18% ADF, 47% NDF, and 1.4% EE (NRC, 1989). Feeding limits are similar to those for brewers dried grains.

2.2 Distillers By-Products

Distillers by-products are obtained by processing residues remaining after removal of alcohol and some water from a yeast-fermented mash (FEED INDUSTRY RED BOOK, 1994). Major distillers by-products are distillers dried grains, distillers dried solubles, distillers dried grains with solubles, and condensed distillers solubles (HATCH, 1991). Solubles produced in the distilling industry may be dried and marketed as distillers dried solubles. Most distilleries add the liquid solubles to the grains and do not produce dried solubles (FEED INDUSTRY RED BOOK, 1994). Distillers grains are commonly used by the feed industry as a component of protein supplements for ruminants. The higher fraction of ruminally undegradable protein relative to soybean meal makes distillers grains particularly attractive in diets for lactating dairy cows.

2.2.1 Distillers Dried Grains

Distillers dried grains (FEED INDUSTRY RED BOOK, 1994) is obtained after the removal of ethyl alcohol by distillation from the yeast fermentation of a grain or grain mixture by separating the resultant coarse grain fraction of the whole stillage and drying by methods employed in the grain distilling industry. The predominant grain shall be declared as the first word in the name; barley (IFN 5-00-518), cereals (IFN 5-02-144), corn (IFN 5-02-842), rye (IFN 5-04-523), sorghum (IFN 5-04-374), and wheat (IFN 5-05-193). Typical nutrient analyses (DM basis) are 23% CP, 17%

ADF, 43% NDF, and 9.8% EE (NRC, 1989). Approximately 50% of the protein is ruminally undegradable compared with 35% in soybean meal (NRC, 1989). The formula feed industry generally limits distillers dried grains to less than 50% of protein supplements and 25% of complete feeds for dairy cattle.

2.2.2 Distillers Dried Grains With Solubles

Distillers dried grains with solubles (FEED INDUSTRY RED BOOK, 1994) is the product obtained after the removal of ethyl alcohol by distillation from the yeast fermentation of a grain or grain mixture by condensing and drying at least three-fourths of the solids of the whole stillage by methods employed in the grain distilling industry. The predominant grain shall be declared as the first word in the name; barley (IFN 5-00-519), cereals (IFN 5-02-146), corn (IFN 5-02-843), rye (IFN 5-04-024), sorghum (IFN 5-04-375), and wheat (IFN 5-05-194). Typical nutrient analyses (DM basis) are 25% CP, 18% ADF, 44% NDF, and 10.3% EE (NRC, 1989). Approximately 50% of the protein is ruminally undegradable compared with 35% in soybean meal (NRC, 1989). Feeding limits are similar to those for distillers dried grains.

2.2.3 Distillers Solubles

Distillers dried solubles (FEED INDUSTRY RED BOOK, 1994) is obtained after the removal of ethyl alcohol by distillation from the yeast fermentation of a grain or grain mixture by condensing the thin stillage fraction and drying it by methods employed in the grain distilling industry. The predominant grain shall be declared as the first word in the name; barley (IFN 5-00-520), cereals (IFN 5-02-147), corn (IFN 5-02-844), rye (IFN 5-02-526), sorghum (IFN 5-04-376), and wheat (IFN 5-05-195). Typical nutrient analyses (DM basis) are 29.7% CP, 7% ADF, 23% NDF, and 9.2% EE (NRC, 1989). Most distilleries add the liquid solubles to the grains and do not produce dried solubles (FEED INDUSTRY RED BOOK, 1994). Condensed distillers solubles (FEED INDUSTRY RED BOOK, 1994) is obtained after the removal of ethyl alcohol by distillation from the yeast fermentation of a grain or grain mixture by condensing the thin stillage fraction to a semi-solid one. The predominant grain shall be declared as the first word in the name; barley (IFN 5-12-210), cereals (IFN 5-12-215), corn (IFN 5-12-211), rye (IFN 5-12-212), sorghum (IFN 5-12-231), and wheat (IFN 5-12-213). Condensed distillers solubles can be marketed as a liquid feed ingredient. Contents of DM from 8 to 26% and CP from 30 to 35% (DM basis) for condensed distillers solubles have been reported in research trials (CHASE, 1991). Michigan State workers found that condensed distillers solubles (8% DM) used as replacement for water at intakes up to 20% of total ration DM did not depress animal performance in finishing cattle (CHASE, 1991). Cornell workers added condensed distillers solubles (26% DM) to rations of early lactating dairy cows at 0, 8, and 16% of total ration DM (CHASE, 1991). Feed intake, milk production, and milk composition were similar for the three rations. Maximum daily intake of condensed distillers solubles was about 15 kg per cow (4 kg per cow of DM).

2.2.4 Distillers Wet Grains

Distillers wet grains (IFN 5-16-149) is defined (FEED INDUSTRY RED BOOK, 1994) as the product obtained after the removal of ethyl alcohol by distillation from the yeast fermentation of a grain or grain mixture. The guaranteed analysis shall include the maximum moisture. Typical nutrient analyses are similar to distillers dried grains (NRC, 1989), except for moisture. Feeding limits are similar to those for brewers wet grains.

2.2.5 Molasses Distillers Solubles

Molasses distillers dried solubles (IFN 4-04-698) is obtained by drying the residue from the yeast fermentation of molasses after the removal of the alcohol by distillation (FEED INDUSTRY RED BOOK, 1994). Molasses distillers condensed solubles (IFN 4-04-697) is obtained by condensing to a syrupy consistency the residue from the yeast fermentation of molasses after the removal of the alcohol by distillation (FEED INDUSTRY RED BOOK, 1994).

2.3 Corn Wet Milling By-Products

The corn wet milling process is used to isolate corn starch for the production of high-fructose sweeteners, corn syrup, ethanol, or use in the food industry (WEIGEL, 1991). Animal feed by-products of this process are corn germ meal, corn gluten meal, corn gluten feed, and condensed fermented corn extractives. The latter will be discussed in more detail because it is a fermentation by-product.

2.3.1 Condensed Fermented Corn Extractives

This feed ingredient (FEED INDUSTRY RED BOOK, 1994; IFN 4-02-890) is obtained by the partial removal of water from the liquid resulting from steeping corn in a water and sulfur dioxide solution which is allowed to ferment by the action of naturally occurring lactic acid producing microorganisms as practiced in the wet milling of corn. It is commonly known as steep liquor and generally consists of 40–50% solids. It is used as replacement for molasses as a liquid feed supplement carrier when cost-effective (WEIGEL, 1991). Fermented corn extractives may also be contained in corn gluten feed or meal.

2.4 By-Products of Fermentation Industries

By-products of various fermentation industries, such as from the production of antibiotics and citric acid, are available for feed purposes. These by-products may contain residual antibiotics or vitamins. Fermentation by-products as defined in the FEED INDUSTRY RED BOOK (1994) are listed below.

2.4.1 Condensed, Extracted Glutamic Acid Fermentation Product

This ingredient (IFN 5-01-595) is a concentrated mixture of the liquor remaining after the extraction of glutamic acid. Solids are mainly derived from the cells of *Corynebacterium lilium* used in the production of glutamic acid.

2.4.2 Extracted Presscake and Meal

Extracted presscake is the filtered and dried mycelium obtained from fermentation. For label identification the source must be indicated as *Penicillium* (IFN 5-07-154), *Streptomyces* (IFN 5-07-155), or citric acid (IFN 5-07-156). Extracted meal is the ground presscake; *Penicillium* (IFN 5-06-162), *Streptomyces* (IFN 5-06-163), or citric acid (IFN 5-06-164).

2.4.3 Dried Extracted Fermentation Solubles

This ingredient is the dried extracted broth obtained from fermentation. For label identification the source must be indicated as *Penicillium* (IFN 5-06-166), *Streptomyces* (IFN 5-06-176), or citric acid (IFN 5-06-165).

2.4.4 Dried Fermentation Extract

This ingredient (IFN 5-06-147) is the dried product resulting from extracting and precipitating by means of non-aqueous solvents or other suitable means, the water-soluble materials from a fermentation conducted for maximum production of enzymes using a non-pathogenic strain of microorganisms in accordance with good manufacturing practice. For label identification, the source must be indicated as *Bacillus subtilis, Aspergillus oryzae, Aspergillus niger*, or as permitted by FDA.

2.4.5 Fermentation Solubles

Dried fermentation solubles is the dried material resulting from drying the water-soluble materials after separation of suspended solids from a fermentation conducted for maximum production of enzymes using a non-pathogenic strain of microorganisms in accordance with good manufacturing practice. For label identification, the source must be indicated as *B. subtilis* (IFN 5-29-779), *A. oryzae* (IFN 5-29-780), *A. niger* (IFN 5-29-781), or as permitted by FDA. Condensed fermentation solubles is the product resulting from the removal of a considerable portion of the liquid by-product resulting from the action of the ferment on the basic medium of grain, molasses, whey, or other media. For label identification the source must be indicated as grain, molasses, or whey.

2.4.6 Undried Extracted Solids and Fermentation Solubles

This ingredient is undried mycelium and extracted broth or the extracted and undried mycelium and broth obtained from fermentation. For label identification the source must be indicated as *Penicillium* (IFN 5-06-172), *Streptomyces* (IFN 5-06-173), citric acid (IFN 5-06-171), or as permitted by FDA.

2.4.7 Dried and Liquid Fermentation Product

Dried fermentation product is derived by culturing on appropriate nutrient media for the production of one or more of the following: enzymes, fermentation substances, or other microbial metabolites, and dried in accordance with approved methods and good manufacturing practice. Protein, fat, fiber, cell count, enzyme activity, or nutrient concentration shall be guaranteed where applicable. For label identification, source must be indicated as *B. subtilis, A. oryzae, A. niger, Lactobacillus acidophilus, Lactobacillus bulgaricus, Lactobacillus bifidus*, or *Streptococcus faecium*. Liquid fermentation product is derived by culturing or fermenting on appropriate liquid nutrient media for the production of one or more of the following: enzymes, fermentation substances, or other microbial metabolites, and stabilized by approved methods in accordance with good manufacturing practice. Percent solids, cell count, enzyme activity, or nutrient metabolite level shall be guaranteed where applicable. For label identification, source must be indicated as *B. subtilis, A. oryzae, A. niger, L. acidophilus, L. bulgaricus*, or *S. faecium*.

2.5 Fermented Ammoniated Condensed Whey

This ingredient (IFN 5-28-223) is defined (FEED INDUSTRY RED BOOK, 1994) as the product produced by the *L. bulgaricus* fermentation of whey with the addition of ammonia. It must contain 35% to 55% CP with no more than 42% equivalent CP from non-protein nitrogen (NPN). It is used as a source of NPN for ruminants. Upper feeding limits for ruminants above 200 kg body weight are 1 to 2 kg/head/day based on its NPN content.

3 Silage Preservation

Ensiled forages form the cornerstone of feeding programs for ruminants, particularly dairy cattle. Preserving forage as silage depends on exclusion of oxygen from the forage mass and reduction of pH through bacterial fermentation. Silage additives, bacterial inoculants and enzymes, have been developed to aid ensiling and reduce storage losses. These additives are used widely in the preservation of silage.

3.1 Ensiling Process

There are four stages of the ensiling process. During Phase 1 (aerobic phase) plant respiration continues after chopping. Plant cells take in oxygen from the surrounding air and give off carbon dioxide. Aerobic bacteria present in the ensiled material begin degradation quickly. Degradation coupled with cell respiration uses up the oxygen present in the forage, and it quickly reaches an anaerobic state. The amount of oxygen trapped in the mass is related to fiber level, moisture content, rate of silo filling and fineness of chop. Provided all oxygen present is converted to carbon dioxide and no additional oxygen enters the mass, mold formation does not occur. Cell respiration causes an initial rise in temperature. Silage temperatures of 25 to 38 °C are optimum for the anaerobic lactic acid producing organisms. Temperatures above 38 °C that can occur with excessive oxygen entrapment and (or) delayed sealing may reduce the nutritive value of ensiled forage.

During Phase 2 (lag phase) plant cell membranes break down, allowing cell juices to become a growth medium for bacteria. Lactic acid producing bacteria begin to dominate the fermentation process during Phase 3 (fermentation phase) after silage pH drops to 5.5–5.7 from 6.5–6.7 in the initial ensiled material. The rate of pH decline determines the duration of Phase 2; Phases 2 and 3 normally merge by day 3. Lactic acid production is dependent on: optimum pH (5.5–5.7), sufficient numbers of viable lactic acid bacteria, adequate available carbohydrate, sufficient moisture, and anaerobic conditions.

Lactic acid production lowers silage pH to 4.4–5.0 in silages with high buffering capacity and low available carbohydrate (primarily legumes). Silages with low buffering capacity and more available carbohydrate (corn, other cereal grain, or grass) will have a final pH of 3.8–4.2. Phase 3 lasts about two weeks, and temperature of the mass gradually declines to 25 °C during this period. Low pH stops bacterial action, and the silage stabilizes in Phase 4 (stable phase). If insufficient lactic acid is formed, production of butyric acid by clostridia and breakdown of protein may occur. This does not occur unless both silage pH and moisture content (above 70%) are high enough to allow clostridial activity. Criteria for a desirable fermentation include: (1) rapid decline in pH, (2) low final pH, (3) rapid rate of lactic acid production, and (4) more than 65% of the total organic acids as lactic acid. Nutrient preservation will be at a maximum when management practices and silage additives direct fermentation to achieve these goals.

3.2 Silage Additives

3.2.1 Microbial Inoculants

The purpose of bacterial inoculants is to rapidly lower pH. This low pH (high acidity) also helps prevent the growth of undesirable organisms which can lower the intake potential and nutritive value of silage. Bacteria that are needed for the fermentation of silage are normally present on plant tissue. Numbers of lactic acid bacteria on the standing crop and in the swath just after cutting are generally low. Higher numbers are found on the chopped forage just prior to ensiling and are apparently due to either growth of lactic acid bacteria as the forage wilts in the field or inoculation during the chopping process. However, under certain harvest conditions sufficient numbers of desirable species of bacteria may not be present in the chopped forage to achieve the best possible fermentation. Air temperature at the time of chopping, wilting time, and average wilting temperature can all affect the number of lactic acid bacteria present on chopped forage. The number of bacteria normally present on chopped forage at the time of ensiling can range from 100 to 100 million bacteria per gram of fresh forage (Muck, 1988). Inoculants have the best chance of improving the ensiling process when numbers of naturally occurring lactic acid bacteria are low.

Criteria which potential organisms should satisfy for use in silage inoculants include: They (1) should have rapid growth rate and ability to compete with and dominate other organisms likely to occur in silage, (2) must be homofermentative, (3) must be acid-tolerant and produce a low final pH, (4) must be able to ferment glucose, fructose and sucrose, (5) should have no action on organic acids, and (6) have low proteolytic activity. European researchers concluded that there is a scarcity of organisms common to silage which meet these criteria. *Lactobacillus plantarum* is the bacterial species most commonly found in commercial silage inoculants, since it grows rapidly and quickly lowers pH through fermentation of sugars to lactic acid. Homofermentative organisms are desired because of their rapid rate of lactic acid production and lower dry matter losses during fermentation. Homofermentative lactic acid bacteria include *Lactobacillus, Streptococcus* and *Pediococcus* species. *Streptococcus* and *Pediococcus* strains grow rapidly and dominate the initial fermentation while the lactobacilli typically dominate below pH 5. Thus, a mixture of these bacterial species has fermentative capability over the pH range found in silage and may offer more potential than inoculants containing only a single genus. Comparison of bacterial species is complicated, since within each species many strains with different fermentative capacities exist. Thus, two inoculants containing different strains of *L. plantarum* may not affect fermentation to the same degree.

Bacterial inoculants have generally improved the ensiling process shown in research trials conducted with several commercial inoculants utilizing laboratory, bunker and tower silos. These bacteria cause a more rapid drop in pH, higher lactic acid content, and slightly lower final pH. This improvement in fermentation results in slightly higher dry matter recoveries (1–2%) from the silo for inoculated silages.

Much of the nitrogen (N) in silage is either non-protein nitrogen (NPN) or protein that is degraded by rumen bacteria. Solubilization of true protein occurs in silage through the action of plant enzymes called proteases. The rate at which these proteases act in silage is higher for legumes and high-moisture silages. Increases in soluble protein content (% of CP) of legume and grass forage during the ensiling process ranged from 33 to 45% and 21 to 30%, respectively, as moisture content increased from 50 to 70% (PITT and SHAVER, 1990). Rapid reduction in silage pH and low final pH help inhibit protein solubilization.

Wisconsin USDA researchers saw a reduction in NPN as a percent of total N on an average from 60 to 55% when inoculants were applied to 70% moisture alfalfa silages stored in bunker silos. Work with laboratory silos also showed a reduction in the NPN fraction for inoculated alfalfa silages. However, this is a modest reduction in the NPN fraction relative to the large amount of NPN present in alfalfa silage. Further, there is little evidence to suggest that this reduction in NPN and soluble protein actually results in less rumen-degradable protein for inoculated silages. The fraction of crude protein that is degraded in the rumen to peptides, amino acids and ammonia ranges from 70 to 85% in silages (SATTER, 1986). Soluble and degradable protein levels are highest for high-moisture, high-protein alfalfa forages. These forages offer the best opportunity for inoculants to improve protein fractions in the silage, but benefits in research trials have been small.

Dry matter intake and milk yield differences between inoculated and untreated alfalfa silages were generally not statistically significant in Wisconsin USDA trials (SATTER et al., 1988). However, when inoculation increased the number of lactic acid bacteria by tenfold or more, a 3% increase in milk yield was observed. This means that if the forage contains a large number of naturally occurring lactic acid bacteria or if the inoculant does not contain enough live lactic acid bacteria, a milk response is highly unlikely. HARRISON (1989) reported a 3.1% increase in milk yield when addition of bacterial inoculants to alfalfa improved silage fermentation in a summary of 15 comparisons. There was no effect on milk yield when the inoculants failed to improve the ensiling process.

Inoculants applied at the rate of 10^5 colony-forming units (CFU) per gram of ensiled forage will likely be most effective when alfalfa is wilted for one day or less or when average wilting temperatures are below 15°C (SATTER et al., 1988). This is because the numbers of naturally occurring lactic acid bacteria were almost always below 10^4 CFU per gram of chopped forage under these conditions. This allows for at least a tenfold rise in bacterial numbers after inoculation.

Inoculation will cost about 60 to 80 cents per ton of fresh forage. Assuming that an average of 11 kg of dry matter from silage is consumed per day over a lactation, bacterial inoculants would cost about 1.5 to 2.0 cents per cow per day. Inoculants must increase the feeding value of alfalfa silage 1.5 to 2.0% to offset the direct cost of the additive. Best chances for success may be with first and fourth cutting hay-crop silage, since this is where numbers of naturally occurring lactic acid bacteria are likely to be lowest due to cooler wilting conditions (SATTER et al., 1988). Benefits to inoculation of corn silage seem less likely, since good fermentation of corn silage generally occurs due to its high content of available carbohydrate and low buffering capacity. However, Kansas State researchers observed better silage fermentation and dry matter recovery from the silo when bacterial inoculants were added to corn and sorghum silages in a summary of 22 trials. Bacterial inoculants may reduce aerobic deterioration of silages on feedout, thereby improving bunk life and silage acceptability.

The following directions are useful: "Apply a minimum of 10^5 CFU live lactic acid bacteria per gram of fresh forage. Check the manufacturers recommended application rates to be sure that this level is being added. Follow the manufacturer's directions for proper on-farm storage and handling to ensure product stability. Use a good quality product from a reputable company. Not all bacterial inoculants are equally effective. Request research data from a series of trials with evidence of proper statistical analyses to substantiate sales claims."

Application of the inoculant as a liquid suspension gives a more even distribution of the bacteria across the forage and allows them to start working more rapidly. Application of the inoculant at the field chopper rather than at the silo blower allows the bacteria in the inoculant to get a head start on the naturally occurring bacteria. This is generally where inoculants are applied when silage is stored in bunker silos.

3.2.2 Enzymes

The principle behind enzyme additives is to break down the complex carbohydrates in the forage into simple sugars that can be readily fermented by the lactic acid bacteria. Enzyme additives containing cellulases, hemicellulases and amylases to break down cellulose, hemicellulose and starch have been developed. *A. oryzae, A. niger* and *B. subtilis* cultures and their fermentation products are included in some silage additives as a source of enzymatic activity. Various combination products containing both enzymes and lactic acid bacteria have also been developed, since in theory these additives should compliment each other with the enzymes providing additional substrate for the added bacteria to ferment, resulting in a more rapid pH drop and lower final pH.

Enzyme additives have the best chance for success when plant sugars are low, provided sufficient numbers of lactic acid bacteria are present. Typical ranges in sugar content for legume, grass and whole plant corn forages are 4–15, 10–20 and 8–30%, respectively (LEIBENSPERGER and PITT, 1988). Minimum initial sugar requirements for complete fermentation exceed average typical sugar contents when moisture content exceeds 60, 75 and 80% for legume, grass and corn silage, respectively (LEIBENSPERGER and PITT, 1988). Enzyme additives would only be expected to improve the ensiling process for direct-cut hay-crop forages or legumes wilted to more than 65% moisture. Researchers are also interested in the potential of enzyme additives containing cellulase and hemicellulase to reduce the neutral detergent fiber (NDF) content of forages in the silo resulting in improved silage digestibility and intake potential, particularly for high-fiber hay-crop silages.

Wisconsin USDA researchers have concluded that addition of cellulase enzyme improved fermentation of alfalfa silages containing 65 to 75% moisture when fermentable substrate was limiting. When both naturally occurring lactic acid bacteria and fermentable substrate were limiting, combined addition of bacterial inoculant and substrate resulted in an improvement of the fermentation. Maine researchers saw improved silage fermentation when enzyme additives containing cellulase or cellulase/xylanase were added to 72% moisture mixed grass–legume forage. Cellulase enzyme reduced silage NDF content and increased intake of dry matter, but had no effect on milk yield. A cellulase/xylanase enzyme mixture added to 70% moisture containing mixed grass–legume forage reduced silage NDF content and increased dry matter intake and milk yield when compared to controls (STOKES, 1992). Wisconsin USDA researchers also reported a reduction in silage NDF content related to the addition of cellulase enzyme, but there was no effect on milk yield. Reduction in NDF content in the silo ranged from 4 to 7 percentage units across Maine and Wisconsin USDA trials for 45 to 50% NDF forages.

Enzyme additives did not improve the ensiling process, dry matter intake or milk yield in four comparisons summarized by HARRISON (1989). Bacteria/enzyme additives improved the ensiling process and dry matter intake, but had no effect on milk yield in four comparisons summarized by HARRISON (1989). Kansas State researchers saw little improvement in fermentation when various enzyme products were added to alfalfa silages.

Enzymes added at rates used in research trials would cost $ 3.00 to $ 3.50 per ton of fresh forage. Enzyme additives are not recommended because of variable results and high cost. Their use should be restricted to situations where substrate supply prevents adequate preservation (direct-cut hay-crop forages or legumes wilted to more than 65% moisture) and sufficient numbers of lactic acid bacteria are present or have been added. Adding enzymes at lower rates or fungal cultures and their fermentation products to bacterial inoculants may increase the cost of the inoculant only 20 to 40 cents per ton of fresh forage, but enzyme activities in these products may not be sufficient for them to be effective. Potential exists for enzyme additives to reduce NDF content of forages in the silo and improve digestibility, intake and milk yield, but this needs further study.

4 Enhancement of the Nutritive Value of Crop Residue

Numerous processing methods, physical (chopping and grinding), physicochemical (steam treatment and ammonia-freeze explosion), chemical (sodium hydroxide, alkaline hydrogen peroxide, urea, and ammonia), and biological (bacterial and fungal), have been evaluated for their ability to enhance the digestibility of ligno-cellulosic residues. Bacterial and fungal treatments have been used to enhance the nutritive value of crop residue for livestock feeding. This practice is of major importance in developing countries where high quality feedstuffs are lacking. This discussion will focus on microbiological treatments.

Fermentation systems with 20 to 30% solids are referred to as solid state fermentation (SSF). This involves use of a moistened substrate where a thin layer of moisture exists on the surface of the particles. Solid state fermentation differs from submerged fermentation (SF) in that there is not enough water used to make a fluid mixture. Solid state fermentation is a cheaper, simpler technology than SF with potential for direct application of the fermented product in animal feeding (REID, 1989). LAUKEVICS et al. (1984) reported that SSF of wheat straw reduced fiber content and increased digestibility. Numerous strains of fungi have been isolated that are capable of degrading ligno-cellulosic residues (AGOSIN and ODIER, 1985). Most of the fungi that degrade lignin are grouped under the common name white rot fungi. Use of white rot fungi for degrading ligno-cellulosic residue has been studied extensively in the wood pulping process (KIRK and FARRELL, 1987). Submerged fermentation as well as SSF with various bacterial and fungal cultures have been shown to improve the nutritive value of banana wastes (BALDENSPERGER and HANNIBAL, 1985), orange peels (NICOLINI et al., 1987), cassava (OFUYA et al., 1990), citrus pulp (RODRIGUEZ et al., 1986), palm oil processing wastes (BAKER and WORGAN, 1981), sugarcane bagasse (ENRIQUEZ and RODRIGUEZ, 1983), straw (HAN et al., 1978), and sugar beets (COCHET et al., 1988). Microbial fermentation techniques have also been used for production of biomass protein (also called single cell protein, SCP) with various non-conventional substrates, such as methanol (HESSELTINE, 1972), corn stover and pulp mill sludge (LEESON et al., 1984; KELLEMS et al., 1981), waste plastics (KARTHIGESAN and BROWN, 1981), sawdust (PAMMENT et al.,

1978), wood pulp waste (CLAYPOOL and CHURCH, 1984), brewery waste (JOHNSON and REMILLARD, 1983), methanol and paraffins (ZIMMERMAN and TEGBE, 1977), whey (TELLER and GODEAU, 1986), and dairy plant wash water (CATON et al., 1989) for conversion to animal feeds.

Technology is becoming available for large-scale SSF systems (1500 kg of substrate; ZADRAZIL et al., 1990). The *in vitro* digestibility of wheat straw incubated with white rot fungi in this system increased from 40% to 54%. In the production of biomass protein, pure cultures of microorganisms are introduced into an inoculum medium containing small amounts of substrate. Enzyme systems are secreted by the microorganisms for degradation of the substrate. The inoculum is then transferred to a fermenter containing nutrients plus substrate for the large-scale production of biomass protein. Following fermentation, the product is steam-treated and then dried for use as livestock feed. The CP content of biomass protein from various substrates ranges from 10% to 75%. Research trials indicate that biomass protein can be used as a partial replacement for conventional protein sources in rations for beef and dairy cattle without significant effects on animal performance. The economics of feeding biomass protein depends on the price of conventional protein supplements. Production of biomass protein may be useful for controlling environmental pollution and providing an alternative protein source in developing countries.

5 Feed Products

5.1 Antibiotics

Antibiotics are used commercially as feed additives for growth promotion and disease prevention and control. They are produced by fermentation processes using fungi and bacteria. Incorporation of antibiotics into feeds is subject to special requirements on use, labeling, and inspection with a feed containing an antibiotic becoming basically a drug in its entirety under the provisions of the FDA (FEED INDUSTRY RED BOOK, 1994). The mode of action for growth promotion has not been well established, but popular concepts include reduction of pathogenic organisms that produce toxins, reduction of microorganisms which compete with the host animal for nutrients, and stimulation or selective inhibition of the growth of microorganisms which synthesize nutrients for the host animal. For example, one class of compounds, ionophores (monensin and lasalocid), are approved for use in beef cattle and dairy replacement heifers to increase feed efficiency and average daily gain through alteration of ruminal volatile fatty acid production. These compounds are also approved for use in the prevention of coccidiosis in poultry and young animals. Antibiotics used commercially for growth promotion are fed at levels ranging from a fraction of a gram to over 100 grams per ton (FEED INDUSTRY RED BOOK, 1994). They are used in feeds for poultry, swine, calves, dairy cattle, and beef cattle. Antibiotics are also used in the prevention and control of poultry and animals diseases. They may be used with other drugs in combinations approved by the FDA. Inclusion rates for disease prevention and control are generally higher than for growth promotion. The animal feed industry continues to develop new uses for antibiotics and new antibiotics for growth promotion and disease prevention and control. Though exciting developments from the standpoint of animal health and productive efficiency are anticipated, the future use of antibiotics in animal feeds will be balanced against growing consumer demand for lesser use of antibiotics and hormones in livestock production systems.

5.2 Yeast

Yeast is available in yeast culture and active dry yeast forms for use in animal feeds. *Saccharomyces cerevisiae* is the most common yeast used in animal feeds. Yeast culture (IFN 7-05-520) is the dried product composed of yeast and the media in which it was grown, dried in such a manner as to preserve the fermenting activity of the yeast (FEED INDUSTRY RED BOOK, 1994). Yeast culture contains yeast cells, yeast metabolites, and the media on which the yeast and metabolites are produced. Active dry yeast (FEED INDUSTRY RED BOOK, 1994; IFN 7-05-524) is yeast which has been dried in such a manner as to preserve a large portion of its fermenting capacity. It must contain no added cereal or filler and must contain no less than 15 billion live yeast cells per gram (FEED INDUSTRY RED BOOK, 1994). Commercially available yeast products are used in feeds for poultry, swine, calves, dairy cattle, and beef cattle. The primary digestive effects of yeast in ruminants have been higher numbers of cellulolytic bacteria, more stable rumen environment, higher ruminal pH and lower lactic acid concentration, and higher ruminal and total tract digestibilities. However, these responses have been variable and inconsistent in research trials. Dairy cows fed yeast produced 1.6 kg more milk per day than control cows in a summary of seven trials (HUTJENS, 1992). Production responses were greater in early lactation cows. Early lactation (2 weeks prepartum to 4 weeks postpartum) appears to be the optimum time to feed yeast as cows switch from low to high energy diets (HUTJENS, 1992). Yeast has been reported to improve average daily gain and feed efficiency in beef cattle, feed efficiency in poultry, feed intake and average daily gain in swine (DAWSON, 1987), and feed intake and average daily gain in calves (FALLON, 1987).

5.3 *Aspergillus oryzae* Fermentation Extract

This fermentation extract (AO) is produced from a selected strain of the enzyme-producing fungus *A. oryzae*, and is commercially available for use in feeds for poultry, swine, calves, dairy cattle, and beef cattle. The primary digestive effects of AO in ruminants have been higher numbers of cellulolytic bacteria, more stable rumen pH, and shifting ruminal volatile fatty acids patterns by a reduction in the proportion of propionate relative to acetate and an increase in butyrate. These responses have been variable and inconsistent in research trials. Dairy cows fed AO produced 1.2 kg more milk per day than control cows in a summary of seven trials (HUTJENS, 1992). Animals in hot environments had lower rectal temperatures when AO was fed, and production responses tended to be greater in trials conducted under heat stress conditions.

5.4 Bacterial Direct-Fed Microbials

The FDA adopted the term direct-fed microbials to refer to a source of viable, naturally occurring microorganisms which includes bacteria, fungi, and yeast. The FDA has accepted 42 organisms as generally recognized as safe (GRAS) and as being appropriate for use in animal feeds (KILMER, 1993). This listing includes the bacterial direct-fed microbial species of *Bacillus, Bacteroides, Bifidobacterium, Lactobacillus, Pediococcus*, and *Streptococcus* (KILMER, 1993). Two of the more common bacteria used in commercially available direct-fed microbial products are specific strains of *Streptococcus faecium* and *Lactobacillus acidophilus*. Several forms of bacterial direct-fed microbials are commercially available for use in poultry, swine, calves, dairy cattle, and beef cattle including powders for liquid incorporation, boluses, pastes, and liquids. Use of bacterial direct-fed microbials is most common in stressed animals, such as newborn calves, incoming beef feedlot cattle, early postpartum dairy cows, and animals that are off-feed. Research data demonstrating efficacy are limited. The mode of action has not been well established, but popular concepts include production of organic acids, hydrogen peroxide, and (or) antibiotics which eliminate undesirable microorganisms in the digestive tract, reduced oxidation/reduction potential which could limit oxygen availability to intestinal pathogens, production of enzymes which could improve nutrient availability, and detoxification of metabolites produced by intestinal pathogens.

5.5 Enzymes

The use of enzymes in animal feeds is gaining popularity (FEED INDUSTRY RED BOOK, 1994). This has been largely stimulated by findings of improved nutritional value of barley for poultry from *beta*-glucanase addition to high barley diets, and improved availability of phytate phosphorus from corn and soybean meal in swine from dietary phytase addition. The former has been shown in research trials to increase average daily gain 7% to 30% and feed efficiency 5% to 22% in broiler chickens fed high barley diets (DUNNE, 1991). The latter has been shown to reduce the need for inorganic phosphorus in swine diets, and reduce fecal phosphorus excreted into the environment (CROMWELL, 1991). Historically, the economics of adding phytase relative to the cost of inorganic phosphorus would determine whether or not it would be used in swine diets. However, environmental regulations on manure phosphorus application to the soil is bringing phytase into swine diets more rapidly around the world, particularly Europe. Other areas where enzymes have shown some promise include baby pig diets (DUNNE, 1991) and as silage additives. Application of enzymes in ruminant diets has been limited, because of the high rate of enzyme production and the possible degradation of supplemental enzymes by the ruminal microorganisms. Recent advances in biotechnology have allowed scientists to genetically engineer microorganisms to produce enzymes in larger quantities and at lower cost than was previously possible (FEED INDUSTRY RED BOOK, 1994). These improved production capabilities for enzymes should increase their use in animal feeds in the future.

6 References

AGOSIN, E., ODIER, E. (1985), Solid state fermentation, lignin degradation, and resulting digestibility of wheat straw fermented by selected white-rot fungi, *J. Microbiol. Biotechnol.* **21,** 397–403.

BAKER, T. W., WORGAN, J. T. (1981), Utilization of palm oil processing effluents as substrates for microbial protein production, *Eur. J. Appl. Microbiol.* **11,** 234–240.

BALDENSPERGER, J., HANNIBAL, L. (1985), Solid state fermentation of banana wastes, *Biotechnol. Lett.* **7,** 743–748.

CATON, J. S., WILLIAMS, J. E., BEAVER, E. E., MAY, T., BELYEA, R. L. (1989), Effects of dairy biomass protein on ruminal fermentation and site and extent of nutrient digestion by lambs, *J. Anim. Sci.* **67,** 2762–2771.

CHASE, L. E. (1991), Feeding distillers grains and hominy feed, in: *Proc. Alternative Feeds for Dairy and Beef Cattle, Natl. Invit. Symp.*, St. Louis, MO (JORDAN, E. R., Ed.), pp. 15–19, Columbia, MO: Coop. Ext., Univ. of Missouri.

CLAYPOOL, D. W., CHURCH, D. C. (1984), Single cell protein from wood pulp waste as a feed supplement for lactating cows, *J. Dairy Sci.* **67,** 216–218.

COCHET, N., NONUS, M., LEBEAULT, J. M. (1988), Solid state fermentation of sugar beet, *Biotechnol. Lett.* **10,** 491–496.

CROMWELL, G. (1991), Feeding phytase to increase the availability of phosphorus in feeds for swine, in: *Proc. 52nd MN Nutr. Conf.*, Bloomington, MN, pp. 189–200, St. Paul, MN: Coop. Ext., Univ. of Minnesota.

DAWSON, K. A. (1987), Mode of action of the yeast culture, Yea-Sacc, in the rumen: A natural fermentation modifier, in: *Biotechnology in the Feed Industry* (LYONS, T. P., Ed.), pp. 119–126, Nicholasville, KY: Alltech Technical Publications.

DUNNE, J. (1991), Enzyme application in animal feed and associated industries, in: *Proc. 52nd MN Nutr. Conf.*, Bloomington, MN, pp. 201–224, St. Paul, MN: Coop. Ext., Univ. of Minnesota.

ENRIQUEZ, A., RODRIGUEZ, R. (1983), High productivity and good nutritive value of cellulolytic bacteria grown on sugarcane bagasse, *Biotechnol. Bioeng.* **25,** 877–880.

FALLON, R. J. (1987), Yeast culture in calf rations, in: *Biotechnology in the Feed Industry* (LYONS, T. P., Ed.), pp. 127–136, Nicholasville, KY: Alltech Technical Publications.

FEED INDUSTRY RED BOOK (1994), (GOIHL, J. H., MCELLHINEY, R. R., Eds.), Eden Prairie, MN: Comm. Marketing, Inc.

HAN, Y. W., YU, P. L., SMITH, S. K. (1978), Alkali treatment and fermentation of straw for animal feed, *Biotechnol. Bioeng.* **20,** 1015–1026.

HARRISON, J. H. (1989), Silage additives and their effect on animal productivity, in: *Proc. 24th Annual Pacific Northwest Animal Nutrition Conference*, pp. 27–35.

HATCH, R. H. (1991), Dry corn milling: The industry, in: *Proc. Alternative Feeds for Dairy and Beef Cattle, Natl. Invit. Symp.*, St. Louis, MO (JORDAN, E. R., Ed.), pp. 11–14, Columbia, MO: Coop. Ext., Univ. of Missouri.

HESSELTINE, C. W. (1972), Biotechnology report: Solid state fermentation, *Biotechnol. Bioeng.* **14,** 517–532.

HUTJENS, M. F. (1992), Selecting feed additives, in: *Large Dairy Herd Management* (VAN HORN, H. H., WILCOX, C. J., Eds.), pp. 309–317, Champaign, IL: Management Services, American Dairy Science Association.

JOHNSON, D. E., REMILLARD, R. L. (1983), Nutrient digestibility of brewers single cell protein, *J. Anim. Sci.* **56,** 735–739.

KARTHIGESAN, J., BROWN, S. B. (1981), Conversion of waste plastic to single-cell protein by means of pyrolysis followed by fermentation, *J. Chem. Technol. Biotechnol.* **31,** 55–65.

KELLEMS, R. O., ASELTINE, M. S., CHURCH, D. C. (1981), Evaluation of single cell protein from pulp mills: Laboratory analysis and *in vivo* digestibility, *J. Anim. Sci.* **53,** 1601–1608.

KILMER, L. H. (1993), Direct-fed microbials and fungal additives for dairy cattle, in: *Proc. Four-State Appl. Nutr. Conf.*, La Crosse, WI, pp. 153–161, Madison, WI: Coop. Ext., Univ. of Wisconsin.

KIRK, T. K., FARRELL, R. L. (1987), Enzymatic combustion: The microbial degradation of lignin, *Annu. Rev. Microbiol.* **41,** 465–505.

LAUKEVICS, J. J., APSITE, A. F., VIESTURS, U. E. (1984), Solid state fermentation of wheat straw to fungal protein, *Biotechnol. Bioeng.* **26,** 1465–1474.

LEESON, S., SUMMERS, J. D., LEE, B. D. (1984), Nutritive value of single-cell protein produced by *Chaetomium cellulolyticum* grown on corn stover and pulp mill sludge, *Anim. Feed Sci. Technol.* **11,** 211–219.

LEIBENSPERGER, R. Y., PITT, R. E. (1988), Modeling the effects of formic acid and molasses on ensilage, *J. Dairy Sci.* **71,** 1220–1231.

MUCK, R. E (1988), Factors influencing silage quality and their implications for management, *J. Dairy Sci.* **71,** 2992–3002.

NRC (National Research Council) (1989), *Nutrient Requirements of Dairy Cattle* (6th rev. Ed.), Washington, DC: National Academy of Sciences.

NICOLINI, L., HUNOLSTEIN, C., CARILLI, A. (1987), Solid state fermentation of orange peel and grape stalks by *Pleurotus ostreatus, Agrocybe aegerita*, and *Armillariella mellea, J. Microbiol. Biotechnol.* **26,** 95–98.

OFUYA, C. O., ADESINA, A. A., UPKONG, E. (1990), Characterization of the solid state fermentation of cassava, *World J. Microb. Biotechnol.* **6,** 422–424.

PAMMENT, N., ROBINSON, C. W., HILTON, J., MOO-YOUNG, M. (1978), Solid state cultivation of *Chaetomium cellulolyticum* on alkali treated sawdust, *Biotechnol. Bioeng.* **20,** 1735–1744.

PITT, R. E., SHAVER, R. D. (1990), Processes in the preservation of hay and silage, in: *Proc. Dairy Feeding Symp.*, Harrisburg, PA (SAILUS, M. A., Ed.), pp. 72–87, Ithaca, NY: Northeast Regional Agricultural Engineering Service.

REID, I. D. (1989), Solid state fermentations for biological delignification, *Enzyme Microbiol. Technol.* **11,** 786–803.

RODRIGUEZ, J. A., DELGADO, G., DANIEL, A. (1986), Optimization of solid state fermentation of citrus dried pulp by *Aspergillus niger* in a packed bed column, *Acta Biotechnol.* **6,** 253–258.

SATTER, L. D. (1986), Protein supply from undegraded dietary protein, *J. Dairy Sci.* **69,** 2734–2749.

SATTER, L. D., MUCK, R. E., WOODFORD, J. A., JONES, B. A., WACEK, C. M. (1988), Inoculant research: What has it shown us?, in: *Proc. WI Forage Council's 12th Forage Prod. and Use Symp.*, Dells, WI, pp. 108–117, Madison, WI: Coop. Ext. and WI Forage Council, Univ. of Wisconsin.

STENGEL, G. (1991), Brewers grains: The industry, in: *Proc. Alternative Feeds for Dairy and Beef Cattle, Natl. Invit. Symp.*, St. Louis, MO (JORDAN, E. R., Ed.), pp. 86–89, Columbia, MO: Coop. Ext., Univ. of Missouri.

STOKES, M. R. (1992), Effects of an enzyme mixture, an inoculant, and their interaction on silage fermentation and dairy production, *J. Dairy Sci.* **75,** 764–773.

TELLER, E., GODEAU, J. M. (1986), Evaluation of nutritive value of single cell protein (Pruteen) for lactating dairy cows, *J. Agric. Sci.* **106,** 593–599.

WEIGEL, J. C. (1991), Wet corn milling: The industry, in: *Proc. Alternative Feeds for Dairy and Beef Cattle, Natl. Invit. Symp.*, St. Louis, MO (JORDAN, E. R., Ed.), pp. 1–3, Columbia, MO: Coop. Ext., Univ. of Missouri.

ZADRAZIL, F., DIEDRICHS, M., JANSEN, H., SCHUCHARDT, F., PARK, J. S. (1990), Large scale solid state fermentation of cereal straw with *Pleurotus* sp., in: *Advances in Biological Treatment of Lignocellulosic Materials* (COUGHLAN, M. P., COLLACO, M. T. A., Eds.), pp. 43–58, London: Applied Science Publ.

ZIMMERMAN, D. R., TEGBE, S. B. (1977), Evaluation of a bacterial single cell protein for young pigs and rats, *J. Anim. Sci.* **46,** 469–475.

Index

M

N

O

P

T

V

W